OXFORD PAPERBACK REFERENCE

A Dictionary of Earth Sciences

Oxford Paperback Reference

The most authoritative and up-to-date reference books for both students and the general reader.

Abbreviations
ABC of Music
Accounting
Archaeology*
Architecture*
Art and Artists
Art Terms*
Astronomy
Bible
Biology
Botany
Business
Card Games
Chemistry
Christian Church
Classical Literature
Classical Mythology*
Colour Medical Dictionary
Colour Science Dictionary
Computing
Dance*
Dates
Earth Sciences
Ecology
Economics
Engineering*
English Etymology
English Grammar
English Language*
English Literature
English Place-Names
Euphemisms
Film*
Finance
First Names
Food and Nutrition
Fowler's Modern English Usage
Geography
King's English
Law
Linguistics
Literary Terms
Mathematics
Medical Dictionary
Medicines*
Modern Quotations
Modern Slang
Music
Nursing
Opera
Operatic Characters*
Philosophy
Physics
Plant-Lore
Political Biography
Politics
Popes
Proverbs
Psychology*
Quotations
Sailing Terms
Saints
Science
Shakespeare
Ships and the Sea
Sociology
Superstitions
Theatre
Twentieth-Century Art*
Twentieth-Century Poetry
Twentieth-Century World History
Weather Facts
Women Writers
Word Games
World Mythology
Writers' Dictionary
Zoology

**forthcoming*

A Dictionary of

Earth Sciences

SECOND EDITION

Edited by
AILSA ALLABY
and MICHAEL ALLABY

Oxford New York
OXFORD UNIVERSITY PRESS

Oxford University Press, Great Clarendon Street, Oxford OX2 6DP

Oxford New York
Athens Auckland Bangkok Bogotá Buenos Aires Calcutta
Cape Town Chennai Dar es Salaam Delhi Florence Hong Kong Istanbul
Karachi Kuala Lumpur Madrid Melbourne Mexico City Mumbai
Nairobi Paris São Paulo Singapore Taipei Tokyo Toronto Warsaw

and associated companies in
Berlin Ibadan

Oxford is a trade mark of Oxford University Press

First published 1990
First issued as an Oxford University Press paperback 1991
Second edition 1999

British Library Cataloguing in Publication Data
Data available

Library of Congress Cataloging in Publication Data
Data available

0-19-280079-5

2 3 4 5 6 7 8 9 10

Typeset by Best-set Typesetter Ltd., Hong Kong
Printed in Great Britain by
Cox & Wyman, Reading, Beckshire

From the Preface to the First Edition

Setting intellectual boundaries that would serve to define 'geology' has never been simple. As long ago as 1830, in his *Principles of Geology*, Charles Lyell expressed the view that geologists should be well versed in chemistry, natural philosophy, mineralogy, zoology, comparative anatomy, and botany. For at least a century and a half those who would study the structure and composition of the Earth have had to familiarize themselves with a wide range of scientific disciplines.

Strictly speaking, the word 'geology' describes all studies of the Earth. Traditionally, however, 'geology' has come to mean the study of rocks.

T. C. Chamberlin used the name 'Earth sciences' to embrace astronomy, cosmogony, and cosmology as well as the traditional disciplines, and Alfred Wegener (originally a meteorologist) also used it, but it was not until the 1960s that it began to gain a wider currency. Within ten years it was widely accepted, used sometimes in the singular, nowadays commonly in the plural. When, in the late summer of 1985, our friends at the Oxford University Press invited us to compile a dictionary of terms used in the topics directly related to studies of the Earth, it was clear that it should be a dictionary of 'Earth sciences'.

We had to begin by defining the term for our own purpose. We examined the way it was used by other authors, assembled a kind of consensus, and determined that our dictionary should include terms from climatology, meteorology, economic geology, engineering geology, geochemistry, geochronology, geomorphology, geophysics, hydrology, mineralogy, oceanography, palaeoclimatology, palaeoecology, palaeogeography, palaeontology, pedology, petrology, the philosophy and history of the Earth sciences including brief biographical notes of important figures, planetary geology, sedimentology, stratigraphy, structural geology, tectonics, and volcanology.

The task of a dictionary is descriptive, not prescriptive. It records words and expressions that are in current use and explains the meanings attached to them, but it does not impose those meanings or seek to dictate what a correct usage should be. As recorders, we express no opinions.

We would emphasize that the book is meant to be used as a dictionary. In no sense is it intended to be a textbook in its own right.

Preface to the Second Edition

Work on the first edition of *The Concise Oxford Dictionary of Earth Sciences* was completed in the summer of 1988, and the book was published in 1990. Many changes and advances have occurred during the years that have passed since the initial compilation and we are glad to have been given the opportunity to take account of them in a second edition.

The revision has been extremely thorough. Every entry from the first edition has been scrutinized and many have been amended to bring them up to date. A few entries have been removed as no longer relevant, but many new definitions have been added.

It is in the nature of dictionaries to grow longer with each revision and we do not apologize for the fact that ours conforms to this rule. Such growth is unavoidable, because the language itself is growing. The introduction of a new word or expression does not mean an earlier term has been discarded, as it were to make room for it. Old words survive and while they remain in use we are bound to define them.

Most of the additions have been generated by advances in planetary exploration and the search to identify new reserves of petroleum and natural gas. Planetary exploration has discovered many new satellites and smaller bodies within the solar system and has revealed previously unknown details about the more familiar planetary satellites. This has required us to augment the number of entries devoted to the solar system. Simply listing satellites and explaining the meaning of 'asteroid' and 'comet' is no longer adequate. Satellites have qualities that may now be summarized, and at least some of the minor bodies have names and known dimensions. In pursuing this exploration, new space vehicles have carried new instruments. We have provided some details about the more important space missions, including some that are still in the planning stage, and have defined some of the devices and techniques used.

Space exploration has an immediate influence on the Earth sciences through the use of satellites for observation. We have listed some of the more important satellites and remote-sensing techniques.

Advances in the methods used in petroleum exploration have led to rapid developments in the relatively new discipline of sequence stratigraphy, which has acquired a vocabulary of its own. It has also focused considerable attention on the study of trace fossils and the branch of taxonomy devoted to them. We have defined the terms most widely used in both sequence stratigraphy and ichnology.

Those may have been the main source areas for new terms requiring definition, but they are not the only ones. The discovery of communities of living organisms that thrive in extreme environments, such as those adjacent to hydrothermal vents, has led to important revisions in the scientific classification of organisms in general, with implications for evolutionary science that have influenced palaeontological ideas. These, too, we have aimed to accommodate.

We have retained the system of cross-referencing used in the first edition, but have made two innovations. We have consigned to appendices material that is best displayed in tables, such as time-scales. This makes the information they contain easier to find and use than it was when they appeared in the main body of the text. We have also added a small number of illustrations. Generally, a dictionary is about words and their uses and should use words to explain the meaning of words. There are occasions, however, when a simple diagram can usefully illustrate essentially visual ideas.

The first edition was compiled with the help of many contributors and advisers. The value of their hard work endures, and we fully acknowledge it, for without it there would have been no dictionary to revise. In preparing the first edition we were greatly helped by Professor Hubert Lamb. His contribution remains, but sadly he died in 1997. He was always friendly and unstinting in the trouble he took on our behalf. He is much missed.

In preparing the new edition we have been assisted by Dr Robin Allaby. His contribution has greatly strengthened the revision and we are very grateful for his help.

We also wish to thank Professor D. H. Tarling and Dr C. D. Gribble. They each scrutinized a long list of entries, revising them where necessary.

Finally, we thank Nigel May, the science librarian at the library of the University of Plymouth, for allowing us to make use of the library facilities.

AILSA ALLABY
MICHAEL ALLABY

Contents

Contributors and Advisers xi

A Dictionary of Earth Sciences **1**

APPENDIX A: Stratigraphic Units as Defined in the North American Stratigraphic Code, 1983 600

APPENDIX B: Time-Scales 602

APPENDIX C: Wind Strength 605

APPENDIX D: SI Units, Conversions, and Multiples 607

Bibliography 609

Contributors and Advisers

Ailsa Allaby
Michael Allaby
Robin Allaby, University of Manchester Institute of Science and Technology
Dr Keith Atkinson, Camborne School of Mines
Dr R. L. Atkinson, Camborne School of Mines
Dr T. C. Atkinson, University of East Anglia
Dr A. V. Bromley, Camborne School of Mines
Denise Crook
J. G. Cruickshank, Department of Agriculture for Northern Ireland, Belfast
Dr P. Francis, Open University; Lunar and Planetary Institute, Houston
Professor K. J. Gregory, Goldsmith's College, University of London
Dr C. D. Gribble, University of Glasgow
Dr Colin Groves,* Australian National University
Dr W. J. R. Harries, University of Plymouth
Professor M. Hart, University of Plymouth
Professor Emeritus H. H. Lamb,* University of East Anglia
John Macadam
Dr R. J. T. Moody, Kingston University
Dr J. Penn, Kingston University
Dr John M. Reynolds, Reynolds Geo-Science Ltd.
Dr D. Rolls, Kingston University
Dr I. Roxburgh
Dr N. A. Rupke, Wolfson College, Oxford
Dr Stuart Scott, University of Plymouth
Dr B. W. Sellwood, University of Reading
Dr P. J. C. Sutcliffe, Kingston University
Professor D. H. Tarling, University of Plymouth
Joan Taylor
Professor S. R. Taylor, Australian National University
Dr R. J. Towse,* Kingston University
Dr I. Tunbridge, University of Plymouth
Dr C. E. Vincent, University of East Anglia
Professor Brian F. Windley, University of Leicester
Andrew Yelland, Birkbeck College, London

* Contributor to *The Oxford Dictionary of Natural History* whose earlier entries have been transferred to this book.

aa *See* LAVA.

AABW *See* ANTARCTIC BOTTOM WATER.

AAC *See* ANTARCTIC CONVERGENCE.

Aalenian A *stage in the European Middle *Jurassic (178–173.5 Ma, Harland et al., 1989). *See also* DOGGER.

AAV *See* AGGREGATE TESTS.

Ab *See* ALKALI FELDSPAR.

abapical A directional term meaning away from the shell *apex.

abaptation The process by which an organism is fitted to its environment as a consequence of the characters it inherits, which have been filtered by *natural selection in previous environments. Because present environments seldom differ greatly from recent past environments, adaptive fitness can resemble *adaptation. In this sense, however, adaptation appears to imply advance planning, or design, which is misleading.

abandoned channel A former stream channel through which water no longer flows (e.g. a *cut-off).

abandonment facies association A *facies association formed under conditions of rising sea level, when *clastic deposition has ceased and sediment is deposited very slowly.

Abbé refractometer *See* REFRACTOMETER.

abiogenesis Development of living organisms from non-living matter; as in the supposed origin of life on Earth, or in the concept of spontaneous generation, which was once held to account for the origin of life but which modern understanding of evolutionary processes (*see* EVOLUTION) has rendered outdated.

abiotic Non-living; devoid of life. *Compare* BIOTIC.

ablation 1. Removal of snow and ice by melting and by direct alteration from the solid to the gaseous phase (sublimation). The rate of loss is controlled chiefly by air temperature, wind velocity, *humidity, rainfall, and *solar radiation. Ablation on snowfields is also influenced by aspect, depth of snow, and the nature of the underlying surface. Ablation *till is the glacial debris that may be released. The ablation zone of a glacier is that area in which losses, including *calving, exceed additions. **2.** Removal of *rock material, especially by wind action.

ablation till *See* ABLATION 1; and TILL.

ablation zone *See* ABLATION.

aboral Away from the mouth; on the opposite side of the body from the mouth.

abrasion (corrasion) The erosive (*see* EROSION) action that occurs when *rock particles of varying size are dragged over or hurled against a surface. Some common agents of abrasion are the *bed load of streams, rock debris embedded in the bases of *glaciers, and *sand and *shingle transported by wind or waves.

absolute age (true age) The age of a geologic phenomenon measured in present Earth years, rather than its age relative to other geologic phenomena (*compare* RELATIVE AGE). The term 'absolute age' has been considered rather misleading, as the means for measuring ages (*radiometric dating, *dendrochronology, *varve analysis) are subject to experimental error and the dates obtained are not precise. The alternative term 'apparent age' has been suggested. *See also* DATING METHODS; and GEOCHRONOLOGY.

absolute humidity *See* HUMIDITY.

absolute plate motion The motion of a lithospheric *plate (*see* LITHOSPHERE) with respect to a fixed frame of reference. Various frames of reference have been used, including those defined by *hot spots, no net torque of all the plates, and palaeomagnetic (*see* PALAEOMAGNETISM) Euler poles (*see* POLE OF ROTATION).

absolute pollen frequency (APF) *Pollen data from sediments, expressed in terms of the absolute numbers for each *species, *genus, or *family, per unit volume of sediment and, where deposition rates are known, per unit time. In certain circumstances this approach gives clearer information than does the traditional way of expressing pollen data as *relative pollen frequencies (RPF). APFs are particularly useful in site comparisons in which one or more high pollen producers vary. For example, when trees first appear in the regional pollen rain their prolific pollen may, in an RPF method, give the impression of declining herbaceous species, whereas examination by an APF method will show constant values for herb species.

absolute porosity *See* POROSITY.

absolute temperature Temperature measured using the *Kelvin scale.

absolute vorticity *See* VORTICITY.

absolute zero *See* KELVIN SCALE.

absorptance The ability of a material to absorb *electromagnetic radiation of a specified wavelength. *See also* ABSORPTANCE BAND.

absorptance band The range of wavelengths of *electromagnetic radiation which are absorbed by a material. *See also* ABSORPTANCE.

absorption The amount of seismic energy lost during transmission, by conversion to heat. The absorption coefficient is the fractional loss of energy over a distance of one *wavelength; hence higher-*frequency signals are attenuated more readily than those of lower frequencies over the same path. Typical values for *rocks range from 0.25 to 0.75 dB per wavelength.

abstraction (extraction) The artificial removal of water from a well, *reservoir, or river.

Abukama-type metamorphism The *recrystallization of *rocks under a high *geothermal gradient so that at any given temperature the pressure is relatively low. The term originally referred to a belt of *metamorphic rocks stretching southwestwards from the Abakuma Plateau in Japan, and characterized by the development of *andalusite and *sillimanite in rocks that were originally *shales (*pelites). This belt lies parallel to, and on the continental side of, a high-pressure metamorphic belt.

abundance zone *See* ACME ZONE.

ABW *See* ARCTIC BOTTOM WATER.

abyssal hills Relatively small topographic features of a dominantly flat, deep-ocean floor, commonly 50–250 m in height and a few kilometres in width. They are most typical of the *Pacific Ocean floor at depths of 3000–6000 m.

abyssal plain Smooth, almost level area of the deep-ocean floor in which the gradient is likely to be as low as 1 : 10 000. The covering sediments are usually thin deposits of a *pelagic ooze or *distal *turbidite.

abyssal storm (benthic storm) A large pulse of energy, possibly transferred from the surface, that accelerates *contour currents on the ocean floor to about 40 cm/s, raising large amounts of fine sediment.

abyssal zone Zone of greatest ocean depth, i.e. below a depth of 2000 m. This zone lies seaward of, and deeper than, the *bathyal zone, and covers approximately 75% of the total ocean floor. It is the most extensive Earth environment, cold, dark, with slow-moving currents (less than a few centimetres per second), supporting *fauna that typically are black or grey, delicately structured, and not streamlined.

Acadian orogeny A phase of mountain building affecting an area from the northern Appalachians in what is now New York State to the Bay of Fundy in maritime Canada (the name refers to the colony of Acadie in that region of French Canada). It occurred in the *Devonian about 380 Ma ago, although the precise date and duration are uncertain, and was most intense east of the Taconic area (*see* TACONIC OROGENY). It was caused by the westward movement of the Avalon *terrane. *See* APPALACHIAN OROGENIC BELT.

Acado-Baltic Province *See* ATLANTIC PROVINCE.

acanthodians *See* ACANTHODII.

Acanthodii (acanthodians) Class of primitive, fossil fish, characterized by the presence of a true bony skeleton (*see* BONE), a *heterocercal tail *fin, a persistent *notochord, *ganoid scales, and stout spines in front of the fins. The acanthodians lived from the *Silurian to the *Permian Period and may be related to ancestors of the more modern bony fish.

Acanthograptidae *See* DENDROIDEA.

Acanthostega *See* *ICHTHYOSTEGA*.

acceleration *Evolution that occurs by increasing the rate of ontogenetic (*see* ONTOGENY) development, so that further stages can be added before growth is completed. This form of *heterochrony was proposed by E. H. Haeckel as one of the principal modes of evolution.

acceleration, gravitational *See* GRAVITATIONAL ACCELERATION.

accelerometer A device whose output is directly proportional to acceleration. Accelerometers are used in the measurement of the motion of a ship, helicopter, or aircraft during *gravity surveys. A *seismometer or moving-coil *geophone can also function as an accelerometer.

accessory, lithic *See* LITHIC FRAGMENT.

accessory mineral A *mineral *phase within a rock whose presence does not affect the root name of the rock. For instance, the root name 'granite' is defined by the presence of *quartz, *alkali feldspar, and *mica. These are the '*essential minerals'. The presence of the mineral *sphene does not affect the root name and hence would be an example of an accessory mineral. *Apatite and *zircon are also common accessory minerals.

accessory plate (sensitive tint) In optical microscopy, a plate used to determine the optical properties of *minerals. *Quartz, *mica, and *gypsum are the common minerals used to determine the slow and fast *vibration directions that relate to the two *refractive indices of an *anisotropic mineral. The terms 'length-fast' and 'length-slow' may then be assigned to a given mineral for identification purposes. A wedge of quartz (quartz wedge) is used to determine the order of *interference colour exhibited by a mineral.

accidental lithic *See* LITHIC FRAGMENT.

accommodation space The space in which sediment may accumulate.

accordion fold *See* CHEVRON FOLD.

accretion 1. Process by which an inorganic body grows in size by the addition of new particles to its exterior. It is the mechanism by which primitive planetary bodies are believed to form as a result of the accumulation of minute, cold, homogeneous particles (homogeneous accretion). An alternative hypothesis is that iron-rich cores accumulated first and were later surrounded by silicate material (heterogeneous accretion). Homogeneous accretion yields a planet that initially has the same composition from centre to surface; heterogeneous accretion yields a planet that has a layered structure from the start. **2.** The accumulation of sediments from any cause, representing an excess of deposition over *erosion. **3.** The addition of continental material to a pre-existing continent, usually at its edge. The use of 'accretion' in this sense has evolved from theories of *nucleation to newer theories of the horizontal addition of *allochthonous *terranes of initially coherent bodies of continental *rock, usually more than 100 km² in area, which can collide, rotate, and fragment as they become sutured to a continent.

accretional heating The heating of bodies orbiting a star due to bombardment by smaller objects, the kinetic energy of the impacting body ($\frac{1}{2}mv^2$, where m is mass and v velocity) being released mainly as heat.

accretionary lapilli Pellets of *ash, ranging in size from 2 mm to 64 mm, which commonly exhibit a concentric ('onion skin') internal structure. The *lapilli are formed by the accretion of very fine ash around condensing water droplets or solid particles, particularly in steam-rich eruptive columns (*see* ERUPTION). Once formed they can be transported and deposited by *pyroclastic fall, *surge, or flow processes.

accretionary levée *See* LAVA LEVÉE.

accretionary prism *See* ACCRETIONARY WEDGE.

accretionary wedge (accretionary prism) A tectonically thickened wedge of *sediment found on the landward side of some *trenches. The accretionary wedge consists of oceanic sediment scraped off the subducting *plate (*see* SUBDUCTION), plus sediment derived from landward and deposited in the trench. Slices of sediment are added to the wedge by *underthrusting and the trench migrates seaward, the continuation of this process producing an *inversion.

accumulated temperature Surplus or deficit of temperature with respect to a defined mean value and expressed as an accumulation over a given period, e.g. a month, season, or year. For example, a datum value of 6 °C is used as a critical temperature for sustained vegetation growth, against which accumulated surpluses or deficits may be measured.

accumulation zone That part of a *glacier where the mean annual gain of *ice, *firn, and snow is greater than the mean annual loss. The zone consists of stratified firn and snow together with ice from frozen meltwater. Its lower boundary is the *equilibrium line.

ACF *See* ACF DIAGRAM; and AUTOCORRELATION.

ACF diagram A three-component, triangular graph used to show how metamorphic *mineral assemblages vary as a function of *rock composition within one *metamorphic facies. Besides SiO_2, the five most abundant oxides found in *metamorphic rocks are Al_2O_3, CaO, FeO, MgO, and K_2O. The three components plotted on ACF diagrams are A (Al_2O_3), C (CaO), and F (FeO + MgO), making the diagrams particularly useful for showing assemblage variations in metamorphosed, *basic, *igneous rocks and impure *limestones. However, each of these components has to be modified slightly to account for the presence of other, minor components in the rock. Such modification leads to: A ($Al_2O_3 - Na_2O - K_2O$); C ($CaO - [(10/3)P_2O_5] - CO_2$); and F ($FeO + MgO - Fe_2O_3 - TiO_2$). The minerals *quartz and *albite are assumed to be present in the rocks and are not shown on the diagram. *Tielines connect minerals which coexist in equilibrium and can thus define triangular areas in which three minerals are in equilibrium in the rock, lines on which two minerals are in equilibrium in the rock, and points at which one mineral is in equilibrium in the rock (in addition to the ubiquitous quartz and albite).

achnelith *See* PELÉ'S HAIR.

achondrite Rare stony *meteorite lacking *chondrules and with low nickel-iron content. It is more coarsely crystalline than a *chondrite. Basaltic achondrites resemble terrestrial *lavas.

achromatic line In the three-dimensional graph which plots quantities of the three *additive primary colours contributing to *pixels against each other, the line which runs at 45° to the axes. Pixels which plot close to this line will not be strongly coloured and may be subject to *decorrelation stretching.

acicular Pointed or needle-shaped.

acid According to the Brønsted–Lowry theory, a substance that in solution liberates hydrogen *ions or protons. The Lewis theory states that it is a substance that acts as an electron-pair acceptor. An acid reacts with a *base to give a salt and water (neutralization), and has a *pH of less than 7.

acidophile An *extremophile (domain *Archaea) that thrives in environments where the pH is below 5.0.

acid rain Precipitation with a pH of less than about 5.0, which is the value produced when naturally occurring carbon dioxide, sulphate, and nitrogen oxides dissolve into cloud droplets. The effects of increased acidity on surface waters, soils, and vegetation are complex.

acid rock *Igneous rock containing more than about 60% *Silica (SiO_2) by weight, most of the silica being in the form of silicate minerals, but with the excess of about 10% as free *quartz. Typical acid rocks are *granites, *granodiorites, and *rhyolites. *Compare* BASIC ROCK; and INTERMEDIATE ROCK. *See also* ALKALINE ROCK.

acid soil *Soil having a *pH less than 7.0. Degrees of soil acidity are recognized. Soil is regarded as 'very acid' when the reaction is less than pH 5.0. The *USDA lists five standard ranges of soil acidity (less than pH 4.5, extremely acid; 4.5–5.0, very strongly acid; 5.1–5.5, strongly acid; 5.6–6.0, medium acid; and 6.1–6.5, slightly acid). Surface *soil horizons of acid *brown earths have a reaction of pH 5.0 or less.

acme zone (peak zone, flood zone, epibole, abundance zone) An *informal term for a body of *strata containing the maximum abundance of a particular *taxon occurring within the stratigraphic range of that taxon, and after which the *zone is named.

acoustic impedance (*Z*) The product of density (ρ) and the acoustic velocity (v) for a given rock mass; $Z = \rho v$. The *reflection coefficient for an interface is governed by the contrast in the acoustic impedances of the two adjacent *rock masses.

acquired characteristics Characteristics that are acquired in the lifetime of an organism, according to early evolutionary theorists such as *Lamarck. Lamarck further suggested that traits acquired in one generation in response to environmental stimuli would be inherited by the next generation. Thus over several generations a particular type of organism would become better adapted (*see* ADAPTATION) to its environment. The kinds of acquisition envisaged by Lamarck and their heritability are now discredited, although there has been a recent revival of some aspects of Lamarckism in modified form.

acritarchs A diverse, perhaps unrelated group of organisms, which are organic-walled, hollow structures, 20–150 mm in diameter, ranging from *Precambrian to *Recent times. They are found in marine strata, although some non-marine examples are reported from Recent beds. Acritarchs are used in *correlation and to distinguish on-shore from off-shore *sediments.

Acrothoracica *See* CIRRIPEDIA.

acrozone *See* RANGE ZONE.

actinium series *See* DECAY SERIES.

actinolite A member of the *amphiboles, *$Ca_2(Mg,Fe)_5[Si_4O_{11}]_2(OH,F)_2$ with the ratio Fe/Fe + Mg = 0.9 to 0.5, belonging to the *tremolite–*ferroactinolite series of Ca-rich amphiboles; sp. gr. 3.0–3.4; *hardness 5–6; *monoclinic; light greenish-grey to dark green; white *streak; *vitreous *lustre; habit *acicular, often fibrous and felted; *cleavage *prismatic, good {110}; occurs widely in low- to medium-grade *schists and some *igneous rocks. The asbestiform variety is called *nephrite and such felted forms were used in the past for insulation and fire-resistant materials, but the development of asbestosis in workers has severely restricted their use.

Actinopterygii (ray-finned fish) A subclass of the *Osteichthyes (bony fish, *see* BONE), comprising the ray-finned fish, which include the majority of living bony fish of sea and fresh water. The *fins are composed of a membranous web of skin supported by a varying number of spines and soft rays. They appeared first during the *Devonian.

activation analysis *See* NEUTRON ACTIVATION ANALYSIS.

activation energy (energy of activation) The energy that must be delivered to a system in order to increase the incidence within it of reactive molecules, thus initiating a reaction.

active geophysical methods Geophysical exploration methods which require an artificial signal to be generated. For example, exploration seismology, some *electromagnetic techniques, *electrical resistivity, *remote sensing, and *induced polarization are said to be active geophysical methods. The term is contrasted with *passive geophysical methods.

active layer Seasonally thawed surface layer between a few centimetres and about 3 m thick, lying above the permanently frozen ground in a periglacial environment. It may be subject to considerable expansion on freezing, especially if silt-sized particles dominate, with important engineering implications. *See also* MOLLISOLS; and PERMAFROST.

active margin (seismic margin) The margin of a continent that is also a *plate margin. The alternative term, 'Pacific-type margin', indicates the range of features (e.g. *earthquakes, andesitic (*see* ANDESITE) volcanic chains, offshore oceanic *trenches, and young fold mountains) which may be associated with active margins. Some authors distinguish an 'Andino-type margin', involving an oceanic and a continental plate, from a 'Japan-type margin', involving an oceanic plate and an *island arc. The term 'Mediterranean-type margin' is also in use, although to a lesser extent, to signify the coincidence of continental edges and plate margins in a *collision zone.

active methods *See* ACTIVE GEOPHYSICAL METHODS.

active pool The part of a *biogeochemical cycle in which the nutrient element under consideration exchanges rapidly between the biotic and abiotic components. Usually the active pool is smaller than the *reservoir pool, and it is sometimes referred to as the 'exchange' or 'cycling' pool.

active remote sensing *Remote sensing which is based on the illumination of a scene by use of artificial radiation. An example is *radar. *Compare* PASSIVE REMOTE SENSING.

activity A broadly used term which refers to the rate or extent of a change associated with some substance or system. For example, it may be the tendency of a metal high in the electromotive series to replace another metal lower in the series, e.g. magnesium displacing copper from most of its compounds. It may also be used to describe the rate of decay of atoms by radioactivity.

activity coefficient (γ) The ratio of chemical activity (i.e. the effective concentration, *a*) of a component in a solution, to the actual mole fraction (X) present in solution: ($\gamma = a/X$). Values for activities are determined experimentally in a number of ways, including measuring the ratio of the *vapour pressure (p) of a known concentration of the substance in solution to the vapour pressure (p^*) of the pure substance: $a = p/p^*$. In an ideal solution the activity coefficient = 1, and the activity of the component is equal to its mole fraction. In general, the greater the amount of dissolved material, the lower the activity coefficients of each of the species present.

Actonian A *stage of the *Ordovician in the Upper *Caradoc, underlain by the *Marshbrookian and overlain by the *Onnian.

actual evapotranspiration (*AE*) The amount of water that evaporates from the surface and is transpired by plants if the total amount of water is limited. *Compare* POTENTIAL EVAPOTRANSPIRATION.

actualism The theory that present-day processes provide a sufficient explanation for past geomorphological phenomena, although the rate of activity of these processes may have varied. The theory was first clearly expressed in 1749 by G. L. L. *Buffon (1707–88), and was the essential principle of *uniformitarianism as presented in 1830 by C. *Lyell (1797–1875).

acuity The ability of a human to discern spatial variation in a scene.

ACV *See* AGGREGATE TESTS.

Adam The postulated male ancestor for all modern humans, who lived in Africa between about 100 000 and 200 000 years ago. 'Adam' is based on a change in the human Y chromosome that occurred at that time in one descendant of Adam and is now present in all human males, except for some Africans. *See also* MITOCHONDRIAL EVE.

adamantine Of mineral *lustre, brilliant, like a polished diamond.

adamellite A rock of granitic composition (*see* GRANITE) characterized by the presence of *quartz, *plagioclase feldspar, and potassic feldspar (*see* ALKALI FELDSPAR) accompanied by *biotite and/or *hornblende. The two feldspar types occur in approximately equal proportions, the plagioclase composition lying within the oligoclase range. The name is derived from the type locality of Adamello in the Tyrol where granites of this type were originally defined. In Britain the best-known example occurs at Shap Fell in Cumbria.

Adams–Williamson equation Equation describing a fundamental relationship between seismic velocities (v_p and v_ω), the *gravitational acceleration (g), and the adiabatic change in density ($d\rho$ within the *Earth (assuming only hydrostatic pressure) as a function of radium (dr):

$$d\rho = \frac{g\rho}{dr\,v_\rho^2 - (4/3)v_\omega^2}$$

This equation is directly applicable to the lower *mantle and outer *core, but is invalid where the composition is variable, the pressure is not hydrostatic, or the increase in pressure is not adiabatic.

adapical A directional term meaning towards the shell *apex.

adaptation 1. Generally, the adjustments that occur in animals in respect of their environments. The adjustments may occur by *natural selection, as individuals with favourable genetic traits breed more prolifically than those lacking these traits (genotypic adaptation), or they may involve non-genetic changes in individuals, such as physiological modification (e.g. acclimatization) or behavioural changes (phenotypic adaptation). *Compare* ABAPTATION. **2.** In an evolutionary sense, that which fits an organism both generally and specifically to exploit a given environmental zone.

adaptive radiation 1. A burst of evolution, with rapid divergence from a single ancestral form, resulting in the exploitation of an array of habitats. The term is applied at many *taxonomic levels, e.g. the radiation of the mammals at the base of the *Cenozoic refers to *orders, whereas the radiation of 'Darwin's finches' in the Galápagos Islands resulted in a proliferation of *species. **2.** Term used synonymously with '*cladogenesis' by some authors.

adaptive zone The adaptive specialization(s) that fit the *taxon to its environment, e.g. feeding habits.

addition rule (Weiss zone law) With reference to crystallographic notation, the rule stating that the indices (*see* MILLER INDEXES) of two *crystal faces in the same *zone always add up to the indices of a face bevelling the edge lying between them. The rule may be used to index faces on a *stereogram, or faces at the intersection of two zones.

additive primary colours The spectral colours red, green, and blue, which, when mixed together by projection through filters, can be used to produce all other colours. None of the primary colours can be produced by combinations of the other two. *See also* SUBTRACTIVE PRIMARY COLOURS.

adductor muscles *See* MUSCLE SCAR.

Adelaidean A *stage of the Upper *Proterozoic of south-eastern Australia, underlain by the *Carpentarian and overlain by the Hawker (*Cambrian).

Adelaidean orogeny A late *Proterozoic and *Ordovician phase of mountain building, affecting what is now southern Australia, in which *sedimentary rocks of the Adelaidean System were raised by severe thrusting and overfolding, first in the south and later along the northern margin of the system.

adhesion ripples *See* ADHESION WARTS.

adhesion warts (adhesion ripples) A *sedimentary structure consisting of an irregular, wart-like or blistered, *sand surface, formed by the wind blowing dry sand over a moist surface. The warts tend to be slightly asymmetrical, with steeper sides in the up-wind direction.

adiabatic Applied to the changes in temperature, pressure, and volume in a *parcel of air that occur as a consequence of the vertical movement of the air, and without any exchange of energy with the surrounding air. *See also* DRY ADIABATIC LAPSE RATE; and SATURATED ADIABATIC LAPSE RATE.

adit Horizontal or nearly horizontal tunnel from the surface into a mine, for entry, drainage, or exploration.

admission The substitution of a *trace element for a major element with a similar *ionic radius but a higher *valency during the crystallization of a *magma, e.g. the substitution of Li^{+} for Mg^{2+} in the *pyroxenes, *amphiboles, and *micas.

adobe A silty *clay, often calcareous, found in dry, desert-lake basins. This fine-grained *sediment is usually deposited by desert floods which have eroded wind-blown *loess deposits. The term is of Spanish origin.

adoral On the same side of the body as the mouth.

Adrastea (Jupiter XV) A jovian satellite (a *moom) that orbits within the main ring of Jupiter; it and *Metis may be the source of the material comprising the ring. Both are considered too small to suffer tidal disruption, but eventually their orbits will decay. Adrastea is one of the smallest satellites in the solar system. It was discovered in 1979 by David Jewitt. Its diameter is 20 km ($\pm$20) (23 $\times$ 20 $\times$ 15 km); mass 1.91 $\times$ 10^{6} kg; mean distance from Jupiter 129 000 km.

adsorption The attachment of an ion, molecule, or compound to the charged surface of a particle, usually of *clay or *humus, from where it may be subsequently replaced or exchanged. Ions carrying positive charges (e.g. those of calcium, magnesium, sodium, and potassium) become attached to, or adsorbed by, negatively charged surfaces (e.g. those of clay or humus).

adsorption complex Various materials of the soil, mainly *clay and *humus and to a lesser degree other particles, capable of adsorbing ions and molecules.

adularia *See* ALKALI FELDSPAR.

advection The horizontal transfer of heat by means of a moving gas (usually air).

adventive cone *See* PARASITIC CONE.

AE *See* ACTUAL EVAPOTRANSPIRATION.

aedifichnia A category of *trace fossils that comprises structures in full relief that were constructed by organisms from raw materials, e.g. mud nests of wasps, caddis fly cases, spiders' larders consisting of concentrations of insects, insect remains, and spiders.

aegirine *Pyroxene mineral, $NaFe^{3+}Si_2O_6$; sp. gr. 3.5; *hardness 6; *monoclinic; greenish-black or brown; occurs as fairly short, *prismatic crystals in *igneous and *metamorphic rocks. A variety intermediate in composition between aegirine and augite is called 'aegirine–augite'. *See also* AUGITE; and CLINOPYROXENE.

Aegyptopithecus zeuxis A genus and species of early *catarrhine primates, known from abundant remains, including several nearly complete skulls, from the early *Oligocene of the Jebel al-Qatrani Formation, Fayum, Egypt. The size of a small, living monkey, it had a long tail and could jump from branch to branch. It possessed the dental and some of the cranial characteristics of living catarrhines, but lacked many of the other cranial and most of the postcranial diagnostic features, and so represents a time when catarrhines had separated from other primates, but remained more primitive than living hominoids (*Hominoidea) or Old World monkeys and it could have been ancestral to living catarrhines.

aeolian processes (eolian processes) The erosion, transport, and deposition of material due to the action of the wind at or near the Earth's surface. Aeolian processes are at their most effective when the vegetation cover is discontinuous or absent.

aeolianite General term for the sedimentary products of wind (*aeolian) deposition.

Aeolis Quadrangle A region of Mars formed in the Late *Noachian or Early *Hesperian Epoch, containing both extensional and compressional landforms and *valles, some of which may be outflow channels, but some of which may be tectonic rift features.

aerial photograph A photograph taken from an aircraft. In hydrology, false-colour infrared photographs are used to determine the wetness and temperature of soils and to detect *springs.

aerial photography The taking of aerial photographs of rock exposures and of the ground surface for purposes of geologic interpretation. The photographs may be taken vertically, or at a high-oblique or low-oblique angle, and may be assembled like a mosaic to provide a picture of a large area. Stereoscopic cameras (two cameras within a single body) may be used to produce pairs of pictures that provide three-dimensional pictures when observed through a stereoscopic viewer. *See* PHOTOGEOLOGY.

aerobic 1. Of an environment: one in which air (oxygen) is present. In the case of a depositional environment, one with more than 1 ml of dissolved oxygen per litre of water. *Compare* ANAEROBIC; and DYSAEROBIC. **2.** Of an organism: one requiring the presence of oxygen for growth, i.e. an aerobe. **3.** Of a process: one that occurs only in the presence of oxygen.

aerodynamic roughness Uneven flow of air caused by irregularities in the surface (which may be of a solid, or of air of different density) over which the flow takes place.

aerological diagram Diagram to demonstrate variations with height of the physical characteristics of the atmosphere, particularly its temperature, pressure, and *humidity.

aeromagnetic survey Survey of the Earth's magnetic field, based on data from *magnetometers towed behind aircraft or suspended below helicopters. These instruments measure the total intensity of the *geomagnetic field or, occasionally, components of this field. The resulting measurements can then be compared with theoretical models for the value of the field and the differences (*magnetic anomalies) can be interpreted in terms of changes in the magnetic properties of the rocks below the survey line or grid. The magnetometers are usually flown with other instrumentation, e.g. *radiometric and electromagnetic, at the lowest practicable constant height above the ground. Usually the *magnetometer is housed in a 'bird' towed behind the aircraft, or in a wing-tip pod, or in a 'stinger' in the tail. In cases where the magnetometer is on board, in-board coil systems compensate for the aircraft's own magnetic field.

Aeronian A *stage of the Lower *Silurian (*Llandovery Period) underlain by the *Rhuddanian and overlain by the *Telychian.

aerosol Colloidal substance, either natural or manmade, that is suspended in the air because the small size (0.01–10 μm) of its particles makes them fall slowly. Aerosols in the *troposphere are usually removed by *precipitation and their *residence time is measured in days or weeks. Aerosols that are carried into the *stratosphere usually remain there much longer. Tropospheric aerosols may act as *Aitken nuclei but the general effect of aerosols is to absorb, reflect, or scatter radiation. Stratospheric aerosols, mainly sulphate particles resulting from volcanic *eruptions, may reduce *insolation significantly. About 30% of tropospheric dust particles are the result of human activities. *See* ATMOSPHERIC STRUCTURE; MIE SCATTERING; RAYLEIGH SCATTERING; and VOLCANIC DUST.

Aëtosauria Mainly *Triassic group of primitive thecodontian ('tooth-in-socket') reptiles (*see* THECODONTIA). They resembled heavily armoured crocodiles, and appear to have been specialized herbivores or possibly omnivores. They grew up to 3 m long, and their armour plating comprised rows of bony *plates.

AFC *See* ASSIMILATION-FRACTIONAL CRYSTALLIZATION.

AF demagnetization *See* ALTERNATING MAGNETIC FIELD DEMAGNETIZATION.

AFM diagram A three-component, triangular graph used to show how metamorphic *mineral assemblages vary as a function of *rock composition within one *metamorphic facies. Besides SiO_2, the five most abundant oxides found in *metamorphic rocks are Al_2O_3, CaO, FeO, MgO, and K_2O. The three components plotted on AFM diagrams are derived from a tetragonal diagram, with species Al_2O_3, K_2O, FeO, and MgO, and are ideal for showing mineral assemblage variations as a function of the composition of *pelites. Mineral and rock compositions plotting within this diagram are projected on to the Al_2O_3–FeO–MgO face from either the *muscovite or K-feldspar point on the Al_2O_3–FeO edge. The components of the diagram are thus A (Al_2O_3), F (FeO), and M (MgO), with the projection geometry being accommodated on specially scaled axes. Each of these components has to be modified slightly to account for the presence of other, minor components in the rock, leading to: A (Al_2O_3 − $3K_2O$); F (FeO − TiO_2 − Fe_2O_3); and M (MgO). The minerals *quartz and *albite are assumed to be present in the rocks and are not shown on the diagram. As in *ACF diagrams, *tielines connect minerals which coexist in equilibrium.

AFMAG EM system Audio-Frequency *Ma*gnetic *E*lectro*M*agnetic method, which uses natural electromagnetic (EM) fields (*sferics) in the audio-frequency range (1–1000 Hz) generated by thunderstorms to investigate lateral changes in the *resistivity of the Earth's surface.

African Plate One of the present-day major lithospheric *plates, consisting of the continental mass of Africa surrounded, except to the north, by *oceanic crust and oceanic *ridges. To the north, a complex picture of collision and *subduction zones and *transform faults has been postulated for the boundary with the *Eurasian Plate and various minor plates, e.g. the *Aegean Plate. The northern part of the African Plate also contains remnants of the oceanic crust of *Tethys. To the north-east the Red Sea is interpreted as an actively forming ocean, at the young stage of the *Wilson cycle, while the E. African *rifts, partially defining what is called by some the 'Somali Plate' to the east, may be at the embryonic stage of ocean development, or possibly a stillborn ocean.

aftershock A seismic event that occurs after an *earthquake, usually within days or weeks. Although often of small *magnitude, aftershocks can be more destructive as buildings and structures have already been weakened.

Aftonian The earliest (1.3–0.9 Ma) of four *interglacial *stages in N. America, following the *Nebraskan glacial episode, and approximately equivalent to the *Donau/Günz interglacial of Alpine terminology. Climatically it was marked by mild summers and winters warmer than those in present-day N. America.

Agassiz, Jean Louis Rodolphe (1807–73) A Swiss geologist who worked initially on fossil fish, Agassiz is better known for his *glacial theory (1837). He met *Buckland in 1840, and persuaded him that *drift deposits in Britain were evidence of a glacial epoch. In 1846 he moved to the USA to become professor of zoology and geology at Harvard, where he founded the Museum of Comparative Zoology (1859).

agate (mocha stone) Variety of chalcedonic silica (SiO_2) that is *cryptocrystalline. It is similar to *chalcedony except that impurities of iron and manganese may give it a distinct colour banding which is frequently precipitated in concentric zones. Moss agate contains delicate, fern-like, dendritic patterns. Agates may be cut and polished as decorative stones.

age 1. The interval of geologic time equivalent to the *chronostratigraphic unit '*stage'. Ages are subdivisions of *epochs and may themselves be subdivided into *chrons. An age takes its name from the corresponding stage, so like the stage name it carries the suffix '-ian' (or sometimes '-an'); the term 'age' is capitalized when used in this formal sense, e.g. '*Oxfordian Age'. **2.** An *informal term to denote a time span marked by some specific feature, e.g. '*Villefranchian mammalian age'.

ageostrophic wind The vector difference between the *geostrophic and the actual winds.

agglomerate Coarse-grained volcanic rock with rounded to subangular fragments. These fragments are mainly larger than 2 cm in size, but the mixture of fragments is typically ill sorted and the *matrix may be fine grained. An agglomerate may be the product of a volcanic explosion and therefore a *pyroclastic rock, but often the term 'agglomerate' is applied to brecciated volcanic rocks of uncertain origin. Those deposits may range from vent *breccias to debris from mudflow or lahar deposits.

agglutinate A constituent of lunar soils comprising glass-bonded *aggregates, which consist of *glasses and rock and mineral fragments welded together by glass. These aggregates form during the impact of micrometeorites into lunar soils. Their abundance in a lunar soil is an index of exposure to micrometeorite bombardment, and hence to soil maturity. The average size of agglutinates in mature soils varies, but tends toward a mean of 60 μm.

aggradation The general accumulation of unconsolidated sediments on a surface, which thereby raise its level. A large range of mechanisms may be involved, including *fluvial, *aeolian, marine, and *slope processes.

aggregate 1. In the building and construction industry, a mixture of mineral substances (bulk *minerals), e.g. sand, gravel, crushed *rock, stone, slag, and other materials (e.g. colliery spoil, pulverized fuel ash) which, when cemented, forms *concrete, mastic, mortar, plaster, etc. Uncemented, it can be used as road-making material, railway ballasts, filter beds, and in some manufacturing processes as flux. In road-making, aggregate mixed with *bitumen is called 'coated stone', and different physical characteristics are required for the different layers comprising the road *pavement. Fine aggregate is less than 6.35 mm in diameter, coarse aggregate greater than 6.35 mm. *See* AGGREGATE TESTS; and PAVEMENT. **2.** Group of soil particles adhering together in a cluster; the smallest structural unit, or ped, of soil. Aggregates join together to make up the major structural soil units.

aggregate abrasion value *See* AGGREGATE TESTS.

aggregate crushing value *See* AGGREGATE TESTS.

aggregate impact value *See* AGGREGATE TESTS.

aggregate tests Specific tests used to determine the suitability of *aggregates for special purposes. There are tests for: (*a*) shape and texture (the angularity number), to determine whether particles have a large angle of friction with good bonding properties; (*b*) size and grading, to determine whether particles will pack well; (*c*) moisture content, to discover whether materials absorb so much water that freeze–thaw action might cause the break-up of structures; (*d*) rock density, which may affect the economics of an operation; (*e*) strength, determined by subjecting the rock to hammering in a standard test and measuring the percentage of fine material produced (the aggregate impact value, or AIV); (*f*) resistance to crushing (the aggregate crushing value, or ACV), measured in a similar manner; (*g*) resistance to abrasion, measured by standard equipment to give the aggregate abrasion value (AAV)—the lower the AAV, the more resistant the rock; and (*h*) resistance to polishing, measured in the laboratory to give the polished stone value (PSV)—the higher the PSV, the greater the resistance to polishing and therefore skidding, and the more valuable the material.

aggregation 1. Process in which soil particles coalesce and adhere to form soil aggregates. The process is encouraged by the presence of bonding agents such as organic substances, *clay, iron oxides, and ions (e.g. calcium and magnesium). **2.** Progressive attachment of particles (e.g. ice or snow) or droplets around a nucleus, thereby causing its growth.

Aglaophyton major *See* RHYNIA.

Agnatha (phylum *Chordata, subphylum *Vertebrata) Superclass of jawless, fish-like vertebrates, with sucker-like mouths, including the extant lampreys, slime-eels and hagfish, and some of the earliest primitive vertebrates, with heavily armoured forms, e.g. **Cephalaspis* (*see also* OSTEOSTRACI), *Pteraspis* (*see* HETEROSTRACI), and *Jamoytius* (*see* ANASPIDA). They appeared first during the *Ordovician.

Agnostida An order of *Trilobita that lived from the Lower *Cambrian to Upper *Ordovician. Most were blind, lacking sutures, and typically are found rolled up. They had a subequal *cephalon and *pygidium. There were 2 suborders. They are important stratigraphic markers.

agric horizon Mineral-soil diagnostic horizon formed from an accumulation of *clay, *silt, and *humus, which has moved down from an overlying, cultivated soil layer. It is a *soil horizon created by agricultural management, and is identified by its near-surface position, and by *colloids accumulated in the pores of the soil.

agrichnion (pl. agrichnia) A *trace fossil comprising a burrow that formed the permanent dwelling of an organism and was used to trap or culture smaller organisms for food or use them in chemosymbiosis.

Agricola, Georgius (Georg Bauer) (1494–1555) The author of works on 'geology' and mineral classification, and of the first comprehensive record of mining, *De Re Metallica* (1556). Using Roman sources and contemporary German knowledge, his books became basic reference material for two centuries.

agrometeorology The study of the relationship between conditions in the surface layers of the atmosphere and those in the surface of the Earth, as this affects agriculture.

Agulhas current Part of the large-scale circulation of the southern Indian Ocean. It is a surface-water current that flows off the east coast of southern Africa between latitudes 25° S and 40° S in a south-westerly direction. Flow velocity varies seasonally between 0.2 and 0.6 m/s.

ahermatypic Applied to corals that lack zooxanthellae (symbiotic unicellular *algae) and that are not *reef-forming.

Aiportian The final *stage of the *Serpukhovian Epoch, underlain by the *Chokierian.

airborne dust analysis Sampling and determination of airborne particles. This technique requires size segregation of the particles and a device for collection during updrafts in order to obtain only local particles. Modern equipment sucks dust directly off vegetation for analysis.

airborne gravity survey A regional *gravity survey undertaken from the air. Such surveys are now rapid and precise because of the development of *gravimeters capable of being compensated for changes in the motion and flight path of an aircraft, particularly a helicopter.

airgun A seismic source which discharges a bubble of highly compressed air into water. Airguns are most commonly used in marine seismic exploration, but can also be used as a down-hole seismic source.

air-lift pump A device composed of two pipes, one inside the other. Air is blown down the inner pipe, which is slightly shorter than the outer pipe. The result of this is to push an air–water mixture up the gap between the two pipes. This is a useful pump for obtaining samples from very small diameter boreholes.

air mass (airmass) Large body of air (sometimes of oceanic or continental proportions) identified primarily by an approximately constant wet-bulb-potential temperature (i.e. the lowest temperature to which the air can be cooled by the evaporation of water into it). The temperature and *humidity characteristics of an air mass, which are roughly the same within the one air mass at a particular latitude and height, are modified by and modify the atmospheric environment through which the air mass passes.

air wave A sound wave which travels through the air from a seismic shot. The speed of such a wave is approximately 330 m/s.

Airy, George Biddell (1801–92) A Cambridge astronomer and mathematician, Airy became Astronomer Royal in 1835. He investigated planetary motion and tides, and studied the Earth and its density, using gravity measurements. His name is used to describe one version of the theory of *isostasy. His wide-ranging advice to the government on scientific issues created, for the first time, the role of a professional scientific civil servant.

Airy model A model to account for *isostasy which in the *lithosphere assumes a constant density ($\rho_c = 2670\ \text{kg/m}^3$), but in which topographic elevations (h) are compensated by the presence of 'roots' replacing high-density *mantle rocks ($\rho_m = 3300\ \text{kg/m}^3$) by lower-density lithospheric rocks. The depth of the root (d) is equal to $h\rho_c/(\rho - \rho_c)$. *See also* PRATT MODEL.

Airy phase When a high-frequency seismic wave is superimposed on a low-frequency ground wave, the two frequencies gradually approach one another until they merge, at which point they form a single wave with a relatively large amplitude, called the 'Airy phase'.

Aitken nuclei counter Device for the estimation of the concentration of particles with radii of more than 0.001 μm in a sample of air. Air is made to expand in a chamber: this causes it to cool. Water vapour in it condenses on to particles, forming a mist whose opacity allows estimation of the number of particles present. *See also* AITKEN NUCLEUS.

Aitken nucleus Suspended, atmospheric, solid particle with a radius of less than 0.2 μm. Most Aitken nuclei are about 0.5 μm. On average, their concentration varies from less than 1000/cm³ over oceans to 150 000/cm³ in urban areas. *See* AITKEN NUCLEI COUNTER; and NUCLEUS.

AIV *See* AGGREGATE TESTS.

AIW *See* ANTARCTIC INTERMEDIATE WATER.

åkermanite *See* MELILITE.

aklé French term for a network of sand *dunes found especially in the western Sahara. The basic unit of the network is a sinuous ridge, at right angles to the wind, made up of crescent-shaped sections which alternately face the wind (linguoid) and back to the wind (barchanoid). Aklé patterns require winds from one direction, and a large quantity of sand.

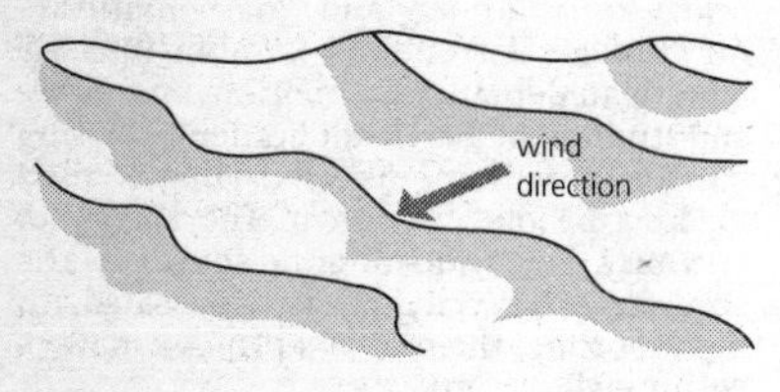

Aklé dune

alabaster *See* GYPSUM.

alar The first lateral protosepta (*see* SEPTUM) on either side of the *cardinal septum. The term is used in descriptions of the septal development of the rugose corals (*Rugosa), and may also be applied to *fossulae which occur in a similar position.

alas A *thermokarst depression with relatively steep sides and a flat floor, which may be occupied by a lake. Alases are well developed in Siberia (the word 'alas' is of Yakutian origin) where they can occupy 40–50% of the land surface.

Alaska current Oceanic water boundary current produced by the deflection of the *N. Pacific current by the N. American continent. It flows in a north-westward direction along the south-eastern margin of Alaska. The Alaska current is also called the *Aleutian current in some texts.

A-layer The seismic layer corresponding to the *crust of the Earth. It varies in thickness from a few kilometres to 70–90 km. The base lies on the *mantle and this boundary is the *Mohorovičić discontinuity.

albedo The proportion of *insolation that is reflected back from the Earth, from the tops of the clouds, and from the atmosphere, without heating the receiving surface. It averages about 30%, but varies widely according to the substance and texture of the surface, and the angle and wavelength of the incident radiation. The value for green grass and forest is 8–27% (over 30% for yellowing deciduous forest in autumn); for cities and rock surfaces, 12–18% (over 40% for chalk and light-coloured rock and buildings); for sand up to 40%; for fresh, flat snow up to 90%; and for calm water only 2% in the case of vertically incident radiation but up to 78% where there is a low angle of incidence. The albedo for cloud surfaces averages 55%, but can be up to 80% for thick *stratocumulus.

Alberta low Storms common in Alberta, Canada, and associated with heavy rain and snow. The storms form as a result of *cyclone regeneration after passage over the Canadian Rockies: as they move eastwards they bring very cold conditions, with blizzards.

Albertian A *series of the Middle *Cambrian of N. America, equivalent to the *St David's.

Albian *Stage (112–97 Ma) in the *Cretaceous, underlain by the *Aptian, and overlain by the *Cenomanian. It is known to contain a great variety of *molluscs, with the *gastropods in particular being useful *zonal indicators between continents. The Gault and Speeton Clays of England are Albian.

albic Applied to an almost white soil in which there is little *clay or oxides coating the sandy or silty particles. The albic *horizon lies at or below the surface.

albite *See* ALKALI FELDSPAR; and PLAGIOCLASE FELDSPAR.

albite–epidote–amphibolite facies A set of metamorphic *mineral assemblages that is produced by the *metamorphism of a wide range of initial rock types under the same metamorphic conditions, and is typically characterized by the development of the mineral assemblage *albite–*epidote–*hornblende in rocks of *basic *igneous composition such as *basalts. Other rocks of contrasting composition, e.g. *shales or *limestones, would each develop their own specific mineral assemblage, even though they are all being metamorphosed under the same conditions. The variation of mineral assemblage with starting rock composition reflects a particular range of pressure, temperature, and $P(H_2O)$. Experimental studies of mineral P–T stability fields indicate that the *facies represents a range of low-pressure, moderate-temperature conditions. *See* AMPHIBOLITE.

albite twin The *plagioclase feldspars, particularly *albite ($NaAlSi_3O_8$), are frequently twinned on the albite law (*see* twin-law) where the *twin-plane and composition plane is (010). This twinning is often repeated to give a series of fine lamellae, seen in the hand specimens as striations (particularly on the basal plane); such twinning is usually called '*polysynthetic' or '*lamellar' twinning.

albitization The partial or complete replacement of pre-existing *plagioclase or *alkali feldspar by albite. There are a number of ways in which this can be achieved. A common process involves the residual water-rich vapour released during the final stages of crystallization of a *granite body. This vapour, which can carry high concentrations of Na^+ in solution, rises through the granite body and reacts with the feldspars present in the granite, converting them to albite which is stable under the lower temperature vapour-rich conditions. A typical reaction that partially or completely replaces plagioclase would be: $CaAl_2Si_2O_8 + 4SiO_2 + 2Na^+ \rightarrow 2NaAlSi_3O_8 + Ca^{2+}$; anorthite + quartz + sodium (in aqueous solution) → albite + calcium (in aqueous solution). This type of reaction, where a rock simmers in its own juices, is termed a '*deuteric reaction'. Another way in which albitization can be achieved is by the reaction of ocean-floor *basalts with sea water in thermal circulation cells within the basalt layer of the *oceanic crust.

alcove A steep-sided hollow eroded by a stream from an exposed rock face.

alcrete *See* DURICRUST.

Aldingan A *stage in the Lower *Tertiary of south-eastern Australia, underlain by the *Johannian, overlain by the *Janjukian, and roughly contemporaneous with the *Bartonian and *Priabonian Stages.

alete *See* SPORE.

Aleutian current (Sub-arctic current) The oceanic current that flows westwards south of the Aleutian Islands and parallel to, but north of, the *N. Pacific current. The water mass is a mixture of water from the *Kuroshio and *Oyashio currents. *See also* ALASKA CURRENT.

Aleutian low Region of the N. Pacific, near the Aleutian Islands, where the average value of atmospheric pressure is low, owing to the frequency of low-pressure systems (cyclones) moving into and occupying the region. Any one of these systems, when present on an individual day, may be called 'an Aleutian low'. Some of them are intense, others much less so. The term is the Pacific equivalent of '*Iceland low', used in the Atlantic.

Aleutian Trench The oceanic *trench which marks the boundary between the *N. American Plate and the *Pacific Plate. The *subduction of the Pacific Plate changes from normal to oblique from west to east along the trench, with the boundary becoming a *transform fault before subduction continues in the *Kuril Trench. Towards the eastern end of the Aleutian Trench there is an increasingly wide *accretionary wedge, and an absence of andesitic *volcanoes.

Alexandrian A *series of the Lower *Silurian of N. America equivalent to the Lower–Middle *Llandovery.

alexandrite *See* CHRYSOBERYL.

Alfisols (grey-brown podzolics) An order of mineral soils that have *clay-enriched or *argillic B *horizons; are alkaline to intermediate in reaction, with the *base saturation in the B horizon more than 35%; are usually derived from base-rich parent materials; and are drier than –15 bars moisture potential for at least three months when plants could grow.

alga (pl. algae) Common (non-*taxonomic) name for a relatively simple type of plant which is never differentiated into root, stem, and leaves; which contains chlorophyll *a* as the primary photosynthetic pigment; which has no true vascular (water-conducting) system; and in which there is no sterile layer of cells surrounding the reproductive organs. The algae range in form from single cells (*Protista) to plants many metres in length; algae can be found in most habitats on Earth, although the majority occur in freshwater or marine environments. *See* BACILLARIOPHYCEAE; CHAROPHYCEAE; CHLOROPHYCEAE; CHRYSOPHYCEAE; DINOPHYCEAE; PHAEOPHYCEAE; and RHODOPHYCEAE.

algal bloom Sudden growth of algae in an aquatic ecosystem. It can occur naturally in spring or early summer when primary production exceeds consumption by aquatic herbivores (*see* PRIMARY PRODUCTIVITY). Algal blooms may also be induced by nutrient enrichment of waters due to pollution.

algal limestone *See* LEIGHTON-PENDEXTER CLASSIFICATION.

algal mat A sheet-like accumulation of blue-green algae (*Cyanobacteria) developed in shallow marine *subtidal to *supratidal environments, as well as in lakes and swamps. The algae cover the *sediment surface, and will in turn trap sediment to produce a laminated alternation of dark, organic-rich algal layers and organic-poor sediment layers. *See also* STROMATOLITE.

alginite *See* COAL MACERAL.

Algonkian A *Precambrian *system (Van Eysinga, 1975) of equivalent time period to the *Proterozoic.

aliasing A distortion in the frequency of sampled data produced by insufficient sampling per wavelength, which can result in spurious frequencies. When the sampling rate is too low to represent the wave-form accurately, then aliasing will occur. To avoid aliasing, the sampling frequency should be at least twice that of the highest-frequency component contained within the sampled wave-form. Alternatively, an anti-alias filter can be applied, which removes frequency components above the *Nyquist frequency.

alkali–aggregate reaction A chemical reaction that can lead to damage in *concrete structures. Free lime (CaO) in *cement reacts with CO_2 in the atmosphere to precipitate $CaCO_3$ around the cement grains. This protects them from *weathering and also gives an alkalinity level (*pH higher than 7) which helps to protect steel from corrosion. If the aggregate contains soluble *silica, however, new minerals may precipitate by reaction between the aggregate and the cement. These may absorb water, causing the concrete to swell and eventually crack. Water entering these cracks may cause rusting of reinforcement bars and repeated wetting and drying may eventually destroy a structure.

alkali basalt A fine-grained, dark-coloured, volcanic rock characterized by *phenocrysts of *olivine, titanium-rich *augite, *plagioclase and iron oxides. For similar SiO_2 concentrations, alkali basalts have a higher content of Na_2O and K_2O than other *basalt types such as *tholeiites. They are also characterized by the development of *modal *nepheline in their *groundmass (only seen with the highest powered lens on a petrological microscope) and normative nepheline (Ne) in their *CIPW norms. Alkali basalts are typically found on updomed and rifted *continental crust, and on oceanic islands such as Hawaii and Ascension Island.

alkalic *See* ALKALINE.

alkali-calcic series *See* CALC-ALKALINE.

alkalic series *See* CALC-ALKALINE.

alkali feldspar A group of *silicate minerals that contain the alkali metal elements potassium and sodium. The normal feldspar minerals (including the calcium-bearing varieties) can be plotted on a chemical basis into a triangle which has $KAlSi_3O_8$ (potassium feldspar, sanidine, orthoclase (Or), or microcline), $NaAlSi_3O_8$ (sodium feldspar, albite, or Ab), and $CaAl_2Si_2O_8$ (calcium feldspar, anorthite, or An) at the three apices. The alkali feldspars are represented by the edge of the triangle joining $KAlSi_3O_8$ and $NaAlSi_3O_8$ and these minerals may also contain up to 10% by weight of the third phase ($CaAl_2Si_2O_8$). At high temperatures the alkali feldspars show complete *solid solution between the potassium and sodium *end-members, but as the temperature drops unmixing occurs and potassium feldspar and sodium feldspar separate out to produce a perthitic texture. Depending upon the final temperature, a range of perthites may result, from coarse (*perthite), representing perthites formed during a large drop in temperature, to fine (*microperthite), and finally to very fine (*cryptoperthite), representing perthites invisible to the naked eye and often invisible under the microscope, but observed by *X-ray diffraction (XRD) techniques. If the amount of potassium exceeds that of sodium, then potassium feldspar is the host and sodium feldspar occurs within the host mineral as *blebs, irregular patches, etc. In the alkali feldspars, perthitic textures occur in the compositional range $Or_{85}Ab_{15}$ to $Or_{15}Ab_{85}$ (or Or_{85} to Or_{15}). K-feldspar ($KAlSi_3O_8$) is the general name for the *monoclinic, potassium-rich end-member: sp. gr. 2.6; *hardness 6; white, sometimes with a reddish tint; *vitreous *lustre; crystalline, *prismatic, with simple twins (*see* CRYSTAL TWINNING). It is an *essential constituent of *acid *igneous rocks and *arkoses and is used in the manufacture of glazes, porcelain, and pottery. Microcline has the same physical properties and composition as orthoclase, but is *triclinic and is characterized by 'cross-hatched' twinning. It is greyish-white, but bright green in the variety known as 'amazonstone' ('amazonite'). Anorthoclase is very similar to microcline, but the amount of sodium exceeds that of potassium. Crystal twinning is common particularly along the *pericline and albite laws. Sanidine is the high-temperature variety of orthoclase and the inversion temperature is at 900 °C. It occurs in quickly cooled lavas. Adularia is a variety of microcline, but with up to 10% sodium substituting for potassium. It may show an opalescent play of colours to give a variety known as 'moonstone'. Albite ($NaAlSi_3O_8$) is the sodium-rich end-member of both the alkali feldspars and the *plagioclase feldspars. The semi-precious moonstone, with its characteristic bluish sheen or *schiller, is an example of a perthitic alkali feldspar.

alkaline (alkalic) 1. Having a *pH greater than 7. **2.** *See* alkaline rock.

alkaline rock *Igneous rock containing a relatively high concentration of the alkali (lithium, sodium, potassium, rubidium, caesium, and francium) and alkaline earth metals (magnesium, calcium, strontium, barium, and radium). Both silica-saturated and silica-undersaturated varieties exist, expressed in the presence of *alkali feldspars and *feldspathoids respectively. Alkali *ferromagnesian minerals are usually present, and their identity depends on the composition of the rock. Igneous rocks of the alkaline suite span the composition range from *basic to *acid, and may be *intrusive or *extrusive.

alkaline soil Soil with a *pH greater than 7.0. Degrees of soil alkalinity are recognized. The *USDA lists soils with pH 7.4–7.8 as mildly alkaline; 7.9–8.4 as moderately alkaline; 8.5–9.0 as strongly alkaline; and more than 9.0 as very strongly alkaline. Soil is not regarded as highly alkaline unless the reaction is between 8.0 and 10.0. The full range of the pH scale (0–14) is not used in soils, as the reaction of most soils is between pH 3.5 and pH 10.0. A *base saturation of 100% indicates a pH of about 7.0 or higher.

alkaliphile An *extremophile (domain *Archaea) that thrives in environments where the pH is above 9.0.

allanite (orthite) *Mineral, with the formula $(Ca,Ce,Y,La,Th)_2(AlFe)_3Si_3O_{12}(OH)$; sp. gr. 3.4–4.2; *hardness 5.0–6.5; *monoclinic; light brown to black; pitchy to *submetallic *lustre; faintly radioactive; *crystals normally *prismatic, often *tabular, sometimes *massive; *cleavage imperfect {001}; often occurs as an *accessory mineral in granitic rocks, *syenites, *gneisses, and *skarns.

Alleghanian orogeny A phase of mountain building, that began in the Early *Carboniferous and was completed by the end of the *Permian, caused by the collision between N. America and Africa. It formed part of the general WSW to ENE *Hercynian belt. The orogeny affected the Lower *Palaeozoic *basement and Lower Permian strata along the western margin of the southern and central parts of the Appalachian Mountains extending from what is now Pennsylvania to Alabama, with effects as far north as New Brunswick and Newfoundland. *See* APPALACHIAN OROGENIC BELT.

allele Common shortening of the term 'allelomorph'. One of two or more forms of a *gene arising by mutation and occupying the same relative position (locus) on homologous *chromosomes.

allelomorph Term that is commonly shortened to '*allele'.

Allen's rule A corollary to *Bergmann's rule and *Gloger's rule, holding that a race of warm-blooded species in a cold climate typically has shorter protruding body parts (nose, ears, tail, and legs) relative to body size than another race of the same species in a warm climate. This is because long protruding parts emit more body heat, and so are disadvantageous in a cool environment, but advantageous in a warm environment. The idea is disputed, critics pointing to many other adaptations for heat conservation which probably are more important, notably fat layers, feathers, fur, and behavioural adaptations to avoid extreme temperatures.

Allerød Late glacial (i.e. late *Devensian) period marking a prolonged warmer oscillation or *interstadial during the general phase of ice retreat in NW Europe. *Radiocarbon dating suggests it lasted from about 12 000 BP to 10 800 BP. Pollen records for the NW European area indicate a cool temperate flora with birch (*Betula* species) widespread, in marked contrast to the preceding and following, colder, *Dryas, phases.

allochem The collective term for particles (grains) which form the framework in mechanically deposited *limestones. In the limestone classification of *Folk, allochems are often found together with a *carbonate mud *matrix (*micrite) and may subsequently have *pore spaces filled by sparry *calcite *cement (*sparite). Common allochems include skeletal fragments (*bioclasts), *ooids, *peloids, and *intraclasts.

allochemical A *limestone defined by the *Folk classification as comprising *allochems with either a sparry *calcite *cement (*sparite), or a *microcrystalline *calcite (*micrite) *matrix. Limestones lacking allochems are defined by Folk's classification of limestones as *orthochemical limestones or *autochthonous *reef rocks.

allochthon A body of rock that has been transported to its present position, usually over a considerable distance. *See* ALLOCHTHONOUS.

allochthonous Not indigenous; acquired. In the Earth sciences the term is applied to geologic units that originated at a distance from their present position. Such displacement may be due to lateral thrusting and overfolding, or to gravity gliding. *Compare* AUTOCHTHONOUS.

allochthonous terrane *See* TERRANE.

alloclast A *clast produced by subterranean, igneous processes that break up pre-existing volcanic rocks. *Compare* AUTOCLAST; EPICLAST; and HYDROCLAST.

allocyclic mechanisms Events responsible for the accumulation of sediments that are external to the sedimentary system itself (e.g. sea-level changes, tectonic activity, or climate). *Compare* AUTOCYCLIC MECHANISMS.

allodapic Applied to materials deposited by turbidity (*see* TURBIDITY CURRENT) or *mass flow, particularly used in relation to *limestones deposited by mass flow.

alloformation *See* ALLOSTRATIGRAPHIC UNITS.

allogenic Applied to minerals, or other components of a rock, that have been derived from pre-existing rocks and transported some distance to form part of the present unit; e.g. *quartz grains in a *sandstone. *Compare* AUTHIGENIC.

allogenic stream Stream originating outside a particular area and whose continuation is inconsistent with its new surroundings. Type examples are the Nile and Indus, whose discharges are sufficient to carry them through arid regions, and the Yugoslav Neretva, which is large enough to pass over permeable limestone.

allogroup *See* ALLOSTRATIGRAPHIC UNITS.

allomember *See* ALLOSTRATIGRAPHIC UNITS.

allometry Differential rate of growth such that the size of one part (or more) of the body changes in proportion to another part, or to the whole body, but at a constant exponential rate. For example, the antlers of the extinct *Irish elk (*Megaloceros giganteus*), the largest of all deer, grew 2.5 times faster than the rest of its body to reach an adult span of up to 3.5 m in the largest individuals. Allometry may in other cases be negative, leading to comparatively smaller parts.

allopatric speciation Formation of new *species from the ancestral species as a result of the geographic separation or fragmentation of the breeding population. Genetic divergence in the newly isolated daughter populations ultimately leads to new species; divergence may be gradual or, according to punctuationist models, very rapid. *See also* PUNCTUATED EQUILIBRIUM.

allopatry The occurrence of *species in different geographic regions. When closely related species are separated, differences between them that minimized their competition for food, shelter, or other resources, usually decrease (i.e. the characteristics converge). The process is called character displacement and may be morphological or ecological.

allophane (kandite) *Clay mineral of the *kaolinite group, $Al_2Si_2O_5(OH)_2$ whitish; amorphous, non-crystalline; occurs along *faults or *joint planes in a variety of rocks.

allostratigraphic units Allogroups, alloformations, and allomembers; these are subdivisions of sedimentary structures that are the subject of *allostratigraphy.

allostratigraphy The study of sedimentary strata that can be defined and identified from the discontinuities bounding them, and that can be mapped.

allotriomorphic *See* ANHEDRAL.

alluvial Applied to the environments, action, and products of rivers or streams. Alluvial deposits (alluvium) are *clastic, *detrital materials transported by a stream or river and deposited as the river floodplain. The term is also applied to surface flow, as in *alluvial fans, *bajadas, etc.

alluvial cone *See* ALLUVIAL FAN.

alluvial fan (alluvial cone) Mass of sediment deposited at some point along a stream course at which there is a sharp decrease in gradient, e.g. between a mountain range and a plain. Essentially, a fan is the terrestrial equivalent of a river-delta formation.

alluvium An *alluvial deposit.

almandine Member of the *garnet group of *minerals, $Fe_3Al_2(SiO_4)_3$; sp. gr. 4.25; *hardness 6.5–7.5; *cubic; red, brown-red, or black; greasy to vitreous *lustre; most common *crystals are dodecahedra, and many are irregular grains; widely distributed in *metamorphic and *igneous rocks, and in beach *sands and *placers. Transparent crystals are used as *gemstones, and the mineral is useful in general as an abrasive.

alnöite An *intrusive, *basic, *igneous, *carbonatite rock, distinctive in possessing primary *calcite, and consisting of *melilite (1/3); *biotite (1/3); and *pyroxene, calcite, and *olivine (1/3). *Feldspar is not present in the rock, its place being taken by the mineral melilite which has the general formula: $X_2YZ_2O_7$; where X = Ca, Na; Y = Mg, Al; Z = Si, Al. The type location for this rock is Alnö island off the coast of Sweden.

alpha decay Certain radionuclides (radioactive *nuclides) decay by the spontaneous emission of alpha particles from their nuclei. The alpha particle is composed of two protons and two neutrons and has a charge of +2. It also has an appreciable mass and its ejection from the nuclide creates a certain amount of recoil energy in the nucleus. The total energy (E_x) created by alpha decay is, therefore, the sum of the kinetic energy of the particle, the recoil energy given to the new *nucleus and the total energy of any emitted *gamma rays. *See also* RADIOACTIVE DECAY.

alpha diversity Diversity among members of a species within a single population.

alpha-mesohaline water *See* HALINITY.

alpha–proton–X-ray spectrometer (APXS) A set of instruments carried on Russian *Vega and *Phobos missions and by *Sojourner*, the rover vehicle carried on the 1997 *Mars Pathfinder mission, that measures the elemental chemistry of surface materials. The sensor head of the instrument contains curium, as a source of alpha particles, an alpha particle detector, a proton detector, and an X-ray detector. The head is placed in contact with a sample and remains there for 10 hours. Alpha particles of known energy bombard the sample. Scattered alpha particles, protons from alpha–proton reactions, and X-rays produced by excitation by the alpha particles of the atomic structure of the sample are measured by the detectors. The energy spectrum of detections by all three instruments is then recorded and transmitted to Earth.

alpine glow At sunset, beginning as the Sun nears the horizon, mountains exposed to direct sunlight in the east, particularly if snow-covered, assume a series of colours changing from yellow-orange to a rosy pink, which finally becomes purplish. The same series of colours in reverse order is seen on mountains in the west at sunrise.

Alpine–Himalayan orogeny Period of mountain building that affected both northern and southern margins of the ancient *Tethyan ocean. It began in the *Triassic, but reached its high point during the Late *Oligocene and *Miocene. The Alps are an obvious testament to this orogeny, while the gentle folds of northern France, and the Weald and London Basin in England, reflect its outer effects.

Alportian *See* SERPUKHOVIAN.

alteration A change produced in a rock by chemical or physical action.

alteration halo A border of minerals produced by *hydrothermal *alteration in the rock surrounding a *vein.

alternating current The current output, with a sinusoidal wave-form, from an alternator or dynamo.

alternating-magnetic-field demagnetization (AF demagnetization, thermal cleaning) A common method for demagnetizing (*see* DEMAGNETIZATION) rock samples that is widely used in *palaeomagnetism and *archaeomagnetism because of its simplicity and because it produces no chemical change in the samples. It can cause problems associated with *anhysteretic and *rotational remanences, and is only fully suitable for *magnetite-bearing rock samples.

altiplanation Process of relief reduction or planation (i.e. the smoothing of the surface) under periglacial conditions. Two mechanisms are involved: destruction of upstanding relief features by *gelifraction or *nivation, and accumulation of debris in depressions or as terraces. In many areas only partial altiplanation has been achieved, with the emergence of altiplanation terraces, such as those of Cox Tor on Dartmoor, England.

altocumulus From the Latin *altum* (height) and *cumulus* (heap). A genus of cloud composed largely of water droplets, and consisting of grey-white sheets, or banded layers and rolls, which may also be broken up into cells. Sometimes it has a banded appearance, occasionally giving a mackerel-sky effect; this is probably associated with strong vertical wind shear in middle altitudes. *See also* CLOUD CLASSIFICATION.

Altonian A *stage in the Upper *Tertiary of New Zealand, underlain by the *Otaian, overlain by the *Clifdenian, and roughly contemporaneous with the upper *Burdigalian Stage.

altostratus From the Latin *altum* (height) and *stratus* (spread out). A genus of cloud consisting of greyish sheets or layers; the cloud may be striated, fibrous, or uniform. It may be composed of ice crystals as well as water droplets. *See also* CLOUD CLASSIFICATION.

alumstone *See* ALUNITE.

alunite (alumstone) *Mineral, $KAl_3(SO_4)_2(OH)_6$; sp. gr. 2.6–2.8; *hardness 3.5–4.0; *trigonal; white, sometimes grey to reddish; white *streak; *vitreous *lustre; *crystals rare, *habit *massive; *cleavage basal {0001}, distinct; *fracture uneven, *conchoidal; slightly astringent taste. Occurs as a *secondary mineral in volcanic rocks containing potassic *feldspars altered by sulphuric-acid solutions. It is difficult to distinguish from *dolomite, *anhydrite, and *magnesite.

alveolus *See* BELEMNITIDA.

A/m *See* AMPERES PER METRE.

Amalthea (Jupiter V) The jovian satellite with the closest orbit to Jupiter. Its surface colour is reddish, apparently because of sulphur emitted from *Io. Its diameter is 189 km (262 × 146 × 134 km), the irregular shape suggesting a rigid body. Its mass is 7.17×10^{18} kg; mean distance from Jupiter 181 000 km. It radiates more heat than it receives from the Sun.

Amarassian *See* KAZANIAN; and TATARIAN.

Amazonian A division of *areological time, lasting from 1.80 Gy to the present in the Hartmann–Tanaka Model and 3.55 Gy to the present in the Neukum–Wise Model, and divided into three epochs: Lower Amazonian (1.80–0.70 or 3.55–2.50 Gy); Middle Amazonian (0.70–0.25 or 2.50–0.70 Gy); and Upper Amazonian (0.25–0.00 or 0.70–0.00 Gy).

amazonite *See* ALKALI FELDSPAR.

amazonstone *See* ALKALI FELDSPAR.

amb *See* AMBULACRUM.

amber Fossil conifer resin which is brittle and hard, translucent to transparent, and yellow to brown in colour. It is found in *sediments or on the shore and takes a fine polish.

ambient pressure Atmospheric pressure in the surrounding air.

ambient temperature The dry-bulb temperature prevailing in the surrounding air.

ambitus The outline or edge of an echinoid (*Echinoidea) when seen from above or below. Usually it is the place where the *plates of the *test are at their widest.

ambulacral Applied to those areas of the body of an *echinoderm that bear *tube feet.

ambulacral groove *See* AMBULACRUM.

ambulacrum (amb) In *Echinodermata, an area of the body surface (covered in most classes by calcitic (*see* CALCITE) *plates), that overlies one of the radial canals of the internal water vascular system, and bears the *tube feet. In some echinoderms, e.g. Asteroidea, *Blastoidea, and

*Crinoidea, the ambulacrum is marked by a deep linear depression, the ambulacral groove. Typically, echinoderms have five ambulacral areas, or a multiple of five. *See* ECHINOIDEA.

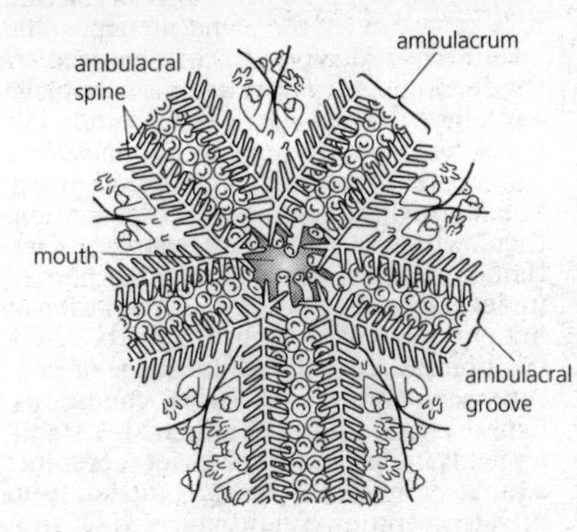

Ambulacrum

Ambulocetus natans The most completely known early cetacean, described in 1994 by J. G. M. Thewissen, S. T. Hussain, and M. Arif, from Lower to Middle *Eocene beds in Pakistan. It is known by parts from most of the skeleton, showing that it had a long neck, relatively long hind limbs, and five separate (hoofed) digits on each limb. It was the size of a sea lion.

amensalism An interaction of species populations, in which one population is inhibited while the other (the amensal) is unaffected.

American Province *See* PACIFIC PROVINCE.

Amersfoort An *interstadial in the last *glaciation of the Netherlands during the early *Devensian (somewhere between 60 000 and 70 000 Ma). The July temperature (based on floral evidence) was perhaps 15–20 °C.

amesite *See* CHAMOSITE.

amethyst *See* QUARTZ.

amino acid Organic compound containing an acidic carboxyl (COOH) group and a basic amino (NH_2) group. They constitute the fundamental building blocks of peptides and proteins and are classified either (*a*) as neutral, basic, or acidic (according to their *pH), or (*b*) as non-polar, polar, or charged (according to their electrical configuration).

amino group The chemical group $-NH_2$.

ammonites Family of *Ammonoidea.

Ammonoidea (ammonoids; phylum *Mollusca, class *Cephalopoda) Subclass of cephalopods which generally have *planispirally coiled, septate shells (*see* SEPTUM). Characteristically the shells are tightly coiled and planispiral, although some are coiled loosely or spirally; the *protoconch is globular; the shells may be either *involute or *evolute. Some forms have marked *ventral *keels; ribs and nodes may also be present. The *siphuncle is variable but mainly ventral in position. *Sutures are often very complex. *Camaral deposits are absent. The Ammonoidea were probably tetrabranchiate (four-gilled) cephalopods. They constitute the largest cephalopod subclass, with 163 families including the ammonites, in which the suture lines form very complex patterns; the ceratites, in which part of the suture line is frilled; and the goniatites, with relatively simple suture lines. They range in age from *Devonian to Upper *Cretaceous. All members are now extinct. *See also* APERTURE; APTYCHUS; FORAMEN; PHRAGMOCONE; and VENTER.

ammonoids *See* AMMONOIDEA.

amnion *See* AMNIOTIC.

amniotic Applied to a type of development typical of higher *vertebrates (*reptiles, *birds, and *mammals), in which the amnion (a protective membrane) surrounds the embryo in a bag of ('amniotic') fluid. Evolutionarily, the amnion is *primitively associated with a shell and is capable of gaseous exchange; its development thus enabled eggs to be laid on dry land for the first time in vertebrate evolution. *Compare* ANAMNIOTIC.

amorphous cloud Continuous cover of low, featureless cloud (e.g. *nimbostratus), often producing rain.

amosite *See* ANTHOPHYLLITE; and ASBESTOS.

amperes per metre (A/m) The SI unit of magnetic moment per unit volume. 1 A/m = 10^3 gauss.

Ampferer subduction *See* A-SUBDUCTION.

Amphibia (amphibians) Class that appeared first in the *Devonian, having evolved from rhipidistian (lobe-finned) fish (*see* RHIPIDISTIA). They flourished in the *Carboniferous and *Permian. During the *Triassic some forms, e.g. *Mastodontosaurus*, grew to 6 m long, and the first modern types were established. Today the amphibians are represented by just three groups, of which the Urodela (salamanders) and Anura (frogs and toads) are the best known (the third group, the caecilians (Apoda), are worm-like and burrowing). Most amphibians are found in damp environments and they occur on all continents except Antarctica.

amphibians *See* AMPHIBIA.

amphiboles A group of *minerals possessing double chains of silicon–oxygen $[SiO_4]$ tetrahedra with a composition of $[Si_4O_{11}]_n$ running parallel to the *crystallographic axis; i.e. parallel to the *prism zone of a crystal. The double chains are held together by monovalent, divalent, or trivalent cations, of which Na^+, Ca^{2+}, Mg^{2+}, Fe^{2+}, Al^{3+}, and Fe^{3+} are the most important; hydroxyl ions also occur. There are three main groups of amphibole minerals: (*a*) calcium-poor amphiboles with the general formula $X_2Y_5[Z_4O_{11}]_2(OH,F)_2$, where X = Mg or Fe^{2+}, Y = Mg, Fe^{2+}, Fe^{3+}, Al^{3+}, etc., and Z = Si or Al; (*b*) the calcium-rich amphiboles with the general formula $AX_2Y_5[Z_4O_{11}]_2(OH,F)_2$, where A = Na, X = Ca, Y = Mg, Fe^{2+}, Fe^{3+}, Al, etc., and Z = Si or Al; and (*c*) the alkali amphiboles in which Na > Ca and with the general formula $AX_2Y_5[Z_4O_{11}]_2(OH,F)_2$, where A = Na or K, X = Na (or Na and Ca), Y = Mg, Fe^{2+}, Fe^{3+}, Al, etc., and Z = Si or Al. Calcium-poor amphiboles include the *orthorhombic amphiboles (called the orthoamphiboles) and include *anthophyllite and gedrite, but the other two groups are *monoclinic and include the common *hornblendes *tremolite and *actinolite, as well as the sodium-rich varieties such as *glaucophane and *riebeckite. Amphiboles are common rock-forming silicate minerals that occur in *intermediate and *alkaline *igneous rocks and also in many regional *metamorphic rock types.

amphibolite A medium-grained, dark-coloured, regional *metamorphic rock composed of *hornblende and *plagioclase with minor *epidote, *sphene, *biotite, and *quartz. The rock may show a well developed planar or linear alignment of elongate hornblende crystals as a result of suffering deformation at the same time as *regional metamorphism. These alignments define *fabrics within the rock known as *schistosity and *lineation respectively. Amphibolites are formed by medium-grade *metamorphism of *basic *igneous rocks such as *basalts, both *extrusive and *intrusive types.

amphibolite facies A set of metamorphic *mineral assemblages produced by the *metamorphism of a wide range of starting rock types under the same metamorphic conditions and typically characterized by the development of the mineral assemblage andesine (*plagioclase)–*hornblende in rocks of *basic *igneous composition. Other rocks of contrasting composition, e.g. *shales or *limestones, would each develop their own specific mineral assemblage, even though they are all being metamorphosed under the same conditions. The variation of mineral assemblage with starting rock composition reflects a particular range of pressure, temperature and $P(H_2O)$ conditions. Experimental studies of mineral P–T stability fields indicate that the facies represents a range of moderate-pressure, moderate- to high-temperature conditions.

amphicoelous Applied to the condition in which the central part of each *vertebra (the centrum) is concave on both anterior and posterior faces. *Compare* HETEROCOELUS.

Amphicyonidae *See* CARNIVORA.

amphidetic Applied to a bivalve (*Bivalvia) ligament located around the *umbones in both anterior and posterior areas. *Compare* OPISTHODETIC.

amphidromic point The centre of an amphidromic or tidal system; a no-tide or nodal point around which the crest of a standing wave or a high-water level rotates once in each tidal cycle. The tidal range increases progressively with increasing distance from the central point. The high water rotates anticlockwise around the central point in the northern hemisphere and clockwise in the southern hemisphere.

Amphineura (phylum *Mollusca) Class of elongate, bilaterally symmetrical, marine molluscs, first appearing in the Upper *Cambrian, in which the shell, if present, consists of seven or eight overlapping, calcareous plates on the dorsal surface. The class comprises the *Polyplacophora (chitons), and the Aplacophora which have no fossil record.

amphitheatre Flat-topped, steep-walled, depression, shaped like a horseshoe and resembling an ancient Greek theatre. It may be the result of glacial erosion, forming a *cirque, or the collapse of a volcano. In the case of Mount St Helens, the catastrophic collapse of the summit and northern slopes of the cone and the subsequent *pyroclastic flow eruption on 18 May 1980 created a north-facing amphitheatre enclosing a small active vent, later occupied by a small *dacite dome.

amplitude 1. Of a *fold, the height of the maximum displacement of a folded layer measured from a median trace which passes through the *inflexion points of adjacent limbs. **2.** Of a wave, the distance between the highest point of the wave and the position of zero displacement.

ampulla An inflatable sac at the end of each *tube-foot in the echinoids (*Echinoidea). It controls water pressure in the tube-foot when the muscles in its walls are contracted.

amygdale (amygdule; adj. amygdaloidal) Spheroidal, ellipsoidal, or (literally) 'almond-shaped' cavity or *vesicle within a *lava, filled with secondary minerals, e.g. *calcite, *quartz, or *zeolites.

amygdaloidal *See* AMYGDALE.

amygdule *See* AMYGDALE.

An *See* ALKALI FELDSPAR.

anabatic wind A wind that blows up a slope, often gently, and usually when the sloping ground surface has been warmed by the Sun. *Compare* KATABATIC WIND.

anabranching channel A type of *distributary river channel that separates from its trunk stream and may flow parallel to it for several kilometres before rejoining it. The anabranching channel remains undivided, and so differs from an anastomosing channel which has major distributaries that branch and then rejoin it.

anacline A common condition of the interarea in a strophic brachiopod (*Brachiopoda) shell. It describes the situation where the interarea is at an angle of less than 90° to the plane of *commissure. *Compare* APSACLINE.

anaerobic 1. Of an environment: one in which air (oxygen) is absent. In the case of a depositional environment, one with 0.0–0.1 ml of dissolved oxygen per litre of water. *Compare* AEROBIC; and DYSAEROBIC. **2.** Of an organism: one able to grow only in the absence of oxygen, i.e. an anaerobe. **3.** Of a process: one that can occur only in the absence of oxygen.

anafront Front (warm or cold) at which there is upward movement of the warm-sector air, commonly producing clouds and precipitation. *Compare* KATAFRONT.

anagenesis In the original sense, refers to evolutionary advance. Now it is often applied more widely, to virtually all sorts of evolutionary change, along a single, unbranching lineage.

analcime (analcite) *Mineral, $Na(AlSi_2O_6)H_2O$; sp. gr. 2.2–2.3; *hardness 5.0–5.5; *cubic; can be colourless or white, with grey, red, and greenish tints, sometimes pigmented with iron oxides; *vitreous *lustre; *crystals polyhedra and tetragonal trioctahedra, also found in granular aggregates; *cleavage poor, cubic; occurs in rocks of low-temperature *igneous origin that are rich in soda and water (e.g. *nepheline syenites), and also volcanic *tuffs.

analcite *See* ANALCIME.

analcite-basanite *See* BASANITE.

analcitite An *igneous rock in which the abundant *plagioclases have been replaced by the *mineral *analcime during late-stage deuteric activity (*see* DEUTERIC reaction). *Magma-derived, late-stage aqueous fluids penetrate along *cleavages and *fractures in primary plagioclase within the crystalline rock body and react with it, converting the plagioclase to analcime which is stable in the presence of a late-stage fluid.

analog data (analogue data) Data which are recorded continuously, as opposed to digital data which are recorded by discrete (digital) sampling.

analogous structures Features with similar functions which have developed independently in unrelated taxonomic groups, in response to a similar way of life, or similar method of locomotion, or similar food source, etc. Thus the wings of birds and insects are analogous.

analogue image In *remote sensing, an image in which continuous variation in the scene being sensed is represented by continuous variation in image tone, such as the image produced from photosensitive chemicals in a photographic film. *Compare* DIGITAL IMAGE.

analyser A piece of *Polaroid in a transmitted- or *reflected-light microscope which may be inserted into the light path above the *mineral section. When the analyser is out of the path of light through the microscope observations are made in *plane-polarized light (PPL); when the analyser is in the path of light, observations are made in crossed polarized light, or *crossed polars (crossed nicols, xpols, or XPL). *See* NICOL PRISM.

anamniotic Applied to a type of development typical of lower *vertebrates (*fish and *amphibians). The egg lacks a shell and protective embryonic membranes: consequently it must be laid in water or in a suitably damp environment. *Compare* AMNIOTIC.

Ananke (Jupiter XII) One of the lesser satellites of *Jupiter, with a diameter of 20 km; its orbit is *retrograde.

Anapsida Group of *reptiles characterized by a skull that lacks apertures in the temple regions behind the eyes. This was the condition in the earliest reptiles, the *Cotylosauria or 'stem reptiles'. Four groups flourished in the late *Palaeozoic: the *Captorhinomorpha; the *Mesosauria; the *Pareiasauridae; and the *Procolophonia, which survived into the *Triassic. The only living representatives are the turtles (Chelonia).

Anaspida (anaspids; superclass *Agnatha) Extinct order of fish-like *vertebrates. The body, including the head region, was covered in dermal scales (*see* DERMAL BONE), and was less flattened than in the *Osteostraci. The tail was hypocercal (tilting downwards). Anaspids were small (up to 15 cm in length). Genera included *Jaymoytius*, *Pharyngolepis*, and *Pterygolepis*. Anaspids ranged from the Late *Silurian to the Late *Devonian.

anastomosing channel *See* ANABRANCHING CHANNEL.

anatase (octahedrite) *Mineral, TiO_2; sp. gr. 3.9, *hardness 5.5–6.0; *tetragonal; yellow, or shades of brown, blue, or black; white *streak; *vitreous *lustre; *crystals normally *bipyramidal octahedra, but sometimes *tabular; *cleavage perfect, basal, and *pyramidal; occurs as an *accessory mineral in *metamorphic and *igneous rocks, also in *hydrothermal veins and *granite *pegmatites.

anatexis The partial or incomplete melting of a rock in response to an increase in temperature (at constant pressure), or a drop in pressure (at constant temperature). Melting takes place along grain boundaries, and the *melt can either be extracted from the partially molten rock system, or can remain within the system. Typical examples of anatexis would be the generation of granitic melts (*see* GRANITE) by partially melting aluminous crustal rocks, and the generation of *basalts by partially melting *mantle *peridotite.

Anatolepis heintzi Possibly one of the earliest jawless fish, known from a number of associated scales and fragments discovered in Lower *Ordovician (*Arenig) strata at Spitzbergen.

anchialine Applied to flooded caves lying deep below the ground surface. They form in limestone environments and from volcanic tubes.

anchimetamorphism A *metamorphic grade in sedimentary rocks where changes due to *diagenesis are overtaken by the very earliest phases of *metamorphism.

anchor 1. Rock anchors are long bolts or cables with one end *grouted into a *drill hole and with a plate and nut on the exposed end. These can carry considerable *loads, although slow *failure of the rock will lessen the support. **2.** Soil anchors may be used in sediments where the material is strong enough to provide sufficient reaction to the load. Holes must be drilled and the anchor installed and grouted quickly, as soil around the hole may crumble and reduce the strength of the bond.

ancient biomolecule Any molecule produced by an organism during its lifetime which persists post mortem. Identification and characterization of an ancient biomolecule may allow inferences to be drawn about the organism in which it was produced, i.e. the type of organism or evolutionary relationships. The types of molecule studied include nucleic acids, lipids, proteins, and complex polysaccharides such as lignin.

Ancient Cratered Terrain *See* MARTIAN TERRAIN UNITS.

andalusite (chiastolite) A member of the Al_2SiO_5 (or $Al_2O_3.SiO_2$) group of *polymorphic minerals; the other members of this group are *kyanite and *sillimanite. The mineral *mullite ($3Al_2O_3.2SiO_2$) is related to this group of minerals; sp. gr. 3.2; *hardness 6.5–7.5; *orthorhombic; pink or red, sometimes grey-brown and green, transparent varieties used as *gemstones; *vitreous *lustre; *crystals *prismatic with a pseudotetragonal form which has a characteristic cross-section, can also be *massive; *cleavage at right angles, distinct {110}; occurs principally in thermally *metamorphosed rocks that show alteration to aggregates of white *mica, in low-grade, regionally metamorphosed rocks, and also in *granites and *pegmatites together with *corundum, *tourmaline, *topaz, and other minerals. The name is derived from Andalusia in Spain.

Andean orogenic belt Part of the active Cordillera Mountain Belt on the eastern margin of the Pacific Ocean, the Andes extend approximately 10 000 km from the Caribbean Sea to the Scotia Sea. They developed during the *Mesozoic and *Cenozoic Eras, largely as a result of the *subduction of normal oceanic *lithosphere. Unlike the N. American Cordillera, very few *allochthonous *terranes have been recognized, and there have been no collisions of major continental masses as in the Himalayas. These differences have led to the concept of a distinctive 'Andean-type orogeny'.

andesine *See* PLAGIOCLASE FELDSPAR.

andesite Fine-grained volcanic rock (named after the Andes mountains) characterized by the presence of *plagioclase feldspars of the *oligoclase–*andesine end of the range, and of some combination of *augite, *orthopyroxene, and *hornblende. In chemical and mineralogical terms andesites are similar to *diorites. The feldspars frequently occur as strongly zoned *phenocrysts. Glassy andesites are rare. An increase in the *quartz percentage marks the gradation of andesites into *dacites. An increase in the percentages of *alkali feldspars marks the change from andesites into *trachyandesites, some of which (e.g. mugearite and hawaiite) exhibit basaltic features, including a higher *olivine content. *See* BASALT.

andhis Local term, used in the north-west of the Indian subcontinent, for dust storms accompanying violent squalls, caused by strong convection.

Andino-type margin *See* ACTIVE MARGIN.

Andisols An order of volcanic soils, deep and light in texture, that contain compounds of iron and aluminium.

andradite Member of the *garnet group of *minerals, $Ca_3Fe_2(SiO_4)_3$; sp. gr. 3.75; *hardness 6.5–7.5; *cubic; yellow, greenish, or brownish-red, to black; greasy to *vitreous *lustre; most *crystals dodecahedral; occurs widely in *metamorphic and *igneous rocks often as a result of *metasomatism, also in beach *sands and *alluvial *placers. Transparent crystals are used as *gemstones, and the mineral can be used as an abrasive.

anemometer An instrument for measuring the speed of flow in a fluid. Meteorological anemometers measure wind speed. Most do so either from the rotation of a wind-driven turbine ('rotating cups anemometer') or from the wind pressure through a tube aligned by a vane to point into the wind ('pressure-tube anemometer').

anemometry The measurement of the speed of flow in a fluid.

aneroid barometer Instrument for measuring atmospheric pressure as represented by the height (or thickness) of a sealed, flexible, metal, concertina-shaped drum from which air has been partially evacuated. Variations in the height of the drum are transmitted to a pressure scale either mechanically, through a series of levers and a pointer, or electronically.

Angara A continental mass of Asia, China, and the Far East that existed during the *Palaeozoic. In the latest Palaeozoic it became joined to *Euramerica along the line of the present Ural Mountains.

Angaraland Name given by the Austrian geologist Eduard *Suess (1831–1914) to a small *shield in north-central Siberia, where *Precambrian rocks are exposed, and which was considered to be the nucleus for subsequently developed structural features in Asia.

angiophyte A plant belonging to the evolutionary line leading to the *angiosperms. Angiophytes may have begun to diversify in the Late *Triassic, become rarer during the *Jurassic, but radiated vigorously in the Early *Cretaceous.

angiosperm A flowering plant, distinguished by producing seeds that are enclosed fully by fruits. Angiosperms are the most highly evolved of plants, and appear first in rocks of Lower *Cretaceous age. All of them are included in the division Anthophyta.

Angiospermae Formerly, 1 of the 2

divisions of the Spermatophyta (seed plants), the other being the Gymnospermae. The term is now used only informally (*angiosperm).

angle of draw The angle between the end of an underground working and the point on the ground surface to which *subsidence, due to that working, may extend. Draw usually proceeds at an angle of 65–75° to the horizontal.

angle of incidence (incident angle) The angle between an incident ray or wave and the normal to a reflecting or refracting surface. The angle of incidence is related to the *angle of refraction by *Snell's law. *See also* CRITICAL ANGLE.

angle of internal friction (friction angle) A measure of the ability of a unit of rock or soil to withstand a *shear stress. It is the angle (ϕ), measured between the normal force (N) and resultant force (R), that is attained when failure just occurs in response to a shearing stress (S). Its tangent (S/N) is the coefficient of sliding friction. Its value is determined experimentally.

angle of reflection The angle between a reflected ray or wave and the normal to a reflecting surface.

angle of refraction The angle between a refracted ray or wave and the normal to a refracting surface. The angle of refraction is related to the *angle of incidence by *Snell's law. A ray or wave entering a medium in which its speed is higher than in the medium it is leaving will deviate away from the normal, whereas when it enters a medium of lower speed it will deviate towards the normal.

angle of repose Maximum angle at which unconsolidated material can stand. Various factors determine this angle, including particle size and angularity, the degree of interlocking between particles, and *pore-water pressure. A typical angle of repose for coarse scree is 32–36°.

angle of shearing resistance (internal angle of friction, angle of frictional resistance; ϕ) Approximate angle of repose for clean *sand; it reduces with moisture content and is zero for a sheared, saturated *clay. For rocks, a rough determination of the angle may be made by placing two small blocks of material together and inclining them until the top one slides.

anglesite *Secondary lead mineral, $PbSO_4$; sp. gr. 6.3–6.4; *hardness 3; *orthorhombic; normally colourless to white, but sometimes yellow, grey, or coloured by a bluish tinge; white *streak; *adamantine *lustre; *crystals *tabular, *prismatic, or pyramidal, or occurring as *massive, granular, or compact aggregations; normally occurs in the oxidized zone of lead deposits, and often surrounds a core of *galena.

Anglian 1. A *glacial *stage of the middle *Pleistocene in Britain. **2.** A middle-*Pleistocene, cold-climate series of deposits. There were a number of glacial advances in East Anglia (England), which are difficult to understand. Near Lowestoft there are two *till sheets separated by stratified sands. The lower, North Sea Drift, contains Scandinavian *erratics, and the upper, thicker till, contains erratics of *Jurassic or *Cretaceous material. The highest terraces of the Thames may belong to this period and perhaps could also be correlated with the *Elsterian deposits of Europe.

ångstrom (Å) Unit of length equal to 10^{-10} m, used in measuring *electromagnetic radiation including visible light and X-rays. Replaced in SI units by the nanometre (10 Å = 1 nm).

angularity number *See* AGGREGATE TESTS.

angular momentum The momentum of a body rotating in a plane around a point. It is formally the product of the mass of the body, the radius of the orbit, and the square of the angular velocity ($mr\omega^2$). Because of its rotation about its axis, the Earth has rotational angular momentum, and orbital angular momentum on account of its annual revolution around the Sun. The angular momentum of the Earth–Moon system (3.45×10^{34} rad/kg/m^2/s) is the sum of the rotational angular momentum of the Earth and the Moon's orbital angular momentum, and is high compared to that of the other terrestrial planets.

angular shear strain A measure of the angular rotation from two mutually perpendicular reference axes following simple *shear. The shear strain is given as the tangent of angular shear ψ, and may be positive or negative with respective clockwise or anticlockwise rotations in relation to the reference perpendicular.

angular unconformity A *discordant surface of contact between the deposits of two episodes of sedimentation in which the older, underlying *strata have undergone folding, uplift, and *erosion before the deposition of the younger *sediments, so that the younger strata truncate the older. *See also* UNCONFORMITY.

anhedral (allotriomorphic) A morphological term referring to grains in *igneous rocks which have no regular crystalline shape. Anhedral forms are developed when a *crystal's free growth in a *melt is inhibited by the presence of surrounding crystals. The shape of the growing crystal is thus controlled by the arrangement and orientation of the surrounding pre-existing crystals. *Compare* EUHEDRAL.

anhydrite Mineral sulphate, $CaSO_4$, found in *evaporite deposits; sp. gr. 2.9–3.0; *hardness 3.0–3.5; *orthorhombic; normally white to colourless, sometimes with a bluish tinge, but occasionally grey and reddish; white *streak; *vitreous to pearly *lustre; *crystals rare, the mineral being usually *massive, granular, and fibrous; three *cleavages at right angles, perfect {010}, good {100}, {001}; occurs with *gypsum and *halite, deposited directly from sea water at temperatures in excess of 42 °C, can form from the dehydration of gypsum; may occur as a *cap rock above *salt domes, and as a minor *gangue mineral in *hydrothermal veins. It is used as a raw material for cements.

anhysteretic magnetization The magnetization acquired when a direct magnetic field is applied to a specimen which is simultaneously held in an alternating magnetic field. *See* ALTERNATING MAGNETIC FIELD DEMAGNETIZATION.

Animalia (Metazoa) Multicellular organisms that develop from embryos; one of the three kingdoms of multicellular organisms (the other two being *Fungi and *Plantae). The kingdom includes all animals other than protozoons (some of which are colonial); *Porifera (sponges) are sometimes excluded because their structure differs markedly from that of other animals. Animals first appeared in the *Precambrian, the Porifera from one kind of *protist forebear, and all other animals from another (or possibly more than one other) protist. The oldest fossils are burrows of a *coelomate in rocks rather less than 700 Ma old.

Animikian A *system of the *Palaeoproterozoic dated at about 2225–1700 Ma (Harland et al., 1989).

anion A negative ion, i.e. an atom, or complex of atoms, that has gained one or more electrons and thereby carries a negative electric charge, e.g. Cl^-, OH^-, and SO_4^{2-}. So-called because when an electric current is passed through a conducting solution the negative ions present in the solution are attracted to the anode (the positive electrode). *Compare* CATION.

Anisian 1. A Middle *Triassic *age, preceded by the *Scythian Epoch, followed by the *Ladinian Age, and dated at 241.1–239.5 Ma (Harland et al., 1989). **2.** The name of the corresponding European *stage, which is roughly contemporaneous with the Guan Ling (China) and *Etalian (New Zealand).

Anisograptidae *See* GRAPTOLOIDEA.

anisometric growth A change in the ratio between the sizes of two parts of an organism during its *ontogeny, such that if an organism grows anisometrically its shape will change.

anisomyarian *See* MUSCLE SCAR.

anisotropic 1. Applied to substances whose optical or other physical properties vary according to the direction from which they are observed. In optical mineralogy the term is applied to *minerals other than *cubic that split the light passing through them into two *vibration directions, each with a different velocity (provided the light is not travelling along an *optic axis), as a result of the light being doubly refracted (*see* DOUBLE REFRACTION). In addition, each beam is refracted differently for different colours of light. **2.** In engineering geology, anisotropy refers to a rock whose engineering properties vary with direction. For example, *schist, a highly anisotropic rock, has a compressive strength that varies depending upon the orientation of the *foliation to the applied load (*see* COMPRESSIVE STRESS).

anisotropic meter Any instrument that measures differences of particular parameters in different directions. Specific instruments measure *electrical conductivity, *seismic waves, and low- and high-field magnetic *susceptibility.

anisotropy The variations in the physical properties of a medium that depend on the direction in which they are measured. *Compare* INHOMOGENEITY; and ISOTROPY.

ankerite *Mineral, $Ca(Mg.Fe)(CO_3)_2$; sp. gr. 2.9–3.2; *hardness 3.5–4.0; *trigonal;

yellowish-brown, sometimes white, yellow, or grey; white *streak; *vitreous *lustre; *crystals can be rhombohedral, but ankerite also occurs *massive and granular; *cleavage rhombohedral, perfect {1011}, faint {0221}; occurs as a *gangue mineral with iron ores, and as fillings associated with coal seams, and in an environment similar to that of *dolomite. It is a form of ferroan dolomite where ferrous iron substitutes for Mg in a *solid solution series from dolomite ($CaMg(CO_3)_2$) to ankerite ($(CaMg_{0.75}Fe_{0.25})(CO_3)_2$). Ankerite often forms as a *cement by diagenetic reaction (*see* DIAGENESIS) from ferroan *calcite at burial depths of about 2.5 km. It is named after the Austrian mineralogist M. J. Anker. *See also* CARBONATITE.

***Ankylosaurus* (ankylosaurs)** Genus of *Cretaceous *dinosaurs, heavily armoured with bony plates, rather in the manner of armadillos. They had small heads with insignificant teeth, and some forms were toothless.

ankylosaurs *See ANKYLOSAURUS.*

annealing *See* HOT WORKING.

Annelida Phylum of *coelomate worms which possess a definite head and good *metameric segmentation. The body is elongate, and each segment has bristles. Their fossils are found in rocks dating from the *Cambrian, and possible fossil annelid worms are known from *Precambrian sediments in S. Australia. They are represented today by the earthworms, sandworms, and leeches.

annual snow-line *See* FIRN LINE.

annulus (planetary) A ring-like structure used, for example, to describe a discrete ring of ejecta around a martian *rampart crater.

anode A positive electrode. *See* ANION.

anomaly *See* GEOBOTANICAL ANOMALY; GEOCHEMICAL ANOMALY; GRAVITY ANOMALY; HEAT FLOW ANOMALY; and MAGNETIC ANOMALY.

anomphalous *See* IMPERFORATE.

anorogenic Applied to **1.** a feature that was not formed as a result of tectonic disturbance, i.e. in an interval between orogenic events, and **2.** such an interval.

anorthite *See* ALKALI FELDSPAR.

anorthoclase *See* ALKALI FELDSPAR.

anorthosite *Plutonic rock composed almost entirely of *plagioclase feldspar (90% or more) which may be oligoclase, andesine, labradorite, or bytownite. These rocks occur most frequently in *Precambrian *shield areas as non-stratiform intrusions or in layered complexes. Anorthosite also constitutes the *lunar highlands where it is largely composed of the calcium *end-member anorthite.

anoxic The condition of oxygen deficiency or absence of oxygen. Anoxic *sediments and anoxic bottom waters are commonly produced where there is a deficiency of oxygen due to very high organic productivity and a lack of oxygen replenishment to the water or sediment, as in the case of stagnation or stratification of the water body.

Antarctic air Very cold, continental polar *air mass, which originates over the frozen surface of Antarctica and the surrounding pack-ice, and over the coldest waters of the *Antarctic Ocean. As this air moves northward over warmer waters it becomes convectively unstable.

Antarctic bottom water (AABW) A dense bottom-water mass, formed in the Weddell and Ross Seas, which moves in an easterly direction around Antarctica under the influence of the deep-reaching, wind-driven, surface *Antarctic Circumpolar Current (West Wind Drift). It is typified by a *salinity of 34.66 parts per thousand and a low temperature (−2 °C to −0.4 °C). The high salinity and dense nature is caused by the removal of pure water as sea ice.

Antarctic Circumpolar Current (West Wind Drift) The largest and most important ocean current in the southern hemisphere. It flows in an eastward direction around Antarctica, and occupies a wide tract of water in the S. *Pacific, S. *Atlantic, and *Indian Oceans. There is very little separation between surface- and *bottom-water circulation within this area. The current is remarkably constant and is characterized by low *salinity (less than 34.7 parts per thousand, and cold waters (−1 to 5 °C). It is the only current which flows right around the world.

Antarctic convergence (AAC) A convergence line in the seas that circle Antarctica between latitudes 50° S and 60° S. It is where the cold waters from the Antarctic region meet and sink beneath the warm waters from the middle latitudes, so forming the *Antarctic intermediate water. This convergence line is now called the Antarctic polar front.

Antarctic front Frontal boundary between cold Antarctic air and warmer air, usually lying to the north of it.

Antarctic intermediate water (AIW) A water mass formed at the surface near to the *Antarctic convergence, at about 50° S. It is typified by a low *salinity (33.8 parts per thousand) and low temperature (2.2 °C). As it spreads northwards it sinks to depths of 900 m and can be traced even in the North Atlantic at 25° N.

Antarctic meteorites *Meteorite finds from *ablation zones on Antarctic glaciers, which represent about 25% of all the meteorites so far discovered. They vary in composition but are usually quite fresh and may have fallen from one million years ago to the present day. They are therefore of great importance in the study of meteorites. Antarctic meteorites include the first meteorites known to have been derived from the *Moon and *Mars.

Antarctic Ocean (Southern Ocean) Term used to describe the oceanic waters surrounding Antarctica. It extends northwards to about 40° S latitude, the limit of the northward drift of ice from the Antarctic region, where there is a marked change in water temperature and *salinity. The Antarctic Ocean is typified by low water temperatures (–1.8 °C to 10 °C).

Antarctic Plate A present-day major lithospheric plate that extends beyond the Antarctic continent to the surrounding *constructive plate margins.

Antarctic polar current A surface current which flows in a westward direction around Antarctica under the influence of easterly winds blowing off the ice cap.

Antarctic polar front *See* ANTARCTIC CONVERGENCE.

ante- From the Latin *ante*, meaning 'before', a prefix meaning 'preceding' or 'previous'.

antecedent drainage The hypothesis that a stream crossing a geologic structure may be older than that structure. Such a stream is believed to maintain its course across a developing *fold or *fault. This is a favoured explanation for streams crossing structures in geologically active areas, as do the Arun in the Himalayas, and the Grand Canyon section of the Colorado in the Rockies.

Anthocyathea *See* IRREGULARES.

anthophyllite *End-member $(Mg,Fe)_2(Mg,Fe)_5[Si_4O_{11}]_2(OH,F)_2$, along with gedrite, of the calcium-poor, orthorhombic *amphiboles. In anthophyllite Mg > Fe, whereas in gedrite Mg < Fe. It is found exclusively in *metamorphic rocks, where it may occur with *cordierite. Amosite is the asbestiform variety of anthophyllite.

Anthophyta *See* ANGIOSPERM.

Anthozoa (sea-anemones, corals, sea pens; phylum *Cnidaria) A class of exclusively polyploid, marine cnidarians. They probably first appeared in the *Ordovician although there are possible records for some groups in the *Cambrian. They are solitary or colonial and usually sedentary. The oral end is expanded as an oral disc with a central mouth that has one or more rings of hollow tentacles. Anthozoans have a well-developed stomodeum (gullet) leading from the mouth to the enteron (gastrovascular cavity). The interior of the enteron is divided by *mesenteries (infoldings of the gut wall). Those members of the class that secrete hard, calcareous skeletons are important in the geologic record from the *Palaeozoic onwards, and at some levels form true coral *reefs. In the Palaeozoic they are often associated with other organisms, e.g. stromatoporoids (*Stromatoporoidea), to produce organic build-ups or reef mounds. *See* HETEROCORALLIA; OCTOCORALLIA; RUGOSA; SCLERACTINIA; TABULATA; and ZOANTHARIA.

anthracite A type of *coal, relatively hard, jet black, with a metallic *lustre, sub-*conchoidal fracture, unbanded, with less than 10% *volatiles and more than 90% carbon. It burns with intense heat and a nonluminous flame.

Anthropogene *See* QUATERNARY.

anthropogenic Applied to substances, processes, etc. of human origin, or that result from human activity.

anthropogeomorphology The study of those land-forms and processes that are a direct result of human activity, including accelerated *erosion, channelized river channels (i.e. rivers made to flow along fixed, sometimes concrete-lined, channels), the melting of *permafrost, and ground subsidence due to the extraction of water or minerals. Particular examples include the Norfolk Broads, England, which are essentially flooded peat quarries, and the Zuider Zee, whose damming has had a major impact on the coastal morphology of the Netherlands.

Anthropoidea (Simiiformes; cohort Unguiculata, order Primates) Suborder comprising the monkeys, apes, and humans. Monkeys and apes have a common ancestor and diverged in the *Oligocene. The dryopithecines of the succeeding *Miocene were undoubted apes. The traditional palaeontological view is that these Miocene apes gave rise in turn to three new lines, one leading to the gibbons, another to the great apes, the third to humans. However, on the basis of anatomical characteristics and genetic criteria, it has long been maintained that ancestors of both the gibbons and the orang-utan diverged from the ancestral line of the advanced Primates at an early date and that only subsequently did that line split to give one group comprising humans and another comprising the gorilla and the chimpanzee; recent palaeontological evidence now tends to support this second view.

anti- From the Greek *anti*, meaning 'against', a prefix meaning 'against' (in the sense of opposed to), 'opposite', or 'preventing'.

Antian The Lower *Pleistocene, temperate, marine deposits that form the upper section of a tripartite division of deposits revealed in a *borehole at Ludham, in eastern England. *See also* BAVENTIAN; LUDHAMIAN; PASTONIAN; and THURNIAN.

anticlinal trap A *fold structure with an arch of non-porous rock overlying porous strata (*reservoir rock), providing a trap in which oil, gas, or water may accumulate. In Middle East oilfields, large, upright folds occur in thick, competent *limestones which extend for many kilometres; fracturing along the crests of the *anticlines increases *permeability in the reservoir rocks. *See* PETROLEUM; and NATURAL GAS. *Compare* FAULT TRAP; REEF TRAP; STRATIGRAPHIC TRAP; STRUCTURAL TRAP; and UNCONFORMITY TRAP.

anticlinal valley Valley developed along the axis of an *anticline. Inward-facing *escarpments are developed where the upper beds of the anticline are relatively resistant, while softer lower materials occupy the valley floor. It is a common land-form in gently folded strata of varying resistance, such as occur in southern Britain.

anticline Arch-shaped fold in rocks, closing upwards, with the oldest rocks in the core.

anticlinorium A regional antiformal structure (*see* ANTIFORM) composed of a series of smaller, higher-order *anticlines and *synclines, some of which may be small enough to be viewed in *outcrop.

anticoincidence circuit A device to minimize errors that may occur when measurements are made to date radiocarbon samples. These measurements must be extremely accurate due to the very low level of activity (*see* RADIOCARBON DATING). The error quoted on a radiocarbon age determination is solely an error in counting statistics. Such errors may rise from spurious counts generated by contamination of the sample, cosmic activity detected by the counter, and radioactive contaminants in the equipment being used. Initially the counter was shielded by surrounding it with large amounts of iron, lead, distilled mercury, or paraffin wax mixed with boric acid. An anticoincidence circuit is an alternative to material absorbers, and consists of a series of tangentially placed *Geiger tubes operated in anticoincidence (i.e. they do not require input signals to arrive within specified intervals in order to be activated). These are positioned within an iron shield, and around the central counting chamber. Radiation from outside, or from within the shield, is detected by this ring of geiger tubes and can be discounted. Special counters have now been developed in which the anticoincidence counters are built into the same tube as the main counter, so that the same gas is used in the whole system. The wall of such a counter usually consists of a polystyrene foil covered on both sides with aluminium. This is then surrounded by a ring of wires forming the *anode for the anticoincidence circuit.

anticyclogenesis Process whereby an *anticyclone or a ridge of high pressure is formed and developed.

anticyclolysis Process whereby an *anticyclone or ridge of high pressure is dissipated or weakened.

anticyclone Area or system of high atmospheric pressure that has a characteristic pattern of air circulation, with subsiding air and horizontal divergence of the air near the surface in its central region. Winds are generally light because of small pressure gradients; they flow clockwise in the northern hemisphere and anticlockwise in the southern hemisphere. A *temperature inversion is common at the base of the air subsidence, and this restricts the vertical development of cloud. Weather conditions are generally settled. Cold anticyclones form over continents in winter and over polar areas at any time, accompanied by strong inversions: in the clear air, pronounced frosts and very cold surface conditions result. Warm anticyclones (so called because of the warm, subsided air aloft) over land areas typically bring spells of settled and often warm weather. *See also* ANTICYCLONIC GLOOM.

anticyclonic gloom Condition of low visibility associated with *anticyclones accompanied, in the colder months of the year, by well-developed *temperature inversions that can trap dust and other pollutants, and often have *radiation fog in the lower layers. The stability of the high-pressure system can make the resultant reduced visibility very persistent and may establish *smog conditions.

antidune The *sediment *bedform generated by fast, shallow *flows of water with a *Froude number greater than 0.8. Antidunes form beneath *standing waves of water that periodically steepen, migrate, and then break upstream. The antidune bedform is characterized by shallow *foresets (*see* CROSS-STRATIFICATION) which *dip upstream at an angle of about 10°. Their preservation potential is low, but they can be identified by low-angle (less than 10°) foresets, dipping up-current. Normally antidunes show a close association with upper-flow-regime *plane beds.

antiferromagnetic Applied to a *ferromagnetic (in the wide sense) substance in which the magnetic lattices are magnetized in exactly equal and opposite directions. Such a substance does not have an external magnetic field in its pure form, but a distorted lattice may result in a *parasitic magnetization.

antiform Arch-shaped rock structure which, by definition, closes (i.e. arches) upward, but in which it may not be possible to determine the oldest rocks. It is frequently observed in complex orogenic regions.

antigorite *See* SERPENTINE.

antimonite *See* STIBNITE.

antimony, native Metallic element, Sb; soft, whitish; *crystals *tabular *hexagonal; *cleavage basal; occurs native in association with *stibnite (antimonite).

antimony glance *See* STIBNITE.

antiperthite *See* HYPERSOLVUS GRANITE; and PLAGIOCLASE FELDSPAR.

antithetic fault A *fault, usually one of a set, which in vertical section shows a sense of *slip opposite to that of the major fault from which it originates. The term derives from the Greek word *antithethemi*, meaning 'set against'.

antitrade Upper wind in low latitudes that flows counter (i.e. poleward) to the *trade wind below.

anvil Name commonly applied to the tops of those *cumulonimbus clouds that spread in the vicinity of the *tropopause into a characteristic anvil-like shape with fibrous edges. The rising column of air in the cumulonimbus cloud is checked by the stable stratification of the air above the tropopause.

apatite A widely distributed *phosphate *mineral, with the formula $Ca_5(PO_4)_3(F,Cl,OH)$; sp. gr. 3.1–3.3; *hardness 5; *hexagonal; usually a shade of green or grey-green, but may also be white, brown, yellow, bluish, or red; white *streak; *vitreous *lustre; *crystals commonly hexagonal *prisms, and often *tabular, also occurs *massive, and granular; *cleavage, basal {0001}, imperfect prismatic {1010}; found as an *accessory mineral in *igneous rock, in *pegmatites and high-temperature *hydrothermal veins, and in *metamorphic rocks. It is the principal constituent of fossil bones (*see also* COLLOPHANE). Apatite is widely used as a phosphate fertilizer, and for the production of phosphoric acid and various other chemicals.

Apatosaurus One of several gigantic *saurischian *dinosaurs recorded from the Upper *Jurassic. *Apatosaurus* was a sauropod, a quadruped with a long neck, the total weight of which reached 30 tonnes. The name *Apatosaurus* is a senior synonym of *Brontosaurus* (i.e. it was the earlier of the two names). Animals of 22 m in length have been recorded from the Morrison Formation of N. America.

ape A name originally (in medieval times) applied to the Barbary macaque (*Macaca sylvanus*) of North Africa (as were the Latin word *simia* and Greek *pithecus*) and, by extension, applied to other primates as these were made known in Europe. As long-tailed monkeys ('tailed apes', or cercopitheci) became better known, 'ape' came to mean primarily 'tailless ape', and today commonly denotes a member of the *Hominoidea, comprising lesser apes (gibbons) and great apes (orang-utan, gorilla, chimpanzee, and, in some usages, human).

aperture 1. (window) The portion of a data record which is selected for specifying operators for use on the data set, e.g. operators such as *autocorrelation functions and *filters are applied by apertures on *seismic records. **2.** In a mollusc (*Mollusca) shell, the opening through which the soft parts of the animal emerge. It is often a simple, circular opening, but is modified in some genera. In gastropods (*Gastropoda), where the aperture is circular or elliptical it is said to be 'holostomatous' or 'entire'; where it is notched (to accommodate a *siphon) it is said to be 'siphonostomatous'. In cephalopods (*Cephalopoda), the aperture may be indented or notched at the ventral margin by a hyponomic sinus, which houses the *hyponome. Some compressed ammonites (*Ammonoidea) possess a pair of lateral shell extensions (lappets) on either side of the aperture.

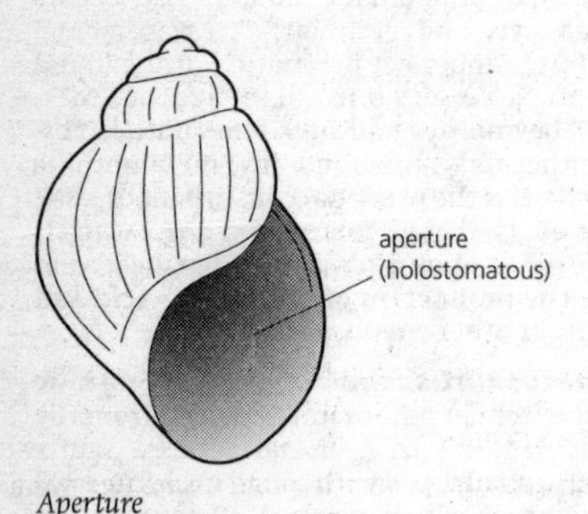

Aperture

apex The first-formed end of a shell, which is usually pointed. The term is most commonly applied to gastropod (*Gastropoda) shells.

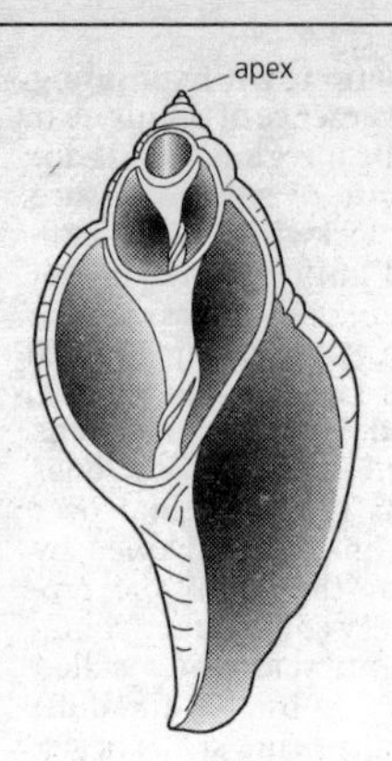

Apex

APF *See* ABSOLUTE POLLEN FREQUENCY.

aphanitic An *igneous rock *texture characterized by mineral grains which are too small to be identified without a petrological microscope. These extremely fine-grained, crystalline *fabrics are formed when a *magma solidifies in response to a very rapid loss of heat and dissolved gases. Emplacement in high-level *dykes or eruption on to the surface can result in the development of aphanitic fabrics.

Aphebian A *stage of the lowermost *Proterozoic of Canada, overlain by the *Helikian.

aphelion Point in the Earth's elliptical orbit at which the planet is farthest from the Sun. *See also* PERIHELION.

aphyric An *igneous rock *texture characterized by a fine-grained *aphanitic *groundmass and by an absence of any *phenocrysts. Aphyric texture forms by the rapid crystallization of *melts lacking large suspended *crystals, and thus these melts must have been very close to their *liquidus temperature (the temperature at which initial crystallization takes place in a cooling melt).

aplite A light-coloured, fine-grained, equigranular *igneous rock composed of *subhedral to *anhedral grains of *quartz and *alkali feldspar, and found as late-stage veins in *granite bodies. The quartz–alkali feldspar composition corresponds to the lowest temperature *melts in granite *magma systems, suggesting that they are residual melts formed by the differentiation of granite *magma (*see* MAGMATIC DIF-

FERENTIATION). The lack of any hydrous minerals and the fine *grain size points to the aplites crystallizing from dry residual melts.

apodeme *See* EXOSKELETON.

apogee The point in the orbit of the Moon (or any artificial satellite), that is most distant from the Earth. *Compare* PERIGEE.

Apollo **1.** A *solar system asteroid (No. 1862), diameter 1.6 km; approximate mass 2×10^{12} kg; rotational period 3.063 hours, orbital period 1.81 years. Its orbit crosses that of earth. **2.** The name of the *NASA manned lunar programme that ran from 1963 to 1972.

apomorph Evolutionarily advanced ('derived') character state. The long neck of the giraffe is apomorphic; the short neck of its ancestor is *plesiomorphic.

apomorphic Applied to features possessed by a group of biological organisms that distinguish those organisms from others descended from the same ancestor. The term means 'new-featured' and refers to 'derived' characters that have appeared during the course of evolution.

apophysis (pl. apophyses) **1.** An irregular or sheet-like *vein or *dyke which originates from a larger *igneous rock body. **2.** *See* VERTEBRA.

Appalachian orogenic belt A 3200 km long, *Palaeozoic, *orogenic belt extending from Newfoundland to Alabama and interpreted as the consequence of the closure of the *Iapetus Ocean. Extensions of the belt along *strike include the Caledonian orogenic belt (*see* CALEDONIAN OROGENY) in north-western Europe, now separated by the Atlantic Ocean. The deformation in the Appalachian belt ranges from late *Precambrian to *Permian, divided into four major *orogenies (*Avalonian, *Taconic, *Acadian, and *Alleghanian), with the transport of *thrust-*nappes predominantly northwestwards. *Seismic reflection profiling by *COCORP has supported, at least for the southern Appalachians, recent interpretations of a thin-skinned tectonic style in which a few kilometres' thickness of highly deformed material has been thrust westward for at least 200 km.

apparent age *See* ABSOLUTE AGE.

apparent cohesion Cohesion of grains due to surface tension in the surrounding pore water.

apparent conductivity (σ_a) The inverse of *apparent resistivity; the units are *siemens/metre.

apparent dip The dip of the trace of a plane, which is measured from the horizontal in any section non-perpendicular to the *strike of the plane. *Compare* TRUE DIP. *See also* DIP, ANGLE OF.

apparent resistivity (ρ_a) A measurement of resistivity which is calculated as the product of the measured resistance (R) and a *geomagnetic factor (K_g) such that $\rho_a = K_g R$, in units of ohms/m. It is important to note that the apparent resistivity is not an explicit measurement of the electrical resistivity of a material and should not be interpreted as being diagnostic of a given material.

apparent velocity (v_a) The velocity which a wave-front exhibits along a line of *geophones. If the wavefront approaches the geophone *array at an angle θ, then the true velocity of the wavefront, v, is given by $v = v_a \cos\theta$. In *refraction seismology, v_a is the reciprocal of the gradient of a straight-line segment of a *time–distance graph.

apparent wavelength (λ_a) If a wave train approaches a line of *geophones at an angle θ, then the distance between corresponding points on successive cycles of the wave form as detected by those geophones is an apparent and not a true wavelength. True wavelength $\lambda = \lambda_a \sin\theta$.

apsacline A condition of the interarea of a strophic brachiopod (*Brachiopoda) shell where the inclination of the *cardinal area relative to the plane of *commissure is 90–180°. It is one of the most commonly occurring conditions. *Compare* ANACLINE.

aptation A character that suits its possessor to its *environment; it may be an *abaptation, *adaptation, or *exaptation.

Aptian **1.** An Early *Cretaceous *age, preceded by the *Barremian, followed by the *Albian, and dated at 124.5–112 Ma (Harland et al., 1989). **2.** The name of the corresponding European *stage, which is named after Apt in France. *See also* NEOCOMIAN.

aptychus A calcitic plate associated with *Mesozoic *ammonites. Normally these plates occur in pairs. Aptychi are shaped like bivalves (*Bivalvia) and have an ornamented outer surface. Since they have been found inside the aperture of ammonoid shells they were originally interpreted as *opercula, protecting the ammonoid body when it was withdrawn into the shell. It has now been shown that aptychi are probably the lower jaws of ammonoids and that the horny upper jaw is rarely preserved.

APXS *See* ALPHA-PROTON-X-RAY SPECTROMETER.

aquamarine *See* BERYL.

aquiclude (aquifuge) A rock with very low values of hydraulic conductivity which, although it may be saturated with *groundwater, is almost impermeable with respect to groundwater flow. Such rocks will act as boundaries to *aquifers and may form confining strata. *See* PERMEABILITY.

aquic moisture regime The moisture balance of humid climates and soils, where annual precipitation exceeds the combined actual evaporation and transpiration, and where the soil moisture status is normally above *field capacity.

aquifer A body of permeable rock, for example, unconsolidated gravel or sand *stratum, that is capable of storing significant quantities of water, is underlain by impermeable material, and through which *groundwater moves. An unconfined aquifer is one in which the *water table defines the upper water limit. A confined aquifer is sealed above and below by impermeable material. A perched aquifer is an unconfined groundwater body supported by a small impermeable or slowly permeable unit. *See* PERMEABILITY.

aquifer test *See* PUMPING TEST.

aquifuge *See* AQUICLUDE.

Aquitanian 1. The earliest *age in the *Miocene Epoch, preceded by the *Chattian (*Oligocene), followed by the *Burdigalian, and with a lower boundary set at 23.3 Ma (Harland et al., 1989). **2.** The name of the corresponding European *stage, which is roughly contemporaneous with the upper *Zemorrian and lower *Saucesian (N. America), parts of the *Otaian (New Zealand), and the upper *Janjukian and lower *Longfordian (Australia). The *stratotype is in the Aquitanian Basin, France. The Aquitanian is itself characterized by the appearance of the planktonic *foraminiferid *Globigerinoides primordia*.

aquitard A *rock with low values of *hydraulic conductivity, which allows some movement of water through it, but at rates of flow lower than those of adjacent *aquifers. *Compare* AQUICLUDE.

Arabian Plate One of the present-day minor lithospheric *plates, separated from the *African Plate by the spreading *Red Sea, the continuation of the *Carlsberg Ridge into the Gulf of Aden, and the Dead Sea transform system, while the boundary with the *Indo-Australian Plate is the Owen Fracture Zone; the plate is colliding with the Iran Plate.

arachnid *See* ARACHNIDA.

Arachnida (arachnids: mites, scorpions, spiders, etc.) Class of terrestrial chelicerates (*Chelicerata) which have book lungs or tracheae derived from gills, indicating their aquatic derivation. Of recent terrestrial animals, the arachnids are probably the oldest known class, scorpions having been recorded from the *Silurian Period. A Silurian scorpion, *Palaeophonus nuncius*, was perhaps the first terrestrial animal. The first fossil spiders are known from the *Devonian. The class is extremely diverse, but except in the mites the body is in two portions: the prosoma (anterior portion) which bears the four pairs of legs, the eyes, the pedipalps (second pair of appendages), and the chelicerae (first pair of appendages, usually pincer-like); and the opisthosoma (posterior portion) which contains most of the internal organs and glands. The two portions may be broadly jointed, or connected by a stalk or pedicel. The prosoma has a dorsal shield or carapace, and the opisthosoma is segmented in most orders, but not in spiders and mites and only very weakly in harvestmen. The number of eyes varies, and can be as many as 12 in some scorpions.

arachnid structure (arachnoids) A strange pattern of ridges, gathered in braids and belts, which merge with radial and concentric ridges to give a 'spiders and cobwebs' appearance on the surface of Venus, particularly in the area between Sedna Planita and Bell Regio (about 43° N, 19° E). The ridges are large, 100–200 km long and up to 20 km wide.

arachnoids *See* ARACHNID STRUCTURE.

aragonite *Mineral, $CaCO_3$; sp. gr. 2.9; *hardness 3.5–4.0; *orthorhombic; colourless, white, grey, or yellowish; white *streak; *vitreous *lustre; *crystals normally *prismatic, often *acicular, sometimes *tabular and pseudo-hexagonal; also occurs fibrous and stalactitic; *cleavage imperfect {010}; occurs in hot springs and in association with *gypsum, also in veins and cavities, and in the oxidized zone of ore deposits with other *secondary minerals. Aragonite is a *polymorph of *calcite, from which it is distinguished by its lack of cleavage and its higher specific gravity. Calcite is the more stable form of $CaCO_3$, and many fossil shells that were made originally of aragonite have either converted to calcite or undergone replacement by some other mineral. Present-day mollusc shells are formed of aragonite crystals. The name is derived from the Aragon province of Spain. *See also* CARBONATES.

aragonite mud A fine *carbonate mud, with particles less than 4 μm, composed mainly of *aragonite needles. The aragonite is generally believed to have been deposited from the break-up of calcareous green algae (*Chlorophyta).

Aratauran *See* HERANGI.

arborescent *See* DENDRITIC.

arc *See* ISLAND ARC.

archae- (arche-) Prefix, from the Greek *arkhaios* ('ancient'), itself derived from *arkhe* ('beginning'). It adds the meaning 'ancient', with the implication 'first', to words to which it is attached.

Archaea The *domain comprising what were formerly known as the archaebacteria. What used to be the kingdom Archaebacteria has been split into two kingdoms: *Crenarchaeota and *Euryarchaeota. The domain Archaea contains the *phenotypes: *methanogens, *sulphate-reducing organisms, and *extremophiles.

Archaean (Archaeozoic, Azoic) One of the three subdivisions of the *Precambrian, lasting from about 4000 to about 2500 Ma ago.

archaebacteria (domain *Archaea) Organisms belonging to the kingdoms *Crenarchaeota and *Euryarchaeota; formerly these were grouped together as the kingdom Archaebacteria.

Archaeocalamites radiatus First described by A. *Brongniart in 1828, this species is one of the earliest recorded equisetaleans or 'horse-tails', a group of plants noted for the jointed nature of the stem. At each joint there is a ring of short branches. *See* SPHENOPSIDA.

Archaeoceti (ancient whales; cohort Mutica, order Cetacea) An extinct suborder comprising the oldest and most primitive cetaceans, which flourished in the *Eocene and may have originated in Africa. Most were comparable in size with modern porpoises, had an elongated snout, and nostrils on top of the skull. The brain case was long and low. The front teeth were peg-like, the cheek teeth *heterodont and characteristic of primitive carnivores. There were 44 teeth in all. The hind legs in most were reduced to vestiges, but in some early genera (**Ambulocetus*, **Basilosaurus*) still protruded from the body wall. They were fish-eating carnivores that had adopted an aquatic life to which they were more highly adapted than, for example, modern seals. The term archaeocete really means any primitive cetacean and probably does not designate a natural, *monophyletic group.

Archaeocyatha Extinct phylum of reef-forming organisms known only from the *Cambrian. They were cup-like, usually 10–30 mm in diameter and up to 50 mm in height. In some respects they were similar to both *sponges and *corals, and may represent a true advance in evolutionary *grade over the former. It is possible that they lived in a symbiotic relationship with some *trilobites. The cause of their extinction is not known. There were two classes: *Regulares and *Irregulares.

Archaeogastropoda An order of *Gastropoda which appears in the Lower *Cambrian and also includes the extant *Patella vulgata* (limpet). Gastropods may be subdivided according to their respiratory structures; archaeogastropods, the most primitive gastropods, possess just two gills. Some forms possess a marginal slit near the *aperture to facilitate the removal of exhaled water and wastes. This may extend posteriorly for some distance but eventually it is filled with shell material as the animal grows and extends its shell. The filled slit (selenizone) is usually completely plugged. In some species (e.g. *Haliotis*) the selenizone is represented by a linearly arranged series of apertures (tremata, sing. trema).

archaeomagnetism The study of the magnetic properties of objects and materials from an archaeological context. Such studies include *magnetic dating, reconstruction of objects and structures, sourcing artefacts, past firing temperatures, etc.

Archaeopteris Early *progymnosperm found first in the *Frasnian stage of the

*Devonian Period. It was identified by its fronds, and is the earliest known representative of the Archaeopteridales.

Archaeopteryx lithographica Only five specimens of this species, the first bird, are known. *Archaeopteryx* is recorded solely from the Lithographic Limestone of the Solnhofen region of Bavaria, Germany. It was first described by H. von Meyer in 1861 and is of Middle *Kimmeridgian or Upper *Jurassic age. Recent work on this species by several palaeontologists tends to support the theory that the birds, through *Archaeopteryx*, evolved from *coelurosaur *dinosaurs similar to *Compsognathus*. The species *A. lithographica* possesses several primitive characters such as teeth, as well as specialized features such as feathers and hollow bones. It is a good example of a connecting species which exhibits a mosaic of evolutionary features.

archaeopyle *See* DINOPHYCEAE.

Archaeosperma arnoldii The earliest (*Devonian) seed-like structure.

Archaeosphaeroides A coccoid cyanophyte (*see* CYANOBACTERIA) from the *Fig Tree cherts of S. Africa, which are *Archaean and date from perhaps 3000 Ma ago.

Archaeozoic *See* ARCHAEAN; and CRYPTOZOIC.

archaic *sapiens* Humans (i.e. *Homo sapiens*) that share features with earlier (non-*Homo*) species, relating mainly to cranial capacity and the robustness of the teeth and skeleton, that are lost in anatomically modern humans. The archaic species are classified as *Homo sapiens*, but not as belonging to the subspecies of modern humans, *Homo sapiens sapiens*.

arche- Alternative spelling for the prefix *archae-.

archetype A hypothetical ancestral form in which all the basic characteristics of a taxonomic group occur, although they are not specialized in any one direction. Thus the modern primitive mollusc *Neopilina* is perhaps close to the molluscan arche-type.

Archie's law An empirical law which relates, for a clay-free sediment, the electrical *resistivity ρ of a porous rock containing water and cement to the fraction of the pore space that is filled with water. $\rho = \rho o^{-m} s^{-n}$, where ρo is the resistivity of the water and s is the fraction of pore space filled with water. The exponent n is usually about 2.0 if one-third of the pore space is filled with water, and exponent m usually varies between 1.3 for unconsolidated sediments to approaching 2.0 for a well-cemented sediment. In the oil-bearing rocks, the remaining pore space is considered to be filled with either oil or gas and the law is usually expressed as $S_w = (R_w/R_t)^{0.5}\phi$, where S_w is the fraction of pore space filled by water with a resistivity R_w in a rock with a true resistivity of R_t and a *porosity ϕ.

arching 1. In an underground *excavation, a small inward movement of the sides, roof, and floor of the cavity caused by *in situ* stress around the excavation. This reduces the permeability of the rock in the immediate vicinity of the excavation. **2.** Masonry or steel support in underground workings.

archipelago Group of islands; a sea containing many scattered small islands.

architecture of sandbodies The large-scale form and arrangement of *sandstone beds. *See* SANDBODY.

archosaur *See* ARCHOSAURIA.

Archosauria (archosaurs) Subclass of *diapsid reptiles, including the *crocodiles, *dinosaurs, *pterosaurs, and *thecodontians. Thecodontians were ancestral to the other groups and appeared at the base of the *Triassic. The term 'archosaur' is from the Greek words *arkhi* ('chief' or 'leading') and *saura* ('lizard').

arctic air Very cold *air mass, generally formed north of the Arctic Circle. As air from this source moves southwards, it cools the regions in which it arrives; but being itself heated in the process it becomes convectively unstable. Polar *lows sometimes form and the accompanying wintry precipitation is often heavy.

arctic bottom water (ABW) Cold, dense water that sinks, to a maximum depth of about 6000 m, from the subpolar *gyre in the Greenland and Norwegian Seas, fills the basins of those seas, then spills as an intermittent southward flow through narrow channels in the ridge between Scotland, Iceland, and Greenland.

arctic front Frontal boundary between cold, arctic air and warmer *air masses, usually lying to the south of it. Many depressions originate on it. In north-western Canada in winter, for example, the frontal zone incorporates cold, dry, continental polar air and modified maritime arctic air from the Gulf of Alaska to the north of continental tropical air.

Arctic Ocean Smallest and shallowest of the major oceanic areas, the shallowness being caused by the surrounding wide continental shelves (up to 1700 km wide). For much of the year the surface is covered by floating pack-ice.

arctic sea smoke (frost smoke) *Fog appearing in very cold air from the arctic-ice or frozen-land regions, when it comes over the warmer water of open parts of the *Arctic Ocean. The rapid heating induces convection currents which rise in the air: these carry moisture upwards from the water surface, and this becomes visible as the moisture quickly condenses again in the very cold surrounding air. Thus a fog of rising columns of condensing water vapour is formed. The fog is usually fairly shallow, wispy, and smoke like. This, and its common occurrence in coastal seas around cold land masses (e.g. Labrador, Greenland, and Norway), gave rise to the name. Similar steam fogs may be seen in winter over the open water of rivers when the air is 10 °C or more colder than the water.

arc-trench gap (fore-arc) The region between an oceanic *trench and the adjacent volcanic *island arc. The arc-trench gap is at least 100 km wide in nearly all cases, and up to 570 km at the eastern end of the Aleutian arc. The width of the arc-trench gap increases with time through the growth of an *accretionary wedge and the development of a *fore-arc basin.

arcus From the Latin *arcus* ('arch'). A cloud feature having a rolled appearance, with fragmented edges on the leading surface of *cumulonimbus and occasionally *cumulus. When well developed, the feature has a prominent arch-like form. *See also* CLOUD CLASSIFICATION.

Ardipithecus ramidus The earliest known member of the human lineage, discovered in 1993 by Tim White, Gen Suwa, and Berhane Asfaw at Aramis, Ethiopia, and dated to 4.4 million years BP. The canine teeth are somewhat reduced from the primitive ape-like condition, but not so much as in **Australopithecus*; the *enamel on the teeth is thin; the deciduous molars are intermediate between those of a human and a chimpanzee. The postcranial skeleton indicates that it was, at least to some degree, bipedal.

Arduino, Giovanni (1714–95) Arduino was a Venetian mining engineer, who devised a classification of the rocks of northern Italy, later adopted by T. O. *Bergman, A. G. *Werner, etc. He distinguished between the Primary (mountain mica-slates), Secondary (mountain limestones, with marine fossils), and Tertiary rocks (fossiliferous valley sediments).

areal erosion Erosion by an *ice sheet of an area too large to be visible as an erosion feature and identifiable only through mapping at a continental scale.

arenaceous Sandy, or sand-like in appearance or texture. The term is applied to *clastic *sedimentary rocks with a grain size of 0.0625–2.00 mm. Three main groups of are-

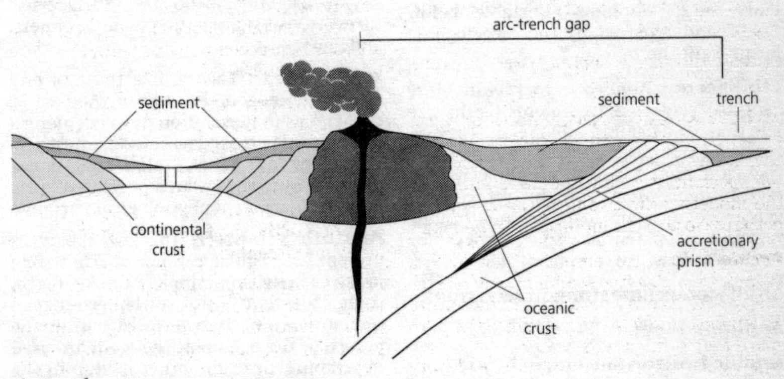

Arc-trench gap

naceous rocks are recognized: quartz sandstones (*quartzites), which contain 95% *quartz; *arkoses, which have greater than 25% *feldspar; and *greywackes, which essentially are poorly sorted sediments with rock (lithic) fragments in a mud *matrix.

Arenicolites An *ichnogenus of *domichnia.

Arenig A *series (493–476.1 Ma) of the Lower *Ordovician, underlain by the *Tremadoc and overlain by the *Llanvirn.

arenite *See* DOTT CLASSIFICATION; and SANDSTONE.

areology The scientific study of Mars. Derived from the name of the Greek god Ares, known to the Romans as Mars, the god of war, and *logos* (the Greek for word or discourse). It is analogous to geology as the study of the Earth.

arête Knife-edged, steep-sided ridge found in upland areas that have been or are being glaciated, and formed by the meeting of adjacent *cirque headwalls. It may be diversified by 'gendarmes' (abrupt rock pinnacles that have resisted frost shattering).

arfvedsonite A sodium- and iron-rich *amphibole.

argentite (silver glance) Ag_2S, *ore mineral for silver; sp. gr. 7.2–7.4; *hardness 2.0–2.5 (can be cut with a knife); *cubic; black and opaque; slimy black *streak; *metallic *lustre; *crystals commonly cubes or octahedra, but can be *massive; *cleavage poor, cubic; occurs in *hydrothermal veins in association with native silver and as a weathering product of primary silver sulphides; unstable below 179 °C and replaced by the *monoclinic form, acanthite.

argillaceous Applied to rocks which are *silt- to *clay-sized *sediments (grain size less than 0.0625 mm in diameter). They account for more than 50% of *sedimentary rocks and most have a very high *clay mineral content. Many contain a high percentage of organic material and can be regarded as potential *source rocks for *hydrocarbons.

argillaceous limestone *See* MARLSTONE.

argillans *See* CUTAN.

argillic horizon Sub-surface B *soil horizon that is identified by the illuvial (*see* ILLUVIATION) accumulation of silicate *clays. The amount of clay necessary is defined in comparison with the quantity in the overlying eluvial (*see* ELUVIATION) horizon, but is at least 20% more. *Cutans or clay skins may be used to help identify an argillic horizon. It is at least one-tenth as thick as the overlying horizons.

argillite (lutite) A well-compacted, non-fissile rock, containing *clay- and *silt-sized particles; more indurated than a *mudstone.

argon-40 *See* POTASSIUM–ARGON METHOD.

aridic moisture regime The moisture balance of arid climates and soils, where the annual precipitation is less than the potential evaporation and transpiration, and where soil moisture status is normally less than *field capacity.

Aridisol An order of soils found in arid environments. These soils have very little organic matter in their surface *horizons, but may contain calcium carbonate or *gypsum, and/or soluble-salt accumulations.

aridity index Indication of moisture deficit. All climatic classifications include arid categories, defined either by quantitative or, more usually, by mainly subjective criteria. C. W. Thornthwaite first used the term 'aridity index' and calculated it as 100 times the water deficit, divided by the potential evaporation. *See also* RADIATION INDEX OF DRYNESS; and THORNTHWAITE CLIMATE CLASSIFICATION.

Ariel (Uranus I) One of the major satellites of *Uranus, with a radius of 581.1 × 577.9 × 577.7 km; mass 13.53×10^{20} kg; mean density 1670 kg/m^3; albedo 0.34. Surface features are visible, including fissures, craters, and flows.

aristogenesis Theory, like those of entelechy, nomogenesis, and orthogenesis, that claims that evolution proceeds along a determined path. Today, however, most people accept that natural selection does not direct evolution towards any particular kind of organism or physiological attribute.

Aristotle's lantern The jaw apparatus present in *regular echinoids. Five strong jaws, each with a single *calcite tooth, come together into a lantern-shaped structure which is suspended within the mouth of the echinoid. The teeth are used in scraping algae and other food from the sea floor.

arkose *Arenaceous rock containing *quartz and 25% or more of *feldspar. The feldspar is easily destroyed during chemical change or transportation, implying that arkoses were deposited rapidly under fairly arid environmental conditions. Most were deposited near to land, probably in close proximity to a granitic area.

arkosic arenite A *sandstone comprising more than 25% *feldspar and with more feldspar than rock fragments, with less than 15% mud *matrix. The matrix is defined as material less than 30 μm in grain size. *See* DOTT CLASSIFICATION.

arkosic wacke (feldspathic wacke, feldspathic greywacke) A *sandstone comprising more than 5% *feldspar, and more feldspar than rock fragments, with more than 15% *matrix. The matrix is defined as material less than 30 μm in grain size. *See* DOTT CLASSIFICATION.

arls A collective name for *marl, *sarl, and *smarl.

Arnsbergian A *stage of the *Serpukhovian Epoch, underlain by the *Pendleian and overlain by the *Chokierian.

Arowhanan *See* RAUKUMARA.

array A geometrical distribution or pattern. A line of *geophones or *shot points constitutes an array, as does a line of electrodes for a *resistivity survey. *See also* ELECTRODE CONFIGURATION.

arrow worms *See* CHAETOGNATHA.

arroyo Gully found along valley floors in an arid or semi-arid region and possessing steep or vertical walls cut in fine-grained cohesive sediments. The floor is flat and usually sandy. Found especially in the south-western United States, parts of India, S. Africa, and around the Mediterranean.

arsenopyrite (known historically as mispickel) *Mineral, FeAsS; sp. gr. 5.9–6.2; *hardness 5.5–6.0; *monoclinic; silver-grey to white, often with a tarnish; dark greyish or black *streak; *metallic *lustre; *crystals *prismatic, often striated, can also be *massive and granular; *cleavage distinct {101}; occurs in high- to moderate-temperature mineral veins in association with *gold, ores of tin, tungsten, *galena, and *quartz, and also disseminated in *limestones, *dolomites, *gneisses, and *pegmatites. It is the principal source of arsenic compounds used for pest control, for the manufacture of dyes and chemicals, and in leather treatment.

artesian water *Groundwater that is confined in an *aquifer, but which may overflow on to the land surface via artificial *boreholes or, sometimes, natural *springs, because of the high *hydraulic head that may be developed in a confined aquifer. Artesian conditions are common when the aquifer has a *synclinal form. The London Basin, England, provided artesian water during the nineteenth century from a chalk aquifer sealed by clays. The term is derived from the Artois region of north-western France.

artesian well (overflowing well) A well that flows at the surface without pumping, because it is sunk into a confined aquifer whose *hydraulic head (sometimes called the potentiometric or piezometric head) lies above ground level. *See* AQUIFER; and ARTESIAN WATER.

Arthrodira (arthrodiriformes; class *Placodermi) Group or order of fossil (*Devonian) fish with a body covered with bony plates, including a heavily armoured *head shield, the gills opening between head and body armour.

Arthropoda A highly diverse phylum of jointed-limbed animals, which includes the crustaceans, arachnids, and insects as the major components, as well as the classes Symphyla, Pauropoda, Chilopoda (centipedes), Diplopoda (millipedes), and the extinct *Trilobita and eurypterids (*see* CHELICERATA; and MEROSTOMATA). Arthropods first appeared in the *Cambrian, already well diversified with such forms as the trilobites, trilobitoids, *Ostracoda, and crabs present, implying an earlier, hidden history, reaching back into the Precambrian. Embryological evidence shows that they are derived either from primitive *Polychaeta, *Annelida, or from ancestors common to both. Arthropods share with annelid worms a metamerically segmented body (*see* METAMERIC SEGMENTATION), at least in the embryo, a dorsal heart, a dorsal anterior brain, and a ventral nerve cord that has segmental, ganglionic swellings. The limbs of all arthropods are paired, jointed, and segmental, and the body has a chitinous *exoskeleton. Primitively, the limbs and cuticular plates correspond to the metameric segmentation of the body, but in many groups there is considerable loss and/or fusion of segments.

arthropods *See* ARTHROPODA.

Articulata 1. (phylum *Brachiopoda) A class of brachiopods, existing from the Lower *Cambrian to the present day, in which the shell is calcareous and comprises valves hinged by teeth in one valve and sockets in the other. The *pedicle is of a horny material. Their main radiation was in the early *Ordovician; of the 7 orders present in the *Palaeozoic, 3 are still extant. **2. (sea lilies; subphylum *Crinozoa, class *Crinoidea)** Subclass of sea lilies (crinoids) in which the basal plates are small or strongly reduced. Articulation between the radial and brachial plates, and in the majority of brachials, is muscular, with a well-developed fulcral ridge. The arms are always *uniserial. All post-*Palaeozoic crinoids belong to the Articulata.

artificial freezing Method of controlling *groundwater and improving strength of ground by pumping a refrigerant, e.g. calcium chloride or liquid nitrogen, through tubes in the ground; as the ground freezes around closely spaced tubes, a continuous frozen zone may be formed. This is an expensive technique, but suitable for mixed *strata as the results are more predictable than *dewatering or *grouting.

artificial rain Rain, or increased rain, produced by seeding clouds artificially with 'dry ice' (frozen carbon dioxide), silver iodide, or other appropriate particles, which act as condensation nuclei.

artificial recharge A process whereby the amount of water in an *aquifer is supplemented by engineered, as opposed to natural, means. Artificial recharge may be through *boreholes, purpose-built ponds, or simply by diverting more water on to the surface *catchment of the aquifer. Artificial recharge may be implemented as part of a conjunctive use scheme.

Artinskian 1. An *age in the *Permian Epoch, preceded by the *Sakmarian, followed by the *Kungurian, and dated at 268.8–259.7 Ma (Harland et al., 1989). **2.** the name of the corresponding eastern European *stage, which is roughly contemporaneous with the upper *Rotliegende (western Europe), lower/middle *Leonardian (N. America), and the Bitaunian (New Zealand).

Artiodactyla (artiodactyls; cohort *Ferungulata) Even-toed ungulates, an order of mammals that includes the living camels, pigs, and ruminants. Descended from the *Condylarthra, they underwent a spectacular burst of *adaptive radiation in *Eocene and early *Oligocene times, largely replacing the initially more numerous *Perissodactyla. The ankle bone (astragalus) is specialized in artiodactyls to give better spring. The axis of the foot is paraxonic (passing between the third and fourth digits). In primitive, four-toed types, e.g. the pig, the first digit is absent; in advanced forms the second and fifth digits are also reduced or lost. Early forms had an unspecialized dentition, but in the course of evolution the upper incisors were lost in some species, the lower incisors biting against the hardened gum of the upper jaw, an adaptation to a herbivorous diet.

Arundian A *stage of the *Visean Epoch, underlain by the *Chadian and overlain by the *Holkerian.

Asaphida An order of *Trilobita that lived from the Upper *Cambrian to *Silurian. There were 6 suborders.

asbestos Fibrous varieties of *amphibole and *serpentine, including chrysotile (fibrous serpentine), *actinolite (asbestos proper), amosite (a variety of *anthophyllite), and crocidolite (blue asbestos; a variety of *riebeckite). Ancient civilizations referred to asbestos cloth as *amianthus* (from the Greek word meaning 'undefiled'), because it could be cleaned by throwing it into a fire and all varieties of asbestos have great heat-resistant properties; varieties with fibres long enough to be spun and woven are used commercially for heat-resistant cements, cladding, and insulation material, and for asbestos corrugated sheets. However, many of its uses, particularly as brake pads for cars and as pipe and ceiling insulation materials, have been banned, because the inhalation of asbestos dust (small, airborne, needle-like fibres) can cause serious lung diseases (e.g. asbestosis) and contribute to pneumoconiosis. In spite of this, the annual world production of asbestos has remained constant at more than 3 million tonnes.

Asbian A *stage of the *Visean Epoch, underlain by the *Holkerian and overlain by the *Brigantian.

aseismic Free from *earthquakes.

aseismic margin *See* PASSIVE MARGIN.

aseismic ridge Long, linear, inactive, volcanic, topographic feature found in many deep-ocean basins. An example is the Walvis Ridge in the south-eastern Atlantic, which extends for 3000 km and in places reaches a height of 2 km above the deep-ocean floor.

ash *Tephra less than 2 mm in size.

ash-cloud surge *See* SURGE.

ash cone *See* SCORIA CONE.

ash-flow *See* PYROCLASTIC FLOW.

Ashgill A *series (443.1–439 Ma) of the Upper *Ordovician, underlain by the *Caradoc and overlain by the *Llandovery (*Silurian).

asiderite *See* STONY METEORITE.

asphalt Brown or black, solid or semi-solid, bituminous substance made almost entirely of carbon and hydrogen. It melts between 65 and 95 °C and is soluble in carbon disulphide. It is formed by the evaporation of volatile *hydrocarbons and occurs in oil-bearing rocks, e.g. in Trinidad, or as a residue from *petroleum refining.

asphaltite *See* GILSONITE.

assay The analysis of *minerals and mine products to determine the concentrations of their components.

Asselian 1. An *age in the Early *Permian Epoch, preceded by the *Stephanian, *Gzelian, and Noginskian (*Carboniferous), followed by the *Sakmarian, and with an initial boundary (the Carbo-Permian boundary) dated at 290 Ma (Harland et al., 1989). **2.** The name of the corresponding eastern European *stage which, because of lower boundary uncertainties, has also been considered as part of the Carboniferous system. It is roughly contemporaneous with the lower *Wolfcampian and lower *Rotliegende (western Europe), and with the lower Somoholoan (New Zealand). *See also* SAKMARIAN.

assemblage zone (coenozone, faunizone) *Biostratigraphic unit or level of *strata characterized by a particular assemblage of animals or plants. An assemblage zone is named after one or more of the distinguishing *fossils present, which are chosen without regard for their total time ranges, so that the assemblage is of purely environmental significance. *Compare* CONCURRENT RANGE ZONE.

assimilation In *petrology, the processes of melting and solution by which wall rock is incorporated into *magma. Partial or complete melting may occur at *contacts and at depth in detached blocks or *xenoliths. Assimilation causing changes in composition of the original magma may lead to hybrid or contaminated rocks. The term does not imply any particular mechanism. *See* PARTIAL MELTING.

assimilation-fractional crystallization (AFC) An important process in *igneous *petrology whereby *melts with widely differing isotopic and *trace elements can be produced. When a primitive *magma, such as a *basalt, invades crustal rocks, portions of the *country rock may become detached and included in the magma as *xenoliths. Because of the high temperature and thermal capacity of the basalt, it is capable of melting a proportion of the country rock. In doing so it loses some of its own heat and thus a proportion of the magma crystallizes. The composition of the resulting magma is determined by the relative amounts of magma and country rock initially present; the rates at which assimilation and crystallization proceed; and by the *partition coefficients of the various elements between solid and liquid.

astatic magnetometer *See* MAGNETOMETER.

Asteriacites An *ichnogenus of *cubichnia.

asteroid A small rocky or metallic body orbiting the *Sun in the asteroid belt between the orbits of *Mars and *Jupiter (from about 2.2 to 3.2 AU (*see* ASTRONOMICAL UNIT), with one family (Trojans) at 5 AU occupying two of the jovian *Lagrangian Points). Two families (Apollo and Aten, as well as some Amor objects) have orbits that cross *Earth's orbit. The largest asteroid is 1 Ceres (diameter 987 ± 150 km). About 30 exceed 200 km in diameter and over 3000 have been identified. *Meteorites are probable samples of the asteroid belt. *See* APOLLO (1862); CASTALIA (4769); CERES (1); CHIRON (2060); EROS (433); GASPRA (951); GEOGRAPHOS (1620); ICARUS (1566); IDA (243); JUNO (3); MCAULIFFE (3352); MATHILDE (253); MIMISTROBELL (3840); NEREUS (4660); PALLAS (2); SHIPKA (2530); TOUTATIS (4179); and VESTA (4).

Asterosoma A mounded, lobed, somewhat star-shaped *trace fossil. The central tube was occupied by an organism; the surrounding, detrital lobes are the result of feeding processes.

Asteroxylon An early lycopsid (*Lycopsida) plant that occurs in the Rhynie Chert, Aberdeenshire, Scotland (Lower *Devonian). It has dichotomous rhizomes and the aerial extensions bear leaf-like enations (outgrowths).

Asterozoa (starfish, brittle stars; phylum *Echinodermata) Subphylum whose members have *radial symmetry of projecting rays, and a star-shaped body. *See also* AMBULACRUM.

asthenosphere The weak zone within the upper *mantle, underlying the *lithosphere, where the mantle rocks deform by plastic flow in response to applied stresses of the order of 100 MPa. It is commonly considered to be coincident with the upper-mantle seismic *low-velocity zone, but this is probably valid only for the oceanic sectors of the mantle. Viscosity is of the order of 10^{21-22} poise, i.e. the same as the underlying mantle, but it is much more 'fluid' than the overlying lithosphere. Originally it was recognized as a possible explanation for *isostatic behaviour, and it is generally recognized as a mantle zone within which convective motions take place. The depth of *earthquake foci in *subduction zones suggests that descending convection limbs penetrate to 700 km, just above the upper mantle–lower mantle boundary. Rising limbs of asthenospheric mantle convection are located under surface spreading centres (*mid-oceanic ridges).

astogenetic heterochrony Among colonial animals, *heterochrony that affects the colony as a whole. *Compare* MOSAIC HETEROCHRONY; and ONTOGENETIC HETEROCHRONY.

astraeoid A condition that occurs in massive *corals. Massive corals are composed of closely packed *corallites and the individuals become polygonal in shape. In the astraeoid condition the walls of the corallites become reduced or lost but the *septa remain entire. *See* COMPOUND CORALS.

astragalus The ankle bone.

astrobleme Literally, 'star-wound'; a terrestrial crater formed by *meteorite, *asteroidal, or *cometary impact. The term is generally used for large craters (more than 10 km diameter), but is not in common usage.

astrogeology *See* PLANETARY GEOLOGY.

astronomical unit (AU) The average distance between the *Earth and *Sun; a unit of measurement equal to 149 597 870 km.

A-subduction The movement of one continental lithospheric *plate under another (*subduction) in a *collision zone, with the separation of part or all of the upper *crust from the lower crust and *mantle. The denser material subducts normally whilst the buoyant material can underthrust or overthrust the crust of the overriding plate. A-subduction, named after O. Ampferer, has also been called 'delamination', and is contrasted with B-subduction, named after Hugo *Benioff, in which oceanic *lithosphere subducts. A-subduction is postulated to involve *shortening of a maximum of only a few hundred kilometres, whereas B-subduction can recycle thousands of kilometres of *oceanic crust and upper mantle.

asymmetrical fold A *fold in which the *axial plane is inclined relative to the median plane, and adjacent limbs *dip in opposite directions.

asymmetric valley Valley that has one side steeper than the other, the opposing sides standing at significantly contrasting angles. This may be due to geologic structure, or to variation in the nature and intensity of erosional processes. Such valleys are common in past and present *periglacial environments, where aspect has a significant effect on the nature of frost-based processes.

Atdabanian A *stage of the Lower *Cambrian (*Caerfai Epoch), underlain by the *Tommotian and overlain by the *Lenian, and dated at 560–553.7 Ma (Harland et al., 1989).

Athabasca *See* TAR SAND.

Atlantic Period in post-glacial times (i.e. post- *Devensian or *Flandrian) from about 7500 to 5000 BP which, according to pollen evidence, was warmer than the present, and moist; with oceanic climatic conditions prevailing throughout north-western Europe. It corresponds to *Pollen Zone VIIa, which throughout north-western Europe is characterized by the most thermophilous (warmth-loving) species found in post-glacial pollen records. The climatic optimum of the post-glacial, or current Flandrian Interglacial, is dated to the early Atlantic period. *Compare* BOREAL. *See* POLLEN ANALYSIS.

Atlantic conveyor A system of ocean currents that plays a major role in the transport of heat from low to high latitudes and, therefore, in global climates. It is driven by the convective overturning of water near the edge of the northern sea ice, where cold, saline water sinks to the ocean floor and travels south as the *North Atlantic Deep Water, eventually to the Southern Ocean. Its place is taken at the surface by warmer water flowing northward.

Atlantic Ocean One of the main oceanic areas of the world. It is relatively shallow, having an average depth of 3310 m; and it is the warmest (average temperature 3.73 °C) and most saline (average *salinity 34.9 parts per thousand) of the major oceans.

Atlantic Province (Acado-Baltic Province, European Province) A subdivision of the early *Cambrian olenellid *trilobite *fauna. The trilobite faunas of the early Cambrian can be divided into two main regional groups: the olenellid fauna found in north-western Europe and N. America, and the redlichiids in Asia, Australia, and N. Africa. The olenellid fauna is also subdivided into two provinces: the Atlantic Province on the southern and eastern flank of the *Iapetus Ocean, and the *Pacific (or American) Province on the northern and western margins. The names Atlantic Province and Pacific Province have also been applied to the *Ordovician trilobite and *graptolite faunas in the same areas.

Atlantic-type coast (transverse-type coast) A coast characterized by subsidences and fractures that cut across the grain of the folded mountain formations inland. Typically, such a coast borders a relatively young ocean that is widening because of *sea-floor spreading. This type of coast was first recognized by Eduard *Suess. *Compare* PACIFIC-TYPE COAST.

Atlantic-type margin *See* PASSIVE MARGIN.

Atlas (Saturn XV) One of the lesser satellites of *Saturn, discovered in 1980 by *Voyager 1, with a radius measuring 18.5 × 17.2 × 13.5 km; visual albedo 0.9.

atlas vertebra *See* VERTEBRA.

atmometer An instrument for measuring evaporation. It is normally in the form of an open-ended glass tube from which water can evaporate.

atmophile Applied to the gaseous elements most typical of, and concentrated in, the Earth's atmosphere; e.g. H, C, N, O, I, and inert gases. They may occur in an uncombined state or combined as, for example, in water (H_2O), carbon dioxide (CO_2) and methane (CH_4). *Compare* BIOPHILE; LITHOPHILE; CHALCOPHILE; and SIDEROPHILE.

atmosphere 1. Air surrounding the Earth. The atmosphere has no precise upper limit, but for all practical purposes the absolute top can be regarded as being at about 200 km. The density of the atmosphere decreases rapidly with height, and about three-quarters of the mass of the atmosphere is contained within the lowest major layer, the *troposphere, whose depth varies between about 10 km and 17 km, being generally smaller further from the equator. **2.** Unit of pressure (abbreviation: atm.). Its value is approximately the average pressure of the atmosphere at sea level, the figure adopted being the pressure at sea level in the International Standard Atmosphere (760 mm of mercury, or 1013.25 mb). In SI units, 1 atm. = 101 325 Pa. *See also* ATMOSPHERIC STRUCTURE.

atmospheric pollution Solid and gaseous contaminants in the atmosphere which occur as dust, smoke, or sulphur dioxide and other gases, particularly from the combustion of fossil fuels and certain industrial processes. Air pollution is most marked in urban areas. *See also* PHOTOCHEMICAL SMOG.

atmospheric pressure Downward force exerted by the weight of the overlying *atmosphere, expressed per unit area in a given horizontal cross-section. Pressure varies throughout the atmosphere, owing to the distribution of mass; there are small diurnal variations of partly tidal and partly thermal origin, as well as bigger changes associated with the passage of depressions and anticyclones. Atmospheric pressure is measured in millibars (mb), 1 mb being equal to 100 kilopascals (kPa). Measurements are usually made with a mercury barometer. The overall global average pressure at sea level is 1013.25 mb, but as air is readily compressible, pressure decreases exponentially with altitude.

atmospheric shimmer The effect observed when light passes through moving air masses which have differing *refractive indices. Results in effects such as star twinkling and ultimately limits the resolution of any *remote sensing system.

atmospheric structure The broadly horizontal layering of the *atmosphere, the layers being distinguished by differences in the rate of change of temperature with height, which either favour or discourage the development of vertical exchanges (convection). From the surface of the Earth upwards the layers are: (*a*) the *troposphere, in which convection is often prominent, especially over warm regions, extending to the *tropopause at a somewhat variable height, generally about 11 km over middle and higher latitudes and 17 km near the equator; (*b*) the *stratosphere, in which there is much less vertical motion, and which extends from the tropopause to about 50 km at the *stratopause; (*c*) the *mesosphere, in which there is once again more convection, extending from the stratopause to a height of about 80 km at the *mesopause; and (*d*) the *thermosphere, extending from the mesopause to the effective limit of the atmosphere, at about 200 km.

atmospheric 'window' The range of wavelengths (about 8.5–11 μm) at which radiation is only slightly absorbed by water vapour. Terrestrial radiation within this range may escape into space unless it is absorbed by cloud (water droplets can absorb in this range). *See also* GREENHOUSE EFFECT; and TERRESTRIAL RADIATION.

Atokan (Derryan) A *series in the *Pennsylvanian of N. America, underlain by the *Morrowan and overlain by the *Desmoinesian Series. It is roughly contemporaneous with the Vereiskian and Kashirskian Stages of the *Moscovian Series.

atoll Ring-shaped organic *reef that encloses or almost encloses a *lagoon, and which is surrounded by the open sea. The reef may be built of *coral and/or calcareous *algae. An atoll is built on an existing structure such as an extinct, submerged volcano.

atomic absorption spectrometry Instrumental technique for the chemical analysis of material. The basic theory was known as long ago as the mid-nineteenth century, and utilizes the observation that certain wavelengths of radiation emitted by excited atoms are strongly absorbed by unexcited atoms of the same element (emission *spectrum). The radiation will be reduced if it passes through an area containing such unexcited atoms. This reduction can be measured by a detector, and gives the *concentration of the element.

atomic number Chemical elements are composed of atoms, all of which are held together by electrical charges. Every atom has a relatively heavy *nucleus that is composed of protons (with a positive charge) and neutrons (neutral particles). Orbiting the nucleus there are a number of exceptionally light *electrons, whose negative charges balance the positive charge provided by the protons. The number of protons in the structure provides the atomic number; atoms having the same atomic number belong to the same chemical element.

attenuation The reduction in amplitude or energy of a signal. Attenuation of *seismic waves (seismic attenuation) occurs as a result of spherical divergence, *absorption, energy losses at interfaces through *reflection and *refraction, and by internal scatterers. For electromagnetic waves *see* SKIN DEPTH.

Atterberg limits Series of thresholds which are observed when the water content of a soil is steadily changed. The 'contraction limit' occurs when sufficient water is added to a dry soil for contraction cracks to close. The addition of further water leads to plastic deformation at the 'plastic limit'. The 'liquid limit' occurs when just enough water is then added for the soil to behave like a liquid. Knowledge of these limits is important for understanding and predicting hillslope failure. The difference in percentage water content between the liquid limit and the plastic limit is called the 'plasticity index'.

attitude **1.** Of a bed or other planar feature, the disposition with respect to the horizontal and compass bearings; these are obtained by measuring the *dip and *strike of the bed respectively. **2.** Of a *fold, the overall disposition; this is defined by measuring the dip and strike of the *axial plane, and the *trend and *plunge of the *hinge line.

aubrite An *enstatite *pyroxene-rich, *achondritic *meteorite that is poor in calcium. The enstatite, which often forms large *crystals, has very low FeO/(FeO + MgO) ratios, Aubrites may contain up to 1.2% elemental silicon included within rare Fe–Ni grains, and appear to have close affinities with the similarly highly reduced enstatite *chondrites.

augen-gneiss A medium- to coarse-grained, banded, *regional metamorphic rock composed mainly of *quartz and *feldspar with *hornblende and *mica also present in variable quantities, and characterized by large ovoidal *megacrysts of feldspar known as 'augens' (derived from the German for 'eyes'). The banding is due to variations in the modal proportions (*see* MODAL ANALYSIS) of the *mineral *phases constituting the rock. The best-developed augen-gneisses are formed by high-grade *metamorphism of aluminous sediments. *See also* GNEISS.

auger Tool used primarily for soil sampling, but also for sampling *peat and other *unconsolidated sediments. The simplest and most universal form has a screw head to bore the soil or sediment. Alternative auger heads are available for more specialized needs. Standard augers sample to one metre depth, but extension rods can be attached enabling sampling at deeper levels.

augite An important member of the *pyroxene group of silicate minerals, occupying a field of composition within the *calc-alkaline pyroxenes between the *diopsides and the *pigeonite field, and with the formula $Ca(Mg,Fe)Si_2O_6$ (Na may substitute for Ca and Al for Si; and a large increase in the amount of Na and Fe^{3+} (for Fe^{2+}) gives aegirine augite $(Na,Ca)(Fe^{3+},Mg,Fe^{2+})Si_2O_6$, an important pyroxene in *alkaline rocks); sp. gr. 3.3; *hardness 6.0; *monoclinic; greenish-black; *crystals short, *prismatic; a common constituent of *basic igneous rocks but may also occur in high-grade *metamorphic rocks. *See also* CLINOPYROXENE.

augite-minette *See* MINETTE.

Aulacocerida *See* BELEMNITIDA.

aulacogen A long-lived, *sediment-filled *graben oriented at a high angle to either a neighbouring modern ocean or a neighbouring *orogenic belt. The sediments in an aulacogen are largely characterized by a lack of major deformation, although late *strike-slip movement on the boundary *faults is known to deform the sediments extensively. Aulacogens have been interpreted as forming within failed rifts (*see* RIFT VALLEY) of *triple junctions and thus are taken to indicate *plate tectonic activity. They have been identified as far back as the early *Proterozoic and are of world-wide distribution.

aulodont Applied to *echinoids with a lantern (*see* ARISTOTLE'S LANTERN) characterized by teeth which are longitudinally grooved and broadly U-shaped in cross-section.

aureole The luminous white or bluish disc, surrounded by a brown ring, sometimes observed directly surrounding the Sun or Moon. The term is also used to describe the bright area with no definite boundary commonly seen surrounding the Sun in a clear sky. *See also* CORONA.

aurora Illumination of the sky, sometimes in brilliant colours, as a result of high-speed solar particles entering the *ionosphere (at a height of 100–130 km) and releasing electrons from air molecules by excitation. The re-establishment of molecules leads to the emission of light, especially red- and green-coloured light, e.g. in arcs or bands over large areas. The effect is called 'aurora borealis' or 'northern lights' in the northern hemisphere and 'aurora australis' or 'southern lights' in the southern hemisphere. Such atmospheric disturbances occur in relation to disturbances on the Sun in the course of the sunspot cycle.

Australian faunal realm Region distinguished by a unique *marsupial fauna, including herbivores, carnivores, and insectivores. These evolved in isolation from the placental mammals which now dominate the other continental faunas. In addition to marsupials there are also very primitive mammals (monotremes): the spiny anteater and the platypus; and small rodents which are relatively recent (probably *Miocene) immigrants.

australopithecines Literally, 'southern apes', early members of the human lineage that lived from about 4 to about 1 million years ago in Africa. The so-called 'robust australopithecines' are nowadays placed in a separate genus, *Paranthropus*. The other ('gracile') australopithecines are also a very diverse group of species, some very primitive and perhaps ancestral to all later hominins, others probably specialized side-lines. The species usually recognized are *Australopithecus anamensis* (3.9–4.1 million years BP) and *A. afarensis* (3.75–3.0 million years BP) from East Africa, *A. bahrelghazali* (about 3.4–3.0 million years BP) from West Africa, and *A. africanus* (about 3.0–2.4 million years BP) from South Africa. Probably some of these species should be placed in different genera.

Australopithecus *See* AUSTRALOPITHECINES.

autapomorphy The possession of an *apomorphic *character state that is unique to a particular species or lineage in the group under consideration.

authigenic Applied to materials (*minerals, *cements, etc.) that formed in the rock of which they are a part during, or soon after, its deposition. *Compare* ALLOGENIC; and ALLOCHEMICAL.

autobrecciated lava A viscous, commonly silica-rich, *lava flow with a congealed crust which has been broken up and fragmented by the continued movement of molten lava within the flow interior. Stressed and deformed by the movement, the crust may fracture in a brittle manner, producing angular, smooth-faced blocks able to weld together if they are hot enough but which otherwise become incorporated into the moving interior of the flow.

autochthonous Indigenous; applied to a material (e.g. *dripstone, *coal) that was formed in its present position. No significant transport has been involved. *Compare* ALLOCHTHONOUS.

autoclast A *clast formed by friction from flowing *lava or autobrecciation (*see* AUTOBRECCIATED LAVA). *Compare* ALLOCLAST; EPICLAST; and HYDROCLAST.

autocorrelation The *correlation of a wave-form with itself; a special case of *cross-correlation. The autocorrelation function (ACF) is especially useful in the identification of *multiples within a *seismic record. While the ACF contains all the amplitude and frequency information of the original wave-form, it possesses no phase information.

autocyclic mechanisms Events responsible for the accumulation of sediments that are part of the sedimentary system itself (e.g. size and configuration of a river channel). *Compare* ALLOCYCLIC MECHANISMS.

automatic point counter An electronic control panel designed to count the number of times a particular *mineral is recorded on multiple, closely spaced traverses of a *thin section viewed down a *polarizing microscope. The control panel, which contains a bank of recording buttons, each assigned by the user to a particular mineral, is linked to a thin-section holder mounted on to the *stage of a polarizing microscope. The holder is designed to move the thin section by a predetermined distance (variable from 0.001 to 1 mm) in one direction, after the user has pressed the control panel button assigned to the particular mineral observed under the *eyepiece *cross-wires. At the end of a complete traverse, the holder automatically resets by a predetermined distance to start the next traverse. On completion of all the traverses, the control panel shows the number of times each mineral was recorded and can recalculate this as a percentage of the total number of points. This gives the percentage volume of each mineral recorded in the rock.

automatic weather station Meteorological station that records measurements of *atmospheric pressure, temperature, *humidity, and wind, and details of weather conditions, and transmits them automatically to a central base.

autosuspension In an active *turbidity current, a feedback mechanism whereby: turbulence maintains the suspended load; the suspended load causes the high density of the suspension; the high density of the suspension causes the flow; the flow causes turbulence.

autotheca One of the three types of graptolite *thecae, possibly containing the female zooid. *See* GRAPTOLITHINA; GRAPTOLOIDEA; and DENDROIDEA. *Compare* BITHECA; and STOLOTHECA.

Autunian A lesser-used stratigraphic name for the lowest of three divisions of the *Permian of Europe, the others being the *Saxonian (mid) and *Thuringian (upper). Together, the Autunian and Saxonian form the Lower Permian.

autunite *Secondary mineral, $Ca(UO_2)_2(PO_4)_2.10–12H_2O$; sp. gr. 3.1; *hardness 2.0–2.5; *tetragonal; usually bright lemon-yellow to greenish-yellow; yellow *streak; *vitreous *lustre; *crystals *tabular, square, forming foliated, scaly masses; *cleavage perfect, basal; radioactive; fluorescent; occurs with *torbernite in the oxidized zone of mineral veins.

auxiliary reference section *See* HYPOSTRATOTYPE.

available nutrients Any elements or compounds in the soil solution that can be absorbed readily into plant roots and that function as nutrients to growing plants. The available amount is usually much less than the total amount of that plant nutrient in the soil.

available relief The part of a landscape that is higher than the floors of the main valleys It is therefore available for destruction by the agents of *erosion, controlled by the local base level. It is measured by the vertical distance between hilltops and valley floors.

available water In soil, the water that can be absorbed readily by plant roots. It is usually taken to be water held in the soil under a pressure of 0.3 to about 15 bars.

avalanche *See* MASS-WASTING.

avalanche wind A blast of often very destructive air ahead of a descending avalanche.

Avalonian orogeny An episode of mountain building, named after the Avalon Peninsula of Newfoundland, that occurred about 650–500 Ma ago (*Cambrian to *Ordovician) as a result of rifting (*see* RIFT) and *volcanicity associated with the opening of the Atlantic. Its rocks are found intermittently from Georgia to Newfoundland. *See* APPALACHIAN OROGENIC BELT.

average In statistics, a summary of the data using a single value. The data may be summarized by the *mean, *median, or *mode values. *See also* VARIANCE.

average velocity (time-averaged velocity, v^-) The ratio of a given depth divided by the travel time to that depth, usually assuming straight ray paths and parallel layering. Thus $v^- = z_n/t_n$, where z_n is the depth of the top n layers and t_n the single-travel time through those n layers. Also, $v^- = \Sigma z_i/\Sigma t_i$ where z_i and t_i are the thickness of and single-travel time through the ith layer respectively, and Σ is the total thickness and total travel time respectively of the overlying layers.

Aves (birds; subphylum *Vertebrata, superclass *Gnathostomata) The class that comprises all the birds. The late *Jurassic **Archaeopteryx lithographica* is still the best-known *Mesozoic bird, but others have been described since the 1980s: **Noguerornis*, from the lowermost *Cretaceous of Spain; the slightly later and more advanced *Concornis* and **Iberomesornis*, known by complete skeletons, also from Spain; and, also early Cretaceous, *Sinornis* and *Cathayornis* from China. There were also some curious, specialized, Late Cretaceous birds, such as the flightless *Mononykus* from Central Asia, in which the forelimbs were reduced to stubby claws, and *Hesperornis*, a diving form. All these early birds had teeth and long, bony tails. Birds arose from within the theropod dinosaurs (*Theropoda) and so should properly be classified as a subgroup of them; those closest were the Dromaeosauridae (the family which includes the famous *Velociraptor* of *Jurassic Park*).

Avicenna (Abu Ali al-Husayn Ibn Abdallal Ibn Sina) (980–1037) Avicenna was an Arab physician and philosopher. His 'geological' ideas were published in *Liber De Mineralibus*, which was attributed to Aristotle, and influential up to about 1500. He wrote about earthquakes, erosion of valleys, the deposition of sediments, etc.

Avogadro constant (Avogadro number) The number of molecules, atoms, or *ions in one *mole of a substance: 6.02252×10^{23} per mol. It is derived from the number of atoms of the pure isotope ^{12}C in 12 grams of that substance and is the reciprocal of atomic mass in grams.

avulsion Lateral displacement of a stream from its main channel into a new course across its *floodplain. Normally it is a result of the instability caused by *channel *aggradation. The avulsion of a stream into an adjacent valley may explain some cases of apparent *river capture.

axial modulus (ϕ) The ratio of longitudinal stress to uniaxial longitudinal strain, i.e. where there is no lateral strain. This is a special form of *Young's modulus.

axial plane The plane that bisects the angle between the two limbs of a *fold.

axial plane cleavage The *cleavage within a *fold which is systematically oriented with respect to the *axial plane, and which lies parallel or sub-parallel to the axial plane, particularly in the *hinge region. Away from this region the cleavage commonly deviates from its parallelism and forms a symmetric, convergent, or divergent fan about the *axial surface.

axial ratio (intercept ratio) In the study of crystals, the position of a *crystal face in space is given by the intercepts the face or plane makes on three (or four) imaginary lines, called 'crystallographic axes'. The X-ray crystallographer can measure the 'unit-cell' dimensions in ångstrom units (Å), and the axial ratios express the relative, and not the absolute, lengths of the cell edges corresponding to the crystallographic axes. These ratios (or '*parameters') are often expressed reciprocally as 'indices', e.g. *Miller indices.

	Crystallographic Axes		
	a (x)	*b* (y)	*c* (z)
Intercepts of crystal face DEF in ångstroms, on *a*, *b*, and *c* axes, measured from origin.	OD 20 Å	OE 10 Å	OF 40 Å
If *b* intercept is made equal to 1 the axial ratio is obtained for the crystal.	*20*/10 = 2	*10*/10 = :1	*40*/10 = :4
Indices are obtained by dividing the intercepts of face DEF *into* those of the parametral plane, which is a face of the unit form with intercepts (111).	*1*/2	*1*/1	*1*/4
Miller's indices of face DEF are obtained by removing fractions.	2	4	1

Note that if face DEF is selected as the *parametral plane, then its indices would be 2/2, 1/1, 4/4 = 111.

axial rift *See* MEDIAN VALLEY.

axial surface The surface which joins the *hinge lines on adjacent folded surfaces. Where it is planar it is called the *axial plane.

axial tilt The angle by which the rotational axis of the Earth differs from a right angle to the orbital plane; this angle varies between 21.5° and 24.5° over a cycle of 40 000 years and at present is about 23.5°. *See* MILANKOVICH CYCLES.

axial trace A line that marks the intersection of the *axial surface of a *fold with any other plane or surface. In practical terms it is the intersection of the axial surface with the surface of the Earth, or with a vertical profile through a fold sequence.

axial trough *See* MEDIAN VALLEY.

axinite An uncommon borosilicate *mineral with variable composition, generally given as $(Ca,Mn,Fe^{2+})Al_2(BO_3)Si_4O_{12}(OH)$; sp. gr. 3.26–3.36; *hardness 7; violet-brown; *crystals normally *triclinic; occurs mainly as a product of *contact metamorphism of calcium-rich *sedimentary rocks (e.g. *contact zones of Cornish *granites).

axiolitic structure Elongate fibres of *alkali feldspar intergrown with *cristobalite which have nucleated and grown outwards from both sides of a linear fracture within rhyolitic *glass. The fibres, which can be seen clearly only by using a petrological microscope, grow in the solid state during devitrification of the *rhyolite glass, the linear fracture acting as the axis and nucleation point for *crystals defining the structure.

axis of rotation The line about which rotation occurs. The Earth's axis of rotation is the line between the North and South geographic Poles, which precesses (*see* CHANDLER WOBBLE). Tectonic *plates rotate about *Euler poles.

axis of symmetry *See* CRYSTALLOGRAPHIC AXES; and CRYSTAL SYMMETRY.

axis vertebra *See* VERTEBRA.

azimuth The angle made by a line on the surface of the Earth with magnetic meridian. In *radar terminology, the direction at right angles to the direction of radar propagation.

azimuth resolution In *radar terminology, the width of ground area illuminated by each pulse of *electromagnetic radiation. The azimuth increases with increasing distance from the radar. The azimuth resolution and the *slant-range resolution govern the resolution of a radar.

azimuthal distribution The spread of directional data measured from features such as cross-beds (*see* CROSS-STRATIFICATION), *ripples, and oriented *fossils, used to determine the direction of current flow. Azimuthal distributions may be displayed on a *rose diagram, or treated statistically to give vector mean and variance data.

Azoic *See* CRYPTOZOIC.

Azores high Semi-permanent anticyclonic region with subsiding air over the *Atlantic Ocean at around 30° N latitude. Movement of the system poleward in summer has a major impact upon the climate of Europe. The aridity of the Sahara Desert and the adjacent Mediterranean region is due to the subsidence of air in this high-pressure system. *See* AIR MASS; and ANTICYCLONE.

azurite *Secondary mineral, $Cu_3(CO_3)_2(OH)_2$; sp. gr. 3.7–3.9; *hardness 3.5–4.0; *monoclinic; various shades of deep azure blue; light-blue *streak; *vitreous *lustre; *crystals often *tabular or short *prisms, and radiating aggregates; *cleavage *prismatic or *pinacoidal; occurs in the oxidized zone of copper deposits, associated with *malachite but less widespread; soluble in nitric or hydrochloric acid, with effervescence. It is a minor ore of copper.

B *See* BAR.

Bacillariophyceae (diatoms) Class of unicellular *algae, usually occurring singly, but may be colonial or filamentous. Cell size ranges from 5 to 2000 μm. The cell wall (*frustule) is impregnated with silica and consists of two *valves, one of which overlaps the other like the lid on a box. The frustule is commonly delicately ornamented and pierced by tiny holes (punctae) which may be covered by porous sieve membranes. There are two *orders. The Centrales (centric diatoms) are circular, with *radial symmetry, and are predominantly marine. The Pennales (pennate diatoms) are elliptical, with *bilateral symmetry, and dominate freshwater environments. The frustules have formed an important constituent of deep-sea deposits since the *Cretaceous. The oldest known diatom is usually taken to be *Pyxidicula bollensis* from the *Jurassic. *See* DIATOMACEOUS EARTH; DIATOMITE; and DIATOM OOZE.

back-arc basin The zone of thickened sedimentation and extensional *tectonics which lies behind an *island arc. Sedimentation in back-arc basins can be very varied, ranging from *pelagic through *turbidites to *alluvial fans, although volcaniclastics (*see* CLAST) are common. Various convective systems have been postulated to explain extension, which is oriented parallel to the compression in a *destructive margin. Back-arc basins have been classified into *retro-arc basins and *inter-arc basins. Examples include the Japan Sea (between Japan and Korea) and the basins west of the *Marianas Trench.

back-arc spreading The formation of the *oceanic crust of the *marginal basins, which is thought to occur by a process similar to that of normal oceanic crust, but involving a convective system developed over a *subducting *lithospheric plate. In most known instances the injection of new oceanic crust appears to be diffuse, rather than concentrated into *spreading ridges.

backing Anticlockwise shift of the direction of the wind. The reverse change is called veering.

backreef Area behind, or to the landward of, a *reef. This zone usually includes a *lagoon between the reef and the land.

backscatter When a surface is illuminated by a radar beam, a portion of the energy is reflected back to the antenna ('specular reflection'). A further portion is scattered back in the same way as light is scattered from non-reflective surfaces. The proportion scattered is controlled by factors such as the roughness and *dielectric properties of the surface, and the *wavelength of the incident beam.

backshore The part of a *beach that is above the level of normal high *spring tides. This zone is usually dry; only when exceptionally high tides or storms occur does wave action influence this part of a beach.

backswamp Area of low, ill-drained ground on a *floodplain away from the main channel. It stands slightly lower than adjacent *alluvial fans extending from the valley sides, and is below natural *levées that rise towards the main channel. It is a site of slow accumulation of silts and clays.

back thrust A *thrust in which displacement is in an opposite direction to that of the main thrust propagation. Back thrusts are thought to form as a result of *layer-parallel shortening in a late stage of thrust sequences.

backwash Seaward return of water down a *beach. The process is affected by wave height and *frequency, and by beach properties such as gradient and permeability. The general effect is to steepen the beach profile.

Bacteria The *domain comprising the kingdom *Eubacteria.

bacterial chemosynthesis *See* CHEMOSYNTHESIS.

badlands Term originally applied to the intricately eroded plateau country of S. Dakota, Nebraska, and N. Dakota, but now widened to refer to any barren terrain that has been similarly intensively dissected. It is most common in areas of infrequent but intense rainfall and little vegetation cover.

bafflestone An *autochthonous *carbonate rock, with the original components organically bound during deposition. The organisms acted as baffles to permit the deposition of finer *matrix material. *See* EMBRY and CLOVAN CLASSIFICATION.

baguio The name given to a *tropical cyclone that forms in the vicinity of Indonesia and the Philippines (Baguio is the name of a town in Luzon, Philippines).

bahada *See* BAJADA.

Bairnsdalian A *stage in the Upper *Tertiary of south-eastern Australia, underlain by the *Balcomian, overlain by the *Mitchellian, and roughly contemporaneous with the *Serravallian Stage.

Bai-u season The principal rainy season of the south-east monsoon in southern and central Japan.

bajada (bahada) Extensive, gently sloping plain of unconsolidated rock debris resting against the foot of a mountain front in a semi-arid environment. Typically it is made of a number of coalescing *alluvial fans laid down by *ephemeral streams as their gradients lessen on leaving the mountain zone. Material is also supplied by the *weathering of the mountain front. Alternatively, it may comprise the alluvial accumulation on the lower part of a *pediment.

Bajocian A *stage in the European Middle *Jurassic (173.5–166.1 Ma, Harland et al., 1989). *See also* DOGGER.

balanced sections A method for reconstructing graphically the original, pre-deformational geometry of folded (*see* FOLD) or *thrust terrains from their present appearance. Balanced sections allow the amount of regional *shortening to be measured and the sequence of thrusting to be deduced. Sections are usually made parallel to the main axis of compression and originate from a reference line within undeformed strata. Three principal methods are used to balance different properties: line-length balancing; areal balancing; and *strain balancing.

Balanidae (acorn barnacles; order Thoracica, suborder Balanomorpha) Family of radially symmetrical, balanomorph (*sessile symmetrical) barnacles, including the familiar barnacles of the genus *Balanus* exposed at low tide and characteristic of the *littoral zone. *Balanus* has a fossil record extending back to the *Eocene.

Balcombian A *stage in the Upper *Tertiary of south-eastern Australia, underlain by the *Batesfordian, overlain by the *Bairnsdalian, and roughly contemporaneous with the upper *Burdigalian and *Langhian Stages.

Balfour A *series in the New Zealand Upper *Triassic, underlain by the *Gore, overlain by the *Herangi (*Jurassic), and comprising the Oretian, Otamitan, Warepan, and Otapirian *Stages.

ball and pillow structure A *sedimentary structure occurring on the base of some *sandstones which are interbedded with *mudstones, and characterized by globular protrusions and isolated pillows of sandstone found in the underlying mudstone. These structures form by the differential settling of the unconsolidated sand into less dense mud below.

ball clay A *sedimentary, usually *lacustrine, *kaolinitic *clay, derived from the intense *weathering of granitic and other rocks to give a unique clay sediment with both plasticity and strength. Ball clays were originally cut into cubes but by the time they had been hauled from Bovey Tracey, Devon, UK, to Stoke-on-Trent where they were to be used, the cubes had been turned into balls as they jostled together in barges and carts, hence the name.

balloon sounding Use of lighter-than-air balloons to establish wind conditions in the upper air. Usually the balloons are tracked by radar, and instruments may be attached to the balloons to record temperature and humidity at given pressure levels. *See also* RADIOSONDE; and RAWINSONDE.

Baltica (Baltoscandia) The continental mass of north-western Europe (including most of what are now the UK, Scandinavia, European Russia, and Central Europe) that formed the south-eastern margin of the *Iapetus Ocean. During the *Caledonian orogenic event, this continental mass was brought into juxtaposition against N. America and Greenland by *subduction of the Iapetus Ocean during the *Silurian and Early *Devonian.

Baltoscandia *See* BALTICA.

band In *remote sensing, the range of wavelengths which are examined.

banded iron formation (BIF) Finely banded, siliceous, *hematite deposits found in *Precambrian rocks, forming *stratiform units often several hundred metres thick and persistent over 150 km or more. They probably formed by chemical–organic processes during sedimentation in stable, shallow basins with little detritus, and so are *syngenetic deposits. In their enriched form (40–60% iron) BIFs are mined *opencast. They include the world's most important sources of iron ore, e.g. at Hammersley, Western Australia; Lake Superior, USA; Labrador, Canada; Ukraine, and Brazil.

band filter A *filter that either passes over (band-pass filter) a discrete range of *frequencies with minimal *attenuation, or rejects (band-reject filter) those frequencies by substantial attenuation. Band-reject filters are the inverse of band-pass filters.

band-pass filter *See* BAND FILTER.

band-reject filter *See* BAND FILTER.

band silicate *See* INOSILICATE.

bankfull flow Maximum amount of *discharge (usually measured in m^3/s) that a stream *channel can carry without overflowing. Its frequency of occurrence varies between streams, from a few times each year to once every few years. The water height at bankfull discharge is referred to as the 'bankfull stage'.

bankfull stage *See* BANKFULL FLOW.

banner cloud Motionless, flag-like cloud, commonly of lenticular (lens-like) shape, forming to the lee (eddy zone) side of a hill or mountain peak. The cloud extends downwind in a strong current of humid air. Many distinctive mountain peaks (e.g. the Matterhorn and Table Mountain) are associated with a characteristic banner cloud. *See also* LEE WAVE.

bar 1. (b) Unit of pressure approximately equal to one atmosphere ($14 lb/in^2$), and precisely equal to 10^5 Pa ($10^5 N/m^2$) in SI units. The pressure of the atmosphere at sea level on average is very approximately one bar, or about 1013 millibars (the bar commonly being divided into one thousand millibars, mb). **2.** Geomorphologic term: (*a*) Low ridge of sand or shingle laid down by marine aggradation in shallow water adjacent to a coastline. There are several varieties: a bay bar joins the two flanks of a bay and may enclose a lagoon; an offshore or barrier bar runs parallel to a coastline and up to 40 km distant. (*b*) Rocky obstruction across a glaciated valley. *See* glacial stairway. (*c*) Lobate river *bedform, typically constructed of gravel, often regularly spaced, and forming a riffle or shallow section. (*d*) Point bar: a low crescentic shoal on the convex side (inside) of a river bend, consisting of material that has been eroded from an outside bend, either opposite or upstream. Point-bar deposits consist of relatively coarse materials, often showing an upstream *dip.

Baragwanathia longifolia One of the earliest known vascular plants (*see* TRACHEOPHYTA). During the *Devonian and *Carboniferous, the lycopods (club mosses) reached the peak in their diversification. *B. longifolia* was an early representative of the family Drepanophycaceae (class *Lycopsida) and is known from the Lower Devonian.

barat Local term for a fierce north-westerly wind common from December to February on the northern coast of Celebes.

barchan (adj. barchanoid) Crescent-shaped mobile *dune in a sand desert in which the wind blows predominantly from one direction. The dune moves by the erosion of sand from the windward slope, and its accumulation on the steeper lee or slip slope which stands at about 32°. Average velocities of dune movement are 10–20 m/yr.

Barchan

barchanoid *See* AKLÉ; and BARCHAN.

barite (baryte) A *mineral, $BaSO_4$, which may form a *solid solution series with *celestite ($SrSO_4$); sp. gr. 4.3–4.6; *hardness 3.0–3.5; *orthorhombic; colourless to white, often tinged yellow, brown, blue, green, and red; white *streak; *vitreous *lustre; *crystals commonly *tabular, *prismatic, but can be fibrous, lamellar, and often granular; *cleavage perfect {001}, present {210}, {010}. Occurs as a vein filling, as a *gangue mineral with ores of lead, copper, zinc, silver, iron, and nickel, associated with *calcite, *quartz, *fluorite, *dolomite, and *siderite, and as a low-temperature mineral which also occurs as a replacement for *limestone, and as a *cement in *sandstone. Insoluble in acid. It is used as a weighting agent in drilling muds, in the chemical industry, in the manufacture of rubber, paper, and high-quality paints, and as an X-ray absorbent.

barkevikite An alkali (sodium and potassium), iron-rich *amphibole, whose colour is very distinctive in *thin section.

barnacles *See* BALANIDAE; and CIRRIPEDIA.

baroclinic **1.** Applied to an atmospheric condition in which isobaric and constant-density surfaces are not parallel, e.g. in a frontal zone. **2.** Applied to a state in the ocean in which the surfaces of constant pressure intersect surfaces of constant density. In this situation, the water density gradient depends on water properties (temperature and *salinity) as well as pressure (depth). This can be contrasted with the *barotropic situation.

baroduric Capable of withstanding high pressures.

barograph *Barometer that gives a continuous recording of air pressure. It is based on an *aneroid instrument with levers attached to the vacuum chambers, and records a trace on a chart mounted around a clock drum.

barometer Instrument for the measurement of *atmospheric pressure. The usual type is a mercury barometer, in which the atmosphere's pressure on a small reservoir of mercury supports a column of mercury in a vacuum tube the open end of which is below the surface of the mercury in the reservoir. The column is on average about 76 cm (30 inches) high. Readings must be corrected to compensate for pressure variation due to gravitational anomalies and for thermal expansion or contraction of the mercury; therefore correction to a standard temperature is necessary. *See also* ANEROID BAROMETER; FORTIN BAROMETER; and KEW BAROMETER.

barothermograph Device for the continuous measurement of both pressure and temperature, on a revolving chart.

barotropic Applied to a state in a water mass in which the surfaces of constant pressure are parallel to the surfaces of constant density. In this situation the density gradient depends on depth only, as in an isothermal freshwater lake. *Compare* BAROCLINIC.

barred basin A partially restricted *sedimentary basin, where free movement of waters is impeded by the presence of a rock sill or sediment barrier. This restriction often results in *anoxic or oxygen-poor waters, or, in arid areas, in *evaporite deposition.

Barrell, Joseph (1869–1919) Barrell was a teacher of mining engineering at Yale University, who developed (in 1917) a chronology for the *Phanerozoic, based on rates of sedimentation and *radiometric dating. He proposed that the anomalously low age of the Earth based on sedimentation rates could be explained by long gaps in the sedimentary record, called *diastems, and that sediments formed only when *subsidence occurred. Barrell was the first person to use the terms *lithosphere and *asthenosphere, based on his observations of *isostasy.

barrel trend *See* BOW TREND.

Barremian **1.** An Early *Cretaceous *age, preceded by the Neocomian *Epoch (but *see* NEOCOMIAN), followed by the *Aptian Age, and dated at 131.8–124.5 Ma (Harland et al., 1989). **2.** The name of the corresponding European *stage, for which the *stratotype is at Angles, France.

barren interzone An *interval zone that is devoid of *fossils.

barren intrazone A measurable breadth of *strata, occurring within a *biostratigraphic unit but itself containing no *fossils.

barrier General term for a depositional feature standing on the seaward side of a coastline. *See* BARRIER BAR; BARRIER BEACH; BARRIER ISLAND; BARRIER REEF; BAYHEAD BARRIER; and BAYMOUTH BARRIER.

barrier bar A major *longshore bar of gravel or sand whose surface is below mean still water level. Normally it is formed off a depositional coast of low gradient and with ample unconsolidated sediment. *See* BARRIER.

barrier beach A relatively small, shingle feature that protects a steep coast. *See* BARRIER.

barrier island An elongated ridge that may extend from a few hundred metres to 100 km along a coast forming a segmented *barrier-bar complex, and found between two tidal inlets. Barrier-island systems have a *lagoonal area on their landward side, and often have wind-blown dunes and vegetation on the exposed (seaward) side of the *barrier. There are three main hypotheses to explain the origin of barrier islands: (*a*) the building up of submarine bars; (*b*) spit progradation parallel to the coast and segmentation by inlets; and (*c*) submergence of subaerial coastal beach ridges by a rise in sea level. Barrier islands are most common in areas of low tidal range.

barrier reef *Reef trending parallel to, but separated by a *lagoon from, a shore. The reef-building organisms build up the structure to approximately the low tide level. One of the finest examples is the Great Barrier Reef which lies off the north-eastern coast of Australia: it extends for about 1900 km and is 30–160 km in width.

Barrovian-type metamorphism A sequence of *regional metamorphic *mineral reactions recorded by the successive mineral assemblages seen in *metapelites (metamorphosed or sandy shales) from the Barrovian terrain around Glen Esk in north-eastern Scotland and characteristic of medium regional metamorphic gradients of temperature and pressure. George Barrow (1853–1932) in 1912 was the first to recognize this sequence of metamorphic mineral assemblages. In *pelites, Barrovian-type metamorphism is marked by the development of a sequence of *index minerals, starting with *chlorite in the lowest-grade rocks, and passing upgrade through *biotite, *garnet, and *kyanite, to *sillimanite in the highest-grade rocks (*see* BARROW'S ZONES).

Barrow's zones The original subdivision by George Barrow (1853–1932) of the sequence of mineral changes seen in rocks of pelitic composition in the Glen Esk region of north-eastern Scotland. Each zone is bounded by two *isograds, each of which marks the appearance, in the direction of increasing *metamorphic grade, of a new *index mineral, and is named after the index mineral seen on the lower-grade boundary. (For instance, the kyanite zone has the kyanite isograd as its low-grade boundary and the sillimanite isograd as its high-grade boundary.) Within each zone, no additional minerals appear in the *pelites, the constant mineral assemblage representing equilibrium over a range of metamorphic conditions. *See* BARROVIAN-TYPE METAMORPHISM. *Compare* BUCHAN METAMORPHIC ZONES.

Bartonian **1.** An *age in the Middle *Eocene, preceded by the *Lutetian, followed by the *Priabonian (Late Eocene) and with an upper boundary dated at 38.6 Ma (Harland et al., 1989). **2.** The name of the corresponding European *stage, which is roughly contemporaneous with most of the upper *Narizian (N. America), upper *Bortonian and *Kaiatan (New Zealand), and part of the *Aldingan (Australia). It was originally considered to be a lateral equivalent of the Priabonian Stage (Upper Eocene). (The name should not be confused with *Bortonian.)

baryte *See* BARITE.

basal conglomerate **1.** A *conglomerate deposited at the base of a sedimentary sequence, e.g. at the base of a *channel-fill deposit. **2.** A conglomerate deposited above an *unconformity surface.

basal sliding The process by which a temperate *glacier moves over its bed. It involves three mechanisms: relatively rapid creep in the basal layers; pressure melting, whereby *ice under pressure melts on the up-glacier side of a small obstacle and the released water freezes on the down-glacier side; and slippage over a layer of water at the bed.

basalt A dark-coloured, fine-grained, *extrusive, *igneous rock composed of *plagioclase feldspar, *pyroxene, and *magnetite, with or without *olivine, and containing not more than 53 wt.% SiO_2. Many basalts contain *phenocrysts of olivine, plagioclase feldspar and pyroxene. Basalts are divided into two main types, *alkali basalts and *tholeiites, with the tholeiites being subdivided into olivine tholeiites, tholeiites, and quartz tholeiites. Petrographically (*see* PETROGRAPHY), alkali basalts have as their *groundmass pyroxene titanaugite (an *augite rich in titanium), whereas tholeiites have pigeonite (a calcium-poor pyroxene). Also, for similar concentrations of SiO_2, alkali basalts have a higher content of Na_2O and K_2O than tholeiites. Basalt flows cover about 70% of the Earth's surface and huge areas of the *terrestrial planets, and are therefore arguably the most important of all crustal rocks. They are formed by *partial melting of *mantle *peridotite. Alkali basalts are typically found on oceanic islands and on the *continental crust in regions of crustal upwarping and rifting. Tholeiites are typically found on the ocean floor and on the stable continental crust where they form large basalt plateaux such as the Deccan Traps of India.

basaltic meteorites *Achondrite *meteorites that have been derived from parent bodies sufficiently evolved to have acquired a basaltic crust (in contrast, for example, to *chondrites, which have derived from relatively primitive bodies). *Eucrites and howardites (similar to eucrites, but more *pyroxene-rich) probably formed through magmatic processes on the same parent body (possibly the *Moon) around 4.5 billion years ago. The *shergottyite–nakhlite–chassignite class of basaltic meteorite has a distinctly different oxygen-isotope composition (*see* OXYGEN-ISOTOPE RATIO), *volatile content, and age (around 1.3 billion years), and originated from a separate parent body, probably the planet *Mars.

basal thrust *See* SOLE THRUST.

basanite An *extrusive, *mafic *igneous rock consisting of a *feldspathoid *mineral (*nepheline, *analcite, or *leucite), *olivine, *plagioclase feldspar, *pyroxene, and minor *accessory minerals. Essentially, the mineralogy is that of a *basalt, with the addition of a feldspathoid mineral and olivine. Varieties of basanites are defined by the type of feldspathoid mineral present to give nepheline-basanite, analcite-basanite, and leucite-basanite. These rocks are found in close association with *alkali basalts. *See also* TRACHYBASALT.

base According to the Brønsted–Lowry theory, a substance that in solution can bind and remove hydrogen *ions or protons. The Lewis theory states that it is a substance that acts as an electron-pair donor. A base reacts with an *acid to give a salt and water (a process called neutralization), and has a *pH greater than 7.

basecourse **1.** The lowest course of masonry. **2.** *See* PAVEMENT.

baseflow (dry-weather flow) In a stream or river, the flow of water derived from the seepage of *groundwater, and/or throughflow into the surface watercourse. At times of peak river flow, baseflow forms only a small proportion of the total flow, but in periods of *drought it may represent nearly 100%, often allowing a stream or river to flow even when no rain has fallen for some time. *See also* INTERFLOW; and SUB-SURFACE FLOW.

baselap The discordant relationship marking the lower boundary of a *depositional sequence, where upper beds lap out over the underlying surface. There are two types of baselap: *onlap, where the overlying beds thin out updip; and *downlap, where younger, initially inclined, overlying beds thin out downdip. *See* DIP.

base level A theoretical plane surface underlying a land mass, denoting the depth below which *erosion would be unable to occur. Sea level provides a base level on a regional scale. Local base levels may be provided by the base of a hillslope, lakes, or by the junction between a tributary and the main river.

basement **1.** Highly folded, *metamorphic or *plutonic rocks, often unconformably (*see* UNCONFORMITY) overlain by relatively undeformed *sedimentary beds (or cover). In this sense, basement is often, though not necessarily, *Precambrian. **2.** The *crust below rocks of interest. In this sense, to a petroleum geologist 'basement' means non-prospective rocks which lie below prospective strata.

base saturation The extent to which the exchange sites of the soil's *adsorption complex are 'saturated' (or occupied) by exchangeable basic cations, or by cations other than hydrogen and aluminium, expressed as a percentage of the total cation-exchange capacity.

base station A station to which reference can be made, to normalize measurements made at out-stations. Geophysical parameters measured at base stations are presumed to be known very accurately and preferably absolutely. For example, in a *gravity survey, the base station is used to determine instrumental *drift and to provide an absolute value to which *gravity anomalies relate. In *magnetic surveying, a continuous-reading *magnetometer may be used at a base station to monitor the *diurnal fluctuations in the Earth's *geomagnetic field and thus aid the interpretation of magnetic survey data.

base surge A turbulent, dilute flow of *ash and either water or steam, which expands radially as a collar-like cloud from the base of a vertically venting eruption column generated by the explosive interaction of *magma and water. The surges are commonly cold and wet, consisting of ash mixed with water at temperatures below 100 °C. With a high magma : water mass ratio the surges can become dry and hot, consisting of ash mixed with steam. Base surges can be very hazardous, as was the one that occurred during the eruption of Taal, Philippines, 1965. Radially expanding basal clouds observed in nuclear explosions are a type of base surge. *See* SURGE.

Bashkirian 1. The earliest *epoch in the *Pennsylvanian, comprising the Kinderscoutian, Marsdenian, and Yeadonian Ages (these are also *stage names in British stratigraphy), and the Cheremshanskian and Melekesskian Ages (stage names in Russian stratigraphy). The Bashkirian Age is preceded by the *Serpukhovian, followed by the *Moscovian, and has its initial boundary (the *Mississippian–Pennsylvanian boundary) dated at 322.8 Ma (Harland et al., 1989). **2.** The name of the corresponding eastern European *series, which is roughly contemporaneous with the *Namurian B and C plus the *Westphalian A (western Europe), and the *Morrowan Series (N. America).

basic 1. Fundamental. **2.** Adjective derived from *base. **3.** *See* BASIC ROCK.

basic rock Rock with a relatively high concentration of iron, magnesium, and calcium, and with 45–53% of *silica by weight. Examples include *gabbro, which is a coarse-grained basic *intrusive rock, and *basalt, which is a fine-grained basic volcanic (*extrusive) rock. *Compare* ACID ROCK; and INTERMEDIATE ROCK. *See also* ALKALINE ROCK.

basic soil A soil with a *pH greater than 7.0. *See* BASE. *Compare* ALKALINE SOIL.

***Basilosaurus* (order Cetacea, suborder *Archaeoceti)** One of the best-known archaeocetes; it grew to approximately 20 m in length and lived during the Upper *Eocene. It was discovered recently that *Basilosaurus* retained small hind limbs, useless for locomotion, but perhaps usable as claspers during copulation.

basin 1. Depression, usually of considerable size, which may be erosional or structural in origin. The converse of a *dome. **2.** *See* PERICLINE.

basin-and-range province 1. Specifically, the structural subdivision of the western USA between the Colorado Plateau to the east and the Sierra Nevada to the west. It is dominated by a series of block-faulted ranges and troughs which dissect it, rising above *bajadas and *alluvial plains which mask down-faulted units. Frequently the ranges are tilted blocks approximately 30–40 km across. Movement occurred initially in *Miocene times, and then during the late *Pliocene to *Recent. **2.** Generally, applied to landscapes having a series of tilted *fault blocks that form long, asymmetric ranges separated by broad *basins.

basin-and-swell sedimentation A form of sedimentation in a region of differential *subsidence, where thin, *condensed, sedimentary sequences are deposited on slowly subsiding highs or 'swells', and thicker, usually muddier sediments accumulate in more rapidly subsiding basins between the swells.

basin modelling The computer simulation of the geological evolution of a sedimentary basin, quantifying certain processes within it, with the aim of predicting the distribution and movement of petroleum within the basin and determining the temperatures and pressures.

Basleoan *See* KAZANIAN; and UFIMIAN.

Batesfordian A *stage in the Upper *Tertiary of south-eastern Australia, underlain by the *Longfordian, overlain by the *Balcombian, and roughly contemporaneous with the mid *Burdigalian Stage.

batholith Large (more than 100 km^2) *igneous intrusion, which may comprise several *plutons amalgamated at depth. Most batholiths are granitic in composition and their genesis is linked with *plate tectonics. Generally, batholiths cut across *country rocks and therefore are discordant in nature.

Bathonian Middle *Jurassic *stage (166.1–161.3 Ma) commonly represented by carbonate sediments in many areas of Europe, but not in the North Sea or Scandinavia. It is known to contain an abundant fauna of invertebrates. *See also* DOGGER.

bathy- From the Greek *bathus* meaning 'deep', a prefix meaning 'deep' as applied to the oceans.

bathyal zone Oceanic zone at depths of 200–2000 m, lying to the seaward of the shallower *neritic zone, and landward of the deeper *abyssal zone. The upper limit of the bathyal zone is marked by the edge of the *continental shelf. In marine ecology, it is the region of the *continental slope and rise. It may be geologically active, and include *trenches and *submarine canyons, with underwater erosion producing avalanches.

bathymetry The measurement of the depth of the ocean floor from the water surface; the oceanic equivalent of topography.

bats *See* CHIROPTERA.

Bauplan (pl. Baupläne) The generalized body plan of an archetypal member of a major taxon.

bauxite A mixture of three hydrates of alumina, mainly *gibbsite, and also *diaspore and *boehmite, and containing impurities of iron, phosphorus, and titanium; colour is variable from dirty white through grey, yellow, brown, and red; sp. gr. 2.0–2.55; *hardness 1–3; it can be compact, *earthy, concretionary (*see* CONCRETION), *pisolitic, or *oolitic. Bauxite results from the tropical weathering of aluminium silicate rocks under good surface drainage to yield *clay minerals which are subsequently desilicated. *Minerals associated with the alumina hydrates in bauxites and *laterites (ferruginous bauxites) include *goethite and lepidocrocite, *hematite, and the clay minerals *kaolinite and *halloysite. Bauxite is the main ore of aluminium and to be commercially exploited should contain more than 25–30% aluminium oxide. The main constraint is the amount of available alumina which can be extracted by the Bayer or similar process. It is named after Les Baux de Provence, in southern France; the major producers are Australia and Brazil.

Baventian The Lower *Pleistocene, cold-stage, marine silts and marine clays, that form part of the tripartite division of deposits revealed in a *borehole at Ludham, in eastern England. *See also* ANTIAN; LUDHAMIAN; PASTONIAN; and THURNIAN.

bay bar *See* BAR.

Bayesian In statistics, applied to the re-evaluation of probabilities based on empirical observation.

bayhead barrier A *barrier beach protecting the head of a bay, but separated from it by a *lagoon. *See* BARRIER.

bayhead beach Sand or shingle *beach in the low-energy environment at the head of a bay. It is typical of irregular coastlines in which bays and promontories alternate.

baymouth barrier A *barrier that partially encloses a bay at its entrance.

Bazin's average velocity equation In 1897, when considering flow in open *channels, H. E. Bazin (1829–1917) proposed that the Chezy discharge coefficient (C) could be related to the *hydraulic radius (r) and a channel roughness coefficient (k_1) by the formula: $C = 157.6/[1 + (k_1/r^{1/2})]$. Other more complex formulae have also been proposed. *See* CHEZY'S FORMULA.

beach An accumulation of sand and gravel found at the landward margin of the sea or a lake. The upper and lower limits approximate to the position of the highest and lowest tidal-water levels. The angle of slope and the *sedimentary structures of a beach are related to the grain size of the beach materials, and to the nature of wave activity and other sedimentary processes active in the area.

beach cusp One of a series of regularly spaced crescent-shaped structures forming local relief along a *beach. The horns or 'headlands' of the cusp are composed of coarse *sand or gravel, and point seaward down the beach. The intervening troughs or 'bays' are made up of finer sand. Beach cusps are usually several centimetres high, although larger examples have been described. The size and spacing of cusps appears to be related to the nature of the waves breaking on the beach.

beach drift The zig-zag progression of *sand and other debris along a *beach. Particles are driven obliquely up a beach by the *swash and are then returned down the steepest gradient of the beach by the *backwash. The combination of these two movements gives the zig-zag progression. *See also* LONGSHORE DRIFT.

beach rock *Cemented beach *sand deposit that develops within the intertidal zone by the precipitation of needle-like crystals of *aragonite in the *pore space between the grains. The cementation process is relatively rapid, taking as little as 10 years for a lithified rock to develop. The precipitation of the cement is favoured by a warm climate, and may be aided by algal or bacterial action.

beaded lightning *See* PEARL-NECKLACE LIGHTNING.

beak *See* SHELL BEAK.

beardworms *See* POGONOPHORA.

bearing capacity Maximum *load per unit area a surface can support in safety without shear *failure.

Beaufort scale Scale of values, from 0 to 12, for describing wind strength, as defined by Admiral Beaufort in the nineteenth century. Each wind force is recognized by its common effects on objects in the landscape (dust, flags, trees, etc.) and on people in the open, or on the state of the sea surface. *See also* SAFFIR/SIMPSON SCALE; and FUJITA TORNADO INTENSITY SCALE.

Beche, Henry Thomas de la (1796–1855) Founder of the British Geological Survey, the Museum of Practical Geology, the Mining Records Office, and the School of Mines, all of which are in Britain. De la Beche was a careful observer and skilled cartographer and artist, who emphasized the importance of stratigraphy and pioneered the reconstruction of ancient environments.

Becke, Friedrich Johann Karl (1855–1931) A Czechoslovakian mineralogist and petrologist from the University of Prague, Becke developed a method for determining the relative *refraction of light in microscopy, later named for him. He also did important work on metamorphic *recrystallization, developing a descriptive terminology and a classification of metamorphic *facies. *See* BECKE LINE TEST.

Becke line test In transmitted-light microscopy, a comparative test used to determine the approximate *refractive index of a *mineral. In *plane-polarized light, when the substage iris diaphragm is partly closed to accentuate grain boundaries, a thin line of light, the 'Becke white line', appears. If the microscope tube is then racked up, or the microscope *stage racked down, the light line will move into the medium with the higher refractive index. The medium could be either an adjacent mineral of known refractive index or, more commonly, the mounting medium, usually a cold-setting resin or Canada balsam (refractive index 1.54).

bed *See* BEDDING; and STRATUM.

bedding Layering of sheet-like units, called beds or strata (the terms are not synonymous, *see* STRATUM); a bed or stratum being the smallest distinguishable division within the classification of stratified sedimentary rocks. Cross-bedding is a type of stratification in which some sediment layers have an inclined attitude in relation to those immediately above or below. *See* CROSS-STRATIFICATION. *Compare* LAMINATION.

bedding foliation A *foliation parallel to the *bedding which is commonly found in limbs of tight and *isoclinal folds, where the *axial plane is parallel or sub-parallel to the limbs.

bedding plane 1. Well-defined, planar surface that separates one bed from another in *sedimentary rock. Each plane marks a break in deposition. **2.** *See* VOIDS.

bedform The shape of the surface of a bed of granular *sediment, produced by the *flow of air or water over the sediment. The nature of the bedform depends upon the flow strength and depth, and upon sediment *grain size. For fine to medium *sand, the typical sequence of bedforms produced under conditions of constant depth and increasing strength of unidirectional flow is: no movement; *ripples; *sand waves; *dunes; and an upper-flow-regime *plane bed. In coarse sand a lower-flow-regime plane bed develops first, then ripples, followed by sand waves, then dunes, and an upper-flow-regime plane bed. At higher-strength flows, the upper flow regime plane bed is replaced by *antidunes.

bed load (traction carpet, traction load) The coarser fraction of a river's total *sediment load, which is carried along the bed by sliding, rolling, and *saltation. It constitutes 5–10% of the total load.

bed roughness The surface relief developed at the base of a flowing fluid, comprising *bedforms (form roughness), and particles projecting from the sediment carpet (*grain roughness). 'Bed roughness' is the quantifiable factor which expresses the frictional effect that the bed exerts on the *flow. Surfaces whose bed-roughness elements do not project through the viscous sub-layer at the base of the flow are said to be smooth. Surfaces whose bed-roughness elements project through the sub-layer are said to be rough.

Beekmantownian A *stage of the *Ordovician in the *Canadian *Series of N. America.

Beestonian 1. A cold period during the Middle *Pleistocene. **2.** Arctic, freshwater, bed deposits of *sand and silts represented at Beeston, Norfolk, England. Some gravels in Norfolk, Suffolk, and Essex, England, have also been ascribed to this period.

belemnites *See* BELEMNITIDA.

Belemnitida (belemnites; class *Cephalopoda, subclass Coleoidea) One order of extinct cephalopods in which the shell is internal and composed of a *phragmocone, *rostrum, and pro-ostracum (*see* SKELETAL MATERIAL). Belemnites appear in the *Jurassic, continue through the *Cretaceous, and a few persist into the *Eocene. The other order, Aulacocerida (*Carboniferous to Jurassic) may have retained a body chamber, but in the Belemnitida this is reduced to the pro-ostracum. The most posterior portion of the shell is known as the 'guard' (*rostrum). This is a bullet-shaped cylinder made up of radiating needles of *calcite with a conical cavity (alveolus) in its anterior end into which fits the phragmocone, a conical, aragonitic (*see* ARAGONITE), septate (*see* SEPTUM) structure, cut by a tiny *siphuncle, that is *homologous to the external shell of other cephalopods. The pro-ostracum is a tongue-like, anterior projection from the phragmocone and perhaps protected the anterior part of the body.

Belinda (Uranus XIV) One of the lesser satellites of *Uranus, with a diameter of 34 km. It was discovered in 1986.

bellerophontiform Applied to the shape of gastropod (*Gastropoda) shells which resemble those of the genus *Bellerophon*. The coiling is isotrophic (i.e. two faces of the shell are symmetrical in respect of a median plane perpendicular to the axis).

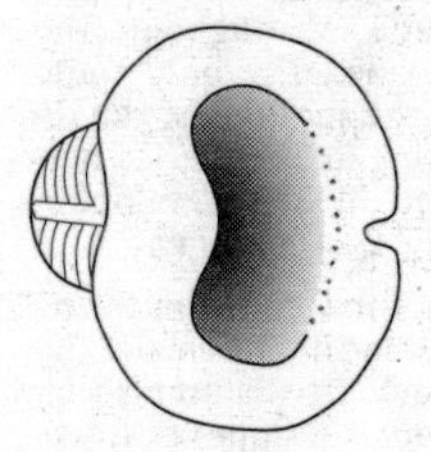

Bellerophontiform

bell pit In mining, an obsolete method for winning *ore or *coal from shallow deposits. Material was extracted and dragged to a central shaft, leaving a bell-shaped excavation. The term comes from ironstone working in Derbyshire, England.

bell trend *See* DIRTYING-UP TREND.

bench 1. Horizontal step along which material is worked in an open pit. **2.** Any narrow, flat surface in solid rock.

Bendigonian A *stage of the Lower *Ordovician of Australia, underlain by the *Lancefieldian and overlain by the *Chewtonian.

beneficiate To increase the grade of ore after grinding by a concentration process, such as *froth flotation or *gravity, or by *magnetic, *electrostatic, or other methods. The term also implies the elimination of waste material. *See* FLOTATION SEPARATION.

Benguela current Oceanic water current that flows northward along the west coast of southern Africa between latitude 15° S and 35° S. It is distinguished by an area of cold upwelling water, and is a relatively weak current, flowing at less than 0.25 m/s.

Benioff, Hugo (1899–1968) An American seismologist, and professor at the California Institute of Technology from 1950 to 1964, Benioff designed seismographs and other instruments used to study *earthquakes. In 1954 he published a cross-section showing *seismicity beneath the Kamchatka Peninsula, which demonstrated that the depths of earthquakes increased with distance away from the trench. This earthquake zone has been named after him. *See* BENIOFF ZONE.

Benioff zone (Wadati–Benioff zone) One of the zones of deep *earthquake *hypocentres whose existence was first demonstrated in 1927 by the Japanese seismologist Kiyoo Wadati. The zones were mapped in the 1940s and 50s by Hugo *Benioff. They dip from near-surface to a maximum depth of approximately 700 km and are associated with oceanic *trenches, *island arcs, volcanic chains, and young *fold mountains, and are thought to indicate active *subduction. *See* SUBDUCTION ZONE.

benmoreite An *extrusive *igneous rock consisting of *anorthoclase, sodic *plagioclase, ferroaugite (an iron-rich *augite), and iron-rich *olivine, and found as a member of the *alkali basalt *magma series. With increasing differentiation (*see* MAGMATIC DIFFERENTIATION) shown by increasing SiO_2 content, the series is alkali basalt–hawaiite–mugearite–benmoreite–*trachyte. The type locality after which the rock is named is Ben More on the Isle of Mull, Scotland.

Bennettitales Extinct *gymnosperm order ranging from the *Triassic into the *Cretaceous. They resembled *cycads, and possessed reproductive structures which must have looked more like flowers than cones.

Benson's flood peak formula *See* FLOOD PEAK FORMULAE.

benthic storm *See* ABYSSAL STORM.

benthos (adj. benthic) In freshwater and marine *ecosystems, the collection of organisms attached to or resting on the bottom sediments (i.e. *epifaunal), and those which bore or burrow into the sediments (i.e. *infaunal).

bentonite *Montmorillonite-rich *clay formed by the breakdown and alteration of volcanic *ash and volcanic *tuffs.

Bergen School The name used to distinguish the group of meteorologists (Vilhelm *Bjerknes, his son Jacob Bjerknes, H. Solberg, and Tor Bergeron) who, working at the Bergen Geophysical Institute, in Norway, between 1917 and 1920, established the existence and role of *fronts and *air masses in the atmosphere.

Bergeron theory (Bergeron–Findeisen theory) A theory, proposed around 1930 by T. Bergeron, and subsequently developed by W. Findeisen, that provides a mechanism for the growth of raindrops in ice/water cloud. It is based on the differential values for saturation vapour pressure over ice and supercooled-water surfaces. At cloud temperatures of –12 to –30 °C air can be saturated over ice but not over water particles, so evaporation can take place from the water droplets, and ice particles can grow by sublimation at the expense of water droplets. When they are large enough, the ice particles can fall from the cloud, melting as they pass through lower, warmer air. The process depends on there being a mixture of ice and water, and so may operate in mid- and high-latitude cloud but not in all clouds, e.g. not in tropical clouds which are at temperatures above freezing. *See also* COLLISION THEORY; and ICE NUCLEUS.

Bergman, Torbern Olof (1735–84) Professor of chemistry at Uppsala, Sweden, Bergman made contributions to mineralogy. He was a diluvialist, believing that the Earth had an aqueous origin; some of his work on the formation of rocks was later developed and extended by *Werner. *See* DILUVIALISM.

Bergmann's rule The idea, proposed by C. Bergmann in 1847, that the size of *homoiothermic animals in a single, closely-related, evolutionary line increases along a gradient from warm to cold temperatures, i.e. that races of species from cold climates tend to be composed of individuals physically larger than those of races from warm climates. This is because the surface-area : body-weight ratio decreases as body weight increases. Thus a large body loses proportionately less heat than a small one. This is advantageous in a cold climate but disadvantageous in a warm one. *See also* ALLEN'S RULE; and GLOGER'S RULE.

bergschrund A wide and deep *crevasse found between a *cirque glacier and its *headwall. It forms when the glacier has developed to the stage at which it pulls away from the rock slope on its upper side. A series of small bergschrunds ('bergschrund crevasses') may form instead of the single feature.

berg wind Local wind, generically of the *Föhn type, which blows offshore in S. Africa.

Beringia Area comprising the Bering Strait and adjacent Siberia and Alaska. At various times in the late *Mesozoic and in the *Cenozoic, the strait was dry land and so provided an important migration route for plants and animals between the Palaearctic and Neoarctic biogeographical regions.

Bering land bridge The intermittent land connection between Siberia and Alaska that operated throughout the *Cenozoic. This provided the only route into N. America for the mammals, the direct route from Europe to N. America having been interrupted by the developing Atlantic Ocean. *See also* LAND BRIDGE.

berm Ridge or nearly flat platform at the rear of a *beach and standing just above the mean high-water mark. Its distinguishing feature is a marked break of slope at the seaward edge.

Berman balance *See* DENSITY DETERMINATION.

Bermuda high Anticyclonic cell in the Bermuda region as a westward extension to, or displacement of, the *Azores high-pressure system.

Bernoulli, Daniel (1700–82) A Swiss mathematician (one of 11 eminent mathematicians his family produced over four generations), whose most important work was in the field of hydrodynamics. In his book *Hydrodynamica* (1738), he showed that the pressure within a flowing fluid depends inversely on its velocity (the greater the velocity, the lower the pressure). This is now known as Bernoulli's principle (*see* BERNOULLI EQUATION). Bernoulli was born at Groningen, the Netherlands, and educated at Basel, Switzerland, where his father had been appointed professor of mathematics on the death of his brother (Daniel's uncle) who held the post previously. Daniel obtained his master's degree at the age of 16 and his doctorate, on the action of the lungs, at 21. In 1725 he was appointed professor of mathematics at St Petersburg Academy, Russia, but left Russia in 1732. In 1733 he became professor of anatomy and botany at the University of Basel; in 1750 he became professor of natural philosophy, a post he held until his retirement in 1777. He died in Basel.

Bernoulli equation An equation that describes the conservation of energy in the steady flow of an ideal, frictionless, incompressible fluid. It states that: $p_1/p_2 + gz + (v^2/2)$ is constant along any stream line, where p_1 is the fluid pressure, p_2 is the mass density of the fluid, v is the fluid velocity, g is the *acceleration due to gravity, and z is the vertical height above a datum level.

Berriasian A *stage of the European Lower *Cretaceous, dated at 145.6–140.7 Ma (Harland et al., 1989), for which the *stratotype is at Berrias, France. *See also* NEOCOMIAN.

Bertrand lens An accessory lens which may be inserted into the light path above the *analyser in a transmitted-light microscope. When determining *vibration directions or *interference figures using parallel or convergent polarized light, the Bertrand lens is inserted to bring the image of the interference figures into focus. Alternatively, if the Bertrand lens is absent, the *eyepiece may be removed and the vibration directions or interference figures observed by looking down the microscope tube. The lens was first used in 1878 by E. Bertrand who adapted an original (1844) design by G. B. Amici.

beryl *Accessory mineral, $Be_3Al_2Si_6O_{18}$; sp. gr. 2.6–2.8; *hardness 7.5–8.0; *hexagonal; normally green, sometimes blue, yellow, or pink, and translucent to transparent; *vitreous *lustre; *crystals hexagonal *prisms, often striated, also occurs *massive; *cleavage perfect basal {001}; occurs extensively in cavities in *granites, *pegmatites, *mica *schists and *gneisses, and associated with *rutile. It is an *ore mineral for beryllium. Transparent green varieties are emeralds, bluish-green are aquamarine, and pink are morganite.

beta decay Some unstable atoms decay by emitting a negatively charged beta particle (negatron) from the *nucleus, often accompanied by the emission of radiant energy (*gamma rays). Beta decay may be regarded as the alteration of a neutron into a proton and an *electron. As a result of beta decay the *atomic number of the atom is increased by one, while the neutron number is decreased by one.

beta diagram A stereographic diagram which represents the *trend and *plunge of the line produced by the intersection of two planes. In its application in structural geology, the *dip and *strike of a *fold surface are recorded as a *great circle, and two great circles representing both limbs of a fold intersect along a line called a β-axis (in this case the fold axis).

beta-mesohaline water *See* HALINITY.

bevelled cliff A sea cliff whose upper part has been trimmed to a relatively low angle by *Quaternary *periglacial processes, while the lower part is still steep as a result of recent marine activity. Such cliffs are common in south-west England. *See* COASTAL PROCESSES.

Bianca (Uranus VIII) One of the lesser satellites of *Uranus, with a diameter of 22 km. It was discovered in 1986.

biaxial interference figure *See* INTERFERENCE FIGURE.

Bibymalagasia An order of mammals described in 1994 by R. D. E. MacPhee, based on the enigmatic genus (*see* ENIGMATIC TAXON) **Plesiorycteropus*, known from subRecent fossil material from Madagascar. The order is distinguished by having large, perforating, transarcual canals in the neural arches of the lumbar, posterior thoracic, and anterior sacral vertebrae (*see* VERTEBRA); a posteriomedial process on the *astragalus, with a ventral groove for flexor tendons; and large ischial expansions (*see* ISCHIUM). *Plesiorycter-*

pus was formerly assigned to the order Tubulidentata.

biconical Applied to a gastropod (*Gastropoda) shell where the shell is in the shape of two cones with the bases in contact. The axial whorls make one cone, the last-formed whorl or living chamber the other cone.

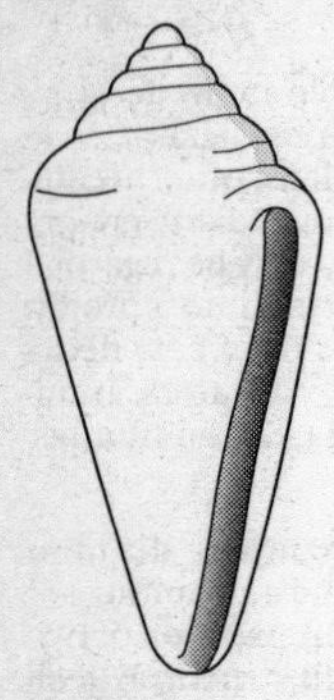

Biconical

BIF *See* BANDED IRON FORMATION.

bifurcation In a phylogenetic (*see* PHYLOGENY) tree, the dichotomous forking of an ancestral *branch which indicates a speciation event.

bifurcation ratio Dimensionless number denoting the ratio between the number of streams of one order (*see* STREAM ORDER) and those of the next-higher order in a drainage network. It may be a useful measure of proneness to flooding: the higher the bifurcation ratio, the greater the probability of flooding.

big bang theory The current explanation for the origin of the universe, in which it expands and evolves from an initial very high-temperature condition about 15–20 billion years ago. The expansion time is given from the reciprocal of the Hubble constant (the rate at which galaxies are receding). All all-pervasive background radiation of 3 K is considered to be residual from the big bang and is the strongest supporting evidence for the theory.

bilateral symmetry The condition, found in many organisms, where one half of the body or structure is the mirror image of the other. Sometimes bilateral symmetry is superimposed on another kind of symmetry, e.g. in some echinoids (*Echinoidea) where basic pentameral (five-sided) symmetry has a bilateral symmetry superimposed upon it.

billow clouds Parallel rolls of cloud with distinct clear areas between the cloud bands, often associated with the cloud variety *undulatus.

bilophodonty In some mammals, a condition in which the 4 cusps on the molar teeth are joined by 2 transverse ridges. *See also* LOPHODONT.

bimodal distribution A distribution of data characterized by two distinct populations. For example, a bimodal grain size will be characterized by two *particle size modes. A bimodal palaeocurrent distribution will exhibit two main current directions (not necessarily opposing directions, which would be termed a 'bipolar' distribution).

bin In a range of data broken into a series of equal intervals, a single interval.

binary system 1. Chemical system of two components, e.g. $MgO–SiO_2$. **2.** *See* STAR PAIR.

bindstone An *autochthonous *carbonate rock, with original components organically bound during deposition. The organisms encrust and bind the finer *matrix *sediment together, but the binding organisms need not necessarily be preserved, as in the case of *stromatolites. *See* EMBRY AND CLOVAN CLASSIFICATION.

Bingham fluid A viscous fluid that possesses a yield strength which must be exceeded before the fluid will flow. Most *lava flows are examples of Bingham fluids. When an initial *shear stress is applied to a fluid lava (e.g. by increasing the slope angle) it will not begin to flow immediately. The slope angle, and therefore shear stress, must be increased until the yield strength of the fluid is exceeded, after which flow will occur. This contrasts with a Newtonian fluid, which has zero yield strength and will flow on any slope.

binomial distribution In statistics, a discrete *probability distribution of the number of outcomes of a particular kind occurring for a set number of trials, where one of two outcomes is possible, each trial is independent, and the probability of a particular outcome is constant.

bio- **1.** From the Greek *bios* meaning 'human life', a prefix associating the word to which it is attached with living organisms or processes. **2.** Prefix for *allochems composed of *bioclastic material. *See* FOLK LIMESTONE CLASSIFICATION.

biochemical oxygen demand *See* BIOLOGICAL OXYGEN DEMAND.

biochron The length of time represented by a biostratigraphic *zone.

biochronology Measurement of units of geologic time by means of biological events. Biochronologists often derive their correlations from widespread and distinctive events in the biological history of the world, based on the first and last appearances of organisms.

bioclast A single, often broken, shell or *fossil fragment.

biocoenosis *See* LIFE ASSEMBLAGE.

biofacies Rock unit, or an association of rock units, characterized by the presence of a *fossil assemblage that is restricted to that particular *facies and that is typical of a specific environment. *Compare* TAPHOFACIES.

biogenic Applied to material, processes, or activities of living or once-living organisms.

biogenesis Principle that a living organism can arise only from another living organism, a principle contrasting with concepts such as that of the spontaneous generation of living from non-living matter. The term is currently more often used to refer to the formation from or by living organisms of any substance, e.g. coal, chalk, chemicals, etc.

biogenetic law Law formulated by the embryologist E. K. von Baer (1792–1876) stating that the early stages of development in animal species resemble one another, the species diverging more and more as development proceeds.

biogenic deposit The formation of rocks, traces (*see* TRACE FOSSIL), or structures as a result of the activities of living organisms.

biogeochemical cycle Movement of chemical elements from organism to physical environment to organism, in a more or less circular pathway. They are termed 'nutrient cycles' if the elements concerned are essential to life. An element may be solid, liquid, or gaseous, or form different chemical compounds, in the various parts of the cycle. Amounts in the inorganic *reservoir pools are usually greater than those in the *active pools. Exchange between the system components is achieved by physical processes (e.g. *weathering) and/or biological processes (e.g. protein synthesis and decomposition). The latter form the vital negative-feedback mechanisms that regulate the cycles. Cycles may be described as varying from perfect to imperfect. A perfect cycle (e.g. the nitrogen cycle) has a readily accessible abiotic, usually gaseous, reservoir and many negative-feedback controls. By contrast, the phosphorus cycle, which has a sedimentary reservoir accessed only by slow-moving physical processes, has few biological feedback mechanisms. Human activities can disrupt these cycles, leading to pollution. Theoretically, perfect cycles are more resilient than imperfect cycles.

biogeochemical exploration *See* GEOBOTANICAL EXPLORATION.

biogeochemistry Science concerned with the effects of living things on subsurface geology; or with the distribution and fixation of chemical elements in the biosphere. Its principles are applied to the systematic collection and analysis of plants in the exploration for mineral deposits. It is also the study of the chemistry of organic sediments and of the chemical composition of fossils and fossil fuels. *See also* GEOBOTANICAL EXPLORATION.

biogeocoenosis *See* ECOSYSTEM.

biogeography The scientific study of the past and present geographical distribution of plants and animals at different taxonomic levels. Modern biogeography also lays great stress on the ecological character of the world vegetation types, and on the evolving relationship between humans and their environment. *See also* PALAEOBIOGEOGRAPHY.

bioglyph Ornamentation on the side of a *trace fossil burrow.

bioherm A build-up of largely *in situ* organisms that produces a *reef or mound of organic origin.

biohorizon **1.** An interface between *strata at which some *biostratigraphic change has occurred. **2.** A biostratigraphic *marker bed.

bioimmuration A type of *fossilization in which soft-bodied encrusting organisms are overgrown by other encrusting organisms. As the soft-bodied organisms disappear, they leave natural moulds on the underside of the layer above them.

biointermediate elements Elements that are partially depleted in surface waters as a result of biological activity. Four such elements are known: Ba, Ca, C, and Ra. *Compare* BIOLIMITING ELEMENTS; and BIOUNLIMITING ELEMENTS.

biolimiting elements A few elements (N, P, and Si) which are almost totally depleted in surface waters, relative to deep water, by biological activity. As these elements are essential to living organisms, the depletion limits further biological production until the scarce elements are replaced, e.g. by *upwelling. *Compare* BIOINTERMEDIATE ELEMENTS; and BIOUNLIMITING ELEMENTS.

biolithite A *carbonate rock formed of organisms that grew and remained in place, comprising a rigid framework of organisms, together with associated debris. A *reef represents a typical biolithite. *See* FOLK LIMESTONE CLASSIFICATION.

biological oxygen demand (BOD, biochemical oxygen demand) Indicator of the polluting capacity of an effluent, where pollution is caused by the take-up of dissolved oxygen by micro-organisms that decompose the organic material present in the effluent. It is measured as the weight (mg) of oxygen used by 1 litre of sample effluent stored in darkness at 20 °C for five days.

biomagnetism The influence of the magnetic field on living organisms, which is strong in some species. Certain *Spirillum* bacteria orient themselves along the geomagnetic lines of force. Higher organisms, e.g. snails, bees, birds, porpoises, and possibly humans, appear able to sense and utilize *geomagnetic field directions for purposes of orientation. Strong magnetic fields may be detrimental to health in humans, but the effects of medium strength fields have not been studied. Magnetic fields are also used for scanning biological organisms.

biome *See* COMMUNITY.

biomicrite A *limestone consisting of *bioclasts set in a *micrite *matrix. It is the product of a poorly sorted accumulation of shell fragments and mud. *See* FOLK LIMESTONE CLASSIFICATION.

biomineralization The incorporation of inorganic compounds, such as salts, into biological structures, often to lend them hardness or rigidity. Biomineralization first occurred in, and defines, the *Cambrian period about 590 million years ago, in *Brachiopoda, *Trilobita, *Ostracoda, and *Graptolithina. In vertebrates, *hydroxyapatite usually occurs, in invertebrates inorganic minerals are more varied: *calcite and *aragonite (a harder, less stable form of calcite) are common, permeating chitin (*see* SKELETAL MATERIAL) to form the hard *exoskeletons of *Arthropoda and also forming the calcareous material of shells; in *Radiolaria and some *Porifera, the skeleton is made of opaline silica; Radiolarians occasionally have a strontium sulphate instead of siliceous skeleton.

biophile Applied to those elements required by, or found in, living plants and animals, including C, H, O, N, P, S, Cl, I, Br, Ca, Mg, K, Na, V, Fe, Mn, and Cu. *Compare* ATMOPHILE; CHALCOPHILE; LITHOPHILE; and SIDEROPHILE.

biosparite A *limestone consisting of *bioclasts together with a sparry *calcite *cement (*sparite). It is the product of an accumulation of clean-washed, mud-free, shell debris, with diagenetic cement growth (*see* DIAGENESIS) in *pore spaces. *See* FOLK LIMESTONE CLASSIFICATION.

biosphere The part of the Earth's environment in which living organisms are found, and with which they interact to produce a steady-state system, effectively a whole-planet *ecosystem. Sometimes it is termed 'ecosphere' to emphasize the interconnection of the *biotic and *abiotic components.

biostratigraphic interval zone *See* INTERVAL ZONE.

biostratigraphic unit A unit of *strata characterized by a particular content of *fossils; these were deposited at the same time as the *sediments and distinguish the unit from adjacent strata. A biostratigraphic unit may be of *chronostratigraphic or environmental significance.

biostratigraphic zone *See* ZONE.

biostratigraphy Branch of stratigraphy that involves the use of *fossil plants and animals in the dating and correlation of the stratigraphic sequences of rock in which they are discovered. A *zone is the fundamental division recognized by biostratigraphers.

biostrome A layered, sheet-like accumulation of *in situ* organisms. It differs from a *bioherm in its geometry, lacking a mound-like or *reef-like form.

biota The living organisms occupying a place together, e.g. marine biota, terrestrial biota.

biotic Applied to the living components of the *biosphere or of an *ecosystem, as distinct from the non-living, *abiotic, physical and chemical components.

biotic index Scheme for grading river quality according to the diversity and abundance of the river fauna present, with particular reference to 'key species' with known pollution tolerances. The scale ranges from 10 (clean water with diverse fauna) to 0 (grossly polluted water with no fauna or with only a few *anaerobic organisms).

biotite An important rock-forming silicate *mineral; a member of the *mica group $K_2(Mg,Fe)_6[Si_3AlO_{10}]_2(OH,F)_4$, in which the Mg/Fe ratio is less than 2:1; biotite forms a series with *phlogopite (of similar composition, but with Mg/Fe greater than 2:1) and may have Fe^{3+}, Ti^{4+}, and Al^{3+} also present in the formula; sp. gr. 2.7–3.3; *hardness 2–3; *monoclinic; black, dark brown, or greenish-black; *vitreous to *sub-metallic *lustre; *crystals *tabular and pseudo-hexagonal, with lamellar aggregates and flakes; *cleavage perfect basal {001}; occurs in *granites, *syenites, and *diorites, in *schists and *gneisses, in contact *metamorphic rocks, and often in *sedimentary rocks.

bioturbation The disruption of *sediment by organisms, seen either as a complete churning of the sediment that has destroyed depositional *sedimentary structures, or in the form of discrete and clearly recognizable *burrows, *trails, and traces (*see* TRACE FOSSIL).

biounlimiting elements Elements (e.g. B, Mg, Sr, and S) that show no measurable depletion in surface waters compared to deep waters as a result of biological activity. *Compare* BIOLIMITING ELEMENTS; and BIO-INTERMEDIATE ELEMENTS.

biozone 1. The total, global, time-stratigraphic range of a given *taxon, i.e. all the rocks laid down in the time interval during which that taxon existed. **2.** A shortened form of 'biostratigraphic zone'. *See* ZONE.

bipolar distribution *See* BIMODAL DISTRIBUTION.

bipyramid In crystallography, a pyramid defines a *crystal face which cuts the vertical '*c*' (or '*z*') axis and one or two of the horizontal '*a*' (or '*x*') and '*b*' (or '*y*') axes. In general, the prefix 'bi-' refers to the repetition of faces about a plane of symmetry; 'di-' usually suggests the presence of an axis of symmetry about which the pyramid may be rotated.

biramous Applied to limbs that are two-branched, e.g. in many *Arthropoda.

bird 1. *See* AVES. **2.** *See* AEROMAGNETIC SURVEY.

birdseye fabric (fenestral fabric) A *fabric characterized by irregular, *equant cavities in *sediment, several centimetres across, often filled with sparry *calcite (*sparite). It is thought to form as the result of the entrapment of gas, and desiccation within *intertidal to *supratidal muddy *carbonate sedimentary environments. *See* FENESTRAE.

bireflectance In reflected-light or ore microscopy, the ability of a *mineral to reflect different quantities of plane-polarized light in different orientations on rotation of the *stage. This property is often related to reflection *pleochroism and is diagnostic of some minerals, e.g. *covellite.

birefringence (double refraction) In optical mineralogy, the ability of *anisotropic *minerals to split *plane-polarized light into two rays as it passes through them. The rays vibrate at right angles to each other and travel along different path lengths within the mineral, depending on the degree of birefringence encountered. As a result, when the two sets of rays emerge from the mineral one set has travelled farther than the other and is retarded. The retardation shows itself as a range of interference colours when the *analyser is inserted, and may be determined by reference to an interference colour chart. *See* MICHEL–LÉVY CHART.

birefringence chart *See* MICHEL–LÉVY CHART.

BIRPS Acronym for the British Institutions Reflection Profiling Syndicate. It is mainly concerned with the deep lithosphere.

Birrimian orogeny A phase of mountain building, affecting *Proterozoic *greenstone belts in an area of what is now W. Africa.

bischofite A hydrated *evaporite mineral, $MgCl_2.6H_2O$, deposited in environments where the *relative humidity is about 30%.

bise Local term for a northerly winter wind affecting mountainous regions of southern France (at Languedoc the 'bise noire' is associated with heavy cloud).

biserial 1. Applied to those graptoloids (*Graptoloidea) in which *thecae occur side by side on each side of the *stipe. **2.** In crinoids (*Crinoidea), applied to the side-by-side condition of the brachial plates (*see* BRACHIA) in the arms. **3.** *See* TRISERIAL.

bismuth, native Metallic element, Bi; sp. gr. 9.7; *hardness 2.5; greyish-white ; soft or as *rhombohedral *crystals; occurs native in veins and *granite *pegmatites associated with tin, silver, cobalt, and nickel mineralization.

bismuthinite *Sulphide mineral, Bi_2S_3; sp. gr. 6.5; *hardness 2; greyish-white; *metallic lustre; *massive; occurs in mineral veins associated with *granites and tin mineralization.

bistatic radar A technique for studying the electrical properties of a planetary surface by the reflection of radio waves. If the transmitter and receiver are located at different places, the term 'bistatic radar' is used; if in the same place, this is called 'monostatic radar'. In the study of the *Moon with bistatic radar during the Apollo missions, the transmitter was located on the command module orbiting the Moon, while the receiver was located on the Earth.

bisulcate Applied to shells that possess two grooves or depressions; the ventral valves of brachiopods (*Brachiopoda) are commonly bisulcate. The term is also applied to the *venter, which has twin grooves, in some ammonoids (*Ammonoidea).

bit 1. The cutting part of a *drill stem, which may either break or crush the rocks as the drill stem rotates. **2.** In computing technology, a binary digit or element of information.

Bitaunian *See* ARTKINSKIAN.

bitheca One of the three types of graptolite *thecae, possibly containing the male zooid. *Compare* AUTOTHECA; and STOLOTHECA. *See* GRAPTOLITHINA; and DENDROIDEA.

bitter lake A saline lake which is rich in sulphates and carbonates, dominated by high concentrations of sodium sulphate.

bitumen Naturally occurring, inflammable, solid or semi-solid *hydrocarbons, black or dark brown in colour, with characteristic pitch odour, and burning with a smoky flame. Group name for *asphalts, mineral waxes, and related substances.

bituminous coal A type of coal between *lignite and *anthracite in *rank, with 18–35% *volatiles, which softens and swells on combustion, either caking or non-caking, and burns with a smoky flame. It is used in town gas and coke manufacture.

bivalves *See* BIVALVIA.

Bivalvia (bivalves; Pelecypoda, Lamellibranchia; phylum *Mollusca) A class of molluscs in which the body is laterally compressed and is enclosed between two oval or elongated valves. The valves are united dorsally by a toothed *hinge and in most species the valves are *bilaterally symmetrical along the plane of junction (*commissure) between them. The valves are opened by a horny, elastic ligament and closed by the action of one or two adductor muscles. Large, modified, ciliated (*see* CILIUM) gills are involved in food collection and bivalves are entirely aquatic. They are adapted to various modes of life, e.g. boring, burrowing, free-living, and sessile, and these modes of life are generally reflected in the shape of the shell which is modified in various ways. Bivalves first appear in the Lower *Cambrian and are generally of limited abundance in the *Palaeozoic. They become more abundant from the *Mesozoic and they now form the second largest molluscan class, with more than 20 000 species.

Bjerknes, Vilhelm Frimann Koren (1862–1951) A Norwegian meteorologist who was born and died in Oslo. As a student, he helped his father, who was professor of mathematics at the Christiania University (now the University of Oslo). He himself held professorships at Stockholm and Leipzig before returning to Norway in 1917 to found the Bergen Geophysical Institute (*see* BERGEN SCHOOL). During the First World War, Bjerknes established a series of weather stations throughout Norway. Information from these allowed Bjerknes and his colleagues, who included his son Jakob and

Tor Harold Percival Bergeron (1891–1977), to develop their theory of air masses bounded by fronts.

black body A body which absorbs electromagnetic energy perfectly. If the body remains at constant temperature then it also radiates *electromagnetic radiation perfectly in equilibrium with that which it absorbs. *See also* STEFAN–BOLTZMANN LAW.

black earth *See* CHERNOZEM.

Blackett, Patrick Maynard Stuart (1897–1974) A British physicist at Imperial College, London, Blackett worked on *magnetometers during the Second World War, and subsequently developed an instrument capable of detecting very small magnetic fields. This led him to the serious study of *palaeomagnetism. He tried, without success, to demonstrate that magnetism is a property of massive rotating bodies.

black ice Type of glazed frost (e.g. on roads or the superstructures of ships), caused when water falls on a surface that is below freezing temperature. The thin sheet of ice is rather dark in appearance and, unlike white hoar frost or rime, may be hard to see. *See also* GLAZE.

black jack *See* SPHALERITE.

Blackriverian A *stage of the *Ordovician in the Middle *Champlainian *Series of N. America.

black shale A *mudstone with high concentrations of organic material, which is deposited in *euxinic environments. Such sediments form important *hydrocarbon *source rocks.

'black smoker' *See* HYDROTHERMAL VENT.

blade A term describing grain shape; bladed grains have ratios of less than 2:3 for both the intermediate : long diameter and the short : intermediate diameter. *See* PARTICLE SHAPE; and ZINGG SHAPE CLASSIFICATION.

bladed In *mineralogy, describes the flat and elongated form or shape of a *mineral, rather like a knife blade or lath, e.g. *kyanite.

Blake *See* BRUNHES.

blanket bog Ombrogenous bog community, typical of flat or moderately sloping areas in very wet, oceanic climates with high humidity. In the British Isles, blanket bogs are widespread on the Pennine summits, in north-west Scotland, and in parts of Ireland.

-blast A suffix applied to metamorphic terms and used to indicate the *in situ* growth of crystals during metamorphic *recrystallization. For example, the term 'porphyroblast' refers to a large, well-formed crystal which grew *in situ* as a rock was metamorphosed.

blasto- A prefix applied to textural terms and used to indicate the partial or complete replacement of a *textural element in the original rock by crystals which have grown during *metamorphic *recrystallization. For example, the term 'blastoporphyritic' refers to a porphyritic *igneous rock (*see* PORPHYRY) which has undergone metamorphism, and in which the *phenocrysts remain as recognizable textural relicts even though each has been replaced by an aggregate of new metamorphic minerals.

blastoporphyritic *See* BLASTO-.

Blastozoa (phylum *Echinodermata) Subphylum of extinct, stalked echinoderms, ranging from the *Silurian to the *Permian. The stem (rarely preserved) was surmounted by a bud-like *theca with five prominent petaloid *ambulacra showing a well-developed *pentameral symmetry. The theca had 13 major plates arranged in three circlets. The ambulacra each had a central food groove, with delicate food gathering appendages called *brachioles to either side. Underlying the ambulacra were the characteristic hydrospires, thought to have functioned in respiration. These were thin-walled, calcareous infoldings (more rarely simple tubes), forming water conduits, that communicated with the outer environment via marginal pores in the ambulacra and five apertures around the mouth (the spiracles). Blastozoa are not common as fossils, except locally in shallow-water calcareous rocks, often associated with *reef *limestones.

blast ratio The ratio of the mass of explosive used and the amount of rock broken by an explosion, usually given as tonnes of rock to one kilogram of explosive.

B-layer The uppermost part of the *Earth's *mantle, mostly above the first major phase change at about 370 km depth and below the *Mohorovičić discontinuity.

bleb A small, rounded inclusion of one mineral within a large mass of another.

bleicherde Ashy-grey soil that forms the *leached layer in a *podzol soil.

blind hole A *borehole from which all the *drilling fluid has escaped into the surrounding rocks.

blind pores *See* EFFECTIVE POROSITY (2).

blind zone A layer of rock that cannot be detected by *seismic refraction methods, usually because it has a seismic velocity lower than that of the overlying layer(s).

blizzard A storm of blowing snow with high winds and low temperatures. Blizzards are a notable climatic feature of the northern and central parts of the USA in winter, and are related to depression tracks. In the USA, a blizzard is defined by the National Oceanic and Atmospheric Administration (NOAA) as a storm with winds of at least 56 km/h, temperatures below −6.7 °C, and enough falling or blowing snow to reduce visibility to less than 0.4 km. In a severe blizzard, wind speed is at least 72.5 km/h, temperatures below −12.2 °C, and visibility is close to zero.

block-and-ash deposit The deposit formed by a *nuée ardente.

blockfield (felsenmeer) Spread of coarse, angular, frost-shattered rock debris resting on a level or gently sloping upland surface, and found in present or former periglacial environments. *See also* FROST WEDGING.

block glide The sliding movement of a large block of rock over a surface that has been lubricated. Alternatively, the block may be carried downslope by the *plastic deformation of underlying material.

blocking In synoptic meteorology, the establishment in the mid-latitudes of a high-pressure system that interrupts or diverts for a considerable period the typically eastward movement of *depressions and other synoptic features in the zonal flow. Over western Europe, for example, blocking often forces depressions to move northward toward Scandinavia or southward over southern France and Spain.

blocking anticyclone *See* BLOCKING HIGH.

blocking high (blocking anticyclone) In synoptic meteorology, an *anticyclone with deep circulation so placed as to interrupt or divert the eastward movement of the typical succession of low-pressure features (*depressions) in the zonal flow of mid-latitudes. At such times the usual mid-latitude zonal flow is diverted into more meridional flow around the high-pressure area, where settled weather can then result. *See also* AIR MASS.

blocking temperature (T_B) The temperature at which a *thermal remanent magnetization becomes locked in a rock for geologic times.

blocking volume (V_B) The grain volume at which a *chemical remanent magnetization becomes locked in a rock for geologic times.

block volume The natural size of a block of rock bounded by *joint (or other discontinuity) planes. Blocks are defined by their length of side as: very large (more than 2 m); large (600 mm–2 m); medium (200–600 mm); small (60–200 mm); and very small (less than 60 mm).

blocky lava A surface flow of hot, molten *lava covered in a carapace of crystalline, angular blocks which tend to be smoothly faceted and may have dimensions up to several metres. The blocks, which have the same composition as the flow interior, are formed by fragmentation of the chilled flow surface as lava continues to move within the flow interior. Blocky lava morphology is usually confined to lavas of high *viscosity and intermediate to high *silica contents.

bloedite A non-marine *evaporite mineral, $Na_2SO_4.MgSO_4.4H_2O$.

blood rain Phenomenon of reddish coloured rainfall, caused by dust particles that have been lifted up from arid areas and carried long distances by the winds before they are washed out in precipitation. Saharan red dust sometimes occurs in rainfall over parts of Europe, even as far north as Finland.

blow-hole *See* COASTAL PROCESSES.

blow-out Wind-eroded section of a sand *dune that has been largely stabilized by vegetation. *Erosion results from a break in the vegetation cover, due typically to overgrazing or to recreational pressure. A parabolic dune may result.

blue A quarrymen's term, applied to stone and meaning 'hard'.

blue-green algae *See* CYANOBACTERIA.

Blue John *See* FLUORITE.

blue Moon (blue Sun) Occasional appearance of the Moon or Sun when partly obscured by large particles in the atmosphere, as in dust storms, or following forest fires or great volcanic explosions. When the Moon or Sun is viewed through dust or smoke trails it usually appears to be very white, but when the suspended particles are predominantly of one size it sometimes appears blue, at other times green or orange. The phenomenon is attributed to diffraction, although no full explanation seems to be known. The smaller the particles, the more the colour of the Sun or Moon tends to the blue end of the spectrum. The phenomenon is believed to be more common in China than elsewhere. (In the neighbourhood of the 1883 Krakatoa volcanic explosion the Sun was seen as an azure blue sphere.)

blue Sun *See* BLUE MOON.

blueschist *Metamorphic rock that has undergone *regional metamorphism at low temperatures and high pressures. Blueschists contain abundant blue *glaucophane, which is a blue *amphibole. Blueschists are usually associated with *destructive plate boundary environments.

blueschist facies *See* GLAUCOPHANE-SCHIST FACIES.

bocca An Italian word meaning 'opening' or 'mouth', used originally by Italian volcanologists when referring to an opening through which *lava is effusing. The term is used extensively to describe secondary effusion points on the surface of active compound lava flows, as well as primary effusion points at the base of cinder cones lying on major eruptive fissure lines.

BOD *See* BIOLOGICAL OXYGEN DEMAND.

Bode, Johann Elert (1747–1826) A German astronomer who popularized the theory, known later as Bode's law, that there is a simple arithmetical relationship between the distances from the *Sun to the planets of the solar system. After the discovery of *Neptune, which did not conform with the 'law', the idea fell into disrepute.

Bode's law *See* TITIUS–BODE LAW.

body chamber The last-formed chamber of a cephalopod (*Cephalopoda) shell, which houses the animal's body.

body wave A seismic wave which propagates through the body of a medium; *P-waves and *S-waves are body waves. These contrast with *surface waves.

boehmite An aluminium hydrate γ-AlO(OH), boehmite forms a continuous series with its *polymorph *diaspore α-AlO(OH) and is a constituent of *bauxite and *laterite; sp. gr. 3.0; *hardness 4; *orthorhombic; usually white; normally occurs as microscopic grains, or *pisolitic aggregates; *cleavage good {010}; occurs extensively in the weathering profiles of rocks together with *gibbsite, diaspore, and *kaolinite. It is used for aluminium production and is named after the nineteenth-century German chemist J. B. Böhm.

Bogen structure *See* COAL MACERAL.

boghead coal *Sapropelic coal with a greasy, dull *lustre and no lamination; in appearance and combustible properties it resembles *cannel coal but with a brown or yellow *streak. It has a high percentage of algal matter and *volatiles, and when distilled gives a high yield of tar and oil.

bog iron ore An impure, limonitic deposit (*see* LIMONITE), usually porous, and probably formed as a result of bacterial action in swampy conditions. It is a low-grade iron *ore which was used extensively in early iron smelting.

bolide Large *meteor that explodes in passing through the Earth's atmosphere. The term is sometimes used synonymously with *fire-ball, but some people reserve 'bolide' for an exploding meteor, and 'fire-ball' for a less-bright object.

Bolindian A *stage of the Upper *Ordovician of Australia, underlain by the *Eastonian.

Bølling Relatively warm period that occurred towards the end of the last (*Devensian) glaciation in north-western Europe, and which is named after the type site in Denmark. The event took place about 13 000–12 000 radiocarbon years BP. *See also* DRYAS.

bolson *Intermontane basin extending from the divide of one block-faulted mountain to the divide of the adjacent mountain. The surface form is made up of mountain front, *bajada, *pediment, and *playa. It is classically described for the *basin-and-range province of the western USA.

Boltwood, Bertram Borden (1870–1927) Professor of radiochemistry at Yale University, in 1907 Boltwood made early determinations of the age of the *Earth using uranium : lead ratios. He was able to show a long geologic time-scale of up to 2000 Ma. *See* RADIOMETRIC DATING.

Boltzmann constant (*k*) The ratio of the ideal gas constant to the *Avogadro constant: $k = R/N_A = 1.3805 \times 10^{-23}$ J/K, where R is the ideal gas constant (equal to 8.3 J/mol/K), and N_A is the *Avogadro constant.

bomb *See* VOLCANIC BOMB; and VOLCANO.

bomb sag The deformation of primary, unconsolidated, volcaniclastic bedding structures by the impact of a large ballistic block. The block, which can be a *volcanic bomb or a fragment of crystalline *country rock, is ejected from its source vent during a period of violent explosive activity. The asymmetry of the deformation structure it produces can be used to locate the position of the source vent.

bone The skeletal tissue of *vertebrates. Bone is composed of about 70 per cent inorganic calcium salts, mostly *hydroxyapatite but carbonate, citrate, and fluoride amines are also present. The organic component is mostly made up of the structural protein *collagen.

bony fish *See* OSTEICHTHYES.

Boomer A marine seismic source in which capacitor plates are highly charged and then allowed to discharge via a transducer in the water. Eddy currents induced in the transducer plates force the plates apart, thus producing a low-pressure region between the plates. Water implodes into this region, generating a pressure wave. The Boomer (the name is a commercial trademark) is used in high-frequency marine surveys as a low-energy (less than 1 kJ), high-resolution seismic source. The term 'boomer' is also used colloquially to describe a high-amplitude, low-frequency reflection event associated with a large and distinctive reflector.

bootstrapping In statistics, an approach to estimating the robustness or degree of error of a statistic when the distribution from which the statistic is drawn is unknown. An estimate of the distribution is constructed based on the data of the sample being analysed, often by repeatedly random sampling from the data itself and recalculating the statistic in order to produce a statistic distribution.

bora Regional term for a cold and typically very dry wind from the north-east, blowing down from the mountains on the eastern side of the Adriatic Sea. They are most common in winter on northern Adriatic coasts. The wind is probably a consequence of continental high pressure in central Europe with low pressure to the south in the Mediterranean. It is often accompanied by much precipitation when associated with *depressions in the Adriatic. In other areas it is used as a generic term for cold squalls moving downhill from uplands.

borax *Evaporite mineral, $Na_2B_4O_7.10H_2O$; sp. gr. 1.7; *hardness 2.0–2.5; *monoclinic; white, sometimes greyish or tinged blue; white *streak; *vitreous to *resinous *lustre; *cleavage good {100}, {110}; occurs as a sediment of saline lakes, and in association with *halite, *sulphates, *carbonates, and other borates such as *colemanite ($Ca_2B_6O_{11}.5H_2O$) and ulexite ($NaCaB_5O_9.8H_2O$). Soluble in water. It is a source mineral for boric compounds.

bord and pillar *See* PILLAR AND STALL.

bore Very rapid rise of the tide, in which the advancing flood waters form a wave with an abrupt front. Bores occur in certain shallow *estuaries and river mouths where there is a large tidal range, and suitably funnel-shaped regions, e.g. the Amazon, the Bay of Fundy, the Tsing Kiang River in China, and the Rivers Severn, Trent, and Ouse in England.

boreal Pertaining to the north (from Boreas, the Greek god of the north wind).

Boreal Period in post-glacial (i.e. post-*Devensian or *Flandrian) times from about 8800 to 7500 BP, which preceded the climatic optimum of early *Atlantic times. Pollen records typically show an increasing abundance of thermophilous (warmth-loving) tree species and also indicate the drier, more continental conditions that characterized the ensuing Atlantic period. The early Boreal corresponds to late *Pollen Zone V; otherwise the Boreal is linked with *Pollen Zone VI, which is sometimes subdivided to give Zones VIa, VIb, and VIc, according to the most abundant tree pollen represented. For Britain, the Boreal period is significant as the last period in post-glacial times in which Britain was joined to mainland Europe by a *land bridge across the Dover Strait. *See also* POLLEN ANALYSIS.

boreal climate The climate associated with the boreal ('taiga') forest zone of Eurasia, where it extends to 65°–70° N in the west and 50° N in the east, and with

N. America, where it extends from the fringe of the tundra southwards to 55° N in the east. Winters are long and cold, with temperatures below 6 °C for 6–9 months, and summers are short, with temperatures averaging more than 10 °C. Precipitation, as snow in winter, typically amounts to 380–635 mm per annum.

Boreal realm The name used to denote that a fauna or flora has northern affinities. At certain times in the *Cenozoic this implies a degree of coldness, while in the *Mesozoic it is less specific, being used in comparative terms with the *Tethyan Realm.

borehole (well, dug well) A hole drilled into rock, usually by *rotary methods, to enable an assessment to be made of the characteristics of the rock itself and of contained fluids, e.g. *groundwater, *natural gas, or *petroleum. The size may range from a few tens of millimetres in diameter to over 300 mm, and boreholes can be drilled at any angle to depths of several kilometres. Rock samples may be recovered from below the surface, the rock may be examined by down-hole geophysical or other methods, or *pumping tests may be run. Boreholes may also be used as water, gas, or oil production wells. A completed borehole is frequently termed a *well. *See* BOREHOLE LOGGING; and WELL LOGGING, GEOPHYSICAL.

borehole effect The distortion on a *well log resulting from changes in the diameter of the hole or in the thickness of the *invaded zone.

borehole logging The recording of geophysical and geologic data recovered from a *borehole. Geophysical borehole logs can be used in conjunction with drill cuttings, sidewall samples, and *cores to establish the geologic succession in the well. This information can then be interpreted geologically. The logs can be used to estimate formation *porosity, *jointing, fluid properties, hydrocarbon saturation, and formation pressures. Water wells are often logged by devices to detect water flows, levels of inflow, and such water properties as temperature and electrical conductivity.

borehole sonde *See* SONDE; and WELL LOGGING.

boring (noun) A *trace fossil that penetrates the hard surface of an organism or rock. The word may be used to describe the drill holes made by natacid gastropods (*Gastropoda) into the shells of prey, or the short, tube-like extensions to larger borings made by the *siphons of certain bivalves (*Bivalvia), e.g. *Pholas*.

bornhardt Rounded, often isolated hill developed in massive rock and found in the humid tropics. Its shape is controlled by large-scale *exfoliation or sheeting *joints. Such hills are sometimes called 'sugar-loaf' hills, after the granitic dome of the Sugar Loaf, Rio de Janeiro, Brazil.

bornite (peacock ore) Common and important copper *ore mineral, Cu_5FeS_4; sp. gr. 5.0–5.1; *hardness 3; *cubic; reddish-brown to various shades of purplish-blue, the colours iridescent on tarnished surfaces; pale grey *streak; *metallic *lustre; *crystals rough cubes, rhombododecahedra, can also be *massive; *cleavage poor {111}; occurs in *hydrothermal veins, in zones of secondary enrichment, and in various *syngenetic copper ores, *porphyry copper deposits, *skarns, etc.; soluble in nitric acid.

bort (bortz) A compact variety of *diamond, occurring naturally in granular *aggregates and used as an abrasive in drill bits, saws, and gem-cutting wheels. It occurs in *ultrabasic, *igneous, *breccia pipes, particularly in southern Africa.

Bortonian A *stage in the Lower *Tertiary of New Zealand, underlain by the *Porangan, overlain by the *Kaiatan, and roughly contemporaneous with the upper *Lutetian and lower *Bartonian Stages. (The name should not be confused with *Bartonian.)

bortz *See* BORT.

boss Discordant *igneous intrusion that has roughly vertical sides and a subcircular outcrop, and an area of less than 25 km^2.

Bothriocidaroida (phylum *Echinodermata, class *Echinoidea) Small order of echinoderms which have five double columns of perforate plates and five columns of single imperforate plates. Traditionally they were regarded as the ancestral stock of all echinoids, but this is now considered unlikely. There are only three genera, *Bothriocidaris* (the earliest echinoid), *Neobothriocidaris*, and *Unibothriocidaris*, ranging in age from Middle *Ordovician to Upper *Silurian.

botryoidal In mineralogy, describes the form or shape of a *mineral that occurs as spheroidal aggregations, often as a result of concentric growth patterns during its formation, e.g. *azurite.

bottleneck A severe reduction in population size, often leading to a *founder effect. Bottleneck events are commonly followed by rapid population expansion in which the rate of loss of new lineages is greatly reduced, giving rise to *star phylogenies.

bottomset beds **1.** The part of a cross-bedded set of *sediments that forms at the base of the downcurrent or lee-side of a *dune- or *ripple-form structure. **2.** Offshore clays formed at the base of a prograding deltaic sequence. *See also* DELTA; and PROGRADATION.

bottom water The water mass that lies at the deepest part of the water column in the ocean. It is relatively dense and cold, e.g. the North Atlantic bottom water has a temperature of 1–2 °C. Bottom-water circulation is slow-moving, is greatly influenced by sea-bed topography, and is driven by differences in water density (*thermohaline circulation).

boudin *See* BOUDINAGE.

boudinage Minor structural feature in which *competent *strata resemble a series of sausages ('boudins') that form by the stretching of the *competent units which, unlike *incompetent ones, cannot deform plastically. Initially, local thickening and thinning occurs ('pinch-and-swell' structures); as deformation proceeds, complete separation between boudins may take place.

Bouguer, Pierre (1698–1758) A French mathematician, Bouguer was a member of an expedition to Ecuador, sent to study the shape of the Earth. He made gravity measurements in the Andes, showing how the pull of gravity diminishes with altitude, and has given his name to corrections used in calculating gravitational anomalies. *See also* BOUGUER ANOMALY; BOUGUER CORRECTION; and GRAVITY ANOMALY.

Bouguer anomaly The gravitational attraction remaining after correcting the measured vertical component of *gravitational acceleration at a point for: (*a*) the theoretical gravity at that point, usually using the *International Gravity Reference Field; (*b*) the *free-air correction; (*c*) the *Bouguer correction; and (*d*) the *topographic elevation correction, usually correcting to sea level. This anomaly is the fundamental gravity anomaly, reflecting all variations in density away from that expected for a homogeneous Earth.

Bouguer correction The correction applied to a measurement of *gravitational acceleration, which allows for the attraction of rock between the observing station and some reference height, usually sea level. The correction is $0.4185\rho h$, where ρ is the *density of the rock in kg/m^3 and h is the difference in height between the two points in metres.

Bouguer gravity map A map of the gravitational field over an area after correction for theoretical gravity, *free-air, solid rock (*Bouguer correction), and topography; i.e. a map showing *Bouguer gravity anomalies.

boulder *See* PARTICLE SIZE.

boulder clay Glacial deposit consisting of boulders of varying size in a clay-dominated matrix, and laid down beneath a valley glacier or ice sheet. Typically it is unstratified and unsorted, and characterized by rock types derived from the country crossed by the depositing glacier. *See also* TILL.

Bouma sequence Idealized sequence of sedimentary structures observed in *turbidity-current deposits. It is named after the geologist, A. H. Bouma, who first emphasized its generality (*Sedimentology of Some Flysch Deposits*, Elsevier, Amsterdam, 1962). The lowest unit, A, a *massive or *graded *sand, is overlain progressively by the B (lower division of parallel *lamination), C (ripple or convolute laminations), D (upper division of plane parallel laminations), and E (*pelagic shale) units. Examples showing the entire sequence are not common. The sequence can be interpreted in terms of deposition under waning current conditions.

boundary current Northward- or southward-directed ocean-water current which flows parallel and close to a continental margin. Such currents are caused by the deflection of eastward- and westward-flowing currents by the continental land masses. Boundary currents on the western margins of ocean basins, such as the *Gulf Stream and the *Kuroshio current, are deep, narrow, fast-moving currents; while currents along the eastern boundaries, such as the *Canaries current and the *California current, tend to be relatively shallow, broad, diffuse and slow-moving.

boundary layer **1.** At the interface between a solid surface and a fluid, a thin fluid layer that is static because of friction between molecules of the fluid and the solid. **2.** General term used to describe the atmospheric layer up to about 100 m above the ground surface, where the air flow is largely conditioned by the frictional effects of the surface. Mean velocities in the boundary layer are typically less than the free-stream values. *See also* PLANETARY BOUNDARY LAYER.

boundary-stratotype A specified rock section (*see* STRATOTYPE) within which the time-line ('golden spike', boundary zonal) occurs that marks the standard demarcation between *chronostratigraphic units. In practice such time-lines are usually based on either the appearance or the disappearance of a key *species (*see* INDEX FOSSIL) or other *taxon. Associated *faunas and *sediments may transgress zonal boundaries. The term *boundary stratotype' has also been used in the sense of the time-line itself.

boundary wave A seismic wave which travels along a boundary between two media of contrasting properties, rather than propagating through the bodies of the media (*compare* BODY WAVE). Where the boundary is a free surface (e.g. ground–air, sea bed–sea water) these waves are called *surface waves and give rise to *ground roll and *mud roll respectively. Boundary waves are of two types: *Love waves and *Rayleigh waves.

boundary zonal *See* BOUNDARY STRATOTYPE.

boundstone A general term for *autochthonous *carbonate deposits in which the sediments are bound during deposition by organisms such as corals (*Anthozoa) or *algae. Boundstones are further subdivided into *bafflestone, *bindstone, and *framestone. *See also* EMBRY AND CLOVAN CLASSIFICATION.

bournonite *Sulphide mineral, $CuPbSbS_3$; sp. gr. 6; *hardness 3; grey; *submetallic *lustre; *massive or rarely as cogwheel *aggregates; occurs in *hydrothermal veins associated with copper and lead mineralization.

Bowen, Norman Levi (1887–1956) A Canadian geochemist and experimental petrologist at the Carnegie Institution, Washington DC, Bowen's classic work was on the chemistry of *igneous rocks, published in *The Evolution of the Igneous Rocks* (1928). Later (1940) he turned to the study of *metamorphism in *limestones and *dolomites. *See* BOWEN'S REACTION PRINCIPLE.

Bowen's ratio The ratio of sensible heat to latent heat transport from the ground to the atmosphere, which is generally calculated from the ratio of the vertical gradients of vapour pressure and temperature. It is used in the assessment of evaporation.

Bowen's reaction principle A concept, first propounded in 1928 by Norman *Bowen, which explains how *minerals can respond to the changing equilibrium conditions when a *magma is cooled, by either a continuous, diffusion-controlled exchange of elements with the magma or discontinuous melting of the mineral. In a *continuous exchange or reaction, *solid-solution minerals such as *feldspar adjust their composition during cooling by a continuous diffusion of elements between magma and mineral, whilst in a *discontinuous reaction, minerals such as *olivine undergo melting at a specific temperature during cooling (the peritectic point) at the same time as a new mineral in equilibrium with the magma begins to crystallize (in this case *pyroxene). Bowen suggested a series of these reactions that might take place during the cooling of a *tholeiitic basalt magma, the so-called Bowen's reaction series, but pointed out that the series was a simplification of very complex reactions and could be misleading if taken at face value. The specific reaction series for tholeiitic magmas was never intended to become a general reaction series for all magmas.

Bowen's reaction series *See* BOWEN'S REACTION PRINCIPLE.

bow shock The shock wave in front of an object moving through a gas or liquid medium. The most familiar *solar-system example is the shock-wave front between the magnetic field of a planet and the *solar wind. It is the outer boundary of a planetary *magnetosphere, where the solar wind is slowed from supersonic to subsonic velocities. It forms upwind of the *magnetopause (the boundary where the planet is shielded from the *solar wind).

bow-tie reflection A concave-upwards depression in a reflector with a curvature greater than that of an incident wavefront is represented on a *brute stack by a 'bow-tie' event. This results from there being three reflection points on the reflector for each surface location. A high degree of reflector curvature is required for each 'bow-tie' event to feature on normal incidence traces, with a reduced degree of curvature needed for a similar effect on offset traces. Consequently, 'bow-tie' events are more likely on long-offset traces and at greater depth within the *seismic section. Curvature of the reflector out of the plane of the section can cause off-section 'bow-tie' *ghosts.

bow trend (barrel trend, symmetrical trend) In the reading from a *wireline log (*see* WELL LOGGING) an expression from a *gamma-ray sonde showing a gradual decrease, then increase in gamma radiation, indicating an increase followed by a decrease in clastic sedimentation within a basin.

boxcar trend (cylindrical trend) In the reading from a *wireline log (*see* WELL LOGGING) an expression from a *gamma-ray sonde showing no change and indicating a river channel bed, *turbidite, or wind-blown sand.

box classification A remote sensing *classification system in which a rectangular box is placed around *digital number parameter space in a graph defining a certain *training area (such as road). If the digital number values of the different *bands plot to within that box then the pixel is assigned to that classification. *See also* MINIMUM-DISTANCE-TO-MEANS CLASSIFICATION; and MAXIMUM-LIKELIHOOD CLASSIFICATION.

box fold A rectangular, *conjugate fold in which the inter-limb angles approximate to right angles. The fold has two angular *hinges and three limbs.

BP Initials which stand for 'before present' and relate to dates before the present day (taken conventionally to be 1950). The term should not be confused with 'BC', which relates to dates prior to the birth of Christ.

brachia (sing. brachium) **1.** In brachiopods (*Brachiopoda), a pair of feathery structures forming part of the lophophore. The brachia are twisted up within the shell in various ways and serve to filter food particles from the water. **2.** In crinoids (*Crinoidea), long, plated, flexible arms made up of a series of articulated *ossicles containing a water vessel from which the *tube-feet arise. When feeding the tube-feet are extended; at other times they are covered by 'brachial plates'. Small arms coming from the main brachia are called 'pinnules' and are made up of 'pinnular plates'.

brachial plates *See* BRACHIA.

brachidium The calcified support for the lophophore in brachiopods (*Brachiopoda). It may be of different shapes in different groups (loops, spires, etc.). Some groups do not possess this calcified structure and the support for their lophophores is entirely hydrostatic.

brachiole The food-gathering arm of a blastoid or a cystoid (phylum *Echinodermata, subphylum *Blastozoa). The arms are *biserial and are simple versions of the arms that occur in crinoids (*Crinoidea). They are rarely preserved.

Brachiopoda (lampshells) A phylum of solitary, benthic, marine, bivalved, *coelomate, invertebrate animals that have existed from the Lower *Cambrian to the present day. Brachiopods are commonly attached posteriorly to the sea bed by a stalk (*pedicle), but may be secondarily cemented, or free-living (e.g. the fossil form **Productus* which, like many productids, was spinose, thick-shelled, and lived partly buried in the mud of the sea bed). Usually they consist of two unequal *valves: a larger pedicle (ventral) valve and a brachial (dorsal) valve, lined by reduplications (*mantle lobes) of the body wall which enclose the large mantle cavity. They are *bilaterally symmetrical about the posterior–anterior mid-line of the valves (i.e. through the valves, *compare* BIVALVIA). The characteristic feeding and respiratory organ, the lophophore, surrounds the mouth and is covered by ciliated tentacles. It may be a simple horseshoe but more often forms two ciliated arms or brachia that project through the gape (thus giving the phylum its name). The alimentary canal is divided into oesophagus, stomach, and intestine, with or without an anus. The nervous system consists of a circum-oesophogeal ring with a small aggregation of nerve cells on the ventral side. The excretory organs are one or two pairs of nephridia (excretory tubules) also acting as gonoducts (for the release of eggs and sperm). The circulatory system is open, with the contractile vesicle (heart) near the stomach. Considerably more than 3000 fossil species are known and about 100 are alive today; these are widely distributed and occur at all depths.

Brachiopods are divided into three classes: *Lingulata, *Inarticulata, and *Articulata.

Brachiosaurus *See* SAUROPODA.

brachium *See* BRACHIA.

brachydont Applied to teeth in which the crowns are low or short and the roots well developed, with narrow canals.

Bradfordian *See* CHAUTAUQUAN.

bradyseism A slow *earthquake; i.e. gradual, differential motions of parts of the Earth's *crust that do not suddenly release seismic energy.

bradytelic evolution One of three types of evolutionary rates within groups. In this instance the rate is exceedingly slow, and is manifest by slowly evolving lineages which survive much longer than would normally be expected. Living fossils, e.g. the *coelacanth, represent the slow end of the bradytelic range; while the better-known groups that have remained relatively stable with time include opossums and crocodiles.

Bragg, William Lawrence (1890–1970) An English physicist who, with his father William Henry Bragg, evolved the technique of X-ray crystallography, used to determine the atomic structure of crystals. He was an early supporter of the concept of *continental drift, and is thought to have been responsible, in 1922, for arranging the translation into English and publication of *Wegener's work. *See also* BRAGG'S LAW.

Bragg equation *See* BRAGG'S LAW.

Bragg's law In crystallography, the law that describes how an X-ray beam is reflected or diffracted in a crystal lattice, given by the Bragg equation $n\lambda - 2d\sin\theta$ where n is any integer, λ is the wavelength of the incident-beam X-ray, d is the spacing between crystal planes (d spacing), and θ is the angle between the crystal plane and the diffracted beam (the Bragg angle). Used in X-ray diffraction for the identification of minerals.

braided stream (braided channel, braided river) Stream whose plan form consists of a number of small channels separated by *bars. The latter may be vegetated and stable (e.g. the eyots (small islets) in the River Thames, England), or barren and unstable (as at glacial margins, where rapid changes occur).

branch The graphical representation of an evolutionary relationship in a phylogenetic (*see* PHYLOGENY) tree.

branchial arch A single gill arch which supports a single pair of gill slits. Together the gill arches form the branchial basket which is part of the visceral skeleton.

branchial basket *See* BRANCHIAL ARCH.

branching decay (dual decay) The decay of an isotope by different methods to two or more different end-members. For example, ^{40}K decays either to ^{40}Ar (12%) by positron emission and *electron capture, or to ^{40}Ca (88%) by emission of a negative beta particle.

Brandenburg One of a series of *moraines which mark the southern limit of *Weichselian ice, extending some 500 km across the N. German Plain. A Russian equivalent extends a further 2000 km into European Russia.

Brandon An episode in the *Upton Warren Interstadial, which occurred within the *Würm (*Weichsel) Glacial.

braunerde *See* BROWN EARTH.

braunite Oxide *mineral Mn_2O_3, but with about 10% by weight of silica, so the mineral may be considered to have the composition $3Mn_2O_3.MnSiO_3$; sp. gr. 4.7–4.8; *hardness 6.0–6.5; *tetragonal; brownish-black to steel-grey; brown-black *streak; *submetallic *lustre; *crystals pyramidal, can also occur *massive or granular; occurs in *hydrothermal veins together with other manganese oxides, as a *secondary mineral, and as a result of the *metamorphism of manganese-bearing sediments. It is very similar to other manganese minerals, but is soluble in hydrochloric acid, when it leaves a residue of silica. It is an ore of manganese and is used in steel-making.

Bravais lattice A lattice is a framework, resembling a three-dimensional, periodic array of points, on which a *crystal is built. The smallest array which can be repeated is the 'unit cell'. In 1850, M. A. Bravais showed that identical points can be arranged spatially to produce 14 types of regular pattern. These 14 space lattices are known as 'Bravais lattices'.

Bravais law *See* BRAVAIS RULE.

Bravais rule (Bravais law) *Crystals are formed by the repetition in three dimensions of a unit-cell structure defined by *lattice points in space. The Bravais rule (proposed by M. A. Bravais) states that the density of these lattice points, or spacing between lattice planes, is proportional to the relative importance of crystal forms. It follows that the faces most likely to be found on a crystal are those parallel to lattice planes of greatest point density.

Brazil current Warm water, forming the oceanic *boundary current that flows southward along the Brazilian coastal margin. It is marked by its high *salinity (36.0–37.0 parts per thousand), its low velocity, and its shallowness (100–200 m) in comparison to its northern hemisphere equivalent, the *Gulf Stream.

bread-crust bomb A ballistic mass of *lava, usually greater than 64 mm in size, ejected from a source vent. The exterior of the lava rapidly chills to form a fine-grained, glassy crust, which is then cracked open by continued vesiculation and the expansion of the hot lava interior. The cracking of the crust resembles that sometimes seen on loaves of bread baked in the traditional manner.

break The onset of an event, in particular the *first break. A sudden onset of new energy giving rise to a noticeably different amplitude event. A time break is the shot-instant time mark on a *seismic record. *See also* UPHOLE TIME.

breaker *Wave that is collapsing or breaking as a result of the wave approaching the shore and reaching shallower water. The decreasing water depth causes the wavelength and speed to decrease and the wave height to increase. Consequently wave steepness increases and the wave becomes unstable: it breaks when the wave height is about 0.8 times the water depth. Several types of breaker have been described, e.g. spilling breakers (in which the wave breaks forward, with broken water spilling down the front of the wave), and plunging breakers (in which the wave crest curls over a large air pocket and falls vertically into the trough).

breccia Coarse, *clastic, *sedimentary rock, the constituent clasts of which are angular. 'Breccia' literally means 'rubble' and implies a rock deposited very close to the source area. The term may also be applied to angular volcanic rocks from a volcanic vent (vent breccia).

breccio-conglomerate A *conglomerate with *clasts which are both rounded and angular; it is intermediate in character between a *breccia and a *conglomerate.

brecciola From the Italian, literally meaning 'little *breccia', normally-graded beds of small, angular, *limestone *clasts.

breeze Relatively light wind, often of convective origin. The term also includes particular local air movements, e.g. mountain, land, and sea breezes. *See also* BEAUFORT SCALE.

breviconic Applied to a *cephalopod conch when it is short and expands quickly. The *aperture is quite wide.

brick Shaped block of baked *clay used for construction. Usually obloid in shape with plane faces and parallel edges; fine uniform texture with no fissures, air bubbles, or pebbles; some are deep-salmon colour from the iron content of the clay. Small proportions of $CaCO_3$ and *silt may be added for strength. The type of clay preferred has a low water content, to limit shrinkage on drying, and is high in organic matter to reduce fuel costs, e.g. the *Jurassic Oxford Clay of England. Furnace, or refractory, bricks are made from *kaolinite, etc.

brickearth Fine-grained silty deposit occurring in south-eastern England. Of complex origin, it probably resulted from the reworking of *loess, either by hillslope washing or by redeposition in standing water.

Brigantian The final *stage of the *Visean Epoch.

bright spot A characteristic phase reversal on a reflection event, usually indicating the presence of gas. The *phase reversal is caused by a marked decrease in the *acoustic impedence of gas-filled material which causes a strong negative reflection from the top of the gas-filled unit. Often this is accompanied by a *flat spot below.

brine A concentrated solution of inorganic salts, formed by the partial evaporation of saline waters. *See also* GEOTHERMAL BRINE.

Brioverian A *stage of the Upper *Proterozoic of Brittany, from about 1000 to 580 Ma ago, underlain by the *Pentevrian.

bristlecone pine *See* *PINUS LONGAEVA*.

British classification (of particle size) *See* PARTICLE SIZE.

brittle *See* TENACITY.

brittle behaviour The manner in which *competent rocks lose their internal cohesion along certain surfaces when the *elastic limit is exceeded under an applied *stress. Such behaviour gives rise to *fractures (either *faults or *joints) and is common in upper crustal regions where temperatures and pressures are relatively low.

Brøgger, Waldemar Christofer (1851–1940) A Norwegian geologist and petrologist, Brøgger was a professor at the University of Oslo from 1890 to 1917. He had interests in many fields, including the geologic mapping of Norway and studies of *pegmatites, but his main work was concerned with the theory of *magmatic differentiation.

Brongniart, Alexandre (1770–1847) Brongniart was professor of mineralogy in the Paris Museum of Natural History, and collaborated with *Cuvier in his work on the mapping of the Paris basin. He demonstrated that lower taxonomic orders of animals were found lower in the *stratigraphic column, and hence gave evidence of progression in the succession of life.

Brontosaurus *See* APATOSAURUS.

bronzite Old-fashioned name for an *orthopyroxene (an *orthorhombic *pyroxene mineral) with a composition in the range $(Mg_{1.4}Fe_{0.6})Si_2O_6$ to $(Mg_{1.8}Fe_{0.2})Si_2O_6$; occurs in *basic and *ultrabasic *igneous rocks and some rare *metamorphic rocks.

brookite *Mineral, TiO_2; sp. gr. 4.1; *hardness 5.5–6.0; *orthorhombic; reddish-brown to brownish-black; white *streak; *metallic *lustre; *crystals normally *platy and *tabular; *cleavage imperfect prismatic; occurs as an *accessory mineral in *igneous and *metamorphic rocks, and in *hydrothermal veins; Brookite, *anatase, and *rutile are all *polymorphs of titanium dioxide. It is named after the British mineralogist H. J. Brooke.

Brørup (Loopstedt) An *interstadial that occurred during the last (*Weichselian) glacial. Named from a place in Jutland, it dates from about 60 000 years BP and is perhaps the equivalent of the *Chelford Interstadial of the British Isles. Estimated mean July temperatures for this period are between 15 and 20 °C.

brown clay *See* RED CLAY.

brown coal Brown to brownish-black *fossil fuel of low *rank, intermediate between *peat and *lignite. It is soft, with a dull lustre and earthy texture, and contains the remains of plants; has a high moisture content (45–66%), and decomposes and darkens on dehydration. Its ash content is 1–5%, and its calorific value varies from low in soft, high-moisture types to high in those resembling *lignites. It comes mainly from east and central Europe, where it is mined from *open-cast pits.

brown earth Freely draining, and only slightly horizonated, *soil-profile type. It has a *mull *humus in the surface *horizon and very little differentiation of horizons below. Brown earths are well weathered and slightly *leached soils, with a *cambic horizon in the middle part of the profile (also known as braunerde and now included in the *Inceptisols).

brown forest soil Little used *soil-profile term that has been applied to both acid *brown earths and *brown podzolics.

Brownian motion Random movement of small particles dispersed in a colloidal solution or suspension.

brown podzolic soil Freely draining, *leached *soil profile that has developed acid surface *horizons, a *mor surface *humus, and a clearly visible enrichment of translocated iron oxide in the middle, or B, horizon. The profile has been leached to the early stage of *podzolization, identified by the movement down-profile of iron and aluminium compounds.

brucite *Mineral, $Mg(OH)_2$; sp. gr. 2.4; *hardness 2.5; *trigonal; white, shades of pale grey, blue, or green; white *streak; pearly to waxy *lustre; *crystals normally broad and *tabular, but also occurs fibrous, *massive, and foliated; *cleavage perfect basal {0001}; occurs in *metamorphosed *limestone, also in *hydrothermal veins together with *calcite, *talc, and *serpentine; soluble in hydrochloric acid. It is named after the nineteenth century American mineralogist A. Bruce.

Brückner cycle Tendency for the cyclical recurrence of runs of wet years at about 35-year intervals, with warmer and drier years in between, reported in 1890 by A. Brückner. Much attention was paid to it for a time, but its operation is obscured by the greater magnitude of the irregular year-to-year variations, and by cyclical recurrences of other periods.

Brunhes The final normal *polarity chron in the *Quaternary, preceded by the Matuyama, and *radiometrically dated from 0.78 Ma to the present. The Brunhes chron contains several brief *reversed *polarity subchrons of which the Blake (*c*.110 000 years BP), Lake Mungo (*c*.30 000 years BP), Laschamp (*c*.18 000 years BP), and Gothernburg (*c*.13 000 years BP) are the best, but still poorly, established.

Bruun rule The idea, proposed in 1962 by P. Bruun, that a marine shoreline maintains an equilibrium profile with a depth and slope determined by the current and wave regime.

Bryophyta (bryophytes) A division of plants which for most authors includes the mosses and liverworts. Although bryophytes lack differentiated water-conducting vessels and rely largely or entirely on water absorbed from rain falling on the plants or from a moist atmosphere, they may have simple water-conducting cells in some larger species. They lack true roots, but possess root-like rhizoids which anchor them to a substrate and which can absorb water and minerals. The plants all show an alternation of generations, with a green vegetative gametophyte (the familiar moss or liverwort plant) and a sporophyte which typically takes the form of a (usually stalked) capsule and which is partially or wholly parasitic on the gametophyte. Most bryophytes are land plants and are found world-wide in a range of habitats. They are known from *Devonian rocks, the earliest fossil bryophyte being a compression of a thalloid liverwort, *Pallavicinites devonicus*. The earliest fossil moss is *Musites polytrichaceus* from the Upper *Carboniferous of France. There is no evidence to link bryophytes with either the green algae (*Chlorophyceae) or the more advanced *pteridophytes.

Bryozoa (moss-animals) Phylum of small, aquatic, colonial animals, related to the *Brachiopoda; many colonies possess a well-developed, *calcite skeleton which comprises microscopic, box-like divisions, each housing an individual animal possessing ciliated tentacles and a *coelom. Food is collected by the tentacles which surround the mouth and are borne on a ridge called the 'lophophore'. Reproduction takes place by (*a*) asexual budding, and (*b*) the release of larvae which give rise to new colonies. There is a possible bryozoan fossil from the Late *Cambrian, and bryozoans have occurred abundantly from the *Ordovician to the present day. Branched colonies are common fossils in some rocks. They were important *reef builders and binders in the *Phanerozoic, and underwent several great radiations.

B-subduction *See* A-SUBDUCTION.

bubble pulse When an explosion or *air-gun is triggered in water, the resulting gas bubble oscillates with decreasing energy with each oscillation generating a pressure pulse, known as the 'bubble pulse'. Unwanted bubble pulses can be prevented by setting off the source sufficiently close to the surface to allow the bubble to vent before it collapses, a method which, although spectacular, is inefficient. Where an *array of tuned airguns is fired at depth, the individual airgun bubble pulses destructively interfere to provide the desired source signature. A knowledge of the bubble pulse period, T, is thus required for each source and this can be calculated using the Rayleigh-Willis formula such that: $T = (0.0452Q^{1/3})/((D/0.3048) + 33)^{5/6}$, where Q is the source energy in joules and D is the water depth in metres to the bubble centre.

bubnoff unit A standard measure for describing the rates of geologic and geomorphologic erosional processes. One bubnoff unit (B) is equal to the removal of one micrometre per year, or one millimetre of surface material per thousand years. The unit is named after S. von Bubnoff (1888–1957).

Buchan metamorphic zones Areas in which a series of mineral assemblages occur, with a *biotite zone surrounded by zones of *cordierite, *andalusite, and *sillimanite, in increasing order of metamorphic grade. The zones are named after the region of north-east Scotland (north of Aberdeen) in which they occur. *Compare* BARROW'S ZONES.

Buchan spells Several periods of the year, 'more or less well defined', when the normal seasonal rise or fall of temperature is halted or reversed for a time, e.g. 9–14 May is a cold period and 3–14 December a warm period. They are named after the Scottish meteorologist Alexander Buchan (1829–1907), whose analysis of temperature records for Scottish stations in the middle of the nineteenth century revealed them. Six cold and three warm periods of this type were identified by Buchan in 1869 from his examination of records covering several years. *See also* SINGULARITY.

Buckland, William (1784–1856) The first

reader in Geology and Mineralogy at the University of Oxford, Buckland was appointed Dean of Westminster in 1846. He developed the English school of historical geology with its emphasis on the progressive (but not evolutionary) nature of Earth history. Initially, he was a diluvialist and his *Reliquiae Diluvianae* (1823) is based on studies of fossil deposits in Kirkdale Cavern and other caves. He subsequently abandoned *diluvialism in favour of Agassiz's *glacial theory which he was the first to introduce into British geology. Buckland's *Bridgewater Treatise* (1836) was not only an attempt to reconcile geology and *natural theology but also an up-to-date manual of historical geology. *See also* CATASTROPHISM.

buckle folding The sideways deflection from a median line of *competent layers in a less competent matrix, due to mechanical 'buckling' instabilities set up by *layer-parallel shortening. A regular dominant wavelength is created as a function of layer thickness, layer *rheology, and contrasts in competence.

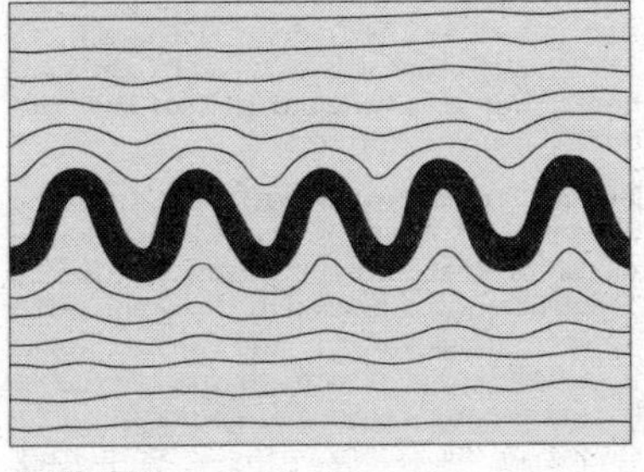

Buckle folding

Buffon, Georges Louis Leclerc, Comte de (1707–88) Buffon was an extremely influential French natural philosopher whose extensive works included one of the last speculative cosmogonies: *Les Époques de la Nature* (1778). Beginning with the formation of the Earth, it covered seven epochs in analogy to the seven days of Creation. He proposed that the Earth was hot at its creation and, from the rate of cooling, calculated its age to be more than 100 000 years.

Buganda-Toro-Kibalian orogeny A phase of mountain building that occurred about 2075–1700 Ma ago, affecting what is now an area of Central and East Africa stretching from northern Zambia to Katanga Province in eastern Congo, and the borders of Uganda. Thrusting in a northwestward direction produced a belt now trending SW–NE and forming the northeastern section of the Kibaride belt.

bulb of pressure Bulb-shaped lines of equal vertical stress below a footing or foundation.

Bulitian A *stage in the Lower *Tertiary of the west coast of N. America, underlain by the *Ynezian, overlain by the *Penutian, and roughly contemporaneous with the upper *Thanetian Stage.

bulk composition of Earth (whole Earth composition) Information on the composition of the Earth as a whole has been deduced from: (*a*) cosmochemical models, which assume that the compositions of all members of the *solar system are related, and that the bulk composition of the Earth can be inferred from the abundances of non-volatile elements in the Sun and some primitive *meteorites (*see* CHONDRITIC EARTH MODEL); and (*b*) geophysical evidence, e.g. *seismic data, *density determinations, and *magnetic surveys. The three layers, *core, *mantle, and *crust, of which the *Earth is formed differ markedly in their composition. The core and mantle make up more than 99% of the Earth's mass but their compositions can be only inferred, unlike that of the crust (*see* CRUSTAL ABUNDANCE OF ELEMENTS). The density and *magnetic field of the Earth, information from seismic surveys, and the existence of iron–nickel meteorites lead to the conclusion that the core is predominantly iron, with a small proportion of lower-density element(s). Nickel is largely excluded because it would make the core too dense. The nature of the light component is controversial, but may be sulphur, carbon, oxygen, silicon, and potassium. From seismic evidence the mantle appears to be composed of *dunite, *peridotite, and *eclogite, rocks similar in composition to *chondritic meteorites. The upper mantle is probably formed from dense *silicates of iron and magnesium, with silicon and magnesium oxides becoming commoner with depth. Information regarding the bulk composition of the Earth is fundamental to resolving such questions as the relationship between the Earth and the *Moon. Perhaps the most critical single feature of the Earth's bulk composition is its content of K, U, and Th, since the radioactive isotopes of these elements control radioactive heat production, and therefore the thermal and geologic history of the Earth. *See also* COSMIC ABUNDANCE OF ELEMENTS; METEORITIC ABUNDANCE OF ELEMENTS; and SOLAR ABUNDANCE OF ELEMENTS.

bulk density Mass per unit volume of soil (sampled as a clod or core), dried to constant weight at 105 °C.

bulking 1. Increase in volume of material when it absorbs water; dry *sand or *clay may swell as much as 50%. **2.** Increase in volume of solid rock or soil when broken.

bulk minerals *See* AGGREGATE.

bulk modulus (incompressibility modulus, *K*) The stress : strain ratio. When a simple hydrostatic pressure p is applied to a cube the resulting volume strain is equal to the ratio of the change in volume δV to the original volume V. Thus: K = (volume stress p)/(volume strain $\delta V/V$).

Bullard, Edward Crisp (1907–80) A geophysicist of the University of Cambridge, England, Bullard worked on many of the geophysical techniques used to gather data in support of *plate tectonic theory, including studies of gravity, heat flow, and *palaeomagnetism. He was an early supporter of plate tectonic theory, publishing in 1964 a computer-generated map showing the matching of *continental shelves on either side of the Atlantic.

bumpiness Unevenness experienced in the flight of an aircraft caused by convection or turbulence in conditions of atmospheric instability or over orographic barriers. *See also* CLEAR-AIR TURBULENCE.

Buntsandstein *See* TATARIAN.

buoyancy Condition arising from the difference between the density of a given *parcel of air or gas and that of the surrounding air. It is this that makes a hydrogen- or helium-filled balloon or airship float or rise through the atmosphere. Also, if an air parcel is warmer than its surroundings, the density difference implies an upwards directed force acting upon that parcel of air and it will rise, with positive buoyancy. When a parcel of air is colder than its surroundings the reverse condition causes it to sink, with downward (negative) buoyancy. *See* THERMAL.

buran Regional wind in Russia and central Asia which blows strongly from the north-east in both winter and summer. In winter it is very cold and is associated with much snow, when the wind is also termed 'purga'.

Burdigalian 1. An *age in the Early *Miocene, preceded by the *Aquitanian and followed by the *Langhian. **2.** The name of the corresponding European *stage, dated radiometrically as about 21.5–16.3 Ma ago, which is roughly contemporaneous with the upper *Saucesian (N. America), upper *Otaian and *Altonian (New Zealand), and the upper *Longfordian and *Batesfordian (Australia). The type section is at Le Coquillat, France. Mammals, including elephants, are important in the stratigraphic definition of this stage.

Burgess Shale A *horizon from the *Cambrian of British Columbia which has yielded an exceptionally preserved fauna. Originally discovered by C. D. Walcott in 1909, it has been redescribed by H. B. Whittington and other authors (1967–8). Apparently the fauna was deposited in deep water on or near a submarine fan. *Arthropods of various types account for more than 30% of the fauna, but other groups are also represented, many of them bizarre. It is probable that in terms of its diversity (140 species in 119 genera) this was a typical Cambrian fauna, but it is unusual in that the soft-bodied forms are preserved.

burial metamorphism A term first used by D. S. Coombs in 1961 to describe metamorphic *recrystallization during *epeirogenic as opposed to *orogenic earth movements. Sediments and volcanic rocks in a developing *basin gradually become buried during sagging of the crust in response to the weight of the accumulating rock column above, so that temperatures, even at great depth, are much lower than those experienced during plate *collision, when forcible depression of the rock mass to regions of much higher temperature and pressure cause metamorphic changes characteristic of *regional metamorphism. *See* METAMORPHIC ROCK.

buried soil Soil covered by an *alluvial, *colluvial, *aeolian, glacial, or organic deposit, and that is a product of a former period of *pedogenesis. In US usage a buried soil is defined as lying beneath 300–500 mm if the covering layer is more than half the thickness of the buried soil, otherwise beneath more than 500 mm.

buried topography An *overstepped erosion surface (*unconformity) which shows topographic relief of hills and valleys. Such relief has a strong effect on the thickness and *attitudes of the overstepping strata.

burner reactor Type of nuclear reactor using ^{235}U as fuel. *Enriched uranium is used to increase efficiency. To prolong the *fission reaction, fast neutrons are slowed down with a moderator (graphite or heavy water) and the rate of reaction is adjusted by control rods (boron) which can absorb neutrons. Some neutrons react to form plutonium, but in smaller amounts than the original uranium, hence the term burner reactor. *Compare* FAST BREEDER REACTOR.

Burnet, Thomas (1635?–1715) A natural philosopher whose *Sacred Theory of the Earth* (1680, 1689) was an early diluvialist work. He tried to correlate the seven days of creation with Earth history, describing the Earth as a giant shell from which flood waters gushed when it was broken by God in the Deluge. The broken fragments of the crust formed mountains. *See* DILUVIALISM.

burrow A *trace fossil formed by an animal during feeding, migration, or in the creation of a resting place. Burrows are formed in soft *sediments and may occur on the surface or be the result of subsurface activities.

'burst of monsoon' Applied to the onset of a marked change in weather conditions in the Indian subcontinent and south-east Asia, associated with the arrival of humid south-westerly winds which displace the hot, dry, pre-monsoon regime. The changed surface-level wind pattern is related to the establishment of a high-level easterly *jet stream.

Burzyan A *system of the Middle *Proterozoic, extending from about 1675 Ma to about 1475 Ma ago (Harland et al., 1989), of western Russian origin.

butte Small, isolated, flat-topped hill resulting from the erosion of near-horizontal strata. The diameter of the caprock is less than the height of the land-form above the surrounding country. Buttes are commonly found in semi-arid regions dominated by *duricrust horizons. *See also* MESA.

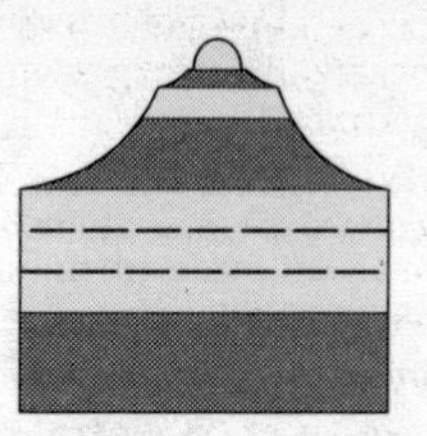

Butte

Buys Ballot's law Law, enunciated by Prof. Buys Ballot of Utrecht in 1857, that in the northern hemisphere the winds blow anticlockwise around centres of low pressure and clockwise around centres of high pressure. In the southern hemisphere both these tendencies are reversed.

by-product In mining, subsidiary material worked from *ore deposits in which other materials are dominant. In some cases the revenue from the by-products can exceed that from the major type, e.g. gold from *porphyry copper deposits. *Compare* CO-PRODUCT.

byssate Applied to the condition found in some bivalves of being attached to the substrate or to objects by strands of byssus (usually tough, horny threads).

byssus *See* BYSSATE.

byte A group of 8 *bits of digital data in binary form. A byte can represent values of 0–225.

bytownite *See* PLAGIOCLASE FELDSPAR.

b-zone A biostratigraphic *zone distinguished by *benthonic *fossils, e.g *brachiopods and *trilobites. The term was proposed by T. G. Miller in 1965. *Compare* P-ZONE.

^{14}C *See* RADIOMETRIC DATING.

cable drilling Drill in which a weight is repeatedly lifted and dropped to create a hole. Only shallow holes can be made in this manner. *See also* ROTARY DRILLING.

cadicone *See* INVOLUTE.

Caerfai Epoch (Early Cambrian Epoch) The earliest *epoch of the *Cambrian *Period in the *chronostratigraphic scale adopted by W. B. Harland et al., 1989, lasting approximately from 570 to 536 Ma ago and comprising the *Tommotian, *Atdabanian, and *Lenian *Ages. This epoch has also been termed the Comley Epoch by some authors, and the rocks deposited during this time the Comley *Series, underlain by the *Ediacaran (*Precambrian) and overlain by the *St David's. *Compare* WAUCOBAN (US usage).

Cainozoic *See* CENOZOIC.

cairngorm *See* QUARTZ.

cake *See* MUD-CAKE.

Calabrian Early *Pleistocene *stage which began about 1.8 Ma ago. It lasted about 1.3 Ma and is noted for major evolutionary changes in mammalian faunas. *See also* CASTLECLIFFIAN; and QUATERNARY.

calamine *See* SMITHSONITE.

Calamites cistiiformes Jointed-stemmed calamitids were an important component of *Carboniferous swampland floras. The generic name *Calamites* was first given to the ridged and furrowed casts of the pith cavity, which are commonly encountered fossils (*see* FORM GENERA). Unlike their smaller modern counterparts (**Equisetum* species) some Carboniferous species grew to 18 m tall. *Calamites cistiiformes* was the first representative of the family Calamitaceae. It was first described by Stur in 1877. *See also* SPHENOPSIDA.

calc-alkaline The name given to a suite of rocks comprising the volcanic association *basalt–*andesite–*dacite–*rhyolite, or the *plutonic association *gabbro–*diorite–*granodiorite–*granite. The suite is defined chemically using a graph depicting the variation of CaO and (Na_2O + K_2O) with SiO_2 (wt. %), and is classified as calc-alkaline if the SiO_2 value at the intersection of the CaO and (Na_2O + K_2O) trends is between 56 and 61 SiO_2 wt. %. (Less than 51 SiO_2 wt. % and the rock suite is classed as belonging to the alkalic series, 51–56 SiO_2 wt. % to the alkali-calcic series, and more than 61 SiO_2 wt. % to the calcic series.) Calc-alkaline rocks are typically developed on the continental side of *plate *subduction zones, well-known examples occurring in the Andes of S. America and in Japan.

Calcarea (Calcispongea; phylum *Porifera) A class of sponge, ranging from *Cambrian to Recent, in which the skeleton is made entirely of calcareous *spicules which are commonly of a tuning-fork shape. The sponges are sometimes associated with *reefs in the *Jurassic, or they may form widespread sponge beds.

calcarenite Calcareous sediment in which a high percentage of the *clasts can be of *quartz within a calcareous *matrix or, more generally, a *clastic *limestone in which both clasts and matrix are calcareous. Calcarenites are characterized by a particle range of 0.006–2.00 mm in diameter.

calcareous ooze Deep-sea, fine-grained, *pelagic deposit containing more than 30% calcium carbonate. The calcium carbonate is derived from the skeletal material of various *planktonic animals and plants, e.g. *foraminiferan *tests and *coccoliths (which are calcitic), and pteropod tests (which are aragonitic). Calcareous ooze is the most extensive deposit on the ocean floor but is restricted to water depths less than about 3500 m. *See also* CARBONATE-COMPENSATION DEPTH.

calcareous soil Soil that contains enough free calcium carbonate to effervesce visibly, releasing carbon-dioxide gas, when treated with cold 0.1 *N* hydrochloric acid, and which could also be regarded as a *basic or *alkaline soil.

calceolid *See* SOLITARY CORALS.

Calcichordata *See* HOMALOZOA.

calcichordates A group of early *Palaeozoic animals with calcareous *exoskeletons, interpreted by some as *carpoid echinoderms (*Echinodermata), and by others as ancestors of the *Chordata.

calcic horizon Mineral *soil horizon with evidence of secondary calcium-carbonate deposition which is more than 150 mm thick, with a calcium-carbonate content of more than 15% by weight, and with 5% more calcium carbonate than is in the parent material or horizons below it.

calcic series *See* CALC-ALKALINE.

calcification Process of redeposition of secondary calcium carbonate from other parts of the *soil profile which, if sufficiently concentrated, may develop into a calcrete, *caliche, kunkar, or *calcic horizon (all similar and usually comprising more than 15% by weight of calcium carbonate in more than 150 mm soil thickness). Calcification involves limited upward and lateral movement of calcium salts in solution, and downward movement in less-wet periods. When occasionally *leaching is deeper, some but not all of it may redissolve.

calcilutite A fine-grained *limestone consisting of *silt and clay-sized *carbonate particles (less than 63 μm in *grain size).

calcirudite A coarse-grained *limestone consisting of limestone *clasts, more than 2 mm in diameter, set in a *carbonate *matrix. The clasts may be angular or rounded.

calcisiltite A fine-grained *limestone consisting of *silt-sized (but not *clay-sized) *carbonate particles (more than 2 but less than 63 μm sizes).

calcispheres Small *calcite spheres, up to 500 μm in diameter, commonly found in *Palaeozoic *limestones and believed to be of algal origin. They consist of a *micrite wall enclosing an interior which is hollow or filled with sparry-calcite (*sparite).

Calcispongea *See* CALCAREA.

calcite Very common, widespread, rock-forming carbonate *mineral, one of two *polymorphs of $CaCO_3$, the other being *aragonite; sp. gr. 2.7; *hardness 3; *trigonal; usually colourless or white, but may be shades of yellow, grey, green, red, or even brown or black; white *streak; *vitreous *lustre; *crystals common, often *tabular, *prismatic, or rhombohedral, but fibrous aggregates and granular masses may also occur; *cleavage perfect rhombohedral $\{10\bar{1}1\}$; cleavage rhombs exhibit *double refraction; a major constituent of calcareous *sedimentary rocks, e.g. *marbles. Calcite can be precipitated from sea water and is a common constituent of invertebrate shells, and late-stage hydrothermal solutions (*see* HYDROTHERMAL ACTIVITY) may precipitate calcite in cavities in some *igneous rocks such as *basalts. It is soluble in dilute hydrochloric acid. It is used as a flux, in cement-making and fertilizers, and as a building stone. *See also* CARBONATES.

Calcite compensation depth (CCD) The depth in the sea at which the rate of dissolution of solid calcium carbonate equals the rate of supply. Surface ocean waters are usually saturated with calcium carbonate, so calcareous materials are not dissolved. At mid-depths the lower temperature and higher CO_2 content of seawater cause slow dissolution of calcareous material. Below about 4500 m waters are rich in dissolved CO_2 and able to dissolve calcium carbonate readily. Carbonate-rich sediments are common in waters less than 3500 m depth, but are completely absent below about 6000 m. *See also* CALCAREOUS OOZE.

calcium feldspar *See* ALKALI FELDSPAR.

calcrete *See* CALICHE.

calcrete uranium In arid climates prolonged evaporation may cause hard surface *crusts to form; if $CaCO_3$ predominates the crust is called *caliche (calcrete). Where the bedrock contains high levels of uranium the calcrete is locally uraniferous and may constitute a workable *ore, e.g. in western Australia and Namibia.

calcsilicate A group of *minerals whose bulk composition consists of calcium silicates. Common calcsilicates include the minerals *wollastonite ($CaSiO_3$), the calcium *garnet grossularite, and the *pyroxene *diopside ($CaMgSi_2O_6$). Calcsilicates are commonly formed by the *metamorphism of *limestones and *dolomites.

calc-sinter *See* TRAVERTINE.

caldera A roughly circular topographic and structural depression, varying in diameter from about 1 to 100 km (but up to 70 km in some martian examples), and formed by the foundering and collapse of a *magma chamber roof into its underlying magma body (e.g. Crater Lake, Oregon, formed by the eruption of Mt Mazama about 6000 years ago). Caldera collapse is commonly preceded or accompanied by rapid explosive evacuation of magma from the chamber in the form of surface *pyroclastic flows. This leaves the chamber roof unsupported by magma pressure and collapse follows. Slumping and erosion of the caldera walls may enlarge the topographic rim of the depression well beyond the structural rim. Later injection of magma into the chamber can cause doming of the caldera floor to create a resurgent caldera. Tobu caldera, Sumatra, almost 100 km in longest dimension, is the largest terrestrial caldera.

Caledonian orogeny Major mountain-building episode which took place during the Lower *Palaeozoic Era. The orogeny affected Greenland, Ireland, Scotland, and Scandinavia, and was associated with the closure of the *Iapetus Ocean between the old continents of *Laurentia, *Gondwana, and *Baltica.

calibration graph A plot of the line intensities (relative to the *internal standard) of a set of samples with known intensities, used in the calculation of the concentration of elements (e.g. in emission spectrometry) from the line intensity of the sample being examined.

caliche (calcrete) *Carbonate *horizon (the K horizon) formed in a soil in a semi-arid region, under conditions of sparse rainfall (20–60 mm/yr) and a mean annual temperature of about 18 °C, normally by the precipitation of calcium carbonate carried in solution. The *soil profile develops over several thousand years, initially in the form of nodules (glaebules), more mature caliches taking a massive, laminar form. It may become *cemented and *indurated on exposure, when it gives rise to a tabular landscape. *See also* CALCIC; DURICRUST; DURIPAN; and PETROCALCIC.

calichnia A type of *trace fossil of a structure that was made for breeding purposes.

California bearing ratio (CBR) **1.** Strength tests of *subgrades and construction material; also used in the design of flexible *pavements to meet specifications. **2.** Measure of soil resistance to penetration under controlled conditions. *See* PENETRATION TESTS.

California current Southward-flowing, eastern *boundary current carrying cool water from the North Pacific current to join the North *Equatorial current. This slow-moving, diffuse water mass flows down the west coast of N. America.

caliper log The record produced by a spring-loaded caliper that continually adjusts itself to the size of a *borehole as it is pulled to the surface. The log is a direct measure of the diameter of the borehole as a function of depth.

Callisto (Jupiter IV) The third largest of the *Galilean satellites, with the lowest density, the body is believed not to have differentiated into a core and mantle, consisting throughout of a mixture of rock and ice. Its surface is the darkest of the Galileans (albedo 0.20; though twice as bright as that of the *Moon) and Callisto is the most heavily cratered body in the solar system. There is believed to be an almost complete absence of geologic activity at the surface, which has an age of about 4 billion years and shows no sign of any extensive resurfacing. The surface is of dirty ice and there are no large mountains. Surface craters and rings are shallow; the largest structures are Valhalla, a bright patch about 600 × 3000 km, and Asgard, a ring about 1600 km across. The surface temperature is about –45 °C. Callisto was discovered on 7 January 1610 by Galileo. Its diameter is 4806 km; mass 1.077×10^{23} kg; density 1851 kg/m^3; surface gravity 0.127 (Earth = 1); mean distance from Jupiter 1.883×10^6 km; mean distance from Sun 5.203 AU; orbital period 16.68902 days; rotational period 16.68902 days.

Callovian A *stage in the European Middle *Jurassic dated at 161.3–157.1 Ma (Harland et al., 1989). *See also* DOGGER.

calm Condition of general lack of wind, indicated by a wind speed of less than 1 knot (0.5 m/s). *See also* BEAUFORT SCALE.

calving Process whereby portions of a glacier's leading edge break off as icebergs into an adjacent body of water.

calvus From the Latin *calvus* meaning 'bald' or 'stripped'; a species of *cumulonimbus cloud in which upper protrusions form a more amorphous mass than appears from the cumuliform outlines. *See also* CLOUD CLASSIFICATION.

calyce The skeletal surface of a coral that in life was in contact with the basal ectoderm of the polyp. It can be quite variable in shape and dimensions, but is commonly saucer- or cup-shaped.

Calypso (Saturn XIV) One of the lesser

satellites of *Saturn, discovered in 1980 by *Voyager 1, with a radius measuring 15 × 8 × 8 km; visual albedo 0.6.

Calyptoptomatida (phylum *Mollusca) Extinct *Palaeozoic class of marine molluscs, with *bilateral symmetry, in which the shell shape is variable: most are subtrigonal in cross-section. The walls are generally thick, and composed of laminated calcium carbonate. The apertural end is the widest; the narrowest is the closed apex, which may be pointed or blunt. The juvenile portion is chambered with *imperforate *septa. Externally the shells are smooth, or ornamented with growth lines and/or longitudinal and transverse ridges. The *aperture is usually protected by an *operculum possessing paired muscle scars. Individuals are 1–150 mm long. The class ranges in age from Lower *Cambrian to Middle *Permian, and contains two orders.

calyx 1. Cup-shaped, plated body of *pelmatozoan *echinoderms. The calyx is made up of several rows of plates. If there is no stem, a centrodorsal plate is succeeded by a ring of five basal plates and then by a further ring of radial plates above them. Above these two rows of plates come the plates making up the arms. Where these two rows of plates are present the calyx is said to be 'monocyclic'. In some species a third row of plates, the infrabasals, is present beneath the basals and the calyx is then said to be 'dicyclic'. In some cases radial plates may be compounded (i.e. split transversely into two plates) and then, depending on their position in the calyx, they may be known as 'infraradial' or 'supraradial plates'. **2.** Bowl-shaped depression at the top of a calcareous coral skeleton, usually formed by the upper edges of the *septa.

calyx drilling Method of *core drilling performed by the rotation of a steel cylinder and cutting with chilled shot, about 2.4 mm diameter, which cuts a formation core. Circulating water carries the cuttings up to a basket-like space at the top of the *core barrel. The core is wedged into the barrel and pulled up one barrel-length at a time. Shafts up to 2 m diameter can be drilled, and holes to a depth of more than 300 m.

camara *See* CAMERA.

camaral *See* CAMERA.

cambering Consistent *dip of strata towards local valley floors, in conflict with the general regional dip. It is well displayed in the English Midlands, where *ironstone beds above clays are cambered down as much as 30 m below their original level. It is probably due to large-scale structural disturbance when *permafrost thawed and when the plastic clays allowed overlying massive beds to flow towards valleys.

cambic horizon Weakly developed mineral *soil horizon of the middle part (B horizon) of *soil profiles, and one that has few distinguishing morphological characteristics except for evidence of *weathering and sometimes of gleying. It is found in *brown earths and *gleys.

Cambrian The first of six periods of the *Palaeozoic Era, which began about 570 Ma ago and ended about 510 Ma ago. Sediments deposited during the period include the first organisms with mineralized skeletons. Common fossils include *brachiopods, *trilobites, *ostracods, and, late in the period, *graptolites. Trilobites are important in the stratigraphic subdivision of the period.

camera (camara) (adj. cameral or camaral) One of the chambers within a chambered mollusc, e.g. a nautiloid (*Nautiloidea) or ammonoid (*Ammonoidea). In living cephalopods (*Cephalopoda), e.g. *Nautilus*, the chambers contain gas at pressures ranging from about 0.3 to 1 atm (30–100 kPa). Some of the chambers may contain 'cameral fluid' which can be extracted by the *siphuncle, allowing the animal to adjust its density according to the depth at which it is living.

cameral *See* CAMERA.

cameral fluid *See* CAMERA.

Camerata (class Crinoidea) Subclass of echinoderms (*Echinodermata), with thecal plates typically united by rigid structures composed of numerous small, polygonal plates, and some lower brachial plates, all the plates firmly sutured together. The *tegmen is plated, some of the plates roofing over the mouth and *ambulacral grooves. The Camerata are known from the Lower *Ordovician to Upper *Permian.

camptonite A dark-coloured, medium-grained *igneous rock composed of *plagioclase feldspar, barkevikite (a sodium-bearing *amphibole) and/or *augite. Camptonites are one variety of a group of *intrusive rocks collectively known as *lamprophyres which characteristically form *dykes. The original camptonite was named from Campton Falls in New Hampshire, USA.

camouflage Way in which a *trace element can substitute for a common element of the same *valency and nearly the same *ionic radius in a crystal lattice, e.g. Hf^{4+} for Zr^{4+}. The trace element is said to be camouflaged by the more common element. *Compare* CAPTURE.

Campanian A *stage in the European *Cretaceous (83–74 Ma, Harland et al., 1989). *See also* SENONIAN.

Campbell–Stokes sunshine recorder Device that utilizes a glass sphere to concentrate the Sun's rays on to a calibrated paper, where the resulting burns register the time and duration of sunshine. The instrument must be pre-set for a given latitude.

Canada balsam resin A naturally occurring resin, distilled from the bark of *Abies balsomea* (the balsam fir) and other N. American *Abies* species, which, when heated to 160 °C, becomes liquid and is used to cement *mineral or *rock chips to glass slides as part of the process of preparing *thin sections. Its use has been largely superseded by warm- or cold-setting epoxy resins which combine low viscosity (about 100 centipoise) with high shear strength (about 11.7×10^6 Pa), good adhesion, and a *refractive index of 1.54.

Canadian A *series of the Lower *Ordovician of N. America, equivalent to the *Tremadoc and *Arenig.

'canali' *See* MARTIAN CANALS.

canalizing selection The elimination of *genotypes that render developing individuals sensitive to environmental fluctuations. Genetic differences may be revealed in organisms by placing them in a stressful environment, or if a severe *mutation stresses the developmental system.

Canaries current Cool, oceanic water current which flows south along the continental margin of Spain, Portugal, and W. Africa. This slow-moving eastern *boundary current is the cause of frequent sea fogs off north-west Spain and Portugal. The water is further cooled by *upwelling of cold water off W. Africa.

cancrinite A member of the *feldspathoid group of silicate *minerals with a complex formula, approximately $(Na,Ca,K)_{6-8}[AlSiO_4]_6(CO_3,SO_4,Cl)_{1-2}.1\text{–}5H_2O$, it may be an alteration product of *nepheline, and is related to vishnevite, the sulphate-rich variety; normally *massive; occurs in nepheline syenites.

Canidae *See* CARNIVORA.

Caniformia *See* CARNIVORA.

cannel coal Fine-grained, tough, compact and uniform *bituminous coal, which is dark grey to black with a dull, greasy *lustre and *conchoidal fracture. It has a high percentage of *volatiles, high ash content, and burns with a smoky, luminous flame. It is a *sapropelic coal formed mainly of *lycopod spores and *algae with fossils of water-dwelling animals. It is found chiefly in Lancashire, England; the local name was 'cannel' meaning 'candle coal'.

cannel shale *See* TORBANITE.

cannonball bomb *See* VOLCANIC BOMB.

canyon A deep, steep-sided gorge cut by a river, generally into bedrock. *See also* SUBMARINE CANYON.

cap *See* PILEUS.

capacity (of stream) Maximum load of solid particles that a stream is capable of carrying. It is largely a function of particle size, in that a decrease in the size of particles involves an increase in the total load that can be transported. Ultimately a heavily loaded stream merges imperceptibly with a mud flow.

capillarity *See* CAPILLARY ACTION.

capillary action (capillarity) The process by which *soil moisture may move in any direction through the fine (i.e. capillary) pores of the soil, under the influence of *surface tension forces between the water and the soil particles. It is analogous to the rise of water in a cylindrical glass capillary tube whose end is placed in a bath of water in the laboratory, except that the pores of the soil are neither cylindrical, nor smooth-walled, nor clean. *See also* CAPILLARY MOISTURE.

capillary fringe *See* CAPILLARY ZONE.

capillary moisture (capillary water) Moisture that is left in the soil, along with hygroscopic moisture and water vapour, after the gravitational water has drained off. Capillary moisture is held by *surface tension (known in the USA as 'water potential') as a film of moisture on the surface of soil particles and *peds, and as minute bodies of water filling part of the pore space between particles. Curved water surfaces or *menisci* (singular: *meniscus*) form bridges across the pores at the boundaries between their water-filled and air-filled parts. Capillary moisture may move through the soil under the influence of surface tension forces (*see* CAPILLARY ACTION) and is available for removal by plant roots.

capillary water *See* CAPILLARY MOISTURE.

capillary wave Water wave whose wavelength is less than 1.7 cm and in which the primary restoring force is the *surface tension of the water. The slightest of breezes may cause slight 'puckering' of the water surface, and the capillary waves so produced will be smoothed and flattened by the effects of surface tension.

capillary zone (capillary fringe) The zone immediately above the *water table, into which water may be drawn upward as a consequence of *capillary action. A typical height for the capillary fringe in clay with a pore radius of 0.0005 mm might be 3 m, compared with less than 10 cm in a fine sand with a pore radius of 0.02 mm.

capillatus From the Latin *capillatus* meaning 'with hair', a species of *cumulonimbus cloud with fibrous cirriform appearance in its upper parts. It is often associated with an anvil or other protrusion in which wisps of cloud trail from the tip to give a hair-like appearance. Typically this cloud type brings squalls with showers or thunderstorms. *See also* CLOUD CLASSIFICATION.

Capitanian A *stage of the *Zechstein Epoch, preceded by the *Wordian and followed by the *Longtanian.

cap rock Layer of hard, impervious rock which lies immediately above a *source rock and which, because of its impervious nature, acts as a barrier, preventing the migration of hydrocarbons or water up-sequence. Such impervious rocks include clay-rich *sandstones, *limestones, and *evaporite deposits associated with salt *diapirs. Where it lies above an oil trap it prevents the upward migration of *petroleum. Above a *salt dome, the layer is usually composed of *anhydrite and *gypsum. In coal mining the term refers to a sandstone layer above shale overlying a coal seam.

Captorhinomorpha (order *Cotylosauria) Suborder of *reptiles, which appeared in the *Carboniferous and became extinct after the early *Permian. The suborder includes the earliest known reptiles, retaining some *amphibian features, but distinctly reptilian in the form of their skulls, and differing from amphibians in the structure of the pelvic girdle. Because they were ancestral to all later reptile groups, the cotylosaurs are known as 'stem reptiles'.

capture 1. Substitution in a crystal lattice of a trace element for a common element with lower *valency or larger *ionic radius, e.g. Ba^{++} for K^{+}. There is often a higher concentration of captured trace elements relative to common elements in the mineral than in the liquid from which it crystallized. *Compare* CAMOUFLAGE. **2.** *See* RIVER CAPTURE.

capuliform Applied to a cap-shaped gastropod (*Gastropoda) shell, similar to those of the genus *Capulus*.

Caradoc A *series (463.9–443.1 Ma) of the Upper *Ordovician, underlain by the *Llandeilo and overlain by the *Ashgill.

carbon Non-metallic element, chemical symbol C, which is unique in the number of compounds it is able to form that contain chains or rings of carbon atoms. This ability to form large, complex molecules in which other elements are bonded to carbon atoms is exploited by all living organisms. The discipline of organic chemistry is essentially the study of cyclic carbon compounds. Carbon is extracted from gaseous carbon dioxide by plants during *photosynthesis, is incorporated in living matter, and when organic matter decomposes its carbon is oxidized and so returned to the atmosphere as carbon dioxide. Pure carbon occurs naturally as *diamond, *graphite, fullerene, and as the amorphous carbon black. *Charcoal, produced by the destructive distillation of organic matter, is also a pure form of carbon. In the Earth sciences, carbon is also important in the form of carbonates, as in *limestones.

carbon-14 *See* RADIO-CARBON DATING.

carbonaceous chondrite Dull, black *stony meteorite, with little or no metal and abundant carbon; iron occurs as sulphide, silicate, or oxide. Carbonaceous *chondrites show very little *metamorphism but display evidence of chemical alteration by water, which continued after their formation, suggesting the parent body was rocky material mixed with ice. They contain a varied suite of organic compounds, including amino acids and a high content of inert gases. Carbonaceous chondrites have very primitive compositions, comparable to that of the *Sun's atmosphere and the nebula from which the *solar system formed. *See also* METEORITE. *Compare* ACHONDRITE.

carbonate lump *See* INTRACLAST.

carbonates 1. Group of *minerals found mostly in *limestones and *dolomites. *Calcite ($CaCO_3$) is the most abundant and most important. *Aragonite has the same formula as calcite but is less stable and shells composed of aragonite change to calcite through geological time. Dolomite (or pearl-spar) is the magnesium-bearing carbonate commonly found as a rock-forming mineral, $CaMg(CO_3)_2$. **2.** The term 'carbonate' is frequently used with reference to those *sedimentary rocks with 95% or more of either *calcite or *dolomite, and is synonymous with *limestone.

carbonation Chemical *weathering process involving a reaction between dilute carbonic acid, derived from the solution in water of free atmospheric and soil–air carbon dioxide, and a mineral. The best-known example is the reaction between *limestone (calcium carbonate) and carbonated water (carbonic acid) which yields calcium and bicarbonate ions in solution:

$$CaCO_3 + H^+ + HCO_3^- \rightarrow Ca^+ + 2HCO_3^-.$$

carbonatite Unusual *igneous rock, rich in *calcite and other *carbonate minerals (including *dolomite and *ankerite), which is considered to be *mantle-derived. Carbonatites occur as intruded masses, *dykes, and as *cone sheets, and rarely as *lavas and *tephra; and are found in association with *alkali-rich igneous rocks, notably those erupted by the volcanoes of the East African Rift. Rare elements, including the REEs (*rare earth elements), barium, niobium, thorium, and phosphorus, are often enriched in comparison with many crustal and mantle-derived igneous rocks.

carbon 'burning' 'Burning' (*see* NUCLEOSYNTHESIS) that occurs in large stars, due to the intense heat of the stellar core, following *hydrogen 'burning' and *helium 'burning'. During this stage of stellar evolution ^{12}C nuclei undergo nuclear reactions at temperatures around 8×10^8 K to produce elements such as ^{20}Ne, ^{23}Na, ^{23}Mg, and ^{24}Mg. *See* OXYGEN 'BURNING'; and SILICON 'BURNING'.

carbon cycle The movement of carbon through the surface, interior, and *atmosphere of the Earth. Carbon exists in atmospheric gases, in dissolved *ions in the *hydrosphere, and in solids as a major component of organic matter and *sedimentary rocks, and is widely distributed. Inorganic exchange is mainly between the atmosphere and hydrosphere. The major movement of carbon results from *photosynthesis and respiration, with exchange between the *biosphere, atmosphere and hydrosphere. Rates of exchange are very small, but over geologic time they have concentrated large amounts of carbon in the *lithosphere, mainly as *limestones and *fossil fuels. This carbon was probably present as CO_2 in the primordial atmosphere. The burning of *fossil fuels and the release of CO_2 from soil air through the clearance of tropical forests may eventually change the balance of the carbon cycle, although the climatic effects may be partly mitigated by the buffering action of the oceans; it is estimated that about 200 billion tonnes of CO_2 have been added to the atmosphere in this way since 1850. *See* GREENHOUSE EFFECT.

carbon dating *See* RADIO-CARBON DATING.

Carboniferous Penultimate period of the *Palaeozoic Era, preceded by the *Devonian and followed by the *Permian. It began about 362.5 Ma ago and ended about 290 Ma ago. In Europe the lower part of the *system is termed the *Dinantian. It is divided into two *series and is characterized by marine *limestones with a rich *coral–*brachiopod fauna. In contrast the upper part, the *Silesian, which is subdivided into three series, is noted for the deposition of terrestrial and freshwater sediments. The vast forests of the Upper Carboniferous gave rise to the rich coal measures of south Wales, England, Scotland, and many other areas worldwide. N. American geologists subdivide the Carboniferous System into two subsystems. Of these the lower (362.5–322.8 Ma ago) is named the *Mississippian and is the equivalent of the Dinantian sub-System plus the lower part of the Silesian sub-System. The upper sub-system, the *Pennsylvanian (322.8–290 Ma ago), is the equivalent of most of the Silesian.

carbon isotopes Natural carbon is composed of three *isotopes: ^{12}C making up about 98.9%; ^{13}C about 1.1%; and ^{14}C whose amount is negligible, but which is detectable because it is radioactive. The relative abundance of these isotopes varies and the study of this variation is an important tool in geologic research, especially *radiometric dating. Carbon-isotope dating is a method of radiometric age-dating using the amount of the heavy, radioactive isotope carbon-14 remaining in organic matter. Carbon-14 has a *half-life of 5730 ± 40 years and the amount of the isotope present can be used to date materials up to about 50 000 years old (*see* RADIOCARBON DATING). In diagenetic studies (*see* DIAGENESIS), measurement of the ratio of carbon-13 to carbon-12 allows the recognition of *carbonate precipitated from a variety of different sources.

carbonization *See* FOSSILIZATION.

cardinalia A collective term that describes the structures found at the posterior end of the interior of the brachial valve of a brachiopod (*Brachiopoda). In its simplest form, as a cardinal process, it is merely the site of a series of muscle bases. In other species it may be much more complex.

cardinal septum One of the initial *septa (vertical plates within the coral skeleton) which occur in rugose corals (*Rugosa). When the coral is very young a single proseptum develops in the *calyce. This soon separates into two, one called the 'cardinal septum', the other the 'counter-cardinal septum'. Other septa then follow to make up the arrangement of septa.

cardinal tooth Large *hinge tooth present in some bivalves which is immediately below the *umbo. More than one may be present on each *valve.

Caribbean current Warm, ocean water current which flows westward through the Caribbean Sea, passes into the *Florida current, and thus contributes to the *Gulf Stream. Its flow velocity averages 0.38–0.43 m/s.

Caribbean Plate One of the present-day minor lithospheric *plates, that is subducting beneath the Antilles in the east, with the *Cocos Plate subducting beneath it on the western side of Central America. The northern and southern boundaries are both *faults: the Bartlett Fault separates it from the *N. American Plate and the Bocono Fault from the *S. American Plate.

carina *See* CONODONTOPHORIDA; and KEEL.

Carlsbad twin The *mineral *orthoclase feldspar forms *crystals which may twin according to several laws. The most common of these is the Carlsbad law, with a *composition plane on 010 and with the vertical '*c*' (or '*z*') axis as twin axis. It is commonly an *interpenetrant (penetration) twin.

Carlsberg Ridge The slow-spreading oceanic *ridge which separates the *African Plate and the *Indo-Australian Plate.

Carme (Jupiter XI) One of the lesser satellites of *Jupiter, with a diameter of 30 km; its orbit is *retrograde.

carnallite *Mineral, $KMgCl_3.6H_2O$; sp. gr. 1.6; *hardness 1–2; *orthorhombic; normally white, occasionally yellowish and reddish; greasy *lustre; *massive and granular, the pseudo-hexagonal *crystals being rare; no *cleavage; occurs in *evaporite deposits and is soluble in water; bitter taste. It is used as a fertilizer, and is named after the nineteenth-century mining engineer R. V. Carnall.

carnassial In many *Carnivora, a modification of premolar or molar teeth, commonly the lower first molar and the upper last premolar, giving them a scissor-like shearing action used for cutting flesh.

Carnian (Karnian) 1. A *Triassic *age, preceded by the *Ladinian, followed by the *Norian, and dated at 235–223.4 Ma (Harland et al., 1989). **2.** The name of the corresponding European *stage, roughly contemporaneous with the Banan (China), and Oretian and Otamitan (New Zealand).

Carnivora (cohort *Ferungulata, superorder Ferae) An order that comprises the modern carnivorous placental mammals and their immediate ancestors. It used to be divided into two suborders, the Fissipedia (mainly land-dwelling) and Pinnipedia (seals, sea lions, walrus), but a more modern classification is into Caniformia (dog-like) and Feliformia (cat-like), with the 'pinnipeds' belonging to the former. The carnivores are descended from a single stock of the probably insectivorous, placental mammals of the early *Cretaceous, the change being reflected in their dentition. Strong incisors for biting, and canines for piercing, were retained from the insectivorous forms, but in general carnivores acquired modified cheek teeth (carnassials) specialized for shearing. These subsequently became reduced in those carnivores which adopted a herbivorous diet. Hoofs have rarely developed, as claws are used for seizing prey, and digits are never greatly elongated (and, apart from the pollex and hallux, they are not reduced). The first true carnivores were the weasel-like Miacidae of the *Palaeocene, which had diverged by the end of the *Eocene to give the Canidae (dogs) and Mustelidae (weasels and their allies) as one branch and the Viverridae (Old World civets) and Felidae (cats) as another. According to some authors, the Mustelidae later branched again to give the Phocidae (seals); and the Canidae diversified widely to produce such forms as the Amphicyonidae ('dog-bears'), Otariidae (sea lions), Procyonidae (raccoons and pandas), and, ultimately, Ursidae (bears); but molecular studies seem to indicate that the Phocidae and Otariidae are descended from a single ancestor which was related to the mustelid–ursid–procyonid stem. Finally, the Hyaenidae (hyenas) emerged in the late *Miocene from viverrid stock; this is the youngest of the carnivore families.

carnosaur *Saurischian (lizard-hipped), carnivorous *dinosaur. Carnosaurs were bipedal, powerfully built, and possessed large, dagger-like teeth. **Tyrannosaurus rex*, of the Upper *Cretaceous, represented the culmination of the carnosaur line.

carnotite *Mineral $K_2(UO_2)_2(VO_4)_2.1-3H_2O$; sp. gr. 4.5; *hardness 2.0–2.5; *monoclinic; normally bright yellow to greenish-yellow; pearly *lustre; *cleavage perfect, basal {001}; strongly radioactive; occurs in the weathered zone of sedimentary *uranium ores, especially in *sandstones enriched with organic matter, normally as coatings and masses in rocks. It is an important ore for uranium.

carpal One of the bones of the wrist that articulate with the digits.

Carpentarian A *stage of the Lower–Middle *Proterozoic of south-eastern Australia, underlain by the *Nullaginian and overlain by the *Adelaidean.

Carpenter, William Benjamin (1813–85) A surgeon and comparative anatomist, who used microscopy to investigate the structure of fossil shells, especially *foraminifera and *crinoids. He took part in several scientific voyages, including the **Challenger* expedition, as a result of which he became interested in marine physics and developed a theory about oceanic circulation.

carpoids (phylum Echinodermata) Informal collective term describing the homalozoan classes Homoiostelea, Homostelea, and Stylophora. *See* HOMALOZOA.

carrier element 1. Major element in a *mineral, for which a trace amount of another element may substitute, e.g. Mg (the carrier element) replaced by Ni (the *trace element) in *olivine. **2.** Inactive material, isotopic with a radioactive transmutation product, which is added to act as a carrier for active material in subsequent chemical reactions. Carrier elements are used in analyses of small samples, as the mass of radioactive material produced in a nuclear reaction is usually too small to be suitable for ordinary analytical procedures such as precipitation and filtration.

Cartesian projection A technique for the mapping of space in which every plane in the area being mapped is projected on to a plane in the map and every line on to a line. Each point in the area under study is identified by three values, representing its location in relation to three mutually perpendicular axes (the Cartesian coordinates of the point). These coordinates are transformed mathematically into a homogeneous set of four coordinates which can then be plotted to produce a graphic representation (a map). The word 'Cartesian' is derived from the name of René Descartes (1596–1650).

cartilage In vertebrates, flexible skeletal tissue formed from groups of rounded cells lying in a matrix containing *collagen fibres. It forms most of the skeleton of embryos and in adults is retained at the ends of bones, in intervertebral discs, and in the pinna of the ear; in *Elasmobranchii, calcified cartilage rather than true *bone provides the entire skeleton.

cartilaginous fish Fish in which the skeleton, including the skull and jaws, consists entirely of *cartilage and never, even in the adult stage, comprises bony tissue. Sharks and rays (*Chondrichthyes) have a cartilaginous skeleton, as have other lower vertebrates, e.g. the jawless lampreys and hagfish (*Agnatha).

cascading system In geomorphology, a type of dynamic system characterized by the transfer of mass and energy along a chain of component subsystems, such that the output from one subsystem becomes the input for the adjacent subsystem. An example is the *valley glacier, where the inputs of snowfall and rock debris from the slopes above, and potential energy (derived from elevation) are cascaded through a sequence of climatic environments with a progressive reduction in mass and dissipation of energy, the output from the glacier being sediment and water which form the input to the *proglacial subsystem.

casing Steel tubes, usually screwed together, which line a *borehole to prevent caving-in of the side walls. The gap between the casing and the natural rock is often filled with concrete. Casings are used mainly where the borehole passes through clays.

Cassadagian *See* CHAUTAUQUAN.

Cassini A mission to *Saturn, launched in 1997, and operated jointly by *NASA and *ESA. *See also* HUYGENS.

Cassini Division *See* RESONANCE.

cassiterite (tinstone) *Mineral, SnO_2; sp. gr. 6.8–7.1; *hardness 6–7; *tetragonal; usually reddish-brown to nearly black, but it can be yellowish and ruby; white to grey *streak; *adamantine lustre; *crystals often pyramidal and *prismatic, may also be *massive and granular; *cleavage prismatic {100}, {110}; occurs typically in high-temperature *hydrothermal veins, with *granites and *pegmatites; associated minerals: *topaz, *quartz, *tourmaline, *mica, *chlorite, and high-temperature metallic *ores; also world-wide in *alluvial deposits because of its resistance to chemical and physical attack. Cassiterite is the only important ore for tin.

cast The preserved sediment infill of an impression or mould made in the top of a bed of soft *sediment (e.g. *flute cast, *load cast, *tool cast, casts of footprints or *trails).

Castalia A *solar system near-Earth asteroid (No. 4769), measuring 1.8 × 0.8 km; approximate mass 10^{11} kg; orbital period 0.41 years. It is a double-lobed object, each lobe about 0.75 km in diameter.

castellanus From the Latin *castellum*, meaning 'castle', a cloud species commonly associated with the upper parts of *altocumulus, *stratocumulus, *cirrus, and *cirrocumulus. Cumuliform, turreted protrusions extend in linear fashion from the cloud top, producing a crenulate form. *See also* CLOUD CLASSIFICATION.

Castlecliffian A *series in the *Quaternary of New Zealand, underlain by the *Nukumaruan, overlain by the *Holocene, and roughly contemporaneous with the uppermost Calabrian and Emilian *Stages, and the subsequent Upper *Pleistocene Series.

castle koppie *See* KOPPIE.

Castlemainian A *stage of the Lower *Ordovician of Australia, underlain by the *Chewtonian and overlain by the *Yapeenian.

CAT *See* CLEAR-AIR TURBULENCE.

cataclasis *See* CATACLASITE.

cataclasite Rock that has been deformed by the process of shearing and *granulation (cataclasis). Cataclasites are the products of dislocation *metamorphism and *tectonism. *See also* MYLONITE.

catagenesis Following *diagenesis, in which sedimentary material is compressed and undergoes chemical changes, a phase in the formation of *petroleum and *natural gas during which continuing sedimentation and subsidence produce temperatures of 50–150 °C and *kerogen is produced. *Compare* METAGENESIS.

catarrhine In primates, applied to nostrils that are close together and open downwards. Old World monkeys, apes, and humans have catarrhine nostrils.

catastrophic evolution (catastrophic speciation) Theory proposing that environmental stress can lead to the sudden rearrangement of chromosomes, which in self-fertilizing organisms may then give rise sympatrically to a new species (*see* SYMPATRIC EVOLUTION). Recent research suggests that at best this explanation applies only to some special cases.

catastrophism Theory that associates past geologic change with sudden, catastrophic happenings. Early geologists, including *Cuvier, *Buckland, and *Sedgwick, claimed that catastrophism was a sound scientific theory. Although it met with considerable scorn in more recent times, many modern geologists would describe themselves as 'neocatastrophists'.

catchment The area from which a surface watercourse or a *groundwater system derives its water. Catchments are separated by *divides. A surface catchment area may overlie an *aquifer system, but may be unconnected with the aquifer rock itself if there are intervening impermeable *aquicludes. In US usage, a catchment is often termed a 'watershed'.

catena Topographic sequence of soils, of the same age and usually on the same parent material, that is repeated across larger landscape transects. Individual *soil-profile types are related to site conditions and to position on a slope. The term was introduced in E. Africa in the 1930s, and is mainly applicable in certain non-glaciated landscapes, particularly those with small, hilly relief, e.g. *loess areas.

cateniform Applied to the form of the *corallum which occurs in the tabulate corals (*Tabulata). The *corallites are elongated and joined side by side to make fence-like structures. In end section they resemble chain links, hence the name (from the Latin *catena* meaning 'chain'). *See* COMPOUND CORALS.

cathode *See* CATION.

cathodoluminescence The luminescence induced by the bombardment of *minerals in polished *thin section with electrons. The technique is particularly useful in the identification of mineral *cements and overgrowths produced in successive episodes, or for distinguishing between particles and *authigenic overgrowths. The particles and each separate generation of cement will be of slightly different chemical composition, and so will luminesce with slightly different colours. *See* QUARTZ OVERGROWTH.

cation A positive *ion, i.e. an atom, or complex of atoms, that has lost one or more electrons and is left with an overall positive electric charge, e.g. Na^+, Mg^{2+}, NH_4^+. The name is derived from the fact that when an electric current is passed through a conducting solution the positive ions present in the solution move towards the cathode (the negative electrode). *Compare* ANION.

cation exchange Process in which cations in solution are exchanged with cations held on the exchange sites of mineral and organic matter, particularly on the surfaces of *colloids of *clay and *humus.

cation-exchange capacity (CEC) The total amount of exchangeable *cations that a particular material or soil can adsorb at a given *pH. Exchangeable cations are held mainly on the surface of *colloids of *clay and *humus, and are measured in milligram-equivalents per 100 g of material or soil.

cation ordering The phenomenon, the extent of which is temperature dependent, in which a *cation shows preference for the site it occupies because it provides greater chemical stability.

cat's eye *See* RIEBECKITE.

caudal Pertaining to the tail.

caudal vertebra *See* VERTEBRA.

cauldron-subsidence Collapse of a volcanic *crater due to the evacuation of a large *magma chamber (cauldron) and marked by a *ring fracture, or *ring-dyke. There are many examples of cauldron-subsidence (e.g. Glen Coe, Scotland), although the various mechanisms of their formation are not necessarily identical.

Cautleyan A *stage of the *Ordovician in the *Ashgill, underlain by the *Pusgillian and overlain by the *Rawtheyan.

cavate **1.** Applied to *spores where the *exine layers are separated by a cavity. **2.** Applied to dinoflagellate (*Dinophyceae) cysts where there is a space between the periphragm and the endophragm.

cavern porosity *See* CHOQUETTE AND PRAY CLASSIFICATION.

cavitation The collapse of a region because it is at a lower pressure than the ambient environment. The collapse of a gas bubble, in water, causes an *implosion that may be detectable as a *seismic energy source. If the hydrostatic pressure in a rock exceeds the hydrostatic pressure of the *drilling fluid, this can cause a collapse of the walls around the hole, or a 'blow-out' of the drill stem. Erosion by cavitation occurs in waterfalls, rapids, and subglacial river channels, and can produce pot-holes.

cay Small, flat, marine island formed from coral-reef material or *sand. The term is applied, for example, to the low-lying, sparsely vegetated islands off the coast of southern Florida.

Cayugan A *series of the *Silurian of N. America equivalent to the Lower *Ludlow to terminal *Pridoli.

Cazenovian *See* ERIAN.

CBR *See* CALIFORNIA BEARING RATIO.

CCD **1.** *See* CALCITE COMPENSATION DEPTH. **2.** *See* CHARGE-COUPLED DEVICE.

CCL *See* CONVECTIVE-CONDENSATION LEVEL.

^{14}C dating *See* RADIOCARBON DATING.

CDP *See* COMMON DEPTH POINT.

CDP stack *See* COMMON-DEPTH-POINT STACK.

CEC *See* CATION-EXCHANGE CAPACITY.

Celastrophyllum circinerve One of the earliest flowering plants known to palaeobotanists. Flowering plants of the family Celastraceae pre-date most others. They began with *C. circinerve* in the Early *Cretaceous, and survive today in the form of *Euonymus europaeus*, the spindle tree.

celerity Velocity with which a wave advances. The celerity (c) of an ideal wave is related to its wavelength (λ) and frequency (f) by the wave equation $c = f\lambda$. The wave frequency (f) is the number of waves (n) passing a point in unit time (t), i.e. $f = n/t$. In deep water the wave celerity may be calculated by the equation: $c = (g\lambda/2\pi)^{1/2} = 1.25\sqrt{\lambda}$ where λ is the wavelength in metres and g is the *acceleration due to gravity (9.81 m/s^2). The speed of shallow water waves may be calculated by the equation: $c = (gd)^{1/2} = 3.13\sqrt{d}$, where d is the depth of water in metres.

celestite *Mineral $SrSO_4$ which may form a *solid solution series with *barite; sp. gr. 3.9–4.0; *hardness 3.0–3.5; *orthorhombic; faint blue to colourless, sometimes stained red; white *streak; *vitreous *lustre; *crystals normally very *tabular, resembling barite, but can also occur fibrous and granular; *cleavage perfect basal {001}, present {210}, {010}; often fluorescent; occurs in *sedimentary rocks, particularly *dolomites, as a cavity-lining in association with barite, *gypsum, *anhydrite, and *halite, with gypsum and anhydrite in *evaporite deposits, also as a *gangue mineral in hydrothermal veins with *galena and *sphalerite, and it will form *concretionary masses in clays and marls. It is the main *ore mineral for strontium and strontium compounds.

cement **1.** Manufactured powder made from limestone and clay which sets to a solid mass when mixed with water. Commercial cements have to fulfil certain defined standards. Combined with *aggregate it forms concrete. **2.** Material, e.g. *calcite, that fills open *pore space in fragmental and organic *sediments.

cementation Process by which *sedimentary rock particles or fragments are cemented together after deposition. Cementing materials are deposited from the mineral-rich waters that percolate through the open *pore space of the rock. The percentage of cement depends on the amount of pore space and on the mud content within a given rock.

cemented Applied to massive, infilled, and *indurated mineral soil: such soil has a hard and often brittle consistency because soil particles are joined together by cementing substances, e.g. calcium carbonate, silica, iron and aluminium oxide, or *humus. Cemented soil usually appears as a highly distinctive and resistant horizon. *Compare* CEMENT; and CEMENTATION.

Cenomanian **1.** A Late *Cretaceous *age, preceded by the *Albian, followed by the *Turonian, and dated at 97.0–90.4 Ma (Harland et al., 1989). **2.** The name of the corresponding European *stage for which the *type locality is near Le Mans, France.

Cenozoic (Cainozoic, Kainozoic) *Era of geologic time extending from about 65 million years ago to the present. It includes the *Tertiary and *Quaternary Periods: the so-called ages of mammals and man. *Molluscs and *microfossils are used in the stratigraphic subdivision of the era. The *Alpine–Himalayan orogeny reached its climax during this period of geologic time.

Centrales *See* BACILLARIOPHYCEAE.

Central European Sea *See* PARATETHYS.

central limit theorem In statistics, the theorem stating of a series of data sets drawn from any *probability distribution, that the distribution of the *means of those data sets will follow a normal distribution.

central vent volcano A point source of the Earth's surface through which *lavas, *pyroclastics, and gas are erupted. The eruptive products have their thickest accumulation around the point source and build up a low-angle shield or higher-angle cone topography constituting the volcanic pile. Later eruptions may occur from fissures within the volcanic pile, which radiate and are fed from the point source. Continuous release of gas (explosive or quiet) from the point source ensures the existence of an open vent at the top of the volcanic pile. This vent can often be enlarged by the peripheral collapse of its walls. *See also* VOLCANO.

centre of curvature The point at which arcs of the circle which forms the surfaces of a typical parallel-concentric *fold have a common origin. The radii of such arcs represent the constant *orthogonal thickness of layers typical of this class of fold.

centre of symmetry *See* CRYSTAL SYMMETRY.

centric diatoms *See* BACILLARIOPHYCEAE.

centrifugal pump *See* PUMP.

centripetal drainage pattern *See* DRAINAGE PATTERN.

Centroceratida (class *Cephalopoda, subclass *Nautiloidea) Order of generally *evolute, *gyroconic cephalopods in which the *siphuncle is sub-central in position. Some forms are *nautiliconic. *Sutures are trilobed. The order ranges in age from Lower *Devonian to Upper *Jurassic.

centrum *See* VERTEBRA.

Cephalaspida *See* OSTEOSTRACI.

***Cephalaspis* (order *Osteostraci)** One of the best known osteostracans. The bony head shield tapered posteriorly and laterally to form two 'horns'. The body segments were freed from the head shield, allowing greater mobility than in earlier genera.

cephalic Pertaining to the head.

cephalic spine A spine occurring on the *cephalon (head) of a trilobite (*Trilobita).

cephalic suture In trilobites (*Trilobita), the structure that includes the facial and ventral cephalic sutures. The facial sutures situated on the dorsal part of the *cephalon are of three main types: protoparian, where the suture runs from the anterior border of the cephalon to a position anterior to the *genal angle; opisthoparian, where the suture runs from the anterior margin to cut the posterior margin posterior to the genal angle; and marginal, where the suture runs along the edge and is not visible on the upper surface. In one family the facial suture runs directly through the genal angle and is 'gonatoparian'. On the ventral side of the cephalon the facial sutures are joined and continue as the ventral cephalic suture. An elongated rostral plate (*see* ROSTRUM) in the anterior mid-line of the ventral surface has a rostral suture anterior to it and a hypostomal suture posterior to it. In some genera the lateral ventral sutures are combined as a single median suture.

Cephalochordata (Acrania) Subphylum of *Chordata, containing only *Amphioxus* (lancelet), probably the most primitive of living chordates, although some soft-bodied *Cambrian fossils are dubiously referred to the group. Cephalochordates have a *notochord extending into the head, gill slits, and segmented muscle blocks, and are thought to resemble the ancestors of vertebrates.

cephalon The anterior or head region of a trilobite (*Trilobita), which consists of at least five fused segments and is generally semicircular in shape. The glabella (the raised median part of the cephalon) is very variable in size and structure and it is probable that in life the stomach lay beneath it. Between the lateral margins of the glabella and the facial suture (*see* CEPHALIC SUTURE) is the fixed cheek (fixigena) and between the facial suture and the lateral margin is the free cheek (librigena). The glabella and the fixed cheeks together are known as the 'cranidium'. Above the visual surface of the eye is the palpebral lobe. In *Cambrian trilobites the eyes are commonly connected to the glabella by narrow ocular ridges. The glabella may be indented by pairs of furrows and a transverse furrow separates the posterior occipital segment from the remainder of the glabella. In some families (e.g. the Trinucleidae) the border of the cephalon is developed into an extensive, pitted, cephalic fringe. The dorsal *exoskeleton of the cephalon also continues on to the ventral surface as the 'doublure'.

Cephalopoda (cephalopods) Literally 'head-foot', a class of *Mollusca, exclusively marine, related to the *Bivalvia and *Gastropoda. The class includes the *Nautiloidea (nautiloids), Sepioidea (cuttlefish), Teuthoidea (squids), Octopoida (octopuses), and the extinct *Ammonoidea (goniatites, ceratites, ammonites), and *Belemnitida (belemnites). The earliest forms belonged to the Nautiloidea and date from the Upper *Cambrian.

ceratites *See* AMMONOIDEA.

ceratoid *See* SOLITARY CORALS.

Ceratopsia Horned, *ornithischian (bird-hipped) *dinosaurs, which had beak-like jaws, from the Upper *Cretaceous. The head accounted for about one-third of the total length of the body because of the development of a large, bony frill which protected the neck and shoulders. *Triceratops* is perhaps the best-known member of the group. It was 5–6 m long and had three forward-projecting horns, one over each eye and the third over the nose.

Ceres The largest *solar system asteroid (No. 1), diameter 974 km, approximate mass 10^{21} kg; rotation period 9.078 hours; orbital period 4.6 years. It was discovered in 1801 by G. Piazzi. It accounts for almost half the total estimated mass of all the asteroids.

cerioid Applied to those corals in which the individuals comprising the colony are packed together and *corallites are polygonal in section, each individual corallite retaining its wall. *See* COMPOUND CORALS.

cerussite *Mineral, $PbCO_3$; sp. gr. 6.4–6.6; *hardness 3.0–3.5; *orthorhombic; usually white or grey; white *streak; *adamantine lustre; crystals often *prismatic or *tabular, also *acicular, but can also occur granular, *massive, and compact; *cleavage good {110}, {021}; of *secondary origin in the oxidized zone of lead veins, associated with *anglesite, *galena, *smithsonite, *pyromorphite, and *sphalerite; soluble (with effervescence) in warm, dilute nitric acid. It is an *ore mineral of lead.

cervical vertebra *See* VERTEBRA.

c.g.s. system A set of units of measurement derived from the metric system and based on the centimetre, gram, and second. It has now been largely replaced by the SI (Système International d'Unités) system.

Chadian A *stage of the *Visean Epoch, dated at 349.5–345 Ma (Harland et al., 1989).

Chaetognatha Phylum comprising the arrow worms, first encountered in the fossil record in *Carboniferous rocks.

chain-former Class of element oxides which form chain structures in *silicate *melts, e.g. SiO_2, $KAlO_2$, $NaAlO_2$, $Ca_{1/2}AlO_2$, and $Mg_{1/2}AlO_2$.

chain lightning *See* PEARL-NECKLACE LIGHTNING.

chain-modifier Class of element oxides which modify or disrupt chain structures in *silicate *melts, e.g. CaO, K_2O, Na_2O, MgO, FeO, TiO, and Al_2O_3.

chain silicate *See* INOSILICATE.

chalcedony The group name for *cryptocrystalline varieties of silica composed of minute crystals of *quartz with submicroscopic pores and with composition ranging from SiO_2 to $SiO_2.nH_2O$ and including the *minerals *agate, *chert, *opal, *onyx, *jasper, and *flint; sp. gr. 2.50–2.67; habits variable from stalactitic to massive; commonly white, greyish-white, or grey, and occasionally yellow.

chalcophile Applied to elements having a strong affinity for sulphur, which concentrate in sulphides and are typical of the Earth's *mantle rather than its *core. Commonly found in *sulphide minerals and *ores. Typical chalcophile elements are Cu, Zn, Pb, As, and Sb. *Compare* ATMOPHILE; BIOPHILE; LITHOPHILE; and SIDEROPHILE.

chalcopyrite (copper pyrites) Most common copper mineral, $CuFeS_2$; sp. gr. 4.1–4.3; *hardness 3.5–4.0; *tetragonal; brass-yellow, often with an iridescent tarnish; greenish-black *streak; *metallic *lustre; sometimes *massive, crystals usually *tetrahedra; *cleavage imperfect {011}; *primary mineral found in *igneous rocks and *hydrothermal veins, in association with *pyrite, *pyrrhotine, *cassiterite, *sphalerite, *galena, *calcite, and *quartz, an important mineral in *porphyry copper deposits, and also occurs in quartz *diorite, and in *pegmatites, crystalline *schists, *porphyry copper deposits, *syngenetic copper ores and *skarns, and contact *metamorphic zones. It is deeper in colour than pyrite, more brittle and harder than *gold, is soluble in nitric acid, and its alteration products are *secondary copper minerals. It is a major *ore mineral for copper.

chalcocite (copper glance) *Mineral, Cu_2S; sp. gr. 5.5–5.8; *hardness 2.5–3.0; *orthorhombic; dark lead-grey tarnishing to black; black *streak; *metallic *lustre; *prismatic and *tabular *crystals are rare, it is usually found *massive, or as powdery coatings; *cleavage prismatic, poor; can occur in *hydrothermal veins in a *primary state, but more usually found in the zones of *supergene enrichment of copper-ore bodies. It is an important *ore mineral for copper.

chalk Porous, fine-grained rock, predominantly composed of the calcareous skeletons of micro-organisms, e.g. *coccolithophores and *foraminifera. The Chalk Formations of the Upper *Cretaceous of

Europe form the White Cliffs of Dover and the cliffs south of Calais.

***Challenger* expedition (1872–5)** The first expedition to explore the deep oceans was led by John *Murray, in the British naval ship HMS *Challenger*. With a staff of biologists, chemists, and geologists, the expedition surveyed the Atlantic, Indian, Antarctic, and Pacific Oceans, taking soundings and collecting specimens in dredges. The extent of the *Mid-Atlantic Ridge was first demonstrated by the crew of the *Challenger*.

chalybdite *See* SIDERITE.

chalybeate Applied to natural waters containing iron.

Chamberlin, Thomas Chrowder (1843–1928) Professor of geology at Chicago University and Head of the Glacial Division of the US Geological Survey, in which capacity he mapped the ice deposits of Wisconsin, Chamberlin developed a theory of successive episodes of mountain building, around an ancient continental *craton. With Moulton, he proposed the *planetismal hypothesis of the formation of the Earth.

Chambers, Robert (1802–71) The author of *Chambers's Encyclopaedia*, who in 1844 published anonymously a book called *Vestiges of a Natural History of Creation*, in which he revived the idea of evolution first proposed by *Lamarck 30 years earlier. The book's popularity and notoriety refocused attention on this issue and so paved the way, among the general public, for *Darwin's *Origin of Species*.

chamosite Silicate *mineral (*see* SILICATES) and member of the *chlorite group and particularly of the septichlorite group which also includes the minerals amesite, greenalite, and cronstedite; composition $Fe^{2+}{}_{10}Al_2[Si_3AlO_{10}]_2(OH)_{16}$; soft but massive; occurs in sedimentary ironstones where it may form by alteration from the original *siderite ore, and is found in association with *chert and *clay.

Chamovnicheskian A *stage in the *Kasimovian Epoch, preceded by the *Krevyakinskian and followed by the *Dorogomilovskian.

Champlainian A *series of the Middle *Ordovician of N. America, generally equivalent to the *Llanvirn and Middle *Caradoc.

Chandler wobble The free oscillation of the *Earth's pole of rotation. This wobble in the Earth's rotation has a 435-day periodicity and appears to have a decay time of the order of 40 years. The cause (excitation) is unknown: atmospheric effects appear to be on too short a time scale; *earthquake activity has been proposed, but has not been established. The major cause appears to be related to the Earth's *core and its magnetic coupling with the lower *mantle.

Changxingian (Tatarian, Tartarian) 1. The final *age in the Late *Permian *Period (*Zechstein Epoch), preceded by the *Longtanian, followed by the *Griesbachian Age (*Triassic), and dated at 254–248 Ma (Harland et al., 1989). **2.** The corresponding eastern European *stage which, because of correlation problems, has been included in the Triassic *System. It is roughly contemporaneous with the Bundsandstein (western Europe), upper Amarassian (New Zealand), and upper *Ochoan (N. America).

channel 1. The preferred linear route along which surface water and *groundwater flow is usually concentrated (although water can flow across wide, flat surfaces as sheet flow). It is commonly a linear, concave-based depression (e.g. river channel, submarine fan channel). The geometry may be sinuous, anastomosing, or straight, and with a widely variable width-to-depth ratio. *See* BRAIDED STREAM; and MEANDER. **2.** A narrow sea-way connecting two wider bodies of water (e.g. the English Channel). **3.** In remote sensing, the range of wavelengths recorded by a single detector to form an image.

channel and vug porosity *See* CHOQUETTE AND PRAY CLASSIFICATION.

Channel Deposits *See* MARTIAN TERRAIN UNITS.

channel fill The *sediment infill of a *channel, produced either by the accretion of sediment transported by water flowing through the channel, or by the infilling of an abandoned channel. *See also* MULTISTOREY SANDBODY.

channel wave An *elastic wave which travels through a layer with a lower velocity than that of the surrounding layers. The wave tends to be confined to that layer because of repeated internal reflection and the refraction of escaping waves back towards the channel owing to the increasing velocities away from the channel. *See also* HIDDEN LAYER.

chaos A state of disorder which is governed by simple and precise laws, but where the outcome is unpredictable and may change greatly with slight variations in starting conditions. Most real systems, such as weather patterns and satellite orbits, display chaotic behaviour. *See also* FRACTAL.

char Solid carbonaceous residue of high calorific value, derived from incomplete burning of organic material. It may be formed into briquettes and burned for fuel; if pure it can be used as a filter medium. Charcoal is made from wood or bone; coke, another char, is derived from *coal. *See* PYROLYSIS.

character Any detectable attribute or property of the *phenotype of an organism. Defined heritable differences in the character may exist between individuals within a species.

character states Particular versions of a *character. Thus, the character 'horns' may have the character states 'straight', 'curly', etc. The proper elucidation of character states and their *polarity is one of the major concerns of *cladistic analysis.

charcoal *See* CHAR.

charge 1. (explosive) The combination of detonator and main explosive. The effective energy released is a function of the nature and weight of the main explosive material, the type of detonator used, and how the charge is fired. **2. (electrical)** A source of electric field forces; the transfer of such charge through a conducting medium is measured as electric current. Electric charge comprises whole-number multiples of electronic charge, of which the electron constitutes a negative charge.

chargeability (*M*) One of the units of measurement of *induced polarization in the *time domain. True chargeability is the ratio of the over- or secondary voltage, V_s, to the observed voltage, V_0, applied by way of an *electrode array so that $M = V_s/V_0$, expressed as a percentage or as millivolts per volt; this quantity is independent of topographic effects and of electrode geometry and is thus a good measure of induced polarization. In reality, what is measured is the apparent chargeability (M_a) which is the area (A) beneath the voltage-time decay curve over a defined time interval (t_1 to t_2) and normalized by the supposed steady-state primary voltage, V_p, such that $M_a = A/V_p = (1/V_p) \times \int_{t1}^{t2}$ of $V(t)dt$, in units of mVs/V.

charge-coupled device (CCD) A light-sensitive semiconductor used to enhance images obtained from faint objects. It accumulates and temporarily stores charge at particular *pixel locations when struck by photons. The pixels can then be moved, allowing them to be pieced together to form an image.

charged-body potential method *See* MISE-À-LA-MASSE METHOD.

Charnian A *stage of the Upper *Proterozoic of Charnwood Forest, England. The Charnwood sequence contains impressions of very early *Metaphyta or *Metazoa.

Charniodiscus *See* EDIACARAN FOSSILS.

charnockite A light-coloured, medium- to coarse-grained *igneous rock containing *quartz and microcline feldspar (*see* ALKALI FELDSPAR) as major components with, in order of decreasing abundance, oligoclase feldspar (*see* PLAGIOCLASE FELDSPAR), *hypersthene, *biotite, and *magnetite. Despite its low abundance, hypersthene (a calcium-poor, iron-rich *pyroxene) is a distinctive feature of this rock. Rocks with a charnockite mineralogy can also be formed by the *metamorphism of quartzo-feldspathic rock under dry *granulite facies conditions. The type location for the rock is in Tamil Nadu (formerly Madras), India.

charnockitic gneiss *See* GRANULITE.

Charon The satellite of *Pluto, discovered in 1979, which orbits at a mean distance of 19 405 km, with an orbital period of 6.387 days. Its equatorial radius is 586 km; mass 1.7×10^{21} kg; mean density 1800 kg/m^3; surface gravity 0.21 (Earth = 1); albedo 0.375. Some consider Charon and Pluto to comprise a double-planet system.

Charophyceae (charophytes) A class (in some classifications a division: Charophyta) of *algae that in some ways resemble bryophytes (*Bryophyta). They occur in fresh and brackish water and their calcified fructifications are ornamented with spiral striae. Neglected by geologists until 1959, they are now used as stratigraphic markers in *Cenozoic strata. *See* GYROGONITE.

Charophyta *See* CHAROPHYCEAE.

Charpentier, Jean de (1786–1855) A Swiss superintendent of mines, Charpentier made extensive field studies in the Alps. Using evidence of erratic boulders and *moraines, he hypothesized that Swiss glaciers had once been much more extensive. His ideas were taken up and developed by *Agassiz.

chart datum Datum, or plane, to which depth measurements on a chart are referred.

chasma Originally, a very large canyon on Mars, generally of structural origin. Such structures are distinguished from smaller, sinuous channels (valles, *see* VALLIS) probably of fluvial origin, and from fossae (*see* FOSSA) which are linear depressions analogous to terrestrial *grabens. The term is now used for any similar large valley on a planetary or *satellite surface.

chattermark Small (less than 5 mm) crescentic scar typically found on the surface of rocks, of rock particles, and of rounded beach pebbles. It is a percussion fracture, produced when particles are thrown together in wind or water environments.

Chattian 1. The final *age in the *Oligocene Epoch, preceded by the *Rupelian (Stampian), followed by the *Aquitanian, and dated at 29.3–23.3 Ma (Harland et al., 1989). **2.** The name of the corresponding European *stage, which is roughly contemporaneous with the upper *Zemorrian (N. America), upper *Whaingaroan, *Duntroonian, and *Waitakian (New Zealand), and most of the *Janjukian (Australia). It is dated by means of *sea-floor spreading rates, small mammals, and *plankton.

Chautauquan A *series in the Upper *Devonian of N. America, underlain by the *Senecan, overlain by the *Kinderhookian (*Mississippian), and comprising the Cassadagian and Bradfordian Stages. It is roughly contemporaneous with the *Famennian (Upper Devonian) and, possibly, the Hastarian (*Tournaisian) Stages of Europe.

Chazyan A *stage of the *Ordovician in the Middle *Champlainian *Series of N. America.

Chebotarev sequence An idealized sequence of chemical changes in *groundwater. As groundwater moves through rock its chemical composition normally changes. In general, the longer groundwater remains in contact with the *aquifer rocks the greater the amount of material it will take into solution. Changes in composition also occur with increasing depth of travel, as bicarbonate *anions, which dominate in many shallow groundwaters, give way to sulphate and then chloride anions, and calcium is exchanged for sodium.

Cheirolepis trailli Early representative of the primitive *bony fish, the *palaeoniscids. It is known from the Middle *Devonian.

chelation Equilibrium reaction between a metallic ion and an organic molecule in which more than one bond links the two components. The metallic ion is termed the complexing agent, the chelating organic molecule the *ligand. Chelation is a naturally occurring mechanism in soils, useful since it removes heavy-metal ions that are in solution in simple inorganic form, in which state they may be directly toxic to plants or may interfere with the uptake of essential nutrients. Heavy-metal toxicity will tend to be reduced by the application of organic material.

Chelford 1. An *interstadial that occurred 65 000–60 000 years BP, during the *Devensian glaciation. **2.** Sections of *alluvial *sands and organic muds containing tree remains from the Chelford Interstadial, overlaid and underlaid by *till, that are exposed in sand pits between Chelford and Congleton, in Cheshire, England.

chelicerae *See* ARACHNIDA; ARTHROPODA; and CHELICERATA.

Chelicerata (phylum *Arthropoda) Subphylum comprising a diverse group of animals which possess an anterior prosoma (fused head and thorax), a posterior opisthosoma (abdomen), and a pair of jointed pincers. The subphylum includes the spiders, mites, and scorpions, as well as the king-crab (*Limulus*) and the eurypterids (*Palaeozoic water scorpions).

Cheltenhamian A *stage in the Upper *Tertiary of south-eastern Australia, underlain by the *Mitchellian, overlain by the *Kalimnan, and roughly contemporaneous with the mid *Zanclian (Tabianian) Stage.

chemical demagnetization The washing of permeable sediments in weak acid, usually 10% hydrochloric, in order to remove *cement with its associated *remanent magnetism. *See* DEMAGNETIZATION.

chemical oxygen demand (COD) Indicator of water or effluent quality, which measures oxygen demand by chemical (as distinct from biological) means, using potassium dichromate as the oxidizing agent. *Oxidation takes two hours and the method is thus much quicker than an assessment of biological oxygen demand (BOD), which takes five days. Since the BOD : COD ratio is fairly constant for a given effluent, COD is used more frequently than BOD for the routine monitoring of an effluent once this ratio has been determined.

chemical potential *See* WATER POTENTIAL.

chemical remanent magnetization (crystalline remanent magnetization, CRM) The magnetization acquired as *ferromagnetic minerals grow through their *blocking volume.

chemical weathering The action of a set of chemical processes operating at the atomic and molecular levels to break down and re-form *rocks and *minerals. The results of chemical weathering are frequently new substances of reduced particle size, greater plasticity, lower density, and increased volume, compared with the original materials. Some of the important processes are *solution, *hydration, *hydrolysis, *oxidation, *reduction, and *carbonation.

chemocline In a *meromictic lake, the transition between the upper *mixolimnion and lower *monimolimnion layers, marked by a change from aerobic to anaerobic conditions.

chemosymbiosis A symbiotic association between a multicelled animal which provides a protected environment and *bacteria that oxidize hydrogen sulphide, using the energy this releases to fix carbon and synthesize carbohydrates and enzymes which sustain the metabolism of the host.

chemosynthesis The pathway by which bacteria in *hydrothermal vent communities synthesize complex organic molecules from hydrogen sulphide gas and dissolved carbon dioxide: $4H_2S + CO_2 + O_2 \rightarrow CH_2O + 4S + 3H_2O$.

Chemungian (Cohoktonian) *See* SENECAN.

chenier Beach ridge, or sandy, linear mound that is built on a marsh area. It is at least 150 m broad, up to 3 m high, and up to 50 km long, and is typical of the Gulf Coast of America. A chenier is formed by the reworking of river-derived materials by waves. There are usually muddy, marshy zones to the front and rear of the chenier.

chenier plain Area of marine *aggradation consisting of sandy ridges (*cheniers) separated by clay-rich depressions. Occasionally the ridges may be vegetated, as on the Gulf Coast of the USA. Chenier plains may be large: that on the north coast of S. America is about 2250 km long and up to 30 km wide.

Cheremshanskian A *stage in the *Bashkirian Epoch, preceded by the *Yeadonian and followed by the *Melekesskian.

chernozem (black earth) Freely draining, dark coloured *soil profile whose name is the Russian word for 'black earth'. Chernozems are associated with grassland vegetation in temperate climates, and identified by the deep and even distribution of *humus and of exchangeable *cations (calcium and magnesium) through the profile (now included in *Mollisols).

chert 1. Chalcedonic (*see* CHALCEDONY) variety of *cryptocrystalline silica, SiO_2, that occurs as nodules or irregular masses in a sedimentary environment, often in association with black *shales and *spilites. **2.** A fine-grained rock consisting of beds of cryptocrystalline silica, usually of *biogenic, volcanogenic, or diagenetic (*see* DIAGENESIS) origin.

Chesterian A *series in the *Mississippian of N. America, underlain by the *Meramecian and overlain by the *Morrowan. It is roughly contemporaneous with the Brigantian Stage of the *Visean Series and the *Serpukhovian Series. The Springerian is an alternative name for the upper part of the Chesterian.

chevron fold (accordion fold, zig-zag fold) A type of *fold which shows characteristically long, planar limbs with a short, angular *hinge zone. Ideal chevron folds have interlimb angles of 60°. Chevron folds occur in sequences of regularly bedded layers of alternating *competent and *incompetent material which deform by *flexural slip and ductile flow respectively.

chevron marks A type of *sole structure characterized by a linear pattern of small, V-shaped ridges formed by the dragging of an object over the surface of a viscous mud. The V-shapes close in the down-current direction, enabling the chevron marks to be used as a palaeocurrent indicator. *See* PALAEOCURRENT ANALYSIS.

Chewtonian A *stage of the Lower *Ordovician of Australia, underlain by the *Bendigonian and overlain by the *Castlemainian.

Chezy's formula An empirical formula relating river *discharge (Q) to *channel dimensions and water surface slope. $Q = AC\sqrt{(rS)}$, where A is the cross-sectional area of the river, C is the Chezy discharge coefficient, r is the *hydraulic radius, and S is the slope of the water surface. This formula is useful for extending river-flow rating curves.

chiastolite *See* ANDALUSITE.

chickenwire structure A closely packed, nodular mass of *gypsum or *anhydrite, with thin, interconnected fingers of mud between. It is commonly found in *sabkha salt deposits.

Chicxulub A small town on the Yucatan Peninsula, Mexico, that has been identified as lying within the crater formed by a major impact event dated at about 65 Ma ago. The crater is about 195 km wide, has a multi-ring structure, and could have been made by an object about 12 km in diameter, the impact believed to be the event associated with the mass extinction marking the end of the *Cretaceous Period.

Chile Rise The oceanic *ridge which separates the *Nazca Plate and the *Antarctic Plate.

Chile saltpetre *See* SODA NITRE.

chilidial plates *See* NOTOTHYRIUM.

chilidium *See* NOTOTHYRIUM.

chilled edge *See* CHILLED MARGIN.

chilled margin (chilled edge) A fine-grained or glassy carapace found around crystalline *magma bodies. The fine grain size is produced by the rapid loss of heat from the external surface of the magma body, causing severe undercooling of the magma which generates many crystal-nucleation sites producing small crystals, or suppression of nucleation sites altogether, producing glass.

china clay (kaolin) *See* CHINASTONE.

chinastone A light-coloured, hydrothermally altered, *igneous rock composed of *quartz, strongly kaolinized orthoclase (*see* ALKALI FELDSPAR), fresh *euhedral albite (*see* PLAGIOCLASE FELDSPAR), and minor *muscovite, *topaz, and *fluorite. The rock is essentially a *granite which is in an arrested state of alteration by the process of *kaolinization. The orthoclase, which shows strong kaolinization, can be veined by secondary quartz and fluorite, both products of the kaolinization process. Superb examples are found in the St Austell granite of Cornwall, England, where the product of granite kaolinization, china clay (kaolin), is extracted for use in the ceramic, paper, pharmaceutical, and other industries. *See also* KAOLINITE.

chine Precipitous ravine found along eroding coastlines. It is typically developed in the soft *Mesozoic and *Cenozoic sediments of southern England. It results from the rapid incision that occurs when a stream responds by down-cutting to the rejuvenating effect of coastline retreat.

chinook Warm, dry, westerly wind of the *Föhn type which blows on the eastern side of the Rocky Mountains. The quick onset of the wind and the sudden large rise in temperature is associated in spring with rapid melting of the snow.

chip sampling *See* SAMPLING METHODS.

Chiron A *solar system asteroid (No. 2060), *comet (95P), or minor planet, diameter 180 km (148–208 km); approximate mass 4×10^{18} kg (2×10^{18}–10^{19} kg); rotational period 5.9 hours; orbital period 50.7 years; *perihelion date 14 February 1996; perihelion distance 8.46 AU. It is in a chaotic eccentric orbit near *Saturn and *Uranus. It was discovered in 1977 by Charles Kowal.

Chiroptera (bats; class *Mammalia) Order comprising the only true flying mammals, possessing features parallel to those of birds, e.g. active metabolism and economy of weight. Insectivores possibly ancestral to the bats are known from the *Palaeocene. The first undoubted bats, however, are preserved in Middle *Eocene deposits in both Europe and N. America. Differentiation of the modern lineages was far advanced by the *Eocene–*Oligocene transition.

chi-squared test (χ^2) In statistics, a *hypothesis test used to determine the goodness of fit of a particular data set with that expected from a theoretical distribution. The test statistic is a function of the difference between observed and expected values which is compared to the chi-squared distribution. The chi-squared distribution is a distribution of sample variance based on a single parameter, the *degrees of freedom.

chitin *See* SKELETAL MATERIAL.

Chitinodendron franconianum Single-celled organisms are known from *Precambrian strata, but the first *protozoans with an external skeleton appear in the Upper *Cambrian. *C. franconianum* is a primitive allogrominid 'foraminiferid with an external 'chitinoid' membrane. It is from the Upper *Cambrian of Wisconsin, USA.

chloralgal Applied to an association of green algae (*Chlorophyta), living in sea water more saline than corals could tolerate, that forms a characteristic calcareous sediment. *Compare* CHLOROZOAN; and FORAMOL.

chlorinity Measure of the chloride content, by mass, of *sea water. It is defined as the amount of chlorine, in grams, in 1 kg of sea water (bromine and iodine are assumed to have been replaced by chlorine). Chlorinity and *salinity are both measures of the saltiness of sea water. The relationship can be expressed mathematically, as salinity is equivalent to 1.80655 times the chlorinity. There is a constant ratio of dissolved chloride to total dissolved salts in all sea water.

chlorite Important group of *phyllosilicate (sheet silicate) *minerals with the general composition $(Mg,Fe,Al)_6[(SiAl)_4O_{10}](OH)_8$ and related to the *micas; sp. gr. 2.6–3.3; soft and green; platy or tabular *habit; occur in low-grade *metamorphic rocks of *greenschist facies and as an alteration product of *ferromagnesian minerals in *igneous rocks. The group includes specific minerals, e.g. clinochlore, delessite, penninite, and thuringite. *Chamosite and greenalite are septichlorites which are chemically similar to the chlorites.

chloritoid (ottrelite) A member of the *nesosilicates with the formula $(Fe^{2+},Mg)(Al,Fe^{3+})Al_3O_2[SiO_4]_2(OH)_4$ and an important metamorphic *index mineral; sp. gr. 3.51–3.80; *hardness 6.5; *monoclinic or *triclinic; *crystals *tabular, pseudohexagonal; dark green to black; occurs in *regionally metamorphosed *pelites with a high $Fe^{3+} : Fe^{2+}$ ratio at low metamorphic grades; it develops at the same time as *biotite and changes to *staurolite with increasing temperature and pressure.

Chlorophyta (green algae) Division of *algae which are typically green in colour. In common with higher land plants, green algae include chlorophylls *a* and *b* among their principal pigments, have cellulose as the main constituent of cell walls, and form food reserves of starch. Consequently it is believed that the ancestors of land plants must have belonged to this group. The organisms take many forms, ranging from unicellular to relatively complex multicellular plants. They are found today mainly in freshwater habitats and their distribution is cosmopolitan. They are known from the *Precambrian onwards, and the earliest *eukaryotes were probably of this division. Marine, lime-secreting green algae have contributed to algal *limestone *reefs since the *Cambrian. Fossil genera include **Palaeoporella* (similar to modern *Halimeda*), and *Coelosphaeridium* (similar to modern *Acetabularia*). *See also* CHAROPHYCEAE.

chlorozoan Applied to an association of calcareous green algae (*Chlorophyta), *hermatypic corals, and molluscs (*Mollusca) that lives in low latitudes, in seas where the temperature is always more than 20 °C and salinity 32–40 per mille, and produces a characteristic carbonate sediment. *Compare* CHLORALGAL; and FORAMOL.

Choanichthyes (*Sarcopterygii) Term used by some zoologists to group together *Crossopterygii (lobe-finned fish) and *Dipnoi (lungfish). Members of this group or subclass were thought to share functional lungs, external and internal nares (nostrils), and narrow-based, paired fins with fleshy lobes.

Chokierian A *stage of the *Serpukhovian Epoch, underlain by the *Arnsbergian and overlain by the *Aiportian.

Chondrichthyes Class of vertebrate animals characterized by a cartilaginous endoskeleton, a skin covered by *placoid scales, the structure of their fin rays, and the absence of a bony *operculum, lungs, and swim bladder. It includes the subclasses *Elasmobranchii (sharks, rays) and Holocephali (ghostfish). The group extends back to the Upper *Devonian. *See also* CARTILAGINOUS FISH.

chondrite *Stony meteorite; the majority of chondrites are characterized by the presence of *chondrules, and they constitute about 86% of meteorite *falls. The principal minerals they contain are *olivine, *pyroxene, *plagioclase feldspar, *troilite, and the iron-nickel minerals kamacite and taenite. Chondrites are grouped according to their petrological type (texture, crystal structure, etc.) into six classes. On the basis of their chemical composition they fall into five main groups: enstatite chondrites (highly reduced, iron in metallic form); high-Fe (H) chondrites; low-Fe (L) chondrites; low-Fe/low-metal (LL) chondrites (some iron in *silicate form); and carbonaceous chondrites (relatively high *oxidation levels, containing *volatiles). C1 chondrites are often used as a chemical model for the *bulk composition of the Earth, but there are notable discrepancies, e.g. in the proportions of volatile elements.

chondrite model (chondritic Earth model) Hypothesis that the *bulk composition of the Earth is close to that of *carbonaceous chondrites.

Chondrites With **Zoophycus*, an *ichnoguild of many-branched, radial, *trace fossils probably made by a worm that moved back and forth through the sediment, each branch of its burrow exploring a new area.

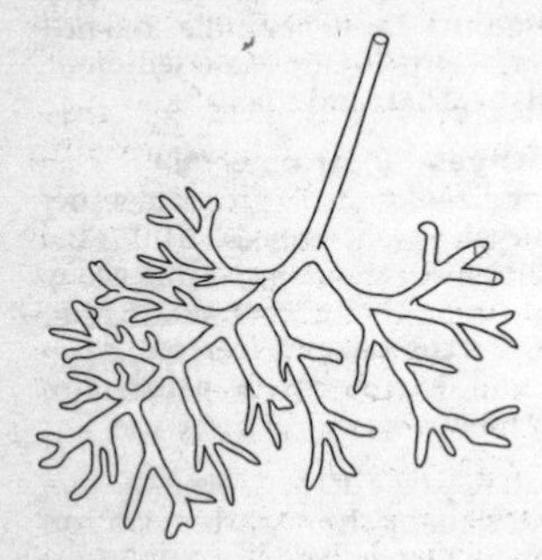

Chondrites

chondritic Earth model *See* CHONDRITE MODEL.

chondritic unfractionated reservoir (CHUR) Undifferentiated material from which *chondrites are believed to have formed and in which elemental abundances are essentially the same as those in the solar atmosphere (*see* METEORITIC ABUNDANCE OF ELEMENTS; and SOLAR ABUNDANCE OF ELEMENTS). CHUR therefore provides a starting point for discussion of the chemical (and in particular the isotopic) evolution of the *solar system.

chondrocranium *See* CRANIUM.

chondrodite Member of the humite group of minerals. The group includes humite, clinohumite, and norbergite; general formula is $n\text{Mg}_2[\text{SiO}_4]\text{Mg(OH,F)}_2$, with $n = 1$ in norbergite, 2 in chondrodite, 3 in humite, and 4 in clinohumite; related to the *olivine group; occurs in contact metamorphic zones (*see* CONTACT METAMORPHISM) of *limestones and *skarns.

chondrophore In certain genera of *bivalves which possess *desmodont dentition, an internal process with a depressed surface that supports the ligament.

chondrosteans *See* CHONDROSTEI.

Chondrostei (chondrostean fish) Group of *bony fish belonging to the subclass *Actinopterygii. Often ranked as a superorder, the chondrosteans have a partly cartilaginous skeleton (*see* CARTILAGE), a *heterocercal tail, a spiracle, and an intestinal spiral valve. The group includes the extant orders Acipenseriformes and Polypteriformes, and the extinct order Palaeonisciformes (*palaeoniscids, or 'ancient fish').

chondrule Small (0.1–2.0 mm), glassy, spherical to subspherical droplets, diagnostic of chondritic *meteorites. They are thought to have been produced by the melting and sudden quenching of pre-existing silicate material. *See* CHONDRITE.

Choquette and Pray classification A widely used, descriptive classification of *porosity types developed in *carbonates. Primary porosity types are classified as: interparticle–pores between the grains; framework–pore spaces between the rigid framework of carbonate skeletons, as in a reef; shelter–pore spaces preserved when curved shell fragments or irregularly shaped grains prevent intergranular spaces from becoming completely occupied by the mud *matrix; intraparticle–pores within the skeletal material which do not become filled with diagenetic (*see* DIAGENESIS) *cement; and *fenestral–pores within carbonate muds provided by *fenestrae. Secondary porosity types, produced by diagenetic and tectonic (*see* TECTONISM) effects, are classified as: intercrystalline–pores produced by *dolomitization or by preferential dissolution of mineral cements; moldic–pore spaces produced by the dissolution of grains, e.g. *ooids, shell fragments; channel and *vug–both produced by the dissolution of the rock by formation waters to generate linear (channel) or patchy (vug) cavities; cavern–the large-scale dissolution of the rock to yield man-sized or larger pores; and fracture–formed by tectonic stresses within the rock.

Chordata (chordates) Large phylum comprising the animals that possess a rod of flexible tissue (*notochord), which is protected in higher forms by a vertebral column (*see* VERTEBRA). The phylum includes the *Craniata, the *Urochordata, and the *Cephalochordata. The first chordates and the earliest *vertebrates (Craniata) are both found in *Cambrian rocks, *See* *PIKAIA*.

chroma One of the three variables of colour (hue and value being the other two). Chroma measures the strength, wavelength purity, or saturation of colour. *See* MUNSELL COLOUR.

chromatid One of the two daughter strands of a *chromosome that has undergone division. Chromatids are joined together by a single centromere, usually positioned in the centre of the pair as they lie beside one another. When the centromere divides during the third stage (anaphase) of cell division (mitosis or meiosis), the sister chromatids become separate chromosomes.

chromatography Analytical technique for the separation of the components of complex mixtures, based on their repetitive distribution between a mobile phase (of a gas or liquid) and a stationary phase (of solids or liquid-coated solids). The distribution of the different component molecules between the two phases is dependent on the method of chromatography used (e.g. *gel-filtration, or *ion-exchange), and on the movement of the mobile phase (which results in the differential migration and therefore separation of the components along the stationary phase).

chrome diopside *See* DIOPSIDE.

chromite Member of the chromite group of *minerals in the *spinel group, $Fe^{2+}Cr_2O_4$, along with magnesiochromite ($MgCr_2O_4$); sp. gr. 5.1; black; *sub-metallic *lustre; *crystals *cubic, but normally it is *massive or a granular *aggregate; occurs in *basic and *ultrabasic rocks. It is the primary source of chromium.

chromosome A protein thread composed of DNA and histone, usually associated with RNA, occurring in the nucleus of a cell. Although chromosomes are found in all animals and plants, bacteria and viruses contain structures that lack protein and contain only DNA or RNA: these are not chromosomes, though they serve a similar function. Chromosomes occur in pairs. Each species tends to have a characteristic number of pairs of chromosomes (23 in humans), found in most nucleated cells within most organisms. The presence of pairs of homologous chromosomes is referred to as the diploid state and is normal for the sexual phase of an organism. Gametes (reproductive cells), and cells of the gametophyte (gamete-producing phase) of plants, however, have only one member of each pair in their nuclei (the haploid state).

chron 1. A small unit of geologic time, equivalent to the *chronostratigraphic unit *chronozone, usually based on *fossil zonation (*see* BIOCHRON). When used formally (e.g. Gilbert Chron) the initial letter is capitalized. **2.** A single time interval of constant polarity of the *geomagnetic field (*polarity chron).

chronohorizon *See* MARKER BED.

chronomere A term for any *geologic-time unit. *See* CHRONOSTRATIGRAPHY.

chronometry Measurement of time, which in a geologic sense is associated with the measurement of *absolute ages.

chronosequence Sequence of related soils that differ in their degree of profile development because of differences in their age. Chronosequences can be found in evolving landscapes such as those produced by deglaciation, volcanic activity, wind deposits, or sedimentation.

chronosome A sedimentary rock unit bounded by planes defined by their time of formation.

chronospecies (evolutionary species) According to one view of *evolution (that of *phyletic gradualism), a new organism may be derived from its ancestor by a process of slow, steady evolutionary change. Conceivably, the descended organism might not be regarded as a member of the same species as its ancestor, in which case it would constitute a new species, in particular a chronospecies.

chronostratic scale An abbreviation for *chronostratigraphic scale.

chronostratigraphic correlation chart A graphic display summarizing the *stratigraphy of a particular area. The vertical scale represents geologic time, the horizontal scale represents distance, and the time-range and geographic extent of known *stratigraphic units are plotted against these two variables. A range of information, obtained from *seismic sections, *well-logs, outcrops, etc., is shown on the chart. This may include *lithology, bounding *unconformities at the top or bottom of *depositional sequences (*see* BASELAP; DOWNLAP; ONLAP; and TOPLAP), the position of known *lithostratigraphic units, *facies changes, and the location of wells. *See also* SEISMIC STRATIGRAPHY.

chronostratigraphic horizon *See* MARKER BED.

chronostratigraphic scale (chronostratic scale) Ideally, a time scale in which the sequence of *geologic-time units and their corresponding time-rock divisions (*chronostratigraphic units) are defined by standard and internationally agreed reference points within *boundary stratotypes. In practice, boundary stratotypes have been agreed only for parts of the scale and the chronostratigraphic units within it continue to be defined by *biostratigraphic means. *See* STANDARD STRATIGRAPHIC SCALE; and STRATIGRAPHIC SCALE.

chronostratigraphic unit (time-stratigraphic unit, time-rock unit) The sequence of rocks formed during a discrete and specified interval of geologic time. Chronostratigraphic units are ranked, according to the length of time they record, into *erathems (the longest), *systems, *series, *stages (the basic working unit), and *chronozones (the shortest). Each unit comprises a number of units of lower rank, e.g. a system would consist of a number of series, and, similarly, a number of stages would constitute a series. All the rocks formed anywhere in the world, regardless of *lithology or local thickness, would be referred to the chronostratigraphic unit appropriate to their time of formation, e.g. all rocks laid down in the *Cambrian Period belong to the Cambrian System. In the traditional *stratigraphic scales, however, note that chronostratigraphic units, and the *geologic-time units to which they correspond, have been defined on the basis of a type section (*see* STRATOTYPE), so historically it is the chronostratigraphic unit that has determined the geologic-time unit and not vice versa. *See* CHRONOSTRATIGRAPHY; and STRATIGRAPHIC NOMENCLATURE.

chronostratigraphy Branch of stratigraphy linked to the concept of time. In chronostratigraphy, intervals of geologic time are referred to as chronomeres. These may be of unequal duration. Intervals of geologic time are given formal names and grouped within a Chronomeric Standard hierarchy. The formal terms are: *eon, *era, *period, *epoch, *age, and *chron. The last four of these are the equivalent of system, series, stage and chronozone in the Stratomeric Standard hierarchy. The formal terms are written with initial capital letters when accompanied by the proper names of the intervals to which they refer. Some geologists hold that the term '*chronostratigraphy' is synonymous with '*biostratigraphy', but most agree that the two branches are separate.

chronozone The lowest-ranking *chronostratigraphic unit. The duration of a chronozone is defined by a *type section which may be based on a *biostratigraphic zone (in which case it would include all rocks laid down during that time interval regardless of *fossil content or rock type), or it may be defined on the basis of the time span of an existing *lithostratigraphic unit. With increasing knowledge, chronozone boundaries may be found not to coincide with the boundaries of the *stages to which they belong, and stages may more precisely be divided into substages. The *geologic-time unit equivalent to a chronozone is a *chron.

chrysoberyl *Mineral, $BeAl_2O_4$; sp. gr. 3.5–3.8; *hardness 8.5; *orthorhombic; various shades of green or greenish-yellow; white *streak; *vitreous *lustre; *crystal form *tabular; *cleavage *prismatic {110}; occurs in *granites, *pegmatites, and *mica *schists, also in *alluvial *sands and gravels. The emerald-green variety is used as a *gemstone (alexandrite).

chrysocolla A hydrated silicate of copper, with the formula $CuSiO_2.2H_2O$; sp. gr. 2.0; *hardness variable; greenish-blue; variable *lustre; normally forms from the weathering of copper-rich mineral deposits, and when sufficiently hard it can be cut and polished to make jewellery.

Chrysophyceae (golden algae, golden-brown algae) Class of predominantly unicellular *algae in which the chloroplasts contain large amounts of the pigment fucoxanthin, giving the algae their brown colour. Many are flagellated, having one flagellum of the tinsel type, with or without a second flagellum of the whiplash type. The silicoflagellates (20–100 μm) form a siliceous skeleton of rings, rods, and spines. They are marine forms with a fossil record going back to the Early *Cretaceous, and are valuable palaeoclimatic indicators. In other chrysophyceans, which are found mainly in freshwater habitats, only the cell walls of the cysts or resting spores contain silica.

chrysotile *See* ASBESTOS; and SERPENTINE.

CHUR *See* CHONDRITIC UNFRACTIONATED RESERVOIR.

Churchillian orogeny A *Proterozoic phase of mountain building that affected an area in what are now northern Saskatchewan and northern Manitoba, Canada. It was probably caused by the collision of a microcontinent (*see* MICROPLATE) within a subducting oceanic *plate, moving northwards, with a *fore-arc, and may have started about 1900 Ma ago and ended about 1850 Ma ago.

cilia *See* CILIUM.

cilium (pl. cilia) Short, hair-like appendage, normally 2–10 μm long and about 0.5 μm in diameter, usually found in large numbers on those cells that have any at all. In certain *protozoa, cilia function in locomotion and/or feeding. They generate currents in the fluid surrounding the cell by beating in a co-ordinated manner.

Cincinnatian A *series of the Upper *Ordovician of N. America, generally equivalent to the Middle *Caradoc to terminal *Ashgill.

cinder cone *See* SCORIA CONE.

cingulum *See* DINOPHYCEAE.

cinnabar The commonest mercury *mineral, HgS; sp. gr. 8.0–8.2; *hardness 2.0–2.5; *trigonal; scarlet-red to brownish-red; vermilion *streak; *adamantine *lustre; *crystals rhombohedral or thick, *tabular plates, but can occur *massive or granular; *cleavage perfect *prismatic $\{10\bar{1}\bar{1}\}$ sedimentary rocks, in fractures in areas of volcanic activity, and around hot springs, associated with *pyrite, stibnite, and *realgar. It is the only major ore of mercury.

CIPW norm calculation Combination of the oxide components of an *igneous rock into a set of water-free, standard, *mineral compounds (termed 'normative constituents') according to a rigidly prescribed order that originally was thought to be the order of mineral crystallization in most *magmas. The calculation, developed in the late nineteenth century by the American petrologists W. Cross, J. P. Iddings, L. V. Pirsson, and H. W. Washington, provides a basis for comparing and classifying rock types independently of their modal mineral assemblages. Rocks of similar chemistry may develop contrasting modal mineral assemblages if they crystallize under contrasting pressure, P_{total}, and $P(H_2O)$ conditions. Use of the CIPW norm calculation eliminates these effects, allowing a comparison based on an ideal mineral assemblage controlled only by the original magma chemistry.

circalittoral zone The area of the *continental-shelf sea bed that lies below the zone of periodic tidal exposure. It is approximately equivalent to the *sublittoral zone. *See also* LITTORAL ZONE.

circularity index 1. The ratio between the area of an inscribed circle to the area of a circumscribed circle fitted to the outline of the crest of a crater (*see* CRATER 2). For fresh lunar craters, the average circularity index reaches a maximum value of about 0.85–0.90 for craters about 10 km in diameter. **2.** *See* DRAINAGE BASIN SHAPE INDEX.

circularity ratio *See* DRAINAGE BASIN SHAPE INDEX.

circular polarization In optical *mineralogy, the polarization that may occur if light passing through a *mineral emerges with a velocity that is uniform in all directions, thus producing a spherical wave front. If the mineral splits the light into two rays, one of them may have circular polarization; if the light is unsplit, all of it may have circular polarization.

circum-oral canal (ring canal) Part of the water-vascular system in echinoderms (*Echinodermata). The *madreporite allows water into a canal that feeds the circum-oral canal, located close to the upper surface of *Aristotle's lantern. Five radial water vessels extend from the circum-oral canal up the centre of each *ambulacrum and from these vessels the *tube-feet, used in locomotion and respiration, occur at intervals.

circulation index *See* ZONAL FLOW.

cirque (corrie, cwm) Half-open, steep-sided hollow in a mountain region that has been or is being glaciated. Its form is due to a combination of glacial scouring (which deepens the floor and often produces a reversed long-profile gradient), and glacial erosion by basal sapping and *gelifraction (which acts on the cirque's head and side walls).

cirque glacier A relatively small body of *ice, *firn, and snow, occupying an armchair-shaped hollow in bedrock. It is generally wide in relation to its length. It is actively supplied by drifting snow and therefore shows vigorous behaviour, involving rotational sliding.

cirri Plural of *cirrus. *See also* CIRRIPEDIA.

Cirripedia (cirripedes; barnacles; phylum *Arthropoda, subphylum *Crustacea) Class of crustaceans comprising the familiar barnacles that settle on rocks, submerged timbers, corals, shells and the undersides of ships. They are entirely marine. The name literally means 'comb-foot'. After a free-swimming larval stage the bivalved cypris larva attaches itself to the substratum by means of cement secreted from glands in the first antennae. The body rotates and the thoracic appendages (usually six) are modified to form cirri (flexible feeding appendages) that point upwards or sideways through the gape in the carapace. The carapace persists in the adult barnacle as an inner mantle which is covered externally by calcareous plates. In the sessile (unstalked) barnacles (e.g. **Balanus*), the plates are large and heavy, making the organism very well adapted to the high-energy conditions of a rocky, intertidal zone. Two orders of cirripedes have fossil records: the Thoracica and Acrothoracica. The Thoracica includes stalked forms (e.g. the modern *Lepas* or goose barnacle) and unstalked forms (e.g. *Balanus*). The stalked form is thought to be the more primitive. The ancestral form of the cirripedes probably resembled the cypris larva. The undoubted fossil record of the Thoracica began in the Upper *Silurian with *Cyprolepis* (a stalked form), but fragments and disarticulated plates of *Cambrian–*Ordovician age have also been assigned to this group. The Acrothoracica are the smallest cirripedes. They bore into calcareous material such as shells and corals. Their distinctive burrows have been recorded as *trace fossils from *Devonian rocks.

cirrocumulus Cloud type comprising shallow, high-level cloud in stable air. The form is sheeted or layered with small-scale billows, or ripples. As freezing of the super-cooled droplets of young cloud occurs, so the fine-patterned features of the cloud become more diffuse. *See also* CLOUD CLASSIFICATION.

cirrostratus Cloud type comprising a semi-transparent veil of fibrous or smooth appearance, often covering the sky, and identified by *halo phenomena. When associated with a cold front, the cloud sheets have distinct edges. *See also* CLOUD CLASSIFICATION.

cirrus (plural cirri) From the Latin *cirrus*, meaning a tuft or lock of hair. **1.** A cloud type comprising high-level, banded clouds in fibrous filaments aligned approximately along their line of movement. *See also* CLOUD CLASSIFICATION. **2.** In certain ciliate *protozoa, an organelle formed by the fusion of a group of *cilia, which usually functions in locomotion or feeding.

citrine *See* QUARTZ.

clade Term derived from the Greek *klados*, a 'twig' or 'branch'. In *cladistics, or phylogenetic systematics, it refers to a lineage branch that results from splitting in an earlier lineage. A split produces two distinct new *taxa, each of which is represented as a clade, or branch, in a phylogenetic diagram.

cladism *See* CLADISTICS.

cladistic analysis The method of analysis which aims to discover *clades and their interrelationships. For each taxon of the group being analysed, *character states are ordered by their *polarity, to find which taxa are united by most *derived states. Only the sharing by two taxa of derived character states is evidence that they belong to the same clade.

cladistics (cladism, phylogenetic systematics) Special *taxonomic system, founded by W. *Hennig (1966), and applied to the study of evolutionary relationships. It proposes that common origin can be demonstrated by the shared possession of derived characters, characters in any group being either *primitive or *derived. In the branching diagrams (*cladograms) used to portray these relationships, it is assumed that *cladogenesis, or splitting of an evolutionary lineage, always creates two equal daughter *taxa: the branching is *dichotomous. Thus each pair of daughter taxa constitutes a *monophyletic group with a common stem taxon, unique to the group, and a parent taxon always gives rise to two daughter taxa which must be given different names from each other and from the parent, so the parent species ceases to exist. A cladogram is therefore synonymous with a classification. A shortcoming of the method would seem to be that usually it takes no account of the time dimension.

cladogenesis In *cladistics, the derivation of new taxa that occurs through the branching of ancestral lineages, each such split forming two (possibly more) equal sister taxa that are often considered taxonomically separate from the ancestral taxon, though this is no longer considered obligatory.

cladogram Diagram that delineates the branching sequences in an evolutionary tree.

Cladoselache Late *Devonian, shark-like fish recorded from Europe and N. America. *Cladoselache* ranged between 0.5 and 1.2 m in length and is noted for the presence of a large ventral fin. Numerous specimens of this genus have been collected from the Cleveland Shales, Upper Devonian, in N. America.

Cladoselachiformes (class *Chondrichthyes, subclass *Elasmobranchii) Order of fossil sharks with an elongate body, the two dorsal fins each with a spine, e.g. **Cladoselache*. They lived from the *Devonian to the *Carboniferous.

Clapeyron equation Equation that relates the temperature (T) at which a mineralogical *phase change occurs, with associated change in volume (ΔV), to the pressure. $\Delta H dT/dp = T\Delta V$, where the thermal gradient as a function of pressure is dT/dp (usually assumed to be adiabatic within the Earth) and ΔH is the heat of fusion.

Clapeyron–Clausius equation Equation that relates the pressure and temperature at which a *phase change occurs in a closed system. $dP/dT = \Delta H_v/T\Delta V$, where P is the pressure, T the absolute temperature, ΔH_v the heat absorbed per mole during the phase change, and ΔV the change in molar volume.

clarain *See* COAL LITHOTYPE.

Clarence A *series in the Lower *Cretaceous of New Zealand, underlain by the *Taitai, overlain by the *Raukumara, and comprising the Urutawan, Motuan, and Ngaterian *Stages. It is roughly contemporaneous with the upper *Albian and lower *Cenomanian.

Clarke, Frank Wigglesworth (1847–1931) As Chief Chemist to the US Geological Survey from 1884 to 1925, Clarke systematized the collection and chemical analysis of *minerals, *rocks, and *ores, and also did important work on the composition of the Earth's *crust. The first of five editions of his book *The Data of Geochemistry* was published in 1908.

Clarke orbit *See* GEOSTATIONARY ORBIT.

Clarke, William Branwhite (1798–1878) A clergyman and amateur scientist, who emigrated from Britain to Australia in 1839. He is credited with the first discovery of gold in Australia (1841). He took part in the geologic investigation of the New South Wales coalfield, and worked on the sedimentary structure of that State.

classification 1. Any scheme for structuring data that is used to group individuals. In ecological and taxonomic studies numerical classification schemes have been devised, but various hierarchical or non-hierarchical classificatory strategies have also been used. In taxonomy, the fundamental unit is the species. Among living forms species are groups of individuals that look alike and can interbreed, but cannot interbreed with other species. In *palaeontology, where breeding capability cannot be determined, species are defined according to morphological similarities. In formal nomenclature, taxonomists follow the binomial system devised by Linnaeus. In this system each species is defined by two names: the generic (referring to the genus) and the specific (referring to the species). Thus various related species may share a common generic name. Genera (sing. genus) may be combined with others to form families, and related families combined into an order. Orders may be combined into classes, and classes into phyla (sing. phylum) or divisions in the case of *Metaphyta. For example, the brachiopods comprise some 11 orders split between two classes and these two classes are the major subdivisions of the phylum *Brachiopoda. The basic groupings, the phyla, are combined together into kingdoms, e.g. *Plantae (the plants) and *Animalia (animals). Some workers have tackled the uncertainties arising from subjectivity in classification by using numerical methods. In their view, if enough characters were measured and represented by cluster statistics, the distances between clusters could be used as a measure of difference. Even so, the worker has to decide (subjectively) how best to analyse the measurements, and so objectivity is lost. Other workers emphasize those features shared by organisms that show a hierarchical pattern (*see* CLADISTICS). **2.** In *remote sensing, the computer-assisted recognition of surface materials. The process assigns individual *pixels of an image to categories (e.g. vegetation, road) based on spectral characteristics compared to spectral characteristics of known parts of an image (training areas). Assignation of pixels is not always possible when the parameter space of different training areas overlaps. In such cases a *principal component analysis prior to classification may be used to allow better separation of training areas by increasing the overall parameter space. *See also* BOX CLASSIFICATION; MINIMUM-DISTANCE-TO-MEANS CLASSIFICATION; and MAXIMUM-LIKELIHOOD CLASSIFICATION.

clast Particle of broken-down rock. These fragments may vary in size from boulders to *silt-sized *grains, and are invariably the products of *erosion followed by deposition in a new setting. *See* CLASTIC ROCK. *See also* BIOCLAST.

clastic (fragmental) Applied to the texture of fragmental *sedimentary rocks.

clastic rock *Sediment composed of fragments of pre-existing rocks (*clasts). Consolidated clastic rocks include *conglomerates, *sandstones, and *shales.

clastogenic flow *See* FIRE-FOUNTAIN.

clathrate A compound in which molecules of one substance, commonly a noble gas, are completely enclosed within the crystal structure of another substance. Typical examples are Kr and Xe encapsulated in *zeolite structures, or Ar, Kr, and Xe trapped in water ice.

Clausius–Clapeyron equation *See* CLAPEYRON–CLAUSIUS EQUATION.

Clavatipollenites A miospore from the *Barremian (Lower *Cretaceous); the oldest known *angiosperm *pollen grain. It is oval and monosulcate (*see* SULCUS).

clay 1. In the Udden–Wentworth scale, particles less than 4 μm in size. *See* PARTICLE SIZE. **2.** In pedology, a soil separate comprising mineral particles less than 2 μm in diameter according to the Atterberg and *USDA classifications. **3.** Class of soil texture, irrespective of particle diameter but usually containing at least 20% by weight of clay particles. *Compare* CLAY MINERALS.

clay dune *See* LUNETTE.

C-layer The lower part of the upper *mantle of the Earth, between about 370 and 720 km depth, within which there are several strong seismic-velocity gradients thought to correspond with mineralogical phase changes.

clay films *See* CUTAN.

clay minerals Members of the *phyllosilicates (sheet silicates) with related chemistry, all are hydrous aluminium silicates with layered structure; layers of $[SiO_4]$ tetrahedra of composition $[Si_4O_{10}]^{4-}$ are joined to Al-O layers (gibbsite-type layers) or (Mg,Fe)-O layers (brucite-type layers). 1:1 sheet silicates have one Si-O layer coupled to one brucite or gibbsite layer and include the *serpentine group and the *kaolinite or kandite group of clays; 2:1 sheet silicates have two Si-O layers joined to one brucite or gibbsite layer and include the *smectite and *illite groups of clays, *bentonite and *montmorillonite, as well as *talc and the *mica group; 2:2 sheet silicates have two Si-O layers joined to two brucite or gibbsite layers and include the *chlorite group. It is difficult to distinguish clay minerals by hand or under the microscope, so sophisticated techniques of *X-ray diffraction and *scanning electron microscopy (SEM) are used to determine the precise clay mineral under investigation.

clay pan *See* PAN.

clayskins *See* CUTAN.

claystone A compacted, non-fissile, fine-grained, *sedimentary rock composed predominantly of clay-sized (less than 4 μm grain size) particles. *Compare* MUDSTONE.

cleaning, magnetic *See* DEMAGNETIZATION.

cleaning-up trend (tunnel trend) In the reading from a *wireline log (*see* WELL LOGGING) an expression from a *gamma-ray sonde showing a progressive upward decrease in the gamma reading, indicating a change in the clay-mineral content.

clear-air turbulence (CAT) The variable pattern of up- and down-draughts, or turbulence, sometimes occurring in the absence of any cloud. It is caused by strong wind shear, especially associated with *jet streams in the upper *troposphere and lower *stratosphere. The phenomenon is significant for aircraft.

clear ice (glaze) Layer of transparent ice formed on objects near the ground, or on aircraft in flight, by freezing rain.

cleat System of joints in most coal seams, along which the coal parts. There are usually two systems at right angles, one better developed and with a more shiny surface than the other. The orientation and intensity of cleats may influence the direction of mine workings.

cleavage 1. In *minerals, cleavage is evident when crystals split along planes of weakness inherent in the structure of their atomic lattices. Cleavage is described by an adjective, e.g. good, poor, etc., and by referring to its crystallographic direction, plane, and degree of perfection, the resulting digits being contained in braces ({}) to distinguish them from descriptions of crystals. *See* MILLER INDICES. **2.** The formation of a set of fractures along closely spaced, parallel surfaces in a rock (the term is usually applied to low-grade *metamorphic rocks) by the alignment of various mineralogical and structural elements during *metamorphism and deformation, e.g. in *slates, where cleavage is due to a parallel arrangement of minerals. The *fabric generally gives rise to a preferred direction of fracturing, broadly analogous to mineral cleavage. Rock cleavages may be divided into two groups: (*a*) continuous cleavages, e.g. '*slaty cleavage' (synonymous with *schistosity and *foliation in high-grade *metamorphic rocks, *see* METAMORPHIC GRADE) which, with further deformation, may be superimposed and cross-cut by a secondary crenulation cleavage; (*b*) spaced cleavages, either crenulation or disjunctive, e.g. *fracture cleavage. Crenulation cleavages form by the microfolding of a pre-existing *anisotropic fabric. Disjunctive cleavages require no such primary fabric. *Compare* FOLIATION.

cleavage refraction The change in orientation (analogous to wave *refraction) of *cleavage planes as they are traced into and through layers of varying *lithology and therefore differing *competence. Angles between the cleavage plane and the bedding are largest where the lithology is most competent; in *graded beds there is a gradational change in orientation.

Clementine A lunar mapping mission on behalf of *NASA and the US Department of Defense, launched in January 1994. It achieved lunar orbit, completed its photographic tasks, and left orbit on 3 May, heading for asteroid *Geographos. On 7 May a housekeeping computer malfunction led to the depletion of the attitude-control propellant and the abandonment of the asteroid mission. Clementine entered an Earth orbit that passed repeatedly through the *Van Allen belts, from which the spacecraft continued to transmit data from its sensors.

Clifdenian A *stage in the Upper *Tertiary of New Zealand, underlain by the *Altonian, overlain by the *Lillburnian, and roughly contemporaneous with the *Langhian Stage.

CLIMAP *See* CLIMATE–LEAF ANALYSIS MULTIVARIATE PROGRAM.

CLIMAPP *See* CLIMATE/LONG-RANGED INVESTIGATION MAPPING AND PREDICTIONS PROJECT.

climate classification The grouping of climates into broad types according to their shared characteristics. There are three principal approaches to the task. (*a*) Generic classification is based on levels of temperature and aridity as these relate to vegetation boundaries. Aridity is usually expressed as 'effective precipitation', which is calculated as the ratio of rainfall to temperature. Climatic types are defined by the response of flora to them. The *Köppen system, with its modifications, uses this approach. (*b*) Classifications based on the moisture budget and 'potential evapotranspiration' (i.e. the maximum moisture that will be transferred from the ground to the atmosphere, provided that sufficient moisture is available) which do not rely on vegetation boundaries. The *Thornthwaite system uses this approach. (*c*) Genetic (i.e. pertaining to its origin) classification, based on factors related to the atmospheric circulation of major winds and *air masses, and on other factors that cause climate, is used in the systems of H. Flohn (1950) and A. N. *Strahler.

Climate–Leaf Analysis Multivariate Program (CLIMAP) An approach to the estimation of mean annual temperatures in the past based on a suite of 29 characters found in the leaves of dicotyledonous plants known to have been present at the site. *Compare* LEAF MARGIN ANALYSIS.

Climate/Long-ranged Investigation Mapping and Predictions Project (CLIMAPP) An integrated project to study the climatic history of the *Quaternary, conducted by a team of scientists engaged in Earth and ocean research. Since 1971 the administrative base for the project has been at Columbia University, New York.

climatic geomorphology That branch of *geomorphology which deals with the effects of climate on geomorphological processes and consequently on the character of land-forms. It has included the identification of climatically controlled zones and has attempted to define provinces with distinctive denudational processes.

climatic optimum Period of highest prevailing temperatures since the last ice age, in most parts of the world about 4000–8000 years ago.

climatic station Place where the basic climatic elements are regularly observed and recorded.

climatic zone Region or zone characterized by a generally consistent climate. Climatic zones approximate to distinct latitude belts around the Earth. The principal ones are the: humid tropical; subtropical arid and semi-arid; humid temperate; boreal (northern hemisphere) or subarctic/sub-antarctic; and polar zones.

Climatiiformes One of the larger orders of fossil fish belonging to the class *Acanthodii. They possessed bony jaws and body skeleton, *ganoid scales, a *heterocercal tail, and stout spines located before the dorsal, anal, pectoral, and pelvic fins. One of the better-known forms, *Climatius*, carried five additional pairs of fins along the belly and reached a length of only 8 cm. These 'spiny sharks' may actually be halfway between the sharks and the bony fish (*Osteichthyes).

climatostratigraphy The study of geologic–climatic units, which are climatic episodes defined in *Quaternary rocks. They are similar to *chronostratigraphic units (i.e. they are inferential), but their boundaries are *diachronous. They record the effects of climate (i.e. the type of biota, soils, etc.), but not the climate itself, and therefore complications arise when planning boundaries. It would be difficult to envisage a stratigraphic scheme which did not recognize the influence of climate in the Quaternary.

climax trace fossil A *trace fossil produced by a member of a climax community.

climbing-ripple cross-lamination *See* RIPPLE-DRIFT CROSS-LAMINATION.

climosequence Sequence of *soil profiles, usually on the same parent material, that differ from each other in their profile development because of local or site differences in climatic conditions. Climosequences can be found along mountain slopes in certain highland areas.

cline Gradual change in *gene frequencies or *character states within a *species, across its geographic distribution.

clino- Prefix derived from the Greek *klino*, meaning 'sloping' or 'inclined'.

clinochlore *See* CHLORITE.

clinoform A sloping depositional surface of a major morphological feature giving seismic expression, e.g. a *delta front.

clinohumite *See* CHONDRODITE.

clinometer An instrument for measuring the angle between an inclined surface and the horizontal, using a pendulum or spirit level and a calibrated circular scale. A clinometer may be incorporated into a magnetic compass, as a 'compass clinometer'. Clinometers are used to measure the inclination of bedding, *cleavage, *lineations, *fault planes, etc.

clinopyroxene (cpx) A group of *pyroxene minerals which crystallize in the *monoclinic crystal system, including the calcium-bearing clinopyroxenes and the sodium-bearing pyroxenes. Important members of the group include *augite, *diopside, *pigeonite, and *aegirine.

clinosequence Sequence of soils in which *soil-profile development is related to the angle of slope of the soil surface. Clinosequences can be found on land-forms such as *escarpments and *drumlins with varying surface angles of slope.

clinothem Rock units generated by *strata which gently *prograde seawards into deeper water.

clinozoisite A member of the *epidote group of *minerals, $Ca_2Al_3(SiO_4)_3OH$; sp. gr.

3.3; *hardness 6.5; greyish-white to yellowish-white; crystallizes in the *monoclinic system (unlike *zoisite which is *orthorhombic) hence the prefix 'clino-'; *massive; occurs as an alteration product of *plagioclase feldspars in *igneous and *metamorphic rocks, particularly *anorthosites, where the rock is said to be 'saussuritized' when feldspars are hydrothermally altered to zoisite or clinozoisite. *See* SAUSSURITIZATION.

clint Joint-bounded surface unit of a hard, almost horizontally bedded *limestone. The diameter varies typically from 1 to 2 m. It may be shallowly dissected by smooth gutters, or runnels, called *grikes. It is a repeating component of a limestone pavement (an extensive *bedding-plane surface exposed by erosion).

Clinton ironstone Red, fossiliferous iron ore from the Middle *Silurian Clinton Formation of east central USA. It often occurs in lenses and may be *oolitic in texture. The *ore mineral is usually *hematite.

clipped trace A wave-form display in which amplitudes greater than a certain level are not shown. For example, clipped *seismic traces have a truncated wiggle with a flat top instead of the usual rounded wiggle.

clitter Local term for the gently sloping spread of coarse, often angular rock debris surrounding many of the tors of Dartmoor, south-west England. It was produced when *blockfields developed on the Dartmoor granite during the periglacial phases of the *Pleistocene and the exposure of *tors, but the rock debris is now stable and largely vegetated.

clod Compact, coherent block of soil, found *in situ* when soil is broken up by digging or ploughing. Clods are of varied sizes.

closed form The *crystal faces of a *mineral which occupy repeat positions in relation to its symmetry elements can be grouped together as forms. A closed form is one which totally encloses a space, e.g. a cube or tetrahedron. *Compare* OPEN FORM.

close fold A *fold in which the inter-limb angle is between 30° and 70° (as defined by M. J. Fleuty (1964), in a classification of folds based on their degree of tightness).

closure age The time at which diffusion within a mineral or rock of a *daughter product of a *radioactive decay system becomes negligible compared to accumulation, so that the system becomes 'closed', usually due to a fall in temperature.

cloud amount The proportion of the sky seen to be covered by cloud. Nowadays it is commonly measured in 'oktas' or eighths of the sky covered, but sometimes it is quoted as a percentage, or as tenths.

cloud base The undersurface of cloud, representing the *condensation level of water droplets or ice.

cloudburst Term popularly applied to brief but exceptionally heavy precipitation, of shower or thunderstorm type.

cloud classification Clouds have been classified by various systems according to form, altitude, and the physical processes generating them. The World Meteorological Organization (*International Cloud Atlas*, 1956) classifies 10 genera in three major groups (cumulus or heap clouds, stratus or sheet clouds, and cirrus or fibrous clouds) by criteria essentially based on cloud form. Some of the genera are subdivided according to variations in internal shape and structure, to give 14 species. Additional or supplementary features, e.g. transparency, arrangement, and characteristics of growth, are defined by Latin names as variants and accessory types of cloud. The cloud genera, with their abbreviations, are: cirrus (Ci), cirrocumulus (Cc), cirrostratus (Cs), altocumulus (Ac), altostratus (As), nimbostratus (Ns), stratocumulus (Sc), stratus (St), cumulus (Cu), and cumulonimbus (Cb). Clouds can also be referred to, according to their composition, as water or ice clouds; combinations of these are called mixed cloud.

cloud discharge Electrical discharge in a thunder cloud. *See also* LIGHTNING.

cloud droplet The liquid component of clouds, occurring as water droplets with an average size of 10 μm. In a non-rain cloud the droplets are suspended at a near-constant level, because air friction approximately balances the gravitational force. *See also* RAINDROP.

cloud seeding Process of introducing nuclei, e.g. silver-iodide crystals or solid carbon dioxide (dry ice), into clouds composed of supercooled water droplets, in an attempt to induce precipitation. Dry ice introduced (at −80 °C) from the air into cloud lowers the air temperature so that (particularly at temperatures below −40 °C) some of the supercooled water droplets are converted into ice crystals which then grow by collisions with further droplets. Silver iodide (which has a crystal structure similar to that of ice), introduced from the air or ground, is the substance most commonly used in seeding: its crystals act as ice nuclei. Other substances, e.g. common salt or fine water droplets, may also be used to encourage coalescence. Natural seeding may be significant in cases where ice crystals from a high 'releaser' cloud (e.g. *altostratus or *cirrostratus) fall into a supercooled water 'spender' cloud (e.g. *nimbostratus) and encourage ice-crystal growth.

cloud street Band (or bands) of (usually) *cumulus cloud parallel to the wind direction in a sky otherwise more or less clear of cloud. Streets may form in an *air mass at a sharply demarcated convection layer, with separate thermals producing parallel streets.

cluster analysis In statistics, the classification of observations into subsets based on a criterion of similarity.

CMP (common mid point) *See* COMMON DEPTH POINT.

Cnidaria (Coelenterata) Phylum comprising the sea anemones, jellyfish, and corals, which is known from the late *Precambrian. The *Ediacaran fauna from southern Australia, from between 680 and 580 Ma ago, includes clearly identifiable jellyfish and their allies. Corals occur for the first time in *Ordovician rocks. Cnidarians are all aquatic and mostly marine. The body plan is characterized by a single internal cavity, the enteron (or gut), and the body wall is composed of two layers of cells separated by a gelatinous mass (the mesogloea). There are two basic body shapes, the sessile polyp (e.g. sea anemones, corals) or the free-swimming medusa (e.g. jellyfish). The mouth is surrounded by tentacles armed with stinging cells (nematocysts) that are unique to cnidarians. Typically, the body approaches *radial symmetry, though some forms are *bilaterally symmetrical. *See* ANTHOZOA; HYDROZOA; and SCYPHOZOA.

co- *See* COM-.

co-adaptation Development and maintenance of advantageous genetic traits, so that mutual relationships can persist. Predator–prey, and flower–pollinator relationships often exhibit examples of co-adaptation, which is an aspect of *co-evolution.

Coahuilan *See* COMANCHEAN.

coal Carbon-rich mineral deposit formed from the remains of fossil plants. These are deposited initially as *peat, but burial and increase in temperatures at depth bring about physical and chemical changes. The process of 'coalification' results in the production of coals of different ranks ('coal series'), from peat, through the bituminous coals and lignite, to anthracite. Each rank marks a reduction in the percentage of *volatiles and moisture, and an increase in the percentage of carbon. They are termed 'woody' or 'humic' coals if formed from fragments of trees or bushes. If the major constituents of coal are pollen grains and/or finely divided plant debris, the term 'sapropelic coal' is used.

coalescence Process of cloud water droplet enlargement, as larger drops (more than 19 μm) fall and collide with smaller drops. Ice crystals in frozen cloud tops may also fall and grow by coalescence with drops below, ultimately melting into raindrops. In cumuliform cloud, turbulence may encourage this process of drop growth by the upthrust of small droplets to overtake and coalesce with other droplets. Repetition of the process can eventually produce rain-sized droplets. Factors favouring coalescent growth are high moisture content, water cloud of large vertical extent, and strong upward turbulence. *See also* BERGERON THEORY; and COLLISION THEORY.

coalification *See* COAL.

coal lithotype Type of *coal, the nature of which depends on original plant structure. Clarain has a semi-bright, shiny *lustre, is finely laminated, with smooth or irregular fracture, and is banded parallel to bedding. Durain is grey to brownish-black, banded, dull, with a granular and rough surface; harder than vitrain and more common. Fusain is sooty black, with silky lustre; it is fibrous and friable like charcoal, and occurrences are usually thin and impersistent. Vitrain is black, with brilliant, glassy lustre, conchoidal fracture, and cubic *cleavage. It is clean and structureless, and occurs in thin bands or lenses.

coal maceral Elementary and microscopic constituent of *coal. There are a number of different types. Alginite is formed from algal remains; typical of *boghead coal. Collinite derives from cell infillings, is structureless and falls within the vitrinite *coal maceral group. Cutinite consists of plant *cuticles, usually the hard outer coat of the epidermal cells of leaves. Fusinite comes from woody material; it has high reflectivity and microhardness (its individual grains are hard), well-defined cellular structure (Bogen structure), and is high in carbon. Micrinite has high reflectivity, medium hardness, is opaque, granular, with grain size less than 10 μm, and has no cell structure. Resinite is formed of small ellipsoids or spindles of resin, a cell-infilling material. Sclerotinite describes variously sized round or oval bodies of irregular structure which may have been fungi or *spores. It is similar in hardness and reflectivity to fusinite. Sporinite is formed of spore *exines, usually flattened parallel to stratification. Telinite is that part of the vitrinite group deriving from cell-wall material.

coal-maceral group One of a particular assemblage of *coal macerals. Exinite (liptinite) is a group consisting of *spores, *cuticles, resins, and waxes, rich in hydrogen and typical of attrital coals. It includes sparain, *cutain, *alginite, *resinite, and liptodetrinite. Inertinite is charcoal-like, the result of bacterial action or forest fires; high in carbon, relatively inert during carbonization. It includes *micrinite, semifusinite, *fusinite, *sclerotinite, and detrinite. Vitrinite is characteristic of vitrain coal derived from *humus; with medium reflectivity. It is subdivided into telinite with visible residual cell-wall material, and collinite which is structureless.

coal series *See* COAL.

coarsening-upward succession A vertical change in a *facies in which the grain size increases with height above the base. *Compare* FINING-UPWARD SUCCESSION.

coastal onlap The deposition and *onlap of coastal non-marine and *littoral deposits further and further inland, due to a relative rise in sea level. If, subsequently, relative sea level falls, the *base level is lowered and *erosion probably occurs at the top of the sequence. At the next relative sea level rise, coastal onlap will recommence but from a lower level. This downward shift in coastal onlap is thus an indication that there has been a relative fall in sea level. *See also* SEISMIC STRATIGRAPHY.

coastal processes The set of mechanisms that operate along a coastline, bringing about various combinations of *erosion and deposition. A cliffed coastline is affected by *slope processes and by wave activity. Both agents give rise to distinctive land-forms, including the *geo, the *bevelled cliff, and the 'blow-hole' (a chamber with a relatively narrow exit at the top of the cliff, from which water and spray are forced when waves are driven against the coast). A low coastline is largely affected by processes in the *surf zone, where most work is done by shoaling and breaking waves, and in the offshore zone, where tidal currents are the chief agents of sediment transfer.

coastal toplap *See* TOPLAP; and SEISMIC STRATIGRAPHY.

coated stone *See* AGGREGATE.

co-axial correlation A graphic method of correlation used in the prediction of storm runoff volumes.

cobalt glance *See* COBALTITE.

cobaltite (cobalt glance) *Mineral, CoAsS; sp. gr. 6.0–6.5; *hardness 5–6; *cubic; from white to dark grey or greyish-black, all with a reddish tinge; grey–black *streak; *metallic *lustre; crystals octahedra, cubes, and pentagonal-dodecahedra, also occurs *massive, and in irregular grains; *cleavage perfect cubic; occurs mainly in *hydrothermal veins and contact *metamorphic zones, and is associated with *chalcopyrite, *sphalerite, *chlorite, *tourmaline, and *apatite. It is one of the principal *ore minerals for cobalt.

cobble *See* PARTICLE SIZE.

Coble creep A form of *diffusion creep in which atoms migrate along grain boundaries. *See also* NABARRO–HERRING CREEP.

coccoliths The microscopic calcareous plates or discs, often oval and commonly intricately patterned and ornamented, that occur as part of the protective covering of a group of the unicellular *algae called *coccolithophorids. Coccoliths are a major component of the modern deep-sea calcareous oozes, and were especially abundant in the *Mesozoic, particularly the *Cretaceous Period, in which they became a major component of the Chalk lithology.

coccolithophorids (class Prymnesiophyceae) A group of unicellular, marine, planktonic *algae which are, at least at some stage in their life cycle, covered in calcareous plates (*coccoliths) embedded in a gelatinous sheath. They are spherical or oval, and less than 20 μm in diameter. They range in age from the Upper *Triassic to *Holocene, although dubious examples have been described from the Upper *Precambrian and *Palaeozoic. Because of their abundance, with large seasonal blooms, and the preservation of coccoliths in sediments, coccolithophorids are believed to play a major role in the global carbon cycle.

Cochiti A normal *polarity subchron which occurs within the *Gilbert reversed *chron (*Pliocene).

COCORP (*CO*nsortium for *CO*ntinental *R*eflection *P*rofiling) A project operated from Cornell University, USA, whose objects are to map the *basement rocks of the USA and to focus on specific geologic problems. In areas of interest the practice is to run long profiles, typically 50–200 km, and to record data to 18–20 seconds *two-way travel time (one second of two-way travel time is equivalent to about 3 km of crystalline basement).

Cocos Plate One of the present-day minor lithospheric *plates that lies beneath the *Pacific Ocean. A remnant of the *Farallon Plate, it is subducting under the *North American Plate and *Caribbean Plate and has constructive boundaries with the *Pacific Plate (along the *E. Pacific Rise) and the *Nazca Plate (along the *Galápagos Rise).

COD *See* CHEMICAL OXYGEN DEMAND.

coefficient of compressibility Amount by which a *stratum of given thickness will compress under increasing load pressure. It is measured as the change in *void ratio per unit increase in pressure.

coefficient of consolidation Factor governing the rate by which compression can occur in a particular soil. The rate and amount of compression in soils varies with the rate at which pore water is lost, and therefore depends on *permeability. The coefficient is defined as: coefficient of permeability × (1 + initial void ratio)/ coefficient of compressibility × density of water.

coefficient of nivosity A measure introduced by Pardé to give an indication of the amount of snow melt contributing to warm-season river flows.

coefficient of sliding friction *See* ANGLE OF INTERNAL FRICTION.

coefficient of variation A measure of variability within a sample, representing a population or a species; it is calculated as the *standard deviation × 100 divided by the mean. Experience shows that within a homogenous population of mature individuals the coefficient of variation rarely exceeds 10.

coelacanth *See* COELACANTHIFORMES.

Coelacanthiformes (class *Osteichthyes, subclass *Crossopterygii) Order of *bony fish thought to have been extinct for 50 million years (since the end of the *Mesozoic Era) until the discovery of the coelacanth (*Latimeria chalumnae*) off S. Africa in 1938. Both fossil and living coelacanths are bulky marine fish with a diphycercal or tri-lobed tail fin; the second dorsal, anal, pectoral, and pelvic fins are very unusual as they are supported by movable stalks or lobes. The much smaller fossil species lived from the *Devonian to the *Cretaceous. They were initially freshwater fish, but in the *Triassic marine representatives also evolved.

Coelenterata Name formerly applied to a phylum comprising both *Cnidaria and Ctenophora. Today, when these two groups are universally separated in different phyla, the term 'Coelenterata' is sometimes used as a synonym for Cnidaria alone.

coelom Principal body cavity in most animals, forming the cavity around the gut in many *Annelida, and in *Echinodermata and *Vertebrata. In *Arthropoda and *Mollusca, the main body cavity is an expanded part of the blood system (a haemocoel) and the coelom is small.

coelomate Applied to animals possessing a coelom.

Coelophysis One of the first flesh-eating *dinosaurs, recorded from the Late *Jurassic of N. America. It was a slender, bipedal *coelurosaur less than 2 m in length, weighing approximately 23 kg. It had a small skull with sharp serrated teeth, a long neck and tail, and hands with long, grasping fingers.

Coelurosauria (coelurosaurs; order *Saurischia, suborder *Theropoda) Infraorder of carnivorous, bipedal *dinosaurs, which were the most persistent of the infraordinal dinosaur groups, extending from *Triassic to *Cretaceous times. The largest measured about 3 m long, although the great majority of these rather slender, agile dinosaurs were much smaller.

coelurosaurs *See* COELUROSAURIA.

Coenopteridales *See* PRE-FERNS.

coenosteum In stromatoporoids (*Stromatoporoidea), the name given to the skeleton. It is a laminated structure and the broad laminae or latilaminae seen in vertical section probably record growth periods in the organism.

coenozone *See* ASSEMBLAGE ZONE.

coercivity, magnetic The magnetic field (direct or alternating) required to reduce the external magnetization of a *ferromagnetic substance to zero. It depends on the composition, grain size, and temperature of the substance.

coesite A variety of *quartz (SiO_2) produced at high pressures (greater than 4 GPa) and found in rocks subjected to impact by large *meteorites.

co-evolution Complementary evolution of closely associated species. The interlocking adaptations of many flowering plants and their pollinating insects provide some striking examples of co-evolution. In a broader sense, predator–prey relationships also involve co-evolution, with an evolutionary advance in the predator, for instance, triggering an evolutionary response in the prey. *See also* CO-ADAPTATION.

cognate lithic *See* LITHIC FRAGMENT.

coherence Two wave trains which are *in phase. Coherence is also an indication of the similarity of two functions and is the *frequency-domain analogy with *correlation in the *time domain.

cohesion Ability of particles to stick together without dependence on interparticle friction. In soils, cohesion is due to the *shearing strength of the *cement or film of water that separates individual grains. In *powder technology, cohesion refers to the forces of attraction by which the particles are held together either by *compaction or a binding substance.

Cohoktonian (Chemungian) *See* SENECAN.

cohort Group of individuals or *taxa of the same age.

coign The top and bottom corners of a *crystal when the crystal is held between thumb and forefinger in order to examine its symmetry elements. However, any three-dimensional 'corner' in a crystal, such as the corners of a cube, may be called a coign.

co-ignimbrite breccia *See* LAG BRECCIA.

coiling In many univalve and bivalve molluscs (*Mollusca) the shells are coiled. The condition is most noticeable among gastropods (*Gastropoda) and cephalopods (*Cephalopoda), where it is obvious that the shell is a hollow cone, coiled up to a greater or lesser extent. These rolled-up cones grow at the apertural end only and form a logarithmic spiral. Since the shell is a hollow cone, coiling about a vertical axis and growing at the apertural end, it is possible to generate a number of shapes. The shape of the tube in section (known as the 'generating curve') when expanding and coiling around and away from the vertical axis in a single plane defines a 'planispirally coiled' shell. If the coiling does not remain in a single plane but moves down the axis (translation), a helically coiled shell results. Where the translation down the axis is in a clockwise direction the coiling is termed 'dextral'; where it is anticlockwise it is termed 'sinistral'. In most cases the coils (whorls) remain in contact during expansion and coiling, but in some cases they do not, resulting in a loosely coiled or 'disjunct' shell.

coke *See* CHAR.

col 1. High pass or saddle in a ridge. It may mark the line of a former stream valley or of a former glacier, and so provides evidence of an early stage in the development of the landscape. **2.** The saddle region of the atmospheric-pressure field between two high- and two low-pressure centres.

colatitude The latitudinal distance from the pole, i.e. 90° – latitude.

cold front Boundary between dense, cold air and the warmer air ahead of it, which the cold air tends to undercut as it advances. The gradient of the upper surface of the cold air may be steep, e.g. 1 : 50. Along this steep front violent upwelling and instability result in high *cumulonimbus cloud with rain and thunder. Weather changes occur with the passage of a cold front, sometimes including a pronounced temperature fall, a rise in pressure, and wind veering (often with squalls) to northerly or north-westerly (in the northern hemisphere). The passage of the front commonly brings clearer, brighter weather, but the unstable air can produce showers.

cold-front clearance Clearing of the sky after the passage of a *cold front, the cold air displacing the warm, moist air of the warm sector as a depression moves away. The process is usually marked by a veering of the wind (in the northern hemisphere, backing in the southern hemisphere), rising pressure, and a fall in temperature. The clearing of the cloud may be partly due to some subsiding motion in the upper layers behind the front.

cold glacier *See* GLACIER.

cold low Basically non-frontal depression, typically associated with circulation in cold *air masses in the mid-*troposphere over north-eastern USA and north-eastern Siberia, though sometimes occurring also over the oceans in air that has emerged from the arctic. Such depressions are marked by more or less concentric isotherms around the core of the low. Such lows may originate along the arctic coast as a result of strong vertical uplift and adiabatic cooling in *occlusions; they do not necessarily influence surface weather conditions, but when they occur over warmer surfaces in middle latitudes, convection develops strongly. *See also* CUT-OFF LOW.

cold pole The area or point in each hemisphere that has the lowest mean temperatures. Verkhoyansk, in north-eastern Siberia is in the region of the northern-hemisphere cold pole, with a January mean of −50 °C and an absolute minimum of −68 °C. Parts of Antarctica have recorded absolute temperatures of around −90 °C.

cold sector Zone of colder air surrounding the narrowing wedge of warm air, the warm sector, in a developing depression. At the *occlusion stage the whole of the surface layer forms a cold sector, with the warm air lifted off the surface.

cold seep Hypersaline brine or a mixture of hydrocarbons that seeps from the ocean floor in a manner similar to that of a *hydrothermal vent, but at about the ambient temperature. Cold seeps are rich in methane and support large and diverse communities sustained by the primary productivity of chemosynthetic bacteria; methane is taken up by the gills of the dominant mussels (*Bathymodiolus*), where it is oxidized by intracellular bacteria to release energy.

cold wave The conditions associated with air of continental polar origin, often dominated by an *anticyclone behind a *cold front, that moves south into central and eastern parts of the USA. Cold waves are defined as a fall of 11 °C or more to a minimum base (−18 °C in northern, central, and northeastern regions) within a 24-hour period. In southern states (Florida, California, and the Gulf Coast) the minimum fall is 9 °C and the base minimum 0 °C.

cold working A mechanism of low-temperature deformation which involves the development of intracrystalline distortions by bending and twin-gliding. Cold working typically produces considerable strain-hardening.

colemanite A hydrated calcium borate, with the formula $Ca_2B_6O_{11}.5H_2O$; sp. gr. 2.4; *hardness 4.5; *monoclinic; crystals short, *prismatic, also *massive; occurs in saline lake deposits in arid regions.

coliform count A count made of the numbers of coliform bacteria present as part of most standard analyses of water intended for potable use. The number of organisms present is normally expressed per 100 ml of water.

collagen Fibrous scleroprotein of high tensile strength which is a major constituent of connective tissue and the organic material in *bone. It may represent up to 6% of the total body weight. When boiled, collagen yields gelatin.

collinite *See* COAL MACERAL.

collision theory Theory to account for the growth of water droplets in cloud to produce raindrops, based on the mechanisms of collision, coalescence, and 'sweeping'. It holds that larger drops, with terminal velocities increasing in proportion to their diameter, fall faster than smaller drops, and collide with them. The probability of collision depends on the spacing of the drops in the cloud (i.e. on the mean free path) and on the relative sizes of droplets. For example, if some drops are up to 50 μm diameter in cloud consisting mainly of droplets smaller than this, collisions can be frequent. Such collisions can lead to coalescence, and an overall increase in size to produce particles of raindrop size. 'Sweeping' is an ancillary process whereby small drops that are swept into the rear of larger drops may be absorbed. These mechanisms are believed to be entirely responsible for rainfall from tropical convection cloud, as well as playing a part in other clouds, including those of mid-latitudes. *See also* BERGERON THEORY.

collision zone A type of *convergent margin in which two continents or *island arcs have collided. The zone may be marked by young fold mountains, a suture, *ophiolites, scattered, mostly shallow *earthquakes, and *faults with relatively short lengths of *outcrop.

colloform banding A texture, often found in certain types of *mineral deposits, where *crystals have grown in a radiating and concentric manner which may reflect underlying geochemical controls. Lead-zinc deposits of sedimentary origin often show colloform banding of *pyrite and *sphalerite (e.g. at Silvermines, Ireland).

colloid **1.** Substance composed of two homogeneous phases, one of which is dispersed in the other. **2.** Soil colloids are substances of very small particle size, either mineral (such as *clay) or organic (such as *humus), which therefore have a large surface area per unit volume. Colloids usually provide surfaces with high *cation exchange capacity, and also exhibit an instability controlled by soil chemistry.

collophane The name often given to a *cryptocrystalline variety of *apatite having the same formula $Ca_5(PO_4)_3.(OH,F,Cl)$, which commonly occurs in phosphatic sediments and as the principal mineral in fossilized bone and fish scales.

colluvial Applied to weathered rock debris that has moved down a hillslope either by creep or by surface wash.

colonization window The period of time during which organisms are able to colonize a substrate in which *trace fossils will be left.

colonnade Sets of regular columns that form the lower part of a thick lava flow showing *columnar *jointing and lie beneath an *entablature.

colorimetric analysis A method of chemical analysis in which reagents are added to a rock solution to form coloured compounds with specific elements. The intensity of the colour, measured on a *spectrophotometer, is proportional to the concentration of the element.

colour *See* PIXEL COLOUR.

colour index The total volume percent of *mafic (coloured) *silicate *minerals in an *igneous rock. The mafic minerals would include the *olivine group, *pyroxene group, and *amphibole group of minerals, as well as *biotite. Colour index is one of many methods used for classifying igneous rocks. If a rock has 60% *modal mafic minerals it has a colour index of 60. Rocks with colour indices between 50 and 90 are termed '*mafic' rocks (e.g. *basalts).

columbite (niobite, tantalite) A *mineral oxide $(Fe,Mn)(Ta,Nb)_2O_6$; when Ta > Nb the mineral is called tantalite, but when Ta < Nb it is called columbite; sp. gr. 5.2–6.5, increasing with increasing Ta content; *hardness 6.0–6.5; *orthorhombic; weakly magnetic; black or brownish-black; very dark red *streak; *sub-metallic *lustre; *crystals normally *tabular and short *prismatic forms, and can also be blade-like; *cleavage pinacoidal; occurs in *granites and *pegmatites in association with high-temperature minerals, e.g. *cassiterite, *wolframite, *tourmaline, *quartz, and *feldspar. It is a major *ore mineral of niobium.

columella 1. In corals, a rod-like structure formed from the swollen axial end of the counter-*septum that may occur as a central structure within the skeleton. **2.** In gastropods (*Gastropoda), a spiral, rod-like structure formed by the fusion of the inner shell surfaces of the whorls where *coiling around the vertical axis is very tight.

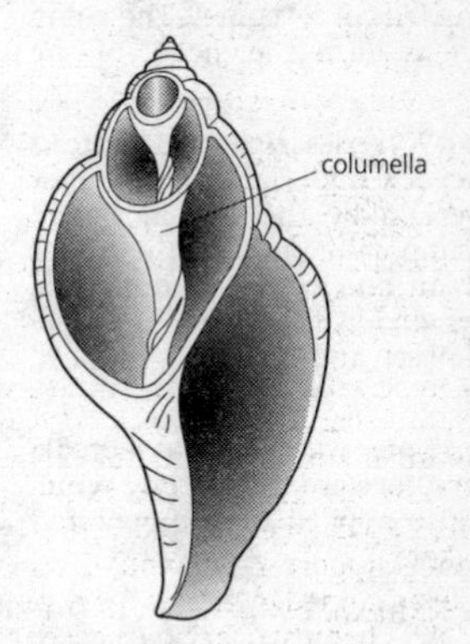

Columella

columnal One of the plates that form the stem (stalk) of crinoids (*Crinoidea). Commonly cylindrical, columnals have a variety of forms and may produce an articulated stem of variable length. *See also* OSSICLE.

columnar Applied to the *habit or shape of a *crystal that is rod like, with parallel faces, giving it the appearance of a column. Rocks such as *basalts and *dolerites may cool and develop a polygonal system of *joints, usually at right angles to the cooling surfaces of the *igneous body; the joints are well exposed by *weathering. This feature is known as 'columnar jointing'.

columnar joint *See* COLUMNAR; and JOINT.

columnar section A simplified graphic representation of the succession of rocks in a particular area, shown in the form of a column with the oldest rocks at the bottom. Each rock unit is distinguished by an appropriate *lithologic symbol, and the thicknesses of the units are drawn to scale. Columnar sections from across a region are often shown side by side to demonstrate lateral variations in *lithology and *stratigraphy.

com- (co-, con-) From the Latin *cum* or *com-*, meaning 'with', a prefix meaning 'with', or 'jointly'. 'Com-' is generally used before words beginning with b, m, and p, and sometimes before words beginning with vowels; 'con-' before words beginning with other consonants; and 'co-' before most words beginning with a vowel and before words beginning with h, gn, l, and r.

coma The diffuse shell of gas, typically about 150 000 km in diameter, which surrounds the nucleus of a *comet. The coma and the nucleus form the 'head' of the comet.

Comanchean A *series in the N. American Lower *Cretaceous, underlain by the Upper *Jurassic, overlain by the *Gulfian, and roughly contemporaneous with the upper *Aptian, *Albian, and lower *Cenomanian. Some authors divide this series into the Coahuilan (lower) and a much reduced Comanchean.

comber Deep-water wave that has a breaking crest blown forward by a strong wind. The term is also applied to a long-period *spilling breaker.

combe rock *See* HEAD.

combination trap Oil, gas, or water trap combining structural and stratigraphic features. *See also* STRUCTURAL TRAP; and STRATIGRAPHIC TRAP.

combined plate margin A *plate margin which has some combination of convergence or divergence with tangential movement, e.g. a *leaky transform fault.

comet A small body composed of meteoric dust and frozen ices (H_2O, CO_2, CO, HCHO) in a highly elliptical or parabolic orbit around the Sun. The average *perihelion distance is less than 1 AU, and the average *aphelion distance is about 10^4 AU. Comets are derived from the *Oort cloud and have average lifetimes of about 100 passages. Comet nucleii are irregular in shape, a few kilometres in diameter and have a low density (100–400 kg/m^3). Due to solar radiation, they emit gas and dust, forming the characteristic tail, when within a few astronomical units of the Sun; the dust composition appears to resemble that of primitive *carbonaceous chondrites.

Comet Nucleus Tour A *NASA mission to map the cores of comets and analyse dust streaming from them. It is due to be launched in 2002, at a cost of $154 million, and will pass comets in 2002, 2006, and 2008.

comfort zone Range of atmospheric temperature and humidity characteristics within which the human body feels and works comfortably and efficiently. The typical zone is delimited by a temperature range of 19–24 °C. Beyond the limits corrective adaptations are necessary to produce bodily comfort.

Comely Epoch *See* CAERFAI EPOCH.

comminution The liberation of valuable *minerals from their *ores by crushing and grinding the ore to a particular grain size so that the residue is a mix of relatively clean particles of ore minerals and waste. Comminution is carried out initially to make excavated material easier to handle and of a specific size. Comminution theory is concerned with the relationships between energy input and particle size from a given feed size.

commissure A line or plane of junction, e.g. between the two valves of the shell in a brachiopod (*Brachiopoda) or bivalve (*Bivalvia).

common canal In graptoloids (*Graptoloidea), a passage into which all the *thecae open (the thecae being short tubes arranged in overlapping series along the *stipe).

common depth point (CDP) In multichannel *reflection profiling, the unique point on an individual reflector from which seismic reflection information is recorded in traces at different *offsets. A set of traces containing information for one CDP is called a 'CDP gather', and the midway position between the shots and their respective detectors for a CDP gather is called the 'common mid point' (CMP). *See also* FOLD.

common-depth-point stack (CDP stack, horizontal stack) The sum of the traces of seismic *reflection data, acquired from a *roll-along survey, which correspond to the same *common depth point but which originate from different seismic *profiles and different *offsets. The technique reduces substantially the amplitude of incoherent noise, *multiples with their different normal *moveouts, diffractions, etc., compared with the seismic reflections.

common lead Lead which differs from primeval lead by the addition of *radiogenic lead resulting from the decay of uranium and thorium (*see* DECAY SERIES), but which occurs in minerals whose U : Pb and Th : Pb ratios are so low that its isotopic composition does not change appreciably with time.

common mid point (CMP) *See* COMMON DEPTH POINT.

common strontium *See* INITIAL STRONTIUM RATIO; and RUBIDIUM–STRONTIUM DATING.

community In ecology, a general term applied to any grouping of populations of different organisms found living together in a particular environment; essentially, the *biotic component of an *ecosystem. The organisms interact (by competition, predation, mutualism, etc.) and give the community a structure. Globally, the climax communities characteristic of particular regional climates are called 'biomes'.

compaction Physical process that reflects the increase in pressures brought upon sediments as a result of deeper and deeper burial. The individual grains or particles of the sediment are packed closer and closer together, thus reducing pore space. Pore waters are forced out and ultimately compaction will result in grain-to-grain suturing. *See also* PRESSURE SOLUTION.

compaction test A field trial to assess the type of equipment required to compact soil, the number of passes needed, and the thickness and moisture content of the soil layer, in order to achieve a high soil density for engineering works.

compass clinometer *See* CLINOMETER.

compensation level *See* PRATT MODEL.

competence 1. The maximum size of rock particle transported by a particular flow of ice, water, or air. Ice has a high competence, because of its viscosity; flowing water has a lower competence, though this increases sharply as velocity increases. Wind has the least competence. **2.** The relative rheological (*see* RHEOLOGY) properties of different rock types that are adjacent to one another. Typically, a competent rock is more viscous than an incompetent rock, more prone to *fracture, and maintains its thickness on deformation. An incompetent rock is usually more ductile than a competent rock and therefore flows more easily.

competent rock *See* COMPETENCE.

complex *See* LITHOSTRATIGRAPHIC UNIT.

complex twins (compound twins) A twin *crystal in which twinning has taken place on two or more *twin laws.

component-stratotype One of the individual sections of *strata constituting a *composite-stratotype.

composite fault-line scarp *See* SCARP.

composite-stratotype A *stratotype (type section) comprising more than one designated standard section of *strata. Such a designation may be made: (*a*) if an entire *lithostratigraphic unit is not completely exposed as a discrete section at any one locality, in which case only one section can be the *holostratotype, the other or others being subsidiary *parastratotypes; or (*b*) in the case of *chronostratigraphic units, when those of higher rank (e.g. *systems) may be composite-stratotypes of constituent lower-rank stratotypes, (e.g. *series).

composite volcano *See* STRATO-VOLCANO.

composition plane (composition surface) The plane along which two halves of a twinned *crystal are joined (*see* CRYSTAL TWINNING). It may or may not be coincident with the reflection or twin plane.

composition surface *See* COMPOSITION PLANE.

compound corals Colonial corals that form a compound *corallum (skeleton).

Where the *corallites are packed so closely as to be polygonal in section the corallum is said to be 'massive'. Other shapes are described as *dendroid, *cateniform, *fasciculate, *phaceloid, *cerioid, and *astraeoid.

compound twins *See* COMPLEX TWINS.

compressed air Air under pressure. This may be used in tunnels to balance the pressure of water in the ground and so control *groundwater. Compressed air is also widely used to drive machinery in mining and tunnelling operations, e.g. in rock *drills and ventilation fans.

compressibility (β) Ability of a substance to decrease in specific volume and increase in density under pressure; the reciprocal of *bulk modulus. In *powder technology, the ratio of the volumes of loose to compact powder.

compressional wave *See* P-WAVE.

compressive stress The stress produced by forces acting on a unit area of rock, which can be resolved into a *normal stress and a *shear stress, acting perpendicularly and parallel to a plane respectively. Compressive stress is measured in pascals (Pa) or kilobars (kb).

con- *See* COM-.

concealed coalfield Coal deposits buried beneath younger strata, e.g. the Kent coalfield, south-eastern England.

concentration 1. In chemistry, the number of molecules or *ions in a given volume of a substance, expressed as *moles of solute per litre of solution (molarity). **2.** In mineral processing, the production of a concentrate from its *ore, or the process of increasing concentration by evaporation, etc.

concentration factor The amount by which an element must be increased above its normal *crustal abundance in an *ore to make it commercially extractable.

concentration-Lagerstätte *See* LAGERSTÄTTE.

concentric fold A *parallel fold in which individual layers maintain a uniform *orthogonal thickness and ideal fold surfaces are arcs of circles sharing a common centre. Due to geometric constraints, the fold profile must change continuously upwards and downwards. *See also* FOLD.

conchoidal fracture Curved fracture pattern characteristic of some siliceous minerals, e.g. *quartz, fine-grained *igneous *rocks, and volcanic glasses, especially *obsidian.

concordant 1. Applied to an *igneous *intrusion that has been emplaced parallel with the structure (*bedding, *foliation, etc.) of the invaded *country rock. *Sills are examples of concordant intrusions. **2.** Applied to a relationship in which adjacent *strata are structurally concurrent with, or parallel to, each other. *Compare* DISCORDANT. *Contrast* CONFORMABLE.

concordant age If the *minerals in a rock have remained geochemically undisturbed since their formation, it is possible for the ages obtained by different radiometric methods (for example, the decay of ^{40}K to ^{40}Ar, ^{87}Rb to ^{87}Sr, ^{238}U to ^{206}Pb, ^{235}U to ^{207}Pb, and ^{232}Th to ^{208}Pb) to agree within the levels imposed by the accuracy of their known half-lives (*see* DECAY CONSTANT). If all the methods employed give a reasonable level of agreement, then the age is concordant. *See* RADIOMETRIC DATING.

Concordia diagram A plot of $^{206}Pb:^{238}U$ against $^{207}Pb:^{235}U$ for concordant samples of various ages should define a single curve, named 'Concordia' by G. W. Wetherill (1956). If the ratios plotted for samples fall below this *concordant age pattern expected for the rock body, this produces a *discordant age pattern. A straight line drawn through two or more such points should intersect the Concordia curve at two points. One of these will give the true age of the rock, while the younger intersect should give the date of any *lead loss (which is what causes the discrepancy).

concrete Building material composed of *cement, *aggregate, and water in varying proportions according to use; when mixed together the material hardens to a rock-like consistency.

concrete dam Dam made of Portland cement or cyclopean concrete (i.e. with large stones). A dense material is important in its construction to reduce percolation and ensure permanence; a relatively dry mix with careful tamping is required.

concrete minimum temperature An alternative to the *grass minimum temperature, producing more uniform results, and relevant to the likely freezing on road surfaces, etc. The temperature is recorded by a standard thermometer exposed in contact with a concrete surface.

concretion 1. Roughly spherical or ellipsoidal body, produced as a result of local early *cementation within a sediment. It is often found with a *fossil as a 'nucleus'. The size ranges from approximately 1 mm to more than 1 m, and concretions are generally monomineralic. **2.** In pedology, the localized concentration of material, e.g. calcium carbonate or iron oxide, in the form of a nodule; such nodules are of various sizes, shapes, and colours.

concretionary *See* NODULAR.

concurrent range zone (overlap zone) A body of *strata which is characterized by the overlapping stratigraphic range of two or more *taxa, selected as diagnostic, and after which the zone is named. Concurrent range zones are very widely used in the time *correlation of strata. **2.** *See* *OPPEL ZONE. *Compare* ASSEMBLAGE ZONE; and PARTIAL RANGE ZONE.

condensation level The atmospheric level at which condensation occurs as a result of convection, the lifting of air (e.g. *orographic lifting), or vertical mixing. *See also* LIFTING CONDENSATION LEVEL; and CONVECTIVE CONDENSATION LEVEL.

condensation nucleus Small particle of an atmospheric impurity (e.g. salt, dust, or smoke), which provides surfaces for the condensation of water. Condensation nuclei vary in size from about 0.1 μm to more than 3 μm. Some nuclei, e.g. salt and acid particles, can encourage condensation at a relative humidity well below 100%. *See also* AITKEN NUCLEUS; and HYGROSCOPIC NUCLEI.

condensation trail (contrail) Broadening trail of water droplets or ice particles behind an aircraft, caused by the release of water vapour by the engines. *See also* DISSIPATION TRAIL.

condensed bed A series of *strata which show a much thinner succession than their lateral equivalents due to greatly reduced sedimentation rates. A condensed bed was formerly known as 'remanié beds'.

condenser In microscopy, a hemispherical lens or series of lenses which helps to direct the optimum amount of light through varying sizes of aperture in objectives of various magnifications. Condensers are usually placed on the microscope below the *stage, between the *polarizer and the *mineral specimen. In the control of illumination, the area of light from the illuminator is normally cut by the condenser to equal the field of view, thus minimizing glare due to interference from marginal light.

conditional instability Atmospheric condition in which otherwise stable air, on being forced to rise (e.g. over an orographic barrier), cools at a rate less than that at which the temperature drops with height in the surrounding air. The rising air therefore becomes warmer than the surrounding air, and so continues to rise. This lesser rate of fall of temperature in the rising air is due to the condensation that occurs, as this is accompanied by a release of the latent heat of condensation. In such cases the instability is thus conditional upon the relative humidity of the rising air. *See also* INSTABILITY.

Condobolinian A *stage in the *Devonian of Australia, underlain by the *Cunninghamian, overlain by the *Hervyan, and roughly contemporaneous with the *Givetian and, possibly, *Frasnian Stages of Europe.

conductance (κ) In a direct current circuit, the reciprocal of the circuit's *resistance; in an alternating current circuit, the circuit's resistance divided by the square of its impedance, the result being the real component of admittance. In both cases, the SI unit is the siemens (S), formerly known as the mhos (reciprocal ohms).

Condylarthra (condylarths; class *Mammalia) Extinct order comprising seven families of primitive *ungulates, including some resembling the original eutherian (*see* EUTHERIA) stock so closely that their classification has presented many difficulties. Condylarths appear to be transitional between insectivores and true ungulates. They ranged from the late *Cretaceous to the latter part of the *Miocene and were a major part of the *Palaeocene fauna. Some forms possessed claws and may have been arboreal (Hyopsodontidae), others became highly specialized, but the development of hoofs and ungulate dentition can be traced to families in the order.

condylarths *See* CONDYLARTHA.

cone of depression The region, shaped like an inverted cone, in which the *water table is drawn down, or depressed, in the vicinity of a *borehole from which *groundwater is being abstracted by pumping.

cone penetrometer Instrument for testing the relative *shear strength of soil. A standard cone is pressed into the ground under known *load and depth of penetration measured. *See* PENETRATION TESTS.

cone sheet A *dyke shaped in cross-section like a cone dipping inwards to a central *pluton. This characteristic form is usually explained by reference to a stress field, shaped like an inverted umbrella, generated by the parent pluton at depth.

Conewangoan (Bradfordian) *See* CHAUTAUQUAN.

confidence interval In statistics, a range of values based on the observed data which are likely to contain the true unknown value for a specified proportion of the time (confidence level) usually expressed as a percentage.

confidence level *See* CONFIDENCE INTERVAL.

confined aquifer *See* AQUIFER.

confining pressure The combined *hydrostatic stress and *lithostatic stress; i.e. the total weight of the interstitial pore water and rock above a specified depth.

confluence Of air flow, convergence of adjacent streamlines, which increases air velocity. *See also* CONVERGENCE.

conformable Applied to a sequence of *strata deposited in an apparently continuous succession. *Contrast* CONCORDANT.

congelifraction *See* FROST WEDGING.

congeliturbation *See* GELITURBATION.

congestus From the Latin *congestus* meaning 'piled up', a species of *cumulus cloud with deep bulging form and an upper part having a cauliflower-like appearance. *See also* CLOUD CLASSIFICATION.

conglomerate Coarse grained (*rudaceous) rock with rounded *clasts that are greater than 2 mm in size. Conglomerates may be termed 'intraformational' if formed of local, recently deposited clasts, or 'extraformational' if the clasts are derived from outside the area of deposition. The term 'polygenetic' is used to describe a conglomerate rock that has been produced under a variety of conditions or processes.

congruent dissolution Transition from a solid substance to a liquid of the same composition.

congruent solution Solution of a double salt (i.e. a salt formed by the crystallization of two or more components, e.g. *dolomite) to yield quantities of component *ions in the same proportions as existed in the solid.

Coniacian A *stage in the European Upper *Cretaceous, for which the *stratotype is at Cognac, France. It is dated (Harland et al., 1989) at 88.5–86.6 Ma. *See also* SENONIAN.

Coniferales In some classifications, the order comprising the conifers. *See* CONIFEROPHYTA.

Coniferophyta (Pinophyta) The biggest division of *gymnosperms, with a long fossil history, comprising trees and shrubs; most are resinous. The leaves are often needle- or scale-like. Fertile parts occur in unisexual cones, variously containing sterile scales. The ovule and seed are naked and borne on a scale. They first appear as fossils in *Carboniferous rocks.

Coniferopsida In some classifications, a subdivision of the *gymnosperms, comprising 4 orders, the oldest of which, the *Cordaitales, appeared in the *Carboniferous. They became extinct at the end of the *Permian, but their place was taken by 2 other coniferopsid orders: the Ginkgoales (ginkgos) and Coniferales (conifers). These are now usually ranked as divisions, the *Ginkgophyta and *Coniferophyta.

cone-in-cone structure A *secondary sedimentary structure consisting of small cones nested one inside another and most commonly made from calcium carbonate. They are believed to form by the growth of fibrous crystals in the sediment while this is still plastic.

conifers *See* CONIFERALES.

conjugate fault set A cross-cutting set of *fault planes which ideally intersect at angles of 60° and 120°, and have both left-handed and right-handed *shear senses. The line of intersection is parallel to the direction of intermediate *principal stress (σ_2). The maximum principal stress bisects the acute angle and the minimum principal stress bisects the obtuse angle.

conjugate fold A set of paired, asymmetric *folds whose *axial planes *dip towards one another. Limbs are commonly straight, and *hinge zones short and angular. Conjugate folds are thought to be formed during the final stages of deformation.

conjunct Applied to the distribution of populations that have overlapping ranges, allowing DNA to be exchanged between them. *Compare* DISJUNCT.

connate water From the Latin *connatus* meaning 'born together', water that has remained trapped in a *sedimentary rock since the original sediments were laid down in that water, prior to *lithification. Connate water may be very old and saline.

Conodontophora The category into which *conodonts were formerly placed. Conodonts are tiny, phosphatic, tooth-shaped *fossils that occur in rocks from the *Cambrian to the *Triassic.

conodonts Small, phosphatic, *fossil teeth, common in rocks from the *Cambrian to *Triassic (and formerly placed in the category *Conodontophora) that belonged to elongated, fish-like animals that were probably chordates (*Chordata), possibly vertebrates, and lived as active predators. Two eyes were located in lobe-shaped structures at the anterior end, a *notochord ran down the length of the worm-shaped body, there were muscular fins at the posterior end, and the feeding apparatus comprised the only hard parts.

conoscopic Applied to the converging light in a transmitted-light microscope, when the addition of a condenser below the microscope stage through which *plane-polarized light is passed converges the light on to the stage to produce *interference figures and optical effects characteristic of individual *minerals. *Compare* ORTHOSCOPIC.

Conrad discontinuity A boundary within the Earth's *continental crust that can be detected seismically at about 10–12 km depth, although exploratory deep drilling has failed to locate it. The boundary separates the crust into a lower, basic layer and an upper, granitic layer.

consequent stream Stream whose course is consequent upon the shape of a newly emerged land surface. Its course shows no necessary relationship with the underlying geologic structure, although older usage tended to restrict the term to a stream flowing in a downdip direction across gently inclined strata.

conservation-Lagerstätte *See* LAGERSTÄTTE.

conservative margin The zone between two lithospheric *plates in which *crust is being neither created nor destroyed, but the plates are sliding tangentially relative to each other along a *transform fault.

consistence (consistency) The resistance of soil to physical impact such as ploughing, digging, or handling. It is controlled by the degree of adhesion between soil particles. It is described when dry as loose, soft, or hard, when moist as loose, friable, or firm, and when wet as sticky, or plastic.

consistency *See* CONSISTENCE.

consistency index In *cladistic analysis, a measure of *homoplasy in a phylogenetic tree (or *cladogram), calculated as the number of steps (i.e. *character state changes) in the cladogram divided by the smallest possible number of steps. The index therefore runs from 0 to 1. A low consistency index (less than 0.5) tends to indicate that much homoplasy has occurred.

consolidation In geology, any process by which loose earth materials become compacted, including *cementation, *diagenesis, *recrystallization, *dehydration, and *metamorphism. In soil mechanics, the term implies a slow reduction in volume and an increased density of saturated soil under *load, e.g. beneath buildings. Rate of consolidation depends on the rate of pore water escape, therefore on the *permeability of the soil.

constant head permeameter *See* PERMEAMETER.

constant offset A constant separation between a geophysical source and a receiver (*see also* OFFSET). Constant-offset profiling (COP) is a specialized method of marine seismic profiling using two ships, one shot-firing and the other recording, which travel along a profile at a constant offset. COP is used for mapping variations in crustal structures over large areas.

constant-separation traversing (CST) An electrical technique in which an *electrode array (commonly a *Wenner or *dipole–dipole array) is moved along a survey line, keeping the electrode separations constant. Similar survey techniques are used in *electromagnetic surveying, where a dual-coil method is employed; in this case constant coil separations used.

constant-velocity gather (**CVG**) In a *seismic velocity analysis, a method that assumes a constant velocity for the entire path of a ray, allowing the *normal move-out (NMO) to be calculated for each *seismic trace as a function of two-way travel time (TWTT). Where the correct velocity has been applied for a given reflector, the NMO will appear horizontal on the CVG; if the velocity applied is too high or too low, it will appear curved. CVG is a necessary precursor for *stacking.

constant-velocity stack (**CVS**) A method similar in principle to a *constant-velocity gather (CVG), in which the *normal move-out is applied as in a CVG and up to ten gathers are stacked (*see* STACKING) and displayed as CVS velocity panels. The correct velocity is that which produces the strongest event for a given reflector. CVS tends not to produce such precise values for velocity as the CVG method.

constructive boundary *See* CONSTRUCTIVE MARGIN.

constructive margin (**constructive boundary**) The zone between two lithospheric *plates which are diverging and consequently where new *crust is being formed. Constructive *plate margins are associated with shallow-focus earthquakes, high heat flow (up to ten times the average), and tholeiitic basalt (*see* THOLEIITE). Constructive plate margins develop oceanic *ridges.

constructive wave Wave that leads to the build-up of a beach, due to the *swash of the wave being more effective in moving material than the *backwash. Usually, constructive waves are associated with low-energy conditions and a gentle offshore gradient.

contact The depositional, *intrusive, or faulted surface along which two different rock types are juxtaposed. The term is especially applicable to situations where *plutonic *igneous rocks intrude into *country rocks; in this context 'contact' also refers to the effect on country rocks of conductive or convective heat transfer (i.e. *contact metamorphism).

contact aureole (**metamorphic aureole**) A region in which *country rocks surrounding an *igneous intrusion have been *recrystallized in response to the heat supplied by the intrusion. The widths of contact aureoles are quite variable and partly depend on the size of the igneous intrusion—the larger the intrusion, the wider the aureole. For intrusions of similar size, an aureole formed in response to simple conduction of heat will, in general, be thinner than an aureole formed in response to the more efficient convective transfer of heat in escaping mineralizing fluids. *See also* CONTACT METAMORPHISM.

contact goniometer *See* GONIOMETRY.

contact metamorphism The *recrystallization of rocks surrounding an *igneous intrusion in response to the heat supplied by the intrusion. Since there is no significant increase in the pressure gradient around an intrusion, recrystallization processes in the surrounding *country rocks are a response only to an increase in the thermal gradient around the intrusion. Hence contact metamorphism is also known as 'thermal metamorphism'. *Metasomatism often takes place during contact metamorphism, unlike *regional metamorphism. *See also* CONTACT AUREOLE.

contact resistance The measured resistance between an earthed electrode and the ground, or between a polarizable electrode and a rock specimen, or between contacts in an electrical circuit.

contact twin Twinned *crystals (*see* CRYSTAL TWINNING), where the two individuals comprising the twins are in contact along a *composition plane.

contained fragments *See* PRINCIPLE OF INCLUDED FRAGMENTS.

contessa del vento Type of cloud typically with a rounded base and bulging upper surface. Sometimes a number of separate discs of this cloud-form extend one above another. Such cloud occurs on the lee side of distinct mountains within an eddy zone. In the case of Mt. Etna the cloud is related to a westerly air stream.

continental crust That portion of the Earth's surface overlying the *Mohorovičić discontinuity, and with an average density of 2700–3000 mg/m^3. The thickness is variable, mostly 30–40 km, except for areas of recent mountain building where the thickness can be 70 km. The crust comprises two layers, but the boundary between them is poorly defined. The upper layer is of granitic composition with surface enrichment in *radiogenic and other *lithophile elements, and has an average density of 2700 mg/m^3. The lower layer was previously considered to be of gabbroic composition but is now thought to be quartz *diorite in *granulite metamorphic grade.

continental drift Hypothesis proposed around 1910 to describe the relative movements of continental masses over the surface of the Earth. A major theorist of continental drift, and certainly the one who gave the hypothesis scientific plausibility, was Alfred *Wegener (1880–1930). His work was based on qualitative data, but has been vindicated in recent years by the development of the *plate tectonics theory, which has provided geologists with a viable mechanism to account for continental movements. *See also* polar wander.

continental freeboard The average level of the sea surface relative to the continents.

continentality A measure of how the climate of a place is affected by its remoteness from the oceans and oceanic air. The difference between the average temperatures prevailing in January and July is most often quoted as an indicator of this.

continental margin Zone that consists of the *continental shelf, *continental slope, and *continental rise. It extends from the shoreline to the deep-ocean floor at a depth of 2000 m. The zone is underlain by *continental crust. Continental margins have been divided into *active margins or *passive margins depending on their coincidence, or otherwise, with *plate margins.

continental rise Smooth-surface accumulations of sediment which form at the base of the *continental slope. The surface of the rise is gently sloping with gradients between 1 : 100 and 1 : 700. The width of the rise varies but is often several hundred kilometres. Two types of deposit lead to the formation of rises, *turbidites laid down by *turbidity currents flowing down the continental slope, and *contourites laid down by *contour currents flowing along the rise at the base of the *continental margin.

continental shelf Gently seaward-sloping surface that extends between the shoreline and the top of the *continental slope at about 150 m depth. The average gradient of the shelf is between 1 : 500 and 1 : 1000 and, although it varies greatly, the average width is approximately 70 km. Five major types of shelves may be recognized: (*a*) those dominated by tidal action; (*b*) those dominated by wave and storm action; (*c*) those dominated by carbonate deposition; (*d*) those subject to modern glaciation in polar areas; and (*e*) those floored by *relict sediments which constitute up to 50% of the total shelf area.

continental-shelf waves *Vorticity waves produced in a *continental-shelf area where there is a sea-bed slope. In the northern hemisphere, if a water column is displaced into shallower water it develops negative relative vorticity, or anticyclonic motion; if displaced into deeper water it will develop positive relative vorticity, or cyclonic motion. The net result is for shelf waves to progress in a poleward direction along the west coasts of continents, or equatorwards along east coasts.

continental slope The relatively steeply sloping surface that extends from the outer edge of a *continental shelf down to the *continental rise. The total relief is substantial, ranging from 1 km to 10 km, but the slope is not precipitous and ranges from 1° to 15° of slope (average 4°). Along many coasts of the world the slope is furrowed by deep submarine canyons, terminating as fan-shaped deposits at the base.

continuation The use of one set of measurements of a potential field (usually gravity or magnetic) over one surface to determine the set of values the field would have over another surface, usually at a different elevation. *See also* UPWARD CONTINUATION; and DOWNWARD CONTINUATION.

continuous distribution Data that yield a continuous spectrum of values. Examples are measurements of the wing lengths of birds, or the weights of mammals, or the heights of plants.

continuous profiling A seismic technique in which the *geophone and shooting patterns provide 100% subsurface coverage and which can be applied in reflection surveys and, with more practical difficulty, in refraction surveys. In a refraction survey the refractor must be monitored, especially if it is not planar, and this may necessitate laying out the geophones in an irregular pattern.

continuous reaction series The continuous change in composition of a *solid-solution *mineral in order to maintain a state of equilibrium with a cooling *magma. The mineral changes its composition by continuously exchanging *cations with the cooling magma in which it floats. The *plagioclase feldspar solid-solution mineral series is a good example of a continuous reaction series. During the cooling of a magma, the plagioclase continuously exchanges Ca and Al for Na and Si in order to maintain an equilibrium composition. The principle of continuous reaction was first presented by Norman *Bowen in 1928. *See also* BOWEN'S REACTION PRINCIPLE.

continuous velocity logging (sonic log) A technique which provides a record of the single travel time a sound wave takes to pass through the sidewall of a *borehole from a *sonde source to a receiver, when these are a constant distance apart and on the same down-hole tool. As the tool is drawn up the borehole to the surface a record of the variations in travel time is plotted as a function of down-hole depth. Measured in units of 10^{-6} feet per second, it provides the inverse of the formation velocity.

CONTOUR A *NASA mission, to be launched in 2002, that will visit three comet nuclei.

contour current Undercurrent typical of the *continental rise which flows along the western boundaries of ocean basins. Such currents occur particularly in regions in which density stratification is strong because of the supply of cold waters originating near the poles. A well-known example is the Western Boundary Undercurrent which hugs the continental rise of eastern N. America. Contour currents are persistent, slow-moving (velocity 5–30 cm/s) flows capable of transporting mud, *silt, and *sand.

contour diagram A stereographic *equal-area net on which orientation data (i.e. the *azimuths of structural features) are plotted as lines or points and then joined to form contours linking areas of equal density of data, thus providing a visual description of the range and concentration of the data. There are several ways to construct such diagrams manually or by computer, and there are statistical tests for evaluating the significance of the densities revealed.

contourites Sediments that have been deposited by *contour currents on the *continental rise. The sediments are thin-bedded *silts, *sands, and muds. The sands are well sorted, laminated, or cross-laminated, with many internal erosion surfaces, and there are concentrations of heavy minerals. Both bottom and top contacts of the beds are sharp, and the beds lack great lateral continuity.

contracting Earth hypothesis The hypothesis stating that mountain building is caused by the contraction of the Earth. Conventionally, the wrinkles that form on a drying apple are used as an analogy for the formation of mountains on Earth. The hypothesis was prevalent in the 19th and early 20th centuries, prior to the discovery of *radiogenic heating within the Earth. The Earth has probably reduced by some 5% in volume since its original formation, but mountains mainly result from *plate tectonic motions.

contraction limit *See* ATTERBERG LIMITS.

contrail *See* CONDENSATION TRAIL.

contrast In *remote sensing, the ratio of energy emitted or reflected by an object to that emitted or reflected by its immediate surroundings.

contrast stretching In *remote sensing, the artificial increase in range of *digital number values of *pixels in an image in order to increase *contrast. *See also* DECORRELATION STRETCHING; and EDGE ENHANCEMENT.

control system In *geomorphology, a *process–response system where human interference changes the natural functioning of the system. For example, the trapping of sediments that occurs where beach *groynes are built may lead to *erosion within the adjacent subsystem and so to a change in the geometry of the overall process–response system.

convection 1. Vertical circulation within a fluid that results from density differences caused by temperature variations. Convection currents occur in the oceans when a water mass that is denser than the water below it sinks and is replaced by lighter, warmer water. **2.** In meteorology, the process in which air, having been warmed close to the ground, rises. The convective uplift of *air parcels is one of the main processes leading to condensation and cloud formation. *See also* DISH-PAN EXPERIMENT; FORCED CONVECTION; HADLEY CELL; INSTABILITY; LEVEL OF FREE CONVECTION; and STABILITY. **3.** Within the Earth, the *radiogenic heat release results in convective motions causing tectonic *plate movements. The location and configuration of the *convective cells is uncertain, but they appear to be *mantle-wide and marked by most heat loss along the *mid-ocean ridges. The difference in temperature between upgoing and down-going convective limbs within the mantle may be only 1–2 °C. In the upper *oceanic crust, heat loss is mainly by convective circulation systems combined with thermal conduction. *See also* CREEP MECHANISMS; GEOTHERMAL GRADIENT; HEAT FLOW; NUSSELT NUMBER; and RAYLEIGH NUMBER.

convection current *See* CONVECTION (1).

convective cell The pattern formed in a fluid when local warming causes part of the fluid to rise, and local cooling causes it to sink again elsewhere. The atmosphere in low latitudes forms such cells, known as *Hadley cells, as warm equatorial air rises, moves away from the equator and cools, and then descends over the subtropics.

convective condensation level The level at which surface air will become saturated when rising by convection. *See also* LIFTING CONDENSATION LEVEL.

convective instability *See* POTENTIAL INSTABILITY.

convergence 1. Situation in which, over a given lapse of time, more air flows into a given region than flows out of it. It is commonly accompanied by confluence of the streamlines, but may be caused by differences of velocity, e.g. where the wind comes against a coast or a mountain wall. Surface friction can produce convergence. *Compare* DIVERGENCE. **2.** The point, line, or region where two oceanic water masses or surface currents meet. This leads to the denser water from one side sinking beneath the lighter water of the other side.

convergent evolution The development of similar external morphology in organisms which are unrelated—except through distant ancestors—as each adapts to a similar way of life. Sharks (fish), dolphins (mammals), and *ichthyosaurs (extinct reptiles) provide good examples of convergence in the aquatic habitat.

convergent margin A boundary between two lithospheric *plates which are moving towards each other. Some such boundaries involve *subduction of *oceanic crust and are called '*destructive' boundaries; others are *collision zones between continents where it is thought that all the oceanic crust has been subducted and that the ocean basin has reached the final stage of the *Wilson cycle.

converted wave *See* S-WAVE.

convex slope *See* SLOPE PROFILE.

convolute In a coiled gastropod (*Gastropoda), applied to the condition in which the outer whorls embrace the inner ones so that the latter are nearly or partly invisible. In nautiloid cephalopods (*Nautiloidea) this type of *coiling is sometimes called 'nautilicone'.

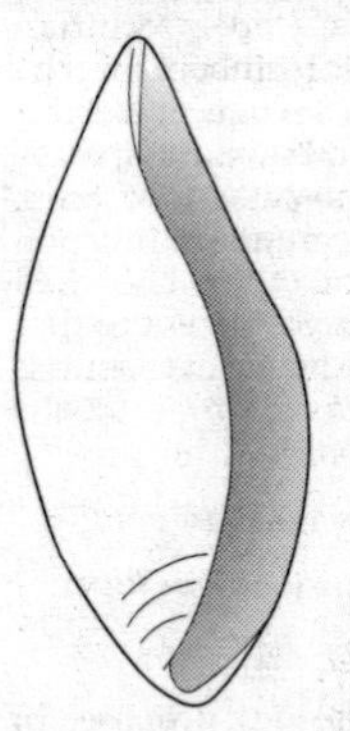

Convolute

convolution A mathematical operation (symbol *) to define the change in the shape of a wave-form caused by its passage through a *filter. If an *impulse response function f is convolved with an input g, then the output, h, is given by: $h = g*f = \Sigma_x^j g(t-j) \times f(j)$.

convolute lamination A *sedimentary structure consisting of a series of upright or overturned *folds whose intensity dies out both upwards and downwards within a single bed. The convolutions are formed by the expulsion of water from rapidly deposited sediments by an external shock, e.g. an earthquake, the effect of large waves, or by the rising and falling of the *water table through the sediment.

Conybeare, William Daniel (1787–1857) An English clergyman, Conybeare is best known as co-author, with William Phillips, of *Outline of the Geology of England and Wales* (1822), one of the most influential textbooks on stratigraphy of the period. He also described and reconstructed *saurian fossils from the Lyme Regis area of England. As a friend and collaborator of William *Buckland, Conybeare was an influential member of the Oxford School of Geology.

Cook, James (1728–1779) A navigator, surveyor and explorer, who captained three expeditions to the Pacific between 1768 and

1779. He surveyed the coasts of New Zealand and eastern Australia, and explored part of the seaboard of Antarctica. He was able to show that the supposed 'Great Southern Continent' did not exist.

Cooksonia hemispherica **(family Rhyniaceae)** Extremely primitive plant, a few centimetres high, from the Upper *Silurian and Lower *Devonian. Members of the family did, however, possess an epidermis (protective outer layer) and stomata (specialized pores) to control the passage of gases. They also had an underground rooting portion, the rhizome, and they branched *dichotomously. *C. hemispherica* is the first representative of the Rhyniaceae, and is known from the Silurian of Europe. *See also* PSILOPHYTALES; and *RHYNIA*.

cooling age *See* FISSION-TRACK DATING.

cooling joint (shrinkage joint) *See* JOINT.

coombe rock *See* HEAD.

coordinated stasis The idea, proposed in 1992 by Gordon Baird, that certain groups of species remain unaltered for tens of millions of years, then experience an episode of rapid extinction and the formation of new species. This resembles *punctuated equilibrium acting at the level of communities and may occur because the species interact so closely they cannot evolve, instead responding to environmental change by moving as a group to a more hospitable location. The fossil record of animals dwelling in ocean-bottom muds in the *Silurian to middle *Devonian appears to support the idea, but it is not accepted by all palaeontologists.

coordination number Number of atoms, ions, molecules, or groups of these that surround a given atom in a complex molecule. In geology the most common groups are oxygen atoms surrounding a *cation, e.g. in silicate minerals (*see* SILICATES) silicon is always surrounded by four oxygen atoms.

COP (constant-offset profiling) *See* CONSTANT OFFSET.

Copernican System *See* LUNAR TIME SCALE.

Cope's rule In 1871, the American palaeontologist Edward Drinker Cope (1840–97) noted a phylogenetic trend towards increased body size in many animal groups, including mammals, reptiles, arthropods, and molluscs. This came to be known as Cope's rule. It remained unchallenged until a study of more than 1000 insect species in 1996 and was finally disproved in 1997, by a study in which David Jablonski made more than 6000 measurements on 1086 species of Late *Cretaceous fossil molluscs spanning 16 million years and found as many lines led to decreased size as increased. Evolutionary lineages show no overall tendency to greater size, but if the extant survivor happens to be larger than its immediate ancestor (e.g. the horse) this coincidence appears to validate Cope's rule.

copper, native A malleable metallic element, Cu; sp. gr. 8.9; *hardness 2.5; *cubic, but normally found *massive, or as *platy flakes infilling cracks or *vesicles in volcanic rocks, or as a result of *weathering of copper-rich *mineral deposits.

copper glance *See* CHALCOCITE.

copper pyrites *See* CHALCOPYRITE.

coppice dune A sand *dune that forms around and in the lee of a clump of vegetation.

co-product Many *ore deposits are worked for more than one element. Where those elements are of almost equal economic significance they are called 'co-products'. *Compare* BY-PRODUCT.

coprolite Fossilized droppings or excreta (i.e. a fossilized faecal pellet). Coprolites may have distinctive shapes or markings which can provide information regarding the structure of the animal's alimentary canal, and analysis of the contents may also reveal its diet.

coquina *Clastic or detrital *limestone that contains a high proportion of coarse shell debris cemented by calcium carbonate.

coral *See* ANTHOZOA; HETEROCORALLIA; HYDROZOA; MILLEPORINA; OCTOCORALLIA; RUGOSA; SCLERACTINIA; STYLASTERINA; TABULATA; and ZOANTHARIA.

coral growth lines All corals that have a calcified outer wall display minute growth lines on their outer surfaces. The carbonate is secreted by symbiotic *algae (zooxanthellae) which, responding to day and night, create a series of diurnal growth increments. Studies on *Devonian isolate corals indicate a 400-day year and, therefore, a 22-hour day. Post-Devonian data confirm a near-linear deceleration of the Earth's rotational velocity towards the present 24-hour day. The number of daily growth lines can therefore be used in a crude way to calibrate the geologic record.

coralline limestone *See* LEIGHTON–PENDEXTER CLASSIFICATION.

corallite The skeleton formed by an individual coral polyp, which may be either solitary or part of a colony.

coralloid Literally, coral-like, but used to describe the circular, curved, or irregular shapes that sometimes result from the chemical precipitation of minerals, e.g. *marcasite.

corallum The skeleton of a colonial coral, made up from individual skeletons (*corallites) each secreted by one polyp. Distinctive corallum shapes are given a variety of names.

Cordaitales Extinct *gymnosperm order, included in the *Coniferopsida, which appeared in early *Carboniferous times and disappeared towards the end of the *Permian. They produced trees up to 30 m high, with strap-like leaves and primitive cones. Some cordaitaleans developed stilt roots, probably lived in swamp habitats, and were analagous to modern mangroves. Fossils of stems, leaves, roots, and cones are locally abundant in coal. The term *Cordaites* is a *form-genus name, and strictly applies to the leaves of cordaitaleans, but it has often been used informally for the whole plant. Other form-genera include *Cordaianthus* for cordaitean cones, *Amelyon* for the roots, and *Dadoxylon*, *Araucarioxylon*, *Mesoxylon*, and *Pennsylvanioxylon*, according to type or age, for stem fragments and wood.

Cordelia (Uranus VI) One of the lesser satellites of *Uranus, with a radius of 13 km. It was discovered in 1986.

cordierite *Silicate mineral $Al_3(Mg,Fe)_2[Si_5AlO_{18}]$ which can be iron-rich and is a member of the *cyclosilicate group; sp. gr. 2.5–2.8; *hardness 7; *orthorhombic; dark blue or greyish-blue, translucent to transparent; *vitreous *lustre; *crystals rare, *prismatic or pseudo-hexagonal, but usually occurs *massive; *cleavage imperfect {010}, parting {001}; occurs in aluminous rocks that have been subjected to thermal or regional *metamorphism, in *hornfels, *schists, and *gneisses, in association with *andalusite, *spinel, *quartz, and *biotite. Fine dark blue examples are used as *gemstones. It is named after the 19th-century French geologist P. L. A. Cordier.

cordillera 1. Broad assemblage of mountain ranges belonging to *orogenic belts of different ages which formed originally at *destructive plate margins. **2.** System of mountain ranges, together with their related plateaux and *intermontane basins. For example, the Cordillera of N. America includes all the mountain ranges and plateaux west of the Great Plains and of the Mexican lowlands. **3.** Subsidiary complex of ranges within a mountain system, e.g. the eastern and western cordillera of the Andes. **4.** Individual range, e.g. the Cordillera Patagonica of the southern Andes.

core 1. The central zone or unit of the Earth. It is composed of iron, with a lighter element, probably sulphur, and accounts for 16% of the Earth's volume and 32% of its mass. The core is separated into inner and outer units. The inner core is a solid with a radius of about 1220 km and the outer core, which does not permit the passage of shear waves (*S-waves), is liquid. Other planets have mass distributions that suggest they possess cores, e.g. *Mars, *Venus, and *Mercury. The *Moon may have a small core. *Saturn has magnetic fields interpreted to indicate a metallic core, probably of liquid hydrogen. **2.** A rock specimen obtained by *drilling.

core barrel The part of *borehole drilling equipment in which the core is collected. Usually the barrel is 50–100 mm in diameter with the drill *bit at one end. Sometimes an inner tube is present, linked by a swivel to prevent its rotation (double core barrel), and even a triple core barrel may be used to aid the recovery of weak rock.

core-logging Geologic description of a *core obtained from a *borehole. This includes lithological, palaeontological, and structural information, often with reference to engineering properties. *See also* ACOUSTIC LOGGING; CONTINUOUS VELOCITY LOG; GEOPHYSICAL WELL-LOGGING; INDUCED POTENTIAL LOG; NEUTRON LOG; RESISTIVITY LOG; and SELF-POTENTIAL LOG.

core recovery Length of *core recovered from a *borehole, compared with depth of hole cored, expressed as a percentage. *See* TOTAL CORE RECOVERY (TRC)

core slicer 1. A diamond-edged cutter that can be used inside a *borehole to cut triangular slices 2–3 cm wide and up to 1 m long from the rock walls. **2.** A diamond-tipped saw blade used to slice specimens from a rock *core drilled in the field.

corestone (woolsack) Rounded boulder, occurring individually or in piles at the ground surface, or in exposed sections. It results from an initial phase of subsurface *chemical weathering, of a *joint-bounded block, followed or accompanied by surface *erosion that exposes the corestone.

core wall *See* DIAPHRAGM WALL.

Cor F *See* CORIOLIS FORCE.

Coriolis force (Cor F) An apparent force acting on moving objects that results from the Earth's rotation. It causes objects in motion, and oceanic and atmospheric currents, to be deflected to the right in the northern hemisphere and to the left in the southern hemisphere. The force is proportional to the speed and latitude of the moving feature, and therefore varies from zero at the equator to a maximum at the poles.

corona 1. Coloured rings of lights, typically from blue inside to red outside, that sometimes appear to surround the Sun or Moon. The effect is created by diffraction of light by spherical water drops in such clouds as *altocumulus. *Compare* HALO. **2.** Concentric zones of one or more *minerals surrounding a core mineral. Coronas can be formed in a number of ways. (*a*) The *discontinuous reaction of minerals with a *magma can be preserved as coronas around the original high-temperature mineral if the cooling rate is fast enough to prevent the reactions going to completion. (*b*) Late-stage fluids may react with an earlier *primary mineral to develop a corona of *secondary minerals. (*c*) Two minerals may undergo sub-*solidus reactions (reactions occurring after the rock has solidified) to maintain equilibrium as a rock mass cools, developing a corona of lower-temperature minerals. These types of *texture are also known as 'reaction rims'. *See also* CRYSTAL ZONING. **3.** One of the large, circular features (150–600 km diameter) of uncertain origin, comprised of up to 10–12 subconcentric ridges and grooves, which surround an inner region of irregular relief, found on the surface of Venus, mainly in a latitudinal belt 55° N–80° N along the borders of Ishtar and Tethus Regio. Most are associated with what appear to be lava flows.

corrasion *See* ABRASION.

correlated progression The hypothesis that evolutionary change of *characters occurs by correlated response, i.e. a change in one character may influence change in another, such that the rate of change of the two characters is not independent. *Compare* MOSAIC EVOLUTION.

correlation 1. In stratigraphy, correlation is the establishment of a correspondence between stratigraphic units. It depends on the similarities that exist in terms of *lithology or *fossil content. Isolated stratigraphic units, or successions, may be either 'correlated', i.e. they were once physically continuous, or time-correlated, i.e. equated in terms of time. **2.** In geostatistics, correlation is a technique used to determine the degree of association between two data sets. **3.** In geophysics, the comparison of one wave-form with another in the *time domain. It is analogous to *coherence in the *frequency domain. *See* AUTOCORRELATION; and CROSS-CORRELATION.

correlation diagram Diagram illustrating probable stratigraphic equivalence from place to place.

correlogram A graph showing the strength of correlations in data at different time intervals and thereby exposing the existence and phases of cycles.

corridor dispersal route As originally defined by the American palaeontologist G. G. *Simpson (1940), a corridor is a migration route that allows more or less uninhibited faunal interchange. Thus many or most of the animals of one *faunal realm can migrate to another one. A dispersal corridor has long existed between western Europe and China via Central Asia.

corrie *See* CIRQUE.

corundum *Mineral, Al_2O_3; sp. gr. 3.9–4.1; *hardness 9; *trigonal; the two main varieties blue and green, but can be yellow, or brown to almost black, and transparent; *adamantine to *vitreous *lustre; *crystals usually rough and barrel-shaped, tapering, and also flat and *tabular; no cleavage, partings {0001}, {$01\bar{1}2$}; occurs in silica-poor rocks such as *nepheline syenites and undersaturated (*see* SILICA SATURATION) alkali *igneous rocks, in *contact aureoles in thermally altered alumina-rich *shales or *limestones, in aluminous *xenoliths found within *basic igneous rocks in association with *spinel, *cordierite, and *orthopyroxene, in metamorphosed *bauxite deposits and in *emery deposits, and in *alluvial deposits because of its hardness and resistance to *abrasion along with *muscovite, *hematite, and *rutile. Flawless crystals are the *gemstones blue sapphire, red ruby, and green oriental emerald. The main use of corundum is based on its hardness. It is made into grinding wheels and discs, emery paper, and powders for grinding and polishing.

Corynexochida An order of *Trilobita that lived from the Lower *Cambrian to Middle *Devonian. The glabella (*see* CEPHALON) is variable, but usually with parallel sides or wider anteriorly. There are 3 suborders.

coset *See* CROSS-STRATIFICATION.

cosmic abundance of elements Based on data from the *Sun and other stars, hydrogen and helium are by far the most abundant elements of the cosmos (e.g. the Sun's atmosphere may contain 70% hydrogen and 28% helium by mass). In general, elements show an exponential decrease in abundance with increasing atomic number (Z), up to about 45; the abundances of heavier elements thereafter appears fairly constant. Other regularities are superimposed on this general pattern: elements of even atomic number are more abundant than adjacent elements of odd atomic number (*Oddo–Harkins rule); there is a pronounced peak at atomic number 26 (the 'iron' peak); and isotopes whose mass numbers are multiples of four, i.e. multiples of the alpha-particle (helium nucleus) mass, e.g. carbon, oxygen, neon, magnesium, silicon, sulphur, and iron, have enhanced abundances. Cosmic abundances of elements constrain the way in which the *solar system evolves and dictate the composition of its members (which include the Earth and ourselves on it). *See also* NUCLEOSYNTHESIS.

cosmic dust Particles with a wide range of masses (10^{-2}–10^{-18} g) and velocities occurring in interplanetary, circumstellar, and interstellar space. They are typically porous with low densities and a 'cluster of grapes' morphology, but many are compact with densities of about 2000 kg/m^3. Most particles collected from the *stratosphere are comprised of layer lattice *silicates, *olivine, or *pyroxene. In the terrestrial neighbourhood, they are mostly derived from *comets and *asteroids.

cosmic radiation Ionizing radiation from space, comprised principally of protons, alpha particles, and 1–2% heavier atomic nucleii, as well as some high-energy photons and electrons. On encountering the Earth's atmosphere, secondary radiation is produced, mainly gamma rays, electrons, pions, and muons. Three sources are identified: (*a*) galactic cosmic rays, from outside the *solar system, with energies in the range 1–10 GeV per nucleon; (*b*) solar cosmic rays, mainly associated with *solar flares, with energies in the range 1–100 MeV per nucleon; and (*c*) *solar wind, with energies of about 1000 eV per nucleon.

cosmic-ray track The interaction of ionizing cosmic rays (*see* cosmic radiation) with mineral surfaces produces solid-state damage which, when etched with acid, is revealed as tracks of varying length. Most tracks are produced by iron group nucleii (nuclear charge (Z) 18–28, VH ions). Track lengths due to galactic cosmic rays may extend to depths of 20 cm. In contrast, *solar flare tracks are mostly less than one millimetre in length. Track densities, typically about 10^{12}/m^2 on unshielded lunar mineral surfaces, provide a measure of *exposure age.

cosmine *See* COSMOID SCALE.

cosmoid scale Type of scale found only in fossil lung-fish (*Dipnoi), and in *Crossopterygii, including the living *coelacanth. The thick scales are composed of layers of vitrodentine (harder than dentine and containing little organic matter), followed by cosmine (a type of *dentine perforated by branching canals), and finally by layers of vascular and laminated bone underneath.

cosmology The study of the origin and evolution of the universe. The current *big bang cosmology derives the observable universe from a singular event 15–20 billion years ago. Previous hypotheses include the steady-state theory, in which the expansion of the universe was due to the continuous creation of matter. The most celebrated of earlier world views was the Ptolemaic system, in which the Earth was the centre of the universe; this was superseded by the Copernican revolution, which displaced the Earth from its central position.

cosmopolitan distribution (pandemic distribution) Distribution of an organism that is world-wide or 'panendemic'. Apart from weeds, commensal animals, and some of the lower groups of cryptogams (plants reproducing by *spores or gametes, rather than seeds), there are relatively few organisms that occur on all six of the widely inhabited continents. *See also* FAUNAL REALM.

Cosmorhaphe A complex grazing *trail found in deep-water *flysch deposits. It is a single, sinuous trail formed by a *deposit feeder which exhibited efficient feeding behaviour in an area of low productivity.

costa 1. In corals, an external projection of a *septum beyond a *corallite wall, forming a rib. **2.** A rib-like thickening of a shell, extending from the *umbo to the margins in brachiopod (*Brachiopoda) and lamellibranch (*Bivalvia) shells. In gastropods (*Gastropoda) the thickening of the shell usually runs either axially or spirally, and thus there may be axial or spiral costae.

Costonian A *stage of the *Ordovician in the Lower *Caradoc, overlain by the *Harnagian.

cotectic curve In a system of three mineral components represented by a block diagram, the *liquidus surface forms a trough which defines the composition of the *melt having the lowest freezing temperature. The cotectic curve is the axis of this trough.

cotectic surface Curved surface in a *quaternary system which defines the temperature range over which two or more solid *phases crystallize simultaneously from a liquid. Analagous to a cotectic line in a *ternary system.

coterminous With shared boundaries. In stratigraphy, applied, for example, to a time line that is common to *stratigraphic units of different *rank.

cotidal line Line joining points at which given tidal levels (such as mean high water or mean low water) occur simultaneously. The lines are shown on certain hydrographic charts. The same information is contained in tide tables, where the data are given as differences from the times of high or low water at a 'standard port'.

Cotylosauria *Anapsid 'stem reptiles' which appeared in the *Carboniferous, having diverged from *labyrinthodont *amphibians. They flourished in the remainder of the *Palaeozoic, but rapidly dwindled to extinction in the *Triassic.

coulée flow A very thick, and relatively short, blocky *lava flow formed from viscous lava, usually *rhyolite or *dacite in composition, which has been extruded from a *volcanic cone or *fissure. The short flows can be up to 100 m thick and hence have a morphology lying between that of a *dome and a flow. (When used in the French literature, the term 'coulée' is applied to lava flows in general.)

Coulomb failure criterion The simplest relationship between *shear stress (τ) acting parallel to a plane and *normal stress (σ) acting perpendicular to the plane at the point of failure. The relationship $\sigma = \sigma_o + \tau\tan\phi$, where σ_0 is the normal stress after failure and ϕ is the angle of failure, implies that σ_2 (*see* PRINCIPAL STRESS AXES) does not affect brittle fracture strength.

coulter counter An electronic measuring device used mainly for determining the grain-size distribution in silt- and clay-sized *sediments (*see* PARTICLE SIZE). The coulter counter measures the volume of the individual particles in a suspension and is particularly useful for determining the size of mud flocculates (*see* FLOCCULATION). It can also be used for sand-sized sediments.

counterpoised beam balance (Walker's steelyard) A long, graduated, horizontal beam which is pivoted near one end on a vertical shaft, permitting the beam (arm) to swing freely. It is used to weigh objects, by attaching them to the short end of the arm and balancing their weights with counterweights which can be moved along the graduated arm, and it can also be used to determine the densities of minerals.

country rock The rock which has been intruded into and/or surrounded by a *plutonic *igneous intrusion.

couple Two equal, parallel forces which act in opposite directions in the same plane but not along the same axis.

coupled substitution The substitution of two or more elements in a crystalline compound, which maintains electrical neutrality. In forming a *solid solution series, ionic size is more important than *ionic charge, as this can be compensated for elsewhere. For instance, when a *plagioclase feldspar solid solution series forms, in passing from albite ($NaAlSi_3O_8$) to anorthite

($CaAl_2Si_2O_8$), Al^{3+} replaces Si^{4+}. This leaves a negative charge that is balanced by the (coupled) substitution of Ca^{2+} for Na^+.

Couvinian *See* EIFELIAN.

covalent bond Bond in which a pair (or pairs) of electrons is shared between two atoms. The bond is often represented by drawing a single line between the symbols of the two atoms that have bonded together. Sometimes the bonding is between atoms of different elements (e.g. hydrogen chloride, H–Cl), and sometimes between atoms of the same element (e.g. fluorine, F–F). The name 'molecule' is used to describe any uncharged particle containing covalently bonded atoms. *See also* HYDROGEN BOND; IONIC BOND; and METALLIC BOND.

covalent compound A compound with *covalent bonds.

covalent radius The atomic radius of an atom (e.g. carbon), determined in a covalent compound by a technique which can be used only for structures in which that atom (carbon) is *covalently bonded. Measurement of the distance between planes of atoms (the 'd-spacing') gives the sum of the radii (the bond length) of the two covalently bonded atoms in two planes. If all the atoms in the compound are of the same element this gives the value of the radius of an atom in each plane (e.g. the bond length in *diamond (pure carbon) is twice the radius of an individual carbon atom). If atoms of different elements are covalently bonded, the bond length of one of the atoms can be derived from the sum of the two different bond lengths only if the bond length of the other atom is known.

covariance In statistics, a measure of the association between two variables. Covariance is calculated as the difference between the average product of corresponding values in the two data sets and the product of the *means of the two data sets. A value of zero indicates no relationship between the data sets. The covariance value can be used to calculate *linear regression and *principal components.

covellite *Mineral, CuS; sp. gr. 4.6; *hardness 2; dark blue; *sub-metallic *lustre, may show an iridescent tarnish; found *massive or in thin plates; occurs in the zone of *secondary enrichment overlying copper-rich mineral deposits.

cow-dung bomb A *volcanic bomb whose name is derived from its characteristic shape.

coxa The uppermost joint of the leg of an insect.

cpx *See* CLINOPYROXENE.

crabs *See* MALACOSTRACA.

crachin Condensation as low cloud or fog, often with drizzle, which is frequent in spring in coastal areas of the Gulf of Tonkin and southern China. The weather results from the *advection of warm air over a cold surface, or the mixing of *air masses at the surface.

cracking Breaking down or decomposition of large, more complex molecules, particularly *hydrocarbons, into small simple molecules, usually by heating. The process can occur naturally, resulting in older deposits containing lighter crude oil than those of younger fields. Commercially, cracking is carried out in oil refineries.

crag Shelly sand.

crag and tail Land-form consisting of a small rocky hill (crag) from which extends a tapering ridge of unconsolidated debris (tail). The crag is a residual feature left by selective glacial erosion, while the tail is drift-deposited by ice on the lee side of the obstacle.

Craniata (Vertebrata; vertebrates; phylum *Chordata) The subphylum of animals that have a bony or cartilaginous skull and a dorsal vertebral column. It includes the fish, *amphibians, *reptiles, *birds, and *mammals, which appear successively in the fossil record, starting in the *Ordovician. *See also* BONE; and CARTILAGE.

cranidium *See* CEPHALON.

cranium (skull) In the axial skeleton of vertebrates, the bony structure that encases the brain. The cranium comprises three parts: dermatocranium; chondrocranium; and splanchnocranium. The dermatocranium includes the roof of the cranium, the area around the orbits, and the jaw. The chondrocranium includes the floor of the cranium. The splanchnocranium gives rise to the visceral skeleton (gill arches and derivatives such as larynx and trachea).

crater 1. General term for a circular, funnel-shaped depression, up to 1 km in diameter, produced by volcanic processes by which gases, *tephra, and *lava are or have been ejected. Several types are recognized: a crater at the summit of a volcanic cone marks the site of *magma degassing and ejection of material; a *maar, often occupied by a lake, results from explosive activity; and a *caldera is a large volcanic depression greater than 1 km in diameter. **2.** Near-circular depression produced by the impact of an extraterrestrial body, e.g. Meteor Crater, Arizona. *Meteorite craters are formed by the explosion outward and upward of material compressed and heated strongly by the energy of impact, and so usually are circular at the time they form. They are characterized by topographically raised rims and by *ejecta blankets which show inverted stratigraphy with respect to the target rocks. *See also* SHATTER CONES.

crater counting *See* CRATER DENSITY STUDIES.

crater density studies (crater counting) The establishment of the relative age of a portion of a planetary or *satellite surface from the observed density of *meteorite impact craters. If a calibration is available for the meteoric flux (e.g. from dated lunar surfaces), *absolute ages may be obtained by this technique. Problems with the method include the identification of primary from secondary craters, and whether the surface is saturated with craters.

Cratered Plains *See* MARTIAN TERRAIN UNITS.

Cratered units *See* MARTIAN TERRAIN UNITS.

crateriform Having a crater-like form: a circular depression with a raised rim, surrounded by an *ejecta blanket.

craton (shield; adj. cratonic) Area of the Earth's *crust, invariably part of a continent, which is no longer affected by orogenic activity. This stability has existed for approximately 1000 Ma. A classic example is the Canadian Shield.

creep 1. Slow downslope movement of the *regolith over hillslopes, due to gravity. The necessary disturbance of the regolith may be due to freezing and thawing; to expansion and contraction (resulting from temperature change or from wetting and drying); to additional weight and lubrication by water; or to the activities of burrowing animals. **2.** *See* CREEP MECHANISMS. **3.** The behaviour of *minerals under low differential *stress over long periods of time. Typically there is an initial stage of transient creep (*primary creep) with viscoelastic strains, which changes progressively to a state of purely viscous strain, until the mineral ruptures in a final (tertiary) stage of creep.

creep mechanisms Mechanisms by which materials deform at the Earth's surface or, more commonly, at depth. These can be: (*a*) cataclastic, in which individual grains or fragments physically rotate or glide past one another (*see* cataclasis); (*b*) dislocation creep, by a gliding motion along crystalline dislocations; (*c*) grain-boundary sliding; (*d*) *recrystallization; and (*e*) *diffusion of individual atoms. Each process is dependent on the stress, temperature, and duration of the stress.

creep strength The strength of a rock which is undergoing long-term *creep processes. It is the threshold value, beyond which creep gives way to permanent rupture, and is virtually synonymous with *fundamental strength.

Crenarchaeota (domain *Archaea) The less derived (*see* APOMORPH) of the two kingdoms of the Archaea, composed principally of extreme *thermophiles and *psychrophiles. Members of the Crenarchaeota show a greater genetic similarity to those belonging to the *domains *Eucaryota and *Bacteria than do those of the *Euryarchaeota.

crenulation cleavage *See* CLEAVAGE.

Creodonta (class *Mammalia) The more ancient of the two placental mammalian carnivorous orders, an extinct order comprising two families (Oxyaenidae and Hyaenodontidae), which appeared in the late *Cretaceous and dwindled to extinction in the *Pliocene. Oxyaenids (e.g. *Oxyaena*) were rather weasel like, although some of them were large (*Patriofelis* was the size of a bear). Hyaenodonts had narrower skulls, longer legs, and well-developed *carnassials, and diversified into forms reminiscent of the dogs, cats, and hyenas. Only the hyaenodonts survived the Eocene, and filled the role of scavengers until they were displaced by the modern hyena. Creodonts were small-brained and slow-moving, and are not closely related to modern carnivores.

creodont-like teeth Teeth resembling those of the *Creodonta, carnivorous mammals of the early *Tertiary (although the first representatives appeared in the late *Cretaceous), in which *carnassials were formed by the molars, rather than premolars and molars as in modern *Carnivora.

crepuscular rays Beams of sunlight made visible by haze in the atmosphere, and seen where rays penetrate gaps in clouds such as *stratocumulus; this effect is called 'Jacob's ladder'. In other cases rays from a low Sun diverge upwards above cumuliform cloud.

crescent-and-mushroom A two-dimensional *outcrop pattern, produced by the superposition of two *fold systems, whose forms resemble alternating crescents and mushrooms. The pattern is sometimes called 'dome–crescent–mushroom'. *See* INTERFERENCE PATTERN.

Cressida (Uranus IX) One of the lesser satellites of *Uranus, with a diameter of 33 km. It was discovered in 1986.

crest *See* CRESTAL PLANE.

crestal plane A plane or surface which contains the highest points (crests) of all beds within a folded sequence.

crest line A line on a folded layer which is common to and links the highest points on that layer. The term is employed most usefully in respect of plunging *folds where the *axial trace and crest line occupy different positions.

Cretaceous Third of the three *periods included in the *Mesozoic Era. It began approximately 145.6 Ma ago and ended about 65 Ma ago. It is noted for the deposition of the chalk of the White Cliffs of Dover, England, and for the mass extinction of many invertebrate and vertebrate stocks. Among these were the *dinosaurs, mosasaurs, *ichthyosaurs, and *plesiosaurs. *See* K/T BOUNDARY EVENT.

crevasse 1. Deep fissure in the surface of a glacier, caused when tensile stress overcomes the shear strength of ice in the brittle upper few metres. Appropriate tensile stresses are typically developed when a glacier moves over a convex-up slope. **2.** A breach in a *levée along the bank of a river through which flood water may flow and produce *crevasse deposits (2) (crevasse splay).

crevasse deposit 1. The gravelly or sandy sediment infill of a *crevasse in glacial ice. **2. (crevasse splay)** The deposit generated by a river crevasse event. It is sheet-like in geometry, thinning away from the side of the breach in the river bank, and characterized by rapidly deposited sands, fining upwards to a muddy top, produced by the waning flow of a flood event.

crevasse splay *See* CREVASSE DEPOSIT.

Crinoidea (crinoids; subphylum Crinozoa; phylum *Echinodermata) The most primitive living class of echinoderms, whose members are either stalked (sea lilies) or unstalked (feather stars). The body is contained within a cup-like *calyx, composed of regularly arranged plates, consisting of a lower dorsal cup which is covered by a dome (the *tegmen). There are usually five plated and branching arms (brachial processes, or *brachia) that articulate freely with the calyx. The upper surface contains the mouth and anus. There are tube feet along each arm with a median food groove between them leading to the mouth. The stem, when present, consists of a column of calcite discs (ossicles or *columnals) each with a central hole (lumen) for extensions of the soft parts. All *Palaeozoic forms were stemmed (sometimes of considerable length), but most modern forms are free swimming. They first arose in the Lower *Ordovician, and fossil crinoids are an important constituent of Palaeozoic *limestones.

crinoids *See* CRINOIDEA.

cristobalite A high temperature form of *quartz SiO_2 which has a stability field at atmospheric pressure above 1470 °C and below 1713 °C, at which latter temperature cristobalite melts and the quartz *liquidus is reached; sp. gr. 2.32; occurs as fine aggregates in cavities in *basalts and in some thermally metamorphosed (*see* CONTACT METAMORPHISM) *sandstones.

critical angle (Θ_c) The *angle of incidence at which a refracted ray grazes the interface between two media whose velocities are V_1 and V_2, such that, according to *Snell's law, $\sin\Theta_c = V_1/V_2$, provided that V_2 is greater than V_1.

critical damping (μ_c) The minimum amount of *damping which will stop oscillation or movement.

critical erosion velocity *See* CRITICAL VELOCITY.

critical flow Critical flow occurs when the flow velocity in a *channel equals the wave velocity generated by a disturbance or obstruction. In this condition the Froude number (Fr) = 1. When the wave velocity exceeds the flow velocity (Fr is less than 1) waves can flow upstream, water can pond behind an obstruction, and the flow is said to be subcritical or tranquil. When Fr is greater than 1 waves cannot be generated upstream and the flow is said to be supercritical, rapid, or shooting. In this condition a standing wave is formed over obstructions in the river bed. In nature, supercritical flow is found only in rapids and waterfalls, but it is often created artificially by *weirs and *flumes with the aim of measuring *discharge.

critical point The temperature above which a gas cannot be liquefied, regardless of the pressure, and becomes a supercritical fluid with the molecular freedom of a gas, but the density of a liquid. The critical point for water is 374 °C; at that temperature the gas will liquefy under a pressure of 0.022 GPa.

critical reflection The reflection, normally at a very large angle, observed when the *angle of incidence and *angle of reflection of a ray incident on a surface are both equal to the *critical angle.

critical velocity (critical erosion velocity) The minimum velocity of a flowing fluid that is required in order to entrain a particle. *See* HJULSTROM RELATIONSHIP.

CRM *See* CHEMICAL REMANENT MAGNETIZATION.

crocidolite *See* ASBESTOS; and RIEBECKITE.

Crocodilia (class *Reptilia) Order of the 'ruling reptiles' (*Archosauria) which includes the crocodiles, alligators, caimans, and gavials (gharials). They are derived from *thecodontian ancestors and are closely related to the *dinosaurs and *pterosaurs. They are first recorded from *Triassic rocks and were the only *archosaurs to survive the *Mesozoic Era, although strictly speaking the birds (*Aves) are also archosaurs.

Croixian A *series of the Upper *Cambrian of N. America, equivalent to the *Merioneth.

Cromerian **1.** A northern European *interglacial *stage dating from about 0.6–0.55 Ma ago. It coincides approximately with the *Günz/Mindel interglacial of the Alps. **2.** Temperate deposits of Middle *Pleistocene age found at West Runton, Norfolk, England. Estuarine *sands and silts, and freshwater peat (Upper Freshwater Bed), with a temperate-forest flora, are succeeded by glacial deposits. It is not possible to correlate them with continental European deposits.

Crommelin A *comet with an orbital period of 27.89 years; *perihelion date 1 September 1984; perihelion distance 0.743 AU.

Cromwell current *See* EQUATORIAL UNDERCURRENT.

cronstedite *See* CHAMOSITE.

cross-bedding *see* BEDDING; and CROSS-STRATIFICATION.

cross-correlation In comparing one wave-form with another, the *correlation of two digital traces (i.e. wave-forms that have been *digitized) which are similar but not the same, with one being delayed in time with respect to the other. The operator slides one trace past the other in small time steps (called delays or lags) and at each step the elements of the traces are multiplied together, term by term, and the products added. The maximum value (almost equal to unity) of this cross-correlation is obtained when the two traces are in closest alignment with each other. A value of −1 means the wave-forms are identically matched but opposite in phase; a value approaching zero indicates low degrees of similarity. The method is extremely useful for detecting wave-forms swamped by noise and, in particular, in the analysis of *Vibroseis records. *See also* AUTOCORRELATION.

cross-cutting relationships *See* LAW OF CROSS-CUTTING RELATIONSHIPS.

cross-dating The matching of tree-ring width patterns and other properties among the trees and fragments of wood from a particular area. This enables the year in which each ring was formed in living trees and recent stumps to be determined accurately, the presence of false rings or the absence of rings in individual specimens being made apparent. By matching ring series from living specimens with those from older (e.g. constructional) timbers, the chronology may be extended backwards in time.

crossed nicols *See* CROSSED POLARS.

crossed polars (crossed nicols, XPL, xpols) The situation in a reflected- or transmitted-light microscope when the pieces of *Polaroid of both the *analyser and sub-stage *polarizer are inserted into the light path through the microscope. The pieces of Polaroid are aligned at right angles to each other so that no light is transmitted to the observer, but a *mineral (or rock) *thin section present on the microscope stage interferes with the light rays into which the light is split by the mineral and produces interference leading to *interference colours. *See also* BIREFRINGENCE.

cross-hairs *See* CROSS-WIRES.

cross-lamination The smallest structure formed of inclined laminations (*see* LAMINA), with thicknesses measured in millimetres. *See* CROSS-STRATIFICATION.

Crossopterygii (class *Osteichthyes) Subclass of *bony fish comprising both fossil and living lobe-finned or tassel-finned fish, including the *Coelacanthiformes and *Rhipidistia. The former are well known from *Palaeozoic and *Mesozoic rocks, but were thought to have become extinct by the end of the *Cretaceous until living specimens were netted this century in the Indian Ocean. The Rhipidistia did become extinct, although not before they gave rise, in the *Devonian, to the *amphibians. The Crossopterygii are characterized by the fact that all fins (except the tail-fin) are based on movable stalks or lobes. The tail fin is either *heterocercal or *diphycercal.

cross-over distance (x_c) The distance on a seismic *refraction survey *time-distance chart at which the *travel times of the *direct and refracted waves are the same. This distance also marks the point when the refracted wave overtakes, and thus arrives before, the direct wave. The cross-over distance is related to the refractor depth, h, and the velocities of the overlying medium and the refractor, V_1 and V_2 respectively, such that $x_c = 2h[(V_2 + V_1)/(V_2 - V_1)]^{1/2}$. The value of x_c will always be greater than twice the refractor depth.

cross set *See* CROSS-STRATIFICATION.

cross-stratification A family of primary *sedimentary structures formed by the migration of the slip-faces of rippled bedforms or of *bars. It is characterized by inclined laminations (*foresets) bounded by planar surfaces (planar or tabular cross-stratification), or by scoop-shaped surfaces (trough cross-stratification). The foresets *dip at the *angle of repose of the sediment on the *ripple slip-face and are oriented in the direction of migration of the ripple (*see* PALAEOCURRENT ANALYSIS). Tabular cross-stratification is produced by the migration of straight-crested, asymmetrical ripples or *sand waves. Trough cross-stratification is generated by the migration of *linguoid ripples or *dunes. The term 'cross-lamination' is applied to cross-stratification formed by the migration of ripples; 'cross-bedding' is used for cross *strata formed by the migration of large-scale forms such as dunes, sand waves, or bars. The term 'cross set' is used to define the cross-stratification preserved between any upper and lower bounding surface. Where the original bedform which produced the cross set is preserved and forms the upper bounding surface to the set, the term 'form set' is used. A number of cross sets preserved within a single bed are called a 'coset'.

cross-well seismic A technique, used in prospecting for crude oil and natural gas, in which a powerful sound is produced at different levels in one well and its vibrations recorded in one or more other wells. The character of the received vibrations provides information about the rock structures between the wells.

cross-wires (cross-hairs) Two very thin, black lines or pieces of wire set at right angles to each other in the *eyepiece of a microscope. They are normally positioned N–S and E–W, parallel to the alignment of the *polarizer and *analyser. Many optical properties, e.g. *extinction angles, are measured with respect to the position of the cross-wires.

Crotonian *See* QUATERNARY.

crotovina *See* KROTOVINA.

crown group In *cladistic analysis, the extant taxa descended from a common ancestor. *Compare* STEM GROUP.

crude oil *See* PETROLEUM.

Crudinian The basal *stage in the *Devonian of Australia, underlain by the *Silurian, overlain by the *Merionsian, and roughly contemporaneous with the *Gedinnian Stage in Europe.

crumb structure Type of soil structure in which the structural units or *peds have a spheroidal or crumb shape. Crumb structure is more often found in porous than granular organo-mineral surface *soil horizons, and provides optimal pore space for soil fertility.

crura *See* CRUS.

crus (pl. crura) Part of the calcified brachial support (*brachidium) in *brachiopods which attach the brachium to the interior *shell beak region of the brachial valve. *See* BRACHIA.

crushing In mineral processing, the breaking up of large particles by compression between two surfaces. Primary crushing produces pieces 100–50 mm in diameter; secondary crushing reduces those to 10 mm.

crust 1. The thin outermost solid layer of the Earth. It represents less than 1% of the Earth's volume, and varies in thickness from approximately 5 km beneath the oceans to approximately 60 km beneath mountain chains. Most of the *terrestrial planets have a solid surface, generally considered to be of different composition to the underlying, higher-density rocks, and regarded as crust. *See also* CRUSTAL ABUNDANCE OF ELEMENTS; CONTINENTAL CRUST; and OCEANIC CRUST. **2.** A surface soil layer, sometimes slightly cemented with calcium carbonate, silica, or iron oxide, which may be from a few millimetres to many tens of millimetres in thickness but which is always harder and more compact than the soil below. Crusts are now being produced by mechanical action or *pedogenesis, mainly in arid environments; less commonly they are relict or fossil features exhumed on the soil surface.

Crustacea (crustaceans; phylum *Arthropoda) Diverse subphylum of mandibulate arthropods, the body usually divided into three parts: head, thorax, and abdomen. In some crustaceans (e.g. crayfish) the head and thorax may be joined to form the cephalothorax. The head bears two pairs of antennae, one pair of *mandibles, and two pairs of *maxillae. The limbs are *biramous, and are adapted for a wide range of functions. Closely placed *setae on the limbs function as filters in filter-feeding species. Respiratory gills are situated on the appendages, but vary greatly in location and number; they are absent only in very small species. In addition to the antennae, sense organs include a pair of compound eyes, and a small, dorsal, median, nauplius eye, comprising three or four closely applied ocelli (clusters of photoreceptors). The nauplius eye, characteristic of crustacean larvae, is absent in many adults; and some groups lack the compound eyes. Mainly marine, but there are many freshwater species, and a relatively small number have invaded the land. Four classes of crustaceans have an important fossil record. The Malacostraca (crabs, lobsters, woodlice, etc., *Cambrian to Recent) includes the earliest crustaceans of the subclass Phyllocarida. The Branchiopoda (similar to modern water fleas, Lower *Devonian to Recent) are valuable *index fossils in non-marine strata. The *Cirripedia (barnacles) occur from Upper *Silurian to Recent, and the *Ostracoda from Lower *Cambrian to Recent. The living class Cephalocarida (e.g. *Hutchinsonella*) is thought to be closest to the ancestral crustacean stock, but the group is without any unequivocal fossil representative.

crustaceans *See* CRUSTACEA.

crustal abundance of elements The Earth's *crust has an average density of 2800 kg/m^3 and a thickness varying from about 30 km below the continents (up to 60 km beneath some mountains) to 5 km beneath the oceans. By mapping the major rock types and averaging their composition the abundance of elements can be estimated. The crust is enriched in *incompatible elements (e.g. K and Rb) as well as *lithophile elements, but a few elements predominate, especially in *silicate minerals, while some *ore metals are rare (e.g. Cu and Sn). Because the crust was formed from material extruded from the Earth's mantle, it is to be expected that the mantle is depleted in 'crustal' components. Oxygen (O) constitutes almost 50% of the Earth's crust by weight and is the most abundant element. Other major elements include: silicon (Si), which is the second most abundant, constituting 27.72% of the crust by weight; aluminium (Al) third; sodium (Na); magnesium (Mg); calcium (Ca); and iron (Fe). Other elements, including such desired metals as gold (Au), silver (Ag), and platinum (Pt), are rare in the crust.

Cruziana The name given to the *tracks made by an animal (some by Trilobita) as it crawled over the sea floor. The tracks are bilobed in appearance, divided along the mid-line, and each half is scored by fine grooves created by the action of the walking limb. *See* REPICHNIA.

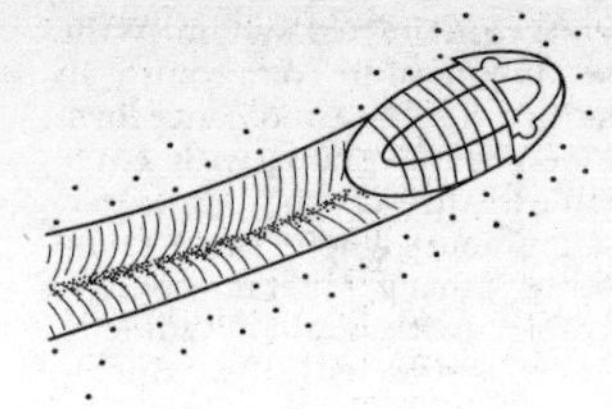

Cruziana

cryergic Applied to the work of ground *ice, and so includes *frost heaving, *frost wedging, and the thaw processes. The term has been used as a synonym for *periglacial.

cryic *See* PERGELIC.

cryogenic Applied to: **1.** features or materials generated by the action of ice; **2.** instruments operating a few degress above absolute zero (0K = –273.15 °C).

cryolite *Mineral, Na_3AlF_6; sp. gr. 3.0; *hardness 2.5; *monoclinic; colourless to white, sometimes brown to reddish; white *streak; *vitreous to greasy *lustre; *crystals rare, but cubic-like and hexahedral, also occurs *massive; no *cleavage, basal parting and poor prismatic parting; occurs in *pegmatites that have been enriched with fluorine, and in association with *siderite, *quartz, *galena, *chalcopyrite, *fluorite, and *cassiterite. The powder becomes almost invisible in water due to its low *refractive index. Synthetic cryolite is used as a flux in the production of aluminium and enamels.

cryonival Applied to the set of geomorphological processes comprising *cryergic and *nival mechanisms.

cryopediment A bench-like land-form, cut indiscriminately across bedrock, and confined to past or present *periglacial environments. Its position on the lower part of a hillslope is the main criterion for distinguishing it from a *cryoplanation terrace. It is the periglacial analogue of the warm desert *pediment.

cryoplanation The reduction of relief to a gently undulating land surface under periglacial conditions. Equivalent to extensive *altiplanation.

cryosphere That part of the Earth where the surface is frozen, comprising the area covered by *ice sheets and *glaciers, *permafrost regions, and sea areas covered by ice, at least in winter.

cryoturbation *See* GELITURBATION.

cryovolcanism Volcanism occurring at low temperatures, where liquid erupts through an overlying crust of ice. The liquid probably consists of brine containing magnesium and sodium sulphates; at extremely low temperatures ammonia may also be present. Cryovolcanism is known to be active on *Enceladus, *Europa, and *Ganymede.

Cryptic A *period of the *Priscoan, dated from about 4550–4150 Ma (Harland et al., 1989).

cryptocrystalline Applied to a very fine aggregate of crystals in an *igneous rock, and to minerals in which the individual crystals are too fine to be distinguished even under a petrological microscope. They will produce a diffraction pattern with X-rays. Rapid cooling of a *magma as it is extruded as a lava often produces cryptocrystalline aggregates of minerals.

cryptodome An uplifted area, approximately dome-shaped, caused by the intrusion of viscous *magma.

Cryptodonta (phylum *Mollusca, class *Bivalvia) Subclass of bivalves most of which have thin, equivalve shells composed of *aragonite. They have an *amphidetic to *opisthodetic external ligament. The *hinge plate is narrow or absent, with most forms having a toothless hinge margin; some members are *taxodont. They have an *infaunal mode of life. They first appeared in the *Ordovician.

cryptoperthite *See* PERTHITE.

Cryptozoic (Archaeozoic, Azoic) A period of time equivalent to the *Archaean, i.e. prior to 2500 Ma (Van Eysinga, 1975).

crystal A homogeneous, ordered solid, having naturally formed plane faces and a limited chemical composition. Crystals have definite geometric forms that reflect the arrangement in *lattices of the atoms of which they are composed. *See* CRYSTAL CLASS; and CRYSTAL SYMMETRY.

crystal class (point group) Crystals are formed by the repetition in three dimensions of a *unit cell structure, defined by lattice points in space. There are only 32 ways to arrange the space lattices in terms of symmetry elements, and these are called the 'crystal classes'.

crystal face One of the relatively flat

surfaces by which a crystal is bounded. Faces are produced naturally during the process of crystal growth. Cut and polished *gemstones are bounded by plane faces which are often produced artificially and which, therefore, are not crystal faces.

crystal-field theory A theory describing the behaviour of elements with partly filled d- or f-orbitals. In *geochemistry it is particularly concerned with the crystal-field (*ligand-field) effects on the first transition metal *ions, i.e. Sc, Ti, V, Cr, Mn, Fe, Co, Ni, and Cu. If the *anions surrounding a *cation are considered as point charges then the electrostatic interactions between them will vary according to: (*a*) ionic distance; (*b*) the strength of the charge; and (*c*) the *coordination number (how anions are distributed around the cation). Some of the five d-orbitals will have maximum electron density nearer to the anion than others. Interactions between cation and anion will affect the energies of the d-orbitals and cause a separation of energy levels between orbitals (crystal-field splitting).

crystal group *See* CRYSTAL SYSTEM.

crystalline A general term, applied to *metamorphic or *igneous rocks formed by the process of crystallization from solid or liquid precursors (although certain *sedimentary rocks, e.g. some *limestones, may also be made up of crystalline grains, cemented by crystalline *cement).

crystalline carbonate A *carbonate rock whose original sedimentary texture has been recrystallized (*see* recrystallization). *See* DUNHAM CLASSIFICATION.

crystalline limestone *See* DUNHAM CLASSIFICATION.

crystalline remanent magnetization *See* CHEMICAL REMANENT MAGNETIZATION.

crystallite A microscopic, often skeletal *crystal which represents the initial form of crystalline material just after *nucleation has taken place in a *magma. Crystallites are usually preserved in volcanic *glass, which represents a quenched magma. *Quenching is the only way in which crystallites can be preserved.

crystalloblastic A metamorphic *texture characterized by the mutual interference of polygonal grains which meet at approximately 120° triple junctions. The texture is produced during solid-state *crystal growth as a result of increasing the pressure on or temperature in the rock system. As the grains grow, they have to compromise in their competition for space and hence form the typical polyhedral grains with triple junctions. The texture can be '*isotropic' (no grain alignment) where there is no directed *stress during *metamorphism, or '*anisotropic' (with grain alignment) where there is a directed stress during metamorphism.

crystallographic axes Axes used to define the position of a *crystal face in space by the intercepts of the face on three (or four) imaginary lines.

crystallography The study of crystals, including their form, structure, habit, and symmetry.

crystal symmetry In well-formed crystals, the symmetrically arranged faces reflect the internal arrangement of atoms. The symmetry of individual crystals is determined by reference to three elements. The plane of symmetry (also called the 'mirror plane' or 'symmetry plane') is a plane by which the crystal may be divided into two halves which are mirror images of each other. The axis of symmetry is a line about which a crystal may be rotated through $360°/n$ until it assumes a congruent position; n may equal 2, 3, 4, or 6 (but not 1), depending on the number of times the congruent position is repeated. These correspond respectively to 2-fold (diad), 3-fold (triad), 4-fold (tetrad), and 6-fold (hexad) axes. The centre of symmetry is a central point which is present when all faces or edges occur in parallel pairs on opposite sides of the crystal. Using these elements of symmetry, crystallographers have recognized 32 *crystal classes and seven *crystal systems. Symmetry is highest (high symmetry) in the *cubic system, where many elements are repeated, and lowest (low symmetry) in the *triclinic system, where only a centre of symmetry may be present (i.e. there may be no plane or axis of symmetry).

crystal system (crystal group) One of the seven systems or groups into which crystals can be placed by reference to their *crystallographic axes. It is possible to arrange the 14 *Bravais lattices into seven systems (groups) each characterized by a primitive (*p*) lattice.

crystal twinning Feature, common to many single crystals, where the crystal *lattice is differently orientated in two or more parts of the crystal, e.g. twin plane, twin axis, etc., and simple or multiple twins. *See* ALBITE TWIN; CARLSBAD TWIN; COMPLEX TWINS; CONTACT TWIN; DEFORMATION TWINNING; GENICULATE TWIN; GROWTH TWINNING; INTERPENETRANT TWIN; MIMETIC TWINS; NORMAL TWINS; PARALLEL TWINS; PERICLINE TWINNING; SWALLOWTAIL TWINNING; TRANSFORMATION TWINNING; TWIN AXIS; TWIN LAW; and TWIN PLANE.

crystal zoning A *texture developed in *solid-solution minerals and characterized optically by changes in the colour or *extinction angle of the mineral from the core to the rim. This optical zoning is a reflection of chemical zoning in the mineral. For example, a *plagioclase can be zoned from a Ca-rich core to an Na-rich rim. Zoning results from the mineral's inability to maintain chemical equilibrium with a *magma during rapid cooling; the zonation represents a frozen picture of the *continuous reaction series for that mineral. Zoning can be of three types, the first two applying mostly to plagioclase feldspars. (*a*) Normal zoning is where the mineral is zoned from a high-temperature core composition to a low-temperature rim composition. (*b*) Reverse zoning is where a mineral is zoned from a low-temperature core composition to a high-temperature rim composition. (*c*) Oscillatory zoning is where the mineral chemistry continuously oscillates between high- and low-temperature compositions going from the core to the rim. *Compare* CORONA.

CST *See* CONSTANT-SEPARATION TRAVERSING.

CTD An instrument for measuring seawater conductivity (from which *salinity can be calculated), temperature, and depth (actually, pressure). A sensor unit is lowered through the water on the end of an electrical conductor cable which transmits the information to indicating and recording units on board ship.

cube A hexahedral crystal shape or form which can be referred to three axes of equal length which intersect at right angles.

cubic (isometric) One of the seven *crystal systems and the one with the highest number of symmetry elements present. It is characterized by four triad axes of symmetry, and requires three *crystallographic axes of equal lengths intersecting at right angles.

cubichnia A category of *trace fossils that mark the temporary resting places of various organisms.

cuesta Asymmetric land-form consisting of a steep scarp slope, and a more gentle dip (or back) slope. It is typical of areas underlain by strata of varying resistance that are dipping gently in one direction, and is intermediate between the flat-topped *mesa and *butte, and the more symmetric ridge form of the *hog's back.

culmination The highest antiformal point (*see* ANTIFORM) of a *crest line along all non-*cylindroidal folds.

cumec *See* DISCHARGE.

cummingtonite A *monoclinic, Mg-rich member $(Mg,Fe)_2(Mg,Fe)_5[Si_4O_{11}]_2(OH,F)_2$ of the calcium-poor *amphiboles, along with grunerite (Mg < Fe) with which it forms a series, similar to *anthophyllite–gedrite; sp. gr. 3.1–3.6; *hardness 5.5; dark green; occurs in amphiboles and some *intermediate *igneous rocks.

cumulate Applied to *igneous *intrusive rocks formed by the accumulation of crystals as a result of *gravity settling. It is typical of layered intrusions and common in

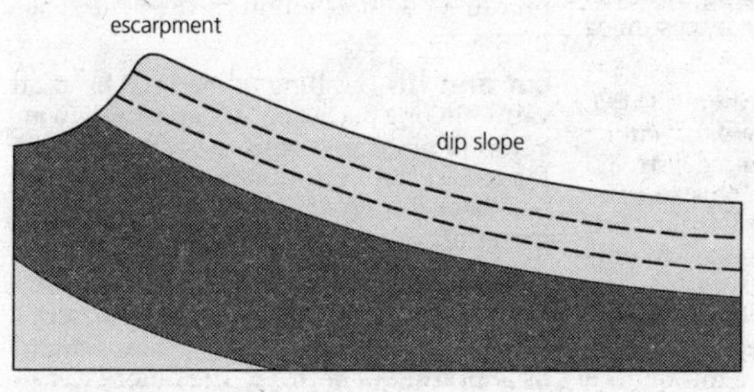

Cuesta

some differentiated *meteorites. The early-formed minerals are called 'cumulus' minerals and show a regular variation in composition with their height in the intrusion. *See* MINERAL LAYERING.

cumulative percentage curve A graphical plot in which size classes are plotted against the percentage frequency of the class plus the sum of the percentages in preceding size classes. When plotted on normal graph paper, the cumulative frequency curve resembles an S-shape. When plotted on a normal-probability scale, the cumulative percentage data appear in a series of straight-line segments, each with a different gradient.

cumulonimbus From the Latin *cumulus* meaning 'heap' and *nimbus* meaning 'rain', a cloud of bulging, dense form, often towering to great height in unstable air. Young clouds have distinctive fibrous or lined features; older, glaciated types, with abundant ice crystals, are lustrous. Typically, the upper parts are spread into anvil or plume features. The cloud base is dark and usually gives rise to precipitation, often with *virga. *See also* CLOUD CLASSIFICATION.

cumulus 1. From the Latin *cumulus* meaning 'heap', a dense, isolated, and clearly defined cloud with vertical growth in bulges or domes, and a flattened, darker base. Sharply outlined, bulging cloud tops indicate vigorous growth. Occasionally a more ragged form occurs. *See also* CLOUD CLASSIFICATION. **2.** *See* CUMULATE.

Cunninghamian A *stage in the *Devonian of Australia, underlain by the *Merionsian, overlain by the *Condobolinian, and roughly contemporaneous with the *Emsian and *Eifelian Stages of Europe.

cupola A small, dome-shaped, satellite intrusion (*see* INTRUSIVE) projecting upwards from the main body of a larger intrusion or *batholith.

cuprite (red copper ore) *Mineral, Cu_2O; sp. gr. 5.8–6.1; *hardness 3.5–4.0; *cubic; red to nearly black; brownish-red *streak; *adamantine to *sub-metallic *lustre; crystals usually octahedral and *acicular, but can be granular and *rhombododecahedral; *cleavage poor {111}; usually occurs in the oxidized zone of copper deposits, as a *secondary mineral, and associated with *malachite and *azurite. It is a minor *ore mineral for copper.

Curie point *See* CURIE TEMPERATURE.

Curie temperature (Curie point) The temperature at which thermal vibrations prevent quantum-mechanical coupling between atoms, thereby destroying any *ferromagnetism (in the wide sense). Typical Curie temperatures are 675 °C for *hematite and 575 °C for *magnetite.

current electrode An electrode by which an electrical current enters or leaves a conducting medium. Two current electrodes are used in a variety of *electrode configurations in *resistivity surveys (*vertical electrical sounding and *constant-separation traversing) and *induced-polarization surveys.

current meter An instrument for measuring the speed of flow in a watercourse. The most common type of current meters relate current speed to the rate at which an impeller is rotated by the flowing water.

Curvolithus A genus of *trace fossils comprising short, horizontal burrows close to the surface.

cuspate foreland Large, triangular area of coastal deposition, which is dominated by many shingle ridges, and is often terminated landward by poorly drained terrain. It is the result of a long episode of local marine aggradation under wave advance from two dominant directions. Dungeness, on the south coast of England, is a typical example.

cuspidate Having a sharp tip or point.

cutan 1. (clay films, clayskins, argillans, tonhäutchens) Deposited skin or coating of material on the surfaces of *peds and stones, which is usually composed of fine, *clay-like soil particles which have been moved down through the soil. **2.** A complex, insoluble biopolymer with water-proofing qualities found in the *cuticle of plants.

cut and fill Levelling of material by excavation in one place and its deposition in an adjacent place, to produce a uniform height for roads, railways, canals, etc. In mining, the back-filling of excavations by waste materials.

cuticle 1. The impervious covering to the outer walls of the epidermal cells of aerial plant organs, composed of *cutin, *cutan, or a mixture of both. **2.** The outer layer of an insect, secreted by epidermal cells; it has a complex structure that varies according to species. *See* SKELETAL MATERIAL.

cutin A complex biopolymer comprising a mixture of fatty-acid derivatives with water-proofing qualities found in the *cuticle of plants.

cutinite *See* COAL MACERAL.

cut-off (ox bow) Section of a river channel that no longer carries the main discharge. Its abandonment results from meander development associated with lateral channel migration across a *floodplain. Channel length is shortened by contact at the neck of a loop, which becomes a cut-off.

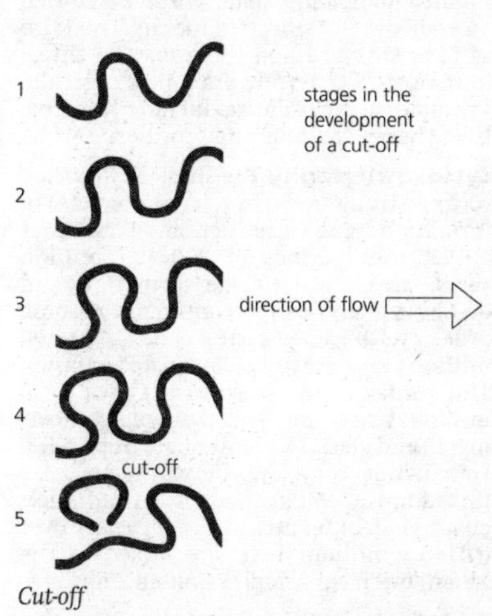

Cut-off

cut-off grade Lowest grade or assay value of ore in a deposit that will recover mining costs; the cut-off grade determines the workable tonnage of an ore.

cut-off high *Anticyclone isolated from the main subtropical belt of high pressure and around which the main flow of the upper westerlies is diverted, causing a blocking situation in middle latitudes.

cut-off low *Cold low in mid-latitudes (occasionally almost in subtropical latitudes) where it is cut off from the main subpolar belt of low pressure. Sometimes a cut-off low occurs with a *cut-off high over the higher latitudes, typically in *blocking situations. In summer, weather associated with such slow-moving lows is unsettled and thundery.

cut-off trench *See* DIAPHRAGM WALL.

cutting bar *See* CUTTING BOOM.

cutting boom (cutting bar) Part of a cutting machine used to undercut or overcut coal or other stratified material. The cutting machine consists of a rubber-tyred base and a cutter bar 3 m long and 380 mm wide, with a chain running along its edge and holding the *bits. It is driven by an electric motor.

Cuvier, Chrétien Frédéric Dagobert ('Georges'), Baron (1769–1832) Professor at the National Museum of Natural History in Paris, where he developed the discipline of comparative anatomy and applied it to the study of fossil quadrupeds in his *Recherches sur les ossements fossiles de quadrupèdes* (1812). Its *Discours préliminaire* (1811), in which he outlined a theory of multiple catastophes, became one of the most influential scientific treatises of the early 19th century. With *Brongniart he produced a map of the Paris basin, having used his fossil studies to work out the stratigraphy. Cuvier opposed Lamarckian evolution, arguing that species were stable and became extinct because of 'revolutions' (changes in sea level, etc.). *See* CATASTROPHISM; DILUVIANISM; and LAMARCKISM.

CVG *See* CONSTANT VELOCITY GATHER.

CVS *See* CONSTANT VELOCITY STACK.

cwm *See* CIRQUE.

cyanobacteria A large and varied group of bacteria which possess chlorophyll *a* (carried on specialized membranes (thylakoids) within the cells) and carry out *photosynthesis in the presence of light and air with concomitant production of oxygen. They do not have chloroplasts. Cyanobacteria were formerly regarded as *algae (division Cyanophyta) and were called 'blue-green algae'. Fossil cyanobacteria have been found in rocks almost 3000 Ma old and they are common as *stromatolite colonies in rocks 2300 Ma old. They are believed to have been the first oxygen-producing organisms and to have been responsible for generating the oxygen in the atmosphere, thus profoundly influencing the subsequent course of evolution. The organisms may be single-celled or filamentous, and may or may not be colonial. Some are capable of a gliding motility when in contact with a solid surface. Many species can carry out the fixation of atmospheric nitrogen.

Cyanophyta *See* CYANOBACTERIA.

cyanophyte Cyanobacterium (**blue-green alga**). *See* CYANOBACTERIA.

cycad *See* CYCADOPHYTA.

Cycadaceae In some classifications, a family comprising the extant members of the order Cycadales, now included in the division *Cycadophyta.

Cycadales In some classifications, an order of *gymnosperms comprising the cycads, now included in the division *Cycadophyta.

Cycadophyta (**cycads**) A division of *gymnosperms comprising plants with leaves and habit similar to those of palm trees, although some species are quite small. Most cycads bear large, coloured, female or male cones. Pollen grains have motile spermatozoa within them, which is a very primitive feature. Formerly they were much more important and, following their appearance in the *Permian, remained important members of the world's *Mesozoic floras. Their reduction was particularly marked in the Late *Cretaceous as they were progressively displaced by *angiosperm trees. The survivors are regarded as 'living fossils'.

cycle of erosion *See* DAVISIAN CYCLE.

cyclic sedimentation A style of sedimentation where the sequence of sedimentary *facies develops in a regular and repeated manner. The cycle of sediments may consist of two or more facies, and the cycles may be symmetrical or asymmetrical (e.g. A → B → C → D, A → B → C → D; or A → B → C → D → C → B → A, A → B → C → D → C → B → A. The cyclicity may be the product of repeated episodes of marine *transgression and *regression linked to climatic cycles (*see* MILANKOVICH CYCLES), or of the lateral shifting of depositional environments, or due to repeated cycles of tectonic activity in source areas.

cycling pool *See* ACTIVE POOL.

cyclogenesis Formation and strengthening of cyclonic air circulation, tending to form or deepen depressions. The process is associated with upper-air divergence over or near the frontal zone.

cyclolysis The processes of dissipation of cyclonic air circulation around a depression, or the weakening of lesser cyclonic features.

cyclone 1. The name given to a *tropical cyclone that develops in the Indian Ocean and Bay of Bengal. Cyclones usually travel north, on tracks that carry them over Bangladesh. **2.** *See* DEPRESSION.

cyclopean concrete *See* CONCRETE DAM.

cyclopel A laminated sediment, formed in a glaciomarine environment, that consists of layers of silt and mud. *Compare* CYCLOPSAM.

cyclopsam A laminated sediment, formed in a glaciomarine environment, that consists of layers of sand and mud. *Compare* CYCLOPEL.

cyclosilicate (**ring silicate, metasilicate**) Applied to the 'ring' structure of linked SiO_4 *tetrahedra in *silicate minerals. The ratio of Si to O is 1 : 3 and there may be three, four, or six SiO_4 tetrahedra linked together. Examples of cyclosilicates include *tourmaline, *beryl, *axinite, and *cordierite.

cyclostratigraphy The study of stratified rocks (*stratigraphy) in relation to cycles of formation and destruction. First-order cycles, over periods of 200–400 million years, are linked to the formation and breaking apart of supercontinents. Second-order cycles (super cycles), lasting 10–100 million years, are linked to *plate tectonics. Third-order cycles (mesothems), over 1–10 million years, are linked to plate movements and glacial and interglacial episodes. Fourth-order cycles (*cyclothems) are linked to the *Milankovich solar radiation curve. Fifth-order cycles (minor cycles), over 0.01–0.2 million years, are linked to the Milankovich curve and astronomic forcing.

cyclothem Unit, or given set of deposits, laid down as a result of either cyclic or rhythmic sedimentation. In a cyclic sequence the cyclothem would represent the whole succession, e.g. 1234321, whereas in a rhythmic sequence the cyclothem units would be repeated, e.g. 12341234. *See* CYCLOSTRATIGRAPHY.

cylindrical *See* SOLITARY CORALS.

cylindrical trend *See* BOXCAR TREND.

cylindroidal fold A *fold which maintains its two-dimensional profile in a third dimension, like a cylinder (as compared to a circle).

cyrenoid Applied to a type of *heterodont *dentition that occurs in bivalve molluscs (*Bivalvia), in which there are three *cardinal teeth present in each valve. *Compare* LUCINOID.

cyrtoconic Applied to the conch of a *cephalopod when it is a curved, tapering cone.

Cystoidea (cystoids; subphylum Blastozoa, phylum *Echinodermata) Extinct group of echinoderms, ranking in some classification schemes as a class or superclass, and ranging from the Lower *Ordovician to the Upper *Devonian. The body is spherical or ovoid, covered with a plated *theca, and was usually attached to the substratum either directly or by a short stalk. The thecal plates are calcitic and perforated by characteristic pore structures that functioned in respiration. The plates are arranged in circlets but vary in number from around 13 to more than 100, giving an irregular appearance. The mouth, on the upper surface, is surrounded by *ambulacral food grooves that in some cystoids were equipped with food gathering appendages called *brachioles. These are rarely preserved. In many cases *radial symmetry is poorly developed.

cystoids *See* CYSTOIDEA.

Cytherean An alternative name for 'venusian'. In Greek mythology, Cythera was an alternative name for Aphrodite, the goddess of love, identified with the Roman Venus. It is derived from Cythera, in Cyprus, near the place where, according to the legend, Aphrodite was born in the sea.

dacite A light-coloured, fine-grained *igneous rock containing 63–70 wt.% SiO_2, as well as *plagioclase feldspar, *alkali feldspar, *quartz, *biotite, and *hornblende as *essential minerals, and *sphene, *apatite, and *magnetite as *accessory minerals. Plagioclase feldspar dominates over alkali feldspar in a 2:1 ratio. Dacites are the volcanic equivalents of *granodiorites and are found as members of the *calc-alkaline series, typically developed as eruptive products on the continental side of oceanic-continental *plate *subduction zones. They are often surprisingly coarse-grained for lavas.

Dactyl *See* IDA.

Dactylodites ottoi A feeding *trace fossil (*fodinichnia) in the form of a J-shaped burrow.

Daedalus *See* SPREITEN.

dagalas *See* KIPUKA.

Dalmation-type coast A *Pacific-type coast that has been partially drowned by a rise in sea level, producing many long, narrow islands lying parallel to the shore.

Dalradian The last, or youngest, stratigraphic unit of the *Precambrian of Scotland and Ireland.

Dalslandian A *stage of the Middle–Upper *Proterozoic, from about 1600–650 Ma ago, and equivalent to the *Jotnian (Van Eysinga, 1975).

Dalslandian orogeny (Gothian orogeny, Gothic orogeny, Sveconorwegian orogeny) An eastward continuation of the *Proterozoic *Grenvillian phase of mountain building that affected what are now southern Sweden and southern Norway. It occurred about 1050–1100 Ma ago and may have been caused by *subduction along the north-western margin of the *Iapetus Ocean, by the closing of an earlier ocean, or by rifting (*see* RIFT) that ended before producing significant *sea-floor spreading.

Daly, Reginald Aldworth (1871–1957) A mathematician who turned to geology and the study of *igneous rocks and *volcanoes, Daly was a professor at Harvard University. In the 1920s he was one of the very few supporters in America of the theory of *continental drift. He proposed that drift was caused by continental 'downsliding' due to gravity, above a molten *mantle which had the properties of glass. It was Daly who first suggested, in 1936, that *submarine canyons may be excavated by suspension currents. *See* TURBIDITY CURRENTS.

damping A slowing down or prevention of oscillation due to the dissipation of the kinetic energy of oscillation. Friction will dampen a mechanical system, and electromagnetic damping uses eddy currents to oppose motion. *See also* CRITICAL DAMPING.

Dana, James Dwight (1813–95) American mineralogist and geologist, responsible for the term 'geosyncline', who proposed that the Earth is contracting as it cools, causing deformation which is concentrated at continental margins on the limbs of geosynclines, thus forming mountains, metamorphic belts, etc.

Danian 1. The earlier of two *ages in the *Palaeocene Epoch, preceded by the *Maastrichtian (Late *Cretaceous), followed by the *Thanetian, and dated at 65–60.5 Ma (Harland et al., 1989). **2.** The name of the corresponding European *stage which, in Denmark, is characterized by chalky *limestone rich in *reef-dwelling organisms. It is roughly contemporaneous with the Montian (Belgium), Danian (*see* (3) below) and lower *Ynezian (N. America), lower *Teurian (New Zealand), and *Wangerripian (Australia). In the past, because of its chalk *facies, some authors considered the Danian to be Upper Cretaceous, and succeeded in the Lower *Tertiary by the Montian. **3.** The basal stage in the Lower Tertiary of the west coast of N. America, overlain by the Ynezian and roughly contemporaneous with just the lower part of the Danian Stage in Europe.

darcy The unit of intrinsic *permeability, used particularly in the oil industry. One darcy is equal to 0.987×10^{-12} m^2.

Darcy's law A description of the relationships among factors that determine *groundwater flow, expressed as an equation. At its simplest, Darcy's law states that $Q = kIA$, where Q is the groundwater flow, k the *hydraulic conductivity of the rock, I the hydraulic gradient (i.e. gradient of *hydraulic head), and A the cross-sectional area through which flow occurs.

d'Arrest A *comet with an orbital period of 6.51 years; *perihelion date 1 August 2001; perihelion distance 1.346 AU.

Darriwilian A *stage of the Middle *Ordovician of Australia, underlain by the *Yapeenian and overlain by the *Gisbornian.

darwin A measure of evolutionary rate (introduced by J. B. S. Haldane in 1949), given in units of change per unit time. *See also* HALDANE.

Darwin, Charles Robert (1809–82) English naturalist who is remembered mainly for his theory of evolution, which he based largely on observations made in 1832–6 during a voyage around the world on HMS *Beagle*, which was engaged on a mapping survey. In 1858, in collaboration with Alfred Russel Wallace (who had reached similar conclusions), he published through the Linnaean Society a short paper, 'A theory of evolution by natural selection'; and in 1859 he published a longer account in his book, *On the Origin of Species by Means of Natural Selection*. In this he presented powerful evidence suggesting that change (evolution) has occurred among species, and proposed natural selection as the mechanism by which it occurs. His theory may be summarized as follows: (*a*) the individuals of a species show variation; (*b*) on average, more offspring are produced than are needed to replace their parents; (*c*) populations cannot expand indefinitely and, on average, population sizes remain stable; (*d*) therefore there must be competition for survival; (*e*) it is the best-adapted variants (the fittest) that survive and reproduce. Since environmental conditions change over long periods of time, a process of natural selection occurs which favours the emergence of different variants and ultimately of new species (the 'origin of species'). This theory is known as Darwinism. The subsequent discovery of chromosomes and genes, and the development of the science of genetics, have led to a better understanding of the ways in which variation may be caused. Modified by this modern knowledge, Darwin's theory is called 'neo-Darwinism'.

Dasycladales (class *Chlorophyceae) An order of green *algae in which the thallus contains a single nucleus, becoming multinucleate prior to reproduction. The thallus consists of an erect axis with branches, the whole sharing radial symmetry. The group appeared in the *Ordovician, where elaborate and ornate forms occurred. They were more diverse in the *Permian and *Jurassic, but became reduced in variety during the *Cretaceous and survive now only in the tropics.

data *See* DATUM.

dating errors *See* ERRORS; and ISOCHRON.

dating methods During the last century geologists constructed a relative time scale based on *correlation of palaeontological and stratigraphic data. Depositional rates of *sediments have also been employed as a dating method, but only recently has absolute dating been made possible through the use of radioactive *isotopes. Of the various methods the last is obviously the most precise, but *fossils, *lithologies, and cross-cutting relationships do enable the geologist to give an approximate *relative age in field studies. *See also* ABSOLUTE AGE; RADIOACTIVE DECAY; RADIOMETRIC DATING; ISOTOPIC DATING; RADIO-CARBON DATING; DENDROCHRONOLOGY; GEOCHRONOLOGY; GEOCHRONOMETRY; and VARVE ANALYSIS.

Datsonian A *stage of the Lower *Ordovician of Australia, underlain by the *Payntonian (Upper *Cambrian) and overlain by the *Warendian.

datum (pl. data) Something that is known or assumed to be true. *See also* DATUM LEVEL; and MARKER BED.

datum level A surface or level which is regarded as a base from which other levels can be counted (i.e. a *datum). For example, sea level is often used as a datum level against which the height of land and depth of the sea bed are measured.

daughter *See* CLADISTICS; RADIOACTIVE DATING; and RADIOACTIVE DECAY.

daughter minerals *See* FLUID INCLUSION.

Davis, William Morris (1850–1934) An American geologist from Harvard University, Davis initiated the study of *geomorphology. He evolved the concept of the cycle of erosion (the *Davisian cycle), and described the role of rivers in the evolution of landscape.

Davisian cycle (cycle of erosion) Orderly series of stages through which land-forms were believed to pass from their initiation following uplift to their final planation by erosion. The main stages were those of youth, when hillslopes were steep and river profiles irregular; of maturity, when river profiles were smoothly concave-up and incision had markedly slowed; and of old age, when the landscape was reduced to a gently undulating surface or *peneplain. This framework for land-form studies has now largely dropped out of use.

day degrees The departure of the average daily temperature from a defined base, e.g. 4 °C, the minimum recognized temperature for the growth of many plant species. The number of day degrees may be totalled to assess the accumulated warmth of a particular year's growing season for crops. *See also* ACCUMULATED TEMPERATURE; AGROMETEOROLOGY; and MONTH DEGREES.

day length *See* EARTH ROTATION.

dB *See* DECIBEL.

D days (geomagnetic) *See* DISTURBED DAYS.

death assemblage (thanatocoenosis) An assemblage of *fossils of organisms that were not associated with one another during their lives. The remains were brought together after death, often by the action of currents.

débâcle Break-up of river ice in spring over northern Eurasia and N. America. The onset of thaw begins in March in low latitudes, and later further north.

debris flow Slow-moving, sediment gravity flow composed of large *clasts supported and carried by a mud-water mixture. Debris flows occur as overland and submarine mudflows, and as submarine deep-sea deposits. The deposits are poorly sorted and internally structureless; typically they have a pebbly, *mudstone texture. Debris-flow deposits cover many thousands of square kilometres of the *abyssal plain after originating on the *continental slope as slumped material.

debris slide A shallow landslide within rock debris, characterized by a displacement along one or several surfaces within a relatively narrow zone. It may take place as a largely unbroken mass, or may be disrupted into several units, each consisting of rock debris.

debrite A deposit formed by a *debris flow.

decay constant *Radioactive decay involves only the *nucleus of the parent atom, and thus the rate of decay is independent of all physical and chemical conditions (e.g. pressure, temperature, etc.). The decay was shown by *Rutherford to follow an exponential law. The fundamental equation describing the rate of disintegration may be written as: $-(dN/dt) = \lambda N$, where λ is the decay constant, representing the probability that an atom will decay in unit time t, and N is the number of radioactive atoms present. It is a fundamental assumption in geochronology that λ is a constant and that the only alteration in the amount of daughter or parent in the system is due to radioactive decay. The constant λ is usually expressed in units of 10^{-10} per year (e.g. ^{235}U is 9.72, ^{40}K is 5.31, ^{87}Rb is 0.139, and ^{238}U is 1.54). The total lifetime of a radioactive parent in a given system cannot be specified; in theory it is infinite. It is a simple matter, however, to specify the time for half of the radioactive parent atoms in a system to decay. This is called the 'half-life' (T), which is related to the decay constant by the expression $T = 0.693/\lambda$. *See also* DECAY CURVE.

decay curve A graphical representation of the exponential rate at which radioactive disintegration occurs (*see* RADIOACTIVE DECAY). If half the parent *nuclide remains after one time increment, one-quarter will remain after the next (identical) time increment, and so on. A plot of the surviving parent atoms against time in half-lives (*see* DECAY CONSTANT) gives a decay curve that approaches the zero line asymptotically. In theory it should never attain zero. The number of surviving parent atoms $N_{(t)}$ at the end of a number of half-lives (n) is simply $N_o/2^n$. Plotted as a function of time t, survivors form a characteristic decay curve, the equation of which is $N_t = N_o e^{-\lambda t}$, where e = 2.718 and λ is the *decay constant.

decay index In *cladistic analysis, the number of additional steps required to dissolve a given *clade.

decay series *Radioactive decay of a parent nuclide through a sequence of radioactive daughter nuclides to a final, stable daughter nuclide. Uranium has three naturally occurring *isotopes: ^{238}U, ^{235}U, and ^{234}U; all are radioactive. Thorium exists mainly as one isotope (^{232}Th) which is also radioactive. In addition, five radioactive isotopes of thorium occur in nature as short-lived, intermediate daughters of ^{238}U, ^{235}U, and ^{232}Th. Each of these isotopes is the parent of a chain (decay series) of radioactive daughters, each ending with stable isotopes of lead. Decay of ^{238}U gives rise to the 'uranium series' which includes ^{234}U as an intermediate daughter and ends in stable ^{206}Pb. The decay of ^{235}U gives rise to the 'actinium series' which ends in stable ^{207}Pb. The decay of ^{232}Th results in the emission of six alpha and four beta particles (*see* ALPHA DECAY; and BETA DECAY) leading to the formation of stable ^{208}Pb.

decibel (dB) One-tenth of a bel (named after Alexander Graham-Bell) and the unit in which two power levels are compared. It is used most commonly in acoustics and in describing electrical signals. The decibel difference (N) between the largest (A_{max}) and smallest (A_{min}) measurable amplitudes is given by $N = 20\log_{10}(A_{max}/A_{min})$. The ratio of values for two power levels, P_1 and P_2, is given by $N = 10\log_{10}(P_1/P_2)$.

declination 1. The angle between magnetic North and true (geographic) North. **2.** The angle between a celestial object and the celestial equator.

declined *See* STIPE.

décollement Literally, 'unsticking', a concept formulated by geologists studying the structure of the Swiss Jura, and dealing with the sliding and buckling of the *Mesozoic cover over the crystalline *basement. *Triassic *evaporites act as the 'lubricant' of materials in the Jura environment.

deconvolution Inverse *filter, or the action of undoing the effect of a previous *convolution process. The convolution of the Earth's *impulse response R (where R is the set of * *reflection coefficients characteristic of particular layered strata) with a wavelet W results in the observed *seismic trace S, such that $S = R{*}W$. A spike signal (*Dirac function) (δ) at time zero is the convolution of the wavelet W with some designed operator D such that $D{*}W = \delta$. Thus if the operator D is applied to the seismic trace S, R can be obtained: $D{*}S = D{*}R{*}W = D{*}W{*}R = \delta{*}R = R$. (It should be noted that convolution of a spike function δ leaves the function unaltered, hence $\delta{*}R = R$).

decorrelation stretching In *remote sensing, a type of *contrast stretching that results in the artificial enhancement in colour of an image. The spread of multispectral data is increased along the natural maximum as identified by *principal component analysis.

decussate Applied to a *texture in which crystals have no preferred orientation, but have grown in a random arrangement.

dedolomite *Limestone formed by the replacement of *dolomite by *calcite in a *carbonate rock. This replacement occurs mainly when a dolomite is flushed through by *meteoric or sulphate-rich waters. A characteristic feature of dedolomite is the presence of dolomite-shaped, rhombohedral crystal forms now occupied by calcite *pseudomorphs.

deepening In meteorology, a lowering of the central pressure in a depression. *Compare* FILLING.

deep scattering layer (DSL) Sound-reflecting layer in ocean waters, consisting of a stratified, dense concentration of zooplankton and fish. Such organism-rich layers, which cause scattering of sound as recorded on an echo sounder, may be 50–200 m thick.

Deep Sea Drilling Programme (DSDP) An international programme initiated in 1963, which resulted in more than 500 *boreholes being drilled in the sea bed of the *Atlantic, *Pacific, and *Indian Oceans and the Mediterranean Sea. Until 1975, the programme was financed mainly by the National Science Foundation of America, but subsequently it has received support from the United Kingdom, France, West Germany, Japan, and the USSR. The managing institution was the Scripps Institute of Oceanography, and the drilling ship used was the *Glomar Challenger*. The programme evolved into the more ambitious *International Programme of Ocean Drilling (IPOD) and now the Ocean Drilling Programme (ODP).

deep-sea fan Fan-shaped body of sediment that accumulates at the lower end of a submarine canyon, either at the foot of the *continental slope or on the *continental rise.

deep-sea trench Narrow, elongate, steep-sided and often rock-walled depression that is 3000 m or more deeper than the adjacent deep-sea floor. Trenches are the deepest parts of the ocean, the greatest known depth being that of the Challenger Deep in the *Marianas Trench where a depth of 11 022 m has been recorded. Trenches usually mark the position of a destructive *plate margin, where the sea floor is slowly being destroyed by *subduction.

Deerparkian (Oriskanyan) *See* ULSTERIAN.

deflation The removal of material from a land surface by *aeolian processes. It is most effective where extensive unconsolidated materials are exposed, e.g. on beaches, and on dry lake and river beds. The very large, enclosed hollows of many *deserts (e.g. the Qattara Depression of the Egyptian Sahara) may be due to deflation.

deflation hollow Enclosed depression produced by wind erosion. It may be found both in hot deserts, where wind may scour a hollow in relatively unconsolidated material, and in more temperate regions, where a protective vegetational cover has been removed from a sand *dune.

deflexed *See* STIPE.

deformation lamellae A feature that may be present in twinned *crystals where rows of atoms are offset in response to directed pressure, giving *lamellar gliding twins. Examples include *sphalerite, *calcite, and *pyroxenes.

deformation twinning (lattice gliding) When *crystals are stressed they may be deformed plastically by gliding or sliding along planes between rows of atoms within the crystal structure. This may take the form of 'translation gliding', whereby one or more rows of atoms may be displaced laterally along the glide plane, or 'twin gliding', where a smaller displacement is taken up by each row within the *lattice. Twins produced in this way are called 'gliding twins' or 'deformation twins'.

de Geer moraine A type of *moraine landscape that consists of a series of separate, narrow ridges trending parallel to a former ice front, and which can form annually. The ridges may be up to 300 m apart and up to 15 m high. They consist typically of a *till core, capped by a layer of partly rounded boulders. This landscape may have formed beneath the grounded part of an *ice sheet that extended into a lake or sea. It is named after an early Swedish geologist.

degrees of freedom 1. When a substance is heated its kinetic energy increases. Kinetic energy is made up from the translation and rotation of particles, and the vibration of atoms which constitute the molecules of a substance. A substance may, therefore, absorb heat energy supplied to it in several ways, and is said to possess a number of degrees of freedom. In general, a molecule consisting of N atoms will have $3N$ degrees of freedom; thus for a diatomic molecule there will be six degrees of freedom: three will be translational, two rotational, and one vibrational. In a *phase diagram, describing, for example, a three-phase system (such as ice–water–vapour), pressure and/or temperature can be altered independently, in an area where only one phase exists, without altering the one-phase condition. Along the line separating two areas, if temperature is altered then pressure must alter accordingly, or vice versa, to maintain the two-phase equilibrium. At a point where three phases are in equilibrium, alteration of either temperature or pressure will cause one phase to disappear. The system thus possesses (*a*) two degrees of freedom in the area; (*b*) one degree of freedom along the line; and (*c*) no degrees of freedom at the point. **2.** In statistics, the number of independent variables involved in calculating a statistic. This value is equal to the difference between the total number of data points under consideration, and the number of restrictions. The number of restrictions is equal to the number of parameters which are the same in both observed data set and theoretical data set, e.g. total cumulative values, means.

dehydration Removal of water, especially from a chemical compound by heat, sometimes in the presence of a catalyst or dehydrating agent, e.g. sulphuric acid. Also refers to removal of water from crystals, oil, etc., commercially by distillation, chemicals, or heat. In *metamorphism, *prograde metamorphic reactions commonly involve dehydration of hydrous minerals.

dehydration curve (vapour pressure curve) The equilibrium curve for dehydration reactions on graphs showing increasing total pressure (vertically) against increasing temperature (horizontally). The slope of the dehydration curve varies with *vapour pressure and depends on the relationship with total pressure. This is important in *prograde metamorphism when many reactions involve dehydration reactions as increasing temperature produces water vapour. *Retrograde metamorphism is more limited, as most of the water has been driven off previously and the breakdown of existing grains, nucleation, and growth are much slower in the absence of water.

Deimos *See* MARS.

delamination 1. An increasingly accepted hypothesis according to which the lower part of the continental or oceanic *lithosphere becomes sufficiently dense, usually by cooling and mineralogical phase changes, that it separates from the upper lithosphere and descends into the *mantle, being replaced by warmer, less dense mantle rock. These physical and chemical changes introduce a new tectonic regime. **2.** *See* A-SUBDUCTION.

Delaware effect The distortion of lines of force by a change of resistivity between adjacent layers as *laterolog or *induction sondes are pulled through a *borehole. The change of *lithology correlates with the onset of the change of signal and not halfway through the change-over.

delayed flow Water that reaches river channels after flowing through subsurface routes and also from *groundwater. It contrasts with *surface runoff; together, surface runoff (quick flow) and delayed flow make up the total of river flow.

delay time A time gap between the shot-instant and the start of recording by a *seismograph to avoid long, blank sections on a record. It is used particularly in *reflection work over deep water where there is no geologic reason to record while the energy is travelling through the water column, and the recording starts just before the expected *first arrival. It is also used in *time-domain *induced polarization surveying to allow for the dissipation of transient voltages which have no direct relation to the *overvoltage.

Delmontian A *stage in the Upper *Tertiary of the west coast of N. America, underlain by the *Mohnian, overlain by the *Repettian, and roughly contemporaneous with the upper *Messinian, *Zanclian (Tabianian), and lower *Piacenzian Stages.

delta A discrete protuberance of *sediment formed where a sediment-laden current enters an open body of water, at which point there is a reduction in the velocity of the current. This results in rapid deposition of the sediment, which forms a body, for example at the mouth of a river where the river discharges into the sea or a lake. There is a characteristic coarsening upwards of sediments. A river provides the sediments to form a delta; but the shape and nature of a delta is controlled by a variety of factors including climate, water discharge, sediment load, rate of subsidence of the sea or lake floor, and the nature of the river-mouth processes (particularly tidal and wave energy). One classification of delta types, based on variations in transport patterns on the delta, subdivides deltas into three classes: (*a*) river-dominated, e.g. the Mississippi and Po; (*b*) wave-dominated, e.g. the Rhône and Nile; (*c*) tide-dominated, e.g. the

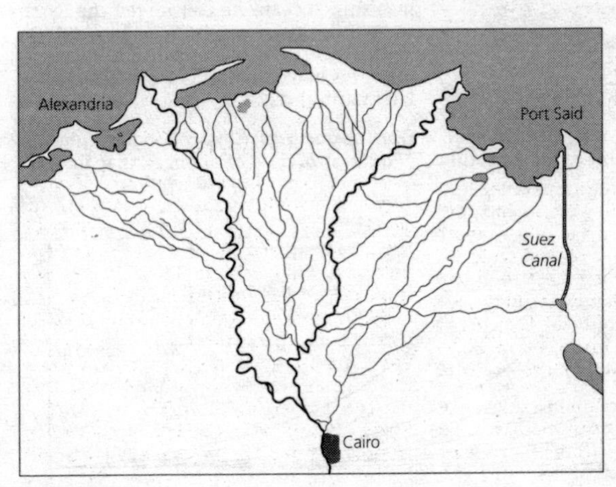

Delta (Nile)

Ganges and Mekong. *See also* GILBERT-TYPE DELTA.

delta front The sloping portion of a *delta, developed offshore from the bar at the mouth and passing at its toe into the *pro-delta. Delta fronts are the site of active, and often rapid, sedimentation, frequently characterized by *growth faulting and *slump structures generated by over-steepening of the sediment pile.

Deltatheridium Small, probably carnivorous mammal from the Upper *Cretaceous. It was once thought to be an insectivore, but the possession of certain dental characters indicate that it was a likely ancestor or sister group to the creodonts (*Creodonta) and *Carnivora.

ΔT method In seismic investigations, a method of velocity analysis associated with the normal *moveout correction (ΔT), which enables the *root-mean-square velocity (v_{rms}) to be calculated down to a particular *reflector for which the normal incidence two-way travel time (t_0) and ΔT are known for a given *offset (x) such that $v_{rms} = x/(2t_0\Delta T)^{1/2}$. Once v_{rms} is known, the depth (z) to the reflector can be calculated using $z = v_{rms}t_0/2$.

delthyrium Subtriangular opening or slit-like notch beneath the apex of the pedicle valve of some *brachiopods, for the passage of the *pedicle.

deltidial plate Calcareous deposit at the side of the *delthyrium which serves to constrict or close the opening.

demagnetization (cleaning, magnetic) Geologic and archaeological samples are usually partially demagnetized in a series of incremental steps to determine their *coercivity and/or *blocking-temperature spectra. *Viscous remanences are more readily removed than either *thermal or *chemical remanence, so it is often possible to isolate the original (primary) magnetization acquired when the samples were first formed. This is known as *alternating magnetic field demagnetization or thermal 'cleaning'. *Chemical demagnetization can also be used on permeable sedimentary samples in which the *cement is usually most readily removed by acid washing, thereby preferentially removing the chemical remanence associated with the cement and isolating the *detrital remanence acquired during deposition. Direct magnetic fields can be applied to reduce the observed remanence to zero, this field corresponding to the effective mean coercitivity of the total remanence. New developments include the use of tuned *microwaves to demagnetize magnetic minerals without much heating and chemical change.

demagnetizer One of the various instruments used to remove gradually the *remanent magnetization of specimens. There are three main types of instrument, based on *alternating magnetic fields, *thermal treatment in zero magnetic field, *chemical treatment, and the use of *microwaves.

de Maillet, Benoît (1656–1738) De Maillet was a French diplomat and traveller, whose posthumous work *Telliamed* (has name spelled backwards) was published in 1748. A cosmogony, covering the creation and prehistory of the Earth, it argued that all rocks and all forms of life had their origin in a diminishing ocean. He considered the Earth to be more than two billion years old. De Maillet's belief in mermaids and mermen made his book an object of easy ridicule by later naturalists.

Demospongea (Demospongiae; phylum *Porifera) A class of sponges that first ap-

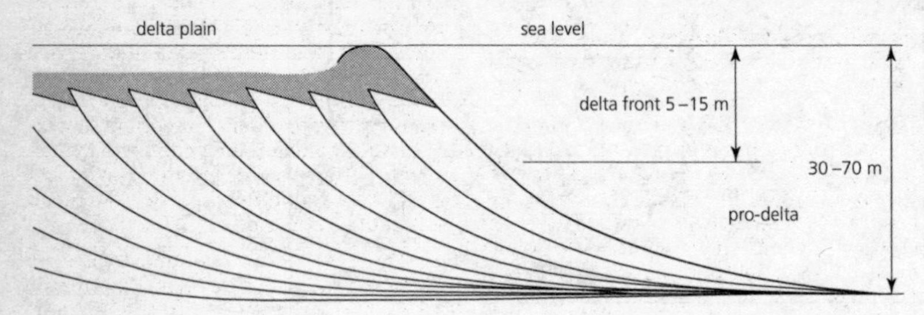

Delta front

pears in the *Cambrian. The soft tissue of the skeleton is supported by siliceous *spicules, each spicule consisting of either a single ray (monaxon) or four rays (tetraxon) diverging at 60° or 120°. Most *fossil species are represented by their spicules only, but they can usually be referred to modern families.

Demospongiae *See* DEMOSPONGEA.

demultiplexing The reordering of a magnetic tape to allow a computer to commence the processing of multiplexed (*see* MULTIPLEXING) data. The data samples are rearranged into trace-sequential format, i.e. all the samples for channel 1 are followed by those for channel 2, . . . , followed by those for channel *n*.

dendritic (arborescent) Applied to the shape or *habit of a *crystal that is deposited or precipitated in tree-like, slender branches, often along narrow joint planes, e.g. manganese oxide.

dendritic drainage A *drainage pattern whose shape resembles the pattern made by the branches of a tree or veins of a leaf, which may develop on homogeneous rock.

dendrochronology (tree-ring analysis) 1. The science of dating by means of tree rings. **2.** All aspects of the study of annual growth layers in wood.

dendroclimatology Branch of *dendrochronology dealing with the relationships between the annual growth increment in tree trunks and climate, and especially with the reconstruction of past climates from dated tree-ring series. It is assumed that by studying present tree-ring patterns in relation to climate, older tree-ring chronologies may be used to indicate the climatic conditions experienced before detailed climate and weather records were kept.

dendrogeomorphology The use of dated tree-ring series to study land-forms and geomorphological processes.

dendrogram A diagram that represents relationships among groups of taxa, with the highest taxon at the base of a vertical line from which lower taxa branch at appropriate levels. There are 2 principal types: (*a*) the *phenogram, which is based solely on similarities in *phenotypes; and (*b*) the *cladogram.

Dendrograptidae *See* DENDROIDEA.

dendrohydrology The use of dated tree-ring series to study hydrological questions, especially relating to the periodicity of river flow and flooding.

dendroid 1. In corals (*see* COMPOUND CORALS), applied to a colony formed by the irregular branching of *corallites. The individual corallites are separated from one another but may be connected by tubules. **2.** In graptolites (*Graptolithina), applied to a bushy colony formed by irregular branching of the *stipes.

Dendroidea (subphylum Stomochordata, class *Graptolithina) An order of graptolites that existed from the Middle *Cambrian to the Lower *Carboniferous. Most lived attached to the sea bed and were upright and bushy in appearance. They are many-branched, with numerous small thecae (*see* THECA), in some connected by *dissepiments. The *stipes bear two kinds of thecae, *autothecae and *bithecae, opening from a continuous, closed *stolotheca. There are concentric sheets of tissue in the lower part of the *rhabdosome. The order contains three families: Acanthograptidae, Dendrograptidae, and Ptilograptidae.

dendroid graptolites *See* DENDROIDEA.

Denekamp (Zelzate) A Middle *Devensian *interstadial from the Netherlands, dating from about 30 000 years BP, when the estimated mean July temperature for the Netherlands was 10 °C. The Denekamp Interstadial, together with the *Moorshoofd and *Hengelo Interstadials, is perhaps the equivalent of the *Upton Warren Interstadial of the British Isles.

dense-medium separation (heavy-medium separation, HMS) The simplest of the *gravity separation processes used to concentrate *minerals before their final grinding and liberation. It is also used to separate *coal from the heavier *shale. Heavy media of suitable density are used, in which lighter minerals float and denser minerals sink. In industry, the medium is usually a thick pulp made from a heavy mineral in water. Dense medium separation is suitable for any *ore which has enough difference in specific gravity between waste and mineral particles to repay the cost of treatment.

density Mass per unit volume, expressed in SI units as kilograms per cubic metre (kg/m^3). It is determined directly, or indirectly using *gravity or *seismic velocity measurements. Typical densities in unconsolidated, wet sediments range between 1200 and 2600 kg/m³, and 1000 to 2000 kg/m³ if dry. Consolidated sediments range between 1600 and 3200 kg/m³. (*See also* NAFE–DRAKE RELATIONSHIPS for seismic determinations of the density of sediments.) *Basic *igneous rocks range between 2300 and 3170 kg/m³, and *metamorphic rocks are mostly in the range 2400–3100 kg/m³, with *eclogite being one of the denser of the commoner metamorphic rocks, between 3200 and 3540 kg/m³. The density of the upper *mantle increases with depth from about 3330 to 4000 kg/m³, below which the lower mantle systematically increases to about 5400 kg/m³ above the *core. The outer core density is about 10 100 to 12 100 kg/m³; the inner core is more uniform, with a density of about 13 000 kg/m³.

density current Current that is produced by differences in density. Where a flow of sea water has a greater density than that surrounding it, the more dense water will dive beneath the less dense water. Density of sea water is affected by temperature, *salinity, and the content of suspended sediment. A turbidity current is a gravity-controlled density current, in which the density contrast is due to the high suspended-sediment content. Most deep, and bottom, ocean currents are density currents.

density–depth profile Within the Earth, this relationship is determined primarily using *seismic waves, as their velocity is dependent on density, complemented by gravitational-field and *free-oscillation measurements. The upper *mantle is complicated by the presence of *phase transitions, particularly at depths of around 300 km and 650 km. Within the lower mantle, the density increase with depth is essentially adiabatic, except in the region immediately overlying the *core. The outer core density (about 10 000 kg/m³) is almost twice that of the lower mantle (about 5000 kg/m³), reflecting a compositional change. The core density increases with depth, with a small jump from the outer to inner core of 12 000 kg/m³ to 13 000 kg/m³.

density determination (specific gravity determination) The determination of the density of a substance. In the case of *minerals several techniques are available. (*a*) Using a Jolly's spring balance, the mineral is first weighed in air and then immersed in water and reweighed. The density is calculated by comparing the two weights. (*b*) The Berman balance is a very sensitive torsion balance used for determining the density of small fragments of minerals, based on the same principle as the Jolly balance. (*c*) A pycnometer is used to determine the density of soils or powdered minerals. It consists of a small bottle fitted with a ground-glass stopper with a capillary opening. The specimen is placed in the pycnometer, weighed, and then the bottle is filled with water and reweighed. (*d*) Heavy liquids can also be used. Liquids which are relatively dense, such as bromoform (sp. gr. 2.89) and methylene iodide (sp. gr. 3.33), may be mixed with acetone (sp. gr. 0.79) to produce a series of liquids of known density. If a mineral is introduced into one of these liquids and it neither rises nor sinks its density is the same as that of the liquid. These heavy liquids are often toxic, and great care must be taken when using them. Gloves and face masks should always be worn. *See also* COUNTERPOISED BEAM BALANCE.

density log A *well-log (record) of the density of the rocks penetrated by a drill hole. Normally it is based on the output from a *gamma-gamma sonde, with compensation for the effects of the *mud cake. It is particularly useful for the identification of *evaporites, and when examined in conjunction with the *neutron-log it can provide an indication of the presence of gas zones.

denticle In many fish, a scale composed of *dentine, with a pulp cavity, which resembles a tooth. In *Elasmobranchii denticles cover the entire body.

dentine *Bone-like substance, lacking cell bodies and consisting mainly of calcium phosphate in a fibrous matrix.

dentition In bivalve *Mollusca, the articulating tooth-and-socket system in the shell *hinge system. Various types may be recognized, e.g. *desmodont, *dysodont, *heterodont, *isodont, *pachydont, *schizodont, and *taxodont.

denudation From the Latin *denudare*, to strip bare, and involving the processes of *weathering, transportation, and *erosion.

denudation chronology Branch of landform studies that deals with the historical development of landscapes by *denudation, especially during pre-*Quaternary time. Evidence for developmental stages is provided by studies of erosion surfaces and their mantling deposits, *drainage patterns, stream long-profiles, and geologic structures.

depleted mantle The residue that remains after a given element has been removed from *peridotite to form a *basalt melt. The *incompatible elements (e.g. Rb, U, and *rare earth elements) are preferentially partitioned into a *melt, and during crustal formation these elements in particular have been removed from the *mantle, leaving the mantle depleted in incompatibles. *Compare* UNDEPLETED MANTLE.

depocentre The site of maximum deposition within a sedimentary basin, where the thickest development of the sedimentary sequence will be found.

deposit feeder An animal that lives on or in the *sediment of the sea floor, and swallows mud rich in organic material in order to obtain nutriment.

deposit gauge Device for collecting solid and liquid atmospheric pollutants.

depositional remanent magnetization (DRM) The magnetization acquired by a sediment during deposition. This is predominantly *detrital remanent magnetization, but used more widely, the term covers all magnetizations acquired at such a time. *See also* POST-DEPOSITIONAL MAGNETIZATION.

depositional sequence A discrete succession of *strata, deposited more or less continuously, and bounded at top and bottom by either an *unconformity surface or the equivalent, correlative *disconformity or conformity surfaces (*see* CONFORMABLE). The ancient stratal surfaces can be traced laterally by seismic *reflection *profiling. A depositional sequence is the basic, operational, *stratigraphic unit in *seismic stratigraphy. Where top and bottom boundaries can be traced to conformity surfaces a depositional sequence becomes of *chronostratigraphic significance, as it represents a specific interval of geologic time during which the particular unit was deposited. *See* BASELAP; DOWNLAP; ONLAP; and TOPLAP.

depositional sequence model One of the two schools of *sequence stratigraphy, devised at Exxon Production Research (EPR), that uses *unconformities and their correlated conformities as the boundaries of genetically related strata. *Compare* GENETIC STRATIGRAPHIC SEQUENCE MODEL.

depositional system An assemblage of *lithofacies (*see also* FACIES, FACIES ASSOCIATION) formed within a particular depositional environment (e.g. *fluvial).

depositional systems tract An association of related *depositional systems, e.g. a systems tract comprising fluvial, deltaic (*see* DELTA), and *continental-shelf systems.

depression 1. Enclosed area of low pressure revealed by the pattern of pressure distribution. Depressions are also described as cyclones, or cyclonic systems, and have a characteristic pattern of wind circulation (anticlockwise around low pressure areas in the northern hemisphere). Mid-latitude depressions are associated with the convergence of polar and tropical *air masses along a frontal zone: this commonly becomes deformed, and each air mass in turn advances over parts near (especially south of) the depression path, bringing first a warm and then a cold front. **2.** The downward convexing of a *crest line in a non-*cylindroidal fold.

depression angle In *radar terminology, the angle between the horizontal plane passing through the radar antenna and the line between the antenna and object. *Compare* LOOK ANGLE.

depth point A subsurface point on a reflector which contributes to a reflection event on a *seismic trace. *See* COMMON DEPTH POINT.

deranged drainage An original *drainage pattern that has been deranged or disturbed by the intervention of external factors such as tectonic activity and glaciation.

derived 1. As applied to *fossils, a term implying that material has been fossilized elsewhere, eroded from its original site, and then incorporated in another, younger, geological horizon. The terms 'reworked' and 'remanié' are sometimes used as alternatives. **2.** *See* APOMORPH. *See also* CLADISTICS.

dermal bone *See* BONE.

dermal denticle *See* PLACOID SCALE.

dermatocranium *See* CRANIUM.

Derryan *See* ATOKAN.

desalination A series of processes whereby salt water is rendered potable.

Desdemona (Uranus X) One of the lesser satellites of *Uranus, with a diameter of 29 km. It was discovered in 1986.

desert A term which has no precise definition. Deserts may be thought of in terms of biomes in which evaporation exceeds the average precipitation. The rate of evaporation varies with temperature, but desert conditions are likely to develop wherever precipitation is less than 250 mm/yr. Typically, precipitation in deserts is very erratic. Plants and animals are either absent or sparsely distributed, and are adapted to long droughts or to a lack of access to free water.

desertification The spread of *desert-like conditions, particularly in arid or semi-arid areas, due to the influence of human activity and climatic change.

desert pavement A thin covering of gravel and stones found in many desert areas, left after erosion by wind and water has removed the finer soil materials.

desert rose A radiating series of petal-shaped *calcite or *gypsum *minerals, sometimes resembling the form of a rose, developed in the early stages of sand *diagenesis in arid regions, and particularly in *sabkhas.

desert varnish Thin, dark surface veneer of iron and manganese oxides which coats exposed rocks, especially in hot deserts. It results from the surface precipitation of minerals released from the rock by *chemical weathering.

desiccation Long-term loss of water associated with regional climatic change.

desiccation cracks (mud-cracks, shrinkage cracks, sun-cracks, syneresis cracks) The polygonal-shaped cracks developed in mud which has dried out in a terrestrial environment. They are most often preserved when loose sand infills the cracks and then buries the desiccated mud surface.

Desmarest, Nicolas (1725–1815) French encyclopaedist and amateur geologist who did field-work and mapping in the Auvergne, where he recognized that *basalt was associated with lava flows. He realized that the hexagonal cracks in basalt were the result of cooling.

desmodont Applied to a type of *hinge condition found in certain *bivalves in which the teeth are very small or lacking, and ridges may have replaced them. The ligament may be supported by a *chondrophore.

Desmoinesian A *series in the *Pennsylvanian of N. America, underlain by the *Atokan, followed by the *Missourian, and roughly contemporaneous with the Podolskian and Myachkovskian Stages of the *Moscovian Series.

Despina (Neptune V) A satellite of *Neptune, with a diameter of 148 km; visual albedo 0.06.

desquamation *See* WEATHERING.

destructive margin The contact between two lithospheric *plates which are moving towards each other and where *oceanic crust is being destroyed by *subduction. Destructive margins, a type of *convergent margin, are marked by shallow- to deep-focus *earthquakes and typically andesitic *volcanicity, and most are also marked by an oceanic *trench. *See* PLATE MARGIN.

destructive wave Relatively high-energy shallow-water wave that causes degradation of a beach by moving more material seawards than landwards, thus having a net erosional effect on the adjacent beach. It is characterized by high frequency, which implies that the *swash is impeded by the backwash from the preceding wave. The backwash is more effective than the swash in moving material, and waves having this character are usually steep and associated with onshore winds. The erosional effect is also enhanced by the near-vertical plunge on breaking. It is favoured by a relatively steeply sloping offshore zone.

detachment fault A low-angle *normal fault, formed due to the gravitational instability of an uplifted block, along which there is considerable horizontal displacement.

detrital Applied to material derived from the mechanical breakdown of rock by the processes of *weathering and *erosion.

detrital remanent magnetization (DRM) The magnetization of a sediment acquired as *ferromagnetic detrital grains become aligned by the *geomagnetic field as they fall through the water and become part of the sediment on the bottom of a lake, pond, river system, or the sea. *See also* POST-DEPOSITIONAL MAGNETIZATION.

deuteric alteration Textural and mineralogical changes occurring within an *igneous rock during the final crystallization stages of the molten rock. It usually occurs at well above room temperature, but some changes can be delayed for a few years after solidification. *See* DEUTERIC REACTION.

deuteric reaction The 'simmering' of an *igneous rock in its own 'juices'. After the last drop of *silicate *melt has crystallized and the rock is, in effect, solid, residual water-rich vapour, concentrated by the crystallization of a large proportion of non-water-bearing minerals, can permeate along crystal boundaries and through crystal fractures and react with the *primary minerals. Reaction of the low-temperature vapour with high-temperature magmatic minerals (*see* MAGMA) is termed 'deuteric alteration'.

Devensian The last glacial stage in Britain, lasting from around 70 000 years BP (possibly earlier) to about 10 000 years BP. It is approximately synchronous with the *Wisconsinian Stage in N. America, the Weichselian Glaciation in northern Europe and the Würm Glaciation in the Alps. The Devensian was preceded by the *Ipswichian Interglacial, and followed by the *Flandrian (i.e. present) Interglacial. *See also* ALLERØD; BØLLING; CHELFORD INTERSTADIAL; DRYAS; LATE-GLACIAL; LOCH LOMOND STADIAL; UPTON WARREN INTERSTADIAL; and WINDERMERE INTERSTADIAL.

deviatoric stress A *stress component in a system which consists of unequal *principal stresses. There are three deviatoric stresses, obtained by subtracting the mean (or *hydrostatic) stress (σ^-) from each principal stress (i.e. $\sigma_1 - \sigma^-$, $\sigma_2 - \sigma^-$, and $\sigma_3 - \sigma^-$). Deviatoric stresses control the degree of body distortion.

devitrification The crystallization of minerals following the melting of a glass; this most commonly occurs in glassy *ignimbrites, where the glass is replaced by *microcrystalline *cristobalite and *alkali feldspar.

Devonian The fourth of the six *periods of the *Palaeozoic Era and the first of the Upper Palaeozoic Sub-era. It began about 408.5 Ma ago and ended about 362.5 Ma ago. In Europe there are both marine and continental *facies present, the latter being commonly known as the *Old Red Sandstone. Although originally described from the type area in Devon, the marine Devonian is subdivided stratigraphically into *stages established in the exceptionally fossiliferous deposits of the Ardennes in Belgium. These stages are the Gedinnian (408.5–401 Ma), Siegennian (401–394 Ma), and Emsian (394–387 Ma) of the Lower Devonian, the Eifelian (387–380 Ma) and *Givetian (380–374 Ma) of the Middle Devonian, and the *Frasnian (374–367 Ma) and Famennian (367–362.5 Ma) of the Upper Devonian. The subdivision of the marine deposits is based on lithologies and the presence of an abundant invertebrate fauna including *goniatites and *spiriferid brachiopods. The continental Old Red Sandstone deposits contain a fauna of jawless fish and plants belonging to the primitive *psilophyte group. As a result of the *Caledonian orogeny of late *Silurian times, much of the British Isles was covered with continental *red-bed facies.

dewatering The removal of *groundwater to reduce flow-rate or diminish pressure. Methods used depend on the *permeability of the ground, proximity of hydrogeologic boundaries (*see* HYDRAULIC BOUNDARY), *storage coefficient of the soil, pressure, and *hydraulic gradient. Methods used include *abstraction by *wells, *electro-osmosis, sumps and drains, vertical drains, or exclusion by *grouting, *compressed air, or freezing techniques (*see* ARTIFICIAL FREEZING). Dewatering is usually undertaken to improve conditions in surface *excavations and to help construction work at or near the surface.

dew-point The temperature, with constant pressure and water-vapour content, to which air must be cooled for saturation to occur. Dew-point may be determined from the dew-point hygrometer which indicates initial condensation on a cooled surface.

dew-point hygrometer *See* DEW-POINT; and HYGROMETER.

dextral coiling *See* COILING.

dextral fault (right lateral-fault) The sense of displacement in *strike-slip fault zones where one block is displaced to the right of the block from which the observation is made. *See* FAULT.

D/H ratio Ratio between deuterium (heavy hydrogen, ^2H) and hydrogen (^1H) in natural waters and other fluids, and in water combined in hydrous *minerals. This ratio yields information about the origin and geologic history of the fluid, and about fluid/rock interactions. *See also* ISOTOPE FRACTIONATION.

diabase *See* DOLERITE.

diachronous Applied to a lithologic unit that differs in age from place to place.

diad *See* CRYSTAL SYMMETRY.

diadochy *See* IONIC SUBSTITUTION.

diagenesis All the changes that take place in a sediment at low temperature and pressure after deposition. With increasing temperature and pressure, diagenesis grades into *metamorphism. Diagenetic processes such as *compaction, dissolution (*see* PRESSURE DISSOLUTION), *cementation, replacement, and *recrystallization are the means by which an unconsolidated, loose *sediment is turned into a *sedimentary rock, e.g. *sand into a *sandstone, or *peat into *coal. *See also* CARBON ISOTOPES; and OXYGEN ISOTOPES.

diagnostic horizon Soil layer containing a combination of characteristics typical of that kind of soil.

diallage A variety of the *pyroxenes *diopside or *augite which has a characteristic lamellar structure due to a parting along {100}.

diamagnetism When a magnetic field is applied to *electrons orbiting a *nucleus, the individual electron spins precess and result in a field in the opposite direction to that applied. All atoms and molecules show this form of magnetization, usually with a negative susceptibility of the order of 10^{-5} or less, but it can have *paramagnetism and/or *ferromagnetism superimposed upon it.

diamict A general term describing both *diamicton and *diamictite.

diamictite A lithified, conglomeratic, *siliciclastic rock which is unsorted, with *sand and/or coarser particles dispersed through a mud *matrix. The term is commonly used today in preference to 'tillite', which has clear genetic connotations.

diamicton The unlithified equivalent of a *diamictite.

diamond Crystalline form of carbon that is the hardest naturally occurring material (*hardness 10 on *Mohs's scale); sp. gr. 3.5; *cubic; white or colourless, sometimes yellow, green, red, and rarely blue or black; crystals octahedral; cleavage perfect {111}; of *igneous origin and frequently associated with *kimberlites.

diamond drilling The drilling of a *borehole using diamond-studded *bits, usually for *core recovery in exploration.

diaphragm wall (core wall, cut-off trench) A nearly impervious wall or core at the base of a dam or similar structure which reduces percolation. *Bentonite suspensions are used in support of deep trench excavations formed for the construction of cast-in-place concrete diaphragm walls.

diaphthoresis *See* RETROGRADE METAMORPHISM.

diapir Upward-directed, dome-like intrusion of a lighter rock mass, e.g. *salt or *granite, into a denser cover. The process is termed diapirism.

diapirism *See* DIAPIR.

diapsid In *Reptilia, describes a skull with two temporal openings behind the eye. This type of skull is characteristic of two subclasses of reptiles: the Lepidosauria, represented today by lizards, snakes, and the *rhynchocephalian *Sphenodon* (tuatara); and the *Archosauria, which includes the extinct (thecodonts, *pterosaurs, *dinosaurs, and the living crocodiles. The earliest known diapsids were lepidosaurs (e.g. *Youngina*) from the Upper *Permian and Lower *Triassic of S. Africa. Formerly, many authorities grouped all such reptiles in the subclass Diapsida, but the arrangement has been largely abandoned because the more primitive and more advanced forms are not clearly related to one another. *Birds are considered to be closely related to the archosaurs. Although the skulls of birds have only one temporal opening, this is apparently derived from the fusion of the two diapsid apertures.

Diapsida *See* DIAPSID.

Diarthrognathus broomi Of the many mammal-like reptiles (*see* SYNAPSIDA) recorded from the *Triassic, this species is one of the closest to mammalian ancestry. It is assigned to the Ictidosauria and is characterized by a number of advanced cranial features. The skull of *Diarthrognathus* is only 4–5 cm long and its dentition is not as specialized as that of certain other mammal-like reptiles from the Upper Triassic.

diaspore An aluminium hydrate α-AlO(OH), forming a continuous series with its *polymorph *boehmite γ-AlO(OH); sp. gr. 3.3–3.5; *hardness 6.5–7.0; white; *massive or foliaceous or as *orthorhombic prismatic crystals; it is a constituent of *bauxite and *laterite deposits along with *emery and *corundum.

diastem A very small break in a *conformable succession of *strata, indicated only by a *bedding plane, and representing a brief interruption in the deposition of *sediments, with little or no *erosion having occurred. Diastems may be very localized, with deposition having occurred elsewhere. *Compare* NON-SEQUENCE.

diastrophism The deformation of the Earth's *crust on a large scale. *Faulting, *folding, and *plate movement can be involved, to produce mountain ranges, *rift valleys, continents, and ocean floor.

diatom *See* BACILLARIOPHYCEAE.

diatomaceous earth (kieselguhr) Deposit composed of *fossil diatoms (*see* BACILLARIOPHYCEAE), which is mined for many industrial uses: as a mild abrasive in metal polishes, as a filtering medium (e.g. in sugar refineries), for insulation of boilers and blast furnaces, etc., and under the name 'kieselguhr' as a vehicle for explosives. Vast deposits of diatomaceous earth (called '*diatomite') are mined at Lompoc, California.

diatomite Diatom-rich *sediment, which has been laid down in a lacustrine or deep-sea environment. The diatom cell wall is made from *silica, therefore the sediment is siliceous. *See* BACILLARIOPHYCEAE; DIATOMACEOUS EARTH; and RADIOLARIAN EARTH.

diatom ooze Soft, siliceous, deep-sea deposit composed of more than 30% (by volume) *diatom cell walls (*see* BACILLARIOPHYCEAE). Diatomaceous *sediments predominate in the high latitudes both around the coast of Antarctica and in the N. Pacific, but in the N. Atlantic the diatom content is overwhelmed by the *terrigenous sediment derived from the adjacent continents. *See also* RADIOLARIAN OOZE.

diatreme Carrot-shaped volcanic vent that has formed by explosive action: it is often filled with coarse, angular fragments injected by gas *fluidization, e.g. *kimberlites. Diatremes typically cut through non-volcanic *basement rocks, and contain fragments derived from great depths.

dichograptid A family of graptolites (order *Graptoloidea), 'dichograptid' being the name given to a fauna of graptoloids occurring in the early *Ordovician.

Dichograptina *See* GRAPTOLOIDEA.

dichotomize To split into two equal parts.

dichroic *See* DICHROISM.

dichroism (adj. dichroic) The property some *minerals have of absorbing more light in one *vibration direction than in another, and consequently of giving different colours in two different *vibration directions. It is a form of *pleochroism and is expressed by giving the colours for each of the two vibration directions.

Dickinsonia *See* EDIACARAN FOSSILS.

dickite *See* KAOLINITE.

Dictyonema flabelliforme One of the best known of all *graptolite species, *D. flabelliforme* is characteristically conical or bell-shaped in outline, with numerous branches connected by horizontal crossbars (*dissepiments). *D. flabelliforme* is a dendroid graptolite which was probably *planktonic in its mode of life. It ranged from the *Cambrian into the *Carboniferous.

diductor muscle *See* DIVARICATOR MUSCLE; and MUSCLE SCAR.

dielectric A material which does not conduct electricity, but in which an applied electric field displaces *charge rather than causing it to flow.

dielectric constant (relative permittivity, symbol ϵ_r) Measure of the polarity of a medium. The force (F) between two electric charges (e) at a distance (d) apart in a vacuum is expressed as: $F = e^2/d^2$. In any other medium: $F = e^2/\epsilon_r d^2$ where ϵ_r is the dielectric constant. Typical values are: 1.0 for air; 1.013 for steam; 15.5 for liquid ammonia; 80.36 for water at 20 °C; usually 3.18 for ice, although at low frequencies it can be up to two orders of magnitude higher; 5–19 for granite; and 3–105 for dry to moist sand. The dielectric constant is temperature dependent, increasing as temperature increases, and strongly frequency dependent with low-frequency (less than 100 Hz) values being up to 30% higher than those of high frequency (more than 100 Hz). It is analogous to the *magnetic permeability. The complex relative permittivity, ϵ^* is given by $\epsilon^* = \epsilon' - j\epsilon''$ where ϵ' and ϵ'' are the real component (relative permittivity) and the imaginary component (*dielectric loss factor) respectively, and $j = \sqrt{-1}$.

dielectric loss factor (ϵ'') Related to the complex *relative permittivity, ϵ'' is a measure of the loss of energy in a *dielectric material through conduction, slow polarization currents, and other dissipative phenomena. The peak value for a dielectric with no direct-current conductivity occurs at the *relaxation frequency, which is temperature related. The maximum value can be used as an important measure of the dielectric properties of rocks and ice.

dielectric permittivity (ϵ_0) The 'absolute' permittivity of a *dielectric, being the ratio of the electric displacement to the electric field strength at the same point. Its value for free space is 8.854185×10^{-12} F/m. The real component of the complex *relative permittivity is an important diagnostic parameter when considering the high- and low-frequency values for zero *dielectric loss. It relates to grain size, packing variations, and density of the material. *See also* DIELECTRIC CONSTANT.

Dienerian *See* SCYTHIAN.

Dietz, Robert Sinclair (1914–95) An American naval oceanographer, Dietz worked on *magnetic anomaly patterns in the north-eastern Pacific, and in 1961 he coined the term *sea-floor spreading for his theory that new ocean floor was created at ocean *ridges. He also worked on lunar cratering, and recognized craters on Earth as being of meteoric origin.

differential settlement Unequal settling of material; gradual downward movement of *foundations due to compression of soil which can lead to damage if *settlement is uneven.

differential stress *See* MOHR STRESS DIAGRAM; and STRESS DIFFERENCE.

differentiation *See* MAGMATIC DIFFERENTIATION.

differentiation index The sum of the normative constituents (*see* CIPW NORM CALCULATION) Q + Ab + Or + Ne + Kp + Lc in an *igneous rock, where Q = *quartz, Ab = *albite, Or = *orthoclase, Ne = *nepheline, Kp = kaliophilite, and Lc = *leucite. The index, defined in 1960 by two American petrologists, Thornton and Tuttle, seeks to quantify the degree of differentiation a rock has undergone. The greater the degree of differentiation, the more enriched the rock is in *felsic minerals and hence the higher the differentiation index.

diffraction The radial scattering of any wave (light, radio, seismic, water, etc.) incident upon an abrupt discontinuity in accordance with *Huygens' principle. A *fault plane, *angular unconformity, small isolated objects (e.g. boulders, fragments of wrecked ships, etc.) will all give rise to the diffraction of incident seismic energy. The quasi-hyperbolic curvature of a seismic diffraction event is related to the velocity within the media through which the diffracted wave travels. In media with slow velocities, the hyperbola is strongly curved, the curvature decreasing as velocity increases.

diffuse reflection The reflection of *electromagnetic radiation from a surface in all directions evenly. *Compare* SPECULAR REFLECTION.

diffusion 1. Movement of molecules or *ions from a region of higher to one of lower solute concentration as a result of their random thermal movement. For example, ions diffuse through a solution or *melt towards growing *crystals as their incorporation into the solid phase reduces their concentration in the immediately adjacent liquid. **2. (in crystals)** (*a*) Self diffusion involves the movement of a unit of a given composition through a crystal *lattice of the same composition. (*b*) Volume diffusion, the movement of atoms or ions through the crystal lattice. It includes simple self diffusion and more complex situations where ions of a certain species migrate through a lattice containing a variety of ions of different sizes or charges and in various configurations. Self diffusion leads only to change in shape or texture, volume diffusion leads to changes in composition.

diffusion coefficient (*D*) In diffusion calculations, the square of the dimension distance divided by time; this varies with the type of particles, the diffusion medium, and temperature.

diffusion-controlled growth *Crystal growth in a *melt or *hydrothermal fluid where growth rate is controlled by volume *diffusion in the liquid. This generally occurs in multi-component systems at large degrees of *supercooling.

diffusion creep Deformation as a result of the migration of atoms along a stress gradient. It usually becomes important as the temperature rises close to the melting temperature. *See also* COBLE CREEP; CREEP MECHANISMS; and NABARRO–HERRING CREEP.

digenite A copper sulphide Cu_9S_5, closely

related to *chalcocite Cu_2S; sp. gr. 5.7; *hardness 3; greyish-blue; *sub-metallic *lustre; occurs as irregular aggregates in association with other copper sulphides, e.g. *chalcopyrite and *bornite, in mineral deposits.

digital image In *remote sensing, an image in which the continuous variation in the scene being sensed is converted to discrete variation in the form of a finite range of integer values (*pixels) each assigned a *digital number. *See also* ANALOGUE IMAGE.

digital number In *remote sensing systems, a variable assigned to a *pixel, usually in the form of a binary integer in the range of 0–255 (i.e. a *byte). The range of energies examined in a remote sensing system is broken into 256 *bins. A single pixel may have several digital number variables corresponding to different *bands recorded.

digitize 1. To translate graphical information into a series of numbers suitable for processing by (digital) computer. For example, topographic detail can be taken from a map and digitized to produce a computer-generated topographic cross-section. The superimposition of an orthogonal coordinate system on to an image and the recording of the data in a machine-readable form assumes that a two-dimensional image is being analysed. It can be done three-dimensionally if *stereophotographs are analysed. Fully automated systems also exist for this type of work. **2.** To convert analog data into digital form by sampling the continuous record at discrete sample intervals. *See also* ALIASING; and SAMPLING FREQUENCY.

dihedron *See* SPHENOID.

dike *See* DYKE.

dilatancy In rocks, the increase in volume during deformation. This is achieved by an increase in pore volume, rotation of grains, grain boundary slippage, and microfracturing.

dilatation (dilation) The process in which an applied *stress leads to a state of *strain involving a change in volume. The change in volume is given by: $\Delta_v = (V_f - V_o)/V_o$, where V_o is the initial volume and V_f the final volume.

dilatational wave *See* P-WAVE.

dilation *See* DILATATION.

diluvialism Theory of the Earth that used the biblical account of the Noachian Deluge to explain major geologic phenomena. Before about 1800 the biblical flood was frequently cited as the cause of fossilization, sedimentation, deposits, stratifications, etc. In the 19th century the theory was extended by *Cuvier, *Buckland, and others, who believed in a series of deluges, which had caused the extinction of species and the breaks between major formations. The term is also sometimes (incorrectly) used to describe *Werner's theories.

dimension stone Building stone quarried and cut as regular blocks according to specified dimensions, e.g. rectangular, columnar, tabular. Common dimension stones are *limestone and *granite; nowadays often cut quite thinly and used as cladding panels on top of existing *concrete structures, etc. *Compare* FREESTONE.

Dimetrodon angelensis Large, specialized, meat-eating *reptile from the Lower *Permian of North America. *Dimetrodon* had a *synapsid-type skull and is grouped amongst the *pelycosaurs. It grew to over 3 m in length and is noted for the presence of a large sail on its back. The skull of this carnivore was high and narrow, with the teeth differentiated and well adapted to predatory habits.

dimorphism 1. The presence of one or more morphological differences that divide a species into two groups. Many examples come from sexual differences of particular traits, such as body size (males are often larger than females), or plumage (male birds are usually more colourful than females). *See also* SEXUAL DIMORPHISM. **2.** *See* POLYMORPHISM.

dim spot In a seismic reflection event, a reduction in amplitude caused by the presence of *hydrocarbons which lessen the contrast between the *acoustic impedance of the overlying material and that within the reservoir. *Compare* BRIGHT SPOT; and FLAT SPOT.

dimyarian *See* MUSCLE SCAR.

Dinantian The Lower *Carboniferous sub-System in western Europe, overlain by the *Silesian and comprising the *Tournaisian and *Visean Series. It is dated at 362.5–332.9 Ma (Harland et al., 1989) and is roughly contemporaneous with the Carboniferous Limestone (Britain), *Kinderhockian, *Osagean, *Meramecian, and lower *Chesterian (N. America).

dinocyst *See* DINOPHYCEAE.

dinoflagellates *See* DINOPHYCEAE.

Dinophyceae A class of Pyrrophyta, comprising *algae that are unicellular and have two flagella (thread-like structures) of unequal length. Most dinoflagellates belong to this class and the cysts (dinocysts) are useful in *biostratigraphy. The organisms have two biological stages. (*a*) In the motile (thecate) stage (*see* theca) the organism may have either a flexible cell wall or a rigid, armoured one, and it maintains itself in the water by active movement of the flagella. The cell surface bears two furrows, each holding one flagellum. The transverse furrow is called the 'cingulum', the longitudinal one the 'sulcus'. The cingulum divides the cell into an anterior epitheca and a posterior hypotheca. The apex of the epitheca is sometimes extended to form an apical horn. (*b*) In the cyst stage the organism is dormant. When encystment occurs a two-layered cyst wall (phragma) is formed. Proximate cysts develop in the cyst wall, in contact with the wall from the motile stage. Chorate cysts develop deeper in the motile cell and are linked to the cell by processes. Cavate cysts are those where the two layers of the wall are separated by cavities. The organism leaves the cyst by an opening (archaeopyle) when conditions change. There are three important orders: Gymnodiniales, Peridiniales, and Dinophysiales.

Dinophysiales *See* DINOPHYCEAE.

dinosaurs Literally the name means 'terrible lizards', but in fact the dinosaurs were not lizards. They were *diapsid reptiles whose closest living relatives are the *crocodilians and *birds. Dinosaurs first appeared in the Middle *Jurassic and produced an astonishing array of different types and sizes before becoming extinct at the end of the *Cretaceous. The two groups of dinosaurs, *Saurischia and *Ornithischia, are not usually thought to be more closely related to each other than to other *archosaurs, so the concept of 'dinosaur' is a heterogeneous one.

Dione (Saturn IV) One of the major satellites of *Saturn, with a radius of 560 km; mass 10.52×10^{20} kg; mean density 1440 kg/m^3; visual albedo 0.7. It was discovered in 1684 by G. D. Cassini.

diopside An important *pyroxene $CaMgSi_2O_6$, and member of the *clinopyroxenes, forming a continuous series with *hedenbergite $CaFeSi_2O_6$; this series may be called diopside solid solution series and denoted di_{ss} (or Di_{ss}); sp. gr. 3.22 (di) to 3.44 (hed); *hardness 5.5–6.0; *monoclinic; normally a pale, dirty green or grey (or, rarely, colourless); *vitreous *lustre; *crystals usually short, *prismatic, columnar grains, but the bulk are irregular grains; *cleavage prismatic good {110}; a common constituent of magmatic, *basic, and *ultrabasic rocks, e.g. *pyroxenites, *peridotites, *gabbros, and diabases, and also occurs in *basalts and *dolerites; it is common in many *metamorphic rocks, particularly metamorphosed *dolomites and calcareous sediments. Chrome diopside is a bright green, chromium-rich (1–2% Cr_2O_3) variety commonly found in *kimberlite pipes.

dioptase A rare mineral, formula $CuSiO_2(OH)_2$; sp. gr. 3.3; *hardness 5; *trigonal; emerald-green, transparent to translucent; *vitreous *lustre; crystals usually short, *prismatic, and terminated by rhombohedra, but also occurs *massive; *cleavage perfect rhombohedral; occurs in the oxidized zone of copper-sulphide deposits.

diorite *Intermediate, coarse-grained *igneous rock with up to 10% *quartz. *Plagioclase feldspars are of the *oligoclase–*andesine varieties; and *ferromagnesian minerals, e.g. *pyroxenes or *hornblende, are present.

dip 1. The angle of inclination of a planar feature measured from a horizontal *datum. The true dip is always measured in a vertical plane perpendicular to the *strike of the plane of either *bedding or *cleavage. The angle of dip measured in any plane not perpendicular to the strike is an apparent dip and will always be less than the true dip. **2. (magnetic)** The *inclination of the Earth's magnetic field from the horizontal. Positive dips are downwards, negative are upwards from the horizontal plane.

dip circle magnetometer *See* MAGNETOMETER.

dip fault A *fault which *dips parallel to the regional dip of the bed it faults.

dip-isogon method A method developed by J. G. Ramsay (1969) for distinguishing three fundamental classes of *folds based on their relative thicknesses and the curvatures of their surfaces, as indicated by the inclination of (dip) *isogons.

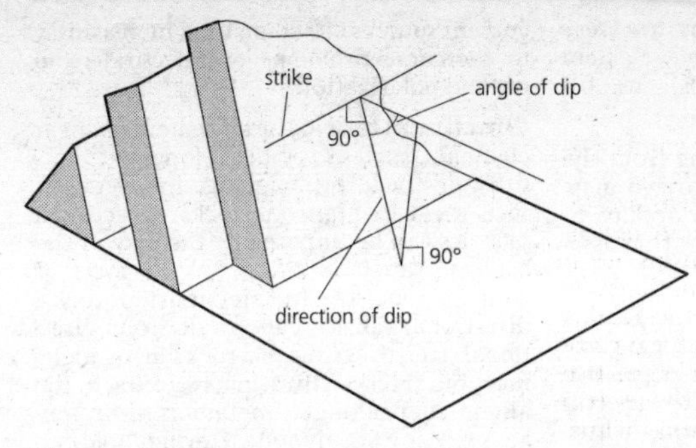

Dip

diphycercal tail Possibly the original type of tail fin in fish, in which the body axis divides the fin into equal *dorsal and *ventral sections.

Diplichnites A *trace fossil (*repichnia) made by an animal as it crawled across a surface.

Diplocraterion A permanent, U-shaped dwelling *burrow created by an animal that lived in the *sediment but was a suspension feeder. The two parallel tubes are circular in section. Some organisms were capable of responding to rapid deposition or erosion. This response is reflected in the presence of concentric laminae on either side, either inside the two arms of the burrow or below the outer curve (or, in the form *Diplocraterion yoyo*, in both positions).

diplograptids A family of graptoloids (order *Graptoloidea) comprising a fauna that spans the period from the *Llanvirn to the lowermost *Silurian.

Diplopleurozoa (phylum *Cnidaria) Extinct, primitive class, known only from the Lower *Cambrian, whose members have a bell-shaped body of elliptical outline, showing distinct *bilateral symmetry. There is a median furrow along the main axis of the body, from which arise numerous, flat, simple segments which reach the periphery and are separated by narrow grooves. The margins are scalloped into short lappets, each bearing a simple tentacle.

Diploporita (subphylum Blastozoa) Extinct class of *cystoids which have *uniserial *brachioles, *theca composed of a large number of irregularly arranged plates, and paired pores (diplopores) present on all plates. They are known from the Lower *Ordovician to Middle *Devonian.

dipmeter log A down-hole geophysical log designed to measure the *dip and dip direction of dipping surfaces in a *borehole. The logging tool consists of four resistivity logging devices set at 90° to one another, and held against the side of the borehole (*see* RESISTIVITY LOG; and RESISTIVITY METHODS). When wound back to the surface they respond instantly to layers of differing electrical resistivity (e.g. a *clay *horizon or porous sands) associated with the bedding, while dipping beds produce a response with a time delay related to the dip of the horizon. Computer processing of the data yields a *tadpole plot of dips and dip directions in the well. It is claimed that dipmeter data can be used to identify tectonic as well as sedimentary structures. In practice, the data are often equivocal and must be interpreted with great caution.

dip move-out *See* MOVEOUT.

Dipneusti *See* DIPNOI.

Dipnoi (Dipneusti; class *Osteichthyes) Literally, 'double breathing'. Often ranked as a subclass, the group includes the extant lung-fish and their fossil relatives (e.g. the Middle *Devonian *Dipterus* and the *Triassic *Ceratodus*). Early forms have an elongated body, a well-ossified internal skeleton, *heterocercal tail, fleshy-lobed fins, and *cosmoid scales. Teeth are absent, but one of the commonest fossils is the broad fan-shaped tooth plate that served for shearing and crushing small invertebrates. Dipnoans first appear in Lower *Devonian rocks and were common in freshwater habitats in the late *Palaeozoic and the *Triassic;

thereafter their fossil remains are very sparse. There are just three surviving genera (*Neoceratodus*, *Protopterus*, and *Lepidosiren*), all tropical.

dipole field The field resulting from the presence of two oppositely polarized magnetic poles. The term is usually applied to the *geomagnetic dipole field or that associated with a *magnetic anomaly.

dip pole A point, usually on the Earth's surface, where the magnetic vector is vertical, i.e. the *inclination is 90°. A positive (downward) inclination is a positive (north) dip pole, and a negative (upward) inclination is a negative (south) dip pole. Many dip poles exist and these are not the same as the two *geomagnetic poles.

dip shooting A field procedure in seismic-reflection studies where, commonly, *split-spread and/or *cross-spread shooting is used to determine the structural *dip of a reflector. The asymmetry of the *normal move-out (NMO) associated with a split-spread shot is diagnostic of a dipping reflector, with the up-dip side having the smaller NMO corrections. The true structural dip can be determined from the dip information provided by a cross-spread shot.

dip-slip fault A fault with a relative displacement parallel to the *dip of the *fault plane. In practice many dip-slip faults have some component of *strike-slip in their displacement. *See* FAULT.

dip slope A geomorphological topographic surface which *dips in the same direction, and often by the same amount, as the *true dip or *apparent dip of the underlying strata. Dip slopes are commonly found in *cuesta and vale topography.

dipyramid *See* BIPYRAMID.

Dirac function ($\delta(t)$) A spike *impulse which has an amplitude at only one instant in time and one unit (however measured) of energy.

direct circulation Circulation system in which lighter air rises, while denser air descends, leading to the conversion of potential energy to kinetic energy. Land and sea breezes are examples of such circulation.

direct current A unidirectional current provided by a source of electrical energy. It is used in *induced polarization surveys and in some electrical *resistivity surveys, although very-low-frequency alternating current sources are being used increasingly to overcome problems of *electrode and ground polarization.

directional fabric The alignment of linear elements, such as crystals, elongate *xenoliths, or *bedding–*cleavage intersections in a rock. In *igneous rocks, directional *fabrics can be imposed by the flow of the *silicate *melt aligning *phenocrysts or elongate xenoliths. In *metamorphic rocks, directional fabrics can result from directional stresses acting on a rock during metamorphic *recrystallization, resulting in the alignment of elongate metamorphic minerals such as *hornblende. During *folding, directional fabrics can result from the intersection of two planar surfaces such as bedding and cleavage to give an 'intersection lineation'.

directional filter In *remote sensing, a type of *spatial-frequency filter which enhances edges within an image which run in selected compass directions.

direct problem *See* FORWARD PROBLEM.

direct wave A seismic wave which travels through the ground directly from the source to the detectors without being reflected off or refracted by a subsurface layer.

dirtying-up trend (bell trend) In the reading from a *wireline log (*see* WELL LOGGING) an expression from a *gamma-ray sonde showing a progressive upward increase in the gamma reading, indicating a change in the clay-mineral content.

discharge A measure of the water flow at a particular point, e.g. a river *gauging station, sewage works, or *groundwater abstraction well. Various units of measurement are in common use depending on the nature of the discharge being measured. River flow may be expressed in cubic metres per second (misleadingly called 'cumecs' in some literature), while *borehole flows may be more conveniently expressed as litres per second.

discharge hydrograph *See* HYDROGRAPH.

discoid *See* SOLITARY CORALS.

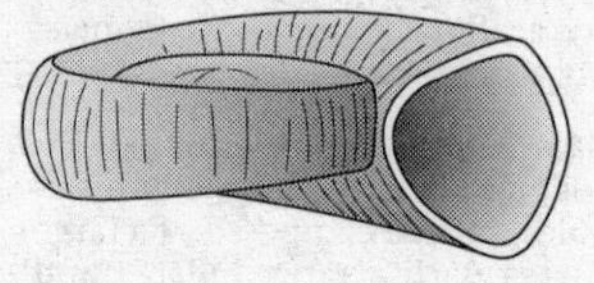

Discoid

disconformity Unconformity such that the beds above and below the surface are parallel. Some disconformity surfaces are highly irregular, whereas others have no obvious relief.

discontinuous reaction series A sequence of mineral reactions each of which takes place at a specific temperature during the cooling of a *magma. A high-temperature mineral will remain in equilibrium with a cooling magma until it reaches its reaction temperature. At this point the mineral completely reacts with the *melt and dissolves away to produce a new mineral in equilibrium with the melt. This second mineral remains in equilibrium with the magma as the magma continues to cool until its reaction temperature is also reached. The process then repeats itself. The principle of discontinuous reaction was first advanced by Norman *Bowen in 1928. In sub-alkaline magmas the discontinuous reaction *olivine–*pyroxene–*amphibole–*biotite is commonly observed.

discontinuity 1. A boundary or layer at depth, marked by a significant change in the speed of transmission of *seismic waves, e.g. *Mohorovičić discontinuity. **2.** In meteorology, sharp changes in temperature, wind, or humidity across a frontal boundary. **3.** A break in sedimentation.

discordant 1. Applied to the cross-cutting relationship of an *igneous intrusion, e.g. a *dyke, when it intersects the *bedding planes or *foliation in the host rock. *See* LAW OF CROSS-CUTTING RELATIONSHIPS. **2.** Applied to a lack of parallelism between the bedding planes (or tectonic *fabric) of one rock unit and others adjacent to it. *Compare* CONCORDANT.

discordant age The situation in which a series of age determinations of a single rock body do not yield the same value (*compare* CONCORDANT AGE). Discordant ages indicate that some transfer of materials has taken place into or out of the rock or mineral, and this transfer could have affected either the parent or daughter *nuclides, or both.

discordant drainage A *drainage pattern that runs across (i.e. is discordant to) the geologic structure.

discordant intrusion *See* BATHOLITH; BOSS; and DYKE.

Discovery A series of missions planned by *NASA, based on small-scale spacecraft that are designed to proceed from development to flight within less than three years for a cost of less than $150 million.

disharmonic fold A *fold which exhibits sharp changes in its geometric properties (i.e. wavelength, symmetry, and shape) when compared with other folds in adjacent layers. It is due to the inter-layering of competent and incompetent beds (*see* COMPETENCE) which buckle on different wavelengths.

dish-pan experiment Method for simulating the convective cell and other global atmospheric motions. A shallow layer of water in a round vessel is heated at the edge, to represent the equator, and cooled at the centre, to represent the polar region. As the dish-pan is rotated, various motions are produced in response to the radial temperature gradient and to the speed of rotation. At lower speeds, the flow pattern is zonal, but with faster rotation large waves develop and incorporate closed circulations. *See also* HADLEY CELL; and ROSSBY WAVE.

dish structure A *sedimentary structure seen in *sandstones, and characterized by repeated horizons comprising a series of concave-upwards, dish-like *laminations. The dishes are commonly lined with thin (0.2–2.0 mm) *clay *laminae and are separated by vertical *pillar structures. Both the dish structures and associated pillars are formed by *dewatering of the pore water from unconsolidated sands.

disjunct Applied to the distribution of populations that have distinct ranges separated from one another by a distance large enough to ensure DNA cannot be exchanged between them. *Compare* CONJUNCT.

disjunct shell *See* COILING.

dislocation 1. The relative displacement to either side of a fracture. **2.** Imperfections developed along a line of atoms within a crystal structure, which may displace or add a row of atoms into the regular structure. Edge dislocations add rows of atoms; screw dislocations displace rows of atoms along a plane.

dislocation creep *See* CREEP MECHANISMS.

dismicrite A *micrite-dominated *limestone disrupted by an abundance of *fenestrae or *birdseye structures. Dismicrites are characteristic of *intertidal to *supratidal environments. *See* FOLK LIMESTONE CLASSIFICATION.

dispersion 1. The spreading of a body of water as it flows. Lateral dispersion is the widening of the path taken by *groundwater as it flows from a known point of origin through a rock *matrix, due to its movement around individual mineral grains within the main rock body. Unless it flows in well-defined fissures or fractures, water does not travel through a rock in a straight line but is forced to flow across a widening front because of the granular nature of the rock matrix. Longitudinal dispersion is the spreading out of a body of water along its own flow path, due to the differences in water velocities in larger and smaller pores of the rock. Both modes of dispersion are normally observed by means of *tracers. Lateral and longitudinal dispersion also occur in river *channels, where they are due to differences in *flow velocity across the channel and between the water surface and the bed, and also to random fluctuations in velocity caused by turbulent eddies. **2.** The process of separating soil particles (as in *aggregates) from each other so that they may react as individual particles. Aggregates or *peds of soil particles are destroyed by dispersion (and their formation is initiated by flocculation. **3.** The distortion of a wave train that occurs when velocity varies with frequency and consequently the *phase velocity is not the same as the *group velocity. There is minimal dispersion for most *body waves (although electromagnetic body waves are subject to considerable dispersion) but it is important for *surface waves, especially *Love waves, particularly in the presence of velocity-layering near the surface. **4.** In *mineral optics, a measure of the difference in *refractive index, determined by direct observation of the biaxial *interference figure using the two extreme wavelengths of the visible spectrum (red and violet). Both *isogyres are observed for red and violet tints. **5.** *See* SWELL.

displaced terrane *See* TERRANE.

displacement The relative movement on either side of a *fault plane; it may be in any direction parallel to the plane. The finite displacement on a fault plane is defined by a straight line connecting the positions of the initial and final points.

displacive transformation *See* POLYMORPHIC TRANSFORMATION.

di$_{ss}$ *See* DIOPSIDE.

disseminated deposit Deposit in which usually fine-grained *ore minerals are scattered throughout the rock. Large, disseminated deposits form important sources of ore, e.g. *porphyry copper deposits.

dissepiment 1. In *corals, one of the small, horizontal, domed plates which form cyst-like enclosures around the edge of a *corallite and which do not extend right across the corallite. **2.** In *graptolites, a strand of chitinous material which connects adjacent branches in a *dendroid colony.

dissepimentarium The part of the interior of a *corallite that is occupied by small plates (dissepiments) which are convex towards the axis of the corallite.

dissipation trail (distrail) Phenomenon of cloud evaporation caused by exhaust heat in the rear of an aircraft flight path. *See also* CONDENSATION TRAIL.

dissolved load The part of a river's total load that is carried in solution. Five *ions normally constitute approximately 90% of the dissolved load: chloride (Cl^-), sulphate (SO_4^{-2}), dissolved bicarbonate (HCO_3^-), sodium (Na^+), and calcium (Ca^{2+}). Generally the load is at its maximum concentration during low-discharge conditions when *groundwater is the main source of flow.

dissolved-oxygen level The concentration of oxygen held in solution in water. Usually it is measured in mg/l (sometimes in $\mu g/m^3$) or expressed as a percentage of the saturation value for a given water temperature. The dissolved-oxygen level is an important first indicator of water quality. In general, oxygen levels decline as pollution increases.

distal Applied to a depositional environment sited at the furthest position from the source area, and generally characterized by fine-grained sediments. *Compare* PROXIMAL.

distribution coefficient *See* PARTITION COEFFICIENT.

distributary channel A natural stream channel that branches from a trunk stream which it may or may not rejoin. It occurs typically on the surface of an *alluvial fan or *delta, where it may be part of a complex, fan-shaped network that distributes the discharge and sediment load of the main channel among many small distributary channels between which a variable assemblage of bays, lakes, *tidal flats, or marshes may exist. The larger distributory channels show *crevasse splays and *levées. *See* ANABRANCHING CHANNEL.

disturbance General term used to denote low-pressure features, e.g. a depression or trough. Disturbances commonly appear as waves in the major air flows in the mid-*troposphere, e.g. the equatorial easterlies, the prevailing westerlies over middle latitudes, and the trade winds.

disturbed days (D days) (geomagnetic) The five days in each month when the disturbance of the geomagnetic field is greatest, determined from observatory records of the *diurnal variation.

Ditomopyge During the *Carboniferous and *Permian, the number of *trilobite genera fell consistently until the extinction of the group at the end of the *Palaeozoic. *Ditomopyge* was one of the last representatives of the family Phillipsiidae. It was comparatively small, with an opisthoparian *suture (*see* CEPHALIC SUTURE), well-developed eyes and a fused *pygidium or tail area. *D. scitula* is known from the Upper *Carboniferous of N. America.

diurnal temperature variation Daily variations in temperature at a particular place, related to the local radiation budget. In mid-latitudes, for example, maximum temperatures usually occur after noon and minimum temperatures in the early morning. The range varies according to location, with high variation in continental areas, and low variation in maritime areas. The diurnal range in equatorial areas exceeds the annual variation in average temperature.

diurnal variation (geomagnetic) Daily variations in the *geomagnetic field, determined from hourly measurements made at observatories, of the horizontal component (*H*), *declination, and vertical component (*Z*) of the field. Most of these variations (Sq variations) are attributed to ionospheric currents induced by solar radiation, but some are associated with lunar effects (Lq variations).

divariant assemblage Two or more metamorphic minerals in equilibrium over a range of pressure and temperature (*see* METAMORPHISM). Varying the pressure or temperature within defined limits will not initiate reactions within the equilibrium assemblage of minerals. Since two independent variables, pressure and temperature, can be changed or varied without upsetting the equilibrium, the assemblage of minerals is said to be in divariant equilibrium.

divaricator muscle (diductor muscle) In *Brachiopoda, the muscle that opens the shell. It is attached at one end to the cardinal process (*see* CARDINALIA) and at the other to the floor of the other valve.

divergence 1. In meteorology, a situation in which, over a given time, more air flows out of a given area than flows in. Some subsiding motion can be expected to take place in the air over this region. **2.** A horizontal flow of water in different directions away from a common centre or line. A particular example of divergence in the oceans is seen in areas of *upwelling. *Compare* CONVERGENCE. **3.** Evolutionarily, genetic segregation and differentiation within a *taxon to the extent that distinct derivative taxa result. The divergence may be at the species, genus, family, order, or higher level (*see* CLASSIFICATION). Thus it is possible to refer for instance to the divergence of *reptiles and *mammals from a stem group, to the divergence of mammal orders, and to the divergence of a breeding population into two related species.

diversification Increase in the *diversity of distinct types in one or more taxonomic categories (i.e. species, genus, etc., *see* CLASSIFICATION). *Phanerozoic, marine invertebrates with well-developed hard parts provide an illustration: their diversity at phylum level remains much the same throughout, whereas at family level there is a peak at the mid-*Palaeozoic, and a trough at the Permo–*Triassic boundary; after this there is a steady increase to a second, higher peak in the *Cenozoic.

diversity Most simply, the *species richness of a *community or area, though it provides a more useful measure of community characteristics when it is combined with an assessment of the relative abundance of species present.

divide The boundary between separate *catchment areas or drainage basins. It is normally marked topographically by high ground. In British usage, a divide is sometimes called a watershed, but watershed has a different sense in US usage.

divining (dowsing, water-witching) The use of hand-held hazel sticks, pendulums, copper wires, etc., to detect the presence of *groundwater. When groundwater is present the hand-held instrument is reputed to move or twitch. There is no conclusive evidence to show that divining can locate groundwater any better than a random sinking of wells, but it is widely used nevertheless.

Dix formula The equation by which the *interval velocity (v_{int}) can be calculated for a zone between two depths on a *seismic section. For two reflectors with reflected-ray travel times t_1 and t_2, and *root-mean-square velocities v_{rms1} and v_{rms2} respectively, then: $v_{int} = [(t_2 v_{rms2}^2 - t_1 v_{rms1}^2)/(t_2 - t_1)]^{1/2}$. The stacking velocities are normally assumed to be true root-mean-square velocities but there may be occasions when it is necessary to correct them for the effects of *dip.

D-layer The lower *mantle, from a depth of about 720 km to the *core boundary at 2886 km depth, in which the seismic velocities increase as a function of adiabatic compression.

DMO *See* MOVEOUT.

Dnepr-Samarovo Glacial deposits occurring in the European USSR which may be the equivalent of the *Riss Stadial of Alpine terminology. They are Lower *Saalian in age.

doctor, the Local term for the W. African *harmattan wind: it refers to the health-giving properties of this wind.

documentation map A map showing the total documentary evidence relating to a proposed engineering site, e.g. mining rights, owners of property, etc.

dodecahedron A *crystal form composed of 12 rhomb-shaped faces, each face intersecting two of the *crystallographic axes and being parallel to the third. It belongs to the *cubic (isometric) system.

Dogger 1. An alternative name for the Middle *Jurassic *Epoch, which is dated at 178.0–157.1 Ma (Harland et al., 1989). It is preceded by the *Lias, followed by the *Malm, and comprises the *Aalenian, *Bajocian, *Bathonian, and *Callovian *Ages. Some authors do not recognize the Aalenian as a separate age, including it instead within an enlarged Bajocian Age. **2.** The name of the corresponding European *series, which is roughly contemporaneous with the uppermost *Herangi and lower *Kawhia (New Zealand). **3.** A *lithostratigraphic unit in the Aalenian of Yorkshire, England.

dog-tooth spar *See* SCALENOHEDRON.

Dokuchaev, Vasily Vasilievich (1840–1903) A Russian soil scientist, Dokuchaev was Director of the Kharkov Institute of Agriculture and Forestry. He studied the formation of soils, especially the *chernozem, developing a soil classification and making soil maps.

doldrums The oceanic equatorial zone, which has low pressure and light, variable winds. The zone moves seasonally north and south of the equator.

dolerite (diabase, microgabbro) A dark-coloured, medium-grained *igneous rock which contains *plagioclase feldspar of labradorite composition and *pyroxene of *augite or titanaugite composition as *essential minerals, and *magnetite, titanomagnetite, or *ilmenite as *accessory minerals. Where *olivine also occurs as an additional mineral, the rock is termed an 'olivine dolerite'. Where *quartz occurs as an additional mineral in the *groundmass, the rock is termed a 'quartz dolerite'. Dolerites are the medium-grained equivalents of *basalts and, like the basalts, can be divided into alkali and tholeiitic types (*see also* ALKALI BASALT; and THOLEIITE). Dolerites are commonly found in shallow level *intrusions such as *dykes, *sills, or *plugs.

Dolgellian A *stage of the Upper *Cambrian, underlain by the *Maentwrogian and dated at 514.1–510 Ma (Harland et al., 1989).

doline (swallow-hole, sink-hole) Steep-sided, enclosed depression in a *limestone region. It is normally located at a site of increased *joint density, which focuses drainage passing vertically through the rock. It enlarges by solution (*carbonation) and by collapse, thus allowing so-called solution dolines to be distinguished from collapse dolines. A shaft may lead from its floor to a cave system.

Dollo, Louis Antoine Marie Joseph (1857–1931) Born in France, Dollo became a professor at the University of Brussels. He was a palaeontologist whose major interest was in the development and evolution of fossil reptiles. He also worked on other fossil vertebrates and modern *marsupials, especially the structure of their limbs.

Dollo's law Evolutionary irreversibility: once regarded as inevitable, but now considered to apply mainly in special cases. The potential for further useful mutation may well be very limited in highly specialized organisms, since only those mutations that will allow the organism to continue in its narrow niche will normally be functionally possible. In such cases there is therefore a self-perpetuating, almost irreversible, evolutionary trend, so much so that it is regarded virtually as a law, 'Dollo's law' (after the palaeontologist Louis *Dollo). The trend results from steady directional selective pressure, or *orthoselection reinforced by specialization, or developmental canalization (*see* CANALIZING SELECTION).

dololithite A *dolomite composed of *detrital dolomite fragments derived from the *weathering and *erosion of pre-existing dolomites.

Dolomieu, Déodat de Gratet de (1750–1801) French explorer and mineralogist, who became Professor of Mineralogy at the Jardin du Roi in Paris. He gave his name to the mineral *dolomite and to the Dolomite Alps.

dolomite 1. (pearlspar) Widely distributed rock-forming *mineral, $CaMg(CO_3)_2$; sp. gr. 2.8–2.9; *hardness 3.5–4.0; *trigonal; usually white or colourless, but can be yellowish and brown; white *streak; *vitreous *lustre; crystals are usually rhombohedral with curved, composite faces, also occurs *massive and granular; *cleavage perfect rhombohedral {1011}; usually *secondary, having formed by the action of magnesium-bearing solutions on *limestones (*dolomitization), also occurs as a *gangue mineral in *hydrothermal veins particularly associated with *galena and *sphalerite. It dissolves very slowly in cold, dilute acid, but effervesces very readily when warmed. It is used as a building stone and in the manufacture of bricks for furnaces. **2. (dolostone)** A *sedimentary rock type, usually formed by the *dolomitization of *limestones, and commonly occurring interbedded with them. Most limestones contain some magnesium carbonate and strictly the term 'dolomite' refers to rocks containing 90% or more of the mineral dolomite (*see* 1). Dolomite that has formed soon after deposition tends to be fine-grained and to have preserved the original *sedimentary structures, whereas *recrystallization in late-diagenetic dolomites produces a coarser-grained rock, a loss of sedimentary structures, and an increases in *porosity. *See also* ANKERITE.

dolomitization The transformation of *limestone to *dolomite (1), by the conversion of $CaCO_3$ to $CaMg(CO_3)_2$. This occurs by the addition of magnesium to the sediment or rock and may take place soon after deposition or at various stages during *diagenesis. Dolomitization is thought to occur in *supratidal areas where sea water is drawn into the sediment and evaporation results in a high ratio of magnesium to calcium in the pore waters. Deeply buried limestone may become dolomitized if a mixture of fresh water and sea water passes through the rock. Dolomitization results in a 13% increase in *porosity, making dolomites important *reservoir rocks.

domain The primary division of living systems. There are three domains: *Archaea, *Bacteria, and *Eucaryota.

domal uplift The upwarping of an area in the form of a dome. Domal uplifts are usually about 1 km high, and 10^2–10^3 km across, and are characterized by negative *gravity anomalies and by alkaline *volcanicity. Domal uplifts may also have a radiating system of *rift valleys, and many are thought to be precursors of continental rifting, leading to the development of an ocean basin, and thus represent the first stage of the *Wilson cycle.

dome 1. *Anticlinal structure which plunges in all directions. **2. (volcanic dome, tholoid)** A mound of viscous *lava, usually *rhyolite in composition, which has grown and built up over a vent. The mound of solid lava is covered by coarse, angular blocks which form by chilling and brecciation of the growing dome's surface. The blocks accumulate around the growing dome to produce a *scree slope of crumble *breccia. Domes can grow by repeated injection of *magma into the dome body (endogenous dome) or by repeated eruption of small volumes of *magma from the surface of the dome (exogenous dome). **3. (salt dome)** A circular or elongate plug, 1–2 km in diameter but extending downwards for many kilometres, formed by the upward movement of buoyant and less dense evaporitic material (commonly *halite) into denser overlying rocks. The diapiric movement (*see* DIAPIR) may be initiated by tectonic thickening. **4.** A special form of *crystal development characterized by two roof-like faces symmetrical about a plane of symmetry. The faces are repeated once only about an *axis of symmetry. **5.** *See* PERICLINE. **6.** *See* ICE DOME.

dome and basin *See* INTERFERENCE PATTERN.

dome–crescent–mushroom *See* CRESCENT-AND-MUSHROOM; and INTERFERENCE PATTERN.

domichnia A category of traces (*see* TRACE FOSSIL) made by animals in the creation of a permanent dwelling structure. The *borings of bivalves (*Bivalvia) such as *Pholas* are included in this category.

Dominian Reef *See* RANDIAN.

Donau A period of glaciation which occurred at about the beginning of the *Pleistocene. It may correspond to the *Nebraskan Stage in N. America. Evidence of it has largely been removed by later glacial episodes.

Donau/Günz Interglacial An *interglacial *stage of the Alpine areas that may be equivalent to the *Waalian of northern Europe and the *Aftonian of N. America.

Doppler radar A device that measures the *Doppler shift in a radar beam reflected from water droplets on either side of a rotating *mesocyclone or *tornado. Angular velocity can be calculated from the extent of the red shift on one side and blue shift on the other.

Doppler shift The apparent change in frequency of waves whose source is moving towards or away from an observer. It was first described by the Austrian physicist Christian Doppler (1803–53).

Dorogomilovskian A *stage in the *Kasimovian Epoch, preceded by the *Chamovnicheskian and followed by the *Klazminskian (*Gzelian Epoch).

dorsal Towards the upper surface of an organism (in vertebrates the side of the animal closest to the spine); the opposite of *ventral.

dorsum The inner margin of an ammonoid (*see* AMMONOIDEA) shell; the opposite margin to the *venter.

dot chart A semi-circular, transparent overlay with points proportional to the distance from a central point. It is used to determine potential effects, particularly for assisting in calculating the *topographic correction in gravity surveys. The elevation and density are determined for each point and assumed to be typical for the area immediately surrounding the point.

Dott classification A widely used classification of *sandstone types which divides sandstones into arenites (less than 15% of rock is mud *matrix) and wackes (more than 15% but less than 75% of the rock is mud matrix). The arenites are subdivided into quartz arenite (more than 95% *quartz grains), arkosic arenite (more than 25% *feldspar and more feldspar than rock fragments), subarkose arenite (5–25% feldspar grains and more feldspar than rock fragments), lithic arenite (more than 25% rock fragments and more rock fragments than feldspar), and sublitharenite (5–25% rock fragments and more rock fragments than feldspar). The wackes are divided into quartz wacke (more than 95% quartz grains), lithic wacke (more than 5% rock fragments and more rock fragments than feldspar), and feldspathic (arkosic) wackes (more than 5% feldspar and more feldspar than rock fragments).

double core barrel *See* CORE BARREL.

double couple (Type II earthquake source) A seismic wave pattern, consisting of four lobes (for *P-waves) and four lobes (for *S-waves) of alternate compression and dilation, which is generated by movement along two *fault planes at right angles to each other. *Compare* SINGLE COUPLE.

double planation In tropical areas, *weathering below the surface at the basal *weathering front is one level of planation, whereas on the land surface weathered material can be modified by wind and water processes and this provides a second level of planation. Climate and local conditions determine the relative rates of development of these two levels of planation, and they can contribute to the production of an *etchplain.

double refraction *See* BIREFRINGENCE.

double sulphides *See* SULPHIDES.

double zig-zag *See* INTERFERENCE PATTERN.

doublure *See* CEPHALON.

down-hole hammer drilling A method of drilling a *borehole using a pneumatic, percussive hammer drill. The rock is fragmented by repeated impaction presented directly to it. This produces a medium-diameter hole, but drilling speed is slow so that operating costs tend to be high. The method is often used in drilling holes for a pre-split surface, since such holes have to be precisely located. *See* PERCUSSION BORING.

downlap A discordant relationship at the lower boundary of a *depositional sequence where younger, initially inclined *strata thin out down-dip across an underlying inclined or horizontal surface. *Compare* ONLAP. *See* BASELAP; and SEISMIC STRATIGRAPHY.

downthrow The relative downward *displacement on one side of a *fault.

Downtonian In British stratigraphy, a *stage at the base of the *Devonian roughly contemporaneous with the lower *Gedinnian Stage. Some authors extend it into the uppermost *Silurian.

downward continuation A technique, which must be used with extreme caution, in which measured values of a potential field (usually gravity or magnetic) at one surface are used to determine the values that field would have at a lower surface. However, the computed field is often erratic and unreliable as a result of the noise contained in the original measurements becoming exaggerated with downward continuation. The method may be useful in resolving anomalies which overlap at the surface where measurements are made, provided the depth to which the field is continued is not below that of the causative bodies themselves. If that depth is exceeded, the computed field may deteriorate completely and become meaningless.

downwelling (sinking) The downward movement of surface waters caused by the convergence of different water masses in the open ocean, or where surface waters flow towards the coast. An example of the latter is found along the Washington–Oregon coast in winter. The *Antarctic convergence zone is a major downwelling region in the Southern Ocean and is the source of *Antarctic intermediate waters.

dowsing *See* DIVINING.

draa High (more than 300 m) sand ridge or chain of *dunes in the Sahara, lying about 0.5–5 km from its nearest neighbour and moving at 2–5 cm per year. It is the largest land-form of the *erg, or sand desert. A star-shaped dune, or 'rhourd', is developed at the site where two draa chains cross.

drag 1. The flexuring of bedding and *cleavage traces along the margins of a *fault plane produced by the *displacement of either side. Cleavage and bedding traces are seen to *dip asymptotically into and out of the plane of faulting. **2.** Ductile deformation of features towards and into *shear zones.

drainage 1. The passage of water over and through the land surface, ultimately towards the sea. *See* DENDRITIC DRAINAGE; DERANGED DRAINAGE; DISCORDANT DRAINAGE; DRAINAGE DENSITY; DRAINAGE PATTERN; INCONSEQUENT DRAINAGE; and SUPERIMPOSED DRAINAGE. **2.** Process of removing the gravitational water from soil, using artificial or natural conditions, such that freely moving water can drain, under gravity, through or off soil. *See* MOLE DRAIN; and TILE DRAIN.

drainage basin morphometry The measurement of the surface form of a drainage basin, and of the arrangement and organization of the associated river network. Properties such as area, shape, gradient, and relief are important elements of form (*see* DRAINAGE BASIN SHAPE INDEX; and DRAINAGE BASIN RELIEF RATIO), while the stream network is investigated through a study of its components and of the ways in which they are related. *See* DRAINAGE NETWORK ANALYSIS.

drainage basin relief ratio An index (Rh) of the relief characteristics of a drainage basin. It is expressed as $Rh = H/L$, where H is the difference in height between the highest and lowest points in the basin and L is the horizontal distance along the longest dimension of the basin parallel to the main stream line. The ratio can provide a measure of the rate of sediment loss from a basin, with which it tends to be positively correlated.

drainage basin shape index A measure of the shape of a drainage basin, normally expressed as the ratio between two dimensions of the basin being considered. One such measure is the circularity index (or ratio), C, expressed as $C = A_b/A_c$, where A_b is the area of the basin and A_c is the area of a circle with the same length of perimeter as the basin. Another index is the form factor, F, expressed as $F = A/L^2$, where A is the area of the basin and L is its length. Such indices may help in forecasting the flood potential of a basin.

drainage density A measure of the average spacing between the streams draining an area. It is obtained by dividing the total length of streams by drainage area. Its magnitude is affected by factors such as the amount of rainfall, permeability of the ground surface, and age. *See* DRAINAGE.

drainage morphometry (morphometric analysis, Horton analysis) Calculation of a range of dimensionless drainage-network relationships (*see* BIFURCATION RATIO for example), based on a system of stream ordering, i.e. the numerical ranking of channel segments within a channel network. Proposed by R. E. Horton in 1945.

drainage network *See* DRAINAGE PATTERN.

drainage-network analysis The study of the way in which the pattern of streams in a drainage basin is organized. Classical work focused on the relations between the components of a network, e.g. on the link between the importance (or 'order') of a stream segment and its frequency. Certain 'laws' of drainage network composition were derived. The modern approach emphasizes the importance of random processes in an explanation of these 'laws' and is more concerned with the density of the drainage network (drainage density = total length of channels/drainage area).

drainage pattern The spatial relationship between individual stream courses in an area. The resulting pattern often reflects the underlying rock type and structure, and several varieties are recognized. A dendritic pattern is the most common, characterized by a randomly branched arrangement. It is not structurally controlled, and is developed on a homogeneous rock, e.g. *clay. A trellis pattern consists of subparallel streams, usually aligned along the geologic *strike, and joined at right angles by tributaries. A rectangular pattern is dominated by right-angled bends, and reflects control by *joints or *faults. A centripetal pattern consists of stream courses converging into a central depression. The drainage network is the drainage pattern viewed geometrically.

drainage-sediment survey *See* STREAM-SEDIMENT ANALYSIS.

drainage wind *See* KATABATIC WIND.

drained test *See* TRIAXIAL COMPRESSION TEST.

dravite *See* TOURMALINE.

drawdown The lowering of the *water-table or *potentiometric surface, normally as a result of the deliberate extraction of *groundwater.

dreikanter *See* VENTIFACT.

Dresbachian A *stage of the *Cambrian in the *Croixian *Series of N. America, overlain by the *Franconian.

driblet cone *See* SPATTER CONE.

drift 1. Any sediment laid down by, or in association with, the activity of glacial ice. The term is often widened to include related submarine and lacustrine deposits. The British Geological Survey has used it to refer to all superficial (i.e. draft) deposits. It was introduced by C. *Lyell (1797–1875), who suggested that glacial deposits were laid down by melting icebergs which drifted across an ice-age sea covering Britain. This old term is now largely superseded by more recent classifications. **2. (Instrumental)** The change in the output of a recording device due to internal factors. Systematic drift can be compensated for by repeat readings at a *base station.

drifter *See* PERCUSSION BORING.

drift map A geologic or geomorphological map which shows the distribution of more recent glacial, *fluvial, fluvioglacial, *alluvial, and marine *sediments. Depending on the distribution and extent of *drift the map may show a combination of solid and drift exposures.

drill hole *See* BOREHOLE.

drilling The process of making a hole, usually into the Earth. Two common methods are *cable drilling and *rotary drilling. *See also* BIT; CALYX DRILLING; DIAMOND DRILLING; DOWN-HOLE HAMMER DRILLING; and PERCUSSION BORING.

drilling bit *See* BIT.

drilling mud *See* MUD, DRILLING.

drill string *See* STRING.

dripstone (flowstone, speleothem) Calcium carbonate rock deposited in caves by the precipitation of *calcite from water as excess dissolved carbon dioxide is diffused into the atmosphere. Dripstone takes various forms, including *stalactites, helictites (having spiral form), curtains, ribbons, and *stalagmites.

drizzle Precipitation of very small (200–500 μm) water droplets generated by *coalescence at the base of cloud such as *stratus. *See also* CRACHIN.

DRM 1. *See* DEPOSITIONAL REMANENT MAGNETIZATION. **2.** *See* DETRITAL REMANENT MAGNETIZATION.

dromaeosaurid Member of a *coelurosaurid branch of the *theropod *dinosaurs, which developed from the genus *Dromaeosaurus* (the 'emu reptile'), known from the early Upper *Cretaceous of Canada. Dromaeosaurids had relatively large brain cases and may have been among the most intelligent reptiles ever to have lived.

drop ball A simple and economical method for breaking up large stones after quarry blasting by dropping a heavy weight from a crane on to the boulder.

dropstone 1. A *clast, released by melting from the base of a floating *ice sheet or *glacier, which falls through the water body to settle in muddy sediment. Dropstones can often be recognized as isolated, coarse clasts set in fine, laminated sediments. **2.** A volcaniclastic bomb (*see* VOLCANIC BOMB), ejected from a *volcano, which falls into water and settles in muddy sediment on the lake or sea floor.

drought A relative term denoting a period during which rainfall is either totally absent or substantially lower than usual for the area in question, so that there is a resulting shortage of water for human use, agriculture, or natural vegetation and fauna.

drought cycle Temporary and repetitive phase of drier conditions in an otherwise favourable environment, e.g. the 22-year drought cycles of N. American grasslands.

drumlin Smooth, streamlined, oval-shaped land-form, one end of which is blunt and the other tapered. Drumlins may occur singly, but more commonly they are found within a large group, called a 'drumlin field' or 'drumlin swarm'. Usually they are composed of *boulder clay, but occasionally they are made of solid rock (hence 'rock drumlin'). They are believed to be formed beneath the outer zone of an expanding *ice sheet, during a major advance: they result from the selective deposition of material that is then streamlined by the moving ice. The long axis of a drumlin lies parallel to the direction of the advance.

drumlin field *See* DRUMLIN.

drumlin swarm *See* DRUMLIN.

druse (adj. drusy) A cavity (*vugh) in an *igneous rock or a *mineral vein into which *euhedral (well-formed) *crystals of the rock or mineral vein project, or the crystals themselves. The cavity represents a volume of late-stage, vapour-rich *magma trapped by the rock crystallizing around it. Crystals can grow freely into this medium and hence crystallize in perfect forms (e.g. smokey quartz in granite) to give well-shaped *crystal faces. The word is German, *Druse* meaning decayed or weathered ore.

drusy *See* DRUSE.

dry adiabatic lapse rate The rate at which dry (i.e. unsaturated) air cools when rising adiabatically through the atmosphere as a result of the utilization of energy in expansion. It is 9.8 °C/km. *See also* INSTABILITY; SATURATED ADIABATIC LAPSE RATE; and STABILITY.

dry air 1. (climatol.) Air with low *relative humidity. **2. (meteorol.)** Unsaturated air.

Dryas Part of the characteristic three-fold *late-glacial sequence of climatic change and associated deposits following the last (*Devensian) ice advance and prior to the onset of the markedly warmer conditions of the current (*Flandrian) *Interglacial. The type sequence was first described for *Allerød in Denmark, and shows upper- and lower-clay deposits rich in remains of *Dryas octopetala* (mountain avens), and between them deposits of lake mud with remains of cool-temperate flora, e.g. tree birches. The colder Dryas phases mark times of cold, tundra-like conditions throughout what is now temperate Europe. The three-fold Dryas–Allerød–Dryas sequence forms *Pollen Zones I, II, and III of the widely accepted late and post-glacial chronology of Europe. The basal, Older, Dryas deposit forms Zone I; the Allerød Zone II; and the Younger Dryas Zone III. In north-western Europe Pollen Zone I is subdivided into a, b, and c. Zone 1b represents a proposed *Bølling *Interstadial, with Zones 1a and 1c referred to as Oldest and Older Dryas respectively.

dry-bulb thermometer Thermometer that registers normal air temperature. It may be used in conjunction with a *wet-bulb thermometer: the *relative humidity can be found by measuring the depression of temperature registered by the wet bulb. *See also* HYGROMETER; and PSYCHROMETER.

dry ice Solid carbon dioxide. It is used in *cloud seeding to cool air in supercooled clouds by *sublimation at low temperatures. This can generate many ice crystals for further ice nucleation.

dry melt *See* MELT.

dry season Period each year during which there is little precipitation. In tropical climates, e.g. over much of India, the dry period is often in the winter season. In very low latitudes, two dry seasons may occur each year, between the northward and southward passage of the equatorial rains. In subtropical, Mediterranean, and west-coast climates, the dry season is in the summer.

dry valley Linear depression that lacks a permanent stream but that shows signs of past water *erosion. It is a common land-form in areas underlain by permeable rock, e.g. the Chalk of southern England. The dry valley was eroded during an episode of surface drainage, perhaps due to *permafrost conditions, to greater precipitation, or to a higher *water table.

dry-weather flow *See* BASEFLOW.

DSDP *See* DEEP SEA DRILLING PROGRAMME.

DSL *See* DEEP SCATTERING LAYER.

d-spacing *See* COVALENT RADIUS.

dual decay *See* BRANCHING DECAY.

ductile behaviour The response to *stress of certain materials which undergo permanent deformation without fracturing. This produces permanent strain marked by smooth variations within the deformed rock. Ductile behaviour is enhanced where high *confining pressures are combined with high temperatures and low rates of strain, conditions characteristic of deeper crustal levels.

dug well *See* BOREHOLE.

dump structure A conical mound of sediment formed from a large amount of debris released when an iceberg overturns or breaks up.

dune A land-form produced by the action of wind on unconsolidated sediment, normally sand. Aeolian dune forms range from small *ripples less than 1 cm in height, to the *draa forms of the Sahara which rise to more than 300 m. Such dunes may be divided into three basic categories: *barchans; longitudinal or 'seif' dunes, which parallel the wind direction; and transverse dunes which are aligned normally to the dominant wind. Transverse dunes are initial forms on sandy coastlines in temperate regions. They migrate inland and may be eroded locally by the wind to form a damp hollow or '*dune slack'. The enclosing crescentic dune is a 'parabolic' dune whose form reverses that of the barchan. *See also* AKLÉ DUNE; COPPICE DUNE; DUNE BEDFORM; and STAR DUNE.

dune bedform ('megaripples') Mounds or ridges of *sand which are asymmetrical, and are produced subaqueously by flowing water. The external morphology is similar to the smaller '*ripple' and larger '*sand wave', with a gently sloping, upstream side (stoss), and a steeper downstream side (lee). The crestline elongation extends transverse to the flow direction and is sinuous or lunate in plan. The height varies between 0.1 m and 2 m, while the wavelength (spacing) between dunes is 1–10 m. Size and growth are limited by water depth and, in general, dune height is less than one-sixth of the flow depth. The down-current migration of dune bedforms leads to the formation of cross-bedding in sediments.

dune slack Flat-bottomed, hollow zone within a sand-dune system that has developed over impervious strata. The slack may result from erosion or *blow-out of the *dune system, and the flat base level is therefore close to or at the permanent *water-table level. Characteristically, dune slacks have rich, marshy flora, with *Salix* species (willows) as typical woody colonizers.

Dunham classification A widely used *limestone classification, proposed by Robert Dunham in 1962, which divides limestones on the basis of their texture and mud content. For limestones which retain their original, depositional texture, the main subdivisions are: lime mudstone (limestone with less than 10% grains in a mud-supported sediment); lime wackestone (limestone with more than 10% grains in a mud-supported sediment); lime packstone (grain-supported limestone with mud *matrix between the grains); lime grainstone (grain-supported limestone with no mud matrix); and lime boundstone (limestone whose original components were bound together (e.g. by corals or algae) during deposition). For limestones in which the depositional texture has been destroyed by *recrystallization, Dunham defines two types: crystalline limestone (recrystallized limestone with a fine texture); and sucrosic limestone (recrystallized limestone with a coarse texture). The original Dunham classification does not subdivide limestones with particles coarser than 2 mm, or differentiate between different types of organically bound limestone. These categories of limestone are defined by Embry and Clovan in their modifications to the Dunham classification. *See* EMBRY AND CLOVAN CLASSIFICATION.

dunite Coarse-grained, *igneous rock, consisting mainly of *olivine. It was first described from the Dun Mountain Range, New Zealand.

Duntroonian A *stage in the Lower *Tertiary of New Zealand, underlain by the *Whaingaroan, overlain by the *Waitakian, and roughly contemporaneous with the mid *Chattian Stage.

duplex A series of *horses bounded by a *roof thrust and a *floor thrust. There are three main types of compressional duplexes: hinterland dipping; foreland dipping; and antiformal stacks. There are also extensional duplexes in normal *dip-slip regimes. Duplex terminology has also been applied to *strike-slip terrains where smaller *en échelon features bound by two continuous, major *fault zones form a strike-slip complex. In strike-slip regimes extensional and compressional duplexes may coexist.

duplicatus From the Latin *duplicatus* meaning 'doubled', a type of cloud with overlapping layers, typified by such genera as *stratocumulus, *altocumulus, *altostratus, *cirrostratus, and *cirrus. *See also* CLOUD CLASSIFICATION.

durain *See* COAL LITHOTYPE.

duricrust Deposit of the *weathering zone, especially in subtropical environments, which may ultimately develop into a hardened mass. A range of types occurs, each distinguished by a dominant mineral: *ferricrete and *alcrete are dominated by sesquioxides of iron and aluminium respectively; silcrete by silica; and *calcrete (caliche) by calcium carbonate.

duripan Mineral diagnostic *soil horizon which is *cemented by silica and so will not slake or fall apart in water or hydrochloric acid. It may contain secondary cement, e.g. carbonates and iron oxide. Where duripans are exposed on the soil surface, they are called duricrust. *Compare* CALICHE.

durophagic Adapted to the eating of hard materials, such as the diet of many benthic dwellers, comprising shelled invertebrates. Skates and rays, for instance, have a tough dentition and protrusible mouth capable of powerful suction which may be used to dislodge shellfish from rock faces.

dust Solid particles, the size of clay and silt particles (*see* PARTICLE SIZE), that can be raised and carried by the wind.

dust-bowl An area of the Great Plains region, USA, where a combination of drought and inappropriate farming practices, especially an expansion of wheat production, led to severe *deflation and soil erosion during the middle 1930s. More generally, any region where deflation of cultivable land occurs.

dust devil (dust whirl, sand pillar) Very localized *whirlwind in a desert area, where strong convection uplifts dust and sand often to a height of a few tens of metres.

Dust storm A wind carrying sufficient dust for visibility to be reduced to less than 1 km.

Dutch cone Type of *cone penetrometer which may be used to give a continuous log of layered sequence in a *soil profile.

Du Toit, James Alexander Logie (1878–1948) Du Toit was a South African geologist who made field studies of the provinces of *Gondwanaland, and found extensive evidence for *continental drift. He published his ideas in *Our Wandering Continents: An Hypothesis of Continental Drift* (1937).

Dutton, Clarence Edward (1841–1912) An officer in the US Army, Dutton was seconded to the US Geological Survey. He introduced the term '*isostasy', a concept for which he saw evidence in the Colorado plateau. Interested also in *volcanology and *seismology, he investigated the Charleston (USA) earthquake of 1886, and in 1904 published a textbook on *seismology.

dyke (dike) *Discordant, or cross-cutting, tabular intrusion. Most dykes are vertical or near vertical, having pushed their way through the overlying *country rock. *See* DYKE SET; DYKE SWARM; and RADIAL DYKE.

dyke set A suite of *dykes whose alignment is parallel or subparallel, reflecting their emplacement from a common source and under a common *stress regime.

dyke swarm A collection of many subvertical *radial dykes around a central *intrusion; or many parallel to sub-parallel *dykes occurring over a large regional area (*dyke set).

dynamic correction *See* STATIC CORRECTION.

dynamic correlation A *cross-correlation process which involves traces of different offsets, and the adding together of the cross-correlations for similar pairs of traces over a number of adjacent *depth points. The cross-correlations for successive differences of offset are squared, displayed, and alignments selected to calculate the residual *normal move-out and *stacking velocity for each alignment. It is a GSI (Geophysical Service Inc.) process.

dynamic equilibrium *See* EQUILIBRIUM.

dynamic metamorphism Fragmentation and *recrystallization of rocks in narrow zones such as *faults or *thrusts where strong deformation has occurred. Rocks are ground to a fine powder in the zone of deformation and, because of their fine *grain size, recrystallize efficiently under the extreme directional stress and release of frictional heat during deformation. The fine-ground powder recrystallizes to a flinty rock which often surrounds fragments of uncrushed *country rock in the deformation zone to form a *mylonite. During extreme deformation all the fragments in the deformation zone are ground down to powder and recrystallize to form a fine-grained, banded rock known as an 'ultramylonite', and rocks can melt to form *pseudotachylyte.

dynamic viscosity In a flowing fluid, a measure of the resistance of the fluid to changing its shape, defined as the ratio of the sheer stress to the rate of deformation sustained across the fluid.

dysaerobic (poikiloaerobic) Applied to a depositional environment with 0.1–1.0 ml of dissolved oxygen per litre of water. *Compare* AEROBIC and ANAEROBIC.

dysodont Applied to a type of hinge *dentition, found in certain *bivalves, where teeth are simple, small, and situated very close to the *dorsal margins of the *valves.

Earth The third planet in the *solar system, outwards from the *Sun. The mean distance of the Earth from the Sun is 149.6×10^6 km. This distance provides the standard 'astronomical unit' (AU) of measurement. The Earth has a mean radius of 6371 km, density of 5517 kg/m^3, and a mass of 5.99×10^{27} g. The *oceanic (5–7 km thick) and *continental (40 km thick) crusts are separated by the *Mohorovičić discontinuity from the silicate *mantle, which extends to the *Gutenberg discontinuity at 2900 km depth, and overlies a molten, iron-rich *core. The oldest rocks are about 3980 million years old, and the Earth formed about 4600 million years ago.

earthflow Flow of unconsolidated material down a hillslope, normally resulting from an increase in pore-water pressure, which reduces the friction between particles. Flow velocities vary from slow, when behaviour is plastic, to rapid, when behaviour is more liquid, and reflect variations in water content. Dry flows may occur when an earthquake shock breaks inter-granular bonds.

earthquake Motion of the Earth. Tectonic earthquakes result from the release of accumulated *strain when brittle failure occurs. This failure coincides with the release of *stress on the rocks that actually break. Earthquakes are usually classified in terms of their depth: shallow are less than 70 km depth; intermediate 70–300 km; and deep more than 300 km. No earthquakes are known below 720 km depth. Earthquakes may also be caused by volcanic activity or induced explosions (e.g. A-bombs) to which the elastic model of tectonic earthquakes does not apply. The energy released is not stored kinetic energy, but chemical/physical energy which imposes a sudden stress that locally exceeds the strength of the rocks and no significant accumulated strain is involved as the rocks yield to the imposed stresses.

earthquake energy The amplitude of a seismic signal is proportional to the energy (E) released in an earthquake. There is no consensus on the actual equation. This is based on the *Richter magnitude but also depends on the frequency of the wave, distance from the source, etc., and is of the form $\log E = aM + b$. The annual energy dissipation by this means is estimated to be more than 10^{11} watts, of which 75% is released in shallow earthquakes and 3% in deep.

earthquake intensity *See* EARTHQUAKE MAGNITUDE.

earthquake magnitude (earthquake intensity) The magnitude of an *earthquake can be estimated from its destructiveness using the *Mercalli scale. As this measure is also dependent on the local geologic and building context, it has been largely replaced by the *Richter scale, based on the amplitude of *seismic waves.

earthquake mechanisms Natural, artificial, or induced events that cause earthquakes. Natural mechanisms include rock falls and slides, spontaneous rock-bursts, volcanic explosions, and tectonic plate motions. Artificial and induced earthquakes can result from explosions (quarry blasts, pressure release below dam sites, nuclear bombs, etc.) or rock-bursts associated with pressure release due to mining, etc. Generally such stress releases are sudden, resulting in the release of seismic energy, but *bradyseisms gradually release stress and thus do not result in an earthquake.

earthquake prediction Most predictions are based on attempts to determine a stress increase prior to rock rupture. This may involve *geodetic measurements to monitor relative motions, changing elevation, etc., or phenomena resulting from stress accumulation (changes in the magnetization, temperature, gas release, etc.), some of which may affect animals. So far, most methods indicate only an increasing probability of seismic activity and cannot be used to predict an actual occurrence, other than the use of *foreshocks, often only minutes prior to a major main *earthquake, but such small earthquakes do not necessarily lead to major activity. Quiescence within an active seismic area can indicate either a gradual increase in stress or that stress release is taking place gradually. *See* FOCAL MECHANISM; and EARTHQUAKE MAGNITUDE.

Earth rotation (day length) Astronomical observations, mainly based on *eclipses of the *Sun, suggest a deceleration of the *Earth's rotation by about 41 seconds of arc per century. Fossil coral growth rings indicate rotation rates corresponding to about 400 days/year some 400 million years ago. Rotation is affected by periodic and irregular events. Most rapid irregular changes are meteorological or oceanic in origin and depend on the mechanical coupling between the *atmosphere, oceans, and solid Earth, but they also affect the Earth's moment of inertia. The major long-term change is due to the slowing of the Earth's rotation by the tidal drag of the *Moon and, to a much lesser extent, by the Sun and planets, most of it attributed to the M_2 *ocean tide. Irregular fluctuations of largely unknown origin also occur, possibly associated with electromagnetic coupling between the Earth's *core and *mantle. *See also* CHANDLER WOBBLE.

Earth tides All terrestrial tides are caused by an imbalance between the centrifugal forces operating on the *Earth and the changing gravitational fields of the *Moon, *Sun, and planets. The *oceanic tide is thus a manifestation of the same phenomenon as that causing internal ('solid' Earth) tidal effects, particularly in the outer, liquid *core of the Earth. The tides are seen at the surface mostly as resonance effects between the tidal components of similar harmonic frequencies as the Earth's *free oscillation periodicities.

earthy Applied to the non-*metallic *lustre of porous aggregates of *minerals, e.g. *clays, *laterites, and *bauxite.

East Australian current Oceanic water current that flows along the east coast of Australia. This narrow (100–200 km wide) current forms the westerly part of the anticyclonic circulation in the S. Pacific. The flow velocity varies in the range 0.3–0.5 m/s. It is an example of a western *boundary current.

easterly wave Type of weak *trough in a tropical easterly airflow, which has a wavelength generally of 2000–4000 km. Such waves occur, for example, over W. Africa and in the Caribbean area where they develop in summer and autumn with a weak or absent *trade-wind inversion, and in the central Pacific, when the equatorial trough is displaced northward. Disturbances in such waves often lead to tropical storms, which vary in intensity.

Eastonian A *stage of the Upper *Ordovician of Australia, underlain by the *Gisbornian and overlain by the *Bolindian.

East Pacific Rise The oceanic *ridge which separates the *Pacific Plate from the *Antarctic, *Nazca, and *Cocos Plates. The East Pacific Rise is a fast-spreading ridge with a maximum *half-spreading rate calculated at 4.4 cm/yr, and its topographic profile is relatively smooth compared with slow-spreading ridges, e.g. the *Mid-Atlantic Ridge.

ebb tide Falling *tide: the phase of the tide between high water and the succeeding low water. *Compare* FLOOD TIDE.

Eburonian A northern European *stage dating from about 1.8 to 0.9 Ma ago, associated with a period of glaciation at about the beginning of the *Pleistocene. It may correspond to the *Donau stage in the Alpine area and the *Nebraskan stage in N. America.

Ecardines Alternative name for the class *Inarticulata of the phylum *Brachiopoda.

ecdysis Periodic shedding of the *exoskeleton by some invertebrates, or of the outer skin by some *Amphibia and *Reptilia.

Echinodermata (echinoderms) Phylum of 'spiny-skinned' invertebrate animals which are entirely marine. They are characterized by an internal skeleton of porous calcite plates; a *pentameral symmetry (although a *bilateral symmetry is often superimposed upon this radial plan, especially in many modern *Echinoidea); and the presence of a water-based vascular system, a complex internal apparatus of fluid-containing tubes and bladders which pass through pores in the skeleton and are seen from the outside as *tube feet. The phylum is varied, and includes *Ophiuroidea (brittle stars), Asteroidea (starfish), *Echinoidea (sea urchins), *Holothuroidea (sea cucumbers), and *Crinoidea (sea lilies); and the extinct Edrioasteroidea, *Blastoidea, *Cystoidea, and members of the subphylum *Homalozoa (*carpoids). *Tribrachidium* from the late *Precambrian of Australia is probably an echinoderm, close to the stock from which the other groups evolved in the *Cambrian and *Ordovician. Echinoderms first appeared in the Lower Cambrian, but of the 20 classes known in the *Palaeozoic, only six survive into the *Mesozoic and on to the present day.

Echinoidea (echinoids; sea urchins, sand dollars, heart urchins; phylum *Echinodermata) Class of free-living echinoderms in which the body is enclosed in a globular, cushion-shaped, discoidal, or heart-shaped *test built of meridionally arranged columns of interlocking, calcareous plates, which bear movable appendages (spines, pedicellariae, and spheridia). The test is composed of 20 vertical rows of plates arranged in five double rows of perforate (*ambulacral) plates and five double rows of imperforate (interambulacral) plates. *Tube feet, connected to the internal water-based vascular system, emerge through the pores of the ambulacra. The apical system on the upper surface consists of five ocular plates and up to five genital plates. In all *regular echinoids the anus is enclosed within the apical system, but in many *irregular echinoids it is in the posterior interambulacrum. The mouth is always on the lower surface and may be central or anterior in position. The class first appeared in the *Ordovician; underwent a great *adaptive radiation in post-*Palaeozoic times, when rigid tests were evolved; experienced a marked reduction in the *Permian and *Triassic; and thereafter resumed its diversification, which has continued until the present day. Fossils of the inner skeletons of these echinoderms are common in both *Mesozoic and *Cenozoic sediments. There are about 125 Palaeozoic, 3670 Mesozoic, 3250 Cenozoic, and more than 900 extant species.

echo-sounding A method for determining water depth by using the elapsed time (t) between the transmission of a pulse of high-frequency sound and the return of the echo from a reflector, e.g. the sea floor. Knowing the velocity of sound in water (v_w), the depth (h) to the reflector is given by $h = v_w t/2$.

eclipse The partial or complete obscuration of one heavenly body by another, as perceived by an observer on one of the bodies. The proper description of an eclipse also refers to the period of time involved.

ecliptic The plane of the orbit of the *Earth around the *Sun. It forms an angle of 23°27′ with the Earth's equator. The orbits of the planets all lie within 3.4° of this plane, except for those of *Pluto (17.2°) and *Mercury (7°).

ecologic reef A name proposed in 1970 by R. J. Dunham to describe a *reef that is a rigid, wave-resistant, sediment-binding structure actively built by organisms. *Compare* STRATIGRAPHIC REEF.

eclogite Very rare, coarse-grained, *igneous rock with a chemical composition similar to that of *basalt, but noted for the presence of the rare, bright-green *pyroxene omphacite, and red, almandine–pyrope *garnets. Eclogites may be basalt metamorphosed by high temperatures and pressures.

eclogite facies A set of metamorphic *mineral assemblages produced by *metamorphism of *basic *igneous rocks under high-pressure, moderate-temperature conditions and typically characterized by the development of the mineral assemblage pyrope *garnet and omphacite (an Na-rich, high-pressure *pyroxene). Other rocks of contrasting composition, e.g. *shales or *limestones, are not known to show mineral assemblages characteristic of the high pressures and moderate temperatures involved. Eclogite facies can form when a slab of *oceanic crust is *subducted into the *mantle. The basic igneous rocks of the slab can then be converted to a typical *eclogite mineral assemblage.

ecology The scientific study of the interrelationships among organisms and between organisms, and between them and all aspects, living and non-living, of their environment. Ernst Heinrich Haeckel (1834–1919) is usually credited with having coined the word 'ecology' in 1869, deriving it from the Greek *oikos*, meaning 'house' or 'dwelling place'.

economic basement Rocks below which the chance of finding economic mineral resources is minimal. For instance, oil seldom occurs in economic quantities below 6–7 km.

ecophenotype *See* ECOPHENOTYPY.

ecophenotypic effects Non-heritable modifications of a *phenotype, produced in response to factors in the environment or habitat, that become preserved in the *fossil.

ecophenotypy The divergence of *phenotypes due to developmental changes induced by local environmental conditions, producing distinct ecophenotypes. Such divergence is not heritable, but when found in the fossil record can be mistaken for speciation.

ecosphere *See* BIOSPHERE.

ecostratigraphy The study of the occurrence and development of *fossil *communities throughout geologic time, as evidenced by *biofacies, with particular reference to its relevance in stratigraphic *correlation and other fields such as biogeography and basin analysis.

ecosystem (ecological system) Term first used by A. G. Tansley (1935) to describe the interdependence of *species in the living world (the biome or *community) with one another and with their non-living (abiotic) environment. Fundamental concepts include the flow of energy via food-chains and food-webs, and the cycling of nutrients biogeochemically (*see* BIOGEOCHEMICAL CYCLE). Ecosystem principles can be applied at all scales—thus principles that apply to an ephemeral pond, for example, apply equally to a lake, an ocean, or the whole planet. In Russian and central European literature 'biogeocoenosis' describes the same concept.

ecotone Narrow and fairly sharply defined transition zone between two or more different communities.

ectocochlear Applied to those cephalopods (*see* CEPHALOPODA) in which the shell is totally external to the body. Cephalopods whose shell is encased in soft tissue are said to be 'endocochlear'.

Ectoprocta 1. (phylum *Bryozoa) The major subphylum of bryozoans, in which the lophophore (feeding and respiratory organ) surrounds the mouth but not the anus (in contrast to the *Entoprocta). Many possess a calcite skeleton. Ectoprocts are mostly, but not exclusively, marine, and have an extensive fossil record from the *Ordovician to the present. **2.** Alternative name for Bryozoa.

ectotherm *See* HOMOIOTHERM.

edaphic Of the soil, or influenced by the soil. Edaphic factors that influence soil organisms are derived from the development of soils and are both physical and biological (e.g. mineral and humus content, and pH).

eddy Motion of a fluid in directions differing from, and at some points contrary to, the direction of the larger-scale current. In air, eddies vary in size from small-scale turbulence (which can transport dust and diffuse pollutants) to large-scale movements (e.g. cyclone and *anticyclone cells) within the general global circulation of the atmosphere.

eddy currents The *alternating current induced in a conductor when it is subjected to a time-varying magnetic field in accordance with *Lenz's law. The eddy currents then generate their own secondary *electromagnetic field which is of considerable importance in *electromagnetic exploration techniques.

eddy viscosity A coefficient relating the average shear stress within a *turbulent flow of water or air to the vertical gradient of velocity. The eddy viscosity depends on the fluid density and distance from the river bed or ground surface. The concept of eddy viscosity is fundamental to the von *Karman–Prandtl description of the velocity profile in *turbulent flow, and is important in determining rates of evaporation or cooling by wind, and the shear stress exerted by rivers on moving particles on their beds.

Edenian A *stage of the *Ordovician in the Lower *Cincinnatian *Series of N. America.

edenite *See* HORNBLENDE.

edentulous 1. Applied to a condition found in some *Bivalvia in which hinge teeth are absent. **2.** In mammals, toothless, either naturally or as a result of tooth loss.

edge 1. A *crystal consists of a three-dimensional stacking of a *unit cell defined by a space lattice. In *crystallography, the three edges of the lattice are labelled *a*, *b*, and *c* (or *x*, *y*, and *z*), and they define both the edges of crystals in the seven *crystal systems and their *crystallographic axes. **2.** In *remote sensing, a boundary between area of different *tones. *See also* EDGE ENHANCEMENT.

edge dislocation *See* DISLOCATION.

edge enhancement The process of increasing the *contrast between adjacent areas of different tones in an image in order to define *edges more clearly.

Ediacara A *series of the Upper *Proterozoic (*Vendian), from about 590–570 Ma ago, underlain by the *Varangian sequences and characterized by the varied assemblages of soft-bodied metazoan *fossils (Ediacara Formation, Australia).

Ediacaran fossils Late *Precambrian

*fossils, from Ediacara, Australia, dated to about 640 Ma. They come from a shallow, littoral, marine environment and the animals appear to have been stranded on mudflats or in tidal pools. About 30 genera are known and include: medusoids (jellyfish), e.g. *Medusina mawsoni* and *Medusinites*; pennatulaceans (soft corals), e.g. *Charniodiscus*; annelid worms, e.g. *Spriggina*; and *Dickinsonia*, up to 1 m long, which may have resembled an anthozoan polyp, but has been variously assigned to the annelids, medusoids, and to a phylum of its own. There are also ovoid or discoid forms of unknown affinity. Since the fossils were first described by M. F. Glaessner in 1961 similar faunal types of equivalent age have been found elsewhere in the world.

Eemian An *interglacial *stage in northern Europe, dating from about 100 000 years BP to about 70 000 years BP, which may be the equivalent of the *Riss/Würm interglacial of the Alpine area and the *Ipswichian of the East Anglian succession.

effective porosity **1.** That proportion of the total *pore space in a rock which is capable of releasing its contained water. *Clay, for example, may have a total porosity of 50% or more, but little if any of the water contained in these pores may be released, because of the retentive forces (e.g. surface tension) that hold it within the rock. **2.** The proportion of the pore space through which *groundwater flow occurs. For example, in fractured rocks the majority of flow occurs in the fractures, and intergranular pore water may be almost static. In porous rocks, some pores may have only one connection with the general pore space ('blind' pores) and so contain only static water.

effective precipitation Net precipitation after losses by evaporation. As higher temperatures increase evaporation, an index of effective precipitation derived from a temperature : precipitation ratio has been used as a criterion for some systems of climate classification (e.g. those of *Köppen in 1936, *Thornthwaite in 1948).

effective stress In soil, pressure between grains at points of contact; it is at equilibrium in saturated soil. Effective stress equals total pressure minus the neutral pressure of water in *pores. During *consolidation, effective stress increases and reaches maximum at complete consolidation before shear *failure occurs.

effective temperature (T_e) The temperature of a planetary surface in the absence of an atmosphere. The effective temperature of Earth is some 35–40 °C lower than the actual Earth surface temperature (T_s): the latter is approximately 15 °C owing to the *greenhouse effect of the Earth's atmosphere.

EGT *See* EUROPEAN GEOTRAVERSE.

Egyptian jasper *See* JASPER.

E_H *See* REDOX POTENTIAL.

Eifelian (Couvinian) **1.** An *age in the Middle *Devonian *Epoch preceded by the *Emsian, followed by the *Givetian, and dated at 386.0–380.8 Ma (Harland et al., 1989). **2.** The name of the corresponding European *stage, which is roughly contemporaneous with the upper *Cunninghamian (Australia) and upper Onesquethawian (N. America).

Eildonian A *stage of the Middle *Silurian of south-eastern Australia, underlain by the *Keilorian and overlain by the *Melbournian.

einkanter *See* VENTIFACT.

ejecta blanket The blanket of debris surrounding an impact *crater. It is composed of material ejected from the crater during its formation, and is laid down with stratigraphy inverted from that of the bedrock. There is typically a star-shaped distribution of ejecta around the crater rim. In addition to rock fragments excavated from the crater, and melted material, surface material from outside the crater may be incorporated by *base surge erosion or excavated by secondary craters caused by large ejected blocks. There is a chaotic size distribution of ejected material. Many Martian craters are surrounded by fluidized ejecta blankets which have flowed across the surface. In contrast, lunar ejecta blankets are due mostly to ballistic sedimentation.

Ekman depth *See* EKMAN SPIRAL.

Ekman spiral Theoretical model to explain the currents that would result from a steady wind blowing over an ocean of unlimited depth and extent. In the northern hemisphere the surface layer of the water would flow at an angle of 45° to the right of the wind direction. Water at increasing depths would flow in directions more to the right, until, at a depth known as the Ekman depth, the water would move in a direction opposite to that of the wind. The Ekman depth varies with latitude but is of the order of 100 m in mid-latitudes. The velocity of the water flow decreases with depth throughout the spiral. In the northern hemisphere, the net water transport is at 90° to the right of the wind direction, and is known as the Ekman transport.

Ekman transport *See* EKMAN SPIRAL.

Elara (Jupiter VII) One of the lesser satellites of *Jupiter, with a diameter of 80 km (±20 km).

Elasmobranchii Subclass of shark-like fish. They have 5–7 gill slits, fairly rigid fins, *placoid scales, a spiracular opening behind the jaws, numerous teeth, and claspers on the ventral fins of the male. Elasmobranchs have a cartilaginous skeleton and their fossil record consists mainly of teeth and fin rays. Well-known fossil genera include *Cladoselache* (*Devonian), and *Ctenacanthus* (Upper Devonian to Lower *Permian). *See* CARTILAGE; and CARTILAGINOUS FISH.

elastic constants For an *isotropic material which obeys *Hooke's law, where *strain is linearly proportional to *stress, there are only two independent elastic constants, those of stress and strain. The elastic properties of such materials are defined by elastic moduli, notably the *bulk modulus, *shear modulus, *Young's modulus, *Lamé's constant, and *Poisson's ratio. *Seismic wave velocities are governed by the elastic moduli and the densities of the media through which they travel.

elastic deformation Temporary deformation, from which material recovers, caused by an applied *stress, such that on release of the stress the body reverts to its former, unstrained condition. In purely elastic materials such deformation is described by a linear stress–strain relationship (*see* HOOKE'S LAW). In rocks, ideal elastic strain is combined with viscous components.

elastic limit (yield point) The point at which a material achieves its maximum elastic *strain and beyond which strain is no longer linearly related to *stress. Beyond the elastic limit flow or rupture induces permanent deformation.

elastic rebound theory Theory which holds that accumulated potential energy, stored as elastic *strains, is released by faulting (i.e. when the material ruptures). Zones adjacent to the *fault plane 'rebound' elastically, leaving them relatively unstrained.

elastic wave An acoustic or *seismic wave.

elastoviscous behaviour The *strain behaviour of materials which are essentially viscous but which deform elastically under *stresses of short duration. Following the release of a longer-term stress the degree of elastic recovery of elastoviscous materials is somewhat less marked than would be observed in purely elastic materials.

E-layer 1. The outer *core of the Earth between about 2886 km and 5156 km depth, mainly defined on *seismic velocities. **2.** The *ionospheric layer at about 100 km height from which 3.5 MHz radio waves are strongly reflected.

elbaite *See* TOURMALINE.

elbow of capture *See* RIVER CAPTURE.

electrical charge *See* CHARGE.

electrical conductivity (σ) The ease with which an electrical current will pass through a material, in units of siemens or reciprocal ohms (mhos) per metre. In materials which are assumed to be *isotropic, conductivity is equal to the inverse of *resistivity (ρ), therefore $\sigma = 1/\rho$. *See also* APPARENT CONDUCTIVITY.

electrical conductivity (in Earth) This is determined at shallow depths by various *electrical and *electromagnetic survey methods. Upper *mantle conductivity is studied mainly by using geomagnetic variations, particularly the *diurnal variation, and *magnetotelluric soundings. Lower mantle conductivities are estimated by the influence on geomagnetic *secular variations.

electrical drilling *See* ELECTRICAL SOUNDING.

electrical sondes Various *well-logging devices for measuring the electrical resistivity (1/conductivity) of the rocks through which a *borehole passes. *See* INDUCTION SONDE; MICROLOG SONDE; and SELF-POTENTIAL SONDE.

electrical sounding (vertical electrical sounding (VES), electrical drilling) In electrical profiling, a technique in which the spacing of the *electrodes or *coils is expanded in order to increase the depth below the surface station from which information is obtained. A variety of *electrode configurations is used for electrical *resistivity sounding and *induced polarization surveys. In resistivity sounding, apparent resistivity data are recorded on a log–log plot as a function of their respective current-electrode separations. *Master curves are often used to provide preliminary interpretation of the curves, and followed by more detailed and sensitive computer analysis. For electromagnetic (EM) sounding the *coil spacing is increased or, in *time-domain EM surveys, the frequency is varied. *Compare* CONSTANT SEPARATION TRAVERSING.

electrical tomography A geophysical technique, sometimes used in archaeology, in which metal electrodes are inserted into the ground and a current passed between them, through the subsoil. The resistivity is measured and many such measurements, taken at different depths and in different directions, allow a three-dimensional image of the site to be constructed.

electrode array *See* ELECTRODE CONFIGURATION.

electrode configuration (electrode array) A geometrical pattern of electrodes used in *electrical sounding, *constant separation traversing, and *induced polarization surveys. Usual configurations comprise two *current electrodes and two *potential electrodes whose separations are known and defined by a *geometric factor. Common configurations include the *dipole–dipole, *Schlumberger, *square, and *Wenner arrays.

electrode potential *See* OXIDATION POTENTIAL; and POTENTIAL ELECTRODE.

electrolyte A chemical compound which, while molten or in solution, is decomposed by the conduction of an electrical current through it. The current is moved by the passage of *ions rather than, as in a metal, free *electrons.

electrolytic conduction The conduction of an electrical current by the movement of *ions, as in an *electrolyte. *Compare* OHMIC CONDUCTION.

electrolytic polarization The dissociation of an *electrolyte by *electrolytic conduction, which drives *ions of constant polarity towards a particularly-charged electrode. For example, if a current is applied to a dilute solution of sulphuric acid, positively charged hydrogen ions move towards the *cathode and negatively charged sulphate ions towards the *anode; after a time the solutions adjacent to the electrodes have an abundance of ions of appropriate charge and the solution is said to be polarized.

electromagnetic methods A range of methods by which magnetic or electrical fields generated in subsurface conductors by artificially induced *eddy currents are measured. The term includes *radar, electromagnetic ground *conductivity, and *VLF, but excludes those in which induction is insignificant, for example, *resistivity and *induced polarization, and those with insignificant depth penetration, for example, microwave sensing. The *AFMAG method, which uses naturally generated electrical signals, is included, but *magnetotelluric methods are not.

electromagnetic radiation (EMR) The range in radiation extending from wavelengths of less than 10^{-12} m to more than 10^{3} m. In order of increasing wavelength are included cosmic ray photons, gamma rays, X-rays, ultraviolet radiation, visible light (violet to red), infrared radiation, microwaves, radio waves, and electric currents.

electromagnetic spectrum The range of frequencies or wavelengths of *electromagnetic radiation.

electromagnetic wave A wave comprising an electrical and a magnetic component at right angles to one another but which are in phase and have the same frequency. The electrical component represents the electrical field strength, E, and the magnetic component the magnetic flux density, B. Electromagnetic waves travel at the speed of light (about 2.998×10^{5} km/s in free space), such that the wave velocity $v = 1/(\mu_0 \epsilon_0)^{1/2}$, where μ_0 and ϵ_0 are the magnetic permeability ($4\pi \times 10^{-7}$ H/m) and *dielectric permittivity of free space, respectively.

electrometer *See* MASS SPECTROMETRY.

electron Elementary particle of mass 9.11 $\times 10^{-31}$ kg and negative electrical charge of 1.602×10^{-19} C (coulombs). Electrons can exist independently, or in groups around the *nucleus of an atom. Experiments show that electrons in an atom may occur at a range of distances from the nucleus but are most likely to exist in certain low-energy orbits or shells, and within these shells there are further subshells, the configuration being such that no two electrons in any one atom have identical properties. When an electron moves from one subshell to another of lower energy, electromagnetic radiation is given off; if an electron moves to a subshell of higher energy, electromagnetic radiation is absorbed. An electron moves about the nucleus in a circular or elliptical orbit and also spins on its axis.

electron capture A mechanism by which a *nucleus can decrease its proton number and increase its neutron number by capturing one of its extranuclear *electrons. The process can be envisaged as the reaction between an extranuclear electron and a proton in the nucleus to form a neutron and a neutrino, the neutrino then being emitted from the nucleus. The event may leave the product nucleus in an excited state, and this is then followed by the emission of a *gamma ray. Removal of an extranuclear electron from the k-shell, or from higher-energy shells, leaves a vacancy that is subsequently filled by other electrons that fall into the vacant position. In the process these electrons emit a series of X-rays that can be detected.

electronegativity **1.** Tendency to form negative *ions, measured by combining *ionization-potential and electron-affinity values for an element to find the degree to which its atoms attract *electrons. **2.** The ability of an atom to attract electrons, usually in non-metallic, acid-forming elements. Elements with sharply contrasting electronegativities tend to form ionic compounds (*see* IONIC BOND), e.g. NaCl, where Na and Cl have electronegativities of 0.9 and 3.0 respectively. Elements with similar electronegativities are likely to form *covalent bonds, e.g. CH_4 (methane), where C and H have electronegativities of 2.5 and 2.1 respectively. *See* VALENCY.

electron-probe microanalyser An instrument used to determine the chemical composition of a 1 μm diameter specimen of *mineral or *glass at the surface of a polished rock or mineral slice. A narrow beam of electrons is focused on to the polished surface of the specimen to cover a 1 μm diameter spot. The electrons excite the atoms in the specimen to emit X-rays, whose wavelength is characteristic of the elements present and whose intensity at a given wavelength is proportional to the relative concentration of the element corresponding to that wavelength. By comparing the X-ray intensity at any one wavelength with that in a standard sample of known composition, the absolute concentration of the element in the sample can be deduced.

electro-osmosis Phenomenon whereby some fine-grained sediments with low *permeability expel *pore water when an electric current is passed through them. This is sometimes exploited to reduce *groundwater by passing currents between *anodes and *cathodes.

electrophoresis The migration, under the influence of an electric field, of charged particles within a stationary liquid. The liquid may be a normal solution or held upon a porous medium (e.g. starch, acrylamide gel, or cellulose acetate). The rate at which migration occurs varies according to the charge on the particle and also its size and shape. The phenomenon is exploited in a variety of analytical and preparative techniques employed in studies of macromolecules.

electropositive element An element whose *electrode potential is more positive than that of the standard hydrogen electrode which is assigned an arbitrary value of zero. Electropositive elements tend to lose *electrons and form positive *ions, e.g. the univalent alkali metals Li^+, Na^+, K^+, etc., and the divalent alkaline-earth metals Be^{2+}, Mg^{2+}, Ca^{2+}. *Compare* ELECTRONEGATIVITY.

Elektro The name of a Russian meteorological satellite that monitors conditions over the Indian Ocean.

eleutherozoan Applied to unattached *echinoderms. Previously 'Eleutherozoa' ranked as a subphylum, but the term is now used only informally. *Compare* PELMATOZOAN.

elevation correction **1.** The correction applied to a measurement of gravitational attraction to allow for the distance to the station from the theoretical reference surface and for the attraction of the rocks in this zone. It is a combination of the *free-air and *Bouguer corrections. **2. (static correction)** The correction which is made to seismic travel-time data to compensate for irregular topography and to reduce the data to a common *datum. Land 'statics' are extremely important in seismic surveying on land. As part of a 'statics' survey, refraction shooting is undertaken to determine the depth and velocity of the weathered zone (the low velocity layer, or LVL). This is done in order to design more effective seismic source parameters (e.g. ideally, the shot depth should be below the weathered layer) thereby optimizing the seismic reflection surveys which are undertaken after an LVL survey. In marine surveys, marine 'statics govern the source–*hydrophone geometry and provide a correction for the finite *offset between the source *arrays and the hydrophone *streamer.

elevation head *See* ELEVATION POTENTIAL ENERGY.

elevation potential energy (elevation head) The energy possessed by a mass, e.g. a body of water, by virtue of its being raised above a particular datum point, usually taken as either sea level or local ground level. The energy may be released when the mass is allowed to fall to a lower level, and may be harnessed, e.g. in the case of water by powering a turbine in a hydroelectric scheme. Elevation head is the energy possessed by a unit weight of water at a point, due to this cause. *See also* BERNOULLI EQUATION; DARCY'S LAW; and PRESSURE HEAD.

Élie de Beaumont, Léonce (1798–1874) A French geologist who made a detailed study of European folded rocks, and concluded that these showed evidence of distinct mountain building episodes. His overly geometric treatment of the directions of mountain systems was criticized by his contemporaries, but his theory that the Earth is cooling, and therefore shrinking, found wider support.

elite In *ichnology, applied (*a*) to a structure constructed within a sediment by living organisms that indicates a high level of biological activity because of its high content of oxygen or organic matter, and (*b*) to a *trace fossil that is more conspicuous than others nearby because some feature of its structure or composition has been emphasized by the processes of *diagenesis.

elliptical polarization In optical mineralogy, light passing through a mineral may be split into two rays. The velocity of one ray is uniform in all directions, to give a spherical wavefront; the other ray varies in velocity with direction, to give an ellipsoidal wavefront. Elliptical sections so produced are used to determine the optical properties of minerals.

ellipticity *See* PRINCIPAL STRAIN RATIO.

El Niño A warm-water current which periodically flows southwards along the coast of Ecuador. It is associated with the *Southern Oscillation (these effects are collectively known as an El Niño–Southern Oscillation, or ENSO, event) and with climatic effects throughout the Pacific region. A similar phenomenon may also occur in the Atlantic. Approximately once every seven years, during the Christmas season (the name refers to the Christ child), prevailing trade winds weaken and the *Equatorial countercurrent strengthens. Warm surface waters, normally driven westward by the wind to form a deep layer off Indonesia, flow eastwards to overlie the cold waters of the *Peru current. In exceptional years, e.g. 1891, 1925, 1953, 1972–3, 1982–3, 1986–7, 1994–5, and 1997–8 the extent to which the upwelling of the nutrient-rich cold waters is inhibited causes the death of a large proportion of the plankton population and a decline in the numbers of surface fish.

elongation index (I_E) The percentage by weight of particles whose long dimension is greater than 1.8 times the mean dimension measured with a standard gauge. The elongation, *n*, is length divided by breadth and the elongation ratio is $1/n$.

elongation ratio *See* ELONGATION INDEX.

ELR *See* ENVIRONMENTAL LAPSE RATE.

Elsasser, Walter Maurice (1904–91) A German-born physicist, Elsasser became a US citizen in 1940. He worked on the structure of the nucleus and on terrestrial magnetism. In 1946, he proposed that a '*self-exciting dynamo' in the Earth's outer *core might be the cause of the magnetic field.

Elsonian orogeny A *Proterozoic phase of mountain building that occurred about 1500–1400 Ma ago, affecting what are now the Central and W. Nain Provinces of the Canadian Shield. It was preceded by the *Hudsonian and succeeded by the *Grenvillian orogenies.

Elsterian A glacial period in northern Europe, dating from about 0.5 Ma to 0.3 Ma ago, that is probably equivalent to the *Mindel glaciation of the Alpine area.

elutriation A method of *grain-size analysis in which the finer *grades are separated from coarser and heavier particles by the use of a rising current of air or water, which carries the light particles upwards and allows the heavier grains to sink. By controlling the velocity of the flow, grains of different sizes can be separated.

eluvial deposit A residual accumulation of *ore minerals that occurs above the *source rock and has experienced no transport. In eluvial deposits concentrations of economic minerals are usually increased above the value of those in the underlying rocks by the removal of soluble elements from the host material, e.g. the tin deposits of Rondonia, Brazil. *Compare* ALLUVIAL DEPOSITS.

eluviation Removal of soil materials in suspension or in solution from surface *horizons, and with partial deposition in the lower horizons of *soil profiles. Removal in solution is called *leaching, and hence the term 'eluviation' is often limited in use to removal in suspension.

elvan *See* QUARTZ PORPHYRY.

embankment dam Earth or gravel wall above ground level to confine water in *reservoirs or channels, to prevent flooding, to restrain *tailings in a pond, or to carry roads and railways. On weak or unconsolidated deposits *load is spread by wide embankments with a core of impervious *concrete or rolled clay and with an outer layer of crushed rock or soil.

Embry and Clovan classification A *limestone classification, based on textural principles, which expands the *Dunham classification to include conglomeratic limestones and different types of organically bound limestone. E. F. Embry and J. E. Clovan proposed their system in 1971. They retain the Dunham terminology of mudstone, wackestone, packstone, and grainstone for limestones with particles less than 2 mm in size, but for limestones containing particles more than 2 mm in size they define two new terms: floatstone (*matrix-supported limestone in which more than 10% of the *clasts are larger than 2 mm in size); and rudstone (clast-supported limestone in which more than 10% of the clasts are coarser than 2 mm in size). For *autochthonous limestones whose original components were organically bound during deposition Embry and Clovan replace the Dunham term 'boundstone' with three new terms: bafflestone (autochthonous limestone whose original components were bound during deposition by organisms which act as baffles, permitting sediment to be trapped in the lee of the baffles); bindstone (autochthonous limestone whose original components were bound during deposition by organisms which encrust and bind, e.g. algae); and framestone (autochthonous limestone whose original components were bound during deposition by organisms which built a rigid framework, e.g. corals (*Anthozoa) in a *reef structure).

embryophytes The first land plants, from which all later terrestrial groups are descended. Gene sequences and comparative morphology suggest land plants form a *monophyletic group, descended from *Charophyceae. Embryophyte fossils occur in rocks of middle *Ordovician (Llanvirn, about 470 Ma) age.

emerald *See* BERYL.

emery A greyish-black variety of *corundum (Al_2O_3) which can be crushed and powdered for use as an abrasive for polishing hard surfaces.

Emilian *See* CASTLECLIFFIAN; and QUATERNARY.

emission spectrum *See* SPECTRUM.

emmissivity The ratio of *exitance of a body to the exitance of a *black body at the same temperature.

EMR *See* ELECTROMAGNETIC RADIATION.

Emsian 1. An *age in the Early *Devonian Epoch preceded by the *Siegenian, followed by the *Eifelian, and dated at 390.4–386.9 Ma (Harland et al., 1989). **2.** The name of the corresponding European *stage, which is roughly contemporaneous with the lower *Cunninghamian (Australia) and part of the Onesquethawian (N. America).

emu Electromagnetic units in the *c.g.s. units (gauss, oersteds, etc.). These have now been replaced by SI units of *ampere per metre, *weber per metre, and *tesla.

En *See* ENSTATITE.

en- From the French *en* meaning 'in', a prefix meaning 'in', 'into', or 'inside'.

enamel Crystals of a calcium phosphate-carbonate salt, containing 2–4% of organic matter, which are formed from the epithelium of the mouth and provide a hard outer coating to *denticles and the exposed part of teeth.

enantiomorphy The property of different

structural forms of the same substance, e.g. certain crystals, of appearing to mirror each other (as do the right and left hands). No rotation or inversion can bring one into coincidence with the other. A good example of this property is found in *quartz.

enargite A *sulphide *mineral Cu_3AsS_4 and *end-member of an *isomorphous series, the other end-member being famatinite Cu_3SbS_4; sp. gr. 4.5; *hardness 3; greyish-black; *metallic *lustre; occurs as irregular grains in association with other sulphides in copper-rich mineral deposits.

enation theory The theory that accounts for the origin of the fern leaf by suggesting that it arose from the development of simple outgrowths (enations). Any such theory has to account for the large, branching fronds of a fern with branching veins (a 'megaphyll') and also for the small leaves with one or two veins ('microphyllus').

Enceladus (Saturn II) A major satellite of *Saturn, discovered in 1789 by Sir William Herschel. It has the brightest surface of any body in the *solar system (albedo more than 0.9), composed of clean, fresh ice (not necessarily water ice). The surface is cratered, but there are also smooth plains and long linear cracks and ridges. The surface appears to be young, probably less than 100 Ma old, indicating it has been geologically active until very recently, and possibly is still active with some kind of water volcanism. This activity may make Enceladus the source of the material comprising the tenuous E ring of Saturn. Enceladus is much too small to be heated by radioactive decay; the heat would have dissipated long ago. The orbit of Enceladus is locked in a 1:2 *resonance with *Dione, which may provide some *tidal heating, but probably not enough to melt water ice. Enceladus is 238 020 km from Saturn; its radius is 249.1 km; mass 0.73×10^{20} kg; mean density 1120 kg/m^3; visual albedo 1.0.

Encke A *comet with an orbital period of 3.3 years; *perihelion date 11 June 1997; perihelion distance 0.339 AU.

endemism Situation in which a species or other taxonomic group (*see* CLASSIFICATION; and TAXONOMY) is restricted to a particular geographic region, due to factors such as isolation or response to soil or climatic conditions. Such a taxon is said to be endemic to that region. The size of the region in this context will usually depend on the status of the taxon: thus a family will be endemic to a much larger area than a species, all other things being equal. Reference is frequently made to 'narrow endemics', i.e. taxa with markedly restricted ranges. Some of these are evolutionary relics, such as the maidenhair tree (*Ginkgo biloba*), the only surviving species of the *Ginkgoales, confined to Chekiang Province, China, where it was discovered in 1758.

endichnia *See* TRACE FOSSIL.

end-member **1.** One of two or more simple substances forming the extreme ends of a *solid solution series, e.g. *albite ($NaAlSi_3O_8$) is one end member, and *anorthite ($CaAl_2Si_2O_8$) the other end-member, in the *plagioclase series. **2.** In palaeontology, one of the two distinct forms between which gradual variation occurs.

end-member textural classification A method of classifying *sediments consisting of a mixture of *grain sizes which defines the sediment type by plotting the relative proportions of three end-members (e.g. *sand, *silt, *clay; or *gravel, sand, mud). There are many end-member classifications that have been proposed, but all are based on the relative proportions of three end-member components.

end moraine *See* MORAINE.

endobiont An organism that lives beneath the surface of a substrate, e.g. the bed of a sea or lake.

endobyssate Applied to the habit of specific bivalves (*Bivalvia) that live in *sediment. In contrast to *epibyssate forms, the byssus (*see* BYSSATE) of these animals is used to anchor the animal within a burrow or boring.

endocochlear *See* ECTOCOCHLEAR.

endochondral bone *See* BONE.

endocone Apically pointed, conical layers of calcareous material that may fill the *siphuncle in some groups of nautiloids (*Nautiloidea). The layers build forward from the rear of the conch.

endogenetic processes Processes which originate below the Earth's surface; particularly applied to Earth movements (by faulting and *earthquakes) and volcanic activity.

endogenous dome *See* DOME (2).

endolith An *ooid that has been infested

by boring micro-organisms (e.g. fungi or *cyanobacteria).

endopunctate *See* PUNCTATE.

endorheic lake A lake that loses water only by evaporation (i.e. no stream flows from it). *Compare* EXORHEIC LAKE.

endoskeleton A skeleton that is contained within the body. In vertebrates, the endoskeleton comprises the axial skeleton and the appendicular skeleton. In *Echinodermata, the skeleton lies beneath the body surface and technically it is therefore an endoskeleton. *Compare* EXOSKELETON.

endotherm *See* HOMOIOTHERM.

en échelon The parallel or subparallel alignment of separate structural features (e.g. tension fractures) which are arranged obliquely to a specific directional axis.

energy budget *See* RADIATION BUDGET.

energy of activation *See* ACTIVATION ENERGY.

engineering geophysics The application of geophysical survey techniques to site investigations for civil engineering purposes. Common investigations include: finding the depth to bedrock for foundation studies; determining the *rippability of material for excavations; measuring the degree of fracturing of rock; the detection of underground fissures, cavities, and mineshafts; the location of pipes; and, in a marine environment, the determination of the strength of sediment infill and the detection of dangerous near-surface gas pockets before drilling with a jack-up rig.

englacial Contained within the interior of a glacier, as opposed to being at its base (subglacial) or on its surface (supraglacial). Normally the term is applied to meltwater or *drift.

enhancement seismograph *See* SEISMOGRAPH.

enigmatic taxon A genus or higher *taxon of restricted diversity, comprising only one or a few species with affinities that are poorly understood. Many species belonging to enigmatic taxa are known from very few specimens; some have not been seen since they were first described.

enriched uranium Uranium containing up to 3% ^{235}U. It is produced industrially for use in some nuclear reactors. Natural uranium contains as little as 0.7% ^{235}U.

ensialic belt An *orogenic belt developed on sialic *continental crust. Ensialic belts are thought to have developed by mechanisms involving little horizontal movement and so are unrelated to *plate tectonics. Many *Proterozoic belts have been classed as ensialic belts, although increasingly these have also been interpreted as *collision zones of *plates.

ENSO event *See* EL NIÑO.

enstatite (En) An important *orthorhombic *pyroxene (*orthopyroxene, or opx) $Mg_2Si_2O_6$ and *end-member of an *isomorphous series, the other end-member being orthoferrosilite (Fs) $Fe_2Si_2O_6$; this series forms the base of a compositional equilateral triangle whose third apex is *wollastonite ($CaSiO_3$) and into which the common pyroxene minerals (*augite, *diopside, *pigeonite) can be plotted, as well as *olivine; sp. gr. 3.2 (En) to 3.96 (Fs); *hardness 5.5; colourless or greyish-white, with a green tinge, or brownish-green; *vitreous *lustre; *crystals are *prismatic or *tabular, but normally forming very irregular grains; *cleavage good prismatic {110}, poor {010}, {100}; occurs in magnesium-rich magmatic rocks (e.g. *peridotites, *gabbros, *norites, and *basalts) and in contact *metasomatic zones associated with these rocks; and in *partial melt envelopes around very high-temperature *basic *igneous intrusions.

entablature An irregular arrangement of *columnar *jointing that lies above a *colonnade in a thick lava flow.

entelechy *See* ARISTOGENESIS.

enterolithic structure Irregular, tight to open folding, developed particularly in *evaporite sequences due to volume changes in the rock brought about by chemical transformations of the salts. Such structures are commonly formed by the swelling of *anhydrite during its *hydration to *gypsum.

enteron *See* ANTHOZOA.

enthalpy (*H*) Heat content per unit mass of a substance measured as the internal energy plus the product of its volume and pressure.

Entisols An order of embryonic mineral soils that have no distinct pedogenic horizons. Representing only the initiation of *soil-profile development, Entisols are common on recent *floodplains, steep eroding slopes, stabilized sand *dunes, and recent deep ash or wind deposits.

Entoprocta (phylum *Bryozoa) Subphylum of freshwater bryozoans which entirely lack a mineralized skeleton. The lophophore surrounds both the anus and the mouth. Many fossil forms are known, but the subphylum is known only from the *Cenozoic. Formerly the Entoprocta was classified as a separate animal phylum, although its many resemblances to the Bryozoa were recognized.

entrainment Process by which air from the environment outside a growing cloud is caught into the rising convective current within the cloud and mixed with the cloudy air. This is significant in that it reduces the buoyancy of the rising current and causes cloud growth by reason of the cooler, drier air which is introduced. This also causes some evaporation of cloud droplets. When very dry air is introduced, entrainment can produce rapid dissipation of the cloud.

entrenched meander *See* MEANDER.

entropy 1. Measure of disorder or unavailable energy in a thermodynamic system; the measure of increasing disorganization of the universe. **2.** *See* LEAST-WORK PRINCIPLE; and LEAST-WORK PROFILE.

environmental geology Study of the problems resulting from natural hazards and human exploitation of the natural environment. The geologic techniques used include those of engineering geology, economic geology, hydrogeology, etc., as applied to waste disposal, water resources, transport, building, mining, and general land use.

environmental lapse rate (ELR) The rate at which the air temperature changes with height in the atmosphere surrounding a cloud or a rising *parcel of air. The overall average rate is a decrease of about 6.5 °C/km, but the rate varies greatly in different regions of the world, in different airstreams, and at different seasons of the year. Where the lapse rate of temperature is negative (temperature increases with height), an inversion is said to exist.

Eobactrites sandbergeri Recorded from the Lower *Ordovician of Czechoslovakia, *E. sandbergeri* is the species thought by some workers to be the earliest representative of the cephalopod molluscs, the *Ammonoidea. *Eobactrites* was a straight-shelled form with no internal deposits related to buoyancy control.

Eocambrian Little-used term employed in the description of sequences of unfossiliferous rocks that were deposited at the end of the *Precambrian.

Eocene *Tertiary *epoch which began at the end of the *Palaeocene (56.5 Ma) and ended at the beginning of the *Oligocene (35.4 Ma). It is noted for the expansion of mammalian stocks (horses, bats, and whales appeared during this epoch), and the local abundance of *nummulites (marine protozoans of the *Foraminiferida). The Eocene Epoch is divided into the *Ypresian, *Lutetian, *Bartonian, and *Priabonian *Ages.

Eocrinoidea (subphylum Blastozoa) Extinct class of *cystoid-like *echinoderms, with *radial symmetry, which range in age from Lower *Cambrian to Middle *Silurian (Lower *Palaeozoic). Their globular or flattened *theca are composed of numerous, irregularly arranged plates that lack the thecal pores typical of cystoids, and their sutural pores (i.e. those at plate margins) are unlike those of cystoids. Food grooves are extended on an exothecal skeleton of erect, unbranched, *biserial, food-gathering appendages (called *brachioles), which are not homologous with the arms of *crinoids. The eocrinoids are the earliest blastozoans and may have included ancestors to other cystoid groups, possibly to crinoids also, but the question is still unresolved.

eocrinoids *See* EOCRINOIDEA.

Eodelphis Known from the Upper *Cretaceous of N. America, *Eodelphis* is a genus of early opossums, *marsupials with long prehensile tails. Opossums are arboreal in habit and are thought to represent the ideal stem stock from which evolved the whole marsupial group. Several species of *Eodelphis* are known from the Milk River Formation (Upper Cretaceous) of Alberta. The teeth of *Eodelphis* are sharply cusped and somewhat primitive in form.

Eoembryophytic An epoch in the evolution of plants, proposed in 1993 by J. Gray, that lasted from the early Llanvirn (about 476 Ma) in the middle *Ordovician to the late Llandovery (about 432 Ma) in the Early Silurian. Fossil tetrads comprising four spores bound by a membrane make their first appearance during this epoch and occur over a wide geographic range.

'Eohippus' *See* EQUIDAE; and *HYRACOTHERIUM*.

eolian *See* AEOLIAN.

eon 1. The largest *geologic-time unit, incorporating a number of *eras. The equivalent *chronostratigraphic unit is the *eonothem. Originally, two eons were proposed in 1930 by G. H. Chadwick. The younger was the *Phanerozoic Eon (time of evident life), comprising the *Cenozoic, *Mesozoic, and *Palaeozoic Eras, and this term is still used. The term suggested for the preceding eon was the *Cryptozoic (time of hidden life). This time has also been called the *Archaeozoic (time of most ancient life), but most commonly has been known simply as the *Precambrian. Three eons have been proposed for Precambrian time: the Priscoan for time before 4000 Ma ago; the *Archaean for 4000–2500 Ma ago; and the *Proterozoic for 2500–590 Ma ago; the term 'Precambrian' is still in frequent use but is *informal. **2.** A time unit of 10^9 present Earth years.

eonothem The *chronostratigraphic unit equivalent to the *geologic-time unit '*eon'. At present it is a little-used term.

Eosimias A fossil primate, known mainly by jaws and teeth, from the Middle *Eocene of southern China, described in 1994 by C. K. Beard, Tao Qi, M. R. Dawson, Banyue Wang, and Chuankuei Li. According to its describers, it has the features of a very primitive member of the Simiiformes (monkeys, apes, and humans) and may stand at the base of this group.

Eosphaera A tiny, sphaeroidal *fossil that is associated with other *microfossils in the Gunflint Chert, a rock, some 2000 million years old, from western Ontario, Canada. It is suggested that these fossils are the remains of the first photosynthetic organisms.

Eotracheophytic An epoch in the evolution of plants, proposed in 1993 by J. Gray, that lasted from the latest Llandovery (about 432 Ma) in the Early *Silurian until the mid-*Lochkovian (about 402 Ma) in the Early *Devonian. During this epoch simple plant spores, dispersed individually, become increasingly common in several plant groups, including early vascular plants, and the earliest undoubted plant megafossils have been dated to this time.

Eötvös effect The gravitational effect due to the vertical component of the *Coriolis force when the *gravimeter is in motion, e.g. while at sea. It depends on the velocity and direction of the motion.

epeiric sea (epicontinental sea) Shallow sea which extends far into the interior of a continent, e.g. Hudson Bay and the Baltic Sea. The term also denotes shallow sea areas that cover the *continental shelf and are partially enclosed, e.g. the North Sea.

epeirogenesis The large-scale upward or downward movements of continental or oceanic areas. Epeirogenic movements should not be confused with the more dynamic mountain-building episodes of an *orogeny.

ephemeral stream A stream which flows only after rain or snow-melt and has no *baseflow component. A desert *wadi may form an ephemeral stream.

epi- From the Greek *epi* meaning 'upon', a prefix meaning 'upon', 'in addition to', or 'above'.

epibenthos The organisms living on the surface of the sea bed or bed of a lake.

epibole *See* ACME ZONE.

epibyssate Applied to animals that use the byssus (*see* BYSSATE) to anchor themselves to rock or seaweed. *Compare* ENDOBYSSATE.

epicentral angle (Δ) The angular distance between an *earthquake *focus and a seismic station.

epicentre The point on the Earth's surface immediately overlying an *earthquake *focus (the *hypocentre).

epichnia *See* TRACE FOSSIL.

epiclast A *clast produced by the chemical or mechanical *weathering, or other surface processes, of volcanic rock. *Compare* ALLOCLAST; AUTOCLAST; and HYDROCLAST.

epicontinental sea *See* EPEIRIC SEA.

epicratonic Applied to processes which are active on the surface of a *craton, and the products of such processes.

epidote (pistacite) A rock-forming *mineral $Ca_2(Al_2Fe^{3+})Si_3O_{12}(OH)$ and a member of the epidote group of minerals, which includes *zoisite, *clinozoisite, and *allanite; sp. gr. 3.4–3.5; *hardness 6.5; *monoclinic; normally pistachio-green, but also various shades of green, yellow, grey, or black;

*vitreous *lustre; *crystals are *prismatic parallelograms, often rod-like and isometric, often showing radial, fibrous, and columnar masses; *cleavage perfect {001}; it occurs in *hydrothermal formations, especially associated with altered, *basic, *igneous rocks, and also in *contact metamorphic zones with *quartz, *chlorite, *calcite, and *sulphides, and can also replace various minerals, such as *amphiboles, which break down under late-stage hydrothermal alteration.

epifaunal Applied to benthic organisms that live on the surface of the sea bed, either attached to objects on the bottom or free-moving. They are characteristic of the *intertidal zone. *Compare* INFAUNAL.

epigene Produced or occurring at the Earth's surface. The term is used especially in relation to the processes of *weathering, *erosion, and deposition. '*Epigenetic drainage' is sometimes used as a synonym for 'superimposed drainage'.

epigenesis The hypothesis that an organism develops by the new appearance of structures and functions. An alternative hypothesis (termed 'preformation') is that the development of an organism occurs by the unfolding and growth of characters already present in the egg at the beginning of development.

epigenetic drainage Superimposed drainage. *See also* DRAINAGE; and EPIGENE.

epigenetic ore Deposit later in origin than the host rock.

epilimnion The upper, warm, circulating water in a thermally stratified lake in summer. Usually it forms a layer that is thin compared to the *hypolimnion.

Epimetheus (Saturn XI) One of the lesser satellites of *Saturn, discovered in 1979 by *Pioneer II, with a radius measuring 69 × 55 × 55 km; mass 0.0055×10^{20} kg; mean density 630 kg/m^3; visual albedo 0.8.

Epiphyton *See* RHODOPHYCEAE.

episodic evolution The fossil record is characterized by extinction events and succeeding phases of rapid evolutionary innovation. The overall picture is thus one of episodic *evolution. However, the term has recently acquired other connotations, and tends to be linked with *punctuated equilibrium.

epitaxy The overgrowth of one *crystal on another such that the structural orientations of the two *minerals are specifically related. The minerals themselves do not have to be closely similar in their structures.

epitheca 1. The outer wall of a *corallite in the subclass *Zoantharia. **2.** *See* DINOPHYCEAE.

epithermal Vein deposit formed within about a kilometre of the Earth's surface by hot (50–200 °C), ascending solutions which often produce shatter zones. Typical minerals are *stibnite, *cinnabar, *gold, and *silver.

epoch One of the intervals of geologic time recommended by the International Subcommission on Stratigraphic Terminology. An epoch is ranked as a third-order time unit, and is the equivalent of the chronostratigraphic unit *series. Several epochs form a *period; several periods an *era. Epochs are themselves subdivided into *ages. When used formally, the initial letter is capitalized, e.g. Early Devonian Epoch.

e-process *See* EQUILIBRIUM PROCESS.

epsilon cross-bedding One of a series of cross-bedding types proposed by J. R. L. Allen in 1963. Epsilon cross-bedding is the type of cross-bedding formed by the lateral accretion of a river *point bar. Other terms (nu, gamma, beta cross-bedding, etc.) proposed in the same scheme are not in current usage.

epsomite (Epsom salts) *Mineral, $MgSO_4.7H_2O$; sp. gr. 1.7; *hardness 2.5; *orthorhombic; colourless to white; *vitreous to earthy *lustre; crystals rare except locally, and show a fibrous structure; *cleavage one direction, perfect; occurs as encrusting masses in caves and old mine workings, and also in the oxidized zones of *pyrite deposits in arid regions. It is readily soluble in water and has a bitter taste. It is named after mineral springs at Epsom, England.

Epsom salts *See* EPSOMITE.

equal-area net (Schmidt–Lamber net) A stereographic net (i.e. a system of coordinates providing a two-dimensional representation of a sphere) used to display structural data in graphic form. The net is divided into 2° blocks of equal area bounded by *great and *small circles, and the data plotted on it refer to the orientation of features. The range of orientations and any tendency towards preferred orientation may be observed.

equant Applied to a *clast whose dimensions are broadly similar on each of its long, intermediate, and short axes. *See* PARTICLE SHAPE.

Equatorial countercurrent *See* EQUATORIAL CURRENT.

Equatorial current Oceanic current which flows in an east–west direction in the equatorial regions of all the oceans. The broad (1000–5000 km) westward-flowing currents (North and South equatorial currents) are separated by a comparatively narrow westward-flowing countercurrent (the Equatorial countercurrent). The flow tends to be limited to the upper 500 m of the water column and the velocity is 0.25–1.0 m/s. The strength and position of the equatorial currents are controlled by the overlying wind system.

equatorial orbit An *orbit of a satellite in which the orbital plane deviates from the equatorial plane of a planet by less than 45°. *Compare* POLAR ORBIT.

equatorial plane *See* PLANE OF PROJECTION.

equatorial trough Shallow low-pressure zone around the equator, where the trade-wind systems tend to converge. *See also* INTERTROPICAL CONVERGENCE ZONE; and DOLDRUMS.

Equatorial undercurrent (Cromwell current) A shallow, subsurface, eastward-flowing current in the central *Pacific Ocean between 1.5° S and 1.5° N. The current is 300 km wide and flows at up to 1.5 m/s at depths of 50–300 m. The surface waters in this area flow in the opposite direction.

Equidae (horses; order *Perissodactyla) Family that includes the modern horses, asses, and zebras (all of which are placed in a single genus *Equus*), and many extinct forms. Sufficient of these are believed to be ancestral to modern equids for the evolution of the family to have been traced in considerable detail. The horses are first represented in the fossil record by **Hyracotherium* ('*Eohippus*') which diverged from a *condylarth predecessor in the *Palaeocene. Numerous evolutionary lines subsequently appeared from this fox-sized prototype. Most of the evolutionary advances occurred in the New World, although the Equidae were to survive the *Pleistocene only in the Old World. The conquistadores reintroduced horses to the Americas in the 16th century. The domestic horse (*E. caballus*) is probably not descended from the only true living wild horse, Przewalski's horse (*E. ferus przewalskii*), but from a progenitor closely related to it.

equifinality The theory that the members of morphologically similar land-forms may each have been produced by a different process or sequence of processes. For example, an armchair-shaped hollow may have been produced by a *cirque glacier, by a rotational landslide, by *nivation, or by spring-sapping. Their present-day similarity may make the origin of such land-forms difficult to determine.

equilibrium In geomorphology, a steady state of balance between the processes acting on a landscape and the resisting Earth materials so that over time the geometric shape of the landscape is little changed. Individual land-forms, such as river profiles and hillslopes, may show similar balance. As it is recognized that a perfect balance cannot exist between input and output of energy, the terms 'quasi-' and 'dynamic' equilibrium are sometimes used. Quasi-equilibrium is an apparent equilibrium which is recognized for short periods of time.

equilibrium flow *See* STEADY FLOW.

equilibrium line Line on a glacier that divides the zone of *ablation from the zone of accumulation.

equilibrium process (e-process) In stellar evolution, the culmination of reactions following a rise in temperature after *silicon 'burning'. The e-process rearranges nucleons to produce the most stable nuclei. *See* NUCLEOSYNTHESIS.

equinoctial gale Term derived from the popular misconception that gales are more frequent at periods close to the equinoxes.

equipotential 1. (line) The two-dimensional locus of all points with equal value for a potential field. **2. (surface)** The three-dimensional distribution of all points of equal value for a potential field. Equipotential lines and surfaces are perpendicular to the lines of force associated with the potential field. For example, sea level is an equipotential surface for the Earth's gravity field.

Equisetites hemingwayi The first known species of the family Equisetaceae, found in the *Carboniferous of Europe, and the direct ancestor of the only extant genus *Equisetum*. *See* SPHENOPSIDA.

Equisetum *See* SPHENOPSIDA.

equivalence A problem of non-uniqueness in the interpretation of *electrical soundings, when identical trough- or peak-shaped apparent resistivity graphs for a three-layer case can be synthesized using different models for the middle layer. Identical peak-shaped graphs can be generated if the transverse resistance, R_t, remains a constant equal to $h\rho$, where h is the thickness of the layer and ρ its true resistivity. Identical trough-shaped graphs can be generated if the longitudinal conductance, G_x, remains a constant equal to h/ρ. Such problems can be resolved if access is available to borehole geologic control and the range of h or ρ can be deduced.

equivalent aperture width The width of a rectangular *aperture in the *time domain with the identical peak amplitude, containing the same amount of energy.

Equus *See* EQUIDAE.

era First-order geologic time unit composed of several *periods. The *Mesozoic Era, for example, is composed of the *Triassic, *Jurassic, and *Cretaceous Periods. When used formally, as above, the initial letter of the term is capitalized.

erathem The *chronostratigraphic unit that is equivalent to the *geologic-time unit *era. An erathem comprises a number of *systems grouped together and takes its name from the corresponding era, e.g. the *Mesozoic Erathem would refer to the rocks laid down during the Mesozoic Era. At present it is a little used term.

Eratosthenian System *See* LUNAR-TIME SCALE.

e-ray *See* EXTRAORDINARY RAY.

erg 1. Sand sea in a hot desert. It is a feature of the Sahara, where sand typically is accumulated in wide shallow basins as *alluvial and *lacustrine deposits derived from adjacent rocky desert during the *Cenozoic. It is often very large: the Grand Erg Oriental covers 196 000 km^2 in Algeria and Tunisia. **2.** Unit of energy or work in the *c.g.s. system. 1 erg = 10^{-7} J.

Erian A *series in the Middle *Devonian of N. America, underlain by the *Ulsterian, overlain by the *Senecan, and comprising the Cazenovian, Tioughniogan, and Taghanician *Stages. It is roughly contemporaneous with the *Givetian Stage in Europe.

Eros A *solar system asteroid (No. 433), measuring 41 × 15 × 14 km; approximate mass 5 × 10^{15} kg; rotational period 5.270 hours; orbital period 1.76 years. Eros is to be orbited in February 1999 by the *Near Earth Asteroid Rendezvous mission.

erosion 1. The part of the overall process of *denudation that includes the physical breaking down, chemical solution and transportation of material. **2.** Movement of soil and rock material by agents such as running water, wind, moving ice, and gravitational *creep (or mass movement). *See* BUBNOFF UNIT.

erosion rate The rate at which geomorphological *processes wear away land surfaces. Rates vary widely, depending on both processes and environments, as shown by the following figures: glacial *abrasion, 1000 B (*bubnoff units); soil *creep (under a temperate, maritime climate), 1–5 B; *solifluction (cold climate), 25–250 B; slope wash, 2–200 B; solutional loss, 2–100 B; sea cliff retreat (rocks of medium hardness), 4000 B; soil erosion resulting from human activity, 2000–8000 B.

erosion surface 1. (planation surface) Gently undulating land surface that cuts indiscriminately across underlying geologic structures and that is the end-product of a long period of *erosion. Different types are produced under different environments. Among the more important are the *peneplain (humid temperate), *etchplain (humid tropical) and *pediplain (semi-arid). Marine processes may also cut a marine erosion surface. **2.** An irregular surface cut into rock or sediment by the eroding action of flowing water, ice, or by the wind.

erratic Glacially transported rock whose lithology shows that it could not have been eroded from the local *country rock. An example is the occurrence near Snowdonia, Wales, of granites derived from Scotland. An 'indicator' is an erratic whose origin can be located precisely.

errors Generally, the deviation of measured values from their true values. Such errors may be random or systematic. Random errors should have a *Gaussian (normal) distribution about the arithmetic mean of the measurements and this should approach the true value as the number of measurements increases. Systematic errors are consistent differences between the true value and a set of measurements, such that their arithmetic mean is displaced from the true value. Thus random errors determine the precision of a set of measurements, while systematic errors limit their accuracy. The use of the *isochron is a common method for minimizing dating errors.

eruption The release of *lava and gas from the Earth's interior on to the Earth's surface and into the atmosphere. Where the gas component is minimal (e.g. in basaltic *magma) lava is released quietly; where the gas component is large (e.g. in *rhyolite magma) lava is explosively fragmented by the expanding gas as it is released, forming high eruption columns of *ash and *pumice above the vent. Such contrasts in the explosivity of eruptions have been used to define several eruption styles, ranging from the quiet Hawaiian type to the violently explosive Phreatoplinian type.

ESA *See* EUROPEAN SPACE AGENCY.

escape velocity The velocity required for atoms or molecules at high altitude to escape from a planet's gravitational field. For example:

Planet	*Escape velocity (km/s)*
Earth	11.2
Moon	2.4
Mercury	4.3
Venus	10.3
Mars	5.0

escarpment *See* SCARP.

escutcheon Depressed area, variable in shape, which is found to the posterior of the *shell beaks of certain bivalves. *Compare* LUNULE.

esker Long, sinuous, steep-sided, narrow-crested ridge which consists of cross-bedded sands and gravels. It is laid down by glacial meltwater either at the retreating edge of an ice sheet, or in a subglacial, or *englacial ice tunnel, or in an ice-walled tunnel.

Eskola, Pentii Elias (1883–1964) A Finnish geologist and mineralogist, Eskola was a professor at the University of Helsinki from 1928 to 1953. His major work was a study of the origin of *metamorphic rocks; he developed a classification of *metamorphic facies, based on typical *minerals.

essential mineral A *primary mineral whose presence in an *igneous rock is essential to defining the root name of that rock. For example, *plagioclase and the *pyroxene *augite are essential to defining the root name '*gabbro' and hence are essential minerals. Where a primary mineral is present but is not essential to the naming of the rock it is termed an '*accessory mineral'.

estuary Semi-enclosed coastal body of water which has a free connection with the open sea and where fresh water, derived from land drainage, is mixed with sea water. Estuaries are often subject to tidal action and where tidal activity is large, ebb and flood tidal currents tend to avoid each other, forming separate channels. In estuaries where tidal activity is small, the invading dense sea water may flow under the lighter fresh water forming a *salt wedge. A positive estuary is one in which surface salinities are lower within the estuary than in the open sea due to freshwater inflow exceeding outflow caused by evaporation. A negative estuary is one in which evaporation exceeds freshwater inflow and therefore hypersaline conditions exist in the estuary. Normally an estuary is the result of valley drowning by the post-glacial rise in sea level. The action of tidal currents on the large amount of available sediment may give rise to a range of mobile bottom forms, including ebb and flood channels, sandbanks, and *sand waves.

Etalian *See* GORE; and ANISIAN.

Etched Plains *See* MARTIAN TERRAIN UNITS.

etch figures (etch marks) Pits (etchings), regular in shape, which may develop in certain directions consistent with the orientation of the symmetry elements for the *crystal as a whole when *crystal faces are treated with suitable chemical reagents. These etch figures may therefore be used to assign crystals to their appropriate *crystal systems.

etch marks *See* ETCH FIGURES.

etchplain Plain produced in a tropical or subtropical environment as a result of a phase of deep *chemical weathering during tectonic stability, followed by one of *erosion in which the weathered debris is stripped away. Etchplain relief reflects differences in the resistance of the bedrock to weathering and involves *double planation.

etesian winds Greek name for north-easterly, easterly, northerly, or north-westerly winds which blow between May and September in the Aegean Sea. The equivalent Turkish term is *meltemi*.

Ethiopian faunal realm Area which corresponds with sub-Saharan Africa, although it is not completely separated from the neighbouring *faunal realms; generally it is taken to include the south-west corner of the Arabian peninsula.

eu- From the Greek *eu* meaning 'well' or 'easily', a prefix meaning 'well', 'good', etc. It is used in ecology to denote, in particular, enrichment or abundance, e.g. '*eutrophic', nutrient-rich; 'euphotic', light-rich.

Eubacteria (domain *Archaea) The single kingdom of the *domain *Bacteria, which contains the true bacteria. There are 11 main groups: purple (photosynthetic); gram positive; *cyanobacteria; green non-sulphur; spirochaetes; flavobacteria; green sulphur, Planctomyces; Chlamydiales; Deinococci; and Thermatogales. The only feature common to all bacteria is their prokaryotic cellular organization. The majority of bacteria are single-celled, and most have a rigid cell wall. Cell division usually occurs by binary fission; mitosis never occurs. Bacteria are almost universal in distribution and may live as saprotrophs, parasites, *symbionts, pathogens, etc. They have many important roles in nature, e.g. as agents of decay and mineralization, and in the recycling of elements (such as nitrogen) in the *biosphere. Bacteria are also important to humans, e.g. as causal agents of certain diseases, as agents of spoilage of food and other commodities, and as useful agents in the industrial production of commodities such as vinegar, antibiotics, and many types of dairy products. The oldest fossils known are of bacteria, from rocks in S. Africa that are apparently 3200 million years old. These must have been heterotrophic bacteria, feeding off organic molecules dissolved in the oceans of that time. The first photosynthetic bacteria, of *anaerobic type, appeared a little later, about 3000 Ma ago.

Eucaryota (Eukarya) The domain which contains the eukaryotic kingdoms, i.e. plants, animals, fungi, and *protists.

eucrite A type of *meteorite of basaltic composition, mainly pigeonite (low-calcium *pyroxene) and *plagioclase feldspar, with a little metallic iron, *troilite, and one or more *silicates. All eucrites seem to have crystallized at or near the surface of the parent body.

Euechinoidea (class *Echinoidea) Subclass of sea-urchins in which the normally rigid *test is composed of five *ambulacra and five *interambulacra, each made up of two columns of plates. The subclass includes both *regular types of echinoid (e.g. *Hemicidaris*) and *irregular types (e.g. *Clypeus*, *Micraster*). They first appeared in the Upper *Triassic.

eugeocline An association of *calc-alkaline volcanics, *greywackes, and *shales. These materials are thought to be related to *island arcs. The word is now little used, unlike the contrasting term '*miogeocline'.

eugeosyncline The part of a *geosyncline that is characterized by the presence of *volcanism and *plutonism. Geosynclinal theory has needed some reinterpretation and modification in the light of the newer, unifying theory of *plate tectonics.

euhaline water *See* HALINITY.

euhedral (idiomorphic) A morphological term referring to grains in *igneous rocks which have a regular crystallographic shape. Euhedral forms are developed when a crystal grows freely in a *melt and is uninhibited by the presence of any surrounding crystals. The shape of the growing crystals will thus be controlled by their own natural crystallographic form. *Compare* ANHEDRAL.

Eukarya *See* EUCARYOTA.

eukaryote (adj. eukaryotic) Organism with cells that have a distinct nucleus, i.e. all *protists, fungi, plants, and animals. The first eukaryotes were almost certainly green algae (*Chlorophyta), and what appear to be their microscopic remains appear in *Precambrian sediments dating from a little less than 1500 Ma ago. *Compare* PROKARYOTE.

Eulerian current measurement A technique for measuring the direction and speed of water movement at a series of fixed points. A current-measuring device is held at a fixed point and as the water flows past its speed is measured. A series of measurements taken at different times or places may be plotted on a map as individual current vectors or streamlines. A number of devices exist for Eulerian current measurement, the most common being the propeller-type current meter.

Euler pole *See* POLE OF ROTATION.

eulite *See* ORTHOPYROXENE.

eulysite A *metamorphic rock consisting of iron and manganese-bearing *silicates, such as the *pyroxene group *minerals *hedenbergite and iron-rich *hypersthene, the *olivine group minerals fayalite and manganese fayalite, and the *garnet group minerals *almandine and *spessartine.

Euparkeria Bipedal *thecodont ('tooth-in-socket') reptile known from the Lower *Triassic. The thecodonts are considered to be ancestors of the *dinosaurs and, apart from a limited number of primitive features in the skull, *Euparkeria* is a likely ancestor for most archosaurian (*see* ARCHOSAURIA) stocks. It was a small reptile, only 60–100 cm in length.

euphotic zone *See* PRIMARY PRODUCTIVITY.

Euramerica The continental mass which resulted from the fusion of north-western Europe and N. America during the *Caledonian orogeny. This cratonic (*see* CRATON) area subsequently fused with *Angara and *Gondwanaland during the Variscan orogenic event to form *Pangaea.

Eurasian Plate One of the present-day major lithospheric *plates, extending from the oceanic *ridges in the *Atlantic Ocean to a currently poorly-identified margin with the *North American Plate in the east. The southern boundary is basically the Alpine–Himalayan *fold belt, involving a collage of *microplates, minor plates, and the *African and *Indo-Australian Plates.

Europa (Jupiter II) The smallest of the *Galilean satellites, and the smoothest object in the solar system, with no feature more than 1 km high. The surface is icy, the ice being about 10–30 km thick, with two types of terrain, one mottled, brown or grey, with small hills, the other comprising large, smooth plains criss-crossed with straight and curved tracks, some thousands of kilometres long, producing a surface resembling that of the Arctic Ocean. There are very few craters. The crust is believed to be no more than 150 km thick and may include a liquid ocean beneath the surface ice. The inner core is believed to be of iron and sulphur beneath a rocky mantle. The satellite was discovered on 7 January 1610, by Galileo. Its diameter is 3130 km; mass 4.8×10^{22} kg; density 2990 kg/m^3; visual albedo 0.64; surface gravity 0.135 (Earth = 1); mean distance from Jupiter 670 900 km; mean distance from Sun 5.203 AU; orbital period 3.551181 days; rotational period 3.551181 days.

European GeoTraverse (EGT) A geological survey of the European continent along a traverse from the North Cape, Norway, to Tunisia with the aim of producing a three-dimensional description of the evolution, nature, structure, physical properties, and dynamics of the continent. The project was proposed in 1980, involved scientists from 14 countries, and was completed in 1992.

European Province *See* ATLANTIC PROVINCE.

European Space Agency (ESA) An organization, founded in May 1975 to replace the European Space Research Organization, that conducts space research and operates its own launch vehicles.

europium anomaly Most *rare earth elements exist in the trivalent state. Europium, however, can exist as the Eu^{2+} *ion and it can proxy for calcium in *plagioclase feldspar during *igneous fractionation. Crystallization of calcic plagioclase thus depletes residual *magmas of europium, relative to the other rare earth elements. The europium anomaly can therefore be used as an indication of the amount of fractionation a magma has undergone. Efforts to explain the marked europium anomaly discovered in *lunar *basalts led to much debate about the geochemistry of the *Moon, with suggestions that the europium might have been lost by volatization, retained selectively in the lunar interior, or, more probably, that melting and differentiation caused it to be depleted in the basalts but retained in the highlands.

Euryapsida Reptiles which have a single, upper temporal opening behind the eye. The group includes the *plesiosaurs, *nothosaurs, and the *placodonts. The *ichthyosaurs had a similarly placed opening, but there were key differences in the arrangement of the bones forming its margins: for this and other reasons, the ichthyosaurs are usually placed in a different subclass from the euryapsids, the Ichthyopterygia. However, like the ichthyosaurs, the euryapsids were typically aquatic types, especially from the mid-*Triassic onwards. Euryapsids first appeared in the *Permian and became extinct at the end of the *Mesozoic.

Euryarchaeota (domain *Archaea) The more derived (*see* APOMORPH) of the two kingdoms of Archaea, comprising a broad range of *phenotypes including *methanogens, *halophiles, and *sulphur-reducing organisms. Members of the Euryarchaeota show less genetic similarity to those belonging to the *domains *Eucaryota and *Bacteria than do those of the *Crenarchaeota.

euryhaline Able to tolerate a wide range of *salinity.

eurypterid *See* CHELICERATA; and MEROSTOMATA.

eurythermal Able to tolerate a wide range of temperature.

eurytopic Able to tolerate a wide range of several factors.

eustasy The world-wide changing of sea level caused either by tectonic movements, or by the growth or decay of glaciers (*glacio-eustasy, or glacio-eustastism).

eutaxitic structure A planar *fabric found in welded *ignimbrites, defined by flattened and elongate *pumice *clasts (*fiamme), usually 10–40 mm in length and set in a lighter matrix of flattened and sintered, *ash-size, glassy shards.

eutectic point *See* EUTECTIC SYSTEM.

eutectic system Mixture of two or more minerals in definite proportions which have crystallized from a *melt or solution simultaneously. The temperature at which this occurs is the eutectic point.

Eutheria (class *Mammalia, subclass Theria) The infraclass that includes all of the placental mammals, and which probably arose during the *Cretaceous. The embryo is retained in the uterus, nourished by means of an allantoic placenta, and born in an advanced stage of development.

Eutracheophytic An epoch in the evolution of plants, proposed in 1993 by J. Gray, that lasted from the late *Lochkovian (about 398 Ma) in the Early *Devonian until the middle *Permian (about 256 Ma). During this time the diversity of spores and plant megafossils increases greatly, with several classic assemblages, and a marked increase in the diversity of vascular plants.

eutrophic Applied to nutrient-rich waters with high *primary productivity. *Compare* OLIGOTROPHIC.

euxinic Applied to an environment in which the circulation of water is restricted, leading to reduced oxygen levels or anaerobic conditions in the water. Such conditions may develop in swamps, *barred basins, stratified lakes, and *fjords. Euxinic sediments are those deposited in such conditions, and are usually black and organic-rich. *Compare* ANOXIC.

evaporation *See* EVAPOTRANSPIRATION.

evaporation pan A broad, shallow, water-filled pan of standard size. For example, the class A pan used by the US Weather Bureau is 122 cm in diameter and 25 cm deep. The amount of water within the pan is monitored to obtain an estimate of evaporation losses.

evaporimeter *See* LYSIMETER.

evaporite General lithologic term applied to *sedimentary rocks which have formed by the precipitation of salts from natural brines during evaporation of marginal salt pans, *lagoons, supratidal flats, and saline lakes. Common rock types are *limestone, *dolomite, *anhydrite, *gypsum, and *rock salt.

evapotranspiration Combined term for water lost as vapour from a soil or open water surface (evaporation) and water lost from the surface of a plant, mainly via the stomata (transpiration). The combined term is often used since in practice it is very difficult to distinguish water vapour from these two sources in water-balance and atmospheric studies.

evapotron An instrument developed in Australia to measure the extent and direction of vertical air eddies which are involved in the vertical transfer of water vapour, and thus to provide a direct measurement of evaporation rates over short periods of time.

event deposit *See* STORM BED.

event stratigraphy A term first proposed by D. V. Ager (1973) for the recognition, study, and *correlation of the effects of significant physical events (e.g. marine *transgressions, volcanic *eruptions, geomagnetic *polarity reversals, climatic changes), or biological events (e.g. extinctions), on the stratigraphic record of whole continents, or even of the entire globe. It is argued that by correlating these effects, as they are evidenced in the sedimentary record, it will be possible to define truly synchronous horizons, thus leading to greater resolution and a more accurate *chronostratigraphic scale. More recently A. Seilacher (1984) has suggested the term 'event stratinomy' for the study of events at the level of individual *beds.

event stratinomy *See* EVENT STRATIGRAPHY.

evolute Applied to coiled *cephalopod conchs, in which all the whorls are exposed. *Compare* INVOLUTE.

evolution Change, with continuity in successive generations of organisms. The phenomenon is amply demonstrated by the fossil record, for the changes over geologic time are sufficient to recognize distinct *eras, for the most part with very different plants and animals. *See also* DARWIN; MACROEVOLUTION; MICROEVOLUTION; NATURAL SELECTION; PHYLETIC EVOLUTION; PHYLETIC GRADUALISM; PHYLOGENY; and PUNCTUATED EQUILIBRIUM.

evolutionary lineage Line of descent of a *taxon from its ancestral taxon. A lineage ultimately extends back through the various taxonomic levels, from the species to the genus, from the genus to the family, from the family to the order, etc. *See also* CLASSIFICATION.

evolutionary rate Amount of evolutionary change that occurs in a given unit of time. This is often difficult to determine, for several reasons. For example, should the unit of time be geologic or biological (the number of generations)? How should morphological change in unrelated groups be compared? In practice it is necessary to adopt a pragmatic approach, such as the number of new genera per million years.

evolutionary species *See* CHRONOSPECIES.

evolutionary trend Steady change in a given adaptive direction, either in an evolutionary lineage or in a particular attribute, e.g. dentition. Such trends are often apparent in unrelated *taxa. Formerly they were attributed to *orthogenesis; now *orthoselection or the contending theory of species selection are invoked.

evolutionary zone *See* LINEAGE-ZONE.

Ewing, Maurice (1906–74) An American geophysicist and oceanographer, Ewing developed offshore *seismic reflection profiling for use in oil prospecting in the 1930s. In the post-war period he made extensive studies of the structure of the floor of the Atlantic ocean, using *seismic refraction, sediment cores, etc. Ewing was instrumental in making the Lamont–Doherty Geological Observatory a leading research centre.

ex- From the Latin *ex* meaning 'out of', a prefix meaning 'out' or 'not having'.

exaerobic Applied to a laminated *biofacies formed in a depositional environment that was *anaerobic, or almost so, and that contains epibenthic (*see* EPIBENTHOS) fossils of macroinvertebrates (e.g. *Brachiopoda, *Bivalvia, *Mollusca).

exaptation A characteristic that opens up a previously unavailable niche to its possessor. The characteristic may have originated as an *adaptation to some other niche (e.g. it is proposed that feathers were an adaptation to thermoregulation, but opened up the possibility of flight to their possessors), or as a neutral mutation.

excavation Hole created in the ground by *drilling, augering (*see* AUGER), boring, blasting, scraping, ripping, or digging, depending on the strength and condition of the rock requiring removal. Excavation may be on the surface, e.g. for buildings, or underground, e.g. for mines and tunnels.

exchangeable ions Charged ions that are adsorbed on to sites (with a charge opposite to that on the ion) on the surface of the *adsorption complex of the soil (mainly *clay and *humus *colloids). Exchangeable ions can replace each other on this surface, and are also available to plants as nutrients. Although *cations (e.g. calcium and magnesium) are the most common, exchanging at negatively-charged sites, some complexes (e.g. sulphate and phosphate) do exchange at positively-charged sites. *See also* ANION-EXCHANGE CAPACITY; CATION-EXCHANGE CAPACITY; and EXCHANGE CAPACITY.

exchange capacity Total ionic charge of the *adsorption complex in the soil that is capable of adsorbing *cations or *anions.

exchange pool *See* ACTIVE POOL.

exfoliation Weakening and separation of the surface layers of rock as a result of *chemical or (possibly) thermal *weathering, or of pressure release due to *erosion. The decomposition of *biotite and hydration of *feldspar in *granite causes swelling that may lead to failure. Expansion and rock fracturing may also result from temperature change (although this is questioned by many geologists), and sheet failure may result from the release of internal stress in massive rocks when an overburden is removed. *See also* THERMOCLASTIS.

exhumed topography An ancient landform or landscape that had been buried beneath younger rocks or sediment and that is exposed by their subsequent *erosion.

exichnia *See* TRACE FOSSIL.

exine Outer, decay-resistant coat of a *pollen grain or *spore, composed of sporopollenin, an inert polymer. The exine is characteristic for different plant families and genera, and sometimes even for different species. Hence it forms the basis for the identification and quantitative analysis of the vegetation composition of *peats and other suitable sedimentary deposits dating back many thousands of years. *See also* PALYNOLOGY.

exinite *See* COAL MACERAL GROUP.

exitance The radiant flux density of *electromagnetic radiation leaving a surface.

exobiology The biology of outer space: a study currently limited to the seeking of evidence for the existence of life beyond the Earth, and speculation on the possible alternative forms of such life.

exocuticle *See* EXOSKELETON.

exogenetic processes A blanket term for those processes which operate on or close to the surface of the Earth and which involve *weathering, *mass movement, *fluvial, *aeolian, *glacial, *periglacial, and *coastal processes. The term is normally used in contrast to the *endogenetic processes, whose origin is within the Earth.

exogenous dome *See* DOME (2).

exorheic lake A lake that has one or more outflow streams. *Compare* ENDORHEIC LAKE.

exoskeleton A general term applied to the hard covering of many *invertebrates. It may be a shell growing at the edges only (by accretion) or a series of plates. The term is most commonly applied to the horny skeleton enclosing the body of all *Arthropoda and secreted by the underlying cellular layer. The exoskeletal material (*cuticle) is composed of a complex glycoprotein, is relatively impermeable to water, and has a high strength-to-weight ratio. It provides insertion sites for muscles (apodemes) and is divided into separate plates that facilitate movements. The plates are connected by thin, untanned cuticle. In trilobites (*Trilobita) and crustaceans (*Crustacea) the cuticle is impregnated with mineral salts (calcium carbonate and calcium phosphate) which give increased strength. The exoskeleton is moulted periodically (*ecdysis) to permit body growth.

exosphere Outer region of the upper atmosphere extending from a base of 500–750 km altitude. The zone has a very low concentration of gases (mostly molecules of oxygen, hydrogen, and helium, although some are ionized). Gases can escape from it into space, as molecular collisions are much reduced because of the low gas density. The exosphere and much of the underlying *ionosphere form part of the *magnetosphere.

exothecal External to the *theca, or *test.

exotherm *See* POIKILOTHERM.

exotic Applied to geological materials derived from an extrabasinal source, e.g. exotic *clasts in a *conglomerate derived from some distant or extraformational source, or an exotic *terrane (i.e. a structurally emplaced *allochthonous unit).

expanding Earth A hypothesis which is now supported mainly by Warren Carey, but which was first proposed by M. R. Mantovani in 1907, and raised again in the 1930s by Hilgenbirg and others. It holds that the diameter of the Earth has increased with time, fragmenting the continents and causing the growth of ocean basins at spreading axes (ridges). A conference to discuss the idea was hosted by Carey at Hobart, Tasmania, in 1956.

expanding spread In seismic refraction shooting, a method in which a *spread of *geophones is used at increased *offset for repeated shots at the same location. This is equivalent to using many spreads of geophones for a single shot. In seismic-reflection surveys it is used to provide information about *root-mean-square velocities and depths to reflectors using the *$T^2 - X^2$ method. *See also* ELECTRICAL SOUNDING.

Explorer 59 *See* INTERNATIONAL SUN–EARTH EXPLORER-C.

explosive charge *See* CHARGE.

exposure age (sun-tan age) The period during which a rock has been exposed at the lunar surface as measured by fission tracks induced by exposure to *solar flare particles, which penetrate to depths of less than 0.5 cm. The average exposure age is less than three million years.

exsiccation Dehydration of an area by a process, e.g. drainage, in the absence of changes in precipitation levels. Draining of marshlands and deforestation are examples of processes that can lead to exsiccation. *See also* DESICCATION.

exsolution Unmixing. Some homogeneous *solid solutions of *minerals are stable only at high temperatures. On cooling these become unstable and one mineral separates from the other at a certain temperature. An intergrowth of two separate minerals may result, e.g. *perthite, an intergrowth of sodium and potassium feldspar. This occurs without loss or addition to the mineral as a whole.

extension A measure of the change in length of a line from its initial unit length. Extension may be positive (elongation) or negative (shortening), depending on whether the length of the line increases or decreases. The simplest type of extension (e) is calculated from: $e = (L_f - L_0)/L_0$, where L_f is the final length and L_0 the original unit length.

external mould *See* FOSSILIZATION.

extinction **1.** In optical mineralogy, a *mineral is said to be in extinction when the *vibration direction of the two rays of a doubly refracting *crystal coincide with the vibration directions of the two pieces of *Polaroid in a thin-section microscope that is parallel to the *polarizer and *analyser so that no light reaches the eye. This phenomenon occurs four times in a complete 360° rotation of the stage. *See* OBLIQUE EXTINCTION; STRAIGHT EXTINCTION; SYMMETRICAL EXTINCTION; and UNDULOSE EXTINCTION. **2.** The elimination of a *taxon. This may take place in several ways. In the simplest case the taxon disappears from the record and is not replaced. Alternatively, one taxon may replace another, the earlier group consequently disappearing. Thus there is a process of either subtraction or substitution. Extinction generally takes place at particular times and places but there are recurring periods when episodes of mass extinction have taken place. Environmental catastrophe, occurring for whatever reason, removes many groups from the environment and ecosystems collapse. Eventually new forms appear and evolution resumes. It would appear that periods of mass extinction control the pattern of evolution.

extraclast A fragment of *carbonate rock derived from the erosion of an exposed ancient *limestone on land outside the depositional basin in which it is found. *Compare* INTRACLAST.

extraction *See* ABSTRACTION.

extraformational *See* CONGLOMERATE.

extraordinary ray (e-ray) In mineral optics, one of the two rays produced when light is passed through a doubly refracting *crystal. The ordinary ray (o-ray) travels with uniform velocity in all directions; the velocity of the extraordinary ray varies with direction. The extraordinary ray is refracted within the *mineral.

extremophile A micro-organism (domain *Archaea) that thrives under extreme environmental conditions of temperature, pH, or salinity. *See also* ACIDOPHILE; ALKALIPHILE; HALOPHILE; HYPERTHERMOPHILE; PSYCHROPHILE; and THERMOPHILE.

extrusion The emission of *magma from a vent or *fissure on to the Earth's surface where it forms a *lava flow. Gas is released quietly, rather than explosively, from the magma at the source vent and consequently there is little associated *pyroclastic activity. Gas may also be vented from a different source, or extrusion may be of lava (magma) which has been degassed previously.

extrusive Applied to all ejected material of volcanic origin. The sense applies to *lavas and flows, rather than to *pyroclastic rocks.

'eye' of storm Central part of a tropical cyclone with light winds, generally clear skies, and a slight, horizontal pressure gradient. The diameter of the 'eye' averages 20 km but in a large cyclone can be 40 km or more. The 'eye' is an area of some air subsidence which produces adiabatic warming.

eyepiece (ocular) An eye lens, a fixed diaphragm with *cross-wires, and a field lens, all contained in a short tube which is inserted into the top of a microscope. The internal construction of the eyepiece may vary depending on whether the focal plane lies above (Huygenian or negative eyepiece) or below (Ramsden or positive eyepiece) the field lens. Most eyepieces are either 5× or 10× magnification.

eyot *See* BRAIDED RIVER.

fabric (petrofabric) The physical arrangement of particles and *minerals in a rock, including its texture and structure, both microscopic and macroscopic.

fabric analysis Analysis of the elements that make up the *fabric of a rock to determine the response of that rock to *stress. In a rock the three-dimensional pattern comprising the distribution, shape, size, and size distribution of *crystals or *grains constitutes the fabric, e.g. bedding in *sedimentary rocks, or metamorphic banding (*see* GNEISS). Fabric analysis is important in engineering because, for example, rock strength parallel to banding can differ markedly from that perpendicular to banding. In soils, fabric is analysed by comparing the strength of an undisturbed sample with its strength when *remoulded; this describes the *sensitivity of the soil.

Fabrosaurus australis Described in 1964 by Ginsburg, *F. australis* is one of the earliest recorded *ornithischian *dinosaurs. Like other 'bird-hipped' dinosaurs, *Fabrosaurus* possessed a predentary bone at the tip of its lower jaw. Its teeth were small and pointed, and were used to grind plant material. *Fabrosaurus* was a biped with short forelimbs, and a long tail to provide good balance.

face pole *See* POLE OF A FACE.

facial suture *See* CEPHALIC SUTURE.

facies 1. Sum total of features that reflect the specific environmental conditions under which a given rock was formed or deposited. The features may be lithologic, sedimentological, or faunal. In a sedimentary facies, *mineral composition, *sedimentary structures, and bedding characteristics are all diagnostic of a specific rock or lithofacies. **2.** *See* METAMORPHIC FACIES.

facies association A group of sedimentary *facies used to define a particular sedimentary environment. For example, all the facies found in a *fluviatile environment may be grouped together to define a fluvial facies association.

facies fossil *Fossil organisms that are restricted to particular *lithologies, reflecting the original environments of deposition.

facies sequence A vertical succession of *facies; such sequences often fine or coarsen upwards, and may be repeated many times in a cyclic manner due to the migration of the facies through time and space.

facing direction (younging) The direction (upward or downward) in which the stratigraphy throughout most of a *fold becomes progressively younger. *Anticlines face upwards, by definition; *antiforms may face upwards or downwards depending on the direction in which the stratigraphy is younging. It may be determined by such 'way-up' criteria as *graded bedding, *sole structures, and *cross-bedding.

faecal pellet Rounded, initially soft particles, mostly 100–500 μm in diameter, which are excreted by organisms. The internal structure of the pellet is usually fine grained. Worms (*Annelida), *gastropods, and *crustaceans produce these pellets in large quantities. Faecal pellets are most likely to accumulate and be preserved in low-energy, muddy environments colonized by an abundant fauna, e.g. *lagoons and *tidal flats. As a type of *fossil excreta, the term is commonly applied to small droppings, often of invertebrate origin, which may make up important parts of some *lithologies. *See* COPROLITE.

failed arm *See* AULACOGEN; and RIFT VALLEY.

failed rift *See* AULACOGEN; and RIFT VALLEY.

failure The process by which a body under *stress loses cohesion and divides into two or more parts, commonly by means of a brittle fracture.

failure strength *See* ULTIMATE STRENGTH.

failure stress envelope On a *Mohr stress diagram, the curve which joins points of *failure at progressive *stress configurations, thus delimiting the field of stable stress configurations (within the envelope) as distinct from the failure field outside the envelope.

falling head permeameter *See* PERMEAMETER.

falls *Meteorites that are seen to fall or are collected immediately, and whose time and locality of impact are accurately recorded. *Compare* FINDS.

fall-stripes *See* VIRGA.

false body *Clay displaying *thixotropy.

false cirrus *See* SPISSATUS.

false colour A term used in *remote sensing techniques to describe the display of data collected in a number of different wavelengths, usually longer or shorter than those perceptible to the naked eye. Typical false-colour images include infrared data, which are often displayed as visible red. Thus green vegetation, which is highly reflective in the infrared, typically appears red on a false-colour image.

famatinite *See* ENARGITE.

Famennian 1. The final *age in the *Devonian Period, preceded by the *Frasnian, followed by the Hastarian Age (*Carboniferous), and dated at 367–362.5 Ma (Harland et al., 1989). **2.** The name of the corresponding European *stage, which is roughly contemporaneous with the upper *Hervyan (Australia) and the *Chautauquan (N. America).

family *See* CLASSIFICATION.

FAMOUS project The Franco-American Mid-Ocean Undersea Study, undertaken on a 50 km length of the Mid-Atlantic Ridge between latitudes 36.5° N and 37° N.

fan cleavage The structural arrangement in which *cleavage planes form upwardly convergent or divergent fans throughout a sequence of folded layers. Generally, cleavage planes are parallel to the *axial plane only in the *hinge region of a *fold; elsewhere the cleavage planes may deviate systematically from a parallel alignment.

fanglomerate Applied generally to *conglomerates and *breccias deposited on *alluvial fans.

fan shooting A simple *seismic refraction method used for delineating subsurface geologic features (e.g. *salt domes, buried valleys, and back-filled mineshafts) by contrasting their *seismic velocities with those of the surrounding materials. *Geophones are set out around a segment of arc in a fan-like *array centred on one or more shot locations. A base line of geophones is used in relation to one of the shot locations to provide a time–distance curve where no subsurface feature is present, thus calibrating the travel times for a given range. Using the fan arrays, travel times of refracted rays are measured to each detector; any ray encountering a zone of anomalously high or low velocity will arrive ahead of or behind the expected travel time for that shot-to-geophone range.

Farallon Plate A present-day minor lithospheric *plate subducting beneath the *North American Plate, the Farallon Plate is the remnant of a plate that was once large. The amount of Farallon *crust which has been subducted can be estimated from the area of Pacific crust which still exists, assuming symmetrical spreading from the Pacific–Farallon *constructive margin. The present-day remnants are also called the *Gorda, *Juan de Fuca, and *Cocos Plates.

far-field barrier (geological barrier) In the disposal of *radioactive waste, a structure with geological and hydrological characteristics which make it impermeable to radionuclides and thus ensure that it will provide permanent containment. *Compare* NEAR-FIELD BARRIER.

fasciculate Applied to any *compound coral where the *corallites are spaced far enough apart to avoid mutual interference. However, the corallites are closely associated or bundled together.

fasciole A groove on the *test of spatangoid echinoids (*Echinoidea) which does not bear large spines or tubercles. Tiny spines situated in these grooves are covered in cilia (*see* CILIUM) that move water and mucus so as to remove extraneous material from the surface of the animal.

fast breeder reactor Nuclear reactor which uses fast neutrons to convert uranium to plutonium and which creates more fuel than it uses. A chain reaction is set up in which ^{238}U, with an initial charge of ^{239}Pu to begin the *fission process, discards fast neutrons each of which induces fission in another nucleus. Excess ^{239}Pu is produced where plutonium production exceeds rate of fission. About 60% of fuel elements in fast breeder reactors is converted to useful energy compared to 0.5–1% in *burner reactors. Plutonium can be recycled, but other radioactive products are waste. *See* RADIOACTIVE WASTE.

fast Fourier transform **(FFT)** An algorithm (e.g. the Cooley–Tukey method) which enables the Fourier transformation of digitized wave-forms to be accomplished more rapidly by computer than would be possible using direct evaluation of the Fourier integral. FFT usually involves iterative techniques. *See also* FOURIER ANALYSIS; and FOURIER TRANSFORM.

fathom Unit of water-depth measurement, originally six feet, equal to 1.83 m.

fatty acid Long-chained, predominantly unbranched, carboxylic acid; it may be saturated or unsaturated. Fatty acids have the general formula $R\text{-}(CH_2)_n\text{-}COOH$, where R represents a hydrocarbon group, e.g. CH_3 or C_2H_5 and n is any whole number between 1 and 16.

fault Approximately plane surface of fracture in a rock body, caused by brittle failure, and along which observable relative displacement has occurred between adjacent blocks. Most faults may be broadly classified according to the direction of slip of adjacent blocks into *dip-slip, *strike-slip, and *oblique-slip varieties. The term 'dip-slip fault' comprises both *normal and *reverse slip faults, and the special cases of low-angle *lag and *thrust faults. Strike-slip faults (wrench, transform, transcurrent) result from horizontal displacement (dextral or sinistral movements), and on a regional scale may involve *transpression and *transtension. *See also* VOIDS.

fault block **(fault slice)** A rock mass which is bound on at least two sides by *fault planes. The block may be uplifted or depressed in relation to adjacent blocks.

fault-block mountains Mountains or ranges that result from the upthrow of large fault blocks and that are separated from others by basins or troughs, producing an upland unit bounded by normal or reversed faults. It is classically developed in the Great Basin, Utah, USA, where a typical block is tilted and bounded by a steep fault scarp on one side and a by a more gentle dip slope on the other. Usually it is dissected by erosion. *See* BASIN-AND-RANGE PROVINCE; and HORST.

fault line *See* FAULT TRACE.

fault-line scarp *See* SCARP.

fault outcrop *See* FAULT TRACE.

fault plane A discrete, planar surface along which there has been appreciable relative displacement of the rock masses on either side.

fault-plane solution The use of the direction of first motion of seismic waves detected at seismic stations in different areas to determine the nature and orientation of the *stress field involved in the initial generation of an *earthquake and hence used to determine the *focal mechanism.

fault scarp *See* SCARP.

fault slice *See* FAULT BLOCK.

fault trace **(fault line, fault outcrop)** A generally linear feature which marks the intersection of a *fault plane with the surface of the Earth. Fault traces are sometimes marked by positive or negative topography and the emergence of *springs.

fault trap Structure in which water, oil, or gas may be trapped on one side of a *fault plane by an impervious horizon thrown above it by a *fault. *Compare* ANTICLINAL TRAP; REEF TRAP; STRATIGRAPHIC TRAP; STRUCTURAL TRAP; and UNCONFORMITY TRAP.

fault zone A region, from metres to kilometres in width, which is bounded by major *faults within which subordinate faults may be arranged variably or systematically. Single fault zones are marked by fault *gouge, *breccias, or *mylonites.

fauna **(adj. faunal, faunistic)** The animal life of a region or geologic period. *Compare* FLORA.

faunal province *See* FAUNAL REALM.

faunal realm **(faunal province, faunal region, zoogeographical region)** Biological division of the Earth's surface (i.e. a large geographical area) containing a fauna more or less peculiar to it. The degree of distinctiveness varies with the region concerned and reflects partly climate and partly the existence of barriers to migration. The number of realms recognized varies from one authority to another, but a minimum of six are recognized: *Australian, *Ethiopian, *Nearctic, *Neotropical, *Oriental, and *Palaearctic.

faunal succession *See* LAW OF FAUNAL SUCCESSION.

faunizone *See* ASSEMBLAGE ZONE.

Faye correction *See* FREE-AIR CORRECTION.

feather angle The angle subtended between a *streamer and the track of the towing vessel when a cross current causes the cable to drift off-line. The feather angle is of considerable importance when the streamer is several kilometres long and multi-fold coverage is to be obtained, because the position of the streamer affects the *common-depth-point coverage.

feather ore *See* JAMESONITE.

Federov stereographic net *See* STEREOGRAM.

feldspars The most important group of rock-forming *silicate *minerals, including the *alkali feldspars $KAlSi_3O_8$ to $NaAlSi_3O_8$ (K-feldspar to *albite) and the *plagioclase feldspars $NaAlSi_3O_8$ to $CaAl_2Si_2O_8$ (*albite to *anorthite). The potassium, sodium, and calcium *end-members can be taken to represent the three apices of an equilateral triangle into which all the alkali feldspars and the plagioclase feldspars can be plotted (usually with the plagioclase feldspars along the base of the triangle and the alkali feldspars along the left-hand side).

feldspathic greywacke *See* ARKOSIC WACKE.

feldspathic wacke *See* ARKOSIC WACKE.

feldspathoid The name for a group of framework *silicate minerals which are similar to feldspars in their structure but contain less silica per formula unit. The group includes the minerals *nepheline (a sodic feldspathoid), *leucite (a potassic feldspathoid), and *sodalite (a sodium, chlorine-bearing feldspathoid). The mineral *analcime, although a *zeolite, is closely related to the feldspathoids and often described with them. Minerals within the feldspathoid group crystallize from silica-deficient (*silica undersaturated) *melts and can occur instead of, or with, *feldspar.

Felidae *See* CARNIVORA.

Feliformia *See* CARNIVORA.

felsenmeer *See* BLOCKFIELD.

felsic A term applied to light-coloured *igneous *minerals and *igneous rocks rich in these minerals. Typical felsic minerals are *quartz, *feldspar, *feldspathoids, *muscovite, and *corundum. The term felsic derives from the two common minerals, *fel*dspar and *si*lica.

felsite A very light-coloured, *aphanitic *igneous rock, with or without *phenocrysts present. The term is used in the field as an initial classification and can refer to devitrified *rhyolite glass (*obsidian), or primary, *cryptocrystalline rhyolite.

felsitic An *igneous texture characterized by an equigranular, *cryptocrystalline aggregate of minerals, usually *quartz and *feldspar, in *rhyolites. The texture is formed during extreme *undercooling of a *magma, a condition which generates numerous *nucleation sites and imposes slow rates of *crystal growth.

femic A little-used term to describe the normative *ferromagnesium *minerals in a rock. The *CIPW normative components *hypersthene (Hy), *diopside (Di), fayalite (Fa), and forsterite (Fo), are all femic components of a rock.

femtoplankton Marine planktonic organisms, 0.02–0.2 μm in size, about which little is known.

femur 1. In *tetrapods, the upper bone of the hind limb. **2.** In *Insecta, the third and usually the largest and most robust segment in the leg.

fence diagram A three-dimensional depiction of an area, resembling an open area surrounded by a 'wall' or 'fence', showing the location and relationships of its sedimentary deposits. The diagram is constructed from several stratigraphic sections drawn in positions corresponding to their actual locations and their strata are joined.

fenestrae Irregular cavities found in muddy *intertidal to *supratidal *carbonate sediments. They take a number of forms: *birdseye fenestrae (irregular, 'birdseye'-shaped cavities, usually 1–5 mm across, formed by gas entrapment in the sediment); laminoid fenestrae (long, thin cavities, parallel to the sediment laminae, formed particularly in algal, laminated muds, and produced by the decay of organic material); and tubular fenestrae (cylindrical, near vertical tubes, formed by burrowing organisms or plant rootlets). Fenestral cavities may become filled with sparry calcite (*sparite). If they remain unfilled, the fenestrae are responsible for the development of *fenestral porosity in the sediment. The term comes from *fenestra* (pl. *fenestrae*), the Latin for an opening or window.

fenestral fabric *See* BIRDSEYE FABRIC; and FENESTRAE.

fenestral porosity *Porosity developed

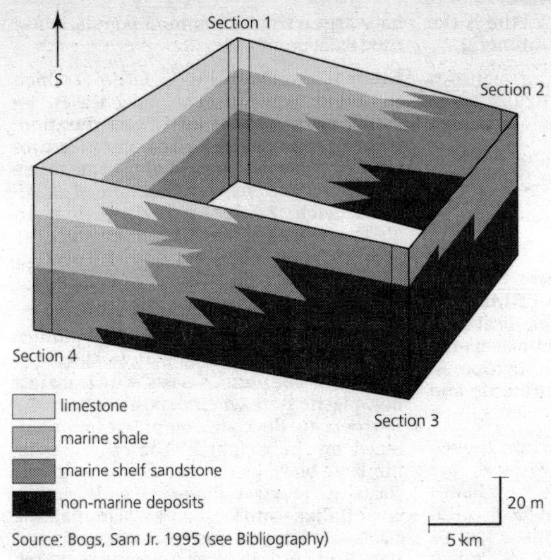

Source: Bogs, Sam Jr. 1995 (see Bibliography)

Fence diagram

in *carbonates due to the presence of *fenestrae. Rocks with fenestral porosity will not form good *reservoir rocks unless the fenestrae are interconnected to permit a good *permeability to be established. *See* CHOQUETTE AND PRAY CLASSIFICATION.

fenestrated Perforated with small openings or transparent areas.

fenite A rock that was first described from the Fen carbonatite intrusion in southern Norway. It is sodium-rich, metasomatic, *orthoclase–*nepheline–*arfvedsonite–*aegirine-bearing, and developed around *carbonatite intrusions. Alkali-rich fluids, migrating away from the crystallizing carbonatite *magma, react with the surrounding rocks, converting them to an assemblage of alkali-bearing *minerals. This process of alkali *metasomatism is given the specific name of 'fenitization' and produces particularly striking results when the *country rocks are *granites.

fenitization *See* FENITE.

Fennoscandian Border Zone (Tornquist Line) An ancient *Caledonian lineation separating the crystalline mass of Scandinavia and Russia from the fragmented crustal blocks of the remainder of north-western Europe. This NW–SE trending structural feature complements the major Caledonian *faults of northern Scotland (Great Glen Fault, Highland Boundary Fault, etc.). The Fennoscandian Border Zone has been active since that time, and in the *Mesozoic formed the boundary between the stable, crystalline *basement of Sweden and the Baltic and the subsiding Danish–Polish Furrow and North Sea Basin to the south-west. It is a complex zone of faults, *horsts, and half-*grabens which are best seen in southern Sweden (Scania), on the Danish island of Bornholm, and in southern Poland.

Fennoscandian uplift The uplift of Fennoscandinavia as a consequence of the inflow of *mantle in order to return to *isostatic balance following unloading of the area as the *Pleistocene ice sheets melted.

ferns *See* PTEROPSIDA.

ferrallization Part of the *leaching process found in tropical soils, by which large amounts of iron and aluminium oxides accumulate in the B *horizon of such soils as krasnozems.

ferricrete *See* DURICRUST.

ferrimagnetism A *ferromagnetic substance in which one of the two antiparallel magnetic lattices is stronger than the other, so that the substance has a net *remanent magnetization even in the absence of an external magnetic field. *Magnetite is the commonest ferrimagnetic rock mineral.

ferro- A prefix attached to an *igneous rock name to indicate a particular abundance of iron-bearing *minerals, or a high whole-rock iron content.

ferroactinolite *See* TREMOLITE.

ferroaugite *See* BENMOREITE.

ferrohastingsite *See* NEPHELINE SYENITE.

ferromagnesian minerals *Silicate *minerals in which *cations of iron and magnesium form *essential chemical components. The term is used to cover such minerals as the *olivines, *pyroxenes, *amphiboles, and the*micas *biotite and *phlogopite.

ferromagnetic In the wide sense, applied to substances in which *electron spins are magnetically coupled by either exchange or super-exchange quantum-mechanical forces. Such materials can acquire a spontaneous magnetization which is much greater than either *diamagnetism or *paramagnetism. There are three types of ferromagnetism: *ferromagnetism (in the strict sense); *ferrimagnetism; and *antiferromagnetism.

ferromagnetism In the strict sense, magnetism occurring in substances in which the *electron spins are coupled by quantum-mechanical exchange forces so that, within a single volume element, all electron-spin vectors are in the same direction. Typical materials are pure iron, nickel, and iron-nickel alloys found in *meteorites and lunar samples. *See also* FERRIMAGNETISM; and ANTIFERROMAGNETISM.

ferrosilite *See* ORTHOPYROXENE.

fertility, soil Condition of a soil relative to the amount and availability to plants of elements necessary for plant growth. Soil fertility is affected by physical elements, e.g. supply of moisture and oxygen, as well as by the supply of chemical plant nutrients.

Ferungulata (class *Mammalia, infraclass *Eutheria) A cohort which is proposed on palaeontological grounds but regarded by some authorities as artificial, that includes the *Carnivora (all of the modern mammalian carnivores), the primitive ungulates (including the elephants and sea cows), the *Perissodactyla (tapirs, rhinoceroses, horses, etc.), and the *Artiodactyla (pigs, camels, cattle, etc.), which are believed to have arisen from a common population in the *Palaeocene.

fetch 1. Length of water surface over which the wind blows in generating waves. Together with wind velocity and duration, this determines wave height. Many features of coastal deposition tend to become orientated normally to the direction of maximum fetch. **2.** Distance over which an airstream has travelled across sea or ocean.

Ffestiniogian *See* MAENTWROGIAN.

FFT *See* FAST FOURIER TRANSFORM.

fiamme Flattened, elongate, *pumice *clasts that are found in many welded *ignimbrites. The pumice clasts, which are in a hot, plastic state when deposited from the *pyroclastic flow, are compressed and flattened by the weight of the overlying ignimbrite body, producing elongated glassy clasts with ragged, flame-like ends (*fiamme* is the Italian word for 'flame'). Many fiamme clasts continue to vesiculate after formation, leading to the development of spherical *vesicles within them.

fiard (fjard, firth) A coastal inlet similar to a *fiord, but with lower relief.

fibratus From the Latin *fibratus* meaning 'fibrous', a species of separate cloud or cloud veil which has rather curved elements, but without hooks. *See also* CLOUD CLASSIFICATION.

fibril Cloud trail observed in *cumulonimbus, where *drizzle-sized droplets are large enough for their terminal velocities to allow them to depart from the main cloud body.

fibrolite *See* SILLIMANITE.

fibrous Applied to the physical *form of a mineral that occurs in fine, thread-like strands which may be parallel or radiating in nature, e.g. *asbestos.

fibula In *tetrapods, the post-axial bone of the lower part of the hind limb.

fiducial point 1. Temperature at which the atmospheric-pressure scale of a particular barometer reads correctly. The temperature at which this is so in latitude 45° is called the standard temperature of that barometer. At other temperatures and other latitudes, corrections must be applied. **2.** Fixed point (indicated by a pointer) that is the zero of the scale of a *Fortin barometer.

field capacity The water content which can be retained by a soil after excess moisture has drained freely away. Usually it is measured as a percentage of the soil volume or of the weight of oven-dry soil. *See also* SOIL MOISTURE CONTENT.

field reversal *See* GEOMAGNETIC FIELD; and POLARITY REVERSAL.

Figtree *See* SWAZIAN.

Filicopsida *See* PTEROPSIDA.

filiform Thread-like; long and slender.

filling Term used in synoptic meteorology to describe an increase in pressure at the centre of a *depression. *See also* DEEPENING.

film water *See* PELLICULAR WATER.

filter 1. A device for removing unwanted components from water. Coarse material may be recovered by the use of simple mesh sieves, but finer material and certain pollutants may require the use of other filters, e.g. of activated carbon or sand. **2.** To discriminate against a portion of information entering a device (the filter), typically by removing unwanted *noise or isolating specific parts of the information (e.g. separating high-frequency from low-frequency data). Filters usually operate within the frequency domain although others exist (e.g. the *velocity filter). Frequency filters have the disadvantage of invariably distorting the signal pulse shape, lengthening the pulse and so causing a phase shift, and displacing peaks and troughs in time. Linear filtering is known as *convolution. *See also* ALIASING; BAND FILTER; SPATIAL-FREQUENCY FILTER; and WIENER FILTER.

filter route Term introduced by the American palaeontologist G. G. *Simpson (1940), to specify a faunal migration route along which the spread of some animals is very likely but the spread of others is improbable. The route thus filters out part of the fauna, but permits the rest to pass. Deserts and mountain ranges provide examples of filter routes.

filtrate *See* MUD FILTRATE.

filtration Essentially, the removal of solid matter from liquids. Filter systems may be physical, mechanical, biological, chemical, or electrokinetic.

fin Appendage of fish and fish-like aquatic animals used for locomotion, steering, and balancing of the body. The skin fold forming the fin membrane is supported by cartilaginous, horny, or bony fin rays, which can be soft and flexible (soft rays) or hard and inflexible (fin spines).

finds *Meteorites identified by their composition and structure but not seen to fall. *Compare* FALLS.

fines 1. Particles of material below a specified size; or fine-grained sediment which settles very slowly. **2.** In *ore processing, material crushed or ground too finely, or ores too powdery, for normal smelting.

fines 10% test A test that is similar to the *aggregate crushing value (ACV) test. A sample of aggregate is subjected to varying loads and the percentage of *fines calculated for each load. A graph of loading versus percentage fines is used to find the loading required for 10% fines.

Fingerlakian (Fingerlakesian) *See* SENECAN.

fining-upward succession A vertical change in a *facies in which the grain size decreases with height above the base. *Compare* coarsening-upward succession.

finite resource *See* NON-RENEWABLE RESOURCE.

finite strain The total amount of *strain which has accumulated incrementally over a period through the addition of many smaller strains. The investigation of present-day rock structures involves the analysis of finite strain states which are evaluated by reference to natural objects whose original shape is known.

fiord (fjord) Long, narrow, deep, U-shaped coastal inlet which usually represents the seaward end of a glaciated valley that has been partially submerged. The water depths often exceed 1000 m except near the mouth where a bar or sill may be present.

fire-ball A very bright *meteor, approximating the average magnitude (brightness) of Venus. *See also* BOLIDE.

fire-fountain A continuous spray of disrupting *magma through a vent to form a persistent fountain of molten magma above the vent. The fountain, which may rise to 200 m, is supported either by the hydrostatic pressure of magma in the upper levels of the main volcanic superstructure, or by expanding gas released from the magma during the *eruption. Fall-out from the column produces a *spatter rampart around the vent and if the accumulation rate is high the molten spatter may coagulate to form a flow of *lava (a 'clastogenic flow').

firn (névé) Snow that has survived a summer melting season. It is an intermediate material in the conversion of snow to glacial ice. Normally it is granular, due to the partial melt.

firn limit *See* FIRN LINE.

firn line (annual snow-line, firn limit) A line on a *glacier marking the upper limit to which winter snowfall melts during the summer *ablation season. It is often clearly marked, and on many glaciers separates hard, blue ice below from snow above.

firn wind (glacier wind) Downhill airflow which develops over a glacier during the day, usually in summer. The greater air density over the glacier than over the surrounding surfaces causes this air to sink.

first arrival *See* FIRST BREAK.

first break (first arrival) The first wave from a discrete seismic-source impulse, naturally or artificially generated, that is recorded at a seismic detector. First breaks are used in *seismic refraction surveying. *See* BREAK.

firth *See* FIARD.

fish An instrument package towed behind a ship so that the measurements are unaffected by the ship and its equipment. A fish commonly contains a *magnetometer and *side-scan sonar.

Fisher, Osmond (1817–1914) An Anglican clergyman, Fisher was the author of the first textbook on *geophysics: *The Physics of the Earth's Crust* (1881). He argued that the Earth had a thin crust, and that convection currents in the fluid interior were the cause of mountain building, *rifts, etc. He postulated that the *Moon had been torn from the Pacific, causing the continents to be pulled apart.

fish-tail bit A drill *bit used for cutting through soft sediments.

fissility Ability of rock materials to split. The term is applied to *shales, flags, *slates, and *schists.

fission Splitting of a heavy atomic nucleus by collision, with the ejection of two or more neutrons, and the release of much energy.

fission hypothesis One of the three classical hypotheses for lunar origin. Proposed by George Darwin in 1879, it derives the *Moon from the silicate *mantle of the *Earth, following *core separation. Although this accounts for the low density of and paucity of metallic iron in the Moon, the process requires about four times the observed *angular momentum of the present Earth–Moon system. Detailed compositional differences between the Moon and the terrestrial mantle, from the Apollo data, appear fatal to the hypothesis.

fission-track dating Charged particles, from the spontaneous fission of ^{238}U in *minerals and in natural and synthetic *glasses, leave a trail of damage (fission tracks, radiation tracks) as they travel through a solid medium. This is the result of the transfer of energy from the particles to the atoms of the medium. These tracks, suitably enlarged by etching, can be seen in some minerals by using a petrological microscope (*see* POLARIZING MICROSCOPE). The number of tracks per unit area is a function of the age of the specimen and its uranium concentration, provided that it cooled rapidly on formation and has not been reheated at a later date. The uranium concentration can be measured by counting tracks produced by fission of ^{235}U caused by irradiation of the specimen with thermal neutrons in a nuclear reactor. Fission-track dates can, by this method, be obtained for minerals such as *micas, *apatite, *sphene, *epidote, and *zircon. The dates obtained are 'cooling ages' and indicate the time elapsed since the temperature dropped below the 50% track retention value; the tracks are known to fade by annealing of solids at elevated temperatures. The method can also be used to date *tektites, volcanic glass, and some archaeological objects.

fissure volcano A linear fracture on the Earth's surface through which *lavas, *pyroclastics, and gas are erupted and effused. The eruptive products accumulate most thickly along the linear fracture and build up an elongate, low-angle shield or higher-angle cone topography, constituting the volcanic pile. In addition to *eruptions from the main fissure, material can be erupted from secondary fissures developed locally within the growing volcanic pile and radiating from the trend of the main fissure. *See also* CENTRAL VENT VOLCANO; and VOLCANO.

fixation (pedol.) Soil process by which certain nutrient chemicals required by plants are changed from a soluble and available form into a much less soluble and almost unavailable form.

fixed-source method A geophysical exploration method in which the source (transmitter) is kept at a fixed position and a detector (receiver) is moved over the survey area to take measurements which are then plotted as profiles or maps. Electromagnetic methods such as *VLF and the *TURAM technique are examples of fixed-source methods. *Compare* MOVING-SOURCE METHOD.

fixigena *See* CEPHALON.

fjard *See* FIARD.

fjord *See* FIORD.

***f-k* space** A means of representing *frequency-domain data in terms of the independent variables *frequency (*f*) and *wavenumber (*k*).

Fladbury A site in Worcestershire, England, where palaeontological and palaeozoological evidence suggests there was a barren, tundra-like landscape during a cold period following the *Upton Warren Interstadial.

flakiness index (I_F) The specification for stone for bituminous surfacing, applied to *aggregate coarser than 6.5 mm. It is expressed as the percentage by weight of particles (in a sample of more than 200) whose smallest dimension is less than 0.6 times the mean dimension.

flame photometry (flame spectrometry) A technique analogous to emission spectrometry, but using a flame to excite electrons, rather than an arc or plasma. It is a simple and straightforward analytical technique that is basically a quantitative version of a 'flame test'. A known weight of sample is dissolved in hydrofluoric acid and either perchloric or sulphuric acids, portions of the solution are added to a flame, and the strength of emission of light of a particular wavelength produced by the potassium in the flame is recorded. This is then compared with those produced by standard solutions. The final results may be affected by sodium concentrations as well as by the sulphuric acid. Perchloric acid, iron, magnesium, aluminium, and calcium also interfere with the potassium emission but their effects may be reduced by buffering and by the removal of interfering *ions.

flame spectrometry *See* FLAME PHOTOMETRY.

flame structure A sedimentary structure in which wavy tongues of mud, with shapes resembling flames, project into the rock above them, which is commonly a *sandstone.

Flandrian The present *interglacial. Evidence suggests that the so-called post-glacial period, the warm phase following the last (*Devensian) ice advance or cold phase, is more appropriately treated as another interglacial of the *Quaternary (or Late *Cenozoic) Ice Age. In Europe, the warmest Flandrian stage occurred during *Atlantic times, about 6000 BP (the Hypsithermal is the equivalent N. American climatic optimum). No consensus view exists as to when the ice advance or extreme cold conditions will prevail once again in high mid-latitudes, nor as to how quickly these conditions will arise. The Flandrian is sometimes referred to alternatively as the *Holocene interglacial.

flank eruption The release of *lava and *pyroclastic material from a source on the slopes of a *volcano, away from its primary *central vent or fissure-vent area. The location of a flank eruption is often controlled by local inflation stresses. These initiate fractures within the volcanic pile which propagate outwards from the primary volcanic conduit to feed lava towards the volcanic flanks. A flank eruption occurs where the lava-filled fractures intersect the slope of the volcano.

flaser bedding A form of *heterolithic bedding characterized by *cross-laminations draped with *silt or *clay. Flaser beds form in environments where *flow strengths fluctuate considerably, thus permitting the transport of *sand in *ripples, followed by low-energy periods when mud can drape the ripples.

flaser gneiss *See* FLASER ROCK.

flaser rock (flaser gneiss) A rock displaying ovoidal *megacrysts enclosed within a fine-grained, streaky, *anisotropic *groundmass. The groundmass is typical of a *mylonite fabric produced by intense deformation. The megacrysts are regions of the original rock which have survived deformation. 'Flaser' is a German dialect word meaning 'irregular vein', or 'knot' (as in wood).

flash flood A brief but powerful surge of water either over a surface ('sheet flood') or down a normally dry stream channel ('stream flood'). Usually it is caused by heavy convectional rainfall of short duration, and is typical of semi-arid and *desert environments.

flat That part of a staircase *thrust plane trajectory which has a horizontal or sub-horizontal orientation. Flats represent areas of décollement (*see* DÉCOLLEMENT PLANE) along weak, layer- or bedding-parallel planes. Although initially horizontal, flats may steepen during later compression. *See* RAMP; and THRUST.

flat bed *See* PLANE BED.

flat-iron 1. A land-form of roughly triangular shape, and with one side (the *dip slope) that is both steep and uniform. It is formed between two adjacent valleys that cut through a *hogback ridge roughly at right angles to its *trend. It is a common land-form along the eastern front of the Rocky Mountains, USA. **2.** Applied to *clasts that have been shaped by glacial *erosion at the base of sliding *ice. Typically, they show a distinctive 'bullet' form, with one end plucked and the other streamlined, and may vary in size from a few centimetres to many metres.

flat spot A characteristic, strong, flat reflection within an otherwise dipping reflection event which is produced by a strong positive reflection from a gas-water interface, typically in a *trap structure. Often a flat spot occurs below a *bright spot.

flattening A change in the shape of an object caused by the application of a *stress, and which may be described by reference to the transformation of a sphere into an oblate ellipsoid as a result of *pure shear. Flattening is the strain state found in S-*tectonites (i.e. tectonites marked by a single, penetrative *foliation).

F-layer The transition zone between the liquid outer *core and the inner solid core of the *Earth. The boundaries are poorly defined but it is at a depth of about 5100 km.

flexural rigidity The flexural rigidity of an elastic sheet is defined as $ET^3/12(1 - \sigma^2)$ where E is *Young's modulus, T is the thickness of the sheet, and σ is *Poisson's ratio. In the Earth, the *lithosphere is usually treated as an elastic sheet that responds to loading by *ice-caps, volcanic piles, etc., from which the thickness of the lithosphere and the *viscosity of the *mantle can be derived.

flexural slip Folding in which there is *slip along the contacts between parallel layers. This discontinuous *simple shear mechanism may occur along *bedding planes or cleavage planes.

flexure 1. The lateral deflection from a datum line of a planar feature as it is shortened. **2.** The form of a *monocline, or a gentle fold whose inter-limb angles are 120–180°.

flight *See* STRING.

Flinn diagram A graphical representation of the full range of three-dimensional *strain states in deformed rocks, with or without a change in their volume. The diagram plots a value for the *principal strain axes Y/Z against X/Y of ellipsoids (*see* STRAIN ELLIPSOID) which result from *dilation and/or distortion of a reference sphere. The data plotted on the diagram are obtained from analyses of strain using *strain markers.

flint (silex) Variety of *chert, which occurs commonly as nodules and bands in *chalk. It is deposited in the porous, permeable structures of *sponge, *diatom, and *echinoid skeletons and also in burrows.

floating point A style of expressing a number that avoids losing significant figures should the number be too large or too small for a given register on a calculating device. For example, 165 400 can be written as 1.654×10^5 in floating point (although computers usually use base-2 rather than base-10). Floating-point amplifiers rather than binary amplifiers are being used increasingly in seismic amplification systems to increase the effective range of digital recordings.

floatstone A coarse-grained *limestone with *matrix-supported *clasts, 10% or more of which are coarser than 2 mm in size. *See* EMBRY AND CLOVAN CLASSIFICATION.

flocculation Process in which *clay and other soil particles adhere to form larger groupings or *aggregates. The reverse of this process is known as *dispersion.

floccus From the Latin *floccus* meaning 'tuft', a species of cloud with a tufted appearance, the lower parts of which are rather ragged, often with *virga. The species is most associated with *cirrus, *cirrocumulus, and *altocumulus. *See also* CLOUD CLASSIFICATION.

flood basalt *Lava of basaltic composition which is erupted from a laterally continuous fissure to form a widely dispersed, *low aspect-ratio, flow sheet. The Roza Member of the mid-*Miocene Columbia River Plateau in the USA is the largest flood *basalt unit in the geologic record, comprising basalt flows with a total volume of more than 1500 km^3 which have travelled up to 300 km from their source fissure. The largest individual lava flows in this member have volumes up to 700 km^3.

flood forecasting A technique which uses the known characteristics of a river basin to predict the timing, discharge, and height of flood peaks resulting from a measured rainfall, usually with the objective of warning populations who may be endangered by the flood. *Compare* FLOOD PREDICTION.

flood-peak formulae A number of methods (including those attributed to Benson, Potter, Morisawa, and Rodda) for the prediction of flood peaks by reference to rainfall intensity and frequency, topography, and orographic, temperature, and other relevant factors.

floodplain The part of a river valley that is made of unconsolidated river-borne sediment, and periodically flooded. It is built up of relatively coarse debris left behind as a stream channel migrates laterally, and of relatively fine sediment deposited when *bankfull discharge is exceeded.

flood prediction (flood forecasting) The study of rainfall patterns, catchment characteristics, and river hydrographs to predict the future average frequency of occurrence of flood events. Flood predictions seek to estimate the probable discharge that, on average, will be exceeded only once in any particular period, hence the use of such terms as '50-year flood' and '100-year flood'. *Compare* FLOOD FORECASTING.

flood tide Rising *tide: the phase of the tide between low water and the next high tide. *Compare* EBB TIDE.

flood zone *See* ACME ZONE.

floor thrust The lowest *thrust surface bounding a *duplex system, which joins the *roof thrust at the leading and trailing edges of the duplex. If it is the lowest regional thrust surface it may be called a *sole thrust.

flora (adj. floral, floristic) All the plant species that make up the vegetation of a given area. The term is also applied to assemblages of fossil plants from a particular geologic time, or from a geographical region in a former geologic time. Examples of all three types of usage, respectively, are: British flora, *Carboniferous flora, and *Gondwana flora. *Compare* FAUNA.

Florida current Part of the *Gulf Stream: it extends from the southern tip of Florida to Cape Hatteras, N. Carolina. It is a fast-flowing (1–3 m/s), narrow (50–75 km wide), and deep current, still evident at depths of 2000 m where velocities of up to 10 cm/s have been measured. It is an example of a western *boundary current.

floristics *See* PHYTOGEOGRAPHY.

flotation separation A concentration process whereby finely ground *ore is dispersed in water containing a flotation reagent which causes selected minerals to become hydrophobic. Aeration and agitation of the suspension allows hydrophobic particles to float while unaffected minerals sink. The floating particles are skimmed off or overflow the flotation cell.

flow *See* BANKFULL FLOW; BASEFLOW; CRITICAL FLOW; FROUDE NUMBER; GROUNDWATER FLOW; INTERFLOW; LAMINAR FLOW; QUICKFLOW; STEADY FLOW; SUB-SURFACE FLOW; SURFACE RUNOFF; TURBULENT FLOW; UNDERFLOW; and UNIFORM FLOW.

flow cleavage A type of *cleavage which is intermediate between a *slaty cleavage and a *schistosity. The term has fallen from favour and several authorities (e.g. C. McA. Powell (1979)) recommend that it should not be used in descriptions of cleavage because it implies the mode of origin of the cleavage.

flower structure A series of convex-upward thrust or *reverse faults found in *transpressional *strike-slip zones. On seismic sections the appearance of the structure is reminiscent of the petals of a flower or the leaves of a palm tree, hence the analogy. Such structures are important in oil exploration as they indicate strike-slip movements and therefore the possible development of pull-apart basins, and also provide potential areas for oil accumulation.

flow folding Folding which results mainly from continuous *simple shear or viscous flow within layers, producing deformations analogous to lamellar flow in liquids. The term 'flow fold' has also been used synonymously with *ptygmatic fold.

flowmeter An instrument for measuring the flow of liquids.

flowstone *See* DRIPSTONE.

flow till Sediments which flow after they have been deposited by *ablation. *See* TILL.

fluid inclusion Usually minute amount of liquid and/or gas trapped in a crystal during crystallization or *recrystallization. There are two ways in which solid phases in fluid inclusions may originate. (*a*) One or more mineral grains may be trapped along with the fluid phase(s) during formation of the fluid inclusion. (*b*) One or more solid phases may form in a fluid inclusion after its initial formation as a result either of a reaction between the fluid and the host mineral, or by precipitation from the fluid upon cooling (in which case the solids are known as daughter minerals). The temperature and composition of the fluid from which the enclosing mineral originated can be estimated from studies of such inclusions.

fluidization Process of passing gas through loose, fine-grained particles causing the mixture to flow like a liquid; this facilitates mixing and chemical reaction. The faster the gas flows the more the mixture expands and movement increases. A bubble phase may form which travels upwards until all the solid particles are transported by the gas. Fluidization is used in coal-fired power stations and may occur naturally in volcanic eruptions producing *pyroclastic flows and surges. It is the phenomenon which enables pyroclastic flows to travel distances in excess of 100 km and to surmount topographic obstacles hundreds of metres high.

flume 1. A short section of artificial channel constructed in a river in order to create a constriction in which *critical flow will be established, allowing the *discharge to be calculated from the water depth. **2.** An experimental channel used for studying relationships between *sediment movement and *flow conditions. There are a number of different flume designs but most flumes are capable of carrying water at variable depths and velocities, either in a unidirectional flow or generating waves. Flume studies have been responsible for establishing the important relationships between the grain size (*see* PARTICLE SIZE) and *erosion velocities and stability fields for the various sediment *bedforms.

fluorescence Kind of luminescence, in which an atom or molecule emits radiation when electrons within it pass back from a higher to their former, lower energy state. The term is restricted to the phenomenon in cases where the interval between absorption and emission is very short (less than 10^{-3} s). *See also* PHOSPHORESCENCE; X-RAY FLUORESCENCE; and X-RAY FLUORESCENCE SPECTROMETRY.

fluorite (fluorspar, Blue John) Mineral, CaF_2; sp. gr. 3.2; *hardness 4; *cubic; often yellow, green, blue, or purple, but can be colourless, pink, red, or black, and often colour banded; white *streak; vitreous *lustre; *crystals often cubes, but can be *octahedra and *rhombdodecahedra, and a mixture of forms; *cleavage perfect {111}; widely distributed in mineral veins alone or as a *gangue mineral with metallic ores, and in association with *quartz, *barite, *calcite, *galena, *cassiterite, *sphalerite, and many other minerals; soluble in sulphuric acid with the evolution of hydrogen sulphide. It is used extensively as a flux in the smelting of iron, in the ceramic industry, and in the chemical industry. The deep-purple, banded variety, Blue John, is used as an ornamental stone.

fluorometer An analytical instrument used mainly in chemical analysis to measure the fluorescent radiation emitted by any particular substance. It works by exposing the substance under investigation to monochromatic radiation.

fluorspar *See* FLUORITE.

flushed zone (invaded zone) The zone in a sediment bordering a drill hole in which the *groundwater has been replaced by *mud filtrate.

fluted moraine A ground *moraine surface which shows streamlined ridges and grooves trending (*see* TREND) at right angles to the ice front. Individual ridges are generally less than 1 km long, and less than 10 m high. They may form as a consequence of high vertical ice pressure forcing a subglacial, plastic *till up into the low-pressure zone downglacier of a large boulder.

flute cast *See* FLUTE MARK.

flute mark A tongue-shaped scour cut into mud by a turbulent flow of water. The tongue is deepest at the up-current end and the flute can thus be used as a *palaeocurrent indicator. If the flute is infilled by *sediment a flute *cast will be preserved in the base of the overlying bed. Although once believed to be diagnostic of *turbidite deposition, flutes can form in any setting where water flows strongly over soft mud.

fluvial Pertaining to a river.

fluvial processes The set of mechanisms that operate as a result of water flow within (and at times beyond) a stream channel, bringing about the *erosion, transfer, and deposition of *sediment. The erosional processes include: the displacement of bed particles through drag and lift forces; *corrasion, the wearing away of bed and banks as mobile sediment is dragged against them; and bank collapse, a consequence of hydraulic activity. Transport processes include the transfer of material in solution and suspension, and by *saltation. Depositional processes act when the immersed weight of a particle is greater than the force driving it down-channel.

fluviatile Applied to *sediments of fluvial (river) origin.

fluxgate magnetometer *Magnetometer based on two parallel solenoids, equally and oppositely wound on high-permeability cores, and driven by a high-frequency alternating current. A signal coil detects any bias arising from the presence of an ambient magnetic field.

fluxoturbidite A poorly graded *sediment, the product of gravity-induced flow in which little turbulent mixing of particles occurs. It is transitional between a *slump and a turbidity flow (*see* TURBIDITY CURRENT).

flyer *See* STRING.

flysch Sedimentary *facies term used to describe a thick succession of redeposited, deep-sea, *clastic material of *synorogenic character.

focal mechanism, earthquake Shallow *earthquakes are considered to occur when prolonged tectonic stress exceeds the local yield strength of rocks, so that brittle *failure occurs suddenly, with associated earthquake and stress drop. Intermediate and deep earthquakes appear to have variable focal-plane mechanisms, although they are predominantly associated with *slip along fracture planes. Other mechanisms include sudden phase transitions, resulting in an implosion as the mineral density increases and its volume decreases. *See* FAULT-PLANE SOLUTION.

focus (of earthquake) *See* HYPOCENTRE.

fodinichnia The excavations formed by *deposit feeders in search of food. The category includes radial traces (*see* TRACE FOSSIL), e.g. **Chondrites*, and U-shaped tubes, e.g. *Rhizocorallium*.

foehn wall *See* FÖHN WALL.

foehn wind *See* FÖHN WIND.

fog Condition of atmospheric obscurity near the ground surface, caused by the suspension of minute water droplets in conditions of near saturation of the air. The formation of fog is aided by any concentration of smoke particles, as these act as condensation nuclei, and may cause fog at levels of humidity below saturation point. Visibility is below 1 km. The cause of condensation of the water droplets may be radiation-cooling of the ground, advection of warm air over a cold ocean or cold ground, or conditions at a *front. *See also* SMOG. *Compare* MIST.

föhn wall (foehn wall) Mass of cap cloud and associated precipitation over windward slopes and parts of leeward slopes on mountain barriers, resulting from the föhn effect (*see* FÖHN WIND).

föhn wind (foehn wind) Generic term for warm, dry winds in the lee of a mountain range. It was originally used in the European Alps. After cooling on the windward ascent at the *saturated adiabatic lapse rate of 0.5 °C/100 m, with the resulting condensation and precipitation, the air descending on the leeward side of the mountain range is warmed through compression at the *dry adiabatic lapse rate of 1 °C/100 m. This produces a warming wind on the lee side, with higher temperatures than occurred in the same air on the upslope side of the mountains.

foid A contraction of the term *feldspathoid, which is applied to any plutonic rock containing up to 60% *modal feldspathoid minerals. For example, a *syenite with significant *nepheline present can be termed a 'foid-bearing syenite', or a 'foid-syenite'. Such terminology is used on the Streckeisen classification of *igneous rocks with a *colour index of less than 90. In practice, the terms 'foid-bearing' and 'foid-' are replaced with the specific feldspathoid mineral name. In the above example, the name 'nepheline-syenite' would apply. Where the modal volume of feldspathoid minerals exceeds 60%, the rock is termed a 'foidolite'.

foidolite *See* FOID.

fold 1. A bend in rock strata or in any planar feature. The feature (e.g. *bedding, *cleavage, or layering) is deflected sideways and the amount and direction of *dip is altered. Four principal regimes are responsible for folding: *layer-parallel or lateral compression; differential vertical subsidence; differential shearing; and *thrusting. In a simple *anticline–*syncline fold pair, an individual fold consists of a curved *hinge zone and two planar limbs. An imaginary *fold axis lies parallel to the hinge zone (line) and marks the intersection of the *axial plane (or surface) with this zone. This basic geometric form gives rise to many fold profiles, including *parallel, *similar, *concentric, *open, and *isoclinal fold types. To define the attitude of a fold accurately, the orientation of both the hinge line and the axial plane have to be measured. Varying orientations of the hinge line and axial plane may give rise to widely differing fold attitudes, thus *vertical, *upright and *inclined (horizontal and plunging, *see* PLUNGE), and *reclined and *recumbent forms may be described. **2.** In *seismic reflection sampling, the number of *offset distances which sample one *common depth point. For example, if one CDP is sampled at 24 offset distances it is referred to as '24-fold' coverage. The signals recorded for the CDP on each separate trace are then summed by *stacking to improve the signal-to-noise ratio.

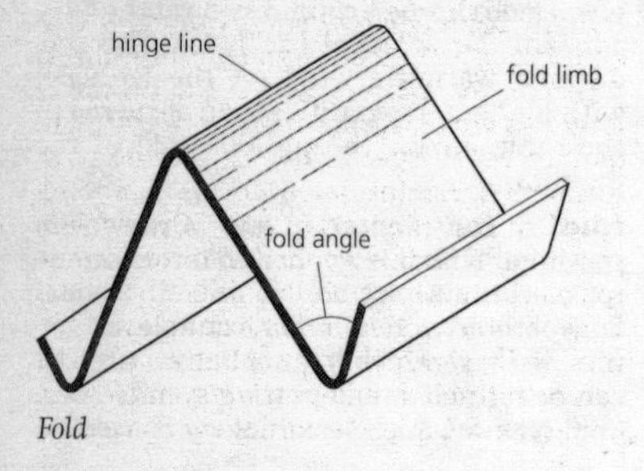

Fold

fold-and-thrust belt A linear or arcuate belt in which compression has produced a combination of *thrusts and *folds. The *dip of the thrust planes decreases with depth, and the belt normally lies against the *foreland of an *orogenic belt. *See also* FOLD BELT.

fold angle (inter-limb angle) The angle between the limbs of a *fold, whose tightness or openness reflects the intensity with which the structure has been deformed. The fold angle can be quantified by measuring the angle of intersection between two tangents through the *inflexion points of opposite *fold limbs.

fold axis A line which lies parallel to the *hinge line and marks the intersection of the *axial plane with the hinge zone.

fold belt A linear or arcuate region characterized by compressional tectonics, including folding. Some authors refer to '*fold-and-thrust belts' in recognition of the importance of thrusting in crustal *shortening.

folding frequency *See* NYQUIST FREQUENCY.

fold limb The generally planar region of a *fold which lies between two adjacent *hinge lines and is confined between the zones of maximum curvature.

fold test The main test in *palaeomagnetism for determining the age of the *remanent magnetization of rocks. Comparison of the scatter of directions of natural remanence from both sides of a *fold, before and after correcting for the tectonic effects of the folding, enable the magnetization to be determined as pre-folding if the scatter is least after correction for folding.

foliation A continuous, sub-planar rock *fabric formed by the preferred orientation of minerals with a generally *platy or *tabular habit. The layers are planar, and parallel but not necessarily to *bedding or *cleavage. The term 'foliation' is commonly applied to textures in high-grade *metamorphic rocks (*see* METAMORPHIC GRADE).

folivorous Leaf-eating.

Folk limestone classification A widely used classification of *carbonates, based on the type of particles and the nature and proportion of the *matrix and/or *cement present. In his original classification Robert L.

Folk defined three main components to *limestones. These are *allochems, comprising various *grains and particles; *micrite (*micro*crystalline *calc*ite* mud matrix); and *sparite (*spar*ry calc*ite* cement). The main allochems are *bioclasts ('bio-', *see* below), *pellets ('pel-'), *intraclasts ('intra-'), and *ooids ('oo-'). In defining a limestone by the Folk classification the rock is named according to the nature of the material filling the spaces between the particles (i.e. micrite matrix or sparite cement), prefixed by an abbreviation to denote the main allochems present: bio- for bioclasts, pel- for pellets, oo- for ooids, and intra- for intraclasts. For example, a limestone comprising pellets in a mud matrix is called a 'pelmicrite', and a limestone comprising shell fragments with a sparry calcite cement is a *'biosparite'. It is possible to combine the prefixes of several allochems where appropriate, e.g. 'oobiosparite'. Limestones which lack allochems, consisting only of micrite are termed 'micrites'. *Dismicrite is used for micrites with fenestral cavities (*see* FENESTRAE). Organically bound limestones, e.g. *reef rocks, or *stromatolites, are termed *biolithites. Folk subsequently modified his classification to include various carbonate textures. Under his textural scheme a limestone with varying proportions of bioclasts, mud matrix and cement would be classified as follows: micrite with less than 1% shell fragments = micrite; micrite with 1–10% shell fragments = fossiliferous micrite; micrite with 10–50% shell fragments = sparse biomicrite; micrite with over 50% shell fragments = packed biomicrite. Limestone with micrite and sparry calcite together with shell fragments = poorly washed biosparite; poorly sorted shell fragments with sparry calcite cement = unsorted biosparite; well sorted shell fragments with sparry calcite = sorted biosparite; rounded and abraded shell fragments with sparry calcite cement = rounded biosparite. Similar textural terms apply for other allochems with micrite and/or sparite. *See* BIOMICRITE; INTRAMICRITE; INTRASPARITE; PELSPARITE; OOSPARITE; and OOMICRITE.

fondaform A flat surface beneath a body of water that gives a seismic reflection.

fondathem A rock unit produced beneath a body of water and giving a seismic reflection.

fool's gold *See* PYRITE.

footwall The *fault block which lies below any inclined *fault surface. *Compare* HANGING WALL.

foram *See* FORAMINIFERIDA.

foramen A pore or opening. Applied to various openings, e.g. the *pedicle foramen in *Brachiopoda through which the *pedicle emerges. In cephalopods (*Cephalopoda) the term is applied to the opening in the *septum through which the siphuncular cord passes.

foramen magnum Opening at the posterior end of the skull through which the spinal cord passes.

foramina A series of openings connecting the various chambers of a foraminiferid (*see* FORAMINIFERIDA) *test.

foraminifera *See* FORAMINIFERIDA.

Foraminiferida (informally foraminifera, foraminiferans, forams (pl.); foraminiferid, foraminifer, foraminiferan, foram (sing.); class Rhizopoda) Order (in some classifications, subclass Foraminifera) of testate, amoeboid *protozoa in which the cell is protected by a *test, consisting of one to many chambers, whose structure and composition is of great importance in foraminifera classification. The three main types are: (*a*) most primitively, a test wall composed of a secreted, chitinous-like, organic material called tectin, which also often forms an underlying layer in the other two types; (*b*) a test formed from agglutinated *sedimentary particles, which may be cemented with an organic, calcareous, or ferric oxide *cement; (*c*) a fully mineralized test, composed of secreted calcareous or siliceous minerals, of which the calcareous types (*aragonite and *calcite) are the most common. The arrangement of multiple chambers may be linear, spiral, cone-like, etc. Numerous fossil foraminifera are known, usually less than 1 mm across; though some, like the *fusilinids (*Carboniferous to *Permian) and *nummulitids (*Eocene to *Oligocene) were appreciably larger (some measured up to 100 mm in diameter). All species live in marine environments. Agglutinated forms predominated in the *Cambrian and *Ordovician, presumably derived from a tectinous ancestor, while forms with fully mineralized tests appeared in the Ordovician and diversified greatly in the *Devonian. The Foraminiferida are important zonal fossils, and some *planktonic varieties can be used for stratigraphic correlation on virtually a world-wide scale. *See* GLOBIGERINA OOZE.

foramol Applied to an association of *Bryozoa, *Foraminiferida, coralline red algae (*Rhodophyceae), and *Mollusca that inhabits seas where the temperature often falls below 15 °C and forms a characteristic sediment. *Compare* CHLORALGAL and CHLOROZOAN.

forams *See* FORAMINIFERIDA.

forced convection Mechanical turbulence, with the development of eddies, in air flowing over an uneven surface.

forced regression *Regression caused by a fall in sea level.

forced regressive systems tract *See* REGRESSIVE SYSTEMS TRACT.

fore-arc *See* ARC-TRENCH GAP.

fore-arc basin The part of the fore-arc (*arc-trench gap) adjacent to the *island arc which is characterized by flat-lying sediments, in contrast to the highly deformed *accretionary wedge adjacent to the oceanic *trench. Fore-arc basins lie behind the topographic high-point of the wedge, which in places forms an outer (sedimentary) island arc. Fore-arc sediments lie unconformably on accretionary wedge material and show progressive shoaling. The source of the material is the adjacent volcanoes and the erosion of uplifted *plutonic–metamorphic basement.

foredeep A basin adjacent to a *craton which is filled with a thick accumulation of sediment derived from an *orogenic belt during uplift. The sediments are typically non-marine to shallow-marine and commonly suffer deformation within a few million years of deposition.

foreland A stable area on the edge of an *orogenic belt; a foreland is usually on the margin of a *craton and is underlain by *continental crust. Many forelands have suffered warping during *orogeny and also carry a superficial *fold-and-thrust belt. The major direction of movement in an orogenic belt is towards the foreland. Where the orogenic belt lies between two stable areas, the other is called the hinterland.

fore reef *Talus slope on the seaward side of a *reef, constantly under attack by waves and currents.

foreset 1. The inclined surface within a cross set (*see* CROSS-STRATIFICATION) produced by the forward movement of the slip-face of a *ripple, *dune, *sandwave or bar. **2.** The slip-face of a *Gilbert-type delta.

foreshock A small *earthquake, sometimes occurring in swarms, that precedes a major earthquake (or volcanic eruption).

foreshore Lower shore zone that lies between the normal high- and low-water marks. The foreshore may either be a plane slope dipping seawards at a low angle, or be marked by the development of longshore bars (ridge-and-runnel topography), depending on the nature of the wave attack.

foreshortening In *radar terminology, the distortion of a radar image caused by shallow surface angles relative to the radar wavefront angle, such that the base of a hill is actually closer to the radar than the top. The resulting image gives the appearance of a shorter slope. *Compare* LAYOVER.

forked lightning Lightning discharge in which luminous branches from the main channel are seen. *See also* SHEET LIGHTNING.

form The overall shape of a crystal. If it is able to grow freely, a crystal develops with a regular pattern of *crystal faces and interfacial angles which are characteristic of a particular *mineral. The study of this regularity of crystal form, and of the internal structure to which it is related, is called '*crystallography'.

formal The name of a *stratigraphic unit is formal when it has been established according to the conventional principles of *stratigraphic nomenclature and is being used in the sense of a proper name, e.g. *Barremian Stage, Late *Devonian *Epoch. In formal use the initial letter is capitalized. *Compare* INFORMAL.

formation The fundamental unit used in lithostratigraphy. Specific features distinguish one rock formation from another. The thickness of the formation is unimportant in its definition, as a given formation may vary within different outcrops. Formations may be subdivided into members and together several formations constitute a group.

formation age Time which has elapsed since a *meteorite formed, obtained by *radiometric dating, assuming no loss of a gaseous daughter isotope has occurred (*see also* GAS RETENTION AGE). Most meteorites give formation ages around 4.5 billion years, although there are exceptions (*see* SHERGOTTYITE/NAKHLITE/CHASSIGNITE METEORITES).

formation evaluation The detailed analysis and interpretation of borehole data, drilling results, geophysical downhole logs, etc., to determine the physical characteristics of the rock *formations through which the drill has penetrated. This is done mainly to ascertain whether or not economic reserves of hydrocarbons are present and, if they are, to determine the most economical and efficient way to extract them. Formation evaluation is an important component of reservoir engineering design.

formation velocity (v_{for}) The uniform seismic velocity of a particular homogeneous rock type. For a rock thickness h and a single-travel time t, $v_{for} = h/t$. *Compare* INTERVAL VELOCITY.

form factor *See* DRAINAGE BASIN SHAPE INDEX.

form-genus Non-phylogenetic, artificial *taxon of convenience. In palaeobotany the disarticulated parts of fossil plants, whose natural affinities are unknown, have been assigned to form-genera on the basis of similarities in morphology. Later discoveries may show different form-genera to have derived from the same plant, e.g. **Stigmaria*-type roots and **Lepidodendron*-type bark. *See also* CORDAITALES; and *CALAMITES CISTIIFORMES*.

form roughness *See* BED ROUGHNESS.

form set *See* CROSS-STRATIFICATION.

Fortin barometer Mercury barometer that requires the accurate setting of the mercury level at a fixed point (scale zero, *see* FIDUCIAL POINT). The reading of the mercury height is then taken by adjusting a vernier scale to the top of the mercury column. *See also* KEW BAROMETER.

Förtsch discontinuity An irregular seismic discontinuity within the upper *continental crust, at 8–11 km depth, usually interpreted as a change from upper granitic rocks to deeper dioritic composition. *See also* CONRAD DISCONTINUITY; and MOHOROVIČIĆ DISCONTINUITY.

forward problem (direct problem, normal problem) The problem of calculating what should be observed for a particular model, e.g. calculating the gravity anomaly that would be observed for a given model of a salt *dome. *Compare* INVERSE PROBLEM.

fossa (pl. fossae) A linear depression on a planetary surface, closely analogous to a terrestrial fault-bounded depression or *graben. The radial fractures with mainly *normal faults (e.g. Tempe, Tantalus Menonia, Claritas fossae, etc.) which surround the Tharsis bulge on Mars, form type examples.

fossil 1. Generally, anything ancient, especially if it is discovered buried below ground (e.g. *fossil fuel, fossil soil). **2.** The remains of a once-living organism, generally taken to be one that lived prior to the end of the last glacial period, i.e. fossils are older than 10 000 years. The term includes skeletons, tracks, impressions, trails, borings and casts. Fossils are usually found in consolidated rock, but not always (e.g. woolly mammoths living 20 000 years ago were recovered from the frozen tundra of Siberia). In its original sense, fossil meant anything dug up from the earth, including ores, precious stones, etc. The modern use of the word dates from the late 17th century. *See also* LIVING FOSSIL; and TRACE FOSSIL.

fossil fuel All deposits of organic material capable of being burnt for fuel; chiefly coal, oil, and gas. These are formed under pressure by alteration or decomposition of plant or animal remains.

fossiliferous micrite *See* FOLK LIMESTONE CLASSIFICATION.

fossilization The process by which a *fossil is formed. It is unusual for organisms to be preserved complete and unaltered; generally, the soft parts decay and the hard parts undergo various degrees of change. *Solution and other chemical action may reduce the tissues to a thin film of carbon; this process is called 'carbonization'. The organism may be flattened by the compaction of sediments to form compressions. Porous structures, e.g. bones and shells, may be made more dense by the deposition of mineral matter by *groundwater; this process is called 'permineralization' or 'petrifaction'. The internal physical structures of some shells may be changed as a result of solution and reprecipitation; in this process ('recrystallization') the original structure may be blurred or lost. Many shells which were originally composed of *aragonite are recrystallized into the more stable mineral *calcite. The solution of an original shell and the simultaneous deposition of another mineral material constitutes 'replacement'; this may occur molecule by molecule, in which case the microstructure is preserved, or *en masse*, where it is not. Common replacement minerals include *silica or iron sulphide, but there are many others. The impression of skeletal remains in surrounding sediments constitutes a 'mould'. Where the external structures are preserved it is called an 'external mould' and where the internal features are preserved it is called an 'internal mould' or 'steinkern'. Filling of a mould cavity by mineral matter may produce a 'natural cast'. Tracks, trails, burrows, and other evidence of organic activity may also be preserved. These are called 'ichnofossils' or *trace fossils.

fossil-Lagerstätte *See* LAGERSTÄTTE.

fossula A gap or depression in the floor or *calyx of a rugose coral (*Rugosa). It may be formed by the absence of septa (*see* SEPTUM) in a particular part of the calyx.

foundation Lowest part of a structure, below ground surface and in contact with natural earth materials, which transmits *load to the soil or rock. In a dam the foundation may include the valley floor and abutments.

founder effect (peripatric speciation) The derivation of a new population (e.g. on an oceanic island) from a single individual or a limited number of immigrants. The founder(s) represent a very small sample of the *gene pool to which it or they formerly belonged. *Natural selection operating on this more restricted genetic variety yields gene combinations different from those found in the ancestral population.

founder lineage In *phylogenetics, an ancestral lineage, often still extant, from which other lineages have risen. The term is usually applied to intraspecific studies of populations and used to describe operational taxonomic units that occur at *internal nodes of a *phylogenetic tree.

fourchite An *intrusive *igneous rock consisting of essential titanaugite (titanium-rich *augite) and *kaersutite (with or without *biotite) set in a light-coloured base of *analcime or *glass. This rock type is a member of the alkali *lamprophyre group of rocks, which includes *camptonites, *monchiquites, and *alnoites.

Fourier analysis The method whereby any periodic function can be broken down into a covergent trigonometric series of the form $f(x) = a_0/2 + \Sigma_\infty^{n=1}(a_n \cos nx + b_n \sin nx)$ where a_n and b_n are constant coefficients. Fourier analysis is the process of determining the *frequency-domain function from a time function (e.g. a seismic-trace waveform). *See also* FOURIER TRANSFORM.

Fourier synthesis The superimposition of sinusoidal waves of known frequency, amplitude, and phase to represent an observed wave-form. It is the process of determining the *time-domain function from a frequency function. *See also* FOURIER TRANSFORM.

Fourier transform The mathematical formulae by which a time function (e.g. a seismic trace) is converted into a *frequency-domain function and vice versa. *See also* FOURIER ANALYSIS; and FOURIER SYNTHESIS.

fractal A geometric entity which has a basic pattern that is repeated at ever decreasing sizes. Fractal patterns are not able to fill spaces and are hence described as having fractal dimension. Fractals occur frequently in nature, such as in forked lightning or in chaotic systems (*see* CHAOS).

fractional crystallization The removal of early formed *crystals from an originally homogeneous *magma (for example, by gravity settling) so that these crystals are prevented from further reaction with the residual *melt. The composition of the remaining melt becomes relatively depleted in some components and enriched in others, resulting in the precipitation of a sequence of different minerals. Fractional crystallization is one of the main processes of *magmatic differentiation.

fractionation *See* MAGMATIC DIFFERENTIATION.

fracto- *See* FRACTUS.

fracture 1. General term applied to any break in a material, but commonly applied to more or less clean breaks in rocks or minerals that are not due to *cleavage or *foliation. **2.** *See* VOIDS.

fracture cleavage A *cleavage (defined originally by Leith in 1905) which resembles very closely spaced, parallel *joints or fractures, but where in fact the 'fractures' are due to a loss of material resulting from *pressure solution. In modern terminology the term would be replaced by 'spaced cleavage'.

fracture porosity A form of *secondary porosity generated by tectonic fracturing (*see* TECTONISM) of the rock. Such porosity can develop in any rock, allowing the development of productive *reservoirs in rocks such as *granites and *gneisses. *See* POROSITY; CHOQUETTE AND PRAY CLASSIFICATION; and RESERVOIR ROCK.

fracture spacing index (I_f) The number of fractures in a one-metre length of drill *core.

fracture zone A linear feature on the deep-sea floor across which the *lithosphere changes abruptly in both age and water depth. Most fracture zones transect oceanic *ridges and are small circles whose radii of curvature depend on the distance from the *pole of rotation of the two lithospheric *plates which are diverging at the ridge. Many fracture zones also contain deep-sea basins.

fractus (fracto-) From the Latin *fractus* meaning 'broken', a species of cloud that has an irregular or ragged form. The term is applied to *cumulus and *stratus. *See also* CLOUD CLASSIFICATION.

fragipan Subsoil *horizon, found deep in a *soil profile and having a high bulk density. It is a dense, brittle, and compact layer, apparently with little or no cementation horizon, associated with *acid soil conditions.

fragmental *See* CLASTIC.

framestone An *autochthonous, organically bound *limestone, where the organisms, e.g. corals (*Anthozoa), form a rigid framework during deposition. *See* EMBRY AND CLOVAN CLASSIFICATION.

framework porosity *See* CHOQUETTE AND PRAY CLASSIFICATION.

Franconian A *stage of the *Cambrian in the *Croixian *Series of N. America, underlain by the *Dresbachian and overlain by the *Trempealeauan.

franklinite A member of the *spinel group of minerals, and an *end-member of the *magnetite series, with composition $ZnFe^{3+}{}_2O_4$, although appreciable amounts of Mn^{2+} and Fe^{2+} are also present; sp. gr. 5.0; *hardness 6; black; *metallic *lustre; *crystals *cubic but normally found as *octahedra or granular aggregates; occurs at Franklin, New Jersey, in association with other zinc minerals in a metamorphosed *limestone.

Frasien *See* FRASNIAN.

Frasnian (Frasien) 1. An *age in the Late *Devonian Epoch, preceded by the *Givetian (Middle Devonian), followed by the *Famennian, and dated at 377.4–367 Ma (Harland et al., 1989). **2.** The name of the corresponding European *stage, which is noted for the culmination of the first major radiation of the single-celled *Foraminiferida and for the origin of important groups of *goniatites, prolecanitids, goniatitids, and clymeniids. It is roughly contemporaneous with the *Senecan (N. America) and part of the *Condobolinian and *Hervyan (Australia).

frazil ice Flowing water ice that forms platelets rather than continuous sheets; often observed in Canadian rivers. The name is derived from the French *fraisil*, meaning cinder.

free-air anomaly The *gravitational acceleration remaining after correction of a measurement for the theoretical gravity, usually the *International Gravity Reference Field, and by the *free-air correction. No correction has been made for the gravitational attraction of rocks between the observing station and the reference datum, which is usually sea level.

free-air correction (Faye correction) The correction applied to a measurement of gravity which allows for the variation of gravity with height above a reference level, usually sea level. This correction assumes there is only air between the station and the reference level, and it is 0.3086 mgal/m. *See also* BOUGUER CORRECTION.

free atmosphere The atmosphere above the level to which the effect of friction reaches (commonly taken as about 500 m).

freeboard The distance between the maximum permitted water level in the reservoir behind a dam and the top of the dam wall. *See also* CONTINENTAL FREEBOARD.

free face *See* SLOPE PROFILE.

free oscillations The harmonics at which any body, e.g. the Earth, tends to vibrate most freely, i.e. resonates. There are two fundamental types: torsional (vibration with motions perpendicular to the Earth's radius); and spheroidal (vibrations that are both radial and tangential to the Earth's surface). The study of such resonances, e.g. those induced by major earthquakes, provides information on the internal nature of the Earth. A major earthquake can make the entire globe vibrate or ring like a bell, and some earthquakes have been so large that sensitive *seismometers have continued to record the oscillations for weeks after the event. The decay of the vibrations gives valuable information about the elastic layering of the Earth, and especially of the *low velocity zone. *Moonquakes produce similar phenomena.

freestone A building stone; poorly jointed *sandstone or *limestone which can be worked easily in any direction. *Compare* DIMENSION STONE.

freezing nuclei Any nuclei, commonly ice crystals but sometimes the suitably shaped crystals of other substances, that when present in clouds at temperatures below 0 °C will cause any supercooled droplets with which they collide to change to ice (in the form of a crystalline growth upon the nucleus). *See also* NUCLEUS. *Compare* ICE NUCLEUS.

freibergite *See* TETRAHEDRITE.

frequency (*f*) The number of complete *wavelengths which pass a given point in a specified time; units are hertz (Hz; one hertz is one cycle per second). The frequency of a periodic wave-form is given by $f = 1/T$, where T is the *period; and by $f = v/\lambda$, where v is the *velocity and λ the wavelength.

frequency domain A reference framework in which measurements are related to frequency rather than to time (as in the *time domain).

fresh water Water containing little or no chloride ion. According to the Venice system, which classifies brackish waters by their percentage chloride content, fresh water contains 0.03% or less of chloride. *Compare* HALINITY.

Fretted and Chaotic Hummocky Terrain *See* MARTIAN TERRAIN UNITS.

friable Applied to the consistency or handling properties of soil, meaning that the soil crumbles easily.

frictional resistance, angle of *See* ANGLE OF SHEARING RESISTANCE.

frictional angle *See* ANGLE OF INTERNAL FRICTION.

frigid *See* PERGELIC.

fringing reef *See* REEF.

Fronian A *stage of the Lower *Silurian, underlain by the *Idwian and overlain by the *Telychian.

front Boundary or boundary region separating *air masses of different origins and characteristics. Temperature gradients in any horizontal surface are large through the front. Different types of front are distinguished according to the nature of the *air masses separated by the front, the direction of the front's advance, and the stage of development. The term was first devised during the First World War by the Norwegian school of meteorologists (headed by Prof. V. *Bjerknes). *See also* ANAFRONT; COLD FRONT; WARM FRONT; KATAFRONT; OCCLUDED FRONT; and POLAR FRONT.

frontal wave Wave-like deformation of the line of a *front between two *air masses. The wave develops from the northward incursion of warm air and usually travels along the front, with colder air ahead and to the rear. Typically, frontal waves occur in sequences, or 'families', of several waves, and develop into *depressions or storm centres travelling more or less eastward as 'secondaries' along the extended cold front to the rear of the original low. The secondaries tend to catch up and merge with the original depression as it slows up in its fully developed stage.

frontal zone Transition zone, sometimes amounting to a discontinuity, that separates adjacent *air masses. Some turbulent mixing takes place. The sloping zone separating a cold wedge under warm air typically extends about 1 km vertically and about 100 km horizontally.

frontogenesis Development and intensification of frontal boundaries between adjacent *air masses.

frontolysis Processes of dissolution or dissipation of a *front. Frontal decay results when different *air masses stagnate together, or move together or in succession along the same track at the same speed, or incorporate air of the same temperature.

frost Condition in which the prevailing temperature is below the freezing point of water (0 °C). This may lead to a deposit of ice crystals on objects, e.g. grass or trees. Such deposits result from condensation when the *dew-point temperature is below freezing. *See also* BLACK ICE.

frost boil *See* INVOLUTION.

frost heave (frost heaving) Upward movement of the ground surface or of individual particles, due to the formation of lenses of ice up to 30 mm thick in the *regolith. It reaches its maximum in silt-dominated material, in which the greatest volume of ice may develop (more than 68% ice by volume). When the total uplift of the surface is measured, it is found to be approximately equal to the sum of the thicknesses of the layers of ice. Surface stones may be heaved by the development of needle-ice columns ('pipkrakes').

frost heave test A laboratory test in which *aggregate or soil is frozen under controlled conditions. A cylinder containing rock aggregate, 150 mm high and 100 mm in diameter is placed in freezing conditions with its base in running water for 250 hours. The *frost heaving must be less than 12 mm.

frost heaving *See* FROST HEAVE.

frost hollow Area (e.g. a valley bottom or a smaller hollow) that is very liable to severe and frequent *frosts as a result of dense, cold air moving downslope (katabatic flow) and collecting there under conditions of radiation cooling (e.g. at night).

frost pull and frost push *Periglacial processes that bring about the upward migration of *clasts through the *regolith. Frost push takes place when an ice lens forms beneath a clast and so pushes it upwards. Frost pull occurs when a clast adheres to ice within a freezing regolith and so is drawn upwards as the ground heaves.

frost-shattering *See* FROST WEDGING.

frost smoke *See* ARCTIC SEA SMOKE.

frost table *See* TJAELE.

frost wedging (congelifraction, frost-shattering, gelifraction, gelivation) Fracturing of rock by the expansionary pressure associated with the freezing of water in planes of weakness or in pore spaces.

Froude number (Fr) A dimensionless number equal to the ratio of water velocity to the speed of a gravity wave, used to assess whether flow in an open channel is critical, tranquil, or shooting. If the Froude number is less than 1, flow is said to be subcritical or slow; if Fr = 1, flow is critical; and if Fr is greater than 1, flow is fast or supercritical.

frustule Silica wall of a diatom. *See* BACILLARIOPHYCEAE.

Fs Orthoferrosilite. *See* ENSTATITE.

fugacity (*f*) A measure of the tendency of a gas to escape or expand, used in calculations of chemical equilibrium. Fugacity (f_i) is the pressure value needed at a given temperature to make the properties of a non-ideal gas satisfy the equation for an ideal gas, i.e. $f_i = \gamma_i P_i$, where γ_i is the fugacity coefficient, and P_i is the partial pressure for the component *i* of the gas. For an ideal gas, $\gamma_i = 1$.

fugichnia The so-called 'escape structures' that mark the response of animals to changes in the rate of deposition or erosion or to predation. The original trace (*see* trace fossil) may be regarded as a permanent dwelling structure, but the presence of fine, crescentic laminae (*spreiten) indicates that the organism made efforts to escape burial or exposure. Thus the presence of spreiten implies Fugichnia.

Fujita Tornado Intensity Scale A standard six-point scale for reporting the intensity of a *tornado by inferring its wind force from the type and extent of the damage it caused. The scale was introduced in 1971 by Tetsuya Theodore Fujita and Allen Pearson. *See* APPENDIX C.

fuller's earth 1. A clay consisting mainly of expanding *smectites such as *montmorillonite used industrially for its absorptive properties. **2.** Capitalized, Fuller's Earth is the stratigraphic name of a *Jurassic clay formation outcropping in southern Britain.

fulvic acid Mixture of uncoloured organic acids that remains soluble in weak acid, alcohol, or water after its extraction from soil.

fumarole Vent in a volcanically active area that emits steam, gas (SO_2, CO_2, etc.), or other *volatile constituents at high temperatures (from 100 °C to 1000 °C). The fumarolic condition has been thought to indicate a late stage in volcanic activity, but may actually precede volcanic *eruptions, e.g. Mt. St Helens, Cascade Range, Washington, in 1980.

fumarolic stage *See* FUMAROLE.

functional morphology The attempted interpretation of the functions of particular organs or structures that occur in various *fossils. In some cases it is difficult to interpret function in groups that have no living examples, but it is often possible to analyse growth and form in invertebrate groups and to relate this growth both to the biology of the organism and to the environment.

fundamental form *See* PARAMETRAL PLANE.

fundamental strength The maximum *stress a material can sustain indefinitely at a given temperature and *confining pressure. The fundamental strength is always less than the breaking strength and *ultimate strength.

Fungi One of the three multicellular kingdoms, along with the *Plantae (plants) and *Animalia (animals). Although resembling plants, Fungi feed by ingesting organic matter, whereas plants are autotrophic and require only inorganic substances as nutrients. As fungi generally lack hard parts they are rarely found as fossils, but thread-like representatives have been found in *Precambrian rocks. They probably left the sea about 400 million years ago, when the first plants colonized the land.

funnel cloud Cloud produced in a low-pressure vortex in the centre of a spiral storm, e.g. a tornado or waterspout.

funnelling The constraining of an airflow by valleys, leading to higher wind speed, convergence, and uplift. Similar effects occur in the air between an advancing front and the face of a mountain barrier.

fusain Fossil charcoal, a *coal lithotype sometimes called 'mother-of-coal', produced by the burning of plant material under airless conditions. This converts the material to almost pure carbon and can preserve small plant parts and cellular structures.

fusiform Spindle-shaped; elongated with tapering ends.

fusilinid One of the so-called 'larger foraminifera' (*Foraminiferida) which usually have a *fusiform or discoid shape. Many genera are differentiated by rapid development and evolution, making them important *index fossils, particularly in *Carboniferous and *Permian rocks.

fusinite *See* COAL MACERAL.

fusion 1. Generally, the melting of a solid substance by heat. **2.** In nuclear fusion, the combining of two light atomic nucleii to form a heavier nucleus with the sudden release of energy, e.g. in the hydrogen bomb.

G *See* GAUSS.

G *See* GRAVITATIONAL CONSTANT.

g *See* GRAVITATIONAL ACCELERATION.

gabbro A coarse-grained, *basic *igneous rock, consisting of *essential calcium-rich *plagioclase feldspar (approximately 60%), *clinopyroxene (*augite or titanaugite), and *orthopyroxene (*hypersthene or *bronzite), plus or minus *olivine with *accessory *magnetite or *ilmenite. Gabbros result from the slow crystallization of *magmas of basaltic composition, and like the *basalts they can be divided into tholeiitic and alkali types (*compare* ALKALI BASALT; and THOLEIITE). Tholeiitic gabbros are characterized by the presence of two pyroxene types (augite and hypersthene) and *interstitial silica-rich *glass, whereas alkali gabbros are characterized by one calcium/titanium-rich pyroxene (titanaugite) and scattered interstitial *feldspathoid minerals. Many large gabbroic *intrusions display mineral layering, testifying to the complex processes taking place within basic *magma chambers. Gabbros are commonly found intruded as ring complexes (e.g. Ardnamurchan and Skye in Scotland), large *lopoliths (Bushveld complex, S. Africa), or layered complexes (Skaergaard in eastern Greenland being the most famous).

gaging station *See* GAUGING STATION.

gahnite *See* SPINEL.

Gaian hypothesis Hypothesis, formulated by James E. Lovelock and Lynn Margulis, that the presence of living organisms on a planet leads to major modifications of the physical and chemical conditions pertaining on the planet, and that subsequent to the establishment of life the climate and major *biogeochemical cycles are mediated by the living organisms themselves.

gaining stream (influent stream) A stream that receives water emerging from a submerged spring or other groundwater seepage which adds to its overall flow.

gal The unit, named after Galileo, for measuring *gravitational acceleration. 1 gal = 1 cm/s^2. The gal has been largely replaced by the *gravity unit.

galactic cosmic rays *See* COSMIC RADIATION.

Galápagos Rise The oceanic *ridge between the *Cocos and *Nazca Plates.

Galatea (Neptune VI) A satellite of *Neptune, with a diameter of 158 km; visual albedo 0.06.

galaxite *See* SPINEL.

gale Wind blowing at more than 30 knots (17 m/s).

galena Mineral, PbS; sp. gr. 7.4–7.6; *hardness 2.5; *cubic; lead-grey; lead-grey *streak; *metallic *lustre; *crystals cubes or *octahedra, and often octahedral twins; *cleavage perfect cubic {100}; widely distributed in *hydrothermal veins and *syngenetic exhalative deposits and as a replacement in *limestones and dolomitic rocks, associated in veins with *sphalerite, *pyrite, *chalcopyrite, *barite, *quartz, *fluorite, and *calcite.

Galilean satellites The four classical *satellites Io, Europa, Ganymede, and Callisto (in order outwards from Jupiter) which were discovered in 1610 by Galileo. The observation that they orbited a body other than the Earth was fatal to the Ptolemaic *cosmology. Ganymede (radius 2638 km) is the largest satellite in the *solar system, larger than both Mercury and Pluto. Europa (radius 1536 km), the smallest of the four, is a little smaller than the Moon (radius 1738 km). They occupy equatorial orbits. There is a regular decrease in density from Io (3550 kg/m^3) to Callisto (1830 kg/m^3), Callisto preserving one of the most heavily cratered surfaces of any *satellite.

Galileo A *NASA spacecraft, launched on 15 October 1989 from the shuttle *Atlantis* on an international deep-space mission, that passed and photographed Venus on 9 February 1990, for a gravity assist. On its journey through the asteroid belt, Galileo encountered *Gaspra on 29 October 1991, and *Ida on 28 August 1993, where it discovered the satellite Dactyl. It reached *Jupiter in December 1995, then entered an orbit that brought it into repeated encounters with the *Galilean satellites. The spacecraft comprises an orbiter and a probe, released in July 1995, as Galileo was still approaching Jupiter, which penetrated the atmosphere and returned data for 61.4 minutes, by which time it had descended to a level where the pressure was 24 bars, 140 km below the 1 bar pressure altitude.

Gallic The middle *epoch of the *Cretaceous Period, dated at 131.8–88.5 Ma (Harland et al., 1989).

gamma A unit of magnetic field strength. 1 gamma = 10^{-5} gauss. It has now been replaced by SI units: 1 gamma = 10^{-9} tesla (i.e. 1 nano tesla, nT).

gamma–gamma sonde A *well-logging instrument package in which a source of gamma radiation, usually $^{27}_{60}Co$ or ^{137}Cs, bombards the wall of the *borehole, and the backscatter of *gamma rays, together with natural radiation, is recorded some 45 cm above the source. The record is known as a *density log, as the backscatter is an exponential function of electron density in the rocks. Mostly the instrument responds to the nearest 10 cm of rock wall. Sometimes two detectors are used at different locations to distinguish between *mud cake and rock.

gamma-ray log *See* GAMMA-RAY SONDE.

gamma rays Electromagnetic radiation, about 10^{-10} to 10^{-14} m in wavelength, similar to, but of shorter wavelength than X-rays, emitted by radioactive substances.

gamma-ray sonde The *well-logging instrument, comprising a scintillometer, used to measure the natural radioactivity of the rocks through which the drill hole passes. Potassium (^{40}K) is the most abundant radioactive element and occurs in *clays (especially the mineral *illite and *micas). The record is a gamma-ray log, expressed in API units (a gravity scale devised by the American Petroleum Institute and applied mainly to measurements of crude oil). The log is particularly useful for delineating the alternation of *clay-rich and clay-poor *lithologies, e.g. *claystones interbedded with *limestones or *sandstones. Conventionally, claystone horizons yield API values of more than 75. High gamma-ray log values will also be recorded from organic-rich *shales which also concentrate other radioactive elements, e.g. uranium and thorium. In addition, *glauconitic sands, volcaniclastic sands, *zircon-rich sands, and clay-matrix-rich sands produce gamma-active sediments. *See also* PHOTON LOG.

gamma-ray spectrometer *See* SCINTILLATION COUNTER.

gamma-ray spectrometry Analytical method used in some branches of chemistry and physics for the measurement of the intensities and energies of gamma radiation. Scintillation or semi-conductor radiation detectors, coupled to various types of electronic circuitry, enable a spectrum to be accumulated. This may be used to identify the gamma-emitting radioisotopes, and their energy intensities can be used to determine concentrations of the corresponding elements. The technique is also used in *remote sensing to determine the abundance of some elements in distant objects, e.g. the surface of the *Moon.

gangue That portion of an ore deposit which is of no commercial value but which cannot be avoided during mining; it is removed during processing as waste. Common gangue minerals are *quartz, *calcite, and *fluorite, *see* ORE MINERAL; OREBODY; and ORE GRADE.

ganister (gannister) A fine-grained, *arenaceous rock that underlies certain coal measures. It is used for its refractory qualities, e.g. to make furnace hearths.

gannister *See* GANISTER.

ganoid scale Type of fish scale with a rhomboid shape, found in some fossil as well as extant *bony fish (e.g. *Polypterus* and *Lepisosteus*). The scale consists of a superficial layer of enamel-like ganoine, a middle layer of *dentine, and a basal layer of vascular bony tissue.

Ganymede (Jupiter III) One of the *Galilean satellites, and the largest jovian satellite; it is bigger than *Mercury and *Pluto. It is believed to have a rock and metal core surrounded by a large mantle of water or water ice, 800–900 km thick, and the surface is of ice, with two types of terrain, one very cratered and dark, the other rather lighter, with many grooves and ridges. These terrains result from tectonic activity, but the details are not known. Both terrains are extensively cratered, the craters being flat, with no ring mountains and central depressions, and suggest the surface is about 3–3.5 Ma old. Ganymede has a magnetic field, embedded within that of Jupiter. Ganymede was discovered in 1610 by Simon Marius and Galileo. Its diameter is 5268 km; mass 1.48×10^{23} kg; mean density

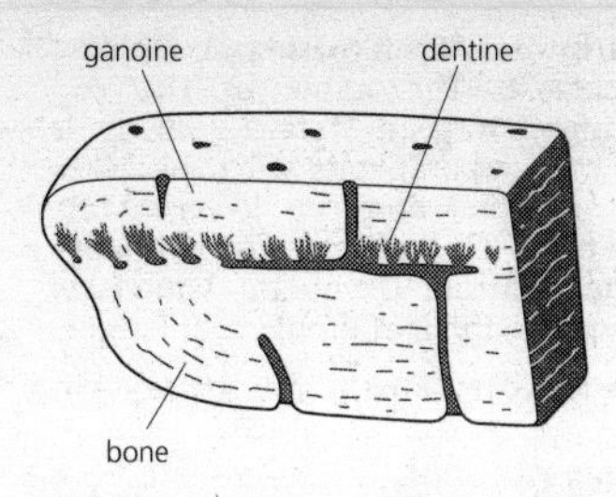

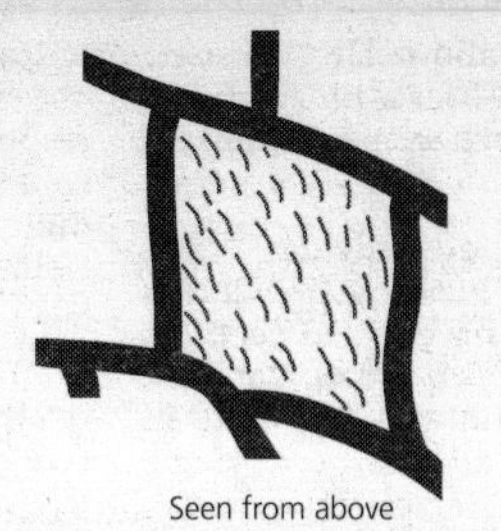

Ganoid scale

1940 kg/m^3; visual albedo 0.42; mean distance from Jupiter 1.07×10^6 km.

gap Transverse valley that cuts through a ridge. It is termed a water gap when occupied by a stream; otherwise it is a wind gap. It may be a relic of an early stage in the development of a drainage pattern.

Gardar rifting An episode of rifting (*see* RIFT) that occurred about 1400–1000 Ma ago, after the *Ketilidian orogeny but probably not causally connected to it.

garnet An important rock-forming *mineral group, with the general formula $X_3Y_2Si_3O_{12}$, where X may be Ca, Mg, Fe^{2+}, or Mn and Y may be Al, Fe^{3+}, or Cr^{3+}; the main minerals are *grossular (X = Mg, Y = Al), *pyrope (X = Mg, Y = Al), *almandine (X = Fe^{2+}, Y = Al), *spessartine (X = Mn, Y = Al), *andradite (X = Ca, Y = Fe^{3+}), and *uvarovite (X = Ca, Y = Cr^{3+}), and there is continuous chemical variation in the group; an unusual variety called hydrogrossular $Ca_3Al_2[SiO_4]_2[SiO_4]_{1-m}(OH)_{4m}$ has hydroxyl ions in the structure and is found in the rare rock type *rodingite; sp. gr. 3.6–4.3; *hardness 7.0–7.5; colour very variable depending on its chemical composition, and can vary from shades of deep red-brown to almost black, green, white, yellow, and brown; usually *vitreous *lustre; *crystals *cubic, with the most common form being *dodecahedra; no *cleavage; found in high-grade *metamorphic and *igneous rocks, in beach *sands, and *alluvial *placers. Transparent pyrope crystals may be used as *gemstones, but garnet is more generally used as an abrasive.

GARP *See* GLOBAL ATMOSPHERIC RESEARCH PROGRAMME.

gas *See* NATURAL GAS.

gas chromatography Analytical technique in which the components of a sample are separated by partitioning between either a mobile gas and a thin layer of non-volatile liquid held on a solid support (gas–liquid chromatography), or between the gas and a solid absorbent as the stationary phase (gas–solid chromatography). Partitioning occurs repeatedly throughout the column, and, as each solute travels at its own rate, a band corresponding to each solute will form. Solutes are eluted (washed out) in increasing order of partition ratio, and enter a detector attached to the column exit. The time of emergence of a peak on the display identifies a component, and the area under each peak is proportional to the component's concentration. Gas chromatography is used mainly in the analysis of *volatile organic compounds.

Gasconadian A *stage of the *Ordovician in the *Canadian *Series of N. America.

gas–liquid chromatography *See* GAS CHROMATOGRAPHY.

Gaspra A *solar system asteroid (No. 951), measuring $19 \times 12 \times 11$ km; approximate mass 10^{16} kg; rotational period 7.042 hours; orbital period 3.29 years. It was imaged by *Galileo in October 1991.

gas-retention age Measurement of the age of *meteorites using the amount of radiogenic argon from the decay of potassium-40, or of helium from uranium and thorium, to indicate the time since the meteorite was last at a temperature at which gas leakage could occur. This is based on the principle that at high temperatures gases diffuse easily through *silicate *lattices, e.g. ^{40}A from ^{40}K diffuses out and is lost at temperatures above about 300 °C. If a meteorite has not suffered reheating its gas-retention age is the same as its *formation age.

gas–solid chromatography *See* GAS CHROMATOGRAPHY.

gastrolith A stone swallowed (e.g. by some reptiles and birds) to break up food and so assist digestion. Such stones acquire a rounding and polish.

Gastropoda (gastropod; phylum *Mollusca) The class of Mollusca that includes snails and slugs. They have a true head, an unsegmented body, and a broad, flat foot. They appear in the *Cambrian and occur in *sedimentary rocks of all ages, occupying a range of aquatic and terrestrial environments. The majority of modern gastropods and all the *fossil forms possess a coiled shell, which is all that is left to the palaeontologist for determining identification. The classification of living forms is based largely upon soft parts, so that similarly shaped shells developed by unrelated groups cause problems of nomenclature.

gauging station (gaging station) A point at which river flow or groundwater levels are measured.

Gault Clay Glutinous marine deposit found in south-eastern England and in France, containing abundant fossil *bivalves, *gastropods, *ammonites, and vertebrates. It is Lower *Cretaceous (*Albian) in age.

Gauss, Karl Friedrich (1777–1855) A German mathematician, Gauss developed the study of spherical geometry, which is essential to the theory of *plate tectonics. His major work, *Theoria Motus Corporum Coelestium* ('Theory of the Motion of Heavenly Bodies') (1809), was on planetary movements. He also helped to initiate the international geomagnetic survey of 1834.

Gauss A normal *polarity chron in the late *Pliocene, preceded by the *Gilbert and followed by the *Matuyama reversed *chrons, and *radiometrically dated at 3.58–2.60 Ma (Harland et al., 1982). The Gauss contains at least two reversed *polarity subchrons: *Mammoth and *Kaena.

gauss (G) The c.g.s. unit of measurement of (*a*) magnetic field and (*b*) magnetic moment per unit volume. It has now been replaced by the *SI units weber/m^2 (Wb/m^2) and tesla (T). 1 G = 10^{-4} T = 10^{-4} Wb/m^2.

Gaussian distribution *See* NORMAL DISTRIBUTION.

GCM *See* GENERAL CIRCULATION MODEL.

Gedinnian 1. The earliest *age in the *Devonian Period, preceded by the *Silurian, followed by the *Siegenian, and dated at 408–401 Ma. **2.** The name of the corresponding European *stage, which is roughly contemporaneous with the *Crudinian (Australia) and the lower/middle Helderbergian (N. America). The boundary (also the Silurian–Devonian boundary) *stratotype section is at Klonk, near Prague.

gedrite *See* AMPHIBOLES; and ANTHOPHYLLITE.

gehlenite *See* MELILITE.

Geiger counter *See* GEIGER–MÜLLER COUNTER.

Geiger–Müller counter (Geiger counter) An instrument for detecting ionizing radiation which is used in general geologic prospecting. It consists of a cylindrical metal cathode with a wire anode along its axis, the whole being enclosed in a thin-walled tube filled with low-pressure inert gas. In operation the cathode carries a charge of about 1000 volts, which is just short of that level needed to produce an electrical discharge across the cathode–anode space. A charged particle or *gamma ray traversing this space collides with atoms of the inert gas, producing positive *ions and negative *electrons. Under the high voltage these are rapidly accelerated towards the cathode and anode, colliding on the way with other gas atoms and producing many more charged particles in a chain reaction. This avalanche arriving at the anode and cathode is registered as a pulse which is amplified to produce a click in a headphone set, or a succession of such pulses which can be expressed as a meter reading in milliroentgens per hour or counts per second. For more accurate surveys (especially from the air) a *scintillation counter is required, which is a more sensitive instrument.

Geikie, Archibald (1835–1924) Director of the British Geological Survey from 1881 to 1901, Geikie made studies of glacial and fluvial *erosion, and attempted to calculate the age of the Earth from rates of denudation. This led to conflict with *Kelvin. Geikie was also one of the first historians of geology, stressing in his *Founders of Geology* (1897, 1905) the importance of the work of his fellow Scotsman *Hutton.

gel Translucent to transparent, jelly-like material formed by the coagulation of a *colloid; a non-homogeneous gelatinous precipitate; or a liquefied mud. A gel is more solid than a *sol and is able to withstand some *shear stress. *Bentonite slurry is used as a gel in *diaphragm walls.

gel-filtration Column-chromatography technique, normally employing polymeric carbohydrate-gel beads of controlled size and porosity as a stationary phase. Mixture components are separated on the basis of their sizes and rates of diffusion into the beads. Smaller molecules tend to diffuse more rapidly into the beads, thereby leaving the mainstream of solvent and so becoming retarded with respect to larger molecules. This method can also be used to determine the molecular weight of an unknown substance.

gelifluction (congelifluction) Flow of water-saturated sheets of rock debris over perennially frozen ground, and on slopes as low as 1°. It is the cold-climate variety of *solifluction and occurs only in the active layer (to a depth of 3 m).

gelifraction *See* FROST WEDGING.

geliturbate *See* GELITURBATION.

geliturbation (congeliturbation, cryoturbation) General term for all frost-based movements of the *regolith, including *frost heaving and *gelifluction. The material disturbed by such movements is called 'geliturbate'.

gelivation *See* FROST WEDGING.

gemstone Naturally occurring mineral that has been artificially polished, faceted, and shaped for decorative purposes. It is normally classified as precious (e.g. *diamond, ruby, and emerald) or semi-precious (e.g. *garnet, *zircon, and *topaz). Many gemstones are hard, clear, and free from natural imperfections.

genal angle The angle where the lateral and posterior margins of a trilobite (*see* TRILOBITA) *cephalon meet.

genal spine A spine that occurs in some species of trilobites (*Trilobita) at the *genal angle.

gendarme *See ARÊTE.*

gene Fundamental physical unit of heredity. It occupies a fixed chromosomal locus, and when transcribed has a specific effect upon the *phenotype. It may mutate, and so yield various *allelic forms. A gene comprises a segment of DNA (in some viruses it is RNA) coding for one function or several related functions. The DNA is usually situated in thread-like *chromosomes, together with protein, within the nucleus; in bacteria and viruses, though, the chromosomes comprise simply a long thread of DNA.

gene flow Movement of *genes within an interbreeding group that results from mating and gene exchange with immigrant individuals. Such an exchange of genes may occur in one direction or both.

gene pool The total number of *genes, or the amount of genetic information, that is possessed by all the reproductive members of a population of sexually reproducing organisms.

genera *See* CLASSIFICATION.

general adaptation Adaptation that fits an organism for life in some broad environmental zone, as opposed to 'special adaptations' which are specializations for a particular way of life. Thus the wing of a bird is a general adaptation, while a particular kind of bill is a special adaptation.

general circulation The term generally used to describe the large-scale circulation of the atmosphere over the globe, or over one hemisphere, with its more or less persistent features (which may be brought to prominence by considering long-term, or even shorter-term, averages), and all the transient features on various scales. Although in its nature it is a matter of the winds, the general circulation may be studied by means of barometric pressure maps because of the intimate relation between pressure and wind.

general circulation model (GCM) A computer simulation of the *general circulation of the atmosphere for purposes of climatological studies. The atmosphere is contained within a notional three-dimensional grid and its condition described by the gas laws calculated at each point of intersection within the grid. Climatic development is then simulated by repeated recalculations of the effect of changes at one intersection on the intersections surrounding it. Present GCM studies are limited by the coarseness of the grids used, which cannot reflect events, such as cloud formation, occurring on a smaller scale, and by ignorance of the detail of certain important processes, such as those affecting energy transfer between the oceans and atmosphere.

generalized reciprocal method (GRM, Palmer method) A method of interpreting seismic refraction profiles over irregular layers, using forward and reverse shooting to obtain matched *time–distance graphs. It is comparable to the *plus–minus method, but less restricted in its application.

generating curve *See* COILING.

Genesis A *NASA project to collect charged particles from the *solar wind and return them to Earth. It is due to be launched in 2001.

genetic drift The random fluctuations of *gene frequencies in a population, such that the genes amongst offspring are not a perfectly representative sampling of the parental genes. Although drift occurs in all populations, its effects are most marked in very small isolated populations, in which it gives rise to the random fixation of alternative *alleles so that the variation originally present within single (ancestral) populations comes to appear as variation between reproductively isolated populations.

genetic stratigraphic sequence model One of the two schools of *sequence stratigraphy, proposed in 1989 by W. E. Galloway ('Genetic stratigraphic sequences in basin analysis. 1: Architecture and genesis of flooding-surface bounded depositional deposits', *Bull. Am. Ass. Petrol. Geol.*, 73, 125–42), based on studies of the coast of the Gulf of Mexico. It uses marine flooding surfaces as the boundaries of strata, because these reorganize the sedimentary system and are easy to identify and correlate. *Compare* DEPOSITIONAL SEQUENCE MODEL.

genetic stratigraphic unit A large-scale set of *facies sequences, which were formed by slow deposition and are bounded by breaks in sedimentation, which can be related to one another for purposes of analysis and interpretation.

geniculate twin (knee twin) A special kind of twinned *crystal in which the *twin-plane has markedly changed the shape of the crystal, rather like a knee or elbow joint. The twin-plane is a reflection plane.

genitus Growth of a new cloud from a mother-cloud, where only a limited part of the mother-cloud is affected by the change. *See also* CLOUD CLASSIFICATION; and MUTATUS.

genotype Genetic constitution of an organism, as opposed to its physical appearance (*phenotype).

genus *See* CLASSIFICATION.

genus-zone *See* TAXON RANGE ZONE.

geo- From the Greek *ge* meaning 'Earth' (a version of 'Gaia'), a prefix meaning 'pertaining to the Earth'.

geo Narrow inlet of a cliffed coastline, which has developed along a major near-vertical *joint or *fault.

geobarometer *Mineral or group of minerals whose existence, coexistence, or element distribution, is stable between known pressure bounds at given temperatures and which is therefore useful as an indicator of the pressure under which a rock equilibriated. For example, the FeS content of the mineral *sphalerite, in equilibrium with *pyrite and *pyrrhotite, is unaffected by temperatures in the range of 300–550 °C, but has been shown to vary according to pressure, and has been used as a geobarometer in regionally metamorphosed terrains.

geobotanical anomaly The marked local concentration, above background levels, of one or more elements in an ecological assemblage or specific plant which may indicate the presence of an ore deposit or a concentration of hydrocarbons.

geobotanical exploration (biogeochemical exploration) Traditionally, the use of indicator plant species or assemblages to detect the possible presence of metal-rich deposits. It is based on the limits-of-tolerance principle, i.e. it assumes that only specialized species can withstand metal-contaminated soils. In practice, plant response may be confusingly more complex (e.g. plants may respond to the low availability of essential nutrients rather than to a high concentration of toxic minerals) which makes such indicators unreliable. In modern use the concept includes the collection and chemical analysis of plant materials or soil layers, especially humus, in which metal ions may accumulate. It is a supplementary rather than a primary prospecting method.

geochemical affinity Affinity of an element for a particular environment.

geochemical anomaly 1. Abnormal concentration of elements in earth materials compared with background levels. **2.** Increase of hydrocarbons in soils.

geochemical cycle A continuous cycle of elements passing through and between the Earth's *lithosphere, *biosphere, *hydrosphere, and *atmosphere. For example, sodium is released from rocks (lithosphere) by *weathering and is transported in solution or suspension to the sea (hydrosphere). Sediments formed in the oceans take up sodium and may be compacted to join the geologic cycle, becoming *sedimentary, *metamorphic, and perhaps ultimately new *igneous rock. *See* BIOGEOCHEMICAL CYCLE.

geochemical differentiation *See* PRIMARY GEOCHEMICAL DIFFERENTIATION; and SECONDARY GEOCHEMICAL DIFFERENTIATION.

geochemical soil survey The process of collecting and analysing unconsolidated soil sediments in order to locate geochemical anomalies in the underlying rock and to use these to find *ore bodies. Where the loose material or soil is stable there is a geochemical balance between the original rock, *weathering, pore water, and biological activity. The soil profile is usually layered, and in a geochemical soil survey the horizon giving the sharpest contrast between background and anomaly is chosen for analysis.

geochemistry Branch of geology concerned with the abundance and distribution of the chemical elements and their *isotopes within the Earth or within solid bodies in the solar system, their circulation in natural systems (the *atmosphere, *hydrosphere, *biosphere, and *lithosphere), and the laws governing this distribution and its evolution.

geochronologic unit *See* GEOLOGIC-TIME UNIT.

geochronology Determination of time intervals on a geologic scale, through either absolute or relative dating methods. Absolute dating methods involve the use of radioactive elements and knowledge of their rates of decay: this yields an actual age in years for a given rock or *fossil. Relative dating involves the use of fossils or sediments to place events and rock sequences in order, and does not provide absolute dates. *See also* DATING METHODS; ABSOLUTE AGE; RELATIVE AGE; PLANKTONIC GEOCHRONOLOGY; and GEOCHRONOMETRY.

geochronometric scale (chronometric scale) A time-scale based on years BP (conventionally before 1950). Subdivisions on the scale are defined by particular units of duration (e.g. 10^6 years, 10^9 years) rather than reference points in actual rock successions. An example of such a subdivision is the placing of the boundary between the *Archaean and the *Proterozoic at 2500 Ma (i.e. 2500×10^6 years) ago.

geochronometry The determination of the length of time intervals. Geochronometric resolutions for zonations based on different organisms may be calculated by dividing the time-span of a series by the number of *zones and the intervals between zones. However, this will give only an approximate measure of time. *See also* DATING METHODS; and GEOCHRONOLOGY.

geode A hollow, rounded body, which has a lining of *mineral *crystals pointing inward, e.g. *quartz or *calcite. The crystals grow into the cavity unimpeded and form perfect crystals which are frequently collected and valued for their beauty.

geodesy The science of measurement of the shape or figure of the Earth and its gravitational field. This science has expanded from topographic and astronomic surveying with the advent of satellite positioning systems, e.g. *GPS and *SPS.

geodetic latitude The latitude as defined by the vertical, relative to the geodetic reference ellipsoid (i.e. the shape that most closely matches mathematically the figure calculated by geodesy). The centre of the reference system is not geocentric.

geodetic measurement Any measurement concerned with the shape figure of the Earth, often involving the *geoid.

geo-electric section A diagrammatic section of stratified layers that is deduced from electrical (resistivity) depth probing or drilling, where layers are identified by their *apparent resistivities. Such sections are useful in detecting *water-table levels and determining whether water is saline or fresh at the water-table.

geognosy *See* WERNER, ABRAHAM GOTTLOB.

Geographos A *solar system asteroid (No. 1620), diameter 2 km; approximate mass 4×10^{12} kg; rotational period 5.222 hours; orbital period 1.39 years. It was to have been visited by the Clementine spacecraft, but a computer malfunction ended the mission.

geoid The gravitational equipotential surface corresponding to mean sea level, including the level at which the sea would stand in a continental area if it were able to do so.

geological barrier *See* FAR-FIELD BARRIER.

Geological Long Range Inclined Asdic (GLORIA) A sonar device, towed behind a ship, that transmits one sound beam to port and another to starboard and is used for studying the sea bed.

geologic cross-section An interpretation of a vertical section through the Earth's surface, most usefully a profile, for which evidence was obtained by geologic and geophysical techniques or from a *geologic map.

geologic map A map which shows the surface distribution of rock types, including their ages and relationships, and also structural features.

geologic map symbols Symbols used on a geologic map to provide a reduced or condensed pictorial representation of data, so that space is conserved.

geologic time-scale A two-fold scale that subdivides all the time since the Earth first came into being into named units of abstract time, and subdivides all the rocks formed since the Earth came into being, into the successions of rock formed during each particular interval of time. The branch of geology that deals with the age relations of rocks is known as *chronostratigraphy. The concept of a geologic time-scale has been evolving for the last century and a half, commencing with a relative time-scale (mainly achieved through *biostratigraphy), to which it has gradually become possible to assign dates (*see* DATING METHODS) which are, nonetheless, subject to constant revision and refinement. Since the first International Geological Congress in Paris in 1878, one of the main objectives of stratigraphers has been the production of a complete and globally accepted *stratigraphic scale to provide a historical framework into which all rocks, anywhere in the world, can be fitted (*see* STANDARD STRATIGRAPHIC SCALE; CHRONOSTRATIGRAPHIC SCALE; UNIFIED STRATIGRAPHIC SCALE). Such a standard scale is still a long way off, but the names for *geologic-time units and *chronostratigraphic units down to the rank of *period/*system are in common use; many *epoch/*series and *age/*stage names are still regionally variable. Appendix B gives an outline geologic time-scale employing currently common names and dates (though they are not necessarily universally accepted).

geologic-time unit (geochronologic unit) A subdivision of geologic time, based on the rock record of the corresponding *chronostratigraphic unit. Each time unit coincides with a particular chronostratigraphic unit and, like them, time units are ranked in order of decreasing duration, each unit comprising a number of units of shorter time interval (e.g. two or more *chrons comprise an *age, two or more ages comprise an *epoch, etc.).

Chronostratigraphic Unit	*Geologic-time Unit*
*eonothem	*eon (longest)
*erathem	*era
*system	*period
*series	*epoch
*stage	*age
*chronozone	*chron

Geologic-time units generally bear the same name as their chronostratigraphic counterparts, but the terms 'lower', 'middle', and 'upper' are changed to 'early', 'middle', and 'late'.

geomagnetic dipole field The best mathematical fit to the observed *geomagnetic field using a single dipole. If not stated, this is usually the geocentric dipole field, which is axial and inclined at about 11.3° relative to the Earth's axis of rotation. The axial geocentric dipole field is the best-fitting dipole, located at the centre of the Earth and aligned along the Earth's axis of rotation, and is usually considered to represent the Earth's magnetic field after averaging out the geomagnetic *secular variation. *See also* NON-DIPOLE FIELD.

geomagnetic equator A line joining points on the Earth's surface where the *inclination of the *geomagnetic field is zero. To avoid *magnetic-anomaly distortions it is usually calculated from harmonics of the geomagnetic field less than 12.

geomagnetic field The *Earth's magnetic field shows variability on all time-scales, ranging from nanoseconds to millions of years. Most *transient variations are of external origin, reflecting interactions between the *solar wind and the Earth's *atmosphere. Longer-term changes, called *secular variations, are of internal origin. The average annual field has been closely defined by satellite observations for 1980. Previous annual patterns were based on geomagnetic observatories, supplemented by field surveys and corrected to some particular time on the basis of observed secular variation in the preceding years. The intensity of the field varies from about 30 μT near the equator to 60 μT near the observed *geomagnetic poles at 73° N 100° W and 68° S 143° E. Most of the field (80%) can be accounted for mathematically by a single microgeocentric dipole inclined at 11.3° to the Earth's rotation pole, known as the geomagnetic dipole, with a magnetic moment of 8.01×10^{22} A/m^2. The remaining field, the micro-non-dipole field, forms 12 main areas, varying by ±1.5 μT. An improved mathematical fit to the observed field was obtained using an inclined dipole offset from the geocentre by 340 km. During this century, the pattern of the field shows a tendency to drift westwards at about 0.2°/year, but this is not considered to be persistent on archaeological time-scales. The field shows an ability to reverse polarity on a geologic time-scale, with three polarity changes per million years during the last 60 million years, but long periods, of about 50 million years, of constant polarity are also known. The geomagnetic field is generally attributed to fluid motions within the outer *core; these carry magnetic lines of force with them (*magnetohydrodynamics), creating a coupled *self-exciting dynamo that allows polarity reversals.

geomagnetic pole The location on the Earth's surface of the axes of the calculated inclined *geomagnetic dipole field. These are not *dip poles.

geomagnetic polarity interval *See* CHRON.

geomagnetic reversal time-scale *See* MAGNETO-STRATIGRAPHIC TIME-SCALE.

geometric distribution *See* LOG-NORMAL.

geometric factor (K_g) A numerical multiplier defined by the geometrical spacings between electrodes, which is used in conjunction with the voltage-to-current (R) ratio measured in *electrical resistivity surveys to give an *apparent resistivity (ρ_a) such that $\rho_a = K_g \times R$. The generalized formula for calculating K_g for a four-electrode configuration is: $K_g = 2\pi(1/C_1P_1 - 1/C_1P_2 - 1/C_2P_1 + 1/C_2P_2)^{-1}$ where C_1 and C_2, and P_1 and P_2 are the current and potential electrode positions respectively. The geometric factor for the three main electrode configurations are: dipole–dipole, $K_g = \pi n(n+1)(n+2)p$; Schlumberger, $K_g = (\pi p^2/q)(1 - q^2/4p^2)$; Wenner, $K_g = 2\pi p$, where p and q are defined for each case.

geomorphic sequence A sequence of strata (*see* STRATUM) formed during a full cycle of rising sea level and its subsequent fall.

geomorphology The scientific study of the land-forms on the Earth's surface and of the processes that have fashioned them. Recently an extraterrestrial aspect has developed, resulting from studies of lunar and planetary surfaces.

geopetal structure A sedimentary *fabric which records the way up at the time of deposition. Geopetal structures are commonly found in cavity fills within *limestones, where the lower part of the cavity has been filled with *sediment and the upper part filled later with *cement.

geophone (seismometer, pickup, 'jug') A rugged device used to detect the arrival of seismic waves by transforming the ground motion into an electrical voltage. *Compare* HYDROPHONE.

geophysics The science concerned with all aspects of the physical properties and processes of the Earth and planetary bodies and their interpretation, including, for example, *seismology, gravity, magnetism, *heat flow, and *geochronology.

geostatic stress *See* LITHOSTATIC STRESS.

Geostationary Operational Environmental Satellite (GOES) A series of meteorological satellites placed in *geostationary orbit by *NASA. GOES-8 was launched in 1994.

geostationary orbit (Clarke orbit) A satellite orbit in which the satellite travels on the equatorial plane in the same direction as the rotation of the Earth at a height of about 36 000 km (more than 5 Earth radii) above the equator. Its orbital period is exactly one *sidereal day and therefore the satellite remains vertically above a fixed spot on the surface of the Earth. At this height it has a view of almost the whole of one hemisphere. The possibility of such an orbit was first suggested by Arthur C. Clarke, for whom the orbit is sometimes named. *Compare* GEOSYNCHRONOUS ORBIT.

geostrophic current An ocean current

that is the product of a balance between *pressure-gradient forces and the *Coriolis force. This produces a current flow along the pressure gradient. Such a current does not flow directly from a region of higher pressure to one of lower pressure (i.e. 'down the slope' of the sea surface) but flows parallel to the gradient. All the major currents in the oceans, such as the *Gulf Stream, are very nearly true geostrophic currents. The *Gulf Stream, for example, can be likened to a river that does not run down a hill but around the hill.

geostrophic wind The wind blowing above the *boundary layer that, by its strength and direction, represents the balance between the *pressure-gradient force, acting directly from the region of higher pressure towards the region of lower pressure, and the *Coriolis force (effect), deflecting moving air to the right (in the northern hemisphere and to the left in the southern hemisphere). When these are in balance the wind flows parallel to the isobars (*see* BUYS BALLOT'S LAW). Air is also subject to a centrifugal force, owing to the curvature of the air's path around a centre of low or high pressure. In boundary layer, air experiences friction with the surface, causing it to flow across the isobars, at an angle of 10–20° over the sea and 25–35° over land (where friction is greater). *See also* GRADIENT WIND.

geosynchronous orbit A satellite orbit around the Earth that has a 24-hour periodicity, so it follows the same path over the Earth's surface every day. *Compare* GEOSTATIONARY ORBIT; and SUN-SYNCHRONOUS ORBIT.

geosyncline Large, downwarp structure generally of considerable extent, which may develop along a *continental margin. The term was introduced originally by James *Hall in 1859, and later an elaborate terminology was developed to describe and interpret the component parts. Many of these have fallen into disuse since the replacement of geosynclinal theory by more unifying theory of *plate tectonics; the terms 'eugeocline' and 'miogeocline' are used today to denote particular rock associations.

geotechnical map Map recording existing geology, and estimating likely conditions in terms helpful to the selection of construction techniques and ground treatment, and to the prediction of the reaction between ground and structure. Included in geotechnical maps are analytical maps, comprehensive maps, interpretive geologic maps, and slope category maps.

geothermal brine Hot, concentrated, saline solution that has circulated through crustal rocks in an area of anomalously high *heat flow and become enriched in substances leached from those rocks (e.g. chlorides of Na, K, and Ca); it often contains dissolved metals, in which case it forms an important intermediary in the deposition of *ore deposits. One of the best documented examples is the brine from *boreholes in the Salton Sea *geothermal field in southern California, discovered in the early 1960s, with temperatures between 300 and 325 °C, density 1021 kg/m^3, and containing appreciable concentrations of Cu, Pb, Zn, and Ag. Hot brine solutions may also form through sea water–rock reactions in hydrothermal systems at *oceanic ridges, e.g. in the median valley of the Red Sea.

geothermal field An area of the Earth characterized by a relatively high *heat flow. The anomolously high rate of heat flow may be due to present, or fairly recent, *orogenic or magmatic activity, or to the *radioactive decay of *isotopes of K, Th, and U where these occur at very high concentrations in crustal *granites (*hot dry rocks). In *sedimentary basins, low thermal-conductivity values for the rocks are balanced by high thermal gradients, thus maintaining a constant heat flow. The high thermal gradients raise the temperature of deep, permeating water. Extraction of the water up deep *boreholes provides surface water at temperatures useful for space heating. *Hot springs and *fumaroles can be important surface manifestations of a geothermal field.

geothermal gradient The increase of temperature with depth. It usually refers to depths below 200 m. In the continents, the gradient is usually between 20 and 40 °C/km, although it can well exceed this in volcanic regions. In the oceans, the depth of penetration of most core barrels is so short that the gradient can be determined over only a few metres and varies considerably. The average geothermal gradient at the surface of the Earth is about 24 °C/km, but it is assumed to decrease with depth as widespread *mantle melting would otherwise occur. The observed gradients are therefore modified to result in an estimated temperature of about 1200 °C at the top of the seismic *low-velocity zone in the upper mantle. Within the mantle, the increase of temperature with depth is considered to be less than 0.1 °C/km greater than the adiabatic increase of 0.33 °C/km. *See also* HEAT FLOW.

geothermic survey A survey measuring the *heat-flow variations within a region.

geothermometer An indicator of the temperature, or range of temperatures, at which a geologic event (e.g. the crystallization of a *magma or the metamorphism of pre-existing rocks) occurred. Apart from the presence or absence of *minerals or mineral assemblages known to be stable within certain temperature ranges, among the most widely used indicators are: (*a*) *stable-isotope distribution, e.g. the ratios of ^{18}O to ^{16}O between different mineral pairs varies according to temperature (*see also* OXYGEN-ISOTOPE ANALYSIS); (*b*) mineral transformations or inversions known to be temperature dependent, e.g. the transition of α quartz to β quartz at 573 °C; (*c*) liquid–vapour homogenization points in *fluid inclusions (subject to certain assumptions, the temperature of formation of a crystal is indicated by the temperature at which the vapour bubble co-existing in the inclusion disappears upon heating); (*d*) unmixing or *exsolution lamellae of mineral pairs below a particular temperature, e.g. chalcopyrite–bornite at 500 °C; (*e*) temperature-dependent element distribution between co-existing minerals, e.g. iron–titanium oxide distribution between the co-existing mineral pairs *magnetite–ulvöspinel and *ilmenite–*hematite (in this instance subject also to oxygen *fugacity).

germanate system Series of compounds which have identical structures to the *silicates, except that *phase transitions take place at lower pressures. Used by early workers to study possible phase transitions in the Earth's *mantle before apparatus capable of attaining realistic geologic pressures was available.

geyser A small opening on the Earth's surface which periodically spouts a fountain of boiling water into the air. The largest fountain height recorded was 500 m, from a now extinct geyser in New Zealand. Water beneath the mouth of a geyser is heated by conduction from surrounding hot rocks, water at the base of the column boiling before that higher in the column. Expanding vapour bubbles rise in the column of water, expelling water at the top and lowering the pressure at the base. This allows the onset of further boiling, the system being self-sustaining, until the entire column of water is blown out of the system as a water spout. The water involved carries a large load of dissolved minerals which precipitate around the mouth of the geyser as *siliceous sinter. The name is from Geysir, about 45 km from the active volcano Hekla, Iceland, and was first used as a technical term in 1847 by the German chemist R. W. von Bunsen, who spelled it 'geysir'. This spelling is still sometimes used.

GHOST *See* GLOBAL HORIZONTAL SOUNDING TECHNIQUE.

ghost A spurious seismic *reflection which occurs when energy is transmitted upwards from a subsurface *shot and then reflected downwards from the surface of the ground or sea. A ghost wave train may interfere with other downward-moving waves, thus modifying their wave-form, and may add a reverberation tail. This type of ghost reflection is one kind of *multiple event. A seismic event recorded from a reflector located outside the plane of the seismic section is called an 'off-section ghost' and may be a problem in an area where the three-dimensional subsurface topography is pronounced but only a two-dimensional seismic survey is being conducted.

ghost stratigraphy The alignment of *country rock *xenoliths within large *granite bodies such that they reflect a continuation of the *stratigraphy and structure of the country rocks surrounding the *intrusion. An old idea, ghost stratigraphy was regarded as evidence favouring the view that granite may form by the replacement of country rocks.

Ghyben–Herzberg relationship Beneath oceanic islands, the thickness (d) of a lens of fresh *groundwater (density ρ_w) overlying sea water (density ρ_m) can be determined if the height above sea level (h) of the top layer of the lens is known and the conditions are static. The relationship is $d = ah = \rho_w/(\rho_m - \rho_w)$ where a is typically about 38.

Giacobini–Zinner A *comet with an orbital period of 6.52 years; *perihelion date 21 November 1998; perihelion distance 0.996 AU.

gibber *See* GIBBER PLAIN.

gibber plain A term used in Australia to describe an extensive plain (normally a *pediplain) that is mantled by loose rock fragments (gibber). These fragments are typically the rubble left from the destruction of a *silcrete *duricrust or from the breakdown of resistant *conglomerates (in which case the gibber consists of *quartz pebbles).

Gibbs free energy *See* GIBBS FUNCTION.

Gibbs function (Gibbs free energy; symbol *G* or *F*) Generally defined in terms of changes in free energy: $\Delta F = \Delta H - T\Delta S$, where H = enthalpy, T = absolute temperature, S = entropy. In a mixture of reactants, if ΔF is less than 0 a reaction may take place spontaneously, whereas if ΔF is positive energy must be supplied to the system for a reaction to occur at all. In geochemistry, the sign and magnitude of ΔF are important. The sign indicates whether a given reaction can occur spontaneously, and the magnitude indicates how far the reaction can go before equilibrium is attained.

gibbsite Mineral, $Al(OH)_3$ that is a constituent of *bauxite; sp. gr. 2.4; *hardness 3; greyish-white; occurs as an alteration product of aluminium silicates in *laterite and bauxite deposits.

giga- from the Greek *gigas* meaning 'giant', a prefix (symbol G) attached to SI units, meaning the unit $\times 10^9$ (e.g. 2 Gm = 2 gigametres = 2×10^9 m).

Gigantoproductus giganteus *See PRODUCTUS GIGANTEUS.*

Gilbert A reversed *polarity chron in the mid *Pliocene, which is followed by the *Gauss normal polarity chron. It started about 5.70 Ma ago and lasted until 3.58 Ma. The Gilbert contains at least four normal *polarity subchrons: *Thvera; *Sidufjall; *Nunivak; and *Cochiti.

Gilbert, Grove Karl (1843–1918) An officer on the US Geological Survey, Gilbert made studies of Meteor Crater, Arizona, suggesting that it was the product of a collision with an *asteroid, and also studied lunar *craters. He investigated crustal movements around Lake Bonneville, developing a theory of crustal isostatic compensation (*see* ISOSTASY). He also distinguished between folded and block-faulted mountains.

Gilbert-type delta A type of river *delta which consists of a wedge-shaped body of *sediment, comprising relatively thin, flat-lying, *topset sediments, long, steeply dipping *foresets which prograde (*see* PROGRADATION) from the river mouth, and thinner, flat-lying, *bottomset or *toeset deposits. Gilbert-type deltas are often developed in lakes, where river water and lake waters are of the same density.

Gilbert, William (Gilberd, William) (1540–1613) A natural philosopher and physician to Elizabeth I, Gilbert studied terrestrial magnetism. In his book *De Magnete Magneticisque Corporibus, et de Magna Magnete Tellure* (1600) he made a firm distinction between magnetism and electricity. He explained the Earth's magnetic field by likening the Earth to a vast spherical magnet.

gilgai Undulating micro-relief of soils which contain large amounts of clay minerals (e.g. *montmorillonite) that swell and shrink considerably on wetting and drying, to an extent that may be sufficient to fracture pipelines or move telegraph or fence poles from the vertical. The word is derived from a settlement in Queensland, Australia, where the soils of swelling clays are especially common. *See also* PATTERNED GROUND.

gilsonite (asphaltite) Solid form of *hydrocarbon which occurs as pure, natural *bitumen; sp. gr. 1.05–1.10; *hardness 2; black; occurs in *veins, *lodes, and sedimentary rocks. On heating it softens rapidly and flows like a liquid. It is used in waterproof coatings, wire insulations, and lacquer.

Ginkgoales Formerly a subdivision of the *Coniferopsida, one of the gymnosperm groups. Its members are now classified as the division *Ginkgophyta.

Ginkgophyta The division of *gymnosperms that includes only the extant *Ginkgo biloba* (maidenhair tree) and its extinct relatives. The first undoubted maidenhairs occur in *Triassic rocks, and in the subsequent *Jurassic Period their distribution was practically world-wide. The surviving species is restricted (in the wild) to China, and its leaves are strikingly similar to fossil *Ginkgo* leaves from the Triassic. The restricted geographical range, the unchanged appearance of the leaves, and the motile male sperms (otherwise known only in living seed plants in the cycads) have together led to the maidenhair being referred to as a 'living fossil'.

Giotto An *ESA mission, launched in 1985, to the comets *Halley and *Grigg–Skjellerup.

gipfelflur A plain or apparent surface

which is made up of the summit levels in a mountainous region. *Gipfelflur* is a German word whose literal meaning is 'peak-plain'.

Gisbornian A *stage of the Middle *Ordovician of Australia, underlain by the *Darriwilian and overlain by the *Eastonian.

GISP (Greenland Ice Sheet Project) A US drilling project that extracts ice cores from the Greenland ice sheet and uses them to obtain atmospheric and palaeoclimatological information. In 1993 drilling reached bedrock at a depth of about 3000 m and is now drilling its second hole (GISP2), about 30 km from *GRIP.

Givetian 1. An *age in the Middle *Devonian Epoch preceded by the *Eifelian, followed by the *Frasnian, and dated at 380.8–377.4 Ma (Harland et al., 1989). **2.** The name of the corresponding European *stage that is zoned on goniatites and spiriferid brachiopods. It is roughly contemporaneous with the lower *Condobolinian (Australia) and *Erian (N. America).

glabella *See* CEPHALON.

glaci- (glacio-) Dominated by glacial ice. The prefix is followed by a term indicating the environment or process that is so dominated, e.g. glaciaquatic (of water derived from a *glacier), glacioeustasy (the theory that changes in sea level result from the growth and decay of *ice sheets), glacifluvial (of sediments or land-forms produced by meltwater streams escaping from a glacier), glacioisostasy (the theory that local flexing of the Earth's *crust occurs as a result of the loading and unloading that takes place as large ice sheets wax and wane), glacilacustrine (pertaining to a lake adjacent to a glacier), and glaciomarine (of sediments laid down in a sea environment near a glacier). *See also* GLACIOTECTONICS.

glacial breach A glacially eroded trough that cuts through a ridge and so breaches a former *watershed. It is formed when the outflow of a *glacier (or *ice sheet) is impeded, its thickness consequently increases, and ultimately a new escape route (the breach) is exploited. This process is called 'glacial diffluence' when a single glacier spills out of its valley, and 'glacial transfluence' when several breaches are formed due to the accumulation of a large ice sheet. The many breaches through the western Highlands of Scotland are due to the accumulation of transfluent ice east of the main watershed.

glacial diffluence *See* GLACIAL BREACH.

glacial diversion Displacement of a preglacial stream by the action of a *glacier. In upland areas the new drainage often follows a *glacial breach. In lowlands, glacial drift may block an existing valley and lead to stream diversion.

glacial drainage channel (meltwater channel) Channel cut by the action of glacial meltwater or by water from an ice-dammed lake. Various types may be recognized, classified by the position of the channel with reference to the *glacier, e.g. ice-marginal, *englacial, or subglacial. Usually these channels are steep-sided and flat-floored, and are unrelated to the present drainage pattern.

glacial limit A line marking the furthest extent of a former glacial advance. It may be identified on the ground through the recognition of features associated with glacial margins, including lateral and terminal *moraines, outwash spreads (*sandur), marginal *meltwater channels, and *proglacial lakes. *See also* TRIM LINE.

glacial period A general term used to describe either a glacial *stage (e.g. *Devensian), or an indeterminate period of glaciation.

glacial plucking (quarrying) The removal of relatively large fragments of bedrock by direct glacial action. The process involves several mechanisms, including the incorporation of rock fragments into the base of the *glacier when it freezes to weakened bedrock, and the removal of bedrock material when fragments already included in the ice are dragged over it.

glacial stairway Long profile of a *glacial trough: it is characterized by alternating rock bars (*riegels) and rock basins, giving the impression of a stairway. The structure is attributed to variations in the erosive power of ice, or to the influence of rock jointing.

glacial theory The theory, developed in the late 1830s and 1840s by Venetz, *de Charpentier, and *Agassiz, that most of Northern Europe, N. America and the north of Asia, had been covered by *ice sheets during a period later termed the *Pleistocene. The hypothesis was used to explain *erosion, and the subsequent deposition of *till or *boulder clay, and the extinction of species such as the *mammoth. Since that time, glacial theory has been developed to include multiple glaciation, and evidence of much older ice ages.

glacial transfluence *See* GLACIAL BREACH.

glacial trough A relatively straight, steep-sided, U-shaped valley that results from glacial *erosion. Its cross profile approximates to a parabola, while its long profile is often irregular, with rock bars (*riegel) and over-deepened rock basins being typical features. The world's largest glacial trough is that of the Lambert Glacier, Antarctica, which is 50 km wide and about 3.4 km deep.

glaciaquatic *See* GLACI-.

glaciated rock knob *See* ROCHE MOUTONNÉE.

glaciation The covering of a landscape or larger region by ice; an ice age.

glacier A large mass of ice, resting on or adjacent to a land surface, and typically showing movement. Glaciers may be classified in several ways. The most useful division (as it relates to work done) is based on temperature, and three categories are recognized. In temperate (or warm) glaciers (e.g. those of the Alps) the ice is at *pressure melting point throughout, except during winter when the top few metres may be well below 0 °C. Movement is largely by basal slip. Polar (or cold) glaciers (e.g. parts of the Antarctic sheet) have temperatures well below the pressure melting point and movement, which is slow, is largely by internal deformation. Subpolar glaciers (e.g. those of Spitzbergen) have temperate interiors and cold margins and so are composite. The morphological classification is based largely on the size, shape, and position of the ice mass and *cirque glaciers, *valley glaciers, and *piedmont glaciers are among the types recognized.

glacier creep The deformation of *glacier ice in response to *stress, by a process involving slippage within and between ice crystals. The rate of creep is dependent on both stress and temperature. When the *shear stress is doubled, the *strain rate increases eight times, and a rise in temperature from −22 °C to 0 °C involves a ten-fold increase in strain rate.

glacier ice *See* ICE.

glacier power A general term for the ability of a *glacier to erode its bed. A distinction is made between total glacier power (basal *shear stress multiplied by average velocity) and effective power (determined by the contribution of *basal sliding to the total velocity). It follows that temperate glaciers may be roughly ten times more powerful than polar and subpolar types.

glacier surge A relatively rapid movement of a valley *glacier, or of an individual *ice stream within a major *ice sheet. The movement may build up over a period ranging from a few months to several years and may be a hundred times faster than the 'normal' velocity. Surging may result from an increase in ice thickness or from excessive basal water.

glacier wind *See* FIRN WIND.

glacifluvial *See* GLACI-.

glacilacustrine *See* GLACI-.

glacio- *See* GLACI-.

glacioeustasy **(glacioeustastism)** The theory that sea levels rise and fall in response to the melting of ice during *interglacials and the accumulation of ice during *glaciations. A major *glacioeustatic oscillation has an amplitude of some 100 m. *See* GLACI-.

glacioisostasy The adjustment of the *lithosphere following the melting of an *ice sheet. The gradual rise of the land enables an estimation to be made of the flexural rigidity of the lithosphere and the viscosity of the *mantle. Adjustments need to be made to gravity readings in previously glaciated areas to allow for this effect, which occurs on a regional scale. Some 1000 m of depression may have occurred in Scandinavia during the last *ice age, where 520 m of recovery has been recorded. The process gives rise to warped shorelines. *See* GLACI-.

glaciology The scientific study of *ice in all its forms. It therefore includes the study of ice in the atmosphere, in lakes, rivers, and oceans, and on and beneath the ground. Commonly, however, it is the study of *glaciers.

glaciomarine *See* GLACI-.

glaciomarine sediment High latitude, deep-ocean sediment, which originated in glaciated land areas and has been transported to the oceans by *glaciers or icebergs. Such sediments may contain large dropstones, transported by and dropped from icebergs, in the midst of fine-grained sediments.

glaciotectonics 1. The study of structures within a *glacier. These may be identified by contorted layers of rock debris. They are most common when a glacier is frozen to its bed and can move only by thrusting and folding. **2.** The study of structures imposed on bedrock by glacial movement and on *drift by both movement and the loss of support that occurs on melting.

glaebules *See* CALICHE.

glass An amorphous, *metastable solid with the atomic structure of a *silicate liquid. Glass can be formed by *quenching a silicate *melt, the short time-scale for cooling or pressure reduction preventing the reorganization of the random liquid structure into an ordered crystalline structure. Since cohesion between atoms in the liquid silicate increases with increasing silica content, melts with high silicate contents are most likely to form glasses. Natural *igneous glasses of *rhyolite composition (70% SiO_2) are termed 'obsidians'. A wide variety of glasses, formed by meteoritic impact into the lunar *regolith, exist on the lunar surface. Shapes include spheres averaging 100 μm in diameter, tear-drops, dumb-bells, etc., typical of rotational shapes assumed by splashed liquids. They do not resemble meteoritic *chondrules. Volcanic glasses, formed by *fire fountains during eruption of *mare basalts, also occur locally on the Moon. Their compositions match those of local surface rocks, soils, or minerals. No *tektite compositions are found.

glass-plate reflector *See* REFLECTOR.

glass shards Angular, glassy particles, less than 2 mm in size, formed either by the explosive magmatic fragmentation of *pumice *vesicle walls, or by the chilling and brittle fragmentation of *magma when it comes into contact with groundwater or surface water. Magmatically formed shards commonly have 'Y' or cuspate shapes and may deform plastically and weld together if they are hot enough when deposited and the overburden load is sufficient. Shards formed by *hydrovolcanic processes have a variety of shapes, ranging from those with curviplanar surfaces and low vesicularity to those with smooth, fluid-form surfaces and moderate vesicularity.

glauconite Although sometimes considered as *clay mineral, glauconite is more accurately a member of the *mica group, with the composition $(K,Ca,Na)_{<2}(Fe^{3+},Al,Mg,Fe^{2+})_4[(Si,Al)_4O_{10}]_2(OH)_4$; sp. gr. 2.4–3.0; *hardness 2; *monoclinic; olive green, yellowish, or blackish green; dull *lustre; granular; occurs in marine sediments as aggregates up to 1 mm in diameter. It is being formed on many modern *continental shelves, at depths from a few tens to hundreds of metres, where the sedimentation rate is low and decaying organic matter is present in a generally oxidizing environment. In many *sandstones glauconite can impart a green colour when abundant, as in the *Cretaceous Greensands of Britain and the eastern USA.

glaucony A green-coloured, marine sedimentary *facies characterized by the presence of *grains of the *glauconite mineral family, which develop on *continental margins and on ocean highs. Glaucony is of variable mineralogy due to the replacement of mineralogically different initial substrates by *authigenic minerals of the glauconite family.

glaucophane An important alkali *amphibole of composition $Na_2(Mg_3Al_2)[Si_8O_{22}](OH)_2$, and *end-member of the glaucophane–riebeckite $(Na_2(Fe^{2+}{}_3Fe^{3+}{}_2)[Si_8O_{22}](OH)_2$ series; sp. gr. 3.0; *hardness 6.0; blue, bluish-black; *prismatic crystals, *fibrous or *granular habit; it is an *essential mineral of *blueschists under conditions of low temperature and high pressure in metamorphosed sediments at *destructive plate margins.

glaucophane-schist facies A set of metamorphic *mineral assemblages produced by *metamorphism of a wide range of starting rock types under the same high-pressure/low-temperature metamorphic conditions and typically characterized by the development of the mineral assemblage *glaucophane–lawsonite–*quartz in rocks of *basic *igneous composition. Other rocks of contrasting composition, for instance *shales or *limestones, would each develop their own specific mineral assemblage, even though they are all being metamorphosed under the same conditions. The variation of mineral assemblage with starting rock composition reflects a particular range of pressure, temperature and $P(H_2O)$ conditions. Experimental studies of mineral *P–T* stability fields indicate that the facies represents high-pressure/low-temperature conditions which can be met during *subduction of *oceanic crust under *continental crust. Because rocks of basic igneous composition within this facies are characterized by the blue-coloured mineral glaucophane, the *facies is sometimes referred to as the 'blueschist facies'.

G-layer The inner, solid *core of the *Earth, extending from a depth of about 5150 km to the Earth's centre at 6371 km depth.

glaze Clear-ice deposit on objects, produced by the freezing of supercooled water droplets on to surfaces at temperatures below 0 °C. *See* CLEAR ICE.

Gleedonian A *stage of the *Silurian, underlain by the *Whitwellian and overlain by the *Gorstian.

gley The product of waterlogged soil conditions, and hence an *anaerobic environment; it encourages the reduction of iron compounds by micro-organisms and often causes mottling of soil into a patchwork of grey and rust colours. The process is known as gleying, or gleyzation (US usage).

gleying *See* GLEY.

gleyzation *See* GLEY.

gliding tectonics *See* GRAVITY TECTONICS.

gliding twins *See* DEFORMATION TWINNING.

glimmerite An *ultrabasic *igneous rock, consisting almost wholly of *essential dark *mica, either *phlogopite or *biotite. These rocks are rather rare, being found among *ultramafic *xenoliths in *kimberlite pipes and within old *basement *gneisses. These occurrences testify to the deep-seated origin of glimmerites, which might be considered as metamorphic (*see* METAMORPHISM) rather than igneous.

Glinka, Konstantin Dimitrievich (1867–1927) A Russian soil scientist who worked at the University of St Petersburg, Glinka was a student of *Dokuchaev and developed and organized his work. He was responsible for the soil surveys of most of European Russia and of Siberia. His book *Soil Science* was first published in 1908.

Global Atmospheric Research Programme (acronym GARP) International project that aims to provide a comprehensive knowledge of atmospheric structure as an aid to prediction. The complex system is based on land and marine weather stations, together with upper-air soundings, satellite remote-sensing, and sensors carried by balloons.

Global Horizontal Sounding Technique (acronym GHOST) Programme for the direct sensing of the atmosphere, using balloons designed to float at various constant-density levels. These are tracked and their various measurement readings are monitored by polar orbiting satellites. Sensors on the balloons record temperature, humidity, and pressure; tracking gives mean winds for each balloon. The project forms part of the *World Weather Watch.

Global Positioning System (GPS) A total of 18 satellites, in different orbits, so that at least four are visible at any time from all points on the Earth's surface. They give a locational accuracy of less than 2 cm on baselines of 1000 km. It is planned to use them for measuring the motions of small blocks of the Earth's *crust, e.g. opposite sides of a fault system, *microplates, etc. (The satellites also have obvious military applications.)

global tectonics The study of the relative movements of large parts of the Earth. Implicit in the phrase 'global tectonics' are the ideas that: (*a*) the movements may not have involved lithospheric *plates (indeed, the earliest supposed plate movements are in the early *Proterozoic); (*b*) energy changes in one part of the Earth have repercussions in other parts; and (*c*) large-scale tectonics have effects on many other systems, e.g. global weather patterns, evolutionary change, and the formation of natural resources.

global warming potential (GWP) A ranking of the absorptive capacity of the principal *greenhouse gases expressed as the atmospheric warming effect of each compared with that of carbon dioxide, which is given a value of 1. The values take account of the wavelengths at which each gas absorbs radiation and its atmospheric residence time. On this scale, the GWP of methane is 11, nitrous oxide 270, CFC-11 4000, and CFC-12 8500.

global water budget *See* WATER BUDGET, GLOBAL.

***Globigerina* ooze** Deep-sea ooze in which at least 30% of the sediment consists of *planktonic foraminifera (*Foraminiferida) including chiefly *Globigerina*. It is the most widespread *pelagic deposit, covering almost 50% of the deep-sea floor, and it covers most of the floor of the western *Indian ocean, the mid-*Atlantic Ocean, and the equatorial and S. *Pacific. Species occurring in this deposit have been used to establish climatological and temperature criteria. *Globorotalia menardii* is supposed to indicate warmer conditions and *Globigerina pachyderma* to indicate colder temperatures. Another foraminiferan, *Globorotalia truncatulinoides* can coil in either a left- or right-handed manner and it is suggested that right *coiling indicates warmer conditions and left coiling colder.

Gloger's rule Individuals of many species of insects, birds, and mammals are darkly pigmented in humid climates and lightly coloured in dry ones. This may well be a camouflage adaptation (moist habitats are usually well vegetated and tend to lack pale colours). There are many exceptions to this so-called rule. *See also* ALLEN'S RULE; and BERGMANN'S RULE.

glomeroporphyritic Applied to those *phenocrysts which have attached together to form clusters set within a finer-grained *groundmass. The phenocryst clusters can form by the aggregation of separate *crystals suspended in a *melt or by partial disaggregation of crystal accumulations on the walls of the *magma chamber.

GLORIA *See* GEOLOGICAL LONG RANGE INCLINED ASDIC.

Glossifungites An assemblage of *trace fossils that includes separate vertical, U-shaped, or sparsely branched *burrows. These may have been temporary or permanent dwelling places, depending on the feeding habits of the animals concerned, i.e. whether they were foragers or suspension feeders. The *Glossifungites* assemblage is characteristic of marine *littoral and shallow *sublittoral environments.

***Glossopteris* flora** The *Permian glacial deposits of S. Africa, Australia, S. America, and Antarctica are succeeded by beds containing a *flora very different from that of N. America and Europe. The flora of the south grew in a cold, wet climate, while that of the north existed under warm conditions. Plants with elongate, tongue-shaped leaves dominated the southern flora, with the genera *Glossopteris* and *Gangamopteris* being among the best known. Of these two, the genus *Glossopteris* gives its name to the flora. *Glossopteris* is characterized by a leaf with a fairly well defined midrib and a reticulate (net-like) venation. *G. indica* is the last species referred to the genus and the family Glossopteridales. It is known from the *Trias of India.

gloup A blow-hole. *See* COASTAL PROCESSES.

glow curve *See* THERMOLUMINESCENCE.

Gnathostomata 1. Superclass comprising all vertebrates with true jaws. **2. (phylum *Echinodermata, class *Echinoidea)** Superorder of echinoids, in which the *periproct is outside the apical system, there are no compound *ambulacral plates, a lantern (*see* ARISTOTLE'S LANTERN) and girdle are present, and the teeth are keeled. It includes the orders Holectypoida (*Jurassic–Recent), and Clypeasteroida (sand dollars, *Palaeocene–Recent).

gneiss (adj. gneissose) General petrological term applied to coarse-grained, banded rocks that formed during high-grade *regional metamorphism. The banding (gneissose banding, or gneissosity) is a result of the separation of dark minerals (e.g. *biotite, *hornblende, and *pyroxenes) and the light-coloured quartzofeldspathic minerals. A gneissose rock may be described more strictly by adding a qualifying prefix, e.g. biotite gneiss, hornblende gneiss, or *pelitic gneiss.

gneissose banding *See* GNEISS.

gneissosity *See* GNEISS.

Gnetales *See* GNETOPHYTA.

Gnetophyta A remarkable and probably artificial division of *gymnosperms that comprises only the Gnetales. Possible fossil material of the group is known from the early *Permian, and there is also some fossil pollen from the *Tertiary. With such a scant record it is not surprising that the evolutionary relationships of Gnetales remain unclear.

gnomon A device that is erected beside a rock or structure of interest to provide a reference for the vertical and colour. It was used during the manned exploration of the Moon.

goaf 1. Waste material. **2.** An area from which *coal has been removed.

gobi An Asian name for *desert pavement.

GOES *See* GEOSTATIONARY OPERATIONAL ENVIRONMENTAL SATELLITE.

goethite A hydrous iron oxide, α-FeO(OH),

which forms a series with lepidocrocite, γ-FeO(OH); sp. gr. 4.0; *hardness 5; reddish-brown; *massive or earthy; occurs as an alteration product of iron-bearing minerals in association with *limonite or *hematite. It is named after J. W. von Goethe (1749–1832).

gold Malleable metallic element, Au; sp. gr. 19; *hardness 2.5; bright yellow, but yellowish-white if silver is also present; occurs native as grains, threads, wire, sponge, and in many other forms, and may be combined with a variety of other metals including silver, copper, palladium, rhodium, or bismuth; occurs in *placer deposits, in residual soils, and in veins associated with a variety of *igneous and *sedimentary rocks.

golden algae *See* CHRYSOPHYCEAE.

golden-brown algae *See* CHRYSOPHYCEAE.

golden spike *See* BOUNDARY STRATOTYPE. *Compare* SILVER SPIKE.

Goldschmidt, Victor Moritz (1888–1947) A Norwegian geochemist, Goldschmidt developed some of the basic techniques of physics for use in *geochemistry, including *X-ray diffraction and spectroscopy. He did important work on *metamorphism and *trace elements, described in his textbook *Geochemistry*, published posthumously in 1954.

Goldschmidt's rules In a system where the two variables, temperature and pressure, are controlled externally, the number of *phases will not usually exceed the number of components. In geology the maximum number of naturally occurring minerals in a rock is equal to the number of components where a given mineral assemblage is stable over a range of temperatures and pressures.

Gomphotheriidae (order *Proboscidea) Extinct family of *mastodons, characterized by the development of multiple accessory tooth cusps and believed to be ancestral to the later mastodons, though not to modern elephants. These long-jawed mastodons or gomphotheres were one of three distinct proboscidean lines established by the *Miocene. They in turn diverged into a variety of descendant lines, several of which had highly specialized lower jaws. The family persisted into the *Pleistocene in both Old and New Worlds. Long snouts appeared in the earliest forms and in the Miocene *Gomphotherium* the lower jaw and premaxilla (front bone of the upper jaw) were very long, both bearing tusks, which may have been used for digging, and probably permitting only a short trunk. In later forms the face became shorter and the trunk presumably longer. There were many genera and species. The evolutionary importance of the gomphotheres lies in the probable parallel between their development and that of the line which led to the true elephants.

gonatoparian suture *See* CEPHALIC SUTURE.

Gondwanaland Former supercontinent of the southern hemisphere from which S. America, Africa, Madagascar, India, Sri Lanka, Australia, New Zealand, and Antarctica are derived. Their earlier connection explains why related groups of plants and animals are found in more than one of the now widely separated southern land masses. Examples include: the *Dipnoi (lung-fish) common to S. America, Africa, and Australia; *marsupial mammals found today in Australia and for most of the *Cenozoic in S. America; and the monkey-puzzle tree (*Araucaria*) common to S. America and Australia.

goniatites *See* AMMONOIDEA.

goniometer *See* GONIOMETRY.

goniometry The technique for measuring the *interfacial angles of *crystals. The instrument used, a goniometer (or contact goniometer) resembles a 180° protractor with a pivoted straight edge at the point of origin. The angle between the normals to *crystal faces is then read directly from the graduated scale. More accurate measurements on very small crystals may be made using an optical system and a telescope (optical goniometry), with the crystal mounted on a rotating *stage fitted with a vernier scale. These instruments are known as 'one-circle' or 'two-circle' reflecting goniometers. The crystal faces are arranged with the crystal edges parallel to the axis of rotation of the goniometer head.

Gorda Plate A minor lithospheric *plate which is subducting under the *North American Plate to the west of Oregon. The *subduction zone is not marked by an oceanic *trench, but inland there is the andesitic volcanic chain of the Cascade Mountains. The Gorda Plate is separated from the *Juan de Fuca Plate by a *transform fault;

both are remnants of the once-large *Farallon Plate.

Gore The basal *series in the New Zealand *Triassic. It is overlain by the *Balfour, and comprises the Malakovian, Etalian, and Kaihikuan *Stages.

Gorstian A *stage of the Upper *Silurian, underlain by the *Gkedonian and overlain by the *Ludfordian and dated at 424–415.1 Ma (Harland et al., 1989).

gossan Near-surface, iron oxide-rich zone overlying a sulphide-bearing *ore deposit, caused by the *oxidation and *leaching of sulphides. Useful in mineral exploration as a visible guide to sulphide mineralization by its yellow or red colour.

Gothenburg *See* BRUNHES.

Gothian A *stage of the Lower *Proterozoic, from about 2100–1600 Ma ago, of the Baltic Shield region, underlain by the *Svecofennian and overlain by the *Jotnian. According to Van Eysinga, 1975, Gothian is synonymous with Karelian.

Gothian orogeny (Gothic orogeny) *See* DALSLANDIAN OROGENY; and GRENVILLIAN OROGENY.

gouge *Clay filling in a mineral vein, or clay material between *fault planes produced by movement along the fault.

GPS *See* GLOBAL POSITIONING SYSTEM.

Grabau, Amadeus William (1870–1946) An American geologist and palaeontologist, Grabau was a professor at Columbia University, New York, and later worked on the geologic survey of China. In 1940 he developed a theory of rhythms in the growth of the Earth's *crust, and of repetitions in mountain building.

graben A downthrown, linear, crustal block, bordered lengthways by normal *faults. The upstanding blocks on either side may have been lowered by erosion. It is a structural feature, which may be of considerable length, caused by the relative lowering of a block between two faults or *fault zones. The faults are commonly high-angle, normal faults which are parallel in their *strike direction. Half-graben are bounded on only one side by one or more faults and are associated mainly with *tilt-block tectonics. Graben with a regional extent and topographic expression can also be called *rift valleys. *Compare* HORST.

grab sampling *See* SAMPLING METHODS.

grade 1. Group of things all of which have the same value. **2.** A balanced condition, especially of a river (river or stream grade) when it has just sufficient energy to transport the load supplied from the drainage basin; a balance between erosion and deposition. The concept has also been applied to hillslopes ('graded slopes') that are stable dynamically and so maintain themselves in the most economical configuration. The term is no longer used widely as it oversimplifies the issues involved. **3.** The fraction of a *sediment falling within a particular size limit, e.g. sand grade, silt grade, and boulder grade. *See* PARTICLE SIZE. **4.** The quality of a mineral ore. **5.** Classification of an ore by the quantity or purity of the mineable metal in an orebody. **6.** In civil engineering, the gradient of a road. **7.** Distinctive functional or structural level of complexity in the organization of an organism. Thus fish, *amphibians, *reptiles, and *mammals represent successive *vertebrate grades. Grades may occur within a single lineage; or the same grade may be achieved independently in different ones (e.g. warm-bloodedness evolved independently in birds and mammals). **8.** *See* METAMORPHIC GRADE.

graded bedding *Sedimentary structure in which there is an upward gradation from coarser to finer material, caused by the deposition of a heterogeneous suspension of particles. The feature may be used to establish the 'way up' or natural succession of *strata.

graded reach A length (reach) of stream channel whose gradient and cross-sectional form have become adjusted to carry just the discharge and sediment load that are normally supplied from upstream. Such a reach is said to be in equilibrium. Early definitions emphasized the smooth long profile of such a reach; it is now realized that a profile may be irregular and the reach still be graded.

graded sediment 1. A sedimentary deposit which is sorted (*see* SORTING) with the coarsest grain size at the base and the finest at the top is termed 'normally graded'; a *sediment which is sorted with the finest at the base and coarsest at the top is termed 'inverse-graded'. **2.** A well-sorted sediment.

graded slope *See* GRADE.

gradient In meteorology, the rate of

change of a function (e.g. pressure or temperature), at right angles to the isolines.

gradient wind Wind that represents the balance among all the forces acting upon moving air. It may be expressed by V in the equation: $G = 2D\omega V \sin\phi \pm DV^2/r + F$, where G is the pressure gradient, D the air's density, ω the angular velocity of the Earth's rotation about its axis, ϕ the latitude, r the radius of curvature of the air's path, F the friction, and V the air's velocity when the forces are in balance. The sign before the second term is + when the motion is around a cyclonic (low-pressure) centre, – when the motion is around an anticyclonic (high-pressure) centre. *See also* GEOSTROPHIC WIND.

grading curve (soil grading curve) Graph of grain size (horizontal logarithmic scale) against percentage distribution (vertical arithmetic scale); a point on the curve indicates the percentage by weight of particles smaller in size than the grain size at the given point.

gradiometer Any instrument that measures the gradient of a potential field rather than its absolute value. *See* GRAVITY METER; and MAGNETOMETER.

gradualism *See* PHYLETIC GRADUALISM.

grain 1. A *detrital mineral or rock fragment (particle), of *sand size. **2.** Quarrying term, for the parting fabric of a rock (i.e. the direction in which a rock is most easily split by a quarry worker).

grain boundary sliding *See* CREEP MECHANISMS.

grain flow The movement of *sediment under gravity where the sediment is supported by direct grain-to-grain contact. This differs from turbidity flow, where the sediment moves under the influence of gravity in a turbulent flow of water.

grain roughness *See* BED ROUGHNESS.

grain shape *See* PARTICLE SHAPE.

grain size, igneous rocks Arbitrarily defined limits for the sizes of *crystals within an *igneous rock. Commonly accepted grain-size ranges are:

Very coarse	>3 cm
Coarse	5 mm to 3 cm
Medium	1–5 mm
Fine	<1 mm
Glassy	no grains present

For sedimentary rocks, *see* PARTICLE SIZE.

grainstone In the Dunham classification, a grainstone is defined as a *limestone consisting of *grain-supported particles without any mud *matrix.

grain-support A sedimentary *fabric in which the particles are in contact with one another and form the mechanical framework of the rock or *sediment.

granite A light-coloured, coarse-grained, *igneous rock, consisting of *essential *quartz (at least 20%), *alkali feldspar, *mica (*biotite and/or *muscovite), with or more commonly without *amphibole, and *accessory *apatite, *magnetite, and *sphene. *Hypersolvus granites are characterized by one type of alkali feldspar, usually *microperthite, whereas subsolvus granites are characterized by two types of alkali feldspar: *microperthite and *albite. Granite can be formed by *partial melting of old *continental crust, on a local scale by *in situ* replacement of continental crust (*granitization), by *fractional crystallization of *basalt *magma, or by a combination of these processes.

granite minimum In the *ternary system *quartz–*albite–*orthoclase, the common low-temperature point marking either the last of a *melt to crystallize after *fractional crystallization has occurred, or, conversely, the first fraction of any solid mixture of *alkali feldspars and quartz to melt upon heating. As this liquid has a very similar composition to many *granites, the temperature point at which it forms is known as the granite minimum.

granitic layer The upper 10–12 km of the *continental crust that overlies the *Conrad discontinuity. The name is based on the common occurrence of *granites near to the surface, and average *seismic velocities and densities are consistent with a *granodioritic composition.

granitization The conversion of crustal rocks to a granitic *mineral assemblage by the action of metasomatic fluids (*see* METASOMATISM) without going through the magmatic stage. The *essential chemical components of *granite are introduced into the solid parent rock, and those elements not required are removed from the solid rock by the percolation of metasomatic fluids along grain margins and the *diffusion of ions through *crystals. 'Granitization' was proposed as a process for generating granites in order to get around the space problem associated with injecting large volumes of granitic *magma into the crust; the idea is no longer much in vogue.

granitoid A term used to encompass the granitic rock types: *alkali-feldspar granite, *granite, *granodiorite, and *tonalite as defined in the classification of rocks by the *International Union of Geological Sciences (IUGS).

granoblastic A *textural term referring to a mosaic of equidimensional *anhedral grains in *metamorphic rocks. If inequant grains such as *micas are present, they are randomly oriented.

granodiorite A coarse-grained *igneous rock consisting of *essential *quartz, *plagioclase feldspar, *alkali feldspar, *biotite, and *hornblende, with *accessory *sphene, *apatite, and *magnetite. Plagioclase is the dominant feldspar in the rock, equal to or greater than two-thirds of the total feldspar present. Where alkali feldspar is completely, or almost completely suppressed, the rock is known as a 'trondhjemite' or, more generally, 'plagiogranite'. Granodiorites are commonly found as *intrusions into the *crust above *subduction zones, and are the *plutonic equivalent of *extrusive *dacites.

granofels A little-used name for a massive *granoblastic *metamorphic rock, consisting of equant grains of *quartz and *feldspar, hence lacking a *foliation. Minor *mafic *minerals consist of *pyroxene and *garnet. The anhydrous mineral assemblages of such rocks are produced under the high-grade, anhydrous *granulite facies conditions found at the base of the *crust.

granophyre A light-coloured, medium-grained, *igneous rock with a *granite mineralogy but characterized by the development of a *granophyric texture.

granophyric (micrographic) Applied to a fine-scale intergrowth of *quartz and either *alkali feldspar or *plagioclase, found as *interstitial, late-stage products in the *groundmass of *granites. The texture is formed by simultaneous and rapid crystallization of the two *phases from the late-stage liquid trapped between the earlier formed crystals.

granular **1.** Applied to the *texture of equigranular *igneous rocks with a *grain size ranging from 0.05 to 10 mm. **2.** Applied to the *form of a *mineral *aggregate when it is composed of grains. Evenly granular mineral aggregates have a saccharoidal *texture.

granulation The comminution, during deformation, of crystals into smaller grains of equal size which are then able to rotate in relation to one another. The process is a common feature of *plastic deformation in *granitoids.

granule A particle between 2 and 4 mm in size. *See* PARTICLE SIZE.

granulestone A *siliclastic rock consisting of *granules.

granulite A coarse-grained, equigranular *metamorphic rock, consisting of *quartz, *feldspar, and the anhydrous *ferromagnesium minerals *pyroxene and *garnet. There is some confusion over the use of the term granulite, different authors using the name in different ways. Consequently, *basic granulites, rich in the ferromagnesium minerals *orthopyroxene and *clinopyroxene, are better termed 'pyroxene gneisses', whilst *acid granulites, rich in quartz and feldspar, are better termed 'charnockitic gneisses'. These rock types are thought to be formed by *metamorphism of deep crustal rocks which have suffered earlier dehydration by the removal of a wet *granite *melt.

granulite facies A set of metamorphic *mineral assemblages produced by *metamorphism of a wide range of starting rock types under the same metamorphic conditions and typically characterized by the development, in rocks of *basic *igneous composition, of the mineral assemblage *clinopyroxene–*plagioclase–*orthopyroxene–*quartz. Other rocks of contrasting composition, for instance *shales or *limestones, would each develop their own specific mineral assemblage, even though they were all being metamorphosed under the same conditions. The variation of mineral assemblage with starting rock composition reflects a particular range of pressure, temperature, and $P(H_2O)$ conditions. Experimental studies of mineral P–T stability fields indicate that the facies represents high-pressure/high-temperature conditions which can be met near the base of the *continental crust.

granulometry The measurement of grain sizes.

grapestone A composite *carbonate

*grain, consisting of an aggregate of carbonate *peloids or other particles, bound together by *algae or micritic cement (*see* MICRITE). The aggregate has an appearance resembling a bunch of grapes, and forms in low-energy, *subtidal areas.

graphic A *texture seen in *granites and *pegmatites and characterized by a coarse intergrowth of *quartz and *alkali feldspar on a scale of tens of microns. The texture can be very coarse when developed in certain pegmatites, the intergrowths developing on a scale of several millimetres, and is easily seen in hand specimen, looking rather like cuneiform (i.e. wedge-shaped) writing. When developed on a microscopic scale, graphic texture is termed *granophyric texture.

graphic log An outline description of a particular sequence of rocks, made in the field and set out in columnar form. It is usual to sketch in each succeeding rock unit with its appropriate *lithologic symbol, and draw thicknesses of units to scale. Subsequent columns contain data concerning: *particle size (shaded in according to a horizontal scale); *sedimentary structures; *fossils; colour; palaeocurrent directions (*see* PALAEOCURRENT ANALYSIS); *weathering; *sorting; and any other salient features gathered under the general heading of 'remarks'. *See also* LITHOLOGIC SYMBOL.

graphite Pure carbon, C; sp. gr. 2.1; *hardness 2; greyish-black; feels soft and greasy; good basal *cleavage; scaly, columnar, granular, or earthy; occurs in veins and may be disseminated through rocks as a result of *metamorphism of original carbon-rich sediments. It is used as a lubricant, electrical conductor, and in the manufacture of crucibles and paint.

graphoglyptid A *trace fossil comprising a system of horizontal tunnels that were used as permanent dwellings and to produce or trap food.

graptolite *See* GRAPTOLITHINA.

Graptolithina (graptolites; phylum *Hemichordata, subphylum Stomochordata) An extinct class of stick-like, colonial, marine organisms that existed from the Middle *Cambrian to the Lower *Carboniferous. Their fossils are used to establish a stratigraphical time-scale for the Lower *Palaeozoic. There were 2 principal orders, *Dendroidea and *Graptoloidea, and a number of minor, short-lived orders.

Graptoloidea (subphylum Stomochordata, class Graptolithina) An order of graptolites that existed from the Lower *Ordovician to Lower *Devonian. The *rhabdosomes had up to 8 *stipes in early forms, but 2 and finally 1 in later forms. The *thecae are of only one type, equivalent to the *autotheca of *Dendroidea, occur on one or both sides of the stipe, and vary quite widely in morphology. In adults, there is always a *sicula bearing a *nema. There are 3 suborders: 1, to which no name has been assigned, comprises a single, *paraphyletic family, Anisograptidae. The other 2 are Dichograptina and Virgellina.

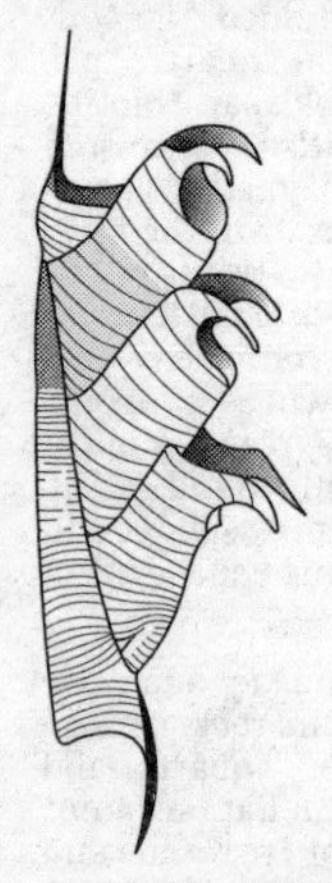

Graptoloidea (Saetograptus chimaera)

grass minimum temperature Minimum temperature recorded in open ground at night by a thermometer whose bulb is exposed over the tips of short grass.

graupel Soft hail, composed of particles resembling small snowballs, and formed by accretion when a snowflake falls through supercooled water droplets which freeze on contact and cover the flake.

gravel In the British classification for *particle sizes, grains with diameters between 2 and 60 mm.

gravimeter An instrument for measuring the *gravitational acceleration. Most field instruments are relative instruments, i.e. they determine the difference in gravitational acceleration between two or more points. Most operate by determining the change in extension of a spring loaded with a constant mass. Laboratory instruments can be absolute and are based on dropping masses, vibrating strings, or oscillating pendulum systems.

gravimetric analysis A method of chemical analysis which involves the quantitative precipitation of an element or elements from solution in a compound whose composition is accurately known and which may be isolated in a pure form prior to weighing. When new electronic techniques of elemental analysis were introduced the term 'wet chemistry' came to be applied to the classical gravimetric techniques.

gravimetry The science of measuring *gravitational *acceleration at different locations.

gravitational acceleration (*g*) Following Newton's law, the force F between two masses m_1 and m_2, separated by a distance r, is given by $F = Gm_1m_2/r^2$, where G is the *gravitational constant. The gravitational acceleration, g, is then given by $g = F/m_1 = Gm_2/r^2$ and is measured in *gravity units.

gravitational constant (*G*) The constant of proportionality relating the gravitational force (F) between two masses (m_1 and m_2) separated by a distance r. $F = Gm_1m_2/r^2$. The value of G is 6.672×10^{-11} Nm^2/kg^2.

gravitational equipotential A surface of equal *gravitational acceleration, e.g. the *geoid.

gravitational field Theoretically, the *gravitational attraction of a unit mass extends to infinity; in practice, this field is the space within which the mass's gravity is an effective force.

gravitational signature *See* SIGNATURE.

gravitational water Water that moves through soil under the influence of gravity and that must be removed before the soil can attain *field capacity.

gravity anomaly The *gravitational acceleration remaining after allowing for other factors. Common anomalies are *Bouguer, *free-air, and *isostatic, but a gravity anomaly is any acceleration remaining after allowing for different gravitational attraction models.

gravity assist A technique used to accelerate or slow spacecraft without consuming fuel by transferring *angular momentum from a planet to the spacecraft. The craft approaches the planet, is accelerated towards it by gravitational force, and as it leaves decelerates to its original speed relative to the planet, but not relative to the Sun. If a spacecraft approaches from behind, travelling in the same direction as the planet, it will accelerate relative to the Sun by acquiring angular momentum; if it approaches from ahead, travelling in the opposite direction to the planet, it will slow relative to the Sun by losing angular momentum to the planet.

gravity corer A sampling tube that penetrates into marine or lake sediments under its own weight. *See also* HYDRAULIC CORER.

gravity gliding *See* GRAVITY SLIDING.

gravity separation A concentration process based on differences in specific gravity between *ore and *gangue minerals. Separation is performed by shaking tables, Humphreys spirals, jigs, hydroclones, *dense media, etc.

gravity settling The settling of heavy *minerals and their accumulation on the floor of a *magma chamber, e.g. gravity-accumulated *chromite in some *ultramafic rocks. *See* FRACTIONAL CRYSTALLIZATION.

gravity sliding (slide, gravity gliding) The movement of rock bodies in response to gravitational instability along particular planes in unstable regions which leads to the formation of *thrust, *nappe, and *slump structures.

gravity survey A survey, usually a profile or grid, undertaken to determine the *gravitational acceleration within an area.

gravity tectonics (gliding tectonics) Mechanism whereby large masses of rocks move down a slope under gravitational force, producing *folding and *faulting of varying extent and complexity.

gravity unit Not an official SI unit, but one commonly used to measure *gravitational acceleration. One gravity unit is equal to 10^{-6} m/s^2, or 0.1 mgal.

graywacke *See* GREYWACKE.

greasy Of *mineral *lustre, having the appearance of an oily coating, due to the scattering of light by a surface that is rough at the microscopic scale.

great circle A line on the surface of a sphere representing the circumference of a circle whose centre is coincident with the centre of that sphere. In making a *stereographic projection, a horizontal (equatorial) projection of the sphere (i.e. as a plane at right angles to the N–S plane) is a primitive circle of given radius, and also a great circle. All other great circles are similar to lines of longitude. A line joining the points on the conceptual sphere at right angles to the horizontal plane will project on a *stereogram as a straight line passing through the centre of the projection, with the N and S poles as its diameter, both of which plot coincidentally in the centre of the horizontal plane. *See* PLANE OF PROJECTION.

Great Interglacial *See* MINDEL/RISS INTERGLACIAL.

Great Red Spot An anticyclonic storm system in the hydrogen–helium atmosphere of the southern hemisphere of *Jupiter. It covers 10° of latitude, is about 12 000 km across (close to the diameter of the Earth), and has been observed for more than 300 years.

green algae *See* CHLOROPHYTA.

greenalite *See* CHAMOSITE; and CHLORITE.

greenhouse effect The effect of heat retention in the lower atmosphere as a result of absorption and reradiation by clouds and gases (e.g. water vapour, carbon dioxide, methane, and chlorofluorocarbons) of long-wave (more than 4 μm) *terrestrial radiation. The insulating effect is analogous to that of greenhouse glass (i.e. it is transparent to incoming short-wave radiation but partly opaque to reradiated long-wave radiation) and alters the balance of incoming and outgoing radiation in the Earth's energy budget. Marked increases in atmospheric carbon dioxide, generated for example by the combustion of *fossil fuels, could result in a global increase of atmospheric temperatures if not offset by other (perhaps natural) changes. *See also* ATMOSPHERIC WINDOW.

greenhouse gas A gas composed of molecules that absorb and reradiate infrared electromagnetic radiation. When present in the atmosphere, therefore, the gas contributes to the *greenhouse effect. On Earth, the principal greenhouse gases are water vapour, carbon dioxide, methane, nitrous oxide, ozone, and certain halocarbon compounds. *See* GLOBAL WARMING POTENTIAL.

greenhouse period A time during which there were no glaciers on Earth. Sea levels were high, ocean waters were not well mixed and tended to be anoxic, and nutrients were recycled on *continental shelves. In *cyclostratigraphy, third-order cycles were predominant. *Compare* ICEHOUSE PERIOD.

Greenland Ice Core Project *See* GRIP.

Greenland Ice Sheet Project *See* GISP.

greensand Term applied to *glauconite-rich *sandstones and calcareous sandstones. The Lower and Upper Greensands of south-east England occur beneath and above the *Gault.

greenschist A low-grade, *regional metamorphic rock, containing abundant *chlorite with *albite, *epidote, and *sericite, and possessing a distinct *cleavage. Greenschists form by *metamorphism of *basic *igneous rocks, the chlorite component being derived from the *ferromagnesium minerals present in the original rock. Chlorite imparts a green colour to the cleaved rock, hence the name 'greenschist'.

greenschist facies A set of metamorphic *mineral assemblages produced by *metamorphism of a wide range of starting rock types under the same metamorphic conditions and typically characterized by the development of the mineral assemblage: *chlorite–*actinolite–*albite–*epidote–*quartz in rocks of *basic *igneous composition. Other rocks of contrasting composition, e.g. *shales or *limestones, would each develop their own specific mineral assemblage, even though they are all being metamorphosed under the same conditions. The variation of mineral assemblage with starting rock composition reflects a particular range of pressure, temperature, and $P(H_2O)$ conditions. Experimental studies of mineral *P–T* stability fields indicates that the facies represents a range of moderate-pressure (about 4–7 kb), moderate-temperature (about 400–500 °C) conditions, usually developed in *orogenic belts associated with continent–continent or ocean–continent plate *collision zones.

greenstone A low-grade, *regional metamorphic rock containing *actinolite, *epidote, and *albite, and lacking a *cleavage. Greenstones form by *metamorphism of *basic *igneous rocks, the actinolite and epidote components being derived from the *ferromagnesium minerals present in the original rock. Actinolite and epidote impart a green colour to the massive rock, hence the name 'greenstone'.

greenstone belt Large geologic formation, up to 250 km across, that is largely of *Archaean age. Greenstone belts are considered to represent ancient volcano-sedimentary *basins bordered and intruded by granitic *plutons. These formations represent an important phase of crustal evolution and currently it is commonly considered that they are remnants of *back-arc basins.

greisen An altered, light-coloured, *igneous rock consisting of white *mica and *quartz. The rock forms by the reaction of crystalline *granite with hot fluorine-rich vapour derived from the crystallizing granite at a deeper level in the *intrusion. The granite mineralogy is unstable in the presence of this vapour and reacts to give the stable mica-quartz assemblage. This process is termed *pneumatolysis. Greisens are found as marginal modifications to granite adjacent to *mineral *veins, as thin veins and *dykes in granite fissures, and as large bodies at the tops and sides of granite intrusions.

Grenvillian orogeny An episode of mountain building that ended about 1000 Ma ago and affected an area extending from what are now Columbia and Mexico, eastern N. America, eastern Greenland, and Scandinavia (where it is known as the *Dalslandian, or Gothian orogeny). It is well exposed in the south-eastern part of the Canadian Shield along a line extending north-east from the northern shore of Lake Huron. It was caused by *plate movements associated with the opening of the Atlantic and leading to the displacement of rocks in a north-westerly direction.

Grenz horizon *See* SUBATLANTIC.

grey-brown podzolic *Soil-profile term describing *eluviated, freely draining soils that have a distinctive *clay-enriched B *horizon. *See also* ALFISOLS.

grey level A calibrated sequence of grey tones, ranging from black to white. In *remote sensing, electromagnetic radiation falling on a photosensitive receiver generates an electrical current proportional to the intensity of the radiation. The receiver is usually tuned to specific wavelength bands and the signal from each receiver is amplified and its intensity classified into different levels, usually 0 (black) to 256. These are the *digital numbers of each of the *pixel units that together make up a remotely sensed frame. In multispectral scanning systems, the digital numbers are classified into grey-scale bands, of which there are usually 10 to 12, but up to 256 shades of grey (brightness) may be used.

greywacke (graywacke) Texturally and mineralogically immature *sandstones that contain more than 15% *clay minerals. They may consist of angular to sub-rounded grains of *quartz and *feldspar, small pebbles, and a fine *matrix of clay minerals, *chlorite, and *carbonate.

grèze litée A bedded hillslope deposit made up of alternate layers of coarse rock fragments (up to 25 mm diameter) and finer material, and laid down in a *periglacial environment. It is probably formed by a combination of *frost wedging, *gelifluction, and the action of meltwater derived from the thawing of frozen ground.

grid reference In Great Britain, a map reference system, overprinted on all Ordnance Survey (OS) maps, which can describe uniquely the location of a point on the ground, on the smallest-scale maps to within 10 m. From a reference point to the west of the Isles of Scilly, the National Grid divides Great Britain (Ireland is excluded) into squares of 100 × 100 km, each designated by two letters. These squares are subdivided into 1 × 1 km squares whose western and southern margins are numbered, and on the smallest-scale maps 100 × 100 m squares are also used. To describe a particular location, the grid reference begins with the identifying letters of the relevant 100 km square (e.g. SH), followed by two digits (e.g. 60) to identify the western edge of the 1 km square and a third digit (e.g. 9) as an estimate of the distance of the point from the western margin of that square. This procedure is called 'easting'. 'Northing' is similar, starting with two digits for the southern edge of the 1 km square (e.g. 54) and a third (e.g. 3) as an estimate of the distance from the southern edge. The full grid reference, consisting of two letters and six digits locates the point to within 100 m (SH 609543 is the grid reference for the summit station of the Snowdon Mountain Railway). On smaller-scale maps, two additional digits, one added to the easting and one to the northing, locate the point to within 10 m.

Griesbachian The first *stage of the *Scythian Epoch, overlain by the *Nammalian.

Griffith cracks *See* GRIFFITH FAILURE CRITERION.

Griffith failure criterion The two-dimensional relationship between *shear stress and *normal stress at the point of failure. The mechanism of failure is based on the formation, propagation, and joining of microscopic 'Griffith cracks' whose leading edges concentrate stress. Failure occurs when a critical stress is reached and the cracks propagate fully.

Griffith–Murrell failure criterion A three-dimensional modification of the *Griffith failure criterion which includes the effect of intermediate *principal stress (σ_2) on failure.

Grigg–Skjellerup A *comet with an orbital period of 5.09 years; *perihelion date 22 July 1992; perihelion distance 0.989 AU.

grike Deep, tapering cleft, normally a solution-widened *joint, cut into the surface of a near-level area of hard *limestone. In north-western Yorkshire, England, it may be 15–60 cm wide at the surface, and commonly 0.5–3 m deep. The deepest grikes form where several joints intersect, and may develop into shallow caves. *See also* CLINT.

GRIP (Greenland Ice Core Project) A European drilling programme that extracts ice cores from the Greenland ice sheet, from which atmospheric and palaeoclimatological data are obtained. It operates at a site about 30 km from the second *GISP site. In 1993, GRIP reached bedrock at about 3000 m.

GRM *See* GENERALIZED RECIPROCAL METHOD; and GYROREMANENT MAGNETIZATION.

groin *See* GROYNE.

Grooved Terrain *See* MARTIAN TERRAIN UNITS.

groove mark A linear groove, cut in a muddy substrate by the dragging of an object through the *sediment by flowing water. The orientation of the groove will be parallel to the current direction. Subsequent infilling of the groove by sediment will result in a groove cast being preserved on the base of the overlying bed.

grossular A member of the *garnet group of minerals, $Ca_3Al_2Si_3O_{12}$; sp. gr. 3.5; *hardness 7; green to yellowish-brown; well-formed crystals or granular; occurs in metamorphosed impure *limestones and as a *detrital mineral, and is used as an abrasive or for gemstones.

ground-control point A point which can be mapped by Cartesian coordinates in two dimensions common to both an image and a topographical map. Such points allow orientation and the correction of distorted images.

ground data (ground information, 'ground truth') Information collected in the field, e.g. during a survey of land resources.

ground frost Condition in which the ground surface has a temperature below 0 °C.

ground ice *See* ICE.

ground information *See* GROUND DATA.

groundmass The finer-grained material of *igneous rocks in which larger crystals (*phenocrysts) and *xenoliths are enclosed. The groundmass of an igneous rock commonly represents that part of the *magma system which has cooled rapidly. Rapid cooling initiates numerous crystal *nucleation sites which limit the size to which crystals can grow. In *lavas, the fine groundmass is formed when the magma is erupted on to the Earth's surface and is cooled rapidly by air convection over the flow. *Compare* MATRIX.

ground moraine *See* MORAINE.

ground range In *radar terminology, the distance between the *nadir and an object. *See also* RANGE.

ground roll A *surface wave, typically a *Rayleigh wave, which propagates along the surface of the ground with a characteristically low frequency and low velocity, but with a relatively high amplitude. Such waves degrade the quality of the *seismic record by masking reflections from the subsurface. In a marine environment, the same type of phenomenon is called 'mud roll'.

ground surge *See* SURGE.

ground truth 1. In *remote sensing, the verification of image interpretation by direct observation of the ground. **2.** *See* GROUND DATA.

groundwater All the water contained in the void space within rocks. The term is generally taken to exclude vadose water (water travelling between the surface and the *water-table). Most groundwater derives from surface sources (*meteoric water); the remainder is either introduced by magmatic processes (*juvenile water) or is *connate water.

groundwater facies A *groundwater of particular character and chemistry. The concept of hydrochemical facies is based on the assumption that the chemical composition of groundwater at any point reflects a tendency towards chemical equilibrium with the matrix rocks under the prevailing conditions. *See* CHEBOTAREV SEQUENCE.

groundwater flow The movement of water through interconnected voids in the *phreatic zone. *See also* DARCY'S LAW.

group 1. A number of *geophones whose output is summed to feed one *seismic channel. A particularly large number of geophones used per channel may be referred to as a 'patch'. *See also* ARRAY. **2.** *See* FORMATION.

group interval The horizontal distance between the mid-points of adjacent *groups of *geophones.

group speed (of wave) In deep-water areas individual waves within a group move forward through the group at a faster speed than the group itself moves. The group form is produced by the interference of waves of different wavelengths (and therefore speeds) as they move forward from their source. New waves form at the rear of the group, lose amplitude as they reach the front of the group, and disappear. The group speed is half that of the individual waves that it comprises.

group velocity (*U*) The speed with which an envelope of a wave train travels. The term is contrasted with the *phase velocity.

grouting Injection of liquid *cement or chemicals into the ground where they set thus impeding or preventing water flow by reducing *permeability and improving the strength of rocks by filling *pores and *fractures. Primary injection holes are spaced at regular intervals with infilling of secondary holes where necessary. The type of grout and its *viscosity depends on rock type because of varying size of pores and fractures and hydraulic resistance. The migration of grout is controlled by permeability.

growan *See* SAPROLITE.

growth band (growth line) A band or line, found in many organisms, that marks growth. Growth lines in *Brachiopoda, for example, record the former position of the *commissure; in *Gastropoda growth may be intermittent and a band of rapid growth may be preserved in the shell.

growth curve A graph that records the changing isotopic levels of ^{207}Pb, ^{206}Pb, and ^{204}Pb brought about by the *radioactive decay of ^{238}U, ^{235}U, and ^{232}Th. Very early in the Earth's cooling history there was no *radiogenic lead. With the passage of time radiogenic ^{206}Pb, ^{207}Pb, and ^{208}Pb accumulated at the sites of uranium and thorium. The accumulation of radiogenic lead from an environment with constant ratios of U to Pb and Th to Pb would be expected to follow a simple growth curve, illustrating the increase of daughter *isotopes with time. The curve would begin at the level of *primordial lead and is normally plotted against axes of $^{207}Pb:^{204}Pb$ (*y*), and $^{206}Pb:^{204}Pb$ (*x*). The curve rises steeply at first, while more ^{207}Pb than ^{206}Pb is produced, but then flattens as ^{235}U becomes depleted following its more rapid decay than that of ^{238}U. Plotted in this way, the growth curve is the reverse of the decay curve. In reality, for any suite of samples, there would be a family of growth curves, each of which is specified by a particular value of the parameter $^{238}U:^{204}Pb$. Each of these curves represents the course along which lead isotope ratios would evolve in a system presently containing a given $^{238}U:^{204}Pb$ ratio. As the equation governing these graphs is only a function of time, systems of the same age but different $^{238}U:^{204}Pb$ ratios will contain lead-isotope ratios lying along straight lines, known as *isochrons. These isochrons will diverge from the assumed *common lead-isotope ratio and will have slopes that are a function of age.

growth fault A *fault occurring as a primary, post-depositional or syn-depositional feature commonly associated with *unconsolidated *sediments recently deposited in a *basin which is undergoing active subsidence. Growth faults are typically *normal and movement occurs at the same time as sedimentation, so the sediment is usually thicker on the down-faulted side. *See* LYSTRIC FAULT.

growth-fibre analysis The analysis of the orientation of *quartz or *calcite *growth fibres; changes in their orientation mark changes in the direction of *extension.

growth fibres Elongate *crystals which lie in a crack or *vein parallel to the direction in which the walls of the fissure have opened (i.e. the direction of *extension).

growth line *See* GROWTH BAND.

growth twinning A type of *crystal twinning that occurs naturally during crystal growth where the energy state near the twin boundary is slightly higher than that of a single crystal. Most of these twinned crystals grow in a natural state by the addition of clusters of atoms a layer at a time, e.g. *aragonite growth twins.

groyne (groin) A breakwater made from rock, concrete, wood, or metal, erected on a beach to inhibit the movement of sand and shingle and to protect against *longshore drift.

Gruneisen ratios (r) A variety of parameters that have been determined to relate crystal lattice energy, pressure, and density. They are used mostly in determining the volume coefficient of thermal expansion for convecting *mantle materials.

grunerite *See* CUMMINGTONITE.

grus *See* SAPROLITE.

Guadalupian A *series in the Upper *Permian of N. America, underlain by the *Leonardian, overlain by the *Ochoan, and roughly contemporaneous with the *Ufimian and lower *Kazanian Stages.

guano The leached residue of profuse accumulations of bird or bat excrement, rich in calcium phosphate. Such deposits are found particularly on arid oceanic islands, and in caves. Guano is worked industrially as a phosphate resource.

guard *See* ROSTRUM.

Guettard, Jean Étienne (1715–86) A French geologist who in 1746 constructed the first mineralogical map of France, using symbols to indicate minerals and rock types. He produced the first mineralogical map of N. America in 1752 (although he never went there), and in 1766 began a survey of France with the aim of producing detailed geologic maps.

guild A group of species that have similar ecological roles, because they require the same resources and obtain them by similar means.

Gulf An *epoch of the *Cretaceous Period, dated at 97–65 Ma (Harland et al., 1989).

Gulfian A *series in the N. American Upper *Cretaceous, underlain by the *Comanchean, and overlain by the Midway (*Palaeocene).

Gulf Stream The most important ocean-current system in the northern hemisphere, which stretches from Florida to north-western Europe. It incorporates several currents: the *Florida Current, the Gulf Stream itself, and an eastern extension, the *North Atlantic Drift. The Florida Current is fast, deep, and narrow, but after passing Cape Hatteras the Gulf Stream becomes less effective at depth and develops a series of large meanders which form, detach, and re-form in a complicated manner. After passing the Grand Banks (off Newfoundland), the flow forms the diffuse, shallow, slower-moving N. Atlantic Drift. The temperature (18–20 °C) and salinity (36 parts per thousand) tend to be seasonally constant, unlike neighbouring coastal water masses.

gully A feature of rain *erosion that develops from the run-off of a violent torrent that bites deeply into topsoil and soft sediments. Gullies can develop on valley sides as valley-side gullies, and also along valley floors as *arroyos. *See also* RILL-WASH.

Gunflint Chert *See* *EOSPHAERA*.

Günz The first of four glacial episodes established by *A. Penck and E. Bruckner in 1909. It is named after an Alpine river, and so the term is really applicable only to its type area, but it has come to be used much more widely. The Günz may correlate with the *Menapian of northern Europe.

Günz/Mindel Interglacial An Alpine *interglacial *stage, that is perhaps the equivalent of the *Cromerian stage of northern Europe.

gust Sharp increase in wind strength close to the ground, caused by mechanical disturbance in an air flow. Gusts may also be generated by temperature lapse rates and by wind shear, e.g. in *clear-air turbulence.

Gutenberg, Beno (1889–1960) A German seismologist who emigrated to the USA in the 1930s, in 1913 Gutenberg used seismic data to calculate the diameter of the *core. In 1926, he established the existence of the low-velocity layer (*see* LOW-VELOCITY ZONE), noting that seismic waves from foci with depths of 50–250 km took longer to arrive than expected. At the California Institute of Technology he continued his seismic studies in collaboration with C. F. *Richter. *See also* GUTENBERG DISCONTINUITY.

Gutenberg discontinuity The seismic-velocity *discontinuity between the Earth's *mantle and *core. The boundary is at a depth of about 2600 km and is thought to have surface irregularities of a few kilometres.

gutter cast An elongate *cast found on the base of a bed and formed by the infilling of a gutter structure. The gutter is a linear to sinuous, U-shaped depression, up to 10 cm wide and of similar depth to its width, formed by fluid scour from helical vortices travelling parallel to the flow direction.

guyot Flat-topped submarine mountain or *seamount, the summit of which lies 1000–2000 m below the ocean surface. The flat top may be a result of marine and/or subaerial erosion.

G-wave *See* LOVE WAVE.

GWP *See* GLOBAL WARMING POTENTIAL.

Gymnodiniales *See* DINOPHYCEAE.

gymnosperm A seed plant in which the ovules are carried naked on the cone scales, in contrast to the *angiosperms, in which they are enclosed by an ovary. Gymnosperms date from the *Carboniferous and subsequently dominated the floras of the world until the *Cretaceous, since when they have been progressively displaced by the angiosperms (flowering plants).

Gymnospermae Formerly, a subdivision of the seed plants (Spermatophyta), but now regarded as a group of plants that are not closely related and '*gymnosperm' is used only informally to describe what are now classified as 4 separate divisions: *Ginkgophyta, *Cycadophyta, *Coniferophyta (Pinophyta), and *Gnetophyta.

gypcrete A gypsiferous (*see* GYPSUM) soil profile developed in arid regions. Gypcretes are formed by the precipitation of $CaSO_4$ from saline waters drawn to the surface by capillary action.

gypsic Applied to a *soil horizon (a gypsic horizon) where secondary *gypsum ($CaSO_4$) has accumulated through more than 150 mm of soil, so that this horizon contains at least 5% more gypsum than the underlying horizon.

gypsum *Evaporite mineral, $CaSO_4.2H_2O$; sp. gr. 2.3; *hardness 1.5–2; *monoclinic; clear white, but sometimes shades of yellow, grey, red, and brown; white *streak; vitreous *lustre; *crystals usually *tabular, often with curved faces, but can occur *massive or granular; *cleavage perfect {010}, good {100}, {011}; occurs in bedded deposits in association with *halite and *anhydrite. It is very insoluble and therefore the first mineral to precipitate from evaporating sea water; usually succeeded by anhydrite and halite. Occasionally it results from the reaction of sulphuric acid on *limestone in volcanic areas. It can also result from the secondary hydration of anhydrite. Selenite is the colourless, transparent form; satin spar is the fibrous variety; alabaster is the fine-grained variety and can be carved.

gyre (ocean gyre) Circular or spiral motion of water, the term usually being applied to a semiclosed current system. A major gyre exists in each of the main ocean basins, centred at about 30° from the equator and displaced towards the western sides of the ocean ('western intensification'). Gyres are generated mainly by surface winds and move clockwise in the northern hemisphere and anticlockwise in the southern hemisphere.

gyrocompass The axis of a spinning gyroscope, within a set of gimbals, and with an unbalanced mass distribution, will precess about true North. Such an instrument can be used to make very precise determination of the direction of true North.

gyroconic Applied to the conch of a *cephalopod when it coils relatively loosely.

gyrogonite *Fossil cast of the female reproductive structure (nucule) of charophytes (stoneworts, *see* CHAROPHYCEAE). Charophyte nucules are small (around 400–600 μm), nut-like objects that have a fossil record commencing in the *Silurian. The various characteristic spiral ornamentations of gyrogonites make them useful *index fossils in some freshwater deposits. *See also* *PSEUDOSYCIDIUM*.

gyroremanent magnetization A magnetic remanence acquired by magnetically anisotropic materials that are rotated within an alternating magnetic field. *See also* ROTATIONAL REMANENCE.

gyttja (nekron mud) Rapidly accumulating, organic, muddy deposit, characteristic of *eutrophic lakes. The precise nature of gyttja varies with the producer organisms involved, which include small *algae or larger aquatic plants.

Gzelian 1. The final *epoch in the *Pennsylvanian, comprising the Klazminskian and Noginskian *Ages (these are also *stage names in eastern European stratigraphy). The Gzelian is preceded by the *Kasimovian Epoch, followed by the *Asselian Age (Early *Permian), and has its upper boundary (the Carbo–Permian boundary) dated at 290 Ma (Harland et al., 1989). **2.** The name of the corresponding eastern European *series, which is roughly contemporaneous with the upper *Virgilian (N. America) and upper *Stephanian (western Europe).

H The horizontal component of the *geomagnetic field.

H_0 *See* HUBBLE PARAMETER.

haar Especially in coastal areas of eastern Scotland and north-eastern England, common term for a sea fog: these are frequent in early summer.

habit The development of an individual crystal, or aggregate of crystals, to produce a particular external shape, with development depending on the conditions obtaining during formation. Individual crystals may possess habits such as acicular (needle-like), tabular (broad and flat), fibrous (hair-like), or prismatic (elongated in one direction). Aggregates of crystals may possess habits such as *botryoidal, *dendritic, or reniform (kidney-shaped).

haboob From the Arabic *habb* meaning 'to blow', a local term for a dust storm in northern Sudan. Typically, the storm is experienced late in the day during summer.

hackly fracture A sharp or jagged *fracture typical of brittle minerals, e.g. in metals such as silver or platinum.

HAD *See* HELIUM ABUNDANCE INTERFEROMETER.

hadal zone The part of the ocean that lies in very deep trenches below the general level of the deep-ocean floor (the *abyssal zone).

hade The angle between a *fault plane, mineral *vein, or *lode and the vertical.

Hadean The first *Precambrian '*system' (Harland et al., 1989), extending from about 4500 Ma to about 3875 Ma ago, and best preserved in lunar rocks.

Hadley cell One of the most fundamental divisions of the global wind circulation, comprising the net ascent of air over the lowest latitudes due to convection, the compensatory downward motion in the subtropical anticyclone belt, and the resulting trade winds which blow toward the meteorological equator (intertropical convergence). It is named after G. Hadley, who in 1735, seeking to account for the trade winds, proposed a single, large-scale convective cell representing a thermally driven, low-latitude atmospheric circulation.

Hadrosauridae (hadrosaurs) Duck-billed *ornithischian *dinosaurs, of possibly amphibious habits, which averaged about 9 m in body length. Several mummies of hadrosaur dinosaurs have been found, from which it was possible to establish that they were unarmoured and that the digits of the forelimbs were connected by a web of skin. Hadrosaurs flourished in the Upper *Cretaceous.

hadrosaurs *See* HADROSAURIDAE.

Hadrynian A *stage of the Upper *Proterozoic of the Canadian Shield region, underlain by the *Helikian.

haematite *See* HEMATITE.

Hagedoorn method *See* PLUS–MINUS METHOD.

hail Form of precipitation comprising ice in the shape of balls or irregular particles (hailstones), whose concentric structure indicates a growth by *coalescence and freezing of supercooled water drops. Hail is usually associated with *cumulonimbus cloud.

hair hygrometer Instrument that indicates *relative humidity, based on the expansion and contraction of a treated human hair.

hairpin A sudden change in a palaeomagnetic *polar wander path, usually taken to indicate a tectonic collision between the tectonic plate concerned and some other plate.

haldane A unit of evolutionary morphological change, equal to one standard deviation per generation, and named for the Scottish physiologist and geneticist J. B. S. Haldane. *See also* DARWIN.

Hale–Bopp A *comet with an orbital period of more than 4000 years; *perihelion date 31 March 1997; perihelion distance 0.914 AU.

half-field prism *See* REFLECTOR.

half-graben *See* GRABEN.

half-life *See* DECAY CONSTANT; and RADIOMETRIC DATING.

half-shadow test (shadow test, Schroeder Van Der Kolk method) A test used to determine the approximate *refractive index of a mineral by immersion in a liquid of known refractive index. A card is inserted below the rotating *stage to cut out half the light. The *condenser is then adjusted to focus or blur the card. If the mineral differs in refractive index from the liquid, one side of the grain will have a distinct shadow. If the shadow is on the same side as the card the mineral has a higher refractive index than the liquid; if it is on the opposite side of the grain the mineral has a lower refractive index than the liquid.

half-space A mathematical model in which only one boundary exists, all others being infinitely far away. The medium under consideration is usually assumed to be perfectly homogeneous and *isotropic.

half-spreading rate *See* SPREADING RATE.

half-width ($w_{1/2}$) Half the width of an anomaly measured between points with half the peak amplitude. It is used when determining depths to the centres of causative bodies in *gravity and *magnetic profiles.

halides A group of minerals which contain halogens, principally chlorine and fluorine. The group includes *halite (NaCl), *sylvite (KCl), and *fluorite (CaF_2). Halides are characterized by ionic bonding and are mainly *cubic in form, soft, and generally light in weight. They frequently occur as precipitates resulting from the evaporation of saline waters.

halinity The extent to which particular water contains chloride. According to the Venice system, brackish waters (which are saline, but less so than sea water) are classified by the chloride they contain and divided into zones. The zones, with their percentage chlorinity (mean values at limits) are: euhaline 1.65–2.2; polyhaline 1.0–1.65; mesohaline 0.3–1.0; alpha-mesohaline 0.55–1.0; beta-mesohaline 0.3–0.55; oligohaline 0.03–0.3; fresh water 0.03 or less.

***Haliomma vetustum* (family Actinommidae)** One of the earliest recorded species of *radiolarians, characterized by a siliceous shell, and a member of a family whose members have a spherical or ellipsoidal shape and no internal spicule. *H. vetustum* is known from the *Ordovician of Europe.

halite (rock salt) Mineral, NaCl; sp. gr. 2.2; *hardness 2.5; *cubic; perfect cubic *cleavage; colourless, white, or shades of yellow, red, and blue; white *streak; vitreous *lustre; crystals usually cubes, often with curved faces, but it can be granular and compact; widely distributed in stratified *evaporite deposits, associated with other water-soluble minerals (e.g. *sylvite) and with *gypsum and *anhydrite of various geologic ages, large masses frequently forming plugs which rise through and arch overlying *sedimentary rocks, thereby forming oil traps; soluble in water; tastes salty. It is widely used as a road dressing in icy weather.

Halley A *comet with an orbital period of 76.1 years; *perihelion date 9 February 1986; perihelion distance 0.587 AU.

Hall, James (1761–1832) Scottish geologist and physicist who was the pioneer of experimental petrology. He was able to demonstrate (1800) that a *basalt *melt will crystallize if cooled slowly, thus giving important support to plutonist theory.

Hall, James (1811–98) American geologist, palaeontologist, and member of the US Geological Survey. A uniformitarian, he made detailed observations of the stratigraphy of the Appalachians. He proposed that the sediments were laid down in a trough, which subsided under their weight, before final uplift and erosion. He opposed catastrophist and contractionist hypotheses of mountain formation. *See* CATASTROPHISM; and CONTRACTING EARTH HYPOTHESIS.

Hall effect A strip of metal or semiconductor carrying an electrical current within a strong, transverse, magnetic field develops a potential at right angles to both the current and the field. This forms the basis for some sensitive *magnetometers. The effect was discovered by Edwin Hall (1855–1938).

Hallian The last of two *stages in the *Pleistocene of the west coast of N. America, underlain by the *Wheelerian, overlain by the *Holocene, and roughly contemporaneous with the Upper Pleistocene Series of southern Europe.

halloysite *See* CLAY MINERALS.

halmyrolysis Early *diagenesis, modification, or decomposition of sediments on the sea floor. For example, both the breakdown of *ferromagnesian minerals, and the growth of *glauconite aggregates in sea-floor sediments, are types of halmyrolytic process.

halo Rainbow-coloured, sometimes white, ring around the Sun or Moon, due to refraction by ice crystals when thin *cirrus cloud obscures the luminary. *Compare* CORONA.

halocline Water layer with a large vertical change in salinity. The halocline is usually well developed in coastal regions where there is much freshwater input from rivers producing surface waters of low *salinity, a zone where salinity increases rapidly with depth (the halocline), and a deeper zone of more saline, denser waters.

halokinesis The study of salt tectonics, which includes the mobilization and flow of subsurface salt, and the subsequent emplacement and resulting structure of salt bodies (e.g. the *Mesozoic salt domes of the North Sea basin). *See also* HALOTECTONISM.

halophile An *extremophile (domain *Archaea) that thrives in extremely saline environments.

halotectonism Movements of subsurface salt that involves tangential compressive stress.

hamada (hammada) A rocky *desert, or desert region, which does not have surficial materials and which consists mainly of boulders and exposed bedrock. Two basic types occur: stony hamada, jaggedly developed across crystalline rocks; and pebbly hamada, cut across sedimentary material and mantled with bedrock fragments.

hammada *See* HAMADA.

hammer chart A circular graticule divided by radial and concentric lines to provide a template with many compartments. It is used when calculating the *terrain correction in gravity surveying.

hammer source In *seismic refraction surveys, a heavy sledge-hammer is swung on to a plate on the ground to provide a seismic impulse and trigger an appropriate recording system. It is used extensively with signal-enhancement *seismographs, when repeat hammer blows can improve the signal-to-noise ratio. It is mainly a source of *P-waves, but can be modified to become a source of *S-waves.

hand lens A hand-held, steel- or plastic-mounted, optical magnifying system usually consisting of two or more optical elements, which is used in the field to provide an enlarged image of rocks, minerals, and fossils, ranging from 5× to 20×. Hand lenses vary in diameter from 0.5 to 1.5 cm, can be carried conveniently in a pocket, and form an essential tool in a geologist's field kit.

hanging valley Tributary valley whose floor is well above that of the adjacent main valley and where there is therefore often a waterfall. It is typical of glaciated uplands, where it may result from glacial widening and/or deepening of the main valley.

hanging wall The *fault block which lies above any inclined fault surface. *See* FOOTWALL.

hardground A term first introduced into geologic literature in 1897, drawn from an oceanographic source but used in a narrower sense to describe a specific *horizon that had initially been lithified a short distance beneath the sea floor. *Lithification resulted from *calcite precipitation around local foci, giving rise to *concretions. The *sediment above and surrounding the concretions remained soft and was consequently bioturbated (*see* BIOTURBATION). The initial *cementation of the hardground may have been triggered by the chemical influence of organisms that dwelt within the sediment. Progressive encroachment of the cemented areas on to the soft sediment gradually resulted in total lithification. 'Incipient' hardgrounds represent periods of arrestation in the cementation process, with the result that nodular and soft-sediment units alternate one with the other. The cemented hardground may itself support a fauna of fixed or cemented organisms and may be bored extensively by bivalves (*Bivalvia). *Pyrite films may line *burrows, which may also be filled by *silica.

hardness **1.** A measure of the ability of water to form a carbonate scale when boiled, or to prevent the sudsing of soap. Permanent hardness is due mainly to dissolved calcium and magnesium sulphate or chloride, the bicarbonate ion causes temporary hardness. Dissolved carbon dioxide and the weathering of carbonate rocks are the main sources of hardness in water. **2.** A physical property of minerals and one of the most useful tests for mineral identification. *Mohs's scale of hardness (H), which ranks minerals by their hardness and thus makes possible a diagnostic test in which one mineral is used to scratch another, was introduced in 1822 and is still the standard used today. Useful tools for determining hardness are the finger nail (H about 2.5) and a penknife (H about 5.5). With a little practice, the hardness of a mineral may be determined by means of a scratch to within one or two points on Mohs's scale. *See also* VICKERS HARDNESS NUMBER.

hardpan Hardened *soil horizon, usually found in the middle or lower parts of the profile, that may be *indurated or *cemented by a variety of possible cementing materials. *See also* CALICHE; DURICRUST; and PAN.

HARI *See* HIGH ASPECT RATIO IGNIMBRITE.

harmattan wind (the doctor) Dry, dusty, north-easterly or easterly wind which occurs in W. Africa north of the equator. Its effect extends from just north of the equator in January, almost to the northern tropic in July. In W. Africa it is known as 'the doctor' because of its invigorating dryness compared with humid tropical air. The harmattan wind stream occasionally extends south of the equator during the northern winter as an upper air wind over the south-westerly monsoon.

harmonic A *frequency which is a whole multiple of the fundamental frequency; thus the second harmonic has twice the fundamental frequency, the third harmonic three times, etc.

harmonic fold A *fold which maintains its geometric form, integral wavelength, and symmetry throughout a sequence of layers. Such folds form where the competent layers comprising the sequence are of similar thickness and evenly spaced, and the contrast in *competence is constant between each layer and the next.

Harnagian A *stage of the *Ordovician in the Lower *Caradoc, underlain by the *Costonian and overlain by the *Soudleyan.

Hastarian A *stage of the *Tournaisian Epoch, dated at 362.5–353.8 Ma (Harland et al., 1989).

hastingsite *See* HORNBLENDE.

Haumurian *See* MATA.

Hauterivian A *stage in the European Lower *Cretaceous dated at 135–131.8 Ma (Harland et al., 1989). *See also* NEOCOMIAN.

Haüy, Abbé René-Just (1743–1822) One of the founders of modern crystallography, Abbé Haüy was Professor of Mineralogy at the Sorbonne. He applied mathematics to mineralogy in order to determine the structure of crystals, proposing a 'law' of rational intercepts. His work, *Traité de crystallographie* (1822), was for a time the standard book on the subject.

haüyne A *feldspathoid and member of the *sodalite group of minerals, formula $(Na,Ca)_{4-8}(Al_6Si_6O_{24})(SO_4,S)_{1-2}$; sp. gr. 2.5; *hardness 5.5; greyish-white; granular; occurs as a constituent of *alkali volcanic rocks, often with *pyrite.

Hawaiian-Emperor chain The Earth's largest volcanic mountain range, extending from the oldest *seamount, dated at 78 Ma, near Kamchatka to an active submarine volcano south-east of Hawaii. The chain is regarded as a plume trace (*see* MANTLE PLUME). It forms two distinct, gentle curves, with a kink dated at approximately 43 Ma and thought to be related to a change in the direction of movement of the *Pacific Plate over a stationary *hot spot.

Hawaiian eruption A type of volcanic *eruption that produces very fluid, mobile, basaltic *lava, characterized by spectacular *fire fountains, with little explosive activity. The lava flows that result are often formed by the accumulation of fire-fountained lava and are termed 'spatter-fed flows'. Hawaiian-type eruptions produce a broad, low-angle cone (a shield *volcano). *Compare* PELÉEAN ERUPTION; PLINIAN ERUPTION; STROMBOLIAN ERUPTION; SURTSEYAN ERUPTION; VESUVIAN ERUPTION; and VULCANIAN ERUPTION.

hawaiite *See* ANDESITE; and BENMOREITE.

Hawker *See* ADELAIDEAN.

haze Atmospheric condition in which visibility is reduced because of the dispersion of light by very small, dry particles, e.g. fine dust.

HDR *See* HOT DRY ROCK.

head 1. In certain areas of Britain, the name given to a sheet of poorly sorted, angular rock debris, mantling a hillslope and deposited by *gelifluction. Similar material is commonly found as a fossil deposit in the extraglacial areas of N. America and Europe. *Coombe rock is similar, but is found on chalk. **2.** *See* BERNOULLI EQUATION; DARCY'S LAW; ELEVATION POTENTIAL ENERGY; HYDRAULIC HEAD; and PRESSURE HEAD.

head cut *See* KNICK POINT.

heading blasting An old method of quarry blasting in which explosive is confined in small tunnels at suitable intervals in the quarry face. For large blasts, several tunnels may be used.

head shield **1.** The term sometimes applied to the *cephalon in trilobites (*Trilobita). **2.** The dorsal head covering of solid bone which occurs in many ostracoderms (*Ostracodermi) (agnathan fish, *see* AGNATHA).

headwall, glacial The steep rock slope at the head of a *cirque or *valley glacier. It is a site of active *erosion, perhaps by *frost wedging.

head wave A *refracted wave which enters and leaves a high-velocity medium at the *critical angle. Usually the term refers to the refracted wave which arrives to give a refraction *first break.

heat capacity The energy required to increase the temperature of a given mass of material by one degree Celsius.

heat flow (heat flux) A measure of the heat being conducted through the surface rocks of the Earth. It is usually determined by taking two or more temperature readings at different depths down a *borehole or sampling tube, and measuring or estimating the thermal conductivities of the rocks in between. On land, measurements are usually taken at depths greater than 200 m to avoid temperature changes associated with climatic oscillations. The largest uncertainties are due to local or regional heat transference by convective circulation. In oceanic sediments the observed heat flow decreases in proportion to $t^{0.5}$, where t is the age of the *oceanic crust. Near the oceanic ridge crests the heat flow averages 100 mW/m^2, decreasing to less than 50 mW/m^2 for crust older than 100 Ma. In the continents, the older crustal areas appear to have a lower heat flow, averaging about 38 mW/m^2, than the younger crust; in *Mesozoic and *Tertiary orogenic areas it is 60–75 mW/m^2. The heat flow in the cratonic shields (*see* CRATON) is almost entirely accounted for by *radiogenic heat production within the surface rocks themselves, implying depletion of radiogenic heat-producing elements in the *crust and upper *mantle beneath these areas. The annual global heat loss by the Earth is somewhat uncertain because of heat loss by convection, but is about (4.1–4.3) $\times$ 10^{13} W (1.3 $\times$ 10^{21} J/year), about 75% of which is lost through the oceanic crust. Internal heat sources within the Earth are: (*a*) radiogenic heat-producing elements; (*b*) gravitational energy released by core formation; (*c*) exothermic chemical reactions; and (*d*) heat remaining from the initial accretion of the Earth, comprising the original heat of the materials, kinetic heat from early meteoritic and *planetismal bombardment, and exothermic chemical reactions. The proportion of each source is not clear, although radiogenic heat is probably the major component; gravitational energy release may have been dominant during core formation.

heat-flow anomaly An area characterized by higher or lower *heat flow than the average for the area.

heat-flow unit (HFU) A unit of 1 cal/cm^2/s, now replaced by SI units: 1 HFU = 41.8 mW/m^2.

heat flux *See* HEAT FLOW.

heat of formation The amount of heat evolved or absorbed during the formation of one *mole of a substance from its constituent elements.

heave 1. The amount of horizontal (lateral) displacement between two sides of a *dip-slip fault, measured at right angles to the *strike of the fault. **2.** Lifting of earth due to frost (*see* FROST HEAVE TEST), overloading, swelling clay, etc. **3.** Upwarp in the floor of a mine due to the floor being too weak to resist the forces resulting from the weight of the overlying rock on the adjacent, supporting pillars.

heavy liquids *See* DENSITY DETERMINATION.

heavy medium separation *See* DENSE MEDIUM SEPARATION.

hedenbergite A member of the *clinopyroxene group of *minerals and the iron-rich *end-member of the *diopside series, with formula $CaFeSi_2O_6$; sp. gr. 3.7; *hardness 6; black; forms short, *prismatic *crystals or lamellar masses; occurs in metamorphosed iron-rich sediments, *skarns and *eulysites, and in some *igneous ferrogabbros and *granophyres along with fayalitic *olivine.

Heezen, Bruce Charles (1924–77) An American oceanographer from Lamont-Doherty Observatory, Heezen is perhaps best known for his maps of the ocean floors, produced in collaboration with Marie Tharp. He also made important studies of the axial ridge systems of the oceans, and in 1952 found the first evidence for *turbidity currents during his studies of the Grand Banks earthquake (1929).

Heim, Albert (1849–1937) A Swiss alpine geologist, Heim did important work concerned with the theory of mountain building, and on the formation of *overthrusts and *nappes in the Alps. He supported the idea of a *contracting Earth. He also studied the mechanics of rock deformation, proposing that rocks can deform plastically under pressure and that the same pressure causes *metamorphism.

Heinrich events The release into the ocean of large amounts of ice, as icebergs, at intervals of 8000–10 000 years, with major consequences for the circulation of ocean water and the global climate. The phenomenon was first described in 1988 by Hartmut Heinrich.

Heinrich layers North Atlantic sediments that have a high ratio of debris carried by ice to *Foraminiferida shells and thus record episodes of major iceberg release or surges in the *Laurentide ice sheet.

Helderbergian *See* ULSTERIAN.

Helene (Saturn XIII) One of the lesser satellites of *Saturn, discovered in 1980 by *Voyager 1, with a radius of 16 km; visual albedo 0.7.

helicitic structure Sigmoidal trails of mineral inclusions in *porphyroblasts formed during metamorphic *recrystallization. The sigmoidal inclusion trails may represent remnants of a pre-existing fold-fabric preserved as the growing *crystal enclosed the *fabric, leaving untouched those minerals it did not require to provide its chemical components. Sigmoidal inclusion trails may also represent relative movement between the growing porphyroblast and the *groundmass or external *fabric. Thus, if the porphyroblast grows under the influence of *shear stress, it rotates during growth, and may include relicts of any parallel external fabric as an S-shaped internal inclusion trail.

helictite *See* DRIPSTONE.

Helikian A *stage of the Middle *Proterozoic of the Canadian Shield region, underlain by the *Aphebian and overlain by the Hadrynian.

Heliopora *See* OCTOCORALLIA.

helium abundance interferometer (HAD) An instrument used in *remote sensing to determine very precisely the relative abundances of helium and hydrogen in planetary atmospheres by means of an *interferometer.

helium 'burning' The fusion of helium in the contracted core of a red giant star at extremely high temperatures, hotter than those reached in the *Sun. It is particularly important because the reaction produces carbon and oxygen, so helping to explain their *cosmic abundance. *See* CARBON 'BURNING'; HYDROGEN 'BURNING'; SILICON 'BURNING'; and NUCLEOSYNTHESIS.

helium clock A method for measuring the passage of geologic time based on the accumulation of helium in the Earth's atmosphere. As one of the rare or 'noble' gases, helium is essentially inert chemically. There is very little of it on Earth and this is attributed to the fact that its low atomic weight allows it to escape from the Earth's gravitational field. The mean *residence time in the atmosphere is only a few million years. No helium in the atmosphere can therefore be residual since the formation of the Earth, but must be the product of *alpha decay in the Earth's crust. Thus a state of equilibrium must exist between the continual loss of helium into outer space and the supply of new *radiogenic helium.

Hellenic Arc The mostly inactive east–west volcanic chain on the *Hellenic Plate which is associated with the *subduction of the *African Plate. Crete forms part of the outer, non-volcanic arc and extinct *volcanoes, including Santorini, form the volcanic arc, approximately 150 km to the north.

Hellenic Plate A *microplate in the eastern Mediterranean which lies in the *collision zone between the *African Plate and the *Eurasian Plate. The southern boundary of the Hellenic Plate is the *Hellenic Trench, north of which there is a rising, non-volcanic arc. The Aegean Sea, further to the north, is undergoing widespread extension and subsidence.

Hellenic Trench A linear depression between the Ionian Basin and Crete which forms the boundary between the *Hellenic Plate and the *African Plate. To the east the boundary is a *transform fault along the Pliny-Strabo Trench. Several hundred kilometres of *oceanic crust are thought to have been subducted under the Hellenic Plate, and the Hellenic Trench may now be the junction between the African *continental margin and the Hellenic Plate.

Helminthoida A two-dimensional grazing trace represented by tightly packed meanders or spiral *trails. The *trace fossil is related to the feeding behaviour of an efficient *sediment-eating organism.

helm wind Local name for a type of lee wave frequent in winter and spring on the western (lee) slope of the Crossfell range in Cumbria, England. Associated with this cold, gusty, north-easterly wind, the helm itself is a thick bank of cloud along the mountain range, together with an outlier of narrow, almost motionless cloud away from the range.

Helvetian *See* SERRAVALLIAN.

hematite (haematite, iron glance, kidney ore, red iron ore, specularite) Iron *mineral, Fe_2O_3; one of the main ores for iron; sp. gr. 4.9–5.3; *hardness 5–6; *trigonal; steel-grey to black, often iridescent, compact varieties dull to bright red; red to reddish-brown *streak; *metallic *lustre; crystals *tabular or *rhombohedral with curved, striated faces, also occurs as columnar, *mammillated, and *botryoidal masses; no *cleavage; widely distributed as an *accessory mineral in *igneous rocks, *hydrothermal *veins, as a rock-forming mineral in *sedimentary rocks, as a *primary mineral, as concretions or a cementing agent, and as a replacement for other minerals. Bedded ores of hematite form huge deposits in the *Precambrian of N. America and elsewhere. *See* BANDED IRON FORMATION (BIF).

hemera A period of geological time determined by the maximum development of a *fossil plant or animal.

hemi- From the Greek *hemi* meaning 'half', a prefix meaning 'half' or 'affecting one half'.

Hemichordata (acorn worms) Phylum first encountered in the Middle *Cambrian *Burgess Shale, British Columbia, Canada. The acorn worms are related to the *chordates, for although they lack a *notochord, the gill slits are very similar to those of primitive *vertebrates.

hemimorphism The property of certain crystals in which there is no element of symmetry present to cause the repetition of upper-hemisphere faces in the lower hemisphere. A good example is *tourmaline, where there are no horizontal *axes of symmetry or *centre of symmetry, but only one vertical axis of three-fold symmetry and three vertical planes of symmetry.

hemimorphite A *silicate mineral $Zn_4[Si_2O_7](OH)_2.H_2O$ which accompanies zinc, iron, and lead sulphides and is associated with *smithsonite $ZnCO_3$, a principal *ore of zinc; sp. gr. 3.4–3.5; *hardness 4–5; *orthorhombic; when *massive, white, grey, yellow, brown, green, or light blue, sometimes colourless; *vitreous *lustre; *crystals can be small, *tabular grains, but normally form aggregates of radial or earthy masses; *cleavage perfect prismatic {110}; occurs in the oxidation zone of lead-sulphide/zinc-sulphide deposits, in *hydrothermal veins near the surface, and as a *primary mineral in hydrothermal veins, associated with sulphides and *fluorite.

hemipelagic sediment (hemipelagite) A deep-sea, muddy *sediment formed close to continental margins by the settling of fine particles, in which biogenic material comprises 5–75% of the total volume and more than 40% of the *terrigenous material is *silt. *Compare* PELAGIC SEDIMENT.

hemipelagite *See* HEMIPELAGIC SEDIMENT.

Hengelo (Hoboken) A Middle *Devensian *interstadial dating from about 40 000 years BP, when estimated July temperatures in the Netherlands were about 10 °C. Together with the *Moershoofd and *Denekamp interstadials, the Hengelo is equivalent to the *Upton Warren Interstadial of the British Isles.

Hennig, Willi (1913–76) A German zoologist who originated *phylogenetics. He was awarded his Ph.D. by the University of Leipzig in 1947 and conducted research on *Drosophila* larvae. His book, *Grundzüge einer Theorie der phylogenetischen Systematik* (1950), made no immediate impact, but when an English translation appeared in 1966, as *Phylogenetic Systematics* (with a second edition in 1979), it suddenly achieved authoritative status (comparable, perhaps, to *The Origin of Species*!). Hennig argued that, as *taxonomy aims to depict relationships and the only objective meaning of 'related' means sharing a common ancestor, taxonomy must be based on *phylogeny. This struck an immediate chord. Hennig coined such terms as *apomorphic, *plesiomorphic, and *sister groups, and offered a redefinition of monophyly (*see* MONOPHYLETIC), which he insisted must be paramount in taxonomy.

Hennig's dilemma Phylogenetic trees constructed from an examination of two or more characters (e.g. two different genes) may result in two contradictory *phylogenies. In such cases no single tree can be constructed using all characters compatibly. (*Cladistics, in which the dilemma may arise, was founded by the German entomologist W. Hennig.)

Herangi A *series in the Lower *Jurassic of New Zealand, underlain by the *Balfour (*Triassic), overlain by the *Kawhia, and comprising the Aratauran and Ururoan *Stages. It is roughly contemporaneous with the *Lias and *Aalenian.

Herbert Smith refractometer *See* REFRACTOMETER.

Hercynian orogeny A major and prolonged episode of mountain building that began in the Late *Devonian and continued throughout the *Carboniferous, affecting a broad belt along an approximately WSW to ENE line from what are now south-western England and north-western Europe to southern Europe and the Iberian Peninsula, and eastern N. America and the Andes. Probably it was caused by northward thrusting and *plutonism along a discontinuous front.

hercynite *See* SPINEL.

Heretaungan A *stage in the Lower *Tertiary of New Zealand, underlain by the *Mangaorapan, overlain by the *Porangan, and roughly contemporaneous with the upper *Ypresian and lower *Lutetian Stages.

hermatypic Applied to *corals that contain zooxanthellae and are reef forming. Modern *scleractinian hermatypic corals are characterized by the presence of vast numbers of symbiotic zooxanthellae (unicellular *dinoflagellates) in their endodermal tissue. They live in waters of normal marine salinity, at depths of up to 90 m, in temperatures above 18 °C, and grow vigorously in strong sunlight.

herringbone cross-bedding A form of cross-bedding (*see* CROSS-STRATIFICATION) in which the *foresets in successive sets are directed in opposite directions, so producing a structure which somewhat resembles the bones of a fish. The bipolar orientation of foresets seen in herringbone cross-bedding is commonly generated by the reversing currents developed in many tidal environments.

Hervyan The uppermost *stage in the *Devonian of Australia, underlain by the *Condobolinian Stage, overlain by the *Carboniferous, and roughly contemporaneous with the upper *Frasnian and *Famennian of Europe.

Hesperian A division of *areological time, lasting from 3.50 Gy to 1.80 Gy in the Hartmann–Tanaka Model and 3.80 Gy to 3.55 Gy in the Neukum–Wise Model, and divided into two epochs: Lower Hesperian (3.50–3.10 or 3.80–3.70 Gy); and Upper Hesperian (3.10–1.80 or 3.70–3.55 Gy).

Hess, Harry Hammond (1906–69) An American geophysicist from Princeton University, Hess made important contributions to the theory of *plate tectonics. He devised the concept of *sea-floor spreading (*see also* DIETZ, ROBERT SINCLAIR), and discovered and named *guyots. His *Essay in Geopoetry* (1960, 1962) was an attempt to link the features of the sea floor in a common hypothesis, in which he proposed that the continents move passively on rafts or rigid plates.

Heterian *See* KAWHIA.

hetero- From the Greek *heteros* meaning 'other', a prefix meaning 'different from'.

heterocercal tail In fish, a tail in which the tip of the vertebral column turns upward, extending into the dorsal lobe of the tail fin; the dorsal lobe is often larger than the ventral lobe. The heterocercal tail is present in many fossil fish, in the sharks (*Chondrichthyes), and in the more primitive bony fish, e.g. the families Acipenseridae and Polyodontidae. In the later, ray-finned fish (*Actinopterygii) the homocercal tail developed, in which the vertebral column stops short of the tail fin, which is supported only by bony rays, giving rise to an apparently symmetrical type. *Compare* DIPHYCERCAL TAIL.

heterochrony Dissociation, during development, of factors of shape, size, and maturity, so that organisms mature in these respects at earlier or later growth stages. This leads to either *paedomorphosis or *recapitulation.

heterocoelus Applied to the condition in which the articulate surface of the *vertebra centrum is saddle shaped, as is the case in birds. *Compare* AMPHICOELUS.

Heterocorallia (subclass *Zoantharia) Small order of corals, known only from *Carboniferous rocks in Europe and Asia, and represented by two genera, *Hexophyllia* and *Heterophyllia*. It is included in the *Zoantharia because its members possess *septa, and because of the microstructure of the calcareous skeleton. The septa are arranged differently from those in other orders. The four original septa are conjoined axially, and new septa are formed attached to these at their axial ends.

heterodont 1. (invertebrate) Applied to a *hinge dentition occurring in the *Bivalvia, where teeth of differing sizes occur in the hinge plate. The teeth are differentiated into *cardinal teeth, which occur beneath the *umbo, and lateral teeth, which occur posteriorly and anteriorly to the umbo. **2. (vertebrate)** Possessing teeth that are differentiated into several forms, e.g. incisors, canines, premolars, and molars (as in mammals). This contrasts with the homodont condition, in which the teeth are all of the same form.

Heterodonta (phylum *Mollusca, class Pelecypoda) Subclass of bivalves (*Bivalvia), which have *heterodont *dentition consisting of a *hinge plate with distinct *cardinal teeth below the *umbo, and lateral teeth posterior to the cardinals. Some forms may be *desmodont. The ligament is *opisthodetic. Shell shape varies according to the mode of life of the organism, most having a crossed-lamellar, *aragonitic shell structure. The *pallial line is entire, or with a sinus. Heterodonta are mainly *infaunal siphon-feeders, the *siphons being well developed. They first appeared in the Middle *Ordovician.

Heterodontosauridae Ornithopod *ornithischian *dinosaurs, known only from the Upper *Triassic of S. Africa. These 'different-toothed lizards' had tusk-like canines.

heterogeneity *See* INHOMOGENEITY.

heterogeneous accretion A model for the accretion of the planetary bodies from the primitive solar nebula (PSN), in which the rate of accretion of solid particles into the planets is slow relative to the rate at which the PSN cools. The consequence is that the surface layer of each body at any one time is in equilibrium with the pressure and temperature conditions prevailing in the nebula, and thus each planet accretes successive 'onion-skin' layers of material with different compositions. According to this model, the layered structure of planets may be partly of primary origin. *Compare* HOMOGENEOUS ACCRETION.

heterogeneous simple shear A *fold mechanism in which folding is due to changes in the amount and orientation of a *simple shear displacement. Layers arranged obliquely to planes of simple shear are rotated passively, producing similar folds of ideal form.

heterolithic bedding A closely interbedded deposit of *sand and mud, generated in environments where current flow varies considerably. The three main types of heterolithic bedding are called *flaser, *wavy, and *lenticular. Flaser bedding is characterized by cross-laminated sands with thin mud drapes over *foresets. Wavy bedding consists of rippled sands with continuous mud drapes over the ripples. Lenticular bedding consists of isolated lenses and *ripples of sand set in a mud *matrix. Heterolithic sediments can be deposited in storm-wave influenced shallow marine environments, river *floodplains, *tidal flats, or *delta-front settings where fluctuating currents or sediment supply permit the deposition of both sand and mud.

heterolithic unconformity *See* NONCONFORMITY.

heteropygous *See* PYGIDIUM.

heterosphere The layers of the atmosphere, above 80 km altitude, through which the chemical composition of the air changes markedly with height, principally as a result of oxygen dissociation. *Compare* HOMOSPHERE.

heterospory *See* SPORE.

Heterostraci (Pteraspida; heterostracans; superclass *Agnatha) Oldest known vertebrate order of jawless, heavily armoured fish-like forms, ranging from the Upper *Cambrian to the *Devonian. The *dermal plates of the body lacked true *bone cells and no internal skeleton has been preserved, so this presumably was *cartilaginous. The anterior part of the body was covered by large dorsal and ventral plates, with smaller plates to the side. There was a single lateral gill opening, eyes at the sides of the head, and the impression of paired nasal sacs on the inner surface of the rostral shield (which protruded in front of the mouth). The rest of the body was covered by scales; the tail was hypocercal (tilted downwards). Heterostracans were usually only a few centimetres long, although some species reached 1.5 m. The body in typical forms, such as the Devonian *Pteraspis*, was rounded, but the group included dorsoventrally flattened, bottom-dwelling forms such as *Drepanaspis*.

heterotopy An evolutionary change in the site at which a particular development occurs. The term was coined by E. H. Haeckel in 1866 to complement *heterochrony.

Hettangian A *stage in the European Lower *Jurassic (208–203.5 Ma, Harland et al., 1989), roughly contemporaneous with the lower Aratauran (New Zealand). *See also* LIAS.

Hexacorallia Alternative name for the order *Scleractinia (stony corals).

Hexactinellida (Hyalospongea; phylum *Porifera) A group of sponges that first appeared in the *Cambrian. They are of normal sponge shape and have skeletal *spicules consisting of opaline *silica. Large spicules (megascleres) and small spicules (microscleres) are connected in various patterns, and this forms the basis for the classification of the group. They were common in the *Palaeozoic but many forms disappeared in the *Permian. They were again important in the *Jurassic and *Cretaceous, but are now confined to much deeper marine waters.

hexad *See* CRYSTAL SYMMETRY.

hexagonal A *crystal system characterized by four *crystallographic axes of which three are horizontal and equal in length with an angle of 120° between them, and a fourth axis which is vertical. The vertical axis is one of six-fold symmetry. The hexagonal system differs from the *trigonal system, which has a vertical axis of three-fold symmetry.

Hexapoda *See* INSECTA.

HFS *See* HIGH-FIELD-STRENGTH ELEMENTS.

HFU *See* HEAT-FLOW UNIT.

hiatus 1. A gap in a sedimentary succession. **2.** The geologic-time interval represented by such a gap.

hidden layer A layer whose presence remains undetected by *seismic refraction methods because (*a*) it is too thin and has insufficient velocity contrast to provide a discernible arrival; or (*b*) it is of lower velocity than the overlying medium. The term is analogous to the *suppressed layer in the interpretation of *electrical resistivity soundings.

hierarchical method 1. The classification of organisms or units into a graded succession or hierarchy. **2.** In stratigraphy, the practice of defining *stratigraphic units in terms of their *rank within a hierarchy and the units of lower rank which they comprise. For example, the geologic *systems are composed of *series, which are themselves composed of *stages. *Compare* TYPOLOGICAL METHOD.

high-alumina basalt An *aphyric, *extrusive *igneous rock of basaltic composition containing a high proportion of *modal *plagioclase, and an alumina content of more than 17 wt.%. Tholeiitic and calc-alkali *basalts (*see* THOLEIITE; and CALCALKALINE) can have alumina contents of more than 17 wt.%, so the term does not refer to a specific basalt type. Rather, the variety of basalt should be specified, e.g. 'high-alumina basalt of calc-alkali type'.

high-aspect-ratio ignimbrite (HARI) An *ignimbrite sheet displaying a value for the ratio of its average thickness (*V*) to its horizontal extent (*H*) of 10^{-2} to 10^{-3}, where *H* is taken as the diameter of a circle with a surface area equal to that of the flow. *Compare* LOW-ASPECT-RATIO IGNIMBRITE.

high-field-strength elements (HFS) Elements of high *valency (greater than 2), e.g. Sn, W, and U, which are not readily incorporated into the *lattices of common rock-forming *silicate minerals. During crystallization of *igneous rocks they are generally incorporated into accessory phases (*see* ACCESSORY MINERAL), e.g. *zircon and *monazite, or else continually concentrated into residual *pegmatitic or *hydrothermal fluids, from which they may be precipitated in economic concentrations.

high-level waste *See* RADIOACTIVE WASTE.

high-pass filter In *remote sensing, a type of *spatial-frequency filter which selectively removes low *spatial-frequency data from an image and increases contrast between high spatial-frequency data. The result is that the lines which make up the edges of objects are more clearly defined.

High-pass filters are used to achieve *edge enhancement.

high-potassium basalt An *igneous rock of basaltic composition characterized by the appearance of *sanidine as well as *labradorite in the *groundmass, and by $K_2O/Na_2O > 1$. Such *basalts are commonly found well inland on the continental side of a *plate *subduction zone. The K_2O content of basalts in these above-subduction-zone environments increases inland as the depth to the subducted oceanic slab increases.

highstand A time during which sea levels are at their highest. *Compare* LOWSTAND.

highstand systems tract (HST) In the *genetic stratigraphic sequence model used in *sequence stratigraphy, a bounding surface formed by maximum marine flooding, where the sea level was high, and stable or falling slowly. *See* SYSTEMS TRACT. *Compare* LOWSTAND SYSTEMS TRACT and TRANSGRESSIVE SYSTEMS TRACT.

high symmetry *See* CRYSTAL SYMMETRY.

Hiller borer *See* PEAT-BORER.

Hiller peat-borer *See* PEAT-BORER.

hill fog General term denoting low cloud that covers high ground.

Himalayan orogenic belt An *orogenic belt interpreted as being the result of the convergence and collision of India with Asia. Because of variable uplift, great relief, high rates of *erosion and upthrusts of mid-crust, most depths within the *orogen are exposed, down to the mid-crust, and the belt is thought to illustrate the later stages of the *Wilson cycle. To the north of the Indus *suture the *Tibetan Plateau has *crust up to 80 km thick: a few authorities interpret this as being due to *A-subduction of the crust of the *Indo-Australian Plate after collision in the *Eocene, but most attribute the thickness to internal thrusting since the early *Mesozoic.

Himalia (Jupiter VI) One of the lesser satellites of *Jupiter, with a diameter of 170 km (±20 km).

hinge 1. The surface region of a *fold about a *hinge line, which occupies the area of maximum curvature. **2.** In *Brachiopoda

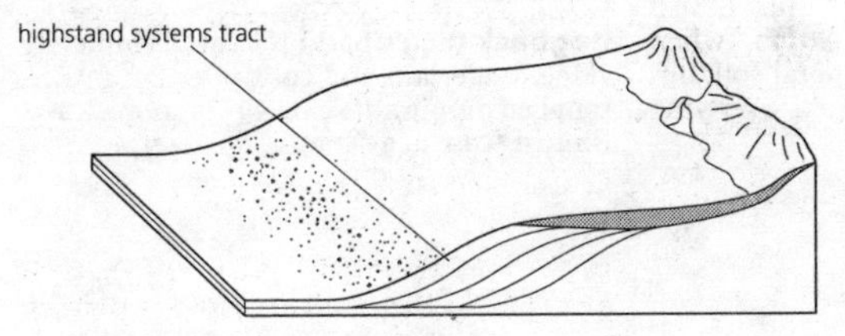

Highstand systems tract

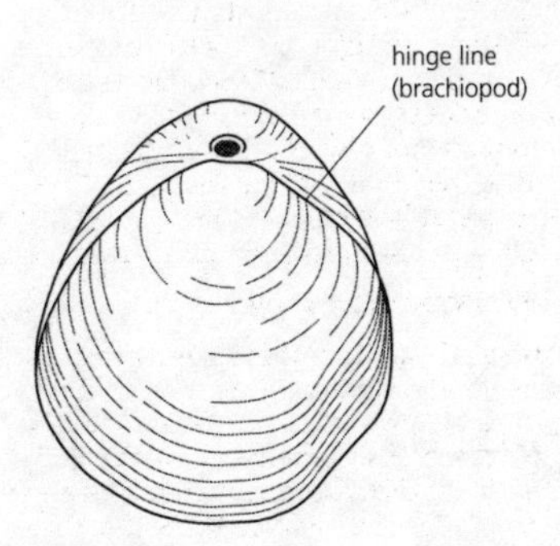

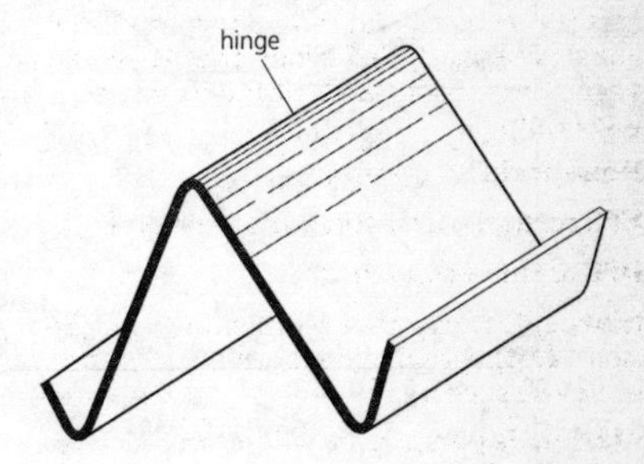

Hinge

and lamellibranchs, the area of the *commissure where the two valves of the shell are permanently in contact. In those articulate brachiopod (*Articulata) shells where the whole width of the posterior part of the commissure remains in contact during opening the shell is said to be 'strophic'. In those brachiopods where the hinge axis passes through the teeth and sockets, and these represent the fulcra, the shell is said to be 'non-strophic'.

hinge line A straight or curved line which joins the points of maximum curvature along the *hinge of a *fold.

hinterland *See* FORELAND.

Hirnantian A *stage of the *Ordovician in the Upper *Ashgill, underlain by the *Rawtheyan and overlain by the *Rhuddanian (*Silurian).

histic epipedon Surface *soil horizon, not less than one metre in depth, high in organic carbon, and saturated with water for some part of the year. *See also* HUMUS (2).

histogram A frequency graph in which the data range is divided into *bins.

Histosols An order of soils composed of organic materials. Histosols must have a thickness of more than 40 cm when overlying unconsolidated mineral soil, but may be of any thickness when overlying rock.

Hjulström effect The contrast between the flow velocity at which a fine-grained cohesive *sediment may be deposited and that at which it will be eroded. F. Hjulström published an important diagram which plots the relationship between grain size and velocity for *erosion, transportation, deposition, and settling velocity. One of the many important consequences of these relationships is that although fine-grained cohesive sediments (fine *silt and *clay) will be deposited only if flow velocities are very low, a very high velocity is required to erode the same sediment, once deposited. This is because of the cohesive nature of the sediment, which makes silt and clay more difficult to erode than pebbly sediment.

hlaup A burst of water through an icecap, triggered by volcanism.

HMS *See* DENSE MEDIUM SEPARATION.

hoar frost Deposit of patterned ('feathers', 'needles', 'spines', etc.) ice crystals on surfaces chilled by radiation cooling. The feature is seen particularly well on vegetation. The ice is derived from the *sublimation of water vapour on surfaces, as well as from frozen dew.

Hoboken *See* HENGELO.

hogback (hog's back) Narrow, symmetric ridge, underlain and controlled by a resistant bed dipping at some 40° or more. It is a limiting case of a *cuesta.

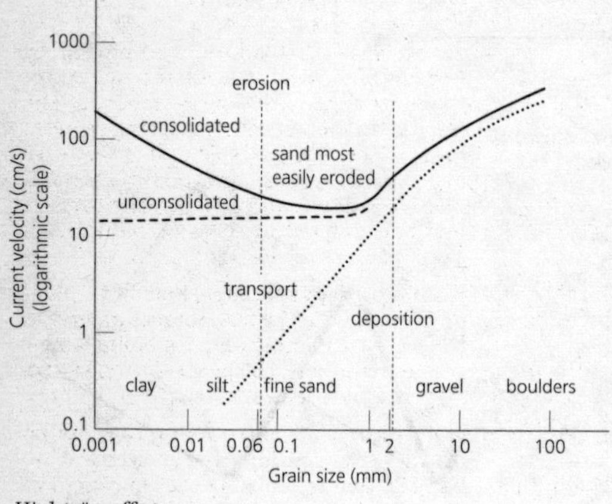

Hjulström effect

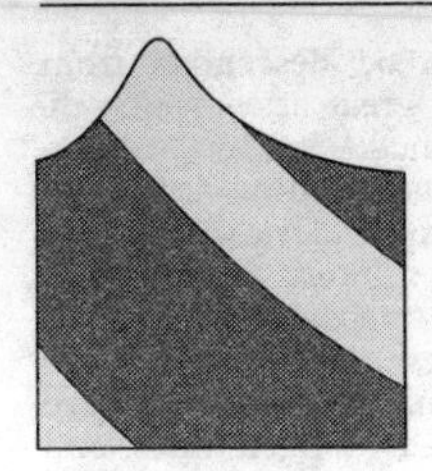

Hogback

hog's back *See* HOGBACK.

Holkerian A *stage of the *Visean Epoch, underlain by the *Arundian and overlain by the *Asbian.

Holmes, Arthur (1890–1965) A British geologist of the Universities of Durham and Edinburgh, Holmes did important work on *radiometric dating and developed the concept of convection currents in the *mantle, caused by heat from radioactive decay. This latter work persuaded him to become an early supporter of *continental drift theory. His *Principles of Physical Geology* (1944, 1965, 1993) has been an important textbook for more than 50 years.

holo- From the Greek *holos* meaning 'complete', a prefix meaning 'whole' or 'complete'.

Holocene *Epoch that covers the last 10 000 years. It is often referred to as Recent or Post-Glacial. *See also* FLANDRIAN.

holochroal *See* TRILOBITE EYE.

holocrystalline A *textural term referring to *igneous rocks or parts of igneous rocks which are composed entirely of crystals.

holohyaline A rarely used *textural term referring to *igneous rocks or parts of igneous rocks which are composed entirely of *glass.

hololeucocratic A rarely used term applied to an *igneous rock with a *colour index less than 5.

holometabolous Applied to a type of development in which distinct larval and adult forms occur.

holomictic Applied to lakes in which all of the water circulates, from the surface to the bed, when the lake cools in winter. *Compare* MEROMICTIC.

holophyletic Applied to a *taxon that includes all descendants of the common ancestor. The term is a special case of *monophyletic.

Holostei Group of marine and freshwater *bony fish including many fossil species, e.g. *Lepidotes* (*Triassic–*Cretaceous), and *Dapedius* (*Jurassic). Holosteans arose at the end of the *Permian from *palaeoniscid ancestors and are particularly abundant in marine Jurassic deposits. The main characteristics shown by the group relate to improvements in swimming and feeding: the use of the air sac to control buoyancy; a reduction in the bony fin rays, now unjointed; the development of a shorter and more mobile jaw; the gradual reduction in scale thickness; and the development of an almost symmetrical tail. Holosteans are represented today only by the garpike (*Lepisosteus*), and the bowfin (*Amia*), both freshwater.

holostomatous *See* APERTURE.

holostratotype The original *stratotype chosen and described to act as the standard reference section, or reference point within a section, when a stratigraphic unit or boundary is first established. *Compare* LECTOSTRATOTYPE; and NEOSTRATOTYPE. *See also* PARASTRATOTYPE; and HYPOSTRATOTYPE.

holosymmetric Within each *crystal system, applied to the class with the highest order of symmetry, where the *crystal has the same symmetry as the lattice.

holothuroid (sea cucumber) *See* HOLOTHUROIDEA.

Holothuroidea (holothuroids; sea cucumbers; subphylum *Echinozoa) Class of worm-like echinoderms which may be free-living or attached. The mouth, surrounded by variably branched tentacles, is at one end of the elongate body, the anus at the other. The calcitic skeleton is not rigid, being reduced to small sclerites (or *spicules) of very variable shape: hooks, anchors, rings, and plates. Fossil spicules of sea cucumbers are first encountered in rocks of *Ordovician age.

holotype (type specimen) Individual plant or animal chosen by taxonomists to serve as the basis for naming and describing a new species or variety. *Compare* LECTOTYPE; NEOTYPE; PARATYPE; and SYNTYPE.

Holsteinian A north European *interglacial period, dating from about 0.3 Ma to 0.25 Ma, that is probably the equivalent of the *Mindel/Riss Interglacial of the Alpine areas and may also be equivalent to the *Hoxnian of East Anglia.

Homalozoa ('carpoids'; phylum *Echinodermata) Extinct subphylum whose members have no trace of *radial symmetry. The *theca is depressed and asymmetrical. Homalozoa are known from Middle *Cambrian to Middle *Devonian rocks. There are three classes: Homostelea, Homoiostelea, and Stylophora. The affinities of the group have generated some controversy. The similarities between the larvae of some echinoderms and some chordates (*Chordata) has led to the suggestion that the phyla are in some way related. Most recently, the Stylophora have been reassigned by R. P. S. Jeffries on morphological grounds to a new subphylum of primitive chordates, the Calcichordata. The matter remains unresolved.

homeomorph An organism which, as a result of *convergent evolution, comes to resemble another to which it may not be closely related.

homeotherm *See* HOMOIOTHERM.

Homerian A *stage of the Middle *Silurian, underlain by the *Sheinwoodian and overlain by the *Gorstian.

Hominidae (order Primates, suborder *Anthropoidea, superfamily *Hominoidea) The mammalian family that includes humans and immediately ancestral forms now extinct. Its members are distinguished from the apes (Pongidae) (*a*) by the possession of a much larger brain, in which the frontal and occipital lobes are especially well developed, allowing more complex behaviour including communication by speech; (*b*) by a fully erect posture facilitated by the positioning of the *foramen magnum beneath the skull so that the head is held upright; (*c*) by a bipedal gait; and (*d*) by the slow rate of post-natal growth and development, which favours complex social organization and the emergence of distinct cultures. The family includes the genera **Paranthropus*, *Australopithecus* (but *see* *AUSTRALOPITHECINES), and **Homo*. Whether a possible ancestral genus **Ramapithecus* should also be included in this family, or with the apes, is a matter for debate.

Hominoidea (order Primates, suborder *Anthropoidea) Superfamily comprising the Hylobatidae (gibbons), Pongidae (great apes), and *Hominidae (humans and their immediate ancestors). The latter two families are believed to be descended from a common stock of 'great apes' which diverged to form distinct Asian and African lines, the African line dividing again 4–6 million years ago into the African apes and the hominids. The hominoids lack tails and cheek pouches; have opposable thumbs (reduced in some species); and differ from the Cercopithecidae in having less-specialized dentition, larger heads, longer limbs, and wider chests which some authorities believe they inherited from ancestral brachiating forms. Today only the Hylobatidae (gibbons) are specialized brachiators (i.e. swing from branches hand over hand).

homo- From the Greek *homos* meaning 'the same', a prefix meaning 'the same' or 'alike'.

homocercal tail *See* HETEROCERCAL TAIL.

homodont *See* HETERODONT.

homogeneous accretion A model for the accretion of planetary bodies from the primitive solar nebula (PSN) in which the rate of accretion is fast relative to the rate at which the PSN cools. The consequence is that each body forms very quickly and consists entirely of material that was in equilibrium with the physical conditions of the PSN over only a very short period of time. According to this model, the layered structures of the planets are entirely of secondary origin. *Compare* HETEROGENEOUS ACCRETION.

homogeneous non-rotational strain *See* PURE SHEAR.

homogeneous nucleation Spontaneous condensation or freezing of water in the atmosphere in the absence of substances to act as nuclei. This is most likely in supercooled air below −40 °C.

homogeneous rotational strain *See* SIMPLE SHEAR.

homogeneous strain The *strain, distributed evenly throughout a body, which is produced when straight lines and parallel lines in an undeformed body remain straight and parallel as the body is deformed. There are three types of homogeneous strain, varying smoothly from one to the next: axially symmetrical extension; axially symmetrical shortening; and plane strain. *See* PURE SHEAR; and SIMPLE SHEAR.

homogenization temperature 1. Temperature at which an *exsolution pair of minerals, e.g. *chalcopyrite–*sphalerite, homogenize to a single crystalline phase. **2.** Temperature at which the liquid and vapour phase in a *fluid inclusion homogenize to monophase liquid or vapour. This gives an indication of the minimum trapping temperature of the fluid inclusion. *See also* GEOTHERMOMETER.

homoiotherm (homeotherm) An organism whose body temperature varies only within narrow limits. It may be regulated by internal mechanisms (i.e. in an endotherm) or by behavioural means (i.e. in an ectotherm), or by some combination of both (e.g. in humans, who light fires and wear thick clothes to keep warm in cold weather and wear light clothing to keep cool in warm weather, but who are also endothermic).

homologous Applied to similar structures in different animals when both are thought to have the same evolutionary origin, although their functions may differ widely, e.g. forelimbs, wings, and flippers in *vertebrates.

homoplasy In the course of *evolution, the appearance of similar structures in different lineages (i.e. not by inheritance from a common ancestor). The term includes *convergent evolution and *parallel evolution. *See also* REVERSAL.

homopycnal flow At a river mouth, a flow of river water that is of the same density as the water in the basin receiving it. This results in intense local mixing of waters, with considerable sedimentation, and is typical of *Gilbert-type deltas. *Compare* HYPERPYCNAL FLOW and HYPOPYCNAL FLOW.

homosphere Atmospheric layer from the Earth's surface to approximately 80 km altitude, where the relative proportions of the various gaseous constituents, excluding water vapour, remain almost constant. *See also* HETEROSPHERE.

homospory *See* SPORE.

homotaxis Literally, 'the same arrangement' (from the Greek *homos* and *taxis*). The term was proposed by T. H. Huxley (1825–95), in an address to the Geological Society of London, to describe *strata from different areas that contain similar lithologic or *fossil successions but are not necessarily of the same age.

Honda–Mrkos–Pajdusakova A *comet with an orbital period of 5.29 years; *perihelion date 17 January 1996; perihelion distance 0.581 AU.

Hooke, Robert (1635–1703) An experimental philosopher, Hooke investigated a very wide range of topics and published extensively. He is responsible for *Hooke's law, which laid the foundations for the scientific study of elasticity. In the field of Earth sciences, he studied gravity and planetary movements, earthquakes, geomagnetism, and the nature of fossils and crystallography.

Hooke's law The law, named after Robert *Hooke, which describes the behaviour of perfectly elastic materials in terms of a straight-line relationship between *stress and *strain in such materials. Strain is directly proportional to the applied stress provided the medium remains elastic (i.e. the *elastic limit is not exceeded); stress is equal to the strain multiplied by a constant of elasticity (*Young's modulus). The law also states that stress divided by strain is a constant. *See also* ELASTIC CONSTANTS.

'hopper' crystals Cubic crystals of *halite or *pseudomorphed (replaced) halite, characterized by sunken depressions on the crystal faces.

horizon 1. An *informal term used in *stratigraphy to denote a plane within a body of *strata. This may be at a boundary of lithological change, or commonly the term may refer to a thin, distinctive *bed within a lithological unit. *See also* BIOHORIZON. **2.** An interface separating two media with different geophysical properties. **3.** In soil, a horizontal layer that can be distinguished from the layers below and (except for the surface layer) above it. Identified by a coding system using a capital letter, sometimes followed by a subscript, such layers are used to diagnose soil types. *See* SOIL HORIZON.

horizontal drilling The drilling of an oil or natural gas well at an angle to the vertical, so the well runs parallel to the formation containing the oil or gas. Production from the resulting well, known as a horizontal hole, is often three to five times greater than that from a vertically drilled well.

horizontal hole *See* HORIZONTAL DRILLING.

horizontal stack *See* COMMON-DEPTH-POINT STACK.

hornblende An important rock-forming *silicate mineral and member of the *amphiboles, particularly the calcium-rich *monoclinic amphiboles; it is the general name given to a group of amphiboles (the hornblende series) with similar properties, which includes hastingsite $Ca_2(Mg_4Al)[Si_7AlO_{22}](OH,F)_2$, tschermakite $Ca_2(Mg_3AL_2)[Si_6Al_2O_{22}](OH,F)_2$, edenite $NaCa_2Mg_5[Si_7AlO_{22}](OH,F)_2$, and pargasite $NaCa_2(Mg_4Al)[Si_6Al_2O_{22}](OH,F)_2$; sp. gr. 3.0–3.5; *hardness 5.0–6.0; black or greenish-black; white *streak with a greenish tint; *vitreous *lustre; *crystals normally *prismatic, *columnar, and occasionally isometric; *cleavages in two directions set at 124°, prismatic {110}; occurs as a common constituent of medium *basic, *igneous rocks (e.g. *syenites, *diorites, and *granodiorites), also in contact *metasomatic zones, and extensively in *metamorphic rocks (e.g. *schists, *gneisses, and *amphibolites).

hornblende–hornfels facies A set of metamorphic mineral assemblages produced by *contact metamorphism of a wide range of starting rock types under the same metamorphic conditions and typically characterized by the development of the mineral assemblage *hornblende–*plagioclase in rocks of *basic *igneous composition. Other rocks of contrasting composition, e.g. *shales and *limestones, would each develop their own specific mineral assemblage, even though they are all being metamorphosed under the same conditions. The variation of mineral assemblage with starting rock composition reflects a particular range of pressure, temperature, and $P(H_2O)$ conditions. Experimental studies of mineral P–T stability fields indicate that the facies represents the low-pressure, moderate-temperature conditions met close to *igneous *intrusions.

hornblendite An *ultrabasic *igneous rock composed of more than 90% *hornblende. Where other *ferromagnesium minerals constitute 10–50% of the rock, the rock can be named according to the other types of ferromagnesium minerals, for example, pyroxene hornblendite, olivine–pyroxene hornblendite.

hornfels (pl. hornfelses) A massive, fine-grained, *granoblastic contact metamorphic rock (*see* CONTACT METAMORPHISM), commonly displaying *conchoidal fracture and splintery debris. Heat from an *igneous *intrusion initiates *recrystallization of surrounding rocks, thereby producing a contact metamorphic hornfels which is usually finer-grained than the *country rocks. Hornfelses formed by recrystallization of *shales, and occurring away from the contact with the igneous body, quite often develop prominent *graphite-rich spots formed by the aggregation of organic material in the original shales.

hornitos Small cones or chimneys of *lava *spatter, often several metres high, found on the surface of *pahoehoe lava flows. Lava spatter is thrown out from the flow interior by the explosive release of primary magmatic gas or by the explosive conversion to steam of trapped pockets of *groundwater beneath the flow.

horotely (adj. horotelic) Normal or average rate of *evolution per million years, of genera within a given *taxonomic group. Thus slowly or rapidly evolving lines may be horotelic during certain episodes in their history.

horse A lenticular or sigmoidal mass of rock which is completely bounded by two or more *thrust faults which rejoin along the *strike and up-*dip. The term may also be used for the analogous structure in strike-slip terrains (*see* STRIKE-SLIP FAULT).

horse latitudes Subtropical latitudes coinciding with a major anticyclonic belt; they are characterized by generally settled weather and light or moderate winds. When sailing ships carrying cargoes of horses were becalmed in these latitudes, horses would sometimes be thrown overboard, mainly to reduce the demand for drinking water.

horsetails *See* SPHENOPSIDA.

horst Upthrown block lying between two steep-angled *fault blocks. *Compare* GRABEN. *See also* RIFT.

Horton analysis *See* DRAINAGE MORPHOMETRY.

Hortonian flow *See* SURFACE RUNOFF.

hot dry rock (HDR) Rocks, usually *granite, which have abnormally high heat production as a result of the decay of *radiogenic elements rather than merely residual heat. Potentially these are a source of geothermal energy. One method of exploiting the heat generated is to fracture the rocks at depth using small, downhole, explosive charges, and then initiate a water circulation system from the surface. When cold water is pumped down it returns considerably warmer, and this energy can be extracted by heat exchangers. *See also* GEOTHERMAL FIELD; and GEOTHERMAL GRADIENT.

hot spot An area of high volcanic activity. Some hot spots, e.g. Iceland, are located on *constructive margins. Others occur within lithospheric *plates, often lying at the end of a chain of progressively older volcanoes, e.g. the *Hawaiian-Emperor Chain. The hot spot is thought to be stationary, or nearly so, and to produce volcanoes intermittently as the plate moves over it. It has been suggested that *mantle plumes lie beneath hot spots.

hot spring A continuous flow of hot water through a small opening on to the Earth's surface. The water is usually *groundwater heated at depth by hot rocks and recycled to the surface by convection. Hot spring waters are rich in dissolved minerals which are often precipitated around the spring mouth.

hot working (annealing, polygonization) A *strain-recovery process in which new, unstrained, polygonal grains develop from and replace highly strained grains at high temperatures.

Howard, Luke (1772–1864) An English meteorologist who was the first person to devise a successful classification system for clouds, which he published in 1803 as a paper, 'On the Modifications of Clouds'. He defined four main and several secondary cloud types, calling the main types 'stratus', 'cumulus', 'cirrus', and 'nimbus'. He became very well known and his classification system forms the basis of the one used today. Howard was trained and earned his living as an apothecary. He lived in London.

Hoxnian 1. An *interglacial period. **2.** A series of temperate-climate deposits, named after Hoxne, Suffolk, England, with a characteristic vegetational sequence that occurs in *tills of the earliest glacial *stages, sometimes filling deep channels. They may be equivalent to the *Holsteinian deposits of continental Europe. Sometimes during the Hoxnian Interglacial the sea rose to well above its present level. There is a correlation between this stage and the Boyn Hill terraces of the Thames Valley, and also with *raised beaches (30 m ordnance datum) found on the Sussex coast.

HREE Abbreviation for 'heavy *rare earth element'. *See* MID-OCEAN-RIDGE BASALT.

HSR *See* HIGHSTAND SYSTEMS TRACT.

Hubble constant *See* HUBBLE PARAMETER.

Hubble parameter (H_0) A measure of the rate at which the recessional speed of galaxies varies with distance, calculated from *Hubble's law; H_0 = 50–100 km/s/Mpc (megaparsec). This was formerly known as the Hubble constant.

Hubble's law As formulated in 1936 by the American astronomer Edwin Powell Hubble (1889–1953) and his colleague Milton Lasell Humason (1891–1972), the distance to a galaxy in light-years is equal to one-hundredth of its red-shift velocity in miles per second.

Hudsonian orogeny A *Proterozoic phase of mountain building, which ended about 1750–1800 Ma ago, and that affected the shield area in what is now Canada. It was preceded by the *Kenoran and succeeded by the *Grenvillian orogenies.

hue A measure of the relative amounts of the *additive primary colours which contribute to the colour of an object. *See* MUNSELL COLOUR.

Hugoniot The relation between pressure and density within the *Earth as derived from *seismic velocities.

Humberian orogeny *See* TACONIC OROGENY.

Humboldt, Friedrich Heinrich Alexander von (1769–1859) A Prussian naturalist and physicist, and great explorer, Humboldt made contributions to the study of volcanoes, vegetation and its relation to climate, ocean currents, and mountain ranges. He made a particular study of *geophysics, aiming to collect and correlate all relevant data, and played an important role in the international geomagnetic survey of 1834, which determined declinations and inclinations world-wide.

Humboldt current *See* PERU CURRENT.

humerus In *tetrapods, the upper bone of the forelimb.

humic acid Mixture of dark-brown organic substances, which can be extracted from soil with dilute alkali and precipitated by acidification to pH 1–2 (in contrast with fulvic acid, which remains soluble in acid solution).

humic coal *See* COAL.

humidity Expression of the moisture content of the atmosphere. Measures of humidity include statements of the total mass of water in 1 m^3 of air (absolute humidity), the mass of water vapour in a given mass of air (specific humidity), *relative humidity, vapour pressure, and the *mixing ratio.

humification The sequence of reactions by which decaying organic material is converted to *humus.

humilis From the Latin *humilis* meaning 'low', a species of shallow *cumulus cloud which typically has a flattened appearance. *See also* CLOUD CLASSIFICATION.

humite *See* CHONDRODITE.

hummocky cross-bedding A form of cross-bedding (*see* CROSS-STRATIFICATION) characterized by cross-laminations which have both concave and convex-upwards forms. The cross sets cut across one another with concave and convex-upwards surfaces, and the cross-beds have an external form of convex-upwards hummocks with wavelengths of about 1–5 m. The structure has been recognized only in ancient sediments and is thought to be the product of storm waves.

hummocky moraine A strongly undulating surface of ground *moraine, with a relative relief of up to 100 m, and showing steep slopes and deep, enclosed depressions. It results from the downwasting (i.e. thinning) of ice which is usually stagnant. Blocks of ice may squeeze debris released from the ice into *crevasses between the blocks.

humus 1. Decomposed organic matter in soils that are *aerobic for part of the year. It is dark brown and amorphous, having lost all trace of the structure and composition of the vegetable and animal matter from which it was derived. **2.** Surface organic *soil horizon that may be divided into types: *mor (acid and layered) or *mull (alkaline and decomposed). It is now known as a '*histic epipedon'.

Huronian 1. A *system of the *Proterozoic, extending from about 2475 to 2225 Ma ago (Harland et al., 1989) and containing three glacial cycles. **2.** A Lower–Middle Proterozoic *stage, extending from about 2600 to 1500 Ma ago (Van Eysinga, 1975).

hurricane 1. The name given to a *tropical cyclone that develops over the N. Atlantic and Caribbean. Hurricanes move westward, then swing north, on tracks that often carry them across inhabited islands and coastal areas of Mexico and the USA. **2.** A wind blowing at more than 120 km/h (75 mph), which is Force 12 on the *Beaufort scale.

Hutton, James (1726–97) Scottish natural philosopher, prominent advocate of Plutonist and uniformitarian theories, who published *The Theory of the Earth* (1795). His ideas were based on the study of crystalline rocks, but he paid little attention to stratigraphy and organic remains. He recognized that unconformities implied that earth movements must have occurred. His work was popularized by *Playfair. *See* PLUTONISM; and UNIFORMITARIANISM.

Huygenian eyepiece *See* EYEPIECE.

Huygens A mission by *NASA and *ESA, carried by *Cassini and launched in 1997, to *Titan.

Huygen's principle Each point of an advancing wave-front can be thought of as a new source of secondary wavelets, so that the envelope tangent to all these wavelets forms a new wave-front.

Hyaenidae *See* CARNIVORA.

Hyakutake A *comet with an orbital period of more than 65 000 years; *perihelion date 1 May 1996; perihelion distance 0.230 AU.

hyaline Translucent or transparent; glass-like.

hyaloclastite An aggregate of fine, glassy debris formed by the sudden contact of hot, coherent *magma and either cold water or water-saturated sediment. Rapid heat loss from the magma to the cold water sets up tensile thermal stress in the magma carapace as it cools, chills, and contracts, causing the glassy, chilled zone to fragment and form a quench-fragmented debris. Thick hyaloclastite deposits form over *basalt flows when they erupt beneath the sea or enter the sea. If the deposit remains in contact with water after its formation, the glassy debris can easily be hydrated to form *palagonite.

hyalopilitic A *texture found in *extrusive *igneous rocks and characterized by a felt of crystals set in a background of *glass. The felts of crystals represent an arrested stage of early crystal growth from a liquid now represented by the background glass.

Hyalospongea *See* HEXACTINELLIDA.

hydration Chemical combination of water with another substance, e.g. the addition of water to a mineral (such as *anhydrite) to produce a hydrous phase (in this case, *gypsum). This usually involves expansion. Hydration may be important in the mechanical *weathering of rocks to produce *clays and economically important minerals such as *kaolin, *talc, and *goethite.

hydraulic boundary Within a *groundwater system, the interface between regions of different hydraulic characteristics such as *porosity, storativity (*see* STORAGE COEFFICIENT), conductivity (*see* PERMEABILITY), or *transmissivity. For example, when modelling groundwater systems or carrying out pumping tests, workers must make judgements regarding the homogeneity of those systems. Distinct and marked changes in the hydraulic characteristics (e.g. where an *aquifer abuts an *aquiclude) may require marking as hydraulic boundaries.

hydraulic conductivity *See* PERMEABILITY.

hydraulic corer (piston corer) A sampling tube that penetrates marine or lake sediments by hydraulic methods, instead of being either a *gravity corer or a *drilling corer. One version is the *Mackereth corer.

hydraulic equivalent *Grains of different density will settle through a fluid at different rates. The concept of hydraulic equivalence is used to relate the size of a mineral grain to the size of a *quartz grain with the same settling velocity. Thus a *magnetite grain (sp. gr. 5.18) of 0.2 mm diameter is the hydraulic equivalent grain size to a 0.5 mm diameter *quartz particle (sp. gr. 2.65).

hydraulic fracture An extensional fracture and/or crack-seal vein produced by the expulsion of fluid from a *sediment that is subjected to a rapid increase in *load pressure and therefore high fluid pressure.

hydraulic fracturing *See* HYDROFRACTURING.

hydraulic geometry A description of the adjustments made by a stream in response to changes in discharge, both at a cross-section and in the downstream direction. Adjustments are made in width, mean depth, mean velocity, slope, frictional resistance, suspended-sediment load, and water-surface gradient. The relationship between discharge and adjustment is expressed as a power function: $y = aQ^b$, where y is the adjusting variable, Q is discharge, and a and b are coefficients.

hydraulic gradient A measure of the change in *groundwater head over a given distance. Maximum flow will normally be in the direction of the maximum fall in head per unit of horizontal distance, i.e. in the direction of the maximum hydraulic gradient. *See* HYDRAULIC HEAD.

hydraulic head In general, the elevation of a water body above a particular datum level. Specifically, the energy possessed by a unit weight of water at any particular point, and measured by the level of water in a *manometer at the laboratory scale, or by water level in a well, *borehole, or *piezometer in the field. The hydraulic head consists of three parts: the elevation head (*see* ELEVATION POTENTIAL ENERGY), defined with reference to a standard level or datum; the *pressure head, defined with reference to atmospheric pressure; and the velocity head. Water invariably flows from points of larger hydraulic head to points of lower head, down the *hydraulic gradient.

hydraulic radius Ratio between the cross-sectional area of a stream channel and the length of the water-channel contact at that cross-section (the wetted perimeter). It is a measure of channel efficiency: the higher the ratio, the more efficient is the channel in transmitting water.

hydric *See* MESIC.

hydrocarbon Naturally occurring organic compound containing carbon and hydrogen. Hydrocarbons may be gaseous, solid, or liquid, and include *natural gas, *bitumen, and *petroleum.

hydrochemistry The study and representation of the chemical composition of waters, normally those of natural occurrence.

hydroclast A *clast produced by a reaction between *magma and water. *Compare* ALLOCLAST; AUTOCLAST; and EPICLAST.

hydrofracturing (hydraulic fracturing) Process of breaking up rocks under pressure by introducing water or other fluids; usually done to increase *permeability in oil, gas, and geothermal reservoirs. The pressure opens joints, cracks, and bedding planes which can be kept open by the introduction of sand, glass beads, or aluminium balls. Hydrofracturing may occur naturally as a result of internal hydraulic overpressures, e.g. in the formation of porphyry deposits. *See* GEOTHERMAL FIELD; and PORPHYRY COPPER.

hydrogen 'burning' A thermonuclear process that produces energy by the combination of hydrogen nucleii, and referred to in stellar evolution as the 'main-stage sequence'. The stage ends when the inner core of hydrogen is exhausted, causing the contraction of the inner part of the star and the expansion of the outer part to produce a red giant. *See* CARBON 'BURNING'; HELIUM 'BURNING'; SILICON 'BURNING'; NUCLEOSYNTHESIS; and THERMONUCLEAR REACTIONS.

hydrogeologic map A geologic map that is specially prepared to emphasize features of hydrogeologic importance. For example, rocks are shown not just according to their age or lithology but also as *aquifers or *aquicludes, and details may be included of *groundwater levels, *springs, and water sources. Unlike many geologic maps, they provide a high degree of interpretation for the user.

hydrogeology The scientific study of the occurrence and flow of *groundwater and its effects on earth materials.

hydrograph A graph showing the plot of water flow in a water course, or the elevation of *groundwater in a *borehole above a particular datum point, against time. The 'unit hydrograph' is the name given to a method of calculation which allows rainfall to be converted to stream flow and so facilitates the prediction of how particular river basins will respond to changing precipitation patterns. The discharge hydrograph shows the flow rate of water against time for a discharging water body. The stage hydrograph shows water level against time.

hydroids *See* HYDROZOA.

hydrologic cycle Representation of the flow of water in various states through the terrestrial and atmospheric environments. Storage points (stages) involve *groundwater and surface water, ice-caps, oceans, and the atmosphere. Exchanges between stages involve evaporation and transpiration from the Earth's surface, condensation to form clouds, and precipitation followed by runoff. *See also* RESIDENCE TIME.

hydrologic modelling (hydrologic simulation) The use of small-scale physical models, mathematical analogues, and computer simulations to characterize the likely behaviour of real hydrologic features and systems.

hydrologic network An integrated array of meteorological, *groundwater-level, and stream-flow measuring stations which in combination give a complete measurement of the *hydrologic cycle for a particular area. In modern practice the various measuring and *gauging stations are sometimes linked by telemetry to a central monitoring unit for use in flood forecasting. *See* FLOOD PREDICTION.

hydrologic regions The smaller units, with fixed boundaries, into which large tracts of country are divided for the purpose of collecting hydrologic data. The establishment of hydrologic regions allows data to be collected on the same basis from year to year and so facilitates historical analysis. Well-identified regions have a common climate and geologic and topographical structure, so that the *hydrologic cycle operates fairly uniformly within each region.

hydrologic simulation *See* HYDROLOGIC MODELLING.

hydrology The study of the *hydrologic (water) cycle. While it involves aspects of geology, oceanography, and meteorology, it emphasizes the study of bodies of surface water on land and how they change with time. *See* HYDROGEOLOGY.

Hydrology and Water Resources Programme A project by the *World Meteorological Organization to promote international collaboration in the evaluation of water resources and the development of hydrological networks and services.

hydrolysate Sediment consisting of undecomposed, finely ground rock and insoluble material derived from weathered primary rocks; typical of *clays, *shales, and *bauxites, with Al, Si, K, and Na as major components.

hydrolysis 1. Reaction between a substance and water in which the substance is split into two or more products. At the points of cleavage the products react with the hydrogen or hydroxyl ions derived from water. **2.** Process of enriching the soil *adsorption complex with hydrogen after *exchangeable metallic ions have been replaced by hydrogen ions. *Compare* WEATHERING.

hydromuscovite *See* ILLITE.

hydrophone A microphone used to detect acoustic (including seismic) waves under water. A number of hydrophones are linked together to form a *streamer.

hydrosphere The whole of that body of water which exists on or close to the surface of the Earth. The hydrosphere formed as the Earth cooled and atmospheric water condensed.

hydrospire *See* BLASTOIDEA.

hydrostatic stress The component of *confining pressure derived from the weight of pore water in the column of rock above a specified level. All *principal stresses are equal and changes in hydrostatic pressure produce changes only in the volume and density of the material. It can be simulated experimentally by enclosing material in a jacket and pumping in liquids to produce equal pressures throughout.

hydrothermal activity Any process associated with *igneous activity involving the action of very hot waters. The waters involved can be derived directly from an igneous *intrusion (i.e. *juvenile water) as a residual fluid formed during the late stages of crystallization of the body, or can be external *groundwater heated during crystallization of the intrusion. The hydrothermal fluids can react with and alter the rocks through which they pass, or can deposit minerals from solution. Hydrothermal reactions include *serpentinization, chloritization (*see* CHLORITE), saussuritization (*see* SAUSSURITE), uralitization (*see* URALITE), and *propylitization; whilst hydrothermal vein and replacement mineral deposits include Cu, Pb, and Zn sulphides. Hydrothermal activity should not be confused with geothermal activity which involves the convection and movement of hot waters but is not necessarily connected with an igneous intrusion (*see* GEOTHERMAL FIELD; and GEOTHERMAL GRADIENT).

hydrothermal mineral A mineral formed by precipitation from a very hot hydrothermal fluid (*see* HYDROTHERMAL ACTIVITY) as it passes down a temperature or pressure gradient. Common hydrothermal minerals precipitated in *veins and cavities are *quartz, *fluorite, *galena, and *sphalerite.

hydrothermal vent A place on the ocean floor, on or adjacent to a mid-ocean *ridge, from which there issues water that has been heated by contact with molten rock, commonly to about 300 °C. The vent water often contains dissolved sulphides. These are oxidized by chemosynthetic bacteria, which fix carbon dioxide and synthesize organic compounds. Near the vents, at temperatures up to 40 °C, there are highly productive communities comprising animals that utilize the organic compounds or live symbiotically with the chemosynthetic bacteria; these organisms support carnivores and detritivores. These communities include beard worms (phylum *Pogonophora) that completely lack a digestive tract, *Munidopsis* crabs (superfamily Galatheoidea), giant clams (e.g. *Calyptogena magnifica*), mussels, acorn worms (class Enteropneusta), and many more. Vent fluids containing high concentrations of iron, manganese, and copper tend to be hot (about 350 °C) and black. They are known as 'black smokers'. 'White smokers' flow more slowly, are cooler, and contain high concentrations of arsenic and zinc. *See* HYDROTHERMAL ACTIVITY, and HYDROTHERMAL MINERAL. *See also* COLD SEEP.

hydrovolcanic processes A sequence of events initiated by the interaction of a body of *magma and water external to the magma system. For example, when a rising body of magma encounters external *groundwater it converts the groundwater to steam which expands rapidly, fragmenting the surrounding rocks and magma to generate a phreatomagmatic eruption (*see also* PHREATIC ACTIVITY). When lava is erupted under water, the water removes heat rapidly from the outer margin of the flow, which is chilled to a glass, and fractured and fragmented during contraction. The fragmented glassy margin of the flow forms a *hyaloclastite envelope around the flow.

hydroxides Applied to *minerals whose chemical composition includes the OH radical, and often to those with water molecules, whose presence gives hydrated oxides. Among the important hydroxide minerals are: *brucite, $Mg(OH)_2$; *gibbsite, $Al(OH)_3$; *diaspore, α-AlO(OH); *boehmite, γ-AlO(OH); *limonite, $FeO(OH).nH_2O$; and *goethite, α-FeO(OH).

hydroxyapatite A hydrated calcium phosphate mineral, which also contains fluoride, chloride, and carbonate calcium salts. It is often formed as a consequence of *biomineralization, producing hard structures, such as *bone.

Hydrozoa (hydroids; phylum *Cnidaria) Class of multicellular, mainly marine animals with cells arranged in two layers, the epidermis and the gastrodermis (endodermis), separated by a gelatinous mesogloea. These enclose a continuous digestive cavity (coelenteron), which communicates directly with the exterior by a single aperture (mouth) and is lined by a gastrodermis. The gastrodermis lacks nematocysts. Hydrozoa are Lower *Cambrian to Recent. *See also* MILLEPORINA; and STYLASTERINA.

hygromagmatophile elements *See* INCOMPATIBLE ELEMENTS.

hygrometer Instrument for measuring atmospheric humidity. Types include the wet-bulb–dry-bulb, dew-point, and *hair hygrometers, and there is one type based on electrical resistance. *See also* PSYCHROMETER.

hygroscopic nucleus Microscopic particle (e.g. of sulphur dioxide, salt, dust, or smoke) in the free air, on which water vapour may condense to form droplets. *Aerosols that are soluble in water (e.g. salt or sulphuric acid) can induce condensation in unsaturated air, e.g. salt nuclei can induce it at a *relative humidity of less than 80%. The size of nuclei may be from 0.001 μm to more than 10 μm (i.e. 'giant' nuclei such as particles of sea-salt). *See also* AITKEN NUCLEI; and NUCLEUS.

hygroscopic water Water absorbed from the atmosphere and held very tightly by the soil particles, so that it is unavailable to plants in amounts sufficient for them to survive. *Compare* CAPILLARY MOISTURE.

hygrothermograph (thermo-hygrograph) An instrument used for the continuous recording of both the temperature and humidity of the air, on separate traces.

Hylonomus lyelli Ancient, probably the oldest-known, *reptile. *Hylonomus* is a stem reptile and a member of the *Captorhinomorpha. It was first found in fossilized tree stumps in the Coal Measures of Nova Scotia. The skull roof of this species was fully ossified and there were no openings behind the eye sockets. *Hylonomus* measured some 25 cm and possessed a long tail.

hyoid A bone or bones developed from the second visceral arch: it or they support the tongue.

Hyolithida (phylum *Mollusca, class *Calyptoptomatida) Order of pyramid-shaped calyptoptomatids which have conical embryonic chambers. The *operculum possesses one or two pairs of *muscle scars which are bilaterally symmetrical. The shell is externally undifferentiated, and internally it is non-septate. The order ranges in age from Lower *Cambrian to Middle *Permian. The main genus, *Hyolithes*, had worldwide distribution during the Cambrian, with over 300 species.

hyp- *See* HYPO-.

hypabyssal Applied to medium-grained, *intrusive *igneous rocks which have crystallized at shallow depth below the Earth's surface. There is, however, no sharply defined or agreed depth limit to the term 'shallow'.

hyper- From the Greek *huper* meaning 'beyond' or 'over', a prefix meaning 'exceeding' or 'greater than normal'.

Hyperion (Saturn VII) One of the major satellites of *Saturn, with a radius measuring 185 × 140 × 113 km; visual albedo 0.19–0.25. It was discovered in 1848 by W. Bond.

hypermorphosis Acceleration of development so that the organism reaches its adult size and form well before the attainment of sexual maturity, and continues to develop into a 'super-adult'.

hyperpycnal flow At a river mouth, a flow of river water that is denser than the water in the basin receiving it. This occurs during floods. The denser water flows beneath the basin water, as a density current, carrying sediment beyond the shore and inhibiting the progradation of a delta. *Compare* HOMOPYCNAL FLOW; and HYPOPYCNAL FLOW.

hypersolvus granite A *granite which has crystallized above the solvus temperature and hence contains only one *alkali feldspar type. The solvus temperature is that temperature below which two alkali feldspars of contrasting composition are in equilibrium. Granites which crystallize below the solvus temperature (subsolvus granites) immediately crystallize two alkali feldspar types, one K-rich, the other Na-rich. The feldspars from either hyper- or subsolvus granites may suffer sub-*solidus *exsolution if cooling is slow enough. This is an equilibrium process which takes place when the rock is entirely solid and involves splitting of the feldspars into K- and Na-rich lamellae by internal *diffusion of *ions, to give perthitic (*see* PERTHITE) or antiperthitic feldspars.

hypersolvus syenite *See* SYENITE.

hypersthene A member of the *orthorhombic *pyroxenes with approximately equal amounts of Fe and Mg in its composition $MgFeSi_2O_6$, it is a *mineral occurring within the series *enstatite to (ortho)*ferrosilite, although the name hypersthene has fallen into disuse (with the more general term orthopyroxene followed by the exact composition being preferred); sp. gr. 3.5; *hardness 5–6; *orthorhombic; green to greenish, or brownish-black; vitreous *lustre; *crystals can be *prismatic or *tabular, but usually irregular grains; *cleavage good, prismatic {110}; occuss in iron-rich, *bafic, *igneous rocks, e.g. noritic *gabbro, *trachytes, and *andesites.

hyperthermic *See* PERGELIC.

hyperthermophile An *extremophile (domain *Archaea) that thrives in environments where the temperature is extremely high, in some cases preferring a temperature of about 105 °C, tolerating 113 °C, and failing to multiply below 90 °C. *Compare* THERMOPHILE.

hypichnia *See* TRACE FOSSIL.

hypidiomorphic fabric *See* HYPIDIOTOPIC FABRIC.

hypidiotopic fabric A rock *texture characterized by the presence of minerals some of which show their *crystal form (i.e. some *euhedral habits). 'Hypidiotopic fabric' refers to *sedimentary rocks; for *igneous or *metamorphic rocks, 'hypidiomorphic fabric' is used.

hypo- (hyp-) From the Greek *hupo* meaning 'under', a prefix meaning 'below', 'slightly', or 'lower than normal'. 'Hypo-' is generally used before words beginning with a consonant, 'hyp-' before words beginning with a vowel.

hypocentre (focus) The actual location, usually within the Earth, of the first motion of an *earthquake, i.e. the location of the focus of the earthquake. The point at the Earth's surface overlying the hypocentre is the *epicentre.

hypocrystalline A *textural term referring to *igneous rocks, or parts of igneous rocks, which contain both *crystals and *glass.

hypogene *Mineral deposit formed by generally ascending solutions in or from below the Earth's *crust; or processes such as *volcanicity operating within the crust.

hypolimnion The lower, cooler, non-circulating water in a thermally stratified lake in summer. If, as often occurs, the *thermocline is below the compensation level, the dissolved oxygen supply of the hypolimnion depletes gradually: replenishment by *photosynthesis and by contact with the atmosphere are prevented. Re-oxygenation is possible only when the thermal stratification breaks down in autumn.

hyponome A tube or funnel occurring in cephalopods (*Cephalopoda), through which water is expelled from the *mantle cavity. It is used for jet propulsion and its presence may cause an embayment or slit in the ventral margin of a cephalopod shell, called the 'hyponomic sinus'.

hyponomic sinus *See* HYPONOME.

hypopycnal flow At a river mouth, a flow of river water that is less dense than the water in the basin receiving it. The river water is buoyant and flows above the basin water. This is typical of the situation where a river enters the sea, because of the difference in density between fresh and salt water. *Compare* HOMOPYCNAL FLOW, and HYPERPYCNAL FLOW.

hypostomal suture *See* CEPHALIC SUTURE.

hypostratotype (reference section, auxiliary reference section) An additional, subordinate *stratotype, selected after the establishment of a stratigraphic unit, and in another region, to supplement the information contained in the original stratotype (the *holostratotype). *Compare* PARASTRATOTYPE. *See also* LECTOSTRATOTYPE; and NEOSTRATOTYPE.

hypotheca *See* DINOPHYCEAE.

hypothermal Applied to a mineral deposit originating at great depth and at temperatures between 300 and 500 °C; below the *mesothermal zone.

hypothesis An idea or concept that provides a basis for arguments or explanations which can be tested by experimentation. In inductive or inferential statistics, the hypothesis is usually stated as the converse of the expected results, i.e. as a null hypothesis

(H_0). This helps workers to avoid reaching a wrong conclusion, since the original hypothesis H_1 will be accepted only if the experimental data depart significantly from the values predicted by the null hypothesis. Working in this negative way carries the risk of rejecting a valid research hypothesis even though it is true (a problem with small data samples); but this is generally considered preferable to the acceptance of a false hypothesis, which would tend to be favoured by working in the positive way.

hypothesis testing In statistics, the comparison of a data set with a particular theory, called a null hypothesis, in order to see if it deviates significantly from this theory, in which case an alternative theory may more appropriately explain the data. A statistic is considered significantly deviant from the null hypothesis when the statistic falls outside the *confidence interval predicted by the null hypothesis for a defined confidence level. An example is the *chi-squared test.

Hypsilophodontidae Family of bipedal (ornithopod), *ornithischian *dinosaurs, ranging from the Upper *Triassic to the Upper *Cretaceous. *Hypsilophodon* itself, from the *Wealden of the Lower Cretaceous, is the most primitive of the ornithopod dinosaurs known to date, despite its relatively late appearance in the *Mesozoic fossil record.

Hypsithermal *See* FLANDRIAN.

hypsographic curve A graphic representation of the elevation and depth of points on the surface of a planet with reference to a datum; on Earth sea level is used as the datum. This shows that on Earth, on average land projects about 850 m above sea level and the average depth of ocean basins is about 3730 m below sea level. *Compare* HYPSOMETRIC CURVE.

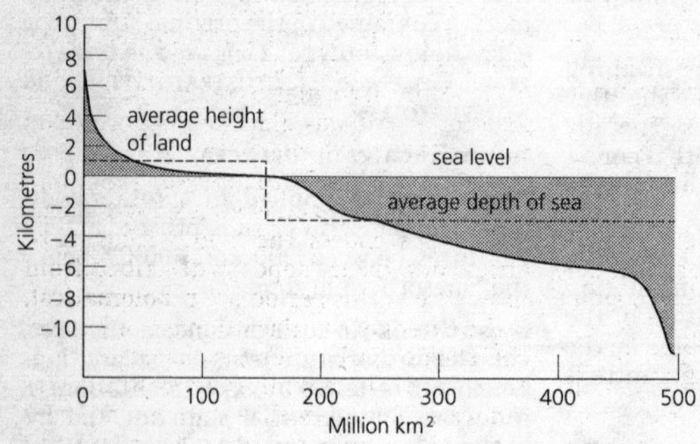

Hypsographic curve

hypsometric curve A graphic representation of the proportion of the surface of a planet that is elevated or depressed in relation to a datum. *Compare* HYPSOGRAPHIC CURVE.

Hyracotherium Known formerly as *Eohippus* (the 'dawn horse'), the earliest known *perissodactyl, which is placed in the family *Equidae. It was only 27 cm high, the size of a fox terrier: it was short-faced with low-crowned cheek teeth, and had four toes on the forefeet, and three on the hind. Abundant in the Early *Eocene of N. America and Europe, it was recently discovered in *Palaeocene deposits in Mongolia. It was a browser dwelling in forest glades, and the likely ancestor for all the horses. Because of its small size, when the fossils were first found, in Europe, they were mistakenly associated with the African hyraxes, hence the name. In America the fossils were identified correctly and the name '*Eohippus*' was adopted, but according to the rules of *taxonomic nomenclature '*Hyracotherium*' takes precedence.

hysteresis loop The curve formed when the intensity of magnetization acquired by a *ferromagnetic material is plotted against the magnitude of a direct magnetic field applied in one direction and then in the opposite direction. This loop is used to define the *coercivity, *saturation magnetization, and *susceptibility of such materials.

I

-ian The preferred suffix that is added to the geographical location of a *type section or *type area to form the name of a *stage or *age, e.g. '*Dolgellian', '*Frasnian', '*Barremian'.

Iapetus (Saturn VIII) One of the major satellites of *Saturn, with a radius of 718 km; mass 15.9 × 10^{20} kg; mean density 1020 kg/m^3; visual albedo 0.05–0.5. It was discovered in 1671 by G. D. Cassini.

Iapetus Ocean (proto-Atlantic) The late *Precambrian and early *Palaeozoic ocean that lay between *Baltica and *Laurentia. The *oceanic crust and upper *mantle of the Iapetus Ocean floor is presumed to have been subducted during the early Palaeozoic, and the ocean to have disappeared completely by the latest *Silurian–early *Devonian (about 400 Ma). The ancient *suture is thought to extend from WSW to ENE across the Solway Firth and Borders Region, Scotland. The *Caledonian *orogenic belt, which removed all trace of the Iapetus Ocean, extends all along the border between the two ancient *cratons, affecting the areas now known as Norway, eastern Greenland, Scotland, northern England, Wales, Ireland, eastern Canada, and the eastern USA.

Iberomesornis The best-known of several Early *Cretaceous birds, from Las Hoyas, central Spain. It is the first bird to show evidence of a perching foot.

Icarus A *solar system asteroid (No. 1566), diameter 1.4 km; approximate mass 10^{12} kg; rotational period 2.273 hours; orbital period 1.12 years. Its highly eccentric orbit crosses the orbit of Earth.

ICE *See* INTERNATIONAL SUN–EARTH EXPLORER-C.

ice 1. Water which has frozen into a crystal lattice. Pure water freezes at 0 °C at 1013.24 mb pressure. The presence of salts in solution depresses the freezing point of water. Liquid water has its maximum density at 4 °C, in consequence of which ice floats on water. With increasing pressure, a series of denser polymorphs of ice forms, each designated by a Roman numeral, ordinary ice being ice I. **2.** Several properties and varieties of ice are important in geomorphological *processes. Expansion on freezing (9.05% in specific volume) generates very high pressures. In an enclosed space in the laboratory the pressure reaches 216 MPa (megapascals) at −22 °C but reaches only about 10% of this when unenclosed, as in nature. The stresses are, however, sufficient to bring about *frost wedging. Such ice I converts into the denser ice III at lower temperatures, but the pressure exerted by it changes little. 'Ground ice' forms when *interstitial water freezes, and this may bring about heaving as well as frost wedging. 'Glacier ice' is a relatively opaque mass of interlocking crystals, and has a density of 0.85–0.91 g/cm^3. '*Regelation ice' is relatively clear and is formed by the freezing of meltwater beneath a temperate *glacier. **3.** In planetary geology other ices are important. Water ice condenses at 160 K at *solar nebular pressures and appears in abundance forming the surfaces of the *Galilean satellites Europa, Ganymede, and Callisto. The *satellites of the *jovian planets are mostly water ice–rock mixtures. Water ice will exist in high-pressure polymorphs (e.g. ice VIII, density 1670 kg/m^3) above about 15–20 kb in satellite interiors. Other possible ices important in satellites (e.g. Titan) include $NH_3.H_2O$, $CH_4.nH_2O$, and $H_2O.CO_2$.

ice ages Periods when ice has accumulated at the poles and the continents have been glaciated repeatedly. Exactly why glaciation occurred is not clear. There are suggestions of a middle *Precambrian glaciation about 2300 Ma in N. America, S. Africa, and Australia. More information exists to suggest that the Earth was glaciated between 950 and 615 Ma, and there are at least two glacial *horizons in Africa, Australia, and Europe. There is good evidence for a glaciation at the end of the *Ordovician in N. Africa, but glacial deposits described from elsewhere at this period are problematical, so the extent of the glaciation is not known. The Permo-Carboniferous glaciation of S. America, S. Africa, India, and Australia was widespread and is well documented. There is no evidence for further glaciation until the *Quaternary. Suggestions have been made for other ice ages during the *Palaeozoic but evidence for them is sparse. The *Pleistocene ice age is the best documented, but there is undoubted evidence of earlier glaciations in the geologic record.

iceberg calving *See* CALVING.

ice blink Whitish appearance seen above the horizon, caused by light reflection over an ice or pack-ice surface.

ice-cap Ice mass less than 50 000 km^2 in area, but large enough to submerge the underlying topography and whose flow behaviour is a consequence of its size and shape. It consists of a central *ice dome together with outlet *glaciers radiating from the periphery.

ice crystal Frozen water composed of crystalline structures, e.g. needles, dendrites, hexagonal columns, and platelets. Ice clouds are composed almost entirely of ice crystals.

ice dome A main component (with the *outlet glacier) of an *ice sheet or *ice-cap. It has a convex surface form, of parabolic shape, and tends to develop symmetrically over a land mass. Its thickness often exceeds 3000 m.

ice field A nearly level field of *ice, whose area may range from about 5 km^2 to near-continental size, formed when the land surface is sufficiently high or uniform for ice to accumulate. It differs from an *ice-cap in that it lacks a domed form and its flow is controlled by the underlying relief.

icehouse period A time during which glaciers reached their maximum extent, ocean waters were well mixed and oxygenated, and sea levels were low. In *cyclostratigraphy, fourth- and fifth-order cycles were predominant. *Compare* GREENHOUSE PERIOD.

Iceland low The region of the N. Atlantic in which the average value of atmospheric pressure is low, owing to the frequency of low-pressure systems (cyclones or '*depressions') moving into and occupying the region. Any one of these systems, when present on an individual day, may be called 'an Iceland low', but the term is used mostly to describe the statistical or climatic feature.

ice nucleus Crystalline, microscopic particle, either of ice or some other substance, which can induce the growth of ice crystals upon it from saturated air in cloud at temperatures of about −25 °C or below. Deposition of water in this way—directly from vapour to form ice, without forming liquid in between—is called 'sublimation'. During growth, splintering of the crystals (e.g. as a result of updrafts) causes large numbers of new nuclei to be formed. *Compare* FREEZING NUCLEI.

ice sheet A large ice mass, with an area usually greater than about 50 000 km^2, made up of *ice domes and *outlet glaciers. The Antarctic ice sheet is the world's largest, with an area of about 11.5×10^6 km^2 and a mean thickness of about 2000 m.

ice shelf The outer part of an *ice-cap or *ice sheet that extends into and over the sea. It typically ends in a cliff that may be 30 m high, and the total ice thickness may be 200 m. Ice wastage is by *calving (the breaking away of ice blocks) and by bottom melt.

ice stream *See* OUTLET GLACIER.

ice wedge Tapering, vertically layered mass of ice, about a metre wide at the top and extending downward for some 3–7 m. It results from contraction cracking of the ground during extreme cold, followed by water penetration from the *active layer and subsequent freezing. It is characteristic of uniform sediment, such as river *alluvium, under *periglacial conditions.

ice-wedge polygon *See* PATTERNED GROUND.

ichnoclast A *trace fossil that has been reworked by a later organism.

ichnocoenosis An assembly of *trace fossils made by members of a single community.

ichnofabric The structure and texture of a sediment that is produced by the activity of living organisms.

ichnofacies A rock or sequence of rocks characterized by its *lithology, inorganic *sedimentary structures, and specific *trace fossils. The traces represent the behaviour of *fossil organisms under particular environmental conditions. The relative abundance of different trace fossil ethotypes is critical to the determination of the correct ichnofacies.

ichnofossil *See* FOSSILIZATION; and TRACE FOSSIL.

ichnogenus A group of *trace fossils that is given a name because the similarity of the traces suggests they were made by closely related species of organisms. Ichnological taxonomy, which applies the principles of biological nomenclature to non-biological material, is governed by the *International Code of Zoological Nomenclature*. Above the level of genus, the Code indicates names should be used formally only to the family level; at higher levels all names are informal. Names of ichnogenera are conventionally written italicized and with a capital initial; ichnogenus is abbreviated as igen. *See* ***DIPLOCRATERION***; ***RHIZOCORALLIUM***; and ***ZOOPHYCUS***.

ichnoguild A group of *ichnospecies that shared particular resources or behaviour in the manner of a *guild.

ichnology The study of the tracks, burrows, and other traces made by living organisms on and within a substrate. If the traces are recent and made by organisms that are still living, the study is called neoichnology. If the organisms are long disappeared and evidence of their presence is preserved as *trace fossils, the study is called palaeoichnology.

ichnospecies A species (trivial) name assigned to *trace fossils within an *ichnogenus and conventionally written italicized and with a lower case initial; ichnospecies is abbreviated as isp.

ichnotaxobase Any morphological feature of a *trace fossil that can be used in classification.

ichnotaxonomy The formal classification of *trace fossils.

Ichthyornis *See* AVES.

Ichthyosauria (class *Reptilia, subclass Ichthyopterygia) Order of so-called 'fish lizards', and the only order in the subclass, the first members of the group date from the *Triassic and were primitive in type. The typical, shark-like form, *Ichthyosaurus*, appeared in the *Jurassic, when the group in general was especially common. Ichthyosaurs disappeared before the end of the *Cretaceous.

Ichthyostega *Amphibians in the form of *Ichthyostega* and *Ichthyostegopsis* first appear in the Upper *Devonian. *Ichthyostega* shows refinements of the skull, and the development of strong limbs. It retains a long, rather fish-like tail, and links with the *crossopterygian fish are confirmed by the presence of teeth with a labyrinthine infolding of the enamel. *Ichthyostega* grew to just under 1 m in length. An unexpected finding about *Ichthyostega* and its relatives, from recently discovered, more complete specimens, is that they had more than the five digits per limb that characterize modern tetrapods: *Ichthyostega* had seven on the hind feet (the forefeet are still unknown), and *Acanthostega* had eight on both fore- and hind feet.

Ichthyostegopsis *See* ICHTHYOSTEGA.

ICP *See* INDUCTIVELY COUPLED PLASMA SPECTROSCOPY.

-id *See* -IDE.

Ida A *solar system asteroid (No. 243), measuring 58 × 23 km; approximate mass 10^{17} kg; rotational period 4.633 hours; orbital period 4.84 years. Images of Ida from *Galileo, taken on 23 August 1993, showed it has a small satellite, later named Dactyl.

Idamean A *stage of the Upper *Cambrian of Australia, underlain by the *Mindyallan and overlain by the Post-Idamean.

-ide (-id, -ides) From the Greek *ides* meaning 'son of', a suffix attached to the name of an element which is a member of a series (e.g. actinide, halide), or to the more electronegative element or radical in a binary compound (e.g. sodium chloride).

-ides *See* -IDE.

idioblastic A *textural term referring to *metamorphic rocks in which the grains display fully developed *crystal forms.

idiomorphic *See* EUHEDRAL.

idiomorphic fabric *See* IDIOTOPIC FABRIC.

idiotopic fabric The *fabric developed in a crystalline *sedimentary rock in which most of the *crystals are *euhedral. In an *igneous or *metamorphic rock the term 'idiomorphic fabric' is used.

idocrase (vesuvianite) A mineral, with the formula $Ca_{10}(Mg,Fe)_2Al_4Si_9O_{34}(OH,F)_4$, closely related to the *garnet group; sp. gr. 3.4; *hardness 6.5; green; normally *massive or granular; occurs in contact-metamorphosed (*see* THERMAL METAMORPHISM) *limestones along with *grossular, *scapolite, and *wollastonite.

Idwian A *stage of the Lower *Silurian, underlain by the *Rhuddanian and overlain by the *Fronian.

I_f *See* FRACTURE SPACING INDEX.

IGC International Geological Congress.

igen. Abbreviation for *ichnogenus.

igneous Applied to one of the three main groups of rock types (igneous, *metamorphic, and *sedimentary), to describe those rocks that have crystallized from a *magma.

ignimbrite A poorly sorted, *pyroclastic rock body formed by deposition from a pumiceous pyroclastic flow. The passage of one pyroclastic flow deposits one ignimbrite flow unit with a number of layers directly related to the components of the flow. Ignimbrite flow units deposited by the passage of successive flows can accumulate into a loosely consolidated, composite sheet. Where the rock-body temperature and the accumulating overburden load are both sufficiently high, sintering and flattening of *groundmass *shards and *pumice *clasts can occur in the lower part of the body, producing a welded zone within the ignimbrite body which is then characterized by the development of a *eutaxitic texture. Hot gases escaping through the ignimbrite deposit can precipitate minerals between the loosely packed shards and pumice in the upper parts of the flow, creating a lithified *sillar horizon in which the glass of the pumice and shards is devitrified. Ignimbrites are found on all scales, from a few hundred metres to more than 100 km long, and from one metre to tens of metres thick. The geometry of the ignimbrite sheet is descried using the aspect ratio. *See* HIGH-ASPECT-RATIO IGNIMBRITE; and LOW-ASPECT-RATIO IGNIMBRITE.

IGRF *See* INTERNATIONAL GEOMAGNETIC REFERENCE FIELD.

Iguanodontidae *Jurassic–*Cretaceous family of bipedal (ornithopod), *ornithischian *dinosaurs, whose best-known representative is *Iguanodon*, remains of which have been found in Europe, Asia, and Africa.

ijolite A medium- to coarse-grained, ultra-alkaline *plutonic *igneous rock, consisting of *essential *nepheline, *aegirine (or aegirine-*augite, or sodic *diopside) with or without melanite *garnet, and *accessory *apatite, *sphene, and *cancrinite. The rock can be considered an undersaturated alkali *syenite (*see* SILICA SATURATION).

ilium In *tetrapods, the dorsal section of the pelvis, which articulates with one or more sacral *vertebrae.

Illinoian The third (0.55–0.4 Ma) of four glacial *stages recognized in N. America. It is represented by deposits from ice moving from the north-east, and in formerly glaciated areas to the west pollen evidence suggests that mean annual temperatures were 2–3 °C cooler than they are now. At the end of this period the climate became warmer and drier. The Illinoian is approximately equivalent to the *Mindel and *Riss glacials of the Alps.

illite (hydromuscovite) Common *clay mineral and important member of the 2 : 1 group of *phyllosilicates (sheet silicates) with the formula $K_{1-1.5}$ $Al_4[Si_{7-6.5}$ $Al_{1-1.5}$ $O_{20}](OH)_4$, which possesses an overall negative charge due to incomplete charge balance; sp. gr. 2.6–2.9; *hardness 1–2; *monoclinic; *crystals form tiny flakes; formed by the *weathering decomposition or *hydrothermal alteration of *muscovite or *feldspar.

illuminator The light source used in a transmitted- or *reflected-light microscope. For most routine work, an incandescent, tungsten filament lamp with a variable rheostat is used. The colour temperature is normally about 2800 K and a pale blue filter may be used to provide a more daylight-like colour temperature. The standard illuminating system may contain additional lenses, diaphragms, and a *polarizer.

illuviation Process of deposition (inwashing) of soil materials, either from suspension or solution, and usually into a lower *soil horizon, after removal from above or from a lateral source.

ilmenite Mineral, $FeTiO_3$; sp. gr. 4.5–5.0; *hardness 5–6; *trigonal; black; black to brownish-red *streak; *sub-metallic *lustre; crystals normally thick and *tabular, but often *massive and compact; no *cleavage; magnetic; occurs as an *accessory mineral in *igneous rocks, e.g. *gabbro and *diorite, in *quartz veins and *pegmatites, in *gneisses, in association with *hematite and *chalcopyrite, and because it is resistant to *weathering it occurs extensively in *alluvial deposits with *magnetite, *monazite, and *rutile. It is used as a source of iron and titanium. The name is derived from that of the Ilmen Mountains, Russia.

image intensifying *See* IMAGING.

imaginary component (quadrature, out-of-phase component) An electromagnetic field induces a secondary field in a conductor and the resultant vector of the two fields can be resolved into two components, one of which is the imaginary component, lagging $\pi/2$ behind the *in-phase component.

imaging The instrumental recording and interpreting of various portions of the electromagnetic spectrum from planetary or *satellite surfaces. This information is obtained mostly in digital form, either from spacecraft or telescopically. A common form is a photographic image in the visible portion of the spectrum. 'Multispectral imaging' consists of simultaneous recording, typically of four or more spectral bands ranging from the visible to infrared wavelengths. 'Image intensifying' is the enhancement of the image by computer processing. *See* REMOTE SENSING.

imaging radar A type of *radar that constructs an image from the echoes reflected back to its antennae.

imaging spectrometer A *remote sensing instrument which records an image in a large number of spectral *channels.

Imbrian A *period of the *Archaean, dated at about 3850–3800 Ma (Harland et al., 1989).

Imbrian System *See* LUNAR TIME SCALE.

imbricate With parts overlapping one another like tiles.

imbricate structure (schuppen structure) Fabric resulting from the stacking of rock fragments, particles, or tectonic units. Some pebble beds show an imbricate structure, with the pebbles leaning in the direction of the current. *See also* BAR.

immersion objective method (oil immersion) A technique commonly used in *reflected-light microscopy, especially when high magnification and high resolution are required. A drop of immersion oil or water, with a *refractive index of about 1.51, is placed on the polished surface of the mineral and the *objective is lowered carefully into the liquid. The advantages of the method include better observations of colour differences, *bireflectance, and *anisotropy.

immobilization Conversion of a chemical compound from an inorganic to an organic form as a result of biological activity; the compound is thereby removed from the reservoir of compounds available to plant roots.

impactite A rock type produced during the impact of a *meteorite on to a planetary or *satellite surface. *See also* SUEVITE.

impactogen *See* RHINE GRABEN.

imperforate Sometimes applied to those gastropod (*Gastropoda) shells which do not possess an *umbilicus. Although the term is much used, 'anomphalous' or 'non-umbilicate' are to be preferred since an umbilicus is not a perforation.

impervious rock Rock which will not permit oil, water, or gas to flow through it. *See also* PERMEABILITY.

implosion A sudden collapse into a zone of very low pressure. In marine *seismic surveying, the discharge of air into water causes water to implode into the bubble space and the impact of the water colliding with itself creates a seismic signal, the *bubble pulse.

impulse response function The specification of the effect of a *filter. For example, if a spike (*Dirac) function is put through a linear filter the output function will appear modified and spread out, this characteristic filter effect being the impulse response function. It is important in *convolution and *deconvolution in data processing.

impunctate *See* PUNCTATE.

Inarticulata (phylum *Brachiopoda) A class of brachiopods, existing from the Lower *Cambrian to the present day, in which the shell is calcareous, but its valves are not hinged by teeth and sockets and the *pedicle is much reduced or absent. There are 3 orders.

Inarticulate brachiopods *See* INARTICULATA.

Inceptisols An order of mineral soils that have one or more *soil horizons in which mineral materials have been weathered or removed. Inceptisols are in the early stages of forming visible horizons, and are only beginning the development of a distinctive *soil profile. The term embraces *brown earths.

incident angle *See* ANGLE OF INCIDENCE.

incipient hardground *See* HARDGROUND.

incised meander *See* MEANDER.

inclination The angle between the horizontal and a magnetic vector. Conventionally, a vector with a magnetic north pole dipping below the horizontal is considered positive, and an upward vector is negative.

inclined extinction *See* OBLIQUE EXTINCTION.

inclined fold A *fold in which the angle of *dip of the *axial plane is between 10° and 80°, and the highest and lowest points on the fold surfaces do not necessarily coincide with the *hinge points. *Compare* OVERFOLD; UPRIGHT FOLD; and RECUMBENT FOLD.

inclinometer, geomagnetic An instrument to measure the *inclination of the *geomagnetic field. It is usually a dip circle, but inclination may also be determined from separate measurements of the horizontal and vertical components of the geomagnetic field.

included fragments *See* PRINCIPLE OF INCLUDED FRAGMENTS.

incompatible elements (hygromagmatophile elements) Elements that, owing to their size, charge, or *valency requirements, are difficult to substitute into the crystal structure of a rock-forming mineral (e.g. the boron *ion is very small and the tungsten ion may have a +6 charge). This results in their being preferentially introduced into a *magma on *partial melting and less likely to crystallize out of it. During the crystallization of *igneous rocks, incompatible elements (e.g. Sn, Li, Rb, Sr, and *rare earth elements) are often concentrated into *pegmatitic or *hydrothermal fluids. During the formation of the *Earth's crust the incompatible elements have been transferred through magmatic processes to the crust from the *mantle, which has consequently become depleted in these elements.

incompetent A relative rheological (*see* RHEOLOGY) term, referring to the ease with which a rock or layer of rock may be deformed. It is applied to materials which are less rigid than competent materials (*see* COMPETENCE) and tend to flow rather than fracture when deformed. Incompetence reflects the inability of a material to transmit *compressive stresses over large distances.

incompressibility modulus *See* BULK MODULUS.

incongruent dissolution Dissolution of a mineral with decomposition or reaction in the presence of a liquid, converting one solid *phase into another, e.g. the conversion of orthoclase (*see* ALKALI FELDSPAR) to *kaolinite:

$$2KAlSi_3O_8 + 11H_2O \rightarrow Al_2Si_2O_5(OH)_4 + 4Si(OH)_4 + 2K^+ + 2OH^-.$$

incongruent melting Melting of a mineral with decomposition or reaction, such that it is replaced by a *melt and a solid *phase of different composition. For example, *orthoclase melts incongruently to give *leucite and a silica-rich liquid.

inconsequent drainage (insequent drainage) A *drainage pattern which bears no apparent relationship to the underlying rock type or structure.

incumbent replacement An evolutionary mechanism, proposed in 1991 by M. L. Rosenzweig and R. D McCord, according to which a well-adapted species (the incumbent) becomes extinct, due to a chance combination of adverse factors, and its vacated niche is occupied by an invading species population.

incus From the Latin *incus* meaning 'anvil', a supplementary cloud feature comprising the flattened, anvil-like shape of the top of *cumulonimbus cloud. *See also* CLOUD CLASSIFICATION.

index ellipsoid *See* INDICATRIX.

index fossil (index species, zone fossil) *Fossil whose presence is chosen to denote the *zone in which it occurs and after which the zone is named. Index fossils are selected for their distinctiveness and/or abundance. To be of use in *biostratigraphy, ideally an index fossil should have a narrow range in time (i.e. to have undergone rapid evolutionary change) but have had a wide geographical distribution. *Trilobites (*Cambrian); *graptolites (*Ordovician and *Silurian); *ammonites (*Jurassic); and *foraminifera (*Cretaceous and *Cenozoic) are among the most notable index fossils.

index mineral A diagnostic *mineral in a *regional metamorphic terrain whose first appearance, going in the direction of increasing *metamorphic grade through the metamorphic sequence, marks the outer limit of a *metamorphic zone. A line marking the first appearance of an index mineral is termed an 'isograd', and represents a line of constant metamorphic grade. To indicate the regional distribution of metamorphic grade the first appearance of index minerals in rocks of *pelitic composition (i.e. *shales) are usually mapped out in the field, as this type of rock is extremely sensitive mineralogically to changes in metamorphic grade.

index species *See* INDEX FOSSIL.

indicated reserve *See* RESERVE.

indicator *See* ERRATIC.

indicatrix (optical indicatrix, index ellipsoid) An ellipsoid which represents geometrically the different *vibration directions in a *mineral and illustrates conceptually the optical features of a *crystal. The origin is a point lying at the centre of the ellipsoid and the axes of the ellipsoid are proportional in length to the *refractive indices of beams of light vibrating at right angles along them. These axes are commonly termed X, Y, and Z, or n_α, n_β, and n_γ, for *orthorhombic, *monoclinic, or *triclinic minerals. *Tetragonal, *hexagonal, and *trigonal minerals are represented by an indicatrix with one principal section circular; and the *cubic minerals are represented by an indicatrix (the isotropic indicatrix) which is a sphere with all axes equal. Measurements of the elliptical plane sections give optical properties which aid specific identification of a mineral. *See* OPTIC AXIS.

Indian Ocean One of the world's major oceans, lying between Africa, India, and Australia. It has a surface area of 77 million km^2 and an average depth of 3872 m. The ocean receives a great deal of sediment from three of the world's major rivers (the Ganges, the Indus, and the Brahmaputra).

Indian summer *See* SINGULARITY.

Indo-Australian Plate One of the present-day major lithospheric *plates, which is having new material added to its south and south-west along the *Carlsberg Ridge and the south-east Indian Rise, but its other margins are the *collision zone of the Himalayan *orogenic belt, *subduction zones (e.g. in the E. Indies), or *transform faults (e.g. the Alpine Fault in New Zealand). It is thought this plate may now be breaking into two separate plates along the line of the 90° E Ridge.

Induan *See* SCYTHIAN.

induced polarization (IP, induced potential, over-voltage, interfacial polarization) An exploration method which uses either the decay of an excitation voltage (*time-domain method) or variations in the Earth's *resistivity at two different but low frequencies (*frequency-domain method). A variety of different *electrode configurations can be used. *See also* CHARGEABILITY.

induced potential *See* INDUCED POLARIZATION.

induced pulsed transient *See* INPUT.

induction The creation of a voltage by changing the magnetic flux such that the amount of voltage induced is directly proportional to the rate of change of the magnetic flux according to Faraday's or Neumann's law (*see also* LENZ'S LAW). In applied geophysics, induction is a fundamental process in *electromagnetic (EM) prospecting; a primary EM field is used to induce a secondary field in any subsurface conductors and the resultant of the two fields is measured. The strength of the secondary field, which is a direct function of the electrical *conductivity of the ground, can then be determined.

induction log A *well-logging system which uses *electromagnetic *induction methods to measure electrical *conductivity and *resistivity in the adjacent *formations.

induction sonde A variety of *electromagnetic (EM) instrument for producing and measuring eddy currents within the rocks surrounding a *borehole. The strength of the eddy currents is proportional to the conductivity (1/resistivity) of the surrounding rocks. Penetration is usually greater than by *laterolog sondes. The main application of the log in hydrocarbon exploration and production is in the estimation of water saturation. Interpretation is affected by the *Delaware effect.

inductively coupled plasma emission spectrometry Technique for the chemical analysis of *trace elements in a wide variety of materials, using a source that depends on the interaction between the magnetic field of an oscillating radio-frequency current and the charged species present in the *plasma. Argon gas is passed through the field producing a toroidal (doughnut-shaped) plasma at temperatures of over 5000 °C. The sample is introduced as an *aerosol into the gas and is vaporized and atomized in the plasma. The source has a *monochromator (a means of selecting a single frequency) and a readout system. It has low limits of detection for most elements, and is a very rapid technique.

induration Process of forming indurated *horizons or *hardpans that have a high bulk density and are hard or brittle. *Cementing materials may be present and responsible for the induration.

industrial mineral Any earth material of economic importance, excluding metal *ores and fuels; e.g. *barite, *fluorite, and china clay (*kaolin).

inertinite *See* COAL MACERAL GROUP.

infaunal Applied to benthic organisms that dig into the sea bed or construct tubes or burrows. They are most common in the subtidal and deeper zones (i.e. the area seaward of the low-water mark). *Compare* EPIFAUNAL.

inferred reserve *See* RESERVE.

inferred tree A *phylogenetic tree based on empirical data pertaining to extant taxa.

infiltration Downward entry of water into soil.

infiltration capacity The maximum rate at which soils and rocks can absorb rainfall. The infiltration capacity tends to decrease as the soil moisture content of the surface layers increases. It also depends upon such factors as grain size and vegetation cover.

infinitesimal strain The extensions and deflections that are required to deform an infinitesimally small reference cube into a *strain parallelepiped. The concept is of use when considering the amount of strain absorbed by a material during each instant of its progressive deformation.

inflexion point 1. The point of no curvature in a *fold. **2.** The point at which the curvature of a fold changes sense between one closure and the next.

influent stream *See* GAINING STREAM.

influx An inflowing of *sediment, fluid, mineralizing solution, or other material.

informal The informal naming of a *stratigraphic unit occurs when (*a*) the unit-term is referred to as an ordinary noun, and not in the context of a proper name, e.g. 'the geologic periods'; and (*b*) when a unit-term (such as *zone, or *formation) is referred to without having been specifically established in classification, e.g. 'a sandstone formation', 'a mineralized zone'. Terms used informally are not capitalized. *Compare* FORMAL. *See* STRATIGRAPHIC NOMENCLATURE.

infrared radiation *Electromagnetic radiation which has a *wavelength between 0.7 μm and 100 μm. *See also* NEAR-INFRARED; MID-INFRARED; REFLECTED INFRARED; and THERMAL INFRARED.

infrared remote sensing Method of distinguishing various types of vegetation, rocks, etc. using either monochrome or coloured infrared film which can be used in conventional cameras. Potentially, *aerial photography using infrared film and coloured filters may be of benefit in preparing *geologic maps. Longer-wavelength infrared can discriminate most rocks, shorter wavelengths can reveal iron oxide, etc.; alteration effects around certain mineral deposits, e.g. *porphyry coppers with their attendant *clay minerals, can also be distinguished, and the technique allows a consideration of plant species affected by the nature of the *soil and rock substrate.

ingrown meander *See* MEANDER.

inhomogeneity (heterogeneity) Irregular spatial variability in physical properties. *Compare* ANISOTROPY.

inhomogeneous strain *Strain, distributed unevenly throughout a deformed body, in which straight lines and parallel lines in the undeformed material become curved and non-parallel on deformation. The mathematical theory describing the geometry of inhomogeneous strain is very complex and it is usual to divide this strain into smaller components of *homogeneous strain.

initial levée *See* LAVA LEVÉE.

initial strontium ratio (common strontium) Strontium has four naturally occurring *isotopes: ^{88}Sr; ^{87}Sr; ^{86}Sr; and ^{84}Sr. Of these, ^{87}Sr is perhaps the most important because it is formed by the natural radioactive decay of the rubidium isotope ^{87}Rb; this decay provides the basis for one of the most important geochronological methods. At its simplest, the initial strontium ratio (common strontium) of a rock is the ratio between the radioactively produced isotope ^{87}Sr and the 'ordinary', non-radiogenic isotope ^{86}Sr at the time when the rock crystallized. In a hypothetical rock containing no rubidium this ratio would remain unchanged for ever, but most rocks contain some rubidium; thus radioactive decay constantly increases the amount of ^{87}Sr, and the ratio of ^{87}Sr to ^{86}Sr constantly increases, at a rate proportional to the amount of rubidium in the rock. The initial strontium ratio of a rock is determined by measuring the present-day $^{87}Sr:^{86}Sr$ ratio in several of its constituent minerals. At the time when it first crystallized each mineral in the rock would have had the same $^{87}Sr:^{86}Sr$ ratio, but each mineral contains a different amount of rubidium, so after any given time its $^{87}Sr:^{86}Sr$ ratio will have increased away from the initial value by an amount exactly determined by the relative proportions of rubidium and strontium in it. Therefore, both the initial strontium ratio and the age of the rock can easily be found by plotting the measured present-day ratios of the constituent minerals on an *isochron diagram. In *petrology, initial strontium ratios are important because they provide information that otherwise would not be available on the chemical composition and ages of the source regions of *igneous rocks. For example, an igneous rock that has a very young radiometric age but a very high initial strontium ratio must have been derived from a source rich in ^{87}Sr. Such a source must have been rich in rubidium and old enough for the ^{87}Sr to have accumulated by the radioactive decay of ^{87}Rb. Young *granites in continental *collision zones (e.g. the Alps) have extremely high (up to 0.8) initial strontium ratios and formed by the melting of old crustal *gneisses. Granites formed by melting of the same rocks at different times can be distinguished immediately by their initial strontium ratios since the source rocks would have accumulated different amounts of ^{87}Sr when melting occurred. Granites formed in *island arcs are derived from young *mantle materials and have strikingly lower (0.704–0.706) initial strontium ratios. Similar arguments can be applied to the source regions of *basalts in the mantle, though the differences between present and initial $^{87}Sr:^{86}Sr$ ratios are very much smaller because the mantle is poor in rubidium; typical ocean *ridge basalts have initial strontium ratios close to 0.703. At the time of its formation, 4.6 billion years BP, the bulk Earth ratio is thought to have been 0.699. *See also* ISOCHRON.

inland sea Extensive body of water that is largely or wholly surrounded by land. Any connection to the open ocean is restricted to one or a few narrow sea passages. Examples of such areas are the Baltic and the Mediterranean Seas.

inlier Structure where older rocks are surrounded completely by newer rocks. It may result from *faulting or *folding followed by *erosion.

inner planet *See* TERRESTRIAL PLANET.

inosilicate (chain silicate, band silicate) Applied to the structure of silicate *minerals where SiO_4 *tetrahedra are linked together into chains by sharing oxygens. Two important groups of rock-forming minerals are included: *pyroxenes, where two of the four oxygens are shared to give a 'single chain' structure with the ratio Si : O = 1 : 3; and *amphiboles, where half the SiO_4 tetrahedra share two oxygens and the other half share three oxygens, to give a 'double chain' or 'band' structure and a ratio Si : O = 4 : 11.

in-phase component (real component) An electromagnetic field induces a secondary field in a conductor and the resultant vector of the two fields can be resolved into two components. One of these is the real component, in phase with the primary field, the other is the *imaginary component (quadrature).

INPUT (INduced PUlsed Transient) An airborne electromagnetic surveying system that comprises a large coil transmitter looped around the aircraft beneath the nose and tail and around each wing-tip. A *bird is trailed behind the aircraft to detect the decaying secondary field at times when the primary field is between source pulses.

Insecta (Hexapoda, insects; phylum *Arthropoda) Class of arthropods that have three pairs of legs and, usually, two pairs of wings borne on the thorax. Typically, there is a single pair of antennae and one pair of compound eyes. Gas exchange takes place through a tracheal system and the gonoducts open at the posterior end of the body. The oldest fossil insects occur in *Devonian rocks, and the first winged representatives are known from *Carboniferous rocks. Dragonflies and beetles were established before the end of the *Palaeozoic; social varieties such as ants and wasps are present in *Cretaceous sediments. The evolution of the flowering plants had a marked influence on insect development, so that many new forms appeared in the Cretaceous and *Tertiary Period. About 950 000 extant species of insects have been described. This is larger than the number of species belonging to all other animal classes combined.

inselberg Steep-sided, isolated hill that stands above adjacent nearly flat plains. It may have a narrow *pediment at its base. Locally, flared or steepened margins occur. It is best developed under a savannah climate. *See also* BORNHARDT.

insequent drainage *See* INCONSEQUENT DRAINAGE.

insolation The amount of incoming solar radiation that is received over a unit area of the Earth's surface. Solar energy received over the planet's surface varies according to season, latitude, transparency of the atmosphere, and aspect or ground slope. On average, equatorial areas receive approximately 2.4 times as much insolation as polar areas.

insolation weathering *See* THERMOCLASTIS.

instability Atmospheric condition in which displaced air tends to maintain its movement away from its original level. This occurs, for example, when rising air cools at the moist-adiabatic lapse rate while a greater *environmental lapse rate allows the air parcel to remain warmer than surrounding air (or even increases the temperature difference) so that continual buoyancy prevails. *See also* CONDITIONAL INSTABILITY; POTENTIAL INSTABILITY; and STABILITY.

instantaneous field of view In *remote sensing, the angle through which a detector is receiving *electromagnetic radiation. It is often expressed as a function of the ground area visible at any one time, which is dependent on height and the angle of radiation reception.

Institute of Space and Astronautical Science (ISAS) The Japanese national research institute dedicated to space and astronautical science under the auspices of the Ministry of Education, Science, and Culture. It was founded in 1981.

intensification In meteorology, the increase of pressure gradient around a pressure system. *See also* WEAKENING.

intensity In *remote sensing, the energy reflected or emitted by a surface.

intensity (earthquake) *See* EARTHQUAKE MAGNITUDE; MERCALLI SCALE; and RICHTER MAGNITUDE SCALE.

intensity-hue-saturation processing In *remote sensing, a form of *contrast stretching in which the visibility of *pixel colour is enhanced. Usually the saturation values of pixels are stretched to fill parameter space.

inter- From the Latin *inter* meaning 'between', a prefix meaning 'between'.

interambulacral *See* INTERAMBULACRUM.

interambulacrum (interamb, adj. interambulacral) In *Echinodermata, that area of the body surface lying between ambulacra. *See* AMBULACRUM.

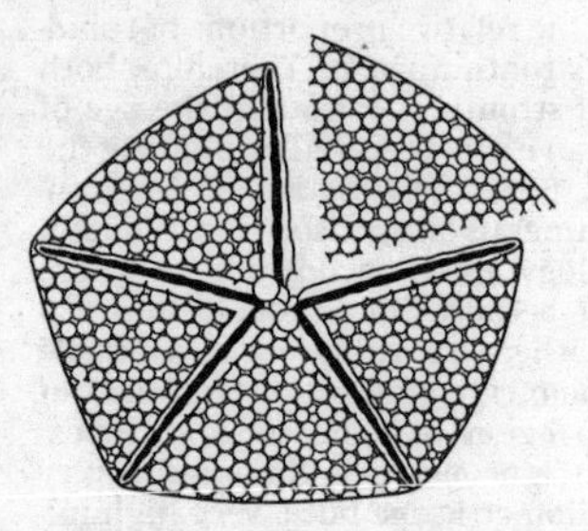

Interambulacrum

inter-arc basin A type of *back-arc basin which is floored by *oceanic crust. The main sediments are turbiditic volcaniclastics derived from the volcanic arc. *See also* INTER-ARC TROUGH.

inter-arc trough A *fore-arc basin developed between an outer, non-volcanic arc and a volcanic arc. *Compare* INTER-ARC BASIN.

interbiohorizon zone *See* INTERVAL ZONE.

interception 1. The capture of rain-water by vegetation from which the water evaporates and is thus prevented from reaching the *water-table and contributing to *surface runoff, *soil moisture, or *groundwater recharge. **2.** The abstraction of groundwater part of the way along its flow path, where otherwise the water might be lost, e.g. as coastal spring discharges.

intercept ratio *See* AXIAL RATIO; and PARAMETER.

intercept time The arrival time of a seismic wave, determined from the intercept of the extrapolation of the refracted straight-line segment of a *time–distance graph at zero *offset.

intercrystalline boundary *See* VOIDS.

intercrystalline porosity *See* CHOQUETTE AND PRAY CLASSIFICATION.

interdigitating (interfingering, interlocking) Applied to a *facies boundary where the line of lithological change between one *sedimentary rock type and the next laterally adjacent rock type is itself broken down into a series of wedge-shaped zig-zags or tongues. Where such interfingering occurs it is a record of the fluctuations in the local depositional environment, implying that both types of *sediment were being laid down at the same time.

interface 1. A device which connects two machines and allows them to communicate with one another. The term is used most commonly in respect of equipment linked to a computer. **2.** The boundary between two substances that have different properties. *See also* HORIZON.

interface-controlled growth *Crystal growth in a *melt or *solution where the rate of crystal growth is controlled by the transport of material across the crystal–liquid interface. This type of growth tends to characterize large *supercooling regimes in one-component systems.

interfacial angle In *crystallography, the angle subtended by the normals to two *crystal faces. It is not the external angle observed or the internal angle between them; it is, however, 180° minus the internal angle. A *goniometer is used to measure interfacial angles.

interfacial polarization *See* INDUCED POLARIZATION.

interference The combination of waves. Constructive interference occurs when peaks add to peaks; destructive interference occurs when a peak coincides with and cancels out a trough.

interference colour chart *See* MICHEL–LÉVY CHART.

interference colours (polarization colours) In *mineral optics, the colours produced when the *analyser is inserted on a thin-section microscope. They are produced as a result of *birefringence (double refraction) whereby one ray of light is retarded relative to the other. The different degrees of retardation give different interference colours. These colours are used in a number of ways as an aid to identification.

interference figure In *crossed-polar mineral optics, the faintly coloured rings and dark curves produced as a result of retardation when convergent polarized light passes through *anisotropic (i.e. *double refracting) *minerals. The black curves or crosses are called 'isogyres'. There are two kinds of interference figures: uniaxial and biaxial. When centred, the uniaxial interference figure resembles a black cross with coloured concentric circles; the biaxial interference figure is more complex, with two curved isogyres and coloured elliptical bands. The coloured bands represent zones of equal retardation and tests using accessory plates give optical properties which are characteristic of individual minerals.

interference pattern A two-dimensional *outcrop pattern resulting from the superimposition of two or more sets of *folds of different generations. The form of the patterns seen depends on the relative *attitudes of the superimposed folds; J. G. Ramsay (1967) recognized four basic types: redundant superposition (in which later folding has not altered the original pattern); dome and basin (egg box); dome–crescent–mushroom (*see* CRESCENT-AND-MUSHROOM); and convergent–divergent (double zigzag).

interferometer An instrument used in *remote sensing that forms and uses interference patterns in radiation to measure the wavelengths of that radiation.

interfingering *See* INTERDIGITATING.

interflow (throughflow) The lateral movement of water through the upper *soil horizons, normally during or following significant precipitation events. Shallow *groundwater or interflow may emerge at the surface at the bottom of slopes and flow across the ground surface for a time. This is known as 'return flow'.

interfluve The elevated part of the landscape that extends between two adjacent valleys. It is normally seen as lying above the steeper slopes of each valley side.

interglacial Period of warmer climate that separates two *glacial periods. Mid-latitude interglacials show a characteristic sequence of vegetation change. *Pollen of heathy *tundra is replaced in the pollen record by abundant herbaceous pollen, which in turn is replaced by that of *boreal and subsequently deciduous forest, including pollen of thermophilous (warmth-loving) species, e.g. *Tilia* (lime). From this peak the sequence reverses as the trend to colder conditions predominates.

intergrade Soil or *soil horizon that has the properties of two genetically different soils or horizons. An intergrade can be regarded as transitional between two distinctive soils or horizons.

intergranular Applied to an *igneous *texture, especially well developed in *basalts, in which the wedge-shaped spaces between a meshwork of lath-shaped crystals, such as *plagioclase, are filled with granules of other minerals.

intergranular displacement The displacement of individual grains within a rock that is undergoing *plastic deformation. Such grain movements induce permanent *strain in the rock.

intergranular pores *See* VOIDS.

inter-limb angle *See* FOLD ANGLE.

interlocking *See* INTERDIGITATING.

intermediate rock *Igneous rock whose chemical composition lies between those of *basic and *acidic rocks, e.g. *andesite. The limits are not fixed rigidly and a number of schemes exist that are based on modal mineralogy and the whole rock chemistry (*see* MODAL ANALYSIS). *Compare* ACID ROCK; and BASIC ROCK. *See also* ALKALINE ROCK.

intermittent stream A stream which ceases to flow in very dry periods. Such streams tend to have permeable beds and during periods of flow water leaking through their beds is added to the local *groundwater. The chalk bournes of southern England provide typical examples of intermittent streams. *See also* LOSING STREAM.

intermontane **1.** Between mountains or mountain ranges. **2.** Applied to basins which are being infilled by sediment eroded from surrounding mountains.

internal angle of friction *See* ANGLE OF SHEARING RESISTANCE.

internal mould *See* FOSSILIZATION.

internal node Within a *phylogenetic tree, a point where two branches join, representing an ancestral species or gene.

internal reflection (**IR**) The reflection of light off *cleavage or *fracture planes just below the surface of certain slightly translucent *ore minerals (e.g. *cassiterite and *sphalerite) and seen as a faint glow when the mineral is viewed under *crossed polars in reflected-light or ore microscopy.

internal standard In many instrumental analytical techniques, the mixing of accurately known amounts of a convenient element or known compound with the sample being considered. For example, in emission spectrometry, an internal standard is employed in order to relate the intensities of the line spectra to the concentration.

internal wave Wave that forms within a water mass at the boundary of two water layers that have different densities. The boundary may be abrupt or gradual, and the slow-moving waves can be detected only by instrumental observations of temperature or *salinity, and by acoustic scattering.

International Cometary Explorer *See* INTERNATIONAL SUN–EARTH EXPLORER-C.

International Geomagnetic Reference Field (**abbreviation IGRF**) The best mathematical fit to the observed *geomagnetic field at any specific time. It is usually evaluated annually.

International Gravity Formula Formula used to determine the *gravitational acceleration at a given latitude (g_ϕ) for a model of the Earth that comprises a rotating, oblate spheroid. $g_\phi = g_0(1 + \alpha \sin^2\phi + \beta \sin^2\phi)$, where g_0 is the value at the equator, of 978.0318 gals, and the constants α and β are 0.0053024 and –0.0000058.

International Gravity Standardization Network A network of stations where the absolute *gravitational acceleration has been established. This enables *gravimeters that are differential instruments to be calibrated, so that they provide absolute values. Such stations are commonly at or near airports.

International Programme of Ocean Drilling (**IPOD**) An international project that evolved from the *Deep Sea Drilling Programme, and was supported financially by the USA, USSR, West Germany, and France. The project has involved scientists from many countries, and the drilling and analysis of many deep *boreholes in the deep sea and *continental shelf areas.

International Sun–Earth Explorer-C (**ISEE-3, ISEE-C, Explorer 59**) One of the three spacecraft comprising a *NASA mission, launched in 1978, to study the relationship between Earth and the Sun, the *solar wind, and *cosmic rays. In 1982, ISEE-3 was removed from its orbit about one of the *Lagrangian libration points, eventually into a heliocentric orbit ahead of Earth that would intersect the comet *Giacobini–Zinner, at which point the spacecraft was renamed International Cometary Explorer (ICE). It traversed the plasma tail of the comet in September 1985.

interparticle porosity *See* CHOQUETTE AND PRAY CLASSIFICATION.

interpenetrant twin (penetration twin) A twinned *crystal where parts of the twin appear to intertwine to give an irregular or indefinable contact surface. It is a special kind of *contact twin, e.g. *quartz Dauphiné twins and interpenetrant twins of *fluorite.

inter-record gap *See* SEISMIC GAP.

intersection cleavage A *cleavage which crosses another planar feature and consequently forms an intersection *lineation with that feature.

intersection lineation *See* DIRECTIONAL FABRIC; INTERSECTION CLEAVAGE; and LINEATION.

intersertal Applied to an *igneous *texture, especially well developed in *basalts, in which the wedge-shaped spaces between a meshwork of lath-shaped crystals, such as *plagioclase, are filled with *glass.

interstade (interstadial) Phase of warmer climate within a glacial period, but of shorter duration (and thought to be less warm) than an *interglacial. Warmth-demanding (thermophilous) species, e.g. *Tilia* (lime), are not represented in the *pollen record, which shows *Boreal affinities. The absence of thermophilous species may, however, be as much a consequence of the shorter time-span of an interstade as of the lack of warmth.

interstadial *See* INTERSTADE.

interstellar clouds Unusually dense patches of gas and dust, usually about 10 000 times more massive than the *Sun, from which stars are believed to form. The clouds break up into many smaller, rotating fragments, which may become stars.

interstellar medium Substance, predominantly hydrogen (with some calcium, sodium, potassium, *hydrocarbons, and cyanogen) found in the space between the stars, mainly in the plane of the Milky Way.

interstitial Pertaining to the spaces (interstices) between sedimentary particles.

intertidal zone Area between mean high-water level and mean low-water level in a coastal region. *See* LITTORAL ZONE.

intertropical confluence Alternative term for the *intertropical convergence zone (ITCZ), preferred by some purists because of the discontinuous occurrence of convergence within it.

intertropical convergence zone (ITCZ, equatorial trough) Low-latitude zone of convergence between *air masses coming from either hemisphere at the boundary between north-easterly and south-easterly trade winds. Low-latitude depressions often form along the zone, which moves latitudinally with the seasons, their occurrence being mainly in the ocean sectors and sometimes leading to tropical hurricanes (or typhoons) when the zone is displaced relatively far from the equator. Over land, continental wind systems, e.g. the south-westerly monsoon and the hot, dry, desert winds in Africa, converge at the zone. *See* INTERTROPICAL CONFLUENCE.

intertropical front Meeting-point of air brought by the trade winds from the circulation of winds of the northern and southern hemispheres. It is not always a sharp *front, but is always a convergence zone. The equatorial rains are associated with this convergence.

interval The time elapsing between two geologic events. *See also* POLARITY INTERVAL.

intervallum The space between the inner and outer walls of *Archaeocyatha.

interval time The difference in *two-way travel time between two reflection events on a *seismic section.

interval velocity (V_{int}) Seismic velocity over a depth interval z. If the rock type is uniform through that depth interval, then V_{int} is equal to the *formation velocity. If the depth interval covers a number of rock beds, then the interval is equal to the *average velocity (V) calculated over the distance z. If z_i is the thickness of the ith interval and t_i is the one-way travel time through it, then $V_{int} = z_i/t_i$. A specific form of V_{int} is given by the *Dix formula, where the interval is defined in terms of the two-way travel time rather than by a discrete difference in depth.

interval zone (interbiohorizon zone, biostratigraphic interval-zone) The unit of *strata lying between the top boundary of a distinctive lower *biostratigraphic unit and the base of an equally distinctive, but different, upper biostratigraphic unit. An interval-zone may or may not contain *fossils.

intortus Twisted or entangled, the name of a species of *cirrus cloud. *See* CLOUD CLASSIFICATION.

intra- From the Latin *intra* meaning 'inside', a prefix meaning 'within' or 'on the inside'. **1.** Prefix used in the *Folk classification to specify a *limestone dominated by *intraclasts. **2.** General prefix for a process or object found or developed within the setting in question, e.g. 'intraformational conglomerate' is a conglomerate derived from within the *formation of deposition.

intraclast A *carbonate fragment of lithified, or partly lithified *sediment, derived from the erosion of nearby sediment and redeposited within the same area (*compare* EXTRACLAST). Such reworked fragments are often amorphous and structureless. Carbonate lumps, formed by the complete micritization (*see* MICRITE) of shell fragments are indistinguishable from true intraclasts. They are, of necessity, classified with intraclasts, although clearly they are not true intraclasts.

intrafolial fold A thinned and modified, tight to *isoclinal fold (*see also* FOLD) which commonly shows a fold *hinge with only vestiges of detached limbs (*see* FOLD LIMB). Such folds are the result of intense deformation which causes *bedding to be transposed along zones of shear and *solution. *See also* SHEAR STRESS.

intraformational *See* CONGLOMERATE.

intramicrite A *limestone consisting of *intraclasts set in a *micrite *matrix. *See* FOLK CLASSIFICATION.

intraparticle porosity *See* CHOQUETTE AND PRAY CLASSIFICATION.

intrasparite A *limestone consisting of *intraclasts cemented together with sparry *calcite (*sparite). *See* FOLK CLASSIFICATION.

intrinsic permeability *See* PERMEABILITY.

intrusion *See* INTRUSIVE.

intrusive Applied to a body of rock, usually *igneous, that is emplaced within pre-existing rocks. Intrusions are classified according to their size, their shape, and their geometrical relationship to the enclosing rocks.

intrusive phonolite *See* TINGUAITE.

invaded zone *See* FLUSHED ZONE.

Inverian A sub-*stage of the *Lewisian, from about 2300–1600 Ma ago (Van Eysinga, 1975). Other authors place this stage within the *Scourian as a metamorphic event.

inverse problem The problem of determining the nature of a physical feature by examining the effects it has on, for example, potential fields, as in *gravity surveying when an anomaly is interpreted to produce a geologic model. *Compare* FORWARD PROBLEM.

inversion A reversal of a particular trend. **1.** A rock sequence in which the younger sediments are at the bottom. Inversion can be caused by *overfolding or *thrusting. It is a major feature of an *accretionary wedge in which progressively younger oceanic and *trench *sediment is *underthrust, so that while each thrust slice is not inverted, each new thrust affects younger sediment, so producing the inversion. In a positive inversion, *normal faults on *passive margins become thrusts on collision. In a negative inversion thrusts become normal faults at the end of an *orogeny. The inversion of sediments takes place during *strike-slip faulting. **2.** During an *orogeny, the uplift that follows subsidence as a reversal of vertical direction. **3.** Seismic velocity usually increases with depth, but occasionally a zone of anomalously low velocity occurs between layers of higher velocities giving rise to a velocity inversion. **4.** *See* polarity reversal. **5.** *See* TEMPERATURE INVERSION.

inversion axis In *crystallography, an *axis of symmetry which can be inverted through 180° about its centre in order to achieve a higher degree of symmetry for the crystal. Thus, an axis of two-fold symmetry on inversion every 90° becomes an axis of four-fold symmetry. An inversion axis also obviates the need for a *centre of symmetry. The notation used to indicate an inversion axis is a 'bar' above the maximum number of repeat positions, e.g. $\bar{2}$ and $\bar{4}$ for a two-fold and four-fold inversion axis respectively.

invertebrate An animal without a backbone; invertebrates make up about 95% of all animal species and are found in every available habitat on Earth.

inverted relief Inverse relationship between a land-form and the underlying geologic structure, as when a hill is developed in a *syncline and a valley in an *anticline. It is a stage beyond the 'normal' or Jura-type relief which is characterized by anticlinal hills and synclinal valleys. The term also denotes the more general case in which, through *erosion, a hill becomes a valley or vice versa.

involute Having edges that roll under or inwards. Applied to those coiled cephalopods (*Cephalopoda) where the final *whorl of the shell envelops earlier ones. The opposite of involute is evolute, where all the previous whorls are exposed. The term 'serpenticone' is sometimes applied to those evolute cephalopods where the shell resembles a coiled snake. Very laterally compressed shells (discus-shaped) are called 'oxycones'; very flat and inflated types are called 'cadicones'; those which are subspherical are called 'sphaericones'.

involution (festooning) Contorted bedding in the near-surface zone of unconsolidated earth material or bedrock. Deformation may be regular, producing festoon-like features (e.g. 'frost boils'), or highly irregular, showing pronounced distortion and twisting. Involution is characteristic of past and present *periglacial zones, and is due basically to ground freezing.

Io (Jupiter I) One of the *Galilean satellites and the most geologically (and especially volcanically) active body in the solar system, its volcanoes being due to heat generated by tidal heating and eruption temperatures reaching more than 1000 K. Io radiates more heat than it receives from the Sun. The *Voyager spacecraft observed 9 volcanic eruptions. It has a metallic core, rock mantle, and a rocky surface covered with sulphur and sulphur compounds, including sulphur dioxide frost. Io was discovered in 1610 by Simon Marius and Galileo. Its equatorial radius is 1821.3 km; mass 8.93×10^{22} kg; mean density 3530 kg/m^3; visual albedo 0.61; mean distance from Jupiter 421 600 km; orbital period 1.769138 days; rotational period 1.769138 days; surface temperature about −143 °C, but with one volcanic feature with a temperature measured as 17 °C.

ion Atom that has acquired an electric charge by the loss (cation; positive charge) or gain (anion; negative charge) of one or more electrons.

ion exchange (IX) Reversible exchange of *ions in a crystal for other ions in solution, without disturbance of the crystal lattice or its electrical neutrality. This occurs by *diffusion, particularly in crystals where weakly bonded ions form one- or two-dimensional channelways. Natural *zeolites are used to capture *anions and *cations from solution. Artificial ion-exchange resins with three-dimensional *hydrocarbon networks are commonly used (e.g. in water softeners; for separating *isotopes; in *desalination; and in the chemical extraction of elements from ores).

ionic bond Bond formed when an *electron is transferred from one atom to another. The atom that loses the electron becomes a positively charged *ion and the atom that gains the electron becomes a negatively charged ion. A strong electrostatic force then bonds the two ions together. The bonding in a sodium-chloride crystal (NaCl) is ionic, the crystal lattice containing Na^+ ions and Cl^- ions. *Compare* covalent bond.

ionic charge The electrical charge of an *ion, created by the gain (negative charge) or loss (positive charge) of one or more *electrons from an atom or group of atoms. Ionic charge is important in determining the strength of bonding in *minerals (e.g. Fe^{3+} makes a stronger bond than Fe^{2+} with similar ions), and also which elements can substitute for each other within a crystal *lattice. *See also* IONIC BOND; IONIC RADIUS; and VALENCY.

ionic potential Measure of the strength of attraction of *ions, expressed as the ratio of *ionic charge (Z) to *ionic radius (r): Z/r.

ionic radius Half the distance between the 'centres' of two *ions of the same element. Although no precise measurement can be made of the size of individual *ions, in practice various techniques (e.g. *X-ray diffraction) can be used to estimate ionic radii in particular crystal structures. Generally, it is found that: (*a*) within the same group of the periodic table ionic radius increases with increasing atomic number; (*b*) for elements of the same period (i.e. same horizontal row) that form positive ions, ionic radius decreases with increasing positive charge (reflecting the greater nuclear attraction on the same number of extranuclear *electrons), for example, $Na^+ = 1.02$, $Mg^{2+} = 0.72$, $Al^{3+} = 0.53$, $Si^{4+} = 0.40$; (*c*) for the same reasons, if an element can exist in different valence states, the higher the positive charge the smaller the ion, for example, $Mn^{2+} = 0.82$, $Mn^{3+} = 0.65$; (*d*) for elements of the same period forming negative ions, the ionic radius increases with increasing negative charge (due to electronic repulsion).

ionic substitution (proxy, diadochy) Replacement or replaceability of one or more kinds of *ion in a *crystal lattice by other kinds of ions of similar size and charge (e.g. in the *olivine series Fe^{2+} and Mg^{2+} substitute for each other).

ionization potential Energy needed to drive an *electron from an atom or molecule without imparting kinetic energy to the electron; this leaves a positive *ion (*cation).

ionosphere The part of the atmosphere that lies above about 80 km altitude, with the highest concentrations of *ions and free *electrons. The most intense concentration is at 100–300 km altitude. Long-distance radio communications use waves that are reflected by certain regions of the ionosphere where there are particular concentrations of ions and free electrons. This allows radio waves to be transmitted around the curved surface of the Earth. Communications satellites make it easier to transmit higher-frequency waves (e.g. television transmissions) around the Earth, but reflection from the ionosphere continues to be used for radio transmissions, being cheaper.

ion pair A positive *ion and a negative ion produced by the transference of an *electron from one atom or molecule to another.

IPOD *See* INTERNATIONAL PROGRAMME OF OCEAN DRILLING.

Ipswichian 1. The temperate, last *Pleistocene *interglacial, named after Ipswich, Suffolk, England. **2.** Late Pleistocene deposits that occur in river valleys, often associated with *terraces. Pollen diagrams compiled from them indicate that the climate was not much different from that of the present day. The deposits can perhaps be correlated with those of the *Eemian of north-western Europe.

IR *See* INTERNAL REFLECTION.

iridescence Physical phenomenon in which fine colours are produced on a surface by the interference of light that is reflected from both the front and back of a thin film.

iridescent clouds High clouds, usually observed within 30° of the Sun, with red, green, yellow, blue, or violet tints along their edges. Tiny cloud particles cause diffraction of light rays, producing the coloured effect.

iridium anomaly The anomalously high (typically 50 p.p.b.) concentrations of iridium, relative to typical crustal abundances of less than 1 p.p.b., observed world-wide in sediments straddling the *Cretaceous–*Tertiary boundary. The anomaly is attributable to fall-out resulting from a massive *asteroidal or *cometary impact, that is thought by many to have been responsible for the mass *extinctions that define the Cretaceous–Tertiary boundary.

Irish elk (*Megaloceras giganteus*) *See* ALLOMETRY.

IRM *See* ISOTHERMAL REMANENCE.

iron, native Native occurrences of the metal, Fe, may be of terrestrial or meteoritic origin; sp. gr. 7.5; hardness 4.5; grey; *massive or granular; malleable; if meteoric may be alloyed with nickel and consists of small fragments that have fallen from the outer atmosphere; terrestrial iron is fairly rare but occurs in *igneous rocks in Greenland and in carbonaceous sedimentary deposits.

iron formation Iron-rich *sedimentary rocks, mostly of *Precambrian age, containing at least 15% iron. The iron occurs as an *oxide, *silicate, *carbonate, or *sulphide, deposited as laminated, deep-water, shelf-sea, and lagoonal *sediments, often associated with *cherts (*see also* BANDED IRON FORMATION). Other iron formations contain iron-rich *ooids, *pellets, and *intraclasts, representing deposits comparable to shallow marine *limestones. The source of the iron in iron formations is the subject of considerable debate; origins from volcanic sources, biochemical precipitation, and the diagenetic (*see* DIAGENESIS) replacement of *limestones are among the suggestions that have been made.

iron glance *See* HEMATITE.

iron meteorite (siderite) *Meteorite composed of iron and nickel (4–30% Ni) with only a small proportion of *silicate minerals. Although very common in museum collections, iron meteorites form only a few per cent of observed meteorite *falls.

iron pan *Indurated *soil horizon, found usually at the top of the B horizon, in which iron oxide is the main *cementing material.

ironstone Iron-rich *sedimentary rock. The source of the iron is primary and/or diagenetic (*see* DIAGENESIS), developed by a number of possible processes. These include: (*a*) replacement of *carbonate particles by *hematite, *siderite, or *chamosite; (*b*) diagenetic development of nodules, and more continuous horizons, of *siderite within *claystone sequences; (*c*) primary deposition of *ooids from Fe/Al-rich gels, and subsequent conversion to *chamosite during shallow burial. *See* PLINTHITE. *See also* MINETTE; and CLINTON IRONSTONE.

irradiance The flux density of *electromagnetic radiation falling onto a surface.

irregular echinoids Echinoids (*Echinoidea) in which a *bilateral symmetry is superimposed on the *radial symmetry and where the apical system no longer contains the *periproct. In some groups, e.g. the spatangoids, the *aboral portions of the ambulacra (*see* AMBULACRUM) terminate above the *ambitus and are situated in recessed, leaf-like 'petals', the mouth is situated far forward, and behind the mouth there is a flattened area (the 'plastron') that bears flat spines and is formed from the posterior interambulacra (*see* INTERAMBULACRUM).

Irregulares (Anthocyathea; phylum *Archaeocyatha) Class of solitary, rarely colonial invertebrate animals found in Lower, Middle, and Upper *Cambrian rocks. The conical cup is from cylindrical to discoid in outer form, often with an irregular outline, having one, or more usually two, porous walls. The *intervallum contains rods and bars, or *septa, always with *dissepiments and commonly with *tabulae. *Compare* REGULARES.

irrigation The process of artificially augmenting the amount of water available to crops. The water may be sprayed directly on to the plants or made available to their root systems through a series of surface channels or ditches.

irrotational wave *See* P-WAVE.

I_s *See* POINT LOAD INDEX.

ISAS *See* INSTITUTE OF SPACE AND ASTRONAUTICAL SCIENCE.

ischium In *tetrapods, the part of the *pelvis that projects backward on the ventral side. In Primates, it bears the weight of the sitting animal.

ISEE-C (ISEE-3) *See* INTERNATIONAL SUN–EARTH EXPLORER-C.

island arc Series of *volcanoes that lies on the continental side of an oceanic *trench of a *lithospheric plate. The *volcanicity, whose products are mainly of intermediate composition, results from the *subduction process; typically it occurs approximately 100 km above the down-going oceanic plate. Island arcs are the sites of strong seismic activity, and have distinctive thermal and magnetic properties. *See also* BENIOFF ZONE; PLATE MOTIONS; and PLATE TECTONICS.

iso- From the Greek *isos* meaning 'equal', a prefix meaning 'equal'.

isobar Line on a weather map connecting points at the same atmospheric pressure. On surface charts the values are 'reduced' to sea level. Such isolines are drawn at a given interval in millibars. Contours of *isobaric surfaces may be drawn to represent surfaces in the upper atmosphere composed of points at the same pressure.

isobaric surface A surface on which any point experiences the same atmospheric pressure. *See also* ISOBAR.

isobase A line drawn on a map linking points at a particular height above sea level (i.e. a contour line) but that lie on surfaces formed at sea level at a specified time in the past.

isobath A line drawn on a map linking points at the same depth below the water surface.

isochore A line on a map which joins points of equal vertical interval between two datum planes.

isochron A line joining points of equal time intervals or ages. In *geochronology the slope of the isochron may be used to determine the age of a suite of rocks. For example, if it is assumed that all the rocks formed from one *magma had the same initial $^{87}Sr:^{86}Sr$ ratio (*see* initial strontium ratio), then there is a simple equation to describe the growth of *radiogenic $^{87}Sr:^{87}Sr = {}^{87}Sr_0 + {}^{87}Rb(e^{\lambda t} - 1)$ where $^{87}Sr_0$ is the number of atoms of the ^{87}Sr *isotopes incorporated into the rock at the time of formation, ^{87}Sr and ^{87}Rb are the numbers of these isotopes after time *t*, and λ is the *decay constant. Because the number of ^{86}Sr is constant, we can derive an equation: $^{87}Sr:^{86}Sr = {}^{87}Sr_0:{}^{86}Sr + ({}^{87}Rb:{}^{86}Sr)(e^{\lambda t} - 1)$ which would give a family of straight lines when plotted on a graph of

$^{87}Sr:^{86}Sr$ (on the *y* axis) against $^{87}Rb:^{86}Sr$ (on the *x* axis). All rock specimens belonging to a co-magmatic suite will plot as points on a straight line called an 'isochron' because all points on the line represent systems having the same age (t) and the same initial $^{87}Sr:^{86}Sr$ ratio. In order to date co-magmatic *igneous rocks by the *whole-rock isochron method, a suite of rocks must be collected which span as wide a range of Rb/Sr ratios as possible, so that the slope of the isochron will be well defined. The age of the suite of rocks is obtained from the slope (m) using the equation $m = e^{\lambda t} - 1$. Isochrons can also be determined in *lead–lead dating by plotting a series of *growth curves. *See also* ISOCHRON MAP.

isochron map 1. A map showing the variation in the difference of the *two-way travel times between two *seismic reflection events. **2.** A map of two-way travel times to the same reflector event marked by *isochrons (lines of equal travel time).

isochronous Applied to events or lithological units that either (*a*) occupy the same time interval (i.e. are of the same duration); or (*b*) occur simultaneously or are of the same age. In the latter sense 'isochronous' is a synonym for 'synchronous'. *Compare* DIACHRONOUS.

isoclinal fold A *fold in which the two limbs are parallel.

isoclinic chart A map showing lines of equal *inclination of the *geomagnetic field.

isoconductivity map A map whose contours link points of equal electrical or thermal *conductivity.

isofrigid *See* PERGELIC.

isogal Line on a map joining places of equal *gravitational acceleration.

isogon A line joining points of equal *dip on the inner and outer bounding surfaces of a folded layer. *See* DIP ISOGON METHOD.

isogonic Applied to lines on a map joining places of equal angle, commonly geomagnetic *declination.

isograd *See* INDEX MINERAL.

isogyre *See* INTERFERENCE FIGURE.

isohel Line connecting points of equal average sunshine duration.

isohyet Line on a climate map connecting points of equal average rainfall.

isohyperthermic *See* PERGELIC.

isomagnetic chart A map of equal magnetic elements, usually geomagnetic intensity or direction.

isomesic *See* PERGELIC.

isometric *See* CUBIC.

isomorphous Applied to two compounds having the same, or nearly the same, crystal *form and containing *ions of approximately the same size or relative size. Isomorphous compounds may show *solid solution. *Compare* ISOTYPIC.

isomyarian *See* MUSCLE SCAR.

isoneph Line on a map joining points of uniform cloud cover.

isopach A contour line joining points of equal thickness in a rock layer. The isopach should represent the true thickness, i.e. corrected for dip effects, but this is rare. Isopachs are essential for estimating the volumes and dispersal of volcanic ashes (i.e. the thickness in centimetres or metres of an air-fall ash around a volcanic vent).

isopach map A subsurface geologic map showing *isopachs in plan view throughout a particular geographic area. Isopach maps are constructed, for example, to enable an estimate to be made of the size and shape of a *petroleum *reservoir, or of the approximate topographic relief of an underlying, older land surface.

isopycnal A line joining points of equal density within a water mass. A three-dimensional surface of equal density is called an isopycnal surface.

isopygous *See* PYGIDIUM.

isoresistivity map A map whose contours link points of equal electrical *apparent resistivity.

isoseismal map A map with lines of either equal seismic intensity or equal frequency of *earthquakes.

isospore *See* SPORE.

isostasy A model for the upper region of the Earth in which differences in elevation are compensated by either low-density roots or lower-density surface rocks. The rigidity of the tectonic *plate allows some departure from this model. *See* ISOSTATIC ANOMALY; AIRY HYPOTHESIS; and PRATT HYPOTHESIS.

isostatic anomaly A *gravity anomaly on a scale of more than 100 km that is associated with areas previously loaded, e.g. by ice (*see* GLACIO-ISOSTASY), lakes, etc., or where recent tectonic activity has loaded the *crust, e.g. mountain formation, volcanic loading, etc. It is generally removed from *Bouguer anomalies as part of the regional gradient, but it can also be calculated. *See* AIRY MODEL; and PRATT MODEL.

isostatic compensation The flexural adjustment of the *lithosphere, increase in topography, or presence of low-density roots that is introduced into a model to account for *isostatic anomalies. The actual compensation depends on the model used for the Earth's lithospheric structure. *See* AIRY MODEL; and PRATT MODEL.

isostructural *See* ISOTYPIC.

isotherm Line on a climate map connecting points of equal average temperature.

isothermal remanence (IRM) The remanence acquired by *ferromagnetic substances when they are placed in direct magnetic fields at room temperature. *See* COERCIVITY; and SATURATION REMANENCE.

isothermic *See* PERGELIC.

isotope One of two or more varieties of a chemical element whose atoms have a common number of protons and electrons (i.e. their atomic number is the same) but which vary in the number of neutrons in their nucleus (i.e. their atomic weight, signified by their mass number, is different). For example, hydrogen exists in the forms $^{1}_{1}H$ (one proton, no neutron), $^{2}_{1}H$ (deuterium: one proton, one neutron), and $^{3}_{1}H$ (tritium: one proton, 2 neutrons). Water in which $^{2}_{1}H$ replaces the more common $^{1}_{1}H$ is known as 'heavy water'. There are 300 naturally occurring isotopes, but only 92 naturally occurring elements, and in nature elements often occur as a mixture of isotopes, with one form being the most common. Isotopes may be produced by various nuclear reactions and the products are frequently radioactive. There are three different ways of specifying an isotope; for example ^{235}U, U-235, and uranium 235 all indicate the isotope of uranium with a mass number of 235. *See also* ISOTOPIC DATING.

isotope dilution An analytical technique used to determine the concentration of an element in a sample by means of a mass spectrometer. The method is based on the determination of the isotopic composition of the element in a mixture. A known quantity of a compound containing an unknown quantity of a particular element is mixed with a *spike (a known weight of a radioactive *isotope of the element). The specific activity (disintegrations per second per kilogram) of the spike is known precisely, so the isotopic composition of the mixture can be used to calculate the amount of the element in the sample. A small amount of the mixture is isolated from the sample, weighed, and its specific activity measured. The concentration of the inactive element in the sample may be estimated by the dilution of the radiotracer. Isotope dilution analysis can be applied to all elements that have two or more naturally occurring isotopes (about 80% of all elements), provided that a spike enriched in one of the isotopes of that element is available. As a technique it has several advantages over other analytical methods. It is free of interference from other elements present and its accuracy is governed by the calibration of the spike solution.

isotope fractionation The separation of *isotopes of an element during naturally occurring processes as a result of the mass differences between their nuclei. Although the chemical properties of the *isotopes of an element are the same, there are differences in their physical properties (e.g. density, vapour pressure, boiling point, and melting point) due to the greater vibrational energy of the lighter isotope. Separation (fractionation) of isotopes will occur during such processes as *evaporation or condensation, melting or crystallization, *diffusion through crystals, and isotopic exchange reactions between water in a *melt and *minerals or mineral pairs. The extent of fractionation is dependent on temperature and is more pronounced the greater the mass difference between isotopes in relation to their individual isotopic mass. Significant fractionation occurs naturally with carbon, oxygen, sulphur, and hydrogen-deuterium. Fractionation ratios and isotopic ratios are useful in determining palaeotemperatures, geologic processes, and the modes of formation of rocks and minerals. *See* D/H RATIO; OXYGEN-ISOTOPE RATIO; OXYGEN-ISOTOPE ANALYSIS; STABLE-ISOTOPE STUDIES; and ISOTOPE GEOCHEMISTRY.

isotope geochemistry The study of the abundance ratios of *isotopes (both *stable and *radioactive) of major and *trace elements in rocks (e.g. Rb/Sr, Pb/U, etc.), to elucidate a number of geologic problems and processes. These include the age relationships of rocks, and the age of the Earth itself (*see* GEOCHRONOLOGY; ISOTOPIC DATING; and RADIOMETRIC DATING); palaeotemperatures and geothermometry (*see* OXYGEN-ISOTOPE ANALYSIS; and GEOTHERMOMETER); and the provenance of natural waters, ore-forming fluids, and *magmas (*see* ISOTOPE FRACTIONATION; SMOW; D/H RATIO; OXYGEN-ISOTOPE RATIO; and STABLE-ISOTOPE STUDIES).

isotope hydrology The use of naturally occurring and introduced *isotopes to date and identify water bodies. Among the most commonly used isotopes are tritium, deuterium, carbon-13, carbon-14, chlorine-36, and oxygen-18.

isotope tracer Radioactive isotope, whose movement can be monitored, that is used to trace the pathways by which individual substances move through an organism, a living system, the *abiotic environment, etc. Non-radioactive chemical analogues of certain substances may be used for the same purpose if their movement can be monitored (e.g. caesium, which can be substituted for potassium).

isotopic dating Means of determining the age of certain materials by reference to the relative abundances of the parent *isotope (which is radioactive) and the daughter isotope (which may or may not be radioactive). If the decay constant (the *half-life or disintegration rate of the parent isotope) and the concentration of the daughter isotope are known, it is possible to calculate an age. *See also* DATING METHODS; RADIOACTIVE DECAY; RADIOCARBON DATING; and RADIOMETRIC DATING.

isotrophic *See* BELLEROPHONTIFORM.

isotropic Applied to substances whose optical or other physical properties are the same from whatever direction they are observed. In thin-section and polished-section microscopy, an isotropic mineral has only one *refractive index or one *reflectance value respectively, whatever the orientation of the mineral as indicated in *crossed polars. The beam of light is not split into two vibration directions, therefore, but passes through or is reflected off the mineral with no change in its optical characteristics.

isotropic indicatrix *See* INDICATRIX.

isotropy The uniformity of physical characteristics of a medium irrespective of the direction in which they are measured. *Compare* ANISOTROPY; and INHOMOGENEITY.

isotypic (isostructural) Applied to a pair of *isomorphous compounds in which the relative sizes of the *ions to each other are the same in each pair, but their absolute sizes are different. *Solid solution is impossible.

isovelocity plot *See* VELOCITY SURVEY.

isovol A line drawn on a map of a coalfield joining points at which coals have similar proportions of *volatiles.

isp. Abbreviation for *ichnospecies.

ISSC International Subcommission on Stratigraphic Classification.

Isuan A *system of the Lower *Archaean, extending from about 3875 Ma to about 3525 Ma ago, and named from the Isuan supercrustal rocks of western Greenland that include the Amitsôq *gneisses.

itabirite The name given in Brazil to *banded iron formations.

ITCZ *See* INTERTROPICAL CONVERGENCE ZONE.

iterative evolution Repeated *evolution of similar or parallel structures in the *development of the same main line. There are many examples of iterative evolution in the fossil record, spanning a wide range of groups. This evolutionary conservatism probably is due to the overriding morphogenetic control exerted by certain regulatory *genes.

IUGS International Union of Geological Sciences.

IVD Abbreviation for *ignimbrite veneer deposit. *See* MANTLE BEDDING.

Ivorian A *stage of the *Tournaisian, dated at 353.8–349.5 Ma (Harland et al., 1989).

I-wave *See* SEISMIC-WAVE MODES.

IX *See* ion exchange.

J

Jaccard's index (Jaccard's coefficient) In *biogeography, an index of faunal resemblance between two regions. It is calculated as $C/N_1 + N_2 - C$, where C is the number of taxa shared between a pair of regions and N_1 and N_2 are the number of species in each of the two regions.

jacobsite *See* MAGNETITE.

Jacob's staff A calibrated rod used in the field to measure the thickness of *strata. It is usually marked off in 10 cm and 1 m intervals.

jade *See* JADEITE.

jadeite Isolated, rare *mineral of the *clinopyroxenes with the composition $NaAlSi_2O_6$; sp. gr. 3.2–3.4; *hardness 6; *granular crystals or *massive; jade is the precious variety of jadeite, although the term includes the *amphibole mineral nephrite; it occurs with *albite in *blueschists (*glaucophane schists) under conditions of low temperature and high pressure at *destructive plate margins.

Jameson, Robert (1774–1854) Professor of Natural History at the University of Edinburgh, who studied with *Werner and actively promoted Wernerian theories to his own students and in the *Edinburgh Philosophical Journal*. He translated Cuvier's *Discours préliminaire* into English (1813), adding Wernerian interpretations.

jamesonite *Sulphide, $Pb_4FeSb_6S_{14}$; sp. gr. 5.6; *hardness 2.5; greyish-black; *metallic *lustre; *massive, fibrous, or as *acicular crystals with a feather-like form ('feather ore'); occurs in vein deposits associated with other antimony sulphides.

Janjukian A *stage in the *Tertiary of south-eastern Australia, underlain by the *Aldingan, overlain by the *Longfordian (*Miocene), and roughly contemporaneous with the *Rupelian and *Chattian (Lower Tertiary) and lowermost *Aquitanian (Upper Tertiary) Stages.

Janus (Saturn X) One of the lesser satellites of *Saturn, with a radius measuring $99.3 \times 95.6 \times 75.6$ km; mass 0.0198×10^{20} kg; mean density 650 kg/m^3; visual albedo 0.8. It was discovered in 1966 by A. Dolfus, Gerard Kuiper, J. Fountain, and S. Larsen.

Japan Trench The oceanic *trench which lies between the northern Japanese islands (an *island arc) bordering the *Eurasian Plate and the *Pacific Plate.

Japan-type margin *See* ACTIVE MARGIN.

Jaramillo A normal *polarity subchron which occurs within the *Matuyama reversed *chron and is dated between 0.98 and 1.05 Ma (Bowen, 1978).

jarosite Member of the *alunite group of *minerals, $KFe_3(SO_4)_2(OH)_6$; sp. gr. 3; *hardness 3; yellowish-brown; *resinous *lustre; massive coatings or fine crystals; occurs as a *secondary mineral, frequently as an alteration product of iron-rich mineral deposits.

jasper A variety of chalcedonic (*see* CHALCEDONY) Silica (SiO_2) which is reddish-brown, opaque, and *cryptocrystalline. It may be banded to give 'Egyptian jasper' ('ribbon jasper'). A contact metamorphosed (*see* THERMAL METAMORPHISM) *shale may also be baked to give 'porcelain jasper' which may be similar in appearance but has a different composition.

jaspillite *Jasper interbedded with *hematite, the name given in Australia to *banded iron formations.

Java Trench The oceanic *trench which marks the outer, deep edge of the E. Indies *subduction zone, where the *Indo-Australian Plate is subducting beneath the *Eurasian Plate. The trench is about 6 km deep along Java, but becomes shallower to the north-west because of progressive in-filling by Bengal Fan *turbidites. There is an outer, non-volcanic arc, formed of an *accretionary wedge, and a *fore-arc basin.

jawless fish *see* AGNATHA.

Jeffreys–Bullen curves The travel times for seismic waves passing through the Earth, including direct, reflected, and refracted waves. They are fundamental to the determination of the density structure of the Earth's interior.

Jeffreys, Sir Harold (1891–1989) A British geophysicist, astronomer, and mathematician who studied *seismic waves and developed a model for the interior structure of the Earth. His tables of *earthquake travel times, calculated in collaboration with K. E. Bullen (1906–76), are still in use. He proposed the *tidal theory for the origin of the solar system.

jet A hard, black, and lustrous form of *lignite, found in isolated masses within organic-rich *shales. Jet is thought to form from the waterlogged debris of driftwood.

jet stream Concentrated, high-speed air flow, generally in a broadly westerly (i.e. west to east) direction. The principal global jets are the polar front and the subtropical jets, at heights of about 10–12 and 12–15 km respectively, and the polar-night or winter jet stream in the upper *stratosphere or *mesosphere at about 50–80 km. The intensity of the jets in narrow bands (the maximum velocity is commonly about 50–100 m/s, but greater speeds are sometimes observed) results from a large poleward increase in pressure gradient with altitude. This is a product of the pole–equator temperature gradient in the air beneath the jet. As pressure decreases more rapidly with height the lower the temperature, so pressure in the colder, polar air masses decreases more rapidly with height than over the regions with warmer air masses. Above the jet, wind speed diminishes as the pressure gradient declines with increasing height, owing to the effects of a different heating pattern in the stratosphere. *See also* POLAR JET STREAM; and SUBTROPICAL JET STREAM.

Johannian A *stage in the Lower *Tertiary of south-eastern Australia, underlain by the *Wangerripian, overlain by the *Aldingan, and roughly contemporaneous with the *Ypresian and *Lutetian Stages.

JOIDES *See* JOINT OCEANOGRAPHIC INSTITUTIONS FOR DEEP EARTH SAMPLING.

joint 1. A discrete brittle fracture in a rock along which there has been little or no movement parallel to the plane of fracture, but slight movement normal to it. Fracture may be caused by shrinkage, due to cooling or desiccation, or to the unloading of superincumbent rocks by *erosion or *tectonism. A group of joints of common origin constitutes a 'joint set' and the joints are usually planar and parallel or sub-parallel in orientation. 'Joint systems' comprise two or more joint sets, which are usually arranged systematically with respect to the *principal stress axes of regional deformation. Cooling joints (shrinkage joints), such as those which split a rock into long prisms or columns to form 'columnar joints', most commonly found in *lavas, are due to differential volume changes in cooling and contracting *magmas. Unloading joints result from erosional unloading of the *crust and form flat-lying, sheet-like joint sets, e.g. those found in granitic rocks. **2.** *See* VOIDS, TYPES OF.

Joint Oceanographic Institutions for Deep Earth Sampling (JOIDES) The original deep-sea drilling programme. This evolved into the *Deep Sea Drilling Programme (DSDP) and then the *International Programme of Ocean Drilling (IPOD).

joint set *See* JOINT.

joint system *See* JOINT.

jökulhlaup Sudden, violent, but short-lived increase in the discharge of a meltwater stream issuing from a *glacier or *ice-cap, sometimes due to volcanic activity beneath. A lake may develop above the heat source; this is subsequently breached to produce a torrent of meltwater, e.g. Lake Grimsvotn on Vatnajökull, Iceland. Flow velocity may reach 7–8 m/s and the discharge may attain 100 000 m^3/s (e.g. the Katlahlaups from Myrdalsjökull, Iceland), comparable to rates of flow of the Amazon river.

Jolly balance *See* DENSITY DETERMINATION.

Joly, John (1857–1933) An Irish physicist and geologist, Joly studied radioactivity in the Earth, showing that it was a source of internal heat and could give rise to convection currents in the interior. He never accepted that radioactivity could be used to determine the age of the Earth, believing that his own calculations, based on the *salinity of the oceans, were more reliable.

Jotnian A *stage of the Middle–Upper *Proterozoic, from about 1600 to 650 Ma ago, of the Baltic Shield region, underlain by the *Gothian and overlain by the *Varegian.

Jotnian orogeny An episode of mountain building that involved the *Proterozoic *sedimentary rocks of the Baltic Shield in what is now Scandinavia.

jovian Adjective derived from Jupiter and used in descriptions of the planet. In planetary geology it is not usual to spell the adjective with an initial capital letter.

jovian planet **(outer planet)** A general term used to refer to the four outer, giant, gaseous planets of the *solar system—*Jupiter, *Saturn, *Uranus, and *Neptune; it contrasts them to the small, rocky, inner or *terrestrial planets—*Mercury, *Venus, *Earth, and *Mars.

jovian satellites *See* ADRASTEA (JUPITER XV); AMALTHEA (JUPITER V); ANANKE (JUPITER XII); CALLISTO (JUPITER IV); CARME (JUPITER XI); ELARA (JUPITER VII); EUROPA (JUPITER II); GANYMEDE (JUPITER III); HIMALIA (JUPITER VI); IO (JUPITER I); LEDA (JUPITER XIII); LYSITHEA (JUPITER X); METIS (JUPITER XVI); PASIPHAE (JUPITER VIII); SINOPE (JUPITER IX); and THEBE (JUPITER XIV).

Juan de Fuca Plate The small lithospheric *plate in the north-east *Pacific Ocean which is being subducted slowly under the *North American Plate, giving rise to the generally andesitic volcanic chain from northern California to southern British Columbia. The Juan de Fuca Plate is a remnant of the *Farallon Plate. The Juan de Fuca Ridge is offset from the East Pacific Rise by the San Andreas Fault.

jug The colloquial name for a *geophone.

Juliet **(Uranus XI)** One of the lesser satellites of *Uranus, with a diameter of 42 km. It was discovered in 1986.

Juno A *solar system asteroid (No. 3), diameter 268 km; approximate mass 2×10^{19} kg; rotational period 7.21 hours; orbital period 4.36 years. It was discovered in 1804 by K. Harding.

Jupiter The fifth and largest planet in the *solar system, distant 5.203 AU from the *Sun. It has a radius of 71 900 km, and a mass 318 times and a volume 1403 times that of the *Earth. Its density is 1310 kg/m^3 and it is comprised mainly of hydrogen and helium. The atmosphere is 0.9H–0.1He (with traces of H_2O, CH_4, and NH_3) which grades down into a liquid shell, overlying a zone of metallic hydrogen. In the centre is a small rock–ice core of about ten Earth masses. Jupiter has at least 16 *satellites, including the four *Galilean satellites.

Jurassic One of the three *Mesozoic *periods: it lasted from 208 to 145.6 Ma, following the *Triassic and preceding the *Cretaceous. The Jurassic Period is subdivided into 11 *stages (*Hettangian, *Sinemurian, *Pliensbachian, *Toarcian, *Aalenian, *Bajocian, *Bathonian, *Callovian, *Oxfordian, *Kimmeridgian, and *Tithonian), with *clays, calcareous *sandstones, and *limestones being the most common rock types. *Brachiopods, *bivalves, and *ammonites are abundant fossils, along with many other invertebrate stocks. *Reptiles flourished on land and in the sea, but mammals were relatively insignificant and presumed to have been predominantly nocturnal. The first birds, including **Archaeopteryx*, appeared in the Late Jurassic.

Jura-type relief *See* INVERTED RELIEF.

Juvavic *See* NORIAN.

juvenile **1.** Applied to volcanic material derived directly from a *magma. Material derived from the surrounding wall rocks is termed 'foreign'. **2.** *See* JUVENILE WATER.

juvenile water Original water, formed as a result of magmatic processes. Juvenile water has never been in the atmosphere. Magmatic water can form in very large quantities. A *magma body with a density of 2.5, an assumed water content of 5% by weight, a thickness of 1 km, and an area of 10 km^2 contains some 1.25×10^9 m^3 of water. *See also* GROUNDWATER.

J-wave *See* SEISMIC-WAVE MODES.

Kaena A reversed *polarity subchron which occurs at about 2.87 ± 0.03 Ma within the *Gauss normal *polarity chron.

kaersutite A member of the alkali *amphiboles with the composition $(Na,K)Ca_2(Mg,Fe)_4Ti[Si_6Al_2O_{22}](OH)_2$ in a group which includes katophorite $Na(Na,Ca)(Mg,Fe^{2+})_4Fe^{3+}[Si_7AlO_{22}](OH)_2$ and oxyhornblende $NaCa_2(Mg,Fe,Fe^{3+},Al,Ti)_5[Si_6Al_2O_{22}](OH,O)_2$; sp. gr. 3.2–3.5; *hardness 5.0–6.0; small *euhedral *crystals; dark brown to black; this rare group of amphiboles occurs in *intermediate alkali *igneous rocks.

Kaiatan A *stage in the Lower *Tertiary of New Zealand, underlain by the *Bortonian, overlain by the *Runangan, and roughly contemporaneous with the *Bartonian Stage.

Kaihikuan *See* GORE; and LADINIAN.

kainite *Evaporite mineral, $MgSO_4.KCl.3H_2O$; sp. gr. 2.1; *hardness 2.5–3.0; *monoclinic; variable in colour, from white through yellow to reddish; vitreous *lustre; crystals rare, usually forms granular masses; occurs widely in salt deposits in association with *halite, *carnallite, etc. It is used as a fertilizer and as a source of potassium salts.

Kainozoic *See* CENOZOIC.

Kalb light line In optical *mineralogy, a test made on *polishing relief, which is visually similar to the *Becke line test, but different in origin. The microscope is focused clearly on the boundary between the two mineral grains. The microscope tube is raised (or the *stage lowered) and as the specimen begins to go out of focus a 'line of light' will move towards the softer mineral.

Kalimnan A *stage in the Upper *Tertiary of south-eastern Australia, underlain by the *Cheltenhamian, overlain by the *Yatalan, and roughly contemporaneous with the upper *Zanclian (Tabianian) and lower *Piacenzian Stages.

kaliophilite A potassium-rich *feldspathoidal mineral $KAlSiO_4$ that is closely related to *kalsilite.

kalsilite A *feldspathoidal mineral $KAlSiO_4$ and *end-member of a series with *nepheline $NaAlSiO_4$ with which it forms limited *solid solution; sp. gr. 2.6; *hardness 5.5; has been reported from the *groundmass of potassium-rich *lava flows.

Kama *See* KAZANIAN.

kame Steep-sided mound composed of bedded sand and gravel which often shows signs of marginal slumping. It is a land-form of glacial deposition, associated with stagnant ice whose removal by melting causes the collapse.

kame delta Flat-topped mound of stratified sand and gravel laid down in standing water at an ice margin. Subsequent ice melt leading to loss of support brings about collapse at the ice-contact margin.

kame terrace Continuous valley-side land-form consisting of stratified sand and gravel whose outer edge typically shows collapse features. It is laid down by meltwater at the junction between an ice mass and the valley wall.

kandite *See* ALLOPHANE; and CLAY MINERALS.

Kansan I and II The second (0.9–0.7 Ma) of four glacial *stages occurring in N. America, during which *isotope evidence suggests that the climate was less extreme than during the *Nebraskan episode, although the Kansan glaciation extended further south. The Kansan is approximately equivalent to the *Günz glacial of the Alps.

kaolin (china clay) *See* CHINASTONE; KAOLINITE; and KAOLINITIZATION.

kaolinite (dickite, nacrite, China clay, kaolin) A very important group of *clay minerals belonging to the 1:1 group of *phyllosilicates (sheet silicates), and with the general formula $Al_4[Si_4O_{10}](OH)_8$, kaolinite represents the final product from the *chemical weathering of feldspars to give clays; sp. gr. 2.6–2.7; *hardness 2.0–2.5; *monoclinic; white, greyish, or stained a variety of colours; dull earthy *lustre; on a microscopic scale crystals are hexagonal plates, but it is usually *massive; *cleavage

{001}; occurs as a *secondary mineral produced by the alteration of aluminosilicates, especially *alkali feldspars; is widely distributed in *igneous rocks, *gneisses, and *pegmatites, and in *sedimentary rocks due to the action of acidic solutions on a wide variety of rocks by processes of weathering and low-temperature *hydrothermal reactions. It is distinguished by its plastic feel, but normally it has to be identified by optical and physical tests. It is extensively used when pure as a cheap, general-purpose filler and coating material for paper, in ceramics, and also in chemicals and paints. It is distinctive in soils for its physical stability during wetting and drying and for its small *cation-exchange capacity. Kaolinite dominates the clay minerals present in certain acid and very old soils, but not in *Oxisols, some Latosols, or soils formed in arid climates. The name is derived from the Chinese *kau-ling* ('high ridge'), referring to the hill from which the first samples were taken.

kaolinitization (kaolinization) High-temperature hydrothermal alteration (*see* HYDROTHERMAL ACTIVITY) and replacement of *feldspars, to varying degrees, to form a fine-grained aggregate of the mineral *kaolinite. Kaolinitization in *granites can be so complete that the rock is reduced to a rotten, friable condition, *quartz being the only mineral to survive the process unscathed. In this condition, the altered granite can be easily broken down by a high-pressure water jet and the kaolinite settled out from suspension in settling pools to produce china clay (kaolin) concentrates. *See also* CHINASTONE.

kaolinization *See* KAOLINITIZATION.

Kapitean A *stage in the Upper *Tertiary of New Zealand, underlain by the *Tongaporutuan, overlain by the *Opoitian, and roughly contemporaneous with the uppermost *Messinian and lower *Zanclian (Tabianian) Stages.

Karatau A *period of the *Riphean sub-era, following the *Yurmatian, succeeded by the *Sturtian, and dated at about 1100–825 Ma (Harland et al., 1989).

Karatavian A *stage of the Upper *Proterozoic, from about 1000 to 700 Ma ago, underlain by the *Yurmatian (Yurmatin) and overlain by the *Vendian *Series (Van Eysinga, 1975).

Karelian *See* GOTHIAN.

Karelian orogeny An *Archaean episode of mountain building affecting the Baltic Shield along a NW–SE line in what is now central Finland. It began about 2000 Ma ago and ended about 1900 Ma ago.

Karginsky *See* VALDAYAN/ZYRYANKA.

von Karman–Prandtl equation An equation describing the logarithmic variation of water velocity within a channel from zero flow at the stream bed to a maximum velocity at the water surface. Originally developed in aerodynamics, the equation also describes the profile of wind velocity above the ground. *See* EDDY VISCOSITY.

K–Ar method *See* POTASSIUM–ARGON METHOD.

Karnian *See* CARNIAN.

karren German term describing the group of solutional features developed at the surface of an outcrop of hard *limestone. Forms range from runnels (shallow troughs) a few millimetres deep to fissures (*grikes) extending several metres into the limestone. The word is sometimes used as a synonym of *lapiés.

karst 1. Area of the Dinaric Alps of Yugoslavia underlain by *limestone. **2.** Any region underlain by limestone and characterized by a set of land-forms resulting largely from the action of *carbonation. **3.** By extension, regions in which other processes produce similar types of land-forms, e.g. *thermokarst in the *periglacial environment.

karstic aquifer An *aquifer within a *karst limestone rock matrix. Such aquifers are normally characterized by large void spaces, relatively high values for hydraulic conductivity (*see* PERMEABILITY), flat *water-tables, and extensive networks of solution channels within which *Darcy's law is not obeyed and flow may be *turbulent.

Kashirskian A *stage in the *Moscovian Epoch, preceded by the *Vereiskian and followed by the *Podolskian.

Kasimovian 1. An *epoch in the *Pennsylvanian, dated at 303–295.1 Ma (Harland et al. 1989) comprising the Krevyakinskian, Chamovnicheskian, and Dorogomilovskian Ages (these are also *stage names in eastern European stratigraphy). The Kasimovian is preceded by the *Moscovian and followed by the *Gzelian Epochs. **2.** The name of the corresponding eastern European *series, which is roughly contemporaneous with part of the *Stephanian (western Europe), and the *Missourian and lower *Virgilian (N. America).

katabatic wind (drainage wind, mountain breeze) Generic term for the wind that occurs when cold, dense air, chilled by radiation cooling, usually at night, moves downslope gravitationally beneath warmer, less dense air. The occurrence is frequent and widespread in, for example, the fjords of Norway, and as an outblowing wind over ice-covered surfaces in Antarctica and Greenland, where the wind may be extremely strong near the coasts and less severe in many mountain regions. *Compare* ANABATIC WIND.

katafront A weak frontal condition in which warm-sector air sinks relative to colder air. The term was coined by T. Bergeron. *Compare* ANAFRONT.

katophorite *See* KAERSUTITE.

Kawhia A *series in the *Jurassic of New Zealand, underlain by the *Herangi, overlain by the *Oteke, and comprising the Temaikan, Heterian, and most of the Ohauan, *Stages. It is roughly contemporaneous with the *Bajocian, *Bathonian, *Callovian, *Oxfordian, and *Kimmeridgian.

Kazanian 1. An *age in the Late *Permian, which is preceded by the *Ufimian, followed by the *Tatarian, and has its upper boundary dated at 253 Ma. Some authors extend the Kazanian to incorporate the Ufimian Age, referring to this longer time-span as the Kazanian (or Kama) Age. **2.** The name of the corresponding eastern European *stage, which is roughly contemporaneous with the upper *Zechstein (western Europe), upper *Guadalupian and lower *Ochoan (N. America), and the upper Basleoan and lower Amarassian (New Zealand).

kb *See* KILOBAR.

K-band A range of radar frequencies, 10–36 GHz, commonly used in doppler radar systems and for scanning vegetation cover.

keel 1. (carina) An external, longitudinal ridge situated on the *venter of an ammonoid (*Ammonoidea). **2.** In flying birds, a ridge projecting forward from the sternum and serving for the attachment of the enlarged pectoral muscles.

Keewatinian A *stage of the Upper *Archaean of New Zealand, overlain by the *Laurentian.

KEI *See* KEY EVOLUTIONARY INNOVATION.

Keilorian A *stage of the Lower *Silurian of south-eastern Australia, underlain by the *Bolindian (*Ordovician) and overlain by the *Eildonian.

kelly The driving chuck of a *rotary drill system through which the rotation is transmitted to the drill stem.

kelp Brown seaweeds that grow below the low-tide level. Large brown *algae, e.g. *Laminaria* species, which anchor themselves firmly to the sea bed are typical. Kelp accumulations are important sources of iodine and potash.

kelvin *See* KELVIN SCALE.

Kelvin, Lord *See* THOMSON, WILLIAM.

Kelvin scale A scale of temperature proposed by Kelvin (William *Thomson), which does not include negative values. The unit of the scale is the kelvin (K). The base of the scale, absolute zero, is the lowest possible temperature for all substances at which no molecule possesses any heat energy. The *triple point of water is given as 273.16 K.

Kenoran orogeny The name, no longer much in use, given to a *Proterozoic phase of mountain building affecting the shield area in what is now the Lake Superior region of Canada; it was followed by the *Hudsonian orogeny.

kenyte A type of *mafic *phonolite characterized by abundant large *phenocrysts of *anorthoclase, smaller phenocrysts of *nepheline, with or without *olivine, and with alkali *pyroxene, set in a glassy *groundmass (*see* GLASS). The type locality is Mount Kenya in E. Africa.

Kepler's laws of planetary motion (1) The orbits of the planets are ellipses with the Sun at a common focus. (2) The line joining a planet to the Sun sweeps out equal areas in equal times. (3) The squares of the periodic times are proportional to the cubes of the mean distances from the Sun. These laws were formulated by the German astronomer Johannes Kepler (1571–1630) and published during the period 1609–19.

keratophyre A fine-grained, *igneous rock consisting of *albite or *oligoclase as the major component, accompanied by *chlorite, *epidote, or *calcite. *Phenocrysts of unaltered *augite are sometimes present. The *mineral assemblage is almost entirely secondary in origin, indicating that the original rock was a *tholeiitic *andesite, the volcanic equivalent of plagiogranite. Keratophyres are found on the ocean floor, and in *ophiolite complexes where they are associated with *spilites. Like that of the spilites, the mineral assemblage of keratophyres may result from low-grade, ocean-floor *metamorphism.

kernal function (*K*) A mathematical function for *resistivity and depth, from which electrical layering can be computed from *apparent resistivity data.

kerogen Solid, bituminous substance formed of fossil organic material in *oil shales, which can yield oil by destructive distillation.

kersantite A type of *lamprophyre, characterized by essential *biotite and *plagioclase feldspar. If *augite is present the rock is termed an 'augite-kersanite'.

Ketilidian orogeny A *Proterozoic phase of mountain building that occurred about 1800–1600 Ma ago, affecting what is now Greenland. *See also* GARDAR RIFTING.

kettle hole (kettle) Depression in the surface of glacial drift (especially ablation or kettle moraine), resulting from the melting of an included ice mass. It may be filled with water to form a small lake ('kettle lake').

kettle lake *See* KETTLE HOLE.

Kew barometer Barometer in which the scale markings are adjusted to take account of the changes of the mercury level in the cistern, so eliminating the need for adjustment of the cistern to the *fiducial point as is required in the *Fortin barometer.

Keweenawan 1. A *stage of the Upper *Proterozoic of New Zealand, underlain by the *Huronian (Van Eysinga, 1975). **2.** A Lower-Middle *Proterozoic *system.

key bed *See* MARKER BED.

key evolutionary innovation (KEI) A newly evolved feature that preadapts the *clade possessing it for new ecological opportunities.

khamsin Hot, dry, dusty wind originating in N. Africa and blowing across Egypt, usually between April and June, ahead of *depressions which move eastward or north-eastward in the Mediterranean Sea or across N. Africa, with high pressure to the east. The term is also applied to very strong southerly or south-westerly winds in the Red Sea.

Kibalian orogeny *See* BUGANDO-TORO-KIBALIAN OROGENY.

kidney ore *See* HEMATITE.

kieselguhr *See* DIATOMACEOUS EARTH.

kill curve A graph, devised by D. M. Raup, on which the number of taxa becoming extinct is plotted against the time intervals between extinction events. It shows, for example, that an extinction of 30% of species occurs, on average, about every 10 Ma.

kilobar (kb) Unit of pressure, equal to 1000 bars (986.923 atmospheres, 10^8 N/m^2, or 10^8 Pa).

Kimberella quadrata A fossil of *Precambrian age, known from Australia and the coast of the White Sea, Russia, that is believed to have been a bilaterally symmetrical animal, possibly resembling a mollusc.

kimberlite A brecciated (*see* BRECCIA), potassic, *ultrabasic *igneous rock, consisting of *megacrysts of *olivine, *enstatite, Cr-rich *diopside, *phlogopite, pyrope-almandine *garnet, and Mg-rich *ilmenite, in a fine-grained *groundmass of *serpentine, phlogopite, *carbonates, *perovskite, and *chlorite, in varying proportions. Kimberlites contain abundant *xenoliths derived from the *mantle and, as such, they are extremely useful for investigating mantle mineralogy and chemistry. Many of the mantle-derived xenoliths have been brought up from such great depth that they contain high-pressure minerals, the best known being *diamond, the high-pressure form of carbon. Many of the megacrysts of enstatite, Cr-diopside and garnet are likewise thought to be derived from the upper mantle through which the kimberlite ascended, and as such should be termed '*xenocrysts'. Near the surface, kimberlite is found in clusters of pipe-like bodies called *diatremes. Mine workings show that they coalesce at depth and link up with *dykes of non-fragmental kimberlite.

Kimmeridgian A *stage in the European Upper *Jurassic (154.7–152.1 Ma), preceded by the *Oxfordian, and succeeded (depending on *biostratigraphic criteria) by the

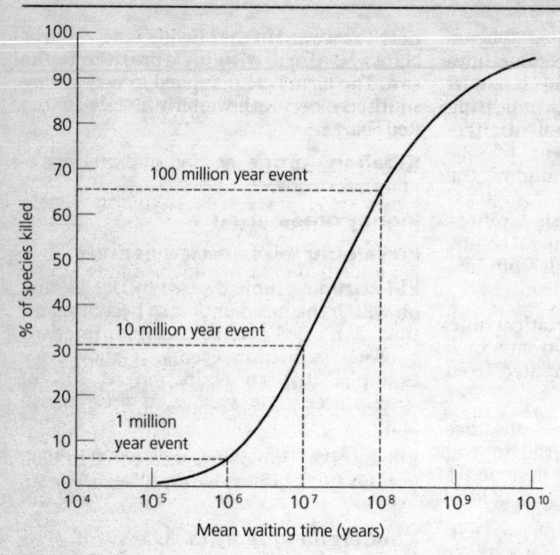

Source: Skelton, P. *Evolution: A biological and palaeontological approach*. Addison-Wesley & OU, p. 797.

Kill curve

*Tithonian (Tethyan) or *Volgian (Boreal) Stages. *See also* MALM.

Kinderhookian The basal *series in the *Mississippian of N. America, underlain by the Bradfordian (*Devonian) and overlain by the *Osagean, and roughly contemporaneous with the Hastarian Stage of the *Tournaisian Series of Europe.

Kinderscoutian A *stage in the *Bashkirian Epoch, overlain by the *Marsdenian.

kinematic viscosity In a flowing fluid, the ratio of *dynamic viscosity to density, an important factor in determining the amount of turbulence in the flow.

K-index In geomagnetism, the maximum values for the horizontal intensity (H) and the *declination ($\tan^{-1} X/Y$), each on a range 0 to 9, for a three-hourly interval.

king crab *See* CHELICERATA.

kink band An asymmetric, linear zone of deformation characterized by short *fold limbs and very small *hinge zones. Kink bands commonly occur as *conjugate sets (*see* CONJUGATE FOLD).

kipuka A Hawaiian term describing an 'island' of land completely surrounded and cut off by *lava. Kipukas are also known by the term 'dagalas' in Italy.

Kirkfieldian A *stage of the *Ordovician in the Upper *Champlainian *Series of N. America.

Kirkwood Gaps *See* RESONANCE.

Klazminskian A *stage in the *Gzelian Epoch, preceded by the *Dorogomilovskian (*Kasimovian Epoch) and followed by the *Noginskian.

klippe In geology, a *tectonic *outlier produced by the *erosion or *gravity-gliding of one or more *nappes. The front portions of the nappes become detached to produce the klippe structure.

knee twin *See* GENICULATE TWIN.

knick point (headcut) Abrupt change of gradient in the generally smooth long-profile of a stream, and typically separating two concave-up segments. It is often attributed to a fall in base level: this, it is said, initiates a knick point which then slowly travels upstream. It may alternatively be due to a change in rock type or load size, or to tributary entry.

knob and kettle (sag and swell topography) The landscape sometimes found on a recent terminal *moraine complex and consisting of a hummocky mound (the 'knob') alternating with a depression (the 'kettle'). The 'kettle' results from the melting of a block of ice enclosed in the *drift.

Knobby Terrain *See* MARTIAN TERRAIN UNITS.

knock and lochan Term descriptive of a glaciated landscape of low relief which is made up of ice-moulded hillocks and intervening *lochans (small lakes) eroded along zones of rock weakness. It is especially well developed in the Lewisian *gneiss area of the coastal lowlands of north-western Scotland.

knot Unit of speed equal to one nautical mile per hour (0.515 m/s). It is still used in many countries as a measure of wind speed and current velocity (as well as for the speed of ships and aircraft).

Kohoutek A *comet with an orbital period of 6.24 years; *perihelion date 28 December 1973; perihelion distance 1.571 AU.

komatiite An *extrusive *igneous rock of *peridotite composition, dominated by *essential magnesium *olivine accompanied by lesser amounts of aluminous *clinopyroxene and *chromite, and found as *lava flows and shallow *sills within *Archaean and *Proterozoic rock successions. Many komatiite flows were erupted above their *liquidus temperature and may have had temperatures in excess of 1600°C. Because their melting temperature is much higher than that of *basalts, *sediments, etc., they were quite capable of melting the rocks over which they flowed. Flows commonly show an upper zone containing magnesium *olivine or aluminous *clinopyroxene crystals with *spinifex texture, indicating extreme *undercooling of the upper part of the flow by contact with the atmosphere on extrusion. Komatiites represent the only known examples of peridotitic *magma.

kona storm Storm conditions associated with very strong southerly winds over the Hawaiian islands, bringing heavy rainfall. The winds blow in conjunction with the passage of a *depression to the north of the islands.

Königsberger ratio (Q, Q_n) Originally the ratio (Q) of the intensity of *natural remanent magnetization (NRM) of an *igneous rock to the intensity of magnetization induced at room temperature in a magnetic field the same as that in which the original NRM was acquired. Now it usually means the ratio (Q_n) of the intensity of NRM to that induced in a magnetic field of 50 μT at room temperature.

kopje *See* KOPPIE.

Köppen climate classification System devised in 1918 by Wladimir Peter *Köppen (1846–1940), with modifications that were completed in 1936, by which climates are divided into six broad groups according to the major vegetation types associated with them, broadly determined by critical temperature and the seasonality of precipitation. For example, a summer temperature of 10°C defines the poleward limit of tree growth; a winter temperature of 18°C is critical for certain tropical plants; a temperature of −3°C indicates some period of regular snow cover. The groups are: (A) tropical rainy climates with temperatures in the coldest month higher than 18°C; (B) arid climates; (C) warm, temperate, rainy climates in which temperatures in the coldest month are between 3°C and 18°C, and in the warmest month higher than 10°C; (D) rainy climates typical of boreal forest, in which temperatures in the coldest month are lower than −3°C (in US usage modified to 0°C), and in the warmest month higher than 10°C; (E) tundra, in which temperatures in the warmest month are 0–10°C; (F) permanent frost and ice-caps, in which temperatures in the warmest month are below 0°C. Subsets of the main classes (written as capital and lower case letters, e.g. Cs) are: absence of a dry season (f); a dry summer season (s); a dry winter season (w); a monsoon climate, with a dry season and rains at other times (m). Arid climates (B) are subdivided into semi-arid steppe-type (S), and arid desert (W). The temperatures within class (B) are indicated as: mean annual temperature higher than 18°C (h); mean annual temperature lower than 18°C with the warmest month higher than 18°C (k); mean annual temperature and warmest month both lower than 18°C (k′). Some criticism of the system centres on its arbitrary criteria of temperature associated with fixed boundaries, and the failure of data on temperature and precipitation to account fully for the effectiveness of the precipitation. *See also* STRAHLER CLIMATE CLASSIFICATION; and THORNTHWAITE CLIMATE CLASSIFICATION.

Köppen, Wladimir Peter (1846–1940) A meteorologist who developed the climate classification system that bears his name. He was born in St Petersburg, Russia, of German parents, studied at the universities of Heidelberg and Leipzig, and during 1872–3 worked in the Russian meteorological service. In 1875 he moved to Hamburg, Germany, where he headed a new division of the Deutsche Seewarte, formed to issue weather forecasts for the land and sea areas of northern Germany. He was able to devote himself entirely to research from 1879. He died in Graz, Austria.

koppie (kopje) An Afrikaans word for 'hill', and applied to a land-form widely described in Africa. It is similar to a *tor, being a steep-sided, isolated land-form, about the size of a house, and best developed on granitic *outcrops. It often shows an angular, castellated outline, when it is called a 'castle koppie'. A koppie may be a late stage in the destruction of a *bornhardt.

Korangan *See* TAITAI.

kosava Local wind in the Danube valley. The term is also used in a generic sense to refer to a type of *ravine wind.

Kotlassia prima Amongst the *amphibians, a few genera referred to the Seymouriamorpha show many developments that suggest they were evolving towards the reptilian condition. Some palaeontologists believe that these animals have attained the 'half-way' stage in the evolution of the first *reptiles. Others, however, including Panchen, claim that *Kotlassia prima* and its relatives retain a lateral-line system and a tadpole stage in the early part of their life cycle. *Kotlassia prima*, known from the Upper *Permian, is one of the last survivors of this important stock. *See also* *SEYMOURIA*.

KREEP volcanism The term for possible pre-mare volcanic activity in the lunar crust, producing 'KREEP basalts' with high concentrations of Th, U, K, REE, P, and other incompatible elements. They are elusive as a rock type and only a few lunar specimens have been identified so far. The Apennine Bench formation, in the Imbrium Basin, has been proposed to be composed of such material. An alternative explanation is that both it and KREEP basalts are derived from impact-generated melts in basin-forming collisions.

Krevyakinskian A *stage in the *Kasimovian Epoch, preceded by the *Myachkovskian (*Moscovian Epoch) and followed by the *Chamovnicheskian.

krotovina (crotovina) Animal burrow that has been filled with organic or mineral material from another *horizon.

***K*-selection** *Natural selection of those organisms that breed in such a way as to maximize their competitive ability, the strategy of equilibrium species. Most typically it is a response to stable environmental resources. This implies selection for low birth rates, high survival rates among offspring, and prolonged development. *K* represents the carrying capacity of the environment for species populations showing an S-shaped population-growth curve. *Compare* R-SELECTION.

K/T boundary event The impact of an asteroid, about 10–11 km in diameter, that struck the Earth about 65 Ma ago near Chicxulub, on the coast of northern Yucatan, Mexico. Initial evidence for the impact was an iridium-enriched layer of clays, found in many parts of the world. The Chicxulub crater, now filled with sediments, is about 180 km in diameter. Many scientists have concluded that the impact caused the mass extinction at the end of the *Cretaceous. This is not accepted by all, and how the impact caused the extinction is not known.

Kuehneosaurus latus First of the eolacertalians, flying lizards which flourished during the Late *Triassic. Essentially these were of ancestral stock, but the outward extension of the ribs indicate a specialized adaptation towards gliding.

Kuenen, Philip Henry (1902–72) A Dutch geologist and one of the founders of modern sedimentology, Kuenen's participation in the Snellius deep-sea expedition to the Moluccas (1929–30) directed his attention to marine geology. One of the very few experimental geologists, he demonstrated the efficacy of high-density *turbidity currents in transporting sand to deep-sea environments. In 1950 Kuenen, together with the Italian geologist C. I. Migliorini, interpreted the graded *sandstone beds of ancient *flysch sequences as *turbidites. The study of their *sedimentary structures has contributed much to *palaeogeography. His most widely known book in the English language is the early classic on *Marine Geology* (1950).

Kuiper, Gerard (or Gerald) Peter (1905–73) A Dutch-born American astronomer who discovered *Miranda and

*Nereid, was a member of the team which discovered *Janus, and who made many important studies of planetary and lunar features and the origin of the *solar system. He also predicted, in 1948, that carbon dioxide is one of the major constituents of the martian atmosphere and discovered that *Titan has a methane atmosphere. He suggested that material left over from the formation of the solar system might form a belt, now called the *Kuiper belt, beyond the orbit of *Pluto. Kuiper was born at Harenskarspel and educated at Leiden. He migrated to the USA in 1933 and became a US citizen in 1937. He worked at the Yerkes Observatory, University of Chicago, and was its director from 1947 to 1949 and again from 1957 to 1960. From 1960 until his death he was director of the Lunar and Planetary Laboratory, University of Arizona. The International Astronomical Union named a crater on Mercury for him.

Kuiper belt A region of bodies, made mainly from ice, that lies just beyond the orbit of *Pluto and supplies the *solar system with short-period comets, e.g. *Halley. At its outer margins the Kuiper belt extends into the *Oort cloud. The existence of the Kuiper belt was proposed by Gerard P. *Kuiper, in 1988 its role as the source of short-period comets was suggested, and in the early 1990s the first objects belonging to the Kuiper belt were discovered.

Kula Plate A formerly large lithospheric *plate in the north-western Pacific that met the *Pacific and *Farallon plates at a *triple junction. It was subducted during the *Jurassic, *Cretaceous, and *Cenozoic, only fragments now remaining as *terranes in the North American *cordillera.

Kungurian 1. The final *age in the Early *Permian Epoch, preceded by the *Artinskian, followed by the *Ufimian (Late Permian), and dated at 259.7–256.1 Ma (Harland et al., 1989). **2.** The name of the corresponding eastern European *stage, which is roughly contemporaneous with the Weissliegende (western Europe), upper *Leonardian (N. America), and Tae Weian (New Zealand). Boundary and correlation uncertainties have led to some rocks being assigned to either the Artinskian or Ufimian Stages.

kunkar *See* CALCIFICATION.

Kuril Trench The oceanic *trench which marks part of the *destructive plate margin between the *Pacific and *N. American Plates. The Kuril Trench is backed by the Kuril *Island Arc, running from Kamchatka to northernmost Japan.

Kuroshio current Oceanic surface current that flows northwards from the Philippines, along the Japanese coast, and then out into the N. Pacific. It is an example of a western *boundary current: fast-flowing (up to 3 m/s), narrow (less than 80 km), and relatively deep. It is second in strength only to the Gulf Stream. The warm water transports heat polewards. The volume transport is variable but is normally about 4×10^7 m^3/s.

kurtosis The degree of peakedness in a distribution curve. If kurtosis (*K*) is one, the curve is said to be mesokurtic; if *K* is greater than 1 it is leptokurtic; if *K* is less than 1 it is platykurtic.

Kutorginida (phylum *Brachiopoda) Order of brachiopods with biconvex, calcareous shells, and cardinal areas in both valves. No teeth or sockets are found. In the past this group of Lower *Cambrian fossils has been assigned to both the *Inarticulata and *Articulata classes, but it is suspected to have been derived independently of these classes and in the future may come to be regarded as a distinct class.

K-wave *See* SEISMIC WAVE MODES.

kyanite An important metamorphic *index mineral and one of the three Al_2SiO_5 mineral *polymorphs, the other two being *sillimanite and *andalusite; sp. gr. 3.5–3.6; *hardness 4–7; *triclinic; light to darker blue, sometimes green or yellow, occasionally black; *vitreous *lustre; *crystals usually long, columnar, often flattened and bladed, with occasional stellate intergrowths; *cleavage pinacoidal, good {100}, {010}, parting {001}; occurs in regional *metamorphic rocks that have been subjected to high pressures and low to moderate temperatures, and may be found in *schists or *gneisses associated with *staurolite and *garnet.

L

Labrador current Oceanic current that brings cold arctic waters southwards into the N. Atlantic along the western margin of Greenland. Frequently it carries icebergs southwards, concentrating them in the area to the east of the Grand Banks in late spring to early summer. Fog banks often occur off the coast of Newfoundland where the Labrador current meets the *Gulf Stream.

labradorite *See* PLAGIOCLASE FELDSPAR.

Labyrinthodontia Subclass of the *Amphibia, primitive in character, and ranging in length from a few centimetres to several metres. They were the *Palaeozoic and *Triassic amphibians. There are no known connections with other amphibians, but they show affinities with *crossopterygian fish and with *reptiles.

laccolith *Concordant, lenticular *pluton, which is circular or elliptical in plan. The floor essentially is flat, but the roof is distinctly domed. *Compare* LOPOLITH.

lacunosus From the Latin *lacunosus* meaning 'with holes', a variety of cloud usually associated with the genera *altocumulus and *cirrocumulus. Thin layers or sheets of cloud display a fairly regular set of holes with frilled edges, so forming a net. *See also* CLOUD CLASSIFICATION.

lacus From the Latin *lacus*, meaning 'lake', a term that has been used to name small, isolated patches of *mare basalt on the lunar surface (e.g. Lacus Somniorum, the 'Lake of Dreams', north of Mare Serenitatis).

lacustrine Pertaining to a lake.

LAD *See* LAST-APPEARANCE DATUM.

Ladinian **1.** A Middle *Triassic *age, preceded by the *Anisian, followed by the *Carnian, and dated at 239.5–235.0 Ma (Harland et al., 1989). **2.** The name of the corresponding European *stage, which is roughly contemporaneous with the Fa Lang (China) and Kaihikuan (New Zealand). It should not be confused with 'Landenian', a Lower *Eocene stage.

lag A *normal, *dip-slip fault, dipping less than 45°, on which there has been a relative downward displacement of the *hanging wall. *See also* FAULT.

lag breccia (co-ignimbrite breccia) A coarse, *lithic-rich deposit accumulated synchronously with the formation of *ignimbrite, but confined to a proximal, near-vent location. As the outer margins of a *pyroclastic *eruption column collapse, forming a pyroclastic flow moving under gravity away from the *vent area, *clasts which are too large to be supported by the column or flow accumulate as a lag *breccia around the vent. Because they are deposited as the ignimbrite forms, they are now called 'co-ignimbrite' breccias to distinguish them from other types of breccias.

lag deposit A coarse-grained residue left behind after finer particles have been transported away, due to the inability of the transporting medium to move the coarser particles.

Lagerstätte (pl. Lagerstätten) A sedimentary deposit that is of value because of the *fossils it contains. A fossil-Lagerstätte contains an abundance of fossils; a conservation-Lagerstätte is a stratum in which fossils are exceptionally well preserved; a concentration-Lagerstätte is a large accumulation of fossils of particular species, such as a shell bed. The term was introduced by Adolf Seilacher, from the German *Lager*, 'stratum', and *Stätte*, 'place'.

lagoon Coastal body of shallow water, characterized by a restricted connection with the sea. The water body is retained behind a *reef or islands.

Lagrangian current measurement A technique for measuring water movements by tracing the path of a water particle over a long time interval. A device is released into the water and allowed to drift passively with it. Measurements can be made, e.g. by following and plotting the progressive position of a *neutrally buoyant float for subsurface currents, or a drift pole or buoy for surface-water movements.

Lagrangian points The locations where the gravitational attraction of two orbiting bodies is equal. Objects located at such points remain fixed so that they are favoured positions for future space stations. There are five such points in the Earth–Moon system; two of them, L^4 and L^5, lie 60° ahead of and behind the lunar orbit. The Trojan family of *asteroids occupy positions close to two of the jovian Lagrangian points. Lagrangian points are named after the French astronomer Comte J. L. Lagrange (1736–1813).

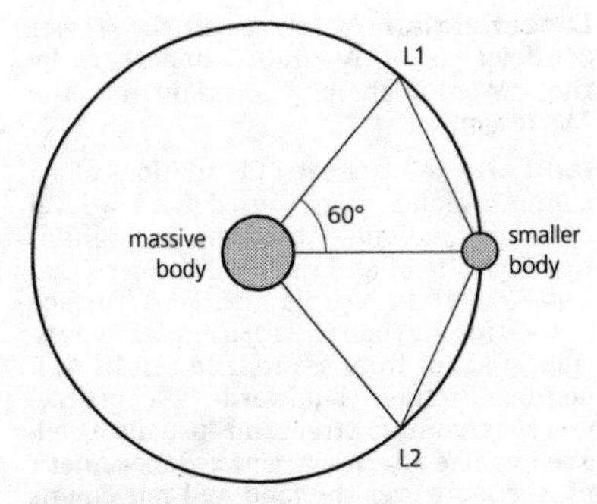

Lagrangian points

lahar (mudflow) Catastrophic mudflow on the flank of a *volcano. It is a notable feature of the volcanic areas of Indonesia, especially Java. Lahars are the cause of most volcanic fatalities. They may reach more than 100 km from the source volcano, when confined along pre-existing valleys.

Lake Mungo *See* BRUNHES.

Lamarckism The theory of evolution propounded by Lamarck.

Lamarck, Jean Baptiste Pierre Antoine de Monet, Chevalier de (1744–1829) A French naturalist who, in 1809, proposed the first formal theory of evolution. He advanced the theory that evolutionary change may occur by the inheritance of characteristics acquired during the lifetime of the individual. For example, fossil evidence suggests that the ancestors of the giraffe had short necks: Lamarck proposed that competition for food encouraged them to stretch upward in order to browse among higher vegetation, causing their necks to lengthen, and that this lengthening was passed on to their descendants. Over millions of years the minute increases from each generation to the next culminated in the long-necked form. It is interesting to note that the theory of the inheritance of acquired characteristics did not hold a central position in Lamarck's own writings. His cardinal point was that evolution is a directional, creative process in which life climbs a ladder from simple to complex organisms. He believed the inheritance of acquired characteristics provided a mechanism for this evolution. Lamarck explained that this progress of life up the ladder of complexity is complicated by organisms being diverted by the requirements of local environments; thus cacti have reduced leaves (and giraffes have long necks). *Compare* DARWIN, CHARLES ROBERT.

Lambeophyllum profundum One of the earliest recorded rugose corals from the Upper *Ordovician of N. America. The majority of early *solitary corals were rather small and horn-shaped. They belonged to the subclass *Rugosa and were noted for the presence of well-defined radiating partitions, the *septa.

Lambertian reflector A perfectly *diffuse reflector which reflects *electromagnetic radiation equally in all directions.

Lamb, Hubert Horace (1913–97) A British climatologist, who was among the first scientists to draw attention to the variability of climates and the social and economic effects of climate change. He studied geography at Cambridge University and in 1936 joined the staff of the Meteorological Office. After refusing to work on the meteorology of spraying poisonous gas, he was transferred to the Irish Meteorological Office, but resigned in 1945, after a disagreement with the director, and returned to the UK Meteorological Office in 1946. In the following years he travelled to Antarctica and in 1954 was transferred to the climatology department of the Meteorological Office. He left this post in 1971 to establish the Climatic Research Unit at the University of East Anglia and was its director until his retirement in 1977.

Lamb's dust-veil index Index, devised by H. H. *Lamb, of the amount of finely divided material suspended in the atmosphere after great volcanic eruptions, and of the duration of an effective veil intercepting the Sun's radiation. It can be calculated from estimates of the amount of solid matter thrown up, or from the reduction of intensity of the solar beam, or from the reduction of temperatures prevailing at the surface of the Earth. The latitude of the volcano affects the index value also, because the maximum extent of the veil over the Earth varies, being greatest after eruptions in low latitudes. (The great eruption of Krakatoa in Indonesia in 1883, which ejected about 17 km^3 of particulate matter into the atmosphere, where it remained for three years, gave an index value of 1000.)

Lamé constant (λ) **1.** An *elastic constant equal to the *bulk modulus minus two-thirds of the shear modulus. *See also* POISSON'S RATIO. **2.** *See* SHEAR MODULUS.

lamellar In *crystallography, applied to a special kind of *crystal twinning (also called 'multiple' or 'repeated' twinning) where there are many twin individuals, each one having a tabular or plate-like appearance. Lamellar twinning is common within the *plagioclase feldspars, in places where two adjoining twin slabs or lamellae are mutually reversed with respect to each other and every alternate twin 'plate' or 'slab' has an identical atomic structure.

Lamellibranchia *See* BIVALVIA.

lamina The finest sedimentary layer, less than 1 cm in thickness. *Strata more than 1 cm thick are termed 'beds' (*see* BEDDING).

laminar flow Type of flow (normally in water) characterized by the movement of fluid particles parallel to each other, with no transverse movement or mixing. Velocity increases steadily away from the bed. Laminar flow in channels is found only at low velocities and adjacent to smooth surfaces. It is almost ubiquitous in soil moisture and *groundwater (except in *karstic aquifers).

laminate Comprising layers (laminae, sing. lamina) of material.

lamination Very fine stratification, composed of discrete layers (laminae, sing. lamina) of *sediment, a millimetre or so in thickness. Lamination is a smaller-scale feature than *bedding.

laminite A finely *laminated *sediment or *sedimentary rock.

laminoid fenestrae *See* FENESTRAE.

lamprophyre A dark-coloured, strongly *porphyritic, *intrusive *igneous rock, containing abundant *euhedral *phenocrysts of *biotite and/or *amphibole which can be accompanied by phenocrysts of *olivine, *diopside, *apatite, or opaque oxides, set in a *mafic, *felsic, or glassy *groundmass. There are no felsic phenocrysts present in this rock type. The lamprophyres are subdivided into a number of types on the basis of their most abundant mafic component and the presence and type of feldspar in the groundmass. Lamprophyre types include *minette, *kerstantite, *alnöite, *vogesite, *camptonite, and *monchiquite, and are found intruded as *dykes and *sills.

lamp shells *See* BRACHIOPODA.

Lancefieldian A *stage of the Lower *Ordovician of Australia, underlain by the *Warendian and overlain by the *Bendigonian.

land and sea breezes Circulations of air common along coasts, caused by a low-level pressure gradient due to the differential heating of land and sea. On summer days, solar radiation warms the land surface more strongly than the adjacent sea: a pressure gradient from sea to land results in a gentle, cooling, landward 'sea breeze' whose maximum strength is usually developed by late afternoon. Upward movement of warm air over the land and movement toward the sea at greater height, followed by subsidence, produces a shallow convection cell. At night and in early morning cooler land and relatively warmer sea produce a reverse-flow convection cell, with a seaward 'land breeze'. The horizontal extent of well-developed land and sea breezes is typically limited to about 40 km from the coast, but associated air movements can often be detected over a much wider coastal belt.

land bridge Connection between two land masses, especially continents, e.g. the Bering land bridge linking Alaska and Siberia across the Bering Strait, that allows migration of plants and animals from one land mass to the other. Before the widespread acceptance of *continental drift, the existence of former land bridges was often invoked to explain faunal and floral similarities between continents now widely separated. On a smaller scale, the term may be applied to land connections that have now been removed by recent tectonics or the *Flandrian *transgression (e.g. between northern France and south-eastern England).

landfill *See* MADE GROUND.

Landsat A series of *satellites, mainly carrying *multispectral scanners, and more recently also carrying *thematic mappers. They are primarily designed for scanning the vegetation cover of the land surface of the Earth and evaluating the effectiveness of satellite-based scanning systems for routine monitoring. The recordings are available on magnetic tape for computer analysis or in photographic form (usually known as Landsat images). Landsat is complemented by *Seasat.

landslide *See* MASS-WASTING.

Langhian 1. That interval in the *Miocene Epoch, dated at 16.3–14.2 Ma (Harland et al., 1989) which is preceded by the *Burdigalian and followed by the *Serravallian Ages. Most authors subdivide the Langhian into the Early Langhian Age (Early Miocene) and Late Langhian Age (Middle Miocene). **2.** The name of the corresponding European *stage, which is roughly contemporaneous with the *Relizian and lower *Luisian (N. America), *Clifdenian (New Zealand), and *Balcombian (Australia). The *stratotype is found between Cessole and Case dei Rossi, northern Italy.

lanthanide *See* RARE EARTH ELEMENT.

lapiés Variety of *karren, consisting of shallow, straight grooves incised by solution into a sloping surface of *limestone. It may constitute a dense, sub-parallel network which develops rapidly upon exposure of the limestone.

lapilli *Pyroclastic fragments or *tephra ranging in size from 2 to 64 mm. Lapilli may be composed of primary magmatic material (e.g. *pumice), accessory lithic material, accidental lithic material, or accretions of wet *ash-size material (*accretionary lapilli). The size-term 'lapilli' is usually applied to the lithological *clast type to give descriptive terms, e.g. 'pumice lapilli' or 'accessory lithic lapilli'.

lapilli-tuff *See* TUFF.

Laplace, Pierre Simon, Marquis de (1749–1827) A French mathematician and physicist, Laplace is best known for his *nebular hypothesis of the formation of the solar system, published in *Exposition du système du monde* (1796). He also worked on planetary motions, the theory of tides, etc.

Laplace's equation The sum of the rates of change of a potential field gradient in three perpendicular directions is equal to zero. For a potential function $U(x,y,z)$, Laplace's equation states: $\nabla^2 U = \delta^2 U/\delta x^2 + \delta^2 U/\delta y^2 + \delta^2 U/\delta z^2 = 0$, where ∇ is the Laplace operator.

lapout The lateral termination of a stratum at the limit of its original deposition. There are two types: *baselap and *toplap, baselap being further divided into *onlap and *downlap.

lappets *See* APERTURE.

lapse rate Rate of decrease of temperature per unit height in the atmosphere. In the *troposphere the average rate is approximately 6.5 °C per 1000 m. *See also* ADIABATIC; and ENVIRONMENTAL LAPSE RATE.

lapse time Estimate of the time interval between the ending of *nucleosynthesis and the consolidation of *meteorite parent bodies. Lapse times may be calculated from the detection in meteoritic material of the daughter *isotopes (*see* DAUGHTER NUCLIDE) of short-lived radionuclides which are now extinct, but which survived until meteoritic condensation. Times of about 10^8 years have been deduced.

Lapworth, Charles (1842–1920) An English geologist, Lapworth was responsible for ending the controversy between *Murchison and *Sedgwick over the Lower *Palaeozoic, by proposing in 1879 that the area in dispute should form a third system, the *Ordovician. He also published a classification of *graptolites (1873), and worked on the structure of the Scottish Highlands, identifying the Moine thrust.

Laramide-Columbian orogeny A late *Cretaceous to early *Eocene mountain-building episode, affecting an area extending from what are now the south-western USA to northern S. America, caused by thrusting associated with the *subduction of the *Farallon Plate beneath the *North American Plate.

large-aperture seismic array (LASA) An *array of *geophones in Montana, USA, which is used to detect seismic events and to distinguish between those caused by nuclear explosions and those caused by *earthquakes.

large-ion lithophile (LIL) Element of large *ionic radius and with a *valency of 1 or 2 (e.g. Rb^{2+}, Pb^{2+}, and Ba^{2+}), which during *igneous fractionation tends to be concentrated in silicic *melts, and from which LILs are incorporated mainly into potassium *silicates such as the *alkali feldspars and *micas. *See also* FRACTIONAL CRYSTALLIZATION; and INCOMPATIBLE ELEMENTS.

LARI *See* LOW-ASPECT-RATIO IGNIMBRITE.

Larissa (Neptune VII) A satellite of *Neptune, measuring 208 × 178 km; visual albedo 0.06.

larvikite A coarse-grained, *intrusive *igneous rock that consists of essential potash *feldspar, *oligoclase feldspar, titaniferous *augite, and sodic *amphibole, together with accessory *biotite, *quartz, or *nepheline (depending on the *silica saturation), *magnetite, *zircon, and *apatite. The two feldspar types occur in equal proportions, making the rock a type of *monzonite. Larvikite was first described by *Brögger in 1890 from the Larvik district of southern Norway. Larvikite can be an impressive ornamental stone.

LASA *See* LARGE-APERTURE SEISMIC ARRAY.

Laschamp *See* BRUNHES.

laser An acronym for *l*ight *a*mplification by *s*timulated *e*mission of *r*adiation, a device for emitting a single, intense beam of coherent, monochromatic light (i.e. light at a single wavelength). *See also* MASER.

laser altimeter An instrument that measures the distance from an orbiting spacecraft to a *satellite surface by measuring the time of travel of a reflected *laser beam. Accuracy is within a few metres. Since the position of the spacecraft is known, elevations on the surface can be obtained.

laser ranging The establishment of precise Earth–Moon distances by aiming *laser beams telescopically, from the Earth at reflecting mirrors placed on the lunar surface by the Apollo 11–15 missions, and measuring the return time to the receipt of the reflections. By making such measurements from different terrestrial sites relative *plate motions can be determined.

last-appearance datum (LAD) The last recorded occurrence of a key taxon in biological history.

Late-Devensian Interstadial *See* WINDERMERE INTERSTADIAL.

late-glacial Term usually applied to the time between the first rise of the temperature curve after the last minimum of the *Devensian glacial stage and the very rapid rise of temperature that marks the beginning of the post-glacial, or *Flandrian period. The late Devensian extends from about 15 000 to 10 000 BP and shows a characteristic three-fold climatic and hence depositional sequence, from cold, Older *Dryas deposits, to warmer *Allerød, to colder, Younger Dryas. In Europe there is some evidence for an additional warmer phase, the *Bølling Interstadial, dated at about 13 000 BP, i.e. during Older Dryas times. Some authorities consider that this partial warming sequence may be characteristic of the late-glacial times of all major glacial advances.

latent heat of transition The heat required to activate a *phase change from a solid to a liquid or from a liquid to a gas, i.e. to a higher energy state (e.g. latent heat of melting), measured in joules per mole (J/mol). Latent heat is released by reactions in the reverse direction (e.g. latent heat of crystallization).

lateral accretion deposit Inclined layers of *sediment, deposited laterally rather than in horizontal *strata, particularly by the lateral outbuilding *sediment on the surface of a river *point bar. The inclined surfaces thus record the progressive migration of the point bar. The *dip of the lateral accretion deposit can be used to determine the size of the point bar and *channel geometry. *See also* EPSILON CROSS-BEDDING.

lateral dispersion *See* DISPERSION.

lateral moraine *See* MORAINE.

laterite Weathering product of rock, composed mainly of hydrated iron and aluminium oxides and hydroxides, and *clay minerals, but also containing some silica. It is related to *bauxites and is formed in humid, tropical settings by the *weathering of such rocks as *basalts. *See also* PLINTHITE.

laterolog sonde An electrical sonde for measuring the electrical resistivity of rocks within a *borehole (resistivity logging). Short sondes (40 cm or less) normally measure the resistivity of the *mud cake and *invaded (flushed) zone. Longer laterologs measure the resistivity of the outer edges of the invaded zone and of the uninvaded rocks. Major differences in the resistivity values measured by shallow and deep sondes will occur in *reservoir rocks containing appreciable quantities of hydrocarbons, so permitting the estimation of the thickness of hydrocarbon reservoirs. Interpretation of long laterologs is affected by the *Delaware effect. *See also* MICROLOG.

latite A *porphyritic, *extrusive *igneous rock containing *euhedral *phenocrysts of calcium-rich *plagioclase feldspar, accompanied by lesser amounts of *biotite, magnesium *clinopyroxene, magnesium *orthopyroxene, iron-titanium oxides, and *olivine phenocrysts set in a *trachytic *groundmass dominated by *alkali feldspar. Plagioclase accounts for 40–60% of the total feldspar content of the rock. Latites are the volcanic equivalents of *monozonites and, as such, they are members of the *calc-alkaline *magma series. The term is sometimes used to describe potassium-rich *andesites.

latite-andesite A *porphyritic, *extrusive *igneous rock containing *phenocrysts of zoned *plagioclase feldspar, *augite (*clinopyroxene), *orthopyroxene, and some *olivine set in a *groundmass of *andesine plagioclase and *alkali feldspar, *pyroxene, and iron oxides. Plagioclase accounts for 60–90% of the total feldspar content. This rock is transitionary to a true *andesite, which is characterized by plagioclase forming more than 90% of the total feldspar content of the rock.

latitude correction The correction made to *gravity readings to allow for the difference in *gravitational acceleration as a function of distance from the equator. *See* INTERNATIONAL GRAVITY FORMULA.

lattice A regular, three-dimensional framework which indicates the ordered arrangement of atoms in *crystals. The smallest complete lattice is known as the 'unit cell' and it may be repeated many times to form a complete crystal. The shape of the unit cell varies according to the arrangement of the points of the lattice in space, i.e. a 'space lattice'.

lattice energy The work required to decompose a crystal lattice of one *mole of a substance into elementary structural units, and displace them so there is a large distance between them.

lattice gliding *See* DEFORMATION TWINNING.

laumontite Member of the *zeolite group of minerals, $CaAl_2Si_4O_{12}.4H_2O$; sp. gr. 2.3; *hardness 3.5; whitish; square *prismatic crystals; occurs as a *secondary mineral in veins and cavities within volcanic rocks and some associated mineral deposits, and may also occur within the zeolite *facies of *metamorphic rocks.

Laurasia The northern continental mass produced in the early *Mesozoic by the initial rifting of *Pangaea along the line of the N. Atlantic Ocean and *Tethys. Laurasia included what was to become N. America, Greenland, Europe, and Asia, while the large, southern, continental mass (called *Gondwanaland) was later to divide into S. America, Africa, India, Australia, and Antarctica.

Laurentia (Laurentian Shield) The *Precambrian shield (*see* CRATON) of central eastern Canada. The name is derived from the St Lawrence River and has been applied to a series of *granites, *gneisses and *metasediments that are older than 2500 Ma. The Laurentian Shield forms the ancient 'core' of Canada, around which younger *orogenic belts have been accreted.

Laurentian A *stage of the uppermost *Archaean of New Zealand, underlain by the *Keewatinian.

Laurentian Shield *See* LAURENTIA.

Laurentide ice sheet An area of continental ice that lay over the eastern part of Canada during the *Pleistocene glaciations. The centre of the ice mass may have originated in or near northern Quebec, Labrador, and Newfoundland, and spread out to the south and west. At its maximum spread it may have covered an area of 13×10^6 km^2.

lava Molten rock, normally a *silicate, erupted by a *volcano. It may be *vesicular, glassy, or *porphyritic in texture, and varies between *acidic and *basic in composition. Its behaviour on extrusion and its relief-forming capacity depend largely on its *viscosity, which is affected by silica content, temperature, and amount of dissolved gases and solids. Generally, the less viscous the lava the faster the flow, and the more viscous the lava the greater the tendency towards explosive eruption. Two varieties of basaltic lava surface are recognized: 'aa', a jagged, stony clinker, bristling with sharp points; and 'pahoehoe', characterized by a smooth, ropy appearance. Andesitic and rhyolitic lavas tend to have 'blocky' surfaces, characterized by smooth-faceted blocks, 1–5 m in diameter.

lava blister The surficial swelling of a plastic *lava flow crust in response to the puffing up of gas or vapour from beneath the flow. Blisters may also form through hydrostatic or *artesian forces in the lava. They are usually 1–150 m in diameter, with a maximum height of 30 m, and are hollow. *Compare* TUMULUS.

lava channel The depression between two parallel and closely spaced retaining walls (*lava levées) in which a flow of *lava is confined. The retaining walls are composed of cooled blocks of lava from the flow itself. As the lava flows, the height of the flow surface pulsates and lava may overflow its retaining walls; this adds a coating of chilled lava to the walls, thereby increasing their height and the effectiveness with which they constrain the flow.

lava lake Molten *lava, usually basaltic in composition, held in a depression (e.g. a large *caldera or *crater) over a magmatic vent. A lava pond may form where a lava flow fills a topographic depression. Large lava lakes found in lava-flooded calderas or craters often display surface features indicating the presence of numerous thermal convection cells within the lake. Lava lakes may be sustained for many years (e.g. Kilauea Crater, Hawaii, which has been an important tourist attraction, and at present there is a continuous lava pond at Pu'u O'o, Hawaii).

lava levée One of the scoriaceous retaining walls on either side of a *lava channel. Four types are recognized: initial levées form when the yield strength of the lava permits the margins of a flow to remain stationary and crystallize while the central part of the flow continues to move; accretionary levées are formed near *boccas where piles of surface clinker are accreted to the flow margins; rubble levées form when the sides of a flow expand outwards by the avalanching of surface and marginal crystalline clinker; and overflow levées form when lava repeatedly overflows existing rubble levées.

lava tube A hollow passage beneath the surface of a solidified *lava flow, formed by the withdrawal of molten lava from a former distributory tunnel, which fed lava below the stationary surface of the flow to the advancing flow margin, sometimes transporting the lava over long distances. Tubes range in width from less than one metre to more than 30 m and in height up to 15 m. They can be tens of kilometres long, e.g. in Victoria, Australia. Most tubes are developed in flows displaying *pahoehoe surface morphology, although they can develop in flows with *aa surface morphology. Mount Etna, Sicily, has examples of both.

law of constancy of interfacial angles In all *crystals of the same substance, the angles between corresponding faces have the same value when measured at the same temperature. This concept was first proposed by *Steno in 1669 and was formulated as a law by Romé de l'Isle in 1772.

law of constant proportions Pure substances always contain the same elements in the same proportion by weight.

law of correlation of facies *See* WALTHER'S LAW.

law of cross-cutting relationships An *igneous rock, *fault, or other geologic feature must be younger than any rock across which it cuts.

law of faunal succession The principle, first recognized at the beginning of the 19th century by William *Smith, that different *strata each contain particular assemblages of *fossils by which the rocks may be identified and *correlated over long distances; and that these fossil forms succeed one another in a definite and habitual order. This law, together with the *law of superposition of strata, enables the relative age of a rock to be deduced from its content of fossil faunas and floras.

law of Haüy *See* LAW OF RATIONAL RATIOS OF INTERCEPTS.

law of original horizontality *Sedimentary rocks are always deposited as horizontal, or nearly horizontal, *strata, although these may be disturbed by later earth movements. The law was first proposed in the mid-17th century by Nicolaus *Steno.

law of rational indices *See* LAW OF RATIONAL RATIOS OF INTERCEPTS.

law of rational ratios of intercepts (law of rational indices, law of Haüy) *Crystal faces cut the *crystallographic axes in simple, whole-number ratios. These ratios may be measured with reference to a unit plane having a ratio of 1 : 1 : 1 for three crystallographic axes and 1 : 1 : 1 : 1 for four crystallographic axes.

law of superposition of strata (principle of superposition) *Strata are deposited sequentially, so that in an undisturbed sedimentary succession each layer of rock is younger than the layer beneath it. Subsequent earth movements may overturn and invert this sequence. The law was first proposed in the 17th century by Nicolaus *Steno.

Laxfordian A sub-*stage of the *Lewisian, from about 1600–1100 Ma ago (Van Eysinga, 1975), named from Loch Laxford, north-western Scotland.

Laxfordian orogeny A mountain-building episode that occurred about 1800–1600 Ma ago, and which is recorded in north-west trending *folds in the Lewisian *gneisses of what is now the extreme north-west of Scotland. It is a continuation of the *Proterozoic *Ketilidian and *Nagssugtoqidian *orogenies of Greenland, and may be a continuation of the *Scourian orogeny which immediately preceded it.

layer cloud One of the principal forms of cloud, with flattened, sheet-like appearance and of limited vertical extent. Common types of cloud exhibiting this form are: (*a*) low-level layer clouds, e.g. *fog and *stratus; and (*b*) multi-layered clouds, e.g. *altostratus, *cirrostratus, and *nimbostratus. *See also* CLOUD CLASSIFICATION.

Layered Deposits *See* MARTIAN TERRAIN UNITS.

layered silicate *See* PHYLLOSILICATE.

layer-parallel shortening The *homogeneous strain which is developed in a layered rock when shortening occurs parallel to the layering. Buckling instabilities, which would normally form *folds, are restricted in such a case, and the layer shortens and thickens as a result.

layover In *radar terminology, the distortion of a radar image caused by steep surface angles relative to the radar wavefront angle, such that the top of a hill is actually closer to the radar than the base. The resulting image gives the appearance of hills or mountains leaning toward the radar. *Compare* FORESHORTENING.

Lazarus taxon A taxon that disappears from the fossil record close to an extinction horizon, but reappears again much later in the sequence.

lazurite *Mineral $Na_8[(Si,Al)_6O_{24}](S,SO_4)$ belonging to the *sodalite group of minerals and with a similar composition to *haüyne and *nosean; sp. gr. 2.3–2.4; *hardness 5.5; *cubic; intense, deep, azure-blue, violet, light blue, or greenish-blue; *vitreous *lustre; usually compact, *massive; *cleavage imperfect, rhombododecahedral; occurs in the contact zone of *alkaline, *igneous rocks, in association with *carbonate rocks, and in alkaline *lavas. It is used to make ornaments.

L-band Radar frequency band between 390 and 1550 μHz, used for radar scanning of the Earth's surface.

leachate Solution formed when water percolates through a permeable medium. In some cases the leachate may be toxic or carry bacteria when derived from solid waste. In mining, leaching of waste tips can produce a mineral-rich leachate which is collected for further processing, as in heap leaching of *porphyry copper and gold deposits.

leaching Removal of soil materials in solution.

lead–lead dating A dating method based on the comparison of *isotopes of lead (Pb). Throughout geologic time the isotopic composition of *common lead in the Earth has evolved from that of *primordial lead by the addition of *radiogenic leads (^{206}Pb, ^{207}Pb, and ^{208}Pb) derived from uranium and thorium decay (*see* DECAY SERIES). ^{204}Pb is not derived from any radioactive parent and appears to be a standard against which the other lead values can be compared. It is normally assumed that in any small part of the Earth's *crust and underlying *mantle which, at the time of formation, contained primordial lead together with uranium and thorium, no radiogenic lead could have been present. With the passage of time atoms of radiogenic ^{206}Pb, ^{207}Pb, and ^{208}Pb gradually replaced uranium and thorium atoms. If, at one instant in time, all the lead in the area under discussion was removed in solution and deposited as a lead ore, then this would preserve a record of the isotopic balance of lead at the time. Given that this ore mineral would not contain any uranium or thorium, it would be preserved as a unique point on the lead growth curve. Using the Holmes–Houtermans model, and plotting $^{207}Pb:^{204}Pb$ against $^{206}Pb:^{204}Pb$, a series of *growth curves would be obtained based on the different isotopic ratios. These curves can then be used to plot *isochrons, the slope of which determines the age of the particular lead assemblage.

lead loss Loss of daughter lead *nuclides during the *radioactive decay of uranium to lead, in which ^{238}U decays to ^{206}Pb, and ^{235}U decays to ^{207}Pb (*see* DECAY SERIES). These lead *isotopes, together with that produced by the ^{232}Th to ^{208}Pb series, are moderated by the presence of non-*radiogenic ^{204}Pb in any calculations of age. After correcting for the original lead in any analysis, if the mineral being investigated has remained a closed system, the ^{235}U : ^{207}Pb and the ^{238}U : ^{206}Pb ages should be concordant. On a graph of ^{235}U : ^{207}Pb against ^{238}U : ^{206}Pb the loci of all *concordant ages define a curve called concordia (*see* CONCORDIA DIAGRAM). If they do not agree, they are discordant and the ratios do not fall on the concordia curve. Because daughter atoms tend to escape from the system, especially when heated or otherwise disturbed, *discordant ages are generally on the young side. Because ^{207}Pb and ^{206}Pb are chemically identical, they are not fractionated by natural processes. As a result of this any lead loss from a mineral is in the same isotopic proportion as that in which it occurs in the mineral. The lead loss would be the same for all parts of the rock body, and the plots of ^{238}U : ^{206}Pb against ^{235}U : ^{207}Pb fall on a straight line below the concordia curve. The two points where this straight line intersects the concordia curve give the age of the rock (higher value) and the time of the lead loss (lower value). In some systems there is not a single time of lead loss but a more continuous process of *diffusion. The lower of the two intersects described above may then be further depressed by this continuous diffusion and the otherwise straight-line relationship may be lost especially at the lower value end.

Leaf Margin Analysis (LMA) A technique for estimating mean annual temperatures in the past from the proportion of dicotyledonous plant species present that had untoothed leaf margins, based on the observation of a strong positive relationship between warmth and plants with smooth-edged leaves. In addition to this univariate approach there is also a multivariate technique. *See* CLIMATE–LEAF ANALYSIS MULTIVARIATE PROGRAM.

leaky transform fault A type of *transform fault in which there is production of basaltic *magma along the *fault plane, indicating some separation of the blocks on each side of the fault. Such transform faults do not exactly follow arcs of small circles about the *pole of rotation of the relevant plate pair. An example is the Azores Fracture Zone, which forms the boundary between the *Eurasian and *African Plates from the *Mid-Atlantic Ridge into the Mediterranean. Leaky transform faults are one kind of *combined plate margin.

least-work principle The theory that geomorphological *processes always operate in such a way as to achieve the work that has to be done with a minimum expenditure of energy (and maximum *entropy). This is typically achieved by the adoption of a certain profile or shape (e.g. a river *meander may be that shape best suited for carrying the discharge and sediment with the least loss of energy). *See* LEAST-WORK PROFILE.

least-work profile That profile whose gradient is just sufficient for the associated geomorphological *process to occur with the minimum possible expenditure of energy. An example is a river long profile, whose concave-up form is the shape best suited for the transfer of increasing quantities of water and sediment in accordance with the *least-work principle. Such a profile expresses a state of high *entropy.

lebensspuren Biologically formed, *sedimentary structures found in *sediments, including *tracks, *trails, *burrows, *borings, faecal casts, and *coprolites.

lectostratotype A *stratotype, chosen after the original establishment of a *stratigraphic unit, that is designed to serve as the standard in the absence of a satisfactory original stratotype (i.e. a *holostratotype). A lectostratotype may be selected from outside the *type area. *Compare* NEOSTRATOTYPE. *See also* PARASTRATOTYPE; and HYPOSTRATOTYPE.

lectotype One of a collection of *syntypes which, subsequent to publication of the original description, is chosen and designated through published papers to serve as the 'type specimen'. *Compare* HOLOTYPE; NEOTYPE; and PARATYPE.

Leda (Jupiter XIII) One of the lesser satellites of *Jupiter, with a diameter of 10 km.

lee depression Non-frontal *depression that develops on the lee side of an upland barrier across the airflow as a result of contraction leading to cyclonic curvature. Dynamic processes are responsible for the low-pressure system rather than wave development along a *front. Such depressions are common, for example in winter on the southern lee side of the Alps.

lee waves Air waves in the lee of a mountain barrier, where a stable layer of air, after displacement by movement over the barrier, returns to its original level. This process results in a series of stationary ('standing') waves extending downwind from the lee side of the barrier. Clouds often form along the wave crests in lenticular form: they may appear stationary, due to condensation of water vapour at the upward side, caused by the upward air movement, and evaporation on the downward side of the wave. The wavelength can be up to 40 km and the wave amplitude is most pronounced in the intermediate levels of the airstream. Circular air motion (in the vertical plane) beneath the wave crests may reverse wind direction locally within the general air flow. This phenomenon is termed a 'rotor'. In addition to stable air at an intermediate level, lee-wave formation requires a constant wind of at least 15 knots. Well-known wave clouds on the lee side of barriers include the Sierra wave of the Sierra Nevada, California, the helm wave of Cumbria, England, and the moazagotl of Silesia.

left lateral fault *See* SINISTRAL FAULT.

Lehmann, Inge (1888–1993) A Danish geophysicist, Miss Lehmann was Director of the Seismological Department of the Royal Danish Geodetic Institute from 1928 to 1953. In 1936 she was able to demonstrate, from studies of the refraction of *P-waves, that the Earth has a solid inner *core.

Lehmann, Johann Gottlob (1719–67) A German mining engineer who moved to Russia, Lehmann worked on the *stratigraphy of mountain rocks, and distinguished between Primary (non-fossiliferous and metal-bearing) and Secondary sediments. His ideas were afterwards extended and developed by *Werner.

Leighton–Pendexter classification A classification for *limestones and *dolomites proposed in 1962 by W. M. Leighton and C. Pendexter, now used infrequently. The classification defines *carbonate rocks by the percentage of *grains, *micrite, *cement, and voids present. Limestones with more than 50% grains are named as 'limestones', prefixed by the name of the main grain type (e.g. 'skeletal limestone', 'pellet limestone'); limestones with 10–50% grains are termed 'micritic limestones', prefixed by the name of the dominant grain type (e.g. 'skeletal micritic limestone'); limestones with less than 10% grains are termed micritic limestones; limestones built by organic frame-builders are termed 'coralline limestones' or 'algal limestones' according to the type of frame-builder. Leighton and Pendexter also define carbonates with variable amounts of dolomite, *calcite, and impurities on a triangular classification.

Leitz–Jelley refractometer *See* REFRACTOMETER.

Lenian A *stage of the *Caerfai Epoch, underlain by the *Atdabanian and overlain by the *Solvan and dated at 553.7–536 Ma (Harland et al., 1989).

lenticular bedding A form of *heterolithic *sediment characterized by the presence of isolated *sand *ripples and lenses, set in a mud *matrix. The ripples may be symmetrical or asymmetrical. Such sediments form in low-energy, muddy environments which suffer episodic higher flows, e.g. areas of *continental shelf lying in water depths affected by storm *waves, or the low-energy zones on *intertidal flats.

lenticularis From the Latin *lenticularis* meaning 'biconvex' or 'lens-shaped', a form of cloud consisting of clearly defined, elongated lenses. The form is typical of *lee-wave clouds and may affect such clouds as *stratocumulus, *altocumulus, and *cirrocumulus. *See also* CLOUD CLASSIFICATION.

Lenz's law When a magnetic field and an electrical circuit are moved in relation to one another, an electric current is induced in the circuit such that it forms a magnetic field opposing the motion.

Leonardian A *series in the Lower *Permian of N. America, underlain by the *Wolfcampian, overlain by the *Guadalupian, and roughly contemporaneous with the *Artinskian and *Kungurian Stages. In some areas it is zoned by the use of *fusilinid foraminiferids. It is the N. American equivalent of the *Rotliegende. *Red-bed localities of *Wolfcampian–Leonardian age in Texas and Colorado have yielded many vertebrate fossils.

Le Pichon, Xavier (1937–) A French marine geologist from the Brittany Oceanographic Centre at Brest, who worked in the USA in the late 1960s, Le Pichon studied the mechanisms of *sea-floor spreading. He worked out the geometry of *plate movements on a sphere, and was the first to recognize that there are six major tectonic plates.

lepidocrocite *See* GOETHITE.

Lepidodendron selaginoides Important species of *Palaeozoic plant, characterized by a *dichotomous branching, by diamond-shaped leaf scars, and by large cones. During the Upper *Carboniferous, *Lepidodendron* species flourished on several continents and the trunks grew up to 30 m before branching. *See also* LYCOPSIDA.

lepidolite Lithium-bearing *mica with composition $K_2(Li,Al)_{5-6}$ $[Si_{6-7}Al_{2-1}$ $O_{20}]$ $(OH,F)_4$; sp. gr. 2.8–2.9; *hardness 2.5–4.0; *monoclinic; *crystals as small flakes; perfect basal {001} *cleavage; lilac to greyish; occurs in late-stage *pegmatites associated with other lithium minerals, such as *spodumene, and minerals of a *pneumatolytic origin, such as *tourmaline and *topaz. Lepidolite is an important *ore mineral of lithium.

lepidomelane An iron-rich variety of *biotite.

Lepidophloios kilpatrickense The earliest known representative of the important *Palaeozoic family of plants, the Lepidodendraceae. Closely related to *Lepidodendron, Lepidophloios kilpatrickense* is distinguished by its internal anatomy. It is recorded from the Upper *Carboniferous of Scotland. *See also* LYCOPSIDA.

Lepidosauria *See* DIAPSID TYPE.

lepisphere A *microcrystalline, blade-shaped crystal of a metastable variety of *quartz, composed of *cristobalite with interlayered lattices of tridymite, aggregates of which often occur during the transformation of *opal into quartz *chert.

Lepospondyli Group of small *Palaeozoic *amphibians, distinguished by the possession of *vertebrae with spool-shaped centra below the neural arches. Each centrum (the massive part of the vertebra) was perforated lengthwise, providing a channel for the *notochord. The lepospondylous, or 'husk', vertebra also occurs in living amphibians, but its evolutionary derivation is unclear.

leptograptid A type of Middle to Upper *Ordovician graptoloid (*Graptoloidea, suborder Didymograptina) characterized by *biramous, slender, flexuous *stipes which are either horizontal or reflexed.

leptokurtic *See* KURTOSIS.

leste Regional wind affecting N. Africa and Madeira, which blows ahead of a low-pressure area and brings hot, dry conditions.

leucite An important *feldspathoid mineral $KAlSi_2O_6$; sp. gr. 2.5; *hardness 5.5–6.0; *tetragonal; normally white or grey; *vitreous *lustre; *crystals form very characteristic tetragonal trioctohedra; *cleavage imperfect {110}; occurs in potassium-rich rocks which may be silica deficient, such as leucite basanites, leucite tephrites, lamproites, and leucitophyres or *leucitites.

leucite–basanite *See* BASANITE.

leucitite An *extrusive, undersaturated (*see* SILICA SATURATION), *mafic *igneous rock consisting of *essential *leucite and *augite, with *accessory iron oxide and *apatite. When *olivine or *nepheline are also present, the rock is termed an olivine-leucitite or a nepheline-leucitite respectively. Leucitites are allied to *phonolites, with which they are commonly associated in the field.

leuco- A prefix attached to a rock name to indicate a lighter than usual colour for the particular rock type. For example, a *gabbro lighter in colour than a normal gabbro owing to the presence of a larger amount of feldspar than usual would be called a 'leucogabbro'.

leucocratic Applied to a rock with a *colour index between 5 and 30.

leucosome A coarse-grained, quartzo-feldspathic vein, varying in thickness from a few centimetres to a metre or two, and found as a high-grade metamorphic product (*see* METAMORPHISM) in rocks of pelitic (*see* PELITE) to psammitic (*see* PSAMMITE) composition. The finer material between the leucosomes is enriched in *mafic and aluminous minerals. Leucosomes may represent a low-melting-point liquid segregated from the *metasediment during high-grade metamorphism to give rise to what is, in effect, a *migmatite *fabric.

levanter Local wind from the east experienced in the Straits of Gibraltar and associated with standing waves in the lee of the Rock of Gibraltar. The wind is especially prevalent in late summer and autumn and brings high humidity.

leveche Local wind affecting south-eastern Spain, especially in summer, and similar to such other hot, dry, dusty winds of tropical continental origin as the *scirocco and *khamsin, which blow in the Mediterranean region.

levée 1. Raised embankment of a river,

showing a gentle slope away from the channel. It results from periodic overbank flooding, when coarser sediment is immediately deposited due to a reduction in velocity. This may lead to a situation in which the river flows well above the level of its outer *floodplain. **2.** *See* LAVA LEVÉE.

level of compensation *See* PRATT MODEL.

Lewisian A *stage of the *Proterozoic of north-western Scotland, from about 2600–1100 Ma ago, named from the Outer Hebridean Isle of Lewis.

lherzolite A two-*pyroxene- and *olivine-bearing, coarse-grained, *ultrabasic rock consisting of essential magnesium-rich olivine, chromium-magnesium *clinopyroxene (Cr-diopside), magnesium *orthopyroxene (*enstatite), and *garnet or *spinel, which is found as *xenoliths in *basalts and as a component of Alpine *ultramafic bodies. A *mantle origin for many lherzolites is attested to by their metamorphic (*see* METAMORPHISM) texture, their high-pressure *mineral assemblage, and their ultrabasic composition. Using evidence from experimental *petrology, lherzolites are interpreted as examples of undepleted mantle (i.e. mantle which has not undergone partial melting to generate basalt *magma). The name is taken from the type locality, at Lherz in the French Pyrenees.

Lias 1. The earliest of the *Jurassic *epochs, followed by the *Dogger. The Lias is dated at 208–178 Ma (Harland et al., 1989), and comprises the *Hettangian, *Sinemurian, *Pliensbachian, and *Toarcian *Ages. **2.** The corresponding European *series, roughly contemporaneous with most of the *Herangi (New Zealand). Blue-grey *shales and muddy *limestones are typical of the Lias, with *Dactylioceras* and *Gryphaea* among the more important *fossils found in the extensive outcrops of England and France.

Libby, Willard Frank (1908–80) Professor of chemistry at the University of California and director of the Institute of Geophysics, Libby developed a method for dating *sediments and artefacts using *carbon-14. The method is described in his book *Radiocarbon Dating*, first published in 1952.

libeccio Local south-westerly wind which brings stormy conditions, especially in winter, to the central Mediterranean.

libration The slow oscillation of a satellite, as seen from the planet around which it revolves. One libration is due to a parallax effect: e.g., because of the rotation of the *Earth, more of the eastern limb of the *Moon is visible at moonrise, while more of the western limb can be seen at moonset. In the Earth–Moon system, a second libration is that of lunar longitude, with a monthly period, since the Moon's revolution around the Earth sometimes exceeds and sometimes lags behind its axial rotation, and the third lunar libration is of latitude, due to the 6° inclination of the lunar equator to its orbital plane, so that more of the polar regions become visible when the Moon is north or south of the *ecliptic.

librigena *See* CEPHALON.

Lichida An order of *Trilobita that lived from the Middle *Cambrian to Middle *Devonian. They were medium-sized to very large, usually spiny, and in many the *pygidium is larger than the *cephalon. There were 3 suborders.

life assemblage (biocoenosis) A *fossil community that is interpreted as representing a former living community. Most assemblages interpreted as life assemblages represent only a small fraction of a former community.

lifting condensation level Level at which air becomes saturated when forced to rise. (The level can be determined on a *tephigram, where a dry-adiabat line intersects a line of saturation mixing ratio through the *dew-point temperature.)

ligand Atom, *ion, or molecule that acts as the electron-donor partner in one or more coordination bonds. A heterocyclic ring is formed if the ligand is an organic compound, and the product is termed a chelate. *See also* CHELATION; and COORDINATION NUMBER.

light absorption Most of the light entering the oceans is absorbed within the first 100 m of water. The depth of light penetration is governed by the amount of material in the water that absorbs and scatters light, e.g. suspended organic material. The absorption of light varies with differing wavelengths of light, blue light penetrating more deeply than red light.

lignite Poor, brown *coal with a visibly woody structure, relatively unaltered, between *peat and *bituminous coal in *rank. It is non-coking, burns with a long, smoky flame, and contains up to 40%

moisture when mined, although this can be reduced to 10–15% by drying in air. It is found in *Tertiary basins in Britain, and also in continental Europe where it is economically important.

Likhvin A series of sediments in the European USSR, perhaps dating from the *Mindel/Riss Interglacial, and the equivalent of the *Holsteinian or the *Hoxnian.

LIL *See* LARGE-ION LITHOPHILE.

Lillburnian A *stage in the Upper *Tertiary of New Zealand, underlain by the *Clifdenian, overlain by the *Waiauan, and roughly contemporaneous with the *Serravallian Stage.

lime Compounds, mostly of calcium carbonates but also other basic (alkaline) substances, used to correct soil acidity and occasionally as a fertilizer to supply magnesium.

lime boundstone *See* DUNHAM CLASSIFICATION.

lime grainstone *See* DUNHAM CLASSIFICATION.

lime mud A general term for *carbonate *sediment composed of particles up to 62 μm in size. Lime muds are found in a wide range of depositional settings, ranging from *pelagic to *intertidal. The mud comes from various sources; it may be derived from microfauna, calcareous *algae, and from the mechanical or biological breakdown of coarser particles or, more problematically, it may be formed by chemical/biochemical precipitation (this is as yet unobserved in modern marine settings).

lime mudstone *See* DUNHAM CLASSIFICATION.

lime packstone *See* DUNHAM CLASSIFICATION.

limestone Sedimentary type of rock composed mainly of *calcite and/or *dolomite, which is often of organic, chemical, or *detrital origin. *See* FOLK'S CLASSIFICATION; DUNHAM'S CLASSIFICATION; and LEIGHTON–PENDEXTER'S CLASSIFICATION.

limestone pavement *See* CLINT.

lime wackestone *See* DUNHAM CLASSIFICATION.

limit-equilibrium analysis In rock and soil mechanics, the study of the point at which a material has reached the limit of its stability using the concept of yield criteria and the associated flow rule in the stress–strain relationship, e.g. where toppling occurs in a steeply jointed rock slope, where a soil becomes plastic, etc.

limonite Mineral, $FeO(OH).nH_2O$; sp. gr. 2.7–4.3; *hardness 4.0–5.3; yellowish-brown to reddish-brown; normally *earthy *lustre; usually amorphous; occurs as a *secondary mineral from the weathering of iron in rocks and mineral deposits, and may accumulate to give iron-rich mineral deposits.

limpet *See* ARCHAEOGASTROPODA; and 'PATELLA' BEACH.

Limulus *See* CHELICERATA.

Lindgren, Waldemar (1860–1936) A Swedish-born officer of the US Geological Survey, and later professor at the Massachusetts Institute of Technology, Lindgren is best known for his classification of ore deposits according to their genesis (1913). He was the first to propose the hydrothermal theory of ore deposition.

lineage-zone (phylozone, evolutionary zone, morphogenetic zone) A unit of *strata containing a clearly defined portion of an evolutionary lineage, marked above and below by some distinct and specified change in form.

linear regression In statistics, a comparison of two sets of data to see if there is a linear relationship.

linear sand ridge Submarine sand mound typical of shallow sea and wide *continental-shelf areas, 3–10 m high, 1–2 km wide, which may extend for tens of kilometres across the shelf and have an average spacing of about 3 km. Such ridges have been described from the North Sea (off Norfolk, England) and the eastern seaboard of the USA. They are the product of storm and tidal action.

lineation 1. Any linear feature that appears on the surface of a rock. Lineation may be formed during deformation by the parallel alignment of minerals, fossils, or pebbles; by parallel crenulation *cleavages; or by striations and grooves resulting from the movement of a rock over a plane, e.g. a *fault surface (*see* SLICKENSIDE), or *flexural slip during folding. An intersection lineation is caused by the crossing of any two planes, e.g. cleavage and *bedding. **2.** Lineations are a series of parallel lines on a rock surface, formed by tectonic processes, by the transportation and deposition of sand under upper-flow-regime plane-bed conditions, or by the movement of glacial ice over the rock surface.

line scanner In *radar terminology, an imaging device which uses photoelectronic detectors combined with a rotating mirror which sweeps across the ground surface taking in image data a line at a time in order to produce a *raster.

line spectrum *See* SPECTRUM.

line squall Stormy conditions, with sudden changes of wind and with heavy cloud, often involving thunderstorms, and a temperature fall, often associated with the passage of a *cold front; it defines the line of the cloud and wind structure. Warm air is overrun by cold air to produce the squall line.

Lingulata (phylum *Brachiopoda) A class of brachiopods, existing from the Lower *Cambrian to the present day, in which the shell is chitinophosphatic, the valves are not hinged by teeth and sockets, and the *pedicle is thick.

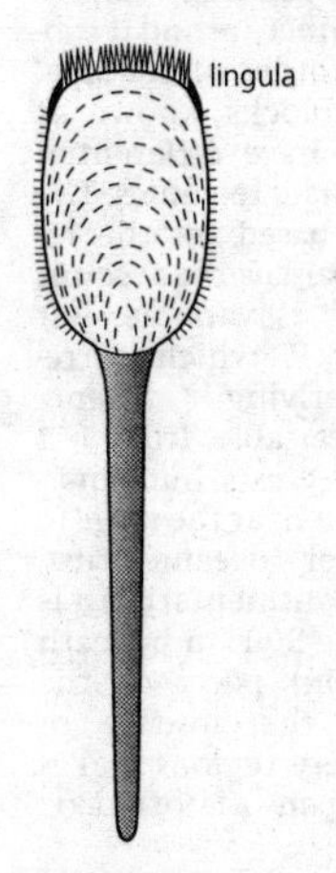

Lingulata

Lingulella viridis One of the earliest recorded *inarticulate brachiopods, and very similar to extant species of the genus *Lingula*, *L. viridis* is described from the lower *Cambrian of England. It is one of several species referred to the cosmopolitan genus *Lingulella*. *Lingula* has been cited as an example of *bradytelic evolution.

Lingulida (class *Inarticulata) Order of *Brachiopoda, with *valves usually of calcium phosphate with some layers of organic material. The valves may be finely *punctate or *impunctate. The *pedicle emerges between the valves posteriorly. Lingulida are usually marine, but some are tolerant of reduced salinity. They appeared first in the Lower *Cambrian. There are two superfamilies: the Lingulaceae (which includes ***Lingulella* and the extant *Lingula*); and the Trimerellaceae, ranging from the *Ordovician to the *Silurian, with a biconvex, probably aragonitic shell, with internal muscle platforms (e.g. *Trimerella*, *Dinobolus*).

linguoid Applied to tongue-shaped asymmetrical *ripples which have a highly sinuous crest and a strong three-dimensional form. *See also* AKLÉ.

liptinite *See* COAL MACERAL GROUP.

liquefaction The process of becoming or making a liquid by heating, cooling, or a change in pressure. In soils, the temporary transformation of material to a fluid state due to a sudden decrease in shearing resistance caused by a collapse of the structure associated with a temporary increase in *pore fluid pressure.

liquid limit *See* ATTERBERG LIMITS.

liquidus In a *temperature–composition diagram, the liquidus is the locus of points marking the boundary above which the *phases are all liquid. In a *binary system it is represented by a line, in a *ternary system by a curved surface, and in a *quaternary system by a volume. Between the liquidus and the *solidus, both liquid and solid phases are present. *See* PHASE DIAGRAM.

listric fault *See* LYSTRIC FAULT.

lithic arenite A *sandstone containing less than 15% mud *matrix, and with a *grain composition comprising more than 25% rock fragments, and more rock fragments than *feldspar present. *See* DOTT CLASSIFICATION.

lithic fragment The dense or crystalline components of a *pyroclastic deposit. Three types are recognized. Cognate lithics are fragments of non-vesiculated, juvenile, magmatic material, e.g. dense, angular, glass fragments. Accessory lithics are fragments of *country rocks ejected explosively during *eruption. Accidental lithics are *clasts picked up locally by pyroclastic flows and surges. Lithics range from blocks to *ash-size fragments and are usually angular, but may be rounded by in-vent abrasion during eruption.

lithic greywacke (lithic wacke) A *sandstone containing more than 15% but less than 75%, mud *matrix, and with a *grain composition comprising more than 5% rock fragments and with more rock fragments than *feldspar present. *See* DOTT CLASSIFICATION.

lithic wacke *See* LITHIC GREYWACKE.

lithification The process of changing unconsolidated *sediment into rock. This involves cementation (*see* CEMENT) of the *grains, but not necessarily burial alteration or *compaction.

litho- From the Greek *lithos* meaning 'stone', a prefix meaning 'pertaining to rock or stone'.

lithoclast A sedimentary *clast composed of a pre-existing rock type.

lithofacies Rock noted for a distinctive group of characteristics, e.g. composition and grain size.

lithofacies map A map which shows the distribution of different *lithofacies for a particular stratigraphic interval or horizon (*see* STRATIGRAPHIC COLUMN). Such maps may be interpreted to yield a *palaeogeography or environmental interpretation for the mapped horizon.

lithologic symbol A mark, or set of marks, representing a particular textural feature or rock type. For example, conventionally a sandstone is represented by stipple, limestone by a 'brick' emblem. However, as there is not a complete range of generally accepted lithologic symbols, where they are used (e.g. in *columnar sections) it is customary to give a key-panel explaining the meaning of the symbols. Symbols may be idiosyncratic to particular authors but they are often standardized, especially within large oil companies.

lithologic trap *See* STRATIGRAPHIC TRAP.

lithology The description of the macroscopic features of a rock, e.g. its *texture or *petrology.

lithophile Applied to elements with a strong affinity for oxygen and which concentrate in the Earth's *crust in *silicate rather than metal or *sulphide minerals; or to elements with greater free energy of oxidation per gram of oxygen than iron. They occur as oxides, and especially in the *silicate minerals which make up 99% of the crust. Examples of lithophile elements are Al, Ti, Ba, Na, K, Mn, Fe, Ca, and Mg. *Compare* ATMOPHILE; BIOPHILE; CHALCOPHILE; and SIDEROPHILE.

lithophysae Concentric shells of *aphanitic material enclosing a hollow core, which form rounded masses a few centimetres in diameter within glassy or aphanitic, *felsic *igneous rocks. The hollow, lithophysal core is sometimes filled by secondary *silica to form a *geode.

lithosequence A sequence of soils where the changing character of the soil is related to, or caused by, the changing *lithology of *rocks and superficial mineral deposits.

lithosome A mass of rock of approximately uniform character penetrated by tongues of rock from adjacent masses with a different *lithology.

lithosphere The upper (oceanic and continental) layer of the solid *Earth, comprising all crustal rocks and the brittle part of the uppermost *mantle. It is generally considered to deform by brittle fracture and if subjected to stresses of the order of 100 MPa. It comprises numerous blocks, known as tectonic *plates, which have differential motions giving rise to *plate tectonics. The concept was originally based on the requirement for a rigid upper layer to account for *isostasy. Its rigidity is variable, but much greater than 10^{21} P, which corresponds with the underlying *asthenosphere. Its thickness is variable, from 1–2 km at mid-oceanic *ridge crests, but generally increasing from 60 km near the ridge to 120–140 km beneath older *oceanic crust. The thickness beneath *continental crust is uncertain, probably some 300 km beneath the cratonic (*see* CRATON) parts of the continental crust, but the absence of the asthenosphere in these regions makes definition difficult. *Compare* ATMOSPHERE; and HYDROSPHERE.

lithospheric plate *See* PLATE.

lithostatic stress (geostatic stress) The component of *confining pressure derived from the weight of the column of rock above a specified level.

lithostratigraphic unit (rock unit, rock-stratigraphic unit) A body of rock forming a discrete and recognizable unit, of reasonable homogeneity, defined solely on the basis of its lithological characteristics (*see* LITHOLOGY). A lithological unit may be *sedimentary, *igneous, *metamorphic, or a combination of these. As with other *stratigraphic units, lithostratigraphic units are defined according to *type sections. Their boundaries are placed at surfaces of lithologic change, usually sudden but sometimes gradational. As the physical nature of the units reflects depositional environments rather than time spans, the boundaries of lithological units may be *diachronous. Lithostratigraphic units are comparatively local in extent when compared to the world-wide compass of *chronostratigraphic units. They are ranked in decreasing order of magnitude in *supergroups, *groups, *formations, *members, and *beds. A diverse, but distinctive and interrelated body of rock that cannot be subdivided into any other lithostratigraphic unit is termed a 'complex'. *See also* BIOSTRATIGRAPHIC UNIT; CHRONOSTRATIGRAPHIC UNIT; and STRATIGRAPHIC UNIT.

lithostratigraphy Branch of stratigraphy concerned with the description of rock units in terms of their lithological features. It deals with the spatial relations of such rock units, but does not take into consideration (*a*) the evolution of the organisms contained within the units (*biostratigraphy), or (*b*) geologic time (*chronostratigraphy).

Lithothamnion *See* RHODOPHYCEAE.

litter (L-layer) Accumulation of dead plant remains on the soil surface.

Little Ice Age A period between about 1550 and 1860 during which the climate of the middle latitudes became generally harsher and there was a world-wide expansion of *glaciers. The effects have been recorded in the Alps, Norway, and Iceland, where farm land and buildings were destroyed. There were times of especial severity during this period, e.g. the early 1600s when glaciers were particularly active in the Chamonix valley, in the French Alps.

littoral drift *See* BEACH DRIFT; and LONGSHORE DRIFT.

littoral zone 1. The area in shallow fresh water and around lake shores where light penetration extends to the bottom sediments, giving a zone colonized by rooted plants. **2.** In marine ecosystems the shore-area or intertidal zone where periodic exposure and submersion by tides is normal. Since the precise physical limits of tidal range vary constantly a biological definition of this zone, which essentially reflects typical physical conditions rather than rarely experienced events, is generally more useful. Thus in Britain the littoral zone is defined as the region between the upper limit of species of the seaweed *Laminaria*, and the upper limit of *Littorina* species (periwinkles), or of the lichen *Verrucaria*.

living fossil An organism that has persisted, essentially unchanged, since its first appearance. For example, *Lingula* (a *brachiopod) has remained much the same since *Ordovician times and *Sphenodon* (tuatara) since the early *Mesozoic.

Llandeilo A *series (468.6–463.9 Ma) of the Middle *Ordovician, underlain by the *Llanvirn and overlain by the *Caradoc.

lizardite *See* SERPENTINE.

Llandovery A *series (439–430.4 Ma) of the Lower *Silurian, underlain by the *Ashgill (*Ordovician) and overlain by the *Wenlock.

Llanvirn A *series (476.1–468.6) of the Lower *Ordovician, underlain by the *Arenig and overlain by the Llandeilo.

L-layer *See* LITTER.

LMA *See* LEAF MARGIN ANALYSIS.

load Total amount of material carried by a stream or river, or the mass of rock overlying a geologic structure.

load cast A bulbous depression formed on the base of a bed of *sediment. Load casts develop by the differential sinking of the sediment, while still soft, into less dense sediment below. Load 'casts' are not strictly casts, as they do not infill an existing depression as in the case of flute casts (*see* FLUTE MARK).

loadstone *See* MAGNETITE.

loam Class of soil texture composed of *sand, *silt, and *clay, which produces a physical property intermediate between the extremes of the three components.

lobe *See* SUTURE.

lobe fins Paired fins, characteristic of *Crossopterygii, that are borne on fleshy lobes containing an axial skeleton.

lobsters *See* MALACOSTRACA.

local range zone *See* TEILZONE.

local wind Air movements, generally of limited geographical range, characteristic of a particular region and/or area of particular land and atmospheric configuration. Many local winds are especially associated with orographic peculiarities or with particular high- or low-pressure systems, and local names may be applied to winds of a broad, general type. Examples include the *föhn winds, of orographic origin, and the many regional winds of the Mediterranean area, whose origins lie in the tropical continental *air masses of the Sahara.

lochan *See* KNOCK AND LOCHAN.

Lochkovian **1.** The earliest *stage in the *Devonian Period, preceded by the *Silurian, followed by the *Pragian, and dated at 408.5–396.3 Ma (Harland et al., 1989). **2.** The name of the corresponding European *stage.

Loch Lomond stadial Relatively cold period that occurred towards the end of the last (*Devensian) glacial stage in Scotland. The event took place about 11 000–10 000 radiocarbon years BP. It is characterized by the development of small *ice-caps and *cirque glaciers in the Highlands.

Lockportian A *stage of the Upper *Niagaran (*Silurian) of N. America.

lode Mineral vein or system of veins; in Cornwall (UK) especially, refers to productive veins only.

lodestone *See* MAGNETITE.

lodgement till *See* TILL.

loess *Unconsolidated, wind-deposited sediment composed largely of *silt-sized *quartz particles (0.015–0.05 mm diameter) and showing little or no stratification. It occurs widely in the central USA, northern Europe, Russia, China, and Argentina. It can give rise to a rugged topography with steep slopes (up to 70°).

logging *See* WELL LOGGING.

log-normal distribution (geometric distribution) A distribution in which the logarithms of the values have a *Gaussian (normal) distribution. This distribution is common in geologic contexts, e.g. grain sizes.

logs Sedimentological logs are vertically measured records of sedimentary successions, illustrating with symbols the vertical sequence of *lithology, grain size, *sedimentary structure, and *fossil content.

Longfordian A *stage in the Upper *Tertiary of south-eastern Australia, underlain by the *Janjukian (*Oligocene), overlain by the *Batesfordian, and roughly contemporaneous with the upper *Aquitanian and lower *Burdigalian Stages.

longiconic Applied to a cephalopod (*Cephalopoda) shell that is very elongate.

Longisquama insignis One of the first *archosaurs known to have the ability to glide or parachute. Described from Soviet Kirgizstan in the 1970s, *Longisquama* is noted for the development of elongate paired scales along its back; its Latin name means 'remarkable long scale'. It is recorded from sediments of *Triassic age.

longitudinal conductance (*S*) The sum of all the thickness/resistivity ratios of $n-1$ layers which overlie a semi-infinite substratum of resistivity ρ_n, such that $S = h_1/\rho 1 + h_2/\rho_2 + h_3/\rho_3 + \ldots + h_{n-1}/\rho_{n-1}$ (mho), where h_1, h_2, etc. are the depths and ρ_1, ρ_2, etc. the resistivities, of successive layers. A knowledge of h_i/ρ_i for the ith layer when it is sandwiched between two layers of much higher resistivity is of importance in resolving the problem of *equivalence.

longitudinal dispersion *See* DISPERSION.

longitudinal-type coast *See* PACIFIC-TYPE COAST.

Longmyndian A *stage of the Upper *Proterozoic of the Welsh borders, underlain by the *Charnian and overlain by the *Moinian.

long-range forecasting Weather forecasting for periods beyond a few days.

longshore bar Linear ridge of sand whose long axis is parallel to the shore and that is in, or immediately seaward of, the intertidal zone. *See also* RIDGE AND RUNNEL.

longshore current Current that flows parallel to the shore within the zone of breaking waves: it is generated by the oblique approach of waves.

longshore drift (littoral drift) Movement of sand and shingle along the shore. It takes place in two zones. *Beach drift occurs at the upper limit of wave activity, and results from the combined effect of *swash and *backwash when waves approach at an angle. Movement also occurs in the *breaker zone, where currents transport material thrown into suspension.

Longtanian A *stage of the *Zechstein Epoch, preceded by the *Capitanian.

Longvillian A *stage of the *Ordovician in the Middle *Caradoc, underlain by the *Soudleyan and overlain by the *Marshbrookian.

long wave Meander in the flow of the circumpolar upper westerly winds, commonly with a wavelength of about 2000 km and sometimes large amplitude. *See also* DISHPAN EXPERIMENT; and ROSSBY WAVES.

Lonsdale, William (1974–1871) A curator and librarian of the London Geological Society, who studied fossils, and especially *corals and *oolitic limestone. Lonsdale is credited with using his knowledge of fossils to demonstrate that sediments in Devon were of the same age as the *Old Red Sandstone (1837), and thus helping to establish the *Devonian System.

look angle In *radar terminology, the angle between the vertical plane passing through the radar antenna and the line between the antenna and object. *Compare* DEPRESSION ANGLE.

looping A geophysical field method in which a *base station is visited regularly because the survey is designed in loops focused through it. This allows instrumental *drift to be measured, especially in gravity and *geomagnetic investigations, and misties and misclosures to be checked for in seismic work.

lophodont Applied to cheek teeth (*molars), found in some *Mammalia, in which the cusps are fused to form transverse ridges (lophs) that aid the mastication of plant material. *See also* BILOPHODONTY.

lophophore *See* BRACHIOPODA; BRYOZOA; and PTEROBRANCHIA.

lopolith *Concordant *igneous *intrusion that has a sagging, saucer-like form. The shape of small lopoliths may be controlled by their emplacement in folds, but there is no such obvious control for great lopolithic intrusions. *Compare* LACCOLITH.

Lopstedt *See* BRØRUP.

Los Angeles abrasion test A method for measuring abrasion resistance in which the sample and a set of steel spheres are tumbled inside a closed, hollow, steel cylinder, about 700 mm diameter and 500 mm long, which is rotated on a horizontal axis.

losing stream A stream with a permeable bed through which water can seep to the *water-table. *See also* INTERMITTENT STREAM.

Love, Augustus Edward Hugh (1853–1940) An applied mathematician and physicist, Love worked at the University of Oxford. His studies of the theory of elasticity led him to suggest that two types of surface *earthquake waves should exist, one of which has been named for him. *See* LOVE WAVE.

Love wave (SH-wave, Q-wave, L-wave, L_Q-wave, G-wave) A type of *surface wave which occurs when the shear-body-wave velocity in the surface medium is lower than that in the underlying strata. Love waves are characterized by horizontal motion normal to the direction of travel, with no vertical motion. In effect a Love wave is a polarized *shear wave and travels slightly faster than a *Rayleigh wave.

low Common term for a low-pressure system, e.g. a depression.

low-angle fault A fault which *dips less than 45°. If the sense of displacement is normal (*see* NORMAL FAULT) it is called a *thrust, if reverse (*see* REVERSE FAULT) it is called a *lag.

low-aspect-ratio ignimbrite (LARI) An *ignimbrite sheet displaying a value for the ratio of its average thickness (V) to its horizontal extent (H) of 10^{-4} to 10^{-5}, where H is taken as the diameter of a circle whose area is equal to that of the flow. *Compare* HIGH-ASPECT-RATIO IGNIMBRITE.

low-level waste *See* RADIOACTIVE WASTE.

low-potassium tholeiite *See* MID-OCEAN-RIDGE BASALT.

lowstand A time during which sea levels are at their lowest. *Compare* HIGHSTAND.

lowstand systems tract (LST) In the *genetic stratigraphic sequence model used in *sequence stratigraphy, a bounding surface formed either from *allochthonous deposits eroded from the platform margin and slope, or from *autochthonous wedges deposited on the upper slope. *See* SYSTEMS TRACT. *Compare* HIGHSTAND SYSTEMS TRACT; and TRANSGRESSIVE SYSTEMS TRACT. *See also* REGRESSIVE SYSTEMS TRACT.

low symmetry *See* CRYSTAL SYMMETRY.

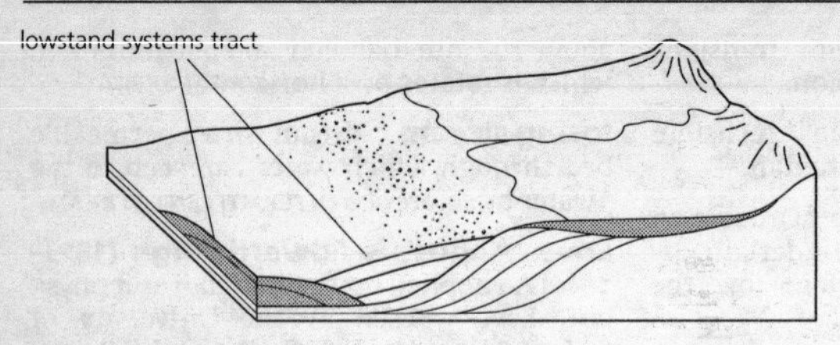

Lowstand systems tract

low-velocity layer *See* ELEVATION CORRECTION.

low-velocity zone (LVZ) The zone within the upper *mantle beneath the oceans within which seismic *P-waves are slowed and *S-waves are slowed and partially absorbed. The top of the zone is some 40–60 km deep near the oceanic spreading *ridges, and this depth increases to 120–160 km beneath the older *oceanic crust. The bottom of the zone is poorly defined, but in the region of 250–300 km in depth. Beneath the continents, a restricted low-velocity zone occurs beneath *crust areas subjected to *orogenesis during the last 600 million years or so, but is not found beneath cratonic (*see* CRATON) areas. It is attributed to the presence of a 0.1% fluid phase and commonly ascribed to the partial melting of mantle rocks at these depths. It is often considered coincident with the *asthenosphere, but probably this is valid only for oceanic areas.

Lq *See* DIURNAL VARIATION (2).

L_Q-wave *See* LOVE WAVE.

LREE Abbreviation for 'light *rare earth element'. *See* MID-OCEAN-RIDGE BASALT.

LST *See* LOWSTAND SYSTEMS TRACT.

L–S-tectonite *See* SHAPE FABRIC.

L-tectonite *See* SHAPE FABRIC.

lucinoid Applied to one of the two types of *hinge-line dentition found in *heterodont bivalves (*Bivalvia). The lucinoid type has two *cardinal teeth in each valve and may have more lateral teeth in the right valve than in the left. *Compare* CYRENOID.

Ludfordian A *stage of the Upper *Silurian underlain by the *Gorstian.

Ludhamian The Lower *Pleistocene, lowermost, temperate-climate, shelly, sand deposit of the tripartite division of the deposits in a *borehole at Ludham, in eastern England. *See also* ANTIAN; BAVENTIAN; PASTONIAN; and THURNIAN.

Ludlow A *series (424–408 Ma) of the Upper *Silurian, underlain by the *Wenlock and overlain by the *Pridoli.

Luisian A *stage in the Upper *Tertiary of the west coast of N. America, underlain by the *Relizian, overlain by the *Mohnian, and roughly contemporaneous with the upper *Langhian and lower *Serravallian Stages.

lumbar vertebra *See* VERTEBRA.

luminance A measure of the *intensity of *electromagnetic radiation emitted by a source in a specified direction.

luminous night clouds *See* NOCTILUCENT CLOUDS.

lumps, carbonate *See* INTRACLAST.

Luna A series of Soviet lunar missions that ran from 1959 to 1976.

lunar From the Latin *luna* (Moon), of, pertaining to, affecting, caused by, or involving the Moon.

Lunar-A A Japanese *ISAS mission, to be launched in 1999, comprising an orbiter and an instrument package to be delivered to the surface.

Lunar Highlands *See* TERRA.

lunar magnetic variation *See* DIURNAL VARIATION (2).

Lunar Orbiter A series of *NASA lunar mapping missions that ran from 1966 to 1967.

lunar time-scale *See* APPENDIX B.

lunate Half-moon shaped.

lunette (clay dune) An accumulation of *aeolian sediment, consisting of clay pellets the size of sand grains, found around the margins of some salt lakes.

lunule 1. In *Bivalvia, a depressed plane or curved area along the *hinge line, anterior to the *umbo. *Compare* ESCUTCHEON. **2.** 'Keyhole-like' perforations in the *tests of many flat clypeasteroid (sand dollar) *Echinoidea.

lustre Of *minerals, the ability to reflect light. The quality and nature of the reflection may be a useful diagnostic aid. The term 'lustre' is often used in conjunction with the qualifying terms '*adamantine', '*metallic', '*resinous', '*waxy', '*pearly', '*silky', 'greasy', and '*vitreous'. The intensity of the reflection is qualified by descriptions such as 'shiny', 'dull' or '*splendent'.

lustre mottling A spotted, shimmering appearance seen on the surface of some calcareous *sandstones, caused by the reflection of light from coarse mineral *cement crystals which enclose *detrital *quartz grains.

Lutetian 1. An *age in the Middle *Eocene, preceded by the *Ypresian (Early Eocene), followed by the *Bartonian, and with a lower boundary dated at 50.0 Ma (Harland et al., 1989). **2.** The name of the corresponding European *stage, which is roughly contemporaneous with the upper *Ulatizian and lower *Narizian (N. America), the upper *Heretaungan, *Porangan, and lower *Bortonian (New Zealand), and the upper *Johannian and lower *Aldingan (Australia).

lutite *See* ARGILLITE.

luxullianite *See* TOURMALINE; and TOURMALINIZATION.

LVL *See* ELEVATION CORRECTION.

LVZ *See* LOW VELOCITY ZONE.

L-wave *See* LOVE WAVE.

lycopods *See* LYCOPSIDA.

Lycopsida (lycopods) Class of the *Pteridophytina, represented today by relatively small plants, the club mosses, although in the *Carboniferous they included large trees which dominated the coal-swamp vegetation of the time. The oldest lycopods date from the *Devonian and derived from the most primitive of vascular plants, the *psilophytes. Lycopods are characterized by *dichotomizing branches; spores borne in spore cases (sporangia) on the upper side of the fertile leaves (sporophylls), which are sometimes organized into cones; and 'microphyllous' leaves, i.e. with a single strand of vascular tissue proceeding unbrokenly (without a 'leaf gap') from the vascular tissue of the stem. The leaves of some lycopods were long and grass like. When they were shed, the leaves left distinctive, spirally arranged leaf scars on the trunk of the lycopod tree. Fragments of the bark are common as fossils in coal balls and the roof shales of Carboniferous coal seams. Prominent genera included **Lepidodendron*, **Lepidophloios*, and *Sigillaria* (in which the spiral leaf-scar pattern fell into discrete vertical rows).

Lyell, Charles (1797–1875) Author of the influential *Principles of Geology*, 12 editions of which were published between 1830 and 1875. He was an extreme uniformitarian, emphasizing a great age for the Earth. He thus influenced *Darwin by allowing the time needed for evolutionary changes to occur. *See* UNIFORMITARIANISM; and DRIFT.

lysimeter (evaporimeter) Device for the direct estimation of *evapotranspiration. Typically it comprises a vegetated block of soil of volume 0.5–1 m^3, to which the amount of water added is known, and from which the amount lost as run-off or percolation may be measured. Recording the changing weight of the soil-vegetation system (keeping vegetation change due to growth static or monitored) reveals the amount of water retained by the system, and thus, by difference, the amount lost as evapotranspiration. For geographic comparisons an easily standardized, short, grass vegetation cover is used. For water-budget experiments, vegetation cover may be varied to simulate different crop types or semi-natural communities.

Lysithea (Jupiter X) One of the lesser satellites of *Jupiter, with a diameter of 24 km.

lysocline The ocean depth at which the rate at which calcium carbonate dissolves increases markedly; this may mark the upper boundary of the bottom water.

lystric fault (listric fault) A curved extensional *fault which characteristically flattens at depth into a *décollement horizon. The *hanging wall rotates down along the concave-up fault surface, in some cases forming a *rollover anticlinal structure. Lystric faults are commonly found in extensional regimes (*see* EXTENSION).

m- *See* MILLI-.

M- *See* MEGA-.

M *See* MAGNETIC MOMENT.

Ma Abbreviation meaning 'million years'.

McAuliffe A *solar system asteroid (No. 3352); diameter 2–5 km; orbital period 2.57 years. Its orbit crosses that of *Mars. McAuliffe is to be visited in 1999 by the *New Millennium Deep Space-1 spacecraft.

maar *Crater, often occupied by a shallow lake, produced by an explosive volcanic *eruption. Normally it is surrounded by a low rampart or ring of ejected material. Typically, maars are formed by the explosive interaction of volcanic *magma with *groundwater. The ejected material is a mixture of *country rock and highly fragmented *ash.

Maastrichtian (Maestrichtian) Final stage (74–65 Ma) of the *Cretaceous in Europe. The *stratotype is described from the Maastricht area of Holland. Throughout western Europe, the Maastrichtian is characterized by chalk *limestones. *See also* SENONIAN.

mackerel sky Pattern of wavy *cirrocumulus (or *altocumulus) cloud with holes which produces an overall resemblance to the body markings of mackerel. *See also* VERTEBRATUS.

Mackereth corer A form of *hydraulic corer commonly used to obtain lake-sediment cores.

macroclimate The climatic character of a large region.

macroevolution *Evolution above the species level, i.e. the development of new species, genera, families, orders, etc. There is no agreement as to whether macroevolution results from the accumulation of small changes due to *microevolution, or whether macroevolution is uncoupled from microevolution.

macropygous *See* PYGIDIUM.

macrotidal Applied to coastal areas where the *tidal range is in excess of 4 m. Tidal currents dominate the processes active in macrotidal areas, e.g. the coast of the British Isles.

maculose Applied to the texture of contact-metamorphic (*see* CONTACT METAMORPHISM) rocks that display spots, each spot representing a fine-grained aggregate of *minerals.

Madagascar A large island lying in the *Indian Ocean off the eastern coast of Africa. It became separated from the main mass of Africa at a time when India was migrating northwards across the Indian Ocean, about 95 Ma. Madagascar is famous for the occurrence of primitive prosimians (lemurs) which survive there because of a lack of mammalian predators. Elsewhere, prosimians have been greatly reduced, probably as a result of competition with true monkeys and apes following their appearance in the *Oligocene.

made ground (made land) Area of dry land that has been constructed by people, generally through the reclamation of marshes, lakes, or shorelines. An artificial fill (landfill) is used, consisting of natural materials, refuse, etc.

Madelung constant The sum of the mutual potential coulombic attraction energies of the *ions in a *crystal. The constant is used in the equation for *lattice energy.

Madreporaria *See* SCLERACTINIA.

madreporite In *Echinodermata, a porous, button-shaped process on the *aboral surface of the body; through its perforations the water-vascular system is connected to the water outside.

Maentwrogian A *stage of the Upper *Cambrian, underlain by the *Menevian and overlain by the *Dolgellian and dated at 517.2–514.1 Ma (Harland et al., 1989). The Maentwrogian is known locally as the Ffestiniogian.

maerl The Breton name for a mixture of carbonate-rich (skeletal) sand and seaweed used as an agricultural dressing. *See also* MARL.

Maestrichtian *See* MAASTRICHTIAN.

maestro Local north-westerly wind of the Adriatic Sea which affects the western coasts, especially in summer. The term is also applied to north-westerly winds in the Ionian Sea and to winds off the coasts of Corsica and Sardinia.

mafic Applied to any *igneous rock which has a high proportion of *pyroxene and *olivine, such that its *colour index is between 50 and 90 (i.e. it is dark coloured).

Magellan A radar mapper, launched by *NASA in 1989 to explore *Venus.

magma A hot, *silicate, *carbonate, or *sulphide *melt containing dissolved *volatiles and suspended *crystals, which is generated by partial melting of the Earth's *crust or *mantle and is the raw material for all *igneous processes. The melt component of silicate magmas, by far the most common magma type, comprises a disordered mixture of single Si–O tetrahedra and chains, branching chains, and rings of Si–O tetrahedra, between which are located randomly positioned *cations (e.g. Ca^{2+}, Mg^{2+}, Fe^{2+}, and Na^{+}) and *anions (e.g. OH^{-}, F^{-}, Cl^{-}, and S^{-}) loosely co-ordinated with the oxygens in the silicate tetrahedra. The greater the silica content of the magma, the more chains and rings of silicate tetrahedra there are to impede each other and hence increase the *viscosity of the magma. The pressure regime and composition of the magma control which minerals nucleate and crystallize from a magma when it cools.

magma chamber A region, postulated to exist below the Earth's surface, in which *magma is received from a source region in the deep *crust or upper *mantle, stored, and from which it moves to the Earth's surface at the site of a *volcano. The existence of 'magma chambers' is often invoked by geochemists to provide a location for processes such as *fractional crystallization, because the chemistry of *lavas is explicable only in terms of such processes. When magma moves rapidly from the chamber, as in a *pyroclastic flow *eruption, the unsupported chamber roof may collapse to produce a *caldera at the surface. The diameter of the caldera can be used to estimate the diameter of the underlying magma chamber; diameters of up to 40 km are not uncommon for terrestrial subvolcanic magma chambers.

magmatic differentiation (magmatic fractionation) Formation of a variety of rock types from an initial single parental *magma. No specific mechanism is implied by the term. *See also* FRACTIONAL CRYSTALLIZATION.

magmatic fractionation *See* MAGMATIC DIFFERENTIATION.

magmatic-segregation deposit Concentration of particular minerals in different parts of a *magma chamber during consolidation, by *gravity settling, filter pressing, flow, *fractional crystallization, liquid immiscibility, or gas transference; for example, the accumulation of heavy minerals such as *chromite and *magnetite by gravity settling.

magnesiochromite *See* CHROMITE; and SPINEL.

magnesioferrite *See* MAGNETITE.

magnesite *Carbonate, $MgCO_3$ and *end-member of a *solid solution series with *siderite, $FeCo_3$; sp. gr. 3.0; *hardness 4; whitish; *earthy *lustre; compact or granular; occurs as an alteration product of *serpentines, *dolomites, or *limestones and may form as a chemical precipitate. It is mined commercially and used in the production of magnesium compounds, refractory products, and special cements.

magnesium-sulphate soundness test A test identical to the *sodium-sulphate soundness test, but in which magnesium sulphate is used instead of sodium sulphate.

magnetic age *See* MAGNETIC ANOMALY PATTERN; CHRON; and MAGNETIC DATING.

magnetic anomaly Any magnetic field remaining after allowance for some particular model. Normally this is the field remaining after allowance for the *International Geomagnetic Reference Field, but sometimes it is the field remaining after further removal of either *regional or near-surface anomalies.

magnetic anomaly pattern (magnetic age) Phenomenon, discovered originally in the north-eastern Pacific Ocean, of linear *magnetic anomalies which lie parallel to oceanic *ridges where spreading has occurred. The magnetic striping, which is determined by instrumental measurements, results from repeated *reversals of the Earth's magnetic field (*see* GEOMAGNETIC FIELD): it exists as corresponding patterns on either side of an oceanic spreading centre (*see* MID-OCEAN RIDGE). These can be correlated with the *magnetostratigraphic time-scale (polarity time-scale), so assigning a magnetic age to some of the individual magnetic anomalies. *See also* SEA-FLOOR SPREADING.

magnetic dating The use of magnetic properties for the age assessment of archaeological and geologic materials. It uses the *natural remanent magnetization acquired at a specific time. Its direction can be compared with time-scales of geomagnetic *secular variations, *polarity reversal sequences, or *polar wander paths. Its intensity can be converted to the strength of the field in which it was acquired and so compared with the time-scales of intensity changes of the *geomagnetic field. *Viscous remanent magnetization may sometimes be used to establish how long the sample has been in a specific position.

magnetic domain A volume of a *ferromagnetic material in which all *electron spins are aligned in the same direction. In most naturally occurring minerals a magnetic domain is about 1 μm in diameter.

magnetic fabric determination The determination of the *petrofabric using magnetic methods. Generally it is the determination of the *magnetic susceptibility ellipsoid, as this is dependent on the shape of *magnetite grains or the crystalline alignment of *hematite grains. The magnetic fabric may also be due to *paramagnetic and diamagnetic materials. Some instruments may be sensitive to the *anisotropy of electrical conductivity in the absence of *ferromagnetic materials.

magnetic field *See* GEOMAGNETIC FIELD.

magnetic flux The magnetic induction, perpendicular to the surface area of a nearby body, multiplied by the area of the body.

magnetic gradiometry A geophysical technique, sometimes used in archaeology, that measures the magnetic properties of subsoil materials.

magnetic induction (B) The vector, B, which is equal to the magnetic permeability of a material (μ) multiplied by the magnetic field intensity vector (H) applied to it.

magnetic moment (M) The measured total intensity of magnetization of a sample, irrespective of the volume or weight of the sample itself. It is measured in units of A/m^2.

magnetic orientation The sensing by certain organisms of the direction of the *geomagnetic field and their use of it for purposes of orientation. *See* BIOMAGNETISM. *See also* MAGNETIC SAMPLING.

magnetic permeability *See* PERMEABILITY.

magnetic profile A series of determinations of the intensity of the *geomagnetic field along a traverse, usually oriented north–south.

magnetic quiet zone Any area showing few or no *magnetic anomalies. Magnetic quiet zones are commonly oceanic areas where magnetic anomalies appear to be absent because either the *geomagnetic field did not change *polarity during the formation of the *oceanic crust or the original magnetization has been destroyed by the effects of thermal blanketing by later sediments.

magnetic sampling The collection of samples in order to study their magnetic properties in the laboratory. The samples may be unoriented if they are only for rock-magnetic properties, e.g. *magnetic susceptibility, *magnetic intensity. For most *archaeomagnetic and *palaeomagnetic purposes samples are collected using a small portable drill and oriented using Sun, magnetic, or surveying methods, or *gyrocompasses, and often utilizing special orientation devices to ensure precision.

magnetic separator A concentrator in which *ore particles pass through a magnetic field or series of fields. The magnets may be permanent or electromagnets, and may work wet or dry.

magnetic signature *See* SIGNATURE.

magnetic storm Major disturbance of the Earth's magnetic field resulting from the passage of high-speed charged solar particles, following a solar flare. The particles are deflected towards the regions of the Earth's magnetic poles.

magnetic survey A survey along a profile or grid using a *magnetometer to determine the strength of the *geomagnetic field at particular points.

magnetic susceptibility The constant of proportionality (κ) between an applied field (H) and the *magnetic moment induced (J), i.e. $J = \kappa H$; κ is measured in dimensionless SI units. *See also* SUSCEPTIBILITY METER.

magnetic variations Periodic and irregular changes in the *geomagnetic field, usually subdivided into brief *transient variations and long-term *secular variations.

magnetite (loadstone, lodestone) An important rock-forming iron *oxide $Fe^{2+}Fe^{3+}{}_2O_4$ ($FeO.Fe_2O_3$) and member of the *spinel group of *minerals; belongs to the magnetite series along with magnesioferrite ($MgFe^{3+}{}_2O_4$), *franklinite, jacobsite ($MnFe^{3+}{}_2O_4$), and trevorite ($NiFe^{3+}{}_2O_4$); sp. gr. 5.2 (to 4.6); *hardness 5.5–6.5; *cubic; normally iron-black with a blue tarnish; black streak; *sub-metallic *lustre; crystals often *octahedral and *rhombdodecahedral, but it is also *massive and granular; *cleavage poor, octahedral; strongly magnetic; found extensively in magmatic rocks associated with *basic rocks, in *pegmatites, in contact *metasomatic zones (especially in *limestones in *regionally metamorphosed rocks due to the dehydration of iron hydroxides), and also in black sands associated with gold gravels.

magnetochronology Geochronological system that is based on geomagnetic *polarity reversals. As no theory yet specifies the existence and duration of separate geomagnetic/polarity intervals, a true magnetochronology is said not to exist. *See also* MAGNETOSTRATIGRAPHIC TIME-SCALE.

magnetogram A recording of the time variations of the *geomagnetic field, usually in analogue form. *See* MAGNETOGRAPH.

magnetograph An instrument for producing a record of changes of geomagnetic elements as a function of time. *See* MAGNETOGRAM.

magnetohydrodynamics The science of relating magnetic fields, mainly mathematically, within a moving conducting medium. It is mainly applicable to the Earth's *core, where the *geomagnetic field is generated by the motion of magnetic lines of force that are 'frozen' within a moving, electrically conducting medium, but it is also applicable to all systems involving the fluid motion of electrically conducting materials within a magnetic field.

magnetometer An instrument for measuring either the direction or the intensity of a magnetic field. Common field instruments include the *fluxgate type and *nuclear-precession types. The fluxgate magnetometer can also be used as a gradiometer and it is therefore most sensitive to shallow magnetic sources. Observatory instruments include the dip-circle and astatic types.

magnetopause The boundary within the *magnetosheath, between the *solar wind and the region closer to the planet which is shielded from the *solar wind.

magnetosheath The region between the *bow shock, where *solar wind particles are slowed from supersonic to subsonic velocities, and the *magnetopause.

magnetosphere The space around a planet in which ionized particles are affected by the planet's magnetic field. The Earth's magnetosphere reaches far beyond the *atmosphere. In the magnetosphere, charged particles are concentrated at altitudes of about 3000 km and 16 000 km. The charged particles oscillate between the northern and southern hemispheres. The outer boundary of the magnetosphere is sharp and well defined, extending to about 10 Earth radii on the sunlit side of the Earth and to perhaps 40 Earth radii on the dark side; but the boundary changes its position in response to solar activity, being depressed by the *solar wind. *See also* EXOSPHERE; and IONOSPHERE.

magnetostratigraphic time-scale (polarity time-scale, geomagnetic reversal time-scale, reversal time-scale) A time-scale based on the periodic *polarity reversals in the Earth's *geomagnetic field. Magnetic minerals within a rock retain an orientation induced by the field at the time the rock was formed (*see* NATURAL REMANENT MAGNETISM). Provided they include suitable minerals, *strata from all over the world thus contain a record of the *normal (as at present) or reversed state of the geomagnetic field at the time of their formation. This reversal pattern has been correlated between different successions of rocks to produce a sequence that, when combined with a *dating method such as *potassium–argon dating, has given a time-scale measured in units of normal or reversed polarity. The scale was first established in detail for the last 4.5 Ma using data from terrestrial, mainly *extrusive, rocks; it has now been extended back to the Upper *Jurassic by means of the *magnetic-anomaly patterns in *oceanic crust. The terms proposed by the *ISSC for geochronologic units of the magnetostratigraphic time-scale are *polarity superchron, *polarity chron (replacing the earlier term 'epoch'), and *polarity subchron (replacing the earlier term 'event'). The corresponding terms proposed for rocks deposited during those *polarity intervals are polarity superchronozone, polarity chronozone, and polarity subchronozone.

magnetostratigraphy Branch of stratigraphy based on geomagnetic *polarity reversals.

magnetotelluric sounding The use of changing components of the *geomagnetic field to study the electrical conductivity of the rocks within the Earth. The geomagnetic-field changes induce eddy currents within the Earth that are in proportion to the electrical conductivity of the rocks concerned, but they can readily be separated from the inducing currents because of their phase difference. It is generally applied to studies of the lower *crust and upper *mantle.

magnetozone A unit of rock of the same magnetic polarity (character). *See also* CHRON.

magnitude, earthquake *See* EARTHQUAKE MAGNITUDE; MERCALLI SCALE; and RICHTER MAGNITUDE SCALE.

Magnolia One of the most ancient of flowering plants, noted for the relatively primitive structure of its flower. *Magnolia* leaves are first recorded from Dakota, USA, in the mid-*Cretaceous. *Magnolia paeopetala* is a reconstruction, by E. E. Leppik, of a whole flower based on the earliest *angiosperm petal ever found (also from the Dakota fossil flora).

magnon In a magnetic material, a quantum of spin-wave energy.

main-stage sequence *See* HYDROGEN BURNING; and NUCLEOSYNTHESIS.

majanna A dry, level plain with a surface encrusted with salt.

malachite Mineral, $Cu_2CO_3(OH)_2$; sp. gr. 3.9–4.0; *hardness 3.5–4.0; *monoclinic; bright green; pale green *streak; in fibrous condition with a silky *lustre, crystals with *adamantine or *vitreous lustre; crystals very rare, and it is usually found as *botryoidal, encrusting masses with bands of varying colour, and frequently in a fibrous, radiating *habit; common *secondary mineral in the oxidized zone of copper deposits, associated with *azurite, native *copper, and *cuprite; soluble with effervescence in dilute hydrochloric acid. It is often used for ornaments and pigments, and as an *ore mineral for copper. *See* SPESSARTITE.

Malacostraca (phylum *Arthropoda, subphylum *Crustacea) One of eight classes of Crustacea, the malacostracans appear in the *Cambrian. There are usually eight pairs of *biramous, thoracic limbs, some forms have pincer-like appendages, and the limbs may be modified for swimming, feeding, or other purposes. The group is diverse and includes crabs, lobsters, and the shrimps, and it has quite a good *fossil record.

Malakovian *See* GORE.

malleolus A small prominence on the *distal end of the *tibia and of the *fibula; in humans they form the 'knobs' of the ankle.

Mallet, Robert (1810–81) A civil engineer from Dublin, Mallet became interested in *earthquakes and *volcanoes, making a detailed study of the Neapolitan earthquake of 1857. He investigated the velocity of *seismic waves, and compiled an earthquake catalogue from which he was able to make a seismic map of the world.

malleus In *Mammalia, the outer *ossicle of the middle ear, which is derived from the articular bone of ancestral vertebrates.

Malm 1. An alternative name for the Late *Jurassic *Epoch, preceded by the *Dogger, followed by the *Neocomian (*Cretaceous), and dated at 157.1–145.6 Ma (Harland et al., 1989). It comprises the *Oxfordian, *Kimmeridgian, and *Tithonian (*see also* VOLGIAN) *Ages. **2.** The name of the corresponding European *series, which is roughly contemporaneous with the upper *Kawhia and *Oteke (New Zealand).

Malvernian A *stage of the Upper *Proterozoic of the Welsh borders and English Midlands, underlain by the *Monian and overlain by the *Uriconian.

mamma (mammatus) From the Latin *mamma* meaning 'udder', a cloud feature consisting of projections from the basal surface of *altocumulus, *altostratus, *stratocumulus, *cumulonimbus, *cirrus, and *cirrocumulus. *See also* CLOUD CLASSIFICATION.

Mammalia (mammals; phylum *Chordata) Class of *homoiothermic animals in which the head is supported by a flexible neck, typically with seven *vertebrae, articulating through two *occipital condyles. The lower jaw is composed of one bone only, the dentary, which articulates directly with the skull; the middle ear contains three small bones, two of which are derived from bones in the lower jaw of *reptiles. Typically, mammalian teeth are of differing forms (*heterodont), they are set in sockets (thecodont), and milk teeth are shed and replaced by a second set. A hard *palate separates the nasal cavity from the mouth. Except in *monotremes, the egg is small and develops in the uterus; and the young are fed milk secreted by mammae (which give the class its name). The skin has at least a few hairs. Many mammalian features were present in *therapsids (mammal-like reptiles) during the *Triassic. Mammals are believed to have appeared first toward the end of the Triassic and to have diversified rapidly from the end of the *Mesozoic, 100 Ma later, following the mass *extinction which marks the *Mesozoic–*Tertiary boundary 65 Ma ago.

mammal-like reptiles *See* SYNAPSIDA.

mammatus *See* MAMMA.

mammillary (mammillated) Applied to the physical *habit of a mineral which has grown from radiating crystals to give curved or rounded surfaces.

mammillated *See* MAMMILLARY.

mammillated topography Hill relief with a streamlined, rounded, and smoothed appearance, normally resulting from the scouring action of an *ice sheet, as in the Adirondack Mountains, eastern USA. However, some crystalline rocks may support a pseudo-mammillated surface formed by non-glacial processes, as on Ben Lomond, Tasmania.

Mammoth A reversed *polarity subchron which occurs within the *Gauss normal *polarity chron.

mammoth *See* MAMMUTHUS.

***Mammuthus* (mammoth)** Line of *Pleistocene elephants adapted to *steppe and *tundra habitats. The tusks were elongated and strongly curved, and the skull was shorter and higher than that of other elephants. The woolly mammoth was adapted to arctic environments. The largest mammoth, indeed the largest *proboscidean of all time, was *Mammuthus armeniacus* of Eurasia, which stood about 4.5 m at the shoulder.

Mammutidae Extinct family of mastodons, comprising the one genus *Mammut* (*Mastodon*) of elephant-like animals, which diverged from the evolutionary line leading to the modern elephants. It was long-lived, extending from the Lower *Miocene to the Recent, and it survived in Africa, N. America, and Eurasia at least until the end of the *Pleistocene. Mastodons were shorter and heavier in build than elephants. *Mammut* (or *Mastodon*) species had short, high skulls, longer jaws than elephants, and usually tusks in both upper and lower jaws, the upper tusks often being large and curving outward and upward. There were never more than two teeth in use at a time, never more than a vestige of the lower incisors, and the molars were low-crowned, simple, and lacked cement. It is believed that the evolution of the mastodons paralleled that of the gomphotheres (*Gomphotheriidae).

mandible In *Crustacea, *Insecta, and Myriapoda (centipedes, millipedes, etc.), one of the pair of mouth-parts most commonly used for seizing and cutting food. In birds, specifically the lower jaw but the term is also used to denote the two parts of the bill of a bird, as upper and lower mandibles.

manganese nodule Concretion of iron and manganese oxides which also contains copper, nickel, and cobalt. Nodules are variable in size, shape, and composition, and are layered internally. The average composition is: manganese 30%; iron 24%; nickel 1%; copper 0.5%; cobalt 0.5%. Manganese nodules are widely distributed on the sea floor of every ocean and in some temperate lakes. They are found in areas of negligible sedimentation and/or strong bottom currents, e.g. on the N. Pacific *abyssal plain at depths of 3500–4500 m, and on the shallow, current-swept Blake Plateau off the east coast of the USA. Submarine mining of these deposits is now considered viable to recover copper and other scarce metals. Growth of manganese nodules apparently reached a peak in the early *Tertiary period.

manganite Mineral, $MnO_2Mn(OH)_2$; sp. gr. 4.2–4.4; *hardness 3–4; *monoclinic; dark steel-grey to black; reddish-brown to black *streak; *sub-metallic *lustre; crystals *prismatic with striated faces, often as bundles or radiating aggregates; *cleavage perfect, pinacoidal; occurs in deposits precipitated in sedimentary conditions in which there is an oxygen deficiency, and in low-temperature *hydrothermal veins in association with *barite and *goethite; soluble in hydrochloric acid with the evolution of chlorine. It is often used in steel-making.

Mangaorapan A *stage in the Lower *Tertiary of New Zealand, underlain by the *Waipawan, overlain by the *Heretaungan, and roughly contemporaneous with the upper *Ypresian Stage.

Mangaotanian *See* RAUKUMARA.

Mangapanian A *stage in the Upper *Tertiary of New Zealand, underlain by the *Waipipian, overlain by the *Nukumaruan, and roughly contemporaneous with part of the *Piacenzian Stage.

mangrove swamp Characteristic vegetation of tropical, muddy coasts, and typically associated with river mouths where the water is shallow and the load of suspended sediment is high. The aerial roots of the mangrove trees trap the sediment, favouring the gradual seaward extension of the land area.

manometer An instrument that can be used for the direct measurement of vacuum, positive pressure, and differential pressure. It usually consists of two interconnected tubes filled with fluid, e.g. water or mercury. A difference in fluid levels in the two tubes records the pressure. It can also be modified to measure flow.

Mantell, Gideon Algernon (1790–1852) A Sussex surgeon, whose hobby was the study of fossils from the Chalk and Weald, on which he became an expert. Mantell discovered *iguanodon and other fossil *dinosaurs which, with the remainder of his collection, were donated to the British Museum (Natural History). He also published a number of popular works on geology.

mantle 1. Zone lying between the Earth's *crust and *core, approximately 2300 km thick. The mantle is probably similar in composition to *garnet *peridotite and it represents about 84% of the Earth's volume and 68% of its mass. A mantle is present in most *terrestrial planets and the *Moon, but is of a different composition in each case. **2. (pallium)** In *Brachiopoda and *Mollusca, a layer of tissue that covers the body and is responsible for the secretion of the shell.

mantle array A term used in isotope geochemistry to describe a graphical plot of $^{144}Nd : ^{143}Nd$ against $^{87}Sr : ^{86}Sr$ for *igneous rocks. Rocks which have been derived from the *mantle tend to plot on a straight line; those that show evidence of crustal contamination tend to fall off the line, often in systematic ways.

mantle bedding The maintenance of a uniform thickness of *pyroclastic material over all but the steepest topographic features. Fall deposits are the main group of pyroclastic rocks which commonly display mantle bedding, although *ignimbrite veneer deposits may also display it.

mantle convection The transfer of heat by movement of material within the *mantle. Material derived from within the mantle is added to the *lithosphere at *constructive margins and cool lithosphere descends at *subduction zones, thus some return flow of material must take place at depth. The return flow may involve only material at shallow depth below the lithosphere, or material through the whole mantle, or possibly two or more layers of convection cells transferring heat between the layers by conduction. *Hot spots may overlie isolated *plumes of material rising from the *core–mantle interface.

mantled gneiss dome A *dome of granitic *migmatites and *gneisses surrounded by a 'mantle' of metasediments which characteristically show outward-dipping parallel *fabrics and *foliations. The feature is thought to be produced by gravitational instabilities due to a density inversion in which dense rocks overlie a less dense, *granite–*gneiss core.

mantle lobe *See* BRACHIOPODA.

mantle plume Localized, hot, buoyant material which is hypothesized to be rising through the *mantle. Mantle plumes are thought by some geologists to rise beneath *hot spots, causing *domal uplift. Plumes may originate near the mantle–core boundary, and it has been suggested that they generally have the form of a cylinder with a radius of about 150 km.

manto A horizontal, bedded *ore deposit; the term may describe a sedimentary *stratum, or a replacement, strata-bound deposit.

map projection The representation on a plane surface of part or all of the surface of the Earth or a celestial body. Various projections have been used in the Earth sciences because of their particular suitability for certain work, e.g. equal-area projections for the global distributions of different sedimentary deposits, polar stereographic projection for the relationships within *Gondwanaland, Mercator's projection for the absolute movements of the lithospheric *plates, and conical projections for the Himalayan–Alpine *fold belt. Projections based on Mercator's are most frequently used because of their general familiarity, unless the information portrayed is peculiarly distorted.

marble Non-foliated, metamorphosed *limestone which is produced by *recrystallization and is hard enough to take a polish. The hardest and most attractive marbles have been used in statuary and for building since antiquity and are still quarried, e.g. from the Carrara quarry which supplied Michelangelo. The statue of Abraham Lincoln at the Lincoln Memorial, Washington, is made from marble quarried in Georgia, USA. Marbles may be variously coloured or banded, depending on their chemical and mineralogical composition (mostly *calcite), e.g. Carrara marble is pure white, but Siena marble, quarried in Tuscany, has red mottling.

marcasite Mineral, FeS_2; sp. gr. 4.8–4.9; *hardness 6.0–6.5; *orthorhombic; pale bronze-yellow; greyish-black *streak; *metallic *lustre; crystals *tabular and spear-like, but it is also *massive, stalactitic, and radiating; *cleavage {101}; occurs in low-temperature mineral *veins together with zinc and lead ores, and in *sedimentary rocks, especially *limestone, *chalk, and *clay. It has been used in the production of sulphuric acid. *Compare* PYRITE.

mare (pl. maria) The Latin word for 'sea', originally used by Galileo in 1610 to denote the smooth, grey areas on the *Moon, visible to the naked eye, that are now known to be plains of basaltic lava filling circular basins excavated by *planetismal impact. *Basalt lavas, mainly in maria, cover 17% of the surface of the Moon, mostly on the near side. The term is also applied to low-lying areas on *Mars.

mare ridge *See* WRINKLE RIDGE.

mares' tails Popular name for tufted *cirrus clouds with *virga (precipitation trails) seen below each cloud.

margarite A dioctahedral *mica with the general composition $Ca_2Al_4(Si_4Al_4O_{20})(OH,F)_4$, found in fine-grained, low-grade, regional *metamorphic rocks (*see* METAMORPHIC GRADE). Calcium is the main *cation found between the individual *silicate sheets of the mica structure. Margarite has the typical appearance of mica, but is much harder than common mica types. *Cleavage sheets are also less elastic than those of the common micas and hence margarite is a member of the 'brittle mica' group. At one time this mineral was thought to be a rare component of metamorphic rocks but recent *X-ray diffraction work has demonstrated that it can be abundant in low-grade metapelites (*see* META-). It occurs with *corundum in *emery deposits and in *mica schists along with *tourmaline and *staurolite.

marginal basin 1. A synonym for *back-arc basin. **2.** A basin developed along a *passive margin.

marginal sea 1. A semi-enclosed body of water adjacent to, and widely open to, the ocean (e.g. the Gulf of Mexico, Caribbean Sea, and Gulf of California). **2.** *See* PERICONTINENTAL SEA.

marginal suture *See* CEPHALIC SUTURE.

marialite *See* SCAPOLITE.

Marianas Trench The oceanic *trench which marks the *destructive margin between the *Pacific and *Philippine Plates. The trench is about 11 km deep, and there is evidence of little or no *accretionary wedge. The *subduction zone steepens at depth, becoming almost vertical, with *hypocentres down to approximately 700 km.

marine platform *See* SHORE PLATFORM.

Mariner A series of *NASA planetary missions exploring *Venus (Mariners 2 (1962), 5 (1967), and 10 (1973)) and *Mars (Mariners 4 (1964), 6 (1969), 7 (1969), 9 (1971)); Mariner 10 also travelled to *Mercury. There was no Mariner 1.

maritime air *Air mass with properties of temperature and humidity derived either from a source region in, or from long passage over, the oceans.

maritime climate General term applied to a climate much modified by oceanic influences. Typical characteristics include relatively small diurnal and seasonal temperature variation, and increased precipitation due to more moist air.

marker bed (key bed, chronohorizon, chronostratigraphic horizon, datum) A thin bed of a distinctive character, often widely distributed and capable of being recognized and traced over a large geographical area. Marker beds are taken to be the results of either (geologically speaking) very short episodes of deposition (e.g. a thin *coal seam), or of almost instantaneous events (e.g. turbidity deposits (*see* TURBIDITY CURRENT), or *bentonite beds derived from volcanic *ash falls). Such beds are thus very valuable as time markers in stratigraphic *correlation.

Markov chain In statistics, a set of sequential observations in which the probability of one member of the sequence occurring conditional on all the preceding members occurring is equal to the probability of that member occurring conditional only on the immediately preceding member occurring.

marl A *pelagic or *hemipelagic sediment (an *arl), typically found interbedded with purer oozes in beds up to 1.5 m thick, with a composition intermediate between a non-biogenic sediment and a calcareous or siliceous ooze. It is 30% clay and 70% microfossils, at least 15% of its volume being siliceous microfossils. *Compare* SARL; and SMARL.

marlstone A semi-lithified *marl. A fully lithified marl is termed an argillaceous *limestone.

Mars (adj. martian) The fourth planet in the *solar system, 1.524 AU from the *Sun. Its radius is 3390 km, its density 3940 kg/m^3, and the inclination of equator to orbit is 25.1°. It has a small *atmosphere (7 mb) of CO_2. The polar caps are of water ice with seasonal solid CO_2. The northern-hemisphere crust is mainly basaltic plains and *volcanoes; the southern an ancient cratered terrain (*see* CRATER). The Tharsis Bulge is an uplifted or volcanic plateau. Large canyons exist and there is evidence of former water *erosion. Some *basaltic meteorites are derived from Mars. It has two small *satellites, Phobos and Deimos, probably captured *asteroids.

Mars 96 A Russian mission to *Mars, launched in 1996, that was intended to place a vehicle in orbit and land instruments on the surface. The launch failed, the rocket and its payload falling back to Earth.

Marsdenian A *stage of the *Bashkirian Epoch, preceded by the *Kinderscoutian and followed by the Yeadonian.

Mars Global Surveyor A *NASA mission to *Mars, launched in 1996, which placed a vehicle in orbit about the planet.

Marshbrookian A *stage of the *Ordovician in the Upper *Caradoc, underlain by the *Longvillian and overlain by the *Actonian.

Mars Observer A *NASA mission to *Mars that was launched in 1992. It should have orbited the planet, surveying and mapping, but communication with the spacecraft was lost before it could be manœuvred into orbit and the mission was abandoned.

Mars Pathfinder A *NASA environmental survey mission to *Mars that was launched in 1996 and soft-landed successfully in July 1997. After landing Pathfinder was renamed the Sagan Memorial Station, in honour of Carl Sagan. Sojourner, a small rover, travelled short distances from the main lander, exploring rocks selected by controllers on Earth from images transmitted by the lander.

marsquake The martian equivalent of an *earthquake or *moonquake. None were positively detected by the Viking Lander *seismometer.

Marsupialia (subclass Theria, infraclass Metatheria) An order that comprises some 250 species of living marsupials and many extinct forms. It is sometimes divided into three suborders (Polyprotodonta, which includes the opossum-like insectivorous, carnivorous, and omnivorous forms; Diprotodontia, containing the phalangers, kangaroos, and other forms evolved from an opossum-like stock, but differing structurally from the polyprotodonts; and Caenolestoidea (classed by others as a superfamily), containing a small group of 'opossum rats'), but nowadays it is usual to divide the marsupials into several orders, often allocated to two cohorts: Ameridelphia and Australidelphia. In this scheme, the name Marsupialia would cease to be used formally. Marsupials are characterized principally by their method of reproduction. The egg is yolky and has a thin shell protecting it from maternal antigens. Placental development is usually very limited and except in the Peramelemorpha the allantois serves no nutritional function, but uterine milk may be taken up by the yolk sac. Within 10–12 days of the breaking of the shell, the embryo (whose fore limbs and associated neural development, mouth, and olfactory system have developed precociously) is born. It crawls into the pouch (marsupium) and attaches itself to a teat, its lips growing around the teat, which injects milk without choking the embryo. In the later stages of its development an offspring may receive high-fat, low-protein milk from one teat while a newer embryo receives high-protein, low-fat milk from another. Marsupials also differ from placentals in their dentition, in the possession of an inflected angular process to the jaw, and in the presence of two marsupial bones which articulate with the pubes. Marsupials and placental mammals apparently diverged from a common ancestor in the *Cretaceous. The first marsupials were similar in general form to the opossums of America. In Australia the marsupials radiated to produce a wide array of adaptive types, while in S. America they filled the insectivorous and carnivorous niches for much of the *Cenozoic, while placentals occupied the herbivorous niches.

marsupials *See* Marsupialia.

Mars Surveyor Four *NASA missions to *Mars, some consisting of two launches. Mars Surveyor '98 comprises an orbiter, to be launched in 1998, and a lander, to be launched in 1999. Mars Surveyor 2001, to be launched in 2001, will place a vehicle in orbit and send a lander with a rover to the surface. Mars Surveyors 2003 and 2005 will each send a lander and rover to Mars, one to be launched in 2003, the other in 2005.

martian Of, or pertaining to, or about, the planet Mars.

martian canals ('canali') Optical illusions, produced by telescopic viewing of Mars with a resolution of poorer than about 100 km, first reported by Schiaparelli ('canali' is the Italian for 'channels') and especially championed by Percival Lowell (1855–1916). These observers produced maps of the martian surface showing interconnected networks of canals, implying the presence of intelligent life on Mars. The intelligence which devised the canals was, however, on the terrestrial side of the telescope.

martian terrain units Polar units include Layered Deposits and Etched Plains; Volcanic units include Volcanic Constructs and Volcanic Plains; Cratered units include Ancient Cratered Terrain and Cratered Plains; Modified units include Knobby Terrain, Fretted and Chaotic Hummocky Terrain, Channel Deposits, and Grooved Terrain.

mascon Abbreviation for mass concentration. There are about ten large-scale positive *gravity anomalies on the *Moon, associated with the large circular maria (*see* MARE). The *basalt filling the multi-ringed impact basins only partly accounts for the excess mascons per unit area, and central subsurface uplift of denser *mantle material into the *feldspathic *lunar highland crust during the impact is required to provide the observed values of up to $+2.2 \times 10^{-3}$ m/s^2 (+220 mgal) observed in the Mare Imbrium.

maser An acronym for *m*icrowave *a*mplification by *s*timulated *e*mission of *r*adiation, a device resembling a *laser but emitting radiation at microwave frequencies.

mass balance 1. Generally, a term used in comparisons of the inputs and outputs of processes. **2.** The balance of elements in ocean water, which is assumed to be constant, i.e. influx and removal of elements occurs at the same rate. Influxes include river water, elements released from sediment pore fluid, and melting ice. Reaction between sea water and rocks is probably insignificant. The removal of elements from sea water occurs by precipitation of chemical sediments, *ion exchange, and the burial of pore fluids. **3.** The relation between the input and output of a *glacier. Input (accumulation) is dominated by snow precipitation and output (*ablation) by surface melting. The difference between accumulation and ablation for a glacier over a year is the net (mass) balance. A positive net balance implies that a glacier is growing; a negative net balance that it is shrinking; and a zero net balance that it is stable.

mass extinction *See* EXTINCTION.

mass flow A down-slope slide of *sediment which moves under the force of gravity. Mass flows include rockfalls (accumulations of *scree), slumps and slides (where masses of sediment move downslope along discrete *shear planes), debris flows (in which ill-sorted masses of sediment move downslope due to the loss of internal strength of the sediment mass), liquefied sediment flows, *grain flows, and turbidity flows.

massif A very large topographic or structural feature, usually of greater rigidity than the surrounding rock. The term 'massif' is used of older crystalline blocks in an *orogenic belt.

massive Lacking in any form or structure, e.g. massive beds are those without internal grading (*see* GRADE), and lacking *sedimentary structures. Applied to a compact *mineral which has no distinguishing *crystal form.

massive sulphide deposit Rich mass of metallic *sulphide minerals, with little *gangue, which are relatively easy to mine. Typical examples are copper–nickel sulphides and copper–zinc–lead sulphides.

mass movement *See* MASS-WASTING.

mass number The sum of the protons and neutrons in the *nucleus of an atom.

mass spectrometry Technique that allows the measurement of atomic and molecular masses. Material is vaporized in a vacuum, ionized, and then passed first through a strongly accelerating electric potential and then through a powerful magnetic field. This serves to separate the ions in order of their charge : mass ratio; the ions are detected, commonly by means of an electrometer which measures the force between charges and hence the electric potential. Mass spectrometry is used in the *radiometric dating of rocks, and in *isotope geochemistry.

mass-wasting (mass movement) General term for the transfer of Earth material down hillslopes. It includes four main categories: flow, slide, fall (*see* ROCK FALL), and *creep. Of these, creep is the most important if least spectacular. It is the result of gravity acting on material that has lost cohesion, typically as a result of an increase in water content. An avalanche is a rapid and often destructive flow of rock or snow. A slide (or landslide) is a comparatively rapid displacement of Earth material over one or more failure surfaces which may be curved or planar. Failure on an arcuate surface is typical of clays, and gives rise to rotational slides such as those of Folkestone Warren, England.

master curve One of a set of theoretical curves, calculated for known models, against which a field curve can be matched. If the two fit very closely the model is considered to apply reasonably well to the field situation and the curve is known as the master curve. Master curves were used extensively in *electrical resistivity depth sounding, but are being replaced by microcomputer curve-matching which is much more accurate and more sensitive to real-life situations.

mastodon *See* GOMPHOTHERIIDAE; and MAMMUTIDAE.

Mata A *series in the Upper *Cretaceous of New Zealand, underlain by the *Raukumara Series and overlain by the *Teurian *Stage. It comprises the Piripauan and Haumurian Stages, which are roughly contemporaneous with the upper *Santonian, *Campanian, and *Maastrichtian.

Mathilde A *solar system asteroid (No. 253), measuring 50 × 53 × 57 km; approximate mass 10^{17} kg; density 1.3 g/cm^3; rotational period 417.7 hours; orbital period 4.32 years. It was imaged by the *Near Earth Asteroid Rendezvous mission in June 1997, revealing 5 craters more than 20 km across, the largest being 30 km wide and 6 km deep.

matrix Lithologic or petrographic term denoting the interstitial material lying between larger crystals, fragments, or particles. It is the background material of small particles in which larger particles and fragments occur. The term is applied to *sedimentary rocks; the *igneous equivalent is *groundmass, although 'matrix' is also commonly used of igneous rocks.

matrix-support *See* MUD-SUPPORT.

maturation The evolution of *hydrocarbons through the increasing pressure and temperature associated with burial. In the immature stage the product is gas; as it matures the development of heavy oils is succeeded by medium, and finally light oils. If the temperature exceeds approximately 100 °C, however, dry gas is the sole product and the stage of incipient *metamorphism is reached. *See also* NATURAL GAS; and PETROLEUM.

Matuyama, Motonori (1884–1958) Professor of theoretical geology at Kyoto Imperial University, Matuyama worked on gravity surveys, and magnetism in *basalts. In 1929 he showed that the polarity of *remanent magnetism in some recent basalts depended on the age of the rock, concluding that the Earth's magnetic field must undergo periodic reversals.

Matuyama A reversed *polarity chron at

the end of the *Pliocene and the beginning of the *Pleistocene. It is preceded by the *Gauss and followed by the *Brunhes normal-polarity chrons, and is *radiometrically dated as occurring between 2.60 and 0.78 Ma. The Matuyama contains at least three normal *polarity subchrons: *Réunion, *Olduvai, and *Jaramillo.

Maury, Matthew Fontaine (1806–73) An American oceanographer and naval officer, Maury studied the winds and currents of the N. *Atlantic and *Indian Oceans, and produced the first bathymetric chart of the N. Atlantic. His *Physical Geography of the Sea* (1855) is said to be the first oceanographic textbook.

maximum-likelihood classification A *remote sensing *classification system in which unknown *pixels are assigned to classes using contours of probability around *training areas using the maximum-likelihood statistic. *See also* BOX CLASSIFICATION; and MINIMUM-DISTANCE-TO-MEANS CLASSIFICATION.

maximum-likelihood tree In *phylogenetics, a tree-building method that uses the maximum-likelihood statistical estimator to calculate the *topology with the highest probability of being correct under assumed rates of *character change.

maximum-parsimony tree The method for selecting a *phylogenetic tree from all possible tree *topologies that requires the smallest number of substitutions.

maximum thermometer Thermometer that records the highest temperature to which it has been exposed (e.g. by allowing a rise of mercury past a restriction in the tube, but preventing the mercury's return on contraction). It is most commonly used to record maximum daily temperatures.

Maysvillian A *stage of the *Ordovician in the Middle *Cincinnatian *Series of N. America.

mean In statistics, the *average as calculated by the sum of each data point divided by the total number of data points. *See also* VARIANCE.

meander The sinuous trace of a stream channel whose length is normally equal to or greater than 1.5 times the down-valley (or straight-line) distance. It is best developed in cohesive *floodplain *alluvium. The relationships between its geometric properties vary little with size: e.g. meander wavelength (the straight-line distance between two points at similar positions (e.g. outward extremities of curve) on two successive curves along the trace) is normally 10–14 times the channel width, irrespective of size. Meander origin is uncertain, but the sinuous curve may be that shape best fitted for the transfer of channelled flow in accordance with the *least work principle. Over time, a meander may move laterally and/or vertically. The process of sideways movement is known as 'meander migration'; it involves the deposition of point *bars on the inner sides of bends and erosion on the outer, and is limited to a tract of floodplain called the 'meander belt'. The migration of two adjacent, concave bands may narrow the floodplain between them, and the restriction is a 'meander neck'. This widens out to form a bulbous feature, the 'meander core', around which the river swings. The surface of a core may show 'meander scrolls', which are low, curved ridges of relatively coarse material lying parallel to the main channel and deposited by the stream. An 'incised meander' results from downcutting, and two types are found. (*a*) If incision is fairly slow, and sideways movement occurs, the result is an 'ingrown meander'. The slope down which the stream has migrated during incision is called a 'slip-off slope'. (*b*) When incision is rapid, with mainly vertical erosion, the consequence is an 'entrenched meander'.

meander belt *See* MEANDER.

meander core *See* MEANDER.

meander migration *See* MEANDER.

meander neck *See* MEANDER.

meander scroll *See* MEANDER.

meander wavelength *See* MEANDER.

mean sea level The average height of the surface of the sea, for all stages of the tide, over a long (usually 19-year) period, being determined from hourly readings of tidal height.

mean square *See* VARIANCE.

measure A *lithostratigraphic term that (in the past) has been used both formally (e.g. Coal Measures, Culm Measures, *see* FORMAL) and informally (*see* INFORMAL) to denote a succession of coal-bearing *strata.

mechanical weathering The *in situ* breakdown of *rocks and minerals by a set of disintegration processes that do not involve any chemical alteration. The chief mechanisms are: crystal growth, including *gelifraction and *salt weathering; *hydration shattering; insolation weathering (*thermoclastis); and *pressure release.

medial moraine *See* MORAINE.

median In statistics, the *average as defined by the central value in a data set when the data set is ordered by value.

median filter A spatial filter which reduces noise in an image by substituting the digital number value of each *pixel in an image for the median value calculated from surrounding classes using contours of probability around *training areas using the maximum-likelihood statistic. *See also* BOX CLASSIFICATION and SPATIAL-FREQUENCY FILTER.

median network A specialist type of *maximum-parsimony tree method for phylogenetic reconstruction in which all the most parsimonious trees are represented in a reticulate grid in a three-dimensional perspective. Median networks are applied to data where the degree of *homoplasy is very high and the number of informative sites very low (e.g. with data derived from populations).

median suture *See* CEPHALIC SUTURE.

median valley (axial rift, axial trough) The valley which lies along the axis of some oceanic *ridges. From observations from submarines and the interpretation of *seismic records of *earthquakes in the median valley, the inner floor has been found to have scattered volcanoes and *black smokers, with *normal faults dipping towards the inner floor on each side. These faults raise slices of old floor to the *ridge crest, from where it is lowered by further normal faults which *dip away from the ridge axis. Median valleys develop on slower-spreading ridges (e.g. *Mid-Atlantic and *Carlsberg) and are up to 3 km deep.

mediocris From the Latin *mediocris* meaning 'medium', a species of *cumulus with limited vertical development and characterized by very slight projections on the upper surface. *See also* CLOUD CLASSIFICATION.

Mediterranean climate Distinctive climatic type, which occurs around latitude 35° N and 35° S, and which is associated with warm-temperate west coasts. Summers generally are hot and dry, winters mild to cool and rainy. The climate is strongly influenced by westerly airstreams in winter, and subtropical high pressure in summer. In the type area, the Mediterranean basin, there is a variety of climatic regimes owing to the complex configuration of seas and mountainous peninsulas in the 3000 km incursion into Eurasia. Annual rainfall is broadly 500–900 mm, but less in more continental locations. Other areas of this type include the coasts of Chile in corresponding latitudes, southern California, south-western Africa, and south-western Australia. Cool ocean currents offshore bring lower temperatures to parts of the Chilean and Californian coasts, but rainfall and temperatures are much affected by the differences of slope and elevation inland.

Mediterranean-type margin *See* ACTIVE MARGIN.

Mediterranean water A water mass, formed in the arid eastern Mediterranean, flows westward, sinking in the Algero-Ligurian and Alboran basins to a depth of approximately 500 m due to its high *salinity (36.5–39.1 parts per thousand). This dense water flows into the *Atlantic Ocean through the relatively shallow Straits of Gibraltar at a depth below 150 m, while above it lighter, Atlantic water flows eastward into the Mediterranean Sea. The Mediterranean water in the Atlantic then sinks to about 1000 m, where it forms a clearly identifiable water mass.

Medusina mawsoni An early jellyfish. During the uppermost *Precambrian, beach deposits laid down in the Ediacara Hills of southern Australia were covered by a layer of fine muddy sediments. The result was the unique preservation of numerous soft-bodied organisms including jellyfish such as *Medusina mawsoni* and *Medusinites*. *See* EDIACARAN FOSSILS.

Medusinites *See medusina mawsoni*; and EDIACARAN FOSSILS.

mega- (M-) From the Greek *megas* meaning 'great', a prefix meaning 'very large'. Attached to SI units, it denotes the unit $\times 10^6$.

megabreccia A *breccia in which individual *clasts may be more than 1 km in their longest dimension.

megacryst Applied to the texture of any *igneous or *metamorphic rock which contains large, usually *euhedral *crystals set in a finer-grained *groundmass. The term has no genetic connotation, unlike 'phenocryst' which implies crystallization from a *magma, and 'porphyroblast' which implies solid-state *recrystallization during *metamorphism.

***Meganeura* (giant dragonfly; order Odonata, suborder Meganisoptera, family Meganeuridae)** Genus of gigantic fossil dragonfly-like insects, from the Upper *Carboniferous. They were probably the largest insects ever to have lived. *M. gracilipes* had a wing-span of 70 cm, and *M. monyi* a wing-span of 60–70 cm.

megaphyll *See* ENATION THEORY.

megaregolith The fractured and *brecciated zone of rock formed on the *lunar highland crust due to the intense early bombardment during the period from 4.4 billion years ago, when the crust was forming, to 3.85 billion years ago, when the Imbrium and Orientale collisions marked the close of the massive basin-forming events. Estimates of the depth of the megaregolith extend from 1 km to 25 km, the latter estimate coinciding with an observed seismic boundary.

megaripple *See* DUNE BEDFORM.

megasclere *See* HEXACTINELLIDA.

megasequence A stratigraphic sequence of rocks formed by sedimentation in a tectonically produced extensional basin, commonly at a *constructive margin.

megasporangia *See* SPORE.

megaspore *See* SPORE.

megathermal Applied to the seasonal patterns of river flow in certain equatorial regions, where the main characteristics are high temperatures and evaporation rates all year round, so that river flow maxima reflect the months of maximum precipitation.

megathermal climate High-temperature climate type, known more commonly in Europe (e.g. in the *Köppenclassification) as humid subtropical or tropical, with the coldest monthly mean temperature above 18 °C. The term is also defined in the *Thornthwaite classification by potential evapotranspiration and moisture-budget criteria.

meionite *See* SCAPOLITE.

mela- A prefix attached to the name of an *igneous rock when the rock has a darker colour than is usual. For example, a *syenite which is darker than normal is called a 'melasyenite'.

mélange A mappable body of rock composed of broken rock fragments, of all sizes and many origins, in a sheared *matrix. A mélange with a chaotic nature, initially of sedimentary origin, is called an '*olistostrome'. Tectonic mélanges are thought to form in *subduction zones at shallow depth.

melanocratic Applied to *igneous rocks whose *colour index is between 60 and 90. The high colour index of such rocks is due to the presence of a high proportion of *ferromagnesian minerals.

melanosome Dark-coloured bands rich in *mafic and aluminous minerals, e.g. *biotite, *sillimanite, and *garnet, which are found between multiple coarse-grained quartzofeldspathic veins in regionally metamorphosed (*see* REGIONAL METAMORPHISM) *pelites and *psammites. The melanosomes represent planar regions which have undergone extreme shortening by dissolution (or melting), and where removal of the *quartz and *feldspar components of the original rock along non-penetrative *cleavage planes developed as a result of high *shear stresses imposed during the regional metamorphic event. The mafic and aluminous components of the melanosome represent the undissolved or unmelted mineral residue left over from the original rock.

Melbournian A *stage of the Upper *Silurian of south-eastern Australia, underlain by the *Eildonian.

Melekesskian A *stage in the *Bashkirian Epoch, preceded by the *Cheremshanskian and followed by the *Vereiskian (*Moscovian Epoch).

melilite A member of the melilite group of *minerals, that has the general formula $(Mg,Al)(Ca,Na)_2(Si,Al)_2O_7$. There is a *solid solution series between gehlenite (aluminium-rich) and åkermanite (magnesium rich); melilite occupies an intermediate position, with some substitution of sodium for calcium, and of aluminium for silicon or magnesium. Sp. gr. 2.9; *hardness 5.5; white or greenish-white; granular or in *tabular *crystals; occurs in *basic lavas, metamorphosed *limestones, and furnace slags.

melilitite A dark-coloured, *extrusive, *ultrabasic, *alkaline *igneous rock containing *essential *melilite, *olivine, and *nepheline. The active alkaline *volcano Nyiragongo in Zaire is famous for its melilitite *lavas which are found associated with other ultra-alkaline lavas, e.g. *leucitites and *nephelinites.

melt Liquid part of *silicate *magma. Magma may consist solely of melt, or of melt and suspended crystals. 'Wet' melt contains dissolved *volatile constituents, mainly water. Conversely, anhydrous melts, without volatiles, are referred to as 'dry' melts.

melteigite An undersaturated, *intrusive *igneous rock containing essential *nepheline and alkali *pyroxene (*aegirine or aegerine-*augite) and displaying a *colour index of 70–90. With decreasing *mafic *mineral content and increasing nepheline content melteigites pass into *ijolites and then into *urtites. All three rock types are essentially *feldspar-free, undersaturated *syenites (*see* SILICA SATURATION) and are well displayed in the Fen intrusive complex of southern Norway.

meltemi *See* ETESIAN WINDS.

meltout till *See* TILL.

meltwater channel *See* GLACIAL DRAINAGE CHANNEL.

MEM *See* MICRO-EROSION METER.

member *See* FORMATION.

membrane stress The stress on a tectonic *plate resulting from its motion over an *Earth with a non-spherical shape. It is largely restricted to plates with a latitudinal motion, i.e. moving away from, or towards, the equatorial bulge and therefore adjusting to the new curvature.

Menapian A glacial *stage (0.9–0.8 Ma) of the northern European plain, perhaps equivalent to the *Günz glacial of the Alps and the *Kansan of N. America.

mendip Hill or ridge that has been buried by younger rocks and subsequently exposed by *erosion. The type example is provided by the Mendip Hills, England, described by W. M. Davis in 1912, but today the term is used only very rarely.

Menevian A *stage of the Middle *Cambrian, underlain by the *Solvan and overlain by the *Maentwrogian and dated at 530.2–517.2 Ma (Harland et al., 1989).

Meramecian A *series in the *Mississippian of N. America, underlain by the *Osagean, overlain by the *Chesterian, and roughly contemporaneous with the Arundian, Holkerian, and Asbian Stages of the *Visean Series of Europe.

Mercalli, Giuseppe (1815–1914) An Italian professor of natural sciences who studied the major *earthquake districts of Italy. Mercalli is best known for his scale of earthquake intensity, first formulated in 1897, and refined several times. *See also* MERCALLI SCALE.

Mercalli scale An *earthquake-intensity scale based on direct observation. It ranges from scales I (not felt) and II (felt by most people at rest), to VII (difficult for a person to remain standing), X (most structures destroyed), and XII (total devastation). It is not a reliable scale for energy release because the extent of destruction depends on the local geology, type of buildings, etc. *See also* RICHTER MAGNITUDE SCALE.

Mercury The innermost planet of the *solar system, 0.387 AU from the Sun, its distance from Earth varying from 77.3×10^6 km to 221.9×10^6 km. It has a tenuous atmosphere, with a surface pressure of about 10^{-15} bar, composed of oxygen (42%), sodium (29%), hydrogen (22%), helium (6%), potassium (0.5%), and possibly trace amounts of argon, carbon dioxide, water, nitrogen, xenon, krypton, and neon (the proportions are approximate). Its diameter is 4880 km; volume 6.085×10^{10} km^3; mass 0.3302×10^{24} kg; mean density 5427 kg/m^3 (making it the densest of the planets); surface gravity 3.7 (Earth = 1); visual albedo 0.11; average surface temperature 440 K (on the sunward side ranging from 590 K to 725 K).

Mercury Orbiter An *ESA mission, to be launched in 2006, to place a vehicle in orbit about Mercury.

meridional circulation Air circulation in a cell, or general flow to or from different latitudes, usually with a marked south-to-north or north-to-south component. *See also* HADLEY CELL; and ZONAL FLOW.

Merioneth A *series (517.2–510 Ma ago) of the Upper *Cambrian, underlain by the *St David's and overlain by the *Tremadoc (*Ordovician).

Merionsian A *stage in the Lower *Devonian of Australia, underlain by the *Crudinian, overlain by the *Cunninghamian, and roughly contemporaneous with the *Siegenian of Europe.

meromictic Applied to lakes whose waters are stratified permanently, usually because of some chemical difference (e.g. contrasting salinities, and hence densities) between *epilimnion and *hypolimnion waters. *Compare* HOLOMICTIC.

Merostomata (phylum *Arthropoda) Class which includes the king crabs (horseshoe crabs) and the extinct eurypterids or 'water scorpions'. Eurypterids were usually 10–20 cm in length, but some were much larger, e.g. *Pterygotus* which grew up to 2 m long. They ranged from the *Ordovician to the end of the *Palaeozoic. The horseshoe crabs also appeared in the Lower Palaeozoic and have survived to the present.

mesa Flat-topped hill of limited extent, but wider than a *butte and normally underlain by near-horizontally bedded sediments. A land-form similar to a mesa but larger is called a 'plateau'.

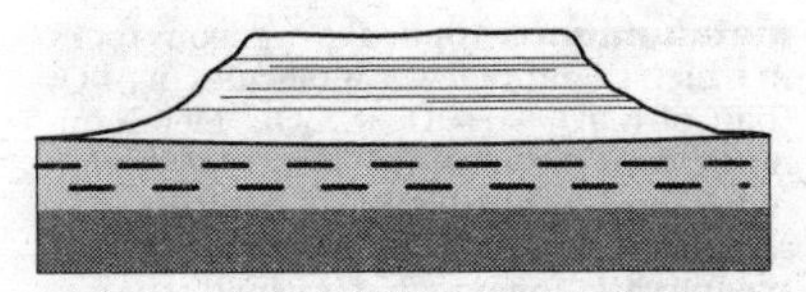

Mesa

mesentery In corals, a radial infolding of the fleshy body wall, forming a partition in the gut. In this central cavity the mesenteries alternate with the skeletal *septa.

mesic Applied to an environment that is neither extremely wet (hydric) nor extremely dry (xeric). *See also* PERGELIC.

mesoclimate General term applied to the characteristics of a relatively small region, e.g. a valley or urban area.

mesocyclone A region of rapidly rotating air, up to 10 km in diameter, inside a large *cumulonimbus cloud. Rotation commences near the middle of the cloud and extends downwards. If it continues beneath the cloud base and reaches to the ground it is a *tornado.

mesohaline water *See* HALINITY.

mesokurtic *See* KURTOSIS.

mesopause In the atmosphere, the inversion at about 80 km height, which separates the *mesosphere from the *thermosphere above. The first 10 km of the mesopause are almost isothermal. The mesopause is liable to be marked by *noctilucent cloud composed of ice crystals on meteoric dust. *See also* ATMOSPHERIC STRUCTURE.

Mesoproterozoic The middle part of the *Proterozoic, lasting from about 1600–1000 Ma.

Mesosauria Order of *anapsid reptiles, with just one family, the Mesosauridae. The mesosaurs (e.g. *Mesosaurus*) are known only from late *Carboniferous or early *Permian rocks in S. America and S. Africa. They were adapted to life in fresh water, lightly built, and up to one metre in length.

mesosphere 1. Upper-atmospheric layer above the *stratopause (at 50 km) through which temperature decreases with height up to about 80 km, where temperatures reach a minimum of about –90 °C. This level is the *mesopause, an inversion above which temperatures rise again. *See also* ATMOSPHERIC STRUCTURE. **2.** That part of the *Earth underlying the *asthenosphere. The term is no longer in current use in this sense.

mesothem *See* CYCLOSTRATIGRAPHY.

mesothermal 1. Applied to the seasonal patterns of river discharge of certain warm, subtropical, and temperate areas. **2.** Applied to a *mineral deposit formed by hot, ascending solutions at about 200–300 °C at moderate depths in the Earth's *crust.

mesothermal climate Climatic type with moderate temperatures, known most commonly in Europe, e.g. in the *Köppen classification, as a warm-temperate rainy climate having a coldest month with temperatures of –3 °C to +18 °C and a warmest month above +10 °C. Such climates are found typically in latitudes 30–45°. The *Thornthwaite classification defines this type according to *evapotranspiration and moisture budget.

mesotidal Applied to coastal areas where the *tidal range is 2–4 m. Tidal action and wave activity both tend to be important in such areas.

mesotype A term applied to *igneous rocks displaying a *colour index between 30 and 60.

Mesozoic The middle of three *eras that constitute the *Phanerozoic period of time. The Mesozoic (literally 'middle life') was preceded by the *Palaeozoic Era and followed by the *Cenozoic Era. It began with the *Triassic approximately 245 Ma ago and ended around 65 Ma at the start of the *Tertiary. The Mesozoic comprises the Triassic, *Jurassic, and *Cretaceous Periods.

Messinian 1. The final *age in the *Miocene *Epoch, preceded by the *Tortonian, followed by the *Zanclian (Tabianian), and with its upper boundary dated at 5.2 Ma (Harland et al., 1989). **2.** The name of the corresponding European *stage, which is roughly contemporaneous with the upper *Mohnian and lower *Delmontian (N. America), upper *Tongaporutuan and lower *Kapitean (New Zealand), and part of the *Mitchellian (Australia). The *neostratotype is found between Caltanisetta and Enna, Sicily. It is an important stage, marked by the presence of thick *evaporite deposits in the Mediterranean. These may indicate that the Straits of Gibraltar were closed approximately 6.5–5.1 Ma ago and that the Mediterranean was reduced to a series of evaporite basins.

meta- From the Greek *meta* meaning 'with' or 'after', a prefix implying change and meaning 'behind', 'after', or 'beyond'. The prefix may be attached to the name of any rock which has undergone *metamorphism. For instance, a *basalt which has been metamorphosed may be termed a metabasalt, and a pelitic rock (*see* PELITE) which has suffered metamorphism may be termed a metapelite.

metabasalt *See* META-.

metacarpal In the forelimb of a *tetrapod, one of the rod-like bones that articulate proximally with the carpus (wrist) and distally with the phalanges (digits). In humans, the metacarpals occupy the palm region of the hand.

metacryst *See* PORPHYROBLAST.

metagenesis Following *catagenesis, a phase in the formation of *petroleum and *natural gas during which the temperature rises above 150 °C and may exceed 200 °C, causing destruction of *kerogen and release of gas (mainly methane); at the higher temperatures petroleum is destroyed, leaving only natural gas.

metal factor (MF) A *frequency-domain measure of *induced polarization where *apparent resistivities are measured at two low frequencies, usually a factor of 10 apart in magnitude (ρ_{dc} and ρ_{ac}); $MF = 2\pi 10^5(\rho_{dc} - \rho_{ac})/\rho_{ac}^2$.

metalimnion In a thermally stratified lake, the layer of water between the *epilimnion and *hypolimnion.

metallic Of mineral *lustre, having the sheen characteristic of a metal.

metallic bond A special type of *covalent bond found in such solids as native *copper. Every atom shares its *electrons with each of its neighbours in turn. The sharing atoms must be able to move and so in metals they appear to be positive *ions moving in a sea of electrons.

metallogenesis Study of the origin of *ore deposits and of the interdependence in time and space of this process with other geologic processes such as *tectonics.

metallogenic province An area of characteristic mineralizing activity, or a particular association of *mineral deposits. It may contain several episodes of mineralization.

metaluminous Applied to *igneous rocks in which there are fewer molecules of Al_2O_3 than of ($CaO + Na_2O + K_2O$). 'Metaluminous', together with the terms 'peraluminous' and '*peralkaline', describe the alumina saturation of an igneous rock, portraying variations in what is usually the second most abundant oxide component in the rock (after silica).

metameric segmentation The repetition of organs and tissues at intervals along the body of an animal.

metamorphic aureole *See* CONTACT AUREOLE.

metamorphic differentiation Separation of an originally homogeneous rock into bands of contrasting mineralogy during high-grade *metamorphism (*see* METAMORPHIC GRADE). The bands of contrasting mineralogy may be produced either by solid state *diffusion of elements during the metamorphic event, or by local *metasomatism.

metamorphic facies A set of metamorphic *mineral assemblages derived from rocks of contrasting composition which have been subjected to the same grade or conditions of *metamorphism. For example, a metamorphosed *shale, *basic *lava, and *limestone which are found adjacent to each other in a metamorphic *terrain must have been subjected to the same grade of metamorphism, yet each of these contrasting rock types displays a different metamorphic mineral assemblage, the individual assemblages reflecting both the starting rock composition and the grade of metamorphism. Since all three rocks have been subject to the same grade of metamorphism, the set of contrasting mineral assemblages reflects only the contrasts in rock composition, and thus constitutes a metamorphic *facies. Any change in the mineral assemblage observed in a particular rock composition represents a mineralogical response to changing metamorphic conditions and thus would define a new facies. The metamorphic conditions represented by a particular facies can be deduced from experimental studies of the overlapping pressure–temperature stability fields of mineral assemblages in that facies. However, the definition of a facies is purely descriptive and based entirely on the mineral assemblages observed. The concept of metamorphic facies was first proposed by the Finnish petrologist P. E. *Eskola in 1920 after he had compared the mineral assemblages in similar rock compositions from Oslo, Norway, and Orijarvi, Finland, two areas which had been subjected to contrasting metamorphic conditions.

metamorphic grade A measure of the relative intensity of *metamorphism. In pelitic rocks (*see* PELITE), an increase of metamorphic grade is marked by a progressive dehydration of the mineral assemblages present whilst in *limestones and impure limestones an increase in grade is marked by progressive decarbonation of the mineral assemblages present. In general, mineral assemblages stable at progressively higher pressures and temperatures characterize an increase in metamorphic grade (e.g. *garnet grade).

metamorphic rock An aggregate of *minerals formed by the *recrystallization of pre-existing rocks in response to a change of pressure, temperature, or *volatile content. Metamorphic rocks can generally be divided into four types: (*a*) regional metamorphic rocks, formed in response to changes leading to high temperature and high pressure (*shearing stress and hydrostatic pressure: *see* HYDROSTATIC STRESS) accompanying orogenic events (*see* OROGENY; REGIONAL METAMORPHISM); (*b*) contact metamorphic rocks, formed in response to changes leading to high temperature (with low pressure) around an *igneous *intrusion (*see* THERMAL METAMORPHISM); (*c*) cataclastic or dynamic metamorphic rocks, formed in response to an increase in directed pressure (shearing stress) particularly in *fault and *thrust zones (*see* CATACLASITE); and (*d*) burial metamorphism, formed in response to changes leading to high pressure (with low temperature).

metamorphic zone The area between two successive *isograds in a metamorphic terrain. The zone is named after the lower-grade isograd (*see* METAMORPHIC GRADE). For example, the area lying between the *garnet isograd and the *kyanite isograd for pelitic rocks in a *regional metamorphic terrain is termed the garnet zone. The zone defines a set of pressure, temperature, $P(H_2O)$, and $P(CO_2)$, over which there were no mineralogical reactions in the particular rock composition concerned. Normally, the chemical composition of *pelites is used as a marker.

metamorphism The process of changing the characteristics of a rock in response to changes in temperature, pressure, or *volatile content. Most metamorphic changes do not include bulk chemical changes, but merely the crystallization of new mineral *phases. These isochemical changes cause major textural changes. *Compare* METASOMATISM. *See also* BARROVIAN-TYPE METAMORPHISM; BARROW'S ZONES; BURIAL METAMORPHISM; DYNAMIC METAMORPHISM; REGIONAL METAMORPHISM; THERMAL METAMORPHISM; and METAMORPHIC GRADE.

metapelite *See* META-.

Metaphyta *See* PLANTAE.

metaquartzite *See* QUARTZITE.

metasediment *See* META-.

metaseptum (pl. metasepta) *See* SEPTUM.

metasilicate *See* CYCLOSILICATE.

metasomatism Kind of *metamorphism that involves the introduction of chemical constituents into a rock, or their removal from it (or both), via a *volatile phase. Complete *mineral transformations may occur, but the original rock *texture may remain (resulting in a *pseudomorph).

metaspecies Ancestral species. They are not *monophyletic (in the strictest sense) so do not, and cannot, conform to the canon that taxa must always be monophyletic.

metastable Applied to a *phase which is apparently stable but is capable of reaction if disturbed, a state usually due to the slowness of a system to attain equilibrium. The term is applied, for example, to supersaturated solutions. A phase is said to be metastable if it exists in the same temperature range in which another phase with lower *vapour pressure is stable. Many minerals that occur at room temperature and pressure are metastable (e.g. *diamond) and most *metamorphic rocks formed under high temperatures and pressures are metastable at the surface under normal temperatures and pressures. A metastable system (mineral assemblage) is in a temporary state of equilibrium and requires only a minimal disturbance to initiate a change to a state of true equilibrium. A *gabbro mineral assemblage which crystallized at high temperatures and moderate pressures, and is thus in equilibrium under these conditions, would be in a metastable state when exposed by *erosion to the low-temperature, low-pressure environment at the Earth's surface. Rainwater falling on the exposed mineral assemblage would act as the disturbance or catalyst initiating the reaction of the gabbro mineral assemblage to a low-temperature, low-pressure, *chlorite–*clay mineral assemblage, thus re-establishing the equilibrium.

metatarsal In the hind limb of a *tetrapod, one of the rod-like bones that articulate proximally with the tarsus (ankle) and distally with the phalanges (toes).

Metazoa *See* ANIMALIA.

metazoan A multicellular animal. *See* ANIMALIA.

meteor Transient, incandescent trail of a *meteoroid entering the *Earth's *atmosphere. All the material burns up before reaching the ground. It is popularly referred to as a 'shooting star'. A very bright meteor is called a fireball. *Compare* METEORITES.

meteoritic abundance of elements The relative abundances of elements contained within chondritic *meteorites (*see* CHONDRITE), and in particular within C1 *carbonaceous chondrites, are believed to provide the best estimate of the composition of the primitive material in the *solar system from which the planets formed. Comparison of C1 carbonaceous chondrites with the *solar abundances of elements (obtained by spectroscopic studies) show that, for non-*volatile elements, the abundances are almost identical. On the other hand, the composition of ordinary chondrites is believed to be similar to the *bulk composition of the Earth.

Element	*Average chondrite (wt. %)*
O	33.24
Fe	27.24
Si	17.10
Mg	14.29
S	1.93
Ni	1.64
Ca	1.27
Al	1.22
Na	0.64
Cr	0.29
Mn	0.25
P	0.11
Co	0.09
K	0.08
Ti	0.06

meteoric water Water of atmospheric origin which reaches the Earth's *crust from above, either as rainfall or as seepage from surface-water bodies. *See also* GROUNDWATER.

meteorites Small, extraterrestrial bodies, most of which probably originate in the asteroid belt, that enter the *Earth's *atmosphere and land on the surface. Most are only a few centimetres in size. Meteorites are classified into four main groups according to their composition and structure as: *chondrites; *achondrites; *stony-irons; and *irons. *See also* ANTARCTIC METEORITE; AUBRITE; BASALTIC METEORITES; CARBONACEOUS CHONDRITE; EUCRITE; SHERGOTTYITE/NAKHLITE/CHASSIGNITE METEORITES; STONY METEORITE; and TEKTITE.

meteoroid Small, extraterrestrial body, within the *solar system, that may enter the *Earth's *atmosphere if its orbit around the *Sun crosses that of the *Earth. Strictly, at the point where a meteoroid enters the atmosphere it is referred to as a *meteor; if it reaches the Earth's surface it is known as a *meteorite.

Meteosat A series of six meteorological satellites built by Aerospatiale and launched by the European Space Agency (ESA) that monitor meteorological conditions over Europe. The first was launched in 1977. Two, known as MOP2 and MOP3, were still active in 1997.

methane hydrates *See* NATURAL GAS HYDRATES.

methanogen A single-celled organism, belonging to domain *Archaea, that produces *methane gas as a product of its *metabolism.

Metis (Jupiter XVI) The innermost of the known jovian planets, discovered in 1979. It orbits inside the main ring of *Jupiter (it is a moom); it and Adrastea may be the source of the material comprising the ring. Metis has a diameter of 40 (±20) km; mass 9.56×10^{16} kg; mean distance from Jupiter 128 000 km.

MF *See* METAL FACTOR.

M-fold The ideal form of a *minor fold which occurs in the *hinge region of a major *antiform (a *synform is distinguished by W-fold forms). In practice it is difficult to recognize ideal M-folds and they are less informative than *Z-fold and *S-fold configurations.

Miacidae *See* CARNIVORA.

miarolitic Applied to a texture found in shallow, level, *plutonic, *igneous rocks and characterized by void spaces or pockets with outlines shaped by neighbouring crystals. The texture is primary representing entrapment of gas segregated by vesiculation (*see* VESICLE) of the late-stage *magma during the final stages of crystallization of the *intrusive body.

micas An important group of *phyllosilicates (sheet silicates) with a 2:1 atomic structure (*see* CLAY MINERALS) and including the important *minerals *muscovite, *biotite, and *phlogopite; the group is characterized by the silicon oxygen tetrahedral layers of composition $[Si_4O_{10}]_n$ and a general composition may be written $(K,Na)_2Y_6[Z_4O_{10}]_2(OH,F)_4$, where Y = Mg, Fe, Fe^{3+}, or Al, and Z = Si or Al; the micas include the common micas already mentioned, plus *glauconite, *lepidolite, and *zinnwaldite, the brittle micas (*see* MARGARITE), and also the related minerals *talc, *stilpnomelane, and *pyrophyllite.

Michel–Lévy chart (birefringence chart, interference colour chart) A chart of standard colours used in measuring *birefringence. In crossed-polarized light, *thin sections of *anisotropic *minerals with a standard thickness of 0.03 mm (30 μm) will give a series of *interference colours, depending on the amount of birefringence (double refraction). A number of sections are examined in different orientations and the maximum interference colour matched with the equivalent colour on the chart. From the point representing 30 μm thickness on the appropriate colour a radiating line is followed to the edge of the chart to give a reading for birefringence. The accessory *quartz wedge may also be used to determine birefringence.

Michell, John (?1724–93) An astronomer and experimental philosopher from Cambridge, Michell made the first scientific investigations of *seismic waves. He studied *earthquake phenomena, especially the Lisbon earthquake of 1755. He proposed a theory for the motion of seismic waves, estimating their velocity, and showed how to determine the *epicentre of an earthquake.

Micraster A genus of Upper *Cretaceous echinoids (*Echinoidea) that is common in France and England and occurs in much of Europe. Much has been written about evolution in this genus, starting with A. R. Rowe in 1899, and adaptations in the genus either to burrowing or to varying substrates have also been suggested.

micrinite *See* COAL MACERAL.

micrite 1. (lime mud) Microcrystalline *calcite, with a grain size finer than 4 μm. This may originate from *biogenic sources in the form of aragonitic calcareous *algae, by bioerosion and physical erosion of coarser *carbonate fragments, inorganic precipitation from carbonate saturated waters, and biochemical precipitation by algal activity. **2.** *Limestones consisting mainly of lithified *lime mud. *See* FOLK CLASSIFICATION; and LEIGHTON–PENDEXTER CLASSIFICATION.

-micrite *See* FOLK LIMESTONE CLASSIFICATION.

micrite envelope A dark-coloured, fine-grained, *carbonate exterior found around carbonate skeletal fragments. It is produced by the alteration of the original mineralogy of the fragment to *micrite, commonly caused by the boring into the skeletal material of certain *algae. In some micrite envelopes the tiny, tube-like borings of the algae are preserved. Some micrite envelopes are formed by coatings of filamentous algae over the grain. If the development of the micrite envelope proceeds to an advanced stage, the primary structure of the skeletal fragment will be totally replaced by micrite, leaving an *amorphous *peloid, or lump (*see* INTRACLAST).

micritic limestone *See* LEIGHTON–PENDEXTER CLASSIFICATION.

micritization The formation of *micrite by the boring into skeletal *carbonate particles by blue-green algae (*Cyanobacteria), and the subsequent precipitation of micrite within the borings. Micritization can also occur if filamentous endolithic algae (i.e. those within the sediment) coat the grain. *See* MICRITE ENVELOPE.

micro- 1. **(μ)** From the Greek *mikros* meaning 'small', a prefix meaning 'extremely small'. Attached to SI units it denotes the unit $\times 10^{-6}$. **2.** In Earth sciences, micro- is a prefix applied in the strict sense to very fine *igneous textures. Individual particles are likely to be below the resolution of the naked eye so they can be resolved only with the aid of a petrological microscope (*see* POLARIZING MICROSCOPE). *Porphyritic texture which can only be resolved by using such a microscope is termed 'microporphyritic' texture and the *phenocrysts are 'microphenocrysts'. **3.** The prefix may also be applied to igneous rock names. For example, a *syenite with a *grain size below that normally accepted for a syenite but above that expected of its fine-grained equivalent, a *trachyte, may be termed a microsyenite.

microbialite A sedimentary body formed on the bed of a lake from the remains of benthic (*see* BENTHOS) communities of algae and *cyanobacteria.

microclimate The atmospheric characteristics prevailing within a small space, usually in the layer near the ground that is affected by the ground surface. Special influences include the impact of vegetation cover on humidity (by *evapotranspiration) and on temperature and winds. *See also* URBAN CLIMATES.

microcline *See* ALKALI FELDSPAR.

microcontinent *See* MICROPLATE.

microcraters *See* 'ZAP PITS'.

microcrystalline Applied to a crystalline texture which is so fine grained that the individual *crystals are too small to be observed by the naked eye but may be distinguished under the microscope.

microdiorite A medium-grained *igneous rock characterized by the *mineral assemblage and chemical composition of *diorite.

microearthquake *See* MICROSEISM.

micro-erosion meter (MEM) A device for measuring the rate at which an exposed rock surface is lowered, perhaps by *weathering. It consists of a gauge which records the extension of a probe. For measurement purposes the unit is placed on three studs, already set in the rock surface, and which provide a reference level. The extension of the probe is then a measure of *erosion.

micro-evolution Evolutionary change within species, which results from the differential survival of the constituent individuals in response to natural selection. The genetic variability on which selection operates arises from mutation and sexual reshuffling of *gene combinations in each generation.

microfabric Structure or arrangement of *mineral and organic particles on a microscopic scale, seen only with a microscope.

microfossil Any *fossil that is best studied by means of a microscope. Material may include dissociated fragments of larger organisms, whole organisms of microscopic size, or embryonic forms of larger fossil organisms. Various groups, e.g. *acritarchs, *Foraminiferida, *Ostracoda, and *Conodontophora, are studied and used as stratigraphic markers. *See* MARKER BED; and MICROPALAEONTOLOGY.

microfracture *See* VOIDS.

microgabbro *See* DOLERITE.

microgranite A medium-grained (1–5 mm grain diameter) *igneous rock characterized by the *mineral assemblage and chemical composition of *granite.

microgranodiorite A medium-grained (1–5 mm grain diameter) *igneous rock characterized by the *mineral assemblage and chemical composition of *granodiorite.

micrographic *See* GRANOPHYRIC.

microlite An extremely small *crystal, usually found embedded in a glassy *groundmass (*see* GLASS) and resolvable only by using high power magnification on a petrological microscope (*see* POLARIZING MICROSCOPE). Such crystals represent the frozen initial stages of crystal *nucleation and growth, and are usually preserved in rapidly chilled *lavas.

microlog The record produced by a small electrical sonde, 3–5 cm between the electrodes, that is usually combined with a caliper sonde (to measure *borehole diameter), the caliper maintaining the microlog against the borehole wall. The microlog mainly records the resistivity and thickness of the *mud cake, and the log is used to assist interpretation of *laterologs. *See also* CALIPER LOG.

micropalaeontology The study of *microfossils. Since the time of A. D. d'Orbigny (1802–57), who is credited with founding the discipline, many thousands of papers have been written on various microfossil groups and many thousands of species have been described. Commercial, or applied, micropalaeontology began in 1877 when the age of strata in a well in Austria was determined by means of *Foraminiferida, since when it has become a powerful tool in geologic investigations. Many oil companies either have their own laboratories devoted to micropalaeontology or employ consultants in this field.

microperthite *See* PERTHITE.

microphenocryst *See* MICRO-.

microphyll *See* ENATION THEORY.

microplate Any small lithospheric *plate. To be classed as a microplate, any present-day, small fragment of *lithosphere should have identifiable *plate margins, though adherence to this requirement is not strict (*see also* TERRANE). Microplates with *continental crust are also 'microcontinents', though the reverse is not necessarily true: Japan is a microcontinent but a part of the *Eurasian Plate, and the Rockall microcontinent has long ceased its independent movement. Microplates, especially those that have continental crust, are considered by many geologists to be important in the formation of several *orogenic belts, e.g. the Cordillera of western N. America. Most authors suggest that the break-up of *Gondwanaland produced many microplates whose subsequent collision with and *accretion to Eurasia may explain the complexities of the Alpine–Himalayan belt, with the strain being taken up partly by the slipping, rotation, and slicing up of microplates and partly by deformation of other terranes.

microporphyritic *See* MICRO-.

micropulsations *See* PULSATIONS.

micropygous *See* PYGIDIUM.

microsclere *See* HEXACTINELLIDA.

microseism (microearthquake) A small *earthquake, usually with a *Richter magnitude less than 2.

microspar Fine, crystalline *calcite with crystals 4–10 μm in size, formed by the *recrystallization of *micrite. Microspar is not synonymous with *microsparite, and the term is reserved solely for neomorphosed micrite. *See* NEOMORPHISM.

microsparite Sparry calcite in 5–20 μm range. *Compare* MICROSPAR.

microsporangium *See* SPORE.

microspore *See* SPORE.

microstylolite A complex, irregular, *pressure-dissolution surface, lined with insoluble residues. It is developed particularly in *carbonate rocks.

microsyenite A medium-grained *igneous rock characterized by the *mineral assemblage and chemical composition of *syenite.

microtektite *See* TEKTITES.

microthermal Applied to the seasonal patterns of river discharge in areas where at least one month has a mean temperature below −30 °C.

microthermal climate Low-temperature climate of short summers, defined in the *Köppen classification as having mean winter temperatures of less than −3 °C. Examples include the cold *boreal forest climate types in continental interiors, and along some eastern seaboards in latitudes 40–65°. The term is also applied in the *Thornthwaite classification according to potential-evapotranspiration and moisture-budget criteria.

microtidal Applied to coastal areas in which the tidal range is less than 2 m. Wave action dominates the processes active in microtidal areas, e.g. the Mediterranean Sea and the Gulf of Mexico.

microwave *Electromagnetic radiation which has a wavelength between 100 μm and 30 cm and frequencies between 1 GHz and 300 GHz. Microwaves lie between *infrared and *radiowaves. *See also* PASSIVE MICROWAVE.

microwave demagnetization The use of *microwaves to excite *magnons, thereby causing loss of magnetization without significant heating and concomitant chemical change.

Mid-Atlantic Ridge The oceanic *ridge which separates the *North and *South American Plates from the *African and

*Eurasian Plates. It is a slow-spreading ridge, with rugged topography and a well-developed *median valley.

mid-infrared *Infrared radiation which has a wavelength between 8 μm and 14 μm.

mid-ocean ridge A long, linear, elevated, volcanic structure often lying along the middle of the ocean floor. Such ridges tend to occupy central positions because the oceans have formed by the symmetrical spreading of two lithospheric *plates from the ridge sites. Ocean ridges occur in all the Earth's oceans, but may be offset from a central position, e.g. the E. Pacific ridge, where one side of the oceanic crust is being consumed along a *subduction zone. At mid-ocean ridges, ocean floor is being formed. At the centre there is a rift valley, formed as discrete segments, bordered by high mountains on both sides. At a fast-spreading ridge (opening at up to 15 cm a year) the crust is smoother, with flat lavas flowing from fissures, than at a slow-spreading ridge (about 2 cm a year), where the median valley contains a chain of small volcanoes linked by fissure eruptions.

mid-ocean-ridge basalt (MORB) A type of tholeiitic *basalt (*see* THOLEIITE), erupted from mid-ocean-*ridge *constructive-plate margins; it is one of the most abundant of all rocks and covers much of the Earth's surface. It is characterized by very low concentrations of K_2O and TiO_2; low iron, P_2O_5, Ba, Rb, Sr, Pb, Th, U, and Zr; and high CaO. When the concentration of each *rare-earth element in the basalt is divided by its mean concentration in *chondrite meteorites (a standard for comparison), this type of basalt shows a progressive lowering of the ratios for the light rare-earth elements (LREEs), compared to the ratios for the heavy rare-earth elements (HREEs). MORB is said to be LREE depleted, a reflection of the chemically depleted nature of the *mantle source regions from which they are derived. Since leaving their source region in the mantle, these basalts, often termed low-potassium *tholeiites, have not been contaminated by passing through any *continental crust and therefore retain the chemical signature of the mantle from which they were derived. MORBs thus provide an insight into the composition of the sub-oceanic mantle.

midstand systems tract *See* REGRESSIVE SYSTEMS TRACT.

Mie scattering Scattering of electromagnetic radiation, mainly in a forward direction, by spherical particles. The theory was proposed by G. Mie in 1908. *See also* RAYLEIGH SCATTERING.

migmatite A coarse-grained, heterogeneous mixed rock consisting of: (*a*) a high-grade metamorphic component with a *gneissose texture (*see* METAMORPHIC GRADE); and (*b*) an *igneous component with a *granite mineralogy and a foliated or unfoliated texture (*see* FOLIATION). Migmatites are found in high-grade metamorphic terrains where a sequence from high-grade metamorphic rocks through migmatites to granite bodies is often seen in the field. The granite component is thought to form by *partial melting of the rock during extreme *metamorphism. Migmatites may thus be a record of the initial stages in the generation of large bodies of granite magma and, as such, they would represent the high-temperature boundary between metamorphic and igneous rocks. Migmatites have an attractive appearance, often being marked with irregular small stripes or patches of contrasting shades ranging from almost white to dark grey, and are widely used as building stone, sometimes being polished for ornament.

migmatization The process whereby a rock undergoes *partial melting during extreme *metamorphism, producing a *migmatite.

migration A method of reconstructing a *seismic section so that dipping reflection events are repositioned to lie beneath their true surface locations and at corrected vertical two-way travel times. There are several migration methods. These include, for example, wave-equation, dip-moveout, ray-trace, finite-difference, wavefront-common-envelope, diffraction, and *frequency-domain.

migration route Link between two biogeographical regions that permits the interchange of plants and/or animals. Various types are recognized in the literature: for example, G. G. *Simpson's (1940) '*corridors', '*filters', and '*sweepstakes routes' are widely referred to in connection with mammalian, and more recently reptilian, migrations.

Mikulino A local name, in Russia, for sediments of *Eemian age.

Milankovich, Milutin (1879–1958) A Serbian mathematician and physicist from the University of Belgrade, Milankovich made important studies of solar radiation. He concluded that there were periodic changes in the amount of radiation received on Earth, caused by eccentricities in the Earth's orbit. These led to cyclical climatic changes, causing glacial periods, etc. *See also* MILANKOVICH CYCLES.

Milankovich cycles Cyclical changes in the rotation and orbit of the Earth that *Milankovich correlated to climatic effects. There are three cycles: changes in the eccentricity of the Earth's orbit, altering the distance between Earth and the Sun at *aphelion and *perihelion, with a period of about 100 000 years; variations in the tilt of the Earth's rotational axis (obliquity of the ecliptic), with a period of about 40 000 years; and a movement (wobble) in the angle by which the axis of the Earth's rotation is tilted in respect of the orbital plane, altering the seasons at which aphelion and perihelion occur (precession of the equinoxes), with a period of about 21 000 years. Climatic changes associated with Milankovich cycles may be recorded in *cyclic sedimentation.

Milazzian *See* QUATERNARY.

milioline winding *Coiling which occurs in the suborder Miliolina of the *Foraminiferida. Coiling starts with a *planispirally coiled *proloculus and may continue in this way. Usually, however, tubular chambers are added lengthwise around the growth axis. It is common to find chambers added at 140° or 120°, leaving either three or five chambers visible on the outside of a *test.

Milleporina (millepores; massive hydrocorals; class *Hydrozoa) Order of *reef-forming *Cnidaria whose members build massive, calcareous *exoskeletons which have pores through which the polyps protrude. Milleporina are known from the Upper *Cretaceous to *Recent. *See also* STYLASTERINA.

Miller, William Hallowes (1801–80) A British mineralogist from the University of Cambridge, Miller developed a classification system for crystallography, based on crystal axes. His method was published in *Treatise on Crystallography* (1839). *See* MILLER INDICES.

Miller–Bravais indices In the *hexagonal and *trigonal *crystal system, there are four *crystallographic axes. M. A. Bravais adapted the *Miller notation to express the intercepts of *crystal faces on these four axes using the modified symbols h, k, i, l; these symbols are the reciprocals of the ratios of the intercepts on the a_1, a_2, a_3, and c (or x, u, y, and z) axes respectively.

Miller indexes *See* MILLER INDICES.

Miller indices (Miller indexes) One of the methods of notation devised to express the intercepts of *crystal faces on *crystallographic axes, proposed by W. H. *Miller. The symbols h, k, and l represent whole numbers which are the reciprocals of intercepts along the a, b, and c (or x, y, and z) crystallographic axes respectively. A face which is parallel to a crystallographic axis is indicated by the symbol o. If a crystal face intercepts the negative end of an axis, a bar (negative sign) is placed above the appropriate symbol. A *prism will have the general notation of 100 or 110, a *pinacoid 001, and a *pyramid 101 or 111. When the indices describe a crystal face they are conventionally written without brackets; when they describe crystal *form, they are enclosed in brackets.

millerite Mineral, NiS; sp. gr. 5.2–5.6; *hardness 3.0–3.5; *trigonal; brass-yellow and opaque; greenish-black *streak; *metallic *lustre; crystals usually long, slender, *acicular, and in radiating groups; *cleavage perfect, rhombohedral; occurs as tufts of radiating fibres in cavities, and as replacement for other nickel minerals, also in veins carrying nickel minerals and other *sulphides, and around some volcanoes as sublimation products. It is a minor ore of nickel and named after the British mineralogist W. H. *Miller.

milli- (m-) From the Latin *mille* meaning 'one thousand', a prefix meaning 'one-thousandth' (e.g. a 'milli-equivalent' is one-thousandth of an equivalent weight). Attached to SI units it denotes the unit $\times 10^{-3}$.

millibar *See* BAR.

milligal One-thousandth of a gal; equivalent to 10 *gravity units.

Millstone Grit Series *See* NAMURIAN.

Milne, John (1850–1913) A British mining engineer who became professor of geology and mining in Tokyo. Milne became interested in *earthquakes, developing and testing *seismographs, and recording waves produced by explosions. He was the first to show that distant earthquakes could be recorded. He also made a study of earthquake-proof buildings.

Mimas (Saturn I) One of the major satellites of *Saturn, with a radius of 198.8 km; mass 0.375×10^{20} kg; mean density 1140 kg/m^3; visual albedo 0.5. It was discovered in 1789 by Sir William Herschel.

mimetic twins Crystals which, on being twinned, appear to have a higher order of *crystal symmetry than is the case. Examples are commonly found in *aragonite and the *zeolite group of minerals.

mimetite *See* VANADINITE.

Mimistrobell A *solar system asteroid (No. 3840), with an orbital period of 3.38 years. It is to be visited in September 2006 by the Rosetta spacecraft.

Mindel The second of four glacial episodes, taking its name from an Alpine river, and established by A. *Penck and E. Bruckner in 1909. It may be equivalent to the *Elsterian of northern Europe.

Mindel/Riss Interglacial (Great Interglacial) An Alpine *interglacial *stage that is possibly the equivalent of the *Hoxnian of East Anglia, England, or the *Holsteinian of northern Europe.

Mindyallan A *stage of the Upper *Cambrian of Australia, overlain by the *Idamean.

mineral Usually inorganic substance which occurs naturally, and typically has a crystalline structure whose characteristics of *hardness, *lustre, colour, *cleavage, fracture, and relative density can be used to identify it. Each mineral has a characteristic chemical composition. *Rocks are composed of minerals. More loosely, certain organic substances obtained by mining are sometimes termed 'minerals'.

mineralization Conversion of organic tissues to an inorganic state as a result of decomposition by soil micro-organisms.

mineral layering The concentration of one *mineral type or a combination of mineral types into layers varying in thickness from a few centimetres to 2–3 m and commonly found in large *intrusive *igneous rock bodies. The sudden appearance or disappearance of a particular mineral marks the boundary between the mineral layers. Mineral layering is also known by the term 'phase layering'. *Compare* CUMULATE.

mineralogy The scientific study of *minerals, comprising *crystallography, mineral chemistry, economic mineralogy, and determinative mineralogy (concerned mainly with physical properties).

mineral saturation index An index showing whether a water will tend to dissolve or precipitate a particular mineral. Its value is negative when the mineral may be dissolved, positive when it may be precipitated, and zero when the water and mineral are at chemical equilibrium. The saturation index (SI) is calculated by comparing the chemical activities of the dissolved *ions of the mineral (ion activity product, IAP) with their solubility product (K_{sp}). In equation form, $SI = \log(IAP/K_{sp})$.

mineral soil Soil composed principally of mineral matter, in which the characteristics of the soil are determined more by the mineral than by the organic content.

minette A type of *lamprophyre, characterized by essential *biotite and *orthoclase feldspar (*see* ESSENTIAL MINERAL). If *augite is present the rock is termed an 'augite-minette'.

minette ironstone Iron *ore from Alsace-Lorraine; *limonite (iron oxide) is the main iron mineral with some *siderite (iron carbonate).

minimum-distance-to-means classification A *remote sensing *classification system in which the mean point in digital parameter space is calculated for *pixels of known classes, and unknown pixels are then assigned to the class which is arithmetically closest when digital number values of the different *bands are plotted. *See also* BOX CLASSIFICATION; and MAXIMUM-LIKELIHOOD CLASSIFICATION.

minimum melting curve A univariant *solidus curve in pressure–temperature space that defines the initial melting temperature of a solid as a function of pressure under water-saturated conditions. Minimum melting temperatures are always achieved under water-saturated conditions.

minimum temperature The lowest temperature recorded diurnally, monthly, seasonally, or annually, or the lowest temperature of the entire record. Daily air-temperature minima are recorded by the screen minimum thermometer. *See also* GRASS MINIMUM TEMPERATURE.

minimum thermometer Thermometer

that records the lowest temperature to which it has been exposed, e.g. by allowing a fall of mercury past a restriction in the tube, but preventing the mercury's return on expansion. It is most commonly used to record minimum daily temperatures.

minor fold A *fold which is generally in *harmonic accordance with the geometry of a major fold system and which has a characteristic *M-fold, *S-fold, or *Z-fold profile in specific areas of an *asymmetrical fold.

mio- From the Greek *meion* meaning 'less', a prefix meaning 'less' (e.g. *Miocene, 'mio' plus 'cene' (from the Greek *kainos*, 'new'), meaning 'less new').

Miocene Fourth of the five *epochs of the *Tertiary Period, extending from the end of the *Oligocene, 23.3 Ma ago, to the beginning of the *Pliocene, 5.2 Ma ago. Many mammals with a more modern appearance evolved during this epoch, including deer, pigs, and several elephant stocks. The Miocene comprises the *Aquitanian, *Burdigalian, Early and Late *Langhian, *Serravallian, *Tortonian, and *Messinian *Ages.

miogeocline An association chiefly of *carbonates, *shales, and clean *sandstones, with an absence of volcanics. These sediments are thought to have formed in shallow water on a continental margin.

miogeosyncline An obsolete term for that part of a *geosyncline characterized both by sediments deposited in shallow water and by the absence of volcanics. In the classification of geosynclines, miogeosynclines had a relatively thin sedimentary pile and were developed closer to the *craton than the volcanic *eugeosyncline. The geosynclinal theory of the formation of *orogenic belts has been superseded by various *plate tectonic models, and the term 'miogeosyncline' has been replaced by '*miogeocline' which is a purely descriptive word having no connection with geosynclinal theory.

mirage Optical effect in which major vertical variation in temperature of the lower atmosphere produces differential refraction of light, resulting for example in raised images and in gaps, which may give the appearance of a water surface.

Miranda (Uranus V) One of the major satellites of *Uranus, discovered by Gerard *Kuiper in 1948. Its radius is 240 × 234.2 × 232.9 km; mass 0.659 × 10^{20} kg; mean density 1200 kg/m^3; albedo 0.27; gravity 0.01 (Earth = 1); distance from Uranus 129 850 km. The surface temperature is about 43 K. Miranda is composed of rocky material and water ice in approximately equal proportions. The first pictures of the surface were provided in 1986 by *Voyager 2 as it passed close to the satellite to gain a gravity assist to take it to *Neptune. The surface of Miranda comprises areas of rolling, cratered terrain, with grooves, valleys, and cliffs, one more than 15 km high. Within this terrain there are three coronae. These are square, with rounded corners, up to about 260 km across, and contain dark and bright patches and sets of parallel ridges and grooves. It is now believed that the crust of Miranda was pulled apart by internal forces as the interior was evolving and the coronae formed above major upwellings of partially melted ice, and tidal distortions by Uranus caused heating.

mirror plane *See* CRYSTAL SYMMETRY.

mise-à-la-masse method (charged-body potential method) A form of *constant separation traversing in which one current electrode is placed directly into a conductive *ore body. The other three electrodes of the *array are on the surface, as in normal *electrode configurations. This method enables the subsurface extent of the conductor to be established more readily than with other electrical methods.

misfit stream (underfit stream) Stream that is too small to have cut the valley it currently occupies. Applies particularly to a meandering stream whose dimensions are much smaller than those of the meandering valley through which it flows.

mispickel *See* ARSENOPYRITE.

Mississippian 1. The Early *Carboniferous subperiod which is followed by the *Pennsylvanian and comprises the *Tournaisian, *Visean, and *Serpukhovian *Epochs. It is dated at 362.5–322.8 Ma (Harland et al., 1989). **2.** The name of the corresponding N. American sub-system which comprises the *Kinderhookian, *Osagean, *Meramecian, and *Chesterian *Series. It is roughly contemporaneous with the *Dinantian plus the *Namurian A of western Europe.

Missourian A *series in the *Pennsylvanian of N. America, underlain by the *Desmoinesian, overlain by the *Virgilian, and roughly contemporaneous with the Krevyakinskian and lower Chamovnicheskian Stages of the *Kasimovian Series.

mist Surface-layer atmospheric condition in which visibility is reduced by very fine, suspended water droplets. In synoptic meteorology, the *relative humidity in a mist condition is more than 95% and overall visibility is at least 1 km. *See also* HAZE; and FOG.

mistral Strong, cold, northerly wind that blows offshore with great frequency along the Mediterranean coast from northern Spain to northern Italy, and that is particularly frequent in the lower Rhône valley. The wind may persist for several days, and is best developed when a *depression is forming in the Gulf of Genoa to the east of a ridge of high pressure. The airstream that feeds the mistral is commonly derived from polar air of maritime origin. In the Rhône valley and similar areas of occurrence, the airflow is strengthened by *Katabatic and funnelling effects producing speeds of up to 75 knots, compared with the typical 40 knots experienced along the coast.

Mitchellian A *stage in the Upper *Tertiary of south-eastern Australia, underlain by the *Bairnsdalian, overlain by the *Cheltenhamian, and roughly contemporaneous with the *Tortonian, *Messinian, and lower *Zanclian (Tabianian) Stages.

mitchellite *See* SPINEL.

mites *See* CHELICERATA.

mitochondrial-DNA (mt-DNA) Circular DNA that is found in mitochondria. It is entirely independent of nuclear DNA and, with very few exceptions, is transmitted from females to their offspring with no contribution from the male parent.

mitochondrial Eve The postulated female ancestor of all modern humans, who was an archaic human living in Africa about 140 000 to 280 000 years ago, based on studies of *mitochondrial-DNA from modern populations by a team led by Allan Wilson. The data suggest there may have been a bottleneck associated with the speciation event leading to *Homo sapiens sapiens*. *See also* ADAM.

mixed cloud Cloud containing unfrozen water droplets as well as ice crystals. Typically the condition is found in *cumulonimbus, *nimbostratus, and *altostratus.

mixed pixel In *remote sensing terminology, a *pixel that has a *digital number which represents the average energy emitted or reflected from several different surfaces occurring within that area represented by the pixel.

mixing condensation level The vertical mixing of air, e.g. in the amalgamation of *air masses with different temperatures, leading to conditions of oversaturation and condensation. The mixing condensation level is the lowest level at which condensation can occur.

mixing depth The extent of an atmospheric layer measured from the surface of the Earth, usually a sub-inversion layer, in which convection and turbulence lead to mixing of the air and any pollutants in it.

mixing ratio Ratio of the mass of a given gas (e.g. water vapour) to that of the remaining gas (e.g. dry air) in the mixture. The examples given yield the 'humidity mixing ratio', expressed most conveniently in grams per kilogram of dry air. *See also* SPECIFIC HUMIDITY.

mixolimnion The upper layer of a *meromictic lake, lying above the *chemocline, where the water is mixed by the wind, circulates freely, and has a low density. *Compare* MONIMOLIMNION.

moberg The Icelandic name for a flat-topped mountain produced by the subglacial eruption of a *central vent volcano.

mobile belt A synonym for *orogenic belt, most often used for those earlier (i.e. early *Precambrian) belts where *plate-tectonic models cannot easily be applied.

mocha stone *See* AGATE.

modal analysis The determination of the *mode of a *rock, usually carried out by point counting using an *automatic point counter attached to a *polarizing microscope.

mode 1. The percentage by volume of each of the *minerals which make up an *igneous rock. Occasionally the term is also applied to *metamorphic rocks. **2.** In statistics, the *average as defined by the most frequently occurring value in a data set.

model Representation of reality in which the main features of some aspect of the real world are presented in simplified terms in order to make that aspect easier to comprehend, and often to facilitate the making of predictions.

Modified units *See* MARTIAN TERRAIN UNITS.

Moershoofd (Poperinge) A Middle *Devensian *interstadial from the Netherlands during which the climate was relatively cool, with average July temperatures of 6–7 °C. This division, together with the *Hengelo and *Denekamp interstadials, is perhaps the equivalent of the *Upton Warren Interstadial of the British Isles.

Mohnian A *stage in the Upper *Tertiary of the west coast of N. America, underlain by the *Luisian, overlain by the *Delmontian, and roughly contemporaneous with the upper *Serravallian, *Tortonian, and lower *Messinian Stages.

Moho *See* MOHOROVIČIĆ DISCONTINUITY.

Mohorovičić, Andrija (1857–1936) A Croatian seismologist, Mohorovičić discovered a seismic discontinuity (since named after him, *see* MOHOROVIČIĆ DISCONTINUITY) at the *crust–*mantle boundary, when analysing the results of an earthquake in 1909. He found two sets of *P-wave arrivals with different travel times, and thus was able to calculate the depth of the discontinuity.

Mohorovičić discontinuity (Moho) The seismic discontinuity that marks the boundary between the *Earth's *crust and *mantle. The mantle has *P-wave velocities of about 8.1 km/s, higher than that of the lower-density crustal rocks. The Moho was originally recognized at a depth of some 20 km in Europe by A. *Mohorovičić on the basis of the arrival of *refracted P-waves before *direct P-waves. Previously considered to be very sharp, the discontinuity is now known to be more complex.

Mohr stress diagram A two-dimensional, graphical representation of the relationship between *shear stress, *confining pressure, and the angle of *failure at the point at which a failure in a material occurs. Circles whose centres lie along the axis of *normal stress (*principal stress axes $(\sigma_1 + \sigma_3/2)$) and whose diameters represent the differential stress (principal stress axes $\sigma_1 - \sigma_3$) represent the *stress state, the data for them having been obtained experimentally. Tangents to each circle join the points of failure for different stress states and so delimit the field of stable stress states (within a Mohr stress envelope) from the failure field outside.

Mohs, Friedrich (1773–1839) A German mineralogist, Mohs was a student of *Werner and succeeded him at Freiberg University before moving to become professor of mineralogy in Vienna. In 1812, he developed a decimal scale for the hardness of minerals which is still in use. His *Grundriss der Mineralogie* (English edition called *Outline of Mineralogy*) was published in 1825. *See* MOHS'S SCALE OF HARDNESS.

Mohs's scale of hardness Scale, devised in 1812 by the German mineralogist Friedrich *Mohs, of the scratch *hardness of *minerals, as: 1 *talc; 2 *gypsum; 3 *calcite; 4 *fluorite; 5 *apatite; 6 *orthoclase; 7 *quartz; 8 *topaz; 9 *corundum; and 10 *diamond. The scale is linear up to a hardness of 9 (corundum), but then rises dramatically to 10, with diamond about 10 times harder than corundum. The scale is still used today.

Moinian A *stage of the Upper *Proterozoic of the north-western Highlands of Scotland, underlain by the *Longmyndian and overlain by the *Dalradian (*Cambrian).

moisture balance *See* MOISTURE BUDGET.

moisture budget (moisture balance, water budget) The balance of water fluxes into and out of a defined area over a defined time period, as represented broadly by the equation: precipitation = runoff + evapotranspiration + the change in soil-moisture storage. In mid-latitudes, for example, the annual budget is balanced by a high level of potential *evapotranspiration and utilization of *soil moisture in summer, compensated by a water surplus and recharge of soil moisture in winter when evaporation is less and precipitation is sometimes greater.

moisture index A term based on the computation of an annual moisture budget by C. W. Thornthwaite (1955), and calculated from the *aridity and *humidity indices, as $I_m = 100 \times (S - D)/PE$, where I_m is the moisture index, S is the water surplus in months when precipitation exceeds *evapotranspiration, D is the water deficit in months when evapotranspiration exceeds precipitation, and PE is the potential evaporation.

Mokoiwan *See* TAITAI.

molarity *See* CONCENTRATION.

molasse A term originally used to describe the mainly shallow-marine and non-marine *sediments produced from the *erosion of a mountain belt after the final stage of uplift in an *orogeny. It is now clear that much so-called molasse is not *post-tectonic, but *syntectonic, developed from the erosion of *nappes while uplift and deformation are still progressing; some workers therefore consider that the term should be discontinued. *See* FLYSCH.

moldic porosity (mouldic porosity) A form of *secondary porosity, developed by the preferential dissolution of shell fragments or other particles, to leave empty spaces previously occupied by the particles. *See* POROSITY; and CHOQUETTE AND PRAY CLASSIFICATION.

mole The amount of a substance containing as many elementary units as there are carbon atoms in 12 g of carbon-12 (i.e. the *Avogadro number). The elementary units by which the amount of substance is being measured may be atoms, molecules, *ions, *electrons, radicals, or any other expressly named constituent.

molecular sieve action Technique used in synthetic *zeolites to separate molecules by trapping them in the *crystal lattice. Sizes are selected to suit particular molecules.

mole drain Drain that can be made in soils by pulling a bullet-shaped device through the soil so that the compacted sides of the tunnel maintain that form for several years.

mollic horizon Surface *soil horizon of mineral soil that is dark in colour, and relatively deep, and contains (dry weight) at least 1% organic matter, or 0.6% organic carbon, the determination of either being acceptable. It is the *diagnostic horizon of *Mollisols and is associated with base-rich materials and grassland vegetation.

Mollisols An order of mineral soils that are identified by a deep *mollic surface *soil horizon (well-decomposed and finely distributed organic matter) and base-rich *mineral soil below.

Mollusca (molluscs) A very diverse phylum of *invertebrates which have a common body plan modified in various ways. There is usually a shell, secreted by a series of tissues (called the *mantle), and beneath this there is the body, which contains a space (mantle cavity) in which lie the gills. The body plan is modified to give the major classes of the phylum. One class, the *Monoplacophora, shows signs of internal segmentation. The *Amphineura (chitons) have a *bilaterally symmetrical shell of eight plates. Scaphopoda have tapering, curving shells which are open at both ends. *Bivalvia have the soft parts enclosed between two shells. *Gastropoda usually possess a coiled univalve shell. The most advanced of the molluscs are the *Cephalopoda, which possess either internal or external, chambered shells. Other groups are also assigned to the Mollusca, but only cephalopods, bivalves, and gastropods have a good geologic record.

molluscs *See* MOLLUSCA.

molybdenite Mineral, MoS_2; sp. gr. 4.6–4.8; *hardness 1.0–1.5; *hexagonal; pale bluish to lead-grey; greenish-grey *streak; *metallic *lustre; crystals hexagonal, often *tabular grains, and it also occurs as foliated and scaly masses; *cleavage perfect basal {0001}; quite widespread but seldom occurs in large quantities, found as an *accessory mineral in *granites, *quartz veins, and *pegmatites, in contact *metamorphic zones with *garnets, *pyroxenes, *scheelite, *pyrite, and *tourmaline, and in *veins with scheelite, *wolframite, *cassiterite, and *fluorite. It is a major *ore mineral for molybdenum.

moment of inertia (*I*) The kinematic properties of a rotating body; a measure of the rotational inertia of an object around a specific axis of rotation, in units of kg/m^2. In the *Earth, the principal moment of inertia lies close to the *axis of rotation and passes through the centre of mass of the Earth. Changes in the mass distribution, e.g. ice sheets, seasonal atmospheric changes, etc., cause changes in the location of the moments of inertia.

monadnock Isolated hill or range of hills standing above the general level of a *peneplain and resulting from the *erosion of the surrounding terrain. It may be located on relatively resistant rock or in a *watershed position where erosion is least. It is named after Mount Monadnock, New Hampshire, USA.

monazite Mineral $(Ce,La,Y,Th)PO_4$; sp. gr. 4.9–5.4; *hardness 5.0–5.5; *monoclinic; clove-brown to reddish-brown to orange and green; off-white *streak; resinous to waxy *lustre; crystals small, short, *prismatic to *tabular grains, with larger crystals showing striated faces; *cleavage imperfect basal {001}; found extensively as an *accessory mineral in *granites and *pegmatites, in *gneisses and *carbonatites, and concentrated in *alluvial sands and *placers. It is used as a source of cerium, thorium, and other *rare-earth metals and compounds.

monchiquite A type of *lamprophyre, characterized by *essential *analcite, *barkevikite (an alkali *amphibole), and/or *augite. There is no *feldspar present in this type of lamprophyre.

Monera One of the five kingdoms of life in the classification originally proposed by R. H. Whitaker (1959, 1969), comprising the *prokaryotic Cyanophyta (blue-green algae) and *bacteria (Schizomycophyta). The bacteria were the first forms of life on Earth, dating from at least 3300 million years ago, while the first *cyanophytes appeared about 2600 million years ago.

Monian A *stage of the Middle-Upper *Proterozoic of Anglesey and the Lleyn Peninsula, north Wales, equivalent to the Lower *Torridonian.

monimolimnion The lower layer of a *meromictic lake, lying below the *chemocline, where the water is dense, static, and does not mix with the water above. *Compare* MIXOLIMNION.

mono- From the Greek *monos* meaning 'alone', a prefix meaning 'single' or 'one'.

monochromator In *reflected-light microscopy, an accessory used to observe the optical properties of minerals with incident light of a specific wavelength. Fixed monochromatic interference filters with a band width of 15–50 nm and continuous-spectrum monochromators are available. They are used primarily for *reflectance measurements at specific wavelengths, normally 546 to 589 nm.

monocline A one-limbed flexure on either side of which the *strata are horizontal or *dip uniformly at low angles.

monoclinic Applied to a *crystal system where the *Bravais lattices have three sets of edges of unequal length, two sets of which are not at right angles to one another. They can therefore be referred to three unequal *crystallographic axes a, b, and c (or x, y, and z). The a (or x) axis (clino) is inclined and forms an obtuse angle with the vertical c (or z) axis usually denoted by the symbol β. The b (or y) axis (ortho) is horizontal and at right angles to both the a (x) and c (z) axes.

Monocraterion A single, tube-like, vertical *burrow that has a series of stacked funnel structures that point towards the top surface of the bed. The multiples of funnel structures are a response to increased sedimentation and reflect the upward migration of the animal.

Monocyathea *See* REGULARES.

monodactyl Applied to the condition in which the lateral *metacarpals and *metatarsals are reduced, leaving only one functional digit.

monograptid 1. A member of a suborder of *Graptoloidea, occurring in Lower *Silurian to *Emsian marine rocks, characterized by the possession of a single, *uniserial *stipe. **2.** Applied to the latest of four graptolite faunas occurring in Silurian and *Devonian rocks.

Monograptus **(class *Graptolithina)** The first recorded *monograptid genus, from the Early *Llandoverian (Lower *Silurian). Successive evolutionary trends within the graptolites resulted in the development of the single-branched monograptids. These had only a limited number of thecal cups (*see* THECA) on one side of the branch although the cups were large and exhibited considerable morphological variation. *Compare* DENDROIDEA; and GRAPTOLOIDEA.

monolete *See* SPORE.

monomictic Applied to lakes in which only one seasonal period of free circulation occurs. In cold monomictic lakes, typical of polar latitudes, the seasonal overturn occurs briefly in summer; in other seasons, surface-water temperatures are below 4 °C, which induces density stratification. In warm monomictic lakes, typical of warm temperate or subtropical regions, the seasonal overturn occurs in winter; at other times thermal stratification, with the formation of a distinct *epilimnion, prevents free circulation through the depth of the lake.

monomineralic Applied to *rocks composed of one *mineral type only. Examples would include the *igneous rock *anorthosite (composed entirely of *plagioclase feldspar) and the *metamorphic rock *marble (composed entirely of *calcite).

monomyarian *See* MUSCLE SCAR.

monophyletic If the members of a given *taxon are descended from a common ancestor they are said to be 'monophyletic', e.g. the families within a class would be monophyletic if they were all descended from the same family or lower taxonomic unit. Under the strictest definition they would all have to be descended from a single species. *Compare* POLYPHYLETIC.

Monoplacophora (phylum *Mollusca) Class of primitive, almost *bilaterally symmetrical, univalved molluscs, whose limpet-shaped shell of calcium carbonate consists of an external periostracum (the outer, non-calcareous layer), a prismatic layer, and an internal, nacreous deposit. The internal organs are *metamerically segmented, and the circular foot and radula are ventrally placed. Typically, Monoplacophora have paired *muscle scars. All are marine, benthic filter-feeders. Fossil forms inhabited shallow water, but modern forms occur in deep water. They first appeared in the Lower *Cambrian.

monopodial 1. Type of branching in which lateral branches arise from a definite main, central stem. **2.** With a single axis, an extension growth from the apex.

Monorhina Older name for the group of fish, including the lampreys and related jawless fossils, which possess a single median nostril between the eyes.

monostatic radar *See* BISTATIC RADAR.

monosulcate *See* POLLEN.

Monotremata (class *Mammalia, subclass Prototheria) An order comprising the duck-billed platypus (*Ornithorhynchus anatinus*) and the echidnas or spiny ant-eaters, *Tachyglossus* and *Zaglossus*. There were some extinct forms of which very few are known in detail, though of those some attained large sizes. The echidnas have no fossil record older than the *Pleistocene, but a fossil platypus, *Obdurodon*, is known from the *Miocene and in the early 1990s teeth of an undoubted platypus-like form, **Monotrematum sudamericanum*, were discovered in *Palaeocene deposits in Patagonia. Two *Cretaceous genera, **Steropodon* and *Kollikodon*, are also known. In view of their reptilian affinities they are thought to represent a separate and direct line of descent from the earliest *Mesozoic animals, independent of the line leading to other mammals. They retain many primitive features and and are quite unlike *marsupials or *eutherians (placental mammals).

Monotrematum sudamericanum A *Palaeocene platypus from Patagonia, Argentina, known so far only by its molar teeth. Its presence demonstrates that monotremes (*Monotremata), like marsupials, were part of the *Gondwana fauna.

monotremes *See* MONOTREMATA.

monotypic In *classification, applied to a group that has only a single representative. Thus a phylum may have only a single class or order and might be represented by a single genus and species.

mons (pl. montis, montes) The Latin word for 'mountain' and a term applied to large planetary mountains. The type example is Olympus Mons, the 26 km high volcano on Mars. Other examples include the Maxwell Montes on Venus.

monsoon From the Arabic *mausim* meaning 'season', a seasonal change of wind direction and properties associated with widespread temperature changes over land and water in the subtropics. Seasonal alternations of pressure systems together with shifting upper wind patterns and *jet streams produce seasonal winds, called 'monsoon winds'. The climate of the Indian subcontinent is especially characterized by the monsoon, where a distinct rainy season occurs in the south-westerly monsoon. Other major areas of monsoons are eastern and south-eastern Asia, the west African coast (latitude 5–15° N) and northern Australia.

month degrees The excess of mean monthly temperatures above 6 °C (43 °F), added together and used as accumulated temperature (indicative of conditions for vegetation growth) in some climate classifications (e.g. that of A. Miller, 1951).

Montian *See* DANIAN.

montes On *Mars, a mountain range, e.g. Montes Alpes, Montes Cordillera.

montmorillonite An important *clay mineral of the approximate composition $\{Al_4[Si_3AlO_{10}]_2(OH)_4\}^{2-}.nH_2O$ with some K^+, Na^+, or Ca^+ ions also present; it belongs to the 2:1 group of *phyllosilicates (sheet silicates) and the montmorillonite or *smectite group includes *bentonite; most members of this group are designated expansive clays by engineers, because they can accommodate many water molecules into their structure and they all possess an overall negative charge; sp. gr. variable but

2.0–2.7; *hardness 2; *monoclinic; white to grey with tints of blue, pink, pink-red, and green; dull *lustre; usually occurs in *massive, *microcrystalline aggregates of very fine, scale-like crystals, but in soils as a hydrous *aluminosilicate *clay mineral with a layer-lattice structure (two sheets of *tetrahedral silicon crystals enclosing a sheet of *octahedral aluminium crystals) that expands when water enters between layers, thus making a soil material, typically with variable water content, subject to swelling and contraction; found very extensively, it results from the decomposition of volcanic ashes in marine basins, and also occurs in the weathering crust of *basic *igneous rocks, e.g. *diabases, *basalts, *gabbros, and *peridotites. It is used extensively as an absorbant, for refining out suspended matter, in the textile and chemical industries.

monzodiorite A coarse-grained *igneous rock consisting of *essential *plagioclase feldspar, *orthoclase feldspar, *hornblende, and *biotite, with or without *pyroxene. Plagioclase is the dominant feldspar making up 60–90% of the total feldspar and varying from *oligoclase to *andesine in composition. The presence of the orthoclase feldspar distinguishes this rock from a *diorite.

monzogabbro A coarse-grained *igneous rock consisting of essential *plagioclase feldspar (*see* ESSENTIAL MINERAL), *orthoclase feldspar, and *pyroxene, with or without *biotite. Plagioclase of labradorite composition is the dominant feldspar, making up 60–90% of the total feldspar. The presence of orthoclase feldspar distinguishes this rock from a *gabbro.

monzonite (syenodiorite) A coarse-grained *igneous rock consisting of essential *plagioclase feldspar (*see* ESSENTIAL MINERAL), *orthoclase feldspar, *pyroxene, and *biotite. The plagioclase feldspar and orthoclase feldspar are in roughly equal proportions, the plagioclase forming between 40% and 60% of the total feldspar.

moom A satellite that orbits inside the rings of its planet.

Moon The *Earth's *satellite, with a mass 1/81 that of the Earth, density 3344 kg/m^3, and radius 1738 km. The average Moon–Earth distance is 384 500 km. The Moon has no atmosphere and surface temperature extremes range from 127 to –173 °C. A feldspathic lunar highland crust 60–120 km thick overlies a silicate mantle. Basaltic lavas cover 17% of the surface. There is probably a small iron core of 300–400 km radius (2–3% of lunar volume).

moonquake Seismic activity on the *Moon resulting from internal sources (differential cooling or tidal distortions) or external sources, e.g. meteoritic impact. It is the lunar equivalent of an *earthquake. All moonquakes are less than 2 on the *Richter magnitude scale. About 3000 per year were recorded by the Apollo *seismometers. The lunar seismic signals have very low attenuation and a large degree of wave scattering, consistent with travel through a brecciated crust. Most moonquakes occur at about 1000 km depth and are probably induced by tidal stresses, related to periodic changes in the Earth–Moon distance. No lunar tectonic activity is evident. Rare surface moonquakes are caused by the impacts of *meteorites.

moonstone A variety of adularia. *See* ALKALI FELDSPARS; and PERTHITE.

mor Type of surface *humus *soil horizon that is acid in reaction, lacking in microbial activity except that of fungi, and composed of several layers of organic matter in different degrees of decomposition.

moraine Term originally applied to the ridges of rock debris around Alpine *glaciers. Subsequently its meaning has been widened to include the rock debris deposit. For example, 'ground' moraine may denote an irregularly undulating surface of *till, glacial drift, or *boulder clay; or it may describe the deposit itself. An 'end' or 'terminal' moraine is a ridge of glacially deposited material laid down at the leading edge of an active glacier. Its height is in the range 1–100 m, and it is accumulated by a combination of glacial dumping and pushing. A 'recessional' moraine is morphologically similar and is laid down at the terminus of a glacier during a period of still-stand that interrupts a long period of retreat of the ice margin. A 'lateral' moraine is a ridge of debris at the margin of a valley glacier and largely derived from rock fall. It is a prominent feature of many contemporary Alpine glaciers. A 'medial' moraine results from the merging of lateral moraines when two glaciers converge. A 'washboard' moraine is a single ridge in a closely spaced pattern (perhaps 9–12 per kilometre) and stands some 1–3 m above the adjacent depressions. It is found in the 'end' moraine belt. A 'push' moraine is a morainic ridge made of unconsolidated rock debris and pushed up by the *snout of an advancing glacier. *See also* DE GEER MORAINE; FLUTED MORAINE; and HUMMOCKY MORAINE.

Morarian orogeny What may have been an episode of mountain building that occurred about 1050–730 Ma ago, prior to the *Caledonian orogeny, affecting the Morar area north-west of the Great Glen Fault, in what is now Scotland. Alternatively, the evidence suggesting an *orogeny may be explained as a phase of *pegmatite formation. If it was a genuine orogeny it may mark the start of the *subduction that led to the closing of the *Iapetus Ocean.

MORB *See* MID-OCEAN-RIDGE BASALT.

Morisawa flood peak formula *See* FLOOD PEAK FORMULAE.

morphogen An environmental factor, such as a chemical substance, that establishes a concentration gradient affecting *morphogenesis in organisms exposed to it.

morphogenesis The development of the form and structure of an organism by growth and differentiation.

morphogenetic zone 1. Area that is characterized by a distinctive assemblage of land-forms and that coincides with a major climatic zone. It is believed that the land-forms largely result from the action of a unique combination of surface processes controlled by climate. **2.** *See* LINEAGE-ZONE.

morphological mapping A method of mapping the form of a land surface. It is based on the assumption that land surfaces can be divided into a number of components, each of which has a uniform gradient or curvature, and which are separated by abrupt or gentle changes of slope. The nature of the change of slope is shown on the map by a standard set of symbols. The angle of each component may be measured instrumentally

morphological system In *geomorphology, a theoretical construct consisting of the relationship between the physical properties of a natural *system (geomorphological). For example, the physical dimensions of a beach (angle of slope seaward, average grain size, porosity, and moisture content) may be related to each other in an orderly manner, and so constitute a morphological system, and the geometric properties of a valley-side slope are typically correlated with certain characteristics of soil and vegetation.

morphology Form and structure of individual organisms.

morphometric analysis *See* DRAINAGE MORPHOMETRY.

morphometrics A technique of taxonomic analysis using measurements of the form (*morphology) of organisms and typically involving multivariate statistics.

morphospace In *theoretical morphology, all the possible forms an evolving taxon might take from a given set of initial parameters.

morphospecies A group of biological organisms that differs in some morphological respect from all other groups.

morphotype In taxonomy, a specimen chosen to illustrate a morphological variation within a species population.

Morrowan A *series in the *Pennsylvanian of N. America, underlain by the *Chesterian (*Mississippian), overlain by the *Atokan, and roughly contemporaneous with the *Bashkirian Series.

mortar texture Closely interlocking *crystals of fine *grain size, formed by the crushing and *recrystallization of a rock that was originally coarser grained. It is a texture produced by shearing of the rock during *dynamic metamorphism. In low-grade dynamic metamorphic rocks, deformation occurs primarily along crystal boundaries, producing an envelope of mortar texture around each crystal.

Mortensnes A *stage of the *Varanger Epoch, dated at 600–590 Ma (Harland et al., 1989).

mosaic evolution Differential rates of development of various adaptive attributes within the same evolutionary lineage. For example, a particular *taxon might show greatly different rates of change with respect to the head, body, and limbs. This is a common phenomenon and makes the reconstruction of transitional fossil types very difficult.

mosaic heterochrony *Heterochrony in which a number of heterochronic processes occur simultaneously, so different parts of an organism develop at different rates. *Compare* ASTOGENETIC HETEROCHRONY; and ONTOGENETIC HETEROCHRONY.

mosaic texture Interlocking *crystals of fine to medium *grain size, displaying triple-point contacts between grains. Such a texture can form during metamorphic *recrystallization of a rock (*see* METAMORPHISM) in response to temperature change without any imposed directional stress.

Mosasauridae (mosasaurs; order Squamata, suborder Sauria) A family of marine lizards, fossils of which have been found in Upper *Cretaceous rocks throughout the world. They evolved into many types but all became extinct at the end of the Cretaceous. Mosasaurs were large, some reaching 9 m in length, with a long, slim body, short neck, and long head. The tail was used for swimming and the limbs for steering; the digits may have been webbed. The teeth were set in pits rather than being fused to the jaws. Most mosasaurs fed on fish, although some are believed to have fed on molluscs.

mosasaurs *See* MOSASAURIDAE.

Moscovian 1. An *epoch in the *Pennsylvanian dated at 311.3–303 Ma (Harland et al., 1989), comprising the Vereiskian, Kashirskian, Podolskian, and Myachkovskian *Ages (these are also *stage names in eastern European stratigraphy). The Moscovian is preceded by the *Bashkirian and followed by the *Kasimovian Epochs. **2.** The name of the corresponding eastern European *series, which is roughly contemporaneous with the *Atokan and *Desmoinesian (N. America) and the *Westphalian B, C, and D plus the lowermost *Stephanian (western Europe). **3. (Moskva Drift)** A *drift unit that overlies the deposits of the *Odintsovo *Interstadial, occurring in European Russia. It is equivalent to part of the *Saalian of western Europe.

Moskva Drift *See* MOSCOVIAN (3).

moss agate *See* AGATE.

Mössbauer spectroscopy An analytical spectroscopic technique based on the fact that certain *nuclides, solidly built into a *crystal *lattice, and at temperatures significantly below a certain temperature, emit *gamma rays. The crystal absorbs the recoil momentum, making it negligible because of the large number of atoms in the lattice, and there is virtually no loss of energy by the gamma photons.

mossy Applied to dendritic (branch-like) aggregates of *minerals that are frequently secondary chemical precipitates, such as iron or manganese oxides (e.g. moss agate and mocha stone).

mother cloud Incipient cloud from which a well-defined cloud type can form and develop. *See also* CLOUD CLASSIFICATION.

mother-of-coal *See* FUSAIN.

mottling Patchwork of different colours in *mineral soil (usually orange or rust against a background of grey or blue) which indicates periods of anaerobic conditions.

Motuan *See* CLARENCE.

mould *See* FOSSILIZATION.

moulin A system of tunnels made by meltwater through a glacier.

mouldic porosity *See* MOLDIC POROSITY.

mountain breeze *See* MOUNTAIN WIND.

mountain wind (mountain breeze) Term commonly used to describe a *katabatic wind or breeze and its counterpart, the upslope anabatic wind or breeze occurring on warm days over heated mountain slopes.

moveout (stepout) 1. The difference between the *two-way travel times of reflected energy detected at two *geophone *offset distances. **2. (normal moveout, NMO, ΔT)** At an offset distance x, the difference in two-way travel time between the detected reflection event at x (t_x) and at zero offset (t_o), such that $\Delta T = t_x - t_o \approx x^2/(2V^2t_o)$, where V is the seismic velocity of the medium above the reflector. **3. (dip moveout, DMO, ΔT_d)** In the case of a planar dipping reflector, the difference between the moveout up-dip and down-dip, proportional to the angle of dip θ, such that: $\Delta T_d = 2x\sin\theta/V$, where x is the offset distance from the mid-point.

M_{sat} *See* SATURATION MAGNETIZATION AND MOMENT.

M-shape Characteristic shape shown in an analysis profile of nickel distribution in *Widmanstätten structure in *meteorites. The temperature-dependence of the *diffusion of nickel within the metal *phase away from kamacite (Ni–Fe alloy with 6% Ni) into taenite (Ni–Fe alloy with 30% Ni), and the faster rate of diffusion in kamacite, leads to disequilibrium nickel distribution and the M-shape.

mud, drilling During rotary drilling, the fluid mud that is circulated to cool the drill *bit, convey rock chippings to the surface, and to seal permeable layers and fractures. The drilling mud is usually maintained under pressure to prevent the drill stem being blown out if it penetrates a pocket of pressurized gas. The muds may be oil- or water-based, and are frequently *thixotropic *bentonites, lime, or *barite.

mud cake (cake) The solid clay deposit formed in a *borehole on a permeable layer when the liquid *mud filtrate permeates into the surrounding rocks.

mudcracks *See* DESICCATION CRACKS.

mud drape A layer of mud deposited over a pre-existing morphological feature, e.g. a *bar, *ripple, or *dune.

mud filtrate (filtrate) The liquid part of a drilling *mud that can penetrate into a permeable layer, leaving behind a solid *mud cake and totally or partially replacing the *groundwater in the rocks.

mudflat Area of a coastline where fine-grained silt or sediment and clay is accumulating. Its development is favoured by ample sediment, by sheltered conditions, and by the trapping effect of vegetation. It is an early stage in the development of a *salt marsh.

mudflow 1. Heavily loaded ephemeral stream whose *viscosity increases with evaporation as it flows over a desert fan. **2.** Rapidly moving variety of *earthflow. This is a typical phenomenon of areas underlain by sensitive clays which may liquefy and flow following a shock, perhaps initiated by sliding. **2.** *See* LAHAR.

mud log 1. The record produced by a *micrولog measuring the *resistivity of the *mud cake in a drill hole. **2.** 'Mud log' is also the term used for the record maintained by well site geologists of the cuttings recovered from the hole.

mud mound A build-up of *carbonate *sediment in the form of a bank or mound, dominated by mud. The accumulation of mud occurs by its deposition in the lee of *in situ* organisms, such as corals (*Anthozoa) or *crinoids; by the sweeping of mud into banks by currents; or by the entrapment and precipitation of carbonate mud by *algae and other organisms acting as baffles.

mud roll *See* GROUND ROLL.

mudrock A lithified mud (*see* LITHIFICATION).

mudstone 1. *Argillaceous or clay-bearing *sedimentary rock which is non-plastic and has a *massive or non-foliated appearance. *Compare* CLAYSTONE. **2.** *See* LIME MUD; DUNHAM'S CLASSIFICATION; and EMBRY AND CLOVAN CLASSIFICATION.

mud-support (matrix-support) Texture of *sedimentary rocks in which there is complete separation of larger fragments or *clasts from each other within the finer-grained *matrix. *Compare* GRAIN SUPPORT.

mugearite *See* ANDESITE; and BENMOREITE.

mull Type of surface *humus *soil horizon that is chemically neutral or alkaline in reaction, that is well aerated, and that provides generally favourable conditions for the decomposition of organic matter. Mull humus is well decomposed and intimately mixed with mineral matter.

mullion structure A *rodding structure composed of elements tens of centimetres wide. The origin of mullions and rodding is not known with certainty but they are thought to be a type of *lineation associated with folding at a competent–incompetent interface (*see* COMPETENCE; and INCOMPETENCE).

multichannel seismic reflection A *reflection event which occurs across a number of different seismic channels, data from which can be used to enhance data processing and the subsequent interpretation of seismic sections.

multifurcation In a *phylogenetic tree, the occurrence of a split in an ancestral branch into more than two branches at an *internal node, because the order in which the progenic branches occur cannot be determined.

multilocular *See* UNILOCULAR.

multiple Generally, re-reflected seismic energy. More specifically, multiple *reflection which is detrimental to the data and should be removed by data processing. 'Short-path' multiples are *ghosts and near-surface phenomena; 'long-path' multiples, which are easier to identify and remove, are due to extra reflections deep within the subsurface.

multiple common-depth-point coverage A method of seismic *profiling in which the same part of the subsurface is sampled on different records. *See* FOLD.

multiple twinning *See* LAMELLAR.

multiplexing The transmission of several channels of information over a single channel without interference and the subsequent recording of that information on magnetic tape. The digital data recorded are in the order: channel 1, sample 1; channel 2, sample 1; ... channel *n*, sample 1; channel 1, sample 2; channel 2, sample 2; ... channel *n*, sample 2; etc., which is a very efficient way to record information on magnetic tape. *Demultiplexing is the process by which the tape is made readable by a computer and its data are made available for processing.

multi-ring basin (ringed basin) A large basin excavated by *asteroidal or *planetismal impact on a planetary surface. Several concentric rings of mountains surround the basin. The ratio of the diameters of adjacent rings approximate to $\sqrt{2}:1$. The minimum dimension for rings to occur is about 300 km on the Moon; elsewhere examples of up to 3000 km diameter are known (Valhalla basin on Callisto). The classic example is the lunar Mare Orientale, 920 km diameter, with three mountain rings. Typical relief is 3 km. The energies involved in excavating the basins are of the order of 10^{27} J.

multispectral imaging *See* IMAGING.

multispectral scanner In *remote sensing, a *line scanner which records data in several different *channels.

multispectral scanning systems The simultaneous use of several different sensors to obtain images or spectral information from various portions of the *electromagnetic spectrum. An example was the multispectral scanner on the Landsat Earth-orbiting satellites, which simultaneously recorded green, red, and two infrared spectral bands. Since different materials (e.g. soil, rock, vegetation) reflect differing amounts of light at different wavelengths, they can be identified by their characteristic spectral 'signatures'.

multistate character A *character that can occur in several *character states.

multistory sandbody A series of *sandstone beds, each deposited by the infilling of a river *channel, stacked one above the other with little or no intervening *mudstone. The multistorey body is formed by the repeated and rapid migration of the channel network over the *alluvial plain, so allowing little chance for the preservation of fine *floodplain *sediment.

Multituberculata (class *Mammalia) Extinct order of rather rodent-like mammals, which appeared first in the late *Jurassic in Europe, flourished during the *Cretaceous and *Palaeocene, but became extinct during the *Eocene. Probably they were the first herbivorous mammals, with skull and teeth analogous to those of rodents. Most were small, but some attained the size of modern woodchucks. The limbs sprawled more widely than those of most mammals. The olfactory bulbs were large, which suggests that the animals depended heavily on their sense of smell. The skull was massive, but unlike that of other mammalian groups. The multituberculates appear to have been a side branch from the main line of mammalian evolution, and they are believed not to be related closely to other groups.

multivariate analysis In a statistical analysis, the measurement of several different attributes of each unit of observation.

Munsell colour Soil-colour system, devised originally in the USA, that is based on the three variables of colour—hue, value, and chroma—and given notation such as '10YR 6/4'. It is now used as an international method for reporting soil colour.

Murchison, Roderick Impey (1792–1871) A gentleman of private means who devoted much of his life to science, Murchison turned professional when he became director of the Geological Survey in 1855. His *Silurian System* (1839) was the first stratigraphic system defined by fossil content, rather than *lithology. He helped clarify the *Devonian System and, after a visit to Russia, the *Permian.

Murray, John (1841–1914) A Canadian-born biologist and oceanographer, Murray did most of his work in Britain. He was leader of the **Challenger* expedition and took part in other oceanographic investigations. He completed the publication of the scientific reports of the *Challenger* survey and produced a bathymetric chart of the oceans.

Muschelkalk 1. The seaway (Muschelkalk Sea) that extended across north-western Europe during the *Triassic, from the present-day UK in the west to northern Germany and Poland in the east. It was bordered to the south by the Bohemian *Massif and in the north by the Baltic Shield (*see* BALTICA). **2.** The mid-Triassic, when using the European tripartite divisions of the Triassic System: Bunter, Muschelkalk, and Keuper. The German word *Muschelkalk* means 'shelly limestone'.

muscle scar A depressed or raised marking on the interior of a shell, usually of a brachiopod (*Brachiopoda) or bivalve (*Bivalvia), which was the point of attachment of a muscle. In articulate brachiopods (*Articulata) there are commonly two pairs of muscle scars occurring in the floor of the *dorsal and *ventral valves, one pair formed by the adductor muscles that close the shell and the second pair by the diductors, that open it. There are additional muscles present in inarticulate brachiopods (*Inarticulata) and these also leave scars. In bivalves there are commonly two muscles present (dimyarian); these are the adductor muscles that close the shell against the opening moment of the ligament. In some species the scars are equal in size (isomyarian), in others unequal (anisomyarian) and the smaller scar is always the anterior one. Sometimes the anterior muscle is lost and the remaining muscle may increase in size; in this condition the bivalve is 'monomyarian'. Connecting the two scars on the interior of the valve there is the pallial line, a linear depression marking the inner margin of the thickened *mantle edges that may be marked by an inward deflection at the posterior part, the pallial sinus. This defines the space for the retraction of the *siphons and the place of attachment of the siphonal retractor muscles.

muscovite An important rock-forming *mineral and member of the mica group with composition $K_2Al_4[Si_3AlO_{10}]_2(OH,F)_4$ (*see* MICAS); when K is replaced by Na the mineral is called paragonite and is related to the *phyllosilicates (sheet silicates) *lepidolite and *zinnwaldite (both lithium-bearing micas); sp. gr. 2.8–2.9; *hardness 2.5–3.0; *monoclinic; normally colourless to very pale grey to green or light brown; vitreous to pearly *lustre; crystals *tabular, *hexagonal, and also occur in foliated masses and disseminated flakes; *cleavage basal perfect {001}; widely distributed in *igneous *alkali *granites, in *pegmatites, as a *secondary mineral resulting from the decomposition of *feldspars, and can be a major constituent of *clastic sediments (e.g. *sandstone and *silts).

Muses-C A Japanese *ISAS mission, to be launched in 2002, to land on the surface of asteroid *Nereus and return samples to Earth.

mushroom rock *See* PEDESTAL ROCK.

Mustelidae *See* CARNIVORA.

mutation 1. Process by which a *gene or *chromosome set undergoes a structural change. **2.** Gene or chromosome set that has undergone a structural change. Mutations are the raw material for *evolution: they provide the source of all variation. For mutations to affect subsequent generations, though, they must occur in gametes (reproductive cells) or in cells destined to be gametes, since only then will they be inherited. A mutation that occurs in a body-cell is called a 'somatic' mutation: it is transmitted to all cells derived, by mitosis, from that cell. Most mutations, however, are deleterious; evolution progresses through the few that are favourable.

mutation rate In *phylogenetics, the number of *mutations arising at a single nucleotide site per gene per unit time.

mutatus Form of cloud evolution in which the mother cloud is greatly affected by the development of cloud from it.

Myachkovskian A *stage in the *Moscovian Epoch, preceded by the *Podolskian and followed by the *Krevyakinskian (*Kasimovian Epoch).

mylonite Rock produced in zones of tectonic dislocation, e.g. *fault and shear zones. Mechanical crushing and grinding or *cataclasis produces a rock that has a *foliation which is often crude but which is sometimes very well developed, and that has a much finer grain size than its precursor. Mylonite is well laminated and often hard and splintery. The name literally means 'milled rock' from the Greek *mylon*, a grinding mill. *See* DYNAMIC METAMORPHISM.

myophore A plate or rod-like structure on the inside of the shell of some bivalves (*Bivalvia). It occurs at the centre of the *hinge for the attachment of muscle.

Mytiloida (phylum *Mollusca, class *Bivalvia) Order of *epifaunal, *byssate bivalves which have an equivalve shell that has a highly non-equilateral shape and a prismatic-nacreous microstructure (e.g. *Mytilus edulis*, the common mussel). They have a *dysodont *dentition, the ligament is elongate and *opisthodetic. Mytiloida have an anisomyarian musculature (i.e. the posterior adductor muscle and its scar are larger than the anterior adductor muscle). The pallial line is complete, the *siphons poorly developed. They first appeared in the *Devonian. *See* MUSCLE SCAR.

n *See* NANO-.

N *See* NEWTON.

NA *See* OBJECTIVE.

Nabarro–Herring creep A form of *diffusion creep in which atoms migrate within the *crystal *lattice. *See also* COBLE CREEP.

nacreous clouds (mother-of-pearl clouds) Type of cloud seen occasionally at great altitude (approximately 22–4 km), just before sunrise or after sunset, characterized by iridescent colouring. The cloud is fine and rather lenticular.

nacrite *See* KAOLINITE.

nadir In *remote sensing, the point on the ground directly beneath a remote sensing system.

NADW *See* NORTH ATLANTIC DEEP WATER.

Nafe–Drake relationship An empirical relationship between the *P-wave velocity and density of water-saturated *sediments and *sedimentary rocks. It is commonly used to evaluate the density of sedimentary rocks in shallow seismic surveys, but errors of the order of 100 kg/m^3 are involved. *See also* DENSITY OF ROCKS.

Nagssugtoqidian orogeny A late *Archaean and *Proterozoic phase of mountain building affecting a belt 240–300 km wide in what is now western Greenland. It comprised two major episodes, the first about 2600 Ma ago, the second about 1900–1500 Ma ago, making it approximately contemporaneous with the *Hudsonian, *Laxfordian, and *Svecofennian orogenies.

Naiad (Neptune III) A satellite of *Neptune, with a diameter of 58 km; visual albedo 0.06.

Nammalian A *stage of the *Scythian Epoch, preceded by the *Griesbachian and followed by the *Spathian.

Namurian The lowermost *series in the *Silesian (*Carboniferous) of Europe, underlain by the *Visean, overlain by the *Westphalian, and subdivided into Namurian A, B, and C. It is dated at 333–315 Ma and is contemporaneous with the Millstone Grit Series (western Europe) and the *Serpukhovian and Lower *Bashkirian Series (eastern Europe).

nannofossil *See* NANOFOSSIL.

nano- (n) From the Greek *nanos* meaning 'dwarf', a prefix meaning 'extremely small' (e.g. *nanofossil). Attached to SI units it denotes the unit $\times 10^{-9}$.

nanofossil (nannofossil) A *fossil of the smallest member of the *plankton (the nanoplankton). Nanofossils are of plants and include various forms, e.g. the coccolithophores (*Coccolithophoridae), which are 5–60 μm in size and have been used as stratigraphic indicators in addition to the larger *microfossils.

nanoplankton Marine planktonic organisms 2.0–20 μm in size.

nappe From the French *nappe*, meaning 'cover', a thrusted mass or folded body in which the *fold limbs and axes are approximately horizontal.

Naraoiidae An order of *Trilobita that lived in the Middle *Cambrian. They were uncalcified and had no thoracic segments.

Narizian A *stage in the Lower *Tertiary of the west coast of N. America, underlain by the *Ulatizian, overlain by the *Refugian, and roughly contemporaneous with the upper *Lutetian and *Bartonian Stages.

NASA The National Aeronautic and Space Administration, the agency of the US federal government that was established under the National Aeronautics and Space Act 1958 to plan, direct, and conduct all US aeronautical and space activities, other than those that are primarily military.

naticiform Applied to a gastropod (*Gastropoda) shell that is globular and umbilicate, resembling those of the genus *Natica*. *See* UMBILICUS.

native element An element which occurs in a free state as a *mineral, e.g. *gold, *copper, and *carbon.

natric horizon A mineral *soil horizon that is developed in a subsurface position in the profile, that satisfies the definition of an *argillic horizon, and that also has a columnar structure and more than 15% saturation of the exchangeable *cation sites by sodium.

natrolite Member of the *zeolite group of *minerals, with the formula $Na_2Al_2Si_3O_{10}.2H_2O$; sp. gr. 2.2; *hardness 5.0; white; *massive, fibrous, or crystalline; occurs as a *secondary mineral infilling cavities in *basic volcanic rocks and as an alteration product of *nepheline.

natron lake (soda lake) A saline lake, rich in the sodium carbonate salt natron ($Na_2CO_3.10H_2O$).

'natural break' Concept that the Earth's history has been punctuated by natural world-wide changes that should be discernible in the stratigraphic record.

natural cast *See* FOSSILIZATION.

natural gas Gaseous *hydrocarbons, chiefly methane (CH_4), ethane (C_2H_6), propane (C_3H_8), and butane (C_4H_{10}), trapped in *pore spaces in rocks with or without liquid *petroleum. It has a high heat value and burns without smoke or soot; it provides raw material for the chemical industry for making plastics, detergents, fertilizers, etc. Gas of this composition is also termed 'natural gas' if it occurs as a gas chimney or after production.

natural gas hydrate A type of *clathrate in which a naturally occurring gas, mainly methane, is held within a lattice of water ice. Such hydrates form under conditions of pressure and temperature found in the upper 300–2000 m of marine sediments where the water temperature is at or below 0 °C and the pressure is greater than 4 MPa, often beneath *permafrost. The hydrates can block gas transmission pipes and disturbing them can cause the release of methane (a *greenhouse gas). Should it prove possible to recover the methane they contain, the amount present in sediments, measured as carbon, is estimated to be up to twice that of all other recoverable and non-recoverable fossil fuels combined.

natural remanent magnetism (NRM) The *remanent magnetization of rocks and naturally occurring objects that has been acquired by normal processes, i.e. excluding laboratory-induced magnetizations. In *igneous rocks and fired archaeological materials, NRM is normally of thermal or chemical origin; in *sediments it is usually of depositional origin; and in *sedimentary rocks it is usually of chemical origin.

natural selection ('survival of the fittest') Complex process in which the total environment determines which members of a species survive to reproduce and so pass on their *genes to the next generation. This need not necessarily involve a struggle between organisms.

natural theology (physico-theology) A philosophy which tries to link the study of natural phenomena with the notion of divine providence, stressing that harmony and order in nature are evidence of God's design. It became important in 18th-century England, being associated with the work of John Ray, William Paley, and many others. The 19th-century *Bridgewater Treatises* were the last major exposition of natural theology. *See* BUCKLAND.

nautical mile Length of mile used in ocean navigation, equivalent to one minute (1/60°) change in latitude. It is internationally defined as being equal to 1852 metres.

nautilicone *See* CONVOLUTE.

nautiliconic Applied to the conch of a *cephalopod when it is coiled and highly *involute.

Nautiloidea (phylum *Mollusca, class *Cephalopoda) Subclass of cephalopods which possess a multichambered, external shell composed of calcium carbonate, which is siphunculate (*see* SIPHUNCLE) and may be coiled. The animal lives in the last-formed chamber, the body chamber. The gill structure is tetrabranchiate (four-gilled). Simple *sutures are produced by contact between the internal *septa and the shell wall. The subclass includes the oldest cephalopods, first recorded from Upper *Cambrian rocks. They diversified and became common throughout the *Palaeozoic, but were greatly reduced at the end of this *era. Further diversification occurred in the *Mesozoic, but the group dwindled again in the *Cenozoic. There is only one extant genus, *Nautilus*, which dates from the *Oligocene. *See also* *PLECTRONOCERAS CAMBRIA*.

Nazca Plate The lithospheric *plate which is subducting under the *South American Plate along the line of the *Peru–Chile Trench. Its other *plate margins are the *ridges and *transform faults which separate it from the *Pacific, *Cocos, and *Antarctic Plates. The Nazca Plate is thought to have decreased in size: the former, more extensive plate has also been called the Phoenix Plate.

neap tide Tide of small range occurring every 14 days, near the times of the first and last quarter of the Moon when the Moon, Earth, and Sun are at right angles. The neap tidal range is 10–30% less than the mean tidal range.

NEAR *See* NEAR EARTH ASTEROID RENDEZVOUS.

Nearctic faunal realm The fauna of N. America, south to Mexico. At the order and family level the fauna is essentially the same as that of the *Palaearctic realm, reflecting their former connection via the *Bering *land bridge, but some genera and more especially species are distinctive to the Nearctic.

Near Earth Asteroid Rendezvous (NEAR) The first of the *NASA *Discovery missions, launched on 17 February 1996. The spacecraft is equipped with an X-ray/gamma ray spectrometer, near-infrared imaging spectrograph, multispectral camera with CCD (*see* CHARGE-COUPLED DEVICE) imaging detector, laser altimeter, and magnetometer. A radio science experiment will also be performed to estimate the gravity field of *Eros. On 27 June 1997, the NEAR spacecraft approached to within 1200 km of *Mathilde. The goal of the mission is to reach and achieve orbit around Eros, in February 1999, and spend about one year studying the asteroid, approaching to within 24 km.

near-field barrier In the disposal of *radioactive waste, the containers in which waste is held, the vault in which these are stored, and the immediately surrounding rocks, which prevent the migration of radionuclides, but which provide a level of containment that is expected to fail over some thousands of years. Containment will then be provided by the *far-field barrier.

near-infrared *Infrared radiation which has a wavelength between 0.7 μm and 2.5 μm. Near-infrared is subdivided into very-near infrared (0.7 μm–1 μm) and short-wave infrared (1.0 μm–2.5 μm). Photographic films respond to wavelengths between 0.7 μm and 1.0 μm, hence very-near infrared is also known as photographic infrared.

near-infrared mapping spectrometer (near-infrared mapping spectrophotometer, NIMS) A *remote sensing instrument that makes measurements in the *near-infrared region of the light spectrum, from which the chemical composition, structure, and temperature of planetary and satellite atmospheres can be determined, as well as the mineral composition and geochemistry of planetary and satellite surfaces.

near-infrared mapping spectrophotometer *See* NEAR-INFRARED MAPPING SPECTROMETER.

near-shore current system System of currents caused by wave activity within and adjacent to the *breaker zone. The current system includes: the shoreward mass transport of water; *longshore currents; and seaward-moving *rip currents. This wave-induced current system often has a reversing tidal-current system superimposed on it.

nebulosus From the Latin *nebulosus* meaning 'fog-covered', a form of *stratus or *cirrostratus cloud or cloudsheets which is rather indistinct, and seen as a nebulous layer. *See also* CLOUD CLASSIFICATION.

necrology The scientific study of all the processes affecting dead animal and plant material, including decomposition and *fossilization.

Nectarian A *period in the early *Archaean, marking the time during which intense bombardment of the Earth ceased and dated at about 3975–3900 Ma (Harland et al., 1989).

Nectarian System *See* LUNAR TIME-SCALE.

needle ice *See* PIPKRAKE.

Néel, Louis Eugène Félix (1904–) A French physicist, Néel became Director of the Centre for Nuclear Studies in Grenoble in 1956. Much of his work was in the field of magnetism, including terrestrial magnetism, and he was the first to explain the existence of remanent *palaeomagnetism in rocks.

Néel temperature The temperature at which the magnetic coupling between two magnetic sublattices in an *antiferromagnetic material is overcome by thermal vibrations; if the material is heated above the Néel temperature it no longer exhibits antiferromagnetic behaviour until it has cooled to or below this temperature.

negative inversion *See* INVERSION.

nekton (adj. nektonic) Free-swimming organisms in aquatic ecosystems. Unlike *plankton, they are able to navigate at will. The nekton includes fish, amphibians, and large swimming insects.

nema In graptolites, a long, hollow, rod- or thread-like structure extending beyond the *sicula *See* GRAPTOLOIDEA.

Nematoda (nematodes, eelworms, threadworms, roundworms; phylum Aschelminthes) Class of aschelminth worms (a phylum in some classifications) which vary greatly in size, from about 1 mm to 5 cm. The cuticle has flanges, and may also have ridges or spines. Morphologically they are all similar, but they occur both as parasites in plants and animals, and as free-living forms. They are first known from rocks of *Carboniferous age (e.g. *Scorpiophagus*).

Nemertini (proboscis worms, ribbon worms) Phylum of unsegmented, non-parasitic worms which are bilaterally symmetrical (*see* BILATERAL SYMMETRY) and elongate. Adults are ciliated (*see* CILIUM) and anteriorly possess a thin, retractile tube, the proboscis. The mouth and brain are well developed, the *coelom is indistinct. Most are marine, although freshwater and terrestrial forms occur. Most are not hermaphrodites. The first fossil forms occur in the Middle *Cambrian *Burgess Shales of British Columbia, Canada.

Neobothriocidaris *See* BOTHRIOCIDAROIDA.

Neocomian The earliest *Cretaceous *epoch, followed by the *Barremian *Age. It is dated at 145.6–131.8 Ma (Harland et al., 1989) and comprises the *Berriasian, *Valanginian, and *Hauterivian Ages. Some authors restrict their use of Neocomian to just the Berriasian and Valanginian, whilst others extend it to include the Barremian and *Aptian.

Neogene The later of the two *periods which comprise the *Tertiary sub-Era, preceded by the *Palaeogene, followed by the *Quaternary, and dated at 23.3–1.64 Ma (Harland et al., 1989). The Neogene Period is subdivided into the *Miocene and *Pliocene Epochs.

Neohelikian A sub-*stage of the Upper *Helikian of the Canadian Shield region.

neoichnology *See* ICHNOLOGY.

neomineralization The formation of a new mineral or minerals from pre-existing minerals during *metamorphism.

neomorphism The diagenetic replacement (*see* DIAGENESIS) of a mineral by a different crystal *form of the same mineral. In *limestones neomorphism often results in the replacement of *calcite by coarser calcite crystals, e.g. *micrite may become neomorphically replaced by *microspar (4–10 μm crystals), or pseudospar (10–50 μm crystals). Neomorphic spar is characterized by irregularly shaped crystals, often patchily developed, and passing gradationally into unaltered areas.

Neoproterozoic The most recent part of the *Proterozoic, from about 1000–575 Ma.

neostratotype A *stratotype chosen to supersede the original stratotype (i.e. the *holostratotype), if the first one has been demolished or invalidated. A neostratotype may be selected from outside the *type area. *Compare* LECTOSTRATOTYPE. *See also* HYPOSTRATOTYPE; and PARASTRATOTYPE.

neoteny Slowing down of bodily development, so that sexual maturity is achieved while the organism still looks like a juvenile; this leads to *paedomorphosis. Since the juvenile stages of many organisms are less specialized than the corresponding adult stages, such shifts allow the organisms concerned to switch to new evolutionary pathways. The word comes from the Greek *neos* (meaning 'youthful'). Some features of human evolution have been ascribed to neotony.

Neotethys *See* TETHYS.

Neotropical faunal realm Region which includes S. and Central America, including southern Mexico, the W. Indies, and the Galápagos Islands. Much of the area was isolated for the greater part of the *Tertiary Period, which explains the distinctiveness of the fauna and the survival of ancient forms of mammals, e.g. the marsupials and edentates.

neotype Specimen chosen to act as the 'type' material subsequent to a published original description: this occurs in cases where the original types have been lost, or where they have been suppressed by the International Commission on Zoological Nomenclature (ICZN). *Compare* HOLOTYPE; LECTOTYPE; and PARATYPE.

NEP *See* NEPHELOMETER.

nephanalysis Interpretation of cloud type and amount from satellite pictures in facsimile or digitized form.

nepheline An important member of the *feldspathoid group of *silicate minerals with composition $Na_3(Na,K)[Al_4Si_4O_{16}]$, which approximates to $NaAlSiO_4$ (with some K replacing Na); nepheline is an *end-member of the series with *kalsilite ($KAlSiO_4$); sp. gr. 2.56–2.66; *hardness 5.5–6.0; *hexagonal; colourless, grey, yellowish, brownish, reddish, or greenish; *vitreous to *greasy *lustre; *crystals normally *prismatic, short, and columnar, thick, and *tabular, or coarse, granular aggregates; *cleavage imperfect prismatic {1010}, poor basal {0001}; it is a primary crystallizing mineral of silica-poor *alkaline *igneous rocks in association with *aegirine augite and alkali *amphiboles, and is an *essential mineral in some silica undersaturated rocks (*see* SILICA SATURATION) such as *nepheline syenites and *nephelinites.

nepheline-basanite *See* BASANITE.

nepheline-monzonite An undersaturated *monzonite (*see* SILICA SATURATION) containing essential *nepheline (*see* ESSENTIAL MINERAL) in addition to the monzonite mineral assemblage.

nepheline-syenite A medium- to coarse-grained *igneous rock consisting of essential *alkali feldspar (*see* ESSENTIAL MINERAL), *nepheline, *pyroxene (*aegirine-*augite or sodic *hedenbergite) and *amphibole (ferrohastingsite, $NaCa_2(Fe^{2+})_4AlAl_2Di_6O_{22}(OH)_2$, which is essentially a sodic, iron-rich *hornblende, *barkevikite and/or *arfvedsonite) with accessory *sphene, *apatite, *titanomagnetite, *ilmenite, and *zircon (*see* ACCESSORY MINERAL). The rock is, in effect, a *syenite which is undersaturated in silica (*see* SILICA SATURATION).

nephelinite A fine-grained, often *porphyritic, ultra-alkaline *extrusive *igneous rock consisting of essential *titanaugite (*see* ESSENTIAL MINERAL), *nepheline, and *titanomagnetite, with accessory *apatite, *sphene, and *perovskite (*see* ACCESSORY MINERAL). Where present, *phenocrysts usually consist of the titanaugite, and occasionally nepheline. *Sodalite and *analcite may also be found in the *groundmass. Where *olivine is an additional component, the rock is termed an 'olivine nephelinite'. Nephelinites are essentially super-undersaturated *basalts (*see* SILICA SATURATION) with nepheline taking the place of all the *plagioclase feldspar.

nepheloid layer A body of water with a high concentration of suspended *sediment which occurs near the deep ocean bottom, close to the base of the *continental slope. Nepheloid layers are usually 1–300 m thick, and carry sediment up to 12 μm in size, in concentrations of 0.3–0.01 mg/l. The sediment is suspended and transported by the movement of the oceanic *thermohaline bottom currents. *See* CONTOUR CURRENT.

nephelometer (NEP) An instrument used in *remote sensing that uses a transmitted *laser beam to determine the scattering of light by atmospheric particles.

nephrite *See* JADEITE.

Neptune Ordinarily the eighth planet of the *solar system, but the ninth when *Pluto's eccentric orbit carries it inside the orbit of Neptune. Neptune is at 30.06 AU from the *Sun and between 4305.6×10^6 km and 4687.3×10^6 km from Earth. Its equatorial radius is 24 766 km; polar radius 24 342 km; volume 6254×10^{10} km^3; mass 102.43×10^{24} kg; mean density 1638 kg/m^3; gravity 11 (Earth = 1); visual albedo 0.41; black-body temperature 33.2 K. Neptune has a dense atmosphere, with a surface pressure greater than 100 bars, composed of molecular hydrogen (89%), helium (11%), and traces of methane, with *aerosols of ammonia ice, water ice, ammonia hydrosulphide, and possibly methane ice (similar to that of Uranus). The average surface temperature is about 58 K, and winds blow at up to 200 m/s. Neptune has eight known satellites (*see* NEPTUNIAN SATELLITES).

neptunian dyke (sandstone dyke) A sheet-like body of *sand cutting through bedded *sediment in a manner analogous to an *igneous *dyke. Neptunian dykes form by upward injection of liquefied sand through a fissure, often as a result of *earthquake activity.

neptunian satellites *See* DESPINA (NEPTUNE V); GALATEA (NEPTUNE VI); LARISSA (NEPTUNE VII); NAIAD (NEPTUNE III); NEREID (NEPTUNE II); PROTEUS (NEPTUNE VIII); THALASSA (NEPTUNE IV); and TRITON (NEPTUNE I).

neptunism A theory of the formation of the Earth, popular at the end of the 18th and the beginning of the 19th century, and associated with *Werner. Werner taught that the oldest, primitive rocks, including *granite and *basalt, had crystallized from a primordial ocean. They were followed by Transition and Floetz formations, also precipitated, and then by Alluvia and Volcanic rocks.

Nereid (Neptune II) A satellite of *Neptune, with a diameter of 340 km; visual albedo 0.2.

Nereites A bilobed *trail that winds over a *bedding plane. The trail has a transverse ornament and was formed by a *deposit feeder during its methodical search for food. The term '*Nereites* facies' is given to deep-water deposits characterized by *Nereites* and depth-related genera.

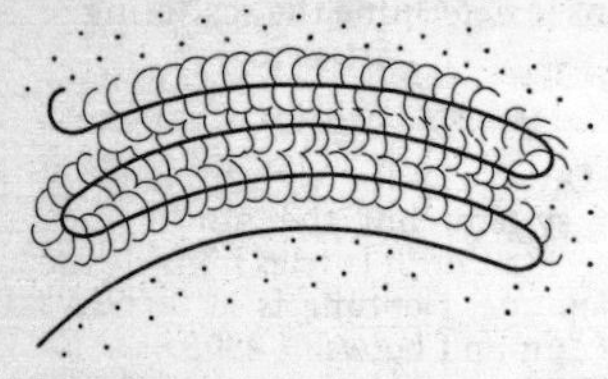

Nereites

Nereus A *solar system asteroid (No. 4660), diameter 2 km; orbital period 1.82 years. It is a near-Earth asteroid, scheduled to be visited by a sample-return mission.

neritic province *See* NERITIC ZONE.

neritic zone (neritic province) The shallow-water, or near-shore, marine zone extending from the low-tide level to a depth of 200 m. This zone covers about 8% of the total ocean floor. It is the area most populated by *benthic organisms, due to the penetration of sunlight to these shallow depths.

nesosilicates (orthosilicates) A group of *silicate minerals characterized by independent SiO_4 *tetrahedra with no shared oxygens. The structure is held together by bonding with interstitial cations. The group includes *olivine, *garnet, *sphene, *zircon, *staurolite, *chloritoid, *topaz, *chondrodite, and the Al_2SiO_5 *polymorphs.

net slip *See* SLIP.

network-former Aluminium (Al), silicon (Si), and often phosphorus (P) atoms, which tend to form network structures in *silicate *melts.

network-modifier Large *cations (e.g. Ca^{2+}, Mg^{2+}, Fe^{2+}, Na^{+}, and K^{+}) occurring as major constituents in *magmas, which usually disrupt and modify the network structures.

neural arch *See* VERTEBRA.

neural spine *See* VERTEBRA.

neutral fold A *fold which closes laterally and is therefore neither *antiformal nor *synformal. Where the *fold axis and *axial plane are inclined vertically, a neutral fold is known as a 'vertical fold'.

neutrally buoyant float (Swallow buoy) A device developed by J. C. Swallow in 1955 for measuring current speed and direction at depth in the oceans. It consists of two aluminium tubes, one containing batteries and sound-generating circuitry, the other containing adjustable weights to allow the device to float and remain at any desired depth. It emits a series of sound pulses which can be tracked by ship, allowing water motions to be measured at selected depths.

neutral soil Soil with a *pH value of 6.6–7.3.

neutron-activation analysis The analysis of a material by bombarding a sample with fast-moving neutrons in the core of a nuclear reactor. Neutrons are added to the nuclei of stable *isotopes to form new radionuclides which decay, producing particles with characteristic energies that can be measured with a *scintillation counter.

neutron–gamma sonde An instrument package for measuring the amount of hydrogen present within the rocks surrounding a *borehole. The *sonde contains a neutron source and a gamma-ray detector. The gamma rays induced in the rocks by neutron bombardment are proportional to the amount of hydrogen present. The device can be used to determine the *porosity of the surrounding rocks.

neutron log The recording of neutron bombardment by a *neutron–neutron sonde being raised through a *borehole. The neutron log is used to estimate formation *porosity and is presented on a scale calibrated in limestone porosity units.

neutron moisture meter (neutron soil-moisture probe) An instrument which uses high-energy (fast) neutrons to detect indirectly the water content of a soil. The fast neutrons are moderated to slow neutrons by the hydrogen atoms in the soil moisture. By measuring the density of the slow-neutron backscatter, an indication is obtained of how much water is present.

neutron–neutron sonde A radioactive instrumental package for lowering down a *borehole. It comprises a neutron source and detector. The number of neutrons backscattered to the detector is proportional to the number of hydrogen atoms within the rocks surrounding the borehole. It can be used to determine the *porosity of the surrounding rocks. *See also* NEUTRON–GAMMA SONDE; and NEUTRON LOG.

neutron soil-moisture probe *See* NEUTRON MOISTURE METER.

névé *See* FIRN.

Newer Drift The deposits marking the maximum extent of the last (*Devensian) glaciation. A morphological distinction occurs between the more-weathered *drift to the south and the less-weathered drift to the north, in the British Isles. The boundaries between the less-weathered Newer Drift and the *Older Drift have been shown to be less distinct than was once supposed.

New Millennium Deep Space-1 A *NASA mission, to be launched in 1998, that will explore the asteroid *McAuliffe and comet *West–Kohoutek–Ikemura.

New Millennium Deep Space-2 A *NASA mission to *Mars, to be launched in 1999.

New Red Sandstone Name given to the terrestrial *red-bed deposits that overlie the *Carboniferous. These deposits are of Permo-Triassic age.

newton (N) The derived SI unit of force, named after Sir Isaac Newton (1642–1727), being the force required to produce an acceleration of 1 m/s^2 in a mass of 1 kg; 1 N = 1 J/m.

Newtonian behaviour The manner in which a fluid flows if it exhibits ideal viscous *strain. The application of a deforming stress induces permanent strain from which there is no recovery when the stress is removed. *Compare* ELASTIC DEFORMATION.

Newtonian fluid *See* BINGHAM FLUID.

Newton's law of gravitation *See* GRAVITATIONAL ACCELERATION.

NEXRAD *See* NEXT GENERATION WEATHER RADAR.

Next Generation Weather Radar (NEXRAD) A network of 175 *Doppler radar installations, completed in 1996, that cover the whole of the USA. They are located at weather stations, airports, and military bases. Each is mounted on a tower and provides three-dimensional images of weather conditions with clear resolution to 200 km and poorer resolution to 320 km.

Ngaterian *See* CLARENCE.

Niagaran A *series of the *Silurian of N. America, equivalent to the Upper *Llandovery to Lower *Ludlow period.

niccolite Metallic mineral, NiAs; sp. gr. 7.8; *hardness 5.0; greyish-red; usually *massive or disseminated; occurs as irregular *aggregates and as complex intergrowths with other *sulphides, associated with *basic *igneous rocks.

Nicol, William (1768–1851) A mineralogist from Edinburgh, Nicol invented the calcite prism named after him and used in polarizing microscopes (*see* NICOL PRISM). His description of a microscope based on his prisms was published in 1829, although the instrument was in use much earlier.

nicol prism Two pieces of optically clear *calcite, cemented by *Canada balsam into the shape of a prism. Light entering the base of the prism is doubly refracted, and when both rays reach the Canada-balsam cement one ray is reflected away from the prism while the other ray continues through the prism. Thus the light emerging from the prism is *plane polarized. These prisms were invented by William *Nicol. Early *polarizing microscopes were fitted with nicol prisms for the *analyser and *polarizer, but modern microscopes are fitted with *Polaroid instead, and, although the term 'crossed nicols' is still in common use, this has been largely replaced by the term '*crossed polars'.

Niggli, Paul (1888–1953) A Swiss crystallographer and petrologist at the University of Zurich, Niggli studied *magmas and *metamorphic rocks, and introduced the use of *phase diagrams in *petrology. He also made a classification of *sedimentary rocks according to grain shape and, in 1928, he and his students attempted to calculate the bulk chemical composition of the Earth. *See* NIGGLI METHOD.

Niggli method The grouping of the oxide components of an *igneous rock according to their chemical similarity: (Al_2O_3 + Cr_2O_3 + *REEs); (FeO + MnO + MgO); (Na_2O + K_2O + Li_2O); and (CaO + BaO + SrO); The oxide groupings are calculated according to *cation percentage, the sum for the rock being normalized to 100. Each group corresponds to an individual Niggli value, each of which can be plotted against silica for a comparison of associated rocks. The method was devised by P. *Niggli.

nimbostratus From the Latin *nimbus*, 'rain', and *stratus*, 'spread out', a dark or grey cloud that obscures the Sun, associated with more or less continuous rainfall which makes the cloud base diffuse. *See also* CLOUD CLASSIFICATION.

NIMS *See* NEAR-INFRARED MAPPING SPECTROMETER.

niobite *See* COLUMBITE.

nitratine *See* SODA NITRE.

nitre (saltpetre) Mineral, KNO_3; sp. gr. 1.9–2.3; *hardness 2; *orthorhombic; white, grey, reddish-brown, or lemon-yellow; *vitreous *lustre; crystals *acicular, usually forms granular crusts and uneven masses; occurs in very arid conditions where vegetation is very sparse, in association with *evaporite minerals of the arid desert type (e.g. *soda nitre, *gypsum, *halite, and occasionally iodates), and little precipitation is required to wash an encrustation of nitre into hollows where it forms saltierras or *massive deposits. It is used as a fertilizer.

nival Applied to the geomorphological (*see* GEOMORPHOLOGY) processes that result from the action of snow.

nivation Complex of surface erosional processes acting under a snow cover. It includes *gelifraction, and the removal of shattered debris by *solifluction and the movement of melted snow. It is an initial process in *cirque development.

NMO *See* MOVEOUT (2).

Noachian A division of *areological time, lasting from 4.60 Gy to 3.50 Gy in the Hartmann–Tanaka Model and 4.60 Gy to 3.80 Gy in the Neukum–Wise Model, and divided into three epochs: Lower Noachian (4.60–3.92 or 4.60–4.50 Gy); Middle Noachian (3.92–3.85 or 4.50–4.30 Gy); and Upper Noachian (3.85–3.50 or 4.30–3.80 Gy).

noctilucent clouds (luminous night clouds) Type of clouds, occurring at 80–85 km altitude, near the upper limit of the *stratosphere, characterized by a blue to yellow hue, and similar in appearance to *cirrostratus. The clouds are seen on summer nights at latitudes between about 50° and 65° in both northern and southern hemispheres; they move rapidly, at speeds up to 300 knots, often in a wave formation.

nocturnal radiation The long-wave radiation from the surface of the Earth at night that is in excess of the incoming radiation from the atmosphere. *See also* ATMOSPHERIC WINDOW; RADIATION BUDGET; and TERRESTRIAL RADIATION.

node 1. A point of zero displacement in a material transmitting *standing waves, produced by destructive *interference between waves propagating in opposite directions. **2.** In a *phylogenetic tree, a representation of an extant (terminal node) or ancestral (internal node) operational taxonomic unit.

nodes The points, diametrically opposite, at which the orbit of a planet intersects the plane of the *ecliptic, or at which the orbit of a *satellite intersects the orbital plane of a planet.

nodular (concretionary) Applied to the *habit of a mineral when its outer surface is rounded or spherical in shape. It may be the result of progressive chemical precipitation, giving a concentric appearance. Minerals such as iron or *manganese nodules form in this way.

nodule A spherical or oval concretion.

Noginskian A *stage in the *Gzelian Epoch, preceded by the *Klazminskian and followed by the *Asselian (*Rotliegendes Epoch of the *Permian).

Noguerornis The earliest known bird after **Archaeopteryx*, from the basal *Cretaceous of Montsec, northern Spain. It was the size of a finch and was the first bird to have a wing that was well developed, as indicated by the elongation of the distal portions of the forelimb and the rigid interlocking of the hand bones.

noise A signal that conveys no useful information (e.g. background sound that makes it difficult to hear a conversation). If the useful signal comprises data that are being recorded, random (white) noise can be reduced by summing the recorded signals; incoherent noise is effectively damped out and the coherent signal is enhanced, thus improving the signal-to-noise ratio.

non-conformity (heterolithic unconformity) An *unconformity in which younger, *sedimentary rocks rest unconformably on older, *igneous or *metamorphic rocks.

non-dipole field The field remaining after removal of the *dipole field (usually the inclined dipole) from the observed *geomagnetic or planetary magnetic-dipole field.

non-frontal depression Low-pressure system that does not develop from a frontal wave as do typical mid-latitude frontal cyclones (or *depressions). Most tropical cyclones are non-frontal. Various conditions can lead to the formation of depressions without frontal characteristics. *See also* COLD LOW; LEE DEPRESSION; POLAR DEPRESSION; and THERMAL LOW.

non-metallic Applied to the *lustre of a *mineral which reflects light from its surface but does not shine like a metal. 'Non-metallic' may be further qualified as 'glassy' or 'vitreous', 'silky', 'resinous', etc.

non-polarizable electrode An electrode whose potential is not affected by the current passing through it, e.g. the porous-pot electrode extensively used in *spontaneous potential measurements which comprises a copper rod immersed in copper sulphate solution that makes ionic contact with the ground through the porous base of the electrolyte reservoir.

non-renewable resource (finite resource) Resource that is concentrated or formed at a rate very much slower than its rate of consumption and thus, for all practical purposes, is non-renewable. *Compare* RENEWABLE RESOURCE.

non-rotational strain *See* PURE SHEAR.

non-selective scattering The *scattering of all wavelengths of *electromagnetic radiation equally in the atmosphere, usually caused by particles which are much larger than the energy wavelengths.

non-sequence A minor break in a *concordant succession of *strata, representing a period during which either no deposition of *sediment occurred or it was subsequently removed. Such periods may be of short duration and localized, but are not necessarily so. *Compare* DIASTEM.

non-steady flow *See* STEADY FLOW.

non-strophic *See* HINGE.

non-spectral hue A *hue which is not present in the spectrum of colours produced by splitting white light with a prism. Non-spectral hues include brown and the pastel colours. *Compare* SPECTRAL HUE.

non-umbilicate *See* IMPERFORATE.

non-uniform flow *See* UNIFORM FLOW.

norbergite *See* CHONDRODITE.

Norian (Juvavic) 1. A Late *Triassic *age, preceded by the *Carnian and followed by the *Rhaetian, and dated at 223.4–209.5 Ma (Harland et al., 1989). **2.** The name of the corresponding European *stage, which is roughly contemporaneous with the Houbachong (China) and Warepan (New Zealand).

norite A coarse-grained, *basic *igneous rock consisting of essential *plagioclase feldspar, *orthopyroxene (*hypersthene or *bronzite), and *clinopyroxene (*augite), with accessory *ilmenite (*see* ACCESSORY MINERAL; and ESSENTIAL MINERAL). Orthopyroxene is dominant over clinopyroxene, and the plagioclase is a calcic type (labradorite or bytownite). Norites, like *gabbros, are found as layers in many large, layered basic intrusions (*see* INTRUSIVE), as well as forming intrusions in their own right.

norm *See* CIPW NORM CALCULATION.

normal distribution (Gaussian distribution) In statistics, a continuous *probability distribution which is asymptotic and symmetrically bell-shaped about the *mean. The normal distribution is widely applied in statistics to model continuous variation. The two parameters of the distribution are the mean and *variance. *See also* CENTRAL LIMIT THEOREM.

normal fault A high-angle (more than 50°), *dip-slip fault on which displacement of the hanging wall is downwards, relative to the *footwall. Normal faults commonly occur in *conjugate fault sets which cause discrete block subsidence and uplift in the form of *grabens and *horsts.

normal field A magnetic field, usually geomagnetic, that has the same *polarity as the present field, i.e. the north magnetic pole either lies in the northern hemisphere or is clearly a continuation of the north-pole *polar wander path.

normal incidence The condition in which a wave-front is parallel to an interface, such that the ray path is perpendicular (normal) to the surface. The *angle of incidence is zero. *See also* SNELL'S LAW.

normalized vegetation index *See* VEGETATION INDEX.

normally consolidated clay Clay that is compacted by exactly the amount to be expected from the pressure exerted by the *overburden; clay which has never been overloaded. *Compare* OVER-CONSOLIDATED CLAY.

normal moveout *See* MOVEOUT.

normal problem *See* FORWARD PROBLEM.

normal stress (σ) The *stress which acts perpendicularly to the plane to which a force has been applied. Normal *compressive stresses (with positive values) inhibit sliding along a plane; normal tensile stresses (with negative values) separate rocks along a plane.

normal travel time The *time–distance graph compiled from *first arrivals, but only where there is no anomalous geologic structure. It is used to calibrate travel times for a given *offset range in *fan-shooting refraction surveys.

normal twins In *crystallography, a twinned *crystal where the twin *axis of rotation is at right angles to the *crystal face.

normal zoning *See* CRYSTAL ZONING.

normative constituents *See* CIPW NORM CALCULATION.

norte (norther) Cold, northerly, local wind affecting the coasts of the Gulf of Mexico, most commonly in winter. Sometimes it brings rainfall.

North American Plate One of the present-day major lithospheric *plates, the N. American Plate extends from the *Mid-Atlantic Ridge in the east to a complex pattern of *subduction zones and *transform faults which form its boundaries with the *Pacific, *Juan de Fuca, *Gorda, *Cocos, *Caribbean, and *South American Plates in the west and south. Along the western side, the N. American Cordillera has been interpreted as a collage of accreted *allochthonous *terranes.

North Atlantic deep water (NADW) A water mass (salinity 34.9–35.03 parts per thousand, temperature 1.0–2.5 °C) that was originally believed to form in an area off the southern tip of Greenland, where winter cooling of saline waters was thought to cause a body of water to sink and spread south. It is now recognized that the main source is in the Norwegian Sea, from which deep water flows over the sills between Scotland, Iceland, and Greenland and cascades into the depths of the Atlantic.

North Atlantic Drift Oceanic surface current in the N. Atlantic, flowing from the Grand Banks off Newfoundland eastwards to north-western Europe and forming a northerly extension of the *Gulf Stream. This diffuse, shallow and relatively warm current has an ameliorating influence on the climate of the coastal regions of north-western Europe.

norther *See* NORTE.

'northern lights' *See* AURORA.

North Pacific current Oceanic surface current in the N. *Pacific that flows eastwards, as an extension of the *Kuroshio current, towards California. It occupies a position in the N. Pacific similar to that in the *Atlantic of the system comprising the *Gulf Stream and *North Atlantic Drift.

nor'wester Convective storm, usually from the north-west, which affects Assam and Bengal between March and May. It is characterized by violent conditions, including a *line squall.

nosean A *feldspathoid mineral and member of the *sodalite group $Na_8[Al_6Si_6O_{24}]SO_4$, where it is closely related to *haüyne; sp. gr. 2.3–2.4; *hardness 5.5–6.0; *cubic; grey, white, or greenish, often with light-blue tinges; *vitreous *lustre; *crystals, when present, *rhombododecahedra, but it is also *massive; *cleavage poor, rhombododecahedral; occurs in extrusive *alkaline undersaturated (*see* SILICA SATURATION) *igneous rocks such as nosean leucitophyres and *phonolites.

Nothosauria (nothosaurs; order *Sauropterygia) Suborder of marine *reptiles that flourished during the *Triassic. They possessed long necks, and their limbs were adapted to swimming. The nothosaurs were replaced by the *plesiosaurs in the Early *Jurassic.

nothosaurs *See* NOTHOSAURIA.

notochord (chorda dorsalis) Somewhat flexible, rod-like structure, composed of disc-like, turgid cells, which extends virtually the entire length of the body of adult and/or larval members of the phylum *Chordata. Lying below the nerve cord, but *dorsal to the intestine, the notochord provides a form of flexible support to the body. In *vertebrates the notochord is replaced wholly or partly by the vertebral column, but it is retained throughout life in *Cephalochordata and *Agnatha.

notothyrium A triangular opening in the posterior margin of the brachial valve of a brachiopod (*Brachiopoda) shell, through which the *pedicle may partly emerge. In some species it is open, in others it is closed by a single plate of shell material, the chilidium, or by two chilidial plates.

nova A new star.

novaculite A dense, fine-grained to *cryptocrystalline *chert.

NRM *See* NATURAL REMANENT MAGNETISM.

nuclear–magnetic log The combined outputs of radioactive and magnetic *sondes used in *well-logging.

nuclear-precession magnetometer A *magnetometer based on the ability of protons to precess about the *geomagnetic field with a frequency dependent on the strength of the ambient magnetic field. A strong field is applied which aligns the protons, and they then precess about the geomagnetic field after the applied field is removed. The instrument has a sensitivity of the order of 1 nT, and is capable of rapid repetition to produce an almost continuous signal.

nuclear waste *See* RADIOACTIVE WASTE.

nucleation 1. Formation of an embryonic *crystallite from a *melt which is followed by the growth of a nucleus to crystal dimensions. *See also* MICROLITE. **2.** Theory according to which *geosynclines developed on the edges of *cratons; the ensuing *orogenic belt then became part of the craton and the products of subsequent erosion filled a new geosynclinal trough which developed on the edge of the enlarged craton. The theory of *plate tectonics, with the emphasis on horizontal movement, modified nucleation in many ways, not least by including oceanic sediment, scraped off the subducting plate (*see* SUBDUCTION), into the new craton. *See* ACCRETION.

nucleic acids Nucleotide polymers, with high relative molecular mass, produced by living cells and found both in the *nucleus and cytoplasm. They occur in two forms, designated DNA and RNA, and may be double- or single-stranded. DNA embodies the genetic code of a cell or organelle, while various forms of RNA function in the transcriptional and translational aspects of protein synthesis.

nucleosynthesis The process by which elements are formed. Modern theories suggest that nucleosynthesis is intimately linked with the stages in the life-cycle of stars (stellar evolution), and that, commencing with hydrogen, heavier elements are created by nuclear fusion of lighter *nuclides at the temperatures and pressures existing in the cores of stars. Because the lighter elements are consumed to produce energy these *thermonuclear reactions are referred to as 'burning', although they have nothing to do with combustion. The stages of stellar evolution conform well with the overall pattern of peaks and troughs in the *cosmic abundance of elements in order of increasing atomic number (Z). During the first and longest (main-sequence) phase, hydrogen (which is by far the most abundant element of the stellar material) is consumed to produce helium (*hydrogen burning). Hydrogen burning is followed in turn by *helium burning, *carbon and *oxygen burning, and *silicon burning, each phase producing heavier elements from lighter ones. The heaviest elements are formed in the last stages in the sequence: the *equilibrium (e) process (coinciding with the 'iron peak' elements, Cr, Mn, Fe, Co, and Ni), followed by the '*slow neutron (s) process producing elements up to Bi (atomic number 83), and finally (in supernova events) the *rapid neutron (r) process producing elements with atomic number greater than 83. *See also* PROTOSTAR.

nucleus (pl. nuclei) 1. The centre of an atom, composed of protons and neutrons and accounting for nearly all of its mass. A proton has a positive electrical charge, equal in magnitude to the negative charge of an *electron; a neutron carries no electrical charge. The nucleus of the hydrogen atom contains a single proton; uranium, the heaviest naturally occurring element, has 92 protons and 142, 143, and 146 neutrons in *isotopes 234, 235, and 238 respectively. **2.** A small, solid particle, e.g. of dust, salt, or smoke on to which water vapour will condense. Such particles are called 'condensation nuclei' and some of them have hygroscopic properties that encourage condensation in unsaturated air. Other nuclei of a suitable shape, e.g. some clay particles such as *kaolinite, probably act as 'freezing nuclei' in the initial stage of ice-crystal formation. *See also* AITKEN NUCLEUS; BERGERON–FINDEISEN THEORY; CONDENSATION NUCLEUS; and ICE NUCLEUS. **3.** The double-membrane-bound organelle containing the *chromosomes, that is found in most non-dividing *eukaryotic cells; it is essential to their long-term survival. It is variously shaped, although it is normally spherical or ovoid. It disappears temporarily during cell division. It is absent from viruses. The *chromosomes, though probably intact, are not visible when the cell is in a resting stage (i.e. not dividing). The nucleus also contains nucleoli, small spherical dense bodies made up of ribosomal RNA and protein, which gives it its integrity. **4.** *See* NUCLEATION.

nucleus number The number of *crystal nucleii formed in a unit volume of *magma in unit time, during the cooling of the magma body.

nuclide A widely used alternative name for an atom. The composition of any nuclide can be given by means of the chemical symbol of the element, the *mass number written as a superscript, and the *atomic number written as a subscript. For example, ${}^{14}_{6}C$ is an atom of carbon (C) having six protons (the atomic number) and 14 nucleons (the mass number).

nuée ardente A French phrase, meaning 'glowing cloud', first used in 1903 by Alfred Lacroix (1863–1948) to describe the complete phenomenon of a basal *pyroclastic flow and its overriding *ash cloud formed by the collapse of a volcanic *dome. Since the pyroclastic flow and ash cloud behave differently from one another during transport and produce contrasting deposits, volcanologists now tend to avoid using the term.

Nukumaruan The basal *series in the *Quaternary of New Zealand, underlain by the *Pliocene, overlain by the *Castlecliffian Series, and roughly contemporaneous with most of the Calabrian *Stage. Some authors consider the Nukumaruan to overlie the *Mangapanian Stage within the uppermost Pliocene and thereby to include the *Tertiary/Quaternary boundary.

Nullaginian A *stage of the Lower *Proterozoic of south-eastern Australia, overlain by the *Carpentarian.

null hypothesis *See* HYPOTHESIS.

null point 1. In nearshore, shoaling waters, each particle-size of *sediment should hypothetically have a null point, where there is no net movement of the particle landward or seaward. It is the point where there is a balance between the component of gravity acting down the slope in a seaward direction and the force on the particle resulting from the difference between the crest and trough velocities of the waves which tends to move the particle landwards. **2.** In an *estuary, the point at which the residual sea water landward flow is balanced by the seaward residual river flow.

numerical aperture *See* OBJECTIVE.

numerical taxonomy The classification of organisms by purely mathematical means. It is based on quantifying observable characteristics of organisms and may be operated at various taxonomic levels to deal with species or higher taxa. It involves the grouping and computation of the similarity of characters; the results are usually displayed graphically, as a *phenogram or dendrogram. There has been discussion as to whether classifications produced in this way are valid and reflect *phylogeny. *See* TAXON; and TAXONOMY.

Nummulites A genus of larger *Foraminiferida, which have a flattened, discoidal or lenticular *test. They lived in warm, shallow, marine waters from the top of the *Palaeocene to the Upper *Oligocene, and in some areas they are numerous enough to be major rock formers. Many species have been described; they occur at different stratigraphic levels and in different *facies associations. *Nummulites planulatus*, which is Cuisian in age, is often found in association with alveolinids and miliolids in a sandy facies; *N. globulus* occurs in the Ilerdian of the Pyrenees in marly and sandy facies; *N. gizahensis* is one of the largest forms and is usually found in sub-reefal *limestones of the Upper *Lutetian of the *Tethyan realm; *N. variolarius* preferred shallow, calm environments and is found with miliolids in sandy limestones of the Middle and Upper Lutetian.

nunatak Rocky summit or mountain range that stands above a surrounding *ice sheet in an area that currently is being glaciated.

Nunivak A normal *polarity subchron which occurs within the *Gilbert reversed *chron.

Nusselt number A dimensionless number relating to the *Rayleigh number. It corresponds to 1 for initial convective motion and increases with an increased probability of convective motions occurring.

nutation Irregularities in the orbital motion of the Earth (or other planets) superimposed on the *precession of its axis. The dominant nutation of the Earth's axis has an 18.6-year periodicity.

nutrient cycle *See* BIOGEOCHEMICAL CYCLE.

Nyquist frequency (folding frequency, f_N) A frequency which is half of the *sampling frequency. The Nyquist interval is the frequency range from zero to f_N, where $f_N = 1(2\Delta t)$ and Δt is the *sampling interval. For example, if the sampling interval Δt is 2 ms (i.e. a sampling frequency of 500 Hz), then $f_N = 1(2 \times 0.002) = 250$ Hz. It is important to know the value of the Nyquist frequency, as frequencies greater than f_N will alias (or 'fold back') and appear as lower frequencies from which they cannot be distinguished.

oasis A fertile spot in the desert, the basis of which is a supply of water that is available throughout the year and which normally originates as *groundwater.

obduction The lateral, sub-horizontal displacement of a *lithospheric plate on to a *continental margin at a *destructive plate boundary. It is the opposite of *subduction.

Obdurodon **(order *Monotremata, family Ornithorhynchidae)** A genus of early *Miocene platypus, with two species, known from very well-preserved material from such sites as Riversleigh, Queensland, Australia. They possessed well-developed, functional teeth, unlike the living platypus, and had skull bones that were less fused, from which it can be seen that monotremes (Monotremata), unlike other living mammals, retained *septomaxillae.

Oberon (Uranus IV) One of the major satellites of *Uranus. Its radius is 761.4 km; mass 30.14×10^{20} kg; mean density 1630 kg/m^3; albedo 0.24. The surface is extensively cratered, the craters being surrounded by bright ejecta. Near the centre of Oberon there is a large crater with a bright central outcrop and background of dark material.

Obik Sea *See* URAL SEA.

objective The magnifying lens which is situated at the base of a microscope tube immediately above the specimen. Objectives may be described in terms of lens quality (e.g. achromat, apochromat, or fluorite), magnification (which may be 5×, 10×, 20×, or larger), and numerical aperture (NA) which is a measure of the amount of light converging on the specimen through the lens. Its value is normally between 0.1 and 0.9.

oblate Description of the shape of a *clast which is tabular or disc-shaped in appearance. It is characterized by a ratio of intermediate to long diameters of more than 2/3, and a ratio of short to intermediate diameters of less than 2/3. *See* PARTICLE SHAPE.

oblate uniaxial strain The *strain state obtained when a reference sphere is shortened along its *z*-axis and extended by an equal amount along its *x*- and *y*-axes and all intermediate axes lying in their plane and perpendicular to the *z*-axis. Such strain alters the sphere into an oblate ellipsoid.

oblique extinction (inclined extinction) In optical mineralogy, the *extinction of a mineral (i.e. it becomes dark) in *crossed polars with its cleavage traces or *crystal boundaries positioned at an oblique angle to the E–W and N–S planes of vibration of the *polarizer and *analyser. Several different angles of extinction may be determined for the same mineral, and the maximum angle may be used for identification purposes.

oblique-slip fault A *fault in which the displacements of the *strike-slip and *dip-slip components have very similar magnitudes; fault movement occurs obliquely across the fault surface.

obliquity of the ecliptic *See* MILANKOVICH CYCLES.

obrusion *See* OBRUTION.

obrution (obrusion) Sudden burial.

obrution deposit A fossil assemblage that has been preserved by the very rapid burial of intact organisms.

obsequent Applied to a land-form whose orientation is opposite to that which may have been expected. For example, an obsequent stream flows against the *dip of underlying strata; and an obsequent *fault-line scarp faces the opposite way to the original fault scarp.

observation well A well that is used to observe changes in *groundwater levels over a period, or more specifically during a pumping test. Pumping does not normally take place from observation wells which are often relatively small in diameter.

obsidian Volcanic *glass of dacitic (*see* DACITE) or rhyolitic (*see* RHYOLITE) composition.

occipital Pertaining to the posterior part of the *cranium.

occipital condyle At the back of the skull, a bony knob which articulates with the first *vertebra. It is absent in fish, and double in *amphibians and *mammals.

occluded front The composite front formed when a *warm front and the sector of warm air behind it is lifted and overtaken by a *cold front. *See also* OCCLUSION.

occlusion Stage in the development of a frontal *depression during which the warm-sector air is gradually lifted from the ground and above the colder air. The word is used to describe an *occluded front. An occlusion may be a 'warm occlusion' such that warmer air follows the frontal system, or a 'cold occlusion', such that colder air follows in the rear. A 'warm occlusion' has characteristics similar to those of a warm front, but there is an upper cold front ahead and aloft which brings showers from *cumulus and *cumulonimbus cloud. A 'cold occlusion' exhibits cold-front features, but is preceded by warm-front cloud types. Occlusions formed in the later stages in the development of travelling depressions commonly cross north-western Europe, especially the 'warm' type in winter and the 'cold' type in summer.

ocean Salt water mass that occupies more than two-thirds of the surface of the Earth (70.8%). The oceans contain $1370 \times 10^6 km^3$ of water; the average depth is 3730 m.

ocean-basin floor The ocean floor in those parts of the oceans that are more than 2000 m deep. It occupies approximately one-third of the *Atlantic and *Indian Ocean floors, and three-quarters of the *Pacific Ocean floor.

ocean current Large-scale water movement in an ocean, arising from three main causes: (*a*) wind stresses acting on the surface of the sea; (*b*) tidal motion caused by the variable attractions of the Sun and Moon; and (*c*) density differences in sea water, caused by differential heating and cooling, *salinity differences, or variations in the suspended-sediment concentration of sea-water masses.

ocean gyre *See* GYRE.

oceanic crust The oceanic rocks which form 65% of the Earth's surface, and are the upper part of the oceanic *lithosphere, overlying the *Mohorovičić seismic discontinuity. The oceanic crust comprises four seismic layers, commencing at an average depth of 4.5 km below sea level. The uppermost layer (*sediments) varies in thickness from being absent over the *oceanic ridges to 2–3 km near the *continental shelves. The other three layers are of remarkably constant thickness and *seismic velocity. Layer 2 is formed by *basaltic lavas with *P-wave velocities of 5 km/s and a *dyke complex just under 2 km thick, overlying a 5 km thick gabbroic layer (layer 3) with a P-wave velocity of 6.7 km/s. Layer 4 is a thin (less than 0.5 km) layer with a P-wave velocity of 7.4 km/s, immediately overlying the *mantle where the P-wave velocity is 8.1 km/s. The total thickness is about 11 km and shows little variation throughout the ocean basins, despite the change in ocean *bathymetry from the ridges to the *trenches. The crust is cut by *fracture zones. It is largely aseismic away from *spreading ridges and *subduction zones. The presence of linear *magnetic anomalies, mostly paralleling the present mid-oceanic ridges, allows the dating of the ocean floors, as does the dating of sediments immediately overlying the *igneous part of the crust and obtained by drilling during the *DSDP and *IPOD projects. The oldest oceanic crust, within the present ocean basins, is less than 200 million years old; it is found in the W. Pacific and the N. Atlantic, and was formed by igneous activity at spreading (accretionary) *plate margins as part of the *sea-floor spreading process. Layer 3 represents the cooled *magma chamber that originally fed the overlying basaltic dykes and lavas.

oceanicity The effects of maritime influences on a climate. *See also* CONTINENTALITY.

oceanic plateau Extensive, topographically high area of an ocean floor that rises to within 2–3 km of the sea surface above the abyssal floor, e.g. the plateau on which Iceland stands, the Galápagos Islands platform, and the Azores platform. Many Pacific plateaux, e.g. the Magellan Rise and Ontong Java Plateau, have a thick covering of carbonate oozes overlying volcanic rocks. The origin of such plateaux is volcanic, but many are now inactive.

oceanic trench *See* TRENCH.

ocean-island basalt (OIB) Quartz *tholeiites, *alkali basalts, and *nephelinites found on *volcanoes which build up from the ocean floor to form the ocean islands away from ocean *ridges. Examples include the Cape Verde Islands, Ascension Island, Tristan da Cunha Island, and Gough Island in the Atlantic, and the *Hawaiian-Emperor *seamount chain in the Pacific. Compared to MORBs (*see* MID-OCEAN-RIDGE BASALT), OIBs are enriched in (*a*) *large-ion lithophile (LIL) elements, (*b*) light *rare-earth elements (LREE) relative to heavy rare-earth elements (HREE), and (*c*) *incompatible elements such as Ti, Ga, Li, Nb, V, Zn, Zr, and Y. Radiogenic-*isotope and *trace-element evidence suggests that OIBs are formed by *partial melting of enriched *mantle.

ocean wave Disturbance of the ocean's surface, seen as an alternate rise and fall of the surface. Ocean waves are of several types: (*a*) wind-generated waves, e.g. sea waves (chaotic wave pattern) and swell (long-period waves); (*b*) catastrophic waves, e.g. *tsunamis, landslide surges, and storm surges; and (*c*) internal waves (subsurface waves at the boundary between two water layers).

Ochoan The final *series in the *Permian of N. America, underlain by the *Guadalupian, overlain by the *Triassic (*see* SCYTHIAN), and roughly contemporaneous with the upper *Kazanian and *Tatarian Stages.

ochre An iron-rich sediment, used as a pigment. *Compare* UMBER.

ochric horizon A light-coloured, mineral *soil horizon, usually at the soil surface, and characteristic of arid-environment soils.

octahedrite *See* ANATASE.

octahedron A three-dimensional, eight-sided form in which all the faces are equilateral triangles (i.e. it resembles a double pyramid). In *crystallography, a *crystal form in which the *crystal faces cut all three *crystallographic axes at equal distances.

Octocorallia (octocorals) A subclass of the class *Anthozoa, poorly known in the fossil record, that ranges from *Ordovician to Recent. The colony is shaped like a flat fan of interconnecting branches from which project tubes with many zooids. *Heliopora* is an important *reef former. The pennatulaceans (Pennatulacea) are also octocorals.

octocorals *See* OCTOCORALLIA.

ocular *See* EYEPIECE.

OD Ordnance datum.

Odderade The third *interstadial within the *Weichselian, which occurred between 70 000 years BP and 60 000 years BP, between the *Brørup and *Moershoofd Interstadials. The vegetational evidence indicates tundra conditions with some temperate boreal elements having been present. After the Odderade Interstadial polar-desert conditions prevailed until the Moershoofd Interstadial.

Oddo–Harkins rule Rule stating that the *cosmic abundance of elements with an even atomic number is greater than that of adjacent elements with an odd atomic number. Consequently, a graph plotting relative atomic abundance against increasing atomic number (*Z*) displays a 'toothed' curve, rather than a smooth line. The reason for this is connected with processes such as helium burning (*see* nucleosynthesis). ${}^{4}_{2}He$ is a basic building block, and so additions produce even numbers, e.g. ${}^{4}_{2}He + {}^{4}_{2}He \rightarrow {}^{8}_{4}Be$; ${}^{8}_{4}Be + {}^{4}_{2}He \rightarrow {}^{12}_{6}C$.

Odintsovo An *interstadial in the Saale Drift of European Russia, marked by a temperate, broadleaved-forest flora.

ODP Ocean Drilling Programme. *See* DEEP SEA DRILLING PROGRAMME.

Oe *See* OERSTED.

oedometer Instrument for testing *consolidation of small samples, including *coefficient of compressibility and *coefficient of consolidation. The sample is compressed in a cylinder allowing only vertical consolidation; the resulting changes in volume over a specific time are measured.

oersted (Oe) A unit of magnetic field intensity, now replaced by SI units. $1 A/m = 4\pi \times 10^{-3}$ Oe.

offlap A conformable sequence of inclined strata, deposited during a marine *regression, in which each stratum is succeeded laterally by progressively younger units, marking the direction in which the sea retreated. *Compare* OVERLAP. *See also* OVERSTEP; and SEISMIC STRATIGRAPHY.

offset The distance between a *geophone, or the centre of a geophone *group, and the *shot position. In-line offset is the displacement within the line of the spread; perpendicular (normal) offset is the distance at right angles between the shot and the line of the spread; reflection offset is the horizontal displacement of a dipping reflection event to its migrated position.

offshore bar *See* BAR.

offshore zone Zone extending seaward from the point of low tide to the depth of wave-base level or to the outer edge of the *continental shelf.

ogive Banded pattern on the surface of a *glacier. Normally it is convex down-glacier due to relatively high velocities at the centre. The banding effect may be due to alternating bands of white ice (containing many air bubbles) and dark ice (few air bubbles), or to variations in longitudinal pressure.

Ohauan *See* KAWHIA; and OTEKE.

Ohm's law The ratio of the voltage (V) applied to a conductor and the electric current (I) caused to flow through it at constant temperature is constant, and is the electrical resistance (R) of the conductor, such that $V/I = R$. At high current densities the law may break down for some materials.

OIB *See* OCEAN-ISLAND BASALT.

oil *See* PETROLEUM.

oil immersion *See* IMMERSION OBJECTIVE.

oil shale Dark grey or black *shale containing organic substances that yield liquid *hydrocarbons on distillation, but that do not contain free *petroleum.

Oka/Demyanka An equivalent in Russia of the *Elsterian (*Mindel), or its *Anglian equivalent, which passes beneath the *Dnepr *drift. Little is known of its occurrence.

okta *See* CLOUD AMOUNT.

Older Drift All the older *drift deposits that mark the maximum extent of ice advance during the last (*Devensian) glaciation in the British Isles. *Compare* NEWER DRIFT.

Older Dryas *See* DRYAS.

Oldest Dryas *See* DRYAS.

Oldham, Richard Dixon (1858–1936) A British seismologist, Oldham worked for the Indian Geological Survey. In 1897 he showed that *P- and *S-waves could be distinguished from each other on *seismograms, and that they travelled through the *Earth's interior. In 1906 he was able to show that the Earth has a fluid *core, through the existence of the S-wave shadow zone, and was able to estimate its size.

Old Red Sandstone Continent The continental facies of the *Devonian in the British Isles. It is characterized by red *sandstones and *conglomerates which were deposited under terrestrial conditions.

Olduvai A normal *polarity subchron which occurs within the *Matuyama reversed *chron. It is dated at 1.76–1.98 Ma (Bowen, 1978).

Olenekian *See* SCYTHIAN.

Olenelloidea *See* REDLICHIIDA.

oligo- From the Greek *oligos* meaning 'small' and *oligoi* meaning 'few', a prefix meaning few or small; in ecology it is often used to denote a lack, e.g. '*oligotrophic' meaning 'nutrient-poor' and '*oligomictic' meaning 'subject to little mixing'.

Oligocene An *epoch (35.4–23.3 Ma) of the *Tertiary Period. It follows the *Eocene and precedes the *Miocene Epochs. The Oligocene Epoch comprises the *Rupelian and *Chattian *Ages.

Oligochaeta (phylum *Annelida) Class of annelid worms which possess very well-developed *metameric segmentation. The segments have bristles, but parapodia (movable, paired lateral appendages) are not present. They are all hermaphrodites; asexual reproduction is predominant in aquatic forms. Eyes and tentacles are absent. A few marine forms occur but most are freshwater or terrestrial. There are 15 families, and the class is first recorded from the Upper *Ordovician.

oligoclase *See* PLAGIOCLASE FELDSPAR.

oligohaline water *See* HALINITY.

oligomictic 1. A *conglomerate containing *clasts of only a few different rock types. *Compare* POLYMICTIC. **2.** Applied to lakes that are thermally almost stable, mixing only rarely. This condition is characteristic of tropical lakes with very high (20–30 °C) surface temperatures.

oligotaxic times A period of low biological diversity among marine organisms, associated with low sea levels (lowstand), sharp temperature gradients, and stronger ocean currents. *Compare* POLYTAXIC TIMES.

oligotrophic Applied to waters poor in nutrient and with low *primary productivity. *Compare* EUTROPHIC.

olisthostrome *See* OLISTOSTROME.

olistolith *See* OLISTOSTROME.

olistostrome (olisthostrome) A sedimentary deposit which consists of a chaotic mass of rock and contains large *clasts composed of material older than the enclosing sedimentary sequence. The clasts may be gigantic and are then called 'olistoliths'. Such deposits are generally formed by *gravity sliding of material, sometimes into oceanic *trenches. Olistostromes have also been called 'sedimentary mélange' (*see* MÉLANGE).

olivine A major rock-forming *mineral group belonging to the *nesosilicates, forming a complete *solid solution series between *forsterite (Mg_2SiO_4) and *fayalite (Fe_2SiO_4); sp. gr. 3.22–4.39 increasing with increasing iron content; *hardness 6–7; *orthorhombic; usually olive-green, but white or yellowish in forsterite, brown or black in fayalite; colourless *streak; *vitreous *lustre; crystals rare, short, *prismatic, usually develops as granular aggregates; *cleavage poor {010}; occurs in silica-poor, *igneous rocks (e.g. *basalt, *gabbro, *troctolite, and *peridotite), extensively with *pyroxene in *dunites, and in *stony meteorites and lunar basalts; alters readily to *serpentine during weathering or *hydrothermal alteration.

olivine dolerite *See* DOLERITE.

omphacite A rare *clinopyroxene of isolated composition $(Ca,Na)(Mg,Fe^{2+},Fe^{3+},A1)[Si_2O_6]$ and with similar properties to *jadeite and *augite; sp. gr. 3.16–3.43; *hardness 5–6; green to dark green; *massive or *granular; occurs in *eclogites at high temperatures and pressures in association with *pyrope garnet.

oncoid *See* ONCOLITE.

oncolite (oncoid, oncolith) Spherical or subspherical particle, up to 5 cm across, produced by the accretion of sedimentary material on to a mobile grain through the action of *algae.

oncolith *See* ONCOLITE.

one-circle reflecting goniometer *See* GONIOMETRY.

Onesquethawian (Onondagan) *See* ULSTERIAN.

one-way travel time The time taken for a seismic wave to travel one way through a medium. In a reflection event it is half the *two-way travel time. *Compare* TRANSIT TIME.

onion weathering *See* SPHEROIDAL WEATHERING.

onlap The progressive spreading of successive beds of *sediment over an increasingly wider area, usually as a result of rising sea level or the gradual subsidence of a land area. Onlap results in an overlapping (*see* OVERLAP) stratigraphical relationship between the sedimentary succession and the underlying *basement rocks. The upper, younger beds 'pinch out' updip against the older, inclined surface on which they rest. *Compare* OFFLAP. *See also* BASELAP; COASTAL ONLAP; DOWNLAP; and TOPLAP.

Onnian A *stage of the *Ordovician in the Upper *Caradoc, underlain by the *Actonian and overlain by the *Pusgillian.

Onondagan (Onesquethawian) *See* ULSTERIAN.

Ontarian A *stage of the Lower *Niagaran (*Silurian) of N. America.

ontogenetic heterochrony In colonial animals, *heterochrony that affects individuals, rather than the colony as a whole. *Compare* ASTOGENETIC HETEROCHRONY and MOSAIC HETEROCHRONY.

ontogeny The development of an individual from fertilization of the egg to adulthood.

Onverwacht *See* SWAZIAN.

onyx (sardonyx) A mixture of chalcedonic (*see* CHALCEDONY) silica SiO_2 and hydrous silica $SiO_2.nH_2O$; occurs as flat, banded varieties with white, grey, brown, red, and black bands.

oo- Prefix in the *Folk classification, for *limestones containing *ooids, e.g. *oosparite.

oobiosparite *See* FOLK LIMESTONE CLASSIFICATION.

ooid (oolith) Sub-spherical, sand-sized, *carbonate particle that has concentric rings of calcium carbonate surrounding a nucleus of another particle. *See also* OOLITE.

oolite Commonly occurring *limestone, consisting largely of *ooids.

oolith *See* OOID.

oolitic Composed, or largely composed of *ooids.

oolitic limestone *See* OOLITE.

oomicrite A *limestone, defined by the Folk classification as comprising *ooids set in a *micrite *matrix.

oomoldic porosity (oomouldic porosity) A form of *secondary porosity developed in oolitic *limestone *(*see* OOLITE), due to the preferential dissolution of the *ooids, leaving a series of empty spaces. *See* POROSITY.

oomouldic porosity *See* OOMOLDIC POROSITY.

^{18}O : ^{16}O ratio *See* OXYGEN-ISOTOPE RATIO.

Oort cloud A spherical zone surrounding, but gravitationally bound to the *solar system, at distances between 20 000 and 100 000 AU from the Sun, which contains about 10^{12} comets. Its existence was first proposed in 1950 by the Dutch astronomer, Jan Hendrik Oort (1900–92). Gravitational perturbations by passing stars or molecular clouds result in comets acquiring orbits within the orbit of *Jupiter.

oosparite A *limestone, defined by the *Folk classification as comprising *ooids together with sparry calcite *cement (*sparite).

ooze A *pelagic mud consisting of the calcareous or siliceous remains of pelagic organisms (e.g. *coccoliths, and *tests of *foraminiferids and diatoms (*Bacillariophyceae)), and *hemipelagic *clay minerals. *Calcareous oozes will accumulate in the deep oceans at depths shallower than the *carbonate compensation depth (CCD). *See also* GLOBIGERINA OOZE; DIATOM OOZE; PTEROPOD OOZE; and RADIOLARIAN OOZE.

opacus From the Latin *opacus* meaning 'shady', a variety of extensive cloud that obscures the Sun or Moon. It is used to describe *stratus, *stratocumulus, *altostratus, and *altocumulus. *See also* CLOUD CLASSIFICATION.

opal Hydrous silica $SiO_2.nH_2O$ associated with the chalcedonic (*see* CHALCEDONY) varieties of silica. A layer of water molecules trapped near the mineral surface causes the *iridescence (*opalescence) which is a diagnostic property of opal; sp. gr. 1.99–2.25; *hardness 5.5–6.5; amorphous; colourless, or milky-white to grey, red, brown, blue, green, to nearly black; resinous *lustre; normally *massive, but can be stalactitic, *botryoidal, and also in veinlets, the various varieties depending on the amount of water contained in the mineral, which can vary from 6% to 10%; no *cleavage; *conchoidal fracture; normally deposited at low temperatures from silica-bearing waters, and occurs as fissure fillings in rocks of any kind, and especially near geysers and hot springs. The variety known as precious opal has a milky-white or sometimes black body colour which exhibits a brilliant play of colours, usually blues, reds, and yellows. The colours can often disappear with the loss of water when the mineral is exposed to air.

opalescence In *mineralogy, a pearly or milky mineral *lustre resembling that of *opal. It results from the reflection and refraction of light from the surface and possibly subsurface layers of the mineral. In *opal, water molecules are trapped in the subsurface layers, giving the opalescent quality.

opaque mineral (ore mineral) In transmitted-light microscopy, a mineral which appears black in *thin section in *plane-polarized light. The term is often used synonymously with 'ore mineral' although neither term is strictly correct; for example, *pyrite is opaque but rarely an ore, and *sphalerite is often an ore but rarely opaque.

open-cast mining (strip mining) Mining in which large strips of land are excavated in order to extract materials without subsurface tunnelling. It may result in short- and long-term environmental damage.

open fold As defined by M. J. Fleuty (1964), a *fold whose inter-limb angle (*see* FOLD ANGLE) is between 90° and 170°. The classification is based on the degree of tightness of folds.

open form In *crystallography, *crystal faces are often described in groups ('forms'); if the form is open at each end (i.e. it does not totally enclose space), it is called an 'open form', e.g. the four prismatic faces of a *tetragonal crystal. *Compare* CLOSED FORM.

open hole A *borehole that has not yet been *cased.

open-pit mining A method of extraction used where the *overburden is limited and easily stripped, but where waste has to be transported to external dumps. It is generally used where deposits are limited laterally but are thicker than in *open-cast mining.

operculum 1. In animals, a lid or cover, sometimes hinged, occurring, for example, in some cylindrical *rugose corals, some *bryozoans, and in *gastropods. **2.** A flap of skin covering the gills in bony fish (*Osteichthyes).

Ophelia (Uranus VII) One of the lesser

satellites of *Uranus, with a diameter of 16 km. It was discovered in 1986.

ophicalcite *See* SERPENTINE.

Ophiocistioidea (ophiocistioids; phylum *Echinodermata, subphylum *Echinozoa) Class of extinct, free-living echinozoans, which possess a low, dome-shaped *test lacking arms. The *peristome is centrally placed on the *ventral surface and is enclosed in a complex jaw system. The *ambulacra are confined to the ventral surface and are composed of three rows of plates. The *interambulacra are narrow and consist of one plate column only. The class is known from the *Ordovician to *Devonian.

ophiocistioids *See* OPHIOCISTIOIDEA.

ophiolite (ophiolite complex) Sequence of rock types, consisting of deep-sea *sediments lying above basaltic *pillow lavas, *dykes, *gabbro, and *ultramafic *peridotite. Some are the remnants of main *oceanic crust, others of crust formed in *back-arc basins.

Ophiomorpha With **Skolithos*, an *ichnoguild of *trace fossils characterized by vertical or steeply inclined *burrows with thick, knobbly walls. The interior of the burrow system is smooth. *Ophiomorpha* ranges from the *Permian to the present day and is associated with arthropods (*Arthropoda). Individual tubes have a diameter of 3–6 cm, and may be branched, with large, bulbous terminations.

ophitic texture Large *crystals of *augite enclosing, either wholly or partially, laths of *plagioclase feldspar, and found within *dolerites and *basalts. The texture is a particular type of *poikilitic texture.

Ophiuroidea (brittle-stars; subphylum *Asterozoa, class Stelleroidea) Subclass of brittle-stars in which a central, rounded disc is sharply demarcated from the long, slender arms. The body cavity of the arms is filled almost completely with axial skeleton. The subclass is known from the *Ordovician to Recent.

Opisthobranchia (phylum *Mollusca, class *Gastropoda) Subclass of marine organisms, which are often 'naked', that includes the sea slugs and pteropods, and ranges from the *Cretaceous to the present day. They have only one gill and are thought to represent an evolutionary grade between the *prosobranch (single-gilled) gastropods and the pulmonate gastropods (i.e. gastropods in which the *mantle cavity functions as a lung).

opisthodetic Applied to the ligament of a bivalve when it is in a position posterior to the umbones (*see* umbo). *Compare* AMPHIDETIC.

opisthogyral *See* OPISTHOGYRATE.

opisthogyrate (opisthogyral) Applied to the umbones (*see* umbo) of bivalves (*Bivalvia) where these are so curved as to point in a posterior direction.

opisthoparian suture *See* CEPHALIC SUTURE.

opisthosoma *See* ARACHNIDA; and CHELICERATA.

Opoitian A *stage in the Upper *Tertiary of New Zealand, underlain by the *Kapitean, overlain by the *Waipipian, and roughly contemporaneous with the upper *Zanclian (Tabianian) Stage.

Oppel, Albert (1831–65) A German geologist and palaeontologist, Oppel made extensive studies of the *Jurassic rocks of Europe. He devised a scheme to divide geologic formations into zones, each based on the period of existence of a single organic species, later to be termed '*index fossils'. *See* OPPEL ZONE.

Oppel zone A *concurrent range zone as originally conceived by Albert *Oppel. The lower boundary of the zone is usually determined by the first appearance of one diagnostic *taxon and the upper boundary by the last appearance of another (either might extend above or below into a region of overlap with the other taxon at the upper or lower end of their range). Other characteristic taxa may be contained within the zone or extend to either side, although long-ranging and slowly evolving lineages are not usually included in the diagnostic assemblage. Not all the significant taxa are required to be present at all levels and in all places. The zone is named after one of the diagnostic species. Oppel zones are thus more flexible and subjective than concurrent range zones in the rigorously applied sense of that term, although the term 'concurrent range zone' is commonly used for what is, in fact, an Oppel zone.

optical continuity In optical *mineralogy, the situation that occurs where two mineral grains are oriented such that their optical properties are consistent and in the same orientation with respect to their fundamental crystallographic properties.

optical emission spectrum *See* SPECTRUM.

optical goniometry *See* GONIOMETRY.

optical indicatrix *See* INDICATRIX.

optic axis In mineral optics, the direction perpendicular to a circular cross-section of an *indicatrix. Sections cut parallel to this circular cross-section are *isotropic.

opx *See* ORTHOPYROXENE.

Or *See* ALKALI FELDSPAR.

o-ray *See* EXTRAORDINARY RAY.

orbicular Disc-shaped, circular, or globular.

orbicular texture Concentric shells of contrasting mineralogy and texture which form spherical masses of diameter 2–15 cm in *basic and *felsic *plutonic *igneous rocks.

orbit 1. The bony socket of the eye. **2.** The path described by a body moving around another under gravitational attraction. *See* EQUATORIAL ORBIT; GEOSTATIONARY ORBIT; GEOSYNCHRONOUS ORBIT; POLAR ORBIT; and SUN-SYNCHRONOUS ORBIT.

orbital forcing The influence on climate and sea level of changes in the orbit and rotation of the Earth. *See* MILANKOVICH CYCLES.

orbit period The time taken by a body to complete a single orbit.

order *See* CLASSIFICATION.

Ordian A *stage of the Lower to Middle *Cambrian of Australia, underlain by the 'Lower Cambrian' and overlain by the *Templetonian.

ordinary ray *See* EXTRAORDINARY RAY.

Ordovician The second (510–439 Ma) of six *periods that constitute the *Palaeozoic Era, named after an ancient Celtic tribe, the Ordovices. The Ordovician follows the *Cambrian and precedes the *Silurian. It is noted for the presence of various rapidly evolving *graptolite genera and of the earliest jawless fish.

ore A *mineral or *rock that can be worked economically.

orebody Accumulation of *minerals, distinct from the host rock, and rich enough in a metal to be worth commercial exploitation.

ore genesis Process by which a mineral deposit forms. Metalliferous mineral deposits may be syngenetic (formed at the same time as the host rocks) or epigenetic (deposited later than the host rocks). Deposits may be classified according to their processes of formation into *igneous, *sedimentary, metamorphic (*see* METAMORPHISM), or *hydrothermal.

ore grade The concentration of an element of interest in a potentially mineable *ore deposit.

ore microscope *See* POLARIZING MICROSCOPE.

ore microscopy *See* REFLECTED-LIGHT MICROSCOPY.

ore mineral A metalliferous *mineral that may be extracted profitably from an *orebody. In mineralogy the term is applied to those minerals that may polish with a *metallic *lustre even when dispersed. *See also* OPAQUE MINERAL.

Oretian *See* BALFOUR; and CARNIAN.

organic soil Soil with a high content of organic matter and water. The term usually refers to *peat. The *USDA defines an organic soil as one with a minimum of 20–30% organic matter, depending on the *clay content.

Oriental faunal realm Area that encompasses India and Asia south of the Himalayan–Tibetan mountain barrier, and the Australasian archipelago, excluding New Guinea and the Celebes. There are marked similarities with the *Ethiopian realm (e.g. both have elephants and rhinoceroses) but there are endemic groups (*see* ENDEMISM), e.g. pandas and gibbons.

orientation survey The first stage in a *geochemical soil survey, concerned with the choice of sampling scheme, type of sample, and analytical method.

original horizontality *See* LAW OF ORIGINAL HORIZONTALITY.

origination The first appearance of a new species. Comparing the rate at which new species appear (origination rate) with the rate at which species go extinct (extinction rate) within an *ecosystem reveals whether the system was stable (origination = extinction), diversifying (origination >

extinction), or losing species (origination < extinction).

Oriskanyan (Deerparkian) *See* ULSTERIAN.

Ornithischia One of two orders of *Mesozoic *dinosaurs, distinguished primarily on the basis of a bird-like pelvis (*see* PELVIC GIRDLE; and PUBIS). Ornithischian dinosaurs were exclusively vegetarian and produced both bipedal forms (*Ornithopoda) and four-footed forms.

ornithischian dinosaur *See* ORNITHISCHIA.

ornithomimid One of the *coelurosaurs, which were *theropod *dinosaurs from the Upper *Cretaceous. They are referred to as the 'ostrich' dinosaurs because of their general build, and their name means 'bird imitators'.

Ornithopoda Suborder of *ornithischian dinosaurs which had an essentially bipedal gait, including several families, e.g. iguanodonts (*Iguanodontidae) and hadrosaurs (*Hadrosauridae). The ornithopods are regarded as the most primitive of the ornithischian suborders.

orogen The body of rock involved in one or more *orogenies. Many authors use the term as a synonym for *orogenic belt.

orogenesis *See* OROGENY.

orogenic belt (mobile belt) A linear or arcuate zone, on a regional scale, which has undergone compressional tectonics. The histories of many orogenic belts have been interpreted using *plate-tectonics models involving the *subduction of oceanic *lithosphere (e.g. the *Andean orogenic belt), the collision of major continental masses (e.g. the *Himalayan orogenic belt), or the accretion of *terranes (e.g. the Cordillera of the western USA and Canada). Orogenic belts were formerly expected to show the phases of the *orogenic cycle, but this has now been replaced by a search for identifiable stages of the *Wilson cycle to see if the belt can be interpreted using a plate-tectonics model.

orogenic cycle An obsolete concept, linked to geosynclinal theory, that several sequential phases were involved in the formation of an *orogenic belt. Many authors produced variations, but most cycles began with a geosynclinal phase (mainly subsidence and sedimentation), followed by an orogenic phase (mainly compression), and ended with a post-orogenic phase (mainly uplift).

orogeny (orogenesis) Mountain building, especially when a belt of the Earth's *crust is compressed by lateral forces to form a chain of mountains. There have been many orogenic episodes in the evolution of the crust, each extending over many millions of years. *See* FOLD BELT; OROGEN; OROGENIC BELT; and OROGENIC CYCLE.

orographic Applied to the rain or cloud caused by the effects of mountains on air streams that cross them. Orographic cloud and rain are produced by the forced uplift of moist air and the consequent condensation when this is cooled to saturation point.

orpiment *Sulphide, As_2S_3; sp. gr. 3.5; *hardness 2.0; yellow or yellowish-orange; foliated or powdery *aggregate; occurs in low-temperature veins and hot-spring deposits, commonly associated with *realgar and *stibnite.

Orthida (orthids; class *Articulata) Extinct order of brachiopods (*Brachiopoda), which have biconvex shells, a straight *hinge line, and well-developed cardinal areas on both valves (*see* CARDINAL TOOTH). The *pedicle opening notches into both *valves and is not usually restricted by plates. The shells are usually *impunctate. Orthida range from the Lower *Cambrian to the Upper *Permian, e.g. *Orthis* (*Ordovician).

orthids *See* ORTHIDA.

orthite *See* ALLANITE.

orthoamphiboles *See* AMPHIBOLES.

orthochemical **1.** The term used in the *Folk classification to describe the *micrite *matrix and sparry calcite (*sparite) *cement in *limestones. **2.** Applied to rocks consisting of micrite without allochemical particles (*see* ALLOCHEM). **3.** Applied to a *carbonate particle formed by direct chemical precipitation.

orthoclase *See* ALKALI FELDSPAR.

orthoconglomerate A *clast-supported *conglomerate containing less than 15% *matrix between the pebbles. The clasts are bound together mainly by mineral *cement.

orthoconic Applied to the conch of a *cephalopod when it is a straight, tapering cone.

orthoferrosilite *See* ENSTATITE.

orthogenesis Evolutionary trends that remain fairly constant over long periods of time and so appear to lead directly from ancestor organisms to their descendants. This was once explained as the result of some internal directing force or 'need' within the organisms themselves. Such metaphysical interpretations have been displaced by the concepts of *orthoselection and species selection.

orthogonal thickness (*t*) The thickness of a folded layer, measured perpendicularly to parallel tangents to the inner and outer fold surfaces.

orthomagmatic Applied to the stage during which the main mass of *silicates crystallizes from a *magma. The stage can be divided into an early orthomagmatic stage, during which anhydrous silicate minerals crystallize, and a late orthomagmatic stage, during which anhydrous and hydroxyl-bearing silicate minerals crystallize.

orthophotograph An aerial photograph in which distortions due to the ground topography, such as elevation and tilt, have been removed, giving the appearance of every object being viewed directly from above, like a map. Distortions are removed by a *photogrammetric system.

orthopyroxene (opx) A series of *pyroxenes which crystallize in the *orthorhombic system. They consist of the *enstatite ($MgSiO_3$) and ferrosilite ($FeSiO_3$) *end-members, with a number of other minerals between, including *bronzite, *hypersthene, and eulite.

orthoquartzite *See* QUARTZITE.

orthorhombic Applied to a *crystal system where the *Bravais lattices have three sets of edges at right angles, but all are of different lengths. They can be referred to three *crystallographic axes, *a*, *b*, and *c* or *x*, *y*, and *z*, of unequal length and at right angles to each other.

orthorhombic amphiboles *See* AMPHIBOLES.

orthoscopic Applied to observations in optical microscopy which use parallel beams of light passing through the atomic structure of a *crystal. This is the normal mode of observation while *conoscopic (i.e. convergent light) observations are used for specialized techniques of *mineral identification.

orthoselection Primary selective pressure of a directional kind, which results in a self-perpetuating evolutionary trend. Species selection has been advanced as an alternative explanation for such trends. *See also* DOLLO'S LAW.

orthosilicates *See* NESOSILICATES.

ortstein *Indurated *soil horizon in the B horizon of *podzols (*Spodosols), in which the *cementing materials are mainly iron oxide and organic matter.

oryctognosy *See* WERNER, ABRAHAM GOTTLOB.

Osagean A *series in the *Mississippian of N. America, underlain by the *Kinderhookian, overlain by the *Meramecian, and roughly contemporaneous with the Ivorian Stage of the *Tournaisian Series and the Chadian Stage of the *Visean Series.

Osborn, Henry Fairfield (1857–1935) An American palaeontologist who taught at Princeton and Columbia Universities, Osborn was an evolutionary theorist who developed the concept of *adaptive radiation. He also arranged the mammalian palaeontology exhibits at the American Museum of Natural History.

oscillation ripple (wave ripple mark) Small ridge of sand formed by wave action on the floor of a sea or lake. Such *ripples, which are usually less than 10 cm in height, commonly display rounded symmetrical troughs and rounded, or sometimes sharp, symmetrical peaks, although many wave-generated ripples are not symmetrical owing to an inequality in forward and backward motion associated with the wave action. The *ripple index of oscillation ripples is usually between 6 and 10.

oscillatory wave Wave that causes a mass of water to move to and fro about a point but not to undergo any appreciable net displacement in the direction of wave advance. The wave-form advances, but the individual water particles move in closed or nearly closed orbits.

oscillatory zoning *See* CRYSTAL ZONING.

osculum *See* PORIFERA.

osmosis The movement of water or of another solvent from a region of low solute concentration to one of higher concentration through a semi-permeable membrane. It is an important mechanism in the uptake of water by plants.

osmotic potential *See* WATER POTENTIAL.

ossicles 1. In *Echinodermata, irregularly *fenestrated calcareous plates, rods, and crosses arranged to form a lattice and bound together by connective tissue; together they comprise the skeleton. **2.** Small bones.

Osteichthyes (bony fish) Fish in which the internal skeleton is largely ossified, and which also have teeth, plates, or scales of *dermal bone. Characteristic features are: the terminal mouth; *homocercal tail; *operculum covering the gills; usually many flat, bony scales embedded in the skin, with epidermis over them; and the presence of a swim bladder (although sometimes this has been lost in the course of later evolution). With more than 25 000 species it is the largest class of *vertebrate animals. The Osteichthyes have a fossil history going back to the Lower *Devonian; they include the *Crossopterygii, the *Dipnoi, the *Chondrostei, the *Holostei, and the *Teleostei.

Osteostraci (Cephalaspida; osteostracans; superclass *Agnatha) Order of extinct, fish-like, jawless *vertebrates, ranging from the *Silurian to the *Devonian. They had somewhat flattened bodies, a broad head covered in a bony shield, dorsally located eyes with an opening between them for the pineal eye, and a single median nostril in front. At the sides and back of the *head shield were areas covered in polygonal *plates, possibly covering sense organs. The internal structure of the head resembled that of the extant lampreys. The body was enclosed in bony scales arranged in a series of vertical rows, and there were either one or two dorsal fins and a *heterocercal tail. The internal skeleton was partly ossified. Osteostracans were usually small, around 30 cm in length; the dorsally placed eyes and flat belly suggest a bottom-dwelling way of life. *See also* CEPHALASPIS.

osteostracans *See* OSTEOSTRACI.

ostia *See* PORIFERA.

Ostracoda (ostracods; subphylum *Crustacea) A class of crustaceans that are typical *Arthropoda but laterally compressed and enclosed within a bivalved carapace. This pair of calcareous valves is an integral part of the epidermis and is closed by a series of muscles that leave scars (*see* MUSCLE SCAR) on the valve interiors. The appendages are typically *biramous but may be modified for digging, swimming, etc. The group includes herbivores, carnivores, and scavengers, and occurs in most aquatic habitats. Most are small (less than 1 mm long) and more than 10 000 species have been described, occurring from the *Cambrian to the present day. The biological *classification of Recent forms is based on the soft-part anatomy and *fossil forms are classified by the nature of the preserved carapaces. Such characters as the nature of the *hinge, the pattern of the muscle scars, as well as overall shape and ornamentation are all used in species determination. Ostracods have considerable stratigraphic use, and are also used to demonstrate variations in *salinity and fluctuations in the positions of shorelines.

Ostracodermi (ostracoderms) The name often used in older textbooks for the fossil, armoured, jawless, agnathan fish of the *Ordovician to early *Carboniferous. Probably the name indicates only a grade of development; some ostracoderms are close to the line of ancestry of the living *Agnatha (lamprey and hagfish), others to the origin of the jawed vertebrates. The informal version of the name survives.

ostracoderms *See* OSTRACODERMI.

ostracum *See* SHELL STRUCTURE; and SKELETAL MATERIAL.

ostrich dinosaur *See* ORNITHOMIMID.

Otaian A stage in the Upper *Tertiary of New Zealand, underlain by the *Waitakian (*Oligocene), overlain by the *Altonian, and roughly contemporaneous with the *Aquitanian and lower *Burdigalian Stages.

Otamitan *See* BALFOUR; and CARNIAN.

Otapirian *See* BALFOUR; and RHAETIAN.

Otariidae *See* CARNIVORA.

Oteke A *series in the Upper *Jurassic of New Zealand, underlain by the *Kawhia and overlain by the Mokoiwian. It comprises the uppermost Ohauan *Stage and the Puaroan Stage, which are roughly contemporaneous with the *Tithonian.

ottrelite *See* CHLORITOID.

outcrop That part of a rock formation which is exposed at the Earth's surface.

outerarc A ridge, or uplifted section, sometimes extending above sea level, found in some *arc-trench gaps.

outer planet *See* JOVIAN PLANET.

outgassing 1. The removal of gas, usually

by heating. **2.** The release of gases by volcanic activity that resulted in the formation of the Earth's atmosphere and hydrosphere.

outgroup In *phylogenetics, a species which is the least related to the species under analysis. The inclusion of a known outgroup allows the identification of *plesiomorphic and *apomorphic *character states, which might otherwise remain unclear, a situation that can give rise to *topological errors. A tree that includes an outgroup is said to be rooted.

outlet glacier A tongue of ice that extends radially from an *ice dome. It may be identified within the dome as a rapidly moving ribbon of ice (an 'ice stream'), while beyond the dome it typically occupies a shallow, irregular depression. The 700 km long Lambert Glacier, Antarctica, is one of the world's largest outlet glaciers.

outlier Area where younger rocks are surrounded completely by older rocks. It may be produced by *erosion, *faulting, or *folding, or any combination of these. *Compare* INLIER.

out-of-phase component *See* IMAGINARY COMPONENT.

outwash 1. The stratified *sands and *gravels deposited at or near to ice margins. **2.** Meltwater escaping from the terminal zone of a *glacier. The resulting streams are typically *braided, and show marked seasonal variations in discharge.

outwash plain (sandur, pl. sandar) Extensive accumulation of rock debris built up by *outwash in front of a *glacier. Its constituents are very coarse close to the ice, and diminish in size further away. Its surface may be dissected by *braided channels. Fossil outwash plains are found at the margins of many *Pleistocene glaciers. The name 'sandur' is Icelandic.

outwelling The enrichment of coastal seas by nutrient-rich estuarine waters.

overbank deposit A *floodplain sediment that lies beyond the limits of the river channel and was left by floodwaters that had overflowed the river banks.

overbreak Excess material removed by blasting, in *excavations. As rock cannot always be cut exactly to straight lines, more must be removed than is required.

overburden 1. Any loose material which overlies bedrock. **2.** In a sedimentary deposit, the upper strata which cover, compress, and consolidate those beneath. **3.** Any barren material, consolidated or loose, that overlies an *ore deposit. The depth and type of overburden may control whether an ore deposit is worked by underground or *open-cast methods. The proportion of overburden thickness to mineral deposit is called the overburden ratio.

overburden ratio *See* overburden.

over-consolidated clay Clay that has been more compacted than would be expected from the existing *overburden, e.g. it has been subjected to pressure from overburden that has subsequently been removed by *erosion. *Compare* NORMALLY CONSOLIDATED CLAY.

overflow channel (spillway) Channel cut by meltwater escaping from a *proglacial lake. Generally it is trough shaped in form, lacks tributaries, and is not integrated with the local drainage pattern. It is often difficult to distinguish from other varieties of *glacial drainage channel.

overflowing well *See* ARTESIAN WELL.

overflow levée *See* LAVA LEVÉE.

overfold (overturned fold) A *fold in which the *axial plane is inclined so that the *fold limbs *dip in the same direction, although not necessarily by the same amount. One limb is thus 'overturned'.

overland flow *See* SURFACE RUNOFF.

overlap An unconformable relationship in which a transgressive sequence (*see* TRANSGRESSION) of progressively younger members of an upper series of strata *onlap and rest upon the underlying (oldest) stratum. Strictly, 'onlap' refers to the process and 'overlap' to the resulting structural relationship, but the two terms are often used synonymously. The opposite of overlap is '*offlap'. *See also* overstep.

overpressure Overpressured zones occur in subsurface *horizons in which the fluid pressure is greater than the normal hydrostatic pressure (*see* hydrostatic stress) for the depth in question. Overpressures develop in rapidly deposited *sediments which have been sealed by impermeable horizons, which prevent the *pore fluids from escaping. Drilling into overpressured horizons can cause hazardous blowouts and well-caving.

overspecialization An old theory which

held that straight-line evolution or *orthogenetic trends might proceed to the point at which the lineage was at an adaptive disadvantage. Overspecialization was therefore considered as one of the causes of extinction. There is no reason to believe, however, that *natural selection would permit evolution to proceed beyond maximum adaptation. More recently, the term has been applied to highly specialized organisms which have proved incapable of responding to environmental change and so have become extinct.

overstep 1. An unconformable relationship in which a younger series of rocks rests upon progressively older rocks, suggesting the series below the *unconformity is tilted. *See also* OFFLAP; and OVERLAP. **2.** The progressive development of *hanging-wall *thrusts.

overthrust A *thrust fault in which the horizontal displacement is large and the *hanging wall is the relatively active element and is thrust over the *footwall.

overturned fold *See* OVERFOLD.

overvoltage *See* INDUCED POLARIZATION.

Owen, Richard (1804–92) An anatomist and palaeontologist, Owen worked on fossil mammals and reptiles, including those brought back from South America by *Darwin. He coined the name '*dinosaur', and reconstructed fossil reptiles such as *Iguanodon. He believed that animals within a major group (e.g. vertebrates) were variations on a single theme, or 'archetype'. His crowning achievement was the founding of the Natural History Museum in South Kensington, London, in 1881.

ox bow *See* CUT-OFF.

Oxfordian A *stage in the European Upper *Jurassic (157.1–154.7 Ma, Harland et al., 1989). Note, however, that the Oxford Clay (Britain) is predominantly of *Callovian age. *See also* MALM.

oxic horizon Mineral subsoil *soil horizon that is at least 30 cm thick and is identified by the almost complete absence of weatherable primary minerals, by the presence of *kaolinite clay, insoluble minerals such as *quartz, hydrated oxides of iron and aluminium, and small amounts of exchangeable bases, and by low *cation-exchange capacity. It is the distinguishing subsoil horizon (B horizon) of an *Oxisol.

oxidation Specifically, a reaction in which oxygen combines with, or hydrogen is removed from, a substance. More generally, any reaction in which an atom loses *electrons. For example, in the reaction between zinc and copper sulphate:

$$Zn + Cu^{2+} + SO_4^{2-} \rightarrow Zn^{2+} + SO_4^{2-} + Cu$$

the zinc has lost two electrons and been oxidized. Conversely the copper has undergone *reduction.

oxidation potential (electrode potential, reduction potential; E$^\theta$) The energy change, measured in volts, required to add or remove *electrons to or from an element or compound. The reference reaction is the removal of electrons from hydrogen in a standard hydrogen half-cell (i.e. H_2(gas) at 1 atm pressure delivered to a 1.0 M solution of H^+ *ions at 25 °C, into which a platinum electrode has been inserted): $H_2 \rightarrow 2H^+ + 2e^-$, This energy change is given the value of zero. The oxidation potential of other species are determined relatively by measuring the potential difference between a half-cell containing an aqueous solution of the oxidized and reduced forms of the test substance, and the standard hydrogen half-cell. For example, for $Fe^{2+} \rightarrow Fe^{3+} + e^-$, $E^\theta = 0.77$, for $Mn^{2+} \rightarrow Mn^{3+} + e^-$, $E^\theta = 1.51$. With decreasing values of oxidation potential, the reduced form of a couple (e.g. Fe^{2+}) will itself reduce the oxidized form of a couple with a higher oxidation potential (e.g. Mn^{3+}). The oxidation potentials obtained under these controlled conditions are called standard electrode potentials, or sometimes standard reduction potentials. *Compare* REDOX POTENTIAL.

oxidation–reduction *See* REDOX REACTION.

oxides A group of minerals in which oxygen is combined with one or more metals to give simple and multiple oxides respectively. Simple oxides include *hematite (Fe_2O_3), *rutile (TiO_2), and zincite (ZnO). Multiple oxides include the *spinels ($MgAl_2O_4$) and hydrated oxides, e.g. *goethite (FeO.OH). Oxides are economically important and are the principal sources of tin (SnO_2), iron (Fe_2O_3, Fe_3O_4), chromium ($FeCr_2O_4$), titanium (TiO_2), manganese (MnO_2), and aluminium ($Al_2O_3.2H_2O$). Oxides are relatively high-temperature minerals occurring in association with a variety of *igneous rocks. They may also form as chemical precipitates in oxidized environments.

Oxisol A *mineral soil, comprising a soil order distinguished by the presence of an *oxic horizon within 2 m of the soil surface, or with *plinthite close to the soil surface, and without a *spodic or *argillic horizon above the oxic horizon.

oxycone *See* INVOLUTE.

oxygen 'burning' A stage of advanced 'burning' in stellar evolution, following *helium 'burning', that takes place at a temperature around 2×10^9 K. ^{16}O nuclei react to form ^{32}S, ^{31}S, ^{31}P, and ^{28}Si. *See* CARBON 'BURNING'; and NUCLEOSYNTHESIS.

oxygen-isotope analysis Method for estimating past ocean temperatures. The ratio of the stable *oxygen isotopes, ^{18}O and ^{16}O, is temperature dependent in water, ^{18}O increasing as temperature falls. Oxygen incorporated in the calcium-carbonate shells of marine organisms will reflect the prevailing $^{18}O:^{16}O$ ratio (*see* OXYGEN-ISOTOPE RATIO). Acidification (to release oxygen) of *fossils of these organisms under carefully controlled conditions can therefore be used, with appropriate calibration, to indicate the record of past ocean temperatures.

oxygen-isotope curve A graphical plot of values for the relative proportions of two *isotopes of oxygen. Oxygen can exist in several isotope forms, but only ^{16}O and ^{18}O are important in the analysis of oxygen isotopes (*see* OXYGEN-ISOTOPE RATIO). In nature, the present-day average ratio of ^{18}O to ^{16}O is about 1 : 500, and measurements are made against this value as a standard. There is now evidence that this ratio changed in ocean waters in a cyclical fashion in succeeding glacial and *interglacial periods (*see* OXYGEN-ISOTOPE ANALYSIS). At the height of the last (*Devensian) glaciation it appears that deep-ocean waters may have been enriched in ^{18}O by about 1.6 parts per thousand. This would be equivalent to a lowering of sea level of about 165 m compared to the present level.

oxygen-isotope ratio ($^{18}O:^{16}O$ ratio) The abundance ratio between two of the three *isotopes of oxygen. They have similar chemical properties because they have the same electronic structure, but because of the differences in mass between their nuclei they have different vibrational frequencies which cause them to behave slightly differently in physico-chemical reactions. These differences can provide information, e.g. regarding the source of water in a past environment or the temperature at which various interactions have taken place. For example, surface waters vary in their oxygen isotopes; light water ($H_2{}^{16}O$) has a higher *vapour pressure than $H_2{}^{18}O$, and therefore is concentrated by evaporation so that fresh water and polar ice are light but sea water is heavy. $CaCO_3$ or SiO_2 are richer in the heavier isotope when precipitated from sea water than when precipitated from fresh water. Moreover, because of meteorological cycles of evaporation/condensation, there is a steady depletion of ^{18}O in sea water towards the poles. *See also* ISOTOPE FRACTIONATION; and OXYGEN-ISOTOPE ANALYSIS.

oxygen isotopes There are three *isotopes of oxygen, ^{16}O, ^{17}O, and ^{18}O. The most important in geology are ^{16}O and ^{18}O, as both these isotopes are found together in *carbonate rocks and *minerals. The $^{18}O:^{16}O$ ratio (*see* OXYGEN-ISOTOPE RATIO) varies with the temperature and chemical composition of the water in which shelly organisms grew, or of the subsurface waters from which diagenetic carbonate *cements crystallized. The $^{18}O:^{16}O$ ratio is therefore a valuable tool for palaeothermometry (*see* OXYGEN-ISOTOPE ANALYSIS) and for diagenetic studies (particularly when used together with a study of $^{13}C:^{12}C$ isotope ratios), provided that there have not been subsequent changes to the isotopic composition. *See* DIAGENESIS.

oxygen-isotope stage One of the glacial or *interglacial *stages revealed by *oxygen-isotope curves. Curves from Atlantic and Pacific deep-sea cores were divided by C. Emiliani into 16 stages, their fluctuations being correlated with ice-sheet growth and decay. N. J. Shackleton and N. D. Opdyke extended these subdivisions in 1973, from their studies of a core from the western Pacific in which 23 stages were recognized. These are presumed to represent a continuous record from about 870 000 years BP. In 1976, J. Van Donk obtained curves from a core in the equatorial Atlantic which yielded 42 isotope stages, representing 21 glacial and 21 interglacial stages.

oxyhornblende *See* KAERSUTITE.

oxyluminescence *See* THERMOLUMINESCENCE.

Oyashio current Western *boundary current in the subpolar gyre of the N. *Pacific. It originates in the Bering Sea and flows south-west off the Kuril Islands to meet the *Kuroshio current east of northern Japan. The current flows at less than 0.5 m/s, it is cold (4–5 °C at 200 m depth), and has a low *salinity (33.7–34.0 parts per thousand).

ozone layer The atmospheric layer at 15–30 km altitude, in which ozone (O_3) is concentrated at 1–10 parts per million. Ozone also occurs in very low concentration at altitudes of 10–15 km and 30–50 km. Generally, atmospheric ozone is produced by the photochemical dissociation of oxygen (O_2), resulting from absorption of ultraviolet solar radiation, to form atoms of oxygen (O). These atoms collide with molecular oxygen (O_2) to form ozone (O_3), which in turn absorbs solar radiation for further dissociation to O and O_2. The ozone layer limits the amount of ultraviolet radiation reaching the ground surface. *See also* ATMOSPHERIC STRUCTURE.

ozonesonde A package of instruments carried by a balloon and used to acquire data from the *ozone layer.

Pa *See* PASCAL.

Pachycephalosauridae 'Bone-headed' *dinosaurs of the Upper *Cretaceous, which were *ornithopods and apparently may have lived in herds in upland regions.

pachydont Applied to a type of *hinge *dentition, found in certain cemented bivalves (*Bivalvia) in which the teeth are heavy, very large, and generally few in number.

Pacific- and Indian-Ocean common water (abbreviation PIOCW) The deep waters of the *Pacific and *Indian Oceans are so similar in character that they are usually classed as one mass, with a mean temperature of 1.5 °C and a *salinity of 34.70 parts per thousand.

Pacific–Antarctic Ridge The *ridge which lies between the *Pacific and *Antarctic Plates and joins the South-East Indian and *East Pacific Rises.

Pacific Ocean The largest of the world's major oceans (179.7×10^6 km^2). It is also the coldest (average 3.36 °C), deepest (4028 m average), and least saline (34.62 parts per thousand) of the ocean basins.

Pacific Plate Although it is the largest present-day lithospheric *plate, the Pacific Plate is shrinking because most of its margins, apart from the *East Pacific Rise and the *Pacific–Antarctic Ridge, are *subduction zones (e.g. the Aleutian, Kuril, Japan, Izu-Bonin, and Marianas subduction zones). The Pacific Plate is the only major plate to consist of nothing but *oceanic crust, parts of which date from the *Triassic Period.

Pacific Province (American Province) The name given to the trilobite (*Trilobita) and graptolite (*Graptolithina) faunas characteristic of the northern and western margins of the *Iapetus Ocean in the Early *Palaeozoic. *See also* ATLANTIC PROVINCE.

Pacific-type coast (longitudinal-type coast) A coast that borders or lies within a mountain chain, so its subsidences and fractures follow the grain of the folding. This type of coast was first recognized by Eduard *Suess. *Compare* ATLANTIC-TYPE COAST; *see also* DALMATIAN-TYPE COAST.

Pacific-type margin *See* ACTIVE MARGIN.

packed biomicrite *See* FOLK LIMESTONE CLASSIFICATION.

packer test (Lugeon test) Test for measuring the *permeability of ground in sections of *boreholes. An inflatable tube ('packer') is lowered down a borehole and expanded so that the sections above and below are isolated. Alternatively, two packers may be used to isolate a certain section. Water is pumped into the section under investigation and leakage can be measured. The rate at which water is absorbed per metre length of hole is measured in units of lugeon, named after the French geologist Maurice Lugeon (1870–1953). One lugeon is approximately equal to 1.0×10^{-5} cm/s permeability.

packstone As defined by the *Dunham classification, a *limestone characterized by a *grain-supported texture, together with a *lime-mud *matrix.

paedomorphosis Evolutionary change that results in the retention of juvenile characters into adult life. It may be the result of *neoteny or *progenesis. It permits an 'escape' from *specialization, and has been invoked to account for the origin of many *taxa, from subspecies to phyla.

pahoehoe *See* LAVA.

paired metamorphic belts Juxtaposed zones of metamorphic rocks; a high pressure–low temperature (*blueschist facies) zone which lies parallel to a low pressure–high temperature (*greenschist facies) zone. Paired metamorphic belts were first recognized in Japan, where the greenschists facies is nearer the *subduction zone, and similar belts in different *Phanerozoic *orogenic belts have been identified and are thought to indicate *destructive margins. Recently it has been suggested that the Japanese belts are not contemporaneous and are separated by a major transcurrent fault.

Palaearctic faunal realm Region coincident with Europe, Asia north of the Himalayan–Tibetan physical barrier, N. Africa, and much of Arabia. The region is similar at the family level, and rather less so at the generic level, to the *Nearctic faunal realm: the *Bering *land bridge connected the two for much of the *Cenozoic Era.

palaeo- (paleo-) From the Greek *palaios* meaning 'ancient', a prefix meaning 'very old' or 'ancient'.

palaeoautecology The study of past populations, usually comprising only one or two species.

palaeobiogeography The scientific study of the geographic distribution of *fossils. *See also* BIOGEOGRAPHY.

palaeobiology The attempt to interpret the biology of *fossil organisms. It is sometimes combined with functional morphological studies in the more general study of *palaeoecology.

palaeobotany The study of *fossil plants.

Palaeocene (Paleocene) The lowest epoch of the *Tertiary Period, about 65–56.5 Ma ago. The name is derived from the Greek *palaios* 'ancient', *eos* 'dawn', and *kainos* 'new', and means 'the old part of the *Eocene' (the subsequent epoch).

palaeoclimatic indicator One of the sources from which evidence concerning past climates can be obtained. For example, glacial, *periglacial, and pluvial deposits provide morphological information related to climate; cave deposits, dunes, and dunefields yield lithologic information; and plants (including pollen), molluscs (*Mollusca), foraminifera (*Foraminiferida), beetles, and ostracods (*Ostracoda) are among the organisms that have been used to derive biotic information.

palaeoclimatology The study of past climates from the traces left behind in the geologic record. It is assumed that the *uniformitarian principle has obtained, but this may not be the case, and the geologic data are always insufficient for palaeoclimatological purposes. Dating methods and *palaeogeographic reconstructions are suspect, and often *fossil faunas and floras cannot be related easily in terms of time. Most glacial *strata can be dated only within wide limits and many climatic criteria leave no mark in the rocks.

palaeocurrent analysis The collection, presentation, and interpretation of directional data measured from *sedimentary structures and textures which were formed under flowing water, the wind, or moving ice. Ranging from the study of a single structure to data collected across a whole *sedimentary basin, palaeocurrents can yield a hierarchy of information, from the local direction of *flow which yielded a single *ripple train, through the direction of migration of a *bar or *channel, to the movement of *sediment through a river or deltaic network (*see* DELTA), and the regional pattern of sediment *provenance and dispersal through a basin.

Palaeodictyon A representative of the *Nereites* *trace fossil assemblage, associated with deep-water *turbidite facies. *Palaeodictyon* was originally thought to be a complex grazing trace but is now thought to represent a permanent or semi-permanent *burrow system. The burrows form a complex, hexagonal pattern with a single, vertical tube reaching towards the bed surface. It is probable that the organism concerned 'farmed' bacteria or algae within the burrows; water currents passed down the vertical tubes to irrigate the burrow system.

palaeoecology The application of ecological concepts to *fossil and sedimentary evidence in order to study the interactions of the Earth's surface, *atmosphere, and *biosphere in pre-historic and geologic times.

palaeoflow The river flow that cut a stream bed that later filled; its direction can be determined from the *cross-stratification it produced.

palaeofluminology The study of ancient stream beds.

palaeogeography The reconstruction of the physical geography of past geologic ages. A palaeogeographical map would normally show the *palaeolatitude of the area under discussion together with the location of inferred shorelines, drainage areas, *continental shelves and depositional environments. At the present time the base map would normally be a reconstruction based on palaeomagnetic data (*see* PALAEOMAGNETISM), although many maps in earlier publications used the present geographical positions of the continents as a foundation. With the advent of *balanced sections in structural geology, many palaeogeographical maps are now being produced that take account of the physical shortening involved and attempt to restore areas to their 'depositional' condition.

Palaeogene The earlier of the two *periods which comprise the *Tertiary sub-Era, preceded by the *Cretaceous, followed by the *Neogene, and dated at 65–23.3 Ma (Harland et al., 1989). The Palaeogene Period is subdivided into the *Palaeocene, *Eocene, and *Oligocene *Epochs.

palaeoguild A *guild that existed in the past.

Palaeohelikian A sub-*stage of the Lower *Helikian of the Canadian Shield region.

palaeohydraulics The analysis of the geometry of ancient *fluviatile features, e.g. preserved channel forms, lateral accretion surfaces, and channel-fill sequences, to estimate the hydraulic parameters which were developed at the time of formation of the channel system.

palaeoichnology *See* ICHNOLOGY.

palaeolatitude The position, relative to the equator, of a geographic or geologic feature at some time in the past. The evidence for the position may come from palaeoclimatic data (*see* PALAEOCLIMATIC INDICATOR; and PALAEOCLIMATOLOGY) or, more normally, a palaeomagnetic determination (*see* PALAEOMAGNETISM).

palaeomagnetic pole The pole position calculated from palaeomagnetic or *archaeomagnetic directions, assuming an *axial geocentric-dipole model of the Earth's *geomagnetic field. It should be calculated on the mean direction of sufficient observations so that past *secular variations of the geomagnetic field will have been averaged out. *See* PALAEOMAGNETISM.

palaeomagnetism Field of *geophysics concerned with the measurement and interpretation of *remanent magnetism or the record of the Earth's past magnetic field (*see* GEOMAGNETIC FIELD). Much valuable information has been obtained concerning *polar wandering and *continental drift.

palaeoniscids A group of primitive fish (*Actinopterygii), most of which were of modest size (about the size of a herring), but possessed thick, rhomboidal, enamel scales. There is great variety in structural detail and they were most abundant in the *Carboniferous. The group diminished rapidly after the early *Triassic and finally became extinct in the *Cretaceous.

palaeontology The study of *fossil *flora and *fauna. Information thus gained may be used to establish ancient environments.

Palaeoproterozoic The earliest part of the *Proterozoic, lasting from about 2500–1600 Ma.

Palaeopterygii Name formerly given to a supposed subclass that included some of the more primitive fish, but whose members are no longer held to be closely related. They were *bony fish with many fossil relatives, including, for example, the still-extant families Acipenseridae (sturgeon), Polyodontidae (paddlefish), and Polypteridae (bichir).

palaeoseismology The study of ancient *earthquakes, primarily by mapping zones of previous earthquake destruction by observation or from historical records.

palaeosol (paleosol) Soil formed during an earlier period of *pedogenesis, and which may have been buried, buried and exhumed, or continuously present on the landscape until the current period of pedogenesis.

palaeosome That part of a composite rock body which appears to be older than an associated element. The term is commonly used with reference to that part of a *migmatite which remained solid during *anatexis.

palaeospecies Group of biological organisms, known only from *fossils, which differs in some respect from all other groups.

palaeosynecology The study of past plant and animal communities, including all the species comprising them.

palaeotaxodont A type of *taxodont dentition in bivalves (*Bivalvia). In early *Palaeozoic forms teeth are formed in two distinct sizes, an anterior row with large teeth and a posterior row with much smaller teeth.

Palaeotaxus rediviva The earliest recorded species of the Taxales (yew) family. Found in the Lower *Jurassic of Europe, it is distinguished by the form of the leaf and the structure of the *cuticle. The Taxales, unlike the conifers, have a single ovule surrounded by a berry-like structure. They bear terminal seeds. *See also* CONIFERALES.

Palaeotethys There is confusion over the relationship between *Tethys (or Neotethys) and Palaeotethys. Some authors refer to the existence of a late *Palaeozoic Palaeotethys which was a gulf at the margin of *Panthalassa between the parts of *Pangaea that were later to separate as *Laurasia and

*Gondwanaland. This ocean, or large parts of it, were consumed in a Palaeocimmerian-Indosinian *subduction zone. In the *Jurassic a new ocean, or oceans, opened along the northern margin of Palaeotethys. This Jurassic-*Cretaceous ocean has been called Neotethys or Tethys (*sensu stricto*).

Palaeozoic The first (570–248 Ma) of the three *eras of the *Phanerozoic. The *Cambrian, *Ordovician, and *Silurian *Periods together form the Lower Palaeozoic sub-era; the *Devonian, *Carboniferous, and *Permian the Upper Palaeozoic sub-era. During the Palaeozoic, two major *orogenies occurred: the *Caledonian during the Lower Palaeozoic, and the *Variscan in late Palaeozoic times. The faunas of the Palaeozoic are noted for the presence of many invertebrate organisms, including trilobites (*Trilobita), graptolites (*Graptolithina), brachiopods (*Brachiopoda), cephalopods (*Cephalopoda), and *corals. By the end of the era, *amphibians and *reptiles were major components of various communities and giant tree-ferns, horsetails, and *cycads gave rise to extensive forests.

palagonite Pale yellow to yellow-brown, hydrated, fragmental, basaltic *glass, formed by the *hydration of *metastable *hyaloclastite and *ion exchange, due to the action of *groundwater percolating in the subaerial environment or to the action of sea water in the case of submarine environments. The high surface area to volume ratio of the *lapilli-size fragments of glass facilitates efficient chemical exchange with water. Much of the original calcium, potassium, and sodium, and some of the aluminium and silicon, are lost, iron is oxidized, and water is gained. Elements such as iron, titanium, magnesium, and aluminium may be either lost or added. The type locality is at Palagonia, Sicily. Some planetary scientists think that the whole surface of *Mars is covered with dust of palagonite origin.

palate The roof of the mouth; it separates the nasal and mouth cavities.

paleo- Alternative spelling for the prefix *palaeo-.

Paleocene *See* PALAEOCENE.

palimpsest A landscape that bears the imprint of two or more sets of geomorphological *processes. For example, much of the Sahel region of Africa shows land-forms resulting from former wet and dry episodes. The word is derived from the Greek *palimpsestos*, 'to rub smooth again', and is also used for a re-used parchment, paper, or ornamental brass whose original writing or engraving has been only partially erased and for a sedimentary structure in which one set of *trace fossils has been partly overwritten by another. *See* RELAXATION.

palingenesis The partial or complete melting of pre-existing rocks to generate new *magma.

palinspastic map A map on which the depiction of folded or faulted strata shows them restored as closely as possible to their original geographical positions prior to folding or faulting. *Balanced sections are used to ensure the palinspastic reconstruction accounts for the entire volume of the materials represented and respects the lengths of lines and the thicknesses of individual layers. The term 'palinspastic' is derived from the Greek *palin* meaning 'again', and *spastikos* meaning 'pulling'.

Pallas The second largest (after *Ceres) *solar system asteroid (No. 2), diameter 538 km; approximate mass 25×10^{19} kg; rotation period 7.811 hours; orbital period 4.61 years. It was discovered in 1802 by H. Olbers.

pallasite A striking and unusual kind of *meteorite in which *olivine and *iron are intergrown in roughly equal proportions; these meteorites are thus intermediate between *stony and *iron meteorites. Only 35 examples are known.

pallial line *See* MUSCLE SCAR.

pallial sinus *See* MUSCLE SCAR.

pallium *See* MANTLE.

Palmer method *See* GENERALIZED RECIPROCAL METHOD.

palsa Mound or ridge, largely made from *peat, containing a perennial ice lens, and found in the damper sites of bogs in *periglacial areas. Widths are in the range 10–30 m, lengths 15–150 m, and heights 1–7 m. Probably palsas are a result of heaving associated with the growth of segregated ice. *See also* PINGO.

paludal Applied to organisms, soils, etc., that are of or associated with a marsh.

palus From the Latin *palus* meaning 'marsh', a term introduced by Giovanni B. Riccioli in 1651 for small patches of lunar *mare *basalt. The Apollo 15 mission landed in Palus Putredinis, the 'Marsh of Decay'.

palustrine Pertaining to a marsh.

palynology The study of living and *fossil *pollen grains, *spores, and certain other *microfossils (e.g. *dinoflagellates and *coccolithophorids). Palynology was developed from *pollen analysis and deals principally with structure, classification, and distribution. It has many applications, e.g. in medicine, archaeology, petroleum exploration, and *palaeoclimatology.

palynomorph A *microfossil, 5–500 μm in size, made from an organic substance that is highly resistant to chemical attack.

pampero Regional storm of the *line-squall type which affects Argentina and Uruguay. Sometimes the storm brings rain, thunder, and lightning, and its passage is marked by a fall in temperature. The storm moves ahead of a south-westerly wind, which follows a *depression.

pan (clay pan) *Soil horizon, usually in the subsoil, that is strongly compacted, *indurated, *cemented, or very high in *clay content.

Pan (Saturn XVIII) One of the minor satellites of *Saturn, with a radius of 10 km; visual albedo 0.5.

Panama Isthmus The narrow neck of land that connects N. America and Mexico with S. America. During the *Mesozoic there was an open marine connection between the *Atlantic and *Pacific Oceans. At the culmination of the Andean–*Laramide orogeny (Late *Cretaceous) a temporary connection between N. and S. America was formed, which explains the similarity between the floras of California and S. America at that time. Some early mammals also migrated southwards across the isthmus. During the early *Cenozoic the land connection was broken but it became reestablished in the *Pliocene, thereby allowing another 14 mammal families to migrate from north to south. Some families, notably the elephants (*see* PROBOSCIDEA), appeared to be unable to migrate southwards, perhaps being constrained by the climatic zones that had to be traversed in such a migration.

pandemic distribution *See* COSMOPOLITAN DISTRIBUTION.

Pandora (Saturn XVII) One of the lesser satellites of *Saturn, discovered in 1980 by *Voyager 1, with a radius measuring 55 × 44 × 31 km; mass 0.0013×10^{20} kg; mean density 420 kg/m^3; visual albedo 0.9.

Pangaea A single supercontinent which came into being in late *Permian times and persisted for about 40 million years before it began to break up at the end of the *Triassic Period; or, in the view of some people, which existed throughout most of the Earth's history prior to the Triassic. It was surrounded by the universal ocean of *Panthalassa.

panning The concentration of heavy minerals by hand in a shallow pan. Loose material is shaken backwards and forwards in water; heavy minerals settle and the *gangue is washed out. It is a simple method for testing in the field.

pannus From the Latin *pannus* meaning 'shred', an accessory cloud term applied to ragged cloud either beneath or attached to another cloud, usually *cumulonimbus, *cumulus, *nimbostratus, or *altostratus. *See also* CLOUD CLASSIFICATION.

panplain (panplane, planplain) Area of very subdued relief that consists of coalesced *floodplains. It is therefore due to lateral stream migration and is a component of a *peneplain. Good examples are found in the Carpentaria region of Australia.

panplane *See* PANPLAIN.

Panthalassa The name given to the vast oceanic area that surrounded *Pangaea when that supercontinent was in existence. *Tethys was a minor arm of this ocean. Once Pangaea began to rift in the *Triassic, the names of the modern oceans are normally applied to the developing ocean basins even though they were, at that time, still very small.

Pantotheria (class *Mammalia, subclass Theria) Extinct infraclass of primitive Middle and Upper *Jurassic mammals, known from N. America, Europe, and E. Africa. They were egg-laying insectivores of shrew-like appearance, and are believed, on the basis of the structure of the jaw and teeth, to include the ancestors of all later placental and *marsupial mammalian groups. By the later Jurassic, they were widely distributed, numerous, and varied in form.

paper shale A dark grey to black *shale composed of thin, parallel laminae that tend to separate on *weathering into tough, slightly flexible sheets reminiscent of sheets of paper.

para- From the Greek *para* meaning 'beside' or 'beyond', a prefix meaning 'beside' or 'beyond'.

Parablastoidea (subphylum Pelmatozoa) A class of echinoderms (*Echinodermata) restricted to the Lower and Middle *Ordovician, which had *biserial *brachioles, well-developed *pentameral symmetry, and *ambulacra composed of *biserial plates.

parabolic dune *See* DUNE.

paraclade A group of *evolutionary lineages; paraclades may be *paraphyletic or *monophyletic, but not *polyphyletic.

paraconglomerate A *conglomerate in which the pebbles are supported by a fine-grained *matrix. The matrix constitutes more than 15% of the rock.

Paracrinoidea (subphylum Pelmatozoa) Class of echinoderms (*Echinodermata) with *uniserial *brachioles, and a box-like *theca composed of numerous *plates tending to *bilateral symmetry. They are restricted to the Middle and Upper *Ordovician.

paracycle In the *genetic stratigraphic sequence model used in *sequence stratigraphy, a sequence of changes in sea level, lasting about 2 million years, and forming part of a hierarchy of sedimentation cycles.

paradigm Essentially, a large-scale and generalized *model that provides a viewpoint from which the real world may be investigated. It differs from most other models, which are abstractions based on data derived from the real world.

paragenesis From the Greek *para* ('beside') and *gen* ('be produced'), a particular assemblage of *minerals all of which formed at the same time. *See also* PARAGENETIC SEQUENCE.

paragenetic sequence The chronological order of crystallization of minerals in a rock or ore deposit. *See also* ORE GENESIS; and PARAGENESIS.

paragonite *See* MUSCOVITE.

parallax The apparent change in position of an object in relation to another when the viewpoint is changed. *See also* STEREOPTIC VISION.

parallel evolution Similar evolutionary development that occurs in lineages of common ancestry. Thus the descendants are as alike as were their ancestors. The nature of the ancestry imposes or directly influences the development of the parallelism.

parallel extinction *See* STRAIGHT EXTINCTION.

parallel fold A *fold in which the *orthogonal thickness of layers remains constant throughout, so that adjacent, bounding, fold surfaces are parallel. Because of their geometry, parallel folds cannot maintain their profiles over long distances; their form changes along the *axial plane.

parallel twins A twinned crystal (*see* CRYSTAL TWINNING) where the twin axis lies in the *composition plane parallel to a *zone axis, e.g. a *Carlsbad twin in *orthoclase feldspar.

paramagnetism Magnetization developed in atoms or *ions that have permanent intrinsic magnetic moments, in the same direction as an applied field. It is usually somewhat larger than that resulting from the *diamagnetism of the material, and may have *ferromagnetism superimposed upon it.

parameter (intercept ratio) In *crystallography, the ratio of the intercepts made by a plane (the parametral plane), parallel to a *crystal face, which intersects the *crystallographic axes, and which has been chosen to define a unit length of intersection along each axis. The *form of which the face is a member is called the 'unit form', 'fundamental form', or 'parametral form'. *See also* AXIAL RATIO.

parametral form *See* PARAMETER.

parametral plane *See* PARAMETER.

paramorph A crystal formed by the conversion of one *mineral polymorph (*see* POLYMORPHISM) to another. The polymorphs of silica provide a good example. If the high-temperature form, *cristobalite, converts to a lower-temperature form, e.g. *tridymite, the tridymite would form a paramorph.

parapatric Applied to species whose habitats are separate but adjoining. *Compare* ALLOPATRIC; and SYMPATRIC.

paraphyletic Of a *taxon, including some but not all descendants of the common ancestor (i.e. not *holophyletic).

Parapithecidae (order Primates, suborder

Simiiformes) An extinct family of primates which lived during the *Eocene and *Oligocene in Egypt; Eocene fossils from Burma are sometimes included in the family in addition. They showed certain similarities in dentition to *Condylarthra, but had short faces and jaws shaped like those of tarsiers.

parasequence In the *genetic stratigraphic sequence model used in *sequence stratigraphy, a relatively conformable sequence of genetically related beds bounded by surfaces formed by marine flooding. A succession of genetically related parasequences forms a parasequence set.

parasequence set *See* PARASEQUENCE.

parasitic cone (adventive cone) A conical mound of ejecta accumulated around an eruptive vent on the lower flanks of a large *volcano. Parasitic cones sometimes grow into large volcanic centres themselves, and may lie on the line of a fissure which radiates a great distance from the main volcanic conduit.

parasitic fold A *fold of small *wavelength and *amplitude which usually occurs in a systematic form superimposed on folds of larger wavelength. Parasitic folds usually show typical S and Z asymmetric profiles on their limbs and M profiles in the *hinge regions.

parasitic magnetization *Ferromagnetic behaviour resulting from the imperfect cancellation of antiparallel magnetic lattices in an *antiferromagnetic substance. *Hematite is a mineral exhibiting this magnetic behaviour.

parastratotype An additional stratigraphic section, designated and described at the time of the establishment of a stratigraphic unit, to augment the definition given by the principal *stratotype (the *holostratotype). The parastratotype is usually selected from within the *type area. *Compare* HYPOSTRATOTYPE. *See also* LECTOSTRATOTYPE; and NEOSTRATOTYPE.

parataxon Artificial classification suggested for certain common organisms of doubtful affinities, or as yet unknown origins, e.g. *fossil *spores, *dinosaur footprints.

Paratethys (Central European Sea) A large, arcuate seaway that, at its maximum development, extended a distance of 4500 km from just north of the Alps to just east of the Aral Sea. By the end of the *Oligocene it had been separated from the Boreal Sea by closure of the Ural, Polish, and Alsace Straits. Towards the end of the *Miocene it became more lagoonal in character and by the *Pliocene was represented only by a series of land-locked lakes: Lake Balaton in Hungary, the Black Sea (rejoined to the Mediterranean Sea by *Quaternary faulting), the Caspian Sea, and the Aral Sea.

paratype Specimen, other than a *holotype, used as the 'type' material by an author at the time of the original description, and designated as such by the author. *Compare* LECTOTYPE; and NEOTYPE.

parcel of air Quantity of air with more or less uniform characteristics.

Pareiasauridae The largest (up to 3 m in length) of the stem reptiles (*cotylosaurs), the pareiasaurs are found in Middle and Upper *Permian rocks in Europe and Africa. They were herbivorous, and perhaps lived semi-aquatically in swamps. They were distinctive among cotylosaurs in that the limbs were rotated in towards the body, thus supporting the body in a more upright fashion.

parent material The original material from which the *soil profile has developed through *pedogenesis, usually to be found at the base of the profile as weathered but otherwise unaltered mineral or organic material.

parent *See* RADIOACTIVE DECAY; and RADIOMETRIC DATING.

pargasite *See* HORNBLENDE.

parivincular In *Bivalvia, applied to a type of ligament that is elongate and cylindrical in shape. It occurs in a position posterior to the *shell beaks.

parsimony In *cladistic analysis, the convention whereby the simplest explanation is preferred, because it requires the fewest conjectures, although the most parsimonious explanation is not always the correct one.

partial melting Incomplete melting of parent rock, characteristically producing a *melt whose chemical composition differs from that of the parent material. It is thought that partial-melting processes play a major role in generating more-defined liquids from less-evolved ones, so that many *basalts may be the result of partial melting in the (*ultrabasic) upper *mantle, and many *granites may have derived partly or completely from the partial melting of *continental crust (*anatexis). Partial melting preferentially enriches melts with *incompatible elements. In a *subduction zone, rocks of *intermediate composition may form (e.g. *andesites). With increasing temperature and pressure, the subducted *oceanic crust (of *basic composition) first undergoes *metamorphism and then begins to melt or release watery fluids; this material rises into the overlying mantle, which may also begin to melt, giving rise to intermediate *magma.

partial pressure In a mixture of gases, the partial pressure of any one is the pressure it would exert alone in the same space as that occupied by the whole. The interaction of the partial pressures of each of the gases equals the total pressure of the mixture.

partial range zone A body of *strata containing the documented lowest occurrence of one taxon and the documented highest occurrence of another taxon, but with no stratigraphic overlap of the taxa. *See also* INTERVAL ZONE. *Compare* CONCURRENT RANGE ZONE.

particle density The mass per unit volume of soil particles, usually expressed in grams per cubic centimetre. *Compare* BULK DENSITY.

particle shape (grain shape) Particle shape is defined by the relative dimensions of the long, intermediate, and short axes of the particle. The ratio of intermediate to long diameter and the ratio of short to intermediate diameter is used to define four shape classes. These are: *oblate (tabular or disc-shaped forms); *prolate (rod-shaped); *bladed; and *equant (cubic or spherical forms).

particle size (grain size) The diameter or volume of the *grains in a *sediment or *sedimentary rock. The particle size can be determined by sieving, by measuring the settling velocity, or (for pebbles, boulders, and cobbles) by direct measurement of individual *clasts. The smaller particles are normally defined by their volume diameters, i.e. the diameter of a sphere with the same volume as the particle. There are numerous particle-size classifications. Two commonly used ones are the Udden–Wentworth scale (Wentworth scale) and the British standard classification. In the Udden–Wentworth scale the sediment *grades are: more than 256 mm, boulder; 64–256 mm, cobble; 2–64 mm, pebble; 62.5–2000 μm, sand; 4–62.5 μm, silt; less than 4 μm, clay. In the British classification: more than 200 mm, boulder; 60–200 mm, cobble; 2–60 mm, gravel; 600–2000 μm, coarse sand; 200–600 μm, medium sand; 60–200 μm, fine sand; 2–60 μm, silt; less than 2 μm, clay. *See* PHI SCALE.

particle velocity The velocity of a particle in a medium affected by the propagation of a seismic wave. It depends on the amplitude of the wave and for weak seismic events may be of the order of 10^{-8} m/s. *Compare* SEISMIC VELOCITY.

parting lineation A series of parallel, linear, low-relief ridges and hollows, spaced a few millimetres apart, seen on *bedding planes in parallel-laminated *sandstones. The *lineation is the preserved trace of primary current lineations formed when *plane-bedded sand is deposited from a high-velocity and usually shallow flow of water. The lineation is oriented parallel to the flow direction. *See* BED FORM.

partition coefficient (distribution coefficient) 1. If a substance is dissolved in two immiscible liquids standing in contact with each other, the substance will partition or distribute itself between them in a constant ratio, called the partition coefficient. The value of this constant is dependent on the temperature, and on the identities of the solute and the solvents. (The number of solute molecules in solvent A divided by the number of solute molecules in solvent B is a constant.) **2.** The ratio of the concentration (by weight) of an element (e.g. Ti) in a crystallizing *mineral to its concentration in the *magma. For example, K_{Ti} = $[Ti]_{min.}/[Ti]_{magma}$, where k_{Ti} is the partition coefficient for Ti, and $[Ti]_{min.}$ and $[Ti]_{magma}$ are the concentrations of Ti in the mineral and magma respectively. The value of k is dependent on temperature, pressure and the composition of crystallizing mineral and magma.

pascal 1. (Pa) The derived SI unit of pressure, equal to 1 N/m^2. **2.** A high-level computer programming language. Both are named after the French mathematician Blaise Pascal (1623–62).

pascichnia One of five groups of *trace fossils established in a behavioural (ethological) classification by A. Seilacher (1953). Grazing traces (pascichnia) result from the distinctive behavioural pattern of an animal during feeding. *Deposit feeders, e.g. **Nereites*, leave a recognizable trace through the exploitation of *sediment for nutrition.

Pasiphae (Jupiter VIII) One of the lesser satellites of *Jupiter, with a diameter of 36 km; its orbit is *retrograde.

pass band The range of frequencies that can be passed through a band-pass filter with negligible *attenuation. *See* BAND FILTER.

passive margin (trailing edge) A continental margin which is not also a *plate margin. Such margins are also known as 'aseismic margins' or 'Atlantic-type margins' and are contrasted with *active margins. Passive margins are characterized by rifted and rotated blocks of usually thick sedimentary sequences. These rocks are often highly prospective for oil and gas, with a variety of traps, including those related to the diapiric (*see* DIAPIRISM) rise of the rock salt formed during the initial separation of the continents.

passive microwave *Electromagnetic radiation with a wavelength between 1 mm and 1 m, which is emitted by all objects at temperatures higher than *absolute zero.

passive remote sensing *Remote sensing which is based on the illumination of a scene by *electromagnetic radiation from a natural source. An example is photography. *Compare* ACTIVE REMOTE SENSING.

Pasteur effect The transition from an anaerobic to an *aerobic life-style, which occurs among certain organisms when the oxygen content of the atmosphere is 1% of that of the present day. The critical point of transition is the 'Pasteur point'. The gradual oxygen enrichment of the Earth's atmosphere during the *Precambrian passed through the Pasteur point approximately 700 Ma ago, resulting in a general transition to an aerobic life-style.

Pasteur pint *See* PASTEUR EFFECT.

Pastonian A Middle *Pleistocene *stage represented by estuarine silts and freshwater peat, revealed as marine clays in the *borehole at Ludham, Norfolk, England. *See also* ANTIAN; BAVENTIAN; LUDHAMIAN; and THURNIAN.

patch *See* GROUP.

patch reef *See* REEF.

'patella' beach Old term referring to a raised beach standing about 3 m above the present high-water mark and found locally on the south coast of England, in southern and western Ireland, and along the Channel coast of France. Traditionally it was given an *interglacial age, but the term is now little used, owing to problems of correlations. It is named after the large number of limpets (*Patella vulgata* is the common limpet) it typically contains.

patellate *See* SOLITARY CORALS.

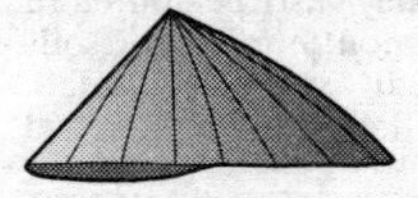

Patellate

Patella vulgata *See* ARCHAEOGASTROPODA; and 'PATELLA' BEACH.

patera An irregular or complex martian structure, with scalloped edges, usually a volcanic *caldera. The classic example is Alba Patera (110° W, 40° N), the volcano of largest areal extent on Mars, but only 3 km high. On the jovian satellite Io, the term is used for a volcanic crater surrounded by irregular flows. The name is derived from a type of shallow, ceremonial dish used by the Romans.

paternoster lake Body of water in a formerly glaciated environment that is aligned with neighbouring lakes, so that it looks like a paternoster in a rosary. It is caused by irregular glacial scouring along a zone of weakness.

patterned ground An assemblage of small-scale, geometric features typically found at the surface of a *regolith that has been disturbed by frost action. The group includes circles, polygons, and nets, which normally occur on level or gently sloping surfaces, and steps and stripes which are found on steeper gradients. Both sorted and non-sorted varieties are recognized. The sorted varieties are typically outlined by coarse, stony material, and so are termed 'stone circles', 'stone polygons', 'stone nets', 'stone steps', and 'stone stripes'. The origin of patterned ground involves a complex interaction of several geomorphological *processes, including ground cracking, frost sorting, *frost heaving, and *mass movement. The 'ice-wedge polygon' is an important member. It is usually 15–30 m in diameter and bounded by *ice wedges up to 3 m wide and about 10 m deep which occupy contraction cracks that form under very low temperatures. The wedges define raised zones (when freezing is active) or depressions (due to thaw). The 'stone garland' is a variety of sorted step, which ends in a stony riser (less than 1 m high) supporting a relatively bare tread (less than 8 m long) upslope. It is found on gradients of 5–15° (in Alaska) and may be due to a combination of *frost pull and frost push, which heave stones to the surface, and mass movement. Patterned ground may also be found in areas underlain by *montmorillonitic soils experiencing markedly seasonal rainfall, where the microrelief forms are called *gilgai.

Pauling's rules Rules relating to chemical bonding based on certain regular features shown by ionically bonded, usually simple solids, and first summarized by L. Pauling (1960).

pavement 1. In geology, bare rock surface resembling a road, e.g. limestone pavement. **2.** The floor of a coal seam. **3.** In roadway construction, any material spread on the subgrade to distribute load and protect against erosion and traffic wear. The types of material used and thickness depend on the type and use of road. Generally the pavement has four layers. (*a*) The wearing course is the top layer of the carriageway. It must be durable, impermeable, skid-resistant, and resistant to polishing. On the most heavily trafficked roads the polished stone value (PSV) (*see* AGGREGATE TESTS) must exceed 60%. The wearing course lies on a surface provided by the basecourse. (*b*) The basecourse is a layer of gravel or crushed rock (*aggregate) of specific dimensions to provide drainage, distribute load on to the road base, and protect against freezing. (*c*) The road base provides the main load-bearing foundation or ballast and consists of irregular-sized rock aggregate, either bonded or loose. (*d*) The subbase is a layer of coarse aggregate below the road base. It provides extra support or drainage. In cold climates large pore spaces are necessary to prevent water rising by capillary action, leading to the collapse of the structure after thawing. (*e*) The subgrade is the rock or subsoil on which the subbase is laid.

Payntonian A *stage of the Upper *Cambrian of Australia, underlain by the Pre-Payntonian and overlain by the *Datsonian *(Ordovician).

paystreak Profitable part of a *mineral deposit. In *alluvial *gold mining, the term often refers to pockets of gold concentrated at or near bedrock.

P-band The radar frequency band between 225 and 390 MHz. It is commonly used in *remote sensing because of its good vegetation penetration.

PcP-wave *See* SEISMIC WAVE MODES.

PE *See* POTENTIAL EVAPOTRANSPIRATION.

peacock ore *See* BORNITE.

peak shear strength *See* SHEAR STRENGTH.

peak zone *See* ACME ZONE.

pearl-necklace lightning (chain lightning, beaded lightning) Infrequent type of discharge, in which variations in luminosity resemble pearls on a string.

pearlspar *See* DOLOMITE.

pearly Of *mineral *lustre, milky, translucent, like a pearl.

peat An organic soil or deposit; in Britain, a soil with an organic *soil horizon at least 40 cm thick. Peat formation occurs when decomposition is slow owing to *anaerobic conditions associated with waterlogging. Decomposition of cellulose and hemicellulose is particularly slow for *Sphagnum* plants, which are characteristic of such sites, and hence among the principal peat-forming plants. Fen and bog peats differ considerably. In fen peats the presence of calcium in the groundwater neutralizes acidity, often leading to the disappearance of plant structure, giving a black, structureless peat. Bog peats, formed in much more acidic waters, vary according to the main plants involved. Species identification of constituents (including animals as well as plants) remains possible after long periods. Recent bog-moss (*Sphagnum*) peat is light in colour, with the structure of the mosses perfectly preserved.

peat-borer Implement designed to extract *peat cores with the minimum of disturbance. The most familiar is the Hiller peat-borer (or corer), which consists of a short screw *auger head to ease penetration of the peat, backed by a chamber which can be opened and closed at the required sample depth, the sharp cutting edge of the chamber assisting detachment of the sample in more consolidated peats. The principal alternatives are the piston sampler, which is particularly good for loose peats, and the Russian borer, which allows easier removal of the complete peat core than is possible with the Hiller borer, but which is more difficult to use in compacted peats since it has no screw auger head.

peat podzol *Podzol *soil profile distinguished by having a surface *mor (peaty) *humus up to a maximum thickness of 30 cm, and usually with an *iron pan at the top of the B *soil horizons. The term occurs in most of the classification systems derived originally from the work of V. V. *Dokuchaev, published in 1886. It has been superseded and podzols now fall within the order *Spodosols.

pebble *See* PARTICLE SIZE.

pectolite *See* WOLLASTONITE.

pectoral girdle In vertebrates (*see* CRANIATA), the skeletal structure that provides support for the fore limbs or fins.

ped Unit of soil structure (e.g. an *aggregate, crumb, granule, or prism) that is formed naturally. *Compare* CLOD.

pedestal rock (mushroom rock) An unstable, mushroom-shaped land-form found typically in arid and semi-arid regions. The undercut base was formerly attributed to wind *abrasion, but is now believed to result from enhanced *chemical weathering at a site where moisture would be retained longest. A famous example is Pedestal Rock, Utah, USA.

pedicel *See* ARACHNIDA.

pedicle A fleshy stalk which attaches most brachiopods (*Brachiopoda) to the sea floor. It emerges either posteriorly from between the two valves of the shell or from a triangular notch (the *delthyrium) in one of them. In some species it issues from a circular *pedicle foramen (in the ventral valve). In other species it is either not developed or atrophied.

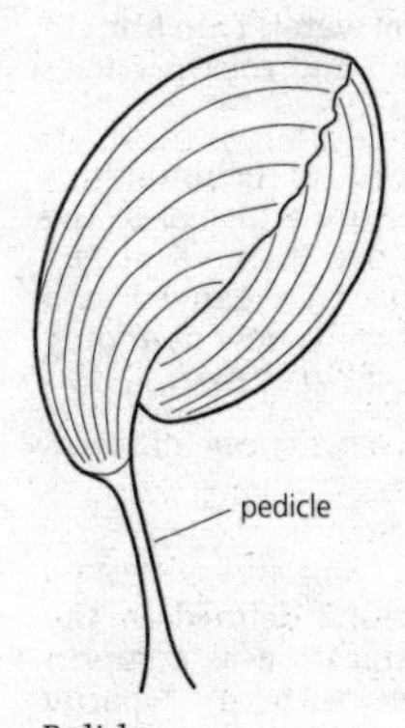

Pedicle

pedicle foramen A rounded, slit-like, or closed opening in the *pedicle valve of a brachiopod (*Brachiopoda) for the passage of the pedicle to the exterior.

pediment (concave slope, waning slope) Surface of low relief, partly covered by a skin of rock debris, that is concave-upward and slopes at a low angle (normally less than 5°) from the base of a mountain zone or *scarp. Classically it is developed and has been investigated in the arid and semi-arid regions of the western USA.

pedion A *crystal form consisting of only one face.

pediplain Extensive plain, best developed in arid and semi-arid regions, showing gently concave or straight-slope profiles and terminated abruptly by uplands. A result of scarp recession rather than of surface lowering, it consists of coalesced *pediments.

pedofacies A *facies, the composition of which can be used to relate the maturity of a *palaeosol formed by *avulsion to its distance from the active river channel.

pedogenesis The natural process of soil formation, including a variety of subsidiary processes such as humification, *weathering, *leaching, and *calcification.

pedology The scientific discipline devoted to the study of the composition, distribution, and formation of soils, as they occur naturally.

pedon Three-dimensional sampling unit of soil, with depth to the *parent material and lateral dimensions great enough to allow the study of all *soil-horizon shapes and *intergrades below the surface.

peel technique Originally a technique for palaeobotanical work, but then refined and now used extensively in carbonate sedimentology and palaeontological work. Calcareous material is etched in a weak solution of hydrochloric acid and differences in relief are produced. After washing, the surface is flooded with acetone, and polyvinylacetate (PVA) sheeting is rolled on to the surface. The acetone softens the sheeting and moulds it to the etched rock surface. After drying the sheeting is peeled from the surface, bringing a thin layer of the surface with it. This 'peel' can then be examined in transmitted light. A series of peels can be taken to reveal and reconstruct buried structures and staining with various chemicals may reveal additional details.

pegmatite Very coarse-grained, *igneous rock, usually of granitic composition, in which the individual *crystals are at least 2.5 cm long. The crystals are often more than 1 m in length, and may be far larger. Crystallization occurs at a late stage, when the *magma is enriched in *volatiles and *trace elements. Pegmatites may concentrate some rare elements (lithium, boron, fluorine, tantalum, niobium, *rare earth elements, and uranium) to economic proportions.

pel- The prefix used in the *Folk limestone classification to denote a *limestone containing *pellets as the major constituent of the particles present.

pelagic In marine ecology, applied to the organisms that inhabit open water, i.e. *plankton, *nekton, and neuston (although neuston are fairly unimportant in such environments).

pelagic ooze Deep-ocean *sediments that accumulate by the settling out of materials from the overlying ocean waters. The dominant constituents are microscopic *pelagic organisms, e.g. the calcareous *globigerina and pteropods, and the siliceous *diatoms and *radiolaria. Minor amounts of fine volcanic, terrigenous, and extraterrestrial debris also contribute to pelagic ooze.

pelagic sediment (pelagite) A sediment formed in the open sea by the slow settling of calcareous and siliceous biogenic particles, biogenic material comprising more than 75% of the volume. *Compare* HEMIPELAGIC SEDIMENT.

pelagite *See* PELAGIC SEDIMENT.

Peléean eruption A very violent type of volcanic activity characterized by the release of *nuées ardentes (glowing clouds of gas-fluidized material). Highly viscous *lavas are typically involved and the *eruptions may be preceded by the growth of a lava dome. Nuées ardentes may be formed by the explosive disruption of a growing dome, or by mechanical collapse of a dome growing on a steep slope. *See* VOLCANO. *Compare* HAWAIIAN ERUPTION; PLINIAN ERUPTION; STROMBOLIAN ERUPTION; VESUVIAN ERUPTION; and VULCANIAN ERUPTION.

Pelecypoda *See* BIVALVIA.

Pelé's hair Thin filaments of basaltic glass formed from chilled *lava spray thrown out from a *volcano during *Hawaiian *eruptions, and named after Pelé, the Hawaiian goddess of volcanoes. The lava is so fluid that on eruption it forms droplets, shaped primarily by surface-tension forces, drawing behind them the long filaments which break to form the Pelé's hair. The filaments may be metres long, and drift downwind for many kilometres. The solidified droplets themselves are known as 'Pelé's tears'. Fragments formed from solidified lava spray are given the general name 'achneliths'.

Pelé's tears *See* PELÉ'S HAIR.

pelite (adj. pelitic) An aluminium-rich, *metamorphic rock formed by the *metamorphism of clay-rich *sedimentary rocks, e.g. *shales and *mudstones. The type of aluminium-bearing *silicate minerals seen in the rock depends on the pressure and temperature of the metamorphism but usually includes one of the *mica group minerals. *Quartz is ubiquitous in pelites.

pelitic *See* PELITE.

pellet limestone *See* LEIGHTON–PENDEXTER CLASSIFICATION.

pellets *Grains of faecal origin, commonly found in *limestones and *phosphorites. Structureless grains resembling pellets which cannot be shown to be of faecal origin are termed 'peloids'. *See* FAECAL PELLET.

pellicular water (film water) Thin films of water that cling to soil and rock particles above the *water-table.

pelmatozoan Applied to echinoderms (*Echinodermata) that are attached to the substratum (e.g. primitive *Crinoidea). Previously the Pelmatozoa were ranked as a subphylum, but the term is now used only informally. *Compare* ELEUTHEROZOAN.

pelmicrite *See* FOLK LIMESTONE CLASSIFICATION.

peloid *See* PELLETS.

pelsparite A *limestone defined in the *Folk limestone classification as comprising pellets together with a *sparite *cement.

pelvic girdle 1. In vertebrates (*see* CRANIATA), the skeletal structure that provides support for the hind limbs or fins. **2.** The part of the abdomen that is surrounded by the bony pelvis.

pelvis In vertebrates, part of the appendicular skeleton that is fused to the sacral vertebrae and provides support for the hind limbs or fins.

Pelycosauria Order of *synapsid reptiles dating from the Upper *Carboniferous and Lower *Permian, e.g. *Varanosaurus*, *Edaphosaurus*, and *Dimetrodon*. Several sported large, sail-like, dorsal fins, and both carnivorous and herbivorous types appeared. They gave rise to and were replaced by the mammal-like reptiles (*Therapsida).

Penck, Albrecht (1858–1945) A German mineralogist from the University of Berlin, Penck's main interest was in *Quaternary glacial land-forms, which he classified according to shape. Some of his work was done in co-operation with his son Walther *Penck.

Penck, Walther (1888–1923) A German geologist, Penck assisted his father Albrecht *Penck in his studies of land-forms. Independently, he worked on the structure of mountain regions, especially the Alps, modifying *Suess's ideas on continental uplift.

pendent Applied to the attitude of the *stipes that hang downwards from the *sicula in graptolites (*Graptolithina). This is the more *primitive condition, from which other graptoloid (*Graptoloidea) stocks arose.

Pendleian The first *stage of the *Serpukhovian Epoch, overlain by the *Arnsbergian.

pene- From the Latin *paene* meaning 'almost', a prefix meaning 'almost' or 'nearly' (e.g. a *peneplain is almost a plain).

peneplain (peneplane) Literally, almost a plain: an extensive area of low relief, dominated by convex-up ('bulging') hillslopes mantled by a continuous *regolith, and by wide, shallow river valleys. Locally, *monadnocks may occur. A peneplain is the end-product of a cycle of erosion, produced by the action of down-wearing over a long period of time, and it is the end-product of the *Davisian cycle.

peneplane *See* PENEPLAIN.

penesaline A level of *salinity intermediate between normal marine and hypersaline, ranging from 72 parts per thousand to 352 parts per thousand). These salinity levels are high enough to be toxic to normal marine organisms, and can be tolerated by only a restricted range of fauna and flora. The characteristic *sediments of penesaline zones are evaporitic *carbonates (*see* EVAPORITE) interbedded with *anhydrite or *gypsum. Penesaline environments are often encountered in the back-*barrier and *backreef zones.

penetration test A standard penetration test of *sand or *silt determines the number of blows of a standard weight from a standard height that are needed to produce a standard penetration. A dynamic penetration test determines the *density of layered *sediments by recording the penetration per blow or series of blows. *See also* CONE PENETROMETER.

penetration twin *See* INTERPENETRANT TWIN.

Pennales *See* BACILLARIOPHYCEAE.

pennate diatoms *See* BACILLARIOPHYCEAE.

Pennatulacea *See* OCTOCORALLIA.

penninite *See* CHLORITE.

Pennsylvanian 1. The Late *Carboniferous subperiod, preceded by the *Mississippian, comprising the *Bashkirian, *Moscovian, *Kasimovian, and *Gzelian *Epochs, and dated at 322.8–290 Ma (Harland et al., 1989). **2.** The name of the corresponding N. American sub-system, comprising the *Morrowan, *Atokan (Derryan), *Desmoinesian, *Missourian, and *Virgilian *Series, and roughly contemporaneous with most of the *Silesian sub-System (i.e. above *Namurian A).

pentadactyl Applied to a limb possessing five digits, or to one modified evolutionarily from an ancestral form which possessed five digits; characteristic of all *tetrapods.

pentagonal dodecahedron *See* PYRITOHEDRON.

pentameral symmetry Five-sided symmetry, found in most members of the phylum *Echinodermata. In some groups this symmetry has a *bilateral symmetry superimposed on it.

Pentamerida (pentamerids; class *Articulata) Extinct order of brachiopods (*Brachiopoda), with thick, strongly biconvex shells. A *spondylium developed in the *pedicle *valve and brachial processes and supporting plates in the brachial (*dorsal) valve, forming an internal calcareous box containing the adductor muscles (which close the valves) and diductor muscles (which open the valves). The shell is *impunctate. Pentamerida first appeared in the Middle *Cambrian and became extinct in the Upper *Devonian.

pentamerids *See* PENTAMERIDA.

pentectic point *See* BOWEN'S REACTION PRINCIPLE.

Pentevrian A *stage of the Lower and Middle *Proterozoic of Brittany, from about 2600 to 1000 Ma ago, overlain by the *Brioverian.

pentlandite An iron and nickel sulphide, with the formula $(FeNi)_9S_8$, and the main *ore mineral for nickel; sp. gr. 4.5–5.0; *hardness 3–4; *cubic; light bronze-yellow; greenish-black *streak; *metallic *lustre; very rarely found in well-formed crystals, usually occurring as irregular grains and inclusions; no *cleavage, parting {111}; occurs with the other *sulphide ores genetically associated with *basic and *ultrabasic rocks, together with *pyrrhotine and *chalcopyrite.

Penutian A *stage in the Lower *Tertiary of the west coast of N. America, underlain by the *Bulitian, overlain by the *Ulatizian, and roughly contemporaneous with the lower *Ypresian Stage.

peperite A mixture of solidified *magma and sediment found along the margin of an *intrusion or at the base of a laval flow.

peralkaline A chemical classification applied to *felsic *igneous rocks in which there are more molecules of $(Na_2O + K_2O)$ than of Al_2O_3. Rocks with such a chemistry crystallize alkali-rich *ferromagnesian minerals such as *aegirine, aegirine-*augite, *barkevikite, and *arfvedsonite. *See also* METALUMINOUS.

peraluminous *See* METALUMINOUS.

peramorphosis Evolutionary change that results in the descendant incorporating all the ontogenetic stages of its ancestor, including the adult stage, in its *ontogeny, so that the adult descendant 'goes beyond' its ancestor. It may occur by *acceleration, *hypermorphosis, or *predisplacement.

perched aquifer *See* AQUIFER.

perched block Glacially transported boulder resting on bedrock where it was deposited by melting ice. Some examples are spectacular and give rise to legends locally.

percolation In *pedology, the downward movement of water through soil, especially through soil that is saturated or close to saturation.

percussion boring Drilling with a chisel-shaped *bit which is repeatedly struck against rock to form a hole. It produces very fine debris, which can be washed away by drilling fluid. If only the drill bit can oscillate up and down to produce the percussive effect, the drilling equipment is called a 'down-the-hole hammer'. If the whole series of drilling rods and bit are oscillated from outside the drill hole, the equipment is called a 'drifter'. *See* DOWNHOLE HAMMER DRILLING.

perennial stream A stream that would normally be expected to flow throughout the year, albeit with only small dry-weather flows in some cases.

pergelic The lowest of the soil-temperature classes for family groupings of soils in the soil taxonomy system, applied to soils in temperate regions. The assessment of soil temperature is based on the mean annual temperature, and on the difference between mean summer and mean winter temperature, measured at a depth of 50 cm or at the surface of the underlying rock, whichever is shallower. In order of ascending temperature, the higher-temperature classes in temperate-region soils are called cryic, frigid, mesic, thermic, and hyperthermic, and in tropical regions the scale from cold to hot is isofrigid, isomesic, isothermic, and isohyperthermic.

pergelisol *See* PERMAFROST.

peri- From the Greek *peri* meaning 'around', a prefix meaning 'around' or 'enveloping'.

periclase Magnesium oxide (MgO), a member of the *oxide group of minerals; sp. gr. 3.5; *hardness 5.5; white, yellow, or green; crystalline or granular; occurs in contact-metamorphosed (*see* THERMAL METAMORPHISM) *limestones and may be altered to *brucite $(Mg(OH)_2)$ by hydration.

periclinal fold *See* PERICLINE.

pericline (periclinal fold) A *fold shown on a map as concentrically arranged contour patterns whose amplitude decreases to zero in a regular fashion in two directions. An *anticlinal pericline is called a 'dome', a *synclinal pericline a 'basin'.

pericline twinning One of the many types of twins (*see* CRYSTAL TWINNING) shown by the *plagioclase feldspars. The twin plane is along the *b* (or *y*) axis and produces polysynthetic *lamellar twinning which results in a series of fine striations on the *cleavage.

pericontinental sea (marginal sea) A sea that surrounds a continent.

periderm The thin, sheet-like material that makes up the skeleton in graptolites (*Graptolithina). Long thought to be *chitin, it is now considered perhaps to be *collagen (a scleroprotein). It is bi-layered, with an inner fusellar layer and outer cortical layer which have distinct structural differences.

Peridiniales *See* DINOPHYCEAE.

peridotite A coarse-grained, *ultrabasic, *igneous rock consisting of *essential magnesium-rich *olivine accompanied by lesser amounts of other *ferromagnesian minerals, e.g. *orthopyroxene (*enstatite-*bronzite), *clinopyroxene (chromium-*diopside), and *chromite, with or without *pyrope *garnet. Peridotites are found in large, layered, *ultrabasic, *intrusive bodies, in *ophiolite complexes, and as *xenoliths brought up in *alkali *basalts and *kimberlites. As well as forming by *crystal fractionation processes in igneous bodies, their mineral chemistry and presence as xenoliths in kimberlites suggest that much of the Earth's *mantle, and therefore of the mass of the Earth, is peridotite. *Meteorites are also composed largely of peridotite, suggesting that peridotites are probably the commonest rock in the *solar system.

peridotite model A model for the composition of the Earth's upper *mantle based on compressional and *S-wave velocities and gravity data. The probable composition is *peridotite, *eclogite, or an intermediate between the two. *Xenoliths in *kimberlites, and *mafic volcanic rocks found at the surface, also suggest this composition.

perigee The point nearest the planet in the orbit of a *satellite, originally defined for the lunar orbit. *Compare* APOGEE.

periglacial Applied strictly to an area adjacent to a contemporary or *Pleistocene *glacier or *ice sheet, but more generally to any environment where the action of freezing and thawing is currently, or was during the Pleistocene, the dominant surface process.

perignathic girdle Continuous or discontinuous ring of internal processes around the *peristome of echinoids (*Echinoidea), to which the muscles which support and control the lantern (the chewing mechanism, *see* ARISTOTLE'S LANTERN) are attached.

perihelion The point in the Earth's elliptical orbit at which the planet is closest to the Sun. During the present epoch this occurs about the beginning of January, when the average amount of incident solar radiation is at a maximum. *See also* APHELION.

period 1. Second-order geologic time unit which is the equivalent of the chronostratigraphic unit '*system'. Periods are subdivided into *epochs; together, several periods constitute an *era. When used formally the initial letter of the term is capitalized, e.g. the *Devonian Period. **2.** (T) The time that elapses between repetitions of the same phase of a wave-form, i.e. the time taken to complete one cycle. For a simple harmonic function, $T = 2\pi/\omega$ where ω is the angular velocity; for a wavetrain of single *frequency f, $T = 1/f = \lambda/V$ where λ is the *wavelength and V the *phase velocity.

periostracum *See* SHELL STRUCTURE; and SKELETAL MATERIAL.

peripatric speciation *See* FOUNDER EFFECT.

periproct In *Echinoidea, the area surrounding the anus and covered in leathery skin in which small, calcitic plates are loosely embedded.

Perissodactyla (cohort *Ferungulata) Odd-toed ungulates. Order that comprises those ungulates in which the number of functional toes is reduced to three or one, and a fourth, if present, is reduced, the weight of the animal being borne by the central digit. The order includes the three suborders Ceratomorpha (tapirs, rhinoceroses, and their extinct relatives); Ancylopoda (or Chalicotheres), which are extinct forms with claw-like extremities in place of hoofs, e.g. *Moropus*; and Hippomorpha (horse-like forms). They appeared in the *Eocene, derived from the *condylarths, and reached their zenith in the mid-*Tertiary, when they were the most numerous of ungulates. So many *fossil remains of them have been found that their evolutionary history is known in greater detail than that of any other mammalian group. Since then the order has declined dramatically, having been displaced by the *artiodactyls, and the group as a whole is moving towards extinction.

peristome **(adj. peristomal)** In *Echinoidea, the area around the mouth, covered in a leathery skin studded with small plates.

peritectic point *See* BOWEN'S REACTION PRINCIPLE.

peritidal Applied to the zone extending from above the level of the highest *tide to below that exposed at the lowest tide, and thus somewhat wider than the *intertidal zone.

perlitic Displaying curved or sub-spherical cracks. The texture is found in glassy or devitrified *igneous rocks and forms by contraction during rapid cooling of the *magma.

perlucidus From the Latin *perlucidus* meaning 'allowing the passage of light', a variety of cloud comprising extensive layers or sheets with holes which allow a view beyond the cloud. The term is applied to *stratocumulus and *altocumulus. *See also* CLOUD CLASSIFICATION.

permafrost **(pergelisol)** Permanently frozen ground which occupies some 26% of the Earth's land surface under thermal conditions where temperatures below 0 °C have persisted for at least two consecutive winters and the intervening summer. Considerable thicknesses may develop, e.g. 600 m on the North Slope of Alaska, and 1400 m in Siberia, but these are partly relict from the last glaciation. Permafrost may contain an unfrozen unit, called 'talik', and may be overlain by an *active layer.

permafrost table The upper limit of *permafrost. *Compare* TJAELE.

permanent wilting percentage **(permanent wilting point, wilting coefficient, wilting point)** The percentage of water remaining in the soil after a specified test plant has wilted under defined conditions, so that it will not recover unless it is given water.

permanent wilting point *See* PERMANENT WILTING PERCENTAGE.

permeability **1. (hydraulic conductivity)** In general, the ability of a rock, *sediment, or soil to permit fluids to flow through it. More precisely, the hydraulic conductivity is the volume flow rate of water through a unit cross-sectional area of a porous medium under the influence of a *hydraulic gradient of unity, at a specified temperature. It is measured in units of m/s or m/day and varies with temperature. Typical values range from 10^{-6} m/day for *clay to 10^{3} m/day for coarse gravel. The magnitude of hydraulic conductivity depends on the properties of both the fluid and the medium. An alternative measure, used in the oil industry, is intrinsic permeability, measured in m^2 (or in industrial units called *darcies), which depends on the properties of the rock alone. **2.** Property of a membrane or other barrier, being the ease with which a substance will diffuse or pass across it. **3.** Capacity of a material to transmit fluids, expressed as hydraulic conductivity. In soils, the ease with which gases, liquids, or plant roots penetrate into or pass through a layer of soil. **4. (magnetic permeability)** The ratio of the *magnetic flux density in a medium to the magnetizing force. In free space (i.e. air) this is a constant, μ_0 equal to $4\pi \times 10^{-7}$.

permeameter A laboratory device for measuring the *hydraulic conductivity of rock and soil samples or the coefficient of *permeability of soil. Two main types are commonly used: those that require movement of water and those that do not, known respectively as falling head and constant head permeameters.

Permian Final *period (290–248 Ma) of the *Palaeozoic *Era, which is named after the central Russian province of Perm. The period is often noted for the widespread continental conditions that prevailed in the northern hemisphere and for the extensive nature of the southern hemisphere glaciation. Many groups of animals and plants, including the rugose corals (*Rugosa), trilobites (*Trilobita), and *blastoid echinoderms (*Blastozoa), vanished at the end of the Permian, in a mass *extinction that was one of the great crises in the history of life. The period is divided into seven *ages: the *Asselian, *Sakmarian, *Artinskian, and *Kungurian Ages in the Early Permian Epoch (290–256.1 Ma); and the *Ufimian, *Kazanian, and *Tatarian Ages in the Late Permian Epoch (256.1–248 Ma).

permineralization *See* FOSSILIZATION.

permitted intrusion The 'passive' intrusion of *magma which ideally produces no tensional forces and involves no forceful emplacement. The intrusion occurs by *stoping, with the melting and assimilation of wall rock.

permittivity *See* DIELECTRIC CONSTANT; and DIELECTRIC PERMITTIVITY.

perovskite A *mineral *oxide $CaTiO_3$; sp. gr. 3.98–4.26; *hardness 5.5; *monoclinic or *orthorhombic; tiny, *cubic *crystals; yellow-brown to black; common *accessory mineral in undersaturated (*see* SILICA SATURATION) *alkaline *igneous rocks found with *nepheline or *melilite. Occasionally occurs in some ultramafic (*see* ULTRABASIC), *plutonic, igneous rocks.

perovskite model A theoretical model of the Earth's composition suggesting that the mineral perovskite ($CaTiO_3$) makes up most of the lower part of the *mantle between 600 and 2800 km depth, at 2000–2500 °C. Perovskite has been made artificially by compressing *olivine, *pyroxene, and *garnet, and has been shown to withstand temperatures of 2000 °C in the laboratory.

persistence Continuation of an *anomaly beyond its expected limits in time or space.

perthite A series of layers that occurs as intergrowths in *alkali feldspars which form at high temperatures and cool slowly, unmixing as they cool. Most alkali feldspars in such slowly cooled rocks (e.g. *plutonic and *igneous) consist of an alkali feldspar host with an exsolved, sodium-rich, *plagioclase feldspar phase that segregated from the alkali feldspar during cooling. Perthitic intergrowths have textures ranging from macroperthite (visible to the naked eye) through microperthite (visible under a microscope) to cryptoperthite (which can be detected only by *X-ray diffraction and other techniques). The nature of the perthite depends upon the ordering of the silicon and aluminium atoms achieved by the mineral lattice as it cools.

Peru–Chile Trench The oceanic *trench which marks the boundary between the subducting *Nazca Plate and the *South American Plate.

Peru current (Humboldt current) Oceanic water flowing northwards along the west coast of Chile and Peru, driven by the westward flow of the S. *Equatorial current (itself driven by south-east trade winds). It is essentially a 'continuity current': water flows into the low-sea-level region left by the S. Equatorial current. This eastern *boundary current is slow-moving, broad, and shallow, and is noted for a prominent area of *upwelling bringing cold bottom waters to the surface. As the near-surface concentrations of nutrient elements are high there is an abundance of marine life associated with this current.

petals *See* IRREGULAR ECHINOIDS.

petrifaction *See* FOSSILIZATION.

petrocalcic horizon *Indurated *calcic *soil horizon that is *cemented by a high concentration of calcium carbonate, often comprising 40% by weight of the mineral material, and that is impenetrable to plant roots or to spade digging.

petrofabric *See* FABRIC.

petrogenesis The origin and evolutionary history of rocks, especially *igneous rocks. *Compare* PETROGRAPHY. *See* PETROLOGY.

petrogenetic grid A diagram that links the stability ranges of metamorphic minerals or mineral assemblages to the conditions of *metamorphism. Experimentally determined mineral or mineral-assemblage stability ranges are commonly plotted as reaction boundaries in pressure–temperature space to produce a petrogenetic grid for a particular range of *rock compositions. The region of overlap of the stability fields of those minerals forming the equilibrium mineral assemblage define the pressure–temperature conditions of the equilibrium. This corresponds to the pressure and temperature of the metamorphism.

petrographic microscope *See* POLARIZING MICROSCOPE.

petrography Systematic description and interpretation of rock *textures and *mineralogy in *thin section and as hand specimens. *Compare* PETROGENESIS. *See* PETROLOGY.

petrogypsic horizon Surface or subsurface *soil horizon cemented by *gypsum so strongly that dry fragments will not slake in water. The *cementation restricts penetration by plant roots. This is a *diagnostic horizon.

petroleum (crude oil) In geology, a term used generally to describe naturally occurring liquid *hydrocarbons formed by the anaerobic decay of organic matter. Oil is rarely found at its original site of formation but migrates to a suitable *structural or *lithological trap. Petroleum is frequently associated with salt water and with gaseous hydrocarbons. *See also* NATURAL GAS; and OIL SHALE.

petroleum geology Study of the mode of origin and conditions of accumulation of *hydrocarbons in the Earth's crust; their exploration, involving the application of the techniques of *geophysics, *geochemistry, *palaeontology, *stratigraphy, and *tectonics; and their evaluation.

petrological microscope *See* POLARIZING MICROSCOPE.

petrology The study of rocks in general, including their occurrence, field relations, structure, origins and history (*petrogenesis), and their *mineralogy and *textures (*petrography). It may usefully be qualified as 'igneous petrology', or 'sedimentary petrology'.

petrophysics The study of the character of rocks using data from down-hole geophysical *well logs. Petrophysical analysis allows an estimate of *porosity, shaliness, rock density, gas content, water saturation, and *lithology, using the suites of *logs normally run in a well.

p-form A smooth, plastically sculptured feature (e.g. a groove, crescentic depression, or pothole), generally less than 20 m in size, cut into an exposed rock surface by the action of a *glacier. It may be attributed to abrasion when it is linear and shows striations, but the origin of other types is uncertain.

PGF *See* PRESSURE-GRADIENT FORCE.

pH The negative logarithm of the hydrogen *ion concentration in a solution, pH = $\log_{10} 1/[H^+]$. If the hydrogen ion concentration of a solution increases, as happens with increasing acidity, the pH will decrease, and vice versa. The pH is measured on a scale of 0–14; a neutral medium (such as pure water) has a pH of 7, numbers above 7 indicate relative alkalinity, numbers below 7 indicate relative acidity. Most pH values in natural systems lie in the range 4–9. Human blood has a pH of 7.4, ocean water 8.1–8.3, water in saline environments may have a pH around 9.0 or higher, while for water in acidic soils it may be as low as 4.0 or less.

phaceloid *See* COMPOUND CORALS.

phacolith A curved, lensoidal *intrusion which is *concordant and follows the arches and troughs of folded *strata.

Phacopida An order of *Trilobita that lived from the Lower *Ordovician to Upper *Devonian. In most, the *cephalic suture is *proparian. There are 3 suborders.

Phaeophyceae (brown algae) Class of *algae which includes no single-celled species; almost all are marine, growing mostly in the intertidal regions. They are the dominant seaweeds in the colder waters of the northern hemisphere. They are typically olive brown or greenish in colour (at least when wet) due to the presence of the pigment fucoxanthin in the chloroplasts. Some fossil plants resembling Fucales (an order of brown seaweeds) have been reported from the *Palaeozoic, but their fossil record is equivocal owing to their lack of calcification.

Phanerozoic Period of geologic time comprising the *Palaeozoic *Mesozoic, and *Cenozoic *Eras, It began approximately 570 Ma ago at the start of the *Cambrian *Period and is marked by the accumulation of *sediments containing the remains of animals with mineralized skeletons. Although the name is derived from the Greek *phaneros* meaning 'visible' and *zoion* meaning 'animal', the term is no longer used in the sense of 'visible life', but merely defines the base of the Cambrian.

phase 1. An individually distinct and homogeneous part of a system. For example, liquid water and water vapour are each single phases; a mixture of the two constitutes a two-phase system. Similarly *minerals crystallizing from a *melt form separate phases within it. A 'phase boundary' is the line marking the contact between two constituent or liquid phases. **2.** A short unit of time, or an episode of development or change, usually within the context of a longer period. The term has been used informally (*see* INFORMAL) in this sense in many branches of the Earth sciences, e.g. 'a phase of *igneous activity', or 'a cold phase' during a generally warmer period.

phase angle The angle, usually termed α, formed by lines joining the centres of the *Earth, the *Sun, and another *solar system object. Aristarchus of Samos (third century BC) measured the Earth–Sun–Moon angle for the first lunar quarter and thus established that the Sun was much more distant than the Moon.

phase diagram Graph to show fields of stability for *phases in a heterogeneous system; variables such as temperature, pressure, and concentration are plotted against each other.

phase layering *See* MINERAL LAYERING.

phase rule Rule stating that for any system in equilibrium, $F = (C - P) + 2$, where F is the number of *degrees of freedom, C the number of components, and P is the number of phases.

phase transitions in the mantle Measurement of the velocity of *seismic waves and determination of the Earth's *moment of inertia indicate that at depths of 80–100 km there appears to be a *low velocity zone and from 400–900 km velocities increase more rapidly than had been expected. These variations in the *mantle may be due to changes in composition or to *phase change, e.g. a transition to a *cubic form of *olivine.

phase velocity (*V*) The speed with which a particular *phase (e.g. peak or trough) of a wave travels. *Compare* GROUP VELOCITY.

phena *See* PHENON.

phenoclast A large and conspicuous *clast set in a generally finer-grained *sediment, e.g. a large *boulder in a fine-grained *conglomerate, or a *granule in *siltstone.

phenocryst Large and often well-formed *crystals set in a finer *groundmass or *matrix. Rocks containing phenocrysts are said to be porphyritic. *Compare* PORPHYROBLAST.

phenogram A tree-like diagram used in analysis to show similarity or dissimilarity among specimens or groups of specimens.

phenon (pl. phena or phenons) A group of organisms that are similar in appearance. They may or may not belong to the same taxon.

phenotype Those observable properties of an organism that are produced by the *genotype in conjunction with the environment. Organisms with the same overall genotype may have different phenotypes because of the effects of the environment and of *gene interaction. Conversely, organisms may have the same phenotype but different genotypes.

Philippine Plate One of the present-day minor lithospheric *plates, the Philippine Plate is surrounded by *subduction zones (the Ryukyu, Philippine, Marianas, and Izu-Bonin), and is splitting along the Marianas Islands, in the *back-arc of the *Marianas Trench-arc system, at about 60 mm/year.

phi scale The expression of *grain sizes on a logarithmic scale. The phi value (ϕ) is related to grain diameter (d) by the expression $\phi = -\log_2 d$. Increasing positive phi values are for grain diameters progressively finer than 1000 μm, and increasing negative values for grain diameters progressively coarser than 1000 μm (e.g. 2ϕ = 250 μm; 1ϕ = 500 μm; 0ϕ = 1000 μm; -1ϕ = 2 mm; -2ϕ = 4 mm).

phlogopite A member of the *mica group of *minerals, phlogopite is the Mg-rich variety of *biotite $K_2(Mg,Fe^{2+})_6[Si_3AlO_{10}]_2(OH,F)_4$ with Mg/Fe more than 2/1; sp. gr. 2.76–2.90; *hardness 2.0–2.5; subtransparent; pale brown; small, tabular *crystals; occurs in metamorphosed, impure, magnesian *limestones by reaction between *dolomite and K-*feldspar or *muscovite, in *kimberlites, and in some *leucite-bearing rocks.

Phobos **1.** A Soviet mission to *Mars, launched in 1988. **2.** One of the two satellites of Mars.

Phocidae *See* CARNIVORA.

Phoebe (Saturn IX) One of the lesser satellites of *Saturn, with a radius measuring 115 × 110 × 105 km; visual albedo 0.06. Its orbit is *retrograde. It was discovered in 1898 by W. Pickering.

Phoenix Plate *See* NAZCA PLATE.

Pholas *See* BORING; and DOMICHNIA.

phonolite A fine-grained, *porphyritic, *extrusive, *igneous rock consisting of *essential *alkali feldspar (*sanidine or *anorthoclase), *nepheline, sodic *pyroxene, and sodic *amphibole (with or without iron-rich *olivine), with *accessory *sphene, *apatite, and *zircon. Where present, *phenocrysts can consist of alkali feldspar, sodic pyroxene, or sodic amphibole. Phonolites are the extrusive equivalent of *nepheline syenites and are found on off-axis ocean islands and in continental regions subjected to anorogenic upwarping and rifting. The name, from the Greek *phone* meaning 'sound', refers to the fact that the rock rings like a bell when struck with a hammer.

phosphates Rock or deposit made up largely of inorganic phosphate, commonly calcium phosphate (e.g. the *minerals *apatite, *autunite, *monazite, *pyromorphite, *torbernite, *turquoise, *vivianite, and *wavellite).

phosphorescence The property of some *minerals of emitting light during exposure to X-rays, ultraviolet light, or cathode radiation and continuing to do so after the exposure has ceased (if light emission ends when the radiation source is switched off, the property is called 'fluorescence'). The colour of the emitted light varies with the wavelength of the radiation to which the mineral is exposed, and is thought to be due to the presence of traces of organic material or *cations within the atomic structure of the mineral.

phosphorite A *sedimentary rock rich in phosphate, usually in the form of carbonate hydroxyl fluorapatite ($Ca_{10}(PO_4CO_3)_6F_{2-3}$). Phosphorites occur as *nodules and crusts formed in oceanic areas where sedimentation rates are very low, by the early, near-surface diagenetic alteration (*see* DIAGENESIS) of *ooids, *pellets, and *bioclasts, or as accumulations of *bones and fish scales in *fluviatile to shallow marine environments. *See also* GUANO.

photo- From the Greek *phos photos* meaning 'light', a prefix meaning 'light'.

photochemical smog Hazy condition of the atmosphere due to the reaction of hydrocarbons with molecules of nitrogen oxide in sunlight which produces complex organic molecules of peroxyacetyl nitrates (PAN). In humid conditions these molecules produce *smog. Such phenomena are common in large urban areas (e.g. the Los Angeles Basin and Athens) where there are stable atmospheric conditions and a high level of hydrocarbon input from incomplete combustion in car engines. Natural photochemical reactions occur in the high atmosphere with the absorption of radiation by oxygen to produce ozone. *See also* OZONE LAYER.

photodisintegration Decomposition of a compound in the presence of light, particularly sunlight.

photodissociation The splitting of a molecule into atoms or other molecules as a result of its absorption of radiation.

photogeology The determination of the overall geology of an area through the interpretation of photographic data (*see* AERIAL PHOTOGRAPHY) by noting variations in colour, tone, geometry, relative relief, and surface texture of land-forms, rock *outcrop boundaries, vegetation patterns, etc.

photogrammetry In *remote sensing, the use of aerial photographs or satellite images to measure distances between ground objects accurately.

photographic infrared *See* NEAR-INFRARED.

photohydrometer An instrument that calculates the size of sediment grains by measuring changes in the intensity of a light beam passed through a column of settling particles.

photometer In a *reflected-light microscope, an attachment which is used to measure the *reflectance of *ore minerals. It consists of a photomultiplier tube that has high sensitivity throughout the visible spectrum. It is used in conjunction with a stabilized light source and high quality *monochromators; its readings conform to Commission on Ore Microscopy (COM) standards.

photometry A general term for the physical measurement of the ultraviolet, visible, and infrared portions of the electromagnetic spectrum.

photon log The record of a scintillometer that has passed through a *borehole. It is identical to a *gamma-ray log, except that the sonde is central within the hole and therefore sensitive to the size of the borehole. (The gamma-ray sonde is held against the borehole wall and is insensitive to the size of the borehole.)

photopolarimeter-radiometer (PPR) A *remote-sensing instrument that measures the intensity and polarization of sunlight in the visible part of the spectrum, from which the temperature and cloud formation in planetary and satellite atmospheres can be determined, as well as some surface details.

photosphere The thin shell of light around the *Sun from which light escapes and within which the solar *spectra originate.

photosymbiosis A symbiotic (*see* SYMBIOSIS) relationship between two organisms (e.g. *Foraminiferida and algae) one of which (in this example the *alga) is capable of *photosynthesis.

photosynthesis Term given to the series

of metabolic reactions that occur in certain autotrophic organisms, whereby organic compounds are synthesized by the reduction of carbon dioxide using energy absorbed by chlorophyll from sunlight. In green plants, where water acts both as a hydrogen donor and as a source of released oxygen, photosynthesis may be summarized by the empirical equation:

$$CO_2 + 2H_2O \xrightarrow[\text{light}]{\text{chlorophyll}} [CH_2O] + H_2O + O_2\uparrow$$

(oxygen being released as a gas). Photosynthetic bacteria are unable to utilize water and therefore do not produce oxygen. Instead they may use hydrogen sulphide (purple and green sulphur bacteria) or organic compounds (purple non-sulphur bacteria) as a source of hydrogen.

phragma *See* DINOPHYCEAE.

Phragmites cliffwoodensis Early representative of the reed-grasses, recorded from the mid-*Cretaceous of New Jersey, USA. It has rather uncertain affinities but if accepted would be the first of the family Gramineae (grasses).

phragmocone The septate shell of a cephalopod (*Cephalopoda). The term is also applied to the septate portion of a belemnite (*Belemnitida) skeleton which, although reduced in size, is *homologous to the external shell of other cephalopods.

phreatic activity Volcanic *eruptions generated by the interaction between hot *magma and surface lake water, sea water, or *groundwater. The water immediately surrounding the magma is heated and volatilized. Its expansion builds up pressure on the envelope of water surrounding it. When the pressure exceeds the confining pressure of the overlying water column the water vapour expands explosively to produce a steam-dominated, phreatic (i.e. subsurface water) eruption. Where significant amounts of magmatic material are ejected in addition to steam the activity is said to be 'phreatomagmatic'.

phreatic zone (zone of saturation) The soil or rock zone below the level of the *water-table, where all voids are saturated. *See also* VADOSE ZONE.

phreatomagmatic activity *See* PHREATIC ACTIVITY.

Phycosiphon An *ichnoguild of structures made by deposit feeders that moved freely some distance below the surface.

phyletic Pertaining to a line of descent.

phyletic evolution Evolutionary change within a lineage, as a result of gradual adjustment to environmental stimuli.

phyletic gradualism Theory holding that *macroevolution is merely the operation of *microevolution over relatively long periods of time. Thus gradual changes will eventually accumulate to the point at which descendants of an ancestral population diverge into separate species, genera, or higher-level *taxa.

phyllic alteration A type of *alteration that is found in *country rocks of copper and molybdenum *porphyry deposits.

phyllite Fine-grained (less than 0.1 mm), low-grade *metamorphic rock, of *pelitic composition, with a well-developed *schistosity, that often has a silky sheen due to the parallel orientation of *phyllosilicate minerals (e.g. *chlorite, *muscovite, and *sericite). *Compare* SLATE; and SCHIST.

phyllonite A *slate-like, *dynamic metamorphic rock formed in *fault zones and containing a penetrative *cleavage orientated parallel to the *fault plane. The cleavage is formed by the *recrystallization and alignment of sheet *silicates during fault movement.

phyllosilicate (sheet silicate, layered silicate) A large group of *silicate minerals which are characterized by possessing layers of $[SiO_4]^{4-}$ tetrahedra linked together to form a flat sheet with the composition $[Si_4O_{10}]_n$. This group includes *micas, *chlorite, *clays, talc, and *serpentine, which are soft minerals and of variable but generally low density. They form at low temperatures and some, particularly the clays, replace *primary minerals as a result of *hydrothermal alteration or *weathering. They are an essential constituent of *argillaceous sedimentary rocks and some low-grade *metamorphic rocks.

phylogenetics The taxonomical classification of organisms on the basis of their degree of evolutionary relatedness.

phylogenetic tree A variety of *dendrogram in which organisms are shown arranged on branches that link them according to their relatedness and evolutionary descent.

phylogenetic zone *See* LINEAGE-ZONE.

phylogeny The evolutionary relationships within and between taxonomic levels, particularly the patterns of lines of descent, often branching, from one organism to another, i.e. the relationships of groups of organisms as reflected by their evolutionary history. *See* TAXONOMY.

phylozone *See* LINEAGE-ZONE.

phylum In animal *taxonomy, one of the major groupings, coming below subkingdom and kingdom, and comprising superclasses, classes, and all lower *taxa.

-phyre A suffix applied to an *igneous rock which is *porphyritic.

physico-theology *See* NATURAL THEOLOGY.

phytogeography (floristics) The study of the geography of plants, particularly their distribution at different taxonomic levels, i.e. family, genus, and species. Patterns of distribution are interpreted in terms of climatic and anthropogenic influence, but above all in terms of earlier continental configurations and migration routes.

phytophagous Feeding on plants.

Piacenzian **1.** The final *age in the *Pliocene *Epoch, preceded by the *Zanclian (Tabianian), followed by the *Calabrian (*Quaternary), and dated at 3.4–1.64 Ma (Harland et al., 1989). **2.** The name of the corresponding European *stage which was originally described as a subdivision of the (then recognized) Astian Stage. It is roughly contemporaneous with the upper *Delmontian, *Repettian, and *Venturan (N. America), *Waipipian, *Mangapanian, and lowermost *Nukumaruan (New Zealand), and upper *Kalimnan and *Yatalan (Australia). The type area is around Castell' Arquato, Italy.

pick To select one feature from among others, e.g. *first arrivals on a seismic refraction record or reflection events in a seismic section. The feature so selected becomes a 'pick'. For example, on a seismic refraction record the selected first arrivals are the picks for each trace, providing *travel-time data for the compilation of a *travel-time graph.

pickup *See* GEOPHONE.

pico- From the Spanish *pico* meaning 'beak', or 'peak' (i.e. a point), a prefix (symbol p) used with SI units to denote the unit $\times 10^{-12}$.

picoplankton Marine planktonic organisms, 0.2–2.0 μm in size, and consisting mainly of *bacteria and *cyanobacteria.

picrite A strongly *porphyritic, *olivine-rich *basalt. Picrites on the islands of Hawaii are formed during high-discharge-rate eruptions and may reflect mobilization of an olivine-rich crystal-cumulate layer near the base of the *magma chamber.

pi diagram A stereographic projection, used in the analysis of the orientation of *folds, in which poles (points representing lines) perpendicular to folded bedding (π poles) are plotted, and the diagram rotated on an overlay to achieve a best-fit *great circle (a π circle). A pole to the plane of the π circle plane records the *plunge and *trend of the *fold axis.

piedmont The tract of country at the foot of a mountain range, e.g. the Po Valley, Italy, at the foot of the Alps. The word is derived from the Italian *piemonte*, meaning 'mountain foot'.

piedmont glacier A lobe of *ice formed when a *valley glacier extends beyond its restraining valley walls and spreads out over the adjacent lowland, or *piedmont zone. Much of the glacier surface is therefore at a low altitude and may show rapid *ablation. An example is the Malaspina Glacier, Alaska.

piercing fold A *fold which develops by the forcible upward movement of mud or salt *diapirs.

piezoelectricity An electric charge induced by a flow of *electrons when pressure is applied at the ends of a polar axis of a crystal which lacks a centre of symmetry (*see* CRYSTAL SYMMETRY) and which has different crystal forms at opposite ends. This property was first detected in 1881 by Pierre and Jacques Curie.

piezometer An *observation well designed to measure the elevation of the *water-table or *hydraulic head of *groundwater at a particular level. The well is normally quite narrow and allows groundwater to enter only at a particular depth, rather than throughout its entire length.

piezometric surface *See* POTENTIOMETRIC SURFACE.

piezoremanent magnetization The magnetization acquired when material is subjected to prolonged pressure. *See also* SHOCK REMANENT MAGNETIZATION.

pigeonite A member of the *clinopyroxenes with a very low calcium content and the composition (Ca,Mg,Fe(Mg,Fe)Si_2O_6; sp. gr. 3.30–3.46; *hardness 6.0; properties similar to *augite, but occurs in rapidly chilled *igneous rocks. In slowly cooled rocks, pigeonite changes to an *orthopyroxene as cooling proceeds.

piggyback basin (thrust-sheet-top basin) A type of *foreland basin formed, in addition to a *foredeep, on top of a thrust sheet by the progressive development of *thrusts by collapse of the *footwall; the newly formed thrust becomes the active thrust surface and older thrusts, with their thrust slices, are carried forward on it passively.

piggyback thrust sequence A *thrust sequence formed by the progressive development of thrusts by collapse of the *footwall; the newly formed thrust becomes the active thrust surface and older thrusts, with their thrust slices, are carried forward on it passively.

Pikaia An early chordate (*Chordata) from the *Burgess Shale that was possibly related to the modern amphioxus.

pile Timber, steel, or concrete sheet or column sunk into loose ground or cast in a *borehole to carry vertical or horizontal loads and provide support under earth or water pressure.

pileus From the Latin *pileus* meaning 'cap', an accessory cloud occurring as a small cap on or above a cumuliform cloud. The cloud is associated with *cumulus or *cumulonimbus. *See also* CLOUD CLASSIFICATION.

pillar and stall (bord and pillar, room and pillar; in Scotland: stoop and room) Method of mining in which large chambers are excavated, leaving pillars of *ore, rock, or *coal to support the roof.

pillar structure Pipe-like, near vertical tubes, with diameters of a few centimetres, found in association with *dish structures, and formed by the upward escape of water from liquefied *sediment.

pillow lava Piles of elongate basaltic *lava pods, having the general appearance of a stacked accumulation of discrete stone pillows, often many hundreds of metres in thickness. Each 'pillow' is surrounded by a chilled, fine-grained, lava skin and sags into the pillows below it. The pillows are rarely more than one metre in diameter and in cross-section each one has a convex upper surface, radial and concentric fractures, and commonly a central cavity or tube which once fed lava to the front of the advancing finger. The morphology indicates that the pillows continued to behave as fluid bodies after the chilled carapace had formed. This provides good evidence of submarine eruption: lava entering water acquires a glassy outer skin as heat is conducted rapidly from the surface. Because water absorbs heat more readily than air, with little increase in its own temperature, the rapid surface cooling allows the molten plastic state of the pillow interior to be maintained longer than it would be in air. Pillows have been observed forming under water from lava entering the sea off Hawaii.

pilotaxitic A felted mass of *acicular or lath-shaped crystals found in fine-grained *igneous rocks; the crystals may be aligned to produce a flow structure.

pinacoid Applied to a *crystal face which cuts the vertical *c* (or *z*) *crystallographic axis and is parallel to the horizontal *a* (or *x*) and *b* (or *y*) axes. A horizontal plane of symmetry (*see* CRYSTAL SYMMETRY) repeats the face at the opposite end of the crystal. The resulting form is a pair of parallel faces with the index (001).

pinch-and-swell *See* BOUDINAGE.

pinger A high-frequency, high-resolution, shallow penetration device used in marine seismic-reflection profiling and offshore *engineering geophysical investigations.

pingo Ice-cored, dome-shaped hill, oval in plan, standing 2–50 m high, and 30–600 m in diameter, developed in an area of *permafrost. The larger examples have breached crests in which ice may be exposed. They are probably due to local freezing of water that has migrated from adjacent uplands, or to the late freezing of the ground beneath a lake. *See also* PALSA.

pinite A fine-grained mixture of *muscovite and *chlorite with some *serpentine or iron oxide, which forms as an alteration product of *cordierite. It is colourless to bluish-green and similar to *mica in structure and composition.

pinnacle reef *See* REEF.

pinnular plates *See* BRACHIA.

pinnules *See* BRACHIA.

Pinus longaeva **(bristlecone pine)** Pine species from California, famous for its longevity and used to develop an exceptionally long, arid-site, tree-ring chronology. The oldest living specimens date back more than 4600 years, but *cross-dating these with remnants of dead bristlecone pines has extended the arid-site chronology to more than 8200 BP. A 5500-year chronology has been developed for bristlecone pines at the upper tree limit. These pines are also used to calibrate the *radiocarbon-dating method to allow for fluctuations in atmospheric $^{14}C : ^{12}C$ ratios as revealed by measuring $^{14}C : ^{12}C$ ratios of individual tree rings in the long, absolutely dated, tree-ring series. *See also* DENDROCHRONOLOGY.

PIOCW *See* PACIFIC AND INDIAN OCEAN COMMON WATER.

Pioneer A series of *NASA spacecraft that conducted explorations of the solar system. Pioneer 10, launched on 2 March 1972, was the first spacecraft to leave the solar system, in 1987. Pioneer 11 was launched on 5 April 1973, for Saturn. Pioneer Venus, launched in 1978 to *Venus, comprised an orbiter and probes sent to the surface.

pipe 1. Nearly vertical, cylindrical body or opening in rock. **2.** In mining, an ore shoot at the intersection of two barren veins. **3.** At Kimberley, South Africa, pipes of diamond-bearing *breccia. **4.** In sedimentology, tube often filled with mud, particularly in *limestones. **5.** In volcanology, vertical channel-ways below a *volcano through which *magma flows towards the surface.

pipette analysis A standard method for measuring the size of small particles. Sediment is stirred into suspension in a measured volume of water in a *sedimentation tube, aliquots of uniform size are withdrawn by pipette at specified intervals and depths, oven-dried, and weighed. The grain diameter (D) is then calculated by $D = \sqrt{C}/\sqrt{(x/t)}$, where C is a constant depending on the particle density and density and viscosity of the fluid, x is the depth (in cm) from which the particles are withdrawn, t is the time elapsed (in seconds), and x/t is the settling velocity.

pipkrake (needle ice) Columnar ice found beneath individual stones or patches of earth in the *periglacial environment. It is a result of the relatively high thermal conductivity of such materials, leading to freezing under them when the temperature falls. Heaving of less than 0.1 m may occur by this process.

Piripauan *See* MATA.

pisoid *See* PISOLITH.

pisolith (adj. pisolitic) (pisoid) A spherical to subspherical, inorganic *carbonate particle, larger than 2 µm in diameter, and in some cases as much as 10 cm in diameter, characterized by an internal concentric lamination. Some pisoliths are said to form in the same manner as *ooids, but others ('vadose pisoliths'), form in subaerial environments in *calcrete profiles, Pisoliths should not be confused with oncoids (*oncolites) which although superficially similar, are of organic origin.

pisolitic *See* PISOLITH.

pistacite *See* EPIDOTE.

piston corer *See* HYDRAULIC CORER.

piston sampler *See* PEAT-BORER.

Pistosaurus grandaevus A likely ancestor for all *plesiosaurs although it is known only by its skull. It bears some similarities to both *nothosaurs and plesiosaurs, and is found in rocks of Middle *Triassic age.

pitch (rake) The angle made by a *lineation with the *strike of the surface on which it occurs.

pitchblende *See* URANINITE.

pitchstone A glassy *igneous rock, rather like *obsidian but with a waxy, resinous *lustre owing to the absorption of water.

pitting The digging of a pit; in Scotland, mining an outcrop by making shallow pits. In exploration, sampling of *alluvial *sediments by shallow (down to 15–20 m) trial pits.

pixel 1. A picture element. Commonly the smallest component of a *multispectral image as determined by a single optic fibre. **2.** In *remote sensing, a single sample of data in an image. A pixel has a spatial attribute corresponding to its location within the image and the area of ground represented, and a spectral attribute corresponding to the *intensity of a particular wavelength. *See also* PIXEL COLOUR.

pixel colour In *remote sensing, the colour of each *pixel. Pixel colour is dependent on three parameters: the intensity, which is the brightness of the colour; the saturation, which is the perpendicular distance to the *achromatic axis (the closer to the achromatic axis the more pastel the colour, the further away the more vivid the colour); and the *hue. *See also* INTENSITY-HUE-SATURATION PROCESSING.

PKP-wave *See* SEISMIC WAVE MODES.

placental mammals *See* EUTHERIA.

placer deposit Deposit of materials (e.g. gold, diamonds, tin, or platinum) that has been concentrated by mechanical action. Placer minerals generally have high density and resistance, and therefore may concentrate during various types of *weathering.

place value The economic importance attached to the location of a mineral deposit. Minerals or metals with a high intrinsic value, e.g. diamonds or gold, have a low place value as transport costs add little to the eventual market price, so they may be worked anywhere on Earth. In contrast, sands and gravels have a high place value and must be worked near the place of use.

placic horizon Subsurface *soil horizon, formed most readily in humid tropical or cold conditions, that is *cemented by iron and organic matter, by iron and manganese, or by iron alone.

Placodermi Class of archaic, jawed (*gnathostome), and heavily armoured fish, which appeared in the *Devonian and which were virtually extinct by the end of that period. They were rather diverse in body form. They all possessed a *head shield formed by bony plates. Pectoral and pelvic fins appear to have been present. Most of them (e.g. *Coccosteus*) were bottom-dwelling fish, with a depressed body terminating in a *heterocercal tail.

Placodontidae *Triassic *euryapsid reptiles, which were specialized shellfish feeders. Some were heavily armoured and strongly resembled turtles; while the more lightly armoured varieties, e.g. *Placodus*, were analogous in general form to the *nothosaurs, except for modifications relating to their molluscan diet.

placoid scale (dermal denticle) Type of scale that comprises the basic unit of the hard skin cover of sharks. It consists of a hard base embedded in the skin and a spiny process (cusp); these are covered by a hard, enamel-like substance (vitrodentine), and the scale projects outwards and backwards. The bulk of the denticle consists of *dentine surrounding a central pulp cavity. Unlike the scales of *bony fish, placoid scales stop growing after they reach a certain size and new denticles are added instead.

plaggen A man-made *soil horizon more than 500 mm deep, resulting from long-continued manuring, often enriched by phosphate.

plagioclase feldspar One of the most important rock-forming silicate *minerals with the general formula $(Na,Ca)(Al)_{1-2}(Si)_{2-3}O_8$. There is a *solid solution series (*see* PLAGIOCLASE SERIES) between the two *end-members albite (Ab) ($NaAlSi_3O_8$) and anorthite (An) ($CaAl_2Si_2O_8$), and the percentage of calcium end-member present is used to subdivide the series into a number of individual minerals: albite (0–10 mol% An); oligoclase (10–30 mol% An); andesine (30–50 mol% An); labradorite (50–70 mol% An); bytownite (70–90 mol% An); and anorthite (90–100 mol% An). Albite; sp. gr. 2.61; *hardness 6.0–6.5; is whitish, *vitreous; *tabular or irregular; with two *cleavages {010} and {001} meeting at almost right angles on the (100) face; and occurs in *acid *igneous rocks and *spilites. Oligoclase; sp. gr. 2.64; *triclinic; cleavage perfect basal {001}, good {010}; is similar to albite except in the percentage An it contains. Andesine; sp. gr. 2.66; triclinic; cleavage perfect basal {001}, good {010}; has properties similar to those of albite but tends to occur in more *intermediate igneous rocks. Labradorite; sp. gr. 2.67; crystals thin and *tabular, flattened parallel to (010); cleavage perfect basal {001}, good {010}; is greyish-white but may show iridescence due to lattice imperfections on cleavage faces and occurs in *basic igneous rocks. Bytownite; sp. gr. 2.72; triclinic; crystals often tabular prismatic but normally form irregular grains; cleavage perfect basal {001}, good {010}; also greyish-white, is a constituent of basic and *ultrabasic igneous rocks. Anorthite; sp. gr. 2.75; triclinic; crystals tabular and prismatic but normally form irregular grains; cleavage perfect basal {001}, good {010}; also greyish-white, occurs in basic and ultrabasic igneous rocks and in metamorphosed *limestones. These plagioclase feldspars cannot be distinguished from one another in hand specimens, but under the microscope their *extinction angles vary; this is a useful property in their identification, together with the nature of the multiple twinning (*see* CRYSTAL TWIN-

NING) which is a very characteristic feature and serves to distinguish them from *alkali feldspars. *Exsolution of potassium-feldspar in a plagioclase feldspar host is called 'antiperthite'.

plagioclase series *Isomorphous series of *feldspars ranging in composition between albite ($NaAlSi_3O_8$) and anorthite ($CaAl_2Si_2O_8$). Between the two *end-members there is a continuous *solid solution series: albite, oligoclase, andesine, labradorite, bytownite, anorthite. Usually, the higher the temperature of formation the more likely the feldspar is to be calcium-rich, whereas at lower temperatures sodium-rich plagioclase tends to form. The preservation of electrical neutrality in the plagioclases is achieved by concomitant substitution of Ca^{2+} and Al^{3+} by Na^{+} and Si^{4+}. *See* PLAGIOCLASE FELDSPAR.

plagiogranite *See* GRANODIORITE.

planar cross-stratification *See* CROSS-STRATIFICATION.

planation surface **1.** Synonym for *erosion surface. **2.** The final stage produced by the erosion of folded *sedimentary rocks, when inverted relief is planed across by erosion.

plane bed (flat bed) A near-horizontal surface of *sand or gravel. Two types of plane bed are found. Upper-stage plane beds are produced by the intense transport of *sediment by high-velocity, shallow flows (upper-flow-regime conditions), and characterized by primary current lineation on the *sediment surface. Lower-stage plane beds are produced only in coarse sands and gravels by flow conditions broadly similar to those which generate current *ripples in finer sand. The lower-stage plane bed exhibits a series of shallow scours on the sediment surface. The accumulation of plane-bedded sediment gives rise to an internal *sedimentary structure of horizontal *lamination.

plane of projection (equatorial plane) The horizontal surface on which points on a spherical projection can be represented in two dimensions. A *stereographic projection can be compared to a slice through the Earth's equator as seen from the N. or S. Pole, and for this reason is often called the 'equatorial plane', and its circumference is called the 'primitive circle'. Since the plane passes through the centre of the sphere its circumference is a *great circle, the angle between the plane of projection and the N–S axis of the sphere being 90°.

plane of symmetry *See* CRYSTAL SYMMETRY.

plane-polarized light (PPL) As light travels, it normally vibrates in all directions at right angles to the line of transmission. If a strongly absorbing crystal, e.g. *polaroid or *tourmaline, is placed in the light path, the rays are strongly absorbed in all directions except one, and the rays of light that emerge are confined to this one plane of vibration; i.e. the light is 'plane polarized'. Polarization may also result from double refraction (*see* NICOL PRISM) or reflection.

plane strain The *strain state which occurs when a reference sphere is extended along its X axis and shortened along its Y axis, leaving its Z axis unchanged. This strain changes the shape from a sphere to a triaxial ellipsoid.

Planet-B A Japanese *ISAS mission to orbit *Mars for one year, to be launched in 1998. Its purpose is to study the martian atmosphere and the effect of the *solar wind on the martian *ionosphere, while returning images of the martian surface and also of Phobos and Deimos.

planetary boundary layer The layer of the lower atmosphere extending to a height of about 500 m above the ground, in which frictional effects of the underlying surface generate *turbulence. *See also* TURBULENT BOUNDARY LAYER.

planetary geology (planetary geoscience, astrogeology) The scientific study of planetary surfaces, employing the techniques of geology. It was first used formally to establish the stratigraphic sequence on the visible face of the *Moon by *Shoemaker and Hackman in 1962, but was used informally by earlier workers, such as G. K. *Gilbert and R. B. Baldwin, to interpret lunar features as lava flows, craters, etc.

planetary geoscience *See* PLANETARY GEOLOGY.

planetismal Small, solid body, a few kilometres in diameter, accreted by gravitational attraction, after condensation of *mineral *phases from the primitive solar nebula. The *solar-system planets are thought to have formed by the further accretion of planetismals.

planetismal hypothesis A cataclysmic

theory, propounded early in the twentieth century, to explain the origin of the *solar system. This was attributed to tidal interactions produced by a close encounter between the *Sun and a star. The immense gaseous tides produced in the Sun pulled matter away, in the direction of the star, to form a long, gaseous filament. This then broke into several parts, each contracting to form a planet.

plane wave A wavefront which has no effective curvature because of its distance from the source. At a very large range from the source, the degree of curvature over a short local distance is negligible and the wavefront can be regarded as planar.

planèze Sloping triangular facet on the flank of a *volcano, underlain and protected by *lava. Its form is a result of the radial dissection by streams of a complex volcanic cone.

planispiral Applied to the condition in which a univalve gastropod (*Gastropoda) shell is coiled (*see* COILING) in a single horizontal plane and the diameter increases away from the axis of coiling. Many cephalopod (*Cephalopoda) shells are also planispiral.

planitia (pl. planitiae) A plain that is low relative to the surrounding terrain. A typical example is Hellas Planitia on *Mars (42° S, 293° W), a probable degraded impact basin, about 2000 × 1500 km.

plankton (adj, planktonic) Minute aquatic organisms that drift with water movements, generally having no locomotive organs. The phytoplankton (plants) comprise mainly *diatoms, which carry out *photosynthesis and form the basis of the aquatic food-chains. The zooplankton (animals) which feed on the diatoms may sometimes show weak locomotory powers. They include *protozoans, small crustaceans (*Crustacea), and in early summer the larval stages of many larger organisms. Plankton are sometimes divided into net-plankton (more than 25 μm diameter) and nanoplankton, which are too small to be caught in a plankton net (*see* NANO-; and NANOFOSSIL).

planktonic geochronology The use of *planktonic organisms, e.g. *globigerinid *foraminiferids or microscopic *algae, to provide a relative dating of *sediments deposited in marine waters. *Radioactive-decay methods applied to planktonic organisms may also yield an absolute date or information on palaeoclimates.

Planolites An *ichnoguild of structures made by deposit feeders that moved freely through shallow burrows.

Plantae (Metaphyta) A kingdom that includes all the plants. The earliest true plants were probably unicellular green algae (*see* ALGA), which first appeared in the *Precambrian. The *Bryophyta (mosses and liverworts) are known from the *Devonian, and the first recorded vascular plants (*Tracheophyta) date from the *Silurian.

plasma A low-density, high-temperature, completely ionized gas, consisting of free atomic nuclei and free electrons. Overall it is electrically neutral. It is sometimes referred to as the 'fourth state of matter'.

plastic deformation A deformation process which proceeds elastically at low *stress values and becomes viscous when a critical stress value is reached. Plastic deformation occurs in rocks under conditions of high pressure and temperature, producing a permanent alteration of their shape but without failure by rupture.

plasticity index *See* ATTERBERG LIMITS.

plastic limit *See* ATTERBERG LIMITS.

plastron 1. The lower, bony shell of a turtle. **2.** *See* IRREGULAR ECHINOIDS.

plate 1. A segment of the *lithosphere, which has little volcanic or seismic activity but is bounded by almost continuous belts (known as *plate margins) of *earthquakes and, in most cases, by volcanic activity and young subsea or subaerial mountain chains. Most Earth scientists consider there are currently seven large, major plates (the *African, *Antarctic, *Eurasian, *Indo-Australian or Indian, *North American, *Pacific, and *South American Plates). There are also several smaller plates (e.g. the *Arabian, *Caribbean, *Cocos, *Nazca, and *Philippine Plates) and an increasingly long list of *microplates (e.g. the *Gorda, *Hellenic, and *Juan de Fuca Plates). The positions of the boundaries of some present-day plates are disputed, particularly within and adjacent to *collision zones, e.g. the Alpine–Himalayan belt, so it is not surprising that very little agreement has been reached about the histories of plates in the geologic past. **2.** A general term applied to plane pieces of *skeletal material usually formed from calcium carbonate. Plates occur in groups of several types, e.g. the *delthyrium in some brachiopods (*Brachiopoda) is closed by a pair of *deltidial plates. **3.** The outer covering of a crinoid (*Crinoidea) body, which consists of a series of rows of plates. **4.** The bony covering, often fused to the ribs, on the upper and lower surfaces of the body of a turtle. The upper surface is the 'carapace', the lower is the 'plastron'.

plateau basalt An extensive, thick, smooth flow or succession of flows of high-temperature, fluid *basalt erupted from fissures, flooding topographic lows, and accumulated to form a plateau. The Deccan Traps in India covers 260 000 km^2 and the Columbia River Plateau basalt, in Washington State, USA, covers 130 000 km^2 and is more than 1800 m thick. Individual flows may have volumes of the order of 100 km^3. The original area of the Thulean Plateau which formed in the north-eastern Atlantic Ocean region 30 Ma ago, was 1.8×10^6 km^2. The lava plains are built up by many thousands of individual flows from numerous, coalescing, *shield-type volcanoes with extremely low angles of slope.

plate bearing test Static test to measure deformability in terms of the theory of elasticity (*see* ELASTIC REBOUND THEORY). Values may be assigned to the ground for *Young's modulus and *Poisson's ratio.

plate boundary *See* PLATE MARGIN.

plate margin (plate boundary) The boundary of one of the *plates that form the upper layer (the *lithosphere) and together cover the surface of the Earth. Plate margins are characterized by a combination of tectonic and topographic features: oceanic *ridges, *Benioff zones, young fold mountains, and *transform faults. Plate margins are of three main types: (*a*) *constructive margins where newly created lithosphere is being added to plates which are moving apart at oceanic ridges; (*b*) *convergent margins which can be either *destructive margins, where one plate is carried down into the *mantle, beneath the bordering plate, at a *subduction zone, or a *collision zone, where two *island arcs or continents, or an arc and a continent, are colliding; or (*c*) *conservative margins, where two plates are moving in opposite directions to each other along a *transform fault. All three margins are seismically active, with volcanic activity at constructive and destructive margins. Some plate margins exhibit features of more than one of the three main types and are known as *combined plate margins. *See* PLATE TECTONICS; and SEA-FLOOR SPREADING.

plate motions The movement of tectonic *plates is expressed in terms of rotations relative to a *Euler pole. Such motions during the last 200 million years have been determined mainly from *magnetic anomaly patterns in the ocean basins. Determination of older motions is based on *palaeomagnetic studies and evidence for continental collisions and separations.

plate stratigraphy The study of sedimentary strata in order to reconstruct the geographic positions and water depths to which they have been subjected as a consequence of the movement of the *plates on which they lie.

plate tectonics The unifying concept that has drawn *continental drift, *sea-floor spreading, seismic activity, crustal structures, and volcanic activity (*see* VOLCANICITY) into a coherent model of how the outer part of the *Earth evolves. The theory proposes a model of the Earth's upper layers in which the colder, brittle, surface rocks form a shell (the *lithosphere) overlying a much less rigid *asthenosphere. The shell comprises several discrete, rigid units (tectonic *plates) each of which has a separate motion relative to the other plates. The *plate margins are most readily defined by present-day *seismicity, which is a consequence of the differential motions of the individual plates. The model is a combination of *continental drift and *sea-floor spreading. New lithospheric plates are constantly forming and separating, and so being enlarged, at *constructive margins (ridges), while the global circumference is conserved by the *subduction and recycling of material into the *mantle at *destructive margins (trenches). This recycling results in andesitic volcanism and the creation of new *continental crust, which has a lower density than the *oceanic crust and is more difficult to subduct. Many features of the Earth's history are explicable within this model which has served as a unifying hypothesis for most of the Earth sciences. Previous mountain systems are now recognized as the sites of earlier subduction, often ending with continental crustal collision: the movement of plates has been used with varying success in interpreting *orogenic belts as far back as the early *Proterozoic. *Plate motions are driven by mantle *convection and are likely to have occurred throughout Earth history, although the resultant surface features are likely to have changed with time. *See* RIDGE-PUSH; and SLAB-PULL.

platform *See* SHELF.

platform conodonts *See* CONODONTOPHORIDA.

platy Applied to *minerals which develop a *crystal form consisting of thin, leaf-like layers. The *mica group of minerals provides a good example; the mineral splits along *cleavage planes which are parallel to rows of alkali atoms in the crystal structure.

platykurtic *See* KURTOSIS.

play The combination of factors that makes possible the accumulation of oil and gas in a particular area.

playa (salina) The lowest part of an *intermontane basin or *bolson, which is frequently flooded by run-off from adjacent highlands or by local rainfall. *Sediments consist largely of *colloids, *clays, and *evaporites, e.g. *halite, *gypsum, and sodium sulphate. The surface is generally flat, with *mudflats and locally small *dunes. The name (*playa* is the Spanish word for 'beach' and *salina* for 'salt-mine') was first applied to the arid basin-and-range province between the Colorado Plateau and Sierra Nevada, in the western USA, but is now used to describe such areas throughout the world.

Playfair, John (1748–1819) Scottish mathematician, who popularized and promoted the plutonist and uniformitarian theories of James *Hutton, rewriting his work to make it more easily readable (1802).

Plectronoceras cambria One of the first *nautiloids, which was small and horn shaped. As its name suggests, the species is discovered in rocks of *Cambrian age, in Europe and Asia.

Pleistocene The first of two *epochs of the *Quaternary, held conventionally to have lasted from approximately 1.64 Ma until the beginning of the *Holocene about 10 000 years ago. The epoch is marked by several glacial and *interglacial episodes in the northern hemisphere.

Pleistocene refugium Favourable area where species have survived periods of glaciation during the *Pleistocene Era. Such species are termed *relicts.

Pleistogene *See* QUATERNARY.

pleochroic halo The dark, strongly pleochroic (*see* PLEOCHROISM) zone often seen around certain radioactive inclusions (e.g. *zircon, *apatite, and *sphene) in some *minerals (e.g. *biotite, *tourmaline, *amphibole, *chlorite, *muscovite, *cordierite, and *fluorite). The halo is created by the interaction of alpha particles with the atoms of the crystal *lattice. Some haloes contain concentric rings with radii of 10–15 μm, the radii indicating the various kinetic energies of the different sets of alpha particles.

pleochroism In optical microscopy, the differential absorption of light in different crystallographic orientations by a coloured mineral when it is rotated on the stage in *plane-polarized light. The mineral may show variations in shade of the same colour or even a different colour. *See also* DICHROISM.

pleonaste *See* SPINEL.

plesiomorph Primitive (of a character state); the opposite of *apomorph.

plesiomorphic Applied to features that are shared by different groups of biological organisms and are inherited from a common ancestor. The term means 'old-featured' and the features to which it is applied were formerly called 'primitive'.

plesion In taxonomy, a group of superfamilies within a suborder.

Plesiorycteropus A genus of apparently burrowing, ant-eating mammals, known only from incomplete material found in subRecent deposits in Madagascar. Individuals weighed about 10 kg or less. *Plesiorycteropus* has been placed in a new order, *Bibymalagasia.

Plesiosauroidea (plesiosaurs) Suborder of aquatic *reptiles which enter the fossil record in the late *Triassic, and which are common in many *Jurassic and *Cretaceous *sediments. In appearance they were likened by *Buckland to 'a snake strung through the body of a turtle', and some grew up to 15 m in length. There were also short-necked types, as well as the swan-necked. The former are defined as the superfamily Plesiosauroidea (e.g. *Plesiosaurus* and *Muraenosaurus*), the latter as the superfamily Pliosauroidea (e.g. *Pliosaurus* and *Trinacromerum*), the two superfamilies making up the suborder.

plesiosaurs *See* PLESIOSAUROIDEA.

pleura *See* PLEURON.

pleural *See* PLEURON.

pleuron (pl. pleura, adj. pleural) The lateral portion of a single thoracic segment in a trilobite (*Trilobita). Each pleuron is indented by an oblique pleural furrow and the outer ends are downturned. The inner edges are attached to a central axial ring, so that each thoracic segment consists of a central axial ring and a pair of laterally placed pleura.

plicate Folded or wrinkled.

Pliensbachian A *stage in the European Lower *Jurassic, roughly contemporaneous with the lower Ururoan (New Zealand) and dated at 194.5–187 Ma (Harland et al., 1989). *See* LIAS.

Plinian eruption An explosive volcanic *eruption of *pyroclastic ejecta forming an eruption column which may be up to 55 km high, dispersing ejecta over an area of 500–5000 km^2. The eruption column has the distinctive, spreading, branched shape of the stone pine (*Pinus pinea*) native to the Mediterranean region. A Plinian eruption may produce thick, airfall, *pumice deposits, or the eruption column may collapse to generate a *pyroclastic flow. This type of eruption is named after Pliny the elder, who died at Pompeii in AD 79 during the eruption of Vesuvius that destroyed the city.

plinthite Portion of *mineral soil containing a large proportion of iron and aluminium oxides, *clay, and *quartz, which has developed through a combination of *leaching and *gleying in well-weathered tropical soils. On drying, it changes irreversibly to an ironstone *hardpan.

Pliocene The last (5.2–1.64 Ma) of the *Tertiary *epochs, comprising the *Zanclian (Tabianian) and *Piacenzian Ages.

plough mark A depression, in some cases many metres deep and tens of metres wide, made on the sea floor by the base of an iceberg that is driven by wind and tide. Many plough marks are found near the edges of *continental shelves in high latitudes, marking the passage of ancient icebergs.

plug *See* VOLCANIC PLUG.

plunge The angle between a line and a horizontal datum plane; the term is commonly used in respect of the inclinations of *fold axes.

plunging breaker Wave that breaks by plunging forward in the direction of motion, so that its crest falls into the preceding trough and encloses a pocket of air. The wave-form is then lost. Typically it occurs when a fairly low wave approaches a steep shingle beach. It is characterized by a smooth forward and under face, and a convex top.

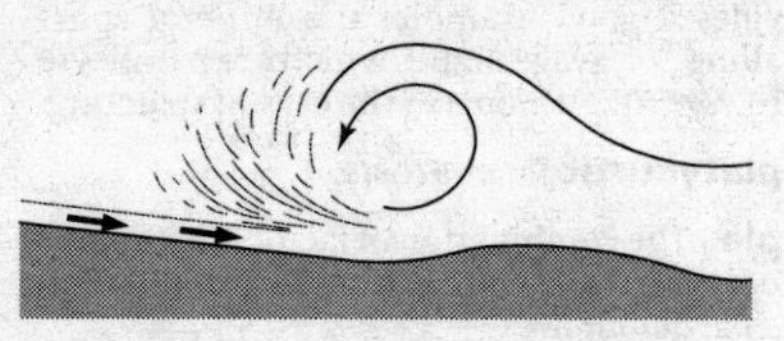

Plunging breaker

plus–minus method (Hagedoorn method) A method of interpreting *seismic refraction profiles over irregular layers whose slope angles do not exceed 10°. It uses both forward and reverse shooting to obtain matched *time–distance graphs. The plus component of the method allows the perpendicular depths to the refractor to be calculated; the minus component is used to determine the *seismic velocity in the refracting medium.

Pluto The ninth and outermost planet of the *solar system, its orbit an average 39.44 AU from the *Sun and highly eccentric, sometimes carrying it inside the orbit of *Neptune. Its distance from Earth ranges from 4293.7×10^6 km to 7533.3×10^6 km. Pluto is the smallest planet (much smaller than the *Moon), with a radius of 1137 km; volume 0.616×10^{10} km^3; mass 0.0125×10^{24} kg; mean density 2050 kg/m^3; surface gravity 0.66 (Earth = 1); visual albedo 0.3; black-body temperature 42.7 K. The atmosphere is very thin, with a surface pressure of about 0.003 bar, and composed of methane and nitrogen. The average surface temperature is about 50 K. It has one satellite, *Charon, so large that some astronomers consider Pluto and Charon a minor double-planet system. Pluto was discovered in 1930 by Clyde Tombaugh.

pluton General term applied to a body of *intrusive *igneous rock, irrespective of its shape, size, or composition.

plutonic A loosely defined term with a number of current usages. Many petrologists use it to describe *igneous rock bodies which have crystallized at great depth, there being no real agreement about what is meant by 'great depth'. Other petrologists use the term to describe any igneous rock that forms in largish *intrusions (i.e. excluding *dykes and *sills). 'Plutonic' has also been used to describe the origin of *magmas and gas derived from near the base of the *crust or in the upper *mantle.

plutonism A theory of the formation of the Earth, prominently advocated by *Hutton, and based on cycles of the growth, decay and regeneration of land masses. The forces involved in these cycles were heat and gravity, the heat deriving from subterranean fires. This generated volcanic rocks and granites, which rose molten from the Earth's interior, to form new land.

pluvial period Prolonged phase of markedly wetter climate that occurs in a normally dry or semi-arid area.

pneumatolysis (adj. pneumatolytic) Changes in rock mineralogy and chemistry that are initiated by the action of a hot, chemically active, gaseous solution derived from a *magma during its final stages of crystallization. The gaseous solution, rich in *volatile elements such as fluorine, chlorine, boron, and hydrogen, is released from the crystallizing interior of the *granite body and rises through cracks and fissures in the crystalline and cooler upper carapace to initiate *greisenization and *tourmalinization of the roof-zone granite.

poached soil *See* PUDDLED SOIL.

Podolskian A *stage in the *Moscovian Epoch, preceded by the *Kashirskian and followed by the *Myachkovskian.

Podopteryx mirabilis Named 'marvellous foot wing', this is the species that is the probable ancestor of the *pterosaurs. Unlike them it was a glider, relying on the large membrane stretched between its hind legs and tail. It is known from the Lower *Triassic of Soviet Kirgizstan.

podsol *See* PODZOL.

podsolization *See* PODZOLIZATION.

podzol (podsol) *Soil profile formed at an advanced stage of *leaching by the process of *podzolization, and identified by its acid *mor *humus, *eluviated and bleached E *soil horizon, and an iron-coloured B horizon, enriched with a variety of translocated materials. *See also* SPODOSOLS.

podzolization (podsolization) An advanced stage of *leaching, podzolization is the process of removal of iron and aluminium compounds, *humus, and *clay minerals from the surface *soil horizons by an organic leachate solution, and the deposition of some of these translocated materials in lower B horizons.

Pogonophora (beard worms) A phylum comprising deep-sea worms, first encountered in early *Cambrian rocks but discovered only in this century. Beard worms bear a superficial resemblance to *Polychaeta and it has recently been proposed that they are actually very highly specialized polychaetes. They live at great depths inside chitinous tubes they secrete for themselves in soft substrates, often in the vicinity of *hydrothermal vents. The body is *coelomate, partly segmented, has 'bristles' (chaetae), and is divided into three parts, the anterior crowned with tentacles. Their most remarkable feature is the complete absence of a gut. This has led to difficulties with classification because it is impossible to distinguish the ventral and dorsal surfaces. The animals are believed to obtain nourishment through a chemosymbiotic association (*see* CHEMOSYMBIOSIS) with bacteria. There are two groups, one found near vents and *cold seeps, the other occurring widely in all oceans.

poikilitic Applied to the *texture produced when several orientated or unorientated crystals are enclosed within a larger crystal in an *igneous rock. The larger crystals have more widely separated nuclei than the enclosed crystals, and may grow faster, thus enclosing surrounding grains. Where *augite encloses a number of *plagioclase feldspar crystals the poikilitic texture is given the specific name of 'ophitic texture'.

poikiloaerobic *See* DYSAEROBIC.

poikiloblast (adj. poikiloblastic) A large crystal which encloses several smaller, orientated or unorientated crystals in a *metamorphic rock.

poikilotherm (exotherm) Organism that regulates its body temperature by behavioural means, e.g. by basking or burrowing. Such animals are often termed 'cold-blooded', although when active their body temperature is little different from that of *homoiothermic ('warm-blooded') animals. Poikilothermy is characteristic of lower vertebrates (fish, *amphibians, and *reptiles), but not of *birds and *mammals.

poikilotopic Applied to the *fabric of a *sedimentary rock in which coarse crystals of *cement enclose a number of smaller, detrital grains.

point bar *See* BAR.

point group *See* CRYSTAL CLASS.

point load index (I_s) The force needed to fracture a sample of rock between conical points: $I_s = P/D^2$, where P is force and D is the distance between the points, both at failure. I_s is related to *uniaxial compressive strength (approximately equal to $I_s \times 24$).

point load tester Equipment for measuring *point load index, consisting of a jack which closes 60° cones across a sample (usually a piece of drill *core) and a gauge to measure pressure just before failing.

point source A source of energy which is considered to originate from a point and to have no effective size in relation to its observed effects, e.g. an explosive shot, a *current electrode in *electrical resistivity sounding, etc.

Poisson distribution In statistics, a discrete *probability distribution which is applied to the number of times an event occurs.

Poisson's ratio (Y) The ratio of latitudinal to longitudinal *strain, which describes the extent to which a material is distorted in a direction perpendicular to an applied *stress:

$$Y = e_{(\text{normal to } \sigma 1)} / e_{(\text{parallel to } \sigma 1)},$$

where Y is Poisson's ratio, e the strain, and σ1 is the maximum *principal stress. It is named after Simeon Poisson (1781–1840).

polar air *Air mass originating in latitudes 50–70°, which may have a maritime or continental source. Maritime polar air has high relative humidity, is warmed in its passage equatorward or over warmer seas, and becomes unstable. Continental polar air is stable in its source region and is associated with very cold surface conditions in winter, e.g. Siberian air moving across Europe.

polar-air depression Non-frontal *depression in the northern hemisphere, which typically results from the southward movement of unstable polar or arctic *maritime air along the eastern side of a large-scale meridional high-pressure ridge. *See also* COLD LOW.

polar climate Climatic type associated with regions inside the Arctic and Antarctic Circles. A gradation of climatic characteristics exists towards the poles, from *tundra conditions to those of perpetual frost. *See also* KÖPPEN CLIMATE CLASSIFICATION.

polar-desert soil *Mineral soil without identifiable *soil horizons, and with almost no surface *humus. It is associated with arid, polar-desert environments where precipitation is less than 130 mm annually, plant cover less than 25%, and thawed, active soil is 20–70 mm deep.

polar front The main boundary line between polar and tropical *air masses along which *depressions develop in mid latitudes, especially over the oceans. The front is in general displaced equatorwards in winter and polewards in summer, though large displacements in either direction take place over shorter periods in individual sectors of the hemisphere.

polar-front jet stream The *jet stream observed in different positions over middle and higher latitudes at heights of 10–13 km, and associated with the boundary zone between polar and tropical air above the *polar front. Maximum velocity averages 60 m/s, but can be twice this in extreme cases. The jet is typically discontinuous but at times can extend almost around the globe. It is more persistent in winter in response to the stronger temperature gradient between north and south than in summer. The jet is related to the development of surface frontal depressions of middle latitudes.

polar glacier *See* GLACIER.

polarimetry The measurement of the degree of polarization of light reflected from a planetary surface. The degree of polarization depends on the composition of the surface material. Polarimetry measurements have been used to obtain information on particle size distribution in the atmosphere of Titan, for example.

polarity The direction of evolutionary change. The polarity of different states of a *character means whether these are *primitive or *derived.

polarity chron The basic time interval in the *magnetostratigraphic time-scale, during which the Earth's *geomagnetic field is constantly, or predominantly, of one polarity, e.g. the *Gauss normal polarity chron and the *Matuyama reversed polarity chron. The duration of polarity chrons is variable, but generally is longer than 0.1 Ma. Polarity chrons may be interrupted by *polarity subchron(s), and grouped to form a *polarity superchron. The term 'polarity chron' has been proposed by the *ISSC to replace *polarity epoch; however, at present both terms are in use. The corresponding *chronostratigraphic unit is *polarity chronozone.

polarity chronozone The polarity *chronostratigraphic unit denoting all the rocks, with or without magnetic minerals (*see* FERROMAGNETIC), formed during a specific *polarity chron.

polarity epoch *See* POLARITY CHRON.

polarity event *See* POLARITY SUBCHRON.

polarity excursion A geomagnetic *polarity subchron.

polarity interval 1. An *informal term for any of the polarity *chronostratigraphic units: *polarity chronozone, *polarity subchronozone, or *polarity superchronozone. **2.** A general, informal term, either spatial or temporal, describing the intervening span between *polarity reversals.

polarity reversal, geomagnetic A change by 180° in the direction of the Earth's magnetic field. *See* GEOMAGNETIC FIELD.

polarity subchron A very short time interval (generally less than 0.1 Ma) of alternate polarity occurring within a *polarity chron, e.g. the *Olduvai normal subchron within the *Matuyama reversed polarity chron. The term 'polarity subchron' has been proposed by the *ISSC to replace '*polarity event'; however, at present both terms are in use. The corresponding polarity *chronostratigraphic unit is *polarity subchronozone. *See also* MAGNETOSTRATIGRAPHIC TIME-SCALE; and POLARITY SUPERCHRON.

polarity subchronozone A polarity *chronostratigraphic unit denoting all the rocks, with or without magnetic minerals (*see* FERROMAGNETIC), formed during a *polarity subchron.

polarity subzone *See* POLARITY ZONE.

polarity superchron The longest polarity time interval in the *magnetostratigraphic time-scale, comprising a number of *polarity chrons, and indicating a period of time (varying between 30 and 100 Ma) during which the polarity of the Earth's *geomagnetic field has a distinct bias. The bias may be towards a *normal field, a reversed field, or an evenly alternating field (called a mixed field), e.g. the present *Cretaceous–*Tertiary–*Quaternary mixed-polarity superchron. The corresponding polarity *chronostratigraphic unit is the *polarity superchronozone.

polarity superchronozone The largest polarity *chronostratigraphic unit, denoting all the rocks, with or without magnetic minerals (*see* FERROMAGNETIC), formed during a specific *polarity superchron.

polarity superzone *See* POLARITY ZONE.

polarity time-scale *See* MAGNETOSTRATIGRAPHIC TIME-SCALE.

polarity transition period The time taken for a change of polarity of the *geomagnetic field. It is thought to take 3000–5000 years for directional changes and about 12 000 years for associated intensity changes.

polarity zone The basic magnetic *lithostratigraphic unit, defined by the measured magnetic polarity of a body of rock. A polarity zone is bounded above and below by a *polarity reversal horizon or a *polarity transition zone. Polarity zones may comprise a number of polarity subzones and be grouped into polarity superzones. When used formally the term is capitalized, e.g. *Gauss Polarity Zone. *See* MAGNETOZONE.

polarization colours *See* INTERFERENCE COLOURS.

polarization, electrode The accumulation of *ions around an electrode, causing the accumulation of a *charge.

polarized radiation *Electromagnetic radiation which is orientated in a single plane.

polarizer A piece of *Polaroid in a transmitted- or *reflected-light microscope which is inserted into the light path between the light source and the *mineral section. Observations are made in *plane-polarized light and the light emerging is confined to either an E–W or N–S plane of vibration.

polarizing microscope A microscope

fitted with an *analyser and *polarizer to allow specimens to be examined in polarized light. The petrological (petrographic) microscope has a light source below the *stage so that light can be transmitted through the specimen. It is used to examine *thin sections of *rocks and *minerals. The ore (reflected light) microscope uses an incident light source, allowing light to be reflected from the polished surface of an *ore (*opaque) mineral. The system of lenses, diaphragms, rotating stage, polarizer, and analyser is broadly similar in both types of microscope, and dual microscopes are available for use with either transmitted or reflected light.

Polaroid The trade name of a plastic sheet impregnated with an organic iodide compound which strongly absorbs light in one *vibration direction and allows it to pass freely in the other vibration direction. The compound, iodocinchonidine sulphate, was first discovered by W. D. Herapath in 1852 and was named 'herapathite' in his honour. In 1928 E. H. Land combined the compound with plastic to produce the Polaroid sheet. It is now used in the *polarizer and *analyser of all *polarizing microscopes to produce *plane-polarized and cross-polarized (*see* CROSSED POLARS) light respectively.

polar orbit An *orbit which is inclined at 45° or more to the equatorial plane. *Compare* EQUATORIAL ORBIT.

polar stratospheric cloud (PSC) A cloud, consisting of ice crystals, which occurs in the *stratosphere late in winter over Antarctica and, less commonly, over the Arctic. The water vapour that freezes may be derived from the dissociation of methane and the cloud forms in the very still, very cold air of a vortex over the pole (the less common occurrence of such clouds in the northern hemisphere is owing to the generally higher temperature of Arctic stratospheric air and the briefer duration of the vortex). The reactions that deplete polar stratospheric ozone in the late winter and early spring take place on the surface of the ice crystals in polar stratospheric clouds.

Polar units *See* MARTIAN TERRAIN UNITS.

polar wander path The successive positions of the *palaeomagnetic pole. The individual pole position is calculated assuming the observed field is that of an *axial geocentric dipole. As the Earth's *axis of rotation is fixed relative to the *ecliptic, the changing positions of the pole, as a function of time, are due to the motion of the tectonic *plate from which rock samples were obtained. The pole path is therefore, more strictly, an apparent polar wander path. Sudden changes in the pole path ('hairpins') are usually caused by continental plate collisions and 'superintervals', between hairpins, correspond to the motion of a plate with little or no collisional interactions with other plates.

polder A low-lying, flat area reclaimed from the sea and protected by embankments or dykes; especially along the Netherlands North Sea coast.

pole of a face (face pole) The normal (line perpendicular) to a *crystal face, and the point at which the normal intersects the *plane of projection. In a *stereographic projection of a *crystal, the crystal is imagined to lie at the centre of a sphere and a pole of each face intersects the surface of the sphere at a point. These points are then projected to either the N. or S. pole of the sphere, on to the horizontal (equatorial) plane to produce an accurate, two-dimensional representation of the faces and *interfacial angles of the crystal.

pole of rotation (Euler pole) A point on the Earth's surface which defines a line through the centre of the Earth about which the relative motion of two *plates may be described.

'Polflucht' ('flight from the poles') A concept that was invoked by *Wegener to explain his ideas about *continental drift. He suggested that a differential gravitational force (the *Eötvös force), caused by the flattening of the Earth at the poles, would cause continental masses to drift towards the equator. This force is now known to be far too weak to cause continental movements.

polished section A specimen of an *ore (*opaque) mineral after it has been prepared for examination under a reflected-light microscope by light reflected from its polished surface. A sample of the mineral is mounted in a cold-setting epoxy resin, sawn to produce a flat surface, and then inverted, ground, and polished in a number of stages using diamond-impregnated fluids and a rotating lap fitted to a polishing machine. Conventional 0.03 mm *thin sections are prepared in the same way to give 'polished thin sections', used in the identification of opaque minerals. Polished sections are also required in the *electron-probe micro-analyser to enable analyses of minerals to be carried out.

polished stone value *See* AGGREGATE TESTS.

polishing relief In the preparation of a *polished section the relief caused when harder *minerals stand out slightly above the surface of softer minerals. Although not desirable, the presence of such relief may help to determine the relative hardness of adjacent minerals. *See* KALB LIGHT LINE.

polje Large, flat-floored depression bounded by steep valley walls and found in a *karst environment. It is classically described for the Dinaric region of Yugoslavia, where it may result from faulting or from solutional processes controlled by a local base level.

pollen Collectively, the mass of microspores (pollen grains) produced by the anthers of a flowering plant (*angiosperm) or the male cones of a *gymnosperm. Different pollen types are described according to their shapes, apertures, etc. Furrows on the surface of the pollen grain are called 'sulci' (sing. sulcus) and monosulcate pollen has a single sulcus. Tricolpate pollens have three, furrow-like, germinal apertures arranged 120° apart and there are many variants of this type.

pollen analysis The study of fossil *pollen and *spore assemblages in *sediments, especially when reconstructing the vegetational history of an area. The outer coat (*exine) of a pollen grain or spore is very characteristic for a given family, genus, or sometimes even species. It is also very resistant to decay, particularly under *anaerobic conditions. Thus virtually all spores and pollen falling on a rapidly accumulating sediment, anaerobic water, or *peat are preserved. Since both pollen and spores are generally widely and easily dispersed, they give a better picture of the surrounding regional vegetation at the time of deposition than do macroscopic plant remains, e.g. fruits and seeds, which tend to reflect only the vegetation of the immediate locality. With careful interpretation, pollen analysis enables examination of climatic change and human influence on vegetation, as well as sediment dating and direct study of vegetation character. The technique has also been applied, more controversially, to the pollen and spore contents of modern and *fossil *soil profiles. *See also* PALYNOLOGY.

pollen-assemblage zone *See* POLLEN ZONE.

pollen diagram Standardized pictorial summary of the *pollen record for a particular location. The vertical axis represents depth, and the proportions or absolute amounts of the various pollen types occurring at different levels are shown by bar histograms or by points on a continuous curve. Conventionally, similar patterns are grouped together on the diagram with arboreal types (i.e. trees) shown first followed in turn by shrubs, herbs, and *spores.

pollen zone (pollen-assemblage zone) Characteristic *pollen-and-*spore assemblage classically considered indicative of a particular type of climate which was assumed to be typical of a fairly extensive geographic region. Changes from one group of pollens to another characteristic assemblage are used to define pollen-zone boundaries. The standard British (Godwin, 1940) and very similar European *post-glacial (i.e. late *Devensian and *Flandrian) chronology recognizes eight major pollen zones, Zones I–III being the characteristic late-glacial sequence: *Older Dryas (I), *Allerød (II), and *Younger Dryas (III). More recently, based on N. American work, the importance of regional variation in the typical zone floras has been acknowledged, and a more flexible approach to pollen-zone definition applied. Pollen-assemblage zones are defined in terms of their pollen and spore profiles alone for a particular site, and initially without reference to or matching with the standard zone models with their strong climatic links. This has enabled local changes, often anthropogenic rather than climatic, to be elucidated more clearly.

poloidal field A magnetic field with radial and tangential components. The *geomagnetic field detected at the Earth's surface is of this form and contrasts with the unmeasurable *toroidal field.

poly- From the Greek *polus* meaning 'many', a prefix meaning 'many'.

polyanions Stable and complex *anions, of various sizes, formed by aluminate and silicate anions linked by shared oxygen *ions.

Polychaeta (bristleworms; phylum *Annelida) Class of annelid worms which possess distinct *metameric segmentation. All have bristly parapodia (movable, paired appendages) on each body segment. Eyes may be present. Males and females occur. Most are marine, although some occur in fresh water and on land. The group first appeared in the *Cambrian, and impressions of polychaete worms are among the fauna of the *Burgess Shale. However, polychaetes are represented in the fossil record largely by *burrows (e.g. **Skolithus*), tubes, and *scolecodont assemblages.

polycyclic landscape (polyphase landscape) A land-form or landscape that has been acted on by the erosional processes associated with two or more partially completed cycles of erosion (*see* DAVISIAN CYCLE). A diagnostic feature is an abrupt break of slope (*see* KNICK POINT) in the profiles of both rivers and hillslopes.

polygenetic *See* CONGLOMERATE.

polygonization *See* HOT WORKING.

polyhaline water *See* HALINITY.

polyhalite Mineral, $K_2Ca_2Mg(SO_4)_4.2H_2O$; sp. gr. 2.8; *hardness 2.5–3.0; *triclinic; normally flesh-pink to brick-red, and translucent; silky to resinous *lustre; usually occurs as fibrous or lamellar masses; *cleavage {100}, parting {010}; occurs in bedded *evaporite deposits and is one of the last minerals to be precipitated from saline waters, due to its high solubility; tastes bitter.

polymetallic sulphide A mineral deposit with three or more metals in commercial quantities; common metals include Cu, Pb, Zn, Fe, Mo, Au, and Ag. It may occur in magmatic, volcanogenic, or hydrothermal environments.

polymetamorphism Repeated episodes of heating and deformation (*metamorphism) acting upon a rock system.

polymictic 1. Applied to lakes (e.g. in high altitudes in the tropics) whose waters are circulating virtually continuously. If periods of stagnation occur they are very short. **2.** Applied to a *conglomerate which contains *clasts of many different rock types. *Compare* OLIGOMICTIC.

polymineralic *See* ROCK.

polymorph One of several possible *crystal forms of an element, compound, or mineral, all of which have the same chemical composition. *See* POLYMORPHISM.

polymorphic Occurring in several different forms.

polymorphic minerals *Minerals which have the same composition, but different atomic lattices, so their physical properties may differ.

polymorphic transformation Change in the atomic structure of a *mineral (but not in its chemical composition) to produce a different form. In reconstructive transformation there are large energy barriers to overcome in order to break down and re-form bonds, and transition tends to be slow, e.g. *graphite to *diamond. In displacive transformation angles are changed between bonds, but no bonds are broken, so little energy is required, e.g. alpha to beta *quartz.

polymorphism Ability of elements or compounds to exist in more than one crystal form, with each having the same chemical composition but different physical properties due to differences in the arrangement of atoms. Examples are *graphite and *diamond (both C); alpha and beta *quartz (both SiO_2); and *calcite (*hexagonal) and *aragonite (*orthorhombic), both forms of $CaCO_3$. The terms 'dimorphism' and 'trimorphism' describe (respectively) two and three different forms.

polypedon (soil individual) Two or more contiguous *pedons, which are all within the defined limits of a single *soil series.

polyphase landscape *See* POLYCYCLIC LANDSCAPE.

polyphyletism (adj. polyphyletic) The occurrence in *taxa of members that have descended via different ancestral lineages. True polyphyletism has traditionally been distinguished from errors of *classification, especially at the higher taxonomic levels, where organisms, as a result of *convergent or *parallel evolution, have been placed wrongly in the same natural group; but modern phyletic taxonomists would hold that any taxon found to be polyphyletic is unnatural, and so an 'error', and must be disbanded.

Polyplacophora (chitons; phylum *Mollusca) Subclass of *Amphineura (or in some classifications ranked as a molluscan class) in which seven or eight *dorsal *plates, generally composed of calcium carbonate, are enclosed by a marginal girdle. The plates articulate with one another and overlap to a varying extent. The anterior and posterior plates differ from the others, the anterior plate often being ribbed. Polyplacophora are entirely marine, and first appeared in the Upper *Cambrian.

porphyrin Heterocyclic derivative of a porphin, which is composed of four linked rings, each containing nitrogen (a tetrapyrrole ring structure). As such it is capable of combining with a variety of metals and so forms part of the structure of many important biological molecules, including haemoproteins, chlorophyll, cytochromes, and vitamin B_{12}.

polysynthetic twin *See* ALBITE TWIN.

polytaxic times A period of high biological diversity among marine organisms, associated with high sea level (highstand), equable climates, little convection in ocean waters, and an abundance of available niches.

polytetrahedron In *crystallography, a *tetrahedron in which negative and positive forms are developed to different degrees in the same crystal, together with the possibility of raised secondary faces on each principal tetrahedron face. In *mineralogy, the term may refer to the ability of SiO_4 tetrahedra in silicate structures to polymerize, giving a variety of structures as a result of oxygen-sharing. These may result in chains (*pyroxenes), double chains (*amphiboles), sheets (*micas), and three-dimensional framework silicates (*quartz and *feldspars).

Pomeranian One of a series of *Weichselian recessional *moraines in northern Germany, which postdates the *Frankfurt series of moraines and predates the Velgart series.

ponente Regional term for a westerly wind that blows in the Mediterranean.

Pongola *See* SWAZIAN.

pool-and-riffle An alternation between a deep zone (the pool) and a shallow zone (the riffle) along the sand and/or gravel bed of a stream. The pool-to-pool spacing is about 5–7 times the width of the channel. The sequence is found in both straight and *meandering channels; in the latter case the pool occurs in the meander bend on the concave side, and the riffle between bends.

poorly washed biosparite *See* FOLK LIMESTONE CLASSIFICATION.

Poperinge *See* MOERSHOOFD.

pop-shooting (secondary blasting) A method of secondary blasting used to break large boulders into pieces of manageable size. A hole is drilled to just beyond the middle of the boulder so that the charge is central. The method is fairly quiet and economical.

pop-up The relatively uplifted section of a *hanging wall which is formed by *back thrusting. The structure is bounded by the back thrust and by the major *thrust from which the back thrust originated.

Porangan A *stage in the Lower *Tertiary of New Zealand, underlain by the *Heretaungan, overlain by the *Bortonian, and roughly contemporaneous with the mid *Lutetian Stage.

porcelain jasper *See* JASPER.

porcellanite A highly siliceous *sedimentary rock with a dull *lustre, porous texture, and *conchoidal fracture, similar in general appearance to porcelain. These rocks are less hard and vitreous than *cherts. Porcellanites include certain *biogenic, siliceous, deep-sea *sediments and some fine-grained, *recrystallized, acid *tuffs.

pore A void surrounded completely by soil or rock materials and created by the packing of mineral and organic particles. Pores can be filled by any proportion of air or water.

pore fluid pressure Pressure of a fluid which fills the *pores in a saturated soil or rock. It can indicate the degree of consolidation of an earthwork, zero pressure indicating complete consolidation. *Compare* PORE-WATER PRESSURE.

pore space The total continuous and interconnecting void space in the bulk volume of soil or rock.

pore-water pressure The pressure exerted on its surroundings by water held in *pore spaces in rock or soil. The pressure is positive when a soil is fully saturated, and is then proportional to the height of the water measured in an open tube (a *piezometer) above the point of interest. A buoyancy effect is achieved and the *shear strength of the soil is reduced. The pressure is zero when the soil voids are filled with air, and is negative when the voids are partly filled with water (in which case surface-tension forces operate to achieve a suction effect and the shear strength of the soil is increased). *Compare* PORE FLUID PRESSURE.

Porifera (Spongiaria, sponges) A phylum of multicellular animals that are not included in the *Animalia. Sponges are sessile, benthic, filter feeders with a bag-like body, a central cavity, and an outer surface pierced with tiny openings (ostia), through which water enters, and an upper, larger opening (osculum), through which it leaves. Many sponges have a skeleton that is either horny, or composed of spongin, or consists of calcareous or siliceous *spicules. They may be fossilized as a complete skeleton or as isolated spicules. Their *fossil record dates from the *Cambrian (or possibly from the *Precambrian) and they are a common source of *biogenic *silica.

porosity Absolute porosity is the total of all void spaces present within a rock, but not all these spaces will be interconnected and thus able to contain and transmit fluids. The *effective porosity is thus defined as the proportion of the rock which consists of interconnected *pores. Porosity is expressed as a percentage of the bulk volume of the rock. *See also* CHOQUETTE AND PRAY CLASSIFICATION; FENESTRAL POROSITY; FRACTURE POROSITY; MOLDIC POROSITY; and OOMOLDIC POROSITY.

porosity and permeability determination *Porosity is measured from rock plugs in a porosimeter. The method involves the extraction of air from the *pores in the rock by a vacuum, and measurement of the bulk volume of the rock. The porosity is calculated by the expression: % porosity = (volume of gas extracted/bulk volume of the rock sample) × 100. Permeability is measured in a *permeameter, by determining the pressure drop ($P_1 - P_2$) from a fluid of known *viscosity (μ) and *flow rate (Q), across a rock sample of known cross-section (A) and length (L). Permeability (K) is then determined by *Darcy's law: $K = Q\mu L/(P_1 - P_2)A$.

porphyritic *See* PHENOCRYST.

porphyroblast (metacryst) A large, well-formed (*euhedral) crystal which grew *in situ* during metamorphic *recrystallization and which is surrounded by a finer-grained *groundmass of other metamorphic crystals.

porphyroclast A large, intact, *mineral fragment or *clast which is surrounded by a finer-grained, crushed *groundmass produced during *dynamic metamorphism. The porphyroclast represents a relict of the original rock which has escaped crushing during the deformation process.

porphyry A medium- to coarse-grained, *intrusive, *felsic, *igneous rock which is conspicuously *porphyritic, containing more than 25% *phenocrysts by volume. The phenocryst *mineral is usually *alkali feldspar. The term can be used as a suffix to a specific name, e.g. *quartz porphyry.

porphyry copper *See* PORPHYRY DEPOSIT (1).

porphyry deposit 1. Porphyry copper occurs as large copper deposits centred around *stocks of *intermediate to *acid, *porphyritic, *igneous rocks. Most occur in *Mesozoic and *Tertiary *orogenic belts. They show concentric zones of *minerals; for example at Bingham, Utah, there is an inner zone of Cu/Mo and an outer zone of Pb/Zn/Ag. The deposits are also characterized by extensive *alteration halos. Most deposits are 3–8 km across and several kilometres deep. They consist of disseminated *chalcopyrite and other *sulphides mined on a large scale from *open pits. The *ore is low grade (less than 1% Cu) but of great economic importance. It was probably formed by a sudden release of *volatiles near the surface, with shattering of the enclosing rocks. **2.** A deposit of molybdenum-bearing ore, usually *molybdenite, associated with rocks of porphyritic texture, and containing minor amounts of copper. It is formed in a similar manner to porphyry copper deposits, but the ore bodies are frequently shaped like inverted cups over the progenitor intrusive, e.g. those in the mineral belt of the USA. **3.** A gold-enriched porphyry copper deposit frequently associated with *island-arc environments. At the present time no true porphyry gold deposits are known, only those with *co-product gold and copper.

porphyry gold *See* PORPHYRY DEPOSIT.

porphyry molybdenum *See* PORPHYRY DEPOSIT.

Portia (Uranus XII) One of the lesser satellites of *Uranus, with a diameter of 55 km. It was discovered in 1986.

Portlandian The youngest *stage of the *Jurassic in Britain, overlain by *sediments of the *Purbeckian Stage and resting in turn on those of the *Kimmeridgian. It is characterized in southern England by *mollusc-rich *limestones. *See also* MALM; TITHONIAN; and VOLGIAN.

positive inversion *See* INVERSION.

post- From the Latin *post* meaning 'after', a prefix meaning 'after', 'behind', or 'later'.

post-depositional remanent magnetization The magnetization acquired by a sedimentary rock after deposition and before undergoing *metamorphism, mostly associated with *chemical remanent magnetization as *ferromagnetic minerals grow during diagenetic alteration (*see* DIAGENESIS), but it also results from the physical rotation of magnetic grains associated with the movement of fluids and gases through the rock.

post-deuteric alteration Changes to the *fabric or composition of an *igneous rock after the completion of *deuteric changes at elevated temperatures.

post-glacial *See* FLANDRIAN; and HOLOCENE.

post-tectonic Applied to a process or event, e.g. the emplacement of *plutons, which occurs after deformation. *Compare* PRE-TECTONIC; and SYNTECTONIC.

postzygapophyses *See* VERTEBRA.

potassium–argon dating (K–Ar method) Geologic dating technique based on the *radioactive decay of potassium (^{40}K) to argon (^{40}Ar). This potassium *isotope has a half-life (*see* DECAY CONSTANT) of 1.3 billion (10^9) years, making this a valuable dating method. The minimum age limit for this dating method is about 250 000 years.

potassium–calcium dating A *radiometric dating method based on the decay of ^{40}K to stable ^{40}Ca. This is not a generally useful technique because ^{40}Ca is the most abundant naturally occurring stable *isotope of calcium (96.94%). The formation of radiogenic ^{40}Ca atoms in a rock or mineral therefore increases its abundance only slightly. The ratio $^{40}Ca:^{44}Ca$ (96.94% : 2.08%) can be used to determine the amount of radiogenic ^{40}Ca present, although the dominance of naturally occurring ^{40}Ca makes it rather insensitive. Furthermore the determination of the isotopic composition of calcium by *mass spectrometry is made difficult by the low efficiency of ionization of calcium atoms in a thermionic source, and by fractionation of isotopes during that process. Because of these disadvantages the $^{40}K:^{40}Ca$ method of dating is really only viable for minerals that are strongly enriched in potassium and depleted in calcium, such as *micas in *pegmatite and *sylvite in *evaporite rocks.

potassium feldspar *See* ALKALI FELDSPAR.

potential electrode An electrode used as a ground contact in a voltage-measuring circuit. In *electrical-resistivity and *induced-polarization surveying, two potential electrodes are used, with a variety of possible *electrode configurations; in *spontaneous potential measurements *non-polarizable (e.g. porous pot) electrodes are commonly used.

potential energy *See* ELEVATION POTENTIAL ENERGY; and HYDRAULIC HEAD.

potential evapotranspiration (*PE*) The amount of water that would evaporate from the surface and be transpired by plants were the supply of water unlimited. It is calculated from the mean monthly temperature, with corrections for day length, and was devised by C. W. *Thornthwaite as part of his system of *climate classification (*see* THORNTHWAITE CLIMATE CLASSIFICATION). From *PE* minus precipitation an approximate index can be calculated of the extent to which the water available for plants falls short of the amount they are capable of transpiring. *Compare* ACTUAL EVAPOTRANSPIRATION.

potential instability (convective instability) Atmospheric condition in which otherwise stable air would become unstable if forced to rise, e.g. over high ground, thereby reaching its saturation point. Large *cumulus with much precipitation often results from the forced uplifting of such air. *See also* INSTABILITY; and STABILITY.

potential reserve *See* RESERVE.

potentiometric surface (piezometric surface) A hypothetical surface defined by the level to which water in a confined *aquifer rises in observation *boreholes. In practice, the potentiometric surface is mapped by interpolation between borehole measurements. As with the *water-table in an unconfined aquifer, the slope of the potentiometric surface defines the *hydraulic gradient and the horizontal direction of *groundwater flow. *See* HYDRAULIC HEAD.

pot-hole An approximately hemispherical depression made in the bedrock of a river channel by stones or boulders that have been spun rapidly by eddies. Each stone is washed away only to be soon replaced by another, so the drilling action is powerful and ceaseless.

Potsdam gravity The value of *gravitational acceleration as measured at Potsdam, eastern Germany, and previously used as a world-wide standard. It is now replaced by the *International Gravity Formula.

Potter's flood-peak formula *See* FLOOD-PEAK FORMULAE.

Poundian A *stage of the *Ediacara Epoch, dated at about 580–570 Ma (Harland et al., 1989).

powder photograph *See* X-RAY POWDER PHOTOGRAPH.

powder technology The study of the properties, behaviour, and uses of finely particulate materials.

powellite *See* TUNGSTATES.

power-law creep *Creep resulting from the movement of *crystal dislocations into systematic patterns, usually polygonal, within a stress field.

***p*-parameter** A measure of the prolateness of an ellipsoid: if p is less than 0.9 the ellipsoid is oblate, if p is greater than 1.1 it is prolate.

p.p.b. Parts per billion (10^9).

PPL *See* PLANE-POLARIZED LIGHT.

p.p.m. Parts per million.

PPR *See* PHOTOPOLARIMETER-RADIOMETER.

p-process *See* PROTON-ADDING PROCESS.

praecipitatio From the Latin *praecipitatio* meaning 'I fall', a supplementary cloud feature in which precipitation from the cloud base falls to the ground. The feature is usually seen with *cumulus, *cumulonimbus, *stratus, *stratocumulus, *altostratus, and *nimbostratus. *See also* CLOUD CLASSIFICATION.

praedichnia *Trace fossils comprising structures that have resulted from predation.

Pragian A *stage of the *Devonian Period, dated at 396.3–390.4 Ma (Harland et al., 1989).

Pratt, John Henry (?1811–1871) Pratt was a mathematician and physicist, and Archdeacon of Calcutta. He attempted to calculate the mass of the Himalayas, after discrepancies were found in the Trigonometrical Survey of India. He proposed that high mountains had a lower density than other parts of the *crust, and went on to develop his own version of the theory of *isostasy. *See* PRATT MODEL.

Pratt model A model for the *lithosphere that accounts for *isostatic anomalies by assuming there is a level of compensation that lies at a constant depth everywhere. Below the level of compensation all rocks have the same density, but above it density decreases as topographic elevation increases. For a column of material anywhere on Earth, the mass lying above the level of compensation will be the same, and $\rho_c h$ a constant, where ρ_c is the density of the crust and h the topographic elevation. If the rocks are on the sea bed, then $\rho_c h + \rho_w d$ is a constant, where ρ_w is the density and d the depth, of sea water. *See also* AIRY MODEL.

pre- From the Latin *prae* meaning 'before', a prefix meaning 'in front of', 'earlier than', 'more important than', or 'better than'.

pre-adaptation Adaptation evolved in one adaptive zone which, quite by chance, proves especially advantageous in an adjacent zone and so allows the organism to radiate into it. No selection for a future environment is implied.

Preboreal The first *Flandrian (*Holocene, or *post-glacial) *stage, a time of rapid forest spread, from about 10 300–9600 BP. 'Preboreal' refers to climatic conditions and in vegetational terms is equivalent to Pollen Zone IV (*see* POLLEN ZONE) of the standard British and European post-glacial pollen chronology.

Precambrian A name now used only informally to describe the *Priscoan, *Archaean, and *Proterozoic, which together comprise the longest period of geologic time that began with the consolidation of the Earth's crust and ended approximately 4000 million years later with the beginning of the *Cambrian Period around 570 Ma ago. The rocks of this period of geologic time are usually altered and few *fossils with hard parts or skeletons have been found within them. Precambrian rocks outcrop extensively in shield areas such as northern Canada and the Baltic Sea.

precession The action of a couple whose axis is perpendicular to the rotational axis (a torque) on a rotating body causing its *axis of rotation to trace out a path about an average position, instead of having a constant alignment, i.e. the axis of rotation itself revolves conically about a central point. The Earth's axis of rotation precesses as a result of several forces, e.g. changes in mass distribution on its surface, changes in the gravitational field due to changes in the relative positions of the Moon, Sun, and planets, etc.

precession of the equinoxes *See* MILANKOVICH CYCLES.

precipitable water The quantity of rainfall that would result from condensation and precipitation of the total moisture in a column of air in the atmosphere. Most atmospheric moisture is contained in the lower atmosphere, below about 5500 m. On average, an atmospheric column of 1 m^2 cross-section contains vapour equivalent to 5–25 mm depth of rainfall. The average *residence time of moisture in the atmosphere is about nine days.

precipitation 1. In meteorology, all the forms in which water (H_2O) falls to the ground as rain, sleet, snow, hail, drizzle, or other more specialized forms, and also the amounts measured. Sometimes precipitation seen falling from clouds evaporates before reaching the ground. **2.** The process of depositing dust or other substances (pollution) from the air. **3.** The deposition of solid particles out of a supersaturated solution.

precipitation-efficiency index Devised in 1931 by C. W. *Thornthwaite, an index based on the ratio of mean monthly rainfall and temperature values to evaporation rates. Summation of monthly values gives an annual precipitation-efficiency index (P-E), which is used to define major climatic regions. *See also* THORNTHWAITE CLIMATE CLASSIFICATION.

precision *See* ERRORS.

predisplacement An alteration in the *ontogeny of a descendant such that some developmental process begins earlier than in its ancestor, and so has progressed further by the time maturity is reached.

pre-ferns A group of plants transitional between the *Psilophytales and the true ferns, having some fern and some psilophyte characters. They had leaves and reproduced by *spores, but had a variety of growth forms. The orders usually included in the group are the Protopteridales and Coenopteridales.

preferred orientation The alignment of inequidimensional *mineral grains in a rock to define a two-dimensional, planar *fabric or a three-dimensional, linear fabric. Alignment may be caused by deformation accompanying *metamorphism, flow of crystal-laden *magma, settling of *crystals in a magma, or settling of flaky minerals in water.

pre-Hadean The earliest *era of the *Priscoan, covering the formation of the Earth and ending around 4550 Ma (Harland et al., 1989).

prehnite A hydrated *silicate and member of the *phyllosilicates (sheet silicates) with composition $Ca_2Al[Si_3AlO_{10}]OH_2$; sp. gr. 2.9; *hardness 6.0; colourless, white, or green; *tabular or in granular *aggregates; has a layered structure similar to the *micas; occurs in association with the *zeolites infilling cavities in *basic *igneous rocks and in contact metamorphosed (*see* THERMAL METAMORPHISM) *limestones. *See also* PREHNITE-PUMPELLYITE FACIES.

prehnite–pumpellyite facies A set of metamorphic *mineral assemblages produced by *metamorphism of a wide range of starting rock types under the same metamorphic conditions and typically characterized by the development of the mineral assemblage prehnite ($Ca_2Al[Si_3AlO_{10}](OH)_2$), pumpellyite ($Ca_2Al_2(Mg,Fe^{2+},Fe^{3+},Al)[SiO_4][Si_2O_7](OH)_2(H_2O.OH)$) and *quartz, with relict *plagioclase and *pyroxene, in rocks of *basic *igneous composition. Other rocks of contrasting composition (e.g. *shales or *limestones) would each develop their own specific mineral assemblage, even though they are all being metamorphosed under the same conditions. The variation of mineral assemblage with starting rock composition reflects a particular range of pressure, temperature, and $P(H_2O)$ conditions. Experimental studies of pressure–temperature stability fields indicates that the *facies represents a range of moderate pressure (2.5–5.0 kb), low temperature (150–300 °C) conditions usually developed during *burial metamorphism in thick sedimentary sequences on *continental margins and in intracontinental basins.

Pre-Imbrian *See* LUNAR TIME-SCALE.

Pre-Nectarian System *See* LUNAR TIME-SCALE.

pre-splitting A method of blasting in which a planar crack is propagated by blasting to determine the final shape of a rock face before holes are drilled for the final blast pattern; firing with a minimum time scatter can then be used. The crack helps to screen the surroundings from ground vibrations during the firing of the main round.

pressure–depth profile In the Earth, the relationship between pressure and depth is controlled by the density of the rocks and the *gravitational acceleration at each point along the profile. Gravitational acceleration is almost constant through most of the *mantle, though it rises slightly close to the *core–mantle boundary, and then decreases linearly to zero at the centre of the Earth. The pressure rises to about 100 GPa at the core–mantle boundary and reaches 375 GPa at the centre of the Earth.

pressure dissolution The process that occurs preferentially at the contact surface of *grains or *crystals as a result of an excess of external pressure relative to the hydraulic pressure of pore fluids. Material in these zones is dissolved and removed, resulting in an increase in *compaction and a decrease in *porosity.

pressure fringe A growth of fibrous *quartz, *calcite, *chlorite, or *muscovite, normal (or sometimes parallel) to the face of a *porphyroblast (usually *pyrite or *magnetite) in a *regional metamorphic rock. During *metamorphism and deformation *minerals dissolved from the region of high pressure, where the *matrix is squeezed against the *crystal, are redeposited in the *pressure shadow zone at the side of the crystal. The minerals nucleate on the face of the porphyroblast and regrow in a fixed crystallographic orientation with respect to the edge of the large porphyroblast. This results in the straight fibres which often intersect at segment junctions. Where the porphyroblast is rotated during the metamorphic event the pressure-fringe minerals grow as curved fibres, thus tracing the stages of detachment between porphyroblast and matrix.

pressure-gradient force (PGF) **1.** The force acting on air due to pressure differences. Horizontal variations in pressure create a tendency for movement from higher to lower pressure: this is only one component of the forces acting on the actual wind, though, so air does not normally flow at right angles across the isobars. Other forces are associated with the rotation of the Earth underneath the moving wind, and with a centrifugal force where the path of the wind is curved. In practice the air moves nearly along the isobars above the friction layer. *See also* CORIOLIS FORCE; and GEOSTROPHIC WIND. **2.** The force acting on a water mass due to pressure differences over distance. Horizontal variations in pressure create a tendency for movement from higher to lower pressure areas. *See also* GEOSTROPHIC CURRENT.

pressure head The potential energy possessed by a unit weight of water at any point when compared with a pressure of one *atmosphere at the same elevation. For *groundwater, it is measured by the depth of submergence between the measurement point and the water level in a *piezometer. *See also* HYDRAULIC HEAD; ELEVATION POTENTIAL ENERGY; and POTENTIOMETRIC SURFACE.

pressure melting The melting of *ice in response to *stress. It comes about because the freezing temperature of water falls as pressure increases, at about 1 °C for every 140 bars ($140 \times 10^5 N/m^2$). The term 'pressure melting point' refers to the temperature at which ice just begins to melt under a given pressure.

pressure ridge A curved, elongated ridge on the surface of a basaltic *lava flow, formed at right angles to the flow direction. It may have been pushed up by the dragging effect of mobile lava below a cooling surface, although collapse may contribute to its detached form.

pressure shadow The area in a *regional metamorphic rock which is protected from deformation by the presence of a relatively rigid *porphyroblast or *porphyroclast. Randomly orientated *quartz and/or *chlorite concentrates in triangular regions next to the faces of the porphyroblast or porphyroclast which are themselves orientated perpendicular to the *schistosity of the surrounding, deformed, metamorphic *fabric. During *metamorphism and deformation quartz and chlorite are dissolved from the region of high pressure, where the matrix is squeezed against the hard, unyielding porphyroblast or porphyroclast, and are redeposited in the no-stress shadow region on either side of the porphyroblast where the deformed fabric wraps around the crystal, so forming the pressure shadow.

pressure-tube anemometer *See* ANEMOMETER.

pressure wave *See* P-WAVE.

pressure welding The suturing together of *grains in a *sedimentary rock as the result of *pressure dissolution taking place along the grain contact. The sutured margin appears as an irregular plane. Pressure welding of grains is particularly common when a *mineral *cement has developed at a late stage, after deep burial of the *sediment, as this results in the main *overburden being supported by the grain-to-grain contacts of the sediment.

pre-tectonic Applied to a process or event which occurs before deformation. *Compare* POST-TECTONIC; and SYNTECTONIC.

prevailing wind In a particular locality, the wind direction that is most frequent over time. For most areas the prevailing wind varies, sometimes quite markedly, according to season. It may also change, or have changed, when the climate changes (as in ice ages).

prezygapophyses *See* VERTEBRA.

Priabonian 1. The final *age in the *Eocene Epoch, preceded by the *Bartonian, followed by the *Rupelian (Stampian), and dated at 38.6–35.4 Ma (Harland et al., 1989). **2.** The name of the corresponding European *stage, which is roughly contemporaneous with the *Refugian (N. America), *Runangan (New Zealand), and part of the *Aldingan (Australia).

Priapulida Phylum comprising the priapus worms, known from *Cambrian times (e.g. *Ottoia* from the *Burgess Shale, Canada), and ranging up to the present.

Pridoli A *series (410–408.5 Ma) of the Upper *Silurian, underlain by the *Ludlow.

primary creep (transient creep) The initial stage of *creep, characterized by viscoelastic *strain, in a material subjected to long-term, low-level *stress.

primary crushing *See* CRUSHING.

primary geochemical differentiation A theory explaining the formation of the Earth's *core, *mantle, and *crust, in which the formation of the nickel–iron core may have been accompanied by partitioning of the elements. It postulates that some of the elements were reduced, alloyed with iron, and concentrated in the core, while the remainder formed the mantle and primitive crust.

primary geochemical dispersion The movement of elements below the Earth's surface by metamorphic, magmatic, or hydrothermal processes (*see* METAMORPHISM; MAGMATIC DIFFERENTIATION; and HYDROTHERMAL ACTIVITY), resulting in the formation of *igneous and *metamorphic rocks.

primary migration First stage in the upward migration of *hydrocarbons within and then out of the *source rock.

primary mineral A *mineral which has crystallized from a *magma. *Essential primary minerals are those primary minerals whose presence is essential for the classification and naming of the rock, while *accessory primary minerals are those primary minerals whose presence does not affect the classification or naming of the rock.

primary porosity *See* POROSITY; and CHOQUETTE AND PRAY CLASSIFICATION.

primary productivity The amount of organic matter synthesized by organisms from inorganic substances. In the oceans, *photosynthesis by phytoplanktonic *algae in the upper 100 m (the euphotic zone) accounts for most primary production. Waters of tropical areas are less productive than those of temperate regions, because in tropical areas the water column undergoes no seasonal vertical mixing. Areas of *upwelling of nutrient-rich deep waters have high productivity.

primary sedimentary structure A structure that forms during or very soon after the deposition of the sediment of which it is composed. *Compare* SECONDARY SEDIMENTARY STRUCTURE.

primary wave *See* P-WAVE.

primitive (evol.) Preserving the character states of an ancestral stage. The term may be used of a character (as a synonym of *plesiomorphic) or, occasionally, of a whole organism.

primitive circle In a *stereographic projection, the circumference of the *plane of projection.

primordial The period of time at, or just before, the final formation of the *Earth.

principal component analysis A *multivariate analysis which maximizes the spread of data by plotting *covariance values on sets of axes in multidimensional space allowing correlations which may have been hidden in the data to be identified. The first principal component corresponds to the first axis in multidimensional space and describes the majority of the spread of the data, subsequent higher order principal component axes are orthogonal to the first axis. Higher order axes display progressively less variation, where the data is less correlated and more representative of statistical noise.

principal point The centre point of an aerial photograph.

principal shock The main *earthquake event with the largest amplitude.

principal strain axes In a strained material, three mutually perpendicular axes (designated *X*, *Y*, and *Z*) which are parallel to the directions of greatest, intermediate, and least elongation, and which describe the state of *strain at any particular point.

principal strain ratio (ellipticity) The ratio (*R*) of the minimum *strain axis to the maximum strain axis. In its simplest, two-dimensional form the long and short axes of initially circular markers are measured and plotted against one another; the slope of the best-fit line through the origin and a number of points is the strain ratio for that plane of measurement. More complex methods are employed to determine *R* minimum, *R* maximum, and *R* tectonic for initially non-spherical strain markers, e.g. *conglomerate *clasts.

principal-stress axes Three mutually perpendicular axes (designated, σ_1, σ_2, and σ_3) which are parallel to the directions of maximum, intermediate, and least principal *stress. Their separate lengths and directions describe the state of stress at a particular point. The stress ellipse contains σ_1 and σ_3; the stress ellipsoid contains all three axes.

principle of contained fragments *See* PRINCIPLE OF INCLUDED FRAGMENTS.

principle of included fragments (principle of contained fragments) The principle that any rock containing fragments of another rock body must be younger than the rock unit from which those fragments were derived.

principle of superposition *See* LAW OF SUPERPOSITION.

principle of uniformitarianism *See* UNIFORMITARIANISM.

Priscoan The first of the three subdivisions of the *Precambrian, lasting from the formation of the Earth to 4000 Ma ago (Harland et al., 1989), and followed by the *Archaean.

prism A *crystal form composed of a number of repeat *crystal faces all of which are parallel to one of the principal *crystallographic axes (usually the vertical *c* (or *z*) axis). *See also* ACCRETIONARY PRISM.

prismatic Applied to elongate *minerals which have well-developed *crystal faces parallel to the *prism.

probability density function In statistics, the mathematical function which allocates probabilities of particular observations occurring. The probability density function may be used to construct a frequency distribution of certain events occurring either discretely, in the form of a *histogram, or continuously.

probability distribution In statistics, the relative frequency distribution of different events occurring, as defined by the *probability density function. Probability distributions may be discrete as in the cases of the *binomial and *Poisson distributions, or continuous as in the case of the *normal distribution.

'Problematica' A named group into which geological structures are placed if their affinity is uncertain but their origin is considered to be organic. They are removed from the 'Problematica' when more certain evidence of their relationships is discovered.

Proboscidea (proboscideans; infraclass *Eutheria, cohort *Ferungulata) Order comprising the elephants and their extinct relatives, e.g. mastodons (*see* MAMMUTIDAE), gomphotheres (*see* GOMPHOTHERIIDAE), and mammoths (*see Mammuthus*). The order was formerly highly successful and occupied the Americas, Eurasia, and Africa. Proboscideans tend toward large size. Since the late *Miocene most have possessed a long trunk: this is developed from the nose and upper lip. Teeth are reduced in number, young adults having three molars in each side of each jaw; these are used one at a time, old teeth being shed and replaced by those behind. The upper incisors are enlarged to form tusks. The jaw muscles are large, and the skull short and high. The *vertebrae and up to 20 ribs carry the weight of the abdomen, which is balanced on the fore limbs by the weight of the head, the hind limbs providing propulsion. The brain is well developed. Parental care of the young is prolonged, and social organization is complex.

proboscidean *See* PROBOSCIDEA.

Procaryotae *See* PROKARYOTE.

Procellarum System *See* LUNAR TIME-SCALE.

processes, geomorphological The set of mechanisms that operate at and near the Earth's land surface, breaking down and transferring rock material and consequently fashioning land-forms at small and medium scales. Processes operating on the surface are termed 'exogenetic'; those originating below the surface 'endogenetic'. Two categories may be recognized: (*a*) zonal processes, which are broadly controlled by climate (e.g. the work of *glaciers); and (*b*) azonal processes, which operate on a world-wide basis (e.g. *fluvial processes). The rate at which such processes operate may be measured by direct field recording or by making use of historical records. *See* EROSION RATE.

process-response system In *geomorphology, a natural *system that is formed by the combination of at least one *morphological and one *cascading system. It therefore shows how form and geomorphological process are related. An example is the coastal process-response system, in which the cascading system of wave energy advances from deep water to the edge of the *swash zone, and is linked to various morphological features of the shallowing zone.

prochoanitic *See* SEPTUM.

Procolophonia Relatively advanced cotylosaurs (*Cotylosauria, 'stem reptiles'), distinguished from the lower types by their shorter jaws and better jaw movement. These anapsids (*Anapsida) lived in the *Triassic and *Permian.

Procyonidae *See* CARNIVORA.

pro-delta The furthest offshore portion of a *delta, lying at the toe of the *delta front, and characterized by a relatively slow rate of fine-grained deposition.

prod mark A type of *tool mark formed by the impact of an object with a soft, muddy surface. Prod marks are characterized by an asymmetrical impact mark, with the steeper margin of the depression being on the down-current side.

production log A *well log which is run inside a casing, primarily to measure the behaviour of fluids within the pipe.

production well A well from which water, gas, or oil is actually to be recovered, as opposed to other wells, e.g. those designed to determine hydraulic characteristics, to recharge an *aquifer, or to act as injection wells to push oil towards the production well.

***Productus giganteus* (*Gigantoproductus giganteus*)** A huge brachiopod (*Brachiopoda) which attained a width of approximately 370 mm. It lived in the warm seas of the Lower *Carboniferous (*Mississippian), adopting a clam-like habit.

Proetida An order of *Trilobita that lived from the *Ordovician to *Permian. The glabella (*see* CEPHALON) was large and clearly defined, often with *genal spines, the thorax had 8–10 segments, and the *pygidium had furrows and was not spiny. The order contained 2 superfamilies.

profiling 1. A method in which an *array is moved progressively along a traverse line to obtain a continuous cover of

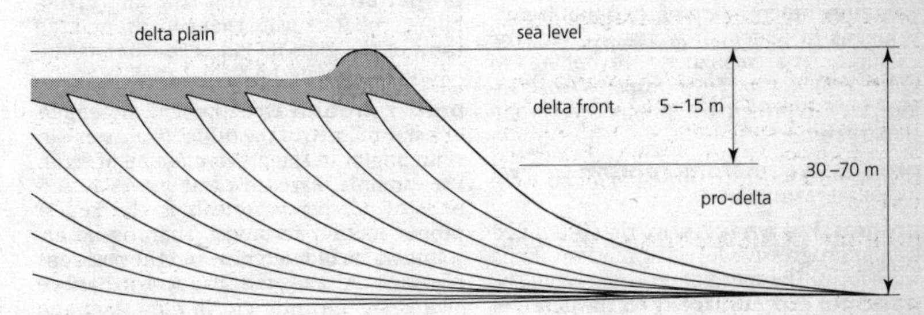

Pro-delta

measurements of the subsurface and so produce a profile. *See* CONSTANT-SEPARATION TRAVERSING. **2.** In reflection seismology, the acquisition of data along a line, which then form a seismic section (*see* SEISMIC RECORD). *See* COMMON DEPTH POINT.

Proganochelys quenstedii **(*Triassochelys quenstedii*)** The oldest known turtle, first described in 1887 from the *Triassic Stubensandstein of Württemburg, Germany, It possessed teeth, and stout ribs on the neck vertebrae.

progenesis The onset of sexual maturity at a younger and smaller stage in development than is usual.

proglacial Applied to the area between a *glacier and adjacent high ground. A proglacial lake is a body of water impounded in such an area and is often inferred for areas of *Pleistocene glaciation from the evidence of strandlines, lake sediments, and *overflow channels.

progradation The outward building of a sedimentary deposit, such as the seaward advance of a *delta or shoreline, or the outbuilding of an *alluvial fan.

prograde metamorphism (progressive metamorphism) The *recrystallization of a rock in response to an increase in the intensity of *metamorphism as this is reflected by an increase of pressure, temperature, and/or $P(H_2O)$. *Compare* RETROGRADE METAMORPHISM.

progressive deformation The accumulation over time of incremental *strain elements within a body as a response to *stress. The accumulated distortions and rotations within the body add up to a final (*finite) strain state.

progressive evolution Steady, long-term improvement of evolutionary grade, which has allowed plants and animals to become ever more independent of the aquatic environment in which they first evolved. For example, the sequence *bryophyte to *pteridophyte to *gymnosperm to *angiosperm represents a progressive evolutionary trend.

progressive metamorphism *See* PROGRADE METAMORPHISM.

progressive wave A wave that is typified by the progressive forward motion of the wave-form. The speed of propagation of the wave-form depends largely on the depth of the water. *See also* CELERITY.

Progymnospermopsida (progymnosperms) The ancestors of the *gymnosperms, which arose in the *Devonian and dwindled to extinction in the latter part of the *Carboniferous. They had trunks with wood resembling that of gymnosperms, but their fertile branches or leaves bore sporangia (*see* SPORE), and their foliage was often fern like. Probably seeds evolved in various different progymnosperms. *See* *ARCHAEOPTERIS*.

progymnosperms *See* PROGYMNOSPERMOPSIDA.

prokaryote An organism, usually unicellular, in which the cells lack a true nucleus, the DNA being present as a loop in the cytoplasm. Other prokaryotic features include the lack of chloroplasts and mitochondria and the possession of small ribosomes. *Compare* EUKARYOTE.

prokaryotic *See* PROKARYOTE.

prolate Applied to *clasts which are rod-shaped, and defined as having a ratio of short to intermediate diameters of more than 2 : 3, and a ratio of intermediate to long diameters of less than 2 : 3.

prolate uniaxial strain The *strain state which occurs when a reference sphere is extended along its X axis and shortened in all other directions perpendicular to X. The resultant strain shape is a prolate ellipsoid.

proloculus The initial chamber of a foraminiferid (*Foraminiferida) *test.

Prometheus (Saturn XVI) One of the lesser satellites of *Saturn, discovered in 1980 by *Voyager 1, with a radius measuring $74 \times 50 \times 34$ km; mass 0.0014×10^{20} kg; mean density 270 kg/m^3; visual albedo 0.6.

promontorium On *Mars, a cape, e.g. Prom. Deville, Prom. Kelvin.

proparian Applied to a trilobite (*Trilobita) *cephalic suture that runs around the front of the glabella (*see* CEPHALON), terminating in front of the *genal angle.

proper motion The apparent movement of a star relative to the other 'fixed' stars, at right angles to the observer's line of sight. For example, Barnard's star moves 10 seconds of arc per year; this is the largest proper motion recorded. There is no absolute frame of reference, and the positions of stars in the familiar constellations change significantly within a few thousand years.

propylitization The process whereby original *plagioclase in an *igneous rock is altered to an *epidote–*sericite–secondary *albite assemblage and original *ferromagnesian minerals are altered to a *chlorite–*calcite–epidote–secondary iron-ore assemblage.

proseptum (pl. prosepta) *See* SEPTUM.

prosobranch gastropods Single-gilled gastropods (*Gastropoda), thought to represent the first stage in the evolutionary development of the group from the ancestral forms, the *Archaeogastropoda.

prosogyral *See* PROSOGYRATE.

prosogyrate (prosogyral) Applied to the umbones (*see* UMBO) of the *Bivalvia, where the beaks are curved so as to point in an anterior direction.

prosoma *See* ARACHNIDA; and CHELICERATA.

protalus rampart Ridge of rock debris less than 10 m high and found near the base of a steep inland face. It consists of frost-shattered debris that has been carried some distance from the face down the steep surface of a basal snow bank.

protaspis The earliest stage recognized in larval trilobite (*Trilobita) development. The larva is small, often spiny, and grows through successive moult stages. Initially it is a small disc but size and segmentation increase with each successive moult.

Proterosuchus Well-known representative of the *Thecodontia, the ancestral archosaurian (*Archosauria) stock. According to some authorities, *Proterosuchus* was the first crocodile. It lived in N. America during the *Triassic.

Proterozoic The most recent (about 2500–575 Ma ago) of the three subdivisions of the *Precambrian.

Proteus (Neptune VIII) A satellite of *Neptune, measuring 436 × 416 × 402 km; visual albedo 0.06.

protist A single-celled, eukaryotic organism (*see* EUKARYOTE) that may resemble an animal or a plant. Animal-like protists include naked and shelled amoebas, foraminiferans, zooflagellates, and ciliates; plant-like protists include dinoflagellates, diatoms, and algae. In a 5-kingdom system of classification protists were grouped as a kingdom, Protista; later some multicellular organisms with protist affinities but previously classed as fungi or plants were transferred into the Protista and the name of the kingdom was changed to Protoctista, although the new name is little used.

Protista *See* PROTIST.

proto- From the Greek *protos* meaning 'first', a prefix meaning 'original' or 'primitive'.

proto-Atlantic *See* IAPETUS OCEAN.

Protoaulopora ramosa Small, colonial coral belonging to the subclass *Tabulata, found in the Upper *Cambrian of Soviet Kazakhstan. With *Bija sibirica* (a doubtful *stony coral) it marks the appearance of an important group of *reef-forming organisms.

Protoceratops andrewsi The first of the horned (*ceratopsian) *dinosaurs, known from the Middle *Cretaceous of Mongolia. It possessed the bony frill and beak characteristic of later forms, but was only 2 m in length and 1.5 t in weight.

protoconch The initial (larval) shell of molluscs (*Mollusca), often retained at the tip of the *spire of the adult shell.

Protoctista *See* PROTIST.

proton-adding process (p-process) Nuclear process in red giant stars at very high temperatures, producing proton-rich, heavier elements. *See also* NUCLEOSYNTHESIS.

proton magnetometer A form of *nuclear-precession magnetometer that is based on the *precession of protons in water or alcohol.

protoparian suture *See* CEPHALIC SUTURE.

protoplanet An individual condensation, representing a very small proportion of the total cloud mass but of similar composition, that occurs during the condensation of an interstellar cloud, from which stars and planets are assumed to have originated. *See* PROTOSTAR.

Protopteridales *See* PRE-FERNS.

protore Rock containing sub-economic material from which economic *mineral deposits may form by geologic concentration processes such as *supergene enrichment. A protore may become profitable with technological advance or change in market value.

protostar Primitive star formed from the

breakup of *interstellar clouds. After a fragment becomes detached, it continues to shrink under the influence of its own gravitation, drawing in more gas and dust and increasing in temperature and pressure. Eventually, the outward pressure associated with the rising temperature balances the inward pressure due to gravitation and collapse ceases, perhaps 10 000 years after separation from the cloud. At this stage, the fragment is called a protostar. When internal temperature exceeds 10^7 K *hydrogen 'burning' begins, marking the transition to a star.

protozoa (sing. protozoon; adj. protozoan) *See* PROTOZOA.

Protozoa Former animal phylum of *eukaryotic, single-celled micro-organisms. In some classifications the Protozoa are grouped with other simple eukaryotic organisms in the kingdom *Protista.

protrusive Applied to the downward extension into *sediment of a vertical or oblique *burrow, frequently reflected in the development of *spreiten distal to the point of entry. In U-shaped burrows the spreiten occur inside the bend of the 'tube'. Protrusive spreiten are found in **Diplocraterion* and **Rhizocorallium*.

proved reserve *See* RESERVE.

provenance The source or origin of *detrital *sediments.

province 1. Region or area of large extent with similar features throughout and capable of being considered as a unit. **2.** In geography, an area of land or sea in one climatic belt.

provinciality The association of species within well-defined biogeographic areas or provinces. Each province contains a distinct assemblage of species, some of which are endemic (i.e. confined to that area only). *See* ENDEMISM.

proximal Applied to a *sediment or sedimentary environment close to the source or origin of the deposit. *Compare* DISTAL.

proximity log The record produced by one version of the *microlog sonde.

proxy *See* IONIC SUBSTITUTION.

psammite A metamorphosed *sandstone, *arkose, or *quartzite, extremely rich in the *mineral *quartz.

PSC *See* POLAR STRATOSPHERIC CLOUD.

psephite A general term used to describe coarse-grained, *detrital, *sedimentary rock, e.g. *gravels, *conglomerates, or *breccias. G. W. Tyrell (1921) suggested the term be restricted to metamorphosed conglomerates and breccias, a suggestion followed by many authors describing *metamorphic rocks.

pseudobreccia An irregularly *recrystallized or partially *dolomitized *limestone, in which the selective growth of coarse crystals gives the rock an apparently fragmented *texture.

pseudoextinction Within an *evolutionary lineage, the disappearance of one taxon caused by the appearance of the next *chronospecies in the series. The extinction is purely taxonomic.

pseudofossil A naturally occurring object that may resemble a *fossil. If there is uncertainty the object may be referred to the *Problematica; it is called a pseudofossil when the resemblance results purely from chance.

pseudo-gravitational field A gravitational field that is transformed to simulate a magnetic field for the purpose of computational analysis.

pseudo-magnetic field A magnetic field that is transformed to simulate *gravitational acceleration for the purpose of computational analysis.

pseudomorph A *secondary mineral or a random aggregate of secondary minerals (monomineralic or polymineralic) which have replaced an earlier mineral but have retained its shape. For example, a random aggregate of *chlorite crystals which have replaced a *euhedral *augite crystal while preserving its shape is said to form a monomineralic pseudomorph of augite. Other examples include *quartz after *fluorite, *serpentine after *olivine, and *limonite after *pyrite, where the original *cubic form is preserved. *See also* METASOMATISM.

pseudonodule A ball-like body of *sandstone, with an internal lamination that is convoluted or upcurled at the edges, set in a bed of *mudstone. The sand ball is the result of the sinking of sand from the base of an overlying bed into the less dense, soft mud below.

pseudopunctate *See* PUNCTATE.

pseudosection (quasi-section) The plotting of data against position along a traverse line, to produce a display of *resistivity or *induced-polarization data in which values are given to the intersection point of 45° lines drawn from mid-points of the current and potential electrode pairs. Depths in the resulting 'section' below a transverse bear no simple relationship to the true geology; a pseudosection shows the variation of the measured parameter with position and with effective depth of penetration, rather than with true depth. It is used widely in displaying *induced-polarization data and *apparent resistivities obtained from *constant-separation traverses with different electrode separations, and apparent conductivities from electromagnetic traverses with different coil separations.

pseudospar *See* NEOMORPHISM.

Pseudosycidium The first of the Charales, a distinct evolutionary line of green *algae which arose during the Upper *Silurian and whose method of reproduction was sexual. Well-developed male (antheridia) and female (oogonia) organs can be identified, the secretion of calcium carbonate around the female organs assisting preservation. *See also* CHAROPHYCEAE; and GYROGONITE.

pseudotachylite A rare, glassy rock produced by frictional melting during extreme *dynamic metamorphism in a *fault or *thrust zone.

psilomelane A hydrated manganese oxide with the possible formula $Ba_3(Mn^{2+}Mn_7^{4+})O_{16}(OH)_4$; sp. gr. 3.7–4.7; *hardness 5–6; grey-black; *sub-metallic or *earthy *lustre; *massive, *botryoidal; occurs as a precipitate in *secondary manganese deposits and may be worked commercially for manganese.

Psilonichnus An *ichnoguild comprising a *biofacies developed on the *backshore of former beaches.

Psilophytales (psilophytes) Primitive *pteridophytes, which were the earliest vascular plants, from the *Silurian and the *Devonian. They had slender, tapering, leafless or scale-bearing stems up to 50 cm high, often with cone-shaped sporangia (*see* SPORE) at the top. Some authorities placed the fossil groups with the present-day psilophytes (*Psilotum*, *Tmesipteris*) in the class Psilopsida, comprising the orders Psilotales (living forms) and Psilophytales (fossil forms). In more recent classifications the members of the Psilophytales have been placed into three separate groups, usually ranked as subdivisions: Rhyniophytina (e.g. **Cooksonia* and **Rhynia*); Trimerophytina (e.g. *Psilophyton* and *Trimerophyton*); and Zosterophyllophytina (e.g. **Zosterophyllum*). The living forms are contained in a fourth subdivision, the Psilophytina.

psilophytes *See* PSILOPHYTALES.

Psilopsida *See* PSILOPHYTALES.

PSV *See* AGGREGATE TESTS.

psychrometer Hygrometer with a *wet-bulb and a *dry-bulb thermometer. *See also* WHIRLING PSYCHROMETER.

psychrophile An *extremophile (domain *Archaea) that thrives in environments where the temperature is low, usually below 15 °C.

Pteraspida *See* HETEROSTRACI.

Pteraspis *See* HETEROSTRACI.

Pteridophytina (pteridophytes) Subdivision of the plant kingdom, comprising the classes *Lycopsida (club mosses), *Sphenopsida (horsetails), Psilopsida (but *see* PSILOPHYTALES), and *Pteropsida (the various families of ferns). They first enter the *fossil record in the *Silurian. They are flowerless plants exhibiting an alternation of two distinct and dissimilar generations. The first is a non-sexual, *spore-bearing, sporophyte generation. It usually appears as a relatively large plant, with stems containing vascular tissue that conducts water and dissolved solutes through the plant, and usually bears the leaves and roots. Spores are produced in sporangia that are either attached to the leaves (as in ferns) or are on specialized scales (sporophylls) grouped into cones (as in horsetails and club mosses), or in the axils of leaves on unspecialized stems (as in some club mosses). The second is a sexual, gametophyte generation, in which the plants generally are relatively small, and without differentiation of stem, leaves, or roots. These plants bear male (antheridia) and female (archegonia) sex organs, together or on separate plants. When the eggs in the archegonia are fertilized by sperms from the antheridia, an embryo results; this can grow into a new sporophyte generation.

pteridophytes *See* PTERIDOPHYTA.

Pteridospermales (seed ferns) Extinct *gymnosperm order, which were the earliest seed plants and flourished in the *Carboniferous, before disappearing in the *Cretaceous. Their foliage was fern-like in appearance, but the fertile leaves bore seeds and *pollen-producing organs. The first plant to be identified (in 1903) as a seed plant rather than a true fern was *Lyginopteridales oldhamia*.

Pterobranchia (phylum *Hemichordata) Class of minute, fixed, colonial, deep-sea organisms which secrete an external cuticular skeleton in which they are housed (e.g. *Rhabdopleura*). Pterobranchs may be the nearest living relatives of the graptolites (*Graptolithina). The mesosoma is small but carries one or more pairs of arms bearing tentacles (the lophophores). The metasoma consists of a long stalk (peduncle) by which the individual is attached. The branchial apparatus is rudimentary.

pteropod ooze Deep-sea ooze in which at least 30% of the sediment consists of the shells of *planktonic small *gastropods (known as pteropods or 'wing-footed' snails). The shells are aragonitic and, as *aragonite solubility increases rapidly with depth, pteropod ooze is restricted to water depths less than 2500 m.

Pteropsida (Filicopsida; ferns) Class of the *Pteridophytina, which comprises all living and extinct ferns. They arose in the *Devonian from the trimerophyte group of *psilophytes (which had developed more elaborate systems of branching, that led on to the formation of true leaves) and made an important contribution to *Carboniferous floras. They are the most advanced, numerous, and varied of the pteridophytes. In most cases they have relatively large, much-divided leaves, and are still significant components of many different plant communities around the globe.

Pterosauria (pterosaurs) *Mesozoic order of flying *reptiles, which were particularly numerous in the *Jurassic, but survived until late *Cretaceous times. Their fossil skeletons suggest that they could not stand upright on land, and so it is assumed that their mode of life involved swooping over the sea to catch fish. Their remains are always associated with marine deposits. *See* *QUETZALCOATLUS NORTHROPI*.

Ptilograptidae *See* DENDROIDEA.

Ptychopariida An order of *Trilobita that lived from the Lower *Cambrian to Upper *Devonian. It was a large, *paraphyletic group, divided into 2 suborders.

ptygmatic fold An irregular, lobate *fold, usually found where single *competent layers are enclosed in a *matrix of low competence. Typically, ptygmatic folds do not maintain their *orthogonal thickness (i.e. they are *similar folds). Characteristically their *axial planes are curved.

Puaroan *See* OTEKE.

pubis In *tetrapods, the anterior, ventral part of the *pelvic girdle. In *ornithischian *dinosaurs the pubis lies alongside the *ischium; in advanced ornithischians there is also a forward prong, the prepubis. In *saurischian dinosaurs the pubis points downwards and forwards from the hip socket.

Puck (Uranus XV) One of the lesser satellites of *Uranus, with a diameter of 77 km. It was discovered in 1985.

puddled soil (poached soil) Soil in which the structure has been destroyed by the physical impact of rain drops, by tillage when wet, or by trampling by animals.

pull-apart basin *See* STRIKE-SLIP FAULT.

pulsations, geomagnetic (micropulsations) Small, almost sinusoidal fluctuations of the *geomagnetic field, usually with durations of seconds to minutes.

pulse length In radar terminology, the total length of an electromagnetic wave emission which is equal to the product of the wavelength, frequency, and time duration of emission.

pumice Extremely vesicular, frothy, natural *glass, having a high (60–75%) *silica content and low density. In some cases it will float on water. Usually, but not always, it is of *pyroclastic origin. *See also* RETICULITE; and VESICLE.

pump A device for moving liquids by adding to the pressure existing within them. For example, a centrifugal pump first increases the velocity of the fluid by the use of impellers; this velocity increase is then converted to an increase in pressure by the use of appropriately orientated guide vanes or the use of a volute casing. Other pump types include multistage turbine pumps, jet pumps, positive-displacement pumps, and suction lifts.

pumpellyite *See* PREHNITE–PUMPELLYITE FACIES.

pumping test (aquifer test) Water may be pumped from one or more wells to determine the particular hydraulic characteristics of an *aquifer or of individual wells. The effect of pumping at known rates is assessed by the use of *observation wells sunk at appropriate places to monitor the height of water in the aquifer or the wells of special interest.

punctae *See* PUNCTATE (2).

punctate 1. Applied to any structure that is marked by pores or by very small, point-like depressions. **2.** Applied to a type of brachiopod (*Brachiopoda) shell structure in which fine pores (punctae) extend from the inner to the outer surface. Three main shell types are recognized: impunctate, where the shell consists of an outer, lamellar layer of shell and an inner, fibrous layer; punctate (or endopunctate), in which the punctae extend through the shell and end beneath the organic periostracum; and pseudopunctate, in which solid rods of *calcite (taleolae) are contained in the fibrous layer.

punctuated equilibrium The theory, first proposed in 1972 by Niles Eldredge and Stephen Jay Gould, that *evolution is characterized by geologically long periods of stability during which little speciation occurs, punctuated by short periods of rapid change, species undergoing most of their morphological changes shortly after breaking from their parent species.

pupaeiform Literally, resembling the pupa of an insect, and applied to a gastropod (*Gastropoda) shell where the shell is an elevated ovoid in which the later-formed *whorls have decreasing radii of curvature.

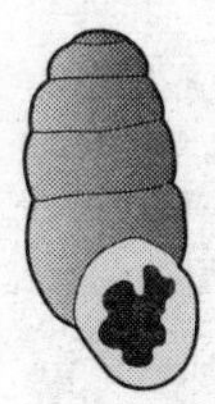

Pupaeiform

Purbeckian A *stage in the British Upper *Jurassic and Lower *Cretaceous, underlain by the *Portlandian and overlain by rocks of the *Wealden Beds. The *type section is found in southern England and consists of sediments deposited under *intertidal to brackish freshwater conditions. *See also* TITHONIAN; and RYAZANIAN.

pure shear (homogeneous non-rotational strain) A flattening *strain in which, during the deformation of a body, the *principal strain axes (*X*, *Y*, and *Z*) remain parallel to their respective *principal stress axes (σ_1, σ_2, and σ_3). *See also* HOMOGENEOUS STRAIN. *Compare* SIMPLE SHEAR.

purga *See* BURAN.

pushbroom system In *remote sensing, an imaging device consisting of a linear array of sensors (*charge-coupled devices) which is swept across the area of observation. Pushbroom systems allow greater resolution of data to be assimilated than do *line scanner systems.

push moraine *See* MORAINE.

push–pull wave *See* P-WAVE.

Pusgillian A *stage of the *Ordovician in the Lower *Ashgill, underlain by the *Onnian and overlain by the *Cautleyan.

puy 1. A volcanic hill in the Auvergne region of France. **2.** Any steep-sided tower of volcanic rock, e.g. the Devil's Tower, Wyoming, USA, and Shiprock, New Mexico, USA. It consists of the resistant central plug or neck of a former *volcano. *See also* VOLCANIC PLUG.

P-wave (compressional wave, dilatational wave, irrotational wave, pressure wave, primary wave, push–pull wave) An elastic *body wave or sound wave. It is the wave most studied in *reflection and *refraction seismology. Particles oscillate about a fixed point, but in the direction of propagation of wave energy. In an *isotropic and homogeneous medium the P-wave velocity (V_p) is given by: $V_p = [\psi/\rho]^{1/2}$, where ψ is the *axial modulus and ρ is the density of the material. P-waves are the fastest of the *seismic waves.

pycnocline Zone in the oceans where water density increases rapidly with depth in response to changes in temperature and salinity. The pycnocline tends to coincide with the *thermocline and *halocline, and separates the surface-water zone of ocean waters from the deep bottom water.

pycnometer (pyknometer) An apparatus used to determine the specific gravity or density of soils, or of *rock or *mineral fragments (*see* DENSITY DETERMINATION). It consists of a small bottle fitted with a ground glass stopper with a capillary opening.

pygidium (adj. pygidial) The posterior part of the *exoskeleton of a trilobite (*Trilobita); it is generally formed by the fusion of several body segments but in some *Cambrian forms consisting of a single segment. Many Cambrian trilobites have small pygidia and are said to be 'micropygous'. Most trilobites are either isopygous, where the *cephalon and pygidium are the same size, or heteropygous, where the pygidium is the smaller. In some cases the pygidium is larger than the head (macropygous).

pyknometer *See* PYCNOMETER.

pyralspite *See* GARNET GROUP.

pyramid A *crystal form composed of a number of non-parallel *crystal faces which meet at a point. The crystallographic notation is frequently (111) or {111} if it is a form, whereby the crystal face intersects all the axes at their unit lengths.

pyrite (fool's gold) *Sulphide mineral, FeS_2; sp. gr. 4.9–5.2; *hardness 6.0–6.5; *cubic; pale brass-yellow, does not tarnish; greenish-black *streak; *metallic *lustre; crystals *cubic, *pyritohedra (pentagonal dodecahedra), *octahedra, or combinations of the two; *cleavage poor basal {001}; occurs with other *sulphide ores genetically associated with *basic and *ultrabasic rocks, and together with *pyrrhotine and *chalcopyrite; very widely distributed in a great variety of environments, and found in *igneous rocks as an *accessory mineral, in *sedimentary rocks (especially black *shales), as nodules in *metamorphic rocks, and common in *hydrothermal veins, in replacement deposits, and contact *metamorphic rocks; alters to iron sulphate and *limonite. It was formerly used widely for the production of sulphuric acid. *Compare* MARCASITE.

pyritohedron (pentagonal dodecahedron) A *crystal form within the *cubic system consisting of a twelve-faced, *closed form (a dodecahedron) with each *crystal face having five sides, but not in the shape of a regular pentagon. It is a common form of the mineral *pyrite.

pyro- From the Greek *pur* meaning 'fire', a prefix meaning 'fire'.

pyrochlore Mineral oxide $(Na,Ca,U)_2(Nb,Ta,Ti)_2O_6(OH,F)$; sp. gr. 4.3–4.5; *hardness 5.0–5.5; *cubic; normally brown to black, green when heated; light brown *streak; vitreous to greasy *lustre; crystals usually *octahedra or irregularly shaped grains; *cleavage octahedral when present; found in granitic rocks and *pegmatites, with or near *alkaline rocks, and in association with *zircon, *apatite, and *carbonatites. The calcium and sodium in the crystal *lattice can be substituted with uranium, thorium, or the *rare-earth elements.

pyroclastic Literally, 'fire-broken'. Applied to volcanic rocks consisting of fragmented particles, generally produced by explosive action.

pyroclastic flow (ash-flow) General term for a hot, high-concentration flow of *pumice or *lithic clasts, entrained and transported in a fluidized *ash matrix. Pyroclastic flows include a wide range of phenomena, from *ignimbrites (large volume, *pumiceous) to block-and-ash flows (small volume, lithic). The flow originates by the gravitational collapse of a dense, turbulent, eruption column at the source vent and moves down-slope as a coherent flow. Fluidization of the ash matrix, which contributes to the high mobility of such avalanches, is achieved by (*a*) the diffusion and release of gas during breakage and the attribution of ash and pumice particles entrained within the flow and (*b*) air ingested and compressed at the front of the advancing flow margin. Where the ash matrix is the dominant component, the term 'ash-flow' is applied by American authors, although British authors prefer to use the term 'pyroclastic flow'.

pyroelectricity The positive and negative electrical charges developed at opposite ends of the polar axis of a *crystal in response to a change in temperature. It is observed in *tourmaline.

pyrogenetic minerals *Minerals which are crystallized from a completely or almost completely anhydrous *magma. These minerals, e.g. *olivine, *pyroxene, and *plagioclase, have anhydrous compositions and are some of the first minerals to crystallize from the magma. If the original magma contained a small amount of water in solution, crystallization of pyrogenetic minerals enriches the residual magma in water.

pyrolite model Model for the composition of upper-*mantle material proposed by Ringwood: one part *basalt to three parts *dunite (mainly *olivine and *pyroxene), *partial melting of which gives basaltic *magma.

pyrolusite Common manganese mineral, MnO_2; sp. gr. 4.5–5.0; *hardness 5–6 for crystals, decreasing to 2 when *massive; *tetragonal; black to bluish-grey; black *streak; *metallic *lustre; crystals rare, *acicular and rod like, but it is usually massive, *cryptocrystalline, and *dendritic on joints and bedding planes; *cleavage *prismatic when present; a *secondary mineral found in the oxidized zone of manganese deposits, in *quartz veins, and as nodules on the sea bed. Synthetic MnO_2 is used for dry batteries, as a decolourizer for glass, and in the manufacture of chemicals.

pyrolysis The heating of organic molecules without oxygen, to produce *hydrocarbons (*char) which have a high calorific value. Fuel produced by this process can be concentrated and stored. Organic waste may be used.

pyrometasomatic deposit (skarn) Deposit formed by *metasomatism, usually in *limestones, at or near *igneous contacts, due to the passage of mineralizing solutions through reactive rocks.

pyromorphite Mineral, $Pb_5(PO_4)_3Cl$; sp. gr. 6.5–7.1; *hardness 3.5–4.0; *hexagonal; various shades of green, yellow, and brown; white *streak; vitreous *lustre; crystals *prismatic and often hollow, or barrel shaped, forming aggregates or crusts, also occurs granular, fibrous or globular; *cleavage prismatic in traces; a *secondary mineral occurring in the oxidized zones of *veins containing lead minerals, often associated with mimetite and *anglesite, and *galena.

pyrope Member of the *garnet group of minerals, $Mg_3Al_2(SiO_4)_3$; sp. gr. 3.51; *hardness 6.0–7.5; *cubic; dark red, pink-red, or black; vitreous *lustre; crystals (dodecahedra, also occurs *massive; occurs in high-grade *metamorphic rocks (e.g. *eclogites) and deep-seated *igneous rocks (e.g. garnet *peridotites), which are often *mantle derived, although the pyrope molecule occurs to a variable extent in most *almandines; it is also distributed in *alluvial deposits. Transparent crystals are used as *gemstones. The name is derived from the Greek *puropos*, meaning 'fiery eyed'.

pyrophyllite Uncommon *silicate mineral $Al_2[Si_4O_{10}](OH)_2$ belonging to the *phyllosilicates (sheet silicates) and with properties similar to *muscovite; sp. gr. 2.65–2.90; *hardness 1–2; occurs as a secondary product from the hydrothermal alteration of *feldspar, and as foliated masses (*see* FOLIATION) in metamorphic *schists. It has been mined as a substitute for *talc.

pyroxene An important group of *inosilicates (chain silicates) comprising the *orthorhombic pyroxenes (*orthopyroxenes) and the *monoclinic pyroxenes (*clinopyroxenes) with the general formula XYZ_2O_6, where X = Mg, Fe, Ca, or Na; Y = Mg, Fe, Fe^{3+}, or Al; and Z = Si (and some Al substitution). The main orthopyroxenes are *enstatite and orthoferrosilite; the main clinopyroxenes include *diopside, *hedenbergite, *augite, *pigeonite, and *jadeite, and also the alkali pyroxenes *aegirine and *aegirine augite (note that *wollastonite, although similar to the pyroxenes, has a different atomic structure); sp. gr. 3.0–3.5; *hardness 5–6; colours variable, but usually dark greens, browns, or black; *vitreous *lustre; *crystals usually short or columnar *prisms; well-marked *cleavage; widely distributed in both *igneous and *metamorphic rocks.

pyroxene gneiss *See* GRANULITE.

pyroxene hornfels facies A set of metamorphic *mineral assemblages (produced by the *metamorphism of a wide range of starting rock types under the same metamorphic conditions), typically characterized by the development of the mineral assemblage *clinopyroxene–*labradorite–*quartz in rocks of *basic *igneous composition. Other rocks of contrasting composition, e.g. *shales or *limestones, would each develop their own specific mineral assemblage, even though they are all being metamorphosed under the same conditions. The variation of mineral assemblage with starting rock composition reflects a particular range of pressure, temperature, and $P(H_2O)$ conditions. Experimental studies of mineral pressure–temperature stability fields indicates that the *facies represents a range of low pressures (0–2 kb) and moderate temperatures (550–750 °C) characteristic of contact metamorphic conditions.

pyroxenite An *ultrabasic, *igneous rock consisting of *essential *clinopyroxene, *orthopyroxene, and *olivine. The *ferromagnesian minerals are magnesium-rich, the pyroxenes dominating olivine, which is usually less than 40% by volume. Different types of pyroxenites are defined using the ratio of clinopyroxene to orthopyroxene to olivine. These include orthopyroxenite, websterite, clinopyroxenite, and olivine-bearing versions of these. Pyroxenites are found as *cumulate horizons within layered *basic intrusions and as components of the sub-oceanic *mantle in obducted fragments of *ocean crust known as *ophiolites.

pyroxenoid *See* WOLLASTONITE.

pyrrhotite Mineral, FeS; sp. gr. 4.6–4.7; *hardness 3.5–4.5; *hexagonal; bronze-yellow but darkens rapidly to reddish-brown on exposure; greyish-black *streak; *metallic *lustre; *crystals rare, and can be *platy or *tabular, but the mineral is usually *massive or granular; no *cleavage; magnetic; occurs in *igneous rocks, e.g. *gabbro or *norite, as disseminated grains, and also in *metamorphic contact zones in association with *chalcopyrite and *pyrite. Pyrrhotite was used formerly for the production of sulphuric acid.

p-zone A biostratigraphic *zone distinguished by *pelagic *fossils, e.g. ammonites (*Ammonoidea) and graptolites (*Graptolithina). The term was proposed in 1965 by T. G. Miller. *Compare* B-ZONE.

Q **1.** *See* KÖNIGSBERGER RATIO. **2.** The ratio of the frequency at the mid-point of a *band-pass filter to the filter width. **3.** The ratio of the peak energy in a waveform to the energy lost by dissipation. As a measure of seismic absorption it has been used effectively in delimiting *Benioff zones using natural *earthquake events (high *Q* implies low absorption and is associated with *crust and deep *mantle; low *Q* implies high absorption and is associated with the sub-crustal *low-velocity layer.

QAPF classification A modal classification scheme for *igneous rocks with a *colour index less than 90, based on the relative proportions of: *quartz to quartz plus total *feldspars (Q); *alkali feldspar to total feldspars (A); *plagioclase to total feldspars (P); and *feldspathoids to feldspathoids plus total feldspar (F).

QAP triangle A three-component, modal classification scheme for *igneous rocks of granitic affinity based on the relative proportions of: *quartz (Q); *alkali feldspar (including albite if the anorthite content is less than 5%) (A); and *plagioclase feldspar (containing more than 5% anorthite (P).

Q days *See* QUIET DAYS.

***Q*-factor** *See* KÖNIGSBERGER RATIO.

Q_n *See* KÖNIGSBERGER RATIO.

quadrature *See* IMAGINARY COMPONENT.

quarrying *See* GLACIAL PLUCKING.

quartz (rock crystal) Widely distributed rock-forming *silicate mineral SiO_2 with the related semi-precious varieties rose quartz (pink), amethyst (purple), cairngorm (dark brown and similar to smoky quartz, although the original cairngorms from the Scottish mountain range were of *topaz), citrine (light brown); sp. gr. 2.65; *hardness 7; *trigonal; commonly colourless or white, but can occur in a variety of colours; *vitreous *lustre; *crystals usually six-sided *prisms terminated by six-faced pyramids, the prisms often striated, also occurs extensively in *massive form; no *cleavage; *conchoidal fracture; found in many *igneous and *metamorphic rocks, extensively in *clastic rocks, and a common *gangue mineral in mineral *veins. *See also* AGATE; CHALCEDONY; COESITE; CRISTOBALITE; FLINT; JASPER; ONYX; OPAL; STISHOVITE; and TRIDYMITE.

quartz arenite *See* DOTT CLASSIFICATION.

quartz dolerite *See* DOLERITE.

quartzite A *metamorphic rock composed mainly of *quartz and usually formed by the *metamorphism of quartz *sandstones. If deformation has accompanied metamorphism the individual quartz crystals are elongate and take on a preferred orientation, defining a planar or linear *fabric within the rock. Unmetamorphosed sedimentary quartzites (*sandstones with a quartz *cement) are known as 'orthoquartzites' to distinguish them from metamorphic quartzites (metaquartzites).

quartz overgrowth The development of *quartz *cement around *detrital *grains, the quartz cement growing in *optical continuity with the grains which they have enclosed.

quartz porphyry (elvan) A *porphyritic *microgranite, *microgranodiorite, or microtonalite.

quartz sandstone *See* ARENACEOUS.

quartz wacke A *sandstone containing more than 15% mud *matrix, with over 95% of the *grains being *quartz. *See* DOTT CLASSIFICATION.

quartz wedge *See* ACCESSORY PLATE.

quasi-equilibrium *See* EQUILIBRIUM.

quasi-section *See* PSEUDO-SECTION.

Quaternary (Anthropogene, Pleistogene) A sub-era of the *Cenozoic *Era, that covers approximately the last 1.64 Ma. The Quaternary comprises the *Pleistocene and *Holocene *Epochs and is noted for numerous major *ice-sheet advances in the northern hemisphere. By the Pleistocene most of the faunas and floras had a modern appearance. Authors differ in their interpretations of the southern European marine sediments. Berggren and Van Couvering (1974) recognize the *Calabrian, *Emilian,

*Sicilian, *Milazzian, *Tyrrhenian, and *Holocene *Stages, whereas Ruggieri (INQUA, 1979) recognizes the *Santernian, Emilian, Sicilian, *Crotonian, Tyrrhenian, and *Versilian Stages. Note that the Emilian of the former was laid down at a later date and represents a significantly shorter time interval than the Emilian of Ruggieri's classification.

quaternary system A *mineral system with four components, for example, the four-component system for *diopside (Di)–*anorthite (An)–*albite (Ab)–*pyroxene (Fs), diagrammatically represented as a tetrahedron. *See* COTECTIC SURFACE; PHASE; and PHASE DIAGRAM.

quenching The sudden chilling of *magma, which may provide a 'snapshot' of the *mineral *phases that were in equilibrium under a particular set of temperature and pressure conditions. Quenching is much used in experimental *petrology to investigate the course of mineral reactions.

Quetzalcoatlus northropi A pterosaur (*Pterosauria) first discovered in 1975 in Texas. It was huge, with a wingspan of 10 m or more, and lived as do vultures, using warm air thermals to soar high over the *Cretaceous plains.

quick clay *Clay which becomes movable or semi-liquid and loses all *shear strength when disturbed, especially when saturated. This may happen naturally due to seismic activity.

quickflow The part of a storm rainfall which moves quickly to a stream channel via *surface runoff or *interflow, and forms a flood wave in the channel.

quiet days (Q days) The geomagnetic intensity and directions for the five least magnetically disturbed days in each month are averaged to determine the quiet-day variations of the *geomagnetic field. *See* DIURNAL VARIATION (GEOMAGNETIC).

quiet zone *See* MAGNETIC QUIET ZONE.

Q-wave *See* LOVE WAVE.

Ra *See* RAYLEIGH NUMBER.

radar Acronym for *ra*dio *d*etection *a*nd *r*anging; the use of electromagnetic energy for the detection of objects which are capable of reflecting it. For example, cloud-detection radars are extensively used in meteorological forecasting and rainfall measurement, while sideways-looking radars are used for topographic mapping. Radar has been used to probe through *ice sheets in order to determine ice thickness and to detect internal reflection events; in polar regions the radar equipment has usually been airborne, but the instruments can be mounted on sledges for surface use. Radar can also be used in arid environments to probe through sand in search of water. Ground radar is being developed for use in engineering site investigations, but has very limited depth penetration where the moisture content is high because of the high *dielectric loss associated with water.

radar altimetry The measurement of the surface relief of a planetary body by recording the travel time of reflected *radar waves. The most spectacular success of the technique has been the establishment of the detailed surface topography of *Venus by the US probes Pioneer Venus 1 and 2 launched in 1978, and the Soviet Venera 15 and 16 launched in 1983. Radar scanning of the Earth's surface from aircraft or satellites provides information on topography, vegetation cover, and *lithology, and is particularly valuable where cloud cover prevents normal surveying. Over the oceans, *Seasat altimetry data provide information on wave height and motion, and on ocean *bathymetry, and assist in definition of the *geoid.

radar cross-section In *radar terminology, the hypothetical area of a perfect *diffuse reflecting surface which would be required to return the same amount of energy to a radar antenna as is observed from a point target. The radar cross-section serves as a measure of the *intensity of the energy that is backscattered from a point target.

radar imaging The use of *radar wavelengths to determine the location, size, and reflectivity characteristics of objects. It is particularly useful in regions with dense cloud cover. *See also* REMOTE SENSING.

radar scattering coefficient The average *radar cross-section value per unit area of a large target. This is the fundamental measurement of the radar properties of a surface, allowing discernment of the texture of the ground.

radar scatterometer A device which measures the electromagnetic energy backscattered from a target as a function of the *depression angle.

radial drainage *Drainage pattern consisting of streams that extend radially from a central zone. It is typical of the patterns developed on freshly constructed landforms (e.g. *volcanoes) and on areas of domed uplift.

radial dykes Vertical to subvertical *dykes which radiate from a central volcanic *plug. Radial dykes associated with the emplacement of *plutons at depth may theoretically lie along the minimum *principal stress axis σ_3 within the stress field that is generated.

radial fault One of a system of *faults which radiate from a central point.

radial fibrous A *mineral *texture seen where crystals are aligned perpendicularly to the curved surface on which they are growing.

radial relief displacement The apparent leaning away from the *centre point of vertical objects in an aerial photograph, due to the conical field of view of the camera lens.

radial symmetry The condition in which the body of an organism is repeated in a circular manner. In *corals the repetition occurs around the mouth, and in some echinoderms (*Echinodermata), where the five rays of the animal are symmetrically placed, the symmetry is also radial.

radiance The *radiant flux density of electromagnetic radiation measured by a *remote sensing detector as it travels through a given solid angle.

radiant flux density The power of *electromagnetic radiation falling on or emanating from a body, measured as watts per square metre.

radiating A crystal form in minerals, where the crystals grow outwards from a central point, frequently diverging to give a concentric pattern of growth.

radiation Any form of emitted wave phenomena, usually the *electromagnetic spectrum, sound, or heat.

radiation budget (energy budget) The difference between the amount of incoming solar radiation and the amount of outgoing terrestrial radiation. The balance is in deficit (i.e. more energy leaves the Earth's surface than reaches it from the Sun) at night. Overall, the highest positive net balance is found in low latitudes.

radiation densimeter An instrument employed to measure the intensity of radiation.

radiation fog Condensation effect over land surfaces on clear nights with light breezes, due to surface radiation cooling. Favoured initial conditions are very humid air, with wet and cold surfaces, e.g. marshes. The fog, most common in winter, is generally cleared by the Sun's warmth in the morning, but thick fog over wet surfaces in winter may persist much longer, particularly if an upper cloud layer screens the Sun.

radiation inversion *Temperature inversion in the lower atmosphere due to radiation cooling of the ground at night. *See also* RADIATION NIGHT.

radiation night A night with clear skies, when terrestrial long-wave radiation cannot be partly returned to the surface by cloud. There is rapid cooling of the air close to the ground, particularly when there is little wind, giving low minimum surface temperatures.

radiation tracks *See* FISSION-TRACK DATING.

radiatus From the Latin *radiatus* meaning 'with rays', a variety of cloud with parallel bands which, due to perspective, appear to meet at the horizon. The effect is seen in *cumulus, *stratocumulus, *altocumulus, *altostratus, and *cirrus. *See also* CLOUD CLASSIFICATION.

radiaxial Radially-axial, applied to *crystals, seen particularly in *calcite, which form a radiating crust of the *mineral within cavities.

radioactive decay Process by which a radioactive 'parent' element loses elementary particles from its nucleus and in doing so becomes a stable 'daughter' element. The rate of decay is constant for a given element and is a very precise and accurate device for the measurement of geologic time. *See also* DECAY CONSTANT; and DECAY CURVE.

radioactive logging *See* DENSITY SONDES; GAMMA–GAMMA SONDE; GAMMA-RAY SONDE; NEUTRON–GAMMA SONDE; and NEUTRON–NEUTRON SONDE.

radioactive survey A survey to measure the natural radioactivity of a region, usually by means of a scintillometer, *spectrometer, or *Geiger counter. It may be a ground survey, or made from a low-flying aircraft, usually a helicopter.

radioactive waste Any discarded substance that is radioactive. Wastes are classified as high-, intermediate-, or low-level according to their level of radioactivity. Low-level waste includes clothing and materials which have been used when handling radioactive sources, e.g. in hospitals. It can be safely buried in trenches 9 m deep beneath a covering of 2 m of clay; no alpha or beta radiation could penetrate the clay cover. Intermediate- and high-level wastes are mainly from the *fission process in nuclear power stations or from military waste. High-level waste is hot and intensely radioactive. It is stored, usually in ponds of water, for up to 50 years, during which time it cools and its short-lived isotopes decay until it can be classed as intermediate-level. It can then be incorporated in a borosilicate glass or synthetic rock (a *synrock), sealed in a container which corrodes at a known rate, and stored in a secure surface or underground facility. After 500 to 1000 years the radioactivity will have decayed sufficiently for the waste to emit no more radiation than many naturally occurring rocks.

radiocarbon dating (^{4}C dating) A dating method for organic material that is applicable to about the last 70 000 years. It relies on the assumed constancy over time of atmospheric $^{14}C:^{12}C$ ratios (now known not to be valid), and the known rate of decay of radioactive carbon, of which half is lost in a period (the 'half-life') of every 5730 ± 30 years. (The earlier 'Libby standard', 5568 years, is still widely used.) In principle, since plants and animals exchange carbon dioxide with the atmosphere constantly, the ^{14}C content of their bodies when alive is a function of the radiocarbon content of the atmosphere. When an organism dies, this exchange ceases and the radiocarbon fixed in the organism decays at the known half-life rate. Comparison of residual ^{14}C activity in fossil organic material with modern standards enables the age of the samples to be calculated. Since the method was first devised it has been realized that the atmospheric ^{14}C content varies, as the cosmic-ray bombardment of the outer atmosphere that generates the ^{14}C varies. Correction for these fluctuations is possible for about the last 8000 years by reference to the ^{14}C contents of long tree-ring series, e.g. those for bristlecone pine (*Pinus longaeva*).

radiogenic Applied to *isotopes produced by the process of radioactive decay.

radiogenic heating The thermal energy released as a result of spontaneous nuclear disintegrations. In the *Earth, the major *isotopes concerned today are of the elements uranium, thorium, and potassium, but various short-lived isotopes may have been important during the early formation of the Earth.

radioimmunoassay A technique for the very precise analysis of proteins, based on the ability of unlabelled proteins to inhibit competitively the binding of labelled protein by specific antibodies (i.e. an immunological reaction). The protein concentration of the unknown sample is determined by comparing the degree of inhibition with that produced by a series of standards containing known amounts of the protein. The technique has been adapted to assay non-proteins.

Radiolaria A large group of marine sarcodinids (*protists), characterized by having a shell with a perforated, membraneous capsule containing the endoplasm, and a siliceous or strontium sulphate skeleton consisting of a lattice shape of variable morphology made up of *spicules, bars, and spines. Radiolarians live mainly in surface waters and the earliest forms are *Cambrian in age. They are used in the biostratigraphic correlation of oceanic sediments, particularly where calcareous *microfossils have been dissolved.

radiolarian earth An unconsolidated or semi-consolidated bed of *radiolarian ooze. It is a relatively rare deposit, as most ancient radiolarian oozes have been compacted into hard *radiolarites.

radiolarian ooze Deep-sea ooze in which at least 30% of the sediment consists of the siliceous radiolarian (*Radiolaria) *tests. Radiolarian-rich oozes occur in the equatorial regions of the *Pacific and *Indian Oceans where the depth exceeds the *carbonate-compensation depth (around 4500 m in the central Pacific). *See also* DIATOM OOZE.

radiolarite A compacted, siliceous, *sedimentary rock composed mainly of the siliceous *tests of the marine zooplankton called *Radiolaria. The sediment from which radiolarites are formed is called a *radiolarian ooze.

radiometer An instrument to measure radiation, but usually a device to monitor infrared radiation. It can be precisely tuned for specific frequencies, but usually has poor spatial resolution.

radiometric dating (radioactive dating) The most precise method of dating rocks, in which the relative percentages of 'parent' and 'daughter' *isotopes of a given radioactive element are estimated. Early methods relied on uranium and thorium minerals (*see* URANIUM–LEAD DATING), but *potassium–argon, *rubidium–strontium, *samarium–neodymium, and *carbon-14–carbon-12 are now of considerable importance. Uranium-238 decays to lead-206 with a half-life of 4.5 billion (10^9) years, rubidium-87 decays to strontium-87 with a half-life of 50.0 billion years, and potassium-40 decays to argon-40 with a half-life of 1.5 billion years. For carbon-14 the half-life is a mere 5730 ± 30 years (*see* RADIOCARBON DATING), and beyond about 70 000 years the amount of carbon-14 remaining in organic matter is beyond accurate measurement. *Compare* INITIAL STRONTIUM RATIO.

radiometry The measurement of incident radiation by a *radiometer. A typical usage was the infrared scanning radiometer on the Apollo 17 lunar mission which used a *thermistor mounted in a small telescope on the orbiting lunar command vehicle to record lunar surface temperatures.

radio occultation The technique of using radio waves transmitted from a spacecraft to probe the atmospheres of planets. The signals received on the Earth as the spacecraft swings behind a planet provide information about both the vertical structures of the planetary *ionospheres and the atmospheric density. Monochromatic radio waves provide more accurate measurements than those obtained using stellar occultations.

radiosonde Instrument, comprising an *aneroid barometer and sensors for temperature and humidity, that is carried aloft to the upper atmosphere by a balloon that ascends at about 5 m/s. Data is transmitted to the surface by radio. *See also* RAWINSONDE.

radius In *tetrapods, the bone of the fore limb on the side of the first digit (in humans, the thumb). *Compare* ULNA.

radius ratio Radius of a *cation divided by that of an *anion. In *silicate minerals, the radius of a cation divided by that of the oxygen anion. Most minerals have bonds that are mainly ionic, therefore radius ratios are most useful in studying mineral structures. *Ions tend to pack closely together in a low-energy, stable state and the most stable arrangements are classified in terms of the cation-to-anion radius ratio. Oxygen is the major anion in most minerals, so the radius ratio of various cations to oxygen is particularly important. This controls the *co-ordination numbers and therefore the structures of silicates, e.g. chains, frameworks, etc.

raft foundation Type of *foundation used for heavy loading, or construction on soft ground, composed of a continuous slab of reinforced *concrete below the entire surface. In Venice, many of the mediaeval raft foundations underneath the large churches are made of wood.

rainbow Phenomenon in which an arc of colours of the spectrum results from the *refraction of the Sun's rays by water droplets in the atmosphere. A primary bow is seen at a limiting angle of 42° to the observer's shadow. A secondary bow, at 51°, may occur; as can multiple bows due to some light being reflected within the raindrops before refraction. The colour separation in the primary bow produces a spectrum with red on the outer edge. The colours are reversed in the secondary (outer) bow. The colour and intensity vary with the size of raindrops, large drops producing bright colours dominated by red.

raindrop A water droplet, formed by condensation of water vapour in a cloud, that is heavy enough to fall from the cloud and large enough to reach the surface of land or sea before evaporating in the unsaturated air beneath the cloud. Droplets reaching the surface range in diameter from about 100 μm in fog to 0.2 mm in drizzle and 5.0 mm in a heavy shower.

rain-gauge Device, usually of copper or polyester, used for measuring rainfall amount. A tapering funnel of standard dimension allows the rain-water to collect in an enclosed bottle or cylinder for subsequent measurement. The gauge is set in open ground with the funnel rim up to 30 cm above the ground surface. Some gauges are calibrated to allow the amount of rainfall to be read directly; with others it must be calculated from the depth of water in the container and the dimensions of the funnel.

rain-making Attempt to induce rainfall by 'seeding' supercooled water clouds. *See also* CLOUD SEEDING.

rain print Small, crater-like pits on the surface of a sedimentary rock, made by rain that fell while the surface was still soft.

rain-shadow The reduction of rainfall to the lee side of a mountain barrier, which results in relatively dry surface conditions, e.g. in the mountains of the south-western USA, where the wetter western slopes of the Coast Range and Sierra Nevada contrast with the desert areas of Nevada and eastern California on the lee side of the mountains.

rain-splash *See* RAIN-WASH.

rain-wash A general term for the transfer of material across the surface and down a hillslope as a result of rainfall. Normally it consists of two components: rain-splash, which is the detachment and subsequent down-slope transfer of small soil particles by raindrop impact; and soil-wash, which is the down-slope movement of material by surface water flow. Rain-wash and *creep are the two main hillslope processes.

raised beach Former beach, now found above the level of the present shoreline as a result of earth movement or of a general fall in sea level. Such beaches are frequently described and correlated in terms of height above present sea level.

rake *See* PITCH.

Ramapithecus Late *Miocene and early *Pliocene ape, known from fragmentary *fossils from E. Africa, south-eastern Europe, and northern India and Pakistan, dating from 14–10 Ma ago, and apparently identical or very similar to the E. African *Kenyapithecus*. *Ramapithecus* is regarded by many as transitional between the true *Miocene apes (the Dryopithecinae) and the later *Hominidae. If this is so, then the human and ape lines diverged prior to the late Miocene, 15–25 Ma ago. More recent evidence, however, suggests that *Ramapithecus* and the related or identical **Sivapithecus* are nearer to the evolutionary line that led to the orang-utan.

rammer 1. Part of the equipment for tunnelling, which pushes into the face. **2.** A tree trunk of suitable dimensions fitted with iron handles and used to ram an *auger into bedrock.

ramose Branched.

ramp 1. That part of a staircase *thrust trajectory which forms the steeply dipping sections between the *flats. Where thrusting occurs in horizontally bedded strata the ramps cut up-section, obliquely to the bedding. A thrust belt may contain several types of ramp, classified as frontal, oblique, and lateral according to their respective perpendicular, oblique, and parallel *strike orientations in relation to the main direction of transport. **2.** *See* SHELF.

rampart craters Martian impact craters with diameters of 5–15 km, with *ejecta sheets extending about one crater radius and terminating in a low ridge or escarpment. The rampart is a primary feature, and not due to secondary modification.

Ramsden eyepiece *See* EYEPIECE.

Randian A *system of the Upper *Archaean, from about 2825–2475 Ma ago, that includes the Ventersdorp, Witwatersrand, and Dominion Reef sequences.

random sampling *See* SAMPLING METHODS.

range In *radar terminology, the distance of radar propagation. In order to avoid interference between consecutive electromagnetic pulses, the interval between pulses must be adequate to allow the return of the previous pulse. Pulse frequency is inversely proportional to the range. Range is measured as either *slant range or *ground range.

Ranger A series of *NASA lunar missions, which ran from 1964 to 1965, sending instruments to the surface.

range zone Unit of *strata defined by the presence and time range of a particular *fossil *taxon. A range zone comprises the entire vertical and horizontal extent of the given organism. Range zones may be local (*teilzone, or local range zone), or the term may be used to refer to the total stratigraphic range of a particular taxon (*taxon range zone). When used formally (*see* FORMAL) the term is capitalized and the qualifying fossil name given in italics, with the generic name capitalized and the specific name in lower-case letters, e.g. the stratigraphic range of the Late *Jurassic *ammonite *Cardioceras cordatum* delimits the *Cardioceras cordatum* Range zone.

rank 1. Category of *stratigraphic unit or *geologic-time unit, classed according to magnitude or duration. **2.** The grade or purity of a substance (referring particularly to *coal).

rapakivi texture Rounded crystals of pink, potassic *feldspar, mantled by white rims of sodic *plagioclase feldspar (occasionally rhythmically zoned with *alkali feldspar), and found as large crystals, up to 4 cm in diameter, in a finer-grained, *igneous *groundmass of granitic composition. The term was applied first to *granites from eastern Finland.

rapid flow (shooting flow) *See* CRITICAL FLOW; and FROUDE NUMBER.

rapid-neutron process (r-process) Neutron-capture chain which takes place on a very short time-scale. It is believed to result from the gravitational collapse of supernovae, leading to a thermonuclear explosion which is capable of forming very large *nuclides in a matter of seconds. *See also* NUCLEOSYNTHESIS.

rare-earth element (REE, lanthanide) One of those elements with atomic numbers between 57 and 71, that have closely similar chemical properties. The ionic radius decreases with increasing atomic number, a phenomenon referred to as the lanthanide contraction. Rare-earth elements occur in minerals only in trace amounts, sometimes replacing Ca^{2+} in *apatite and *hornblende. They tend to become concentrated in the residual fluid of *magmas, and in some *pegmatites the REE cerium replaces the calcium in *epidote to form the mineral *allanite. Lunar rocks, apart from *anorthosite, show considerable enrichment in most of the rare-earth elements relative to the REE *cosmic abundance. *See* EUROPIUM ANOMALY.

raster In *remote sensing, a grid of *pixels used to store and display digitally recorded, remotely sensed images, produced by a series of lines scanned using a *pushbroom, *line scanner, or *radar system at a regular rate of repetition. A separate raster image is used to represent each spectral band.

ratio In *remote sensing, the *digital number value of one band of a multispectral image divided by the digital number value of another band. The ratio allows analysis of relative differences between *channels.

Raukumara A *series in the Upper *Cretaceous of New Zealand, underlain by the *Clarence and overlain by the *Mata. It comprises the Arowhanan, Mangaotanian, and Teratan *Stages which are roughly contemporaneous with the upper *Cenomanian, *Turonian, *Coniacian, and lower *Santonian.

ravinement surface In *sequence stratigraphy, the first surface to have been formed by flooding due to rising sea level, at or close to the shoreline.

ravine wind Wind that passes through an upland barrier along a narrow valley or ravine. The wind is generated by a pressure gradient between the two ends of the valley and the force of the wind is often enhanced as a result of the channelling effect caused by constriction in the valley.

rawinsonde *Radiosonde that is tracked by radio or *radar to observe wind characteristics as well as to record temperature, pressure and humidity.

Rawtheyan A *stage of the *Ordovician in the *Ashgill, underlain by the *Cautleyan and overlain by the *Hirnantian.

Rayleigh criterion In *remote sensing, a method of estimating surface behaviour as either *specular or *diffuse with regard to a particular wavelength of *electromagnetic radiation. A surface is rough if the square root of the mean of the squares of the height of surface irregularities is greater than one-eighth of the wavelength divided by the cosine of the angle of incidence.

Rayleigh, Lord *See* STRUTT, JOHN WILLIAM.

Rayleigh number (Ra) A dimensionless value used to estimate when convection commences in a fluid. The Ra depends on the density and depth of the fluid, the coefficient of thermal expansion, the *gravitational field, the temperature gradient (in excess of the *adiabatic gradient), the thermal diffusivity, and the kinematic *viscosity. Convection usually starts when Ra is 1000 or more, while heat transfer is entirely by conduction when Ra is less than 10.

Rayleigh scattering The scattering of electromagnetic radiation by spherical particles with radii that are less than 10% that of the wavelength of the incident radiation. Such scattering by air molecules produces the blue effect of the sky. Particles such as dust and smoke, that are significantly smaller than 0.4 μm (the wavelength of the blue/violet or lower limit of the visible spectrum) can also scatter visible radiation. Reddish colours at sunset and sunrise result from Rayleigh scattering; these longer wavelengths pass directly through the atmosphere to the observer, while particles in the air scatter out radiation of shorter wavelengths. *See also* MIE SCATTERING.

Rayleigh wave A type of *surface wave which travels along a free *interface. Particle motion is elliptical in a plane perpendicular to the interface and retrograde (at the top of the elliptical orbit movement is in the opposite direction to that in which energy is travelling). Rayleigh waves travel at about 90% of the speed of *S-waves in the same medium. *See* GROUND ROLL.

rays Bright streaks radiating from young lunar craters. They have no surface relief and darken with age, probably due to mixing with the *regolith and to exposure to radiation. They are most probably composed of fine rock powder and glass produced during the impact. The classic example is the spectacular rays emanating from the lunar crater Tycho (85 km diameter), conspicuous through binoculars at full Moon. Hundreds of rays radiate from the crater and some extend across the visible face of the Moon.

reaction time 1. In *geomorphology, the time taken for a *system to react to a sustained change in external conditions. Representative reaction times are difficult to define, because of variations both in the resistance of systems to change and in the magnitude of the external change. For example, a sand-bed river channel reacts more readily to change than does a rock-floored channel. *See also* RELAXATION TIME. **2.** *See* CORONA.

reactivation surface A discontinuity cutting across a *foreset, generated by *erosion or changing flow strength before the resumption of the forward migration of the foreset.

real-aperture radar A *radar system where the *azimuth resolution is determined by the physical length of the antenna, *wavelength, and *range. *Compare* SYNTHETIC-APERTURE RADAR.

real component *See* IN-PHASE COMPONENT.

realgar Minor sulphide ore for arsenic, As_2S_2, associated with *orpiment (As_2S_3) to which realgar changes on exposure; sp. gr. 3.6; *hardness 1.5–2.0 (can be cut with a knife); *monoclinic; red, varying to orange and yellow, transparent to translucent; orange-red *streak; resinous *lustre; crystals rare, short, striated, and *prismatic, also occurs granular, compact, or *massive; *cleavage good *pinacoidal; also found at hot springs and in *limestones and *dolomites; alters to yellow powder.

realistic reaction A chemical reaction which, from criteria based on texture and changing *mineral assemblage, can be convincingly demonstrated to have occurred in a *rock during *metamorphism.

recapitulation of phylogeny Theory, due to E. Haeckel (1834–1919), asserting that *ontogeny (the development of the individual) recapitulates or reflects the *phylogeny (the evolutionary history of the group). The theory as such has been rejected as not of general applicability, von Baer's '*biogenetic law' being sufficient explanation for the observations on which it was based. However, either *hypermorphosis or retardation of sexual maturity can result in individual cases of recapitulation.

Recent *See* HOLOCENE.

recessional moraine *See* MORAINE.

recharge 1. The downward movement of water from the soil to the *water-table. **2.** The volume of water added to the total amount of *groundwater in storage in a given period of time.

recharge area 1. The geographical area of an *aquifer in which there is a downward movement of water towards the *water-table. **2.** The area that acts as a *catchment for any particular aquifer.

reclined *See* STIPE.

reclined fold. As defined by M. J. Fleuty (1964), a dipping *neutral fold in which the *axial plane *dips between 10° and 80° and the *pitch of the *hinge line on the axial plane is more than 80°.

reconstructive transformation *See* POLYMORPHIC TRANSFORMATION.

recovery factor 1. Measure of extraction efficiency. **2.** In mining, the percentage of metal derived from an ore, or *coal from a coal seam, etc. **3.** In *petroleum geology, the percentage of the *in situ* oil that is recoverable. Between 20% and 40% is common for primary recovery techniques, but enhanced recovery by use of water injection, detergents to reduce viscosity, etc., may increase the recovery factor to 75%. Recovery factors for *natural gas can be as high as 90%.

recrystallization 1. The growth of new *mineral grains from pre-existing mineral grains by the solid-state *diffusion of *ions in response to a change in temperature, pressure, or composition of the rock system. **2.** The changing of crystal *fabric or crystal size without an accompanying change in mineral chemistry. **3.** *See* FOSSILIZATION.

rectangular drainage *See* DRAINAGE PATTERN.

rectilinear slope That part (or 'segment') of a hillslope profile that is straight. It is usually the steepest part of the profile. Its gradient varies little in an area of uniform rock types, when it stands at the 'characteristic angle'. More generally, it has been seen (by W. *Penck) as that hillslope profile that develops above a river that is down-cutting at a constant rate.

rectimarginate Applied to a brachiopod (*Brachiopoda) shell where a planar *commissure is present.

recumbent fold A *fold whose *hinge line and *axial plane are horizontal or sub-horizontal. M. J. Fleuty (1964) suggests that the term recumbent fold be restricted to a fold whose axial plane does not *dip more than 10°.

red algae *See* RHODOPHYCEAE.

red-bed copper Conformable copper deposits in *sandstones laid down under terrestrial conditions, usually red in colour. The sands are usually porous and copper minerals, normally *chalcocite, develop in the pores. Such deposits are found in many parts of the world, e.g. in the *Permian of the Urals, and in the *Triassic of central England, Nova Scotia, and the south-western USA.

red beds *Sedimentary rocks, generally *sandstones, which are red due to their *grains being coated with *hematite.

red clay (brown clay) Brown or red, very fine-grained, deep-sea deposit composed of finely divided *clay material that is derived from the land, transported by winds and ocean currents, and deposited far from land in the deepest parts of the ocean basin, especially in mid-latitudes. Red-clay deposits cover about a quarter of the *Atlantic and *Indian ocean floors and almost half the *Pacific ocean floor.

red copper ore *See* CUPRITE.

red edge In *remote sensing, the sharp increase in spectral reflectance of wavelengths in the red and very-*near infrared (700–750 nm) part of the spectrum associated with healthy, green-leaved vegetation. *See also* VEGETATION INDEX.

red iron ore *See* HEMATITE.

Redlichiida An order of *Trilobita that lived in the Lower to Middle *Cambrian. They had large eyes, a large, semicircular *cephalon with strong *genal spines, many small, often spiny, thoracic segments, and a tiny *pygidium. There were 4 suborders. Some, especially members of the suborder Olenelloidea, are important stratigraphic markers.

redox potential (E_H) A scale of values, measured as electric potential in volts, indicating the ability of a substance or solution to cause *reduction or *oxidation reactions under non-standard conditions. The term is sometimes used interchangeably with the term *oxidation potential, but in either case the symbol E^{θ} would refer to standard conditions, while E_H signifies non-standard conditions, usually processes in natural systems such as sea water or soils. The higher the value of E_H, the more oxidizing the conditions. The redox potential is important in *weathering in terms of oxidation and reduction; if the environment will accept *electrons it can precipitate $Fe(OH)_3$, if not the Fe^{2+} *ions will remain in solution. Values in natural environments are closely linked, and vary, with changes in *pH.

redox reaction (oxidation–reduction) Reaction involving the transfer of *electrons from a donor molecule, the reducing agent, to an acceptor molecule, the oxidizing agent.

red podzolic soil *Soil profile formed at an advanced stage of *weathering and *leaching by the process of *podzolization; it is similar in appearance and properties to a *podzol but associated with the greater degree of *chemical weathering and higher iron-oxide concentrations of a humid, tropical environment. *See also* ULTISOLS.

Red Sea An elongate basin, 2000 km long, with shorelines 360 km apart at the widest point and only 28 km apart where it joins the Gulf of Aden. The Red Sea has an inner *median valley, associated with a positive *gravity anomaly, containing *basalts and hot brines, and the sea is thought to be at the young stage of the *Wilson cycle of an ocean.

reduction Chemical reaction in which atoms or molecules either lose oxygen, or gain hydrogen or *electrons. *Compare* OXIDATION.

reduction potential *See* OXIDATION POTENTIAL.

reduction to pole The simplification of the interpretation of *magnetic anomalies by modifying the anomaly pattern to that which it would be in a vertical field, i.e. if the locality were at the north (or south) magnetic pole; induced magnetic effects would then be symmetrical. The anomaly is directly analogous to that of a *gravity anomaly (in which the gravitational force is also vertical).

REE *See* RARE-EARTH ELEMENT.

reef 1. A rigid, wave-resistant build-up constructed by carbonate organisms. Types of reef include patch reefs (small and circular in shape); pinnacle reefs (conical in form); barrier reefs (separated from the coast by a lagoon); fringing reefs (attached to a coast); and atolls (isolated reefs enclosing lagoons). Factors influencing reef growth include: (*a*) water temperature (optimum 25 °C); (*b*) water depth (must be less than 10 m); (*c*) *salinity (normal marine salinity is necessary); (*d*) wave action (intense wave action favours coral growth); and (*e*) turbidity (coral growth requires clear water and an absence of terrigenous suspended sediment). The diversity of species found in a reef will be a function of salinity and water temperature, with stressful conditions resulting in a reduction of species present. **2.** In mining, certain palaeoplacer gold deposits in Australia and South Africa.

reef flat A pavement of naturally

cemented, large skeletal debris and *reef debris to the protected rear (leeward) of a reef crest. Water depth in this zone is shallow, only a few metres at most. Sand shoals and small islands or *cays may be present on the reef flat.

reef front An irregularly sloping ramp extending from the *surf zone to a depth of approximately 100 m on the windward, open-sea side of a *reef. Abundant skeletal growth grades downward into sediment of the *fore-reef zone. The robust, branching form of the coral *Acropora palmata*, common in present-day reefs, typifies this reef area.

reef trap *Stratigraphic oil or gas trap produced by porous *reef *limestones (*reservoir rock) covered by impermeable strata. *Porosity of limestones depends on post-depositional *diagenetic changes. A reef trap may also host lead and zinc mineralization in material deposited from migrating *brines. *Compare* ANTICLINAL TRAP; FAULT TRAP; STRATIGRAPHIC TRAP; STRUCTURAL TRAP; and UNCONFORMITY TRAP.

re-entrant An embayment or recess in the flank of a main valley, often of sufficient length to form a tributary.

reference section *See* HYPOSTRATOTYPE.

reflectance (reflectivity) 1. The ratio of *electromagnetic radiation reflected by a surface to that which is incident upon it. **2.** In *ore microscopy, the amount of incident polarized light which is reflected off the surface of an *opaque mineral. The value of the percentage reflectance (R%) at a specified wavelength is a useful quantitative aid to mineral identification and is determined by: $R\% = $ (intensity of reflected light/intensity of incident light) $\times$ 100.

reflectance spectrometry A number of methods for determining the amount of light reflected from a plane surface. In diffuse *reflectance, this is detected and recorded as a function of wavelength by such instruments as reflectometers, spectroreflectometers, or colorimeters.

reflected infrared In *remote sensing, infrared which is solar-generated *electromagnetic radiation that has been reflected from an object. Characteristically, reflected infrared radiation has a wavelength between 0.7 μm and 3 μm and is therefore *near-infrared. *Compare* THERMAL INFRARED.

reflected-light microscopy (ore microscopy) The scientific study of *ore (*opaque) minerals by means of a *polarizing reflected-light microscope. Systematic observations of *reflectance and *hardness are made in order to identify individual *minerals, and the interpretation of textural relationships and *paragenetic studies may reveal the sequence of mineral formation. It is an important technique in the study of metallic mineral deposits.

reflection The rebounding of an object or wave (light, heat, sound, seismic, etc.) from a surface; the object or wave so reflected. In geophysics, a signal reflected from a *reflector according to *Snell's law. A seismic reflection occurs as a result of a contrast in *acoustic impedances; an electromagnetic reflection occurs because of a contrast in electrical and *dielectrical properties.

reflection coefficient 1. (***R***) The ratio of the amplitude of a reflected ray (A_1) to that of the incoming ray (A_0), such that $R = A_1A_0$. In the case of a normally incident ray, R can be expressed in terms of the *acoustic impedances of the two media above and below the *reflector, Z_1 and Z_2, so that $R = (Z_2 - Z_1)/(Z_2 + Z_1)$. The range of values for R lie between −1 and +1. If R is negative a phase reversal (π) in the wave occurs at the reflector. For water/air R has a typical value of −1; for rocks R has an average value of 0.2 or less. *See also* TRANSMISSION COEFFICIENT. The reflection coefficient can also be expressed in terms of energy (R'), when $R' = R^2$. **2.** (***k***) A ratio of true *resistivities, such that $k = (\rho_2 - \rho_1)/(\rho_2 + \rho_1)$, where ρ_1 and ρ_2 are the true resistivities above and below an *interface.

reflection pleochroism In *ore microscopy, a change in colour exhibited by a mineral on rotation of the *stage in *plane-polarized light. Some minerals, e.g. *covellite, show a strong variation in colour and this is a useful property in their identification.

reflectivity *See* REFLECTANCE.

reflector 1. The surface from which an object or wave (light, heat, sound, seismic, etc.) is reflected. **2.** An *interface which gives rise to a contrast in geophysical properties between the media above and below the boundary. *See* REFLECTION. **3.** A component of an *ore microscope, used in *reflected-light microscopy. There are two types. The glass plate reflector is oriented at 45° to direct the horizontal light source vertically on to the polished surface of the mineral specimen and then to allow the reflected light to pass through it vertically up to the observer. The half-field prism (or mirror system) reflects light downwards through one-half of the aperture of the *objective or lens; the light is then reflected back from the mineral specimen through the other half of the objective, passing behind the prism to the observer.

reflexed *See* STIPE.

reflux theory A proposed mechanism for the continued movement of dense, hypersaline water through a *sediment, causing mineralization or alteration. Reflux is thought to occur by evaporation of salt water producing dense, saline water which sinks through the sediment and displaces lighter waters of normal *salinity. In such a way high concentrations of minerals can be flushed through sediment, so causing effects such as large-scale *dolomitization of buried *limestones.

refolded fold A *fold which, subsequent to the original folding, has undergone one or more further episodes of folding to produce a complex structure in which the folds, or successive folding episodes, are commonly designated chronologically as F_1, F_2, F_3, ... F_n. The *outcrop manifestations of refolding are *interference patterns.

refraction The bending of a ray which travels obliquely from one medium to another, at the *interface separating the two; it is caused by the contrast in velocities with which the ray travels in the two media, and described by *Snell's law. *See also* REFRACTION SURVEY.

refraction survey A field investigation in which seismic *head waves are used to study subsurface geologic structures. Seismic waves travel down from a source to an *interface, where they are critically refracted along the boundary and reradiated back to the surface, and detected by a *geophone *array. The *travel times of the *first breaks are plotted on a *travel-time graph, from which depths to the refractor, its dip, and the velocities of the layers encountered can be calculated. *See* SNELL'S LAW. *See also* CROSS-OVER DISTANCE; and INTERCEPT TIME.

refractive index (*n*) When light travels from air into a substance its velocity is reduced. The light path is also refracted into the substance, and the relationship between the angle of incidence (*i*) and the angle of refraction (*r*) is a constant (*Snell's law). This constant (*n*) is the refractive index of the mineral and is determined by: $\sin i/\sin r = n$. The refractive index is also the ratio of the velocity of light in air (*V*) to the velocity of light in the mineral (*v*): $n = V/v$.

refractometer An instrument used to determine *refractive index. Several types are available. (*a*) The Herbert Smith refractometer measures the *critical angles by total *internal reflection, using a glass hemisphere through which light is directed upwards and reflected from the mineral surface. The reflected beam produces an image in the observing telescope, where a graduated scale measures the critical angle. The refractive index is then calculated by *Snell's law. (*b*) The Abbé refractometer is used primarily for determining the refractive index of liquids. It consists of a pair of glass prisms with a film of liquid between them. The line of the critical angle is measured through a fixed telescope and the refractive index is read off a graduated scale calibrated against a glass plate of known refractive index. (*c*) The Leitz–Jelley refractometer is used to determine the refractive index of small amounts of liquid. It consists of a glass prism cemented to a glass slide which can hold the liquid. A beam of light is directed at right angles to the glass and is refracted by the liquid on to a graduated scale from which the refractive index can be read.

refractory mineral Mineral resistant to decomposition by heat, pressure, or chemical attack. Most commonly applied to heat resistance.

refugia Small isolated areas where extensive changes, most typically due to changing climate, have not occurred. Plants and animals formerly characteristic of the region in general now find a refuge from the new unfavourable conditions in these areas. An example might be a mountain summit projecting above a glaciated lowland region. *See also* RELICT.

Refugian A *stage in the Lower *Tertiary of the west coast of N. America, underlain by the *Narizian, overlain by the *Zemorrian, and roughly contemporaneous with the *Priabonian Stage.

reg 1. Stony desert. **2.** Gravel veneer, normally consisting of small, rounded pebbles, that mantles a Saharan plain and has a gradient as low as 1:5000. The pebble layer may be underlain by a stony soil, or it may be a *lag deposit. *Compare* SERIR.

regelation A process by which water that has been released by *pressure melting beneath a temperate *glacier is refrozen (*see also* BASAL SLIDING; and ICE). The process takes place in a relatively thin zone, the 'regelation layer', which may be only a few centimetres thick. Regelation is associated with the incorporation of bedrock materials in the debris-rich 'sole' of the glacier, which is restricted to the thickness of the regelation layer.

regio (pl. regionis, regiones) A term applied to any feature on a planetary surface that is not clearly defined or understood, usually because of insufficient resolution. Examples include dark regions on *Ganymede and *Iapetus. On *Venus, the term was originally used for radar-bright features such as Beta Regio, which is probably a volcanic construct. Its usage is now extended to cover elevated terrain smaller than continents.

regional field The values in gravity and magnetic surveys that can be attributed to sources within the lower *crust or *basement. These are usually of much longer wavelengths than those associated with near-surface bodies. *See* RESIDUAL GRAVITY MAP.

regional metamorphism The *recrystallization of pre-existing rocks in response to simultaneous changes of temperature, lithostatic pressure, and in many cases *shear stress, occurring in *orogenic belts where lithospheric *plates are converging. The broad areas covered by orogenic belts cause the associated *metamorphism to be developed on a regional scale, hence the name attached to this type of metamorphism. Regional metamorphism can be pre-, syn-, or post-tectonic, depending whether the metamorphic event (or events) is (or are) before, synchronous with, or after the orogenic deformation event (or events). Typical rock *fabrics produced during regional metamorphism are, in order of increasing grain size (reflecting increasing *metamorphic grade), slaty, phyllitic, schistose, and gneissose fabrics. Increase of metamorphic grade in regional *terrains typically produces a *prehnite–pumpellyite–*greenschist–*amphibolite–*granulite facies series. However, each regional metamorphic terrain is characterized by a unique mineral zonal sequence reflecting a particular pressure–temperature gradient during metamorphism.

regional stratigraphic scale *See* STRATIGRAPHIC SCALE.

Regionally Important Geological/Geomorphological Sites (RIGS) A British network of sites selected and conserved by informally constituted groups of volunteers working closely with statutory and voluntary conservation bodies. The scheme began in 1990.

regolith 1. General term for the layer of *unconsolidated (non-cemented), weathered material, including rock fragments, mineral grains, and all other superficial deposits, that rests on unaltered, solid bedrock. It reaches its maximum development in the humid tropics, where depths of several hundreds of metres of weathered rock are found. Its lower limit is the *weathering front. Soil is regolith that often contains organic material and is able to support rooted plants. *Compare* SAPROLITE. **2.** The continuous layer of incoherent fragmental material, produced by *meteorite impact, that typically forms the surface blanket on planets, *satellites, and *asteroids where the atmosphere is thin or lacking. The classic example is the lunar regolith, typically several metres thick, with components ranging from metre-sized blocks to micron-sized dust and glass particles.

regression (marine) The withdrawal of water from parts of the land surface due to a fall in sea level relative to the land. Shallow-water *sediments overlie sediments characteristic of deeper water. *See* OFF-LAP. *Compare* TRANSGRESSION.

regressive systems tract (RST) In the *genetic stratigraphic sequence model used in *sequence stratigraphy, a sigmoid *clinoform produced by *onlap under conditions of rising sea level due to the rapid subsidence of the basin. It is a type of *lowstand systems tract produced under special conditions. If the rate of subsidence was at no time high enough to outpace the rate of sedimentation, thus allowing transgression, the tract is known as a midstand (or forced regressive) systems tract.

regular echinoids Informal term for sea-urchins, of the class *Echinoidea, in which the anus is enclosed within the apical system. The term includes the Perischoechinoidea, Diadematacea, and Echinacea.

Regulares (Monocyathea; phylum *Archaeocyatha) Class of animals that were usually solitary, only rarely colonial, found in Lower and Middle *Cambrian deposits. The conical cup varies from cylindrical to saucer-shaped, and usually consists of two porous walls (a single porous wall in the order Monocyathida). The *intervallum may contain *tabulae alone or with *septa. There is a fan-like divergence of longitudinal pore-rows in the septa. *Dissepiments may be present. Some single-walled monocyathids had a flap (pelta) over the central cavity. *Compare* IRREGULARES.

Reid, Harry Fielding (1849–1944) An American geophysicist, Reid proposed the '*elastic rebound' theory of earthquake motion after studying the 1906 San Francisco earthquake. He was a vigorous opponent of *continental drift theory, describing *Wegener's work as 'pseudo-scientific'.

rejuvenation The marked increase in the rate of *erosion that takes place when a land mass is relatively elevated. Streams respond by incision, with the development of *terraces and *knick points, and finally a *polycyclic landscape emerges.

relative age The position within a time sequence (in the Earth sciences usually the *stratigraphic time-scale) held by an event, *fossil, *mineral, or *rock, compared with others of its kind, e.g. 'an early *Cambrian *trilobite' or a 'Late *Jurassic marine *transgression'. No age in years is implied. *Compare* ABSOLUTE AGE. *See also* DATING METHODS.

relative humidity The water-vapour content of air at a given temperature, expressed as a percentage of the water-vapour content that would be required for saturation at that temperature. Generally the relative humidity decreases during the day, with increase in temperature, and increases at night as the temperature falls.

relative permittivity *See* DIELECTRIC CONSTANT.

relative plate motion The motion of one lithospheric *plate relative to another. This can be described by the *pole of rotation and the angular velocity about this pole.

relative pollen frequency (RPF) Expression of *pollen data from sediments for each species, genus, or family, as a percentage of the total pollen count, or the total tree pollen. It is the traditional and most widely used method for preparing *pollen diagrams. *Compare* ABSOLUTE POLLEN FREQUENCY. *See also* POLLEN ANALYSIS.

relative time-scale *See* DATING METHODS; and GEOCHRONOLOGY.

relative vorticity *See* VORTICITY.

relaxation A term used to describe the 'fading', or loss of topographic relief, of craters on icy *satellites. Such craters are generally shallower than those on rocky satellites, due to viscous flow of the icy crust. Some disappear completely, leaving a discoloured patch or *palimpsest on the surface.

relaxation frequency (f_r) The frequency at which the *dielectric loss factor (ϵ'') reaches a maximum, for a dielectric material that has no static (d.c.) conductivity and that is subjected to an alternating electromagnetic field.

relaxation time 1. The time taken by a disturbed system to reach equilibrium, or the time taken for the magnitude of some parameter to decrease to about 37% of its initial value. For example, the temperature-dependent relaxation time (τ) of a *dielectric is related to the *relaxation frequency (f_r) such that $\tau = 1/(2\pi f_r)$. Physically, it is the time taken for an ionic defect to move within a *crystal *lattice under the influence of an applied alternating electromagnetic field. **2.** In *geomorphology, the time taken for a *system to become adjusted to a sustained change in the nature and/or intensity of external *processes. Such an adjustment normally involves a change in the shape of the land-form or landscape constituting the system. Relaxation times vary. The width of a river channel may adjust in response to an increase in discharge in, perhaps, 10 years, while a glaciated mountain range may require 10^5–10^6 years to lose the imprint of ice.

relict Applied to organisms that have survived while other related ones have become extinct. Often the term refers to *species that have survived periods of unfavourable conditions (e.g. *glacial periods or land submergence) by existing in regions called *refugia, while becoming extinct elsewhere (e.g. some arctic-alpine plants). It may also refer to a surviving species of a group, the other species of which have become extinct (e.g. *coelacanth fish). *See also* RELICT SEDIMENT.

relict sediment *Sediments of the *continental shelf deposited by processes no longer active in the area where the sediments now occur. Relict sediments are remnants from an earlier environment and are now in disequilibrium. Approximately 50% of the present continental shelves are covered by relict sediments deposited during the period of lower sea levels in the *Pleistocene.

relict structure A textural or structural feature inherited from an original *igneous or *sedimentary rock and preserved as a *relict in a low-grade *metamorphic rock which has suffered little or no deformation.

relief In *thin-section microscopy, variations due to the difference in *refractive index between a *mineral and its mounting medium. If the differences are small the mineral appears flat and featureless, with faint outlines. If the differences are large the mineral appears to stand out, with strongly marked outlines and conspicuous *cleavages or fractures. The nature of the relief is determined by the *Becke-line test.

Relizian A *stage in the Upper *Tertiary of the west coast of N. America, underlain by the *Saucesian, overlain by the *Luisian, and roughly contemporaneous with the Lower *Langhian Stage.

remanent magnetization The magnetization remaining after the removal of an externally applied field, and exhibited by *ferromagnetic materials. *See* NATURAL MAGNETIZATION; ISOTHERMAL MAGNETIZATION; GYROMAGNETIZATION; and ANHYSTERITIC MAGNETIZATION.

remanié *See* DERIVED.

remanié beds *See* CONDENSED BED.

remote sensing The gathering of information without actual physical contact with what is being observed. This involves the use of *radars, sonars, spectrosocopy, and the use of airborne and *satellite photography. *See* BISTATIC RADAR; IMAGING; LASER RANGING; POLARIMETRY; RADAR ALTIMETRY; RADIOMETRY; and RADIO OCCULTATION.

remoulding The change that occurs in *clay which has been disturbed and lost *shearing strength but gained *compressibility.

removal time *See* RESIDENCE TIME.

rendzina A *brown earth soil of humid or semi-arid grassland that has developed over calcareous *parent material. Rendzinas may fall within the orders *Inceptisols or *Mollisols.

renewable resource Resource produced as part of the functioning of natural systems at rates comparable with its rate of consumption, e.g. food production by *photosynthesis. Limits to renewable resources are determined by flow rate and such resources can provide a sustained yield. *Compare* NONRENEWABLE RESOURCE.

reniform Kidney-shaped.

repeated twinning *See* LAMELLAR.

Repettian A *stage in the Upper *Tertiary of the west coast of N. America, underlain by the *Delmontian, overlain by the *Venturan, and roughly contemporaneous with part of the *Piacenzian Stage.

repichnia A behavioural category of *trace fossils that result from locomotion. Animals may leave distinct *tracks through walking or crawling across soft *sediment surfaces; repichnia are the fossilized traces of those tracks.

replacement Widely used geologic term denoting a process that involves some kind of transformation. In a petrological sense it refers to the partial or complete alteration of an original mineral to an aggregate of *secondary minerals, by the diffusion of *ions between the solid *phases and an introduced, fluid (usually water-rich) phase. Such diffusion takes place easily when the temperature of the rock system is below the stability limit of an individual mineral and a fluid is present to act as a catalyst to initiate the diffusion reactions. The secondary minerals may be all of one type, or a combination of mineral types. For example, high-temperature, magnesium-rich *olivine can be replaced by an aggregate of secondary *serpentine and *chlorite, while *plagioclase can be replaced by a fine aggregate of white mica (*sericite). *See also* FOSSILIZATION.

reptation A mode of particle transport in which grains are lifted or ejected only weakly and do not rebound or eject other particles when they return to the bed.

Reptilia (reptiles) Large and varied class of *poikilothermic *vertebrates, which arose in the *Carboniferous from *labyrinthodont amphibians. They were the dominant animals of the *Mesozoic world and gave rise to the *birds and *mammals. Reptiles have a body covering of ectodermal scales, sometimes supported by bony scutes. There is no gilled larval phase; development is by *amniote egg, but ovovivipary is common.

Reptiles are air-breathing from hatching onwards.

reptiles *See* REPTILIA.

resequent Applied to a land-form whose orientation is similar to that of the inferred original feature, but which has passed through a complex subsequent history. For example, a resequent *fault-line scarp faces the same way as the original fault scarp.

resequent fault-line scarp *See* RESEQUENT.

reserve Resources of coal, ore, or minerals which can be mined legally and profitably under existing conditions. The indicated reserve is the estimate of ore computed from *boreholes, *outcrops, and developmental data, and projected for a reasonable distance on geologic evidence. An inferred reserve is an estimate based on relationships, character of deposit, and past experience, without actual measurements or samples; it should include the limits between which the deposit may lie. A potential reserve is ore not yet discovered but whose presence is suspected; the term is sometimes used for ore not commercially viable at present time. A proved reserve is a resource reliably established by tunnels, boreholes, or mining.

reservoir 1. A surface body of water whose flow is artificially controlled by means of dams, embankments, or sluice gates in such a way that the water remains static until it is allowed to flow for a specific purpose, e.g. flood control or public water supply. **2.** An underground rock formation with sufficient void space to act as a store for water, natural gas, or oil.

reservoir pool Large and usually *abiotic store of a nutrient in a *biogeochemical cycle. Exchanges between the reservoir pool and the *active pool are typically slow by comparison with exchange within the active pool. Human activity, such as the mining of mineral resources, may profoundly alter this exchange rate, generally releasing an excess into the active pool which can be accommodated only by establishing a new equilibrium. This may in turn produce unfavourable conditions, manifested as chemical pollution, e.g. excess phosphorus in eutrophication, excess sulphur in acid rainfall, and lake acidification.

reservoir rock Any porous rock in which oil, gas, or water may accumulate; usually *sandstone, *limestone, or *dolomite, but sometimes fractured *igneous or *metamorphic rock.

reshabar Regional south-easterly wind affecting mountain slopes in southern Kurdistan (the plateau and mountains in south-eastern Turkey, northern Iraq, northern Syria, and western Iran). The strong, swirling wind is hot and dry in summer but brings cold conditions in winter.

residence time 1. (removal time) The time that a given substance remains in a particular compartment of a *biogeochemical cycle. **2.** The time during which water remains within an *aquifer, lake, river, or other water body before continuing around the *hydrological cycle. The time involved may vary from days for shallow gravel aquifers to millions of years for deep aquifers with very low values for *hydraulic conductivity. Residence times of water in rivers are a few days, while in large lakes residence time ranges up to several decades. Residence times of continental *ice sheets is hundreds of thousands of years, of small *glaciers a few decades. **3.** The average time a particular element of sea water spends in solution between the time it first enters and the time it is removed from the ocean. **4.** The average time that a water molecule or particulate pollutant spends in the atmosphere. For pollutants (e.g. dust from a volcanic *eruption), the residence time may range from a few weeks in the lower *troposphere to several years in the upper *stratosphere, before it is scavenged out by *precipitation. For water molecules the overall average is believed to be 9–10 days.

residual deposit 1. Weathered material remaining *in situ* after soluble constituents have been removed. **2.** *Ore deposit in *clay formed by near-surface oxidation, e.g. *bauxites (aluminium ore), residuál nickel, extensive iron *laterites, and *soil.

residual gravity map Usually a map of the Earth's *gravitational acceleration remaining after allowing for all distorting effects. It is known as a *Bouguer anomaly map when a *regional field (usually a gradient), attributable to gravitational sources within the lower *crust or *basement, has been removed. *See* SMITH'S RULE.

residual shear strength *See* SHEAR STRENGTH.

resinite *See* COAL MACERAL.

resinous Of a mineral *lustre, translucent yellowish to brown.

resistate mineral A mineral which is not readily weathered by chemical attack, e.g. *quartz, *zircon, and *muscovite. The relative ability of minerals to resist *chemical weathering is expressed in the Goldich stability series.

resistivity logging *See* LATEROLOG SONDE.

resistivity methods Geophysical methods in which very-low-frequency or direct electrical current is injected into the ground and its potential distribution is measured in order to obtain information about the Earth's resistivity. Loosely, the term may also include *electromagnetic methods, since *apparent conductivities (σ_a) can be used to derive *apparent resistivities (ρ_a) by: $\sigma_a = 1/\rho_a$. *See* CONSTANT SEPARATION TRAVERSING; ELECTRICAL SOUNDING; ELECTRODE CONFIGURATION; and INDUCED POLARIZATION.

resonance 1. Condition of very large wave amplitude, occurring when the frequency of an external wave-generating force matches and amplifies a natural frequency for waves moving to and fro in an enclosed space such as an *estuary. **2.** The relationship in which the orbital period of one body is related to that of a second by a simple integer fraction (e.g. 1/2, 3/5). Such orbits are common in the *solar system. Well-known examples include the Kirkwood Gaps in the Asteroid Belt and the Cassini Division in Saturn's ring system (a particle moving in the Cassini Division has a period 1/2 that of Mimas, and 1/3 that of Enceladus).

resorption The partial fusion of a *euhedral *phenocryst in a *magma in response to a change in magma temperature, pressure, and/or composition. If the magma is erupted rapidly the partially fused phenocryst can be preserved as a large, *anhedral *crystal with a lobate outline set in a fine-grained *groundmass.

resurgence 1. *See* SPRING. **2.** *See* CALDERA.

resurgent caldera *See* CALDERA.

reticulated Applied to a meshwork of intercalating crystals which may give rise to lattice-like groups of crystals in a mineral.

reticulite A gold-brown, foam-like type of glass produced by *Peléean eruptions and found near Hawaiian volcanoes, often in considerable quantities.

retro-arc basin A type of *back-arc basin which is floored by *continental crust. The main *sediments are fluvial, deltaic, or marine, derived from the uplifted area behind the arc.

retrochoanitic *See* SEPTUM.

retrograde Used in a planetary context to denote a body moving in the opposite sense to that of most *solar-system bodies, i.e. clockwise rather than anticlockwise. The classic example is the retrograde rotation of *Venus. *Triton is in retrograde orbit around *Neptune. At least four of the outer *satellites of *Jupiter, and one of *Saturn, are also in retrograde orbits.

retrograde metamorphism (diaphthoresis, retrogressive metamorphism) The *recrystallization of pre-existing rocks in response to a lowering of *metamorphic grade in the presence of a fluid *phase. After reaching a metamorphic climax, lowering of metamorphic grade does not usually cause retrograde reactions to occur because all the water in the rock system has been expelled at the metamorphic climax, thus preserving high-grade *mineral assemblages. If some water remains in the system, however, or is introduced as the grade decreases, the water can act as a catalyst to initiate retrograde reactions. The reactions produce hydrated mineral types (*see* HYDRATION), in contrast to the dehydration reactions of *prograde metamorphism.

retrogressive metamorphism *See* RETROGRADE METAMORPHISM.

retrosiphonate A condition in some cephalopods (*Cephalopoda) in which the septal (*see* SEPTUM) necks point back, towards the *protoconch.

retrusive Applied to the direction of *spreiten that are extended upwards through the *sediment and are therefore proximal to the point of entry.

return flow *See* INTERFLOW.

return period The frequency, based on statistical analysis of past records, with which a particular environmental hazard may be expected.

Réunion Two normal *polarity subchrons which occur within the *Matuyama reversed *chron.

reverberation (ringing) An oscillatory effect seen on seismic wave-forms and produced by short-path *multiples.

reversal 1. A change of direction, usually by 180°. It commonly refers to a change of *polarity of the *geomagnetic field. **2.** A form of *homoplasy; resemblance between two taxa because one of them has gained a new character, then lost it again, and the other taxa has never gained it. Reversal by character loss is common; there is much doubt about whether it ever occurs by regaining a lost character.

reversal time-scale *See* MAGNETOSTRATIGRAPHIC TIME-SCALE.

reversed field *See* GEOMAGNETIC FIELD; and POLARITY REVERSAL. *Compare* NORMAL FIELD.

reverse fault A low-angle, *dip-slip fault in which the relative displacement of the *hanging wall is upwards. A *thrust is a type of reverse fault.

reverse zoning *See* CRYSTAL ZONING.

reversing dune A *seif dune that has asymmetrical ridges.

revolving storm *See* TROPICAL CYCLONE.

reworked *See* DERIVED.

Reykjanes Ridge The part of the *Mid-Atlantic Ridge to the south-west of Iceland, whose axis, marked by a *median valley, continues into the active *graben across Iceland.

Reynolds number A dimensionless number expressing the balance of viscous and interstitial forces on a small element of moving fluid. The transition from *laminar to *turbulent flow depends on the Reynolds number (R) which is equal to $\rho vd/\eta$, where ρ is the fluid density, v the fluid velocity, d the diameter of the *pore space through which flow occurs, and η the *viscosity. For laminar flow, the Reynolds number is less than 500, while turbulent flow occurs when R is greater than 1000. *Darcy's Law for *groundwater flow is valid for values of R less than about 1–10.

rhabdosome In *Graptolithina, a complete colony.

Rhaetian 1. A Late *Triassic *age, preceded by the *Norian, followed by the *Hettangian (*Lias) and dated at 209.5–208 Ma (Harland et al., 1989). **2.** The name of the corresponding European *stage, which is roughly contemporaneous with the Erchiao (China) and Otapirian (New Zealand). Some authors have placed the Rhaetian in the *Jurassic. Others have questioned its status as a stage, suggesting that it should be a biostratigraphic *zone within the underlying Norian.

Rhea (Saturn V) One of the major satellites of *Saturn, with a radius of 764 km; mass 23.1×10^{20} kg; mean density 1240 kg/m^3; visual albedo 0.7. It was discovered in 1672 by G. D. Cassini.

rheology (adj. rheological) 1. Study of deformation and flow in materials, including their elasticity, *viscosity, and plasticity. **2.** In geology, the study of flow in water, *ice, *magma, and during rock deformation.

Rhine graben The *rift valley which contains the river Rhine and which lies between the Ardennes, the Vosges, and the Black Forest. The uplift occurred in the late *Mesozoic, with rifting in the mid-*Eocene and production of *alkaline *magmas in the *Oligocene. In places, 3 km of *sediment have been deposited. The Rhine graben formed synchronously with the Alpine collision and has been called an 'impactogen', i.e. a collisional rift that forms at the end of the *Wilson Cycle.

Rhipidistia Group of *crossopterygian ('tassel-finned') fish ranging from the *Devonian to the *Permian. They possessed two dorsal fins, lobate or stalked pectoral and pelvic fins, and internal nostrils. Distantly related to the living *coelacanth, they are considered by some to be ancestral to the *Tetrapoda (terrestrial *vertebrates).

Rhizocorallium U-shaped feeding structures excavated during the search for food, and abundant in the *Jurassic, *Rhizocorallium* is typically elongate and is found parallel or slightly oblique to the bedding surface. It may attain a length of 1 m and the parallel tubes usually have a diameter of 2–3 cm. Delicate *spreiten may reveal the direction of excavation. *See* FODINICHNIA.

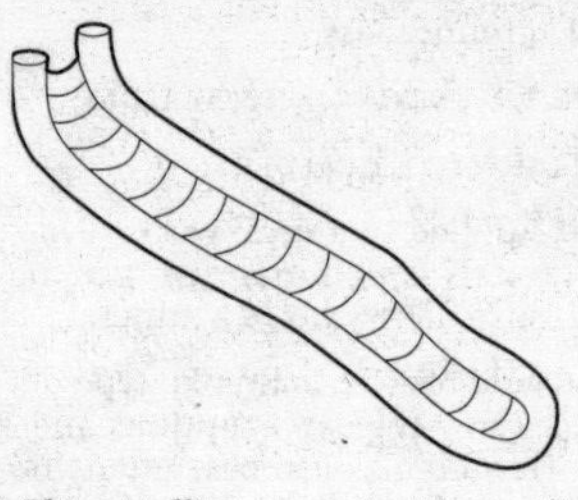

Rhizocorallium

rhodochrosite Mineral, $MnCO_3$; sp. gr. 3.4–3.7; *hardness 3.5–4.5; *trigonal; translucent rose-pink, and sometimes light grey to brown, developing a brown or black crust on exposure; white *streak; vitreous *lustre; crystals rare, but form as *rhombohedra, and rod-like and curved, but it is usually *massive or granular; *cleavage perfect rhombohedral; occurs in *hydrothermal veins containing silver, lead, and copper, and in *metamorphic and *metasomatically altered rocks of sedimentary origin, often a *secondary mineral after manganese oxide; soluble, with effervescence, in hot, dilute hydrochloric acid. It is used in the manufacture of ferromanganese, being added to blast-furnace charges, and in the chemical industry.

rhodonite Member of the *inosilicates with composition $(Mn,Ca)Si_2O_6$ and related to the *pyroxenes but with a different atomic structure, and termed a pyroxenoid (*see also* WOLLASTONITE); sp. gr. 3.57–3.76; *hardness 5.5–6.0; pink to brownish-red; *tabular or *massive; occurs in manganese-rich mineral deposits, e.g. in Franklin, New Jersey.

Rhodophyceae (red algae) A class of marine *algae, most of them red in colour, whose basic shape is filamentous or membranaceous. They tend to occur at greater depths than the green algae (*Chlorophyta) and they are among the oldest groups of *eukaryotic algae, known from the *Cambrian upwards. *Epiphyton* (Cambrian to *Devonian) formed mounds and *Solenopora* (Lower Cambrian to *Cretaceous) formed nodular masses made up of close-packed tubes. The coralline red algae, e.g. *Lithothamnion*, are important rock builders, constructing rigid structures and contributing *lime mud to *sediments.

rhombic dodecahedron *See* RHOMBDODECAHEDRON.

rhombdodecahedron (rhombic dodecahedron) A 12-faced *crystal form (110) which has *cubic *crystal symmetry. Each face is rhomb-shaped and intersects two *crystallographic axes equally and is parallel to the third.

rhombochasm A chasm, rhomboid in shape, that forms deep in the crust through transverse faulting of two blocks.

rhombohedral *See* TRIGONAL.

rhombohedron A six-faced, *closed *crystal form which belongs to the *trigonal system. There are three upper faces and three lower faces arranged about the vertical *c* (or *z*) axis. The three horizontal axes emerge in the middle of each edge. The mineral *calcite frequently occurs in this crystal form.

rhourd A large, star-shaped or pyramidal sand *dune (a 'sand mountain') that may be 100–200 m high and that has been described for the Algerian Sahara. It may form where two zones of sand-laden wind cross one another. *See* DRAA.

Rhuddanian A *stage of the Lower *Silurian, underlain by the *Onnian (*Ordovician) and overlain by the *Aeronian.

Rhynchocephalia ('beak-heads', rhynchocephalians; class *Reptilia) Order of primitive, lizard-like reptiles dating from the *Triassic and often cited as a *living fossil. The order contains only one species, *Sphenodon punctatus*, the tuatara of New Zealand. This survives only in the Bay of Plenty, and it has protected status. The skull is of the primitive *diapsid type, with a fixed quadrate bone. The teeth are fused to the edge of the jaw (acrodont), with a tendency to develop a beak-like structure anteriorly.

rhyncholite One of the beak-like structures which are considered to be the upper jaw structures of *fossil cephalopods (*Cephalopoda). They are approximately rhomboidal in shape, with a slightly concave lower surface. The anterior portion is termed the hood, the posterior portion the shaft. The first examples are found in the *Carboniferous.

Rhynchonellida (rhynchonellids; class *Articulata) Order of brachiopods (*Brachiopoda), with rostrate shells, a functional *pedicle, and a *delthyrium partly restricted by a pair of *deltidial plates. The shell is usually *impunctate. They appeared first in the Middle *Ordovician. The order contains about 250 genera, most of which are extinct.

rhynchonellids *See* RHYNCHONELLIDA.

Rhynia An early vascular plant, first known from the Lower *Devonian Rhynie Chert, Aberdeenshire, Scotland, and described by R. Kidston and W. H. Lang in a series of papers between 1917 and 1921. *Rhynia* was a simple, leafless plant with a creeping, horizontal stem (rhizome) from which the upright, aerial shoots arose. The tips of fertile shoots bore oval-shaped sporangia (*see* SPORE) and the prostrate, horizontal axis was supported by rhizoids rather than true roots. Originally two species were included in the genus: *R. gwynne-vaughanii* (up to 20 cm tall) and *R. major* (20–50 cm). Recent work by David S. Edwards (1986) has shown *R. major* to vary in its branching pattern from *R. gwynne-vaughanii* and to lack the tracheids necessary for it to qualify as a vascular plant. Consequently *R. major* has been transferred to a new genus of uncertain affinity and is now termed *Aglaophyton major*. *See also* COOKSONIA; and PSILOPHYTALES.

Rhyniophytina *See* PSILOPHYTALES.

rhyodacite (toscanite) A fine-grained, *extrusive, *igneous rock characterized by an *adamellite *mineral assemblage and composition. Most rhyodacites are *porphyritic, with *quartz and *plagioclase as common *phenocryst types. The term 'toscanite' was used originally by H. S. Washington in 1897 to describe rocks of rhyodacite composition from Tuscany, Italy; this older term is now little used. Rhyodacites are erupted above subducted *plates and belong to the *calc-alkaline *magma series.

rhyolite A fine-grained, *extrusive, *igneous rock, often with a sugary texture, consisting of *essential *quartz, *alkali feldspar, and one or more *ferromagnesian minerals. Alkali rhyolites are the most common type, being characterized by the ferromagnesian mineral *biotite with or without *pyroxene, and are found in *calc-alkaline *terrains. *Peralkaline rhyolites are characterized by alkali pyroxenes (*aegirine, aegirine–*augite) and alkali *amphiboles (*riebeckite, *arfvedsonite), and are found as *end-members of alkaline *magma series on oceanic islands and rifted *continental crust.

rhythmic sedimentation *See* CYCLOTHEM.

rhythmite A sequence of fine-textured, regularly repeated bands laid down by a sequence of cyclical or rhythmic sedimentation (*see* CYCLOTHEM). Rhythmites are most commonly associated with freshwater environments, but can also be deposited by tidal movements.

ria Drowned river valley in an area of high relief. Classic examples are found in some of the peninsulas of western Europe, notably western Ireland, where they have resulted from the post-glacial rise in sea level.

ribbon bomb *See* VOLCANIC BOMB.

ribbon jasper *See* JASPER.

ribbon lakes *See* TUNNEL VALLEY.

ribbons Straight to sinuous, thin bodies of *sand, with a narrow width in relation to their length. Sand ribbons develop on *sediment-poor, tide-swept shelves (*see* SHELF), oriented parallel to the tidal stream. The term is also used more generally to describe the large-scale geometry of a preserved *sandbody with a width to length ratio in excess of 1 : 100, and a thickness to width ratio greater than 1 : 10.

Richmondian A *stage of the *Ordovician in the Upper *Cincinnatian *Series of N. America.

Richter, Charles Francis (1900–1985) An American physicist and geologist, Richter is best known for his logarithmic scale of *earthquake magnitudes. First proposed in 1927, the scale was later refined with the assistance of *Gutenberg, with whom Richter also co-operated in a study of the world's greatest earthquakes, and other seismological work. *See* RICHTER SCALE.

Richter denudation slope A hillslope that develops at the foot of a cliff that is retreating fairly rapidly, chiefly by *rock fall. The slope has a uniform gradient, is cut across bedrock, and stands at the angle at which the *talus accumulates. With each unit of cliff retreat the related rock fall builds up on older talus, and so the foot of the cliff steadily rises. The Richter slope is revealed when the talus is removed, or it may remain hidden beneath a thick skin of mobile debris. E. Richter described such slopes in the Alps in 1900, and they are named after him.

Richter scale The measurement of the intensity of an *earthquake using the amplitude of *seismic waves. As the amplitude depends on the depth of the earthquake *focus, the distance of the recording station from the focus, the travel path, and local geology at both the source and receiver, such magnitude estimates need to be constrained by several determinations. At any given recording station, the magnitude (M) of a shallow earthquake is given by the equation: $M = \log(A/T) + 1.66 \log \Delta + 3.3$, where A is the maximum amplitude, T is the period, and Δ is the *epicentral angular distance between the earthquake and receiver. For deeper earthquakes, the magnitude is given using 20-second-period *Rayleigh waves by $M = \log(A/T) + af\Delta h + b$, where h is the depth of the focus, and a and b are empirically determined constants for each seismic station.

Ricker pulse A seismic wavelet caused by the passage of a seismic pulse through an ideal viscoelastic medium, where the attenuation is proportional to the square of the frequency.

ridge 1. (wedge) An extension of high pressure from an *anticyclone into a zone where generally lower pressure prevails. **2.** The poleward meanders of the flow of the upper westerly winds over mid latitudes. *See also* long wave. **3.** *See* MID-OCEAN RIDGE. **4.** *See* RIDGE-AND-RAVINE TOPOGRAPHY; and RIDGE AND RUNNEL.

ridge-and-ravine topography A landscape consisting of a monotonous network of branching valleys and intervening low ridges, and which is similar to that of a maturely dissected *peneplain (*see* DAVISIAN CYCLE). The term has been introduced, however, to avoid any genetic implications. It is well displayed in the central Appalachians, USA.

ridge and runnel A series of asymmetrical ridges running parallel to the coast and separated by shallow troughs (runnels) 100–200 m wide. This topography is developed on the foreshore of *mesotidal or *macrotidal beaches. The development of these forms is favoured by moderate wave-energy conditions acting on a flat *beach with an abundant *sediment supply.

ridge crest The highest part of a *ridge, typically 2–3 km above the level of the *abyssal plains. With slow-spreading ridges (e.g. the *Mid-Atlantic Ridge) the crest is split by a *median valley, whereas fast-spreading ridges (e.g. the *East Pacific Rise) have no median valley and the crest has more subdued topography.

ridge-push The hypothetical force, caused by the horizontal spreading of the near-surface *asthenosphere at *constructive margins, which is thought to be one of the two main driving forces for the movement of lithospheric *plates (the other is *slab-pull).

riebeckite A member of the alkali *amphiboles $Na_2(Fe^{2+}{}_3Fe^{3+}{}_2)[Si_4O_{11}]_2(OH,F)_2$ and *end-member of an *isomorphous series with *glaucophane $Na_2(Mg_3Al_2)[Si_4O_{11}]_2(OH,F)_2$; sp. gr. 3.43; *hardness 5; forms either tiny *prismatic crystals or large, *poikilitic, *subhedral, prismatic crystals; dark bluish-green or black; occurs in *alkaline *igneous rocks, especially *granites, in association with *aegirine. Fibrous riebeckite (called crocidolite or blue *asbestos) is formed by *metamorphism of massive *ironstone deposits. When infiltrated with silica it constitutes the semi-precious cat's eye or tiger's eye.

riegel A rock bar that extends across the floor of a *glacial trough. It may be caused by a local reduction in the erosive ability of a *valley glacier or by a local increase in bedrock strength, perhaps due to a reduction in *joint density. It may alternate with a rock basin to give an irregular long profile.

riffle *See* POOL-AND-RIFFLE.

rift 1. A breach or split between two bodies that were once joined. *See* RIFT VALLEY. **2.** In quarrying, a split in *granite, whose plane is oblique or perpendicular to the sheeting.

rift valley An elongate trough, of regional extent, bounded by two or more *faults. Many rifts on land are associated with alkaline *volcanicity and, because their margins are uplifted, many are starved of *clastic *sediments and so contain lakes; the E. African rift system is an outstanding example. Some rifts are thought to be at the embryonic stage of ocean development of the *Wilson cycle, whilst others may become 'failed rifts' (or 'failed arms') and fill with sediment to become *aulacogens. The rift valley developed along the axis of slow-spreading oceanic *ridges is known as the *median valley (or axial rift or axial trough) and is associated with the production of basaltic *magmas. Tibetan rifts form at the end of the Wilson cycle as a result of the northward indentation of India into Asia and the spreading of the thickening Tibetan crust. *Graben (the German word for 'ditch') can be used synonymously for 'rift valley' and also for an infilled, fault-bounded trough of any size, with or without topographic expression.

right lateral fault *See* DEXTRAL FAULT.

rigidity modulus *See* SHEAR MODULUS.

RIGS *See* REGIONALLY IMPORTANT GEOLOGICAL/GEOMORPHOLOGICAL SITES.

rille (rima) A small valley on the *Moon. Three types are recognized. (*a*) Straight rilles are typically 1–5 km wide and hundreds of kilometres long, unrelated to surficial topography, analogous to terrestrial fault *grabens. (*b*) Curved or arcuate rilles are variants of straight rilles, with similar dimensions, and form concentrically to major ringed basins (e.g. Mare Humorum). (*c*) Sinuous and meandering rilles, formed by thermal erosion by flowing *lava. Hadley Rille, 1.2 km wide, 270 m deep, and 135 km long, visited by Apollo 15, is the type example.

rill-wash Eroded material that is concentrated into more or less intermittent trickles and rills on inclined slopes, due to run-off of water.

rima On *Mars, a rille or cleft, e.g. Rima Bradley, Rima Sirsalis.

rime The white deposit of ice that results from *crystal growth on objects that are at a temperature below the freezing point. Supercooled water droplets in fog freeze on contact with such surfaces.

ring canal *See* CIRCUM-ORAL CANAL.

ring-dyke Steeply dipping *dyke of arcuate outcrop formed by the uprise of *magma along a steep conical or cylindrical fracture which bounds central collapsed blocks. *See also* CAULDRON-SUBSIDENCE.

ringed basin *See* MULTI-RING BASIN.

ring fracture A steep-sided, outwardly dipping, fault pattern or fracture, circular or sub-circular in plan view, commonly associated with *ring dykes. Ring fractures surround collapsed volcanic depressions, and are thought to form due to circular *stress trajectories created by a parent *pluton at depth. *See* CALDERA; and CAULDRON SUBSIDENCE.

ringing *See* REVERBERATION.

ring silicate *See* CYCLOSILICATE.

ringwall An archaic term that refers to the ring-like walls enclosing lunar craters or *mare basins, as observed telescopically.

Ringwood's rule A rule stating that: 'Wherever *diadochy in a *crystal is possible between two elements with appreciably different *electronegativities, the element with the lower electronegativity will be preferentially incorporated because it forms a stronger and more *ionic bond.'

rip current Strong, narrow current usually of short duration, flowing seaward from the shore. The presence of a rip current can be detected as a visible band of agitated water flowing seawards, usually as a gap in the line of the incoming waves. Rip currents mark the swift return movement of water piled up on the shore by incoming waves and onshore winds.

Riphean An *era of the Middle *Proterozoic, lasting from about 1675 to 825 Ma ago, that preceded the *Sinian. Russian usage extends the era to about 680 Ma ago.

rippability A measure of the ease with

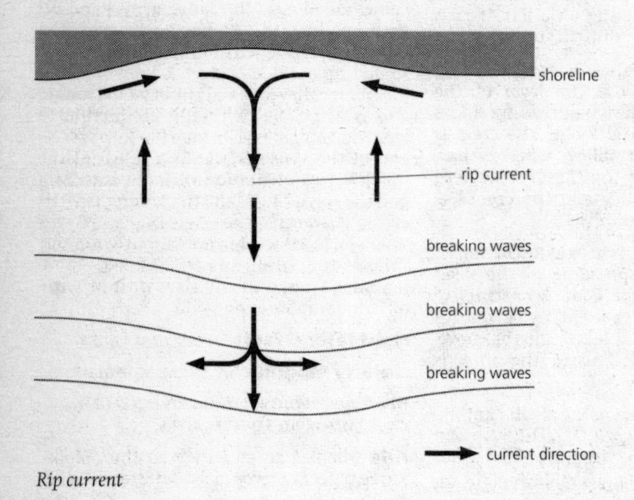

Rip current

which earth materials can be broken by mechanical ripping equipment to facilitate their removal by other equipment. Rippability is related to the *seismic velocity of the material. Rock with a seismic velocity of less than 2 km/s can usually be ripped.

ripple Small-scale ridge of sand produced by flowing water, wind motion, or wave action. The wavelength or spacing of ripple crests is usually less than 50 cm and the heights are less than 20 cm. Ripple form can be described by the wavelength to height ratio, referred to as the *ripple index. The migration of ripples leads to the formation of *cross-lamination in sands. *See also* DUNE BEDFORM.

ripple-drift cross-lamination A form of *cross-stratification (sometimes called climbing-ripple cross-lamination) characterized by set boundaries which *dip in the opposite direction to the *foresets, giving the impression of one set climbing upwards over the underlying set. In some cases the ripple stoss slope (i.e. the gently dipping backslope) is preserved. Ripple-drift cross-lamination forms when there is a rapid rate of net deposition, with the angle at which the sets climb over one another being a function of increasing rate of sedimentation.

ripple index A measure of the symmetry of a ripple form, expressed by the ratio of ripple wavelength to ripple height. Flowing-water ripples (current ripples) have an index of 8–20 and are asymmetric, with a steeper face downstream (lee) and a gentle upstream-facing (stoss) side. The ripple index of wind ripples is 30–70; the structures are much flatter, reaching heights of 1 cm only. Wave-formed ripples (oscillation ripples) have a ripple index of 4–16 and have a more symmetrical profile form.

ripple-symmetry index A measurement of the ratio of the horizontal extent of the stoss side of a *ripple to the horizontal extent of the lee side. Wave ripples have ripple symmetry index values of less than 2.5, whereas current ripples have values of more than 3.0.

ripple train The collective term applied to a series of *ripples which lie one behind the other, as found on a rippled surface.

rip-rap Loose foundation layer of large, irregular, unscreened rock fragments used under water or in soft material for protection and to prevent erosion of dams, sea walls, bluffs, or other structures exposed to wave action. Used extensively in irrigation works and river improvements.

Riss The third of four glacial episodes named after Alpine rivers and established in 1909 by A. *Penck and E. Bruckner. It is perhaps the equivalent of the *Saalian of northern Europe and the *Wolstonian of the East Anglian succession.

Riss/Würm Interglacial An Alpine *interglacial *stage that may be the equivalent of the *Eemian stage of northern Europe or the *Ipswichian of the East Anglian succession. It is the last interglacial and immediately precedes the last glaciation, the *Devensian or in European usage *Weichselian.

river capture Process whereby a stream is able to tap and so capture the discharge of a neighbour. The capturing stream normally extends by headward *erosion along an outcrop of soft rock until it meets and diverts a second, less-favoured, transverse system. A right-angled bend, the 'elbow of capture' is typical of the junction between capturing and captured streams. *See also* AVULSION.

river grade *See* GRADE.

river profile The slope of the long profile of a river, expressed as a graph of distance-from-source against height. It is generally concave-up, and this downstream reduction in gradient may be a consequence of the decreasing energy needed to transfer greater discharge and finer load. In detail, however, it is typically compound, with the profiles of individual segments reflecting the local rock types. It may be broken by *knick points.

river-sediment analysis *See* STREAM-SEDIMENT ANALYSIS.

river terrace (stream terrace) Fragment of a former valley floor that now stands well above the level of the present *floodplain. It is caused by stream incision, which may be due to uplift of the land, to a fall in sea level, or to a change in climate.

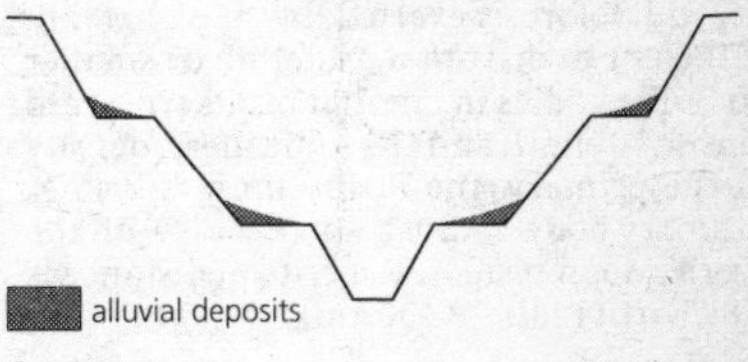

River terrace

river water, major constituents The average composition of river water is very different from that of sea water. The *pH varies greatly from areas of high organic activity, e.g. the equatorial rain forests, to zones of little activity around the poles; the average is usually between 6 and 8. Large rivers, e.g. the Amazon, remove about 10^5 kg of material per km^2 of *catchment area every year. Of this, 2×10^4 kg is dissolved and the rest is carried as solid particles of colloidal debris. The composition of typical river water is:

Ion	*Parts per thousand*	*Percentage of dissolved material*
bicarbonate HCO_3^-	58.5	48.6
calcium Ca^{2+}	15.0	12.5
silicate SiO_2^-	13.1	11.0
sulphate SO_4^{2-}	11.2	9.3
chloride Cl^-	7.8	6.5
sodium Na^+	6.3	5.3
magnesium Mg^{2+}	4.1	3.4
potassium K^+	2.3	2.0
nitrate NO_3^{2+}	1.0	0.8
iron Fe^{2+} or Fe^{3+}	0.7	0.6
TOTAL	120.0	100.0

See also SEA WATER, MAJOR CONSTITUENTS.

RMQ *See* ROCK-MASS QUALITY.

road base *See* PAVEMENT.

roadstone Unconsolidated *aggregate of strong rock which will withstand crushing and abrasion. It is used in the construction of roads.

roadway construction *See* PAVEMENT.

roaring forties Popular maritime term for the prevailing westerly winds which are commonly strong over the oceans in temperate latitudes of the southern hemisphere, particularly between about 40° S and 50° S.

Roche limit The distance within which the tidal forces exerted by a planet are sufficient to disrupt a *satellite or smaller body. For bodies in circular orbits with zero tensile strength and the same mean density as the primary, the Roche limit is 2.46 × (primary-body radius). In the case of the Earth–Moon system, the critical distance is 2.89 Earth radii (18 400 km).

roche moutonnée (glaciated rock knob, stoss-and-lee topography) Mound-like landform of glacial *erosion, consisting of a smoothed, streamlined, up-glacier surface and a broken, shattered, lee flank. It probably results from a combination of abrasion, frost-shattering, and the plucking out of blocks by the *glacier, although crushing has been suggested as a contributory mechanism.

rock A consolidated or unconsolidated aggregate of *minerals or organic matter. The minerals may be all of one type, in which case the rock is 'monomineralic', or of many types, in which case it is 'polymineralic'. The aggregate of minerals can form by: (*a*) accretion or precipitation of grains during Earth surface processes, to give *sedimentary rocks; (*b*) crystallization of *magma to give *igneous rocks; and (*c*) solid-state *recrystallization in response to changes in external conditions (e.g. pressure and temperature) to give *metamorphic rocks. The grain relationships (*textures) of these three rock types contrast. Sedimentary rocks are characterized by one of the following: (i) rounded or angular grains held together by an intergrain precipitate or a fine intergrain mud; (ii) fine aggregates of *clay minerals displaying a preferred orientation of their long axes; (iii) a crystalline aggregate of minerals (e.g. *calcite) displaying straight edges and triple junctions between the grains; (iv) an aggregate of *fossil fragments held together by an interfragment precipitate of *calcite or a fine interfragment mud; or (v) an aggregate of organic material (e.g. *lignite or *coal). All igneous rocks are characterized by an aggregate of minerals displaying an interlocking texture. Metamorphic rocks are characterized by one of the following: (i) a crystalline aggregate of minerals which display a preferred orientation of their long axes; (ii) a crystalline aggregate of equidimensional and randomly oriented non-equidimensional minerals; or (iii) an extremely fine-grained aggregate of sutured, *anhedral, or sometimes elongate minerals.

rock bench *See* VALLEY-SIDE BENCH.

rock bolt A form of support for broken or jointed rock in mines or excavations. The far ends of metal bolts are secured in solid rock, with a face plate and nut at the external end. Rock bolts are usually used in conjunction with steel roof bars, plates, and mesh, combined with *grouting.

rock burst Sudden, often explosive, breaking of rock walls when stressed beyond their limits. It is a hazard encountered in some deep mines (below 1000 m), and may be accompanied by shocks, rock falls, and air concussion. It usually occurs in rocks having high elasticity and strength, so that when a face is exposed the unbalanced *stress is great enough to cause *failure. It can be controlled by systematic *stoping, reduction of open spaces, use of strong supports, and retreating mining methods.

rock crystal *See* QUARTZ.

rock drumlin *See* DRUMLIN.

rock fall The detachment of rock masses of variable size from a steep slope or cliff and their descent, chiefly through the air, by free fall, bounding, or rolling. Movement is very rapid. It is normally caused by *weathering, or by the undercutting of a steep slope by *fluvial or marine processes.

rock fill Waste rock used to back-fill worked-out sections of mines and to support the roof.

rock flour Finely ground rock debris produced chiefly by abrasion beneath a *glacier. It may be removed by meltwater streams, which consequently develop a typically milky appearance.

rock glacier A tongue-like mass of large, angular blocks, finer debris, and ice, found especially in middle latitude Alpine regions (there are about 1000 active examples in the Swiss Alps). It may have a core of ice, in which case it may have originated as a debris-covered glacier, or it may be ice-cemented, when it may be a *permafrost phenomenon.

rock head **1.** Surface between overlying unconsolidated material and solid bedrock below. **2.** Bedrock on which sediments are deposited. **3.** Base of gold-rich gravel where gold concentrates.

Rocklandian A *stage of the *Ordovician in the Middle *Champlainian *Series of N. America.

rock mass A large and indistinct body of solid earth materials, containing features on the scale of *jointing, *folding, *schistosity, etc. The term would not be used to describe a rock only the size of a hand specimen.

rock-mass quality (RMQ) A classification of *rock for engineering purposes, based on the number of major *joints or other discontinuities (planes of weakness) in it, their orientation, and their spacing.

rock mechanics **1.** The study of the physical behaviour of rocks, including crushing, bending, and *shear-strength testing, and also their elasticity, *internal angle of friction, *density, *permeability, and *porosity. *See* ELASTIC DEFORMATION; and ELASTIC LIMIT. **2.** In geology, the study of the mechanics of rock structures, their physical properties, and forces acting on strata. **3.** In engineering, the study of rocks as raw materials, and their behaviour in tunnels, quarries, and mines; and the stability of buildings on rock foundations.

rock pavement *See* RUWARE.

rock-quality designation (RQD) Rough measure of the degree of *jointing or fracture in a *rock mass, measured as a percentage of the drill core in lengths of 10 cm or more. High-quality rock has an RQD of more than 75%, low quality of less than 50%.

rock salt *See* HALITE.

rock slide A landslide made up of fragments and blocks of rock (slide-rock), and which is typical of steep slopes. The failure unit may consist of the *weathered zone, when the form of the unweathered bedrock surface determines the shape of the shear plane. Elsewhere, failure may be defined by major *joints or *bedding planes.

rock : soil ratio (RSR) A measure of the relative proportions of rock and soil. It is a field criterion for separating fresh and weathered rock, which is often difficult to establish, especially in *sedimentary rocks.

rock-stratigraphic unit *See* LITHOSTRATIGRAPHIC UNIT.

rock-structure rating (RSR) The nature and distribution of structural features within a *rock mass which may have a dominant effect on the response of the rock mass to mining. The structure can influence the choice of mining methods and the design of the mine layout.

rock unit *See* LITHOSTRATIGRAPHIC UNIT.

Rodda's flood-peak formula *See* FLOOD-PEAK FORMULAE.

rodding structure A very coarse *lineation of *minerals, rock streaks, lines (striations), and *mullions, which develops in strongly deformed rocks. The rods are cylindrical structures which *strike parallel to the *hinge line.

Rodebaek A glacial *stage in Poland that may be equivalent to the *Amersfoort of the Netherlands. It postdates sediments of *Eemian age.

rodingite A *gabbro or *dolerite which has suffered calcium-*metasomatism to produce a rock consisting of *grossular *garnet and *prehnite, with or without *wollastonite, *diopside, and hydrogrossular.

rogen moraine A field of morainic (*see* MORAINE) ridges that lie at right angles to the direction of a former ice advance. Individual ridges may vary between 10 and 30 m in height, may be more than 1 km in length, and may be 100–300 m apart. They are often linked by cross-ribs. This landscape probably formed beneath a *glacier, but the details of its origin are uncertain. It is named for its fine development around Lake Rogen, Sweden.

roll along A seismic-*reflection field method used to acquire data for a *common-depth-point stack.

rollover The down-warping of the *hanging wall block along a *lystric fault, which occurs when a space opens next to the fault. The resulting structure is usually called a 'rollover anticline'.

rollover anticline *See* ROLLOVER.

roll-type uranium ore C-shaped deposit, most commonly in *sandstones, formed by advancing mineralizing fluids. Deposition takes place where *groundwaters encounter reducing conditions, e.g. where rocks contain organic material. They are usually no more than a few metres thick, and cut across bedding.

Romer, Alfred Sherwood (1894–1973) An American palaeontologist and comparative anatomist, who specialized in the evolution of vertebrates. He obtained his Ph.D. in 1921 from Columbia University, then taught anatomy at the Bellevue Hospital Medical College, New York, and from 1923 to 1934 he was an associate professor at the University of Chicago. In 1934 he was appointed professor of biology at Harvard University, in 1945 director of the biological laboratories, and in 1946 director of the Museum of Comparative Zoology, posts he held until his death.

roof pendant A downward projection of *country rock into a *batholith. The removal of roof rocks by *erosion may leave remnants of roof pendants which resemble *xenoliths in the intruding rock.

roof thrust In a *duplex system, the structurally highest bounding *thrust surface, which joins the *floor thrust at the leading and trailing edges of the duplex.

room and pillar *See* PILLAR AND STALL.

rooted tree A *phylogenetic tree in which the common ancestor is identified, usually by the incorporation of a known outgroup, thus resolving the direction of evolution.

rootlet Strictly, a small plant root, but the term is also commonly used to describe the traces of plant roots preserved in a *fossil *soil or *sediment.

root-mean-square velocity (V_{rms}) The velocity of waves through subsurface media down to the *n*th interface, such that $V_{rms} = (\Sigma V_{int}{}^2 t_i / \Sigma t_i)^{1/2}$, where V_{int} and t_i are the interval velocity and single-travel time through the *i*th interval. V_{rms} is usually derived from normal *moveout measurements from a *T^2–X^2 graph, and is commonly several per cent larger than the corresponding average velocity.

root zone A region from which *nappes appear to be derived, and which has undergone extensive compression. Root zones are characterized by highly flattened and steepened structures.

rope bomb *See* VOLCANIC BOMB.

Rosalind (Uranus XIII) One of the lesser satellites of *Uranus, with a diameter of 29 km. It was discovered in 1986.

rose diagram A circular histogram plot which displays directional data and the frequency of each class. Rose diagrams are commonly used in sedimentary geology to display palaeocurrent data (*see* PALAEOCURRENT ANALYSIS), or the orientation of particles. In structural geology rose diagrams are used to plot the orientation of *joints and *dykes. Wind directions and frequencies can also be plotted on rose diagrams.

rose quartz *See* QUARTZ.

Rosetta An *ESA mission, to be launched in 2003, to comet *Wirtanen.

Rossby waves Named after the Swedish-American meteorologist C.-G. Rossby (1898–1957), Rossby waves are equatorward *troughs and poleward *ridges forming long waves in the circumpolar flow of the upper air, particularly in the mid and upper *troposphere, with a typical wavelength of around 2000 km. Three or four waves usually occur in the circumpolar westerly wind flow over mid latitudes. They may remain stationary (as *standing waves) when wind speed and wavelength have a given relationship. The waves may be initiated by lower winds over mountain barriers, e.g. the Rocky Mountains, or by heating over warm oceans in winter or over land in summer. They are then amplified by vorticity (due to the Earth's rotation) in anticyclonic curvature (in ridges) and in cyclonic curvature (in troughs). Characteristic positions for the main troughs in the upper westerlies over the northern hemisphere are about 70° W and 150° E. The Rossby waves influence the formation of surface *depressions which tend to develop on a *frontal wave ahead of an upper trough. Rossby waves also occur in the oceans.

rostral suture *See* CEPHALIC SUTURE.

rostrum 1. (guard) A massive deposit of fibrous *calcite which makes up part of a belemnite (*Belemnitida) skeleton. **2.** In a brachiopod (*Brachiopoda), a beak-like process formed by the drawing out of the *umbo.

rotary drilling The process of cutting a *borehole in which ground is cut or crushed by a rotating drill system, using a drill *bit turned by means of a *kelly. Engineering and shallow studies usually involve either a mast or an A-frame. Deep drilling is usually from a derrick. *See also* CABLE DRILLING. *Compare* PERCUSSION DRILLING.

rotating-cups anemometer *See* ANEMOMETER.

rotational remanent magnetism (RRM) The magnetization acquired by rotation of a specimen within an alternating magnetic field. The process is unclear, but related to the movement of *magnetic domains with a magnetic field. *See also* GYROREMANENT MAGNETIZATION.

rotational shear *See* SIMPLE SHEAR.

rotational slip (rotational slump) Variety of landslide characterized by movement along a concave-up failure surface. The upper unit of the slump is typically tilted back, and surface water may be retained in the depressed zone. The slump may be caused by basal undercutting, or by the over-steepening of artificial embankments. It is normally found in a uniform, relatively weak material, e.g. *clay.

rotational slump *See* ROTATIONAL SLIP.

rotation of the Earth *See* EARTH ROTATION; and TIDAL FRICTION.

Rotliegende The Lower *Permian *red beds underlying the *Zechstein. Although old, the word is still very much in use, notably in the oil industry, since much of northern Europe's natural gas is obtained from this rock unit.

Rotliegendes The first *epoch of the *Permian Period, dated at 290–256.1 Ma (Harland et al. 1989).

rotor cloud Cloud formed in moist air by condensation in the upper part of an eddy that has been generated beneath the waveform in stable air on the lee side of a mountain barrier. The closed eddy system can result in local reversal of wind direction in the general airstream.

roughness *See* BED ROUGHNESS.

rounded biosparite *See* FOLK LIMESTONE CLASSIFICATION.

roundness index The average radius of curvature of the corners of a particle, divided by the radius of the maximum inscribed circle for a two-dimensional image of the particle, i.e. $(\Sigma r/N)/R$, where Σr is the average radius of curvature at the corners, N is the number of corners, and R is the radius of the largest inscribed circle. In practice, it is used empirically and other techniques are also used. For example, a pebble may be compared with a set of standard silhouettes. The resulting index figure allows inferences to be made about the nature of the depositing process.

Rowe cell Instrument for testing the *consolidation of large *soil or loose *sediment samples. The sample is compressed under hydraulic pressure and drained from below. Pore pressure and change in thickness of the sample can be measured.

RPF *See* RELATIVE POLLEN FREQUENCY.

r-process *See* RAPID-NEUTRON PROCESS.

RQD *See* ROCK-QUALITY DESIGNATION.

RRM *See* ROTATIONAL REMANENT MAGNETISM.

RRR junction A *triple junction formed by the meeting of three *ridges.

r-selection The *natural selection of those organisms that breed in such a way as to maximize their intrinsic rate of increase (*r*) so that when favourable conditions occur (e.g. in a newly formed habitat) the species concerned can rapidly colonize the area. Such an opportunist strategy, based on producing large numbers of seeds, spores, eggs, or offspring most of which perish, is advantageous in rapidly changing environments, as in the early stages of a succession. *Compare* K-SELECTION.

RSR *See* ROCK : SOIL RATIO; and ROCK-STRUCTURE RATING.

RSS Regional stratigraphic scale. *See* STRATIGRAPHIC SCALE.

RST *See* REGRESSIVE SYSTEMS TRACT.

rubble levée *See* LAVA LEVÉE.

rubidium–strontium dating A *radiometric dating method based on the *radioactive decay of ^{87}Rb to ^{87}Sr. Rubidium has two *isotopes (^{85}Rb 72.15%, ^{87}Rb 27.85%), but only ^{87}Rb is radioactive. ^{87}Rb disintegrates in a single step to ^{87}Sr by the emission of a low-energy beta particle (*see* BETA DECAY). Unfortunately this low-energy disintegration makes it very difficult to assess the half-life (*see* DECAY CONSTANT) and two values (5.0×10^{10} years or 4.88×10^{10} years) have been in common use. When a mineral crystallizes, it will usually incorporate both rubidium and strontium *ions and the ratio of Rb to Sr will vary depending on the mineral involved. This initial strontium in the mineral is known as 'common strontium' (*see* INITIAL STRONTIUM RATIO) and is normally in the proportion of ^{88}Sr 82.56%, ^{87}Sr 7.02%, ^{86}Sr 9.86%, and ^{84}Sr 0.56%. Using these proportions it is possible to identify the amount of radiogenic ^{87}Sr present. Originally the above proportions were assumed, but today it is more usual to plot $^{87}Sr:^{86}Sr$ against $^{87}Rb:^{86}Sr$ to produce a straight-line *isochron from which the age of the mineral can be determined. When using the $^{87}Rb:^{87}Sr$ method it is customary to use *whole-rock samples in the analysis, because although ^{87}Sr may leak from one mineral to adjacent minerals over time it usually remains in the system. *Micas and potassium *feldspars are the most suitable minerals for $^{87}Rb:^{87}Sr$ age determinations and the results can commonly be compared with $^{40}K:^{40}Ar$ age determinations from the same sample (*see* POTASSIUM–ARGON DATING). The method has particularly been applied to ancient *metamorphic rocks.

ruby *See* CORUNDUM; and SPINEL.

rudaceous rock General term applied to *sedimentary rocks having a grain size of 2 mm or more.

rudist bivalves An extinct group of coral- and horn-shaped bivalves (*Bivalvia, subclass *Heterodonta). They have a variable *morphology and many species are not easily seen to be bivalves. They are adapted to a sedentary mode of life. Some forms have a larger lower left valve and an upper right valve that forms a flat lid; other forms have a larger lower right valve. The lower valve is often very thick and may have a complex wall structure. There is a tendency to develop a cylindrical and coral-like form. Rudist bivalves are found in *Cretaceous rocks, sometimes forming extensive mounds.

rudite General name given to *rudaceous rocks.

rudstone *See* EMBRY AND CLOVAN CLASSIFICATION.

Rugosa (tetracorals; subclass *Zoantharia) A *Palaeozoic order of solitary and colonial corals which appears in the *Ordovician and becomes extinct in the *Permian. The *corallum contains radial plates (septa) and horizontal plates (tabulae), and sometimes oblique plates (dissepiments). The septa developed in insertion cycles of four and many species retain a degree of *bilateral symmetry. Solitary rugosans seem to have preferred soft substrates since they had no means of cementation and colonial forms seem to have relied for stability on the weight of the skeleton.

rugose Applied to a shell that has a rough or wrinkled texture. The term is commonly used to describe the appearance of the *epitheca of some corals (e.g. *Rugosa), but is also applied to any shell texture where the surface is raised in a wrinkled manner to produce ridges. The word is derived from the Latin *ruga*, meaning wrinkle.

rule of Vs A rule which describes the relationship between the *attitude of a bed and its *outcrop pattern for a given topography. In summary, if the outcrop pattern of a bed forms a V-shape pointing down-gradient, that is the direction in which the bed *dips. The rule is most usefully applied when studying maps on which outcrop patterns show moderately dipping beds traversing valleys.

Rumford, Count *See* THOMPSON, BENJAMIN.

Runangan A *stage in the Lower *Tertiary of New Zealand, underlain by the *Kaiatan, overlain by the *Whaingaroan, and roughly contemporaneous with the *Priabonian Stage.

Runcorn, Stanley Keith (1922–95) A geophysicist and head of the School of Physics at the University of Newcastle, Runcorn was best known for his studies of *polar-wander curves, but he also did work on convection currents in the *mantle. In 1959 he was able to show that the curves for N. America and Europe could be reconciled if it were assumed that the Atlantic Ocean had once been closed.

runnel *See* RIDGE AND RUNNEL.

rupes A term for a straight escarpment. The classic example is Rupes Recta, the 'Straight Wall' in Mare Nubium on the *Moon.

Rupelian (Stampian) 1. The first *age in the *Oligocene Epoch, preceded by the *Priabonian (Late *Eocene), followed by the *Chattian, and dated at 35.4–29.3 Ma (Harland et al., 1989). **2.** The name of the corresponding European *stage, which is roughly contemporaneous with the lower *Zemorrian (N. America), most of the *Whaingaroan (New Zealand), and the upper *Aldingan and lower *Janjukian (Australia).

Rusophycus An *ichnogenus of *cubichnia that are believed to have been formed beneath a thin layer of sand.

Russian borer *See* PEAT-BORER.

Rutherford, Ernest (1871–1937) A New Zealand-born physicist, Rutherford worked at the Universities of Cambridge and Manchester, and at McGill University in Canada. He made important studies of radiation and the structure of the atom, developing the concept of a *half-life for the decay of radioisotopes. This concept led directly to the techniques of *radiometric dating.

rutile An *accessory mineral, TiO_2; iron, niobium, and tantalum may also be present; sp. gr. 4.2–5.6; *hardness 6.0–6.5; *tetragonal; normally reddish-brown, but also yellowish-red, or black; pale brown *streak; adamantine *lustre; crystals normally square *prisms terminated by pyramids; *cleavage prismatic {110} and {100}; occurs in a variety of *igneous rocks, *schists, *gneisses, and *metamorphic *limestones and *quartzites, and becomes concentrated in *alluvial deposits and beach sands. It is an important *ore mineral for titanium and its compounds.

ruware (rock pavement) An area of bare rock with a slightly domed profile that locally *outcrops at the surface of a tropical plain. It is formed when the *weathered profile is stripped from a sound rock surface, and may be seen as the first stage in the emergence of a *dome or *tor.

Ryazanian A *stage in the British Lower *Cretaceous. It is roughly of *Berriasian age and contains the uppermost Purbeck Beds.

Ryukyu Trench The oceanic *trench which forms the boundary between the oceanic *Philippine Plate and the *continental crust of the *Eurasian Plate. The Philippine Plate is subducting obliquely under the Eurasian Plate.

S *See* SIEMENS.

Saalian A northern European glacial *stage dating from about 0.25 to 0.1 Ma that may be equivalent to the *Wolstonian of the East Anglian succession.

sabkha Wide area of coastal flats bordering a *lagoon, where *evaporites, dominated by carbonate–sulphate deposits, are formed. It is named after such an area on the Trucial coast of Arabia.

saccate Applied to *spores and *pollens where a separation of the *exine layers produces an air sac (saccus).

Saccominopsis Known from the *Ordovician, this genus is thought to be the first representative of the very important *protistan group the *fusilinids. *See also* FORAMINIFERIDA.

saccus *See* SACCATE.

saddle *See* SUTURE.

Saffir/Simpson Hurricane Scale A standard scale, introduced in 1955 by meteorologists of the US Weather Bureau, for reporting tropical cyclones. It adds a further five categories to the *Beaufort scale, and includes the surface atmospheric pressure at the centre of the low pressure system and the size of the *storm surge it causes. *See* APPENDIX C.

sag and swell topography *See* KNOB AND KETTLE.

St David's A *series (536–517.2 Ma ago) of the Middle *Cambrian, underlain by the *Caerfai and overlain by the *Merioneth.

Sakigake A Japanese *ISAS mission to comet *Halley, launched in 1985.

Sakmarian 1. An *age in the Early *Permian Epoch, preceded by the *Asselian, followed by the *Artkinsian, and dated at 281.5–268.8 Ma (Harland et al., 1989). Originally it was regarded as the basal Permian age, incorporating the Asselian Age, from which it is now separated. **2.** The name of the corresponding eastern European *stage, which is roughly contemporaneous with parts of the *Rotliegende (western Europe), the upper *Wolfcampian (N. America), and the upper Somoholoan (New Zealand).

salic Applied to the *s*ilicon- and *al*uminium-bearing *CIPW normative components of an *igneous rock. Salic components include the normative *minerals *quartz (Q), *corundum (C), *orthoclase (Or), *albite (Ab), *anorthite (An), *nepheline (Ne), *leucite (Lc), and *kaliophilite (Kp).

salic horizon *Soil horizon, usually below the surface, containing not less than 2% salt and with a figure of 60 or more for the value calculated as the thickness of the horizon in centimetres multiplied by the percentage of salt. It is a *diagnostic horizon.

salina 1. A salt flat or similar place where salt deposits form by evaporation or are found. **2. (solar pond)** A highly saline pond or other body of water where salt crystals are produced by evaporation. The pond may be natural or artificial. **3.** *See* PLAYA.

salination *See* SALINIZATION.

saline giant A thick and extensive salt deposit, produced by the evaporation of a large hypersaline sea. One example of a saline giant is the *Miocene *evaporites of the Mediterranean, which formed by the repeated evaporation of the Mediterranean Sea. Another is the *Permian-aged Zechstein salts of north-western Europe (*see* ZECHSTEIN SEA), which formed as a result of the repeated evaporation of a partially barred marine basin which covered more than 250 000 km^2.

saline-sodic soil Soil that contains more than 15% exchangeable sodium, a saturation extract with a conductivity of more than 0.4 siemens per metre at 25 °C, and in the saturated soil it usually has a *pH of 8.5 or less. Either high concentration of salts or high pH, or both, interfere with the growth of most plants.

saline soil Soil that contains enough soluble salt to reduce its fertility. The lower limit is usually defined as 0.4 siemens per metre.

salinity Measure of the total quantity of dissolved solids in sea water in parts per thousand by weight when all the carbonate has been converted to oxide, the bromide and iodide to chloride, and all the organic matter is completely oxidized. Ocean-water salinity varies in the range 33–8 parts per thousand, with an average of 35 parts per thousand.

salinization (US salination) The process of accumulating soluble salts in soil, usually by an upward capillary movement from a saline *groundwater source.

SALR *See* SATURATED ADIABATIC LAPSE RATE.

salt The product, with water, of the reaction between an *acid and a *base.

saltation Major process of particle transport in either air or water, which involves an initial steep lift followed by travel and then a gentle descent to the bed. An essential requirement for the process is *turbulent flow that can lift particles into the zone of relatively high downstream (downwind) velocity.

salt dome *See* DOME.

salt-dome trap Salt *diapir which has pushed up existing *sediments into a dome structure, and which may result in the trapping of gas, oil, or water in the *pores of the permeable rocks adjacent to and above the salt dome if a suitable *cap rock is present. The rocks ahead of the salt diapir are often severely faulted and may give rise to *fault traps. Oil may also accumulate in the porous top of the salt diapir.

salt fingering A suggested mixing process between layers of saline and less saline water in the ocean. Where warm, saline water overlies cooler, less saline water, e.g. where the saline Mediterranean water flows out into the less saline Atlantic, mixing by this process is thought to take place. The vertical water movements between the water masses of different *salinity occur in small columns or fingers a few milimetres across. These fingers penetrate only a small distance, producing a mixed layer. The mixing process may then be repeated at the two interfaces that are present, and a number of layers may develop.

salt flat Extensive flat surface, found in hot deserts, consisting of salts that have accumulated in a shallow saline lake or *playa. Evaporation then produces a crust of varying hardness.

salt lake A lake with a concentration of mineral salts typically of the order of 100 parts per thousand or greater, dominated by dissolved chlorides, e.g. the Dead Sea, which contains 64 parts per thousand NaCl and 164 parts per thousand $MgCl_2$.

salt marsh Vegetation often found on mud banks formed at river mouths, showing regular zonation reflecting the length of time different areas are inundated by tides. Sea water has a high salt content which produces problems of osmotic pressure for the vegetation, so only plants that are adapted to this environment (halophytes) can survive there.

salt pan A basin in a semi-arid region where chemical precipitates (*evaporites) are deposited due to the concentration by evaporation of natural solutions of salts. The least soluble salts (calcium and magnesium carbonates) precipitate first, on the outside of the pan, followed by sodium and potassium sulphates. Finally, in the centre, sodium and potassium chlorides and magnesium sulphate are deposited. This pattern, slightly distorted through tilting, is seen in Death Valley, California, USA.

saltpetre *See* NITRE.

salt wedge An intrusion of sea water into a tidal *estuary in the form of a wedge along the bed of the estuary. The lighter fresh water from riverine sources overrides the denser salt water from marine sources unless mixing of the water masses is caused by estuarine topography. Salt wedges are found in estuaries where a river discharges through a relatively narrow channel.

samarium–neodymium dating The change from ^{147}Sm to ^{143}Nd is the result of *alpha decay (^{147}Sm has a *half-life of 2.5×10^{11} years) and provides one of the newest methods of dating used in *geochronology. The decay enriches the material in ^{143}Nd relative to the stable isotope ^{144}Nd. The ratio ^{143}Nd : ^{144}Nd that is measured is highly resistant to secondary processes of alteration and *metamorphism. It can be used on terrestrial and extraterrestrial materials and gives valuable secondary information on *petrogenesis in the *crust and *mantle.

sampling frequency (station frequency) The frequency at which a data set is sampled is determined by the number of sampling points per unit distance or unit time, and the sampling frequency is equal to the number of samples (or stations) divided by the record (or traverse) length. For example, if a wave-form is sampled 1000 times in one second the sampling frequency is 1 kHz (and the *Nyquist frequency 500 Hz); if a traverse is 500 m long with 50 stations, the sampling frequency is one per 10 m.

sampling interval (station interval) The distance between points at which measurements are taken or the time which elapses between measurements; it is equal to the traverse (or record) length divided by the number of stations (or samples). For example, a 250 m ground traverse with 25 stations along it has a sampling interval of 10 m; a wave-form might be sampled every two milliseconds, i.e. with a sampling interval of 2 ms (and a *sampling frequency of 500 Hz).

sampling methods Techniques for collecting representative sub-volumes from a large volume of geologic material. The particular sampling method employed depends on the nature of the material being sampled and the kind of information required. Common methods are: (*a*) random sampling, a non-systematic or haphazard distribution of sampling locations; (*b*) systematic sampling, a regularly spaced distribution of sampling locations; (*c*) stratified sampling, a sequentially spaced, vertical distribution of sampling locations based on sampling each *stratigraphic unit in a succession; (*d*) grab sampling, the collection of material using an externally controlled, mechanical grab (a method often used when sampling the sea floor or the surface of another planet); and (*e*) chip sampling, the collection of *borehole drilling chips that are carried to the surface by lubricating mud and are representative of the material traversed by the borehole.

sand 1. In the commonly used Udden–Wentworth scale, particles between 62.5 and 2000 μm. Other classifications exist (*see* PARTICLE SIZE). In pedology, sand is defined as mineral particles of diameter 2.0–0.02 mm in the international system, and as 2.0–0.5 mm diameter particles in the *USDA (American) system. **2.** A class of soil texture. **3.** *See* SHARP SAND.

sandbody A finite unit of *sand (or *sandstone) that usually accumulated in response to one type of depositional process (e.g. as a *channel, beach-bar, or *barrier system). The distribution of the sand and the three-dimensional geometry (*architecture) is controlled largely by the nature of the depositional regime under which it accumulated (i.e. channel sands may be sinuous, and beach or shore-bar sands may be linear and shore-parallel, etc.).

sand line The line on an *electrical log marking the usual *apparent resistivity of a clean *sand.

sand ribbon Longitudinal strip of *sand up to 15 km long, 200 m wide and less than 1 m thick, standing on, and surrounded by, an immobile gravel floor. Sand ribbons are developed on the sea floor of the *continental shelf where there are a paucity of sand, water depths of 20–100 m, and fast-flowing currents. *See* RIBBONS.

sand sheet *See* SHEET SAND.

sandstone (arenite) *Sedimentary rock type, formed of a lithified *sand, comprising grains between 63 μm and 1000 μm in size, bound together with a mud *matrix and a mineral *cement formed during burial *diagenesis. The main constituents are *quartz, *feldspar, *mica, and general rock particles, although the proportions of these may vary widely.

sandstone dyke *See* NEPTUNIAN DYKE.

sandstorm Phenomenon in which *sand and dust particles are uplifted, often to great altitude, by turbulent winds. Visibility is greatly reduced.

sandur (pl. sandar) *See* OUTWASH PLAIN.

sand volcano A conical body of *sand, resembling the form of a small volcano, rarely more than a few metres wide and less than 50 cm high. Internally the sand volcano consists of a massive central plug, surrounded by laminated sand paralleling the external form. Sand volcanoes are formed by the extrusion of liquefied sand through a local vent at the surface. The extrusion usually results from a highly liquefied sand below a confining surface layer.

sand wave Large-scale, transverse ridge of *sand, characteristic of *continental-shelf areas such as the southern North Sea. The external morphology is identical to that of the smaller-scale ripple-and-*dune bedform (megaripple). The wavelength or spacing of sand-wave crests is 30–500 m and the height is 3–15 m. The down-current migration of sand waves leads to the formation of large-scale cross-bedding.

Sangamonian The third (0.55–0.001 Ma) of four *interglacial *stages recognized in mid-continental N. America. It followed the *Illinoian glacial episode and is the approximate equivalent of the *Riss/Würm Interglacial of the Alps. Early warm climates and later cooling climates are represented in well-exposed *pollen spectra.

sanidine *See* ALKALI FELDSPAR.

Santernian *See* QUATERNARY.

Santonian A *stage in the European Upper *Cretaceous (86.6–83 Ma, Harland et al., 1989), for which the *stratotype is at Saintes, France, *See also* SENONIAN.

sapphire *See* CORUNDUM.

saprolite Chemically rotted rock *in situ*. The term is often applied to the lower portion of a *weathering profile. The saprolite on *granite is locally called 'grus' or 'growan', although the latter term may include material broken down by mechanical weathering. *Compare* REGOLITH.

sapropel Organic ooze or sludge accumulated in *anaerobic conditions in shallow lakes, swamps, or on the sea bed. It contains more *hydrocarbons than *peat; when dry it is dull, dark, and tough; it may be a source of oil and gas.

sapropelic coal *See* COAL.

sapropelite A sapropelic *coal, consisting of organic material, particularly *algae, which accumulated in stagnant lake bottoms or the floors of *anoxic shallow seas.

SAR *See* SODIUM ABSORPTION RATIO.

Sarcopterygii Subclass of fleshy-finned fish comprising two main groups (superorders): the coelacanths (*Crossopterygii), and the lungfish (*Dipnoi). The Crossopterygii appeared in the *Devonian, were widely distributed in the *Mesozoic, and are now represented by a single genus, *Latimeria*. One order of the Crossopterygii, the *Rhipidistia, is credited with being the root stock for more advanced vertebrates. The Dipnoi also appeared in the Devonian and are now represented by three species. They inhabit fresh water and have developed organs by which they can breathe atmospheric air.

sardonyx *See* ONYX.

Sargasso Sea Calm centre of the anticyclonic *gyre in the N. *Atlantic. This large eddy of surface water has boundaries demarcated by major current systems such as the *Gulf Stream, *Canary Current, and *North Atlantic Drift. The Sargasso Sea is a large, warm (18 °C), saline (36.5–37.0 parts per thousand) lens of water, which is characterized by an abundance of floating brown seaweed (*Sargassum*).

sarl A *pelagic or *hemipelagic sediment (an *arl), typically found interbedded with purer oozes in beds up to 1.5 m thick, with a composition intermediate between a nonbiogenic sediment and a calcareous or siliceous ooze. It is 30% clay and 70% microfossils, at least 15% of its volume being calcareous microfossils. *Compare* MARL; and SMARL.

saros unit The term for the period of 18 years (inclusive of leap years) and 10.3 days in which the *Earth, *Moon, and *Sun return to the same relative positions, so that solar and lunar *eclipses repeat themselves. Thus the solar eclipse of 30 June 1973, was repeated on 11 July 1991. Each saros period contains about 43 solar and 28 lunar eclipses.

sarsen stone Boulder or block of *sediment consisting of a siliceous *cement binding one of a range of materials, and found widely in south-eastern England, especially in the chalk country. Typical cemented components include angular *quartz grains and rounded to angular *flints (the 'Hertfordshire puddingstone'). Sarsens are of early *Tertiary age, and probably formed as a *duricrust under tropical conditions.

Sartan *See* VALDAYAN/ZYRYANKA.

satellite A minor body orbiting a planet in the *solar system. About 60 are known, divided into three general classes. (*a*) Regular satellites form miniature solar systems and include all the classical major satellites, e.g. the *Galilean satellites. (*b*) Collisional shards are tiny, craggy chunks, probably remnants of larger satellites, e.g. *Amalthea, which is embedded in *Jupiter's planetary ring system. (*c*) Irregular satellites have elongate, highly inclined orbits, mostly far from the planet, suggestive of capture, e.g. the outer satellites of Jupiter. Three bodies, the Earth's *Moon, *Triton (orbiting *Neptune), and *Charon (orbiting *Pluto) do not fit into any of the above classes, and each has to be regarded as a unique case.

satellite photography The production of photographic images, using the visible, infrared, thermal infrared, and other wavebands, by means of satellite-based cameras or sensors. Such images are extensively used in the environmental sciences for the study of the oceans, the atmosphere, and land masses.

satellite sounding Remote sensing of atmospheric properties by an orbiting Earth satellite. For example, indirect sensing is carried out by spectral analysis of outgoing long-wave radiation with infrared photography of clouds and determination of cloud-top height from the temperature variations.

satin spar *See* GYPSUM.

saturated *See* SILICA SATURATION.

saturated adiabatic lapse rate (SALR) The *adiabatic cooling rate of a rising *parcel of air which is saturated (*see* SATURATED AIR), and in which condensation is taking place as it rises, so that the energy release of the latent heat of vaporization moderates the adiabatic cooling. The reduction of the rate of cooling below the *dry adiabatic lapse rate of 9.8 °C/km varies with temperature. This results from the greater energy release by condensation from air at higher temperatures. Thus at a given atmospheric pressure, air at 20 °C may have an SALR as low as 4 °C/km, whereas at –40 °C the SALR may be close to 9 °C/km. The *stability or *instability of the atmosphere at any given time for vertical motion is determined by whether the *environmental lapse rate of temperature within it is less than or greater than the adiabatic lapse rate (i.e. less than or greater than the rate of decrease of temperature of rising parcels of air).

saturated air Air that contains the maximum amount of water vapour that is possible at the given temperature and pressure, i.e. air in which the *relative humidity is 100%.

saturated flow The movement of water through a soil that is temporarily saturated. Most of the loosely held water moves downward, and some moves more slowly laterally.

saturation In *remote sensing: **1.** the maximum digital number value which can be assigned to a *pixel; **2.** a point between the *achromatic line and a pure hue of a *pixel colour corresponding to the relative mixture of hues going to make up a colour.

saturation deficit At a given temperature, the difference between the actual vapour pressure of moist air and the saturation vapour pressure.

saturation magnetization and moment (M_{sat}) The maximum remanent magnetization (*see* REMANENT MAGNETISM) that a material can acquire after being placed in a direct magnetic field. If uncorrected for either volume or weight, the observed magnetization is the saturation magnetic moment.

saturation moisture content (SMC) The maximum amount of water that can be contained in a rock, when all *pore spaces are filled with water; it is expressed as the percentage of the dry weight of the rock.

Saturn The sixth planet in the *solar system, distant 9.52 AU from the *Sun. Its radius is 60 000 km, density 704 kg/m^3, mass 95 × Earth mass, volume 833 × Earth volume, and it has an equatorial inclination to the *ecliptic of 29°. An outer zone of hydrogen and helium is underlain by a zone of metallic hydrogen, around an ice–silicate core. It has 17 known *satellites and is famous for its ring system.

saturnian satellites *See* ATLAS (SATURN XV); CALYPSO (SATURN XIV); DIONE (SATURN IV); ENCELADUS (SATURN II); EPIMETHEUS (SATURN XI); HELENE (SATURN XIII); HYPERION (SATURN VII); IAPETUS (SATURN VIII); JANUS (SATURN X); MIMAS (SATURN I); PAN (SATURN XVIII); PANDORA (SATURN XVII); PHOEBE (SATURN IX); PROMETHEUS (SATURN XVI); RHEA (SATURN V); TELESTO (SATURN XII); TETHYS (SATURN III); and TITAN (SATURN VI).

Saucesian A *stage in the Upper *Tertiary of the west coast of N. America, underlain by the *Zemorrian, overlain by the *Relizian, and roughly contemporaneous with the upper *Aquitanian and *Burdigalian Stages.

saurian Of, or resembling a lizard. Loosely refers to lizard-like animals, and applied to *fossils, life habits, etc. of the extinct *reptiles.

Saurischia The 'lizard-hipped' *dinosaurs, one of the two dinosaur orders. They included bipedal carnivores (*Theropoda) and herbivorous *tetrapods (*Sauropoda). The theropods produced the largest known terrestrial carnivores, and the sauropods yielded the largest known land animals.

saurischian dinosaur *See* SAURISCHIA.

Sauropoda *Jurassic and *Cretaceous quadrupedal *dinosaurs of herbivorous habit. They included *Diplodocus*, from the Upper Jurassic, one skeleton of which, measuring 26.6 m, at one time made *Diplodocus* the longest terrestrial animal known, but larger sauropods have since been discovered. *Brachiosaurus*, from the Late Jurassic of the USA and Tanzania, is estimated to have weighed 80 t, and *Supersaurus* is estimated to have been 15 m tall and possibly 30 m long. **Apatosaurus* (*Brontosaurus*) was also a sauropod, and although shorter than *Diplodocus* its skeleton was more massively built.

Sauropterygia (subclass *Archosauria) An extinct reptilian order, that comprises the two suborders *Plesiosauria (plesiosaurs) and *Nothosauria (nothosaurs).

Saussure, Horace Bénédict de (1740–99) A Swiss naturalist, Saussure made an extensive study of the structure of the Alps, described in the four volumes of *Voyages dans les Alpes* (1779–96). His theory was *neptunian, but with *uniformitarian overtones.

saussuritization The complete or partial alteration of calcium-rich *plagioclase to a fine-grained aggregate of secondary, sodic-rich plagioclase, *epidote, *muscovite, *calcite, *scapolite, and *zeolites. The process commonly takes place during the low-grade *regional metamorphism of *gabbros and *basalts, both of which contain plagioclase as an *essential component.

Saxonian The European stratigraphic term that is equivalent to the Upper *Rotliegende and Weissliegende. *See also* AUTUNIAN.

S-band The radar frequency band between 1550 and 5200 MHz.

scalenohedron A *crystal form consisting of a number (usually 6 or 12) of triangular faces, all with unequal sides. Its *crystal symmetry is either *tetragonal or, more frequently, *hexagonal. Each face cuts the vertical *c* (or *z*) axis and the form is often developed in the mineral *calcite, where it is called 'dog-tooth spar'.

scandent In a graptoloid graptolite (*Graptoloidea), applied to the condition where the *stipes are united back to back with the nema in between.

Scandinavian ice sheet An *ice-cap which developed over Scandinavia during the *Quaternary. It has been suggested, largely on the basis of the degree of downwarp and the partial recovery of the land surface in Scandinavia and the surrounding areas, that the ice was 2600 m thick.

scanning electron microscope (SEM) A microscope that operates by scanning a finely focused beam of *electrons across the specimen. The reflected electron intensity is measured and displayed on a cathode-ray screen to produce an image. The SEM enables magnifications of up to 100 000 times to be made and provides a much better depth of field than a conventional light microscope (which suffers from focus limitations), making the three-dimensional structure of small objects (e.g. *Foraminiferida) spectacularly visible. In geology it is used extensively for *micropalaeontology, diagenetic studies (*see* DIAGENESIS), and *grain textural examination. When coupled with an *electron probe, semi-quantitative determinations of grain chemistry can be made.

scapolite A member of the *feldspathoids, of composition $(Na,Ca,K)_4[Al_3(AlSi)_3Si_6O_{24}](Cl,SO_4,CO_3,OH)$, forming a *solid solution series between the two *end-members marialite (Na and Cl) and meionite (Ca and CO_3); sp. gr. 2.5 (mar) to 2.7 (me); *hardness 5–6; white or pale bluish or greenish; small *prisms or *massive; occurs in some *pegmatites replacing *quartz or *plagioclase, but mainly in *metamorphic or metasomatic rocks (*see* METASOMATISM). Scapolite is found in association with *sphene, *grossular, *diopside, and *epidote.

scar Steep, cliff-like slope of bare rock, developed in the near-horizontally bedded *Carboniferous *limestone of the Yorkshire Dales, England. The steepest and highest scars are normally associated with the *outcrop of the purest and most massively bedded limestone. Often a *scree is formed at the base.

scarp (abbr. of escarpment) Steep slope or cliff found at the margin of a flat or gently sloping area. Many varieties are recognized, and distinguished in terms of origin. A 'fault scarp' results when a *fault displaces the ground surface so that one side stands high. A 'fault-line scarp' is produced by *erosion on one side of an ancient fault: *obsequent and *resequent varieties are recognized. A 'composite fault-line scarp' results from a combination of erosion and faulting. Erosional scarps result from vertical incision, or from the headward enlargement of *pediments.

scarp-and-vale topography A landscape consisting of a roughly parallel sequence of *cuestas (*scarps and *dip slopes) and intervening valleys ('vales'). It is typically found on *uniclinal (homoclinal) structures whose beds show differing *lithological composition and consequently varied resistance to *denudation. It dominates most of lowland Britain, which is characterized by *Mesozoic *sediments dipping gently towards the east and south-east.

scarp-foot knick An abrupt change in gradient that often occurs in semi-arid environments between a *pediment and the adjacent *scarp. It is the boundary between the zones of *scarp retreat and pedimentation. It may be *joint controlled, when it is often so abrupt that a boot may barely be placed in it.

scarp, lobate A fault *scarp characterized by lobes. They are common on the surface of *Mercury, where they are relatively steep, with crest heights from 0.5 to 3 km, lengths from 20 to 500 km, with a broadly lobate outline on a scale of a few to tens of kilometres. They are probably due to *reverse or *thrust faulting resulting from compressive stress. They predate the close of the massive bombardment, and so are probably more than four billion years old. If so, they record an early period of planetary contraction.

scarp retreat The recession of the relatively steep hillslope that terminates a *butte, *mesa, *cuesta, or any elevated, plateau-like surface. Several geomorphological *processes may be involved, including undercutting by an adjacent stream, *spring sapping, *mass movement, *rainwash, and *weathering. Under semi-arid conditions it may give rise to a hillslope that retreats parallel to itself.

scarp slope The relatively steep face of a *scarp. Its steepness is maintained by the *erosion of a relatively weak *stratum that typically underlies the resistant *cap rock that maintains the form of the scarp. Erosion may be achieved by *spring sapping, *sheetwash, and *mass wasting.

scatter diagram In statistics, the diagram obtained when two sets of observations are plotted against each other. Scatter diagrams are usually employed to visualize any correlations that may occur between two sets of observations.

scattering Diffusion of incident radiation by atmospheric particles, e.g. by haze and water droplets as well as by molecules. Such diffused radiation is often refracted many times in passing through the atmosphere. *See also* MIE SCATTERING; and RAYLEIGH SCATTERING.

scavenging The capture and removal by rain or snow of particulate matter in the atmosphere.

scheelite Mineral, $CaWO_4$; sp. gr. 5.9–6.1; *hardness 4.5–5.0; *tetragonal; white, or sometimes shades of yellow, green, brown, or red; white *streak; *vitreous *lustre; *crystals usually *bipyramidal, also occurs *massive or granular; *cleavage good {111}; *fluorescent; occurs in *pegmatite *veins and high-temperature mineral veins, often in association with *wolframite, *cassiterite, *molybdenite, *fluorite, and *topaz, also in contact *metamorphic zones together with *axinite, *garnet, *wollastonite, and metamorphic calcium minerals. It is one of the chief ores of tungsten, and named after the 18th-century Swedish chemist K. W. Scheele.

Scheuchzer, Johann Jacob (1672–1733) A Swiss mathematician and physician, Scheuchzer is best known for his work on fossil fish and plants which he attributed to a universal deluge. He is notorious for finding, in 1726, the fossil remains of *Homo Diluvii Testis* (the man who witnessed the deluge), which was taken as important evidence for the Mosaic Flood, but was later shown by *Cuvier to be a giant salamander.

schiller A metallic, bronze-coloured *lustre, seen for example in moonstone (*see* ALKALI FELDSPAR).

Schindewolf, Otto H. (1896–1971) A German palaeontologist at the University of Tübingen, Schindewolf made extensive studies of ammonites (*Ammonoidea) and their evolutionary history. These studies led him to propose a concept of discontinuous evolution, a theory which is not now generally accepted.

schist A *regional metamorphic rock of pelitic (*see* PELITE) composition which displays a *schistosity. Schists are coarser-grained than *phyllites, having a grain size greater than 1 mm. The minerals defining the schistosity may be *muscovite *mica, *biotite mica, and/or elongate *quartz, depending on the composition and the pressure and temperature of formation. When a *basic *igneous rock is metamorphosed it forms a *hornblende-schist (*amphibolite) or *greenschist if it contains a planar *fabric, or a *greenstone if no fabric is present. Thus in this latter context 'schist' refers to the fabric component and not to the overall rock type.

schistosity The planar alignment of platy *micas and elongate *amphiboles in a *regional metamorphic rock. The alignment of these minerals into a planar *fabric is caused by (*a*) the physical rotation of the mineral grains under the influence of a *shear stress, or (*b*) the syntectonic, metamorphic growth of new minerals with their long axes perpendicular to the principal compressive stress direction.

schizochroal *See* TRILOBITE EYE.

schizodont Applied to a type of *hinge *dentition, found in certain members of the pelecypod order Trigonioida, in which the teeth are large and possess parallel ridges at right angles to the axis of the teeth. The left *valve bears a single tooth.

Schizomycophyta A name formerly given to the *bacteria. *See* EUBACTERIA.

schlieren An *igneous *fabric consisting of streaked-out, linear, planar, or discoidal aggregates of *mafic minerals more densely concentrated than in the host rock, with which they have diffused boundaries. Some schlieren may represent partly remobilized, mafic-rich *xenoliths, streaked out by the flow of the enclosing *magma.

Schlumberger array An *electrode configuration in which the spacing of the two *potential electrodes is less than one-fifth of the distance between the centre of the *array and one *current electrode. *See also* GEOMETRIC FACTOR.

Schmidt hammer An instrument for measuring the compressive strength of a surface by hammering it with a spring-loaded metal piston. It is used to test rock hardness and abrasiveness before drilling.

Schmidt hammer test In *geomorphology, a technique for comparing the surface toughness or hardness of rocks. For example, it has been used to compare the mechanical strength of the surface layers of weathered (*see* WEATHERING) *limestones and *sandstones with that of the unaltered rock beneath. *See* SCHMIDT HAMMER.

Schmidt–Lambert net *See* EQUAL-AREA NET.

schorl A black, opaque variety of *tourmaline which may occur as radiating, needle-like crystals, as seen, for example, in the *granites of Cornwall, Britain.

Schroeder Van Der Kolk method *See* HALF-SHADOW TEST.

schuppen structure *See* IMBRICATE STRUCTURE.

Schwassmann–Wachmann 3 A *comet with an orbital period of 5.35 years; *perihelion date 2 June 2006; perihelion distance 0.933 AU.

scintillation counter (scintillometer) An instrument which measures gamma radiation and is extensively used in airborne and ground radiometric surveys. It utilizes the flash of light emitted when the atoms of a suitable 'phosphor' (e.g. a large sodium iodide crystal 'doped' with thallium) are energized by *gamma rays. The scintillations are detected by the light-sensitive cathode of a photomultiplier tube and are converted by the succession of electrodes in the tube into a stream of electrons which are collected and recorded on a meter. The scintillation counter has now been developed into the gamma-ray spectrometer (*see* GAMMA-RAY SPECTROMETRY) for portable and airborne use. It analyses the complex gamma-ray spectrum of uranium, thorium, and potassium, and indicates the relative gamma-ray contribution of each element to ground gamma-ray emission on a continuous readout.

scintillometer *See* SCINTILLATION COUNTER.

scintillometer survey A geophysical prospecting method using a scintillation detector in which radioactivity causes crystals to emit flashes of light which can be recorded by a photomultiplier tube. It is much more sensitive than methods using a *Geiger counter and can distinguish different types of radiation.

scirocco (sirocco) Regional term for one of the genera of warm winds from south of the Mediterranean. It moves ahead of an eastward-travelling depression and brings hot, dry, dusty conditions to Algeria and the Levant. To the north, where its humidity increases very rapidly as it crosses the sea, it brings moist air to the coast of Europe.

Scleractinia (Madreporaria; stony corals; subclass *Zoantharia) Order of solitary or, more commonly, colonial corals, which always possess an external calcareous skeleton consisting essentially of radial partitions (septa, *see* SEPTUM). Septa develop following the pattern of the radial infoldings in the body wall (*mesenteries) in cycles of 6, 12, 24, 48, etc. The order first appeared in the Middle *Triassic.

sclerotinite *See* COAL MACERAL.

scolecodont The pharyngeal jaws or maxillae of annelid worms (*Annelida), commonly found in paired assemblages. They are usually black and chitinous and are known from most geologic *systems.

scolecoid *See* SOLITARY CORALS.

Scolicia An *ichnoguild of structures made by motile, chemosymbiotic organisms (*see* CHEMOSYMBIOSIS) found in *Tertiary and *Quaternary sands and muddy sands. They were inhabited by *Echinocardium cordatum*, a species of heart urchins, remains of which have been found within them.

scoria Loose, rubbly, basaltic ejecta that accumulate around *Strombolian eruptive volcanic vents, eventually building up as a *scoria cone, whose height may range from a few tens of metres to up to 300 m, and whose slope is determined by the *angle of repose of the loose material. The scoria *clasts range widely in size, and have a light, frothy texture, being full of *vesicles. They are mainly drab grey in colour, although when fresh they may be iridescent, but often the scoria oxidizes by reaction with steam escaping from the vent, when it becomes a deep reddish brown.

scoriaceous Applied to a vesicular *lava or *pyroclastic rock, to describe a frothy, bubbly texture. *See* SCORIA; and VESICLE.

scoria cone (cinder cone, ash cone) Volcanic cone built of *pyroclastic material (cinders and *scoriaceous ejecta) usually of *basaltic or *andesitic composition, dominated by fragments 2–64 mm diameter. These air-fall deposits build a straight-sided cone with slopes of about 30° (the *angle of repose). Complex varieties may occur, including breached or gaping, and cone-within-cone types.

scorpions *See* CHELICERATA.

scour and fill A *sedimentary structure characterized by a concave-upwards *erosion surface cut into the underlying bed by a high-velocity *flow of water, and filled by a *sediment, which is usually coarse, during the waning stage of the flow that cut the scour.

Scourian A sub-*stage of the *Lewisian, from about 2600 to 2300 Ma ago, characterized by Scourian *dykes and *gneiss formation, and named after Scourie, north-west Scotland. *See also* INVERIAN.

Scourian orogeny A mountain-building episode that occurred about 2600 Ma ago, during the *Archaean, prior to the *Laxfordian orogeny but possibly representing its first stages. It is marked by NW–SE trending folds immediately to the south of the Laxford area, affecting part of the Lewisian *gneisses in what is now the extreme north-west of Scotland.

scour lag A coarse-grained *sediment deposited immediately above a scour surface (*see* SCOUR AND FILL).

scree Accumulation of coarse rock debris that rests against the base of an inland cliff. It is added to by the *weathering and release of fragments from the cliff face. Screes are widely found in upland areas that are affected by past or present *periglacial conditions, and in hot, rocky deserts.

screw dislocation *See* DISLOCATION.

scroll bar One of a series of long ridges of sand on a point *bar, lying approximately parallel to the contours.

ScS-wave *See* SEISMIC-WAVE MODES.

scud Popular term for *stratus fractus* (fractostratus), a fragmented low cloud moving quickly beneath rain clouds.

scute In fish, an enlarged, bony, dermal plate or scale.

Scyphozoa (jellyfish; phylum *Cnidaria) Class of marine, mainly *pelagic medusoids, usually with four-part *radial symmetry, in which the polyp stage is reduced or absent. Their fossil record is in general scanty, owing to the absence of hard parts, but jellyfish formed an important component of the *Precambrian *Ediacaran fauna. *See also* *MEDUSINA MAWSONI*.

Scythian (Skythian) Originally defined as a *stage at the base of the *Triassic in the Alps. Now it is generally regarded as a *series divided, according to *ammonite *zones, into four stages. The corresponding Scythian *Epoch is dated at 245–240 Ma (Harland et al., 1989).

sea 1. Large body of usually saline water which is smaller in size than an ocean. **2.** Chaotic waves generated by the action of the wind on the surface layers of the ocean. *See also* OCEAN WAVE; and SWELL.

sea-anemones *See* ANTHOZOA.

sea breeze *See* LAND AND SEA BREEZES.

sea-floor spreading The theory that the ocean floor is created at the spreading (accretionary) *plate margins within the ocean basins. *Igneous rocks rise along conduits from the *mantle, giving rise to volcanic activity in a narrow band along the mid-ocean *ridges. As these cool, the basaltic *lavas and *dykes form the upper part of the *oceanic crust, and the underlying *magma chamber solidifies to form layer 3 of the oceanic crust. The newly formed oceanic crust spreads perpendicularly away from the ridge, probably in response to mantle convective motions (*see* PLATE TECTONICS). As the *basalts originally cooled, they became magnetized by the ambient *geomagnetic field. As this field reverses *polarity, oceanic crust formed at different times is characterized by oceanic *magnetic anomalies that are parallel to the ridge at which they originally formed (Vine and Matthews, 1963). These anomalies allow the dating of the oceanic crust and the determination of its past relative motion. The creation of new ocean floor was implicit in the previous concept of *continental drift, but is mainly characterized by the narrowness of the zone within which the new ocean floor is formed. It is now a fundamental concept within the plate-tectonic theory.

sea fret Popular local term for sea fog, common in spring and summer in Cornwall and on the south, east, and north-east coasts of England.

sea ice In polar regions the surface of the sea freezes, due to the low air and water temperatures: the product is known as 'sea ice'. It exists year-round in the central Arctic and in some Antarctic bays, extending in winter across the entire Arctic and far out to sea around Antarctica. As ice crystals form from sea water, so salt is excluded and eventually returned to the sea. Sea ice therefore contains no salt, except where pockets of sea water become trapped in the ice.

seamount Isolated, submarine mountain rising more than 1000 m above the ocean floor. The sharp, crested summits of seamounts are usually 1000–2000 m below the ocean surface. Seamounts are of volcanic origin. They are increasingly coming under study with modern methods of submarine acoustic imagery, e.g. GLORIA (*Geological LOng Range Inclined Asdic) *side-scan sonar. *See also* GUYOT.

sea pens *See* ANTHOZOA.

seatearth A *clay-rich *fossil soil, found immediately beneath a *coal seam, representing the soil in which the coal-forming vegetation grew.

sea water, major constituents 99.9% of dissolved material in sea water can be accounted for by eleven constituents: Na, Mg, Ca, K, Sr, Cl, SO_4, HCO_3, Br, BO_3, and F, whose relative proportions are almost constant in all oceans regardless of *salinity. The *pH of sea water is maintained at 8.0–8.4 by a buffering system. Because the water is alkaline, most of the Ca^{2+} and HCO_3^- form insoluble $CaCO_3$. Aluminium and iron oxides coagulate to form *colloids because of the high concentration of *electrolytes, and sink to the sea bed, silica is taken up by organisms, and thus Cl, Na, SO_4, Mg, and K, for which no removal mechanism exists, account for almost all the dissolved material, at concentrations much higher than those found in rocks.

Ion	*Parts per thousand by weight*	*Percentage of dissolved material*
chloride, Cl^-	18.980	55.05
sodium, Na^+	10.556	30.61
sulphate, SO_4^{2-}	2.649	7.68
magnesium, Mg^{2+}	1.272	3.69
calcium, Ca^{2+}	0.400	1.16
potassium, K^+	0.380	1.10
bicarbonate, HCO_3^-	0.140	0.41
bromide, Br^-	0.065	0.19
borate, $H_3BO_3^-$	0.026	0.07
strontium, Sr^{2+}	0.008	0.03
fluoride, F^-	0.001	0.00
TOTAL	34.477	99.99

Secchi disc A device used in a simple method for measuring the transparency of water. The disc is 20 cm across, and divided into alternate black and white quadrants. It is lowered into the water on a line until the difference between the black and white areas just ceases to be visible, and this depth is recorded. The Secchi disc provides a convenient method for comparing the transparencies of water at different times.

second arrival The next coherent seismic event to follow the *first break.

secondary blasting *See* POP-SHOOTING.

secondary creep The second stage of *creep, characterized by viscous strains, in a material being deformed as a consequence of long-term, low *stress.

secondary crushing *See* CRUSHING.

secondary depressions *Depressions initiated as part of a 'family' or sequence of wave disturbances along a *cold front; they are in the rear of the first depression of the series. *See also* FRONTAL WAVE.

secondary enrichment *See* SUPERGENE ENRICHMENT.

secondary front Frontal development in the colder air in the rear of a *depression, in the form of a further *trough of low pressure following the main *cold front.

secondary geochemical differentiation A theory concerning the processes which may have taken place in and on the Earth's *crust and in the atmosphere, including the organic reactions leading to the development of the *biosphere.

secondary geochemical dispersion The movement of elements at or just below the Earth's surface, which results from *weathering, *erosion, and deposition.

secondary migration Movement of *hydrocarbons, usually laterally, into and within the *reservoir rock where they accumulate.

secondary mineral A *mineral formed by the subsolidus alteration of a pre-existing *primary mineral in an *igneous rock. Minerals which have crystallized from a *magma are stable only at high temperature and can readily alter to low-temperature, secondary minerals when a fluid, e.g. water, is introduced into the rock system. The fluid acts as a catalyst to initiate the alteration reaction. Most secondary minerals are hydrated *silicates. A typical example is the alteration of primary *olivine to secondary *chlorite and *serpentine.

secondary porosity The forms of *porosity which develop in a rock during and after *consolidation, by the selective dissolution or alteration of mineral *grains and *cements, or by the fracturing of the rock by tectonic processes (*see* TECTONISM). *See* CHOQUETTE AND PRAY CLASSIFICATION.

secondary quartz *Quartz formed as a *cement during *diagenesis.

secondary recovery methods In the petroleum industry, techniques by which recovery has been increased to about 50% since the 1940s, including the injection of natural gas above the oil in a *reservoir to force the oil downwards, or pumping water in below the oil to force it upwards.

secondary sedimentary structure A structure formed by the precipitation of minerals in the pores of a sedimentary rock during or following its consolidation, or by chemical replacement of some of its constituents. *Compare* PRIMARY SEDIMENTARY STRUCTURE.

secondary wave *See* S-WAVE.

second derivative Acceleration due to gravity (g) is the first derivative of the gravity potential field. Where g varies with horizontal distance (due to anomalies), then the gradient in the direction of the variation is the second derivative. Such computations exaggerate noise, as well as highlighting maxima and minima in the gravity field.

secular variation Any long-term variation over a period of the order of 10 to 20 years or longer. In *geomagnetism, a variation that operates on a time-scale greater than one year, but excluding the variation associated with the sunspot cycle.

SEDEX *See* SEDIMENTARY EXHALATIVE PROCESSES.

Sedgwick, Adam (1785–1873) Woodwardian Professor of Geology at Cambridge, where he modernized the teaching of geology. He is best known for unravelling the stratigraphy of North Wales (the *Cambrian System) which resulted in a quarrel with R. I. *Murchison over the boundary between the Cambrian and *Silurian Systems. He made important contributions to structural geology, distinguishing between stratification, jointing, and *slaty cleavage.

sedigraph An instrument that calculates the size of sediment particles by measuring the attenuation of a finely collimated X-ray beam through a settling suspension as a function of height above the base of the *sedimentation tube and time.

sediment Material derived from pre-existing rock, from *biogenic sources, or precipitated by chemical processes, and deposited at, or near, the Earth's surface.

sedimentary basin A subsiding area of the Earth's *crust which permits the net accumulation of *sediment.

sedimentary cycle A cycle which comprises the *weathering of an existing rock, followed by the *erosion of *minerals, their transport and deposition, then burial. First-cycle *sediments are characterized by the presence of less resistant minerals and rock fragments. If this material is reworked through a second cycle, the less resistant minerals will be eliminated, or altered to more stable products. The more sedimentary cycles that a sediment has passed through, the more mature it will become and it will be dominated by well-rounded, resistant minerals (*see* RESISTATE MINERAL).

sedimentary exhalative processes (SEDEX) Processes associated with the upwelling of mineralizing fluids into submarine sedimentary environments, whereby *mineral deposits, usually of base-metal sulphides, are formed. *See* 'BLACK SMOKERS'.

sedimentary mélange *See* OLISTOSTROME.

sedimentary rock Rock formed by the deposition and compression of mineral and rock particles, but often including material of organic origin and exposed by various agencies of *denudation. Sedimentary rocks may be classified as terrigenous (i.e. derived from the breakdown of pre-existing rocks exposed on the land), organic (i.e. produced either directly or indirectly by organic processes such as shell production or *peat formation), chemical (i.e. produced by precipitation from water, e.g. some *carbonates and all *evaporites), or volcanogenic (*pyroclastic, e.g. *tuffs and *bentonites). They may also be described according to their chemical properties and behaviour and their environmental deposition, and each scheme complements the others.

sedimentary structure The external shape, the internal structure, or the forms preserved on bedding surfaces, generated in *sedimentary rocks by sedimentary processes or contemporaneous *biogenic activity. Internal sedimentary structures include: those formed by physical depositional processes (*cross-stratification, flat bedding (*see* PLANE BED), *lamination, and *heterolithic structures); those due to post-depositional deformation (convolute bedding, *slump structures, dish and pillar structures, flame structures, *ball and pillow structures, etc.); those caused by organic disturbance (*bioturbation, *trace fossils); or by post-depositional chemical disturbance (*enterolithic structures, collapse and solution structures, *concretions, etc.) Structures preserved on the tops of beds include: those formed by depositional processes (*ripple marks, primary current lineations); erosional structures (*flutes and scour marks, *see* SCOUR AND LAG); structures caused by the transportation of an object over the bed (*tool marks); and other features such as *desiccation and *syneresis cracks, *sand volcanoes, adhesion ripples and warts, rain prints, and biogenic traces and *trails. Structures preserved on the bases of beds (*sole marks) include *load casts, the casts of flutes, trails and tool marks, and the fill of erosional scours. The external form of sedimentary units (sheet-like, *channel-fill, *reef or mound (*see* MUD MOUND), lenticular, etc.) is a function of the depositional environment and sometimes of post-depositional *compaction.

sedimentation coefficient Measure of the rate of sedimentation of a molecule or particle; it is equal to the velocity per unit centrifugal field (acceleration), and is measured in *Svedberg units.

sedimentation, rate of *See* SUBSIDENCE.

sedimentation tube Apparatus designed to determine rapidly the grain-size distribution of *sand-sized *sediment. The sedimentation-tube method is based on the principle of *Stoke's law that particles of different sizes, shapes, and densities will settle through a column of fluid at different settling velocities. Sediment is introduced into the top of the settling tube, and the time is recorded for the various sediment fractions to settle to the bottom of the tube.

sedimentology The scientific study, interpretation, and classification of sediments, sedimentary processes, and sedimentary rocks.

seed plants The group that comprises the *gymnosperms and *angiosperms, plants distinguished by their production of seeds rather than spores. They arose in the *Devonian from *pteridophyte forebears, probably of a heterosporous character (i.e. bearing microspores and megaspores (*see*

SPORE) on the same plant). These early seed plants were gymnosperms. Angiosperms are not certainly represented in the fossil record until the early *Cretaceous.

seep *See* SPRING.

seepage The slow but often steady flow of water between one water body and another. As a term, it is often used to describe leakage to underlying *aquifers through stream beds or the emergence of *groundwater into a stream channel, but it may also relate to flow between different aquifer units.

seepage velocity The velocity of *groundwater calculated from *Darcy's law. It is not the actual velocity of the water in the pores, but the apparent velocity through the bulk of the porous medium. Actual velocity is higher than seepage velocity by a factor which combines the effects of porosity and the tortuosity of the actual flow path among and around the mineral grains.

seiche A stationary or *standing wave in an enclosed body of water, e.g. a bay or lake. Seiches are usually the product of intense storm activity.

seif dune Linear *dune consisting of curved, sword-like components, and found in hot deserts. Typically it is developed by the elongation of a *barchan arm, and built up by winds blowing from two principal directions.

Seif dune

seism- From the Greek *seismos* meaning 'earthquake', a prefix meaning 'pertaining to earthquakes'.

seismic blind zone A layer which cannot be detected by seismic refraction methods because it is too thin or its velocity is lower than that in overlying strata. *See* HIDDEN LAYER.

seismic gap 1. An area within a known active *earthquake zone within which no significant earthquakes have been recorded. It is not always clear whether this gap represents a zone where gradual motion takes place continually so there is no strain accumulation, or where motion is locked and strain is accumulating. **2. (shot-point gap)** In split-spread seismic *reflection shooting, the distance between the shot point and the nearest *groups of *geophones, which is larger than that between subsequent geophone groups at larger *offsets. **3. (inter-record gap)** A blank space on a magnetic tape which signals the end of one block of seismic data (one seismic record) and heralds the start of another. It is used to facilitate the transfer of magnetically recorded data into a computer system.

seismicity The likelihood that an *earthquake will be felt in a particular area; i.e. earthquakes are frequent in zones of high seismicity (e.g. Japan and California).

seismic margin *See* ACTIVE MARGIN.

seismic moment A measure of the size of an *earthquake, based on the concept of rupture and slippage along a fault plane as a rotational motion about a point on the fault. It is given by: $M_0 = \mu A d$, where μ is a constant varying according to the material, A is the area of the ruptured fault, and d is the average slip.

seismic record (seismogram) The output of a seismic recording system (*seismograph), e.g. a paper or film record, showing the seismic *wiggle traces, usually for a single *shot *spread. In *refraction surveys, many shots into one spread may be summed by a signal-enhancement seismograph to produce a single record. When records are processed they can be placed side by side along a *profile to form a 'seismic section'.

seismic reflection *See* REFLECTION.

seismic-reflection profiling *See* PROFILING.

seismic refraction *See* REFRACTION.

seismic section *See* SEISMIC RECORD.

seismic stratigraphy (seismostratigraphy) The study and interpretation of information obtained by seismic-*reflection *profiling in order to construct subsurface *stratigraphic cross-sections. Analysis of seismic reflections on a seismic section (*see* SEISMIC RECORD) can identify buried stratal surfaces that, when traced laterally and continuously, represent surfaces of synchronous deposition or their correlative *unconformity surfaces. The character of a reflection may vary as the seismic profile moves across a *facies boundary, but the continued presence of the reflection is of *chronostratigraphic significance. More detailed information regarding the age and *lithology of the subsurface strata may be gathered by means of geophysical *well logging. *See also* CHRONOSTRATIGRAPHIC CORRELATION CHART; and DEPOSITIONAL SEQUENCE.

seismic survey The exploration of a subsurface geologic structure by means of *seismic waves which are generated artificially. *Reflection profiling is the most common surveying method, and *refraction surveys are particularly important in land surveys for *static corrections. Small seismic refraction surveys are commonly carried out for *engineering geophysics site investigations.

seismic tomography A range of methods in which the subsurface is divided into a box-grid whose elements are illuminated by seismic rays and the physical character of each element computed. The results are displayed as colour-coded contour maps of subsurface planes. In *borehole *tomography, two holes are used; a source is moved up one while detectors record along the length of the other. By using a succession of *shot positions, most of the box elements in the plane linking the two boreholes are illuminated satisfactorily. Seismic tomography is used to study geologic structures in detail. It has been extended to investigate the *mantle and, increasingly, the effectiveness of extraction techniques in hydrocarbon reservoirs.

seismic velocity The speed with which an elastic wave propagates through a medium. For non-dispersive *body waves, the seismic velocity is equal to both the *phase and *group velocities; for dispersive *surface waves, the seismic velocity is usually taken to be the phase velocity. Seismic velocity is assumed usually to increase with increasing depth and when measured in a vertical direction it may be 10–15% lower than when measured parallel to strata. *See* ANISOTROPY.

seismic wave A packet of elastic strain energy which travels away from a seismic source, e.g. a *shot or *earthquake. *See also* BODY WAVE; and SURFACE WAVE.

seismic-wave modes The conventional notation ascribed to *seismic waves on the basis of their travel times from their *earthquake sources. Letters are used to designate the type of wave along the different portions of its travel path: P and S refer to *P-waves and *S-waves that do not travel through the Earth's *core; K and I refer to P-waves only, which travel through the core and inner core respectively. The letter J refers to S-waves which are generated by the conversion of incident P-waves at the boundary between the inner and outer cores, and which travel only in that mode within the solid inner core. As they propagate towards the Earth's surface, J-waves are reconverted back into P-waves at the inner core boundary. A repetition of the letters (e.g. PP, SS) indicates that the waves have been reflected at the Earth's surface; reflection at the outer edge of the core is indicated by 'c' (e.g. PcP, PcS). For example, the wave PKIKP travels from the surface, through the *mantle to the outer core, into the inner core, and then back to the surface; the wave PPP has travelled only in the *crust and upper mantle, and has been reflected twice off the surface.

seismic zone A region of high *seismicity.

seismogram *See* SEISMIC RECORD.

seismograph A device which records seismic information. Usually the term is used to describe the entire system, including amplifiers and means for filtering data and transferring them to magnetic tape or computer disk, but occasionally it describes only *geophones. An enhancement seismograph can sum successive *hammer impacts or *shots fired into one geophone *spread in order to enhance the signal-to-*noise ratio, and is commonly used in *engineering geophysics site investigations.

seismology The study of elastic (seismic) waves and how they are produced. Global seismology is the study of *seismic waves from *earthquakes (and to a lesser extent nuclear explosions), to investigate the structure of and processes within the Earth. In exploration seismology, artificially generated seismic waves are used in the search for resources (e.g. hydrocarbons, etc.) and the study of the Earth's surface and near-surface. Planetary seismology is the use of seismic waves to investigate the structure of and processes within planets and natural *satellites in the *solar system.

seismometer 1. A device used to detect *seismic waves originating from *earthquakes. **2.** In exploration *seismology, a *geophone.

seismostratigraphy *See* SEISMIC STRATIGRAPHY.

Selene A Japanese *ISAS mission, to be launched in 2003, to place a vehicle in orbit about the Moon.

selenite *See* GYPSUM.

selenizone *See* ARCHAEOGASTROPODA.

selenology The astronomical study of the *Moon. The term is derived from the Greek word *selene*, meaning 'moon'.

self diffusion *See* DIFFUSION (2*a*).

self-exciting dynamo A dynamo that produces a magnetic field around itself whereby motions of an electrical conductor, carrying magnetic lines of force, generates further current, eventually resulting in a stable external magnetic field. It is generally considered that the *geomagnetic field is produced by two self-exciting dynamos, the interaction of which results in *reversals of geomagnetic polarity.

self-potential method *See* SPONTANEOUS-POTENTIAL METHOD.

self-potential sonde (spontaneous potential sonde, SP sonde) A *well-logging instrumental package that measures the electrochemical activity of the rocks within a *borehole. Formations yielding a high and true SP log response include *clay-rich *sediments and *sulphide minerals. The SP method cannot be used for logging wells in offshore areas, where saltwater-based drilling muds are used.

self-reversal The ability of certain *ferromagnetic materials to acquire a thermal remanence in the opposite direction to the ambient magnetic field. Usually this requires the interaction of two or more magnetic lattices, or of minerals with different *Néel or *Curie temperatures.

SEM *See* SCANNING ELECTRON MICROSCOPE.

Senecan A *series in the Upper *Devonian of N. America, underlain by the *Erian, overlain by the *Chautauquan, and comprising the Fingerlakian and Cohoktonian *Stages. It is roughly contemporaneous with the *Frasnian of Europe.

Senonian The final *Cretaceous *epoch which is dated at 88.5–65 Ma (Harland et al., 1989) and comprises the *Coniacian, *Santonian, *Campanian, and *Maastrichtian *Ages. Some authors do not include the Maastrichtian Age within the Senonian.

Sensitive High Resolution Ion Micro-Probe (SHRIMP) A device for dating granite *batholiths *in situ* by analysing zircon grains taken from several samples.

sensitive tint *See* ACCESSORY PLATE.

sensitivity 1. The consistency of *clay as this is affected by *remoulding. The effect depends on the type of clay and the amount of *pore water. In a sensitive clay, *shear strength is decreased dramatically on remoulding when moisture content remains constant. Sensitivity is measured as the ratio of the *unconfined compressive strength to the strength in the remoulded state at the same water content. **2.** In chemical analysis, the smallest change in concentration which can be discriminated by the analytical method.

septa *See* RUGOSA; TABULATA; and SEPTUM.

septarian nodule A *concretion, roughly spheroidal in shape, usually of *clay *ironstone, and characterized by an internal structure of angular blocks separated by radiating mineral-filled blocks. The mineral filling the cracks is usually *calcite. The structure results from the formation of a hard exterior to the nodule due to the development of an aluminous *gel on the exterior, followed by dehydration of the colloidal mass (*see* COLLOID) in the interior, leading to cracking and subsequent mineral infilling of the radiating pattern of cracks.

septomaxilla A bone at the front of the upper jaw in reptiles; monotremes are the only mammals in which this bone maintains its separate identity. *See OBDURODON*.

septum (pl. septa) A cross wall or partition. **1.** A morphological term used particularly with reference to the *Cephalopoda, whose shells are divided internally into a series of chambers (camerae) by septa which are generally concave towards the anterior. There is an opening (foramen) in each septum and this is usually bordered by a collar (septal neck). Septal necks are forward-pointing (prochoanitic) in ammonoids (*Ammonoidea) and backward-pointing (retrochoanitic) in nautiloids (*Nautiloidea). In many ammonoids the septum becomes fluted and there has been much discussion concerning the function of the complex fluting. **2.** The radially arranged *plates that occur in the *corallum of corals. The first-formed septa are called 'prosepta' and are usually larger than the metasepta, which occur between them.

sequence stratigraphy The analysis, within the framework of a *chronostratigraphic scale, of depositional units that are related to one another genetically (i.e. they were formed by similar processes). There are two schools: the Exxon Production Research (EPR) *depositional sequence model and the *genetic stratigraphic sequence model (proposed by W. E. Galloway in 1989), differing mainly in the type of boundaries by which they define the strata.

serac Ice pinnacle found on the surface of a *glacier and resulting from tensional failure in the more rigid upper crust. This is due to the stretching that occurs when a glacier moves over a convex-up slope, spreads out over a plain, or passes round a bend in its valley.

serein The fall of rain from an apparently clear sky. The phenomenon may be explained by the evaporation of cloud particles following the formation of rain droplets, or by the movement of cloud away from the overhead position as the rain approaches the ground.

sericite A white variety of *muscovite or paragonite and member of the *phyllosilicates (sheet silicates) with the formula $K_2Al_4[Si_3AlO_{10}]_2(OH)_2$ formed from the alteration of *feldspar by either *hydrothermal alteration or later-stage *weathering. It appears as small scales or flakes or compact aggregates which give the feldspar a 'cloudy' appearance in *thin section.

series The major subdivision of a *system, and the chronostratigraphic equivalent of an *epoch. It denotes the layers of strata or the body of rock formed during one epoch. A series may itself be divided into *stages. When used formally the initial letter of the term is capitalized, e.g. Lower *Cretaceous Series.

serir Veneer of mixed sand and gravel mantling a Saharan plain and transported originally by *sheetwash and *braided-stream activity. Subsequently it was weathered and modified under more arid conditions.

serpenticone *See* INVOLUTE.

serpentine A group of *minerals belonging to the 1:1 group of *phyllosilicates (sheet silicates) with the composition $Mg_6[Si_4O_{10}](OH)_8$ and including the minerals chrysotile (the asbestiform variety), lizardite, and antigorite; sp. gr. 2.55–2.60; *hardness 2.0–3.5; *monoclinic; chrysotile is fibrous whereas lizardite and antigorite occur as flat *tabular crystals or *massive; various shades of green, also brown, grey-white, or yellow; *greasy to *waxy *lustre, occasionally *silky; formed from altered *olivine and *orthopyroxene. It results from the alteration of ultramafic (*see* ULTRABASIC) rocks either by *hydrothermal action at a late stage or by alteration during *metamorphism, chrysotile forming first and then altering to antigorite; it is a constituent of ophicalcites, a serpentine-*calcite rock derived from the dedolomitization (*see* DEDOLOMITE) of a siliceous *dolomite. It is used extensively as a facing stone and for ornament; chrysotile has been used as a source of commercial *asbestos in Thetford, Canada. Lizardite occurs on the Lizard Peninsula, Cornwall, UK.

serpentine barrens Impoverished vegetation, often dominated by scrub or heath, associated with *serpentine rocks. On *weathering these rocks release an excess of magnesium into the soil, and this often inhibits the development of the natural climax in the areas concerned.

serpentinite Altered rock formed from an *ultrabasic precursor by low temperature and water interaction. Such rocks are compact, variously coloured, and may have considerable ornamental value. They consist mainly of hydroxyl-bearing magnesium silicates formed from original *olivine and *pyroxenes.

serpentinization The process whereby high-temperature *primary *ferromagnesian minerals in an *igneous rock undergo alteration to a member of the *serpentine group of minerals. The process is initiated by the introduction of low-temperature water into the rock system, the water acting as a catalyst for the reaction in which the high-temperature primary ferromagnesian minerals are converted to low-temperature, *secondary, serpentine-group minerals. Serpentinization is extremely common in *ultrabasic rocks, especially those found in *ophiolites where the entire rock may be converted to serpentine-group minerals, forming a *serpentinite rock. Where the original mineral assemblage can be inferred from relict minerals or *pseudomorph textures, the original mineral name can be added to the rock name. Thus a serpentinized *dunite would be an *olivine-serpentinite.

Serpukhovian 1. The final *epoch in the *Mississippian, comprising the Pendleian, Arnsbergian, Chokierian, and Alportian *Ages (these are also *stage names in western European stratigraphy). The Serpukhovian is preceded by the *Visean, followed by the *Bashkirian (*Pennsylvanian), and dated at 332.9–322.8 Ma (Harland et al., 1989). **2.** The name of the corresponding eastern European *series, which is roughly contemporaneous with the upper *Chesterian (N. America) and *Namurian A (western Europe).

Serravallian 1. An *age in the Middle *Miocene, preceded by the *Langhian, followed by the *Tortonian, and dated at 14.2–10.4 Ma (Harland et al., 1989). **2.** The name of the corresponding European *stage, which is roughly contemporaneous with the Helvetian (Europe), upper *Luisian and lower *Mohnian (N. America), *Lillburnian (New Zealand), and *Bairnsdalian (Australia). The *stratotype is in the Scrivia Valley, Italy.

sesquioxides General term for the hydrated oxides and hydroxides of iron and aluminium.

sessile 1. Lacking a stalk. **2.** Attached to a substrate; non-motile.

seston Particulate matter suspended in sea water.

seta Stiff, hair-like or bristle-like structure.

settlement 1. Gradual subsidence of a structure, caused by the compression of soil below *foundation level. Normally a uniform amount of settlement can be accepted but damage may occur when different parts settle disproportionately. *See* DIFFERENTIAL SETTLEMENT. **2.** In mining, the lowering of overlying strata due to extraction of material.

settling lag The period of time it takes for a fine-grained particle to settle through a water body after the cessation of transportation in suspension. The settling time will be dependent on the *particle size and *particle shape, and governed largely by Stokes's law of settling.

sexual dimorphism Phenomenon of morphological differences (besides primary sexual characters) that distinguish the males from the females of a species. For example, male deer often have larger antlers than females, and the males of many birds have differing plumage (often more brightly coloured). Sexual dimorphism is known to have been common in ammonites (*Ammonoidea), and many *fossils originally thought to have represented separate species are now recognized as dimorphs within one species (e.g. the *Jurassic *Kosmoceras jason* and *K. gulielmi* are dimorphs within *K. jason*). *See also* DIMORPHISM.

Seymouria One of the more problematic of fossil *amphibians, possessing a combination of amphibian and reptilian characters so that it has been referred to both groups on different occasions. One of the features separating it and its relatives from other amphibians is the presence of a large, forward-extended optic notch (the primitive site of the ear drum). A medium-sized amphibian, it existed during the Lower *Permian in N. America.

sferics Natural electromagnetic signals, generated in the atmosphere by discharges of lightning, which propagate around the Earth between its surface and the *ionosphere.

S-fold An asymmetrical *parasitic fold whose approximately S-shaped profile, when observed down the *plunge of the *fold axis, indicates its position on the right limb of the major *anticline, but not on the *syncline. *See* MINOR FOLD.

SG The older abbreviation for *specific gravity.

SGCS Standard Global Chronostratigraphic (Geochronologic) Scale. *See* STANDARD STRATIGRAPHIC(AL) SCALE.

shade temperature The temperature of the air, conventionally measured in a standard shelter or screen that protects the thermometer from rain and from direct sunshine, but that allows the free passage of air.

shadow test *See* HALF-SHADOW TEST.

shadow zone 1. A region of the subsurface from which seismic *reflections cannot be detected because their ray-paths do not emerge to the surface. **2.** A zone over the Earth's surface in which *P-waves and/or *S-waves generated by an *earthquake are detected only weakly, or are absent, because of *refraction within the various layers deep within the Earth, and especially in association with the Earth's *core. Seismic signals refracted from the *mantle are usually poorly discernible at a range of up to about 10°. At a range of 103–142° from the earthquake P- and S-waves are observed only very weakly; between 110° and 142° some very weak P-waves can be measured and these are thought to have been refracted from the inner core. Between 142° and 180° from an earthquake no S-waves are recorded because the waves cannot travel through the liquid outer core.

shaft well A well sunk by a mining technique rather than by drilling. Normally it is larger in cross-section than a drilled well.

shale Fine-grained, fissile, *sedimentary rock composed of *clay-sized and *silt-sized particles of unspecified mineral composition. The noun may be qualified by an adjective (e.g. *black shale, *paper shale, and *oil shale).

shale line The line on an *electrical log marking the usual *apparent resistivity for a *shale.

shallowing-upward carbonate cycle A stratigraphic sequence found on platforms, shelves (*see* SHELF), and some former lake beds, where carbonates have been deposited in progressively shallower water. The sequence develops where the rate of carbonate deposition exceeds the rate at which the receiving basin sinks, so the sediment surface repeatedly rises towards the water surface, while at the same time the deposit progrades (*see* PROGRADATION).

shamal Regional north-westerly wind which brings hot, dry conditions in summer to Iraq and the Persian Gulf. It blows with great force during the day.

shape fabric A *fabric in which the orientation of mineral grains is determined by their shape, so the fabric may be modelled by the shapes of various *strain ellipsoids. Three *tectonite fabrics are commonly used to demonstrate shape fabrics. L-tectonites, produced by uniaxial stretching with planar fabrics, correspond to prolate ellipsoids (*see* PROLATE UNIAXIAL STRAIN). S-tectonites, produced by flattening with planar fabrics, correspond to oblate ellipsoids (*see* OBLATE UNIAXIAL STRAIN). L–S-tectonites, produced by stretching and flattening with intermediate fabrics, correspond to triaxial ellipsoids.

sharp sand *Sand composed of angular, not rounded, grains, with little foreign material; it is used in mortar.

shatter cones Striated, conical, nested fracture pattern that has apparently occurred in response to shock waves of the magnitude (20–250 kb) generated by a *meteorite impact. Individual cone shapes may vary from less than a centimetre to several metres in size. The apices of the cones tend to point towards the centre of impact. Shatter cones were first discovered by R. S. *Dietz in the 1940s and have been observed at many postulated impact *crater sites, e.g. the Sudbury Structure, Ontario, Canada; the Ries and Steinheim Basins, southern Germany; and Gosses Bluff, Australia.

shear box Laboratory equipment used to test the *shear strength of soil or rock. The upper and lower halves of a sample are subjected to horizontal shear by lateral pressure, and the effects measured.

shear direction *See* SHEAR PLANE.

shear modulus (μ, Lamé's constant μ, rigidity modulus) The ratio of the shearing stress (τ) to the resultant shear strain component (tan θ in the case of simple shear). It can also be given by $\mu = 1/2E/(1 + \sigma)$ where E is *Young's modulus and σ is *Poisson's ratio.

shear plane A plane that is parallel to the walls of a *shear zone and contains the shear direction (the displacement vector).

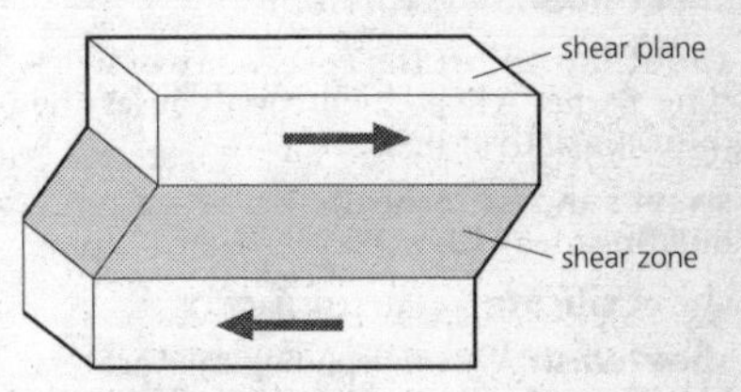

Shear plane and zone

shear strain *See* ANGULAR SHEAR STRAIN.

shear strength The internal resistance of a material to *shear stress; this varies depending on temperature, *confining pressure, shape, size, loading rate, and amount of *pore fluid present. It is measured as the maximum shear stress on an original cross-sectional area that can be sustained. In soils, it is the maximum resistance of a soil to shearing forces under specified conditions. The peak shear strength is the highest stress sustainable just prior to complete *failure of a sample under load; after this, stress cannot be maintained and major strains usually occur by displacement along failure surfaces. The residual shear strength is the ultimate strength along a surface or parting in soil or rock after shearing has occurred. For material not previously sheared there is a rapid decline in strength with increasing shear until the residual shear strength is reached.

shear stress (τ) The *stress which acts parallel to a plane on which a force has been applied. Shear stresses tend to promote sliding along a plane. Conventionally they are designated by the symbol τ; $+\tau$ indicates a left-handed shear sense, $-\tau$ a right-handed shear sense.

shear wave *See* S-WAVE.

shear zone A region, narrow compared to its length, within which rocks have undergone intense deformation. There are two end-members. Brittle shear zones (*faults), marked by a surface of rupture, are a common feature of higher crustal levels. In ductile shear zones deformation is continuous and characterized by high ductile *strains due to the rocks having formed under high temperatures and pressures at deeper crustal levels. Both brittle and ductile shear zones may occur as subparallel or *conjugate sets.

sheep-walk *See* TERRACETTE.

sheet flood *See* FLASH FLOOD.

sheet lightning The appearance of lightning flashes when cloud cover causes the diffusion of their light.

sheet sand (sand sheet) A flat or gently undulating area of sand surrounding a *dune.

sheet silicate *See* PHYLLOSILICATE.

sheetwash A geomorphological *process by which a thin, mobile sheet of water flows over the surface of a hillslope and may transport the surface *regolith. It is important in semi-arid regions, and may also be significant in temperate zones if the vegetation cover has been removed. *See* RAINWASH.

Sheinwoodian A *stage of the Middle *Silurian, underlain by the *Telychian and overlain by the *Whitwellian.

shelf A gently sloping or near-horizontal, shallow, marine platform. Horizontal shelf areas, particularly in *carbonate-dominated areas, are referred to as 'platforms' if they have precipitate margins; more uniformly sloping shallow marine carbonate areas are termed 'ramps'.

shell beak In *Bivalvia, the oldest part of the shell, located near the *hinge.

shell structure Shell growth is similar in *Mollusca and *Brachiopoda. The shell is divided into an outer periostracum, which is organic, and an inner ostracum. Both parts are secreted by the *mantle and shells grow by marginal increments which are commonly visible on the exterior as growth lines. Shell thickening takes place and in some groups accessory or additional shell material may be laid down. External ornament, e.g. ribs, often reflects the shape of the mantle edge or structures contained within it and quite complex patterns of ornament are sometimes produced.

shelly limestone *Carbonate rock that contains a large proportion of shell or shell fragments.

shelter porosity *See* CHOQUETTE AND PRAY CLASSIFICATION.

shergottyite/nakhlite/chassignite meteorites (abbreviation SNC) Association of about a dozen *meteorites, much younger (1.3 billion years) than the majority found (4.6 billion years). They have *igneous *textures and contain iron-rich *silicates and iron oxides, therefore must come from an oxygen-rich environment, and also small amounts of water-bearing minerals. Probably they all come from the same planetary body; the most favoured possibility is the planet *Mars.

Shermanian A *stage of the *Ordovician in the Upper *Champlainian *Series of N. America.

shield 1. A tunnel borer consisting of a conventional shield with thrust rams and erector system. The cutter head and support are inside the shield. This type of machine is usually employed in soils or variable materials, e.g. a sand–rock–gravel sequence. **2.** *See* CRATON.

shield volcano *See* HAWAIIAN ERUPTION.

shingle Beach pebbles, normally well rounded as a result of abrasion, whose diameters are typically 0.75–7.5 cm. They are made of resistant materials such as *flint, which is the dominant constituent of the shingle beaches of south-eastern England. They may also show lateral sorting, e.g. the shingle of the Chesil Beach, Dorset, England, steadily increases in size over 29 km from west to east.

Shipka A *solar system asteroid (No. 2530), with an orbital period of 5.25 years. It is to be visited in October 2008 by the *Rosetta spacecraft.

shoal A mound or other structure raised above the sea bed in shallow water that is composed of, or covered by, unconsolidated material and may be exposed at low water.

shoaling The behaviour of waves as they approach a shelving shore. The waves cease to be symmetrical and sinusoidal and become asymmetrical and solitary. Wavelength and wave velocity decrease, wave height and wave steepness increase, and wave period remains constant.

shoal retreat massif Large *sand accumulation preserved on the *continental shelf during and after a marine *transgression. The massifs represent former *estuary-mouth sand bars (inlet-associated shoals) or former zones of *longshore-drift convergence (cape-associated shoals). In the Middle Atlantic Bight of the east coast of the USA the massifs are up to 70 km long by 20 km wide and can be found offshore from the present Hudson, Delaware, and Chesapeake estuaries, and longshore-drift convergence zones such as Cape Hatteras.

shock metamorphism Changes included in *rocks and *minerals when they are subjected to pressures from shock waves travelling at 10–13 km/s resulting from *meteorite, *cometary, or *asteroidal impact. Pressures ranging to more than 5000 kb are experienced during such events. Materials subjected to less than 100 kb are severely disrupted and fractured; between 100 and 250 kb, framework *silicates exhibit planar deformation features; maskelynite (vitrified *plagioclase feldspar) forms between 250 and 400 kb; *quartz and *feldspar melt from 400 to 600 kb; while rocks subjected to pressures of more than 600 to 700 kb become molten. Peak pressures are sufficient to vaporize material.

shock-remanent magnetization The magnetization acquired as a result of rapidly applied stress while being held in a magnetic field. It is usually attributed to meteoritic impacts. *See also* PIEZOREMANENT MAGNETIZATION.

Shoemaker, Eugene Merle (1928–97) An American geologist who founded the scientific study of impact cratering on Earth, the Moon, and on other solar system planets and their satellites, and who also pioneered the study of near-Earth comets and asteroids, often in collaboration with his wife Carolyn. This led him to recognize the importance of the cratering process in the history of the solar system. In 1952, he became convinced that both Meteor Crater, Arizona, and lunar craters had been formed by impact. In 1956 he was assigned to map the craters formed by nuclear test explosions, and was able to define how ejecta produce crater stratigraphy. Shoemaker was born in Los Angeles and received his first degree from the California Institute of Technology and his Ph.D. from Princeton University in 1960. From 1948 to 1993 he worked for the US Geological Survey, then continuing to serve as scientist emeritus. In the 1960s he established the Astrogeology Branch of the USGS and its centre at Flagstaff, Arizona. Studies of the Moon and other bodies lacking atmospheres led him to propose that their surfaces should be covered with a layer of ejecta he termed 'regolith', and that the age of a surface could be inferred from its cratering. In March 1993, Gene and Carolyn Shoemaker and David Levy discovered a fragmented comet close to Jupiter. The comet, named Shoemaker-Levy 9, collided with Jupiter spectacularly in July 1994. Gene Shoemaker died in a car crash in the Australian outback. He and Carolyn, who survived, had been on their way to study a crater.

shoestring sand An irregular, sinuous *sandbody having a form resembling a shoe-lace, often representing the preserved sandy deposit of a meandering river channel (*see* MEANDER).

shonkinite A dark coloured, coarse-grained, *igneous rock consisting of *essential *diopside (making up about 50% of the rock), *alkali feldspar, and *biotite, with or without *olivine and/or *nepheline. Where olivine or nepheline is abundant the rock name is prefixed with these *mineral names (e.g. olivine-shonkinite). Shonkinites are essentially a type of alkali *syenite.

shooting flow *See* CRITICAL FLOW; and FROUDE NUMBER.

shoreface The subtidal coastal zone between the low-water mark and a depth of about 10–20 m, within which wave action governs the sedimentary processes. Below the lower limit of the shoreface waves do not affect the sea bed. *See* WAVE BASE.

shore platform (marine platform, marine terrace, marine flat, marine bench, wave-cut bench, wave-cut platform) Intertidal bench cut into a land mass by the action of waves and associated processes. It is terminated landward by a sea cliff, and slopes gently seaward at about 1°.

shortening The reduction in length of a line as a result of *strain, e.g. by thickening, folding, thrusting, or the loss of material by solution. Simple shortening may be calculated using the same equation as for *extension (*e*): $e = (L_F - L_O)/L_O$, where L_F is the final length and L_O the original length; if *e* has a negative value when calculated from this equation, the line has been shortened.

short wavelength infrared *See* NEAR-INFRARED.

shot A source of seismic shock waves that are produced for experimental purposes, e.g. by a *hammer, an explosion, an *air-gun, or a *water gun.

shot bounce The noise on a *seismic record that is generated by the physical and mechanical motion of a recording vehicle.

shotcrete A type of *concrete, sometimes reinforced with metal or glass fibre, which is sprayed on to the face of an excavation to form a protective lining and support, usually a few tens of millimetres thick. It is most useful in protecting soft or weak material and can be adapted to suit varying conditions. It is often used in conjunction with other forms of support, e.g. *rock bolts and *arching.

shot depth The depth below ground level of a seismic source. If the source is non-explosive, it is the depth at which the source impulse is generated; if it is explosive, it is the depth to the top of the explosive, or to its mid-point if the length of the explosive material is small in relation to the depth of the hole, or to both top and bottom of the explosive charge (i.e. two shot depths) if it is very large.

shotpoint gap *See* SEISMIC GAP.

shower *Precipitation of short duration, associated with convective clouds which usually do not form a continuous cover of the sky. Intensity varies, e.g. showers may be slight, moderate, or heavy. Hail or snow may form part of the precipitation.

SHRIMP *See* SENSITIVE HIGH RESOLUTION ION MICROPROBE.

shrimps *See* MALACOSTRACA.

shrinkage In *concrete, a measurement of the reduction in dimensions due to the loss of water. Fine-grained *cement has a low coefficient of permeability and behaves in a similar manner to fine-grained soil.

shrinkage cracks *See* DESICCATION CRACKS.

shrinkage joint *See* JOINT.

SH-wave *See* LOVE WAVE.

SI (SI units) *S*ystème *I*nternational d'Unités, an internationally agreed group of units of measurement, used especially in scientific work. The system comprises seven basic and two supplementary units. The basic units are the: metre (m); kilogram (kg); second (s); ampere (A); kelvin (K); mole (mol); and candela (cd). The supplementary units are the: radian (rad); and steradian (sr). A further 18 units are derived from these: becquerel (Bq); coulomb (C); farad (F); gray (Gy); henry (H); hertz (Hz); joule (J); lumen (lm); lux (lx); newton (N); ohm (Ω); pascal (Pa); siemens (S); sievert (Sv); tesla (T); volt (V); watt (W); and weber (Wb).

Siberian high The region of high average pressure in the colder seasons of the year over Siberia. The intensity of the high pressure is increased by the density of the very cold air at the surface over the Siberian plains.

sichelwannen Crescent-shaped *p-forms cut into a flat or sloping rock surface by the action of a *glacier. They are typically 1–10 m long and 5–6 m wide. The horns of the crescents point down-glacier. Their origin is unclear; the action of saturated *till and of subglacial meltwater have both been invoked. The name is German, its literal meaning being 'sickle tubs'.

Sicilian *See* QUATERNARY.

sicula The skeleton of the initial zooid of a graptolite (*Graptolithina) colony.

side-looking airborne radar (SLAR) An airborne *radar system which scans sideways from the flight track and detects the *backscattered radar *reflections from the ground surface in order to produce a radar image of the ground much as *side-scan sonar produces an image of the sea floor. Since different surface materials have different radar reflectance characteristics the radar image can be interpreted in terms of surface features.

sidereal day The time it takes for the Earth to complete one rotation on its axis such that a particular point on the surface returns to its former position in relation to the position of the fixed stars. The sidereal day is 4.09 minutes shorter than the mean solar day, because the movement of the Earth in its solar orbit is imposed on its rotational motion.

sidereal month A period of 27 days, 7 hours, and 43 minutes, which is the average period of revolution of the *Moon around the *Earth, as determined by using a fixed star as a reference point.

siderite (chalybite, spathose iron) Mineral, $FeCO_3$; sp. gr. 3.8–4.0; *hardness 3.5–4.5; *trigonal; grey to grey-brown or yellowish-brown, translucent when pure; white *streak; vitreous *lustre; uneven *fracture; crystals *rhombohedral with curved faces, but also occurs *massive, granular, fibrous, compact, *botryoidal, and *earthy in habit; *cleavage perfect rhombohedral {1011}; widespread in *sedimentary rocks, especially *clays and *shales where it is *concretionary and makes clay into *ironstone, also as a *gangue mineral in *hydrothermal veins together with other metallic ores (e.g. *pyrite, *chalcopyrite, and *galena) and as a replacement mineral in *limestone; dissolves slowly in cold, dilute hydrochloric acid, which effervesces when warmed.

siderolite *See* STONY-IRON METEORITE.

siderophile Applied to elements with a weak affinity for oxygen and sulphur and soluble in molten iron. For example, Ni, Co, Pt, and Ir, which are found in iron *meteorites and are probably concentrated in the Earth's *core. *Compare* ATMOPHILE; BIOPHILE; CHALCOPHILE; and LITHOPHILE.

side-scan sonar A sideways-looking acoustic-survey system which uses the *reflection of high-frequency (30–110 kHz) sound waves from a surface to map the texture of that surface. It is most commonly used to map sea-floor features (sand *ripples, rock *outcrops, pipes, wrecks, etc.) but it can also be suspended vertically from a vessel to investigate vertical surfaces, e.g. the submerged parts of icebergs and major ice fronts.

sidewall corer A device for coring samples from the side of a drill hole. *See also* CORE SLICER.

Sidufjall A normal *polarity subchron which occurs within the *Gilbert reversed *chron.

Siegenian 1. An *age in the Early *Devonian *Epoch, preceded by the *Gedinnian, followed by the *Emsian, and dated at 401–394 Ma. **2.** The name of the corresponding European *stage, which is roughly contemporaneous with the *Merionsian (Australia), and the upper Helderbergian, Deerparkian, and lower Onesquethawian (N. America).

siemens (S) The derived SI unit of electrical conductance in a circuit with a resistance of one ohm, and thus equal to a reciprocal ohm (mho, or Ω^{-1}). It is used in measuring conductivities, where the unit is S/m. The siemens is named after Sir William Siemens (1823–83).

sieve A circular-framed container with a meshed base. The mesh size is accurately machined and the sieve will permit the passage only of particles finer than the mesh size.

sieve deposit A well-sorted, *matrix-free *conglomerate, which forms where the *sediment transported and deposited comprises only *pebble and gravel grades. Sieve deposits are found mainly on *alluvial fans whose source areas consist of well-jointed, resistant *lithologies such as *quartzites.

sieving A method of grain-size analysis (*see* PARTICLE SIZE) in which the sample is passed through a stack of sieves, arranged with the coarsest mesh at the top and finest at the base. The weight of *sediment trapped on each mesh is recorded, and percentage weight plotted either on a histogram or cumulative frequency curve (*see* CUMULATIVE PERCENTAGE CURVE).

sigma-t density (σ-t density) The density of a sea-water sample measured at atmospheric pressure, i.e. at the sea surface. It is defined as density minus 1000, where the density is measured in kg/m³. Thus sea water at 0°C and 35 parts per thousand *salinity has a density of 1028 kg/m³, or a σ-t density, at a pressure of 1 bar, of 28.

signature The characteristics of geophysical anomalies within a region or along a profile. It is often based on *Fourier or power-spectrum analyses of either gravity (gravity signature) or magnetic (magnetic signature) *residual anomalies.

significant wave height The average height of the highest one-third of waves in a given group of waves. This height is usually used as a standard when describing the wave characteristics of a given area.

silcrete *See* DURICRUST.

Silesian The Upper *Carboniferous subsystem in western Europe, underlain by the *Dinantian, and comprising the *Namurian, *Westphalian, and *Stephanian *Series. It is dated at 332.9–290 Ma and is roughly contemporaneous with the uppermost *Mississippian (*Serpukhovian Series) plus the *Pennsylvanian.

silex *See* FLINT.

silica Silicon dioxide (SiO_2) which occurs naturally in three main forms: (*a*) crystalline silica includes the *minerals *quartz, *tridymite, and *cristobalite; (*b*) *cryptocrystalline or very finely crystalline silica includes some *chalcedony, *chert, *jasper, and *flint; and (*c*) amorphous hydrated silica includes *opal, *diatomite, and some chalcedony. *Coesite and stishovite are two high-density *polymorphs of quartz which rarely occur in nature but have been synthesized experimentally.

silica-oversaturated rock *See* SILICA SATURATION.

silica saturation The concentration of silica (SiO_2) in an *igneous rock, relative to the concentration of other chemical constituents in the rock which combine with the silica to form *silicate minerals. On this basis, three classes of igneous rock are recognized: (*a*) silica-oversaturated rocks (e.g. *granite), in which there is more than enough silica to satisfy the requirements of all the major silicate minerals, the free silica appearing as *quartz in the rock; (*b*) silica-saturated rocks (e.g. *diorite), in which there is just enough silica present to satisfy the requirements of all the major silicate minerals, there being neither an excess nor deficiency of silica, resulting in a lack of both quartz and *feldspathoid minerals in the rock; and (*c*) silica-undersaturated rocks (e.g. *nepheline syenite), in which there is not enough silica present to satisfy the requirements of all the major silicate minerals, the silica deficiency being accommodated by the crystallization of feldspathoids (*nepheline, *leucite) in place of *feldspar, the feldspathoids containing less silica in their structure than feldspars.

silicates The most important and abundant group of rock-forming minerals, which can be classified according to the structural arrangement of the fundamental SiO_4 *tetrahedra which are the main building blocks of the group. (*a*) *Nesosilicates have independent SiO_4 *tetrahedra linked by *cations, e.g. *olivine group. (*b*) Sorosilicates have two SiO_4 *tetrahedra sharing one oxygen, e.g. *epidote group. (*c*) *Cyclosilicates have rings of three, four, or six linked SiO_4 *tetrahedra, e.g. *axinite and *tourmaline. (*d*) *Inosilicates (chain silicates) have SiO_4 *tetrahedra linked either into single chains by sharing two oxygens, e.g. *pyroxene group, or into double chains (band silicates) by alternately sharing two or three oxygens, e.g. *amphibole group, (*e*) *Phyllosilicates (sheet silicates) share three oxygens to form a flat sheet, e.g. *mica group. (*f*) Tectosilicates have SiO_4 *tetrahedra linked into a three-dimensional framework by sharing all the oxygens, e.g. *feldspar and *quartz groups.

silica-undersaturated rock *See* SILICA SATURATION.

siliceous ooze Fine-grained *pelagic deposit of the deep-ocean floor with more than 30% siliceous material of organic origin. *Radiolaria and *diatom remains are the major constituents of the siliceous oozes, which tend to occur at depths in excess of 4500 m.

siliceous sinter A *silica-rich precipitate found around the mouth of a *geyser or hot spring whose waters carry large amounts of dissolved minerals which precipitate when the water cools suddenly on exposure to the atmosphere.

silicification The introduction of *cryptocrystalline *silica into a non-siliceous rock via *groundwater or fluids of *igneous origin. The introduced silica either fills *pore spaces in the rock or replaces pre-existing minerals.

siliclastic Applied to a *sediment which comprises particles composed of *silicate minerals and rock fragments, i.e. *mudstones, *sandstones, and *conglomerates.

silicon 'burning' In stellar evolution, the process whereby silicon, with magnesium and sulphur, 'burns' at a temperature of around 3×10^9 K, producing elements of the 'iron peak' (e.g. Cr, Mn, Fe, Co, and Ni). *See* CARBON 'BURNING'; HELIUM 'BURNING'; HYDROGEN 'BURNING'; OXYGEN 'BURNING'; and NUCLEOSYNTHESIS.

silky Applied to the mineral *lustre of some fibrous minerals, e.g. *satin spar.

sill A tabular *igneous *intrusion having *concordant surfaces of contact.

sillar A local Peruvian name for *ignimbrites that have been pervasively altered by *vapour-phase crystallization. The alteration of original *pumice and *glass shards involves deposition of *tridymite, *cristobalite, and *alkali feldspar that occurs as *drusy infills of the *matrix and pumice cavities in the upper parts of ignimbrite cooling units. The gas is derived by diffusion from *juvenile glassy fragments within the flow and from heated *groundwater percolating through the flow. The resulting material makes excellent, lightweight, easily worked building stone, usually white. The city of Arequipa, southern Peru, is renowned for its buildings of sillar.

sillimanite (fibrolite) A *nesosilicate and one of an important group of three *mineral *polymorphs with the same composition Al_2SiO_5, the other two being *andalusite and *kyanite; sp. gr. 3.23; *hardness 6.5–7.5; *orthorhombic; occurs as elongate *prismatic *crystals with a diamond-shaped cross-section or as *fibrous and felted masses; colourless; occurs under conditions of high-grade *metamorphism either in the innermost zones of thermal aureoles (*see* CONTACT AUREOLE) or in high-grade regional metamorphic rocks (*see* REGIONAL METAMORPHISM) formed under high temperatures and pressures. It is named after the American mineralogist B. Silliman.

silt 1. In the commonly used Udden–Wentworth scale, particles between 4 μm and 62.5 μm in size. Other classifications exist. In *pedology, silt refers to mineral soil particles that range in diameter from 0.02 to 0.002 μm in the international system, or from 0.05 to 0.002 μm in the *USDA system. *See* PARTICLE SIZE. **2.** A class of soil texture.

siltstone A lithified *silt, comprising *grains between 4 μm and 62.5 μm in size.

Silurian Third (439–408.5 Ma) of six *periods of the *Palaeozoic *Era. The end of the period is marked by the climax of the *Caledonian orogeny and the filling of several Palaeozoic basins of deposition.

silver, native Malleable metallic element, Ag; sp. gr. 10.5; *hardness 2.5; white; *cubic crystals, thin sheets, or scales; occurs in association with other silver minerals in the oxidized zone of silver-rich mineral deposits and in *hydrothermal vein deposits.

silver glance *See* ARGENTITE.

silver iodide Substance used in the form of fine particles to act as nuclei in *cloud seeding.

silver spike In stratigraphy, a time-line selected as a reference point in a regional succession of *strata. *Compare* GOLDEN SPIKE.

silver thaw Phenomenon resulting from the formation of ice on cold surfaces (e.g. due to rain) during a period of rapid thaw after a severe frost. *See also* GLAZE.

similar fold A *fold in which the *orthogonal thickness changes systematically such that the thickness of the fold remains constant when measured parallel to the *axial surface. Single layers are thicker at the *hinge and thinner in the limb regions. Unlike a *parallel fold, a similar fold can maintain its form throughout a sequence.

simoom Regional wind in desert areas of Africa and Arabia. The wind blows for only a short period at a time and its whirlwind effect carries sand and brings very hot, dry conditions.

simple shear (homogeneous rotational strain, rotational shear) A rotational *strain in which the maximum and minimum strain axes are re-oriented in relation to their original positions. *See also* HOMOGENEOUS STRAIN. *Compare* PURE SHEAR.

Simpson, George Gaylord (1902–84) An American palaeontologist, specializing in mammalian evolution, who also studied mammal migrations. He obtained his Ph.D. from Yale University in 1926, for a thesis on *Mesozoic mammals, and then joined the staff of the American Museum of Natural History, in New York City, becoming curator in 1942. In 1945 he moved, as a professor, to Columbia University. From 1959 to 1970 he was Alexander Agassiz Professor of Vertebrate Palaeontology at the Museum of Comparative Zoology at Harvard University, and in 1967 was appointed professor of geosciences at the University of Arizona. His studies of fossil mammals, especially those of Madagascar, led him initially to oppose the theory of *continental drift. Simpson proposed that the dispersion of species occurred along *sweepstakes dispersal routes.

Sinemurian A *stage in the European Lower *Jurassic (203.5–194.5 Ma, Harland et al., 1989), roughly contemporaneous with the upper Aratauran (New Zealand). *See also* LIAS.

single couple (Type I earthquake source) A *seismic wave pattern, consisting of four lobes (for *P-waves) and two lobes (for *S-waves) of alternate compression and dilation, which is generated by movement along a single *fault plane. *Compare* DOUBLE COUPLE.

single-stage lead *Common lead (Pb) removed at a particular time from a single, constantly evolving, uranium–thorium–lead environment. It can be used to provide the basis for the common-lead method of dating (*see* LEAD–LEAD DATING). After the removal of this lead in solution and its deposition as a lead ore (such as *galena), it is then assumed not to have been altered in isotopic composition by the further addition of *radiogenic products.

singularity In meteorology, a short seasonal episode lasting a few days and commonly occurring about specific dates of the year. One example is the periods of fine, dry, warm weather commonly occurring in September and early October in Britain and central Europe, and known as 'old wives' summer' (or 'Indian summer'), which are due to slow-moving *anticyclones.

Sinian The final sub-era of the *Neoproterozoic, dated at about 825–570 Ma (Harland et al., 1989). The name is derived from a region of central China.

sinistral coiling *See* COILING.

sinistral fault (left lateral fault) The sense of displacement in a *strike-slip fault zone where one block is displaced to the left of the block from which the observation is made.

sink A natural reservoir that can receive energy or materials without undergoing change.

sink-hole *See* DOLINE.

sinking *See* DOWNWELLING.

Sinope (Jupiter IX) One of the lesser satellites of *Jupiter, with a diameter of 28 km; its orbit is *retrograde.

Sinornis An early *Cretaceous bird from Liaoning Province, China. It is one of the earliest to show the advanced avian features of an extensive *synsacrum and complete fusion of the distal *tarsal bones with *metatarsals.

sinus A Latin word meaning 'curve' or 'bay'. **1.** The word 'sinus' was used by Giovanni B. Riccioli in 1651 to designate bay-like features on the lunar maria (*see* MARE). The best known example is Sinus Iridum, the 'Bay of Rainbows', on the north-western margin of the 'Sea of Rains' (Mare Imbrium). **2.** In certain bivalves (*Bivalvia), a recess or embayment in the pallial line; most bivalves with a pallial sinus are burrowers.

siphon In bivalve molluscs (*Bivalvia) and gastropods (*Gastropoda), a tube that connects the mollusc to the world outside, funnelling water towards and away from the gills. In bivalves siphons may occur in pairs.

siphonal canal In some gastropod (*Gastropoda) shells, an indentation or channel that accommodates the *siphon.

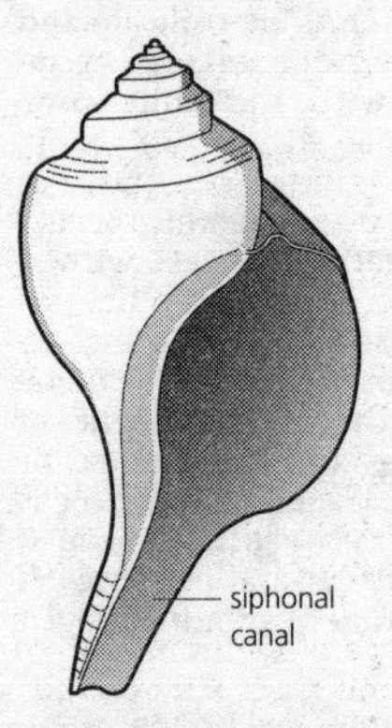

Siphonal canal

siphonate Applied to the *aperture of a gastropod (*Gastropoda) shell when a canal or notch for the *siphon is present.

siphonostomatous *See* APERTURE.

siphuncle A long tube present in those cephalopods (*Cephalopoda) that possess external shells. It runs internally, through all the chambers, and contains the siphuncular cord of body tissue, extending from the visceral mass through a perforation in each *septum of the shell. The siphuncle releases gas into the unoccupied chambers of the shell, making the animal buoyant.

sister groups The twin products of *cladogenesis which, under the rules of *phylogenetic systematics, must be classified at the same rank taxonomically (i.e. the two daughter lineages that result from the splitting of a single parent species).

sister taxa In *phylogenetics, two taxa connected through a single *internal node.

site investigation The geologic examination of a potential development site in order to design the foundations of surface buildings, roads, etc. It includes geophysical surveys, trial pits, and *boreholes.

Sivapithecus Genus of early *hominoid Primates, which probably includes the so-called **Ramapithecus*. They are known from E. Africa, south-eastern Europe, Turkey, Arabia, Pakistan, northern India, and southern China, from the Middle and Late *Miocene 15–8 Ma ago. The genus may include ancestors of great apes and humans, but certainly early relatives of the orang-utan have been identified among species of the genus. Although there are similarities between the teeth of *Sivapithecus* and those of humans, these are probably misleading, and *Sivapithecus* is now regarded as an ape that was ancestral either to the orang-utan alone, or to all living great apes and humans.

skarn A contact metamorphic rock (*see* CONTACT METAMORPHISM) composed of calcium, magnesium, and iron silicates (with or without iron, copper, and manganese sulphides and oxides) which has been derived from *limestone or *dolomite by the metasomatic (*see* METASOMATISM) introduction of large amounts of silicon, aluminium, iron, and magnesium from a nearby *igneous *intrusion, usually a *granite. Many skarns serve as host rocks for economic deposits of *magnetite and copper sulphides.

skeletal limestone *See* LEIGHTON–PENDEXTER CLASSIFICATION.

skeletal material In most vertebrates the skeleton is made from *bone (calcium phosphate); among *invertebrates it is more varied. *Calcite or *aragonite in various forms is common in such groups as the brachiopods (*Brachiopoda) and molluscs (*Mollusca). Normally the invertebrate skeleton is made up from several layers, and often each layer has a distinctive structure. The main, calcified portion of the shell is called the 'ostracum'; an outer layer, the 'periostracum' (made from layers of protein), disappears after death. Calcitic skeletons also occur in corals and *Bryozoa. Echinoderms (*Echinodermata) have skeletons made up of a number of elements. Each element is permeated with living tissue but the hard material is calcite, which forms in optical continuity to form a single *crystal. Chitin (a hydrocarbon related to cellulose) is the principal component of insect cuticle. It has been assumed that chitin impregnated with calcium carbonate forms the exoskeletal material in trilobites (*Trilobita), but although organic material occurs its nature is still unknown. Similarly, recent studies of graptolite (*Graptolithina) skeletal material have shown that the material is not chitin but a scleroprotein (a fibrous, insoluble protein). In some simple animals, e.g. *Radiolaria and some sponges (*Porifera), the skeleton is composed of opaline silica.

skeletal micritic limestone *See* LEIGHTON–PENDEXTER CLASSIFICATION.

skewness An expression of the degree of asymmetry shown on a frequency distribution. A non-skewed distribution is a perfect, normal (*Gaussian) distribution. *Grain-size distributions are skewed positively where larger particles are more numerous than fine ones, or negatively where finer particles exceed coarse within the distribution.

skin depth (z_s) The effective depth of penetration (*m*, in metres) of an *electromagnetic wave of frequency *f* Hz, at which the amplitude of the wave has been attenuated by 1/e (37%) of its surface value. It can be calculated by: $z_s = 503.8/\sqrt{(\sigma f)} = 503.8\sqrt{(\rho/f)}$, where σ is the conductivity (in S/m) and ρ the resistivity (in Ω/m) of the medium.

skip mark A type of *tool mark formed by the intersection at a low angle of an object with a muddy *sediment surface, so that the object skips across the surface, producing a linear series of depressions. *See also* SEDIMENTARY STRUCTURE.

Skolithos With **Ophiomorpha*, an *ichnoguild of cylindrical or U-shaped, vertical or steeply inclined *burrows, permanently occupied by suspension feeders. The burrows lack *spreiten, and were possibly formed in shallow, sublittoral environments. Burrows of this type are recorded from the *Cambrian to Recent.

SKS-wave *See* SEISMIC-WAVE MODES.

skull *See* CRANIUM.

Skythian *See* SCYTHIAN.

sky-view factor The extent of sky observed from a point as a proportion of the total possible sky hemisphere.

slab-pull The force, caused by the sinking of the cold, dense *lithosphere into the *asthenosphere at a *destructive margin, which is hypothesized to be one of the two major driving forces for the movement of *plates (the other is *ridge-push).

slake-durability test A test to estimate the resistance of rocks, particularly *argillaceous rocks, to a combination of wetting and abrasion. Test results are expressed as a slake-durability index for each particular rock.

slaking Breaking up of earth materials when exposed to water or air. The susceptibility of rock to slaking is measured with a *slake-durability test.

slant range In *radar terminology, the distance between the radar platform and an object on the ground.

slant-range resolution The minimum distance by which two objects on the ground must be separated in order to be resolved by a *radar. The minimum distance must be equal to or greater than half the *pulse length of a radar. *See also* AZIMUTH RESOLUTION.

SLAR *See* SIDE-LOOKING AIRBORNE RADAR.

slate Low-grade, *regionally metamorphosed rock, which is highly fissile and fine grained. The *fissility (*slaty cleavage) results from the parallel alignment of numerous fine *phyllosilicate minerals (e.g. *muscovite and *chlorite) induced by compressive tectonic deformation. The smooth, hard, impermeable surface produced when slate is split makes it commercially valuable for roofing, cladding of buildings, and for making such items as billiard-table tops, laboratory benches, and blackboards. *Compare* PHYLLITE; and SCHIST.

slaty cleavage A continuous *cleavage formed by the preferred alignment of homogeneously distributed *phyllosilicates throughout a rock. Slaty cleavage is commonly produced during lower grade *metamorphism. *See* METAMORPHIC GRADE.

sleet *Precipitation in the form of a mixture of rain and melting snow. In N. America the term is applied to ice pellets of less than 5 mm in diameter.

sleeve exploder A seismic source, used in marine investigations, in which a mixture of gases (e.g. propane and oxygen) is detonated inside a rubber sleeve which expands rapidly to accommodate the products of the explosion, thus generating a shock wave in the surrounding water. The exhaust gases are vented to the surface through a pipe to reduce problems caused by a *bubble pulse.

slick Quiescent area on the surface of a water body. The relative smoothness of these areas compared to adjacent waters is usually caused by a thin surface film of oil which changes the *surface tension.

slickenside 1. A *lineation on a *fault or *bedding plane caused by the frictional movement of one rock body against another. The plane may be coated by a *mineral, often *quartz or *calcite, which itself shows striations in the direction of movement. **2.** The polished surface left by the passage of a mud slide. **3.** In soils, the natural crack surfaces produced by swelling and shrinkage in clayey soils that are high in swelling *clays.

slide *See* GRAVITY SLIDING.

slide-rock *See* ROCK SLIDE.

slingram method A dual-coil, *electromagnetic *profiling system in which both the transmitter and receiver can be moved while maintaining a constant distance between them.

sling psychrometer *See* WHIRLING PSYCHROMETER.

slip 1. The relative displacement to either side of a *fault plane of points which were originally coincident. The total displacement (i.e. the sum of the *dip-slip and *strike-slip components) is called the 'net slip'. **2. (translation gliding)** The gliding of intracrystalline zones in relation to one another over distances which are integers of the unit pattern.

slipface The lee face of a sand dune where the surface is at the *angle of repose for sand (30–4°).

slip-off slope *See* MEANDER.

sloc The Gaelic name for long, deep, parallel-sided depressions at the coast, common on the Hebridean islands of Mull and Skye, caused by the erosion of *dykes cutting through the ancient *gneisses.

slope angle The gradient of a stable hillslope. In part this may reflect the *particle-size characteristics of the mantling *regolith. For example, maximum slope angles of about 26° are common in areas where the regolith consists of a mixture of rock rubble and coarse soil particles.

slope processes The set of geomorphological *processes that act on and below the surface of a hillslope, affecting the *regolith and bedrock. Some of the most important are *rainwash, *sheetwash, *weathering, *mass movement, *piping, and linear *erosion by *rill and gully activity. Each process tends to be associated with a distinctive *slope profile.

slope profile The two-dimensional form of a hillslope when measured down the steepest gradient. Traditionally it has been divided into a number of units, each of which reflected a distinctive geomorphological *process. For example, in 1957 L. C. King identified four elements in his ideal profile: a crest (or 'waxing slope' or 'convex slope') dominated by *creep; then a *scarp (or 'free face') affected by *rill activity and *mass movement; followed by a debris (or 'constant') slope where *talus accumulated; and succeeded finally by a *pediment (or 'waning slope') modified by *sheetwash. Subsequently, a nine-unit model has gained some acceptance.

slope stabilization The stability of slopes is important in the design of such excavations as open pits, quarries, and foundations, and in natural slopes forming cliffs, valley sides, and reservoirs, where movement may have serious consequences. Investigations into slope stability include measurements of shape, geologic structures, and soil strengths. Stabilization can be achieved by regrading, grassing, *dewatering, meshing, *grouting, *shotcreting, *rock bolting, or combinations of these.

slow-neutron process (s-process) In a second-generation star, successive neutron capture on a slow time-scale, termed the 's-process', is considered to account for the abundance distribution of the heaviest nuclides. The neutrons are furnished by the basic sequence of energy-producing reactions, i.e. *hydrogen 'burning', *helium 'burning', etc., thus continuing the element-building process beyond Fe. *See also* NUCLEOSYNTHESIS.

slump structure A *sedimentary structure consisting of overturned *folds, formed by the mass sliding of the semi-consolidated *sediment downslope under the influence of gravity.

Smålfjord A *stage of the *Varanger Epoch, dated at 610–600 Ma (Harland et al., 1989).

small circle A circle on the surface of a sphere whose centre is not coincident with the centre of the sphere, so that a plane whose circumference is a small circle does not bisect the sphere. In a *stereographic projection a small circle projects on to the *plane of projection as a circle whose centre lies inside the *primitive circle. If the small-circle plane is horizontal it will project as a circle smaller than the primitive circle but concentric with it. If it is vertical it will project as an arc, concave towards the centre of the primitive circle.

smarl A *pelagic or *hemipelagic sediment (an *arl), typically found interbedded with purer oozes in beds up to 1.5 m thick, with a composition intermediate between a non-biogenic sediment and a calcareous or siliceous ooze. It is 30% clay and 70% microfossils, with siliceous and calcareous microfossils present in approximately equal amounts. *Compare* MARL; and SARL.

SMC *See* SATURATION MOISTURE CONTENT.

smectite A family of *clay minerals that includes *montmorillonite and *bentonite.

Smithian *See* SCYTHIAN.

smithsonite (calamine) Mineral, $ZnCO_3$; sp. gr. 4.4; *hardness 4.5; *trigonal; colour variable, shades of grey, brown, or greyish-white, but green, brown, and yellow types also occur; grey *streak; vitreous *lustre; crystals rare, but when they develop *rhombohedral with curved faces, more usually occurs as *botryoidal and stalactitic masses; *cleavage perfect rhombohedral; occurs in the oxidized zone of zinc ore deposits, commonly associated with *sphalerite, *galena, and *calcite, also as a *replacement in *limestone, and in *hydrothermal veins; soluble in dilute hydrochloric acid, with effervescence. The green variety is used to make ornaments. It is named after the British mineralogist James Smithson (founder of the Smithsonian Institution in Washington, DC).

Smith's rule The formula by which the maximum possible depth of a body of unknown shape can be determined using gravity data. The body can have either a positive or negative density contrast with the surrounding rocks. The maximum depth, d_{max}, is given by: $d_{max} = Ag_{max}/(\delta g/\delta x)$, where $\delta g/\delta x$ is the maximum horizontal gradient of the *gravity anomaly, g_{max} is the value of the anomaly peak, and A varies between 0.86 for a three-dimensional body to 0.65 for an essentially two-dimensional body.

Smith, William (1769–1839) A land surveyor engaged on canal construction whose work led him to see that strata could be identified and correlated by their *fossil content. In 1815 he published the first geologic map and sections of England, showing the sequence of strata.

smog Naturally occurring fog mixed with visible (smoke) and/or invisible pollutants. *See also* PHOTOCHEMICAL SMOG.

smoker *See* HYDROTHERMAL VENT.

SMOW (Standard Mean Ocean Water) A sea-water sample which comprises the international standard for *D : H and *^{18}O : ^{16}O ratios. Differences in isotopic composition are expressed as parts per mille deviations from the isotopic composition of this standard.

smudging The burning of materials (e.g. oil) to produce a smoke layer that reduces the effect of radiation cooling of the air above the ground surface. It is used as a protective measure (e.g. in fruit-growing areas) especially in *frost hollows.

SNC *See* SHERGOTTYITE/NAHKLITE/CHASSIGNITE METEORITES.

Snell's law The ratio of the sine of the *angle of incidence (i) to the sine of the *angle of refraction (r) is a constant for any two *isotropic media bounded by a common *interface. The refractive index n is given by: $n = \sin i/\sin r$, and $n_1 \sin i = n_2 \sin r$, where n_1 and n_2 are the refractive indices of the two media. The law also indicates that a *P-wave incident upon a boundary will be reflected and refracted partly as a P-wave and partly as an *S-wave. The law was formulated by the Dutch astronomer and mathematician Willebrord Snell (1591–1626).

snout The steep, terminal zone of a *glacier. It is usually heavily loaded with debris.

snowblitz theory A theory which proposes that following a bad winter with heavy snowfall, snow persists in lowland areas throughout the summer. This increases the *albedo and thus reduces the amount of solar warming of the ground. More snow is added during the next winter, and more snow may thus accumulate year by year. An *ice-cap may develop, and glaciation may occur after only a few hundred years. Such a sequence of events is more likely in high than in low latitudes.

snowflake The result of the growth of ice crystals in a varied array of shapes. Very low temperatures usually result in small flakes; formation at temperatures near freezing point produces numerous crystals in large flakes.

snow-gauge A device for collecting and measuring snowfall and ice precipitation. The recording is usually made after the snow or ice has melted.

snow grain Small ice particle precipitated as a usually flattish grain.

snow line The lower limit of permanent snow cover. The height of the line varies with latitude; locally it also varies with aspect, because of the relationship to prevailing winds and quantity of snow deposited, and to summer temperatures, etc.

soapstone *See* TALC.

soda lake *See* NATRON LAKE.

sodalite An important group of *silicate minerals belonging to the *feldspathoids and including sodalite $Na_8[Al_6Si_6O_{24}]Cl_2$, *nosean, *haüyne, *cancrinite, *lazurite, and *scapolite; sp. gr. 2.27–2.88; *hardness 5.5–6.0; greyish to bluish, yellow; occurs as *rhombohedra or *massive; *vitreous *lustre; occurs in *nepheline syenites in association with *nepheline and *fluorite, and in metasomatized (*see* METASOMATISM) calcareous rocks near *alkaline *igneous intrusions.

soda nitre (Chile saltpetre, nitratine) Very soluble *mineral, $NaNO_3$; sp. gr. 2.2–2.3;

*hardness 1–2; *trigonal; normally colourless or white, with various darker colours due to impurities; white *streak; *vitreous *lustre; *crystals rhombohedral, also occurs *massive; *cleavage perfect, rhombohedral; occurs in arid regions as surface deposits together with *gypsum, *halite, and other soluble nitrates and *sulphates; deliquescent. It is worked as a source of nitrate.

sodication In soils, an increase in the percentage of exchangeable sodium. Sodium adsorbs on to *cation-exchange sites in the soil, causing soil *aggregates to disperse, which closes soil *pores and renders the soil impermeable to water. *See also* SODIUM-ADSORPTION RATIO.

sodic soil 1. Soil with a sodium content sufficiently high to interfere with the growth of most crop plants. **2.** Soil with more than 15% exchangeable sodium.

sodium-adsorption ratio (SAR) Describes the tendency for sodium *cations to be adsorbed at *cation-exchange sites in soil at the expense of other cations, calculated as the ratio of sodium to calcium and magnesium in the soil; more precisely, it is the amount of sodium divided by the square root of half the sum of the amounts of calcium and magnesium, where *ion concentrations are given in milliequivalents per litre. A low sodium content gives a low SAR value. In practice, allowance must be made for other reactions within the soil that do not involve sodium but that affect concentrations of calcium and magnesium. The SAR value is most likely to be changed by irrigation water.

sodium feldspar *See* ALKALI FELDSPAR.

sodium-sulphate soundness test A method of testing the *weathering resistance, particularly to frost action, of building materials. A sample is soaked in saturated sodium-sulphate solution, drained, and dried. This is repeated and the sample examined for cracks. The method simulates the stresses due to frost action.

SOFAR channel (sound channel) Acronym for the *SO*und *F*ixing *A*nd *R*anging channel, a zone in the oceanic water column at a depth of about 1500 m where the velocity of sound is at a minimum value. Sound passing through the zone is refracted upwards or downwards back into the zone, with little loss of energy, causing sound energy to be trapped in a zone of well-defined depth. The SOFAR channel may be used for the transmission of sound over long distances, exceeding 28 000 km, and can be used to track free-drifting, subsurface, *neutrally buoyant floats.

soil 1. The natural, unconsolidated, mineral and organic material occurring above bedrock on the surface of the Earth; it is a medium for the growth of plants. **2.** In engineering geology, any loose, soft, and deformable material, e.g. unconsolidated *sands and *clays.

soil air The soil atmosphere, comprising the same gases as in the atmosphere above ground, but in different proportions: it occupies the *pore space of the *soil.

soil anchor *See* ANCHOR.

soil association 1. Group of *soils forming a pattern of soil types characteristic of a geographical region. **2.** Mapping unit used to denote the distribution of soil types where the scale of the map does not require or permit the identification of individual soils. *See also* SOIL COMPLEX.

soil-atmosphere survey A method for finding geochemical anomalies, which uses geochemical features of the gases trapped in *soil *pore space. The method can be used only to find *ore minerals which, on *weathering or as a result of their radioactive decay, liberate gases which can readily be analysed (e.g. radon, released by the decay of uranium).

soil borrow The transference of material from elsewhere for refilling excavations, etc.

soil complex Mapping unit used to denote the distribution of *soils: it is more precise than a *soil association, and is used where soils of different types are mixed geographically in such a way that the scale of the map makes it undesirable, or impractical, to show each one separately.

soil conservation The protection of the *soil by careful management, to prevent physical loss by *erosion and to avoid chemical deterioration (i.e. to maintain soil *fertility).

soil formation The action of the combined primary (*weathering and humification) and secondary processes to alter and to rearrange mineral and organic material to form *soil, involving the differentiation of *soil profiles and the formation of loose soil from consolidated rock material. *See also* PEDOGENESIS.

soil grading curve A graph of *grain size plotted on a horizontal logarithmic axis against percentage on an arithmetical vertical axis. A point on the curve gives the percentage by weight of material smaller in size than that at the given point on the graph. *See* GRADING CURVE.

soil horizon A relatively uniform *soil layer which lies at any depth in the *soil profile, which is parallel, or nearly so, with the soil surface, and which is differentiated from adjacent horizons above and below by contrasts in mineral or organic properties. Soil horizons are grouped primarily into O, A, B, and C horizons. O horizons (formerly known as Ao horizons) comprise organic material at the surface. A horizons are surface horizons of mixed organo-mineral composition. Where mineral matter has been lost, the A horizon is sometimes called the E (for eluviated) horizon. Where they are present, B horizons are usually located in the middle of the sequence, and are horizons into which material (mineral and organic) is deposited, thus altering the character of the horizon. C horizons are soil *parent materials, weathered but not otherwise altered by pedogenic processes. The underlying unweathered material is sometimes called the D or R horizon. In addition to these, surface litter may form an L horizon, above a layer of fermented material (F horizon) and, below that, humified material (H horizon), and a mineral crust, often cemented, is sometimes called the K horizon. *See* PEDOGENESIS.

soil individual *See* POLYPEDON.

soil line In *remote sensing, the line which runs at 45° on a plot of *digital number values of red wavelengths of light against digital number values of very-*near infrared wavelengths of light. Soils plot very close to this line, vegetation plots away from this line having a higher tendency to reflect very-near infrared. *See also* VEGETATION INDEX.

soil management A variety of practices and operations with respect to *soil, that aid the production of plants; normally they are planned to allow for sustained yield in the future.

soil mechanics Study of the mechanical properties of loose or unconsolidated particles, especially their composition, shear resistance, and the effects of water. It is applied to *soils to determine their suitability for building sites, mining, etc., and to engineering problems dealing with the stability of *foundations due to mechanical and chemical *weathering of rocks. *See* SHEAR STRENGTH.

soil-moisture content The ratio of the volume of contained water in a *soil compared with the entire soil volume. When a soil is fully saturated, water will drain easily into the underlying unsaturated rock. When such drainage stops, the soil still retains *capillary moisture and is said to contain its *field-capacity moisture content. Further drying of the soil (e.g. by evaporation) creates a soil-moisture deficit, which is the amount of water which must be added to the soil to restore it to field capacity, measured as a depth of *precipitation.

soil-moisture deficit *See* SOIL-MOISTURE CONTENT.

soil-moisture index *See* MOISTURE INDEX.

soil-moisture regime The changing state of soil moisture through the year, which reflects the changing balance of monthly *precipitation and *potential evapotranspiration above the ground surface. When the latter exceeds the former the period is one of soil-moisture deficit in the annual regime.

soil profile A vertical section through all the constituent *horizons of a soil, from the surface to the relatively unaltered *parent material.

soil separates Size divisions of mineral particles (*sand, *silt, and *clay) that comprise the fine earth, each particle being less than 2 mm in diameter.

soil series Basic unit of soil mapping and classification: all *soils in a series have similar profile characteristics and have developed from the same *parent material.

soil structure Grouping of individual *soil particles into secondary units of *aggregates and *peds.

soil survey 1. Systematic examination and mapping of *soil in the field. **2.** Chemical analysis of soil for the detection of geochemical anomalies. *See* GEOCHEMICAL SOIL SURVEY; and ORIENTATION SURVEY.

soil taxonomy The classification of types of soil, in a manner similar to that used for biological classification. In the most widely used system, devised by workers at the US Soil Survey, within the *USDA, soils are divided into 11 orders: *Alfisols, *Andisols, *Aridisols, *Entisols, *Histosols, *Inceptisols, *Mollisols, *Oxisols, *Spodosols, *Ultisols, and *Vertisols. These orders are further subdivided into suborders, great groups, families, and soil series, defined by *diagnostic horizons.

soil variant *Soil sufficiently different in properties from adjacent soils to warrant the use of a new *soil-series name, but occupying a geographical area too small to warrant the issuing of such a new name.

soilwash *See* RAINWASH.

soil-water zone (unsaturated zone, vadose zone) The zone between the ground surface and the *water-table. Water is able to pass through this zone to reach the water-table, but while in the zone it is not given up readily to wells because it is held by *soil or rock particles and capillary forces. *See* GROUNDWATER.

Sojourner *See* MARS PATHFINDER.

sol 1 Colloidal solution (*see* COLLOID) or dispersion of solid particles in a liquid, as in a completely fluid mud. *Compare* QUICK CLAY; and GEL. **2.** One martian day (= 24.7 hours).

sola *See* SOLUM.

solar abundance of elements Studies of the solar spectrum have determined the relative abundances of about 70 elements in the *Sun's atmosphere. Hydrogen and helium predominate, and in general abundance decreases with increasing atomic number. A few exceptions, however, e.g. silicon and iron, have a high abundance, probably correlated with nuclear binding energies and nuclear stability. Spectra originate from the outer layers of the Sun, so their value in determining the total solar abundance of elements should not be overemphasized, but total abundances of elements are much less important than their relative proportions.

solar constant The mean intensity of the solar beam in free space (i.e. before penetrating the Earth's atmosphere) at the average distance of the *Earth from the *Sun. The intensity is not strictly constant for all the wavelengths of radiation involved. The amount of variability is still a subject of debate, but is certainly very small apart from the long-term development in the history of the Sun.

solar cosmic rays *See* COSMIC RADIATION.

solar flare A short-lived, cataclysmic outburst of solar material, driven by magnetic forces, from a relatively small area of the solar surface and generating particles with energies in the range 1–100 MeV which produce track records, e.g. in exposed lunar minerals.

solarimeter Instrument for measuring *solar radiation.

solar magnetic variation *See* DIURNAL VARIATION (2).

solar nebula *See* SOLAR SYSTEM.

solar pond *See* SALINA.

solar radiation Electromagnetic radiated energy from the *Sun. This is the dominant energy input to the *Earth and is intercepted by the *atmosphere and absorbed at the surface. *See also* RADIATION BUDGET.

solar system The system that consists of the central *Sun (G spectral type star), around which orbit nine planets, about 60 *satellites, about 3000 discovered *asteroids, and probably 10^{12} *comets. Most bodies lie close to the plane of the *ecliptic. The age of the solar system, 4.56 billion years obtained from *meteorites, marks the formation of the system from a rotating cloud of dust and gas (the solar nebula).

solar wind General term for the stream of high-energy particles (mainly protons, electrons, and alpha particles) emitted by the *Sun. The particles have velocities of hundreds of kilometres per second and 'wind' strength is thought to be greatest during periods of maximum solar activity. In the neighbourhood of the *Earth the solar wind has velocities in the range of 300–500 km/sec and an average density of 10^7 ions/m^3. *See also* COSMIC RADIATION.

Solenopora *See* RHODOPHYCEAE.

sole structure The term applied to a group of *sedimentary structures found on the base of beds. Sole structures are mostly formed by the scouring action of a current, or by the passage of an implement (*tool) over a muddy substrate, followed by the infilling of the scour by *sands (*scour and fill). Sole structures include *flute marks, produced by turbulent flows of water; *skip, prod, bounce, drag, and *groove marks, caused by the transport of an object over the mud; and *load casts, formed by the sinking of dense sand into underlying layers of less dense mud. Sole structures provide an important means of determining the way-up of beds, and in many cases give an important indication of palaeocurrent direction. *See* PALAEOCURRENT ANALYSIS.

sole thrust (basal thrust) In a *thrust terrain, the lowest regional thrust surface. *See also* FLOOR THRUST.

solfataric activity The quiet escape of hot, sulphur-rich gases from recently emplaced volcanic bodies. The name is derived from the fields of sulphurous gas vents at the Solfatara crater, north of Naples, Italy. When cooled by the atmosphere the escaping gases deposit many minerals, including chlorides, sulphur and *hematite.

solid Applied to a map (e.g. in Britain one published by the Geological Survey) which depicts the *outcrop patterns of rocks unobscured by their cover of recent superficial deposits (e.g. glacial, *alluvial, or marine *sediments).

solid-melt equilibrium The degree of *partial melting which, in any particular rock, is a function of temperature, pressure, and the availability of water, and is usually controlled by the abundance and thermal stability of hydroxyl-bearing minerals, e.g. *muscovite, *biotite, and *hornblende.

solid solution Solid crystalline *phases representing a mixture of two or more *end-members and which may vary in composition within finite limits without the appearance of another phase. A mineral may exhibit solid solution involving atoms of an *isomorphous mineral; for example, magnesium and iron *ions are similar in size, and complete solid solution occurs between *forsterite (Mg_2SiO_4) and *fayalite (Fe_2SiO_4) end-members of the *olivine series, in the most common type of solid solution, known as 'substitutional solid solution'. *See also* IONIC SUBSTITUTION.

solid-state imaging camera (SSI) In *remote sensing, an instrument that operates in the visible spectrum and uses a *charge-coupled device to enhance its images.

solidus The position of points marking the boundary between complete solid and liquid/solid at equilibrium, in a *temperature–composition diagram. In *binary systems the solidus is a straight or curved line, in *ternary systems a flat plane or curved surface.

solifluction (solifluxion) Downhill movement of *regolith that has been saturated with water. It was originally described in *periglacial regions (*see* GELIFLUCTION), but the term was subsequently widened to include all environments. The thick regolith of the humid tropics is particularly prone to solifluction after intense rainfall.

solifluxion *See* SOLIFLUCTION.

solitary corals Those corals where a single *corallite makes up the *corallum. Shape and size range from an extremely low cone with an apical angle of 120° or more (patellate), through discoid (button-shaped), to horn-shaped and slender with an apical angle of 20° or less (ceratoid). A steep, conical shape with an apical angle of 40° is called 'trochoid'; a subcircular, parallel-sided corallite 'cylindrical'; cylindrical corallites that bend crookedly in a worm-like manner are 'scolecoid'; a corallite shaped like the toe of a slipper is 'calceolid'. Other shapes are also described.

solodic soil *Leached, formerly *saline soil, associated with semi-arid tropical environments, in which the A *soil horizon has become slightly acid, and the B horizon is enriched with sodium-saturated *clay. The term was used in soil classification systems derived from early Russian systems based on the work of V. V. *Dokuchaev, but is no longer used in soil classification.

solonetz Mineral soil at a transitional stage of *leaching or a *saline soil during solodization, in semi-arid, tropical environments; it has a sandy, acid, A *soil horizon and a B horizon partially enriched with sodium-clay. The term was used in earlier systems of soil classification. Solonetz soils fall within the order *Aridisols.

solstice The time of most northerly or southerly declination of the sun from the equator. In the northern hemisphere, the summer solstice is around 22 June and the winter solstice around 22 December. These dates are reversed for the southern hemisphere.

solubility product Constant which describes the product of ionic concentrations of any slightly soluble salt. The solubility of a substance is the maximum amount of that substance which can be dissolved in a *solution that is in equilibrium with a solid source of the solute. The main technique for studying solubility involves the use of the solubility product, which is the total number of *ions of each type in a compound that can co-exist in a solution. At a given temperature and at equilibrium the value for the solubility product will always be the same. For example, in a solution of silver chloride, AgCl dissociates to Ag^+ and Cl^-. The solubility product K_{sp} is the product of their concentrations: $K_{sp} = [Ag^+][Cl^-]$ mole²/litre². The solubility of AgCl in moles per litre is equal to the activity of Ag^+ or Cl^- as one mole of AgCl dissolves in water to give one mole of each in solution. The value of K_{sp} for AgCl at 25 °C is $10^{-9.8}$.

solum (pl. sola) The upper part of a *soil profile, above the *parent material, in which processes of soil formation occur, and within which most plant roots and soil animals are found.

solution **1.** A physically homogeneous mixture of two or more substances in which solid, liquid, or gaseous *phases may combine in one of those phases. A constituent of a solution can be separated out by changing its phase, e.g. boiling, condensing, or freezing. Where a solution is formed by dissolving a quantity of one substance in a larger quantity of another, the smaller quantity is called the 'solute', the larger quantity, the 'solvent'. *Compare* COLLOID. **2.** A *weathering process by which weakly bonded ionic components of minerals are detached through the attraction of water molecules (which carry a positive electrical charge at one end and a negative charge at the other, although they are neutral overall), and then carried away from the weathering environment. *Halites and the sulphates and carbonates of magnesium and calcium are especially vulnerable. Solution is usually the first stage of *chemical weathering.

solution channel An elongate void within a rock, which has been enlarged by the *solution action of moving *groundwater on the rock itself. Solution channels are most commonly associated with *carbonate rocks, and groundwater in them can flow as fast as in a river on the ground surface. *See* KARSTIC AQUIFER.

solution cleavage A spaced, usually disjunctive *cleavage which is common in *quartzites and limestones. Solution cleavages commonly contain zones of relatively insoluble minerals indicating that *pressure solution has operated. Some authorities (e.g. C. McA. Powell, 1979) suggest that the term 'solution cleavage' should not be used in descriptions of cleavage as it implies the mode of origin.

solution pipe A cylindrical, near-vertical pipe that is developed at a *joint intersection in a *karst environment. It is caused by a local increase in the rate of *carbonation resulting from enhanced drainage.

Solvan A *stage of the Middle *Cambrian, underlain by the *Lenian and overlain by the *Menevian and dated at 536–530.2 Ma (Harland et al., 1989).

solvus In a geochemical system, a line or surface which separates a homogeneous *solid solution from a field of several *phases which may form by *exsolution or *incongruent melting.

Somali Plate *See* AFRICAN PLATE.

sombric horizon Subsurface *soil horizon of well-drained, mineral, tropical and subtropical soils into which *humus has *leached downward. Base saturation is less than 50%. It is a *diagnostic horizon.

Somoholoan *See* ASSELIAN; and SAKMARIAN.

sonar *See* SIDE-SCAN SONAR; SONIC LOG; and ECHO-SOUNDING.

sonde *See* RADIOSONDE; RAWINSONDE; SELF-POTENTIAL SONDE; SONIC SONDE; and WELL LOGGING.

sonde self-potential *See* SELF-POTENTIAL SONDE.

sonic log *See* CONTINUOUS VELOCITY LOGGING; and SONIC SONDE.

sonic sonde An instrumental package containing two seismic-energy sources and four *geophones which allow the measurement of the *seismic velocities of the rocks in a *borehole as the sonde is pulled through it. The record is a sonic log (velocity log). *See also* WELL SHOOTING.

sonobuoy A disposable, free-floating buoy used in large-scale marine *seismic-refraction surveys. One or more hydrophones are suspended from the buoy and detect the *head waves which are then transmitted back to the firing ship, timed, and recorded. After a certain time the buoys are designed to sink automatically.

sonograph A graphic presentation of reflected sound waves from a sonar scanner.

Sorby, Henry Clifton (1826–1908) An English amateur scientist, Sorby studied estuarial and inland waters of England, but he is best known for developing the study of rocks in *thin sections, using the techniques invented by *Nicol. He was the first to show that individual mineral crystals and grains could be identified using this process. He also used the method to study *meteorite sections.

Sordes pilosus Discovered in 1971, one of the first *pterosaurs known to have been covered in thick fur: the name means 'hairy filth'. The indication is that this reptile and its close relatives were *homoiotherms. It was found in Upper *Jurassic sediments of Chimkent, Soviet Kazakhstan, and was a small, toothed pterosaur with a long tail.

sorosilicate *See* SILICATES.

sorted biosparite *See* FOLK LIMESTONE CLASSIFICATION.

sorting An expression of the range of grain sizes (*see* PARTICLE SIZE) present in a *sediment. A well-sorted sediment is characterized by a narrow range of grain sizes, whereas a poorly sorted sediment contains a wide range of grain sizes.

Soudleyan A *stage of the *Ordovician in the Middle *Caradoc, underlain by the *Harnagian and overlain by the *Longvillian.

sound channel *See* SOFAR CHANNEL.

sounder (radio) An instrument which emits a continuous series of short pulses of electromagnetic energy towards a planetary or *satellite surface. The return signal provides a map of the subsurface electrical conductivity, which can be used to infer subsurface structure, as well as information on topography. This technique was used on the Apollo 17 lunar mission to study subsurface structure.

sound speed The rate at which sound energy moves through a medium. In sea water this is between 1400 and 1550 m/s. In sea water, the speed of sound is a function of temperature, *salinity, and pressure due to depth. At a salinity of 34.85 parts per thousand and a temperature of 0 °C, the speed of sound is 1445 m/s. It increases by approximately 4 m/s for each degree Celsius rise in temperature, by 1.5 m/s for each 1 part per thousand increase in salinity, and by 18 m/s for each 1000 m increase in depth.

source region (for air masses) Extensive areas of essentially uniform surface conditions over land or water, typically of large-scale air *subsidence and lateral divergence, where *air masses develop their initial properties.

source rock 1. *Sediment (usually *shale or *limestone) in which *hydrocarbons originate; it contains more than 5% organic matter and has the potential to generate *petroleum. **2.** Any parent rock from which later sediments are derived.

South American Plate One of the major present-day lithospheric *plates, extending from the *Mid-Atlantic Ridge in the east to the subducting *Nazca Plate in the west, with most of the boundaries with other plates (i.e. *Antarctic, Scotia, *Caribbean, and *North American) being *transform faults.

South-east Pacific Plate A lithospheric *plate which is now coupled with the *Antarctic Plate, but whose oceanic *lithosphere is interpreted to have been subducted under southern Chile and the Antarctic Peninsula.

southerly buster Regional wind in southern and south-eastern Australia, characterized by a rapid shift in direction from north-west to south in the rear of a *cold front. Such winds are especially prevalent between October and March. The change to a southerly wind can bring a great increase in wind speed accompanied by a rapid and marked fall in temperature. Such conditions are akin to *line squalls, and related to the S. American *pamperos.

Southern Ocean *See* ANTARCTIC OCEAN.

southern oscillation A fluctuation of the intertropical atmospheric circulation, in particular in the *Indian and *Pacific Oceans, in which air moves between the SE Pacific subtropical high and the Indonesian equatorial low, driven by the temperature difference between the two areas. The general effect is that when pressure is high over the Pacific Ocean it tends to be low in the Indian Ocean, and vice versa. The phenomenon is strongly linked to *El Niño.

sövite A type of *carbonatite that consists largely of *calcite accompanied by minor *magnetite, and *apatite with or without *phlogopite.

SP *See* SELF-POTENTIAL SONDE.

spaced cleavage *See* FRACTURE CLEAVAGE.

space lattice *See* LATTICE.

spallation 1. (nuclear) A nuclear reaction involving the ejection of many particles from an atomic nucleus, following a collision with a high-energy particle. Both the mass number and the atomic number of the target nucleus are changed by the event. **2. (planetary geol.)** The removal of the surface layers of a rock by the interaction of a compressional shock wave with the surface, caused by micrometeorite impact. *See* 'ZAP PITS'.

spandrels of San Marco An analogy used, in a classic paper by Stephen Jay Gould and Richard Lewontin, to indicate how non-adaptive characters may arise in evolution. Spandrels are the spaces left between the tops of neighbouring arches in churches (in this case, St Mark's cathedral, in Venice); these spaces, not related to the functional architecture, are free to be decorated in a non-functional fashion.

sparite Sparry *calcite *cement. Sparite is the coarse crystalline calcite cement which fills *pore spaces in many *limestones after deposition, formed by the *precipitation of calcite from *carbonate-rich solutions passing through the pore spaces in the *sediment.

-sparite *See* FOLK LIMESTONE CLASSIFICATION.

sparker A seismic source created by the rapid formation of a gas-plasma bubble from the ionization of sea water after the discharge of a high-voltage spark from a comb of electrodes contained within a frame. It provides good resolution (1–2 m) with limited depth penetration (less than 100 m).

sparse biomicrite *See* FOLK LIMESTONE CLASSIFICATION.

spastolith A *grain which, being composed of soft material, became squashed and deformed by mechanical *compaction during burial.

Spathian The final *stage of the *Scythian Epoch, preceded by the *Nammalian.

spathose iron *See* SIDERITE.

spatial frequency In *remote sensing, the frequency of change per unit distance across an image. High spatial frequencies include those changes which occur in very close proximity, such as fine lines, low spatial frequencies include those changes which occur over greater distances, such as broad bands. The ability of the human eye to discern spatial frequency is limited and so selective removal of certain spatial-frequency ranges within an image may result in a more interpretable image with less noise. *See also* SPATIAL-FREQUENCY FILTER.

spatial-frequency filter In *remote sensing, a filter used to enhance the appearance of the spatial distribution of data in an image to make it more interpretable to the human eye. Spatial-frequency filters examine the spatial variations in *digital number of an image and are used to modify the image by selectively suppressing or separating certain *spatial-frequency ranges. Spatial-frequency filters include *directional filter, *high-pass filter, medium-pass filter, and low-pass filter. *See also* MEDIAN FILTER.

spatter Fluid basaltic *pyroclasts which accumulate by fallout from a *Strombolian volcanic *eruption column to form a rampart around the vent. The individual clots are so fluid when they land that they often mould together, like flattened pancakes.

spatter cone (driblet cone) Small (usually 5–20 m high), volcanic cone built from *tephra blown out as clots of relatively fluid basaltic *lava.

spatter-fed flow *See* HAWAIIAN ERUPTION.

specialization Degree of adaptation of an organism to its environment. A high degree of specialization suggests both a narrow habitat or niche and significant inter-specific competition.

species (sing. and pl.) 1. A class of particular chemical individuals all of which are similar, e.g. *ions, atoms, or molecules. **2.** *See* CLASSIFICATION.

species longevity The persistence of species for long periods of time, characterizing for example *Gastropoda and *Bivalvia.

species selection A postulated evolutionary process in which selection acts on an entire species population, rather than individuals. This might occur, for example, as a consequence of the geographical range of a population, which affects the population as a whole and, possibly, its longevity or development.

species zone *See* TAXON RANGE ZONE.

specific gravity (sp. gr.) The ratio of the weight of a substance to the weight of an equal volume of water, expressed as a number. For example, the weight of a given volume of *quartz, with a specific gravity of 2.65, is 2.65 times that of the same volume of water. The average sp. gr. of metallic minerals is about 5. *See also* DENSITY.

specific-gravity determinations In soils, for engineering computations, grains are weighed in a calibrated glass container (*pycnometer), carefully excluding air, to relate the mass of the sample to its volume. The *specific gravity of solid rocks can be determined using a Walker balance, and can be carried out on dry or water-saturated samples. *See also* DENSITY DETERMINATION.

specific humidity The mass of water vapour in a unit mass of air. *See also* HUMIDITY; and MIXING RATIO.

specific retention The ratio of the undrained water to the total water in a rock, the undrained water being water contained in rock voids or *pore spaces, from which it cannot be recovered by drainage or pumping. It is retained against the action of gravity by molecular attraction and *capillarity.

specific yield **1.** The ratio of the water drained from a rock under the influence of gravity, or removed by pumping, to the total volume of the rock voids or *pore space in the drained rock. The difference is caused by the retention of water in the rock, due to molecular attraction and *capillarity. *See* SPECIFIC RETENTION. **2.** The volume of water released by a falling *water-table from a given volume of fully saturated rock.

spectra *See* SPECTRUM.

spectral hue A *hue which is present in the spectrum of colours produced by splitting white light with a prism. Spectral hues include red, green, and blue. *Compare* NON-SPECTRAL HUE.

spectral radiance The *radiance of a specified wavelength of *electromagnetic radiation.

spectrochemical analysis An analytical technique in which a sample is heated to a high temperature, usually in a carbon arc, to produce emission lines whose intensities are proportional to the abundance of elements present. Line intensities may be recorded on photographic plates or measured directly by a photosensitive device, e.g. a photomultiplier. *See also* SPECTRUM.

spectrograph Analytical instrument used mainly for elemental analysis.

spectrometer An instrument which, with associated equipment, furnishes the ratio, or some function of a ratio, of the radiant power of two electromagnetic beams as a function of their spectral wavelength. It requires a source of radiation (in emission spectrometers the sample serves as its own source), a means of distinguishing between different radiation frequencies, and includes narrow band filters, a prism, a diffraction grating, a system of slits to isolate a narrow band of radiation, a sample-containing system, and a photodetector, amplifier, and output device (meter, recorder, VDU, etc.) *See* GAMMA-RAY SPECTROMETRY; and MASS SPECTROMETRY.

spectrophotometer An instrument for determining the intensity of the light absorbed by a compound (usually in *solution). The light absorbed at a given wavelength is proportional to the concentration of the compound in the solution.

spectroradiometer A *spectrometer which measures very narrow wavelengths of the *electromagnetic radiation radiated or reflected by a surface. It is used in *remote sensing to establish the spectral characteristics of a surface material.

spectroscope An instrument used in spectroscopy, whose main features are a slit and collimator, prism, telescope, and counter. A parallel beam of radiation is passed through the prism, so dispersing different wavelengths through different angles of deviation, which can be measured.

spectroscopic binary *See* STAR PAIR.

spectrum (pl. spectra; optical emission spectrum) A series of lines (line spectra), produced as *electrons return to their original energy levels and emit excess energy as infrared, visible, or ultraviolet light of characteristic wavelengths, after atoms have been heated strongly and valence electrons in the outer shell have moved to higher energy levels. Each element has a characteristic line spectrum. The intensity of each line is related to the concentration of the element being excited.

specular *See* SPLENDENT.

specularite *See* HEMATITE.

specular reflection *Reflection of light or a radar beam, as from a mirror or a plane faceted surface such as an angular boulder. *See* BACKSCATTER; *compare* DIFFUSE REFLECTION.

spelean Pertaining to caves.

speleothem *See* DRIPSTONE.

Spermatophyta In former classifications, a division of the plant kingdom comprising 2 subdivisions, the Gymnospermae (*gymnosperms) and Angiospermae (*angiosperms). These are now known informally as *seed plants.

spessartine (spessartite) Member of the *garnet group of *minerals, that has the formula $Mn_3 Al_2 (SiO_4)_3$; sp. gr. 4.18; *hardness 6.5–7.5; *cubic; dark red to orange-yellow or brown; greasy to vitreous *lustre; crystals commonly *dodecahedra; widely distributed in *metamorphic and *igneous rocks, and in beach and river *sands. It is named after the Spessart Mountains, Bavaria, W. Germany.

spessartite 1. A type of *lamprophyre, characterized by *essential *hornblende and *plagioclase feldspar. *Aphyric varieties are called '*malachites'. **2.** *See* SPESSARTINE.

sp. gr. Abbreviation for *specific gravity.

sphaericone *See* INVOLUTE.

sphalerite (black jack, zinc blende) Mineral, ZnS; sp. gr. 3.9–4.1; *hardness 3.5–4.0; *cubic; colour variable, but commonly yellow, brown, or black, and crystals can be transparent to translucent; brownish red to bright yellow or white *streak; resinous to near-metallic *lustre; crystals *tetrahedral or *dodecahedral, with curved faces, but also granular, *fibrous, or *botryoidal; *conchoidal fracture; *cleavage perfect {011}; the most common *ore mineral for zinc metal, Zn–Pb is common in strata-bound veins and massive sulphide deposits, frequently associated with *galena in *hydrothermal veins, and in *limestones where it occurs by *replacement, commonly with *pyrite, *pyrrhotite, and *magnetite; dissolves in concentrated nitric acid with the separation of sulphur.

sphene (titanite) A *nesosilicate $CaTiSiO_4(O,OH,F)$; sp. gr. 3.45–3.55; *hardness 5; *crystals wedge-shaped, occasionally *massive; brown, grey, reddish brown, yellow, or black; *monoclinic; *adamantine to *resinous *lustre; occurs as a primary *accessory mineral in *calc-alkaline *igneous rocks and alkaline igneous rocks in which it may occur as a major constituent along with *apatite, *nepheline, and *aegirine. It is common in *contact metamorphosed *limestones, particularly *skarns.

sphenoid A wedge-shaped, four-faced, *closed form, normally occurring in the *tetragonal or *orthorhombic *crystal systems, but with a special form (the 'monoclinic sphenoid', or 'dihedron') occurring in the *monoclinic system. The terminology for sphenoidal forms is complex and inconsistent, although in general the *crystallographic axes emerge at the centre of the edges where pairs of triangular faces meet.

Sphenopsida (horsetails) Class of the *Pteridophytina that first appeared in the *Devonian and reached the peak of its abundance and diversity during the *Carboniferous, forming a major component of the coal-swamp vegetation. Sphenopsids are characterized by jointed stems with whorls of leaves and branches borne at the joints (or nodes). The internodal part of the stem is vertically ridged and *spores are produced in rings of sporangia arranged in cones, usually at the tips of the fertile shoots. The only living genus, *Equisetum*, is a comparatively small plant (different species ranging between 4 or 5 cm and 12 m), but one of the best-known fossil genera, *Calamites*, included tree-like forms that grew up to 30 m in height. Another common fossil sphenopsid, *Sphenophyllum*, was a slender plant with a ribbed stem only 1–7 mm in diameter but up to several metres in length, that probably scrambled over other vegetation. *See also* *ARCHAEOCALAMITES RADIATUS*; *CALAMITES CISTIIFORMES*; and *EQUISITITES HEMINGWAYI*.

sphericity An expression of how closely the shape of a *grain resembles the shape of a sphere. Sphericity can be determined by examining the relation between the long (L), intermediate (I), and short (S) axes of the particle, the maximum projection sphericity, Ψ, being given by the expression $\Psi = \sqrt[3]{(S^2/LI)}$. For a perfect sphere, $\Psi = 1$. Values less than one relate to increasingly less spherical shapes. *See also* PARTICLE SHAPE; and ROUNDNESS INDEX.

spheroidal oscillation *See* FREE OSCILLATION.

spheroidal weathering (onion weathering) The development of concentric shells of normally chemically weathered material in the outer zone of a joint-bounded mass of rock. The spheroidal or onion-skin appearance results from enhanced *weathering at *joint intersections together with the expansion resulting from chemical change.

spherule A small spherical particle. *Glass spherules resulting from impact-induced melting or volcanic *fire-fountaining are common in the lunar *regolith. A typical diameter is 100 μm. Nickel–iron spherules, usually less than 30 μm in diameter, derived from the impacting *meteorite, are common in impact glasses.

spherulite (adj. spherulitic) A spherical to ellipsoidal aggregate of radiating, fibrous crystals, usually *quartz and *alkali feldspar, found in glassy or *felsitic, *aphanitic *groundmasses of *igneous rocks. Spherulites range in diameter from less than 1 mm to about 1 m and are formed by the devitrification of quenched glassy igneous rocks which are usually silicic in composition.

spicular chert A very fine-grained siliceous *sedimentary rock, the *silica originating from the accumulation of sponge (*Porifera) *spicules. Sponge spicules can be an important source of *biogenic silica in shelf-sea environments. *See* CHERT.

spicule A small needle or spine.

spiculite A *sedimentary rock or *sediment composed largely of sponge (*Porifera) *spicules.

spider diagram A useful plot, used in *igneous *petrology to show variations between two rocks or rock types for a wide range of elements. Usually, one rock is a standard type, e.g. a *mid-oceanic-ridge basalt or a *carbonaceous chondrite meteorite. Data for the test rock are 'normalized' to the standard by dividing the abundance for each element by the abundance in the standard and plotting the quotient, usually in order of atomic number. Elements which have the same abundance in the test rock as in the standard yield a quotient of 1; if the two compositions were identical the plot would be a straight line through 1 on the *y* axis. Elements more abundant in the test rock yield numbers greater than 1; those less abundant yield numbers less than 1. By connecting the points plotted by a single line, a spidery diagram of peaks and troughs is obtained which illustrates immediately systematic differences in composition, e.g. relative enrichment in *incompatible elements.

spiders *See* CHELICERATA.

spike A solution (liquid or gaseous) containing a known concentration of a particular element whose isotopic concentration has been changed by the enrichment of one of its naturally occurring *isotopes. The spike is used in *isotope dilution analysis, being mixed in known proportions with the sample solution, prior to isotopic determination by means of a mass spectrometer (*see* MASS SPECTROMETRY).

spilite A low-grade *metamorphic rock composed of *albite, *chlorite, *actinolite, *sphene, and *calcite, with or without *epidote, *prehnite, and *laumonite, and formed by sea-floor *metasomatism of *mid-oceanic-ridge basalts. Sea water circulating through the *oceanic crust is heated by the cooling *basalt *dykes and *lavas and reacts with them, introducing sodium and water into the rock system and converting the basalt *mineral assemblage into a typical spilite assemblage.

spillway General term for a *glacial drainage channel cut by water during glaciation, and normally including three varieties: (*a*) channels cut by water escaping from a glacially impounded lake (*see* OVERFLOW CHANNEL); (*b*) channels cut by meltwater released from a decaying *glacier (*see* MELTWATER CHANNEL); and (*c*) channels cut by a stream deflected by an advancing glacier. Impressive examples were developed in central Europe (the Urstromtäler of northern Germany) when the Scandinavian *ice sheet diverted streams flowing north from the southern highlands.

spilling breaker (surf wave) Oversteepened wave in which the unstable top of the wave spills down the front of the wave-form as it advances into shallower water. Consequently it gradually diminishes in height until it moves up the beach as *swash.

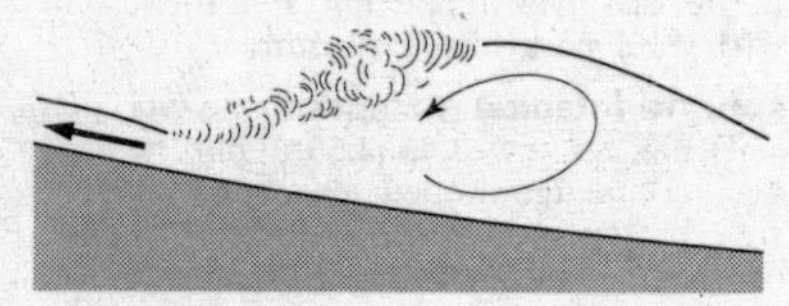

Spilling breaker

spindle bomb *See* VOLCANIC BOMB.

spinal column *See* VERTEBRA.

spinel Important group of non-silicate mineral *oxides, including the subgroups spinel series, *magnetite series, and *chromite series. Members of the spinel series have the general formula XAl_2O_4, where X = Mg in spinel, X = Fe^{2+} in hercynite, X = Zn in gahnite, and X = Mn in galaxite. Complete *solid solution exists between all four *end-members and also magnesiochromite ($MgCr_2O_4$); spinels with appreciable Fe^{2+} (Mg:Fe = 1:3) are called pleonaste and those with Fe^{2+} and Cr^{3+} are called mitchellite; sp. gr. 3.5–4.1; *hardness 7.5–8.0; spinel is dark green and hercynite is dark bluish-green; all occur as small octohedra. Unlike chromite, which occurs in ultramafic (*see* ULTRABASIC) *igneous intrusions, spinel may occur in metamorphic *schists and *gneisses with *sillimanite, *garnet, and *cordierite. In *contact metamorphosed impure *limestones spinel occurs with *chondrodite, *olivine, and *orthopyroxene, and in *emery deposits it occurs with *corundum. Spinel may develop in residual aluminous *xenoliths enclosed in *basic magmatic rocks. Spinel is found in *alluvial deposits and gem quality spinel also occurs. Hercynite occurs in metamorphosed *laterites; gahnite in *granite *pegmatites; and galaxite in manganese vein deposits.

spinifex texture An array of criss-crossing sheafs of subparallel, blade- or plate-like, skeletal, magnesium-rich *olivine or aluminous *pyroxene, between which is found a finer-grained aggregate of devitrified *glass, skeletal pyroxene, and skeletal *chromite. The *texture is usually found as the product of extreme *under-cooling of magnesium-rich *komatiite *lava.

spire In the shell of a gastropod (*Gastropoda), all the *whorls other than the body whorl.

Spiriferida (spiriferids; class *Articulata) Order of *Brachiopoda, whose members have spiral *brachidia, *punctate or *impunctate biconvex shells, and a large body cavity. They first appeared in the Middle *Ordovician, and are last known from the Lower *Jurassic.

spiriferids *See* SPIRIFERIDA.

spissatus From the Latin *spissatus* meaning 'thickened', a species of *cirrus cloud which has sufficient thickness to appear grey even when the cloud is between the Sun and the observer. *See also* CLOUD CLASSIFICATION.

spit Elongated accumulation of sand or gravel projecting from the shore into a water body. *Longshore drift of material is usually responsible for the development of a spit.

splanchnocranium *See* CRANIUM.

splay fault One of a series of branching *synthetic faults near the termination of a major fault which spread the displacement over a large area.

splendent (specular) Applied to the *lustre of a *mineral if it reflects light intensely to give a bright, shining surface. The reflectivity may be related to the high *refractive index of the mineral (e.g. in *gemstones).

SP method *See* SPONTANEOUS-POTENTIAL METHOD.

spodic horizon Subsurface *soil horizon in which organic matter together with aluminium and often iron compounds have accumulated amorphously. It is a *diagnostic horizon in the *USDA Soil Taxonomy.

Spodosols Order of soils in which subsurface *soil horizons contain amorphous materials comprising organic matter and compounds of aluminium and often iron that have accumulated illuvially. Such soils form in acid material, mainly coarse in texture, in humid, cool to temperate climates.

spodumene Unusual, lithium-bearing *pyroxene, $LiAlSi_2O_6$; sp. gr. 3.0–3.2; *hardness 6.5–7.0; *monoclinic; usually greyish-white, often with a greenish or yellowish-green tinge, occasionally violet, transparent to translucent; vitreous *lustre; crystals *prismatic, often striated, etched, and corroded, also occurs *massive and *columnar; *cleavage perfect prismatic {110}, parting {100}; typically occurs in lithium-rich *granites and *pegmatites, associated with *lepidolite, *tourmaline, and *beryl. The green variety (hiddenite) and the lilac variety (kunzite) are used as *gemstones, otherwise it is an *ore mineral of lithium.

spondylium Curved platform for muscle attachment in the *shell beak region of some brachiopods (*Brachiopoda).

sponges *See* DEMOSPONGEA; and PORIFERA.

Spongiaria *See* PORIFERA.

spontaneous potential *See* SELF-POTENTIAL SONDE.

spontaneous-potential method (self-potential (abbreviation SP) method) A method which measures the naturally occurring potential differences between two *non-polarizable electrodes. It is often used in exploration for massive *sulphide and *graphite *ore bodies.

sporangium *See* SPORE.

spore A propagative plant body consisting of a gametophyte enclosed in a non-cellular coat. Spores are enclosed within a capsule (sporangium) and are produced in groups of four (tetrads) when the parent cell divides meiotically. In more primitive plants the spores are identical (isospore) and the condition is called 'homospory'. In more advanced, vascular plants spores of two sizes are produced and the condition is called 'heterospory'. Small, male microspores are contained within a microsporangium; larger, female megaspores within megasporangia. Where spores occur in tetrads the contact surfaces produce a 'trilete' mark on each of the four spores, marking the point for the germination of the prothallus. Less commonly, two contact surfaces are produced, resulting in a 'monolete' marking. Spores which were probably produced singly are 'alete', with no obvious marking.

sporinite *See* COAL MACERAL.

SPOT (Système Probatoire d'Observation de la Terre) A French observation satellite, launched in 1986, which transmits data yielding monochrome, stereoscopic images with a 10 m resolution, superior to those from *Landsat.

spotting Dark-coloured, rounded areas, up to 2 mm in diameter and sometimes larger, found on the surface of *slaty cleavage in low- to medium-grade contact *metamorphic rocks (*see* CONTACT METAMORPHISM) of pelitic composition. Many spots are *graphite-rich, having formed by metamorphic aggregation from organic material originally disseminated through the starting sedimentary *shale. Rocks may also have a spotty appearance when *andalusite is beginning to appear in contact-metamorphosed *pelites.

spread A pattern of *geophone *groups used simultaneously to record data from a single *shot. Examples of spreads include in-line, offset, interlocking, L-spread, reversed, split-spread, and T-spread. *See* ARRAY.

spreading rate The rate, usually in tens of millimetres a year, at which two adjacent lithospheric *plates are separating. The spreading rate varies along a *constructive margin and is at a maximum of 90° from the *pole of rotation. Some authors use 'spreading rate' when 'half spreading rate' (i.e. the rate of movement of a plate from the relevant *ridge) would be more accurate.

spreiten (sing. spreite) From the German *spreiten*, meaning 'to spread out' or 'to extend', sedimentary *laminae that result from the behaviour of an animal during feeding, excavation, or locomotion. They may be U-shaped, sinuous, blade-like, or spiralled, and they are always repeated over a small area. They reflect the intensive working of *sediment for food. Well-defined spreiten are associated with **Diplocraterion*, **Rhizocorallium*, and *Daedalus*. In the latter they are arranged spirally around a single trunk. *See* FUGICHNIA.

Spriggina *See* EDIACARAN FOSSILS.

spring A flow of water above ground level that occurs where the *water-table intercepts the ground surface. Where the flow from a spring is not distinct (i.e. it does not give rise to obvious trickles) but tends to be somewhat dispersed, the flow is more correctly termed a 'seep'. The reappearance of surface water that had been diverted underground in a *karst region is a type of spring known as a 'resurgence'. A major variety is the 'Vauclusian spring', named after the Fontaine de Vaucluse, southern France, and descriptive of the upward emergence of an underground river from a flooded *solution channel.

spring balance A weighing scale, consisting of a pan suspended below a vertical spiral spring, used to measure the weight and often the density of rocks and minerals. When a specimen is placed on the pan the spring stretches along a calibrated scale and a pointer indicates the weight. The Jolly balance (*see* DENSITY DETERMINATION) is a spring balance.

Springerian *See* CHESTERIAN.

spring sapping A set of geomorphological *processes that erode a hillslope around the site where a *spring emerges. The processes may include the collapse of saturated material, surface stream *erosion, and *chemical weathering. It occurs towards the bases of *chalk escarpments in southern England, where its effect may have been enhanced by frost activity under former *periglacial conditions.

spring tide *Tide of greater range than the mean range; the water level rises and falls to the greatest extent from the mean tide level. Spring tides occur about every two weeks, when the Moon is full or new. Tides are at their maximum when the Moon and the Sun are in the same plane as the Earth. *Compare* NEAP TIDE.

s-process *See* SLOW-NEUTRON PROCESS.

SPS A satellite positioning system that uses interaction-geostationary and low, short-period satellite passes to determine the location of stations on the Earth's surface. *See* GLOBAL POSITIONING SYSTEM.

spur A ridge that descends towards a valley floor from the higher ground above. It may be due to an outcrop of resistant rock, or it may develop on the concave side of a winding stream as a result of incision.

Sq variation *See* DIURNAL VARIATION.

squall Short-lived condition with strong winds, which increase by at least 16 knots (30 km/hr). It may include thunder and heavy precipitation. *See also* LINE SQUALLS.

square array An *electrode configuration in which the *current and *potential electrodes are positioned at the four corners of a square of side a. Its *geometric factor is $K_g = 3.41\pi a$ metres.

squeeze-up The extrusion of a small volume of viscous *lava from a crack or opening on the solidified surface of a lava flow, in response to the pressure of fluid lava within the flow interior. Squeeze-ups are generally bulbous or linear in form, range from a few centimetres to several metres in height, and may have vertical grooves down their length.

squeezing ground Weak ground, such as *clay, which has deformed under surrounding loads and has been squeezed into an *excavation as a result of overstressing.

SSI *See* SOLID-STATE IMAGING CAMERA.

SSS *See* STANDARD STRATIGRAPHIC(AL) SCALE.

stability 1. Atmospheric condition in which air that is forced to rise tends to return to its pre-existing position in the absence of the uplifting force. If the *adiabatic *lapse rate of uplifted air is greater than the *environmental lapse rate, then the vertically displaced air will become colder than the surrounding air and as its density increases it will tend to sink back. *See also* INSTABILITY. **2.** In engineering, the resistance of a structure to collapse or sliding, dependent upon the *shearing strength of the material. **3.** In geochemistry, the state of equilibrium towards which a system will move from any other state under the same conditions. **4.** In thermodynamics, the condition when a slight disturbance of temperature, pressure, or composition does not result in the appearance of a new *phase.

stability field Range of temperature and pressure within which a particular mineral or mineral assemblage is stable.

stable isotope Any naturally occurring, non-radiogenic *isotope of an element. Many elements have several stable isotopes.

stable-isotope studies Study of non-radiogenic isotopic ratios of selected elements, e.g. *D : H, $^{18}O:^{16}O$, $^{32}S:^{34}S$, which are fractionated in different proportions during different geologic processes (*see* ISOTOPE FRACTIONATION). Thus natural waters may be 'fingerprinted' by reference to their D : H and $^{18}O:^{16}O$ ratios as being of *meteoric, magmatic, or metamorphic origin; sulphur in *sulphide *ores may be characterized as *sedimentary or *igneous by reference to its $^{32}S:^{34}S$ ratio. *See also* OXYGEN-ISOTOPE ANALYSIS; and OXYGEN-ISOTOPE RATIO.

stack 1. Pillar or block of rock, with near-vertical sides, standing adjacent to a present or former sea cliff. Typically it has been isolated from the main cliff by wave erosion concentrated along steeply inclined *joints or *faults. **2.** The product of *stacking. A 'brute' or 'final' stack is the end product of the standard processing of seismic-*reflection data; the data can be processed further, e.g. by migration programs to produce a 'migration' stack.

stacking The summing of traces from a variety of *seismic records to increase the signal-to-*noise ratio and enhance coherent signals into a composite record (a *stack). *See also* COMMON-DEPTH-POINT STACK; and VERTICAL STACKING.

stacking fault An abnormality in the arrangement of the rows of atoms affecting the structure of a *crystal (e.g. *crystal twinning), caused by changes in the physical and chemical conditions of its immediate surroundings while it was growing.

stacking velocity In seismic investigations, the velocity determined from normal *moveout measurements using *common-depth-point gathers prior to *stacking.

stade (stadial in continental-European usage) A term that is difficult to define with precision, but which refers to a single period of increased cold or advancing ice, which forms a subdivision of a cold *stage within the overall division of a glacial period into periods of cold interspersed with warm, or warmer, periods.

stadial *See* STADE.

staff gauge A graduated pole or board placed in or beside a water course, from which it is possible to measure directly the height of the water surface relative to a known datum elevation.

stage 1. The elevation of the water surface of a river with reference to a fixed datum level. Hence 'rising' and 'falling' stages. **2.** The major subdivision of a *series. A stage is the fourth order unit in chronostratigraphy, the equivalent of *age in terms of geologic time units. It refers to the body of rock accumulated during one age unit. When used formally the initial letter of the term is capitalized, e.g. *Frasnian Stage. **3.** In palaeoclimatology, a climatic, and partly geologic–climatic, term usually defined by a series of *sediments or a sequence of *fossil assemblages and named at a type locality. For example, the *Hoxnian (a temperate stage) is named for organic *interglacial deposits at Hoxne, Suffolk, England. **4.** The degree of development of a land-form or landscape over time, and which traditionally has been described by the terms 'youthful', 'mature', and 'old age' (*see* DAVISIAN CYCLE). The recognition of such stages implies an orderly evolution and this is now seen as unlikely for many parts of the Earth's land surface. **5.** The part of a microscope on which the specimen to be examined is placed. Normally it is flat and may be fixed, as in biological or metallurgical microscopes, or rotating with a 360° calibrated scale as in geologic microscopes. Transmitted-light microscopes have a hole in the centre of the stage through which light passes up to the observer from below. *Reflected-light microscopes have an incident light system, whereby light is directed on to the stage from above and is reflected from the specimen to the observer.

stage hydrograph *See* HYDROGRAPH.

stains and staining techniques Various chemical staining techniques are used to identify *minerals. The procedure followed is to etch the specimen, and then to expose it to a range of organic and/or inorganic compounds which form distinctive coloured complexes with certain minerals. *Feldspars can be identified by etching with hydrofluoric acid and then treating with sodium cobaltinitrate, barium chloride, and potassium acid rhodizonate. *Plagioclases stain red and K-feldspars are stained yellow after this treatment. For *carbonates, a range of stains, including alizarin red S, Feigl's solution, potassium ferricyanide alizarine cyanic green, and titan yellow are used. A number of staining combinations allow the differentiation of *calcite, high-Mg calcite, *dolomite, *anhydrite, and *gypsum.

stalactite Elongated body of *dripstone descending from the roof of a cave in a *karst environment. It is produced by *calcite *precipitation as excess carbon dioxide diffuses from water droplets entering a cave environment.

stalagmite Pinnacle of *dripstone rising from the floor of a cave in a *karst environment. It is produced by the *precipitation of *calcite as excess carbon dioxide diffuses when water droplets strike the floor.

Stampian *See* RUPELIAN.

standard deviation A measure of the normal variation within a set of data. In any given measurement, two-thirds of the samples fall within one standard deviation on either side of the mean, 95% between two standard deviations, and so on; the proportion falls off sharply because of the bell-curve effect. The standard deviation is calculated as the root-mean-square deviation.

Standard Global Chronostratigraphic Scale (abbreviation SGCS) *See* STANDARD STRATIGRAPHIC(AL) SCALE.

standard mean open water *See* SMOW.

Standard Stratigraphic(al) Scale (SSS) (Standard Global Chronostratigraphic Scale, SGCS) Globally standardized *stratigraphic scale whose *chronostratigraphic units will ultimately all be delimited by *boundary stratotypes.

standing wave Type of wave in which the surface oscillates vertically between fixed points called 'nodes', without any forward progression. The crest at one moment becomes the trough at the next and so on. The points of maximum vertical rise and fall are called 'antinodes'. At the nodes particles show no vertical motion but exhibit the maximum horizontal motion. Standing waves may be caused by the meeting of two similar wave groups that are travelling in opposing directions. *See* SEICHE.

stand of the tide Period at high or low water during a tidal cycle when there is little or no change in the height of the *tide. The water level is almost stationary and the tidal currents fall away to zero velocity before reversing.

stannite Comparatively rare mineral, Cu_2FeSnS_4; sp. gr. 4.3–4.5; *hardness 3–4; *tetragonal; steel-grey with an olive-green tinge on fresh surfaces, which become yellowish on exposure; black *streak; *metallic *lustre; crystals rare, *cubes or *tetrahedra, but normally the grains are irregular or *massive; occurs in *hydrothermal tin-*ore deposits and in association with *cassiterite, *chalcopyrite, and *wolframite, also in stanniferous *sphalerite–*galena ores, and associated with sphalerite, *pyrrhotite, and galena.

stapes In *Mammalia, the inner auditory ossicle of the ear, stirrup-shaped because it is pierced by an artery. It is derived from the hyomandibular bone in fish, which connects the cranium and the upper jaw.

star dune A complex *aeolian *dune form characterized by a series of slip faces radiating about a central point, producing a rough star shape. Such dunes are the product of highly variable wind directions and thus have a highly variable palaeocurrent pattern preserved in the dune cross-bedding (*see* CROSS-LAMINATION).

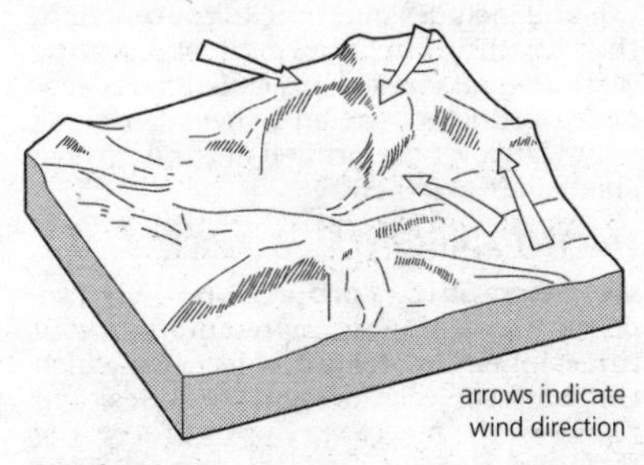

Star dune

Stardust A *NASA mission, due to be launched in 1999, to return samples from the coma of comet *Wild 2.

star pair (binary star; double star) Binary stars are among the commonest stellar systems in our galaxy, accounting for about 50% of all stars. They consist of two stars in orbit around their common centre of gravity. Those which can be resolved by telescope are referred to as visual binaries. Spectroscopic binaries comprise star pairs too close to be resolved visually, but which exhibit *Doppler shifts in spectral lines because of varying velocities of the two stars in the line of sight. Other stars reveal the presence of an invisible companion by changes in their *proper motion. Double stars appear to be visually close to each other, but may be at vastly different distances from the observer.

star phylogeny In a *phylogenetic tree, the occurrence of a multifurcation with many short branches connected at the *internal node. Such topologies are often inferred to represent a recent population expansion event from a common ancestor (the founder lineage). This is often seen in populations which have undergone a *founder effect.

star twinkling *See* ATMOSPHERIC SHIMMER.

stasigenesis Situation in which an evolutionary lineage persists through time without splitting or otherwise changing. So-called '*living fossils' are examples of stasigenesis.

stasis Period of little or no evolutionary change; the 'equilibrium' that alternates with 'punctuations' in the theory of *punctuated equilibrium.

static correction (statics) A correction applied to geophysical data, especially seismic data, to compensate for the effect of irregular topography, differences in the elevation of *shots and *geophones relative to a datum, low-velocity surface layers (weathering correction), and the horizontal geometry of shots and receivers (geophones or *hydrophones), or any correction which applies to the geometry of the source and receiver(s). A static correction provides some form of direct-current shift (e.g. in seismic-*reflection surveys), usually a time element added to or subtracted from the travel times, in contrast to a dynamic correction which involves an operation on the data. *See* ELEVATION CORRECTION; and MOVEOUT.

statics *See* STATIC CORRECTION.

stationary front Condition in which the frontal boundary between cold and warm air is stationary.

station frequency *See* SAMPLING FREQUENCY.

station interval *See* SAMPLING INTERVAL.

staurolite A member of the *nesosilicates and an important metamorphic *index mineral with the approximate composition $(Fe^{2+},Mg)_2(Al,Fe^{3+})_9O_6[Si_4O_{16}](O,OH)_2$; sp. gr. 3.74–3.85; *hardness 7.5; *monoclinic; *crystals *prismatic; shades of brown; occurs in *regionally metamorphosed *schists and *gneisses, such as iron-rich *pelites, with a high Fe^{3+}/Fe^{2+} ratio at moderate grades of *metamorphism and in association with *garnet (*almandine) and *kyanite; it may develop from *chloritoid as the metamorphic grade increases.

steady flow The condition in which flow velocities do not vary with time. This is applicable to both *groundwater and channel flows. In respect of flow to a pumped well, it is sometimes called 'equilibrium' flow. Nonsteady (unsteady or transient) flow changes its velocity and/or direction with time.

steam fog *See* ARCTIC SEA SMOKE.

steatite *See* TALC.

S-tectonite *See* SHAPE FABRIC.

Stefan–Boltzmann law The law stating that energy radiated from a *black body is proportional to the fourth power of its *absolute temperature.

Stegosauridae Suborder of quadrupedal, *ornithischian *dinosaurs, mainly *Jurassic in age, characterized by a double row of plates and spines along the back and tail.

steinkern *See* FOSSILIZATION.

Steinmann trinity *Spilites, *serpentine rocks, and radiolarian *cherts, which, as G. Steinmann observed in 1905, often occur together in mountains such as the Alps, comprising rocks formed as deep-sea sediments.

stem group In *cladistic analysis, those taxa descended from the point where an ancestral taxon split into two *sister groups to the point at which a further split gave rise to an extant *crown group.

stem reptiles *See* COTYLOSAURIA; and CAPTORHINOMORPHA.

Steno (Stenonis), Nicolaus (Niels Stensen) (1638–87) A Danish physician who moved to Florence in 1665. He opposed the prevailing idea that *fossils grew within the Earth, proposing instead that they were organic relics of an earlier period. Steno also distinguished between fossils and inorganic remains such as crystals. He had some conception of stratigraphy, describing strata in Tuscany as being formed sequentially.

stenothermal Unable to tolerate a wide range of temperature.

stenotopic Able to tolerate only a narrow range of several factors.

Stensen, Niels *See* STENO (STENONIS), NICOLAUS.

step faulting The faulting process in which separate *fault blocks are downthrown systematically in one direction, forming a stepped sequence.

Stephanian The uppermost *series in the *Silesian (Upper *Carboniferous) of Europe, underlain by the *Westphalian, dated at 305–290 Ma and roughly contemporaneous with the uppermost *Moscovian, *Kasimovian, and *Gzelian Series. Originally, the lower part of the Stephanian was known as Westphalian E.

stepout *See* MOVEOUT.

stereogram The two-dimensional plot of a *stereographic projection, in which points on the surface of a sphere are represented on a *plane of projection, as points on the Earth's surface might be projected on to a plane representing a slice through the equator by joining these points to either the N. or S. poles. If the plane of projection (equatorial plane) is constructed with *great circles and *small circles drawn at 2° intervals and with an overall diameter of about 20 cm, it is called a 'Wulff stereographic net'. If it is graduated at 10° intervals and includes small circles concentric to the *primitive circle and radii representing vertical great circles, it is called a 'Federov net'. Plotting is done on a sheet of tracing paper laid over the printed net and pivoted about a pin at its centre.

stereographic net *See* STEREOGRAM.

stereographic projection A two-dimensional graphic representation of a three-dimensional solid object, in which the angular relationships of lines and planes of the object are drawn in terms of their relationship to the *great circle formed by the intersection of the equatorial plane with the surface of an imaginary sphere in which the object is contained. Stereographic projections are used widely in structural geology and *crystallography.

stereom (stereome) In *Echinodermata, a mesh, mainly of magnesian *calcite, from which the skeleton is constructed. In *Scleractinia, a secondary structure, composed of bundles of *aragonite crystals arranged transversely, that thickens and strengthens the *epitheca.

stereonet A two-dimensional, circular representation of a sphere in which the lines of longitude and latitude form a system of coordinates (a 'net'), on which projections of *great and *small circles occupy the equatorial plane of a reference sphere. Two types of stereonet are used in the analysis of structural data: an equal-angle net (Wulff net); and an *equal-area net (Schmidt–Lambert net), which is the net preferred for the contouring of data and the evaluation of clusters of data suggesting preferred orientations.

stereophotography The taking of two pictures (stereo pairs), e.g. of a *fossil, from slightly different angles so that when the two images are viewed through a *stereoscope a three-dimensional image is produced, giving greater detail than an ordinary photograph.

stereoptic vision The perception of depth and three dimensions accompanying binocular vision resulting from differences in *parallax producing different images on the retina of each eye.

stereoscope An optical device which allows a pair of overlapping, two-dimensional photographs to be examined with three-dimensional (*stereoptic) vision, thus permitting more detailed interpretation.

Steropodon The earliest Australian mammal, a monotreme known only from opalized jaws and teeth found in *Cretaceous deposits at Lightning Ridge, central Australia (the name is from the Greek *sterope*, meaning 'lightning'). The pattern of the molar teeth suggests that monotremes may be derived from Theria and not from a separate protomammalian stock.

Stettin An end *moraine and a variety of *sediments occurring in Poland and European Russia. Its stratigraphic position is uncertain but it is probably equivalent to the early *Weichselian or *Rodebaek.

stibnite (antimonite, antimony glance) One of the main *ore minerals for antimony, with the formula Sb_2S_3; sp. gr. 4.6; *hardness 2.0–2.5; *orthorhombic; normally lead-grey with a bluish tarnish; *metallic *lustre; crystals *prismatic, columnar, *acicular, and vertically striated; perfect lengthwise *cleavage {010}, imperfect {100}, {110}; occurs mainly in *hydrothermal deposits, at the low-temperature end, often forming sheet-like bodies, associated with *fluorite, *quartz, and *barite, and with sulphides of lead, zinc, and other metals.

stick-slip A discontinuous, jerky pattern of movement along a *fault plane, which is thought to be consistent with *earthquake phenomena.

Stigmaria The *form-genus for the underground axes of Lepidodendrales (*Carboniferous).

Stille, Wilhelm Hans (1876–1976) A German geologist from the Universities of Göttingen and Berlin, Stille's main work was in the field of *orogeny. He believed that mountain building occurred in phases and that the continents were formed from the accretion of mountain belts around ancient *cratons.

stillstand A period of geologic time characterized by unchanging sea levels, i.e. a state neither of *regression nor *transgression.

stilpnomelane A *phyllosilicate (sheet silicate) similar to *biotite with the formula $(K,Na,Ca)_{0-1.4}(Fe^{2+},Fe^{3+},Mg,Al,Mn)_{5.9-8.2}[Si_8O_{20}](OH)_4(OH,F)_{3.6-8.5}$; sp. gr. 2.59–2.96; *hardness 3.0–4.0; properties similar to those of biotite; found in metamorphosed iron- and manganese-rich sedimentary deposits and in some *glaucophane schists.

stipe A branch of a rhabdosome (graptolite colony, *see* GRAPTOLITHINA). A stipe or stipes originate(s) from an initial conical cup (*sicula). The number of stipes can range from one to as many as 64. Their attitudes also vary. In the primitive condition the stipes hang downwards from the sicula and are said to be 'pendent'. If they grow out horizontally from the sicula they are 'horizontal'; if they grow upwards along the nema they are 'scandent'; if they are straight and grow downwards they are 'declined'; if they are curved and slope downwards they are 'deflexed'; if they are straight and grow upwards and outwards they are 'reclined'; and if they are curved and grow upwards and outwards they are 'reflexed'.

stishovite A high-density form of crystalline silica (*see also* QUARTZ) with the same formula SiO_2, but formed at pressures greater than 10 GPa; discovered at some *meteorite impact sites, but rarely in terrestrial rocks; sp. gr. 4.3; minerals possessing a similar atomic lattice may exist in the Earth's upper *mantle.

stock *Igneous *intrusion, approximately circular in plan, that has steep contacts with the *country rocks and a surface area of 20 km^2 or less.

stockwork Mineral deposit formed of a network of small, irregular *veins so closely spaced that it may be mined as a unit.

Stokes's law A law describing the rate at which suspended particles settle, formulated in 1845 by the physicist Sir George Gabriel Stokes (1819–1903). The settling velocity (V) in cm/s is calculated by $V = CD^2$, where C is a constant related to the density and viscosity of the fluid and the density of the suspension and D is the diameter of the particles (assumed to be spheres) in cm.

stolon In colonial invertebrates, the stalk-like structure by which individuals are attached to the substrate.

stolotheca One of the three types of graptolite (*Graptolithina) *thecae, which encloses the main *stolon and the earliest parts of the daughter stolotheca (the *authotheca, and *bitheca).

stomodeum *See* ANTHOZOA.

-stone Suffix for many different kinds of lithified *sediments, e.g. *siltstone, *limestone, *sandstone, *grainstone, *packstone, and *ironstone.

stone canal In *Echinodermata, a canal with walls strengthened by calcareous matter, which connects the *madreporite with the water-vascular system.

stone circle *See* PATTERNED GROUND.

stone garland *See* PATTERNED GROUND.

stone net *See* PATTERNED GROUND.

stone polygon *See* PATTERNED GROUND.

stone steps *See* PATTERNED GROUND.

stone stripes *See* PATTERNED GROUND.

stony-iron meteorite (siderolite) Relatively rare *meteorite type with approximately equal quantities of nickel–iron and basic *silicates, usually *pyroxene and *olivine.

stony meteorite (asiderite) *Meteorite type consisting mainly of rock-forming *silicates (*olivine, *pyroxene, and *plagioclase) with some nickel–iron; more than 90% of meteorites seen to fall are of this type. Stony meteorites are referred to as *chondrites or *achondrites, depending on the presence or absence of *chondrules.

stoop and room *See* PILLAR AND STALL.

stoping 1. The method of emplacement of an *igneous *intrusion in which percolating *magma detaches blocks of *country rock which sink, allowing the magma to move upwards. **2.** In underground mining, the breaking and removal of rock in an *ore body.

storage coefficient (storativity) The volume of water given up per unit horizontal area of an *aquifer and per unit drop of the *water-table or *potentiometric surface. It is a dimensionless ratio and always less than unity. In unconfined aquifers it is equal to the specific yield (*see* SPECIFIC YIELD (2)), but in confined aquifers the storage coefficient depends on elastic compression of the aquifer, and is usually less than 10^{-3}.

storativity *See* STORAGE COEFFICIENT.

storm Common term for gales, squalls, rainstorms, or thunderstorms. It is used specifically for conditions associated with the active areas of low-pressure systems. 'Storm-force winds' are, by definition, strong gales or winds, with speeds exceeding 20.8 m/s.

storm beach Accumulation of coarse beach sediments built above the high-water mark by storm action. Gravel, shell debris, and other coarse materials are thrown into ridge or bank structures by waves during heavy storms.

storm bed A bed of *sediment deposited by a storm event. Storm beds are usually the product of shallow marine wave activity, and are often referred to as 'event deposits', that is, they are the product of a short-lived, high-energy, sedimentary environment.

storm deposit *See* TEMPESTITE.

storm surge Rise or piling-up of water during a storm, as a result of wind stresses acting on the surface of the sea and of atmospheric-pressure differences. If a storm surge occurs at the time of highest *spring tides, flooding of coastal areas may result, as happened in Holland and East Anglia, England, in 1953.

stoss and lee Terms referring to the up-*glacier and down-glacier slopes respectively of a rocky obstacle that has been glaciated. The stoss slope is smoothly abraded, the lee slope roughly plucked. A landscape dominated by such features is said to have 'stoss-and-lee topography'.

stoss-and-lee topography *See* ROCHE MOUTONNÉE; and STOSS AND LEE.

Strahler climate classification A system for describing climates, devised in 1969 by A. N. Strahler, in which world climates are related to the main *air masses that produce them, as: (*a*) equatorial/tropical air masses, producing low-latitude climates; (*b*) tropical and polar air masses, producing mid-latitude climates; and (*c*) polar and arctic air masses, producing high-latitude climates. Subsets of these are based on variations in temperature and *precipitation to give 14 regional types, plus upland (highland) climates which are regarded as a separate category. *See also* KÖPPEN CLIMATE CLASSIFICATION; and THORNTHWAITE CLIMATE CLASSIFICATION.

straight extinction (parallel extinction) In optical *mineralogy, the phenomenon which occurs when the *vibration direction of the light ray is parallel to the *crystal face or *cleavage traces within the mineral. As the mineral *thin section is rotated between *crossed polars, the vibration direction is brought parallel to the plane of the *polarizer. The light passes through the polarizer and is eliminated by the *analyser. *Extinction occurs four times in a 360° rotation of the mineral.

strain The dimensional change in the shape or volume of a body as a result of an applied *stress or stresses. Strain is the ratio of the altered length, area, or volume to its original value, and may be *homogeneous or *inhomogeneous, and involve distortion, *dilation, and rotation. *See* HOOKE'S LAW; POISSON'S RATIO; PURE SHEAR; SIMPLE SHEAR; and SHEAR MODULUS.

strain ellipse A two-dimensional figure used to describe the magnitude and orientation of the maximum and minimum *principal strain axes x and z when a reference circle of unit radius is deformed by *homogeneous strain.

strain ellipsoid A three-dimensional version of the *strain ellipse which describes the magnitude and orientation of the maximum, intermediate, and minimum *principal strain axes when a reference sphere of unit radius is deformed by *homogeneous strain.

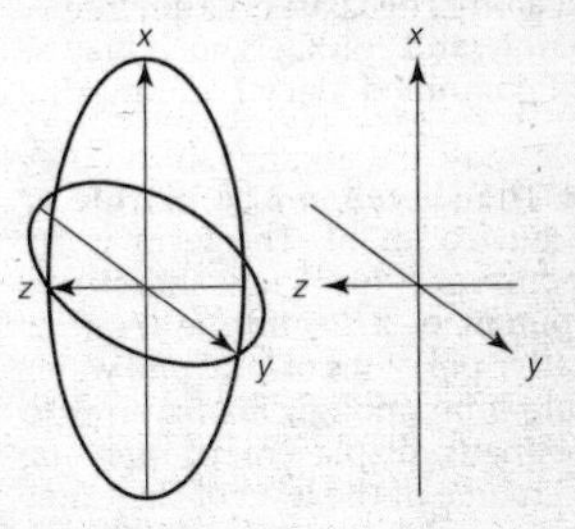

Strain ellipsoid

strain gauge A device for measuring *strain.

strain marker A natural object whose original geometry is known and from which the magnitude and orientation of the *principal strain axes may be determined after deformation. Many objects have been used, e.g. *fossils (*crinoids, *belemnites, *corals, *ammonites, *brachiopods, and *trilobites), rock *textures (i.e. *ooids), *conglomerate pebbles and *clasts, single crystals (e.g. *feldspar *megacrysts), spots in *contact aureoles, volcanic *lapilli, *spherulites, and *xenoliths.

strain parallelepiped The three-dimensional shape produced when a small cube of unit dimensions is deformed.

strain rate The rate of change of the size and shape of a body under an applied *stress. The duration of the stress is very important in determining strain behaviour. Most geologic strain rates are relatively low, with stresses applied over millions of years, leading to a great reduction in yield strength. Evidence for geologically high strain rates is provided by *pseudotachylite glasses whose formation results from 'instantaneous' frictional melting during deformation.

strain–slip cleavage A term synonymous with *crenulation cleavage, but which some authorities (e.g. C. McA. Powell, 1979) have suggested should not be used in cleavage descriptions, owing to the implications it carries of the mode of origin.

strain–time diagram A diagram which illustrates a change in *strain over time for a given *stress. Strain–time diagrams are used to delimit fields of *elastic behaviour, viscous behaviour and *failure, and fields of *primary, *secondary, and *tertiary creep.

strandflat A *shore platform, up to 60 km wide, found along the coasts of Greenland, Iceland, Norway, and Spitzbergen. It may be the result of combined glacial and marine processes.

strandline The shoreline of a marine or *lacustrine environment. The term is applied most commonly to ancient shorelines. The development of a strandline requires that the relative positions of land and water remain stable long enough for features to form. Subsequent displacement may be caused by a change in the level of the water or of the land.

strata *See* STRATUM.

stratified sampling *See* SAMPLING METHODS.

stratiform deposit A mineral deposit that is *concordant with *bedding; usually in sheets but it may be ribbon-like.

stratiformis From the Latin *stratus* meaning 'flattened' or 'spread out' and *forma* meaning 'appearance', a species of cloud consisting of an extensive level sheet or layer, found in *altocumulus, *stratocumulus, and sometimes *cirrocumulus. *See also* CLOUD CLASSIFICATION.

stratigraphic *See* STRATIGRAPHY.

stratigraphic column 1. A succession of rocks laid down during a specified interval of *geologic time. The phrase 'the stratigraphic column' often refers to the whole sequence of *strata deposited throughout geologic time. **2.** A simplified columnar diagram relating a succession of named *lithostratigraphic units from a particular area to the subdivisions of geologic time.

stratigraphic correlation Geologic study concerned with establishing geochronological relationships between different areas, based on geologic investigations of many local successions.

stratigraphic cross-section A section, usually with the vertical scale considerably exaggerated with relation to the horizontal scale, that is designed to show the thicknesses and stratigraphic relationships of successions of named *lithostratigraphic units. For simplicity, the upper part of one of the units is restored to a horizontal position and topography is ignored. Items such as *facies changes, *interdigitating of units, *unconformities, and breaks in succession are shown. The units and their boundaries are related to the subdivisions of *geologic time.

stratigraphic nomenclature The naming of *stratigraphic and *geologic-time units according to established practices and principles. *Formal naming of a stratigraphic unit occurs when the unit is first proposed and described from a type section (*see* STRATOTYPE), which acts thereafter as the standard reference for that unit. Ideally, the name given is binomial, and in the case of *chronostratigraphic and *lithostratigraphic units consists of a preceding geographic name taken from the *type locality (plus lithological description where appropriate), followed by the name of the unit, e.g. Ludlow Series and Elk Point Group. The names of *biostratigraphic units consist of the name of the characteristic *fossil plus the relevant unit term, e.g. *Monograptus uniformis* Range zone. The name chosen for a stratigraphic unit should be unique to that unit. When used as a proper name, as above, the initial letters are capitalized. Except in very special circumstances the first formal name given has priority and is adhered to. In practice many well-known units, e.g. Coal Measures, Millstone Grit, were named long before the present conventions were established, and to avoid confusion these names are preserved in their original form. Geologic-time units generally take their preceding name from that of the corresponding chronostratigraphic unit, plus the name of the unit (*period, *epoch, *age, etc.), e.g. the *Jurassic Period, from the Jurassic System (named after the *type area in the Jura Mountains). The names of *eons and *eras (e.g. *Phanerozoic Eon, *Mesozoic Era) were proposed independently, so that the names for the corresponding *eonothems and *erathems are derived from the time units. *See also* INFORMAL.

stratigraphic reef A name proposed in 1970 by R. J. Dunham to describe a *reef that comprises only a thick mass of pure or nearly pure *carbonate rock. *Compare* ECOLOGIC REEF.

stratigraphic scale A general term to denote a time-scale that incorporates both the traditional elements of the *geologic time-scale as it has evolved over the last century and a half, and, as they are agreed, the reference points of an ideal and globally standardized *chronostratigraphic scale that is defined by *boundary stratotypes. The ideal, globally standardized stratigraphic scale has been termed both the *Standard Stratigraphic Scale (SSS), and the *Standard Global Chronostratigraphic Scale (SGCS). W. B. Harland (1978) has suggested that, for clarity, the older geologic time-scale, which evolved through the designation of type sections (*see* STRATOTYPE) and is gradually being superseded, should be referred to as the Traditional Stratigraphic Scale (TSS). A local stratigraphic scale, from which a standard reference point might be selected later, is termed a Regional Stratigraphic Scale (RSS).

stratigraphic trap (lithologic trap) Oil or gas trap resulting from lithologic variations, e.g. interbedded lenses of *sands and *silts in a deltaic environment. *See* DELTA; NATURAL GAS; and PETROLEUM. *Compare* ANTICLINAL TRAP; FAULT TRAP; REEF TRAP; STRUCTURAL TRAP; and UNCONFORMITY TRAP.

stratigraphic unit A body of rock forming a discrete and definable unit. Such units are determined on the basis of their lithology (*lithostratigraphic units), or their *fossil content (*biostratigraphic units), or their time span (*chronostratigraphic units). It is unlikely that any rock succession will form a unit that accords with all three categories of classification. All stratigraphic units are defined by a *type section. *Geologic-time units are abstract concepts, not actual rock sequences, so do not class as stratigraphic units.

stratigraphy **1.** The branch of the geologic sciences concerned with the study of stratified rocks in terms of time and space. It deals with the correlation of rocks from different localities. Correlation methods may involve the use of *fossils (*biostratigraphy), rock units (*lithostratigraphy), or *geologic-time units or intervals (*chronostratigraphy). **2.** The relative spatial and temporal arrangement of rock strata.

stratocumulus From the Latin *stratus* meaning 'flattened' or 'spread out' and *cumulus* meaning 'heap', a cloud composed of sheets or layers of grey to whitish appearance, typically with dark patches which are not fibrous. *See also* CLOUD CLASSIFICATION.

stratomere A general term for any *chronostratigraphic unit within the Stratomeric Standard hierarchy. *See* CHRONOSTRATIGRAPHY.

stratopause The level that marks the maximum height of the *stratosphere, at around 50 km. After high temperatures in the upper stratosphere (about 0 °C at the stratopause) temperature decreases with increasing altitude in the *mesosphere above. *See also* ATMOSPHERIC STRUCTURE.

stratophenetic classification *See* STRATOPHENETICS.

stratophenetics (stratophenetic classification) In *cladistics, a method for determining the evolutionary relationships among organisms that exist only as fossils. It is based on quantitative assessments of morphological (i.e. phenetic) similarities and geologic age (derived from stratigraphy).

stratosphere The atmospheric layer above the *troposphere, which extends on average from about 10 to 50 km above the Earth's surface. The stratosphere is a major stable layer whose base is marked by the *tropopause, and where temperatures overall average approximately –60 °C. Temperature in the lower stratosphere is isothermal but increases markedly in the upper part, to reach a maximum of about 0 °C at the *stratopause. High stratospheric temperatures result from absorption of ultraviolet radiation (0.20–0.32 μm wavelengths) by ozone concentrated at 15–30 km. Due to the very low air density, even the small amount of ozone concentrated in the upper stratosphere is extremely effective in absorbing radiation, thus giving high temperatures at 50 km. The isothermal condition at the base of the stratospheric inversion layer creates stability, which generally limits vertical extensions of cloud and leads to the lateral spreading of high *cumulonimbus cloud with characteristic anvil heads. *See also* ATMOSPHERIC STRUCTURE.

stratotype (type section) An actual rock succession, chosen at a particular locality (the *type locality) to act as the standard comparison for all other *chronostratigraphic or *lithostratigraphic units of its ilk. Generally, the type section of a *stratigraphic unit should be the rock succession originally so designated and described (the *holostratotype), but circumstances may require amendment (*see* LECTOSTRATOTYPE; and NEOSTRATOTYPE), or amplification (*see* PARASTRATOTYPE; and HYPOSTRATOTYPE).

However, although many simiiar sequences may exist, only a single stratotype at any one time can act as the standard. *See also* BOUNDARY-STRATOTYPE; COMPOSITE-STRATOTYPE; COMPONENT-STRATOTYPE; and TYPE AREA.

strato-volcano (composite volcano) *Volcano built up of layers of *lava alternating with beds of *ash and other *pyroclastics and with material eroded from higher slopes of the cone. Many of the world's highest volcanoes are of this type, and include Mt. Fuji (Japan) and Mt. Egmont (North Island, New Zealand).

stratum (pl. strata) Lithological term applied to rocks that form layers or beds. Unlike 'bed', 'stratum' has no connotation of thickness or extent and although the terms are sometimes used interchangeably they are not synonymous.

stratus From the Latin *stratus* meaning 'flattened' or 'spread out', a cloud of flat, uniform base and of grey appearance, through which the Sun may be outlined clearly when the cloud is not too dense. *See also* CLOUD CLASSIFICATION.

streak The colour of a mineral when in the form of powder, which is usually produced by scratching the solid mineral on an unglazed porcelain plate (streak plate). The colour may be different from the mineral's colour in mass.

streak lightning An electric discharge with a branching appearance of the main channel. It may flash between cloud and air or between cloud and ground.

streak plate *See* STREAK.

streamer A long (up to several kilometres) tube containing a number (sometimes many hundreds) of *hydrophones and filled with oil, designed to be towed by a ship, and used in marine *seismic surveying. It may be balanced to give it neutral buoyancy or have depth controllers to maintain it at a constant depth. Compass devices are attached at intervals to monitor the *feather angle of the cable.

stream flood *See* FLASH FLOOD.

stream grade *See* GRADE.

streamline 1. In a flowing fluid, a hypothetical line which indicates the local direction of flow. *See also* HYDRAULIC GRADIENT; and POTENTIOMETRIC SURFACE. **2.** A shape which allows a body to offer minimum resistance to a fluid through which it moves; to impart such a shape to a body.

stream order Measure of the position of a stream (defined as the reach between successive tributaries) within the hierarchy of the *drainage network. A commonly used approach allocates order '1' to unbranched tributaries, '2' to the stream after the junction of the first tributary, and so on. It is the basis for quantitative analysis of the network.

stream power The rate at which a stream can do work, especially the transport of its load, and measured over a specific length. It is largely a function of channel slope and discharge and is expressed by $\Omega = \gamma Qs$, where Ω is the power, γ is the specific weight of water, Q is the discharge, and s is the slope. Streams tend to adjust their flow and channel geometry in order to minimize their power (*see* LEAST-WORK PRINCIPLE).

stream-sediment analysis (river-sediment analysis; drainage-sediment survey) A technique used in geochemical exploration and analysis, in which semi-mobile and immobile elements are measured from river or stream sediments, although under some circumstances highly mobile elements, e.g. molybdenum, may be used. Anomalies in sediments are not always accompanied by anomalies in the water, there may be seasonal variations in the composition of stream water, and it is easier to collect, carry, and store sediment samples than water samples. Anomalies may occur in the active sediment and in the banks and *flood plains, so sampling of any of these can give satisfactory results. Immobile elements are determined by total-metal analysis; mobile and semi-mobile elements by total- or cold-extractable metal analysis.

stream terrace *See* RIVER TERRACE.

stress A measure of the intensity of a force (F) acting upon a body as a function of its area (A), such that stress = F/A, in units of N/m^2. Stress can be resolved into two important components. Compressive (tensile) stress (σ) acts normal to the surface (*see* NORMAL STRESS) and changes the volume of the body; *shear stress (τ) acts parallel to the surface and changes the shape of the body. *See also* STRAIN.

stress axial cross Three mutually perpendicular *stress axes whose lengths are proportional to the magnitudes of the *principal stresses they represent. The maximum (σ_1) and minimum (σ_3) principal-stress axes define the *stress ellipse; the addition of the intermediate (σ_2) principal-stress axis defines the *stress ellipsoid.

stress difference (σ_d; differential stress) The simple difference between the greatest and least *principal stresses, and used in the construction of *Mohr stress diagrams. The diameter of a Mohr stress circle represents the stress difference for a given combination of σ_3 and σ_1 (*see* STRESS AXIAL CROSS). In *stress–strain diagrams the stress axis is given as σ_d and is usually plotted against percentage *strain. *See* PRINCIPAL-STRESS AXES.

stress field The spatial change in the orientation of *stress throughout a body of material, as opposed to the simple stress at a point. A stress field is represented as a grid of *stress trajectories which may or may not be superimposed.

stress meter Instrument for measuring pressure changes in rocks, that result from mining operations. The stress meter has a steel shaft with a groove containing glycerine. When pressure is exerted the glycerine is squeezed on to a diaphragm which pushes a strain gauge to measure the movement.

stress–strain diagram A diagram which illustrates the change in *strain under an applied *stress, usually as a function of the progressive change in temperature, pressure, and *strain rate. The percentage strain is plotted on the *x* axis, and differential stress (the difference between the maximum (σ_1) and minimum σ_3) *principal-stress axes) on the *y* axis.

stress trajectory A line showing the continuous change in the orientation of a principal *stress throughout a body. Although trajectories may curve, their intersections with other principal stresses remain perpendicular.

strewnfield An area which is associated with a specific group of *tektites and microtektites that can be distinguished according to their age and chemical composition, and which probably represents a particular impact event. Four major strewnfields are known: the Australasian (formed 0.7 Ma ago), which is the largest and covers an area of about 5×10^7 km^2 around Australia and South-east Asia; the Ivory Coast (formed 1.3 Ma ago) covering an area at least 4×10^6 km^2 around and off the coast of W. Africa; the Czechoslovakian (formed 14 Ma ago) without proven associated microtektites as yet; and the N. American strewnfield (formed 34 Ma ago) forming a belt stretching across the Pacific from South-east Asia to the western Atlantic, and of unknown latitudinal extent.

striation Narrow groove or scratch cut in exposed rock by the abrasive action of hard rock fragments embedded in the base of a sliding *glacier. Striation provides a useful clue to the direction of ice movement in formerly glaciated areas.

strike 1. (noun) The compass direction of a horizontal line on an inclined plane. **(verb)** To lie in the direction of such a line. **2. (noun)** The discovery of an economically valuable source of a mineral. **(verb)** To make such a discovery.

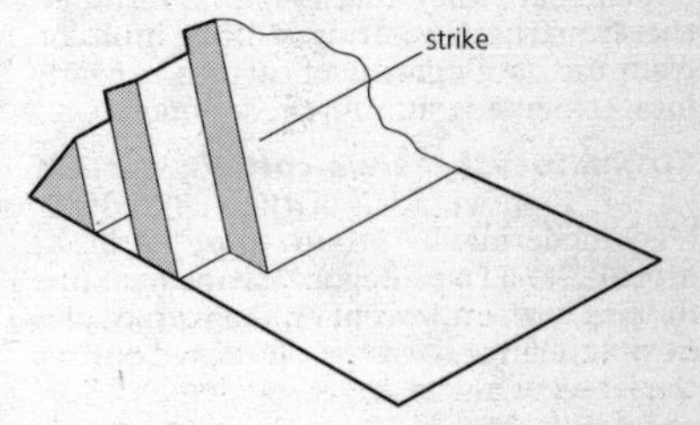

Strike

strike fault A *fault which strikes parallel with the *strike of the layering (i.e. *bedding or *cleavage) in adjacent rocks.

strike ridge An elongated hill developed along the *strike of a bed that is more resistant than its adjacent strata. *See also* CUESTA.

strike-slip fault (wrench fault, tear fault, transcurrent fault) A *fault in which the major displacement is horizontal and parallel to the *strike of a vertical or subvertical *fault plane. Movements along such a fault may be *dextral or *sinistral. Localized zones of deformation due to pressures and tensions across the fault occur at bends in the fault strike and give rise to the formation of pull-apart basins and *grabens, which are rhombic in shape. *Flower structures are also recognized features of such regimes. There are certain geometrical similarities between strike-slip faults and *transform faults, but also significant differences (e.g. displacement along the fault plane is equal and unlimited in a transform fault but not in a strike-slip fault). *See* TRANSPRESSION; and TRANSTENSION.

strike stream *See* SUBSEQUENT STREAM.

strike valley *See* SUBSEQUENT STREAM.

string 1. (flyer) Up to 10 *geophones which are connected together permanently but have only one lead on to the seismic cable. **2. (drill string)** The rods (flights) and tools from the drill collar to the *bit which, when connected together, enable a *borehole to be drilled.

strip mining *See* OPEN-CAST MINING.

stromatactis A series of elongated cavities, with curved or irregular tops and flat bases, filled with *calcite *cements. Stromatactis cavities occur most commonly in carbonate and *mud mounds. They were originally believed to be of organic origin, but currently they are thought to result either from the dewatering of *lime muds or from the development of cavities beneath local cemented crusts on the sea floor.

Stromatocystites walcotti The earliest known representative of the echinoderm (*Echinodermata) group the Edrioasteroidea. It had a pentagonal shape with five distinct 'arms' apparent on the surface of a flexible, many-plated *test. It existed during the Lower *Cambrian.

stromatolite A laminated, mounded structure, built up over long periods of time by successive layers or mats of *cyanobacteria that trapped sedimentary material. Stromatolites are found in shallow marine waters in warmer regions. Some are still in the process of being formed, e.g. those in Shark Bay, Western Australia; *fossil stromatolites dating from the Early *Precambrian are also known, although it is not certain that these were formed by cyanobacteria.

Stromatoporoidea Extinct group that has been attributed to the *Hydrozoa, sponges (*Porifera), foraminifera (*Foraminiferida), *Bryozoa, or *algae, or regarded as a phylum with no modern representatives. Stromatoporoids are calcareous masses built up of horizontal layers (latilaminae) and vertical pillars. The calcareous skeleton is called the *coenosteum. The upper surfaces show a pattern of polygonal markings and may have swellings (mammelons) and stellate grooves (astrorhizae). They are found in *limestones of *Cambrian to *Cretaceous age, often forming *reefs in *Ordovician to *Devonian times.

Strombolian eruption A type of volcanic activity which produces frequent, moderate *eruptions. The *lava is basaltic, but sufficiently viscous for entrapped gases to build up a pressure which is released in continuous small explosions. Lava, flung into the air, falls back to build up a steep-sided cone of interbedded *lava and *tephra. Lava flows are commonly erupted through breaches in the flanks of the cone. *See* VOLCANO. *Compare* HAWAIIAN ERUPTION; PELÉEAN ERUPTION; PLINIAN ERUPTION; SURTSEYAN ERUPTION; VESUVIAN ERUPTION; and VULCANIAN ERUPTION.

strontianite Mineral, $SrCO_3$; sp. gr. 3.7; *hardness 3.5–4.0; *orthorhombic; white to pale green, grey, or pale yellow; white *streak; *vitreous *lustre; crystals *prismatic or *acicular, but also fibrous and *massive; *cleavage prismatic, good {110}; occurs in low-temperature *hydrothermal veins, often in *limestone, and in association with *celestite, *barite, and *calcite; soluble, with effervescence, in dilute hydrochloric acid. It is named after the type locality of Strontian, Highland Region (Argyll), Scotland.

strophic *See* HINGE.

Strophomenida (strophomenids; class *Articulata) The largest order of brachiopods (*Brachiopoda), now extinct, in which one *valve is usually convex and the other flat or concave. There is a straight *hinge line. The *pedicle foramen is filled by one *plate in each valve in adult shells, the *pedicle having degenerated and been lost. Attachment is by cementation of the pedicle valve, with or without tubular spines from the valves. The *shell structure is *pseudopunctate. Strophomenida appeared in the Lower *Ordovician and became extinct in the Lower *Jurassic. The order includes the *Carboniferous **Productus giganteus* (*Gigantoproductus giganteus*), the largest of all brachiopods.

strophomenids *See* Strophomenida.

structural contour map A map on which structural features, e.g. *folds, are represented in three dimensions. The map is read in the same way as a topographic contour map. The contours are based on a single *horizon (e.g. the top of a bed), the position of that horizon being given with reference to a datum plane.

structural trap Trap formed by deformation of porous and non-porous *strata as a result of *folding, *faulting, etc., in which oil, gas, or water may accumulate. *See* NATURAL GAS; PETROLEUM; and POROSITY. *Compare* ANTICLINAL TRAP; FAULT TRAP; REEF TRAP; STRATIGRAPHIC TRAP; and UNCONFORMITY TRAP.

structure grumeleuse A *texture in *limestones, characterized by the presence of *micrite clots completely surrounded by coarser, granular, *calcite or *microspar. This structure is thought to be produced by the selective *recrystallization of the limestone, with larger crystals growing at the expense of smaller ones.

Strutt, John William (Lord Rayleigh) (1842–1919) A mathematician and physicist of Cambridge University, Rayleigh worked on optics, noble gases, wave mechanics, etc. In the field of Earth science, he studied radioactivity in rocks, and gave his name to a type of surface earthquake wave. *See* RAYLEIGH WAVES; RAYLEIGH NUMBER; and RAYLEIGH SCATTERING NUMBER.

Sturtian The penultimate *system of the *Proterozoic, from about 825 to 625 Ma ago (Harland et al., 1989).

Stylasterina (branched hydrocorals; class *Hydrozoa) Order of reef-building *Cnidaria similar to *Milleporina, but without free medusae. Stylasterina are known from the Upper *Cretaceous to Recent.

stylolite An irregular, suture-like contact, produced by *pressure dissolution of rock under deep burial conditions. Stylolites are most commonly found in *limestones, and may be picked out by the concentration of insoluble *clay residues along the stylolite surface. Up to 40% of the original thickness of a limestone sequence can be dissolved through stylolitization.

stylolitization *See* STYLOLITE.

sub- From the Latin *sub* meaning 'under' or 'close to', a prefix meaning 'beneath' or 'lying below'.

Sub-arctic current *See* ALEUTIAN CURRENT.

subarkose A *sandstone characterized by the presence of less than 15% mud *matrix, with between 5% and 25% of the *grains being *feldspar, and there being more feldspar than rock fragments present. *See* DOTT CLASSIFICATION.

Sub-Atlantic A colder, wetter climatic phase which followed more continental *Sub-Boreal times. The change from Sub-Boreal to Sub-Atlantic conditions in Britain is roughly coincident with the transition from Bronze to Iron Age cultures. The Sub-Atlantic marks a period of renewed *peat growth on bog surfaces that in late Sub-Boreal times were sufficiently dry and humified to support heath vegetation, e.g. *Calluna vulgaris* (ling, heather). This renewed peat growth gives a major recurrence surface, the Grenz horizon, which in Britain defines the Zone VIIb/Zone VIII (Sub-Boreal/Sub-Atlantic) boundary of the standard pollen stratigraphy. *See* pollen analysis; and pollen zone.

subbase *See* PAVEMENT.

Sub-Boreal From Scandinavian evidence, a cooler, drier, more continental climate phase that followed the *Atlantic climatic optimum (though with summers still warmer than those at present). In Britain, clear evidence for climatic change (e.g. differing *peat deposits) is lacking, and the *Atlantic/Sub-Boreal boundary is usually taken as the point of marked decline in elm pollen in the pollen stratigraphy (*see* POLLEN ANALYSIS). The reason for the elm decline (which, irrespective of elm species, is characteristic of pollen chronologies throughout Europe) has been the subject of much research. Hypotheses range from colder climate to epidemic disease to anthropogenic causes. The latter is linked to the selective use of elm as fodder for stalled livestock, now thought to have been a characteristic practice of Neolithic, forest-dwelling peoples. In many pollen chronologies this initial elm decline is followed quickly by a general decline in tree pollen caused by temporary forest clearance ('landnam') to provide land for slash-and-burn agriculture. The Sub-Boreal forms *Pollen Zone VIIb, and lasted from about 5000 to 2800 BP.

subcritical reflection In *refraction surveying, *seismic waves which are incident on the refractor *interface at less than the *critical angle of incidence and are weakly reflected to the surface.

subduction The process of consumption of a *lithospheric plate at convergent *plate margins. *See* SUBDUCTION ZONE.

subduction zone The zone, at an angle to the surface of the Earth, down which a lithospheric *plate descends. Most present-day subduction zones extend from *trenches on the ocean floor, from where a zone of *earthquake *hypocentres (called a *Benioff zone) extends, at an angle ranging from near-horizontal to near-vertical, to a depth of up to 700 km. Andesitic *volcanoes form approximately 100 km above the subducting slab, and the presence of andesitic volcanoes in the geologic record is regarded as evidence of an ancient subduction zone and thus of a *destructive plate margin.

subglacial *See* ENGLACIAL.

subgrade *See* PAVEMENT.

subgroup A term that may be used for *formal identification of a distinct and defined assemblage of *formations within an already named *group whose name is in use.

subhedral (hypidiomorphic) An *igneous textural term applied to crystals which are only partly bounded by *crystal faces. The irregular parts of the crystal surface may be caused by partial corrosion of the crystal or by partial intergrowth with other crystals.

sublimate A solid substance that has condensed directly from a gas.

sublimation Direct evaporation from ice. In meteorology, the term is also applied to the reverse process, in which water vapour changes directly to the solid phase. *See also* ABLATION.

sublitharenite A *sandstone characterized by the presence of less than 15% mud *matrix, with between 5% and 25% of the *grains being rock fragments, and there being more rock fragments than *feldspar present. *See* DOTT CLASSIFICATION.

sublittoral zone 1. In freshwater *ecosystems, an alternative name for the limnetic zone. **2.** The sea-shore zone lying immediately below the *littoral (intertidal) zone and extending to a depth of about 200 m or to the edge of the *continental shelf. Red and brown *algae are characteristic of this area. Typical animals include sea anemones and corals on rocky shores, and shrimps, crabs, and flounders on sandy shores. The zone may alternatively be called the subtidal zone. It is approximately equivalent to the *circalittoral zone.

submarine canyon Deep, steep-sided valley cut into the *continental shelf or slope, whose axis slopes seaward at up to 80 m/km. The development of submarine canyons is thought to have originated through erosion by *turbidity currents. However, few turbidity currents have been recorded from the submarine canyons that have been studied.

submersible Underwater vehicle, used for oceanographic investigation or offshore engineering. These small submarines may be manned or unmanned.

sub-metallic Applied to the *lustre of a mineral which is intermediate between *metallic and *non-metallic. Varieties of *chromite or *sphalerite may give a rather dull, metallic appearance on occasions and might therefore be described as sub-metallic. There is no sharp distinction between 'metallic' and 'sub-metallic'.

sub-Plinian eruption *See* VESUVIAN ERUPTION.

subpolar glacier *See* GLACIER.

subsequent stream A stream that follows a line of geologic weakness, such as the *outcrop of a soft bed, a sequence of major *joints, a *fault trace, or the axis of an *anticline. Such a stream tends to extend headwards actively, and may acquire further tributaries through the process of *river capture. It is called a 'strike stream' when its trace follows that geologic *strike, and the associated valley is called a 'strike valley'.

subsidence 1. A progressive depression of the Earth's *crust, which allows *sediment to accumulate and be preserved. Subsidence is caused by *mantle convection and by sediment loading. The subsidence rate will control the proportion of deposited sediment which will be preserved in the subsiding area. Subsidence rates in *sedimentary basins typically vary from 0.3 to 2.5 mm per year. **2.** Sinking or settling of the ground surface due to natural or anthropogenic causes. Surface material with no free side is displaced vertically downwards with little or no horizontal movement. **3.** Local sinking, due to underground mine workings. **4.** Downward movement of air, characteristically gentle (1–10 cm/s) and often in large *anticyclones, which is related to the *divergence in lower layers near the ground surface. Subsidence results from radiation cooling or from convergence of air horizontally in the upper *troposphere. It typically brings settled weather, with evaporation of cloud drops by *adiabatic warming in the subsiding *air mass, causing cloudless skies above the friction layer near the surface, so that in winter fog and low cloud may prevail when moisture is sufficient.

subsoiling The breaking up of subsoils, usually because they are compacted, without inverting them. Subsoiling is usually performed with a chisel-like device that is pulled through the soil.

subsolvus granite An *igneous rock of granitic composition characterized by the presence of two types of *alkali feldspar: a potassium-rich type displaying perthitic (*see* PERTHITE) texture; and a sodium-rich type displaying antiperthitic texture. Subsolvus granites have a high water content, which depresses their *liquidii sufficiently to intersect the subsolidus *solvus surface and cause two extremes of feldspar composition to crystallize from the *melt, instead of one intermediate-composition feldspar as in *hypersolvus granites.

subsolvus syenite *See* SYENITE.

substage A subdivision of a *stage. *See* CHRONOZONE.

sub-surface flow (interflow; throughflow) The flow of water at a shallow depth beneath the ground surface, that occurs when rain falls faster than it can infiltrate downwards. The sub-surface flow re-emerges at the surface at or near the base of ground slopes.

subtidal Applied to that portion of a *tidal-flat environment which lies below the level of mean low water for *spring tides. Normally it is covered by water at all states of the *tide. The word is often used as a general descriptive term for a subaqueous but shallow-marine depositional environment.

subtractive primary colours The colours cyan, magenta, and yellow, which can be subtracted from white light to produce all other colours. *See also* ADDITIVE PRIMARY COLOURS.

subtropical high Surface high-pressure cells, especially prominent and persistent over oceans at around 30° latitude. The *anticyclones develop below the *subtropical jet stream from subsiding air. The development tends to shift equatorward in winter and poleward in summer. The high pressure is weaker over continents in summer. *See also* AZORES HIGH; and BERMUDA HIGH.

subtropical jet stream *Jet stream of subtropical latitudes. The change from a westerly direction in winter to an easterly one in summer influences the surface wind changes that mark the summer *monsoon. The jet is related to a marked temperature gradient in the upper *troposphere. The westerly jet moves equatorward in winter and is associated with subsiding air and settled surface weather. In summer the jet moves poleward. At the seasonal extremes it tends at times to merge with the *polar-front jet.

subzone Division of the fundamental unit (*zone) used in *biostratigraphy. The demarcation of a subzone is based on the *fossil subspecies or assemblage contained within the rock sequence studied.

sucrosic limestone *See* DUNHAM CLASSIFICATION.

Suess, Eduard (1831–1914) Professor of geology in Vienna, Suess published his important work on structural geology, *Das Antlitz der Erde* ('Face of the Earth'), between 1833 and 1909. He studied mountain building, especially the Alps which he believed to have been formed in a *geosyncline, which he named *Tethys. He opposed the concept of *isostasy, arguing that subsidence of the ocean floors had caused what he termed 'eustatic' changes in sea level. *See* OROGENY; and EUSTATIC.

Suess wriggles Small oscillations, mostly lasting 2–3 years, between the ^{14}C determinations and *dendrochronological age determinations of the same piece of wood. These were thought originally to be instrumental, but are now considered to reflect genuine changes in ^{14}C productivity of uncertain cause.

suevite A *breccia of rock fragments in a *matrix of *glass, found within *meteorite impact craters near the site of impact. The shock waves associated with the meteorite impact produce extremely high pressures and temperatures in the rocks for a few microseconds. Near the point of impact, these can brecciate the rock and also melt it. When the melt chills after the passage of the shock wave it produces the glass component known as '*impactite glass'.

suffusion Spreading out of material on the substratum.

Suisei A Japanese *ISAS mission to comet *Halley, launched in 1985.

sulci *See* SULCUS.

sulcus (pl. sulci) Latin for groove or furrow. **1.** A complex region of parallel ridges and furrows on a *satellite surface, particularly well developed on Ganymede (e.g. Uruk Sulcus, an area of bright, grooved terrain bordering Galileo Regio). **2.** A major, rounded depression on the longitudinal mid-line of a brachiopod (*Brachiopoda) shell. It usually occurs in the *ventral *valve and is usually accompanied by a major, rounded, shell elevation (fold) in the other valve. **3.** *See* DINOPHYCEAE. **4.** *See* POLLEN.

sulphates Group of non-silicate *minerals in which the SO_4^{2-} radical is in combination with a number of metal *cations. Examples include *barite ($BaSO_4$), *celestite ($SrSO_4$), *anglesite ($PbSO_4$), and *anhydrite ($CaSO_4$). *Gypsum ($CaSO_4.2H_2O$) is the most common of a number of hydrated sulphates which also occur. Sulphates are normally colourless or white, soft (*hardness about 3), *massive or *earthy, but *tabular when crystalline. They are low-temperature minerals and occur as *gangue minerals in *hydrothermal veins and as chemical precipitates and *evaporites.

sulphides A group of *minerals in which the element sulphur (S) is in combination with one or more metallic elements. Simple sulphides include the common *ore minerals *galena (PbS), *sphalerite (ZnS), and *pyrite (FeS_2). Two metallic *cations may also be present, as in *chalcopyrite ($CuFeS_2$). More complex combinations may also occur to give 'double sulphides' or 'sulpho-salts' in which metallic and metalloid or non-metallic elements are present in combination with sulphur, e.g. *tetrahedrite ($Cu_{12}Sb_4S_{13}$) and *enargite (Cu_3AsS_4 or $3Cu_2S.As_2S_5$).

sulpho-salts *See* SULPHIDES.

sulphur, native Non-metallic element, S; sp. gr. 2.0; *hardness 2.0; yellow; *massive, or *tabular when crystalline; produced by *fumarole volcanic activity and by hot springs, and recovered commercially from bedded sedimentary deposits associated with *gypsum and *salt domes. Most sulphur is now obtained as a by-product of oil-refining, since it is a common contaminant of natural oil.

sumatra Regional squall, usually occurring at night, in the Malacca Strait, accompanied by high winds which veer (*see* VEERING) from southerly to south-westerly and north-westerly. Extensive *cumulonimbus cloud brings heavy rain, with thunder and lightning.

Sun The central star (G spectral type) in the *solar system, 696 000 km in diameter, 333 000 × *Earth mass, 1 300 000 × Earth volume, and with a mean density of 1410 kg/m³. The equator is inclined at 7.25° to the plane of the *ecliptic. It is principally composed of hydrogen and helium. The visible surface is the 'photosphere' (temperature 6000 K). The Si-normalized solar abundances for the terrestrially non-gaseous elements match those of the C1 *carbonaceous chondrites.

suncracks *See* DESICCATION CRACKS.

Sundaland Name commonly given to the unit composed of Malaya, Sumatra, Java, and Borneo, with the intervening small islands; these are linked by the shallow-water (less than 200 m) Sunda shelf, which was exposed during periods of low sea level in the *Pleistocene.

Sundance Sea A shallow marine *embayment that extended over what are now Wyoming and S. Dakota during the late *Callovian and *Oxfordian. The southern edge of this sea (in modern Colorado) was bordered by *tidal flats, the marine connection being northwards through the present-day mid-west and Canada. The *sediments resulting from deposition in the Sundance Sea are characterized by a rich ammonite (*Ammonoidea) fauna (*Quenstedtoceras* and *Cardioceras*), especially the Redwater Shale Member of the Sundance *Formation.

sunshine recorder *See* CAMPBELL–STOKES SUNSHINE RECORDER.

Sun-synchronous orbit A satellite orbit that remains constant in relation to the Sun, passing close to both poles and crossing the meridians at an angle. The orbit, at a height of about 860 km (one-seventh of an Earth radius), takes about 102 minutes and carries the satellite over a different swathe of territory at each pass, so every point on the surface is overflown every 12 hours, at the same solar times each day. *Compare* GEOSYNCHRONOUS ORBIT; *see also* POLAR ORBIT.

sun-tan age *See* EXPOSURE AGE.

super- From the Latin *super* meaning 'on top of', a prefix meaning 'directly over', 'over', or 'above'.

super-adiabatic lapse rate A fall of temperature with increasing altitude, which is greater than the usual *dry adiabatic lapse rate; it occurs in conditions of intense heating over land or sea.

supercell A very large convection cell that forms within a *cumulonimbus by the merging of several smaller convective cells. Inside the supercell air may rise at 45 m/s and the cell may extend to a height of more than 16 km, breaking through the *tropopause. Supercells last much longer than ordinary convective storm cells, producing extremely violent storms, and they can trigger *tornadoes. *See also* MESOCYCLONE.

supercontinent A continental mass which includes several of the *cratons of the present-day continents. The term is used of *Pangaea, *Gondwana, and *Laurasia.

supercooled cloud Cloud containing pure water droplets at temperatures considerably below the nominal freezing temperature of 0 °C. With very pure water (i.e. free from pollutants), supercooling of liquid drops can occur down to around –40 °C; *altocumulus cloud, for example, is usually composed of water droplets at temperatures well below 0 °C. *See also* CLOUD SEEDING.

supercooling (undercooling) The cooling of a liquid to a temperature lower than its normal freezing temperature.

supercritical fluid *See* CRITICAL POINT.

supercritical reflection In *refraction surveying, *seismic waves which are incident on the refractor *interface at greater than the *critical angle of incidence and are strongly reflected to the surface. Such waves travel with the *seismic velocity of the first layer (therefore more slowly than the refracted *head waves).

supergene enrichment (secondary enrichment) Re-precipitation of sulphides and oxides by descending acidic *groundwater which has leached the surface zone of an *ore deposit (*see* GOSSAN); this upgrades the deposits *in situ*, as in *porphyry copper ores.

supergroup A term that may be used for *formal identification of an assemblage of related and adjacent *groups, or related and adjacent *formations and groups.

superimposed drainage (epigenetic drainage) A *drainage pattern that has been established on an earlier surface (perhaps conformable with the immediately underlying strata, and standing well above the present landscape). Subsequently the pattern was lowered by river incision so it now lies across geologic structures to which it bears no relation.

superinterval The time between *hairpins on a palaeomagnetic apparent *polar wander path, usually hundreds of millions of years.

superposition *See* LAW OF SUPERPOSITION.

supersaturation The condition of air in which the humidity is above the level required for saturation at a given temperature (i.e. the *relative humidity is greater than 100%). Supersaturation results when the temperature of air containing no *condensation nucleii falls below its *dew point. *See also* SATURATED AIR.

Supersaurus *See* SAUROPODA.

suppressed layer In *electrical *resistivity depth sounding, a thin layer whose true resistivity is intermediate between those of the layers above and below it, so it may be masked and its effects suppressed. It is analogous to the *hidden layer in *refraction seismology.

supplementary forms *Crystals whose faces have developed in different positions relative to their atomic structure but which nevertheless have the same general *crystal symmetry. They may be distinguished as 'positive' or 'negative' forms, or as 'diploids'.

supra- From the Latin *supra* meaning 'above', 'beyond', or 'earlier in time', a prefix meaning 'above' or 'in a superior position to'.

supraglacial *See* ENGLACIAL.

supralittoral zone The seashore zone immediately above the *littoral fringe and beyond the reach of tidal submergence, though affected by sea spray.

supratidal Applied to that portion of a *tidal flat which lies above the level of mean high water for *spring tides. It is inundated only occasionally by exceptional *tides or by tides augmented by a *storm surge.

surf Breaking waves in the area between the shoreline and the outermost limit of breaking waves.

surface inversion A *temperature inversion in the lower atmospheric layers, extending upwards from the Earth's surface. The condition results, for example, from radiation cooling of the ground and the air above, or from *advection of warm air over cold surfaces.

surface runoff (overland flow, Hortonian flow) The flow across the land surface of water that accumulates on the surface when the rainfall rate exceeds the *infiltration capacity of the *soil. The rate of infiltration, and therefore the possibility of surface runoff, is determined by such factors as soil type, vegetation, and the presence of shallow, relatively impermeable, *soil horizons. Saturated overland flow can occur when a temporary rise of the *water-table inhibits infiltration and causes flow over the surface.

surface tension (γ) Fluid surfaces may take on the behaviour of a stretched elastic membrane as a result of the tendency of a liquid surface to contract. The surface tension of a liquid is given as the tension across a unit length of the fluid surface. Surface tension is temperature-dependent and is closely associated with *capillarity.

surface wave A *seismic wave which propagates along the surface of a medium rather than through it, e.g. *Love waves and *Rayleigh waves. *Compare* BODY WAVE.

surface wind The wind close to the Earth's surface, the velocity of which is usually measured at a standard height of 10 m. Surface-wind velocity is reduced by the frictional effect of the underlying surface. The actual wind is a balance of *pressure-gradient force, *Coriolis force, and frictional effects.

surf wave *See* SPILLING BREAKER.

surge An expanded, turbulent, dilute flow of gas and *pyroclasts. Three main types are currently recognized. *Base surges, which are cold and wet, are generated during phreatomagmatic or *phreatic eruptions; ground surges, which are hot and dry, are generated from the head of pyroclastic flows; and ash-cloud surges, which are also hot and dry, are generated from the overriding gas and *ash cloud above pyroclastic flows.

Surtseyan eruption A high-energy volcanic *eruption which occurs when sea or lake water floods into the top of an active open vent, producing an eruption column up to 20 km high and pyroclastic *clasts with extreme fragmentation and moderate dispersal. The term was first used to describe the activity at the new *volcano, Surtsey, which built up on the ocean floor south of Iceland in 1963.

Surveyor A series of *NASA lunar lander missions that ran from 1966 to 1968.

survivorship curve Graphic description of the survival of individuals in a population, from birth to the maximum age attained by any one member. Usually it is plotted as the logarithm of the number of survivors as a function of age. If a population has a constant mortality rate the graph will be a straight line. The technique may also be used to plot the survivorship of whole populations, *species, genera, or higher *taxa. *See also* COHORT.

susceptibility, magnetic *See* MAGNETIC SUSCEPTIBILITY.

susceptibility meter An instrument for measuring the *magnetic susceptibility of a sample. It may be low field (less than 10 mT), intermediate (10–100 mT), or high field (more than 100 mT).

suspect terrane An area or region that is suspected of being a *terrane, but whose boundary *faults have not been identified.

suspended load The part of the total load of a stream that is carried in suspension. It is made up of relatively fine particles that settle at a lower rate than the upward velocity of water eddies. Its highest concentration is in the zone of greatest turbulence, near the bed. It reaches a maximum in shallow streams of high velocity.

sutural angle *See* SUTURE.

suture 1. A linear belt of highly deformed rocks, including tectonic *mélanges, lenses of *ophiolites, deep-sea *sediments, and usually *blueschists, which is interpreted as the boundary between two collided continents or *island arcs. The location of a suture between collided masses has often led to controversy, and the recognition that *collision zones are in some cases a mosaic of jumbled, sliced, and rotated *terranes has led to the realization that sutures may be diffuse, rather than a narrow belt as was formerly thought. **2.** The line marking the junction between the septa (*see* SEPTUM) and the external wall of a cephalopod (*Cephalopoda) shell that is visible when the shell has been preserved as an internal mould. In some cephalopods the suture lines are simple curves but in ammonoids (*Ammonoidea) the suture becomes crenulate; bends in the suture line that point anteriorly are called 'saddles', those pointing posteriorly 'lobes'. In gastropods (*Gastropoda), the suture is the line of junction between two *whorls of the shell; the angle the line makes with the horizontal is the 'sutural angle'. **3.** *See* CEPHALIC SUTURE.

Svecofennian A *stage of the Lower *Proterozoic, from about 2600 to 2100 Ma ago, overlain by the *Gothian (equivalent to the Karelian according to Van Eysinga, 1975) from the Baltic Shield region.

Svecofennian orogeny An early to middle *Proterozoic mountain-building episode that affected the Baltic Shield, in what are now Sweden and southern Finland, and that occurred approximately 1700 Ma ago, at approximately the same time as the *Hudsonian and *Laxfordian *orogenies.

Sveconorwegian orogeny *See* DALSLANDIAN OROGENY.

Svedberg unit (S) The unit of measurement in which *sedimentation coefficients are expressed. It is equal to 10^{-13} seconds, and is written with no space between the number and the symbol, e.g. 64S.

swale 1. A long, narrow depression, approximately parallel to the shoreline, between two ridges on a beach. **2.** A depression in otherwise level ground. **3.** A shallow depression in the undulating surface of a ground *moraine, caused by uneven deposition by the glacier.

swaley cross-bedding A form of *hummocky cross-bedding in which there are few or no hummocks, but *swales are preserved.

Swallow buoy *See* NEUTRALLY BUOYANT FLOAT.

swallow hole *See* DOLINE.

swallowtail twinning A type of *crystal twinning in which one crystal divides into two along a *twin plane, giving a twinned crystal in the shape of a 'V' or swallowtail. The mineral *gypsum may develop this twinned form.

swash The turbulent uprush of water that occurs when a wave breaks on a beach. It is an important mechanism for transporting sand and shingle landwards.

S-wave (secondary wave, shear wave, transverse wave) An elastic *body wave in which particles oscillate about a fixed point but in a direction perpendicular to the direction of propagation of the wave energy. S-waves cannot travel through a fluid, since a fluid cannot support shear. In an *isotropic and homogeneous medium, S-wave velocity (V_s) is given by: $V_s = \sqrt{(\mu/\rho)}$, where μ is the *shear modulus and ρ the density of the material. S-waves travel at about half the speed of *P-waves in a given medium, and can be polarized into SV- and SH-waves in which particle motion is restricted to the vertical and horizontal planes respectively. S-waves can also be generated from the non-normal incidence of P-waves on to an *interface, when they are known as 'converted waves' and are mostly SV-waves. S-waves can also be converted into P-waves at an interface.

Swazian A sub-*era of the *Archaean (Harland et al., 1989), from about 3525 to 2825 Ma ago, that includes the Onverwacht, Figtree, and Pongola sequences of S. Africa.

sweepstakes dispersal route Term coined by G. G. *Simpson in 1940 to describe a possible route of faunal interchange which is unlikely to be used by most animals, but which will, by chance, be used by some. It requires a major barrier that is occasionally crossed. Which groups cross and when they cross are determined virtually at random.

swell Long-period waves that have built up sufficient energy to move away from the area where wind stresses created them. The waves assume a uniform pattern and move even through areas where winds are weak or absent. The longer-period waves move faster than shorter-period waves, so the waves spread out as they move away from the storm (dispersion). Swell waves generated south of New Zealand have been recorded arriving on the coast of Alaska.

swelling coefficient The amount of swelling pressure generated by the hydration of some *clay minerals, or when water converts to ice. It is measured by an *oedometer.

swirl An enigmatic, light-coloured marking on a *satellite or planetary surface. Swirls have no topographic relief. Examples are known from *Mercury and the *Moon (Reiner Gamma). They appear to be antipodal to major impact structures. The Reiner Gamma swirl is associated with strong surface magnetization. It has been suggested that some may be cometary impact sites.

syenite A *saturated, coarse-grained, *igneous rock consisting of *essential *alkali feldspar and *ferromagnesian minerals (*biotite, *hornblende, *arfvedsonite, *aegirine-*augite, and/or aegirine) and *accessory *apatite, *zircon, and iron oxides. The feldspar constitutes more than 65% of the rock. Hypersolvus syenites are characterized by one type of alkali feldspar, usually potassium-rich and displaying perthitic (*see* PERTHITE) texture. Subsolvus syenites are characterized by two types of alkali feldspar, a potassium-rich type displaying perthitic texture and a sodium-rich type displaying antiperthitic texture. Syenites, which are the *plutonic equivalents of *trachytes, are found as ring complexes and as discrete *intrusions on the stable *continental crust and in the cores of some off-axis, ocean-island *volcanoes.

syenodiorite *See* MONZONITE.

syenogabbro A coarse-grained *igneous rock consisting of *essential *alkali feldspar, calcium-rich *plagioclase, and *ferromagnesian minerals (*augite and *biotite), and *accessory *apatite. The two feldspar types are in equal proportions. The rock is mineralogically half-way between a *gabbro and a *syenite.

syenoids *Igneous rocks of *syenite affinity in which *feldspathoidal minerals take the place of *alkali feldspar. *Ijolite is a syenoid.

sylvite Mineral, KCl; sp. gr. 2.0; *hardness 2; *cubic; colourless to white, but sometimes shades of blue, yellow, or red; vitreous *lustre; crystals usually *cubes, often in combination with *octahedra; *cleavage perfect cubic; occurs in bedded *evaporite deposits, but is one of the last minerals to precipitate because of its solubility in water; tastes much more bitter than *halite. It is used extensively as a fertilizer.

symbiosis A general term describing the situation in which dissimilar organisms live together in close association. As originally defined, the term embraces all types of mutualistic and parasitic relationships. In modern use it is often restricted to mutually beneficial species interactions.

symmetrical extinction In optical *mineralogy, the phenomenon which occurs when the *vibration direction of the light bisects the angles between two sets of *cleavages (as seen in basal sections of *pyroxenes and *amphiboles). A special form of symmetrical *extinction may also occur in twinned crystals (*see* CRYSTAL TWINNING) of *feldspar, and this may be used to determine their composition.

symmetrical fold A *fold in which the limbs are of equal length.

symmetrical trend *See* BOW TREND.

Symmetrodonta (infraclass *Pantotheria) Extinct order of *mammals which lived during the late *Jurassic and early *Cretaceous; they may have appeared toward the end of the *Triassic, making them among the earliest mammals known. They had molar teeth with three cusps arranged in symmetrical triangles. They were very small and probably were predators. They are believed to be ancestral to the marsupial and placental mammals. *See also* MARSUPIALIA; and EUTHERIA.

symmetry plane *See* CRYSTAL SYMMETRY.

sympatric evolution The development of new *taxa from the ancestral taxon, within the same geographic range; it is geographically possible for interbreeding to occur between the potential new taxa, but for some reason this does not happen. Because of the difficulty of envisaging what the reasons might be, until recently few authorities accepted the reality of sympatric evolution, except for certain special kinds of organism; but recent studies have shown that *chromosomal mutation can set up a partial barrier to interbreeding, sufficient to permit sympatric speciation.

sympatry The occurrence of *species together in the same area. The differences between closely related species usually increase (diverge) when they occur together, in a process called character displacement, which may be morphological or ecological.

symplesiomorphy The possession of a *character state that is *primitive (*plesiomorphic) and shared between two or more taxa. Shared possession of a symplesiomorph character state is not evidence that the taxa in question are related.

syn- From the Greek *sun* meaning 'with', a prefix meaning 'together', 'with', or 'resembling'.

synaeresis The process of subaqueous shrinkage of *clays by the loss of *pore water. The loss of pore waters occurs either because of the change in the volume of some *clay minerals as a result of *salinity changes, or by *flocculation of the clays. The shrinkage results in the production of *synaeresis cracks.

synaeresis cracks Irregular, radiating, lenticular-shaped cracks, found on bedding surfaces and often resembling the form of a bird's foot. These cracks form by subaqueous shrinkage rather than desiccation, and are therefore not an indication of subaerial exposure. *Compare* DESICCATION CRACKS.

synapomorphy The possession of *apomorphic features by two or more taxa in common (i.e. the features are shared, derived). If the two groups share a *character state that is not the *primitive one, it is plausible that they are related evolutionarily, and only synapomorphic character states can be used as evidence that taxa are related. *Phylogenetic trees are built by discovering groups united by synapomorphies.

Synapsida (mammal-like reptiles; class *Reptilia) Subclass of reptiles which includes the pelycosaurs (*Pelycosauria) and therapsids (*Therapsida). The pelycosaurs appeared in the Upper *Carboniferous and disappeared in mid-*Permian times, displaced by the therapsids to which they had given rise. The therapsids flourished in the latter part of the Permian and in the *Triassic, but dwindled to *extinction in the early *Jurassic. The therapsids are the ancestors of the *mammals, and share in common with them a synapsid skull (i.e. a skull with one temporal opening).

synclinal ridge An elongated hill underlain by a *syncline whose axis trends parallel with it. Its upstanding nature may result from the relative strength of a compressed downfold compared with the tension-induced weakness of adjacent *anticlines, but other explanations have been proposed. *See also* INVERTED RELIEF.

syncline A basin or trough-shaped *fold whose upper component *strata are younger than those below.

synclinorium A regional synformal structure (*see* SYNFORM) which consists of a series of smaller, higher-order *anticlines and *synclines, some of which may be small enough to be viewed in *outcrop.

syneresis cracks *See* DESICCATION CRACKS.

synform A basin or trough-shaped *fold whose younger *strata may be above or below older ones.

syngenetic ore An *ore deposit formed simultaneously with the host rock and by similar processes, e.g. bedded *ironstone, *magmatic segregation deposits.

synkinematic *See* SYNOROGENIC.

synneusis The drifting together and mutual attachment of crystals suspended in a *magma. *Phenocrysts may cluster together in this manner to form *glomeroporphyritic aggregates.

synodic month The average time between successive new *Moons, which is equal to 29 days, 12 hours, and 44 minutes. This period is curiously correlated with the average length of the human female menstrual cycle (29.5 days), while the average length of human pregnancy (266 days) equals nine synodic months (265.8 days).

synoptic meteorology The presentation of the current weather elements of an extensive area at a particular time. Sea-level and upper-level synoptic charts display weather conditions by symbols at selected synoptic stations. 'Synoptic' is from the Greek *sunoptikos*, meaning 'seen together'.

synorogenic (synkinematic; syntectonic) Applied to a process or event (e.g. *recrystallization of *metamorphic rock, or the emplacement of *plutons) which occurs at the same time as deformation. Synorogenic *sediments need not show contemporaneous deformation, though the sedimentation will be the result of deformation and uplift in the *orogenic belt. Many authors use 'synkinematic', 'syntectonic', and 'synorogenic' interchangeably; others restrict 'synkinematic' to rocks showing evidence of deformation which happened simultaneously with their formation (e.g. rotated *porphyroblasts in a metamorphic rock). 'Synorogenic' is often restricted to events or processes which happen simultaneously with a major phase in *orogeny.

synrhabdosome An association of graptolite (*see* GRAPTOLITHINA) rhabdosomes (colonies) that are radially arranged. They appear to be life associations.

synrock An artificially produced substance, based on petroleum, and used for protection, *grouting, etc.

synsacrum In birds, a structure in the pelvic girdle formed by the fusion of certain vertebrae.

syntaxial growth A mineral *cement which has grown around a *grain in such a way that the *crystal overgrowth is in *optical continuity with the grain.

syntectonic *See* SYNOROGENIC.

syntexis A general term applied to the processes of *magma-rock reactions. *Anatexis and *assimilation can be regarded as end-products of the various process of syntexis. These might include fracturing of *country rock during intrusion, engulfing of country-rock *xenoliths in the magma, and thermal and/or fluid infiltration of xenoliths to initiate melting or solid-state reaction.

synthem A major *stratigraphic unit that is *unconformity-bounded.

synthetic-aperture radar A *real-aperture radar system used for high altitude and satellite *remote sensing. A long antenna is stimulated resulting in a high resolution in the *azimuth direction by using a *Doppler shift effect to identify *backscattered waves returning from ahead and behind the platform.

synthetic fault A *fault whose sense of displacement is the same as that of the main zone of faulting when seen in vertical section. In an extensional regime (*see* EXTENSION), a synthetic fault mimics the displacement of a *lystric fault by forming in the active *hanging wall.

synthetic seismogram A *seismic record which can be modelled for a layered structure where the individual *formation velocities and thicknesses are known (e.g. from *borehole logs). It can then be compared with an observed *seismogram to test geologic interpretations and geophysical models.

synthetic thrust A *thrust whose sense of displacement is in the direction of the major thrust, but which may occur progressively in the *footwall or *hanging wall. A *backthrust in this sense is an antithetic thrust. *See* ANTITHETIC FAULT.

syntype All specimens in a type series in which no *holotype was designated.

Syringopora fischeri One of the last representatives of the stony, colonial corals belonging to the subclass *Tabulata; they flourished during the Upper *Palaeozoic and are known from the Upper *Permian of Germany. At the end of the Permian both tabulate and *rugose corals became extinct.

system 1. The chronostratigraphic equivalent of the time unit *period. Systems are subdivided into *series, and together several systems constitute an '*erathem'. When used formally the initial letter of the term is capitalized, e.g. the *Devonian System. **2.** In *geomorphology, a natural arrangement of interrelated objects or variables, the whole possessing properties that make it greater than the sum of the individual parts. It normally possesses stability, expressed by a balance between the input and output of energy and matter. This equilibrium may be upset by internal or external change. If the change is modest, the system quickly regains equilibrium; if it is extreme, a new equilibrium is established. A hillslope, for example, receives *precipitation and exports water, slope debris, and the products of *weathering. The form of the profile represents a balance between input and output. A landslide, perhaps induced by an increase in precipitation, would destroy this equilibrium, and in due course a new balance would be established. Several varieties of system are recognized. *See* CASCADING SYSTEM; CONTROL SYSTEM; MORPHOLOGICAL SYSTEM; and PROCESS-RESPONSE SYSTEM.

systematic errors *See* ERRORS.

systematic sampling *See* SAMPLING METHODS.

Système Probatoire d'Observation de la Terre *See* SPOT.

systems tract In *sequence stratigraphy, a three-dimensional depositional unit, defined by its boundaries and internal geometry. Three systems tracts commonly occur within a single cycle of sea-level changes: the *highstand (or highland) systems tract, *lowstand (or lowland) systems tract, and, between them, the *transgressive systems tract. *See also* REGRESSIVE SYSTEMS TRACT.

T *See* TESLA.

T *See* TAYLOR NUMBER.

Tabianian *See* ZANCLIAN.

tabula (pl. tabulae) A horizontal *plate extending across the interior of a *corallite. *See* RUGOSA; and TABULATA.

tabulae *See* TABULA.

tabular Applied to the form of a mineral when it occurs in broad, flat, often rectangular surfaces resembling the top of a table. The form may be due to the presence of a prominent *cleavage direction, as in *barite.

tabular cross-stratification *See* CROSS-STRATIFICATION.

Tabulata (subclass *Zoantharia) A *Palaeozoic order of zoantharians that are always colonial, never solitary, and that possess small *corallites. Tabulae (*see* TABULA) are the major skeletal elements and septa (*see* SEPTUM) are usually absent. The *corallum is built of individual corallites, not always connected to each other, and may be encrusting, may consist of thin sheets, or may be massive. Other forms may be joined side by side to form fence-like structures. Smaller tabulates appear to have lived in deeper waters, whereas the larger tabulates appear in coral–stromatoporoid (*see* STROMATOPOROIDEA) associations.

tabulation The arrangement of the *plates in the walls of a dinoflagellate (*see* DINOPHYCEAE) *test that is armoured with a series of plates.

tachylite (tachylyte) A black volcanic *glass formed by the chilling of basaltic *magma. The black colour is due to the presence of numerous microscopic crystals within the glass. Tachylite can be found in the chilled rind of *pillow lavas, but is extremely susceptible to alteration by sea water and may be converted to *palagonite.

tachylyte *See* TACHYLITE.

tachytely A rate of evolution within a group that is much faster than the average (*horotelic) rate. Such accelerated evolution typically occurs when an organism enters a new *adaptive zone and initiates an *adaptive radiation to fill the available niches.

Taconic orogeny A middle to late *Ordovician episode of mountain building, named after the Taconic Mountains (a N–S component of the Appalachian system to the east of the Hudson River, New York); it affected the whole of the Appalachians, in an area from what are now New England to eastern Canada, and is known locally in Newfoundland as the Humberian. In the south it was caused by the *subduction of the western margin of the *Iapetus Ocean, leading to the closing of the western basin of the Iapetus and the collision between N. America and the Piedmont microcontinent and the subsidence of the *continental shelf further north. *See* APPALACHIAN OROGENIC BELT.

taconite The name given in the Lake Superior region of N. America to *banded iron formations.

tadpole plot A graph of *dipmeter results. The bedding *dip is plotted as a dot on a depth–dip graph, and the dot is given a tail corresponding to the direction of the dip at that point (e.g. the tail points vertically upwards from the dot for a northerly dip direction).

Tae Weian *See* KUNGURIAN.

tafoni A hollow, produced by localized *weathering on a steep face. Rock breakdown typically takes place by granular disintegration or by flaking, and the hollow shows a tendency to grow upwards and backwards.

Taghanician *See* ERIAN.

tagma (pl. tagmata) In arthropods, one of the sections into which the body is divided by differences in size, shape, or function.

tagmosis In metamerically segmented animals (*see* METAMERIC SEGMENTATION), functional specialization that leads to differentiation among the segments and the formation of tagmata (*see* TAGMA).

tailings Fine waste material from a mineral-processing plant, that is too poor for further treatment. It is often stored in a *tailings dam.

tailings dam A dam made from material having sufficient *permeability to allow moisture to drain through over a regulated period. The dam is constructed to retain the water-sodden, fine-grained materials (*tailings) which represent the waste product from a mineral-processing plant.

Taitai A *series in the Lower *Cretaceous of New Zealand, underlain by the *Oteke and overlain by the *Clarence. It comprises the Mokoiwian and Korangan, which are roughly contemporaneous with the *Neocomian Series and the subsequent *Barremian, *Aptian, and lower *Albian *Stages.

talc A member of the 2:1 *phyllosilicates (sheet silicates) with composition $Mg_6[Si_4O_{10}]_2(OH)_4$; sp. gr. 2.58–2.83; *hardness 1 (it has the lowest hardness on *Mohs's scale of hardness); *monoclinic; rare crystals are *tabular, often *massive; white to green; *cleavage perfect {001}; massive talc (soapstone or steatite) can be formed during the low-grade *metamorphism of siliceous *dolomites; and as a *secondary mineral during *hydrothermal alteration of *ultrabasic *igneous rocks along shear planes. It is associated with *serpentization with *serpentine changing to talc and *magnetite by addition of CO_2. It is used extensively as a mineral filler.

talc schist A *regional metamorphic rock composed predominantly of *talc, and displaying a *schistosity. The rock forms by the *metamorphism and deformation of *ultrabasic *igneous rocks in regional *terranes.

taleolae *See* PUNCTATE.

talik *See* PERMAFROST.

talus (scree slope) A sloping mass of coarse rock fragments accumulated at the foot of a cliff or slope.

taluvium A hillslope deposit that consists of a mixture of coarse, rocky rubble and finer particles. The term is a hybrid, derived from '*talus' and 'colluvium' (relatively fine debris that has been moved downhill by slope wash).

Tame Valley An *interglacial in the *Upton Warren Interstadial, within the Würm (*Weichselian) glacial.

tangential longitudinal strain In a *fold, a buckling (*see* BUCKLE FOLDING) *strain distribution in which strain is internal (in contrast to layer-parallel strain), and concentrated in the *hinge zone, leaving the limbs relatively unstrained. The *principal strain axes are layer-parallel and there is a neutral surface of zero strain throughout the fold profile.

tantalite *See* COLUMBITE.

taphichnia *Trace fossils made by animals attempting unsuccessfully to escape.

taphofacies (taphonomic facies) A sedimentary rock unit, or association of units, characterized by the combination of preservational features of the fossils contained within it. *Compare* BIOFACIES.

taphonomic facies *See* TAPHOFACIES.

taphonomic grade The degree of preservation with *taphofacies, from which the depositional environment can be inferred.

taphonomy The study of the transition of all or part of an organism and its traces from the biosphere into the lithosphere (i.e. *fossilization). The term was coined by J. A. Efremov in 1940.

tarsal bone In *tetrapod vertebrates, one of the distal bones of the hind foot that articulate with the digits (metatarsals).

tar sand Oil reservoir where the *volatiles have escaped and the rock has become impregnated with *hydrocarbon residue. Sometimes this occurs in commercial quantities, as at Athabasca, Alberta, Canada; the sands are mined *open-cast and treated with steam and hot water to release the oil.

tarsometatarsus A bone formed by the fusion of the *metatarsals with the *tarsals of the appendicular skeleton, and found in birds (*Aves) and some dinosaurs. The beginnings of this fusion can be seen in some *Theropoda, but it is not complete until the early *Cretaceous (**Sinornis*). *See also* TIBIOTARSUS.

Tatarian *See* CHANGXINGIAN

Taupo *See* ULTRAPLINIAN ERUPTION.

taxa *See* TAXON.

taxodont Applied to a primitive type of *hinge *dentition present in certain bivalves (*Bivalvia), in which teeth and sockets are small, numerous, and arranged in a row on each side of the *shell beak, on both *valves.

taxon (pl. taxa) Group of organisms of any taxonomic rank, e.g. family, genus, or *species. *See* CLASSIFICATION.

taxonomy (adj. taxonomic, taxonomical) The formal *classification of organisms, soils, or any other entities, based on degrees of relatedness among the items being considered.

taxon range zone (genus zone, species zone, total range zone) The total body of *strata characterized by the presence of a specified *taxon, both in geographical extent and time-stratigraphic range. *See also* RANGE ZONE.

Taylor, Frank Bursley (1860–1939) Taylor was a respected American glaciologist, but is better known for his more speculative cosmology, which included the idea that the *Moon was a former comet, captured in the *Cretaceous. He published his ideas on *continental drift three years before *Wegener, but used little evidence to support his theory, and was ignored.

Taylor number (*T*) A dimensionless number measuring the influence of rotation on a convecting system. It depends on the scale of the *convective cell, the rate of rotation, and kinematic *viscosity. If T is equal to or greater than 1, then rotational effects are significant. *See* CONVECTION.

TCR *See* TOTAL CORE RECOVERY.

***T–d* curve** A TEMPERATURE–DEPTH CURVE.

T_e *See* EFFECTIVE TEMPERATURE.

tear fault *See* STRIKE-SLIP FAULT.

tectofacies A *lithofacies formed by *tectonism.

tectonic *See* TECTONISM.

tectonism (adj. tectonic) Deformation within the Earth's *crust, and its consequent structural effects.

tectonite A deformed rock with a mineral *fabric which forms a pervasive *foliation or *lineation, or some combination of both. *See* SHAPE FABRIC.

tectosilicates *See* SILICATES.

teeth 1. *See* DENTITION. **2.** *See* BONE. **3.** *See* BIVALVIA. **4.** *See* MAMMALIA.

tegeminal plates *See* TEGMEN.

tegmen The *ventral surface of the *crinoid body between the arms. It is covered with a non-calcareous integument in which small, calcareous plates (tegeminal plates) are usually set.

Teichichnus A so-called 'wall-like' *trace fossil in which the horizontal, tube-like trace is underlain by upwardly curving laminae (*spreiten). It is recorded from the *Carboniferous of Scotland. The concave, upwardly directed spreiten may be used as 'way-up' or *geopetal structures.

teilchron The time unit defined by the local range of a *species, corresponding to its *teilzone.

teilzone (local range zone, topozone) *Strata containing part of the total stratigraphic range of a *fossil *taxon, occurring in a particular area.

tektites Small fragments (usually 2.5–5.0 cm) of *silica-rich, translucent black *glass, found scattered over large areas (*strewnfields) in particular regions of the Earth. Most tektites exhibit 'splash' shapes (tear-drops, dumb-bells, etc.) indicating a rapid cooling and solidification during flight. They are thought to have formed as ejecta from cometary or *meteorite impacts on to silicate-rich rocks, most likely terrestrial, although this is still a matter of some controversy. The material would have melted on impact, been thrown up into the atmosphere or space, and landed, resolidified, far from the site of origin. The ages of known tektites range from around 0.7 to 35.0 Ma. Microscropic tektites (microtektites) have been recorded from ocean *sediments off the Ivory Coast, off southern Australia, and from the Indian Ocean.

Teleostei (teleost, teleostean) Somewhat loosely defined term (an infraclass or superorder according to some authors) that includes all the living bony fish (*Osteichthyes) with the exception of a few orders of primitive fish. Teleosts arose from holostean (*Holostei) stock in the *Jurassic. A transitional form, *Leptolepis*, from Upper Jurassic marine deposits, was a herring-shaped fish, around 23 cm long, with a *homocercal tail and pelvic fins placed well back on the body. The scales still carried traces of enamel. Teleosts diversified in the *Cretaceous and are now the most abundant *vertebrate group. They exhibit a great variety of form, but are characterized by an internal skeleton entirely of *bone, a reduction in thickness of the scales, and a homocercal tail; typically there is a dorsal swim bladder to control buoyancy, a mobile jaw articulation, and the fin rays may be stiffened into spines.

teleseism An *earthquake whose *epicentre is more than 1000 km from, but detectable at, a recording *seismometer. Earthquakes whose epicentres are closer than this are regarded as 'local' events.

Telesto (Saturn XII) One of the lesser satellites of *Saturn, discovered in 1980 by *Voyager 1, with a radius measuring 15 × 12.5 × 7.5 km; visual albedo 0.5.

telethermal Applied to an *ore deposit far from source of *magma. It has been produced by *hydrothermal activity at low temperatures and shallow depth with little *wall-rock alteration.

Television and Infrared Observation Satellite (TIROS) The first meteorological satellite, launched by the USA on 1 April 1960, and operated by the National Oceanic and Atmospheric Administration (NOAA). It monitored strips of territory, missing the poles. Two successors, both called TIROS-N, were launched later.

telinite *See* COAL MACERAL.

telluric anomaly A perturbation in the flow of a *telluric current. In what is assumed to be an *isotropic homogeneous medium at or near the Earth's surface, telluric current flows uniformly and there are no discernible potential gradients. The presence of a subsurface geologic structure (e.g. a *salt dome) perturbs the line of flow of the telluric current, forming a potential gradient which can be measured (the telluric anomaly) using *non-polarizable electrodes spaced several hundred metres apart.

telluric current A naturally occurring electrical current which flows at or near the Earth's surface over very large areas, with a magnitude of about 10 mV/km. At any point from which it is observed it constantly changes direction and magnitude.

telome theory A hypothesis suggesting that the fern leaf arose from various rearrangements of branching stem systems.

telson The spike-like, terminal segment of the opisthosoma that occurs in certain chelicerates (*Chelicerata), e.g. limulids and eurypterids. The prominent, backward-pointing spine at or near the rear extremity of a trilobite (*Trilobita) is also called a telson, although it may or may not be the equivalent of a true telson.

Telychian A *stage of the Lower *Silurian, underlain by the *Aeronian and overlain by the *Sheinwoodian.

TEM *See* TRANSIENT ELECTROMAGNETIC METHOD.

Temaikan *See* KAWHIA.

Temispack A widely used technique for *basin modelling, developed by the Institut Français du Pétrole.

temperate climate General term applied to the characteristics of mid-latitude climates influenced from time to time by both tropical and polar *air masses. Temperature criteria provide subdivisions into warm, cool, or cold-temperate climates. *See also* CLIMATE CLASSIFICATION; KÖPPEN CLIMATE CLASSIFICATION; and STRAHLER CLIMATE CLASSIFICATION.

temperate glacier *See* GLACIER.

temperature–composition diagram Graphic representation, usually with temperature on the 'y' axis, and composition of a *phase or phases (e.g. an *isomorphous series) on the '*x*' axis, which shows temperature-dependent variations in composition of the solid phase and the associated *melt.

temperature distribution with depth *See* GEOTHERMAL GRADIENT; and HEAT FLOW.

temperature inversion An atmospheric condition in which the typical *lapse rate is reversed and temperature increases vertically through a given layer. In the *troposphere an inversion layer marks conditions of great stability, i.e. a region in which vertical motion is strongly damped, with an absence of turbulence. An inversion acts as a ceiling, preventing further upward convection, and is generally the limit for cloud development. Marked and persistent inversions occur at lower levels, with subsiding air in major anticyclonic cells, such as the *Azores high-pressure zone and cold *anticyclones over continents. *See also* ATMOSPHERIC STRUCTURE; and ENVIRONMENTAL LAPSE RATE.

temperature log A record of the temperature measured down a *borehole.

temperature range Extent of diurnal or seasonal (annual) temperature variation. Generally the highest annual range of temperature is experienced in higher latitudes, especially over 65° N, which reflects the continental influences of N. America and Asia. Tropical land masses tend to have the highest diurnal temperature ranges, which in the equatorial zone are much higher than the annual variations. Annual temperature ranges in general are particularly moderated by maritime influences.

tempestite (storm deposit) Material deposited during a single storm, often on a *continental shelf where the *tidal range is small and prevailing winds are strong.

Templetonian A *stage of the Middle *Cambrian of Australia, underlain by the *Ordian.

Temple–Tuttle A *comet with an orbital period of 32.92 years; *perihelion date 27 February 1998; perihelion distance 0.982 AU.

tenacity The physical response of a mineral to stress, e.g. vibration, crushing, or bending. Soft minerals which are easily flattened, e.g. *copper and *gold, are described as 'malleable'. Minerals such as *pyrite and *fluorite crumble easily and are said to be 'brittle'. Minerals such as *mica are 'flexible', meaning they can be bent but will not return to their original shape.

tennantite *See* TETRAHEDRITE.

tensile strength The strength of a substance (e.g. rock) under tension. It can be measured by loading cylindrical samples under tension until failure occurs, or by compressing sample discs along a diameter to induce tension failure on a diametrical surface. It can be measured in a *point load tester.

tensile stress A *normal stress (negative compressive stress) which pulls apart the material on either side of a plane. Tensile stress greatly weakens rocks, reducing the amount of *shear stress that is needed to produce failure in them.

tensiometer A device used to measure the soil-moisture tension in the unsaturated (*vadose) zone. The pressure is negative and cannot be measured using a conventional open *piezometer. The tensiometer consists of a closed tube with a porous base and a *manometer.

tension crack *See* TENSION FRACTURE.

tension fracture (tension crack, tension gash) A discrete, commonly lens-shaped, rock fracture which forms and propagates perpendicularly to the direction of maximum *extension. Tension fractures often occur in *en échelon arrangements.

tension gash *See* TENSION FRACTURE.

tent rocks (wigwams) Conical pillars of white rock, often tens of metres tall, formed by the erosion of *pyroclastic rocks, either by the isolation of blocks by intersecting drainage channels, or (more commonly) where resistant blocks of lava or other rocks protect the material beneath them as surrounding rocks are eroded.

tepee Fold-like structures developed in *calcrete *soil profiles, tidal areas, and the margins of salt lakes, due to the fluctuations in water levels and changes in chemical *precipitation. Tepee structures are characterized by mineral horizons folded with pointed terminations to the anticlinal closure, resembling the form of a tepee tent.

tephigram Diagram showing the vertical variation in properties of the atmosphere, i.e. temperature and humidity are plotted as a function of pressure, with lines indicating *dry and *saturated adiabatic lapse rates. The changes in temperature and humidity of lifted *air parcels can be compared against the environment (surrounding air) curve, revealing the *stability or *instability of the *air mass and the level at which condensation will occur in the uplifted air. *See also* ENVIRONMENTAL LAPSE RATE.

tephra Collective term applied to all *pyroclastic particles or fragments ejected from a *volcano, irrespective of size, shape, or composition. The term is usually applied to air-fall material, rather than pyroclastic flow deposits.

tephrite An *undersaturated, fine-grained, *extrusive, *igneous rock consisting of *essential calcium-*plagioclase, *nepheline, and *titanaugite, with *accessory *apatite and iron oxide. The rock is nearly always *porphyritic. *See* TRACHYBASALT.

tephrochronology A dating method based on the examination of *tephra (volcanic ejecta); in areas of repeated activity it is often possible to recognize distinctive events within a *pyroclastic succession, and to use such markers for local correlation (*see* MARKER BED). The succession, so established, provides useful data on the history of the *volcano as well as a guide to magmatic and geochemical changes operating below ground. In many cases, well removed from the volcanic activity, thin *ash falls may be found within sedimentary successions.

Such thin *horizons are regarded as *isochronous and can be used in correlation. The ash is usually altered to form bentonitic clays (*see* BENTONITE), and because these clays still contain traces of volcanic material they can be dated by *radiometric means. This is especially important where the intervening *strata contain *fossil remains (e.g. of early humans in Africa) which can be calibrated by using the dates of the ash falls. Good examples of this application in regional *stratigraphy occur in the Lower *Palaeozoic strata of southern Scandinavia and the mid-*Cretaceous *sediments of the western interior of the USA.

Teratan *See* RAUKUMARA.

Terebratulida (terebratulids; class *Articulata) Order of *Brachiopoda with *punctate shells, rounded *hinge lines, functional *pedicle, *deltidial plates, and a lophophore support usually consisting of a pair of *crura and a calcareous loop. The three suborders comprising the order first appeared in the Lower *Devonian; the Centronellidina became extinct at the end of the *Permian, the Terebratulidina and the Terebratellidina are still extant.

terebratulids *See* TEREBRATULIDA.

terminal node In a *phylogenetic tree, the point at the end of a branch representing a progenic taxon.

terminator The line on the visible disc of a planet or *satellite which separates the sunlit and shaded regions, marking sunrise and sunset. Because the Sun angle is very low near the terminator, small details (e.g. lava-flow fronts on the Moon) become clearly visible.

ternary system A mineral system in which there are three components, e.g. *diopside (Di)–*anorthite (An)–*albite (Ab). *See also* COTECTIC CURVE; PHASE; and PHASE DIAGRAM.

terra (pl. terrae) The term proposed by Galileo for the white, high-standing regions of the lunar surface, now generally referred to as the Lunar Highlands. These comprise the heavily cratered primary lunar feldspathic crust, forming 83% of the lunar surface. The term is also used for extensive (continent-sized) elevated areas on planetary surfaces (e.g. Ishtar Terra on *Venus).

terrace A nearly flat portion of a landscape, terminated by a steep edge. It may be produced by any one of a range of processes, so the following varieties are recognized: *altiplanation terrace, *kame terrace, *river terrace, *shore platform, and *solifluction terrace.

terracette (sheep-walk) A small-scale land-form consisting of a long, narrow, stepped feature developed in *unconsolidated material mantling a steep hillslope. It may result from slight slippage in material standing at an angle too great for stability.

terrain (N. American terrane) An area of ground with a particular physical character; an area or region with a characteristic geology, e.g. 'metamorphic terrain'. Many British authors now use the 'terrane' spelling, except for 'terrain correction' (*see* TOPOGRAPHIC CORRECTION) and when used in the sense of such expressions as 'rough terrain', i.e. when not meaning 'displaced *microcontinent'. *Compare* TERRANE.

terrain component The influence of the type and features of an area surrounding a proposed mine or engineering site. *See also* TERRAIN EVALUATION.

terrain correction *See* TOPOGRAPHIC CORRECTION.

terrain evaluation The assessment of a *terrain for its suitability as a mining or industrial site. Consideration must be given to whether it can provide: the space required for surface plant; the area needed for disposal, e.g. *tailings dams; adequate areas for stock-piling; parking space for employees; in remote territories, access to an airfield; adequate roads, railways, or access to sea ports; room for future expansion; low fire risk (e.g. the possibility of clearing adjacent woodland to reduce the risk of forest fires); adequate facilities for staff and visitors, e.g. housing, canteens, recreation facilities; space for concentrating plant and a smelting and refining complex. Consideration must also be given to the effect of the development on the environment, e.g. air pollution.

terrain pattern The type of landscape surrounding a mine or industrial site. *See also* TERRAIN EVALUATION.

terrain unit A sub-region of an area that is under consideration for use as a mine or industrial site, e.g. whether it has forest, exposed bedrock, *talus, bog, etc. *See also* TERRAIN EVALUATION.

terrane 1. A *fault-bounded area or region which is characterized by a *stratigraphy, structural style, and geologic history distinct from those of adjacent areas, and which is not related to those areas by unconformable contacts or *facies changes. This usage has recently become frequent in studies of *orogenic belts in which terranes can have origins as *accretionary wedges, *island arcs, and *microplates. Initially, 'displaced' or 'allochthonous' was used to qualify the word 'terrane' and to distinguish such terranes from merely '*suspect terranes', but all three qualifiers are falling into disuse. **2.** N. American spelling of *terrain, increasingly coming into general usage.

terra rossa European soils, red in colour, that developed on iron-oxide-rich residual material over *limestone. They are deep and ancient, some of them being pre-*Pleistocene. They are now classified as *Inceptisols or *Mollisols.

terrestrial planet (inner planet) The term used to classify the four inner, rocky Earth-like planets (*Mercury, *Venus, *Earth, and *Mars) of the *solar system, which, relative to the *jovian (or outer) planets are smaller, have high densities, are composed of metallic (mainly iron) and silicate phases, and have low contents of gaseous elements.

terrestrial radiation Long-wave electromagnetic radiation (wavelengths 4–100 μm, with a peak at 10 μm) from the Earth's surface and atmosphere. *See also* ATMOSPHERIC 'WINDOW'; GREENHOUSE EFFECT; RADIATION BUDGET; and RADIATION NIGHT.

terrigenous The description of a *siliciclastic *sediment which has been deposited, or formed, on land.

Tertiary First sub-era of the *Cenozoic *Era, which began about 65 Ma ago and lasted until 1.64 Ma. The Tertiary followed the *Mesozoic and comprises five *epochs: the *Palaeocene, *Eocene, *Oligocene, *Miocene, and *Pliocene.

tertiary creep In a material undergoing long-term low stress, the final stage of *creep, characterized by accelerated permanent viscous *strain, leading eventually to failure.

teschenite An *undersaturated, medium- to coarse-grained, *igneous rock consisting of *essential calcium-*plagioclase, *analcime, *titanaugite, and *barkevikite, with *accessory iron oxide. The rock is similar to *essexite but with analcime taking the place of *nepheline. The teschenite of the Lugar sill, in Ayrshire, Scotland, is a well-known British example.

tesla (T) A unit of magnetic field strength. $1\,T = 1\,Wb/m^2 = 10^{-4}$ gauss.

test Protective shell that covers the cells of some *protistans and the soft parts of certain invertebrate animals.

testate Having a shell or *test.

Tethyan realm The faunal province (*see* FAUNAL REALM) based in the region of the *Tethys Sea. Characteristically *Jurassic to *Cretaceous in age, the Tethyan Realm denotes a warm-water, tropical to subtropical fauna and flora. The name may also be applied to warm-water faunas and floras of *Mesozoic age outside the area of Tethys, especially when used in contrast to the *Boreal (northern) and Austral (southern) Realms.

Tethys (Saturn III) One of the major satellites of *Saturn, with a radius of 529.9 km; mass 6.22×10^{20} kg; mean density 1000 kg/m^3; visual albedo 0.9. It was discovered in 1684 by G. D. Cassini.

Tethys Sea (Neotethys) The sea that more or less separated the two great *Mesozoic *supercontinents of *Laurasia (in the north) and *Gondwana (in the south). That *land bridges between the two supercontinents existed for much of the Mesozoic is attested to by the cosmopolitan character of *dinosaur faunas. *See* PALAEOTETHYS.

Tetracorallia Alternative name for the order *Rugosa.

tetracorals *See* RUGOSA.

tetrad *See* CRYSTAL SYMMETRY; and SPORE.

tetragonal One of the seven *crystal systems, with two sets of edges of the same length and a third which is either longer or shorter than the other two. The *lattice may be referred to three *crystallographic axes, a_1, a_2, and c (or x, y, and z) where a_1 and a_2 (or x and y) are equal, and c (or z) may be longer or shorter. All three axes are at right angles.

tetrahedrite An *end-member of the tetrahedrite series from tetrahedrite $(Cu,Fe)_{12}Sb_4S_{13}$ to tennantite $(Cu,Fe)_{12}As_4S_{13}$; if sliver is present in quantity (up to 30%) the mineral is called freibergite; sp. gr. 4.5–5.1; *hardness 3.0–4.5; hardness 4.0; greyish-black; *metallic *lustre; black *streak; normally *massive but *tetrahedral when crystalline; occurs in *hydrothermal veins in association with other copper and iron sulphides.

tetrahedron A four-faced *crystal form in which each face is an equilateral triangle. It may be referred to three equal *crystallographic axes at right angles which join the centres of opposite edges of the crystal. It is a special form within the *cubic system with the symbol (111).

Tetrapoda Vertebrate animals which have four limbs, including the *Amphibia, *Reptilia, *Aves, and *Mammalia.

Teurian The basal *stage in the Lower *Tertiary of New Zealand, overlain by the *Waipawan and roughly contemporaneous with the *Danian and lower *Thanetian Stages.

textural maturity An expression of the *sorting, *matrix content, and *grain angularity in a *sediment. An immature sediment is one with poor sorting, a large proportion of matrix, and angular particles. A texturally mature sediment is matrix poor or matrix free, well sorted, and with well-rounded grains.

texture 1. In *petrology, the sizes and shapes of particles in rock and their mutual interrelationships. **2.** In pedology, the proportions of *sand, *silt, and *clay in the fine earth of a soil sample, which give a distinctive feel to the soil when handled, and which are defined by classes of soil texture.

Thalassa (Neptune IV) A satellite of *Neptune, with a diameter of 80 km; visual albedo 0.06.

Thalassinoides An *ichnoguild of branched, horizontal to slightly inclined *burrows or a burrow system, formed by suspension feeders just below water, at the *sediment interface. They usually occur in dense associations. The dimensions of the burrows may reflect the environmental energy level. *Thalassinoides* burrows are typical of *clastics and *detrital *carbonates, and are abundant in certain *Jurassic and *Tertiary formations. Remains of Bryozoa have been found within them.

thalweg Line joining the lowest points of successive cross-sections, either along a river channel or, more generally, along the valley that it occupies.

thanatocoenosis *See* DEATH ASSEMBLAGE.

Thanetian 1. The later of the two *ages of the *Palaeocene *Epoch, preceded by the *Danian, followed by the *Ypresian, and dated at 60.5–56.5 Ma (Harland et al., 1989). **2.** The name of the corresponding European *stage, which is roughly contemporaneous with the upper *Ynezian and *Bulitian (N. America), upper *Teurian and lower *Waipawan (New Zealand), and the upper *Wangerripian (Australia). It is named after the Thanet Sands in Kent, England.

thaw Onset of melting of ice and snow as the temperature rises above the freezing point.

Thebe (Jupiter XIV) The fourth of the known jovian satellites, discovered in 1979, Thebe has a diameter of 100 km (±20 km) (100 × 90 km); mass 7.77×10^{17} kg; mean distance from *Jupiter 222 000 km.

theca (pl. thecae) 1. Shell or *test. **2.** The cell wall in some *protistan algae, e.g. *diatoms and dinoflagellates (*see* DINOPHYCEAE). **3.** The dense wall at the margin of a *corallite in a scleractinian coral (*Scleractinia). **4.** In graptolites (*Graptolithina), one of the cup-like structures that occur along the *stipe, each of which houses an individual member of the colony. Thecae vary from simple, straight-sided tubular structures to those with an introtorted opening, where the thecal aperture faces downward. There are triangulate thecae with very small apertures and in some species the thecae are long, thin, and separated from one another ('isolate'). In some *monograptids the thecae vary in structure along the length of an individual stipe as well as from one species to another. Thecal shape also varies through time.

Thecodontia (thecodonts) 'Tooth-in-socket' reptiles, and the most primitive order of the *Archosauria ('ruling reptiles'), ranging from the Upper *Permian to the Upper *Triassic. They were ancestral to the *dinosaurs, *pterosaurs, and *crocodiles. One of the first thecodonts was *Thecodontosaurus browni* which flourished during the Early *Triassic. It grew to 2–3 m in length and had a small head and neck. Essentially it was a quadruped, but it could walk on its hind legs.

thecodonts *See* THECODONTIA.

thematic map In *remote sensing, an image which has a *classification overlaid on to it.

thematic mapper An improved *multi-spectral scanner that acquires simultaneous images of a scene in different wavelength bands.

theoretical morphology In evolutionary studies, analysis of the differences between all the forms an organism might have and those which have actually existed. By revealing unfilled regions of *morphospace the technique may suggest constraints.

Therapsida Order of *synapsid *reptiles, ancestral to the *mammals, ranging from the latter part of the *Permian to early *Jurassic times.

thermal Used as a noun, the term refers to air that rises in a column over some localized heat source; it rises through the atmospheric environment as a result of its reduced density. Intense heating over land (and particularly over some types of surface) results in *adiabatic expansion and uplift. When thermals rise to *condensation level, short-lived clouds (e.g. fair-weather *cumulus) may form. Continued strong thermals can result in the growth of the cloud into large cumulus or *cumulonimbus. When the upward growth reaches a stable layer, the cloud top may spread laterally.

thermal cleaning *See* DEMAGNETIZATION.

thermal conductivity (*K*) A parameter used as a measure of how well a material conducts heat (for example, for copper $K = 385$ W/m/K, for air $K = 0.02$ W/m/K). K forms the constant of proportionality in the heat-conduction-rate equation: $dQ/dt = -KA d\theta/dl$, where dQ/dt is the heat conduction rate (in joules per second), A is the cross-sectional area of a uniform sample (in square metres), and $d\theta/dl$ is the temperature gradient along the sample (in kelvin per metre). *Compare* THERMAL RESISTIVITY.

thermal emission The emission of *electromagnetic radiation due to the vibration of molecules as a result of their temperature.

thermal equator The zone of highest mean temperature over the Earth, either in the annual or long-term average or at a given moment. On the long-term average, it is located around 5°N latitude. This position north of the geographical equator results from the generally rather higher temperature of the northern hemisphere as compared with the southern hemisphere; this is because the glaciated Antarctic continent maintains colder summers in the southern hemisphere than does the Arctic, with a much smaller land area, in the northern hemisphere. *See also* COLD POLE.

thermal inertia A measure of the responsiveness of a material to variations in temperature. In *remote sensing it is measured by diurnal changes in temperature. Materials with a high *heat capacity display high thermal inertia, consequently such materials will show small changes in temperature through the diurnal cycle.

thermal infrared *Infrared radiation which has a wavelength between 3.0 μm and 100 μm. At normal environmental temperatures objects emit infrared between these wavelengths; hotter objects, such as fires, emit infrared at wavelengths shorter than thermal infrared. *Compare* REFLECTED INFRARED.

thermal low Basically *non-frontal depression associated with strong daytime heating of land surfaces, mainly in summer, occurring for example in Arizona, Spain, and in northern-Indian low-pressure cells. The occurrence of thermal lows is also influenced by imbalances in the main windstreams in the upper *troposphere.

thermal metamorphism *See* CONTACT METAMORPHISM.

thermal resistivity A measure of how poorly a material conducts heat; the inverse of *thermal conductivity.

thermal wind The vertical geostrophic *wind shear for a given layer of the atmosphere, i.e. the vector that defines the difference between the *geostrophic winds at the bottom and at the top of the layer. It is parallel to the mean *isotherms (*thickness contours) and blows with cold air (low thickness) on the left in the northern hemisphere and on the right in the southern hemisphere.

thermic *See* PERGELIC.

thermistor A semiconductor whose electrical resistance decreases markedly with increasing temperature, and which is used as a sensitive device for measuring temperature.

thermoclastic (insolation weathering) Process of physical *weathering whereby the stresses set up when a rock is alternately heated and cooled become sufficient to cause *failure. It may bring about surface spalling (splintering) because of the low conductivity of rock, and may be particularly effective on polymineralic rocks because of varying coefficients of expansion. Some observational and experimental evidence suggests that the process may be less effective than was believed formerly. *See also* EXFOLIATION.

thermocline Generally, a gradient of temperature change, but applied more particularly to the zone of rapid temperature change between the warm surface waters and cooler waters at depth. In the oceans, this zone of rapid temperature change starts at 10–500 m below the surface and can extend down to more than 1500 m. In polar regions the thermocline is generally absent since the ocean surface is covered with ice in winter and solar radiation is small in summer. In thermally stratified lakes in summer the thermocline separates the warm surface waters (*epilimnion) from the cooler deep waters (*hypolimnion).

thermograph Instrument that gives a continuous record of temperature for a day or for a week. The device uses a helical strip of two metals with differing coefficients of expansion. The resulting opening and closing of the coil operates a pen which produces a line over a calibrated chart on a round clock drum.

thermohaline circulation Vertical circulation induced by the cooling of surface waters in a large water body. This cooling causes convective overturning and consequent mixing of waters. In the oceans, this circulation usually involves temperature and *salinity variations acting together.

thermo-hygrograph *See* HYGROTHERMOGRAPH.

thermokarst *Periglacial land-form assemblage characterized by enclosed depressions (some with standing water) and so presenting a *karst appearance. It is caused by the selective thaw of ground ice associated with thermal erosion by stream and lake water and may reflect climatic changes or human activity.

thermoluminescence A phenomenon whereby certain minerals, when slowly heated to temperatures below their level of incandescence, emit light. Some minerals will do this only when in contact with oxygen (or air) and in this case it is known as oxyluminescence. The light energy emitted is measured using a photomultiplier and recorded as a function of temperature, to produce a 'glow curve'.

thermonuclear reactions Nuclear fusion reactions occurring between various nuclei of light elements at very high temperatures. These are the source of energy generation in stars. *See also* NUCLEOSYNTHESIS.

thermophile An *extremophile (domain *Archaea) that thrives in environments where the temperature is high, typically up to 60 °C. *Compare* HYPERTHERMOPHILE.

thermopile Instrument for the measurement of direct and diffuse short-wave radiation, which utilizes a bank of thermocouples covered by a glass or other transparent dome.

thermoremanent magnetization (TRM) As heated *ferromagnetic materials cool through their *Curie temperature down to room temperature or below, the *remanent magnetization they acquire.

thermosphere The upper zone of the atmosphere, above about 80 km, where solar radiation of the shortest wavelengths is absorbed. In this zone, which includes the *ionosphere, temperature increases with height, but because of the very low atmospheric density there, the heat capacity is minute. *See also* ATMOSPHERIC STRUCTURE.

Theropoda Suborder of *saurischian dinosaurs which consists exclusively of bipedal, carnivorous forms. It includes the *coelurosaurs and the *carnosaurs, and ranged from the Upper *Triassic to the *Cretaceous.

thickness The difference in height in the atmosphere between particular pressure levels, e.g. between 1000 and 500 millibars. Isolines of a given interval on charts are termed 'thickness lines'.

thin section A slice of a mineral or rock which is glued to a glass slide with resin of a particular *refractive index (usually 1.540) and ground down to a thickness of 0.03 mm so that light may be transmitted through it when it is examined under a transmitted-light *polarizing microscope. The mineral or rock chip is ground in successive stages, using *carborundum or other suitable abrasives mixed with water, and when finished the section is covered with a glass cover slip, again attached with resin. This is the standard method for preparing most of the rock-forming minerals for study.

thixotropic mud A type of mud commonly used to cool the drill *bit and remove rock chips during drilling. When the drill motion ceases, the thixotropic properties of the mud prevent rock chips descending and blocking the hole. *See* THIXOTROPY; SOL; and QUICK CLAY.

thixotropy The property possessed by some materials of changing from *gel to liquid under *shearing stress, e.g. when shaken, and returning to the original state when at rest. Some muds of the *smectite group have this property and are useful for flushing *boreholes. The change is completely reversible: no change in water content or composition occurs. On a slope the material will flow downwards as long as its velocity is sufficient to maintain minimum shearing stress.

tholeiite An extraordinarily abundant, fine-grained, *igneous rock consisting of *essential calcium-*plagioclase, subcalcic *augite, and pigeonite, with *interstitial *glass or fine *quartz-*feldspar intergrowths. Tholeiite is a type of *basalt, *oversaturated with silica; it occurs as plateau *lavas on the *continental crust and as the main *extrusive component of the ocean floor. *Compare* MID-OCEAN-RIDGE BASALT.

tholoid *See* DOME.

Thompson, Benjamin (Count Rumford) (1753–1814) An American-born inventor, Rumford is best known for founding the Royal Institution in Britain in 1799, 'to promote useful science'. He made studies of heat, leading to investigations of gunpowder, and improvements in lamps, cookers, and fireplaces.

Thomson, Charles Wyville (1830–82) An Irish naturalist, Thomson became professor of natural science at Edinburgh. He was especially interested in life in the deep oceans, which led him to propose an expedition to make detailed studies of the ocean. The result was the **Challenger* expedition, at the end of which Thomson began the arrangement of the collections and publications of the results.

Thomson, William (Lord Kelvin) (1824–1907) A mathematician and physicist at Glasgow University, Kelvin is best known for his theory of thermodynamics, which he applied to his work on the age of the *Earth. He used studies of the internal heat of the Earth, the age of the *Sun, and *tidal friction, to give an estimate of the Earth's age of about 20 million years. This led to conflict with geologists and evolutionists, who saw evidence that the Earth must be much older. *See* KELVIN SCALE.

Thoracica *See* CIRRIPEDIA.

thoracic vertebra *See* VERTEBRA.

thorax The three segments of the body of an arthropod (*Arthropoda) that lie between the head and the abdomen. Each thoracic segment carries a pair of legs. The three thoracic segments are termed the prothorax, mesothorax, and metathorax. In insects (*Insecta), the mesothorax and metathorax may each carry a pair of wings. *See* TRILOBITOMORPHA.

thorium–lead dating A *radiometric dating method based on the *radioactive decay of ^{232}Th, to yield ^{208}Pb + 6He^{4}, with a half-life (*see* DECAY CONSTANT) of 13 900 million years. The minerals used include *sphene, *zircon, *monazite, *apatite, and other rare U/Th minerals. The method is not totally reliable and is usually employed in conjunction with other methods. In most cases the results are *discordant as a result of lead loss. The ratio of ^{208}Pb : ^{232}Th compared with ^{207}Pb : ^{235}U ratio is particularly useful. The Th–Pb system can also be interpreted by means of *isochron diagrams similar to those used in the *rubidium–strontium method.

Thornthwaite, Charles Warren (1889–1963) An American climatologist who devised the system of climate classification that bears his name. Thornthwaite taught on the faculties of the Universities of Oklahoma (1927–43) and Maryland (1940–6), and at Johns Hopkins University (1946–55) before becoming director of the Laboratory of Climatology at Centerton, New Jersey, and professor of climatology at Drexel Institute of Technology, Philadelphia. He was president of the Section of Meteorology of the American Geophysical Union from 1941 to 1944 and in 1951 he was elected president of the Commission for Climatology of the World Meteorological Organization.

Thornthwaite climate classification System for describing climates, devised in 1931 and revised in 1948 by the American climatologist C. W. *Thornthwaite, that divides climates into groups according to the vegetation characteristic of them, the vegetation being determined by precipitation effectiveness (*P*/*E*, where *P* is the total monthly precipitation, and *E* is the total monthly evaporation). The sum of the monthly *P*/*E* values gives the *P*/*E* index, which is used to define five humidity provinces, with associated vegetation. A *P*/*E* index of more than 127 (wet) indicates rain forest; 64–127 (humid) indicates forest; 32–63 (subhumid) indicates grassland; 16–31 (semi-arid) indicates steppe; less than 16 (arid) indicates desert. In 1948 the system was modified to incorporate a moisture index, which relates the water demand by plants to the available precipitation, by means of an index of *potential evapotranspiration (PE), calculated from measurements of air temperature and day length. In arid regions the moisture index is negative because precipitation is less than the PE. The system also uses an index of thermal efficiency, with accumulated monthly temperatures ranging from 0, giving a frost climate, to more than 127, giving a tropical climate. *See also* KÖPPEN CLIMATE CLASSIFICATION; and STRAHLER CLIMATE CLASSIFICATION.

three-cell model An attempt to represent the atmospheric circulation systems over a hemisphere by three adjoining vertical cells of meridional surface motion, transferring energy from equatorial to polar regions. This concept of heat transfer by meridional circulation has generally been superseded by more modern ideas of travelling waves. *See also* HADLEY CELL; and ROSSBY WAVES.

three-dimensional seismology (3D seismic) A technique, used in prospecting for crude oil and natural gas, in which a set of *geophones is located very precisely to detect vibrations from a series of experimentally induced seismic disturbances over a considerable area. In one survey, covering 49 km^2, 1700 seismic disturbances produced 725 000 geophone traces. The technique has led to the identification of larger reserves than had previously been suspected and increased production.

threshold A value beyond which the physical environment adjusts to a change in processes. An example is provided by a *glacier, where a build-up of ice and snow over a number of years reaches a critical level and once that critical level (or threshold) is exceeded there is a sudden change in the process of *basal sliding and the glacier surges forward.

thrombolites Structures with a clotted microtexture and no internal laminae, built by *cyanobacteria and found in calcareous, sublittoral *facies.

throughfall The part of *precipitation which, having been intercepted by vegetation, then falls on to the ground surface. *See* INTERCEPTION (1).

throughflow *See* INTERFLOW.

throw The vertical displacement of a *dip-slip fault; its measurement is often complicated by the difficulty of accurately identifying markers on either side of the *fault plane.

thrust A low-angle (commonly less than 45°) *reverse fault, with a significant *dip-slip component, in which the *hanging wall overhangs the *footwall. *Synthetic thrust sets form *imbricate fan structures which may be thrust-bound, when they form a *duplex. Single thrusts typically show a 'staircase' trajectory composed of *ramps and *flats.

thrust-sheet-top basin *See* PIGGYBACK BASIN.

thufur An earth hummock, some 0.5 m high and 1–2 m wide, which is found in contemporary and past *periglacial environments. Normally it possesses a core of *sediment, which suggests that a form of differential heaving by ground ice is responsible for its development.

thulite *See* ZOISITE.

thundercloud Popular term for *cumulonimbus, which is associated with *thunderstorms.

thunderstorm A storm of fairly local scale in which strongly developed *cumulonimbus cloud produces thunder and lightning, usually with rain and strong, gusting wind, and often with hail.

Thuringian A European stratigraphic term which was originally used as an equivalent to the Zechstein and was later extended to include the whole Upper *Permian. *See also* AUTUNIAN.

thuringite *See* CHLORITE.

Thurnian The cold-*stage marine *silts from the middle of the threefold subdivision of the deposits of the Ludham *borehole. They are Lower *Pleistocene in age. *See also* ANTIAN; BAVENTIAN; LUDHAMIAN; and PASTONIAN.

Thvera A normal *polarity subchron which occurs within the *Gilbert reversed *chron.

Tibetan Plate Lying between the Indian and *Eurasian Plates, the Tibetan Plate consists of two continental blocks, northern Tibet and southern Tibet, divided by a *suture with *ophiolites. Most of the *Cenozoic deformation of this part of the Indian–Eurasian collision has been to the south of Tibet, in the Himalayas.

Tibetan Plateau An area of Tibet with an elevation of approximately 2 km above sea level, below which the *crust is more than 80 km thick; this has been ascribed to *underthrusting of the crust of the *Indo-Australian Plate (i.e. by *A-subduction) or to horizontal shortening as a result of *thrust *tectonics. The uplift has been both rapid and recent: one estimate is 2500–3000 m since the *Pliocene.

tibia **1.** In a tetrapod vertebrate (*Tetrapoda), the anterior long bone of the lower hind limb (the 'shin' bone). **2.** In an insect (*Insecta), the long and often narrow segment of the leg that articulates proximally with the femur (or 'thigh') and distally with the tarsus (or 'foot').

tibiotarsus The bone formed by the fusion of the *tarsal bones and the *tibia, found in birds and some dinosaurs.

tidal barrage A dam or barrier built across a tidal channel to allow a power-generating plant to operate. The siting of the barrage requires there to be a large *tidal range, in excess of 5 m, and a large water body open to the sea, with a narrow entrance. One barrage and generating plant in operation since 1960 is on the Rance Estuary in Brittany, France. It generates electricity by the use of two-way turbines, as water flows in and out of the *estuary through the barrage. The Severn Estuary and Morecambe Bay are potential sites for tidal barrages in the British Isles.

tidal current An alternating, horizontal movement of water associated with the rise and fall of the *tide, these movements being caused by gravitational forces due to the relative motions of Moon, Sun and Earth. Offshore tidal currents tend to exhibit rotary patterns, while in areas near coasts the currents follow rectilinear paths and reverse periodically (ebb and flow currents). Tidal currents can often reach velocities of 2.5 m/sec near shores.

tidal flat An intertidal sandflat, *mudflat, and marsh area developed in some *lagoons in *mesotidal areas, and in protected bays and estuarine areas along *macrotidal coasts. Extensive tidal flats occur in the Wash (eastern England), Wadden Zee (Holland), and along the North-Sea German coast. Tidal flats also occur in warmer climates, as in the Persian Gulf, where *carbonate and *evaporite deposits develop. In tropical areas tidal flats tend to be colonized by *mangrove swamps.

tidal friction The friction exerted on the Earth because of the *phase lag between the *tides and the gravitational attraction of the Moon, Sun, and planets. It is mostly due to the M_2 ocean tide (M_2 is the principal component of the forces acting in the direction of the Moon). *See also* DAYLENGTH.

tidal heating The generation of heat due to friction produced by the strong tidal forces exerted by a very massive parent body on a body moving about it in an elliptical orbit. The intensity of tidal heating is proportional to the square of the orbital eccentricity, being zero in a circular orbit and reaching a maximum in a parabolic orbit, and inversely proportional to the size of the orbit.

tidal inlet A narrow channel that connects the open sea with a *lagoon. Tidal inlets often occur in *barrier-island systems and are typified by small-scale *deltas at each end of the inlets, resulting from the high-velocity, tidal-reversal currents that flow through the channels.

tidalite A *sediment deposited under the dominant influence of *tidal currents.

tidal range The difference in height between consecutive high and low waters. The tidal range varies from a maximum during *spring tides to a minimum during *neap tides. In tide tables daily high- and low-water heights are given for each geographical locality mentioned.

tidal rhythmite A *rhythmite deposited by the action of tides.

tidal stream The flow of water in and out of estuaries, bays, and other restricted coastal openings associated with the rise and fall of the *tide. Landward (flood) and seaward (ebb) streams or currents often follow different paths in shallow-water areas, so forming ebb/flood avoidance cells and a braided pattern of sandbanks in these coastal openings.

tidal theory Theory of the origin of the *solar system, involving the approach near the Sun of another star. This set up tidal forces, and the instability of the Sun resulted in part of its mass being torn off to form the planets. The theory was proposed by Sir James H. Jeans (1877–1946) and Sir Harold *Jeffreys.

tide 1. The periodic rise and fall of the Earth's oceans, caused by the relative gravitational attraction of the Sun, Moon, and Earth. The effect of the Moon is about twice that of the Sun, giving rise to the *spring–*neap cycle of tides. Variation in tides is caused by: (*a*) changes in the relative positions of the Sun, Moon, and Earth; (*b*) uneven distribution of water on the Earth's surface; and (*c*) variation in the sea-bed topography. Semi-diurnal tides are those with two high and two low waters (period 12 hours and 25 minutes) during a tidal day (24 hours and 50 minutes). Diurnal tides have one high and one low water during a tidal day. **2.** *See* EARTH TIDES.

tieline 1. A line on a metamorphic, composition–assemblage diagram (e.g. an *ACF or *AFM diagram), which joins the compositions of *minerals coexisting in equilibrium. **2.** A *survey line that crosses others so that all lines can be calibrated to the same value.

tie point In *remote sensing, a point on the ground which occurs in two or more images and that can be used to co-register images.

'tiger's eye' *See* RIEBECKITE.

Tiglian A late *Pliocene *stage (about 2 Ma) which *isotope records from the tropical Atlantic- and Pacific-Ocean cores and other climatic indicators suggest may have been a time of increased cooling.

tile drain Short lengths of concrete or ceramic pipe placed end to end to make a drain, which is laid at any appropriate depth and spacing to remove water from the soil but allow water to enter at the joints. Nowadays continuous, slotted, plastic piping is often used.

till Collective term for the group of *sediments laid down by the direct action of glacial ice without the intervention of water. The sediments may be classified in terms of *particle size (they range between *clay-rich and *clast-dominated types, according to source area and travel distance), or grouped by the basic process of debris release: subglacial melt gives rise to 'lodgement till'; surface *ablation gives 'ablation till', followed by *flow till after further movement; and the general thaw of static ice produces 'melt-out till'. *See* TILL FABRIC ANALYSIS.

till fabric analysis The measurement of the orientation and *dip of the *clasts contained within a *till, and the subsequent plotting and analysis of the data. These may indicate the direction of a former glacial advance: clasts within a lodgement till tend to lie parallel to the direction of ice movement.

tillite A lithified deposit of *boulder clay or *till produced by the action of *glaciers. *See* DIAMICTITE.

tilloid General term applied to a chaotic mixture of large blocks set in a *clay-rich *matrix; such mixtures include *mudflows, *landslide deposits, and *tillites.

till plain A smooth plain underlain by *till. It may be well preserved, as in midwestern USA, or may be dissected by later *erosion, as in the lowlands of central and eastern England.

tilt-block tectonics An extensional tectonic style in which crustal blocks are rotated along *normal faults which tend to become nearly horizontal at depth. It is thought to be important on the *passive margin of a rifted continental mass.

tiltmeter An instrument used to measure the change in ground slope. Using laser technology, small changes in the relative levels of three connected but displaced fluid reservoirs are accurately measured to monitor the direction and magnitude of slope changes. Changes of a fraction of a radian can be monitored by this instrument. Tiltmeters are particularly useful on active *volcanoes for measuring the rate of ground inflation or deflation prior to and during an eruptive event. Areas showing large ground inflations are usually correlated with *magma rising into the high-level volcanic plumbing system and may pinpoint the potential site of an eruption.

time-averaged velocity *See* AVERAGE VELOCITY.

time–distance curve A graph showing the arrival times of *seismic waves as a function of *shot–*geophone *offset distance. In seismic-*refraction surveys it is also called a t–d curve or t–x curve. In seismic *reflection surveys, travel times squared (t^2) are plotted against the square of the offset distance (x^2) to produce a *t^2–x^2 graph. In both sets of graphs, the inverse gradient of straight-line segments provides information about the *seismic velocity for the appropriate layer.

time domain A reference framework in which measurements are related to time, rather than to *frequency. *Compare* FREQUENCY DOMAIN.

time plane An imaginary surface linking *interfaces of equal age within a body of *strata.

time-rock unit *See* CHRONOSTRATIGRAPHIC UNIT.

time-scale The subdivision of *geologic time based on one of several different criteria. Time-scales related to the rate of sedimentation (*see* SUBSIDENCE), the rate of *sea-floor spreading, and *radiometric dating are used.

time-stratigraphic unit *See* CHRONOSTRATIGRAPHIC UNIT.

tinguaite An *undersaturated, medium- to coarse-grained, *igneous rock consisting of *essential *alkali feldspar, *nepheline, and *aegirine (with or without sodic *amphibole or *biotite). Tinguaite is the *hypabyssal equivalent of *phonolite. The term 'intrusive phonolite' is now preferred.

tinstone *See* CASSITERITE.

Tioughniogan (Tioughiogan) *See* ERIAN.

TIROS *See* TELEVISION AND INFRARED OBSERVATION SATELLITE.

Titan (Saturn VI) One of the major satellites of *Saturn, with a radius of 2575 km; mass 1345.5×10^{20} kg; mean density 1881 kg/m^3; visual albedo 0.21. It was discovered in 1655 by C. Huygens.

titanaugite A titanium-rich *augite.

Titania (Uranus III) One of the major satellites of *Uranus, and the largest. Its radius is 788.9 km; mass 35.27×10^{20} kg; mean density 1710 kg/m^3; albedo 0.18. The surface has many impact craters surrounded by bright ejecta.

titanite *See* SPHENE.

titanomagnetite A titanium-rich *magnetite; i.e. a magnetite ($Fe^{2+}Fe^{3+}{}_2O_4$) in which Ti^{2+} replaces some of the Fe^{2+}.

Tithonian A post-*Kimmeridgian *stage in the southern European Upper *Jurassic, which is distinguished from the *Volgian (northern Europe) by its Tethyan (*see* Tethys) ammonite (*Ammonoidea) fauna. Earlier British stratigraphers referred to rocks formed during Tithonian time (152.1–145.6 Ma; Harland et al., 1989) as Late Kimmeridgian and *Portlandian; some also included the lowermost *Purbeckian. *See also* MALM.

Titius–Bode law (Bode's law) An empirical arithmetical relationship between the distances of the planets from the Sun. If the Sun–Earth distance is taken as 10, then the distances of Mercury, Venus, Earth Mars, Jupiter, and Saturn are approximately satisfied by the sequence: 4, 4 + 3, 4 + 6, 4 + 12, 4 + 48, and 4 + 96. The more accurate version of the 'law' is given by the function $r_n = AB^n$, where r_n is the distance of the nth planet, A is a constant, and $B = 1.73$. This version is due to Mary Blagg (1858–1944). It is not clear that the 'law' has any fundamental significance, but may be merely a consequence of gravitational and tidal evolution following planetary and satellite formation. The *asteroid belt occurs about 4 + 24, corresponding to the former 'missing planet' in the sequence, that was believed to exist for many years. *Uranus, discovered in 1781, occurs close to the next term, 4 + 192, but the position of *Neptune does not fit the 'law'. The regular satellites of the giant planets fit modifications of the 'law'. The 'law' was discovered in 1766 by J. D. Titius (1729–96) and was popularized by J. E. *Bode (1747–1826).

titrimetric analysis Method of analysis in which a *solution of the substance being determined is treated with a solution of a suitable reagent of exactly known *concentration. The reagent is added to the substance until the amount added is equivalent to the amount of substance to be determined. The point of equivalence (end point) in titrimetric analysis is usually determined by a change in colour of an auxiliary reagent (the indicator), but may also be detected by electrical means (potentiometric, conductometric, titrimetry, etc.).

tjaele (frost table) A frozen surface at the base of the *active layer, which moves downwards as thaw occurs. The tjaele should not be confused with the upper limit of *permafrost, the permafrost table.

toad's eye tin *See* WOOD TIN.

Toarcian A *stage in the European Lower *Jurassic (187–178 Ma; Harland et al., 1989), roughly contemporaneous with the mid-Uroroan (New Zealand). *See also* LIAS.

toeset The asymptotic, basal portion of a *Gilbert-type delta, generally characterized by fine-grained *sediment deposited at the toe of the *prograding *foreset. *See also* TOPSET.

tombolo A *spit that links an island to the mainland or to another island, formed by deposition when waves are refracted round the island.

Tommotian The basal *stage of the *Cambrian *System, underlain by the Poundian (*Precambrian), overlain by the *Atdabanian, and dated at about 570–560 Ma. *See also* CAERFAI EPOCH.

tomography Any technique by which details of a subsurface plane can be represented graphically. *See* SEISMIC TOMOGRAPHY.

tonalite An *oversaturated, coarse-grained, *igneous rock consisting of *essential sodic *plagioclase (An_{27-36}), *quartz, *hornblende, and/or *biotite, with *accessory *apatite, *zircon, and iron oxide.

Tonawandan A *stage of the Middle *Niagaran (*Silurian) of N. America.

Tonga–Kermadec Trench The oceanic *trench in the western *Pacific Ocean which forms part of the boundary between the *Indo-Australian and *Pacific Plates. There is a volcanic gap (i.e. a break in the *island arc) to the north-west of the trench, possibly related to the subduction of the Louisville Rise.

Tongaporutuan A *stage in the Upper *Tertiary of New Zealand, underlain by the *Waiauan, overlain by the *Kapitean, and roughly contemporaneous with the upper *Tortonian and *Messinian Stages.

tonhäutchens *See* CUTAN.

tonstein A compact, *kaolinite-rich *mudstone, which developed as a kaolinitic *palaeosol, and is frequently found as thin bands within *coal seams or resting directly above the coal. Some tonsteins are laterally extensive and are believed to be the product of weathered volcaniclastic *ash.

tool mark The impression made on the surface of a soft bed of *sediment by the impact of an object (tool), or the dragging of an object over the sediment by a current. Tool marks include bounce, *prod, *skip, *groove, and *chevron marks, which develop by differing interaction of the tools with the sediment.

topaz A nesosilicate mineral, $Al_2SiO_4(OH,F)_2$; sp. gr. 3.5–3.6; *hardness 8; *orthorhombic; colourless, pale yellow, pale blue, yellowish, or sometimes pink; often transparent; vitreous *lustre; crystals are *prismatic and often *bipyramidal with the vertical faces striated, but it can also be *massive and granular; *cleavage perfect basal {001}; typically occurs in *granite *pegmatites, *rhyolite, and *quartz veins, and extensively as an *accessory mineral in granites, associated with *fluorite, *tourmaline, *beryl, and *cassiterite, also in *alluvial deposits. It is associated with pneumatolytic action (*see* PNEUMATOLYSIS) and is a constituent of greisen. The original cairngorms (*see* QUARTZ) were topaz crystals. It is named after Topazos Island in the Red Sea.

toplap A discordant relationship in which the upper boundary of a *depositional sequence is marked by the termination of initially inclined beds (e.g. those formed in the *foresets of *deltas). Toplap results mainly from nondeposition, possibly with minor *erosion. Coastal toplap occurs during a *stillstand of sea level. Each unit of *strata dips seaward with its upper, terminal edge wedging back towards the land. Successive units of strata build out laterally towards the sea. *Compare* BASELAP.

topographic correction (terrain correction) A correction applied to observed geophysical values to remove the effects of topography. In gravity studies, it is the correction applied to each individual determination of gravity to allow for the attraction of rocks occurring as hills above the height of the recording station and as valleys below this level. For seismic corrections *see* ELEVATION CORRECTION; and STATIC CORRECTION.

topology In *phylogenetics, the branching pattern of a *phylogenetic tree.

toposequence Sequence of *soils in which distinctive soil characteristics are related to the topographic situation.

topotype In taxonomy, a specimen found in the type locality of a taxon to which it is thought to belong, but that is not necessarily of that *type series.

topozone *See* TEILZONE.

topset beds The upper, near-horizontal layers deposited on a *Gilbert-type delta, generally characterized by the coarsest *sediment found on the prograding (*see* PROGRADATION) *delta. *See also* FORESET; and TOESET.

topsoil 1. The superficial layer of *soil that is moved in cultivation. **2.** The A *soil horizon of a *soil profile. **3.** Any surface layer of soil.

tor Mass of exposed bedrock, standing abruptly above its surroundings, and typically but not exclusively developed on granitic rocks. It may be formed by selective subsurface *weathering followed by the removal of the weathered debris, by differential frost-shattering, or as an end-product of *scarp retreat under semi-arid conditions.

torbanite (cannel shale) A carbonaceous *oil shale that is sapropelic, usually occurs as lenses in *coal seams, and was possibly derived from vegetation-rich mud. *See* SAPROPEL; and SAPROPELIC COAL.

torbernite A *secondary mineral, with the formula $Cu(UO_2)_2(PO_4)_2.8{-}12H_2O$; sp. gr. 3.2; *hardness 2.5; *tetragonal; normally bright emerald-green, but occasionally dark green; pale green *streak; vitreous *lustre; crystals *tabular, square, foliated, scaly aggregates; *cleavage perfect basal {001}; occurs in the oxidized zones of *veins containing uranium and copper minerals, often with *autunite.

tornado A relatively small-scale (about 100 m diameter) 'twisting' or rotating column of air, like a funnel, with high wind speeds and great destructive force over the narrow path of its movement. Such systems are especially frequent in unstable air conditions in the central parts of the USA.

Tornquist Line *See* FENNOSCANDIAN BORDER ZONE.

toroidal field A magnetic field with no radial components, e.g. where the magnetic lines of force lie on an electrically conducting spherical surface such as the surface of the Earth's *core. Such a field is not detectable at the Earth's surface, in contrast to a *poloidal field.

Torridonian A *stage of the Upper *Proterozoic of north-western Scotland, from about 1100 to 600 Ma ago, named from Loch Torridon.

torrid zone General climatic term for the region between the northern and southern tropics, broadly the 'equatorial' zone.

torsion 1. The effect of twisting an object in opposite directions from either end. Torsion tests, using rod-shaped samples, allow stable ductile deformation to be studied up to large *strain states. **2.** In *Gastropoda, the twisting of the body through 180° so that the digestive and nervous systems have a U-shape and the *mantle cavity, anus, gills, and two nephridiopores (excretory openings) occupy an anterior position behind the head.

torsion balance A weighing instrument in which the required weight of a sample is set by rotating a lever to move a needle to the appropriate position on a graduated scale, and then the sample is added to the weighing pan, a little at a time, until it counterbalances the pre-set weight precisely and the needle is brought back to the 'zero' position. This is a useful and rapid technique when large numbers of routine weighings need to be performed in the range of 0.01–5.0 g.

torta A low, fairly flat, approximately symmetrical, volcanic dome formed by successive extrusions on level ground, each increment from the vent thinning the previously erupted material as it pushes it outwards. *Torta* is Spanish for 'pie'; volcanic tortas are common in the Andes.

Tortonian 1. A Late *Miocene *age, preceded by the *Serravallian, followed by the *Messinian, and dated at 10.4–6.7 Ma (Harland et al., 1989). **2.** The name of the corresponding European *stage, which is roughly contemporaneous with part of the *Mohnian (N. America), *Waiauan and lower *Tongaporutuan (New Zealand), and the lower *Mitchellian (Australia). The *stratotype is at Tortona, Italy.

torus The shape of a doughnut; a geometrical surface formed by rotating a circle (or other closed curve) about a circle in its plane without intersecting it.

toscanite *See* RHYODACITE.

total core recovery (TCR) The total length of *core recovered from a *borehole as a percentage of the length of the borehole.

total intensity (*F*) The magnitude of the geomagnetic vector field or of the magnetization of an object.

total internal reflection The *reflection of a *seismic wave from the boundary between the layer in which it is travelling and an adjacent layer. A wave in a low-velocity medium which strikes the inner surface of a boundary with a higher-velocity medium at an angle greater than the *critical angle, and therefore is wholly reflected within the layer, is partially converted into another type of wave at the *interface, rather than being refracted out of the layer. *See* S-WAVE.

total range zone *See* TAXON RANGE ZONE.

total stress *In situ* *stress in a rock body, minus the fluid pressure.

tourmaline A member of the *cyclosilicates and a borosilicate $Na(Mg,Fe^{2+},Mn,Li,Al)_3Al_6(BO_3)_3[Si_6O_{18}](OH,F)_4$. There are three important members of this family of *minerals: dravite $(NaMg_3Al_6(BO_3)_3[Si_6O_{18}](OH,F)_4)$; *schorl $(Na(Fe^{2+},Mn)_3Al_6(BO_3)_3[Si_6O_{18}](OH,F)_4)$; and elbaite $(Na(LiAl)_3Al_6(BO_3)_3[Si_6O_{18}](OH,F)_4)$; sp. gr. 2.9–3.2; *hardness 7.0–7.5; *trigonal; black, bluish, pink, or green, never colourless; elongate *crystals common, also *acicular needles and massive or radiating aggregates; *cleavage good $\{11\bar{2}0\}$ prismatic; occurs in granite *pegmatites, pneumatolytic (*see* PNEUMATOLYSIS) veins, and *granites as elbaite and schorl varieties; it may occur in the rock luxullianite formed by pneumatolytic action after boron has been introduced, where it will occur with *topaz, *spodumene, *cassiterite, *fluorite, and *apatite. Dravite variety occurs in metamorphosed impure *limestones and rarely in some *basic *igneous rocks; tourmaline is a common detrital 'heavy' mineral in sedimentary rocks. Good multicoloured crystals can be used as gemstones.

tourmalinization A pneumatolytic (*see* PNEUMATOLYSIS) modification of a pre-existing *igneous rock in which boron-rich, late-stage fluids react with the *mineral assemblage of the *primary rock (usually *granite), eventually producing a rock consisting of *tourmaline and *quartz. Intermediate stages in the process, showing *alkali feldspar partly replaced by rosettes of tourmaline, are commonly preserved in granites. Luxullianite, found at Luxulyan, Cornwall, is a fine example.

Tournaisian 1. The earliest *epoch in the *Mississippian, preceded by the *Famennian *Age (*Devonian), followed by the *Visean Epoch, and dated at 362.5–349.5 Ma (Harland et al., 1989). It comprises the Hastarian and Ivorian Ages. **2.** The name of the corresponding European/Russian *series, which is roughly contemporaneous with the lower *Carboniferous Limestone (Britain), and the *Kinderhookian and lower *Osagean (N. America). *See also* DINANTIAN.

tourquoise *See* TURQUOISE.

Toutatis A *solar system asteroid (No. 4179), measuring 4.6 × 2.4 × 1.9 km; approximate mass 10^{13} kg; rotational period irregular; orbital period 1.1 years. It is a double object, its two components probably in contact, one estimated to be 2.5 km in diameter and the other 1.5 km. On 29 September 2004 Toutatis will reach its closest approach to Earth, at a distance of about 1.5 million km.

tower karst A form of karstic morphology (*see* KARST) that is developed mainly in low latitudes, characterized by residual hills of *limestone rising from a flat plain. The hills have near-vertical sides, resembling towers.

T-peg In *geomorphology, a device for measuring the rate of surface *creep. It consists of a metal rod to which a cross-piece is attached. The apparatus is inserted into the *regolith and the cross-piece is levelled. Soil creep causes the cross-piece to tilt, and the degree of inclination is measured with a graduated spirit-level. The relative movement between the ground surface and the depth of insertion can then be inferred.

trace A recorded data-set for one channel. On a seismic-*refraction record made for a 12-channel *seismograph, each channel would provide one wave-form which, when viewed with the others, would give a *seismic record. *See also* WIGGLE TRACE.

trace element 1. An element that occurs in minute but detectable quantities in minerals and rocks, much less than 1%. All elements except the most common rock-forming elements (O, Si, Al, Fe, Ca, Na, K, Mg, and Ti) generally occur as trace elements, except where locally concentrated in their *ores. **2.** In biology, the occurrence in plant or animal tissue of minor amounts of an element which is essential to its growth.

trace-element fractionation The redistribution (fractionation) of elements between solid and liquid *phases caused by heating and partial melting, and which also takes place in crystallization. For example, fractionation of *trace elements apparently occurs in *chondritic meteorites during the formation of *chondrules (small molten droplets produced by heating and melting in outer space which are subsequently *quenched).

trace fossil (ichnofossil) A *biogenic *sedimentary structure formed by the behavioural activity of an animal on or within a given substrate. The study of trace fossils is called 'ichnology'. Traces are most frequent at the *interface between different *lithologies (e.g. *sandstone and *shale), and are classified on various criteria including *morphology and preservation. Of these two, the second is preferred as a toponomic classification (i.e. classification by place of occurrence) and, apart from the processes of preservation, considers the position of the trace within the depositional unit concerned. In 1970, A. Martinsson divided traces into four groups dependent on their relationship to the casting medium: epichnia are surface ridges or grooves; endichnia are tubes or *burrows formed within the casting medium; hypichnia are grooves or ridges preserved on the lower surface of the main body of the casting medium; and exichnia are formed by *bioturbation outside the main body of the casting medium. *See* FOSSILIZATION.

tracer A substance that is used to follow the passage of *groundwater in places where it cannot be observed directly. Typical tracers include fluorescent dyes and salt. The presence of radioactive *isotopes, e.g. tritium and carbon-14, may also be used as tracers in that they allow the age of groundwater to be determined. The presence of small amounts of other substances may also be used to make deductions about the origin and flow path of groundwaters.

Tracheophyta (vascular plants; kingdom *Metaphyta) Division comprising plants that have vascular tissues (xylem and phloem) through which water and nutrients are transported. In many modern classifications this division embraces the former divisions Pteridophyta and Spermatophyta (*see* PTERIDOPHYTINA; and SPERMATOPHYTINA).

trachyandesite *See* ANDESITE.

trachybasalt The fine-grained, *extrusive equivalent of *syenogabbro. When *undersaturated, *feldspathoidal minerals take the place of *alkali feldspar to generate feldspathoidal trachybasalts known as *tephrites (*olivine-free) and *basanites (olivine present). Trachybasalts are found on the stable *continental crust and on some oceanic islands.

trachyte Fine-grained volcanic rock of the *alkaline series of *intermediate rocks. Trachyte is the volcanic equivalent of *syenite.

trachytic texture General *petrographic term applied to volcanic rocks that have a fine mat of orientated *feldspar laths, suggestive of flow. *See* TRACHYTOIDAL.

trachytoidal Applied to a *foliation resulting from a parallel alignment of tabular *feldspars in a coarse-grained rock, e.g. *syenite. When the same alignment occurs in fine-grained rocks, e.g. *trachytes, it is called '*trachytic texture'.

track A *biogenic *sedimentary structure grouped under the Scoyenia assemblage of *trace fossils. The term may refer to a line of vertebrate footprints or to traces left by the limbs of arthropods (*Arthropoda).

traction carpet *See* BED LOAD.

traction load *See* BED LOAD.

trade-wind inversion The inversion of temperature *lapse rate with height over a major zone of the trade-wind belt, which is very significant in tropical meteorology. Moist tropical air, extending up to 2–3 km above the surface, is 'boxed in' or 'trapped' by dry, clear, warmer air above, resulting from *subsidence in subtropical *anticyclones. The inversion forms where the subsiding air meets a surface flow of cooler maritime air. The top of a cloud layer marks the base of the inversion.

trade winds Old maritime term, much used in meteorology, indicating the steadiness of direction of the prevailing tropical easterly winds, which blow from subtropical high-pressure areas in latitudes 30–40° north and south, generally north-easterly in the northern hemisphere and south-easterly in the southern hemisphere. They are most nearly constant in latitudes centred on 15° north and south. Climatic conditions associated with the belt vary from fine anticyclonic weather in the poleward and eastern margins, caused by *subsidence, to stormier conditions near the equator and western margins, caused by less stable, deeper, moist air.

Traditional Stratigraphic Scale (TSS) *See* STRATIGRAPHIC SCALE.

trail 1. An anterior extension of some brachiopod (*Brachiopoda) shells, usually at a large angle to the general plane of the posterior part of the shell. **2.** A *biogenic *sedimentary structure formed by the movement of snails, clams, or perhaps snakes over the *sediment surface, and classified with *tracks under the Scoyenia assemblage of *trace fossils.

trailing edge *See* PASSIVE MARGIN.

training area *See* CLASSIFICATION.

tramontana Local wind in the Mediterranean, which brings dry, cold conditions from the north across the mountains.

tranquil flow (sub-critical flow) *See* CRITICAL FLOW; and FROUDE NUMBER.

transcurrent fault *See* STRIKE-SLIP FAULT.

transfer fault A vertical or sub-vertical *fault which, via *dip-slip and *strike-slip movements, allows the juxtaposition of two *fault zones which have different displacement characteristics. Lateral *ramps are the equivalent in *thrust terrains.

transfluence *See* GLACIAL BREACH.

transformation twinning A type of *crystal twinning caused by a change in the structure of a crystal under different conditions of temperature and pressure (*polymorphism), e.g. the *cristobalite–*tridymite–*quartz series, where the high-temperature form is *hexagonal and the low-temperature form is *trigonal.

transform fault A type of *strike-slip fault in an ocean, occurring at the boundaries of lithospheric *plates, in which the direction of movement of the crustal blocks is reversed (or 'transformed') in comparison with a strike-slip fault on land. For example, at mid-ocean *ridges the offset between adjacent ridge sections is a transform fault; where the displacement is *dextral (right lateral) the motion, due to spreading, is left lateral, and vice versa. Generally, transform faults occur at right angles to the ridge itself and indicate the direction of spreading. The active transform fault extends into an inactive fracture zone.

transgression (marine) An advance of the sea to cover new land areas, due to a rise in the sea level relative to the land. As a result, shallow-water *sediments are overlain by those characteristic of deeper water, e.g. shelf mud is deposited on coastal sand. *Compare* REGRESSION. *See* ONLAP.

transgressive systems tract (TST) In the *genetic stratigraphic sequence model used in *sequence stratigraphy, a bounding surface formed by a rapid rise in sea level. *Compare* HIGHSTAND SYSTEMS TRACT; and LOWSTAND SYSTEMS TRACT.

transient creep *See* PRIMARY CREEP.

transient electromagnetic method (TEM) An electromagnetic surveying method in which the source signal consists of a train of pulses rather than a continuous wave-form. Quasi-transient methods are used for continuous *profiling (e.g. the *INPUT method), while true TEM methods are most commonly used for depth soundings. It is a *time-domain method.

transient flow *See* STEADY FLOW.

transient variation Any short-term variation. In *geomagnetism, aperiodic fluctuations of the *geomagnetic field, usually on scales of microseconds to a few days.

transit time The time taken for a seismic *P-wave to travel through one foot (0.3048 m), measured by a *sonic log in units of μs/foot.

translation *See* COILING.

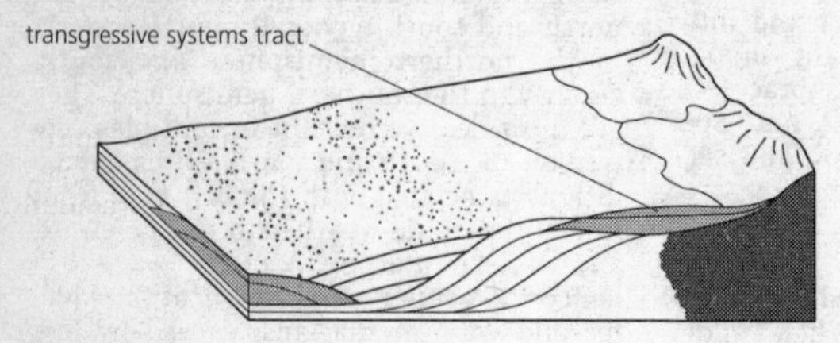

Transgressive systems tract

translation gliding *See* DEFORMATION TWINNING; and SLIP.

translocation In *pedology, the movement of *soil materials in solution or in suspension from one *soil horizon to another.

translucent Applied to a substance (e.g. a mineral) that transmits some light but through which the outlines of objects cannot be seen. This condition is common in varieties of *quartz and *fluorite.

translucidus From the Latin *translucidus* meaning 'transparent', a variety of cloud, occurring in extensive layers or sheet form, which is *translucent, allowing the Sun or Moon, and occasionally some stars, to be visible. This characteristic can be presented by *stratus, *stratocumulus, *altostratus, and *altocumulus. *See also* CLOUD CLASSIFICATION.

transmission coefficient (*T*) The ratio of the amplitude of a transmitted ray (A_2) to that of the incident ray (A_0), such that $T = A_2/A_0$. In the case of normal incidence, and in terms of the *acoustic impedances (Z_1 and Z_2) of the two media above and below the boundary, $T = 2Z_1/(Z_2 + Z_1)$. Also, $T = 1 - R$, where R is the *reflection coefficient. The transmission coefficient can also be expressed in terms of energy (T'), when $T' = 4Z_1Z_2/(Z_2 + Z_1)^2$.

transmissivity (*T*) The rate at which *groundwater is transmitted through a unit width of an *aquifer under a unit *hydraulic gradient. It is often expressed as the product of the *hydraulic conductivity and the full saturated thickness of the aquifer and has units of the form m^3/day/m.

transmittance The ratio of *electromagnetic radiation passing through a material to that incident upon its surface.

transparency Degree to which a substance allows radiation of varying wavelengths to pass through, as opposed to absorbing it. *See also* ABSORPTION.

transparent Applied to a substance (e.g. a mineral) which transmits light and through which the outlines of objects can be seen clearly.

transpiration The removal of moisture from the *soil by plant roots, its translocation up the stem to the leaves, and its evaporation through the stomata. The flow of water through the plant is known as the transpiration stream. It reduces leaf temperatures, and is thought to be important for mineral absorption and translocation within plants. The process imposes a number of environmental requirements upon plants; for example, wilting and desiccation result from an insufficient supply of water.

transpression A tectonic regime which combines both transcurrent *strike-slip movement with oblique compression. *Flower structures are commonly associated with transpressional regimes.

trans-Saharan seaway The marine seaway that, during two intervals in the Late *Cretaceous, extended from *Tethys in the north through what are now Libya, Chad, Niger, and Nigeria, to the newly developing S. Atlantic Ocean. On both occasions ammonite (*Ammonoidea) and ostracod (*Ostracoda) faunas are known to have migrated through this seaway. The first event was in the latest *Cenomanian and earliest *Turonian, and was re-established only in the Late *Campanian to Early *Maastrichtian interval. At the southern end (in present-day Nigeria) there was a structural control in the form of the Benue Trough, but the remainder of the seaway appears to have been controlled only by global sea level change.

transtension A tectonic regime combining transcurrent *strike-slip movement with oblique *extension. Such a regime is associated with oceanic spreading *ridges and *transform faults.

transverse dune *See* DUNE.

transverse-type coast *See* ATLANTIC-TYPE COAST.

transverse wave *See* S-WAVE.

trap *See* ANTICLINAL TRAP; FAULT TRAP; REEF TRAP; STRATIGRAPHIC TRAP; STRUCTURAL TRAP; and UNCONFORMITY TRAP.

trap-door caldera A *caldera in which the floor is hinged on one side.

trapezohedron A *crystal form where the faces are in the shape of a trapezoid with four sides, none of which are parallel. It is developed in the *cubic, *tetragonal, and *hexagonal *crystal systems, to give a *closed form similar in shape to a pyramid. It can be referred to the normal axes of symmetry but frequently lacks a horizontal plane of symmetry.

travel time The time taken for a wave to travel from a source generator (e.g. a *shot or transmitter), through some media, to a detector (*geophone or receiver), at a known *offset. It is used to compute *travel-time curves in seismic-*refraction surveying and *seismic sections in seismic-*refraction surveying. *See also* ONE-WAY TRAVEL TIME; and TWO-WAY TRAVEL TIME. *Compare* TRANSIT TIME.

travel-time curve One of a set of curves which show the *travel times of *P-waves and *S-waves as functions of distance, expressed either in range kilometres, or in degrees (as epicentral angles). The distance from the recording *seismometer to the *epicentre can be determined by measuring the times that elapse between the arrival of the P-wave and S-wave, and the S-wave and *L-wave.

traverse In surveying, a line which connects two points and passes through a series of locations which are to be studied. A traverse may be open-ended and discontinuous, or closed (i.e. it returns to its starting point).

travertine Calcium carbonate deposited by *precipitation from carbonate-saturated waters, particularly from hot springs. Travertine deposits are sometimes *massive, but often display a concentric or fibrous internal structure, sometimes building large, concentric, spherical masses. Travertine is also found in cave deposits in the form of *stalagmites and *stalactites. A porous, sponge-textured form of travertine is referred to as '*tufa' or 'calc-sinter'.

tree-ring analysis *See* DENDROCHRONOLOGY.

trellis drainage pattern *See* DRAINAGE PATTERN.

trema *See* ARCHAEOGASTROPODA.

Tremadoc Oldest (510–493 Ma) of the six *stages that comprise the *Ordovician *Period. *Mudstones and *sandstones of the Tremadoc *Series occur in N. America, Ireland, Wales, England, and Scandinavia. They mark the *continental shelf and slope areas that occur on the southern edge of the northern *Iapetus Ocean.

tremata *See* ARCHAEOGASTROPODA.

tremolite An important member of the *monoclinic calcium-rich *amphiboles $Ca_2(Mg,Fe^{2+})_5[Si_4O_{11}]_2(OH,F)_2$, which forms a series with ferroactinolite; sp. gr. 3.02–3.44; *hardness 5.0–6.0; *monoclinic; white to greyish-white; vitreous *lustre; crystals simple, long, *prismatic, and *acicular, often in thin, radial, rod-like, fibrous, and felted masses; *cleavage perfect *prismatic {110}; widespread *amphibole mineral in *igneous rocks, and also in *metamorphosed, crystalline *limestones, *dolomites, and in *schists and *hornfels. It is named after the Tremola Valley near St. Gotthard, Switzerland. *See also* JADEITE.

Trempealeauan A *stage in N. American usage, spanning the *Cambrian/*Ordovician boundary of other continents.

trench (oceanic trench) An elongate depression of the ocean floor which runs parallel to the *trend of adjacent volcanic islands (*island arc) or continent. Oceanic trenches are up to 11 km deep, typically 50–100 km wide, and may be thousands of kilometres long. In cross-section the trench slopes are usually asymmetric, with a steeper slope on the landward side. Most trenches are associated with *subduction zones.

trend The *azimuth of a geologic feature, commonly of a *fold axis, and written as a compass bearing.

trevorite *See* MAGNETITE.

triad *See* CRYSTAL SYMMETRY.

triangle zone A triangular area bounded by *thrusts, which has been formed by the truncation of an earlier thrust surface by a later *backthrust.

triangular facet *See* VENTIFACT.

Triassic The earliest (245–210 Ma) of the three *periods of the *Mesozoic *Era. As a result of the mass *extinctions of the late *Palaeozoic, Triassic communities contained many new faunal and floral elements. Among these were the ammonites (*Ammonoidea), modern corals, various molluscs (*Mollusca), the *dinosaurs, and certain *gymnosperms.

triaxial cell An instrument for testing the compressive strength of *soils or *rocks. A load is applied to the sample, which is contained in an impermeable membrane surrounded by a fluid. The loading is increased until failure occurs.

triaxial compression test A test for the compressive strength in all directions (*compare* UNIAXIAL COMPRESSION TEST) of a *rock or *soil sample, using a *triaxial cell. Tests in which drainage is prevented are called 'undrained' tests and the strengths obtained are 'undrained' strengths. When *pores are allowed to empty, the tests are called 'drained' tests and the strengths obtained are 'drained' strengths.

triaxial ellipsoid *See* PLANE STRAIN; and SHAPE FABRIC.

triclinic One of the seven *crystal systems, and the one with the lowest *crystal symmetry, characterized by three *crystallographic axes of unequal length, none of which are at right angles to each other. There is a centre of symmetry about which is oriented a pair of parallel faces. The *mineral *plagioclase feldspar is the best-known example.

tricolpate sulci *See* POLLEN.

Triconodonta (class *Mammalia, subclass *Prototheria) Order that includes the earliest of all mammals, living from the late *Triassic until the early *Cretaceous and distributed over the northern continents. Typically the molar teeth each had a row of three sharp conical cusps, the teeth of the upper and lower jaws forming a shearing device. Premolars and molars were differentiated, probably with some replacement, and probably the young were fed on milk secreted by the mothers. Triconodonta may have been *homoiotherms and nocturnal, and possibly they were arboreal. They are believed to have been true carnivores rather than insectivores. *Triconodon*, one of the larger forms, from the Upper *Jurassic, was the size of a modern cat. The order is believed to have evolved from *therapsids independently of the main line of mammalian evolution and to have left no descendants.

tridymite A high-temperature variety of silica (SiO_2) which is stable between 870 and 1470 °C at normal pressures. It occurs as tabular plates in cavities of volcanic rocks and also in *stony meteorites. *See also* QUARTZ.

trigonal (rhombohedral) One of the seven *crystal systems (closely related to the *hexagonal system) in which the unit cell has the shape of a *rhombohedron, consisting of six *crystal faces all of which have two pairs of parallel sides. The crystals may be referred to four *crystallographic axes: one vertical axis of three-fold *crystal symmetry, and three equal, horizontal axes of two-fold symmetry which are separated by an angle of 120°. The minerals *calcite and *quartz always occur in this system.

trilete *See* SPORE.

Trilobita (trilobites; phylum *Arthropoda) The most primitive arthropod class, known from more than 3900 *fossil species. Inhabitants of *Palaeozoic seas, the trilobites appeared first in the early *Cambrian, had their widest distribution and greatest diversity in the Cambrian and *Ordovician Periods, and became extinct in the *Permian. The body was divided into three regions: an anterior *cephalon, comprising at least five, fused segments; a mid-body or *thorax, with a varying number of segments; and a hind region or *pygidium. All three regions were divided by a pair of furrows running the length of the body, giving a trilobite appearance (i.e. a median or axial lobe, flanked on either side by a lateral lobe). The mouth was situated in the middle of the central surface of the cephalon. Paired gill-bearing limbs were attached to the membranaceous, *pleural skeleton. X-ray studies show the eyes to have resembled the compound eyes of living arthropods (*see* TRILOBITE EYE). Trilobites ranged in size from 0.5 mm long planktonic (*see* PLANKTON) forms to those nearly 1 m in length; most species were 3–10 cm long. There were 9 orders: *Redlichiida; *Agnostida; *Naraoiidae; *Corynexochida; *Lichida; *Phacopida; *Ptychopariida; Asaphida; and *Proetida.

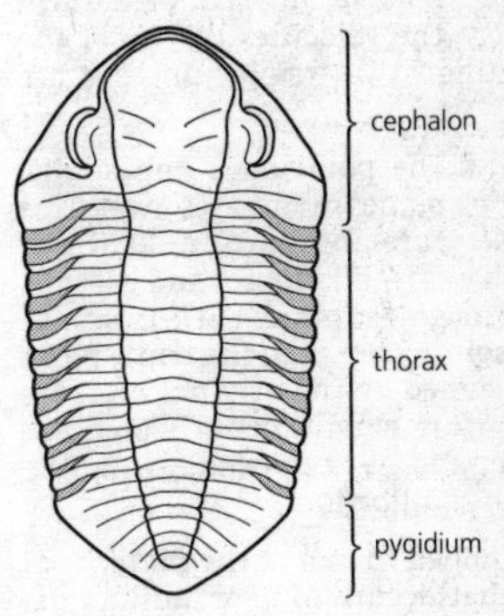

Trilobita

trilobite *See* TRILOBITA.

trilobite eye The eyes of trilobites (*Trilobita) are compound and made up of radially disposed visual units. Most trilobite eyes are 'holochroal' (i.e. many polygonal lenses are in contact with one another and covered by a single cornea). 'Schizochroal' eyes occur in the Phacopina (i.e. the lenses are large, separated from one another, each has its own cornea, and each lens is in two parts); it has been shown that this arrangement produces a sharp focus.

Trimerophytina *See* PSILOPHYTALES.

trim line A boundary (or zone) between frost-shattered and glacially scoured bedrock in an upland region. It marks the junction between glacial and *periglacial activity and so indicates the position of a former *glacial limit.

trimorphism *See* POLYMORPHISM.

triple core barrel *See* CORE BARREL.

triple junction The point where three lithospheric *plates meet. This junction may involve three oceanic *ridges (an 'RRR' junction), thought to have developed from a *domal uplift; or involve some other configuration of ridge ('R'), *transform fault ('F'), and oceanic *trench ('T'). Some types of triple junction are stable, while others evolve geologically rapidly into a different configuration. An example of an RRR junction is the point where the *African, *South American, and *Antarctic Plates meet in the S. Atlantic.

triple-junction method A mathematical method in which a vector diagram is used to calculate the relative velocity of a third *plate, where the velocities of two of the plates meeting at a *triple junction are known.

triple point The point on a *phase diagram at which planes or *phases meet. This is the temperature and pressure at which the three phases (solid, liquid, and gas) of a substance can coexist (*see also* KELVIN SCALE). In a *solid-solution system, the triple point is usually defined by the original pressure and temperature conditions at which the solid, liquid, and gaseous phases of a substance are in equilibrium.

triserial Applied to one of the patterns of chambers that occurs in *Foraminiferida, in which there are three chambers to each *whorl in the *test. In some species there can be a change from a single chamber in each whorl ('uniserial'), through two chambers ('biserial'), to triserial.

tritium clock Tritium (T) is a naturally occurring radioactive *isotope (^{3}H) of hydrogen (^{2}H), having two neutrons and one proton in the nucleus. The isotope decays (half-life = 12.26 years, *see* DECAY CONSTANT) by beta emission (*see* BETA DECAY) to stable helium (^{3}He) and this can be used to measure the age of water samples back to approximately 30 years. Tritium is produced in the upper atmosphere by the interaction of fast cosmic-ray neutrons with stable ^{14}N. The tritium combines with hydrogen and oxygen to form HTO, which is then dispersed throughout the *hydrosphere. The natural production rate of tritium is 15–45 atoms/min/cm^2 of the Earth's surface. In some experiments, artificially produced tritium has been introduced into *groundwater and used to trace and time underground movements. It can also be used to measure and time rates of mixing in oceanic current systems.

Triton (Neptune I) A satellite of *Neptune, with a diameter of 2705.2 km; mass 214.7 × 10^{20} kg; mean density 2054 kg/m^3; visual albedo 0.7. Its orbit is *retrograde.

trochiform Applied to a gastropod (*Gastropoda) shell where the sides of the spire are evenly conical and the base is flat.

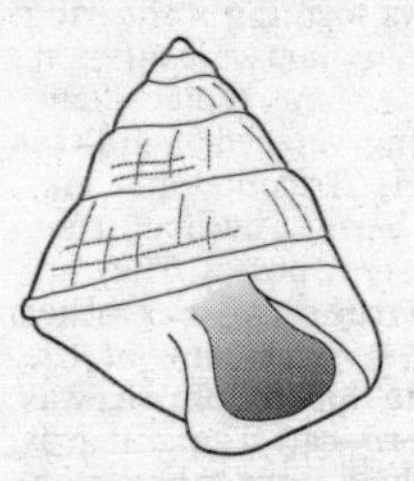

Trochiform

trochoid *See* SOLITARY CORALS.

trochospiral Applied to the growth pattern in *Foraminiferida where material is added to the *test in a helical manner.

troctolite A coarse-grained, *igneous rock consisting of *essential magnesium-rich *olivine and calcium-rich *plagioclase feldspar with *accessory *ilmenite; i.e. it is *gabbro without the *pyroxene component. Troctolites grade into gabbros by a gradual increase in pyroxene content and a gradual decrease in olivine content.

troilite An iron sulphide mineral, FeS; sp. gr. 4.8; *hardness 4.0; greyish-brown; *metallic *lustre; black *streak; *massive or granular; occurs as nodules in *iron meteorites.

trona A mineral, $NaHCO_3.Na_2CO_3.2H_2O$, found in salt lake deposits.

trondhjemite *See* GRANODIORITE.

tropical air *Air masses whose characteristics originate from a tropical source. Broadly, these comprise: (*a*) maritime tropical air, originating over oceans; and (*b*) continental tropical air, originating over land masses, and especially over N. Africa and the Middle East. Movements of these air masses poleward, with modification of their original characteristics, can have marked influence on weather in mid-latitudes.

tropical cyclone (revolving storm) A generally fairly small but intense, closed low-pressure system which develops over tropical oceans. Wind speeds of at least 33 m/s (force 12 on the *Beaufort scale, 64 knots or more) define such storms and distinguish them from less intense systems, e.g. tropical *depressions (of twice or more than twice the diameter) or tropical storms. The atmospheric pressure gradient in such cyclones commonly ranges from about 950 mb at the centre to about 1000 mb at the margins.

Tropical Cyclone Programme (TCP) A project to improve forecasting and warning systems for tropical cyclones. It forms part of *World Weather Watch.

tropopause The boundary separating a lower layer of the atmosphere (*troposphere), in which air temperature generally decreases with height, from the layer above (*stratosphere), in which temperature remains constant or increases with height. The altitude of the tropopause varies according to sea-surface temperature and season, but also over shorter periods, from an average of 10–12 km over the poles (occasionally descending to 8 km or below) to 17 km over the equator. *See also* ATMOSPHERIC STRUCTURE.

troposphere The layer of the atmosphere between the Earth's surface and the *tropopause, within which the air temperature on average decreases with height at a rate of about 6.5 °C/km, though variations that sometimes occur include inversions (temperature increase with height within some limited layer). Most of the atmospheric turbulence and weather features occur in this layer, which contains almost all the atmospheric water vapour and most of the *aerosols in suspension in the atmosphere (although there is also an important aerosol layer at about 22 km). *See also* ATMOSPHERIC STRUCTURE.

trough 1. An extension of low atmospheric pressure from the central regions of a low-pressure system into a zone where generally higher pressure prevails. The term 'trough' is also, and in accordance with this definition, applied to equatorward meanders of the flow of the upper westerly winds over middle latitudes. (The 'equatorial trough', where *trade winds meet, is synonymous with the '*intertropical convergence zone'.) *See also* LONG WAVE. **2.** The lowest point of a *fold surface.

trough cross-stratification *See* CROSS-STRATIFICATION.

trowal Canadian meteorological term for the line of the upper front of an *occlusion and for the region about it where the warm air, having been lifted off the surface, is still relatively low. It is a *trough or valley of warm air that is still undergoing lifting, and is normally marked by clouds and *precipitation.

true age *See* ABSOLUTE AGE.

true dip The *dip which is measured at right angles to the *strike of a plane and which is the maximum value of dip for that plane. *Compare* APPARENT DIP.

true thickness The *orthogonal thickness of a structure or bed, measured at right angles to its surface. Where the structure or bed is inclined, a surface exposure or *borehole may give a value greater than the true thickness. In its simplest form true thickness can be calculated from an *outcrop exposure by multiplying the width of the exposed layer by the sine of the angle of *dip of the layer.

truncated spur A blunt-ended, sloping ridge which descends the flank of a valley. Its abrupt termination is normally due to *erosion by a *glacier which tends to follow a straighter course than the former river.

T_s Symbol used to indicate the actual average temperature at the surface of the Earth, about 15 °C. *See also* EFFECTIVE TEMPERATURE.

tschermakite *See* HORNBLENDE.

TSS Traditional Stratigraphic Scale. *See* STRATIGRAPHIC SCALE.

TST *See* TRANSGRESSIVE SYSTEMS TRACT.

tsunami A seismic sea wave of long period, produced by a submarine *earthquake, underwater volcanic explosion, or massive *gravity slide of sea-bed *sediment. In the open ocean such waves are barely noticeable even though they may be travelling at 700 km/h, but on reaching shallow water they build up to heights of more than 30 m and cause severe damage in coastal areas.

t-test A test to calculate the probability that mean values for a particular measurement are significantly different in two sets of data.

***Tt–d* curve** *See* TIME–DISTANCE CURVE.

tuba From the Latin *tuba* meaning 'trumpet', a supplementary cloud feature of *cumulonimbus or sometimes *cumulus, characterized by a column or cone of cloud projecting from the cloud base. *See also* CLOUD CLASSIFICATION.

tube-feet (podia) In *Echinodermata, hollow appendages connected to the water-vascular system, used in some species for locomotion and in others for feeding.

tubercle The domed, surface structure in trilobites (*Trilobita). Some tubercles appear to be domes covering a space, others may have been the sites of sensory organs.

tubular fenestrae *See* FENESTRAE.

tufa (calc-tufa) *Sedimentary rock formed by the deposition or precipitation of calcium carbonate, or more rarely *silica, as a thin layer around saline springs, or by the encrustations on *stalactites and *stalagmites. *See also* TRAVERTINE.

tuff The compacted (lithified) equivalent of a volcanic *ash deposit, which has been generated and emplaced by *pyroclastic processes or was water lain, and in which the grain size of the pyroclasts is less than 2 mm. Where the proportion of *lapilli-sized pyroclasts exceeds 10% the term 'lapilli-tuff' is used.

tumulus A small mound or dome-like uplift, up to 20 m or more in diameter, on the crust of a *lava flow. Upwarping of the flow crust is caused by the hydrostatic overpressure of lava within the flow interior, or by excess pressure developed due to the difference in rate of flow between the cooler crust and the more fluid lava below it. Unlike a *lava blister, a tumulus is a solid structure.

tundra Treeless plain of the Arctic and Antarctic, characterized by a low, 'grassy' sward. Actually, although grasses are rarely absent, sedges (*Carex* species), rushes (*Juncus* species), and wood rushes (*Luzula* species) are the dominant plants, together with perennial herbs, dwarf woody plants, and various bryophytes (*Bryophyta) and lichens.

Tundra Soil One of the Great Soil Groups, within suborder 1 of the order Zonal Soils of the 1949 *USDA system of soil classification, based originally on the work of V. V. *Dokuchaev, but now superseded. Tundra Soils are now classified as *Inceptisols. They occur on ground that drains poorly (mainly because of *permafrost), and are acid, are 30–60 cm deep, have a high content of organic matter at the surface, and have a microrelief formed by freezing and thawing; their formation, and the decomposition of organic matter, is inhibited by the low temperature.

tungstates A group of non-silicate *minerals in which the WO_4 radical is in combination with a number of metal *cations. Important examples include *wolframite ($(Fe,Mn)WO_4$) and *scheelite ($CaWO_4$). Molybdenum may substitute for tungsten to give related minerals, e.g. the molybdates powellite ($CaMoO_4$) and *wulfenite ($PbMoO_4$).

tunicates *See* UROCHORDATA.

tunnel trend *See* CLEANING-UP TREND.

tunnel valley A valley cut by a subglacial stream escaping from beneath an *ice sheet. It is well developed in Denmark (where such valleys are called 'tunneldale') and in Germany ('Rinnentaler'). Individual examples may be up to 75 km long and up to 100 m deep, with steep sides and flat floors. The long profile may be irregular, as a result of water under pressure being locally forced uphill, and so the valley may now be occupied by a string of lakes ('ribbon lakes').

TURAM method A geophysical *electromagnetic method in which a very long (up to several hundred metres), insulated cable is used, either grounded at both ends or laid out in a large loop, and energized at low frequencies (less than 1 kHz). A receiver coil is moved perpendicularly across the line of the cable and the two orthogonal components of the secondary field are measured.

turbidite Sedimentary deposit laid down by a *turbidity current.

turbidity current A variety of *density current that flows as a result of a density difference created by dispersed *sediment within the body of the current. Such currents occur off *delta fronts, in lakes, and in oceans, and are initiated by the disturbance of sediments on a slope by strong wave action, *earthquake shock, or slumping. Turbidity currents in the oceans are thought to move rapidly (at speeds of up to 7 m/s) down the *continental slope or *submarine canyons along the sea bed, and to deposit originally shallow-water sediments at the foot of the slope or on the *abyssal plain. The ideal sequence of sediments laid down by a waning turbidity current is known as the *Bouma sequence.

turbidity flow *See* GRAIN FLOW.

turbinate Applied to a gastropod (*Gastropoda) shell that is shaped like a spinning top, but with a rounded base.

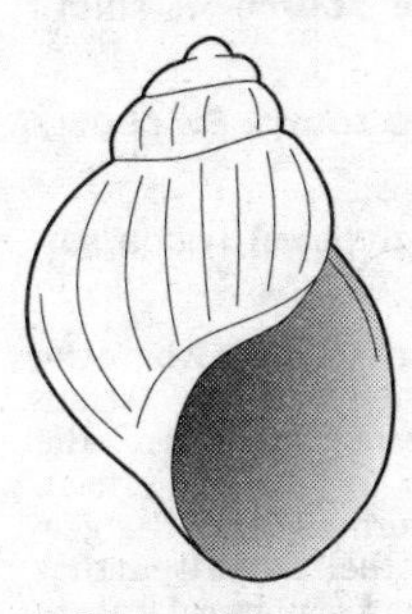

Turbinate

turbulence Disturbed flow in a moving stream of air. It is manifested by variations of wind speed and direction (including vertical components), by vertical exchanges of mass, heat, momentum, water vapour, and any pollutants present, caused by eddies.

turbulent flow Mode of flow occurring in both air and water, and characterized by the superimposition of transverse movements, notably eddies or vortex behaviour, on the general downstream trend. Local upward movement brings about the lifting and removal of particles from stream beds and from the surface of a sandy area. *See* REYNOLDS NUMBER.

turnover rate Measure of the rate of movement of an element through a *biogeochemical cycle. Turnover rate is calculated as the rate of flow into or out of a particular pool, divided by the quantity of the element in that pool. Thus it measures the importance of a particular flux in relation to the pool size. *Compare* TURNOVER TIME.

turnover time Measure of the movement of an element in a *biogeochemical cycle; the reciprocal of *turnover rate. Turnover time is calculated by dividing the quantity of an element present in a particular pool or reservoir by the flux rate for that element into or out of the pool. Turnover time thus describes the time it takes to fill or empty that particular reservoir.

Turonian 1. A Late *Cretaceous *age, preceded by the *Cenomanian, followed by the *Coniacian, and dated at 90.4–88.5 Ma (Harland et al., 1989). **2.** The name of the corresponding European *stage, for which the *type area is between Saumur and Montrichard, France.

turquoise (tourquoise) A phosphate $CuAl_6(PO_4)_4(OH)_8 4\text{–}5H_2O$; sp. gr. 2.60–2.91; hardness 5.0–6.0; *triclinic; sky-blue to blue-green; white or greenish *streak; waxy *lustre; crystals very rare, normally occurs *massive, granular to *cryptocrystalline, or as encrusting masses; *conchoidal fracture; occurs in *veins in association with aluminous, *igneous, or *sedimentary rocks that have undergone *alteration. Finer varieties are used for semi-precious stones.

turreted (turriculate) Applied to a gastropod (*Gastropoda) shell that is very high-spired, with a flat or gently rounded base.

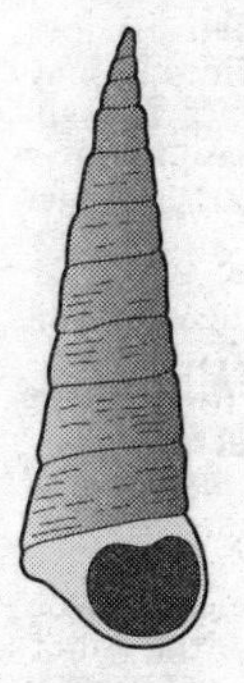

Turreted

turriculate *See* TURRETED.

tuya The name given in British Columbia, Canada, to a flat-topped mountain produced by the subglacial eruption of a *central vent volcano.

Twenhofel, William Henry (1875–1957) An American geologist, Twenhofel was a professor at the University of Wisconsin. He made studies of American *Palaeozoic *sediments, but is best known for his work on the processes of sedimentation, described in his book *Treatise on Sedimentation* (1926, 1932).

twilight Period of half-light caused by scattering and reflection of sunlight in the upper atmosphere at a time when the Sun is some degrees below the horizon.

twin axis An axis of a twinned crystal (*see* CRYSTAL TWINNING), normally at right angles to the *twin plane and frequently related to the *crystallographic axis of the crystal. The relation between the different parts of the crystal may be described with reference to a 180° rotation about the twin axis.

twin gliding *See* DEFORMATION TWINNING.

twinkling In mineral optics, the effect observed in *plane-polarized light when a *thin section of an *anisotropic mineral with widely differing *refractive indices is rotated rapidly on the *stage. The 'twinkling' appearance is caused by the rapid changes in relief. *Calcite is a well-known example.

twin law The law which describes the fundamental elements along or about which a crystal is twinned, in terms of a *twin plane or *twin axis. The various geometric shapes which result from *crystal twinning may also be described with reference to well-known examples, e.g. *orthoclase feldspar may be twinned on the *Carlsbad law and/or the Baveno law.

twin plane The reflection plane which divides a twinned crystal (*see* CRYSTAL TWINNING) such that one half is a mirror image of the other. It is normally parallel to a *crystal face, but in complex multiple twins it may become a highly irregular surface.

two-way travel time The time taken for a *seismic wave to travel from the *shot down to a *reflector or refractor and back to a *geophone at the surface. For finite *offsets, the two-way travel times are affected by normal *moveout; the normal-incidence two-way travel time is measured at zero offset.

***t–x* curve** *See* TIME–DISTANCE CURVE.

t^2–x^2 graph A *time–distance graph, used in *reflection seismology, in which travel times squared (t^2) are plotted as a function of the *offset squared (x^2) to produce straight-line segments each of which represents the reciprocal of the layer *root-mean-squared velocity squared ($v_{rms}{}^2$), and the intercept represents the depth (z) to the reflector. ($v_{rms}{}^2 t^2 = 4z^2 + x^2$.) *See also* *T–D* CURVE; and TIME–DISTANCE CURVE.

Tycho Visually, the most prominent lunar crater, 85 km in diameter, formed about 100 Ma ago. *See* RAYS.

tympanic bone In *Mammalia, the bone supporting the tympanic membrane, derived from the angular bone of the lower jaw and, in many mammals, forming the bulla.

Type I earthquake source *See* SINGLE COUPLE.

Type II earthquake source *See* DOUBLE COUPLE.

type area The geographical region surrounding the *type locality.

type locality The particular site where the type section (*stratotype) for a *stratigraphic unit is located, or where the original type section was first described. Stratigraphic units normally take the geographic component of their name from that of the type locality, or a key physical feature there, e.g. the Kimmeridge Clay from Kimmeridge Bay in Dorset, England. Some early-defined stratigraphic units were named after the *type area, e.g. the *Cambrian System in Wales was named by Adam *Sedgwick (1835) after *Cambria*, the latinized version of the Welsh *cymry* ('fellow countryman') and *Cymru* (Wales).

type section *See* STRATOTYPE; UNIT-STRATOTYPE; TYPE AREA; and TYPE LOCALITY.

type series In taxonomy, all the specimens on which the description of a taxon is based.

type specimen *See* HOLOTYPE.

typhoon The name given to a *tropical cyclone that forms over the Pacific and China Seas.

typological method In stratigraphy, the describing of *stratigraphic units in terms of the kind of rock or *fossil they contain. In practice it is the lower-ranking units, e.g. *zones, that are defined typologically. *Compare* HIERARCHICAL METHOD.

Tyrannosaurus rex Giant, carnivorous *dinosaur which lived during the Upper *Cretaceous in N. America and possibly in Asia. Individuals grew to 12 m in length, 5 m tall, and weighed about 7 tonnes. The apt name means 'king of the tyrant lizards'.

Tyrrhenian *See* QUATERNARY.

Ubendian orogeny A phase of mountain building whose precise dates are uncertain but which probably occurred about 1800–1700 Ma ago, producing what is now a NW–SE belt in southern Tanzania, northern Zambia, and the eastern Congo.

Udden–Wentworth scale *See* PARTICLE SIZE.

Udocanian A *stage of the Lower *Proterozoic, from about 2600 to 2000 Ma ago (Van Eysinga, 1975), overlain by the *Ulcanian.

Ufian *See* UFIMIAN.

Ufimian (Ufian) 1. The earliest *age in the Late *Permian *Epoch, preceded by the *Kungurian (Early Permian), followed by the *Kazanian, and with a lower boundary dated at 256.1 Ma (Harland et al., 1989). **2.** The name of the corresponding eastern European *stage, which is roughly contemporaneous with the lower Zechstein (western Europe), lower *Guadalupian (N. America), and Basleoan (New Zealand). *See also* KAZANIAN.

Uivakian orogeny A phase of mountain building that occurred about 3000 Ma ago, producing what is now an approximately N–S belt in north-eastern Labrador, Canada.

Ulatizan A *stage in the Lower *Tertiary of the west coast of N. America, underlain by the *Penutian, overlain by the *Narizian, and roughly contemporaneous with the upper *Ypresian and lower *Lutetian Stages.

Ulcanian. A *stage of the Lower *Proterozoic, from about 2000 to 1600 Ma ago, underlain by the *Udocanian and overlain by the *Burzyan (Burzyanian).

ulexite *See* BORAX.

ulna In tetrapods (*Tetrapoda), the post-axial bone of the fore limb.

Ulsterian The basal *series in the *Devonian of N. America, underlain by rocks of the *Silurian *System and overlain by the *Erian Series. It comprises the Helderbergian, Deerparkian (Oriskanyan), and Onesquethawian (Onondagan) *Stages, and is roughly contemporaneous with the Lower Devonian Series and *Eifelian Stage (Middle Devonian) of Europe.

ultimate strength (failure strength) The maximum *stress that can be maintained in a body prior to its rupture. It is marked by the highest point on a curve in a *stress–strain diagram.

Ultisols Mineral soils, an order identified by an *argillic B *soil horizon with a *base saturation of less than 35%, and red in colour from iron oxide concentration. Ultisols are *leached, *acid soils, associated with humid subtropical environments. *See also* RED PODZOLIC SOILS.

ultrabasic rock An *igneous rock that consists almost entirely of *ferromagnesian minerals and possesses no free *quartz, and with less than 45% *silica (SiO_2). 'Ultramafic' is a partial synonym.

ultramafic *See* ULTRABASIC.

ultramylonite *See* DYNAMIC METAMORPHISM.

ultraplinian eruption The most extreme type of *Plinian eruption, in which the column of ejecta reaches a height of more than 45 km. Such an eruption, with a column up to 50 km high, occurred near Taupo, North Island, New Zealand, in about AD 186.

ultraviolet radiation *Electromagnetic radiation which has a wavelength between 0.5 nm and 400 nm, located between the visible and X-ray regions of the electromagnetic spectrum. Near ultraviolet occurs at wavelengths between 400 nm and 300 nm, middle ultraviolet between 300 nm and 200 nm, and extreme ultraviolet between 200 nm and 150 nm.

ultraviolet spectrometer (ultraviolet spectrophotometer, UVS) A *spectrometer, used in remote sensing, that measures spectra in the UV waveband from which information can be derived concerning planetary atmosphere and surfaces.

ultraviolet spectrophotometer *See* ULTRAVIOLET SPECTROMETER.

ultraviolet–visual spectrophotometry (UV–Vis spectrophotometry) Method of spectrophotometric analysis wherein the absorbance (or transmission) of a coloured complex is measured at a specific wavelength in the ultraviolet or visible part of the electromagnetic *spectrum. Any metal or *ion which can be made to form a soluble, coloured complex with appropriate absorption maxima may be quantitatively determined by comparing the absorption of the unknown material with that of standards of appropriate *concentration.

Ulysses A *NASA and *ESA probe, launched in 1990, carrying the International Solar Polar Mission to fly over the poles of the Sun, which cannot be seen from Earth, and to study the *solar wind. It travelled first to *Jupiter for a gravity assist.

umber A *mudstone rich in iron and manganese oxides, with *silica, aluminium oxide, and *lime, that is the source of a pigment of the same name. *Compare* OCHRE.

umbilicus The central cavity of a gastropod (*Gastropoda) shell where the later *whorls do not meet centrally; it is sometimes seen basally as an opening, but is also commonly closed by shell material. In cephalopods (*Cephalopoda), an open space that remains in the axis of coiling when successive whorls do not reach the axis.

umbric epipedon Surface *soil horizon similar to a *mollic epipedon but with a *base saturation of less than 50%. It is a *diagnostic horizon.

Umbriel (Uranus II) One of the major satellites of *Uranus. Its mean radius is 584.7 km; mass 11.72×10^{20} kg; mean density 1400 kg/m^3; albedo 0.18. The surface is evenly cratered and darker than the other four major satellites, except for a bright ring 4 km in diameter. The many large craters suggest the surface is old.

umbo (pl. umbones) The first part of a brachiopod (*Brachiopoda) or bivalve (*Bivalvia) shell to be formed. In a brachiopod, the umbo is the posterior part of each *valve; in a bivalve it forms the *dorsal part of the shell.

umbones *See* UMBO.

unavailable water Water that is present in the soil but that cannot be absorbed by plants rapidly enough for their needs because it is held so strongly to the surface of soil particles.

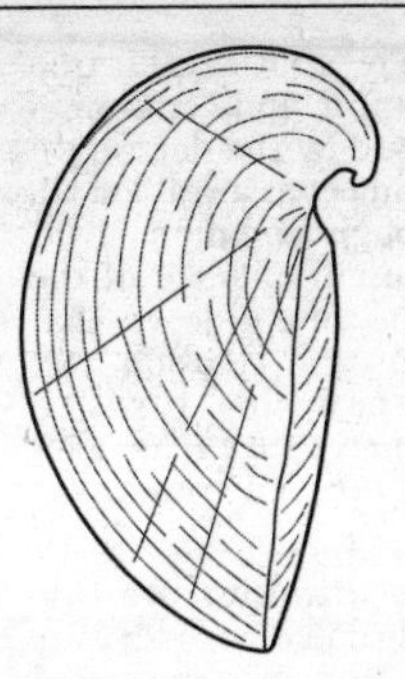

Umbo

uncinus From the Latin *uncinus* meaning 'hooked', a species of *cirrus cloud that is hooked at the end of its filaments and on the upper parts. *See also* CLOUD CLASSIFICATION.

unconfined aquifer *See* AQUIFER.

unconfined compressive strength (uniaxial compressive strength) The strength of a *rock or *soil sample when crushed in one direction (uniaxial) without lateral restraint. *See also* UNIAXIAL COMPRESSION TEST.

unconformity Surface of contact between two groups of unconformable *strata, which represents a hiatus in the geologic record due to a combination of *erosion and a cessation of sedimentation. *Compare* diastem. *See also* ANGULAR UNCONFORMITY; and DISCONFORMITY.

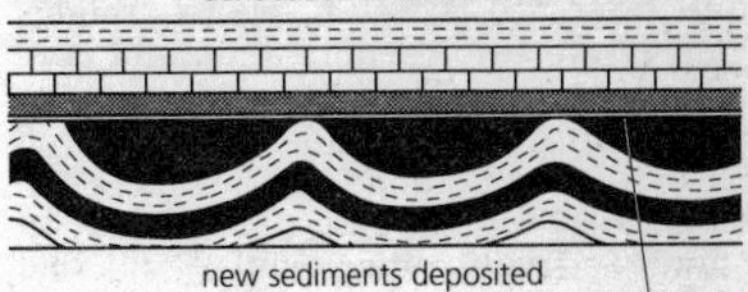

Unconformity

unconformity trap A *stratigraphic trap formed by *folding, uplift, and *erosion of porous *strata, followed by the deposition of later beds which can act as a seal for oil, gas, or water. Although common structures, these traps contain only 4% of the world's oil, perhaps because of losses that occur during uplift and erosion. *See* NATURAL GAS; PETROLEUM; and POROSITY. *Compare* ANTICLINAL TRAP; FAULT TRAP; REEF TRAP; STRATIGRAPHIC TRAP; and STRUCTURAL TRAP.

unconformity-type uranium ore Uranium *ore that is associated with an *unconformity. Uranium is often concentrated in *sedimentary rocks or precipitated from *groundwater, because uranium minerals are readily decomposed under oxidizing conditions. These deposits are all of early *Proterozoic age. In Ontario, uranium is concentrated in the basal bed of the Lorrain Quartzite, which is unconformable on the *Archaean granite, the mineralization being about 1 m thick and covering hundreds of square kilometres.

unconsolidated Applied to particles that are loose and not cemented together.

undaform The approximately level surface beneath a water body, at a level higher than that of the wave bases so its sediments are stirred by wave action and currents, that gives a seismic reflection.

undathem The rock unit produced beneath a water body at a depth higher than the bases of waves, that gives a seismic reflection.

undepleted mantle Primary *mantle material from which *basalt has not been extracted. *Compare* DEPLETED MANTLE.

undercliff A stretch of land that lies parallel with and below a major cliff. A good example is the Undercliff of the Isle of Wight, Britain, between Bonchurch and Blackgang, which is overlooked by a conspicuous cliff in Upper Greensand rocks.

undercooling The state whereby a liquid must be cooled to well below its *solidus temperature before *nucleation and crystallization are initiated. *See* SUPERCOOLING.

underfit stream *See* MISFIT STREAM.

underflow The flow of *groundwater in *alluvial *sediments, parallel to and beneath a river channel. It forms a significant fraction of the total river flow in coarse gravel alluvium.

underplating The addition of material to the underside of a geologic unit, as in the movement of *crust by *A-subduction in a *collision zone so that it underlies the overriding continent. The term has also been used to describe the addition of gabbroic *plutons beneath *lavas and *dykes in a *constructive margin.

undersaturated *See* SILICA SATURATION.

underthrusting The movement of rock such that the lower block moves under a relatively passive upper block, with the plane of contact being gently inclined. Underthrusting is thought to be of major importance in the formation of *accretionary wedges.

undertow The general seaward flow of water beneath individual breaking waves, in contrast to the more localized *rip-current return flow.

undrained test *See* TRIAXIAL COMPRESSION TEST.

undulatus From the Latin *undulatus* meaning 'waved', a variety of cloud whose layers undulate. *See also* CLOUD CLASSIFICATION.

undulose extinction In optical *mineralogy, the irregular or wavy *extinction seen in *thin sections of *mineral grains (e.g. *quartz) on rotation of the *stage. Different parts of the grain go into extinction in different orientations. The phenomenon is thought to result from strain, which may modify crystal orientation slightly.

ungulate **1.** Any hoofed, grazing mammal, which usually is also adapted for running. Hoofed mammals occur in several mammalian groups (*Mammalia), and the term 'ungulate' no longer has any formal taxonomic use. **2.** Hoof-shaped.

uniaxial compression test Test of a *soil or *rock sample for *unconfined compressive strength. An undisturbed sample is loaded from above until it fails. The results depend on the length-to-breadth ratio of the sample and the loading rate.

uniaxial compressive strength *See* UNCONFINED COMPRESSIVE STRENGTH.

uniaxial interference figure *See* INTERFERENCE FIGURE.

Unibothriocidaris *See* BOTHRIOCIDAROIDA.

unicarinate Possessing a single carina (*keel). It may occur as a structure along the *venter of a cephalopod (*Cephalopoda), round each *whorl of a gastropod (*Gastropoda), and may also occur in other groups.

Unified Stratigraphic Time-scale (UTS) A proposed universal time-scale that would combine both absolute and relative time-scales (*see* GEOCHRONOLOGY) that have been deduced by all possible means, including *biostratigraphy, *chronostratigraphy, *stratigraphy, *magnetostratigraphy, and *radiometric dating.

uniform flow 1. In channels, a flow whose velocity and *discharge do not vary along the channel's length. **2.** In *groundwater, a flow whose velocity and direction are the same at all points in the field of flow. Non-uniform flow in channels results from changes in the cross-sectional shape or discharge along the channel's course.

uniformitarianism (actualism) The principle proposed by James *Hutton and paraphrased succinctly as 'the present is the key to the past'. This is a considerable oversimplification, since processes that occurred in historical times may not be occurring now, or may not be observable now, and vice versa.

unilocular Applied to the condition in which the *test of a foraminiferid (*Foraminiferida) consists of a single chamber. Where the test consists of two or more chambers it is said to be 'multilocular'.

uniramous Of an appendage, unbranched.

uniserial 1. Arranged in a single row. **2.** *See* TRISERIAL.

unit cell *See* BRAVAIS LATTICE.

United States Department of Agriculture *See* USDA.

unit form *See* PARAMETER; and PARAMETRAL PLANE.

unit hydrograph *See* HYDROGRAPH.

unit-stratotype The standard succession of *strata, selected from the rock successions within a particular locality (the *type locality), that is designated and described as the standard type section for a particular *chronostratigraphic or *lithostratigraphic unit. The unit-stratotype provides the reference against which all other examples of that stratigraphic unit anywhere in the world may be examined and identified. *See* STRATOTYPE.

unit stress The amount of force applied to a unit area. The term is most commonly used in engineering geology.

univariant assemblage An equilibrium mineral assemblage which, in terms of the *phase rule, has one *degree of freedom; i.e. if equilibrium is to be maintained, a change of any one variable (e.g. pressure) dictates the changes to the values of all the other external variables (temperature and $P(H_2O)$).

unloading joint *See* JOINT.

unsaturated zone *See* VADOSE ZONE; and SOIL-WATER ZONE.

unsorted biosparite *See* FOLK LIMESTONE CLASSIFICATION.

unsteady flow *See* STEADY FLOW.

uphole survey Technique in which seismic sources are energized within a *borehole and arrival times recorded by surface *geophones. It is used particularly to determine weathered-layer velocity.

uphole time The time taken for the first direct wave to travel from the *shot fired at the base of a shot-hole to a *geophone at or near the top of the shot-hole.

upright fold A *fold whose *axial plane or surface is vertical or nearly so.

upslope fog Condensation near the surface in uplifted air on windward slopes.

upthrust Applied to a block that has been moved upwards in a *reverse fault.

Upton Warren A warm *interstadial during the Middle *Devensian, between 42 000 and 43 000 years BP, when faunal and floral evidence suggest the landscape was devoid of trees.

upward continuation The use of one set of measurements of a potential field (usually gravity or magnetic) over one surface to determine the field at a higher surface. Upward continuation is relatively reliable, in that the field is continued into free space where there are no causative bodies to perturb the field further (unlike *downward continuation). The method effectively attenuates high-*wavenumber anomalies due to near-surface features, thus providing a powerful method for examining deeper structures. Upward continuation is used in gravity surveys to determine the nature of the regional gravity pattern over a large area; in magnetic surveys it is particularly useful in tying together aeromagnetic surveys made from different flying altitudes.

upwelling In oceans or larger lakes, a water current, or movement of surface water produced by wind, which brings colder water, loaded with nutrient, to the surface from a lower depth. Ocean upwellings occur off Peru, California, W. Africa, and Namibia, and increase the nutrient content of the surface waters, leading to an abundance of marine and bird life. Upwelling also occurs in the open oceans where surface currents diverge, as deep waters rise to the surface to replace the departing waters, and all along the equator as a result of the effects of the NE and SE *trade winds.

Uralian orogeny A phase of mountain building, affecting what is now a N–S belt in Russia, that occurred at the same time as the *Hercynian orogeny, in the late *Devonian and early *Carboniferous.

uralite *See* URALITIZATION.

uralitization The alteration of *primary *igneous *pyroxene to a fibrous mass of *amphibole, usually *hornblende, during late-stage *hydrothermal activity, or during low-grade *metamorphism of the igneous rock. The amphibole was originally thought to be a distinct *mineral type known as 'uralite'.

Ural Sea (Obik Sea) A *Palaeocene–*Eocene seaway extending from the present-day Caspian Sea in the south to the Arctic Ocean in the north, covering the area immediately east of the Ural Mountains. The area is now covered by the extensive plain across which flows the River Ob.

uranian satellites *See* ARIEL (URANUS I); BELINDA (URANUS XIV); BIANCA (URANUS VIII); CORDELIA (URANUS VI); CRESSIDA (URANUS IX); DESDEMONA (URANUS X); JULIET (URANUS XI); MIRANDA (URANUS V); OBERON (URANUS IV); OPHELIA (URANUS VII); PORTIA (URANUS XII); PUCK (URANUS XV); ROSALIND (URANUS XIII); TITANIA (URANUS III); AND UMBRIEL (URANUS II).

uraninite (pitchblende) Mineral, UO_2; sp. gr. from 8.5 when *massive to 10 for unaltered crystals; *hardness 5–6; *cubic; it is brownish-black or grey-black; brownish-black *streak; sub-metallic, greasy *lustre; crystals very rare, the mineral is usually found as *botryoidal masses; no *cleavage; radioactive; occurs in *pegmatites, and in association with *monazite, *zircon, and *tourmaline, in *hydrothermal veins together with *cassiterite, *pyrite, *chalcopyrite, and *galena, and as a *detrital material in *alluvials. It is a major *ore mineral for uranium metal.

uranium deposit *Ore deposit containing more than 350 p.p.m. uranium. Six types are recognized: pegmatitic (*see* PEGMATITE) together with *disseminated magmatic deposits; hydrothermal veins and *stockworks (*see* HYDROTHERMAL ACTIVITY; and HYDROTHERMAL MINERAL); vein-unconformity deposits (*see* VEIN DEPOSIT); *sandstone deposits; *placer deposits; and phosphatic *limestones and black-*shale deposits.

uranium–lead dating All naturally occurring uranium contains ^{238}U and ^{235}U (in the ratio 137.7 : 1). Both *isotopes are the starting points for complex *decay series that eventually produce stable isotopes of lead. ^{238}U decays to ^{206}Pb (half-life = 4510 Ma, *see* DECAY CONSTANT) by a process of eight *alpha-decay steps and six *beta-decay steps. ^{235}U decays to ^{207}Pb (half-life = 713 Ma) by a similar series of stages that involves seven alpha-decay steps and four beta-decay steps. Also included within this range of methods is that for *thorium–lead dating (^{232}Th to ^{207}Pb; half-life = 13 900 Ma). Uranium–lead dating was applied initially to uranium minerals, e.g. *uraninite and *pitchblende, but as these are rather restricted in occurrence it is more normal to use the mineral *zircon, even though the uranium is present only in trace amounts. The amount of radiogenic lead from all these methods must be distinguished from naturally occurring lead, and this is calculated by using the ratio with ^{204}Pb, which is a stable isotope of the element then, after correcting for original lead, if the mineral has remained in a closed system, the ^{235}U : ^{207}Pb and ^{238}U : ^{206}Pb ages should agree. If this is the case, they are *concordant and the age determined is most probably the actual age of the specimen. These ratios can be plotted to produce a curve, the Concordia curve (*see* CONCORDIA DIAGRAM). If the ages determined using these two methods do not agree, then they do not fall on this curve and are therefore *discordant. This commonly occurs if the system has been heated or otherwise disturbed, causing a loss of some of the lead daughter atoms. Because ^{207}Pb and ^{206}Pb are chemically identical, they are usually lost in the same proportions. The plot of the ratios will then produce a straight line below the Concordia curve. G. W. Wetherill has shown that the two points on the Concordia curve intersected by this straight line will represent the time of initial crystallization and the time of the subsequent lead loss.

uranium series *See* DECAY SERIES.

Uranus The seventh planet in the *solar system, discovered in 1781 by Sir William Herschel, although he described it as a comet. It was named Uranus by J. E. *Bode. Its equatorial radius is 25 559 km and polar radius 24 973 km; volume 6833 km^3; mass 86.83×10^{24} kg; mean density 1318 kg/m^3; visual albedo 0.51; black-body temperature 35.9 K. The inclination of the equator to the plane of the *ecliptic is 97.86°, so the planet is lying on its side (a fact discovered in 1846 by J. Galle). At its closest approach, Uranus is 2581.9×10^6 km from Earth and at its furthest 3157.3×10^6 km. Uranus has an atmosphere, with a surface atmospheric pressure well in excess of 100 bar. The atmosphere is composed of molecular hydrogen (89%) and helium (11%), with *aerosols of methane, ammonia ice, water ice, ammonia hydrosulphide, and possibly methane ice (similar to that of *Neptune). Wind speeds at the surface are 0–200 m/s and the average surface temperature is about 58 K. Two new satellites, so far unnamed, were discovered in 1997 (and designated S/1997U1 and S/1997U2), moving in eccentric orbits at a mean distance of 5.8×10^6 km from the planet (227 Uranian radii), and are estimated to have diameters of 60 km and 80 km, bringing the total number of known satellites to 17 (*see* URANIAN SATELLITES) and it is likely that more remain to be discovered. Except for Titan, the uranian satellites are denser than those of *Jupiter. Oberon and Titania, the two largest, were discovered in 1787 by Sir William Herschel. Umbriel and Ariel were discovered in 1851 by William Lassell. Miranda was discovered in 1948 by Gerard *Kuiper. Ariel, Oberon, and Titania are probably made of water ice, other ices, and silicates. They are believed to be too cold to have a molten core, but on some there are signs of geological activity. The remaining 10 satellites were revealed in images transmitted to Earth from *Voyager 2.

urban climate Modified surface-layer atmospheric conditions caused by the influence of large 'built-up' areas. Changes include pollution, reduction in strong wind speeds toward a city centre, *turbulence of air around buildings, warming of air by the heat output from city structures, and increased evaporation and removal (drainage and runoff) of water.

Urey, Harold Clayton (1893–1981) An American professor of chemistry from California, Urey is best known as the discoverer of deuterium, but he also did important work on the origin and evolution of the universe and of life. In the field of Earth science he developed the use of *oxygen-isotope analysis, used to indicate ancient climatic temperatures.

Uriconian A *stage of the Upper *Proterozoic of Shropshire, England, underlain by the *Malvernian and overlain by the *Charnian.

Urochordata (tunicates; phylum *Chordata) Subphylum comprising the sea-squirts, with reliable *fossils from the *Permian, and possible remains in *Silurian *sediments. Sea-squirts have tadpole-like larvae with *notochords in their tails, a feature that links them with the *chordates.

Ursidae *See* CARNIVORA.

urtite A coarse-grained, *igneous rock consisting of *essential *nepheline (about 85% of the rock) and *ferromagnesian minerals (*aegirine, aegirine-*augite, and soda-iron *amphibole). The rock is named from Lujaur-Urt, in the Kola Peninsula, Russia, and is a type of *undersaturated *syenite.

Ururoan *See* HERANGI.

Urutawan *See* CLARENCE.

USDA (United States Department of Agriculture) The department of the United States federal government that exists to serve the needs of those engaged in agriculture and rural communities. It comprises a number of agencies, each with its own functions, one of which is the US Soil Survey. In 1960, the USDA published *Soil Classification: A Comprehensive System*, prepared by the Soil Survey. This method of classifying soils hierarchically was renamed the US Soil Taxonomy in 1970, was subsequently adopted by the Food and Agriculture Organization of the United Nations, and is now the most widely used taxonomic system. *See* SOIL TAXONOMY.

UTS *See* UNIFIED STRATIGRAPHIC TIME-SCALE.

UV Abbreviation for ultraviolet.

uvala An irregularly shaped hollow in a *karst terrain. It is generally 500–1000 m in diameter and may be 100–200 m deep. It is the result of the coalescence of a number of *dolines.

uvarovite Member of the *garnet group $Ca_3Cr_2Si_3O_{12}$; sp. gr. 3.9; *hardness 7.0–7.5; emerald green; occurs as *cubic crystals and *massive; found in *serpentinite in association with *chromite, and in metamorphosed *limestones and skarn ores.

UVS *See* ULTRAVIOLET SPECTROMETER.

UV–Vis spectrophotometry *See* ULTRAVIOLET–VISUAL SPECTROPHOTOMETRY.

[UVW] *See* ZONE SYMBOL.

V The *vertical component of a magnetic vector.

vadose zone *See* PHREATIC ZONE; and SOIL-WATER ZONE.

Valanginian A *stage in the European Lower *Cretaceous (140.7–135 Ma, Harland et al. 1989) for which the *type locality is at Valangin, Switzerland. *See also* NEOCOMIAN.

Valdayan/Zyryanka A *Weichselian, *loess-like *silt from Siberia, in which there is a threefold division with the Zyryanka *drift at the base, then Karginsky *interstadial *sediments, and finally the Sartan drift.

valency The bonding potential of an atom, measured by the number of hydrogen *ions (valency 1) that the atom could combine with or replace. In an ionic compound the valency (electrovalency) equals the ionic charge on each ion, e.g. in the compound MgO, Mg^{2+} shows a valency of +2, O^{2-} a valency of –2. In a covalent compound the valency (covalency) of an atom is equal to the number of bonds it forms, e.g. in CH_4 carbon has a valency of 4, hydrogen a valency of 1.

valley bulging An upward arching of the bedrock along the axis of a valley. It may not be visible at the ground surface due to subsequent *erosion, but it is revealed by the distortion of the geologic structure. It may be due to *frost heave, or to the compressive forces set up when two opposing valley sides approach each other.

valley glacier A long, relatively narrow ribbon of ice that is confined between valley walls. The Alpine type is fed by a series of *cirque *glaciers that show positive net balances (*see* MASS BALANCE), and is common in the Alps and in the coastal mountains of Alaska, USA. The outlet type is fed by an *ice cap or *ice sheet. The Vatnajökull ice-cap in Iceland feeds several outlet glaciers.

valley side bench (rock bench) A terrace-like land-form standing on the flank of a valley, but lacking a veneer of *alluvium. It may have originated as a true *river terrace from which the alluvium has been stripped by subsequent *erosion, or it may simply result from the exposure of a nearly level *stratum of resistant rock.

valley train An accumulation of fluvioglacial deposits laid down in a valley by meltwaters escaping from a decaying *glacier. The surface slopes quite steeply down-valley, and is incised by shifting *braided streams.

valley wind An *anabatic wind that blows up-valley during the day in otherwise calm conditions, or a *katabatic, down-valley, night wind. *See also* MOUNTAIN WIND; and RAVINE WIND.

vallis A general term for a valley on a planetary or *satellite surface. The largest example is the Valles Marineris system on *Mars. Mostly the term is confined to smaller, sinuous valleys probably of fluvial origin.

valve 1. One of the two halves of the hinged shell of brachiopods (*Brachiopoda), or molluscs (*Mollusca) of the class *Bivalvia. **2.** One half of the cell wall of a *diatom. **3.** A flap or other constriction that can close to ensure that a fluid flows in only one direction.

vanadinite Phosphate mineral $Pb_5(VO_4)_3Cl$ and member of the pyromorphite series with *pyromorphite ($Pb_5(PO_4)_3Cl$) and mimetite ($Pb_5(AsO_4)_3Cl$); sp. gr. 6.88–6.93; *hardness 2.5–3.0; *hexagonal; orange-red, brownish-red, to yellow; white to yellowish *streak; resinous *lustre; crystals sharp, hexagonal *prisms, but occasionally rounded; no *cleavage; found in the oxidized zone of sulphide-ore deposits carrying lead minerals. It can be used as a source of vanadium.

Van Allen belt Two belts of high-energy, charged particles trapped by the *Earth's magnetic field within the *magnetosphere. The inner belt starts at an altitude of about 800 km and reaches a maximum intensity at about 2000 km. The outer belt reaches a maximum intensity at between three and four Earth radii (18 000–25 000 km). Although commonly referred to as 'radiation belts', they contain particles, not radiation. Their presence was predicted by J. A. Van Allen (1914–).

van der Waals force The weak attraction of atoms to each other due to the interaction of *electrons and *nuclei. The resultant linkage is called the van der Waals bond (named after the Dutch physicist Johannes van der Waals, 1837–1923).

vane A device for indicating the direction of wind. The standard exposure for wind vanes is on a mast at 10 m above unobstructed ground.

vane test Test of the *shearing strength of *soil *in situ*. The equipment consists of four thin, rectangular blades which project from a shaft at right angles. This is pressed into the ground and rotated at a uniform rate. The torque required to generate a cylinder of soil determines the shearing strength of the soil.

van't Hoff, Jacobus Henricus (1852–1911) A Dutch chemist, and winner of the 1901 Nobel Prize for chemistry for his work on the relationship between osmotic pressure and vapour pressure, van't Hoff was professor of chemistry, mineralogy, and geology at Amsterdam University and later an honorary professor at the Prussian Academy of Sciences. His work on phase equilibria (*see* PHASE DIAGRAM), involving six components of sea water at a range of temperatures, contributed much to the study of sedimentary processes and in particular provided a theoretical basis for understanding the formation of salt deposits.

vapour-phase crystallization The crystallization of minerals from hot gases escaping through a volcanic body. Cooling of the escaping gases, which carry elements in solution, promotes the crystallization of minerals in rock cavities or the spaces between *pyroclastic *clasts in *ignimbrites. Minerals such as *feldspar and *quartz are common vapour-phase crystallization products.

vapour pressure The pressure exerted by molecules of a substance in the vapour state, at equilibrium with molecules of the same substance in the liquid state, within a closed container. The magnitude of the vapour pressure exerted depends on the temperature and the identity of the liquid; it does not depend on the amount of liquid in the container. The saturated vapour pressure of water at 0 °C is 610 N m^{-2}, rising to 2340 N m^{-2} at 20 °C and 7380 N m^{-2} at 40 °C. *See also* PARTIAL PRESSURE.

vapour-pressure curve *See* DEHYDRATION CURVE.

VAR *See* VARIABLE-AREA DISPLAY.

Varanger A *series of the Upper *Proterozoic (*Vendian), from about 610 to 590 Ma ago.

vardar (vardarac) A type of *ravine wind, which blows in the Moravia–Vardar valley, bringing cold conditions from the north to the Thessaloniki area of Greece.

Varegian A *stage of the Upper *Proterozoic, from about 650 to 580 Ma ago, underlain by the *Jotnian and overlain by the *Caerfai (*Cambrian).

variable-area display (VAR) A way of displaying the wave-form on a *seismic record, in which the positive component of the wave is shown blacked in, and the negative-polarity component is absent, unless the VAR display is in the *wiggle-trace mode.

Variable-area display

variance (mean square) In statistics, a measure of the spread of the data about the mean. In a set of data, the square of the mean variation about the mean value. It is calculated as the mean of the squared data points minus the mean squared; the standard deviation of a data set is the square root of the variance: $s^2 = (\Sigma_{i=1}^{i=n}(x_i - \bar{x})^2)/n$, where s^2 is the variance, x_i is the mean value of the ith measurement of x, $\bar{x}$ is the mean value of x, and n is the number of measurements.

variolitic Applied to a *spherulitic *texture consisting of fine, radiating fibres of *plagioclase or *pyroxene that is found in the glassy, chilled margin of shallow-level, *basic, *igneous *intrusions (*dykes and *sills) or the glassy *groundmass of some quenched (*see* QUENCHING) basaltic *lavas.

varve A banded layer of *silt and *sand deposited annually in lakes, especially near to *ice sheets. The coarse, paler material is deposited in summer; the finer, darker material in winter. One varve consists of one light band and one dark band. Varves can be counted to calculate the age of glacial deposits (varve analysis, also called varve chronology or varve count). Since the pattern of thicknesses of successive varves is often distinctive, correlations can be made between widely separated deposits, using the same principle as that of *dendrochronology.

varve analysis *See* VARVE.

varve chronology *See* VARVE

varve count *See* VARVE

vascular plants *See* TRACHEOPHYTA.

veering A clockwise shift in the direction of the wind. The reverse change is called backing.

Vega Two Soviet missions to *Venus and comet *Halley, both of which were launched in 1984.

vegetation index In *remote sensing, a technique to show vegetation in an image. An example is the normalized vegetation index, which calculates how far *pixels in an image plot from the *soil line by dividing the very-*near infrared *digital number value by the red digital number value of a multispectral data set.

vein A tabular deposit of minerals occupying a fracture, in which particles may grow away from the walls toward the middle.

vein deposit An *ore body that is confined within a sheet-like structure, as a result of magmatic activity or deposition from circulating *groundwater. The deposit is usually narrow but persistent. It may pinch and swell and has well-defined walls.

velocity–depth distribution The variation that occurs in *seismic velocity with increasing depth, from the surface of the Earth to the *core; this reveals marked discontinuities and boundaries (such as the *Mohorovičić discontinuity, *low-velocity zone, and core–*mantle boundary).

velocity log The record of a *sonic sonde while it is being raised through a *borehole.

velocity profile The variation of water velocity with vertical distance from the bed of a river, or of wind velocity with distance from the ground. *See* von Karmann–Prandtl equation.

velocity survey A series of measurements designed to provide information about the variation and distribution of *seismic velocities throughout a series of media, as a function of depth and horizontal distance. A number of velocity analyses along a *seismic section can be displayed to produce an isovelocity plot in which all points of equal velocity are linked by contours. This reveals the velocity structure and any anomalies in the plane of the section, and provides a useful aid to geologic interpretation.

velum An accessory cloud feature of *cumulus or *cumulonimbus, characterized by a widespread veil on or above the upper surface of the cloud. *See also* CLOUD CLASSIFICATION.

vendavale A strong, local, south-westerly wind affecting the Straits of Gibraltar, associated with a *depression, and bringing *squalls and heavy rainfall.

Vendian The ultimate *system of the *Proterozoic, from about 650 Ma ago to the basal *Cambrian (Harland et al., 1989), that includes the *Ediacaran and *Varanger.

Vendobionta A group of fossils of *Ediacaran age that are believed to comprise a *monophyletic group. Adolf Seilacher proposed in 1984 that they be ranked as a kingdom; in 1994 Leo W. Buss and Seilacher proposed they be ranked as a phylum ancestral to the *Cnidaria and lacking nematocysts (cnidae).

Venera A series of Soviet missions to *Venus that ran from 1967 to 1983.

Venice system A system for the classification of brackish water based on the percentage of chloride contained in the water. *See* HALINITY.

Vening-Meinesz, Felix Andries (1887–1966) Professor of *geophysics in the Netherlands, Vening-Meinesz made important contributions to the study of gravity, and was an early advocate of *continental drift. In 1926 he discovered large negative *gravity anomalies over the *Java Trench, and, using *earthquake data to support his theory, suggested that the trenches were the result of buckling of the *crust, caused by convection currents in the Earth's interior. Vening-Meinesz's use of submarines to improve the precision of his gravity measurements was an imaginative and effective innovation.

vent breccia *See* BRECCIA. *Compare* AGGLOMERATE.

vent conglomerate Rounded blocks of juvenile and accessory material infilling a volcanic vent. The blocks are derived from the collapse of the vent and surrounding cone walls during and after explosive *eruptions. The *clasts may be rounded by attrition with other blocks in the turbulent vent conditions prevailing during eruptive activity.

venter The *ventral side of a cephalopod (*Cephalopoda) where the cephalopod is coiled in a plane spiral. It is the periphery or circumference of the shell.

Ventersdorp *See* RANDIAN.

ventifact A pebble that has been faceted by the abrasive action of wind-blown sand and dust. Multiple faceting may reflect pebble movement as underlying sand is disturbed, rather than changes in wind direction. Each facet faces the direction of the abrading wind. A pebble with three facets (triangular facet) is called a 'dreikanter'; one with one facet an 'einkanter'.

ventral Towards the lower or underside of an organism (in a vertebrate the side of the animal furthest from the spine); the opposite of *dorsal.

Venturian The final *stage in the Upper *Tertiary of the west coast of N. America, underlain by the *Repettian, overlain by the *Wheelerian (*Quaternary), and roughly contemporaneous with the uppermost *Piacenzian Stage.

Venus The second planet in the *solar system, orbiting 0.72 AU from the *Sun and $38.2–261.0 \times 10^6$ km from Earth. Its radius is 6052 km; mass 4.869×10^{24} kg; mean density 5204 kg/m^3; surface gravity 8.87 (Earth = 1); visual albedo 0.65; black-body temperature 238.9 K. Venus has a dense atmosphere, with a surface pressure of 92 bars, composed of carbon dioxide (96.5%) and nitrogen (3.5%), with about 150 ppm sulphur dioxide, 70 ppm argon, 20 ppm water, 17 ppm carbon monoxide, 12 parts per million helium, and 7 ppm neon. The surface temperature is 737 K and wind speeds range from 0.3 m/s to 1.0 m/s. Its orbit is *retrograde.

venusian Of, pertaining to, or characteristic of, the planet *Venus. *See also* CYTHEREAN.

veranillo In S. America, the short period of finer weather that occurs during the summer wet season.

verano In tropical America, the drought season that occurs in winter.

Vereiskian A *stage in the *Moscovian Epoch, preceded by the Melekesskian (*Bashkirian Epoch) and followed by the *Kashirskian.

vermiculite Member of the 2 : 1 *phyllosilicates (sheet silicates) with composition $(Mg,Ca)(Mg,Fe^{2+})_5(Al,Fe^{3+})[(Si,Al)_8O_{20}](OH)_4.8H_2O$ and closely related to the *smectite *clay minerals; sp. gr. about 2.3; *hardness about 1.5; *monoclinic; yellow or brown; pearly *lustre; crystals flat and *platy; *cleavage perfect basal {001}; occurs as an *alteration product of *biotite, from the *hydrothermal alteration of biotite and *phlogopite, and from the alteration of *ultrabasic rocks; expands greatly on heating. It is used extensively for insulation and as a lubricant. *See also* CLAY MINERALS.

Verrucano A *Permian formation of *sandstones and *conglomerates with a purplish-red colour.

Versilian *See* QUATERNARY.

vertebra In the axial skeleton of vertebrates, one of a series of bony segments which replace the notochord, forming the vertebral column (or spinal column or backbone), which encases and so protects the spinal cord. Vertebrae differentiate into five types from anterior to posterior: cervical; thoracic; lumbar; sacral; and caudal. Cervical vertebrae facilitate the mobility of the head. The first two vertebrae of the vertebral column, the atlas and axis, are highly specialized cervical vertebrae, the former articulating with the *occipital region of the *cranium. The thoracic vertebrae articulate with the ribs that fuse with the sternum. Lumbar vertebrae are generally larger, with abbreviated ribs fused to the centrum and supporting the posterior coelomic musculature. Sacral vertebrae fuse with the *pelvis, allowing the transfer of force to the appendicular skeleton. Caudal vertebrae are smaller and less specialized, forming the tail of the organism. Six anatomical features are usually recognizable in vertebrae: the centrum is a solid cylinder which surrounds and often replaces the notochord, forming the central body of the vertebra; the neural arch forms a dorsal ring surrounding the spinal cord; a hemal arch grows ventrally on post-anal vertebrae, enclosing blood vessels; neural and hemal spines are anterior–posterior-oriented blades of bone that project dorsally and ventrally respectively; apophyses are bilaterally paired projections to which musculature is usually attached, including prezygapophyses and postzygapophyses, which occur on the anterior and posterior ends of a vertebra respectively and articulate with zygapophyses of adjacent vertebrae; transverse processes are bilaterally paired lateral projections at each side of the neural arch with which the rib articulates.

vertebral column *See* VERTEBRA.

Vertebrata *See* CRANIATA.

vertebrates *See* CRANIATA.

vertebratus A variety of cloud, usually with *cirrus, with cloud elements characterized by a skeletal arrangement, in a form resembling vertebrae. *See also* CLOUD CLASSIFICATION.

vertical component (*V*) The intensity of the *geomagnetic field normal to the local horizontal plane.

vertical electrical sounding (VES) A method for producing information about the subsurface by measuring the *apparent resistivity at increasing *electrode spacings. *See* ELECTRICAL SOUNDING.

vertical fold *See* NEUTRAL FOLD.

vertical seismic profile (VSP) A form of *seismic record obtained by positioning a detector (*hydrophone) in a *borehole at a succession of depths and firing a *shot from a fixed point at the surface. *See also* WALK-AWAY VERTICAL SEISMIC PROFILE.

vertical stacking In seismic investigations, the technique in which the signal-to-*noise ratio is enhanced by summing the signals for a number of *shots into one set of *geophones or *hydrophones at one *offset. It is used extensively in land seismic-*refraction surveys, using signal-enhancement *seismographs which can sum up to 255 shots (e.g. hammer blows, weight-drops, etc.) to produce a good-quality *seismic record, but is of limited value in marine work, where *common-depth-point stacks are preferred.

Vertisols An order of mineral soils that contain more than 30% by weight of swelling clay (e.g. *montmorillonite), and that expand when wet and contract when dry to produce a self-inverting soil and an undulating (*gilgai) microrelief. Vertisols are associated with seasonally wet and dry environments, and are extensive in the tropics.

very-long-baseline interferometry (VLBI) A technique for determining, with precisions of centimetres, the distances between different radiotelescopes, using the *phase difference between radio signals detected by them that come from very distant radio sources (quasars). Some 135 different sources are routinely used.

very-low-frequency method (VLF method) An electromagnetic prospecting method in which permanent, high-power, military transmitters are used as sources of unmodulated carrier waves which induce secondary fields within subsurface conductors that may be many hundreds of kilometres from the source. Various orthogonal components of the secondary field are measured, as a function of the primary field, to obtain information about the nature and position of the subsurface conductors.

very-near infrared *See* NEAR-INFRARED.

VES *See* VERTICAL ELECTRICAL SOUNDING.

vesicle (adj. vesicular) A bubble-shaped cavity in *lava, formed by the expansion of entrapped gases. Such cavities may later become filled with material deposited from solution. Vesicular *basalt (bubbly basalt lava) is basaltic lava containing numerous openings, generally ellipsoidal or cylindrical in shape, formed by the expansion of dissolved gases in the molten rock. *See also* AMYGDALE; and SCORIACEOUS.

vesicular basalt *See* VESICLE.

Vesta The third largest (after *Ceres and *Pallas) *solar system asteroid (No. 4), diameter 526 km; approximate mass 3×10^{20} kg; rotational period 5.342 hours; orbital period 3.63 years. It was imaged in 1995 by the Hubble Space Telescope and appears to have a basaltic crust overlying an *olivine mantle, indicating that differentiation has occurred.

Vesuvian eruption (sub-Plinian eruption) A type of volcanic activity marked by very explosive *eruptions which occur after long periods of dormancy, during which gas pressures in the underlying *magma have built up sufficiently to eject the *plug of solid *lava from the vent. Escaping gases, exsolved in the magma, produce a mobile, frothy lava (*pumice) and clouds of *ash and gases are released into the air. *See* VOLCANO. *Compare* HAWAIIAN ERUPTION; PELÉEAN ERUPTION; PLINIAN ERUPTION; STROMBOLIAN ERUPTION; SURTSEYAN ERUPTION; and VULCANIAN ERUPTION.

vesuvianite *See* IDOCRASE.

VGP *See* VIRTUAL GEOMAGNETIC POLE.

VHN *See* VICKERS HARDNESS NUMBER.

vibration direction Light is an electromagnetic vibration, and can be likened to a transfer of energy by vibrating 'particles' along a path from the source to the receiver. In a ray of light of a single wavelength, the wave is generated by the vibration of 'particles' lying along the path of the ray. When, however, a ray of light passes through an *anisotropic mineral plate, the light is split into two unequal components which vibrate in mutually perpendicular directions, and which are related to the *refractive indices of the mineral in those directions.

vibration, ground Ground vibration due to seismic or other activity (e.g. industrial) may be transmitted to structures, the effects on them varying according to the amplitude of the ground motion, the rock or soil through which the vibration travels, and the length of time for which it persists. *See* VIBROSEIS.

Vibroseis The registered name (trademark) of a device which uses a truck-mounted vibrator plate coupled to the ground to generate a wave train up to seven seconds in duration and comprising a sweep of frequencies. The recorded data from an upsweep or downsweep (increasing or decreasing frequency respectively) are added together and compared with the source input signals to produce a conventional-looking *seismic section. The device is used increasingly in land surveys instead of explosive sources.

vicariance The geographical separation of a species so that two closely related species or a species pair result, one species being the geographical counterpart of the other.

Vickers hardness number (VHN) A quantitative measure of the *hardness of *ore minerals, that is used in their identification. A *diamond is indented into a mineral under a predetermined load (usually 100 g), for a given time (usually 15 s), and the cross-sectional area of the indentation is converted into a number, the Vickers hardness number.

Victoriapithecus The earliest known Old World monkey; together with a related, more poorly known genus, *Prohylobates*, it is now placed in a family, Victoriapithecidae, separate from advanced Old World monkeys. So far, the family is known only from E. and N. Africa. The *bilophodonty characteristic of Old World monkeys is only partially developed in Victoriapithecidae and in other respects, too, they are very primitive.

vidicon In *remote sensing, an imaging device which uses a transparent material whose electrical conductivity changes with varying incidence of *electromagnetic radiation. A sweeping electron beam is used to measure the varying conductivity which is then translated into an image.

Viking Two *NASA spacecraft, launched in 1975, that placed vehicles in orbit and landed instruments on the surface of *Mars, Viking 1 landing on 20 July and Viking 2 on 3 September. Both transmitted substantial amounts of data to Earth, as well as the first colour photographs taken from the martian surface.

Villafranchian A mammalian *age whose base is dated at approximately 3 Ma ago. It lasted an estimated 2 Ma and therefore transgresses the Late *Pliocene/Early *Pleistocene boundary.

virga (fall-stripes) From the Latin *virga* meaning 'rod', a supplementary cloud feature referring to the trails of *precipitation falling from the under surface of cloud but not reaching the ground. This phenomenon is often seen with *cumulus, *cumulonimbus, *altocumulus, *stratocumulus, *cirrocumulus, *nimbostratus, and *altostratus. *See also* CLOUD CLASSIFICATION.

virgella In a graptolite (*Graptolithina), a spine projecting beyond the aperture of the *sicula.

Virgellina *See* GRAPTOLOIDEA.

Virgilian The final *series in the *Pennsylvanian of N. America, underlain by the *Missourian, overlain by the *Wolfcampian (*Permian), and roughly contemporaneous with the upper Chamnovnicheskian and Dorogomilovskian Stages of the *Kasimovian Series plus the *Gzelian Series.

virtual geomagnetic pole The point on the Earth's surface at which a magnetic pole would be located if the observed direction of *remanence at a particular location was due to a magnetic dipole at the centre of the Earth.

viscosity The internal resistance of a substance to flow when a *shear stress is applied. Quantitatively defined, it is the ratio of the shear stress to the *strain rate, in units of pascal seconds (1 Pa s = 10 poise). Resistance to flow is caused essentially by molecular or ionic cohesion. In *magmas, molecular cohesion can be very high, especially if the *silica content is high as in *rhyolite magmas, and a yield strength must be overcome before the magma can flow. The presence of solid crystals increases the effective internal cohesion, and dissolved gas reduces it. In general, basaltic magmas have lower viscosities than rhyolite magmas.

viscous remanent magnetism (VRM) The magnetization acquired, at room temperature, by a sample lying for a period of time in a constant magnetic field. Usually it is exponentially dependent on the duration of the field. Most rock samples acquire this form of magnetization as they lie in the *geomagnetic field after their formation.

Visean 1. An *epoch in the *Mississippian, comprising the Chadian, Arundian, Holkerian, Asbian, and Brigantian *Ages (these are also *stage names in western European stratigraphy). The Visean is preceded by the *Tournaisian, followed by the *Serpukhovian, and dated at 349.5–332.9 Ma (Harland et al., 1989). **2.** The name of the corresponding European *series which has its lower boundary at the base of the Chadian Stage near Clitheroe, Lancashire, England. It is roughly contemporaneous with the upper Carboniferous Limestone Series (Britain), and the upper *Osagean, *Meramecian, and lower *Chesterian (N. America). *See also* DINANTIAN.

vishnevite *See* CANCRINITE.

visible radiation *Electromagnetic radiation which has a wavelength between 380 and 780 nm, visible to the human eye.

visual binary *See* STAR PAIR.

vitrain *See* COAL LITHOTYPE.

vitreous Of a *mineral *lustre, glassy.

vitrinite *See* COAL MACERAL.

vitrophyric Applied to an *igneous *texture in which *phenocrysts are embedded in a glassy *groundmass.

Viverridae *See* CARNIVORA.

vivianite Mineral, $Fe_3(PO_4)_2.8H_2O$; sp. gr. 2.9; *hardness 1.5–2.0; *monoclinic; greyish-blue, occasionally colourless; *vitreous *lustre; crystals *prismatic, rod-like, *acicular, and stellate masses; *cleavage, one, perfect, *pinacoidal; occurs in phosphorus-rich, sedimentary iron ores and *peat bogs, in association with *siderite, in a reducing environment with organic remains present, and in close association with *fossil bones and shells. It is used as a cheap blue pigment.

VLBI *See* VERY-LONG-BASELINE INTERFEROMETRY.

VLF method *See* VERY-LOW-FREQUENCY METHOD.

vogesite A type of *lamprophyre characterized by *essential *hornblende and *orthoclase feldspar.

Vogt, Johan Hermann Lie (1858–1932) A Norwegian geologist, Vogt was professor of metallurgy at the Technical University of Norway. He did important work on the chemistry of *silicates, and on differentiation in cooling *magmas. He also made studies of *ore geology, especially magmatic ores.

void A hole in a rock. If they are interconnected, voids form paths along which water and other fluids may flow. In increasing order of size, the major types of voids are: intercrystalline boundaries; intergranular *pores or spaces between the grains of a *sediment; microfractures or local cracks, usually extending for only a few tens of centimetres and from a few micrometres to 0.1 mm wide; *fractures including *joints, small *faults, and *bedding planes, which are often extensive and may have openings up to a few millimetres wide; fissures formed by solution, *weathering, or local gravitational or *tectonic displacement, and up to about 10 cm wide; and *solution channels, which range up to several metres wide and many hundreds of metres long.

void ratio Measurement of the porosity of rocks and soils. The void ratio (e) = V_v/V_s, where V_v is the volume of air in voids and V_s is the volume of solid particles.

volatile 1. Applied to a substance with a high *vapour pressure, which passes readily into a gaseous *phase. **2.** (*a*) A dissolved element in a silicate *magma which would be gaseous at that temperature except for the confining pressure and solvent nature of the magma (e.g. Cl, F, and S) and that therefore becomes gaseous when the magma reaches the Earth's surface or a zone of reduced pressure. Common volatiles include water vapour, carbon dioxide, sulphur dioxide, hydrochloric acid, and there are many more. The melting temperatures of late, volatile, saturated *melts may be about 600 °C. They form *pegmatites which have large crystals and contain some metals, e.g. lithium, molybdenum, uranium, and tin. (*b*) In *coal, a mixture of combustible gases (hydrogen, carbon monoxide, and methane) with other substances, which is given off when coal is heated without air being present. *Peat contains more than 50% volatiles, *lignites about 45%, *anthracite 10%, and *graphite less than 5%.

volcanic bomb A lump of ejected *lava, more than 32 mm across, that has acquired one of several characteristic shapes during its trajectory to the ground. For example, *bread-crust bombs have a glassy crust, criss-crossed with cracks; a spindle bomb has a tail formed by the twisting or spinning of the blob as it falls; rope or ribbon bombs are twisted strands of solidified lava; and cannonball bombs are formed from solid lumps of lava that have been rounded by abrasion as they bounced to rest.

volcanic cone A conical mound of volcanic ejecta accumulated around an eruptive vent. Cones have outer slope angles of about 30° and are topped by a depression or crater over the site of the vent. The type of material which accumulates to form the cone can be used to name the type of cone. For example, alternate layers of *lava with beds of *ash and other *pyroclastic material characterize a *strato-volcano (composite volcano), and *spatter ejected from a vent during a *Hawaiian-type eruption would accumulate to form a *spatter cone around the vent. *Scoria ejected from a vent during a *Strombolian-type eruption would accumulate to form a *scoria cone around the vent. *See* VOLCANO.

Volcanic Constructs *See* MARTIAN TERRAIN UNITS.

volcanic dome *See* DOME.

volcanic dust Dust, *ash, or other particulate matter commonly suspended in the atmosphere after volcanic *eruptions. After explosive eruptions the dust may be thrown to heights of 20–30 km or more. The fall-out times of dust particles are quite short, a matter of days or weeks, depending on altitude and *precipitation. Volcanogenic *aerosols, usually sulphates, may linger for months, spreading as a long-lived veil in the *stratosphere over much of the Earth.

volcanic-exhalative processes Processes associated with penecontemporaneous *volcanicity that produce sulphide *ore deposits, often lenticular in cross-section and commonly located above mineralized *stockworks (the probable channelways to the sea floor). The Kuroko-type copper deposits are typical of continental-arc volcanic-exhalative sulphides.

volcanic pile *See* CENTRAL VENT VOLCANO.

volcanicity (volcanism, vulcanicity, vulcanism) All the processes associated with the transfer of *magma and *volatiles from the interior of the *Earth to its surface. Current volcanicity is confined to regions of the Earth where lithospheric *plates converge, diverge, or pass over possible *mantle *hot-spots.

volcanism *See* VOLCANICITY.

volcanic neck *See* VOLCANIC PLUG.

Volcanic Plains *See* MARTIAN TERRAIN UNITS.

volcanic plug (volcanic neck) The cylindrical filling of an ancient *volcano which, due to its greater resistance, may be preserved after the volcanic edifice has been eroded away. *See also* PUY.

Volcanic units *See* MARTIAN TERRAIN UNITS.

volcano A naturally occurring vent or fissure at the *Earth's surface through which erupt molten, solid, and gaseous materials. The *viscosity, gas content, and rate of extrusion of the *magma probably determine the shape of the mountain built by the *eruptions. The *magma may reach the surface either through a single channel (*see* CENTRAL VENT VOLCANO), or through a series of vertical fractures (*see* FISSURE VOLCANO). Types of eruptions are named after volcanoes associated with them. *See* HAWAIIAN ERUPTION; PELÉEAN ERUPTION; PLINIAN ERUPTION; STROMBOLIAN ERUPTION; SURTSEYAN ERUPTION; VESUVIAN ERUPTION; and VULCANIAN ERUPTION.

volcano-tectonic depression An extremely large, *caldera- or trough-like, collapse depression, that is usually surrounded by extensive *ignimbrite sheets. One of the largest such depressions is in the

Taupo region of N. Island, New Zealand, and measures 100 × 30 km. Other well-known examples occur in the western USA (e.g. the Yellowstone volcano-tectonic depression) and the Philippines (the Lake Toba volcano-tectonic depression).

Volgian A post-*Kimmeridgian *stage in the northern European Upper *Jurassic which is distinguished from the *Tithonian *to the south) by its Boreal (*see* BOREAL REALM) ammonite (*Ammonoidea) fauna. Some authors equate it broadly with the *Portlandian.

volume diameter *See* PARTICLE SIZE.

volume diffusion *See* DIFFUSION.

volumenometer An apparatus used to determine the density of powdered minerals or of substances which react with the *heavy liquids normally used in the floating equilibrium method. The specimen is placed in a test chamber, graduated with 'upper' and 'lower' marks, and fitted with an airtight lid. The base of the chamber is connected by a flexible U-tube to a mercury reservoir and raising or lowering the reservoir brings mercury into the chamber as far as the 'upper' or 'lower' mark. The volume of compressed air in the chamber is calculated before and after a given weight of a substance is introduced, so the volume of the substance can be determined. The density can then be calculated from the volume and the atomic weight of the substance.

volume scattering In *radar terminology, the scattering of *electromagnetic radiation in the interior of a material, such as a vegetation canopy or soil.

von Karman–Prandtl equation *See* KARMAN–PRANDTL EQUATION.

vorticity The measure of the amount of rotary or circular motion in a water or *air mass about a vertical axis. Two types can be distinguished: 'relative vorticity' is rotation relative to the Earth's surface (it is positive if cyclonic, with the rotation in the same direction as the Earth's direction of rotation, and negative if anticyclonic); 'absolute vorticity' is relative vorticity plus the component of the Earth's rotation about its axis. Absolute vorticity is at a maximum at the poles and zero at the equator.

Voyager Two *NASA spacecraft that were launched in 1977 (Voyager 1 on 5 September and Voyager 2 on 20 August) on a mission to *Jupiter, *Saturn, *Uranus, *Neptune, and out of the *solar system. Voyager 1 passed Jupiter in March 1979 and Saturn in November 1980, its camera being used for the last time on 13 February 1990, to take pictures of the entire solar system, except for Mars, Mercury, and Pluto; Voyager 2 reached Jupiter in July 1979 and Saturn in August 1981. They completed their explorations in 1989.

VRM *See* VISCOUS REMANENT MAGNETISM.

VSP *See* VERTICAL SEISMIC PROFILE; and WALKAWAY VERTICAL SEISMIC PROFILE.

vug (vugh) A cavity in a rock, which may contain a lining of crystalline minerals.

vuggy porosity A form of *secondary porosity in which the *pore spaces are formed by *solution *vugs. *See* POROSITY.

vugh *See* VUG.

Vulcanian eruption An explosive type of volcanic *eruption that occurs when the pressure of entrapped gases in a relatively viscous *magma becomes sufficient to blow off the overlying crust of solidified *lava. A characteristic feature of Vulcanian eruptions is that the material ejected comes from the older rocks in the volcanic edifice and new magma is not erupted. The activity is often long lasting, with the formation of volcanic gas and *ash clouds and the violent ejection of solid angular fragments of all sizes. *See* VOLCANO. *Compare* HAWAIIAN ERUPTION; PELÉEAN ERUPTION; STROMBOLIAN ERUPTION; SURTSEYAN ERUPTION; and VESUVIAN ERUPTION.

vulcanicity *See* VOLCANICITY.

vulcanism 1. A theory of the 18th and 19th centuries, based initially on fieldwork in the French Auvergne and associated with Nicolas *Desmarest and James *Hall. They proposed that volcanic rocks were produced from molten material, and that *volcanoes had formerly existed where *basalts, etc. are now present. **2.** *See* VOLCANICITY.

Waalian An *interglacial *stage that occurred in northern Europe from about 1.3 to 0.9 Ma. It may be equivalent to the *Donau/Günz *Interstadial of the Alpine areas and the *Aftonian of N. America.

wacke A *sandstone which contains between 15% and 75% mud *matrix. *See* DOTT CLASSIFICATION.

wackestone A *limestone defined by the *Dunham classification as consisting of *carbonate particles in a mud-*matrix-supported texture.

wad A variety of *psilomelane, generally with the formula $BaMn_8O_{16}(OH)_4$ but copper and cobalt may substitute for manganese; *hardness 5.5; dull, greyish-black; amorphous or *earthy *aggregates. It is precipitated from water in poorly drained, boggy ground.

Wadati–Benioff zone *See* BENIOFF ZONE.

wadi (ouadi) The Arabic term for an *ephemeral river channel in a *desert area. Flow may occur very occasionally.

Waiauan (Waiaun) A *stage in the Upper *Tertiary of New Zealand, underlain by the *Lillburnian, overlain by the *Tongaporutuan, and roughly contemporaneous with the lower *Tortonian Stage.

Waiaun *See* WAIAUAN.

Waipawan A *stage in the Lower *Tertiary of New Zealand, underlain by the *Teurian, overlain by the *Mangaorapan, and roughly contemporaneous with the upper *Thanetian and lower *Ypresian Stages.

Waipipian A *stage in the Upper *Tertiary of New Zealand, underlain by the *Opoitian, overlain by the *Mangapanian, and roughly contemporaneous with the lower *Piacenzian Stage.

Waitakian A *Stage in the Lower *Tertiary of New Zealand, underlain by the *Duntroonian, overlain by the *Otaian (*Miocene), and roughly contemporaneous with the upper *Chattian Stage.

walkaway vertical seismic profile The *seismic record which is obtained when *hydrophones are positioned down a *borehole to detect *seismic waves originating from successive *shots fired from a surface source with increasing *offset. It provides much information about the geophysical characteristics of *reflectors and about the geologic structures in the vicinity of an existing well.

Walker's steelyard *See* COUNTERPOISED BEAM BALANCE.

Wallace's line An important zoogeographical division which separates the *Oriental and *Australian faunal realms. Alfred Russel Wallace, a zoogeographer and contemporary of Charles *Darwin, first demarcated the boundary, known to this day as 'Wallace's line', between the Oriental faunal realm and the Australian, with its distinctive marsupials (*Marsupialia). This boundary passes east of Java and Bali, northward through the Strait of Makassar (separating Borneo and the Celebes), then extends eastward, south of Mindanao in the Philippines.

wall-rock alteration A reaction of hydrothermal fluids with enclosing rocks, causing changes in mineralogy that are most marked adjacent to the *vein and become less distinct further away. *See* HYDROTHERMAL ACTIVITY; and HYDROTHERMAL MINERAL.

Walther's law (law of correlation of facies) An important statement relating to the manner in which a vertical sedimentary sequence of *facies develops. Walther's law of facies implies that a vertical sequence of facies will be the product of a series of depositional environments which lay laterally adjacent to each other. This law is applicable only to situations where there is no break in the sedimentary sequence.

Waltonian The Red Crag (crags are shelly sands) of Essex and Suffolk, England, which represent part of the Lower *Pleistocene in East Anglia. They mark the first *stage of the British Pleistocene.

Wangerripian The basal *stage in the Lower *Tertiary of south-eastern Australia, overlain by the *Johannian, and roughly contemporaneous with the *Danian and *Thanetian Stages.

Warendian A *stage of the Lower *Ordovician of Australia, underlain by the *Datsonian and overlain by the *Lancefieldian.

Warepan *See* BALFOUR; and NORIAN.

warm front A surface where advancing warm air displaces colder air, e.g. in mid-latitude *depressions where, owing to the convergence of the *air masses and the difference of density between them, the warm air tends to rise over the cold air. Slopes of warm fronts are typically less than 1:100 and the ascent of air is gradual. Stratiform cloud develops in the rising air. High *cirrus cloud followed by lower and thickening *altostratus indicate the approaching front. As the frontal contact with the ground approaches, heavy *nimbostratus and much rain may occur. Passage of the front is marked by a rise of temperature, clearing of precipitation, and (in the northern hemisphere) the wind *veering typically from south or south-easterly to south-westerly.

warm glacier *See* GLACIER.

warm rain Rain, resulting from the coalescence of droplets, in clouds that are unfrozen (i.e. their upper parts are not at freezing level).

warm sector A tongue of relatively warm air of tropical or old polar or maritime origin that appears between colder *air masses. Such a tongue occupies the area between the *warm and *cold fronts in a developing mid-latitude *depression. Within the warm sector, pressure, wind, and temperature remain fairly steady. Cloud and *precipitation depend on the precise condition of the generally stable air; cloud may be produced from *orographic uplift or *fog from passage over a cool sea surface.

washboard moraine *See* MORAINE.

washover delta *See* WASHOVER FAN.

washover fan (washover delta) A fan-shaped body of *sediment that is transported landward by marine waters flowing through or across a coastal barrier such as a *barrier bar or island. Such bodies are formed especially during storms when the barriers are likely to be overtopped.

washplain A nearly flat surface made up of *alluvium mantling a thick layer of deeply weathered (*see* WEATHERING) bedrock and found in a savannah environment. It is washed by seasonally flooding streams which are unable to incise due to the lack of abrasive *bed load and the large volume of *sediment.

water The water molecule is composed of two hydrogen atoms, each of atomic weight 1.00797, combined with an oxygen atom of atomic weight 15.9994. It can occur as liquid, solid, and gas *phases. It is a very powerful solvent which is responsible for the transfer of material on and below the Earth's surface.

water-absorption test A test to determine the moisture content of *soil as a percentage of its dry weight (British Standard 1377: 1967). The sample is weighed, dried in an oven, then reweighed under standard conditions. It is calculated as the moisture content, which is equal to: (weight of the container with wet soil minus the weight of the container with dry soil) divided by (weight of the container with dry soil minus the weight of the container); then multiplied by 100 to express it as a percentage.

water balance (water budget) A method of assessing the size of future water resources in an *aquifer, *catchment area, or geographical region, which involves an evaluation of all the sources of supply or *recharge in comparison with all known *discharges or *abstractions. *See* MOISTURE BALANCE.

water budget *See* WATER BALANCE.

water budget, global The amount of water involved in the *hydrological cycle each year. Average annual *precipitation over the whole globe is about 86 cm, of which 77% falls on the oceans and 23% on land. Evaporation (including transpiration by plants) from the land accounts for 16% of the total precipitation it receives, and 7% of global precipitation returns to the sea as river and *groundwater flows.

water gun A marine seismic source which uses compressed air to drive a piston in order to evacuate a chamber flooded with water. The resulting water jet creates a vacuum in the surrounding water which implodes, causing an acoustic pulse. Since no bubbles are formed there is no *bubble pulse, and consequently the short *shot pulse provides higher resolution than can be attained with an *airgun.

water inventory 1. An inventory to show

how water is consumed, used or otherwise involved in a particular process or place; e.g. within a steel plant or a house. **2.** Globally, approximately 97% of the Earth's water occurs in the oceans. Of the fresh water, 75% is locked up in *ice sheets and *glaciers, almost 25% is *groundwater, and lakes, reservoirs, swamps, river channels, biospheric water, atmospheric water, and soil moisture together account for the remainder. The amount of water at each stage of the *hydrological cycle is called the water storage.

water potential 1. (osmotic potential, chemical potential) The difference between the energy of water in the system being considered and of pure, free water at the same temperature. The water potential of pure water is zero, so that of a solution will be negative. If there is a gradient of water potential between two plant cells, water will diffuse down the gradient until equilibrium is reached. **2.** *See* CAPILLARY MOISTURE.

watershed 1. *See* DIVIDE. **2.** *See* CATCHMENT.

waterspout A *tornado-like vortex that occurs over water and is visible because of the condensation of atmospheric water vapour in the low pressure around the core (not because water is drawn upward from the surface). A waterspout may be a tornado that has moved from over the land, or may form over water, not necessarily beneath a *mesocyclone cloud.

water storage *See* WATER INVENTORY.

water-table The upper surface of *groundwater, or the level below which an unconfined *aquifer is permanently saturated with water. *See also* VADOSE ZONE.

water vapour The gaseous form of water, present in the atmosphere in varying amounts. It is an intermediate stage in the *hydrological cycle. Water vapour in the atmosphere represents 0.01% of the total *water inventory of the Earth. Water molecules have a short *residence time in the atmosphere, on average about nine days, though this varies over a very wide range, from a few minutes upwards. Water enters the atmosphere by evaporation, transpiration, and *sublimation at the Earth's surface. The concentration of water vapour decreases with increasing altitude up to over 20 km, where some increase may occur. Measures of water-vapour concentration include *vapour pressure, the humidity *mixing ratio, and the *absolute and *relative humidity. The physical significance of water vapour in the atmosphere is in its condensation to produce cloud and *precipitation (with release of latent heat of condensation, which often accelerates the uplift of rising air), its absorption and scattering of radiation, etc.

water vascular system *See* ECHINODERMATA.

water velocity The velocity (v m/s) of sound waves in water is dependent on temperature (T °C) and *salinity (S parts per thousand), and can be calculated for any depth (Z) using the formula $v = 1449 + 4.6T - 0.055T^2 + 0.0003T^3 + (1.39 - 0.012T)(S - 35) + 0.017Z$.

water-witching *See* DIVINING.

Waucoban A *series of the Lower *Cambrian of N. America, equivalent to the *Caerfai.

wave A periodic disturbance in a solid, liquid, or gas as energy is transmitted through the medium. (Electromagnetic waves e.g. light, can be transmitted through a vacuum.) In water, waves may occur either at the surface or as *internal waves. The size of the water wave varies from minute *capillary waves to massive *tsunami, while *wave period ranges from a few seconds to several hours.

wave base The depth beneath a water mass below which *wave action ceases to disturb the *sediments. Wave-base depth is approximately equal to half the wavelength of the surface waves.

wave-built terrace A theoretical zone of sedimentation on the outer edge of a *shore platform or, on the larger scale, of a *continental shelf. The depth of its surface below sea level was thought to be determined by the wave base (the depth at which wave activity just fails to disturb *sediment). Modern research has failed to show the existence of such terraces.

wave clouds Clouds occurring in an airstream characterized by *wave motion, usually after passing over a mountain barrier. Lenticular (lens-shaped) clouds and cloudlets appearing stationary in the crests of the waves are a characteristic sight in climates experienced on the lee side of mountain barriers, e.g. in eastern Scotland and Sweden.

wave-cut bench *See* SHORE PLATFORM.

wave-cut platform *See* SHORE PLATFORM.

wave depression A low-pressure area developed at the apex of a developing *wave distortion along a *front. A series of such systems is typical of mid-latitude *depression sequences.

wave diffraction An effect seen as *waves pass through an opening in a breakwater into protected waters. The waves fan out from the opening into the region beyond, but as they do so their height is diminished.

wave equation An equation of *wave mechanics which represents wave displacement (ψ) and wave velocity (v) as a function of space and time (t), where space can be represented by rectangular coordinates (x, y, z), such that: $v^2\psi = (\delta^2\psi/\delta x^2) + \delta^2\psi/\delta y^2 + \delta^2\psi/\delta z^2 = (1/v^2)\delta^2\psi/\delta t^2$; or the equation can be represented in spherical coordinates, such that: $(1/v^2)\delta^2\psi/\delta t^2 = (1/r^2)[\delta(r^2\delta\psi/\delta r)\delta r + (1/\sin\theta)(\delta(\sin\theta \times \delta\psi/\delta\Phi)\delta\theta) + (1/\sin^2\theta)(\delta^2\psi/\delta\theta^2)]$, where r is the radius, θ the *co-latitude, and Φ the longitude.

wave-front The locus of adjacent points with the same *phase in the path of an advancing *wave, e.g. a ripple spreading on the surface of a pond after a stone has been thrown into it.

wavelength 1. **(λ)** The distance on a *wave between successive points that are in *phase, e.g. for a water wave, the wavelength is the distance from one crest to the next. Wavelength is related to the velocity (v) and frequency (f) by: $\lambda = v/f$. The inverse of wavelength is *wavenumber. **2.** Of a *fold system, the distance between one *hinge or *trough and the next. It is rarely possible, or necessary, to measure the wavelengths of folds precisely.

wavellite A *secondary mineral, with the formula $Al_3(PO_4)_2(OH)_3.5H_2O$; sp. gr. 2.3–2.4; *hardness 3.5–4.0; *orthorhombic; normally white, but often greenish, yellow, grey, or brown; vitreous *lustre; normally forms fibrous, radiating aggregates, or hemispherical or globular aggregates; found on *joint surfaces and cavities in rock, especially *slates and *shales, and in association with *limonitic *ore bodies. It is named after the 18th-century British mineralogist W. Wavell, who discovered it.

wavenumber 1. Spatial frequency (k); the number of complete *wave cycles per unit distance, and the inverse of *wavelength (λ), such that $k = 1/\lambda$. **2.** Propagation constant; in electromagnetic theory, wavenumber is defined as $2\pi/\lambda$, and $k^2 = \mu\omega(\epsilon\,\omega + i\sigma)$ for the time factor $e^{-i\omega t}$, where μ is the *magnetic permeability, ω the angular frequency in radians per second, ϵ the *dielectric permittivity, and σ the *electrical conductivity.

wave period The time required for two successive *wave crests to pass a fixed point, or the time for a single wave crest to travel a distance equal to the length of the wave.

wave refraction The process by which the direction of *waves moving in shallow water at an angle to the submarine contours is altered. The part of the wave train travelling in shallower water moves more slowly than that still advancing in deeper water. The lines of the wave crests therefore become more parallel with the submarine contours closer to the coast.

wave-ripple cross-lamination The form of cross-lamination (*see* CROSS-STRATIFICATION) produced by the migration of wave-generated *ripples, or combined flow ripples (i.e. ripples formed by a combination of wave action and unidirectional flow). Wave-ripple cross-lamination is characterized by a variety of distinctive features, including: unidirectional cross-laminae, sometimes with drapes (sand laminae) oriented in the opposite direction; lensoid and complexly interwoven cross-sets; irregular, undulatory bases to cross-sets; and laminae which are discordant with the external ripple form.

wave ripple mark *See* OSCILLATION RIPPLE.

wave spectrum A concept used to describe mathematically the distribution of *wave energy (proportional to the square of the significant wave height) with *wave period. Using this concept, a highly confused pattern of interfering waves can be divided into its constituent wave-forms. The results can be used as an aid to wave forecasting.

wave velocity The velocity at which *waves of energy are transmitted through a medium. It depends on the characteristic properties of the medium—in the case of *seismic waves on its elastic properties and density. The wave velocity (v) is always related to *frequency (f) and *wave length (λ) by the expression $v = f\lambda$.

wavy bedding A form of *heterolithic *sediment characterized by interbedded rippled *sands and mud layers. Wavy beds are commonly found on storm-dominated shelves (*see* SHELF), but also in lakes, *intertidal areas, and other environments where energy levels fluctuate appreciably.

waxing slope *See* SLOPE PROFILE.

waxy Of a mineral *lustre, smooth in appearance, like wax.

weakening In *synoptic meteorology, a decrease in pressure gradient around a pressure system over time. Changes of this kind bring about a weakening of the winds. *See also* INTENSIFICATION.

Wealden The name given to a sequence of freshwater deposits in the British pre-*Aptian *Cretaceous System. In Sussex and Kent the Wealden consists of two major sedimentary units: the Hastings Beds Group and the Weald Clay Group. The former is the lower and more sandy unit, whereas the latter is mud-dominated. The name should not be confused with *Wealdien.

Wealdien *Sand deposits of the *Tithonian *Stage located in France. The name should not be confused with *Wealden.

wearing course *See* PAVEMENT.

weathering The breakdown of *rocks and *minerals at and below the Earth's surface by the action of physical and chemical processes. Essentially it is the response of Earth materials to the low pressures, low temperatures, and presence of air and water that characterize the near-surface environment, but which were not typical of the environment of formation. There are several varieties of rock breakdown (*see also* MECHANICAL WEATHERING). Simple disintegration may occur, resulting in the production of coarse, angular blocks, of peels or skins (the process of 'desquamation'), of *sands, and of *silts. Minerals may be removed in *solution, and *chemical weathering may form new, often easily eroded substances. *See also* CARBONATION; EROSION; FROST WEDGING; HYDRATION; HYDROLYSIS; and THERMOCLASTIS.

weathering correction In *seismic refraction and *seismic reflection surveys, a time correction that is made to allow for the *low-velocity zone or *weathering layer. It allows *travel times to be reduced to a common datum by means of the *static correction.

weathering front The junction between chemically weathered (*see* CHEMICAL WEATHERING) *rock or *regolith and sound rock. Where the front lies between regolith and unweathered bedrock it may be exposed by subsequent *erosion to form an *etchplain.

weathering index A measure of the intensity of *chemical weathering. It consists of a comparison between a mineral or a chemical compound that is relatively stable with one that is readily removed by weathering. For example, the ratio of *quartz to *feldspar is a widely used index. Resistant heavy minerals are also used: the ratio between *zircon and *tourmaline (resistant) and the *amphiboles and *pyroxenes (less resistant) has been employed.

weathering layer A layer at or near the surface, usually marked by a shape transition, in which *seismic velocities are substantially lower than in the geophysical substrate. The geophysical weathering layer may not correlate exactly with a layer of geologically weathered materials; often it is taken to be the zone above the *water-table.

weathering micro-indices Indices used in the grade classification of rock material by microscopic examination in the laboratory. The indices are: I_{mp} (micropetrographic index), the percentage of sound constituents divided by the percentage of unsound constituents; and I_{fr} (microfracture index), the number of microcracks in a 10 mm traverse of *thin section.

weathering profile A vertical section, from the ground surface to unaltered bedrock, which passes through *weathering zones. It is usually best developed in the humid tropics, where depths of 100 m have been recorded but where 30 m is more common. The nature of the profile is a complex response to climatic and geologic controls, and to long-term changes in external conditions.

weathering series A sequence of common *silicate minerals, laid out in the order of their susceptibility to *chemical weathering. A well-known series was suggested by S. S. Goldich in 1938 and runs from *quartz (most resistant), through *muscovite, *alkali feldspar, *biotite, the *plagioclase feldspars, and *olivine (least resistant). This sequence is the reverse of *Bowen's reaction series, which ranks minerals in the order of their crystallization from a *melt. An absolute scale of weathering susceptibility is unlikely to be achieved, partly because of variations in environmental conditions.

weathering zone A distinctive layer of

weathered material that extends roughly parallel to the ground surface. It differs physically, chemically, and mineralogically from the layers above and/or below. A broad distinction may be drawn between the weathering zones in *drift, which are normally distinguished by degrees of oxidation and by carbonate content, and those on bedrock, which are usually separated according to the relative proportions of *corestones and weathered *matrix.

weather report A record of observations of meteorological values and the state of the weather, made at a given place and time.

weather satellite A satellite that senses the state of the atmosphere, e.g. by photographing cloud distribution, or by using infrared photography to record cloud temperature as an indication of the height of the cloud tops.

wedge *See* RIDGE.

wedge-edge trap A *stratigraphic trap in porous beds, in the form of an inclined wedge, which may trap oil, gas, or water at its upper end. *See* NATURAL GAS; PETROLEUM; and POROSITY.

Wegener, Alfred (1880–1930) A German meteorologist and physicist, Wegener is best known for his version of the theory of *continental drift. *Die Entstehung der Kontinente und Ozeane*, published in 1915 (first English language, edition: *The Origin of Continents and Oceans*, 1924). He used a wide range of arguments to support his hypothesis, including palaeontological and palaeoclimatic evidence, the *hypsometric curve, seismic evidence, *polar wandering, and the differences between *continental and *oceanic crust. He found little contemporary support, but has since become regarded as the first to propose the hypothesis in well-argued, scientific terms.

Weichselian *See* DEVENSIAN.

weight drop A seismic-energy impulse produced when a heavy weight is allowed to fall and strike the ground. Weights vary from a few kilograms dropped from 2–3 m, for shallow seismic-*reflection surveys, up to several tonnes dropped from 3–4 m, in large-scale, commercial, seismic-*reflection surveys. In an accelerated weight drop, a mass (usually of no more than a few tens of kilograms) is accelerated towards the ground by compressed air, thus increasing the impact force and energy impulse. *See also* HAMMER; and SHOT.

weighted average The value assessed from a number of samples, where each sample is given a different value of importance according to its reliability.

weir An engineered structure extending across an open water channel. It is possible to calculate the volume of water flowing over a weir from a knowledge of its particular characteristics. Weirs are designed for flow measurement in a variety of shapes but the most common are the suppressed, contracted, V-notch, trapezoid, and broad-crested types. Other weirs may be used to control river water levels or divert flow into channels.

Weissliegende *See* KUNGURIAN; and SAXONIAN.

Weiss zone law *See* ADDITION RULE.

welded ignimbrite *See* IGNIMBRITE.

welded tuff *See* IGNIMBRITE.

well A completed *borehole. The hole may be 'dry' in that it does not produce oil or gas (although containing water), or a producing well.

well injection method A method of solution mining in which *boreholes are used to transport a solvent (usually water) to the working face. The mineral is extracted and the resultant solution lifted to the surface either through an adjacent borehole or an annulus formed by equipping the borehole with concentric casings. The pumping mechanism may be located at the surface or in the borehole. The method is used extensively in *halite mining.

well logging (wireline logging) The lowering of various sensors within an instrumental package (a sonde) down a *borehole. The output of these instruments is then measured to produce a series to well logs as the sonde is wound back to the surface. A wide variety of instruments can be combined: *caliper, *electrical, *radioactive, *temperature, etc. The term 'well logging' is also used to refer to the recording of the nature of drill cuttings encountered during the drilling of the well. *See also* BOREHOLE LOGGING; and MUD LOG.

well-point drainage A method for draining permeable deposits around an excavation that requires small cones of depression. Tubes about 100 mm in diameter, with wire mesh screens, are sunk into the ground and connected by a header pipe to a suction pump at the top. Usually a series of well-points are connected to one header pipe. When these are used in a staged excavation a considerable depth can be drained and the *drawdown restricted, because each well-point acts as a hydrogeologic boundary.

well screen A system of mesh screening or holes designed to allow water to enter a *well or *borehole without undue loss of *head, but to exclude *sand, *silt, and other geologic material. Normally, well screens are required only in relatively unconsolidated materials.

well shooting Technique in which a series of *geophones are deployed down a *borehole and used to determine *seismic velocities, as a function of depth, by measuring arrival times from surface seismic-energy sources. *See also* UPHOLE SURVEY.

Weltian A *system in N. American usage comprising the *Archaean and *Priscoan of Harland et al. (1982), and overlain by the *Xenian.

Wenlock (Wenlockian) A *series of the *Silurian (430.4–424 Ma), underlain by the *Llandovery and overlain by the *Ludlow, throughout Europe. *Reef-dwelling organisms abound on the *bedding planes of the Wenlock *limestone of the Welsh borderlands.

Wenner electrode array An *electrode configuration in which four electrodes are deployed in a line, with equal spacing between the two *potential electrodes, and between each current electrode and its nearest potential electrode. Its *geometric factor (K_g) is $2\pi a$, where a is defined for each case. The Wenner array has five variations, three referred to as the tripotential method with α, β, and γ configurations, one as the Lee partitioning method (which has a fifth electrode at the *array centre acting as a third potential electrode), and one as the Offset Wenner electrode array, which reduces the effects of lateral inhomogeneities. *See* ELECTRODE SEPARATION.

Wentworth scale *See* PARTICLE SIZE.

Werner, Abraham Gottlob (1750–1817) Professor of mineral science at the Mining Academy, Freiberg, who developed a system of mineral classification which he called 'oryctognosy', based on the external characteristics of minerals. He used the term 'geognosy' in preference to geology. The theory of the Earth which he taught became known as *neptunism. He published little but inspired mining students from all over Europe and much of his terminology was adopted even by those who rejected his theory.

Werrikooian The stratigraphic name for the *Pleistocene in south-eastern Australia.

West Australia current The oceanic current that flows north along the western Australian coast. The flow is strong and steady in summer, but is much reduced during the winter months. Low *salinity (34.5 parts per thousand) and low temperature (3–7 °C) typify the waters of this current.

westerlies Popular term for the prevailing, eastward-moving airstreams in the mid-latitudes of the northern and southern hemispheres.

Western Boundary Undercurrent *See* CONTOUR CURRENT.

western intensification The tendency of currents along the western margins of all oceans to be particularly strong, swift, and narrow, flowing northwards in the northern hemisphere and southwards in the southern hemisphere. Currents at the eastern margins of all oceans tend to be slower and more diffuse. *See* GYRE.

West–Kohoutek–Ikemura A *comet with an orbital period of 6.46 years; *perihelion date 1 June 2000; perihelion distance 1.596 AU.

Westphalian A *series in the *Silesian (Upper *Carboniferous) of Europe, subdivided into Westphalian A, B, C, and D (the original Westphalian E is now included in the *Stephanian Series). It is underlain by the *Namurian, overlain by the Stephanian, dated at 315–303 Ma, and is roughly contemporaneous with the Upper *Bashkirian plus most of the *Moscovian Series.

westward drift The apparent westward motion of features of the *geomagnetic field (mostly the *non-dipole component) with time, estimated to be about 0.2° of longitude per year during the last 100 years.

West Wind Drift *See* ANTARCTIC CIRCUMPOLAR CURRENT.

wet-bulb depression The extent to which the temperature recorded by a ventilated *wet-bulb thermometer falls below the dry-bulb air temperature.

wet-bulb thermometer A thermometer, the bulb of which is kept moist by a thin cloth (e.g. muslin) bag connected by a wick to a bath of clean (preferably distilled) water. As long as the air is not saturated, evaporation from the muslin keeps the wet-bulb thermometer at a lower temperature than the dry-bulb thermometer beside it, with which its readings are compared. The depression of the wet-bulb temperature gives a measure of the saturation deficit, and so of the *relative humidity of the air.

'wet chemistry' *See* GRAVIMETRIC ANALYSIS.

wet melt *See* MELT.

wetted perimeter *See* HYDRAULIC RADIUS.

W-fold *See* M-FOLD.

Whaingaroan A *stage in the Lower *Tertiary of New Zealand, underlain by the *Runangan, overlain by the *Duntroonian, and roughly contemporaneous with the *Rupelian and lower *Chattian Stages.

Wheelerian The first of two *stages in the *Pleistocene of the west coast of N. America, underlain by the *Venturian (*Pliocene), overlain by the *Hallian, and roughly contemporaneous with the Lower Pleistocene *Series of southern Europe.

Whewell, William (1794–1866) A mineralogist and moral philosopher of Cambridge, who wrote works on mathematics, *natural theology, and the history and philosophy of science. He also studied the theory and causes of *tides, collecting his information with the assistance of members of the British Association for the Advancement of Science. Whewell is responsible for the introduction of such terms as scientist, *uniformitarianism, and *catastrophism.

whirling psychrometer (sling psychrometer) A *psychrometer with a handle, which allows rapid rotation of mounted *wet- and dry-bulb thermometers to ensure air flow around the bulbs.

whirlwind A spiral wind storm around a low-pressure centre. In arid areas dust may be carried upward several hundred metres.

whitings Extensive patches of white, turbid water consisting of a dense suspension of *aragonite mud. These mud suspensions are seen from time to time in *carbonate-dominated shallow seas, and are mainly due to the disturbance of the muddy sea floor by schools of fish, or by turbulence. It was suspected that some whitings were the result of direct precipitation of aragonite from hypersaline, carbonate-saturated waters, but recently this has been shown not to be the case.

white-out Meteorological condition in which low cloud appears to merge with a snow-covered surface to produce a uniform appearance in which outlines cannot be discerned.

Whiterockian A *stage of the *Ordovician in the Lower *Champlainian *Series of N. America.

'white smoker' *See* HYDROTHERMAL VENT.

Whitwellian A *stage of the *Silurian, underlain by the *Sheinwoodian and overlain by the *Gleedonian.

whole Earth composition *See* BULK COMPOSITION OF EARTH.

whole-rock dating *Igneous and *metamorphic rocks that contain no minerals but have particularly high Rb:Sr ratios (*see* RUBIDIUM–STRONTIUM DATING), or rocks that are too fine-grained for mineral separation, can be analysed as whole-rock samples. Whole-rock samples from different parts of the same body generally differ in rubidium content and the ^{87}Sr:^{86}Sr ratio of each can be plotted as a function of its ^{87}Rb:^{86}Sr ratio in an *isochron diagram. At the time of the initial crystallization different parts of the sample, regardless of rubidium concentration, would have had the same ^{87}Sr:^{86}Sr ratio and hence plot as a horizontal line. With the passage of time ^{87}Rb would be lost and corresponding amounts of radiogenic ^{87}Sr gained. As the ^{87}Sr:^{86}Sr ratio changes in each part of the rock, the slope of the isochron increases progressively, providing a measure of the age of the crystallization. The intercept of the isochron at the ordinate indicates the isotopic composition of common strontium at the beginning of the process.

whorl *See* COILING.

Widmanstätten structure The pattern revealed on polished and etched *iron meteorites, showing a characteristic *texture. It consists of parallel interlocking bands of kamacite (α-iron) and taenite (γ-iron) resulting from the slow cooling of the meteorite. Analysis of the pattern indicates the size and structure of the parent body. It is named after Count Widmanstätten of Vienna who described it in 1804.

Wiechert, Emil (1861–1928) A German physicist, and founder of the Geophysical Institute at Göttingen (1901). Wiechert made important improvements to the *seismograph, which made it possible to distinguish between different types of *seismic waves. He was thus able to identify and name *P- and *S-waves. He also calculated the diameter and density of the Earth's *core from seismic data.

Wiener filter A filter which converts a known input signal into an output signal which, according to a least-squares test, is the one most similar to a desired form of signal output.

Wien's displacement law The law which states that the wavelength of *electromagnetic radiation emitted by a material is inversely proportional to the absolute temperature of that material. As the absolute temperature increases, the wavelength of emitted radiation becomes shorter.

wiggle trace The classic seismic-*trace output from a galvanometer, which illustrates the wave-form of the recorded data. A modern display of *seismic waves is in the form of a *variable-area display with wiggle.

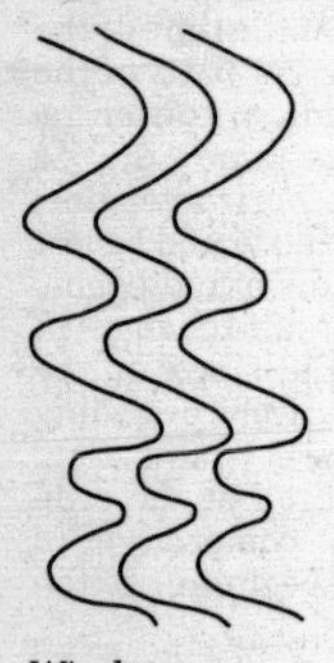

Wiggle trace

wigwams *See* TENT ROCKS.

Wild 2 A *comet with an orbital period of 6.17 years; *perihelion date 6 May 1997; perihelion distance 1.583 AU.

wildflysch Turbiditic, *mass-flow *sediments with numerous ill-sorted *exotic *clasts. In modern terminology wildflysch is better referred to as a form of *diamictite.

willemite Mineral, Zn_2SiO_4; sp. gr. 4.0; *hardness 5.5; *trigonal; greenish-yellow, but can vary to brown or white; vitreous *lustre; crystals can be *prismatic, but it usually forms granular or *massive aggregates; strongly *fluorescent; occurs mainly in the oxidized zones of zinc deposits. It is named after King William I of the Netherlands.

willy-nilly The name given to a *tropical cyclone that forms near western Australia.

Wilson cycle The hypothesis, named after Tuzo *Wilson who proposed it, that an ocean basin has a lifespan with several stages: from opening, through development, to final closing and the destruction of the basin. Six stages, and the *plate tectonic processes involved, have been identified in different parts of the Earth today, and have been postulated for *orogenic belts back to the early *Proterozoic. The earliest (embryonic) stage involves uplift and crustal *extension with the formation of *rift valleys (e.g. the E. African Rifts). The young stage involves further subsidence, plus *sea-floor spreading; the result is a narrow, parallel-sided sea, possibly with *evaporites from intermittent desiccation (e.g. the *Red Sea). At this stage, if the uplift was broadly *domal with a pattern of three radial rifts forming a *triple junction, two of the rifts may widen leaving the third to form an *aulacogen (e.g. the Ethiopian Rift). The next (mature) stage is exemplified by the *Atlantic Ocean, a wide ocean basin flanked by *continental shelves and with the production of new, hot, *oceanic crust along an oceanic *ridge. Eventually this expanding system becomes unstable, and part of the cooled *lithosphere, away from the ridge, sinks into the *asthenosphere, forming an oceanic *trench with an associated *island arc. The shrinking *Pacific Ocean is thought to be at this stage. Further shrinking, with the compression, *metamorphism, and uplift of *accretionary wedges to form young mountain ranges, marks the terminal stage (e.g. the Mediterranean). Finally, all the oceanic crust between the continental masses has subducted and the continents converge in a *collision zone, being joined along a *suture. The suture (e.g. the Indus–Yarlung Zangbo suture in the Himalayas) marks the relic scar between the plates, and the *plate margin finally becomes inactive.

Wilson, John Tuzo (1908–93) Professor of geophysics at the University of Toronto, Tuzo Wilson is best known for his explanation, in 1965, of the *transform faults which offset ocean spreading axes. His ideas were based on studies of linear *magnetic anomalies and *seismicity beneath *ocean crust. He was the first person to use the term *plates, and has also invoked the idea of a *hot spot to explain the evolution of the Hawaiian chain of islands. *See* WILSON CYCLE.

wilting coefficient *See* PERMANENT WILTING PERCENTAGE.

wilting point *See* PERMANENT WILTING PERCENTAGE.

Windermere Interstadial (Late-Devensian Interstadial) A relatively warm period that occurred towards the end of the last (*Devensian) glacial stage in Britain. The event took place about 13 000–11 000 radiocarbon years BP. The Windermere Interstadial includes the *Bølling, *Older Dryas, and *Allerød *chronozones of Scandinavia.

wind noise Electrical *noise generated by *geophones or cables when they are buffeted by the wind; mechanical noise generated when trees and tall structures are moved by the wind and the motion is transmitted to the ground through their roots or foundations.

window *See* APERTURE.

wind rose Quantitative diagram (*rose diagram) representing the relative frequencies of different wind directions and wind speeds at a climatic station over a period of time.

windrow Streak of foam or row of floating debris, aligned in the prevailing wind direction, formed on the surface of a lake or ocean. Where winds blow across a water surface, vertical circulation cells are set up in near-surface waters. These circulation cells are alternately right- and left-handed vortices, and windrows form along the lines of convergence between adjacent cells at the water surface.

wind shear The gradient of horizontal wind velocity with height, which varies according to the rate of change of temperature with altitude. Vertical wind shear can be a cause of cloud formation in the turbulent mixing taking place in a *boundary layer of air moving at different speeds. The shear between the wind at different levels can be expressed as a vector, measuring the difference in speed and direction.

windward 'Upwind', i.e. in the direction from which the wind blows.

wireline Any system which is used down a *borehole and which can be operated at the end of a cable. The term is used with reference to drilling techniques and geophysical logging methods.

wireline logging *See* WELL LOGGING.

Wirtanen A *comet with an orbital period of 5.46 years; *perihelion date 11 June 1997; perihelion distance 0.339 AU.

Wisconsinian The last (80 000–10 000 years ago) of four glacial episodes recognized in N. America. As with previous glacials there were several advances of the ice, and these glacial deposits include sequences of *tills laid down by ice in the western N. American Cordillera. More evidence from areas peripheral to the *ice sheets suggests that temperatures were perhaps 6 °C cooler than they are now. This glacial is perhaps equivalent to the *Würm Glaciation or Glacial of the Alpine areas.

witherite Member of the *carbonate group of minerals, with the formula $BaCO_3$; sp. gr. 4.3; *hardness 3.5; greyish-white; *vitreous *lustre; crystals *tabular and twinned (*see* CRYSTAL TWINNING), but also *massive and granular; occurs as a *gangue mineral in hydrothermal-vein deposits (*see* HYDROTHERMAL MINERAL) associated with *barite and *galena.

within-plate basalt (WPB) *Basalts which are generated within a continental or oceanic *plate, rather than at a *constructive or *destructive plate margin. Continental *flood basalts (e.g. the Columbia River Plateau in the western USA) and *ocean island basalts are types of within-plate basalts. Although each has its own particular geochemical characteristics, they have been identified as a coherent group using their Ti/100; Zr; Y × 3 ratios (against a standard). This effectively separates the WPB group from the *mid-ocean-ridge basalt (MORB), low-potassium *tholeiite, and *calc-alkaline basalt groups.

Witwatersrand *See* RANDIAN.

WMO *See* WORLD METEOROLOGICAL ORGANIZATION.

Wolfcampian The basal *series in the *Permian of N. America, underlain by the

*Virgilian Series (*Pennsylvanian), overlain by the *Leonardian, and roughly contemporaneous with the *Asselian and lower *Sakmarian *Stages.

wolframite *Ore mineral for tungsten, with the formula $(Fe,Mn)WO_4$; sp. gr. 7.0–7.5; *hardness 5.0–5.5; *monoclinic; grey-black to brownish-black; brownish-black *streak; metallic *lustre; crystals normally *tabular, *prismatic, often *bladed, also occurs granular and *massive; *cleavage perfect {010}; occurs in high-temperature *hydrothermal veins, *quartz veins, and *pegmatites, associated with granitic rocks, with *cassiterite, *arsenopyrite, *tourmaline, *scheelite, *galena, *sphalerite, and quartz.

wollastonite Member of the *nesosilicates (chain silicates) $CaSiO_3$ and associated with the *pyroxenes, although it does not possess a pyroxene atomic lattice and is termed a pyroxenoid (along with pectolite ($Ca_2NaH[SiO_3]_3$) and *rhodonite); sp. gr. 2.8–3.1; *hardness 4.5–5.0; *triclinic; white to grey; vitreous to pearly *lustre; crystals *tabular, *prismatic, also occurs *massive and cleavable, or fibrous; *cleavage perfect {100}; occurs in *metamorphosed, siliceous *limestones, and *alkaline, *igneous rocks, and associated with *calcite, *epidote, and *tremolite; soluble in hydrochloric acid with the separation of silica. It is used as a source mineral for rock wool of high strength and with long fibres, and is named after the British mineralogist W. H. Wollaston.

Wolstonian The glaciation which followed the *Hoxnian *Interglacial; a pale, chalky *till (the Gipping Till) occurs in East Anglia, England, and the stone orientation and the included *erratics suggest a movement from the north. The general *stratigraphy needs clarification since it is not certain that the till rests on Hoxnian deposits and there are a number of morainic deposits attributed to this glacial which occur in Norfolk. During this ice advance a large lake, Lake Harrison, was ponded up in the Midlands. The Wolstonian perhaps correlates with the *Saalian.

Wonokan A *stage of the *Ediacara Epoch, dated at about 590–580 Ma (Harland et al., 1989).

wood tin A variety of *cassiterite (SnO_2) with concentric bands and compact and fibrous *texture. Toad's-eye tin shows similar characteristics on a smaller scale. It occurs in *hydrothermal veins associated with *acid *igneous rocks.

Woodward, John (1665–1728) Woodward was a physician with a wide knowledge of *fossils, whose large collection formed the basis for the geologic museum at Cambridge, England. He was a diluvialist, believing that fossils and *sediments had settled out of the ocean of Noah's Deluge in the order of their specific gravities. His *Essay towards a Natural History of the Earth* (1695) influenced both English and continental natural philosophers. *See* DILUVIALISM.

woolsack *See* CORESTONE.

Wordian A *stage of the *Zechstein Epoch, preceded by the *Ufimian and followed by the *Capitanian.

work hardening The effect of a decrease in the ease of deformation of a body undergoing progressive deformation, due to a change in its material configuration, and thought to be caused by the accumulation of defects in the lattice.

World Climate Programme A project established in 1979 by the *World Meteorological Organization for the collection and preservation of climatic data.

World Meteorological Organization (WMO) The United Nations agency responsible for gathering, analysing, and disseminating meteorological data. Its establishment was agreed in 1947 as a successor to the International Meteorological Organization, it came into being in 1951, and later in 1951 it became a UN agency. It has 185 members (179 member states and 6 dependent territories).

World Weather Watch (WWW) A worldwide system for observing, analysing, and forecasting meteorological conditions, established in 1963 under the auspices of the *World Meteorological Organization. It supplies constantly updated weather reports and forecasts to all World Meteorological Organization members, obtaining its data from 4 satellites in *polar orbit and 5 in *geostationary orbit, about 10 000 land observation stations, 7000 weather ships, and 300 moored and drifting buoys. The *Tropical Cyclone Programme is one of the programmes forming part of the WWW. *See also* WORLD CLIMATE PROGRAMME.

World-Wide Standard Seismograph Network (WWSSN) An international network of *seismometer *arrays designed to detect and locate *earthquakes. *See also* LASA.

WPB *See* WITHIN-PLATE BASALT.

wrench fault *See* STRIKE-SLIP FAULT.

wrinkle ridge (mare ridge) A broad, curvilinear swell up to 500 m high and 10 km wide, topped by a narrow, steep-sided ridge up to 200 m high and 2–4 km wide, extending for tens to hundreds of kilometres concentrically near the edges of lunar *mare basins. They are compressional in origin, possibly due to subsidence of mare lavas.

wulfenite Mineral, $PbMoO_4$; sp. gr. 6.5–7.0; *hardness 3; *tetragonal; orange-yellow, olive-green, or brown; white *streak; *resinous *lustre; crystals usually square plates or tablets, sometimes occurs *massive, granular, or *botryoidal; *cleavage good {101}; sub-*conchoidal fracture; occurs as a *secondary mineral in the oxidized zones of *ore minerals of lead and molybdenum, and in association with *anglesite, *cerussite, and *pyromorphite.

Wulff stereographic net *See* STEREOGRAM.

Würm *See* DEVENSIAN.

WWSSN *See* WORLD-WIDE STANDARD SEISMOGRAPHIC NETWORK.

WWW *See* WORLD WEATHER WATCH.

xanthophyllite *See* MICA.

Xenian A *system of the Lower *Proterozoic of N. America, from 2500 to 1600 Ma ago.

xenoblastic A textural term applied to *metamorphic rocks which contain *anhedral *porphyroblasts.

xenocryst A crystal in an *igneous rock which has not crystallized from the *melt but has been introduced into the melt from an external source, e.g. the surrounding *country rocks or a previously crystallized part of the same igneous body. Xenocrysts, which are usually in disequilibrium with the melt, contrast with *phenocrysts which have crystallized in equilibrium from the melt.

xenolith An inclusion or enclave of a preexisting rock in an *igneous rock. Xenoliths are often derived from the *country rocks that have been invaded by the igneous mass, and they frequently show some evidence of reaction, e.g. rounding of their edges and *metamorphism.

xenotime *Accessory mineral, YPO and associated with *monazite (Ce, La, Th) PO_4; sp. gr. 4.4–5.1; *hardness 4–5; *tetragonal; yellowish-brown, greyish-white, or pale yellow; pale brown *streak; resinous to vitreous *lustre; crystals *tetragonal *prisms, very similar to *zircon; *cleavage *prismatic; occurs in granitic and *alkaline, *igneous rocks, as well as in *pegmatites and *gneisses.

xenotopic fabric The *fabric of a crystalline *carbonate rock or *cement, or of an *evaporite deposit, in which most of the crystals are *anhedral.

XPL *See* CROSSED POLARS.

xpols *See* CROSSED POLARS.

X-ray diffraction crystallography A method of analysis in which an X-ray beam of known wavelength is directed at a crystal, and the beam is diffracted by reflections off planes of atoms in the crystal. By recording the angular positions of diffracted beams, the spacing between atomic planes can be determined according to the Bragg equation, $n\lambda = 2d \sin\theta$ (*see* BRAGG'S LAW). The procedure is repeated for various directions in the crystal and a model of its internal structure established. In geology, the technique is used to identify minerals.

X-ray fluorescence (XRF) The secondary X-ray emission that results when *electrons from the outer orbitals of an atom fill vacant inner orbital positions originally created by X- or *gamma-ray excitation. These X-rays are characteristic of excited atoms.

X-ray fluorescence spectrometry An analytical method which can be used to determine the concentration of a wide range of chemical elements, using the intensity of their fluorescent X-rays. An X-ray beam is used to excite atoms in a sample; *electrons near the *nucleus emit secondary or fluorescent X-rays on reversion to their original states. Short-wavelength X-rays are sorted by diffraction in a pure analysing crystal of known d-spacing (*see* COVALENT RADIUS). Since $n\lambda = 2d \sin\theta$ (*see* BRAGG EQUATION), θ can be set to a value and radiation detected for a unique wavelength characteristic of the element being analysed. The intensity of the radiation measured, relative to a standard, is proportional to the concentration of the element. It is an important technique in the geochemical analysis of rocks.

X-ray photography A photographic technique in which X-rays are used that may reveal internal detail (e.g. of a *fossil) which is not visible externally. The technique has the advantage that it is non-destructive.

X-ray powder photograph A photograph produced by monochromatic X-irradiation of a sample of *microcrystalline powder placed at the centre of a circular camera, e.g. a Debye-Scherrer camera. Diffracted X-rays are recorded on a strip of film wrapped around the circumference of the camera. The angular position of the diffracted X-rays on the film gives structural information about the sample. *See also* X-RAY DIFFRACTION CRYSTALLOGRAPHY.

X-ray spectrometer An instrument for

measuring the secondary X-rays emitted when the inner *electrons of an atom (W, Au, etc.) are activated by a primary hard X-ray beam. The content of nearly all elements with an atomic weight greater than that of sodium can be estimated.

XRF *See* X-RAY FLUORESCENCE.

Yapeenian A *stage of the Lower *Ordovician of Australia, underlain by the *Castlemainian and overlain by the *Darriwilian.

yardang A streamlined, wind-sculptured hill ranging in length from metres to kilometres and developed in any bedrock that is at least weakly consolidated. Yardangs are restricted to deserts that have high aridity with minimal plant cover and soil development, and that are dominated by strong, unidirectional winds for most of the year.

Yarmouthian The second (0.7–0.55 Ma) of four *interglacial stages recognized in mid-continental N. America. It follows the *Kansan glacial episode and is equivalent to the upper part of the *Günz/Mindel Interglacial of the Alps. At various times during the Yarmouthian climates were both warmer and cooler than they are now.

Yatalan The final *stage in the *Tertiary in south-eastern Australia, underlain by the *Kalimnan, overlain by the *Werrikooian (*Quaternary), and roughly contemporaneous with the upper *Piacenzian Stage.

yazoo stream A river tributary that may flow for many kilometres on the *floodplain of a trunk stream before finally joining it. The delayed junction is due to the development of *levées by the main stream. They are named after the Yazoo River, Mississippi, USA, which follows the Mississippi River for a considerable distance before breaking through its natural levée.

Yeadonian A *stage in the *Bashkirian Epoch, preceded by the *Marsdenian and followed by the *Cheremshanskian.

yellowcake Concentrated, precipitated, and dried uranium oxide.

yield–depression curve A graph on which the *drawdown is plotted against the yield of a pumped well or *borehole. The resulting plot is invariably curved, and is used to determine the optimum pumping rate for a water supply.

yield point *See* ELASTIC LIMIT.

yield stress The *stress at which the yield strength of a material is exceeded and elastic behaviour gives way to viscous behaviour (*see* VISCOSITY). If continued the stress may lead to the 'failure stress', beyond which *failure occurs.

Ynezian A *stage in the Lower *Tertiary of the west coast of N. America, underlain by the N. American *Danian, overlain by the *Bulitian, and roughly contemporaneous with the upper Danian and lower *Thanetian Stages of Europe.

Younger Dryas *See* DRYAS.

younging *See* FACING DIRECTION.

Young's modulus (*E*) The ratio of longitudinal *stress σ (force F divided by area A, i.e. $\sigma = F/A$) to longitudinal *strain (change in length δL divided by original length L, i.e. $\delta L/L$) in the presence of lateral strain: $E = (F/A)/(\delta L/L)$. If there were no lateral strain, Young's modulus would be equal to the *axial modulus.

Yovian A *system of the Middle *Proterozoic of N. America, from 1600 to 800 Ma ago.

Ypresian **1.** The earliest *age in the *Eocene *Epoch, preceded by the *Thanetian (*Palaeocene), followed by the *Lutetian, and dated at 56.5–50 Ma (Harland et al., 1989). **2.** The name of the corresponding European *stage, named after the Ypres Clay, Belgium. It is roughly contemporaneous with the *Penutian and lower *Ulatizian (N. America), upper *Waipawan, *Mangaorapan and lower *Heretaungan (New Zealand), and the lower *Johannian (Australia).

Yurmatian (Yurmatin) A *system of the Middle *Proterozoic, from about 1375 to 1050 Ma ago (Harland et al., 1989), of western Russian origin.

Z

Z The vertical component of the *geomagnetic field.

Zanclian 1. An *age in the Early *Pliocene, preceded by the *Messinian (*Miocene), followed by the *Piacenzian (Late Pliocene), and dated at 5.2–3.4 Ma (Harland et al., 1989). **2.** The name of the corresponding European *stage, which has a greater correlation potential than the contemporary Tabianian Stage. It is roughly contemporaneous with part of the *Delmontian (N. America), upper *Kapitean and *Opoitian (New Zealand), and the upper *Mitchellian, *Cheltenhamian, and lower *Kalimnan (Australia).

'zap pits' (microcraters) Small craters, ranging in size from less than 1 μm to about 1 cm, on the exposed surfaces of lunar samples, caused by the hypervelocity impact of micrometeorites and dust particles with masses less than about 10^{-3} g; Typical craters consist of a glass-lined pit, a 'halo' zone of fractured material surrounding and underlying the pit, and a spalled zone concentric to the pit. This *spallation may leave the glass-lined pit standing on a pedestal.

Zechstein The final *epoch of the *Permian Period, dated at 256.1–245 Ma (Harland et al., 1989).

Zechstein Sea Name given to an Upper *Permian shallow gulf sea or depositional sequence of rocks that developed in northern Germany and the North Sea Basin. The depositional sequence laid down in this sea consisted of *carbonates and *evaporites. *See also* SALINE GIANT.

Zedian A *system of the Upper *Proterozoic of N. America, from 800 Ma to about 590 Ma ago.

Zelzate *See* DENEKAMP.

Zemorrian A *stage in the *Tertiary of the west coast of N. America, underlain by the *Refugian, overlain by the *Saucesian, and roughly contemporaneous with the *Rupelian and *Chattian (Lower Tertiary), and lower *Aquitanian (Upper Tertiary) Stages.

zeolite facies A set of metamorphic mineral assemblages produced by the *metamorphism of a wide range of starting rock types under the same metamorphic conditions, and typically characterized by the development of the mineral assemblage *smectite–*zeolites (in addition to relict *igneous *plagioclase and *pyroxene) in rocks of *basic igneous composition. Other rocks of contrasting composition, e.g. *shales, would develop their own specific mineral assemblages, even though they are all metamorphosed under the same conditions. The variation of mineral assemblage with starting rock composition reflects a particular range of pressure, temperature, and $P(H_2O)$ conditions. Experimental studies of mineral P–T stability fields indicate that the facies represents a range of conditions involving low pressures (1–4 kb) and low temperatures (300–500 °C), usually developed in (*a*) thick sedimentary sequences on *continental margins, (*b*) rock sequences subjected to hot water-convecting systems, and (*c*) areas of *tectonic burial at the leading edge of *thrust sheets.

zeolites Group of hydrated alumina silicates of sodium, potassium, calcium, and barium, which occur in *geodes, altered *igneous rocks, *hydrothermal veins, and some *sediments. The water molecules are weakly held and the hydration–dehydration reaction has some useful applications, as do the ion-exchange properties of zeolites.

zephyr A prevailing light and warming breeze from the west at the time of the summer *solstice (in the northern hemisphere).

zero-length spring A spring system in which the effective length is zero when measured from a fixed point. This type of spring is commonly used in *gravimeters.

zeugen Mushroom-shaped rock that has been eroded by the abrasive action of wind-blown *sand. The undercutting effect is concentrated near ground level, where sand movement is greatest, and is enhanced in areas of near-horizontal strata when the lowest bed is relatively weak.

Z-fold In a *parasitic fold, an asymmetric fold whose profile is Z-shaped, reflecting its location on the respective limb of a major fold. *See* MINOR FOLD.

zibar A low-relief, rounded, coarse-grained, sand dune with no *slipfaces. Regularly spaced zibars produce an undulating surface on otherwise level ground.

zig-zag fold *See* CHEVRON FOLD.

zinc blende *See* SPHALERITE.

zincite *See* OXIDES.

Zingg diagram A diagram, introduced by T. Zingg in 1935, that is used to plot the relative dimensions of the long, short, and intermediate axes of a particle, allowing its shape to be classified as bladed (*see* BLADE), *oblate, *equant, or *prolate.

zinnwaldite A member of the *mica group of 2:1 *phyllosilicates (sheet silicates) with formula $K_2(Fe^{2+}{}_{2-1},Li_{2-3},Al_2)[Si_{6-7}Al_{2-1}O_{20}](F,OH)_4$; sp. gr. 2.9–3.2; *hardness 2–3; *monoclinic; grey, brown, or even dark green; *vitreous to *pearly *lustre; forms thick, *tabular crystals; *cleavage well marked; occurs in *granites and *pegmatites in association with *scheelite, *wolframite, *cassiterite, and *fluorite. It is pneumatolytic in origin (*see* PNEUMATOLYSIS). It is named after Zinnwald, Saxony, Germany.

zircon Mineral, $ZrSiO_4$; sp. gr. 4.6; *hardness 7.5; *tetragonal; most commonly light brown or reddish-brown, but sometimes grey, yellow, or green; vitreous *lustre; crystals usually square *prisms with bipyramidal terminations; *cleavage *prismatic, indistinct {110} and poor {111}; one of the most widely distributed *accessory minerals in *igneous rocks (e.g. *granite, *syenite, and *pegmatites), when the crystals can be quite large, also occurs in *metamorphic rocks (e.g. *gneisses and *schists), and concentrated in *detrital beach and river *sands. It is the main source of zirconium metal. Common zircon is used extensively in the foundry industry.

Zoantharia (zoantharian corals, Hexacorallia; class *Anthozoa) Subclass of solitary and colonial anthozoans which are many tentacled and have an enteron (gastrovascular cavity) divided by numerous paired *mesenteries. The basal tissues secrete a corallum (cup) made of *aragonite, which has an outer epitheca (skin) and radially arranged septa (*see* SEPTUM). The mesenteries are placed between the septa. The first group appeared in the *Ordovician and is divided into four orders: *Rugosa; *Scleractinia; *Tabulata; and *Heterocorallia. Scleractinian zoantharians (*Mesozoic to Recent) are responsible at the present time for building coral *reefs, restricted to tropical areas. They may build wave-resistant structures, sometimes very thick, and are usually associated with other organisms, e.g. calcareous *algae. Organic build-ups also took place in the *Palaeozoic, and although zoantharians were present they did not always represent the major part of the fauna. Stromatoporoids (*Stromatoporoidea), algae, etc., are associated with zoantharians in these build-ups and it is not always certain whether Palaeozoic reefs were comparable with modern or Mesozoic examples. It is unlikely that Palaeozoic zoantharians were reef-frame builders in the sense that modern scleractinians are, and probable that organisms other than corals formed the wave-resistant structures.

zodiacal light Zone of light apparent at night, after twilight, and again before morning twilight, caused by atmospheric particles scattering sunlight.

zoisite Member of the *epidote group of minerals, with the formula $Ca_2Al_3Si_3O_{12}OH$; sp. gr. 3.3; *hardness 6.5; greyish-white or greenish; crystals *orthorhombic, *prismatic, but normally *massive or *columnar; occurs in regional-metamorphic (*see* REGIONAL METAMORPHISM) rocks. Thulite is a rose-pink variety.

zonal 1. Applied to winds blowing in a mainly west-to-east or east-to-west direction, particularly in describing the main, broad airstreams of the general or large-scale atmospheric circulation, e.g. the zonal westerly winds of middle latitudes. The zonal (or circulation) index is a conventional measurement indicating the strength of the west-to-east airflow over middle latitudes, e.g. in latitudes 35–55° N. Strong westerlies with the pressure systems also in a west-to-east orientation accompany high values of the zonal index. Cellular or meridional airflow (or sometimes weak and chaotic patterns of circulation) accompany low values of the zonal index. **2.** Applied to features (e.g. soils and vegetation) characteristic of a particular region that is approximately bounded by lines of latitude (i.e. a region lying parallel to the equator).

zonal flow The winds that blow in a mainly west-to-east or east-to-west direction, and particularly to the main, broad airstreams of the general or large-scale atmospheric circulation (e.g. the zonal westerly winds of middle latitudes). The zonal (or circulation) index is a conventional measurement indicating the strength of the west-to-east airflow over middle latitudes.

zonal index *See* ZONAL FLOW.

zonal scheme Scheme concerned with the use of numerous *fossils as stratigraphic indicators for the subdivision of a sequence of rocks.

zonation The subdivision of a *stratigraphic unit or units by means of *fossils.

Zond A series of Soviet lunar missions that ran from 1965 to 1970.

zone **1. (biostratigraphic zone)** A unit of rock characterized by a clearly defined *fossil content. To avoid confusion with other types of zone the term '*biozone' (short for 'biostratigraphic zone') is preferred by many authorities, although the term 'biozone' is also used in a different sense. The term is usually qualified to denote the type of zone. *See* ACME ZONE; ASSEMBLAGE ZONE; CONCURRENT RANGE ZONE; LINEAGE ZONE; OPPEL ZONE; RANGE ZONE; SUBZONE; TAXON RANGE ZONE; TEILZONE; ZONULE; and INDEX FOSSIL. **2.** In *crystallography, a set of *crystal faces whose intersecting edges are parallel. They are also parallel to, and may be rotated about, a *zone axis. *See* CRYSTAL ZONING. **3.** *See* METAMORPHIC ZONE.

zone axis An axis of a crystal, which normally passes through the centre of the crystal and may or may not be parallel to the *crystallographic axes, but which is parallel to the edges of *crystal faces that meet in a *zone, so those edges may be rotated about it.

zone fossil *See* INDEX FOSSIL.

zone of aeration *See* SOIL-WATER ZONE.

zone of saturation *See* PHREATIC ZONE.

zone refining **1.** In metallurgy, the small-scale production of high-purity metals using the principle that an impure *melt will deposit pure crystals on solidifying. A rod of the specimen is melted over a very narrow region at one end, and then the molten region is transferred along the rod by means of a moving furnace. Impurities collect in the molten zone and are swept to one end of the metal. **2.** In geology, the same mechanism has been invoked whereby *incompatible *trace elements are partitioned into an advancing *partial melt.

zone symbol ([UVW]) The symbol which defines the position of the *zone axis of a crystal with respect to the *crystallographic reference axes of the mineral. It is enclosed in square brackets to indicate that it refers to a line and not a plane. It may be calculated from the *Miller indices of any two faces in the *zone, or determined by plotting the faces on a *stereographic projection.

zonule Subdivision of a subzone based essentially on the presence of a given species. Several zonules are recognized, for example, in the Upper *Cambrian Franconia Sandstone of Minnesota. Here the *zone was recognized by the presence of the trilobites (*Trilobita) *Ptychaspis* and *Prosaukia*, the subzone by *Prosaukia*, and zonules by *Prosaukia striata* and *P. granulosa*.

zoogeographical region *See* FAUNAL REALM.

zoophycus With **Chondrites*, an *ichnoguild of many-branched, radial, *trace fossils probably made by a worm that moved back and forth through the sediment, each branch of its burrow exploring a new area. They are believed to be *fodinichnia.

Zosterophyllum A Lower *Devonian psilopsid plant (*Psilophytales) whose lower parts underwent H-branching, producing a tufted growth habit. Erect branches were smooth and the sporangia (*see* SPORE) were grouped together on terminal spikes.

Zosterophyllophytina *See* PSILOPHYTALES.

zygapophyses *See* VERTEBRA.

Appendix A: Stratigraphic Units as Defined in the North American Stratigraphic Code, 1983

Material Units

Lithostratigraphic	Lithodemic	Magnetopolarity	Biostratigraphic	Pedostratigraphic	Allostratigraphic
Supergroup	Supersuite				
Group	Suite	Polarity superzone			Allogroup
Formation	*Lithodeme* – Complex	*Polarity zone*	*Biozone* (interval, assemblage, abundance)	*Geosol*	*Alloformation*
Member (lens, tongue)		Polarity subzone Subbiozone realism			Allomember
Bed(s) (flow(s))					

Temporal and Chronostratigraphic Units

Chronostratigraphic	Geochronologic, geochronometric	Polarity chronostratigraphic	Polarity chronologic	Diachronic
Eonothem	Eon	Polarity Superchronozone	Polarity Superchron	
Erathem (supersystem)	Era (superperiod)			
System (subsystem)	*Period* (subperiod)	*Polarity Chronozone*	*Polarity Chron*	*Episode*
Series	Epoch		Diachron	Phase
Stage (substage)	Age (subage)	Polarity Subchronozone	Polarity Subchron	Span
Chronozone	Chron			Cline

(Fundamental units shown in italic)

Appendix B: Time Scales

Geologic Time-Scale

Eon/Eonothem	Era/Erathem	Sub-era	Period/System	Epoch/Series	Starting Ma
	Cenozoic	Quaternary	Pleistogene	Holocene	0.01
				Pleistocene	1.64
		Tertiary	Neogene	Pliocene	5.2
				Miocene	23.3
			Palaeogene	Oligocene	35.4
				Eocene	56.5
				Palaeocene	65
	Mesozoic		Cretaceous		145.6
			Jurassic		208
PHANEROZOIC			Triassic		245

Eon	Era	Sub-era	Period		Age (Ma)
	Palaeozoic	Upper Palaeozoic	Permian		290
			Carboniferous		362.5
			Devonian		408.5
		Lower Palaeozoic	Silurian		439
			Ordovician		510
			Cambrian		570
PRECAMBRIAN: PROTEROZOIC					2500
PRECAMBRIAN: ARCHAEAN					4000
PRECAMBRIAN: PRISCOAN					4600

(After Harland et al., 1989)

Lunar Time-Scale

Stratigraphic mapping of the lunar surface has established the following Systems, in order of increasing age:

System	Events
Copernican	Young ray craters (younger than about one billion years)
Eratosthenian	Older post-mare craters (about 1–3 billion years)
Imbrian	Main mare basalt flooding, preceded by formation of the Imbrium and Orientale basins (3–3.85 billion years)
Pre-Imbrian:	
Nectarian	Formation of 11 major impact basins preceded by the Nectaris basin (3.85–3.92 billion years)
Pre-Nectarian	Formation of about 30 impact basins preceded by the Procellarum basin (before 3.92–4.2? billion years)

The crystallization of the anorthositic lunar highland crust has been dated at 4.44 billion years. This event closely followed the formation of the Moon.
Note that the term 'Procellarum System', to encompass the main period of mare lava flooding, has been abandoned.

Martian (Areological) Time-Scale

Epoch	Absolute age (Gy) (Hartmann–Tanaka Model)	Absolute age (Gy) (Neukum–Wise Model)
Upper Amazonian	0.25–0.00	0.70–0.00
Middle Amazonian	0.70–0.25	2.50–0.70
Lower Amazonian	1.80–0.70	3.55–2.50
Upper Hesperian	3.10–1.80	3.70–3.55
Lower Hesperian	3.50–3.10	3.80–3.70
Upper Noachian	3.85–3.50	4.30–3.80
Middle Noachian	3.92–3.85	4.50–4.30
Lower Noachian	4.60–3.92	4.60–4.50

(These represent two different models published by the University of Arizona Press)

Appendix C: Wind Strength

The Beaufort Scale of Wind Strength

Force 0. 1 mph or less. Calm. The air feels still and smoke rises vertically.

Force 1. 1–3 mph. Light air. Wind vanes and flags do not move, but rising smoke drifts.

Force 2. 4–7 mph. Light breeze. Drifting smoke indicates the wind direction.

Force 3. 8–12 mph. Gentle breeze. Leaves rustle, small twigs move, and flags made from lightweight material stir gently.

Force 4. 13–18 mph. Moderate breeze. Loose leaves and pieces of paper blow about.

Force 5. 19–24 mph. Fresh breeze. Small trees that are in full leaf wave in the wind.

Force 6. 25–31 mph. Strong breeze. It becomes difficult to use an open umbrella.

Force 7. 32–38 mph. Moderate gale. The wind exerts strong pressure on people walking into it.

Force 8. 39–46 mph. Fresh gale. Small twigs are torn from trees.

Force 9. 47–54 mph. Strong gale. Chimneys blown down, slates and tiles torn from roofs.

Force 10. 55–63 mph. Whole gale. Trees are broken or uprooted.

Force 11. 64–75 mph. Storm. Trees are uprooted and blown some distance. Cars are overturned.

Force 12. More than 75 mph. Hurricane. Devastation is widespread. Buildings are destroyed, many trees uprooted. In the original instruction, 'no sail can stand'.

Saffir/Simpson Hurricane Scale

Category	Pressure at centre (inches of mercury)	Wind speed (mph)	Storm surge (feet)
1	28.94	74–95	4–5
2	28.5–28.91	96–110	6–8
3	27.91–28.47	111–130	9–12
4	27.17–27.88	131–155	13–18
5	below 27.17	more than 155	more than 18

Fujita Tornado Intensity Scale

Rating	Wind speed (mph)	Damage expected
F-0	40–72	Light damage
F-1	73–112	Moderate damage
F-2	113–157	Considerable damage
F-3	158–206	Severe damage
F-4	207–260	Devastating damage
F-5	261–318	Incredible damage

Appendix D

SI Units
(Système International d'Unités)

Quantity	Name of unit	Symbol	Equivalent	Reciprocal
length	metre	m	3.281 feet	1 ft = 0.3048 m
mass	kilogram	kg	2.2 pounds	1 lb = 0.454 kg
time	second	s		
electric current	ampere	A		
thermodynamic temperature	kelvin	K	1 °C = 1.8 ° F	1 °C = 1 K
luminous intensity	candela	cd		
amount of substance	mole	mol		

Supplementary Units

Quantity	Unit	Symbol
plane angle	radian	rad
solid angle	steradian	sr

Derived SI units

Quantity	Name of unit	Symbol	Equivalent	Reciprocal
frequency	hertz	Hz		
energy	joule	J	0.2388 calories	1 cal = 4.1868 J
force	newton	N	0.225 pounds force	1 lbf = 4.448 N
power	watt	W	0.00134 horse power	1 hp = 745.7 W
pressure	pascal	Pa	0.00689 pounds force/sq. inch	1 lbf/sq.in = 145 Pa
electric charge	coulomb	C		
electric potential difference	volt	V		
electric resistance	ohm	Ω		
electric conductance	siemens	S		
electric capacitance	farad	F		
magnetic flux	weber	Wb		
inductance	henry	H		
magnetic flux density	tesla	T		
luminous flux	lumen	lm		
illuminance	lux	lx		
absorbed dose	gray	Gy		
activity	becquerel	Bq		
dose equivalent	sievert	Sv		

Multiples used with SI units

Name of multiple	Symbol	Value (multiply by)
atto	a	10^{-18}
femto	f	10^{-15}
pico	p	10^{-12}
nano	n	10^{-9}
micro	μ	10^{-6}
milli	m	10^{-3}
centi	c	10^{-2}
deci	d	10^{-1}
deca	da	10
hecto	h	10^{2}
kilo	k	10^{3}
mega	M	10^{6}
giga	G	10^{9}
tera	T	10^{12}
peta	P	10^{15}
exa	E	10^{18}

Bibliography

Abelson, Philip H. (1997) 'Improved fossil energy technology', *Science*, **276**, 511.

Adams, A. E., MacKenzie, W. S., and Guilford, C. (1984) *Atlas of Sedimentary Rocks Under the Microscope*. Longman, London.

Ager, D. V. (1973) *The Nature of the Stratigraphical Record*. Macmillan, London.

Ahmed, H., Dillon, P. B., Johnstad, S. E., and Johnston, C. D. (1986) 'Northern Viking Graben multilevel three-component walkaway VSPs, a case history', *First Break*, **4**, *10*, 9–27.

Anderton, R., Bridges, P. M., Leeder, M. R., and Sellwood, B. W. (1980) *A Dynamic Stratigraphy of the British Isles*. Allen and Unwin, London.

Attewell, P. B. and Farmer, I. W. (1975) *Principles of Engineering Geology*. Chapman and Hall, London.

Barker, D. S. (1983) *Igneous Rocks*. Prentice-Hall, Englewood Cliffs, New Jersey.

Barnes, John W. (1981) *Basic Geological Mapping*. Geol. Soc. of London Handbook. Open Univ. Press, Milton Keynes, and Halstead Press (John Wiley and Sons), New York, Toronto.

Bates, D. E. B. and Kirkcaldy, J. F. (1976) *Field Geology in Colour*. Blandford Press, Poole.

Battey, M. H. (1981) *Mineralogy for Students*. 2nd edn. Longman, London.

Beatty, J. K. et al. (1982) *The New Solar System*. Sky Pubns., Cambridge, Mass.

Beavis, F. C. (1985) *Engineering Geology*. Blackwell, Oxford.

Beck, A. E. (1981) *Physical Principles of Exploration Methods*. Macmillan Press, London.

Beddow, J. K. (1980) *Particulate Science and Technology*. New York Chemical Publishing Co. Inc., New York.

Best, M. G. (1982) *Igneous and Metamorphic Petrology*. W. H. Freeman, New York.

Billings, Marland P. (1972) *Structural Geology*. 3rd edn. Prentice-Hall, Englewood Cliffs, New Jersey.

Bishop, A. C. (1967) *An Outline of Crystal Morphology*. Hutchinson, London.

Black, Rhona M. (1970) *The Elements of Palaeontology*. Cambridge Univ. Press, Cambridge.

Blatt, H., Middleton, G., and Murray, R. (1980) *Origin of Sedimentary Rocks*. Prentice-Hall, Englewood Cliffs, New Jersey.

Bloom, Arthur L. (1969) *The Surface of the Earth*. Foundations of Earth Science Series, Prentice-Hall, Englewood Cliffs, New Jersey.

—— (1978) *Geomorphology*. Prentice-Hall, Englewood Cliffs, New Jersey.

Blundell, Derek (1996) 'The European GeoTraverse: the anatomy of a continent', *OUGS Jo.*, **17**, *2*, 13–18.

Blyth, F. G. H. and de Freitas, M. H. (1984) *A Geology for Engineers*. 7th edn. Edward Arnold, London.

Boggs, Sam Jr. (1995) *Principles of Sedimentology and Stratigraphy*. 2nd edn. Prentice Hall, Upper Saddle River, New Jersey.

Boillot, G. (1981) *Geology of the Continental Margins*. Longman, Harlow, Essex (first published in French in 1978).

Borradaile, G. J., Bayly, M. B., and Powell, C. McA. (1982) *Atlas of Deformational and Metamorphic Rock Fabrics*. Springer-Verlag, New York.

Bott, M. H. P. (1982) *The Interior of the Earth*. 2nd edn. Edward Arnold, London.

Boucot, A. J. (1984). 'Ecostratigraphy', in Seibold, E. and Meulenkamp, J. D. (eds.)

Stratigraphy Quo Vadis? AAPG Studies in Geology No. 16, IUGS Special Publication No. 14. American Association of Petroleum Geologists, Tulsa, Oklahoma.

Bowen, D. Q. (1978) *Quaternary Geology: A Stratigraphic Framework for Multidisciplinary Work*. Pergamon Press, Oxford.

Boyer, S. E. and Elliot, D. (1982) 'Thrust systems', *Bull. Am. Assocn. of Petroleum Geologists*, **66**, *9*, 1196–1230.

Brady, N. C. (1974) *The Nature and Properties of Soils*. 8th edn. Macmillan, London.

Brasier, M. D. (1980) *Microfossils*. George Allen and Unwin, London.

Bromley, Richard G. (1996) *Trace Fossils: Biology, Taphonomy and Applications*. 2nd edn. Chapman and Hall, London.

Brown, G. C. and Mussett, A. E. (1981) *The Inaccessible Earth*. George Allen and Unwin, London.

—— and Skipsey, E. (1986) *Energy Resources. Geology Supply and Demand*. Open Univ. Press, Milton Keynes, England.

Burns, J. A. and Matthews, M. S. (1986) *Satellites*. Arizona Univ. Press, Tucson.

Buss, Leo W. and Seilacher, Adolf (1994) 'The phylum Vendobionta: a sister group of the Eumetazoa?', *Paleobiology*, **20**, *1*, 1–4.

Butler, R. W. M. (1982) 'The terminology of structures in thrust belts', *J. Struct. Geol.*, **4**, *3*, 239–245.

—— (1987) 'Thrust sequences', *J. Geol. Soc.*, **144**, *4*, 619–34.

Carmichael, I. S. E., Turner, F. J., and Verhoogen, J. (1974) *Igneous Petrology*. McGraw-Hill, New York.

Carr, M. H. (1981) *The Surface of Mars*. Yale Univ. Press, New Haven, Conn.

Cas, R. A. F. and Wright, J. V. (1987) *Volcanic Successions, Modern and Ancient*. George Allen and Unwin, London.

Challinor, John (1971) *The History of British Geology*. David and Charles, Newton Abbot.

Chaloner, W. G. and Macdonald, P. (1980) *Plants Invade the Land*. Roy. Scot. Museum, HMSO, Edinburgh.

Chambers Biographical Encyclopaedia of Scientists (1981) Chambers, Edinburgh.

Chorley, Richard J. (1971) *Introduction to Physical Hydrology*. University Paperbacks (Methuen), London.

Chorley, R. J., Schumm, S. A., and Sugden, D. E. (1984) *Geomorphology*. Methuen, London.

Clarkson, E. N. K. (1979, 1986) *Invertebrate Palaeontology and Evolution*. 1st and 2nd edn. George Allen and Unwin, London.

Clifford, T. N. and Gass, I. G. (1970) *African Magmatism and Tectonics*. Hafner Publishing Co., Darien, Conn.

Cocks, L. R. M. (ed.) (1981) 'Maps of the past', Part V of *Chance, Change and Challenge: The Evolving Earth*. British Museum (Natural History), London, and Cambridge Univ. Press, Cambridge.

Collinson, J. D. and Thompson, D. B. (1982) *Sedimentary Structures*. George Allen and Unwin, London.

Concise Dictionary of American Biography (1964) Charles Scribner's Sons, New York.

Condie, Kent C. (1976) *Plate Tectonics and Crustal Evolution*. Pergamon Press, New York.

Courtney, F. M. and Trudgill, S. T. (1976) *The Soil*. Edward Arnold, London.

Coward, M. P. and Ries, Alison (eds.) (1986) *Collision Tectonics*. Geol. Soc. Special Pubn. 19, Geol. Soc. London.

Cox, Allan and Hart, Robert B. (1986) *Plate Tectonics: How It Works*. Blackwell Scientific Publications, Palo Alto, Calif.

Cox, K. G., Bell, J. D., and Pankhurst, R. J. (1981) *The Interpretation of Igneous Rocks*. George Allen and Unwin, London.

—— Price, N. B., and Harte, B. (1974) *The Practical Study of Crystals, Minerals and Rocks*. McGraw-Hill, Maidenhead, Berks.

Craig, J. R. and Vaughan, D. J. (1981) *Ore Microscopy and Ore Petrography*. John Wiley and Sons, New York.

Cruickshank, James G. (1972) *Soil Geography*. David and Charles, Newton Abbot.

Cummings, A. B. and Given, I. A. (eds.) (1973) *SME Mining Engineering Handbook*, vols. I and II. Society of Mining Engineers, New York.

Curtis, L. F., Courtney, F. M., and Trudgill, S. T. (1976) *Soils in the British Isles*. Longman, London.

Davidson, D. A. (1980) *Soils and Land-use Planning*. Longman, London.

Davis, G. M. (1984) *Structural Geology of Rocks and Regions*. John Wiley and Sons, New York.

Davison, Charles (1927) *The Founders of Seismology*. Cambridge Univ. Press, Cambridge.

Decker, R. W., Wright, T. L., and Stauffer, P. H. (eds.) (1987) *Volcanism in Hawaii*. US Geol. Survey Professional Paper 1350, Washington, DC.

Deer, W. A., Howie, R. A., and Zussman, J. (1966) *An Introduction to the Rock-forming Minerals*. Longman, London.

Derbyshire, E., Gregory, K. J., and Hails, J. R. (1979) *Geomorphological Processes*. Butterworth, London.

Dictionary of National Biography. Oxford Univ. Press, Oxford.

Dineley, David (1979) *Fossils*. Collins, Glasgow.

—— (1984) *Aspects of a Stratigraphic System: The Devonian*. Macmillan, London.

Donahue, Roy L., Miller, Raymond W., and Shickluna, John C. (1977) *Soils: An Introduction to Soils and Plant Growth*. 4th edn. Prentice-Hall, Englewood Cliffs, New Jersey.

Donovan, D. T. (1966) *Stratigraphy: An Introduction to Principles*. George Allen and Unwin, London.

Doyle, James A. and Donoghue, Michael J. (1993) 'Phylogenies and angiosperm diversification', *Paleobiology*, **19**, *2*, 141–67.

Edwards, David S. (1986) '*Aglaophyton major*, a non-vascular land-plant from the Devonian Rhynie Chert', *Bot. Jo. Linnaean Soc.*, **93**, 173–204. Linnaean Society, London.

Edwards, W. N. (1967) *The Early History of Palaeontology*. British Museum (Natural History), London.

Edwards, R. and Atkinson, K. (1986) *Ore Deposit Geology*. Chapman and Hall, London.

Eicher, D. L. (1968) *Geologic Time*. Prentice-Hall, Englewood Cliffs, New Jersey.

Emery, Dominic and Myers, Keith (eds.) (1996) *Sequence Stratigraphy*. Blackwell Science, Oxford.

Evans, A. M. (1987) *An Introduction to Ore Geology*. 2nd edn. (Geoscience Texts, vol. 2). Blackwell, Oxford.

Faul, Henry (1966) *Ages of Rocks, Planets, and Stars*. McGraw-Hill, New York.

—— and Faul, C. (1983) *It Began with a Stone*. John Wiley and Sons, New York.

Faure, G. (1986) *Principles of Isotope Geology*. 2nd edn. John Wiley and Sons, New York.

Fisher, R. U. and Schmincke, H.-U. (1984) *Pyroclastic Rocks*. Springer-Verlag, New York.

Fleuty, M. J. (1964) *Geol. Assoc. Proc.*, **75**, 461–92.

Foth, H. D. and Turk, L. M. (1972) *Fundamentals of Soil Science*. 5th edn. John Wiley and Sons, New York.

Francis, Peter (1976, 1993) *Volcanoes*. Penguin Books, Harmondsworth, Middlesex.

——(1981) *The Planets*. Penguin Books, Harmondsworth.

Fyfe, W. S. (1974) *Geochemistry*. Clarendon Press, Oxford.

Garland, George D. (1971) *Introduction to Geophysics*. W. B. Saunders Co., Philadelphia.

Gass, I. G., Smith, Peter J., and Wilson, R. C. L. (1971) *Understanding the Earth*. Open Univ. Press and Artemis Press, Horsham, Sussex.

Gehrels, T. and Matthews, M. S. (eds.) (1984) *Saturn*. Univ. of Arizona Press, Tucson, Arizona.

Gibbs, A. D. (1983) 'Balanced cross-section construction from seismic sections in areas of extensional tectonics', *J. Struct. Geol.*, **5**, *2*, 153–60.

Gillispie, Charles Coulston (1951) *Genesis and Geology*. Harper and Row, New York.

——(editor in chief) (1970–1980) *The Dictionary of Scientific Biography*. Charles Scribner, New York.

Glass, Billy P. (1982) *Introduction to Planetary Geology*. Cambridge Univ. Press, Cambridge.

Godwin, H. (1940) 'Pollen analysis and forest history of England and Wales', *New Phytology*, **39**, *4*, 370.

——(1975) *History of the British Flora: A Factual Basis for Phytogeography*. 2nd edn. Cambridge Univ. Press, Cambridge.

Goudie, A. (ed.) (1981) *Geomorphological Techniques*. George Allen and Unwin, London.

Greeley, R. (ed.) (1974) *Geological Guide to the Island of Hawaii*. NASA, Washington, DC.

——(1985) *Planetary Landscapes*. George Allen and Unwin, London.

Green, Mott T. (1982) *Geology in Nineteenth Century*. Cornell Univ. Press, Ithaca, New York.

Greensmith, J. T. (1979) *Petrology of the Sedimentary Rocks*. George Allen and Unwin, London.

Gribble, C. D. and Hall, A. J. (1985) *A Practical Introduction to Optical Mineralogy*. George Allen and Unwin, London.

Griffiths, D. H. and King, R. F. (1981) *Applied Geophysics for Geologists and Engineers*. Pergamon Press, Oxford.

Hallam, A. (1973) *A Revolution in the Earth Sciences*. Clarendon Press, Oxford.

——(1983) *Great Geological Controversies*. Oxford Univ. Press, Oxford.

Halstead, L. B. (1982) *Hunting the Past*. Roxby, London.

——and Halstead, Jenny (1981) *Dinosaurs*. Blandford Press, Poole, Dorset.

Hambrey, M. J. and Harland, W. B. (1981) 'The Evolution of Climates', chapter 9 in Cocks, L. R. M. (ed.) *Chance, Change and Challenge: The Evolving Earth*. British Museum (Natural History), London.

Hamilton, E. I. (1965) *Applied Geochronology*. Academic Press, London.

Hamilton, W. R., Woolley, A. R., and Bishop, A. C. (1974) *The Hamlyn Guide to Minerals, Rocks and Fossils*. Hamlyn, London.

Harding, T. P. and Lowell, J. D. (1979) 'Structural styles, their plate tectonic habitats, and hydrocarbon traps in petroleum provinces', *Bull. Am. Assocn. of Petroleum Geologists*, **63**, *7*, 1016.

Harland, W. B. (1971) 'Tectonic transpression in Caledonian Spitzbergen', *Geol. Mag.*, **108**, *1*, 27–42.

——(1975) 'The two geological time scales', *Nature*, **253**, 295–305.

——(1978) *Geochronologic Scales*, in Cohee, G. V., Glaessner, M. F., and Hedberg, H. D. (eds.)

Contributions to the Geologic Time Scale. American Assocn. of Petroleum Geologists (AAPG), Studies in Geology No. 6, Tulsa, Oklahoma.

——Armstrong, R. L., Craig, L. E., Smith, A. G., and Smith, D. G. (1989) *A Geologic Time Scale*. Cambridge Univ. Press, Cambridge.

——Cox, A. V., Llewellyn, P. G., Pickton, C. A. G., Smith, A. G., and Walters, R. (1982) *A Geologic Time Scale*. Cambridge Univ. Press, Cambridge.

——Smith, A. Gilbert, and Wilcock, B. (1964) *The Phanerozoic Time-scale*, vol. 1205, issued as a supplement to *The Quarterly Journal of the Geol. Soc. Lon*. Geological Soc. of London, London.

Harrison, J. E. and Peterman, Z. E. (1982) 'North American Commission on Stratigraphic Nomenclature, Report 9: Adoption of Geochronometric Units for Divisions of Precambrian Time'. *American Assoc. of Petroleum Geologists Bull*., **66**, 6, 801–4.

Hatch, F. H., Wells, A. K., and Wells, M. K. (1972) *Petrology of the Igneous Rocks*. George Allen and Unwin, London.

Hedberg, Hollis D. (ed.) (1976) *International Stratigraphic Guide by International Commission on Stratigraphy*. John Wiley and Sons, Chichester.

Henderson, Paul (1982) *Inorganic Geochemistry*. Pergamon Press, Oxford.

Hills, E. S. (1975) *Elements of Structural Geology*. Chapman and Hall, London.

Hobbs, B. E., Means, W. D., and Williams, P. F. (1976) *An Outline of Structural Geology*. Wiley International Editions, John Wiley and Sons, New York.

Holland, C. H. (1978) *A Guide to Stratigraphic Procedure* (Special Report No. 10). Geological Soc. of London, London.

Holmes, Arthur (1937) *The Age of the Earth*. 2nd edn. Thomas Nelson, London.

Hsü, Kenneth (ed.) (1983) *Mountain Building Processes*. Academic Press, London.

Hubbard, R. J., Pape, J., and Roberts, D. G. (1985) 'Depositional Sequence Mapping as a Technique to Establish Tectonic and Stratigraphic Framework and Evaluate Hydrocarbon Potential on a Passive Continental Margin', in Berg, O. R. and Woolverton, D. G. (eds.) *Seismic Stratigraphy II: An Integrated Approach to Hydrocarbon Exploration*. AAPG Memoir 39, pp. 79–91. American Assocn. of Petroleum Geologists, Tulsa, Oklahoma.

Hunt, Charles B. (1972) *Geology of Soils*. W. H. Freeman, San Francisco.

Hurlbut, C. S. (1971) *Dana's Manual of Mineralogy*. John Wiley and Sons, New York.

Hyndman, D. W. (1985) *Petrology of Igneous and Metamorphic Rocks*. McGraw-Hill, New York.

Institute of Geological Sciences (1974) *Volcanoes*. HMSO, London.

International Subcommission on Stratigraphic Classification of the IUGS Commission on Stratigraphy (1976) *Notes from the International Stratigraphic Guide*. John Wiley and Sons, Chichester.

Kearey, P. and Brooks, M. (1984) *An Introduction to Geophysical Exploration*. Blackwell Scientific Pubns., Oxford.

Keller, Edward A. (1976) *Environmental Geology*. Charles E. Merrill Pub. Co. (Bell and Howell Co.), Columbus, Ohio.

Kenrick, Paul and Crane, Peter R. (1997) 'The origin and early evolution of plants on land', *Nature*, **389**, 33–9.

Kerr, Paul F. (1959) *Optical Mineralogy*. McGraw-Hill, New York.

Khurana, K. K. (1997) *Galileo: A Solar System Explorer*. http://www.igpp.ucla.edu/ssc/galileo/overview.html.

King, Cuchlaine A. M. (1974) *Introduction to Marine Geology and Geomorphology*. Edward Arnold, London.

——(1975) *Introduction to Physical and Biological Oceanography*. Edward Arnold, London.

King, Elbert A. (1976) *Space Geology*. John Wiley and Sons, New York.

Kirkcaldy, J. F. (1967) *Fossils in Colour*. Blandford Press, Poole, Dorset.

Kirsch, H. (1968) *Applied Mineralogy for Engineers, Technologists and Students* (German ed. 1965, translated by K. A. Jones). Chapman and Hall and Science Paperbacks, London.

Knighton, D. (1984) *Fluvial Forms and Processes*. Edward Arnold, London.

Krauskopf, Konrad B. (1979) *Introduction to Geochemistry*. International Student Edition. 2nd edn. McGraw-Hill, Tokyo.

Leeder, M. R. (1982) *Sedimentology*. George Allen and Unwin, London.

Lowe, J. J. and Walker, M. J. C. (1984) *Reconstructing Quaternary Environments*. Longman, London.

McBirney, A. R. (1984) *Igneous Petrology*. Freeman, Cooper and Co., San Francisco.

McClay, K. R. and Price, N. J. (eds.) (1981) *Thrust and Nappe Tectonics*. Geol. Soc. London Special Publication 9.

MacFadden, Bruce J., and Shockey, Bruce J. (1997) 'Ancient feeding ecology and niche differentiation of Pleistocene mammalian herbivores from Tarija, Bolivia: morphological and isotopic evidence', *Paleobiology*, **23**, *1*, 77–100.

McKinnon, William B. (1997) 'Galileo at Jupiter—meetings with remarkable moons', *Nature*, **390**, 23–6.

McLean, A. C. and Gribble, C. D. (1985) *Geology for Civil Engineers*. 2nd edn. (revised by C. D. Gribble). George Allen and Unwin, London.

McQuillin, R., Bacon, M., and Barclay, W. (1984) *An Introduction to Seismic Interpretation*. Graham and Trotman, London.

Mason, Brian (1966) *Principles of Geochemistry*. 3rd edn. John Wiley and Sons, New York.

—— and Moore, Carleton B. (1982) *Principles of Geochemistry*. 4th edn. John Wiley and Sons, New York.

Mason, R. (1978) *Petrology of the Metamorphic Rocks*. George Allen and Unwin, London.

Mather, John D. (1996) 'Exploration for waste disposal: the example of radioactive waste', *OUGS Jo.*, **17**, *2*, 48–52.

Mather, Kirtley F. (ed.) (1967) *Source Book in Geology 1900–50*. Harvard Univ. Press, Cambridge, Mass.

—— and Mason, Shirley L. (eds.) (1967) *A Source Book in Geology 1400–1900*. Harvard Univ. Press, Cambridge, Mass.

Maxwell, J. A. (1968) *Rock and Mineral Analysis*. John Wiley and Sons, New York.

Meadows, Jack (1985) *Space Garbage*. George Philip, London.

Mellanby, Kenneth (1992) *Waste and Pollution*. HarperCollins, London.

Middlemost, Eric A. K. (1985) *Magmas and Magmatic Rocks*. Longman, London.

Mitton, S. (ed.) (1977) *The Cambridge Encyclopaedia of Astronomy*. Jonathan Cape, London.

Miyashiro, Akiho, Aki, Keiiti, and Şengör, Celâl (1982) *Orogeny*. John Wiley and Sons, Chichester (first published in Japanese in 1979).

Mottana, A. (1977) *The Macdonald Encyclopaedia of Rocks and Minerals*. Arnoldo Mondadori Editore, Milan.

Mutch, T. A. (1976) *The Geology of Mars*. Princeton Univ. Press, Princeton, N. J.

Nield, E. W. and Tucker, V. C. T. (1985) *Palaeontology: An Introduction*. Pergamon Press, Oxford.

Nockolds, S. R., Knox, R. W. O'B., and Chinner, G. A. (1978) *Petrology for Students*. Cambridge Univ. Press, Cambridge.

Open University (1971) Course unit S100/25. Open Univ. Press, Milton Keynes.

——(1971) Course Unit S100/22–25.

——(1971) Course Unit S100/26.

——(1972) *Palaeontology and Geological Time*.

——(1972) Course Units *Geochemistry* S2–2.

——(1972) Course Units S2–3.

——(1972) *Internal Processes*, Course Unit S23, Block 4.

——(1974) Course Unit S26.

——(1976) S333 *Techniques Handbook*.

——(1976) 'Porphyry Copper Case Study', Course S333.

——(1976) *Sedimentary Basin Case Study*, Course S333.

——(1976) *Urban Geology*, Course S333.

——(1976) *Lunar Geology Case Study*, Course S333.

——(1977) *Oceanography*, Course Units S334, 3, 5, and 6.

——(1980) Glossary to *Oceanography*, Course S334.

——(1980) *Palaeoclimatology Case Study*, Course S335.

——(1980) *Changing Sea-Levels: A Jurassic Case Study*, Course S333.

——(1980) *Crustal and Mantle Processes: Red Sea Case Study*, Course Unit S336.

——(1981) *Earth Composition*, Course Unit S237, Block 1.

——(1981) *Earth Dynamics*, Course Unit S237, Block 4.

——(1981) *The Evolution of Fish and Amphibians*, Course S364, 'Evolution', Unit 5.

——(1981) *The Evolution of Reptiles, Birds and Mammals*, Course S364, 'Evolution', Unit 6.

——(1981) *Block 2; Earth Structure: Earthquakes, Seismology, and Gravity*.

——(1982) *Handbook*, Course S364, 'Evolution'.

——(1982) *Making Sense of Skulls and Bones*, Course S364, 'Evolution', Audio-Visual 6.

——(1984) *Energy Resources II: nuclear and other options*, Course S238, Block 5, Part II.

Parasnis, D. S. (1986) *Principles of Applied Geophysics*. 4th edn. Chapman and Hall, London.

Palmer, A. R. (1983) 'The decade of North American geology, 1983; geologic time scale', *Geology*, **2**, 503–4.

Park, R. G. (1983) *Foundations of Structural Geology*. Blackie, Glasgow.

——and Tarney, J. (1987) *Evolution of the Lewisian and Comparable High Grade Terrains*. For Geol. Soc. London, Blackwell Scientific Pubns., Oxford.

Parkinson, W. D. (1983) *Introduction to Geomagnetism*. Scottish Academic Press, Edinburgh.

Paterson, M. S. (1978) *Experimental Rock Deformation: The Brittle Field*. Springer-Verlag, New York.

Pauling, L. (1960) *The Nature of the Chemical Bond*. 3rd edn. Cornell Univ. Press, Ithaca, New York.

Peele, R. (1941) *Mining Engineers Handbook*. 3rd edn. 2 vols. John Wiley and Sons, New York.

Pennington, Winifred (1974) *The History of British Vegetation*. English Univ. Press, London.

Pethick, J. (1984) *An Introduction to Coastal Geomorphology*. Edward Arnold, London.

Phillips, F. C. (1955) *The Use of Stereographic Projection in Structural Geology*. Edward Arnold, London.

Phillips, W. R. and Griffen, D. T. (1981) *Optical Mineralogy: The Non-opaque Minerals*. W. H. Freeman, San Francisco.

Pike, R. J. (1980) *Geometric Interpretation of Lunar Craters*. US Geol. Survey Professional Paper 1046-C., Washington, DC.

Porter, Roy (1977) *The Making of Geology*. Cambridge Univ. Press, Cambridge.

Posamentier, Henry W., Summerhayes, Colin P., Haq, Bilal U., and Allen, George P. (eds.) (1993) *Sequence Stratigraphy and Facies Associations*. Special Publication No. 18 of the International Assoc. of Sedimentologists. Blackwell Scientific Publications, Oxford.

Powell, C. McA. (1979) 'A morphological classification of rock cleavage', *Tectonophysics*, **58**, 21–34.

Purnell, Mark (1994) 'Conodonts and the Cambrian origin of vertebrate skeletons', *Jo. Open Univ. Geol. Soc.*, **15**, *2*, 6–9.

Ragan, D. M. (1973, 1984) *Structural Geology: An Introduction to Geometrical Techniques*. 2nd and 3rd edns. John Wiley and Sons, New York.

Ramsay, J. G. (1967) *Folding and Fracturing of Rocks*. McGraw-Hill, New York.

—— and Huber, M. J. (1983) *The Techniques of Modern Structural Geology*, vol. 1 *Strain Analysis*. Academic Press, London.

—— (1987) vol. 2 *Folds and Fractures*. Academic Press, London.

Ravetz, J. R. (1971) *The Roots of Present-day Science*. Open Univ. Press, Milton Keynes.

Raup, David M. and Stanley, Steven M. (1978) *Principles of Paleontology*. 2nd edn. W. H. Freeman, San Francisco.

Read, H. H. (1970) *Rutley's Elements of Mineralogy*. 26th edn. George Allen and Unwin, London.

—— and Watson, Janet (1968) *Introduction to Geology*. 2nd edn. chapter 7, 'Vulcanicity and the Volcanic Association'. Macmillan, London.

Reading, H. G. (ed.) (1978, 1986, 1996) *Sedimentary Environments and Facies*. 1st, 2nd, and 3rd edn. Blackwell Scientific Pubns., Oxford.

Reijes, T. J. A. and Hsü, K. J. (1986) *Manual of Carbonate Sedimentology, A Lexigraphical Approach*. Academic Press, London.

Reineck, H. E. and Singh, I. B. (1980) *Depositional Sedimentary Environments*. 2nd edn. Springer-Verlag, New York.

Reynolds, J. M. (1985) 'Dielectric behaviour of firn and ice from the Antarctic Peninsula', *Journal of Glaciology*, **31**, *109*, 253–62.

Rickards, T. (1984) *Cambridge Illustrated Thesaurus of Physics*. Cambridge Univ. Press, Cambridge.

Roberts, W. L., Rapp, G. R., and Weber, J. (1974) *Encyclopedia of Minerals*. Van Nostrand Reinhold Co., New York.

Rose, A. W., Hawkes, H. E., and Webb, J. S. (1979) *Geochemistry in Mineral Exploration*. 2nd edn. Academic Press, London.

Rosenzweig, Michael L., and McCord, Robert D. (1991) 'Incumbent replacement: evidence for long-term evolutionary progress', *Palaeobiology*, **17**, *3*, 202–13.

Rudwick, M. J. S. (1972) *The Meaning of Fossils: Episodes in the History of Palaeontology*. Macdonald, London.

Rupke, N. A. (1983) *The Great Chain of History: William Buckland and the English School of Geology*. Clarendon Press, Oxford.

Scheidegger, A. E. (1976) *Foundations of Geophysics*. Elsevier, Barking, Essex.

Schneer, Cecil J. (ed.) (1969) *Towards a History of Geology*. MIT Press, Cambridge, Mass.

Sears, D. W. (1978) *The Nature and Origin of Meteorites*. Adam Hilger Ltd, Bristol.

Seilacher, A. (1984) 'Storm beds: Their significance in event stratigraphy', in Seibold, E. and Meulenkamp, J. D. (eds.) *Stratigraphy Quo Vadis?* AAPG Studies in Geology No. 16, IUGS Special Publication No. 14. American Assocn. of Petroleum Geologists, Tulsa, Oklahoma.

Selby, M. J. (1985) *Earth's Changing Surface*. Oxford Univ. Press, Oxford.

Selley, R. C. (1976, 1982) *An Introduction to Sedimentology*. 1st and 2nd edn. Academic Press, London.

——(1978) *Ancient Sedimentary Environments*. 2nd edn. Chapman and Hall, London.

——(1985) *Elements of Petroleum Geology*. W. H. Freeman, San Francisco.

Seyfert, Carl K. and Sirkin, Leslie A. (1979) *Earth History and Plate Tectonics: An Introduction to Historical Geology*. 2nd edn. Harper and Row, New York.

Sharma, P. V. (1986) *Geophysical Methods in Geology*. Elsevier, Amsterdam.

Sheriff, R. E. (1973) *Encyclopedic Dictionary of Exploration Geophysics*. Soc. of Exploration Geophysicists, Tulsa, Oklahoma.

——(1977) 'Limits on resolution of seismic reflections and geologic detail derivable from them', in Charles E. Payton (ed.) *Seismic Stratigraphy—applications to hydrocarbon exploration*. AAPG Memoir 26, pp. 3–14. American Assocn. of Petroleum Geologists, Tulsa, Oklahoma.

——(1985) 'Aspects of Seismic Resolution', in Berg, O. R. and Woolverton, D. G. (eds.) *Seismic Stratigraphy II: An Integrated Approach*. AAPG Memoir 39. American Assocn. of Petroleum Geologists, Tulsa, Oklahoma.

Siebold, E. and Meulenkamp, J. D. (eds.) (1984) *Stratigraphy Quo Vadis?* AAPG Studies in Geology No. 16, IUGS Special Publication No. 14. American Assocn. of Petroleum Geologists, Tulsa, Oklahoma.

Sinclair, John (1969) *Quarrying, Opencast and Alluvial Mining*. Elsevier, Amsterdam.

Skelton, Peter (ed.) (1993) *Evolution: A Biological and Palaeontological Approach*. Addison-Wesley, Harlow, in association with the Open University.

Skinner, Brian J. (1969) *Earth Resources*. Foundations of Earth Science Series. Prentice-Hall, Englewood Cliffs, New Jersey.

Sloan, E. Dendy, Happel, John, and Hnatow, Miguel A. (eds.) (1994) *International Conference on Natural Gas Hydrates: Annals of the New York Academy of Sciences, Vol. 715*. New York Academy of Sciences, New York.

Small, R. J. (1970) *The Study of Landforms*. Cambridge Univ. Press, Cambridge.

Smith, F. Gordon (1963) *Physical Geochemistry*. Addison-Wesley Publishing Co. Inc., London.

Smith, P. J. (1971) Chapter 15 in Gass, I. G., Smith, P. J., and Wilson, R. C. L. (eds.) *Understanding the Earth*. The Artemis Press, Horsham, Sussex.

——(1973) *Topics in Geophysics*. Open Univ. Press, Milton Keynes.

Snelling, N. (ed.) (1985) *Geochronology and the Geological Record*. Geol. Soc. Lond., sponsored by Geol. Soc. Lond. and the IUGS Subcommission on Geochronology.

Stewart, W. N. (1983) *Palaeobotany and the Evolution of Plants*. Cambridge Univ. Press, Cambridge.

Sugden, D. E. and John, B. S. (1976) *Glaciers and Landscape*. Edward Arnold, London.

Suppe, John (1985) *Principles of Structural Geology*. Prentice-Hall, Englewood Cliffs, New Jersey.

Takeuchi, H., Uyeda, S., and Kanamori, H. (1967) *Debate about the Earth*. Freeman, Cooper and Co., San Francisco.

Talwani, Manik and Pitman, Walter C. (eds.) (1978) *Island Arcs, Deep Sea Trenches and Back-arc Basins*. American Geophysical Union, Washington, DC.

Tarling, D. H. (1983) *Palaeomagnetism*. Chapman and Hall, London.

Taylor, Stuart Ross (1975) *Lunar Science; A Post-Apollo View*. Pergamon Press, Elmsford, NY.

—— (1982) *Planetary Science: A Lunar Perspective*. Lunar and Planetary Institute, Houston.

—— and McLennan, S. M. (1985) *The Continental Crust*. Blackwell, Oxford.

Telford, W. M., Geldart, L. P., Sheriff, R. E., and Keys, D. A. (1976) *Applied Geophysics*. Cambridge Univ. Press, Cambridge.

Thomas, Barry (1981) *The Evolution of Plants and Flowers*. Eurobook Ltd (Peter Lowe), London.

—— and Spicer, Robert A. (1987) *The Evolution and Palaeobiology of Land Plants*. Croom Helm, London.

Thornthwaite, C. W. and Mather, J. R. (1955) *The Moisture Balance*. Publications in Climatology, **8**, *1*. Laboratory of Climatology, Centerton, New Jersey.

Thorpe, R. S. and Brown, G. C. (1985) *Field Description of Igneous Rocks*. Geol. Soc. Lon. Handbook. Open Univ. Press, Milton Keynes, and Halstead Press (John Wiley and Son), New York.

Traverse, Alfred (1988) *Paleopalynology*. Unwin Hyman, Boston.

Tucker, Maurice E. (1978) *The Field Description of Sedimentary Rocks*. Geol. Soc. Lon. Handbook Series. Open Univ. Press, Milton Keynes, and Halstead Press (John Wiley and Son), New York.

—— (1981) *Sedimentary Petrology, An Introduction*. Blackwell Scientific Pubns., Oxford.

UNESCO (1976) *Engineering Geological Maps*. UNESCO Press, Paris.

Vail, P. R. et al. (1977) 'Seismic stratigraphy and global changes in sea level', in Charles E. Payton (ed.) *Seismic Stratigraphy; Applications to Hydrocarbon Exploration*. AAPG Memoir 26, pp. 49–213. American Assocn. of Petroleum Geologists, Tulsa, Oklahoma.

Van Eysinga, F. W. B. (compiler) (1975) *Geological Timetable* 3rd edn. Elsevier, Amsterdam.

Vear, Alwyn (1996) 'Basin modelling for petroleum prediction', *OUGS Jo.*, **17**, *2*, 36–41.

Vemura, T. and Mizutani, S. (1979) *Geological Structures*. John Wiley and Sons, New York.

Vernon, R. H. (1983) *Metamorphic Processes*. George Allen and Unwin, London.

Vine, F. J. and Matthews, D. H. (1963) 'Magnetic anomalies over oceanic ridges'. *Nature*, **199**, 947–9. London.

Washburn, A. L. (1979) *Geomorphology: A Study of Periglacial Processes and Environments*. Edward Arnold, London.

Wasson, John T. (1985) *Meteorites: Their Record of Early Solar-system History*. W. H. Freeman, New York.

Watson, Janet (1983) *Geology and Man*. George Allen and Unwin, London.

Wendt, Herbert (1968) *Before the Deluge*. Victor Gollancz, London.

West, R. G. (1977) *Pleistocene Geology and Biology*. 2nd edn. Longman, London.

Weyman, Darrell (1981) *Tectonic Processes*. George Allen and Unwin, London.

Wilf, Peter (1997) 'When are leaves good thermometers? A new case for Leaf Margin Analysis', *Paleobiology*, **23**, *3*, 373–90.

Williams, David R. (1997) *Planetary Missions*. http://nssdc.gsfc.nasa.gov/planetary/projects.html.

—— *Near Earth Asteroid Rendezvous*. http://nssdc.gsfc.nasa.gov/planetary/near.html.

Wills, B. A. (1981) *Mineral Processing Technology*. 2nd edn. Pergamon Press, Oxford.

Wilson, G. (1982) *Introduction to Small-Scale Geological Structures*. George Allen and Unwin, London.

Windley, Brian F. (1977, 1984) *The Evolving Continents*. 1st and 2nd edn. John Wiley and Sons, Chichester.

Wood, Robert Muir (1985) *The Dark Side of the Earth*. George Allen and Unwin, London.

Woodcock, N. H. (1986) 'Strike-slip duplexes', *J. Struct. Geol.*, **8**, 7, 725–35.

York, D. and Farquhar, R. M. (1973) *The Earth's Age and Geochronology*. Pergamon Press, Oxford.

Zeuner, F. E. (1958) *Dating the Past. An Introduction to Geochronology*. Methuen, London.

MORE OXFORD PAPERBACKS

This book is just one of nearly 1000 Oxford Paperbacks currently in print. If you would like details of other Oxford Paperbacks, including titles in the World's Classics, Oxford Reference, Oxford Books, OPUS, Past Masters, Oxford Authors, and Oxford Shakespeare series, please write to:

UK and Europe: Oxford Paperbacks Publicity Manager, Arts and Reference Publicity Department, Oxford University Press, Walton Street, Oxford OX2 6DP.

Customers in UK and Europe will find Oxford Paperbacks available in all good bookshops. But in case of difficulty please send orders to the Cash-with-Order Department, Oxford University Press Distribution Services, Saxon Way West, Corby, Northants NN18 9ES. Tel: 01536 741519; Fax: 01536 746337. Please send a cheque for the total cost of the books, plus £1.75 postage and packing for orders under £20; £2.75 for orders over £20. Customers outside the UK should add 10% of the cost of the books for postage and packing.

USA: Oxford Paperbacks Marketing Manager, Oxford University Press, Inc., 200 Madison Avenue, New York, N.Y. 10016.

Canada: Trade Department, Oxford University Press, 70 Wynford Drive, Don Mills, Ontario M3C 1J9.

Australia: Trade Marketing Manager, Oxford University Press, G.P.O. Box 2784Y, Melbourne 3001, Victoria.

South Africa: Oxford University Press, P.O. Box 1141, Cape Town 8000.

Oxford
Paperback
Reference

CONCISE SCIENCE DICTIONARY

New edition

Authoritative and up to date, this bestselling dictionary is ideal reference for both students and non-scientists. Fully revised for this third edition, with over 1,000 new entries, it provides coverage of biology (including human biology), chemistry, physics, the earth sciences, astronomy, maths and computing.

* **8,500 clear and concise entries**
* **Up-to-date coverage of areas such as molecular biology, genetics, particle physics, cosmology, and fullerene chemistry**
* **Appendices include the periodic table, tables of SI units, and classifications of the plant and animal kingdoms**

'handy and readable . . . for scientists aged nine to ninety'
Nature

'The book will appeal not just to scientists and science students but also to the interested layperson. And it passes the most difficult test of any dictionary—it is well worth browsing through.'
New Scientist